J.M. Gilliss

A catalogue of 16,748 southern stars

deduced by the United States naval observatory from the zone observations made

at Santiago de Chile

J.M. Gilliss

A catalogue of 16,748 southern stars
deduced by the United States naval observatory from the zone observations made at Santiago de Chile

ISBN/EAN: 9783742865236

Manufactured in Europe, USA, Canada, Australia, Japa

Cover: Foto ©berggeist007 / pixelio.de

Manufactured and distributed by brebook publishing software (www.brebook.com)

J.M. Gilliss

A catalogue of 16,748 southern stars

PREFACE.

The reduction of these zones was in progress for more than forty years, and presents three distinct periods: The first extending from the return of the Expedition to the United States, in November, 1852, to the middle of 1866; the second extending from the latter date to January 17, 1891, and the third extending from thence to about the middle of 1894.

During the first period the Expedition had an office of its own, with a small staff of computers engaged in preparing its records for publication, all of whom were under the immediate direction of Capt. JAS. M. GILLISS, U. S. N., until his death on February 9, 1865. In the beginning the work was pushed so rapidly that volumes 1, 2, 3, and 6, entitled "The U. S. Naval Astronomical Expedition to the Southern Hemisphere, during the years 1849–'50–'51–'52," were completed and issued by the House of Representatives in 1855 and 1856 as Executive Document No. 121, Thirty-third Congress, first session. Subsequently the force was reduced to such an extent that toward the end only Messrs. JOHN WIESSNER and GEORGE B. MERRIMAN remained. It was Captain GILLISS's intention that the unpublished fourth and fifth volumes should contain, respectively, the observations made at Santiago with the transit circle, to determine the right ascensions and declinations of 1,963 fixed stars, and the zone observations upon which the present catalogue is founded. At the time of his death much work had been done on these volumes, and before the office of the Expedition was closed the manuscript of the observations to be contained in volume 4 was ready for the printer, and volume 5 was so far advanced that it was erroneously supposed it could be finished by a single computer in about a year.

The computing staff under Captain GILLISS was never large, but there were so many changes in it that the total number of persons who took part in the work at one time or another was considerable. Among those so employed were Lieuts. ROGER N. STEMBEL,* JOEL S. KENNARD, ALBERT N. SMITH, and WILLIAM MITCHELL, of the United States Navy, and Messrs. F. A. P. BARNARD, WM. DIEBITSCH, J. F. FLAGG, JOHN R. GILLISS, WM. HARKNESS, F. G. HESSE, DANIEL MAJOR, GEO. B. MERRIMAN, E. T. PHELPS, and JOHN WIESSNER; but it is not certain that this list is complete.

During most of the second period the papers of the Expedition rested quietly in the archives of the Naval Observatory, because neither money nor computers were available for use upon them. Usually the Observatory had scarcely sufficient force for its own needs, and it was not deemed proper to neglect current work for the sake of

* Since, Rear-Admiral STEMBEL.

advancing the Gilliss reductions, which were only deposited at the Observatory "for safe-keeping" and never constituted any part of the work for which it was directly responsible. Nevertheless Prof. WM. HARKNESS, U. S. N., was nominally in charge of the zones after March 1873, and no opportunity of forwarding them was ever allowed to pass unimproved. The computing done during this period amounted to a total of about fifty-eight months; the persons employed under Professor HARKNESS being Ensign (now Professor) STIMSON J. BROWN, U. S. N., and Messrs. PARKER PHILLIPS, JOHN HEDRICK and EMIL WIESSNER.

The third period was mostly coincident with the removal of the Naval Observatory from its old location in the reservation bounded by the Chesapeake and Ohio Canal, Twenty-third street west, E street north, and Twenty-fifth street west to its present site on Georgetown Heights. In connection with that removal nearly all the principal instruments of the Observatory underwent extensive repairs, and on account of the resulting cessation of observing, the volume of work devolving upon the staff was greatly diminished. Advantage was taken of that circumstance to put Profs. EDGAR FRISBY, U. S. N., and S. J. BROWN, U. S. N., in charge of the Gilliss zones, and their reduction was finally completed by these gentlemen about the middle of the year 1894.

During the forty-two years the reductions were in progress the conditions affecting them changed so much that they were ultimately finished on a plan which would have been impossible in Captain GILLISS's day. To show clearly the nature of that change, and to secure a fairly complete history of the undertaking, it has been thought best that the account of the work done during each of the three periods specified above should be written by the persons who were actually in charge during the respective periods.

HISTORY OF THE FIRST PERIOD.

The only existing account of the work done during the first period is contained in a nearly complete draft of an introduction for the proposed fourth volume of the Expedition, which was written by Captain GILLISS himself. Some parts of it are not strictly applicable to the zone observations, which were to have formed the fifth volume, and much of the detail has since become so familiar that it would now be regarded as superfluous; nevertheless, the document is appended almost entire in order to show exactly the standpoint of the Superintendent of the Expedition about the year 1862.

CAPTAIN GILLISS'S INTRODUCTION.

THE OBSERVATORY.

The building erected for the meridian circle and clock of the Astronomical Expedition occupied an artificial terrace, leveled near the northern crest of Santa Lucia, a rocky eminence near the eastern limits of Santiago de Chile. A description of this hill and of the manner in which the sites were prepared for the observatories has been given, page xxxii, Vol. III; but that progressive changes in the azimuth and level of the meridian circle, to be hereafter discussed, may be the more satisfactorily accounted for, it is proper to repeat here that the stratified porphyritic rocks composing the elevation are columnar, and, at a little distance, may easily be mistaken for basalt. Some of the columns are nearly vertical, and a smaller number lie horizontal, though the larger proportion, as do also the compact strata, stand at every inclination from the vertical toward the west, not one of them toward the east. Partially covered to the depth of an inch or more with decomposed rock and scanty vegetable mold, the eastern face of the hill has an inclination not differing greatly from 45°. The western is absolutely precipitous, and in the afternoon its bare wall of nearly black rock is presented to the direct rays of the sun. From the summit to the north and south extremes the slopes are tolerably regular, that of the northern and longer forming an angle of about 15° with the horizon, though the southern is more abrupt and broken.

Of the two small terraces leveled just below the summit on the northern slope, the lower and longer was selected for the meridian circle. Its elevation above the barometer at our residence, 300 yards distant in a northeast direction, was 170 feet. This was determined from a mean of simultaneous comparisons of two instruments, and many transportations of the same instrument. Assuming the height of the barometer at the level of the sea to be 30.000^{in}, a mean of all our tri-hourly observations during thirty-three months shows the elevation of the cistern of our standard barometer above the sea to have been 1,793 feet, and, consequently, that of the meridian circle 1,963 feet.

Trenches were excavated for the foundations of the observatory on the northeast portion of the terrace, so that the north and east sides of the building were upon the very edges of the rapid slopes, and, except a narrow pathway between the south front and the acclivity leading to the rotary observatory, the only unoccupied ground was to the west. This was a space perhaps 20 by 25 feet, most of which is bare rock in nearly horizontal layers, whose inequalities were filled with coarse sand and finely broken fragments of porphyry. The foundation was built of stone and mortar, to a height of 6 inches above the surface, and the larger portion of it was based upon the native rock.

The observatory building is entirely of wood, 22 feet long, 18 feet wide, and 18 feet high to the eaves, with a pitch to its roof of 5 feet. It was constructed at Washington, and indelibly marked, so that when taken down and transported to Chile, there was no difficulty in putting it together again. Except the roof, its weather-boarding is tongued and grooved, and held in place by screws. The boards of the roof were covered with stout canvas, tightly drawn over, nailed down, and painted. There are meridian openings 20 inches wide in the center of the north and south sides that extend to within 3½ feet of the floor. They are closed by vertical doors opening inward, which are secured by two ordinary door latches connected by an iron rod, and roof-doors that meet at the ridge, and may be elevated by levers and pulleys within the building. To prevent the entrance of drifting rain, these latter doors were made 4 inches wider than the openings, and with half-inch grooves along the under edges that corresponded with similar grooves in the jambs. At the zenith the joint is covered by a small trap, beveled to fit the ridge, and so hinged that it goes up with the first roof-door opened, and at 3 inches from the vertical comes against a steel spring, which throws it down when relieved from the pressure of the door last closed. In each of the gable ends there is a broad door with transom lights.

The rock was laid bare for the circle and clock piers, and, in the former case, was leveled to receive the stone blocks on which the piers were immediately to rest. These blocks are of the same material as the piers; are 3 feet wide and long by 6 inches thick, and are set in hydraulic cement. The piers are of red coarse-grained porphyry, 6 feet 7 inches high above the floor line, 2 feet $1\frac{1}{3}$ inches square at that line, the three external faces battering to a height of 4 feet 7 inches, above which the dimensions of a horizontal section are 1 foot 7 inches each way. The inner faces are vertical, and each pier is composed of three pieces, with hydraulic cement between their joints. Had it been practicable to quarry single blocks of suitable dimensions, there was no machinery at Santiago by which they could have been raised to the site of the observatory.

The clock pier is of timber, 12 inches wide, 4 inches thick, and 8 feet long, of which 2 feet is embedded in masonry below the floor of the observatory. The floor joists are framed and the flooring laid without contact with the piers.

THE MERIDIAN CIRCLE.

In the construction of this instrument many experiments which had been made with meridian circles, as well at Berlin as by eminent artists elsewhere, were availed of to attain for it greater accuracy, simplicity, and strength. To these belong the minute divisions of both circles; modification of the form, materials, and mode of supporting the microscope bearers; the slow motion apparatus; the form of the hanging arms to the counterpoises, so as to permit safer and easier reversal; the application of counterpoise leverage to the reversing carriage, and, finally, in essential modifications of the hanging level. These several points will be more specifically shown in the description.

The instrument was made by Messrs. Pistor & Martins, of Berlin. A representation one-twelfth of the natural size, with some of its parts in detail, is given in Plate I. Together with its reversing carriage and appurtenances, it filled five cases, which reached Santiago safely about January 1, 1850.

The piers were prepared for mounting it by cutting their summits to the same level, and by drilling holes for the several plugs, to which were cemented the columns sustaining the counterpoises; those at *d d* for the Y's and micrometer bearers; those for the frames *k k* serving as fulcra to the slow-motion apparatus; those for the brackets supporting the illuminating lamps, and, finally, the apertures through which light passed to the axis; all, except those for the frames *k k*, which were secured with lead, were cemented in their places with calcined gypsum and iron filings mixed with water.

Its axis is a single cast of bell metal, comprising a hollow cubical box $9\frac{1}{3}$ by $7\frac{2}{3}$ inches, and two conic frustums that terminate in cylinders, one portion of each of which is $2\frac{2}{3}$ inches in diameter by $3\frac{1}{2}$ inches long, and another portion 2 inches in diameter by $3\frac{2}{3}$ inches long. There are broad flanges at the extremities of the larger cylinders, and steel pivots are screwed to a shoulder within the smaller, these pivots being $3\frac{1}{3}$ inches long and $1\frac{2}{3}$ inches in diameter. Thus the entire length of the axis is $40\frac{2}{3}$ inches, and between the bearing points in the Y's $39\frac{1}{3}$ inches. In each of the larger cylinders there is a shallow groove intended as a guide for the friction rollers to the counterpoises. Its telescope is formed of slightly conical tubes of brass, cast with rectangular bases, by which they are firmly screwed to the cubical box of the axis. The object-glass has a clear aperture of 51 lines, with a focal length of 6 feet. Its optical capacity was never directly tested to its limits; but the fact that it showed the color of the companion to *Antares;* brought out distinctly the two very minute stars near *η Argus*, which Sir JOHN HERSCHEL says* "the increased light of the large star had completely obliterated before the conclusion of my observations," even when *η Argus* was almost if not absolutely as brilliant as he ever saw it, and separated double stars of the 9—10 magnitudes whose distance he puts down at ¾ to 1″, are ample evidences of its superior merits. These evidences of its high qualities were not the results of special trials, but presented themselves, as will be seen in subsequent pages, in the regular course of observations, and when the instrument was fully illuminated.

The eye end is furnished with a smaller tube that slides within the main telescope, for adjustment to celestial focus, and admits of slight rotary motion for rectification of the vertical line of the transit system of wires. There are two wire systems at the focus. One comprises seven fixed vertical wires, at intervals of about seventeen seconds of time, a fixed horizontal wire that crosses the center of the field, and a single vertical wire, which is moved by a micrometer screw. The other system comprises seven horizontal wires at intervals of about four minutes of arc, the system being movable by a micrometer screw, one revolution of which is equal to 27.9″. When the instrument reached Chile, there were only five horizontal wires, the two others being subsequently added to facilitate our zone observations by embracing a broader field. The systems inserted by the artists were exquisitely delicate spider fibers, most uniformly placed; but these were accidentally destroyed at a very early period of the observations, as were several later ones, and though Lieutenant MAURY had the goodness to send a supply of the best material from Washington, no systems comparable with the original were ever perfected.

Four positive and a collimating eyepiece accompanied the instrument. The former magnified from 75 to 300 times, and the latter was a simple lens of low power, with a diagonal reflector of highly-polished plate glass placed in advance of it. By means of racks and pinions placed on opposite sides to the micrometer screws moving the vertical and horizontal systems, the eyepiece can be slid in its dovetailed slides until perpendicular to any wire of the system. As constructed by Messrs. PISTOR & MARTINS, the fulcrum for the pinion moving the eyepiece vertically was a slender prolongation of the diaphragm holding the horizontal system of wires; and any movement of the eyepiece, up or down, caused these wires to spring. The position of the fulcrum at that time is shown in fig. 1. It was then altered, as shown in fig. 2, but as there was not sufficient bearing to retain the pinion absolutely in the same plane as the rack, a stout shoulder was finally soldered on, as shown in fig. 3.

Light is admitted through one extremity of the axis, and is reflected down the tube by an annulus placed diagonally in the cubical box, and also through an aperture of the sliding tube, just in advance of the reticles. In the first case the rays are concentrated by a lens placed in the outer extremity of the pivot, and we have a bright field with dark lines; and in the second the aperture is simply glazed, and we have a

* Results of Astronomical Observations made during the years 1834–38 at the Cape of Good Hope, p. 37.

dark field with bright lines. The magnifying power most generally used in observations was 96 times.

The apparatus for communicating slow motion in altitude is: First, a tail-piece, *a a*, that clasps a flange of one of the axis cylinders, to which it may be clamped from either side by an endless screw rod; and second, a stationary pinion in the lower extremity of the tailpiece, that works in a toothed wheel, permanently placed on a fine screw below and parallel with the pinion. On turning the pinion by the Hook's-joint rods, seen in the drawing, looped to the clamping rods, the fine screw passes farther through, or out of the lower end of the tailpiece, according to the direction of the motion.

There is a circle for each arm. They were packed in strong cases, with cross braces, to prevent warping of the wood, and a consequent springing of the metal. When in place, each is held by the pressure of a metallic disk, 5½ inches in diameter, through which and between the radial bars long screws pass into the flanges of the larger cylinders. Being lifted to their places at the same time, the pinching screws were uniformly tightened, the circles being clamped only with sufficient force to prevent

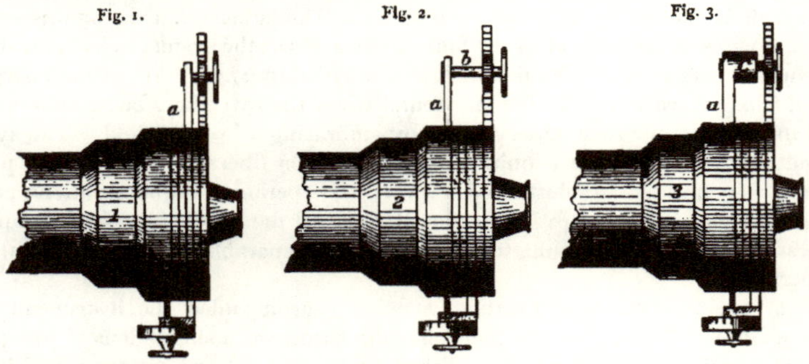

their probable disturbance upon the axis. They are 38⅛ inches in diameter from outside to outside, and both themselves and the radial bars are well proportioned single casts, of somewhat more than ordinary massiveness. Each is divided into spaces of 2′ on an inlaid band of silver, 37 inches in diameter, though only one of them is numbered. The numbering extends from 0° to 360°, and the 0° and 180° divisions were placed perpendicular to the optical axis of the telescope.

Thus it will be seen that, with the exception of the lateral tailpiece and clamp-rod *a a*, the instrument is symmetrically constructed throughout, and this additional weight on one extremity of the axis is counterpoised by the introduction of lead within the corresponding cylinder of the opposite arm.

Usually only one circle is minutely divided, and the reading is by four micrometer microscopes attached to the arms of tubes that intersect each other at right angles, the extremities of the tubes being connected by others, and the frame thus formed centering upon the extremity of the axis. Disturbance in the level of these bearers is almost a necessary consequence of movement of the telescope, and the observer is compelled to examine the level attached to them with every observation. Moreover, it is well known that there are different temperatures at different elevations within the observing room, and that side currents are often generated on opening the meridian doors, which also affect different portions of the circle and microscope bearers unequally. Both the late eminent astronomer at Altona, Counsellor SCHUMACHER, and the artists thought that these objections would be, in a great measure, obviated by having the two circles finely divided, and each of them read by two microscopes, placed on horizontal bearers. The artists proposed, however, to construct the instrument so that all four microscopes might be applied to either circle should it be preferred.

In order to secure solidity of construction and regular and *central* expansion of the supports for the pivots and microscope bearers, such form was given to the plug *d* and to the massive perforated disk *c*, that the expansion would act from the center outward, and therefore be a minimum in its prejudicial effect. Both the disks and the plugs are of red metal and, though separated by thick washers of the same metal, the former are secured to the latter by three strong and equidistant screws. On the successful fastening of the plugs to the pier depends not only the stability of the transit axis, but likewise the immobility of the microscopes, and therefore every precaution was taken to use specially prepared gypsum, and to allow two days for the cement to dry before putting up the disks. Should not the outer faces of the plugs have been placed absolutely parallel, the defect may be remedied by filing down one or more of the washers; otherwise the lines of the microscope bearers will form angles with the planes of the circles. Parallelism of the disks, therefore, is an essential requisite.

The Y's, also, and their adjusting screws, are of red metal. One of them may be moved in azimuth, the other in altitude. They are firmly screwed to the disks. To insure that the pivots rest always on the same points of the Y's, on the side of each and next the pier there is a small steel plate, whose upper edge is beveled outwardly. These are held to the Y's by spiral springs, which yield if the axis descend slightly to one side, but are strong enough to press it in place when resting in the counterpoise arms. As the extremities of the pivots are slightly oval, the beveled edges to the plates prevent all possible injury from inadvertent manipulation. To each Y there is also a hinged lever, with a weight at one end, intended to serve as a guide in adjustment of the pressure of the instrument upon its pivots.

The microscope arms are metallic tubes, protected from the sudden influences of temperature by wooden cases, so contrived as to leave a stratum of air between the tube and wood, and not to cramp the former should the wood warp. They are attached to the dovetailed periphery of the disks by pinching screws, and may be placed at any angles around the circles simply by loosening the screws.

The micrometers are of the ordinary English box form, with reading scales on one side of the field of view. Their screws move two parallel spider lines placed about 8″ from each other, one revolution of the screw representing 1′, and each division 0.5″. There is a marked zero point at the center of the scale, which, together with the numbering on the micrometer head, admits of the usual adjustment. The reading scale is a toothed line, each fifth division of which is indicated by a deep indentation. The graduated planes of the circles are illuminated by light thrown perpendicularly on them from white reflectors, placed at angles of 45° in sliding tubes that fit on the object ends of the microscopes. Two of the microscopes carry pointers—simple slips of silver, with a horizontal line engraved upon each.

The brass frames *k k*, serving as fulcra for the slow-motion apparatus, were next secured to the piers. Should the screw be turned backward, dead motion is prevented by a strong steel spring *f*, which holds a flat-headed steel pin that slides freely through one fork of the frame and presses the fine screw against the other fork. The action of this spring may be released by means of the lever *g*, and to avoid injury to the slow-motion apparatus, when lowering the axis into the Y's, the lower end of the tail-piece is provided with a roof-shaped cover.

Flush with the inner faces of the piers massive brass columns serve as fulcra to the supporting hooks, levers, and counterpoises, which relieve the Y's from any desired portion of the weight of the instrument. The supporting hooks, which are of steel, rest, by strong adjusting screws, upon the extremities of levers of the same metal, and are so adjusted that the friction rollers upon which the axis rests fall directly into the grooves upon its cylinders. When removing the instrument from the piers, it is not necessary to have reference to the counterpoise appendages.

A reversing carriage is provided for the latter purpose. This differs from the ordinary construction by the addition of a system of levers and weights somewhat similar to that made by the same artists for the transit instrument in the prime-vertical of the Washington Observatory. The levers act upon the lower extremity of the elevating screw, and diminish the power to be applied to the crank—a desirable result with an instrument of such massive proportions. There is a stop on the upper part of the carriage frame, which insures motion of the reversing arms through precisely 180°, and a projecting bar beneath the carriage encounters a stop placed on the floor, where the instrument is accurately over its Y's. The carriage travels upon an iron railway permanently screwed to the floor. Before raising out the circle for reversal, it is necessary to lift the hinged levers over the pivots, throw back the spring *f* to the slow-motion apparatus, and slide the illuminating tubes of the microscopes from the faces of the circles. The reversal may then be effected in the dark as safely as by daylight. At the first trial the floor joists were found to be of unequal strength, and the great weight transferred to the carriage (four men are required to lift the instrument) caused such settling on the weaker side that the circle was replaced with difficulty. And although the flooring was removed, and props placed under the joists, much care was always essential in lifting out and restoring the circle. The elevating screw lifted the instrument only high enough for the smaller axis cylinders to clear the microscopes, and when the illuminating tubes were slid back, the distance between two of them on the same side did not exceed that between the outer faces of the disks holding the circles upon the axis more than fifteen-hundredths of an inch.

There are two lamps for illumination of the wire systems. These are supported by brackets that retain them at one foot from the piers. Light from them passes to the axis of the telescope through a burnished brass tube in the perforated piers, and its intensity in the field is modified at will by the interposition of a graduated shaded screen of oiled paper in a vertical toothed sector that turns upon a pivot between the disks *c c* and piers. Motion is communicated to this sector by means of a second whose teeth play into it, and the handles *l l*. In order that the currents of air which the heat of the lamps generate and their effects upon the piers should be as nearly alike as possible, it was recommended that both should be lighted whenever the instrument was in use; and, to avoid the direct influence of their heat on the piers, disks of mahogany *m m*, 1 foot 7 inches in diameter, were interposed between them and the piers. The wooden disks were separated from the piers by others of thick felt; and, to cut off radiant heat that might pass the piers, other mahogany disks *n n*, of the same diameter as the circles, were sent with the instrument to be placed on the inner faces of the piers. But they were found to be somewhat inconvenient; and as the lamps had no sensible effect on a thermometer hung from the counterpoise lever near the circle, they were taken down after a day or two.

There is a hanging level *i i* for horizontal adjustment of the axis. Its tube is inclosed within the brass cylinder, a slip of plate glass inserted in the top of the horizontal cylinder, permitting the bubble and its scale to be seen. There is a reserve chamber at one extremity of the tube, so that risk of leakage is avoided on expansion of the inclosed sulphuric ether, and the bubble may be made of suitable length under all changes of temperature. To obviate the effect of heat from the person during manipulation, both the cylinder and its hollow conical arms are covered with cloth, and the former is provided with convenient handles of wood. The frames for the slow-motion apparatus serve as guides to its arms, and insure readings in the same vertical plane on reversal; but its inverted Y's do not rest over the bearing points of the pivots to the axis. When not in use, the level hung upon a pair of arms put up for it in one corner of the observatory.

The circles are not touched by the observer at any time, but the telescope is directed to the object by means of a long metallic hook covered with chamois leather, and which is furnished with a handle of wood.

The observing chair is made according to the model of those at the Washington Observatory, with two movable cushions that serve alternately as seat and for support of the observer's back at every altitude on each side of the zenith. Its feet were furnished with grooved rollers that rested upon the railway laid down for the reversing carriage.

Earnest and cordial thanks are due to the late counsellor SCHUMACHER and to the present eminent Director of the Royal Observatory at Berlin for their supervision and examination of the circle whilst in the hands of the artists. When completed, it was minutely inspected by Professor ENCKE, and his opinions thereupon were kindly communicated, as follows:

"CERTIFICATE.—The undersigned having yesterday examined the meridian circle ordered for the United States of America, from the establishment of Messrs. PISTOR & MARTINS, takes pleasure in stating that the instrument is worthy of the fame of its makers.

"Experience derived from the use of our meridian circle suggested several improvements, which have been applied for the purpose of combining both simplicity and solidity. The following results have been obtained, viz: Improvement of the microscope bearers, both in the manner of securing them and casing them in wood;

increased strength in the clamp and greater perfection of the slow-motion apparatus; in the levers and counterpoises, so as to facilitate and render secure the process of reversal; in the construction of the reversing carriage; in the arrangement of the index finder; in the suspension level.

"The eyepiece was provided with slides for vertical and horizontal motion, in accordance with instructions, and for the complication of this arrangement the makers do not hold themselves accountable.

"I did not examine the qualities of the object glass, because of the unfavorable position of the instrument; but, from my experience of former instruments of like capacity from the same makers, I entertain no doubt of its excellence.

"I regard the increased strength of the radial bars an essential improvement, experience with meridian circles having shown them to be, generally, too slight.

"J. F. ENCKE, *Director Berlin Observatory.*

"BERLIN, *August* 1, 1849."

THE CLOCK.

This instrument was made for the United States exploring expedition by MOLYNEUX of London; and, as it was intended for use with the invariable-pendulum apparatus, it was furnished with a tripod frame on which to mount it. It is full jeweled, has a dead-beat escapement and a mercurial pendulum, of which the glass cylinder is hermetically sealed within a slightly larger one of brass. Thus, no alteration of its imperfect compensation could be effected without a risk that it was scarcely prudent to incur in a country where the disturbances by earthquakes were continually much greater than the changes of rate under fluctuations of temperature.

The original pendulum spring having been accidentally broken before its shipment to Chile, the clock was sent to New York, where it was thoroughly cleaned, and a new spring applied, by Mr. ARTHUR STEWART. When unpacked at Santiago, its case was screwed immovably against the wooden pier, placed to the north and nearly in a line with the inner face of the east circle pier. As the case is short, its bottom was several inches above the floor of the observatory, and the pier had been so placed that the face of the clock was to the southwest and within 3 feet of the observer when looking south. During our absence from the observatory, and within a day or two after mounting it, the pendulum spring snapped near its center of length, and the cylinder was found at the bottom of the case. Fortunately, the distance it had fallen through was small, and no harm appeared to have been done. A new spring was made from part of a chronometer mainspring, by the most skillful repairer of watches at Santiago, and, soon after, the clock was again set going; but there was a constant tendency to augment its time of vibration which was never overcome.

From November to March, at Santiago, both inclusive, the mean daily range of thermometers shielded from both reflected and radiant heat exceeds 21° Fahrenheit, and during the other seven months it is about 18°. Within the observatory the extremes between day and night temperatures were far greater, for the thin board building, sheltered during the day from the action of the prevailing wind, is exposed, on the surface of a nearly black rock, to the direct rays of the sun, and at night is fully open to the cold drafts descending from the snow peaks along the valley of the Mapocho. Indeed, it would be safe to say that the highest temperature within the *closed* observatory during cloudless days and the lowest temperature of the *open* observatory on clear nights usually differed 30°.

If to this cause of disturbance we add that of frequent earthquakes, which were almost uniformly perceptible by us as two distinct tremors, of which the second followed the first after an interval of five to ten seconds, it will be perceived how much additional labor was imposed on the observer to secure moderate confidence for his results.

Thus, the rate by day will differ from the rate at night, and as the fluctuations are necessarily unequal, the rates derived from diurnally determined errors will, consequently, vary from day to day. Therefore it was important to keep the clock going as nearly as possible to sidereal time, so that if the observations at night did not afford the necessary data for determining the night rate, that for the 24 hours might be used in reducing the zone observations without involving sensible error.

The dial was illuminated, at night, from a lantern placed at such a distance that all the rays of light passing through its lens were received upon the dial.

THE BAROMETER.

Our equipment comprised a standard barometer, differing very little from the form described by Prof. F. R. HASSLER in Senate Document No. 225, second session Twenty-seventh Congress; a mountain barometer, constructed by TROUGHTON & SIMMS; a steel tube siphon barometer, made under the direction of Lieut. M. F. MAURY, and an aneroid. On unpacking the case containing them, the mercury of the mountain barometer was found to be thoroughly black from oxidation and not trustworthy; and the steel tube one very evidently contained air. Of the mercurial barometers, only the Hassler standard had reached its destination in perfect order. Subsequently Lieutenant Maury sent out a second steel tube instrument for us, but it fared no better than the first. Of necessity, therefore, we were obliged to rely upon the standard for the datum requisite in calculating astronomical refractions.

This last instrument will be found described at length in Volume VI, pp. xxxvi–xxxviii. Here it need only be repeated that the internal capacity of its tube is sixty-five (0.65^{in}) hundredths of an inch in diameter, and that the height of its column of mercury is read to 0.002^{in}. It was suspended in the office of the expedition about 900 feet northeast of and 170 feet below the observatory, the doors of the two rooms being visible, and, in fact, during the still hours of the night they were within speaking distance of each other.

It has already been stated, page v, that the height of the barometer cistern above the sea, as derived from all the observations, was 1,793 feet and that the difference of level between its cistern and the observatory had been determined as well by simultaneous observations with the previously compared aneroid as by transportation of the latter many times. The uniform correction for difference of elevations applied to the barometric readings at the office is -0.150^{in}.

THERMOMETERS.

The thermometers used for ascertaining the temperature of the external air near the observatory, as well as that for the mercurial column of the barometer, were made by TAGLIABUE & RONKETTI, at New York. Their divided scales are about 7¾ inches long, each division 1°, and the range from — 5° to + 135°. The external thermometer was inclosed within a glass cylinder, having apertures both above and below the bulb, and, for convenience of reading, was suspended near the south meridian opening. The bulb of the other instrument was immersed in the mercury of the barometer cistern, and temporarily tied to its tube.

EXAMINATIONS AND ADJUSTMENTS.

Value of the level divisions.

The value of each level division given by the makers was 1.07″. To ascertain what confidence should be placed in this determination, the level was suspended by its arms to two of the radius bars of the circle, and the slow-motion screw to the latter was turned until the bubble indicated a horizontal position of the radii. By means of the same slow-motion screw the bubble was then moved through any required number of divisions on the tube, and the corresponding changes in the circle were determined from the mean of the four micrometer microscopes.

The divisions of the level are apparently equal, and occupy the central portion of the tube, the numbers extending from 0 to 75, inclusive.

Observations for value of level divisions.

FEBRUARY 5, 1850.

No.	Circle readings.	North end of bubble.	South end of bubble.	Middle point of bubble.	Whole difference circle reading.	Whole difference divisions of level.	Value of one division.	Remarks.
	′ ″	div.	div.	div.	″	div.	″	
1	16 25.8	44.0	*76.8	60.40				Temperature from 88.8° to 86.7°.
2	15 58.0	17.4	49.5	33.45	27.8	26.95	1.032	
3	16 20.5	39.0	71.4	55.20	22.5	21.75	1.034	
4	15 57.4	17.7	50.0	33.85	23.1	21.35	1.082	
5	15 37.2	1.7	34.5	18.10	20.2	15.75	1.289	
6	15 45.2	9.7	42.0	25.85	8.0	7.75	1.032	
7	15 52.6	15.7	48.0	31.85	7.4	6.00	1.233	
8	15 56.7	22.0	54.7	38.35	4.1	6.50	0.631	
9	15 51.5	16.0	48.8	32.40	5.2	5.95	0.875	
10	16 0.0	23.5	56.4	39.95	8.5	7.55	1.126	
11	16 6.4	30.7	63.7	47.20	6.4	7.25	0.883	
12	15 58.0	23.4	56.3	39.85	8.4	7.35	1.143	
13	16 16.0	39.0	71.8	55.40	18.0	15.55	1.158	
14	15 57.5	18.8	51.7	35.25	18.5	20.15	0.918	
15	15 43.8	9.0	42.2	25.60	13.7	9.65	1.420	
16	15 35.6	2.1	35.2	18.65	8.2	6.95	1.180	
17	16 14.3	39.0	72.7	55.85	38.7	37.20	1.040	
18	16 6.4	30.9	64.0	47.45	7.9	8.40	0.940	
19	15 53.7	20.6	53.9	37.25	12.7	10.20	1.245	Sum $\frac{19.779''}{19} = 1.041''$.
20	15 47.9	9.3	42.8	26.05	5.8	11.20	0.518	

* The bubble being beyond the scale (1.8 div.) is estimated.

There being great discrepancy in the results, and the final mean differing materially from the value given by Messrs. PISTOR & MARTINS, the following additional series was observed on a subsequent day.

FEBRUARY 8, 1850.

No.	Circle readings.	North end of bubble.	South end of bubble.	Middle point of bubble.	Whole difference circle reading.	Whole difference divisions of level.	Value of one division.	Remarks.
	° ′ ″	div.	div.	div.	″	div.	″	
1	288 00 01.5	40.9	* 75.7	58.30				Temperature from 90.3° to 83.5°.
2	287 59 47.0	32.2	61.9	47.05	14.5	11.25	1.289	
3	35.0	23.4	52.8	38.10	12.0	8.95	1.341	
4	25.5	16.5	46.2	31.35	9.5	6.75	1.407	
5	12.2	5.7	35.6	20.65	13.3	10.70	1.243	
6	6.4	1.2	31.1	16.15	5.8	4.50	1.289	
7	21.0	14.6	44.9	29.75	14.6	13.60	1.074	
8	12.8	8.4	38.7	23.55	8.2	6.20	1.323	
9	2.0	—* 1.0	29.7	15.35	10.8	8.20	1.317	
10	13.2	9.8	40.7	25.25	11.2	9.90	1.131	
11	21.8	18.0	49.0	33.50	8.6	8.25	1.042	
12	29.7	24.5	55.7	40.10	7.9	6.60	1.197	
13	41.2	33.5	64.7	49.10	11.5	9.00	1.278	
14	53.5	44.6	* 76.8	60.70	12.3	11.60	1.060	
15	42.0	33.5	65.5	49.50	11.5	11.20	1.027	
16	52.7	43.8	75.0	59.40	10.7	9.90	1.081	
17	3.3	2.0	34.3	18.15	49.4	41.25	1.198	
18	10.0	7.6	40.0	23.80	6.7	5.65	1.186	
19	15.0	12.3	44.8	28.55	5.0	4.75	1.052	
20	23.4	20.7	53.5	37.10	8.4	8.55	0.982	
21	34.6	30.1	63.0	46.55	11.2	9.45	1.185	
22	42.5	36.1	69.2	52.65	7.9	6.10	1.295	
23	49.5	41.0	74.5	57.75	7.0	5.10	1.373	
24	43.0	35.8	69.1	52.45	6.5	5.30	1.227	
25	36.2	30.8	64.2	47.50	6.8	4.95	1.374	
26	30.6	25.8	59.5	42.65	5.6	4.85	1.155	
27	24.4	20.6	54.2	37.40	6.2	5.25	1.180	
28	10.8	10.2	43.9	27.05	13.6	10.35	1.314	
29	3.0	3.4	37.0	20.20	7.8	6.85	1.139	
30	11.1	9.8	44.0	26.90	8.1	6.70	1.209	
31	17.1	16.3	50.8	33.55	6.0	6.65	0.902	
32	10.6	10.5	45.0	27.75	6.5	5.80	1.121	
33	6.6	6.7	41.0	23.85	4.0	3.90	1.026	
34	10.5	10.9	45.6	28.25	3.9	4.40	0.886	
35	18.7	17.3	52.1	34.70	8.2	6.45	1.271	Sum $\frac{42.275''}{36} = 1.174''$.
36	25.5	23.7	58.4	41.05	6.8	6.35	1.070	
37	287 59 30.4	28.3	63.3	45.80	4.9	4.75	1.031	

* Beyond scale; estimated.

Allowing equal weight to every result in each of the two series, the value of one level division is 1.129″, or 0.06″ greater than given by the makers. The value adopted for the reduction of observations is $1^{\text{div.}} = 1.1''$.

It has been stated that the level hung within the bearing points of the pivots in the Y's. Therefore it could not be availed of to determine either the inequality or the irregularity of form of the pivots, if any existed. But with a view to ascertain the care probably bestowed in finishing this portion of the instrument, two series of level readings were made at every 10° elevation of the telescope, with the clamp end of the axis both west and east. The first series was made between noon and $3^h\ 30^m$ p. m. of February 2, 1850, and the second between $1^h\ 30^m$ and $4^h\ 50^m$ p. m. of February 4, both the pivots and the bearing points of the Y's having been carefully wiped prior to commencing observations. From 169° to 190°, and from 349° to 10°, the telescope tube prevents the use of the level.

The numbers given in columns E. and W. of the subjoined table are in every case the means of three readings, between each of which the level was agitated, and the bubble permitted to come to rest.

Circle readings.	CLAMP WEST.						CLAMP EAST.					
	Temperature of observatory.	Level direct.		Level reversed.		East end high.	Temperature of observatory.	Level direct.		Level reversed.		East end high.
		E.	W.	E.	W.			E.	W.	E.	W.	
°	°	div.	div.	div.	div.	div.	°	div.	div.	div.	div.	div.
10	83.5	20.33	17.30	10.80	26.83	3.25	84.7	6.50	22.60	12.45	16.65	5.08
20		20.67	17.40	11.83	26.23	2.78		6.53	22.50	12.53	16.63	5.02
30		20.93	17.27	11.87	26.33	2.73		6.47	22.57	12.03	17.00	5.27
40		20.60	17.47	11.90	26.17	2.79		6.30	22.70	11.60	17.40	5.55
50		20.40	17.63	11.90	26.03	2.84		6.37	22.57	11.50	17.44	5.54
60	83.5	20.30	17.70	11.85	26.15	2.93	84.3	6.40	22.50	11.50	17.40	5.50
70		20.10	17.90	11.87	26.15	3.02		6.40	22.50	11.50	17.40	5.50
80		20.33	17.63	11.80	26.15	2.91		6.30	22.60	11.45	17.45	5.58
90		19.97	18.00	11.77	26.20	3.12		5.67	23.10	11.42	17.35	5.84
100		20.40	17.53	11.87	26.05	2.83		5.73	22.97	11.40	17.30	5.79
110		20.40	17.47	11.95	26.03	2.79		6.10	22.60	11.40	17.30	5.60
120	83.8	20.17	17.67	11.85	26.00	2.91	83.7	6.00	22.67	11.40	17.30	5.64
130		20.40	17.43	11.63	26.20	2.90		6.10	22.43	11.27	17.27	5.58
140		20.40	17.40	11.93	25.87	2.74		5.90	22.60	11.20	17.30	5.70
150		20.40	17.40	10.95	26.85	3.23		5.80	22.50	11.05	17.15	5.70
160		20.40	17.40	10.95	26.85	3.23		5.80	22.60	11.20	17.20	5.70
169	84.4	19.90	18.00	11.23	26.67	3.39	82.8	6.47	22.03	11.40	17.10	5.32
190		20.50	17.47	10.93	27.03	3.27		6.40	21.83	11.00	17.25	5.42
200		20.47	17.47	11.00	26.95	3.24		6.30	21.70	10.85	17.15	5.43
210		20.47	17.43	11.33	26.57	3.05		6.40	21.63	10.87	17.17	5.38
220		20.00	18.00	11.17	26.83	3.42		6.33	21.70	10.87	17.17	5.42
230		20.00	18.00	10.75	27.25	3.63		6.37	21.50	10.80	17.07	5.35
240	85.2	20.03	17.97	10.70	27.30	3.64	81.6	5.90	21.63	10.73	16.80	5.45
250		20.10	17.90	11.10	26.90	3.40		5.80	21.70	10.65	16.85	5.53

Circle readings.	CLAMP WEST.						CLAMP EAST.					
	Temperature of observatory.	Level direct.		Level reversed.		East end high.	Temperature of observatory.	Level direct.		Level reversed.		East end high.
		E.	W.	E.	W.			E.	W.	E.	W.	
°	°	div.	div.	div.	div.	div.	°	div.	div.	div.	div.	div.
260		20.23	17.83	10.77	27.30	3.53		5.80	21.70	10.55	16.87	5.56
270		20.30	17.80	11.00	27.10	3.40		5.87	21.70	11.00	16.55	5.35
280		20.90	17.33	10.83	27.40	3.25		6.00	21.43	11.10	16.33	5.17
290		20.47	17.73	10.60	27.60	3.57		5.97	21.30	11.05	16.25	5.13
300	86.0	20.80	17.60	11.00	27.40	3.30	81.1	5.57	21.43	10.93	16.10	5.26
310		20.83	17.60	10.90	27.55	3.36		5.60	21.53	10.95	16.18	5.29
320		21.00	17.43	10.63	27.80	3.40		5.63	21.30	10.67	16.25	5.31
330		20.40	18.20	10.80	27.83	3.71		5.47	21.50	10.53	16.47	5.49
340		20.80	17.90	10.65	28.05	3.63		5.40	21.40	10.80	16.00	5.30
349	86.0	20.60	18.20	10.87	27.93	3.66	80.2	5.50	21.40	10.47	16.43	5.47

At a later period it was satisfactorily ascertained that the eastern pivot was being constantly and almost uniformly elevated, as though the cement between the joints of the west pier, or the plaster securing its adjuncts, shrank more than those of its fellow. Such result was a probable consequence of the rapid erection of the piers, and at first the changes of level attracted little attention; but when months of hot and dry weather had passed and the plaster surrounding the plugs to which the Y's are screwed appeared indurated as the porphyry, the unceasing disturbance could no longer escape attention or be thus explained.

With time, the error accumulated to inconveniently great amounts, and within the ten months terminating with March, 1851, the support under the eastern pivot was lowered 49.3″, or on an average very nearly 5″ per month. Throughout the following month earthquakes of unusual violence rendered such repeated adjustments necessary that the amounts of alterations made were not always recorded; but it is quite certain that the eastern extremity of the axis was never *elevated* by the screw. During most of the winter (of the southern hemisphere) and until August, 1851, there was little disturbance of the horizontal line. In the last-named month a movement amounting to 5½″ took place in the opposite direction; but, beginning with the spring and until the close of the following autumn, the same uplifting of the eastern pier ensued, so uniform in its monthly amount that the change of error from this cause could be calculated to within a second or two. Of necessity the Y was lowered 45″ more in seven months, and then, during the remainder of our sojourn as throughout the preceding winter, the axis fluctuated about a horizontal line, with a tendency upward, of its western pivot.*

Indications of this progressive change are evident on inspection of the numbers in columns 7 and 13 of the preceding table, and, therefore, in the absence of data to

* When Dr. Moesta, the present director of the *Observatorio Nacional*, last wrote to me on the subject (in 1854), the phenomenon was still going on, and it would seem as if the native rock stratum on which the piers are based is slowly tilting from east to west.

show the actual amount of disturbance between the first and last levelings, it will be proper to divide the results into three groups each, viz, from 10° to 110°, from 120° to 240°, and from 250° to the end, and assume that the mean of each of these groups is the absolute level error with which to compare the separate levelings. In this way the numbers in columns 4 and 5 of the following table have been obtained.

Circle reading.	Mean level.		Difference from mean.		Mean errors.
	First series.	Second series.	First series.	Second series.	
°	div.	div.	div.	div.	div.
10			+0.34	—0.36	—0.02
20			— .13	.42	.28
30			.18	— .17	— .18
40			.12	+ .11	.00
50			— .07	.10	+ .02
60	2.91	5.44	+ .02	.06	.04
70			.11	.06	.09
80			.00	.14	.07
90			+ .21	.40	.32
100			— .08	.35	.14
110			.12	.16	+ .02
120			.31	.13	— .09
130			.32	.07	.13
140			— .48	19	— .15
150			+ .01	.19	+ .10
160			.01	+ .19	+ .10
169	3.22	5.51	.17	— .19	— .01
190			.05	.09	.02
200			+ .02	.08	.03
210			— .17	.13	— .15
220			+ .20	.09	+ .06
230			.41	.16	.13
240			+ .42	— .06	.18
250			— .07	+ .18	.06
260			+ .06	+ .21	+ .14
270			— .07	.00	— .04
280			— .22	— .18	.20
290			+ .10	.22	.06
300	3.47	5.35	— .17	.09	.13
310			.11	.06	.09
320			— .07	— .04	— .06
330			+ .24	+ .14	+ .19
340			.16	.05	.11
349			+0.19	+0.12	+0.16

Although the progression of the signs in the last column appears to indicate an irregularity which follows a sort of law, it may be wholly accidental, and if the bands bearing on the Y's are as nearly cylindrical, the numbers show conclusively that in the reduction of all ordinary observations the form of the pivots may be regarded perfect.

THE MICROSCOPES.

February 4, 1850.—Two microscopes were attached to each pier, and the arms being placed approximately in a horizontal position, the telescope was turned until the middle horizontal wire occulted its image seen in a basin of mercury placed vertically beneath the object glass. By means of the adjusting tools and whilst the circles were in this position, micrometer A, the northwest one, B, the southwest, C, the southeast, and D, the northeast, were made to read accurately 180°, 0°, 180°, 0°.

The runs of the microscopes were next examined. These were found to have been adjusted by the makers, so that at the then temperature and distances of the microscopes and circles, two revolutions of each micrometer head accurately corresponded with the intervals of the single circle divisions in the field of view. But as changes of temperature and other causes produce continual variations in the value of the micrometer revolution, so that two revolutions sometimes represent more and sometimes less than the two-minute spaces into which the circles are divided, a correction thereupon becomes necessary whenever we wish to reduce the measured to the true distance between the micrometer zero and the nearest circle division. The following table shows the amount of the error for *one* revolution of each micrometer head, as found at different subsequent periods, together with the mean of the four values of the ascertained error, the corresponding circle reading at which the determination was made and the temperature at the time. The result for each date is a mean of not less than three separate measurements with each micrometer.

As the angles measured by the micrometers at observations never amount to one entire revolution or sixty seconds of a great circle, the corrections for run applicable to any measurements are diminished proportions of the errors determined for given dates; or, if r represent the measured angle, and X the error of run, the applicable correction $= \frac{r\,X}{60''}$.

Errors of run of each microscope.

Date.		Circle reading.	Temp.	A.	B.	C.	D.	Mean.
1850.		° ′	°	″	″	″	″	″
Mar.	22	305 14	65.0	+ 0.34	— 1.19	+ 1.12	+ 0.19	+ 0.12
Apr.	1	28 49	64.2	0.62	— 0.25	1.43	+ 0.31	0.53
	5	349 08	66.0	0.67	0.00	1.17	— 0.33	0.38
	13	336 57	55.7	0.63	— 0.25	1.50	0.00	0.47
	17	336 57	63.0	0.60	+ 0.40	1.50	0.00	0.63
	19	336 57	57.3	0.75	— 0.17	1.37	0.00	0.49
	29	26 46	54.0	1.25	0.00	1.50	+ 0.50	0.81
May	1	26 46	58.0	0.70	+ 0.10	1.33	0.00	0.53
	7	335 23	62.5	+ 0.75	— 0.10	1.65	— 0.05	0.56
	31	345 57	43.0	— 0.25	0.20	0.95	0.12	0.10
June	5	349 08	52.3	0.30	— 0.26	0.90	0.16	0.05
	19	43 57	60.0	0.20	0.00	0.88	0.14	0.14
July	5	287 57	55.0	0.28	0.00	0.86	0.18	0.10
	20	347 39	43.2	— 0.44	+ 0.31	+ 1.06	— 0.25	+ 0.17

Errors of run of each microscope—Continued.

Date.	Circle reading.	Temp.	A.	B.	C.	D.	Mean.
	° ′	°	″	″	″	″	″
1850.							
July 26	27 06	46.7	0.00	0.00	+ 1.42	− 0.17	+ 0.31
Aug. 6	46 38	49.0	+ 0.10	− 0.06	1.00	+ 0.06	0.28
14	347 39	51.2	+ 0.06	0.12	1.20	0.00	0.29
31	312 10	43.2	0.00	0.27	1.25	0.00	0.25
Sept. 2	356 58	49.5	− 0.45	− 0.05	1.30	− 0.22	0.15
8	332 47	53.2	0.26	0.00	1.27	0.10	0.23
19	321 00	. .	0.12	0.00	1.18	0.16	0.23
25	327 37	49.4	− 0.06	0.00	1.24	− 0.12	0.27
Oct. 18	323 04	56.3	0.00	+ 0.10	1.25	0.00	0.29
20	353 22	71.0	+ 0.30	0.10	1.46	+ 0.26	0.53
27	354 04	65.5	+ 0.05	0.18	1.37	0.50	0.53
Nov. 1	302 56	56.5	− 1.40	0.30	2.75	0.10	0.44
5	354 31	73.8	1.24	0.16	1.36	0.24	0.13
20	353 47	72.0	1.10	+ 0.10	+ 1.88	0.16	+ 0.26
Dec. 8	289 57	78.1	0.70	− 0.26	− 0.26	0.20	− 0.26
9	355 20	55.3	0.71	0.27	0.28	0.17	0.27
12	355 20	62.0	1.00	0.27	0.27	0.13	0.35
13	343 04	68.0	0.78	0.25	0.15	0.18	0.25
31	344 46	70.6	− 0.56	− 0.30	− 0.12	+ 0.14	− 0.21
1851.							
Jan. 6	313 54	76.1	− 0.66	− 0.28	− 0.20	+ 0.24	− 0.23
14	319 36	61.3	0.33	0.87	+ 0.13	− 0.37	0.18
21	344 23	79.0	0.34	0.80	0.00	+ 0.40	0.19
27	344 48	69.4	0.30	0.76	0.00	0.38	0.17
Feb. 5	345 25	71.0	− 0.16	0.66	− 0.08	0.10	0.20
10	287 58	69.2	0.00	0.66	0.00	0.16	0.13
24	319 36	57.7	+ 0.22	0.70	− 0.17	0.12	0.13
27	22 25	65.7	+ 0.16	0.80	0.10	0.10	0.11
Mar. 7	25 12	59.0	0.00	− 0.46	0.20	0.25	0.10
16	25 12	56.5	− 0.20	+ 0.24	0.40	0.32	− 0.01
19	68 25	57.0	0.17	0.31	0.27	0.30	+ 0.04
29	340 32	59.9	− 0.10	+ 0.35	− 0.36	0.18	+ 0.02
Apr. 15	313 53	49.0	0.00	− 0.84	+ 0.27	+ 0.10	− 0.12
28	298 53	42.2	+ 0.16	0.76	0.33	− 0.18	0.11
May 4	20 22	58.5	− 0.10	0.98	0.40	− 0.26	0.24
12	333 51	48.0	0.00	1.09	0.44	+ 0.19	0.12
18	336 57	56.6	0.00	0.90	0.36	0.18	0.09
26	349 08	43.8	+ 0.06	1.00	0.28	0.08	0.15
June 8	334 31	51.0	0.16	0.86	0.37	0.18	0.04
13	55 50	48.9	+ 0.12	0.80	0.30	0.10	0.07
Aug. 17	19 14	49.1	0.00	1.00	0.28	0.37	0.09
31	23 46	48.0	− 0.17	0.76	0.18	0.22	0.13
Sept. 7	14 14	52.0	− 0.08	1.07	0.22	0.10	0.21
14	3 41	47.0	0.00	1.16	0.38	0.18	0.15
Oct. 5	312 10	47.0	+ 0.28	0.86	0.34	0.30	0.02
12	29 41	42.6	+ 0.18	− 0.80	+ 0.26	+ 0.16	− 0.05

Errors of run of each microscope—Continued.

Date.		Circle reading.	Temp.	A.	B.	C.	D.	Mean.
1851.		° ′	°	″	″	″	″	″
Oct.	17	336 24	60.4	+ 0.33	− 0.97	+ 0.41	+ 0.32	+ 0.02
Nov.	1	38 12	52.7	0.00	+ 0.18	0.38	0.28	0.21
	9	310 22	46.5	0.00	0.25	0.16	0.40	0.20
	16	9 47	53.5	− 0.08	0.46	0.08	0.20	0.17
	30	20 18	63.0	+ 0.16	0.52	0.24	0.32	0.31
Dec.	7	311 23	58.0	0.12	0.60	0.28	0.18	0.30
	22	334 56	64.5	+ 0.20	0.75	0.37	0.25	0.39
	26	343 04	60.0	0.00	0.50	0.45	0.30	0.31
	27	37 02	65.0	− 0.07	0.47	0.12	0.35	0.22
	31	43 50	60.5	− 0.12	+ 0.42	+ 0.10	+ 0.42	+ 0.21
1852.								
Jan.	8	334 35	52.5	− 0.40	+ 0.75	+ 0.50	− 0.20	+ 0.16
	12	334 35	53.0	− 0.37	0.70	0.44	0.16	0.15
	16	334 35	58.8	+ 0.10	0.95	0.40	0.18	0.32
	23	303 02	58.0	− 0.35	0.80	+ 0.55	− 0.13	0.22
	24	51 02	68.8	+ 0.30	1.25	− 0.10	+ 0.23	0.42
	27	302 38	60.3	− 0.08	0.38	− 0.33	0.33	0.08
	28	302 33	59.3	0.00	1.30	+ 0.30	+ 0.60	0.55
	29	302 27	54.0	0.00	0.68	0.42	0.00	0.28
	30	302 23	57.6	− 0.05	0.65	0.20	+ 0.40	0.30
	31	302 18	62.8	0.00	0.45	+ 0.50	0.25	0.30
Feb.	2	302 09	65.5	0.00	1.15	0.00	+ 0.37	0.38
	4	302 01	60.0	− 0.12	0.97	+ 0.43	0.00	0.32
	10	301 43	63.7	+ 0.63	1.28	− 0.25	− 0.43	0.31
Mar.	1	301 34	64.8	+ 0.23	1.05	0.00	− 0.37	0.23
	6	350 27	65.6	0.00	1.10	0.00	+ 0.34	0.36
	14	334 35	63.4	+ 0.20	1.00	+ 0.02	− 0.30	0.23
Apr.	26	352 39	51.0	0.30	0.60	0.19	0.25	0.21
May	30	350 26	69.0	0.35	0.55	0.23	− 0.35	0.20
June	2	350 26	64.0	0.20	0.70	0.43	+ 0.17	0.38
	16	180 00	51.4	0.33	0.60	0.18	− 0.23	0.22
July	5	334 35	51.0	0.35	0.60	+ 0.25	− 0.30	0.23
	7	334 35	60.0	+ 0.07	0.33	− 0.32	+ 0.30	0.10
	29	334 35	57.0	− 0.08	0.73	+ 0.20	0.35	0.30
Aug.	7	25 35	64.9	+ 0.18	0.68	0.45	+ 0.16	0.37
	11	180 00	53.0	0.32	0.57	0.23	− 0.26	0.22
	21	25 35	50.0	0.28	0.63	0.19	0.28	0.20
	31	25 35	48.0	+ 0.20	0.76	0.10	0.25	0.20
Sept.	3	25 35	56.1	− 0.07	0.70	0.23	− 0.10	0.19
	7	25 35	54.1	− 0.01	+ 0.66	+ 0.40	0.00	+ 0.26

Examination of the column "Mean" shows that the errors of run for an entire revolution of the micrometer head exceeded half a second of arc in a few instances only, and these were in the year 1850. The values of the errors generally represent portions of celestial space which are far within the optical capacity of the telescope to exhibit

INTERVALS OF THE WIRE SYSTEMS.

The Vertical System.

The wire systems introduced by Messrs. PISTOR & MARTINS were delicate and equal-sized spider threads, inserted at very uniform intervals. Those of the vertical system, *as seen* when looking to the south with the clamp arm of the telescope to the west, were numbered A, B, C, D, E, F, G; A being on the extreme right. Their distances from each other were determined from the transits of circumpolar stars. Very generally the transits of *β Hydri*, or *β Chamæleontis*, or both, were employed, though it occasionally became necessary to resort to the zone observations when accident caused derangement of the system, and a proper number of transits of these stars were not available. In a few instances the results of many transits of a quick moving star have been incorporated with others from circumpolar stars, weight being given the former inversely as their polar distances increased. Unfortunately, with every care, the derangements were of frequent occurrence; and, as the only facility for replacing broken wires was an ordinary reading glass held in the hand, the corresponding intervals of the restored series are not always so uniform as is desirable. When the first breakage took place, May 22, 1850, it was found impossible to obtain a spider that would spin a fiber which would bear the least tension; indeed, the insects invariably cut off the threads and fell to the floor rather than spin a line by which to descend.

After vainly experimenting with more than twenty insects of different species and sizes, there was no alternative but to substitute some other material until a supply could be obtained from the Observatory at Washington. The latter reached Santiago about the close of July of the same year. Meanwhile, the broken fibers were replaced with unspun threads of the finest India white silk, which formed no very bad substitute.

The determination of the value of the equatorial intervals between the wires which are given in the subjoined table was made in the following manner: At every complete transit observation selected for the purpose, the difference between each wire and the mean of the seven wires was taken, and either the mean of all the results relating to the same wire was supposed to apply to the mean of the polar distances of the star on the several days of observation, or, when the period embraced by the observations was long, each transit was computed independently. Then,

$$\text{Equatorial intervals in time} = \frac{\sin \text{ star's } P.\,D. \times \sin \text{ observed int.}}{15 \sin 1''}.$$

Equatorial intervals from mean.

Dates. From—	To—	A.	B.	C.	D.	E.	F.	G.
		s.	s.	s.	s.	s.	s.	s.
1850. Feb. 6 . .	1850. May 22 .	+50.901	+33.966	+16.978	−0.121	−16.923	−33.897	−50.906
May 22 . .	June 14 .	50.935	33.975	17.005	−0.147	16.919	33.895	50.953
June 19 . .	July 26 .	51.194	33.719	16.875	0.000	17.357	33.469	50.965
July 26 . .	Aug. 23 .	50.773	33.835	17.139	+0.023	17.033	33.933	50.804
Aug. 24 . .	Dec. 27 .	50.686	33.727	17.068	−0.031	16.993	33.586	50.867
Dec. 27 . .	1851. July 9 .	50.863	33.874	16.923	+0.056	16.943	33.927	50.849
1851. July 9 . .	Nov. 9 .	50.970	33.775	16.973	−0.039	16.954	33.944	50.860
Nov. 11 . .	1852. Jan. 10 .	50.896	33.871	16.946	0.037	16.948	33.924	50.876
1852. Jan. 10 . .	Apr. 23 .	51.234	33.761	17.126	0.053	17.135	34.041	50.903
Apr. 24 . .	June 3 .	51.261	33.831	17.127	0.067	17.172	34.051	50.931
June 4 . .	Sept. 8 .	+50.934	+33.866	+17.187	−0.015	−17.128	−34.002	−50.839

The Horizontal System.

When the meridian circle arrived, its horizontal system consisted of one stationary wire across the center of the diaphragm holding the vertical series, and five nearly equidistant and parallel ones inserted in a second diaphragm, movable perpendicular to the first by means of a micrometer screw. The interval between each pair of these horizontal wires was about 4¼′, or 17′ between the two extreme ones. With the telescope directed to the south and its clamp arm to the west, *as seen* in the field of view, they are numbered from below upwards, 1, 2, 3, 4, 5, number 3 occulting the stationary horizontal wire at $0^{rev.}$ of the micrometer head. By means of the circles, their absolute distances from each other were measured in the following manner:

The telescope being turned upon the trough of mercury, and wire 3 brought to perfect coincidence with the stationary wire by means of the slow-motion screw, these wires were brought directly over their images seen in the mercury, and a reading of the four microscopes, at this nadir point of the circles, was carefully made. The telescope and circles were then slowly moved by the tangent screw until the image of wire 1 was beneath wire 3 seen directly, and the four microscopes were again read. The images of 2, 4, and 5 were afterwards successively brought under 3, and the corresponding microscope readings of the circles were noted for each.

Conformably with a well-known law of optics, the angles through which the telescope had been turned for the several measurements are precisely equal to half the celestial arcs subtended by the respective wires. The earlier results obtained in this manner are given in the subjoined table, the measurements of the first and second series having been made on February 4, 1850, and of the third and fourth, eight days later.

Intervals of horizontal wires.

Series.	Micrometer readings at nadir point.	Circle readings at coincidence of 3 and 4.	Half angle.	Circle readings at coincidence of 3 and 5.	Half angle.	Circle readings at coincidence of 3 and 2.	Half angle.	Circle readings at coincidence of 3 and 1.	Half angle.
	″	° ′ ″	′ ″	° ′ ″	′ ″	° ′ ″	′ ″	° ′ ″	′ ″
1	A − 0.5	179 57 56.5	2 3.0	179 55 49.0	4 10.5	180 2 2.5	2 3.0	180 4 12.5	4 13.0
	B − 0.0	54.0	6.0	48.5	11.5	2.8	2.8	12.6	12.6
	C − 3.5	50.0	6.5	43.5	13.0	1.3	4.8	11.8	15.3
	D − 2.5	49.0	8.5	45.5	12.0	1.3	3.8	10.0	12.5
2	A − 4.0	50.2	5.8	45.5	10.5	7.0	11.0	11.7	15.7
	B − 2.7	50.2	7.1	44.3	13.0	6.0	8.7	11.8	14.5
	C − 4.8	46.0	9.2	39.8	15.4	2.0	6.8	7.5	12.3
	D − 1.0	49.0	10.0	42.8	16.2	3.3	4.3	11.2	12.2
3	A + 5.8	59.5	6.3	53.0	12.8	15.0	9.2	23.0	17.2
	B + 6.8	59.0	7.8	53.5	13.3	15.0	8.2	24.0	17.2
	C + 5.8	59.0	6.8	53.0	12.8	15.0	9.2	23.5	17.7
	D + 3.6	56.5	7.1	51.0	12.6	12.0	8.4	20.0	16.4
4	A + 5.8	58.0	7.8	51.2	14.6	15.0	9.2	21.5	15.7
	B + 6.8	59.5	7.3	52.0	14.8	15.0	8.2	21.5	14.7
	C + 5.8	58.5	7.3	51.0	14.8	15.5	9.7	21.8	16.0
	D + 3.6	179 57 57.0	2 6.6	179 55 49.2	4 14.4	180 2 13.6	2 10.0	180 4 18.0	4 14.4
Sums			113.1	. . . . . .	212.2	. . . . . .	117.3	. . . . . .	237.4
Means			2 7.07	. . . . . .	4 13.26	. . . . . .	2 7.33	. . . . . .	4 14.84
Intervals			4 14.14	. . . . . .	8 26.52	. . . . . .	4 14.66	. . . . . .	8 29.68

There being too great discordance between the results of the separate series to warrant their adoption as final, two others were made on March 4, as follows:

Intervals of horizontal wires.

Series.	Micrometer readings at nadir point.	Circle reading at coincidence of 3 and 4.	Half angle.	Circle reading at coincidence of 3 and 5.	Half angle.	Circle reading at coincidence of 3 and 2.	Half angle.	Circle reading at coincidence of 3 and 1.	Half angle.
	″	° ′ ″	′ ″	° ′ ″	′ ″	° ′ ″	′ ″	° ′ ″	′ ″
5	A + 18.5	179 58 12.5	2 6.0	179 56 5.0	4 13.5	180 2 26.5	2 8.0	180 4 34.5	4 16.0
	B + 34.0	28.5	5.5	19.7	14.3	42.5	8.5	49.5	15.5
	C + 51.5	45.0	6.5	36.5	15.0	60.0	8.5	67.5	16.0
	D + 40.0	32.5	7.5	24.2	15.8	46.5	6.5	54.5	14.5
6	A + 18.5	11.5	7.0	5.5	13.0	27.0	8.5	34.5	16.0
	B + 35.5	28.0	7.5	20.2	15.3	42.7	7.2	49.5	14.0
	C + 52.5	44.0	8.5	37.5	15.0	59.7	7.2	67.5	15.0
	D + 39.0	179 58 32.5	6.5	179 56 24.2	4 14.8	180 2 47.0	2 8.0	180 4 54.5	4 15.5
Sums			55.0	. . . . . .	116.7	. . . . . .	62.4	. . . . . .	122.5
Means			2 6.88	. . . . . .	4 14.59	. . . . . .	2 7.80	. . . . . .	4 15.31
Intervals			4 13.76	. . . . . .	8 29.08	. . . . . .	4 15.60	. . . . . .	8 30.62
Adopted intervals			4 14.00	. . . . . .	8 27.40	. . . . . .	4 14.98	. . . . . .	8 30.00

The *adopted* intervals are used with all observations to the 22d of May, 1850, inclusive. On turning the telescope object glass downward after the zone work for that night something was heard to fall within it, and, on examination, the stationary horizontal wire, together with Nos. 4 and 5, were found to have been broken by it. They were replaced by fibers of the India silk, before spoken of, and the intervals again measured in the same manner as the preceding. On the 29th of July following, a new system was inserted of the spider's thread received from Washington, and the distances between them were again ascertained. These last wires continued undisturbed until the 27th of December, 1850, at which time advantage was taken of necessary repairs making to the eyepiece to insert two others, one below No. 1, the second above No. 5. The intervals then determined, together with all the subsequent measurements to the same end, are given in the subjoined table. In it wire No. 1 of the early nomenclature has become No. 2, and No. 5 of the old is No. 6 of the enlarged system. For the purpose of exhibiting at one view the intervals used in reducing the whole number of zones, the results of all the measurements are embraced in the table, it being understood that No. 3 coincided with the stationary wire until the 27th of December, 1850; No. 4 with it after that date:

Adopted intervals of horizontal wires.

Period.		1.	2.	3.	5.	6.	7.
From—	To—						
		′ ″	′ ″	′ ″	′ ″	′ ″	′ ″
1850, Feb. 4 . .	1850, May 22 . .	. . . .	8 30.00	4 14.98	4 14.00	8 27.40	. . . .
May 22 . .	July 29 . .	. . . .	8 25.32	4 9.64	4 19.24	8 38.08	. . . .
July 29 . .	Dec. 27 . .	. . . .	8 26.26	4 14.24	4 15.66	8 31.28	. . . .
Dec. 27 . .	1851, July 14 . .	12 26.83	8 29.72	4 16.10	4 17.17	8 28.87	12 35.00
1851, July 14 . .	1852, Jan. 11 . .	12 29.70	8 23.13	4 13.70	4 14.17	8 30.67	12 36.30
1852, Jan. 12 . .	Apr. 23 . .	12 35.46	8 29.48	4 15.81	4 14.23	8 25.44	12 36.90
Apr. 24 . .	Sept. 13 . .	12 33.47	8 29.37	4 15.20	4 13.43	8 25.73	12 37.53

Value of the Micrometer.

The drumhead to the micrometer screw moving the horizontal system is about an inch in diameter, and is divided into one hundred equal parts. To determine the value of each of these parts, it was found from a mean of ten trials on the 4th of March, 1850, that the screw head made 36 revolutions and forty-three hundredths of a revolution whilst moving the horizontal system from coincidence of wire 1 (2 of the preceding table), and the stationary wire, to coincidence of wire 5 (6 of the table) and the stationary. Whence

$$1 \text{ Rev.} = \frac{8'\ 30.00'' + 8'\ 27.40''}{36.43} = 27.927''$$

$$1 \text{ Div.} = \frac{27.927''}{100} = 0.2793''$$

This screw was accidentally injured in September, 1850, when it became necessary to use in its place the one which moved a vertical wire, intended for collimating purposes, etc. The latter had been cut in the same die and properly admitted of the substitution. Subsequently Messrs. PISTOR & MARTINS sent other screws with matrices, for the insertion of which the old screw holes were drilled out. These worked satisfactorily to the close of the observations in September, 1852.

Supposing that they would probably have a different thread, the number of revolutions were counted whilst the micrometer head moved the diaphragm from coincidence of wire 3 with the stationary one, until No. 5 covered the same wire, and were found from a mean of 10 measures to be 18 revolutions and twenty-five hundredths of a revolution. The measures were made January 21, 1852. They give for the value of a revolution 27.942″, a result differing so little from the preceding that the latter (27.927″), being the measurement of a larger arc, has been adopted in the reduction of all the observations.

EXPLANATION OF THE PRINTED OBSERVATIONS.

On the left-hand page the *first* column contains the numbers for convenience of reference to the foot notes.

The *second* column contains the month and day, the latter supposed to commence with apparent noon at Santiago.

The *third* column contains the name of the object observed. With respect to the moon and planets, the limbs observed are always given, except in the few observations of the planet *Neptune*, whose disk was so small that we could more accurately estimate the bisection of its center. NP, SP, NF, SF, following the name of the planet, indicate north preceding, south preceding, north following, or south following limbs, respectively. Preference has been given to the designation of stars in the following order:

1. The Greek or italic letter of Bayer, with the name of the constellation to which it belongs, as given in Baily's Flamsteed.
2. Flamsteed's number, with the name of the constellation.
3. The number in the Catalogue of the British Association.
4. The number in Lacaille's Catalogue of 9,766 stars (L.), reduced at the expense of the British Association.
5. The number in the Histoire Céleste (H. C.) of Lalande, as published by the British Association.
6. The number of the Zone of Bessel (B. Z.), as given in the volumes of the Königsberg observations.

S. P., following the name of a star, indicates that it was observed below the pole.

The *fourth* column contains the estimated magnitude of the star observed. In order that the estimations of brightness by the several observers might be comparable, Lieutenant MACRAE and myself selected as standards well-known stars from the British Association Catalogue, ascending and descending in the scale of brightness to the extreme magnitudes embraced in it. The smallest star visible in the telescope of the meridian circle, when illuminated for ordinary work and the night was favorable, was set down as 12th magnitude. In like manner the judgment of Mr. PHELPS was based on the estimates of Lieutenant MACRAE.

The seven columns next following contain the seconds and decimals shown by the clock at which the object passed the several wires of the vertical system. The uniform practice in observing was to take a second from the clock face before the transit over each wire, and preserve the count by listening to the beats and estimating the fraction of the second from the previous beat to the instant of bisection of the wire. The minutes were usually recorded at the middle and last wires. Subsequent alterations in the hours and minutes of the recorded transits are very seldom if ever stated, but corrections of five seconds, or multiples of that number of seconds, occasioned by erroneous reading of the clock face, are always mentioned in the foot notes. No other changes have been permitted.

The *twelfth* column contains the observed transit reduced to the mean of the seven wires. When the object has not been observed at all the wires, the recorded times of transit have been reduced to the imaginary line passing through the mean of the seven by the equatorial intervals given on page XXIII. For stars more than 10° distant from the pole the correction to an imperfect transit was found by adding together the equatorial numbers for the observed wires, regard being had to their signs, then dividing by the number of wires and multiplying the quotient by the secant of the star's declination. The product was applied to the mean of the observed wires.

For stars within 10° of the pole

$$\text{Sin correction} = \text{Sin equatorial numbers} \times \text{Sec. Declination.}$$

Broken observations of the moon were reduced by the formula which Mr. Airy gives in the Greenwich observations:

$$\text{Correction} = \text{Equat. numbers} + \frac{3600 + \text{I}}{3600} \times \frac{\text{Sin. moon's geo. Z. D.}}{\text{Sin. moon's app. Z. D.}} \times \text{Sec. moon's geo. Dec.},$$

in which I represents in seconds of time the moon's increase of right ascension in one hour of longitude, as found in the tables of moon culminating stars of the Nautical Almanac.

For a planet

$$\text{Correction} = \text{Equatorial numbers} + \frac{3600 + \text{I}}{3600} \times \text{Sec. Declination.}$$

I being the hourly increase of right ascension in seconds of time, as given in the Nautical Almanac.

The *thirteenth* column contains the correction to be applied to the observed transit, on account of the inclination of the instrumental axis to the horizon. The error of level was usually determined both before and after the observations of every night, and from not less than three readings of the bubble in each position of the tube. The results so ascertained will be found at the bottom of the page, and the adopted error for the night in a table at the end of this introduction.

Besides the change apparently of a secular character, which has already been adverted to, there was a slight diurnal movement of the axis occasioned by the action of the sun upon the exposed ends of the metamorphic porphyry, whose prolonged strata composed the foundation of the instrument, and also by constant disturbances from earthquakes. The extent of the diurnal fluctuation within the hours embraced by the observations is shown at the foot of the page, but as it was usually small, and was influenced by both the number of cloudless hours and the direction and force of the atmospheric current during sunlight, no available law can be deduced.

The error of the level of the axis is regarded *positive* when the eastern end is too high, and the numerical correction in seconds of time applicable to the observed transit is

$$\text{Correction} = \frac{i}{15} \cos \text{Z. D. Sec. } \delta$$

in which i represents the inclination determined by the level, Z. D. the zenith distance, and δ the declination of the object, considered negative when south or below the pole.

The *fourteenth* column contains the correction applicable to the observed transit on account of the deviation of the line of collimation of the telescope from the plane described by the middle wire of the vertical system. The amount of this deviation was ascertained each night by the collimating eye piece and a basin of mercury, and generally both before and after the observations for the horizontality of the axis. For the purpose of comparing the errors of collimation obtainable by different methods, during the earlier portions of the work the instrument was also reversed after observing the transit of a circumpolar star over three wires. But from want of rigidity in the floor of the observatory, the method by reversal was attended with so much risk and trouble that it was abandoned after two or three trials had proved satisfactorily that the result derivable from the use of the level with the collimating eye piece and mercury was fully reliable.

The observations for level being completed, the telescope was pointed object glass downward, and being clamped in this position the slow-motion key was turned until the image of the stationary horizontal wire, seen in the basin of mercury, was occulted by the same wire seen directly. If the axis of the telescope was truly horizontal, and there was no error of collimation, then the image of the middle vertical wire was occulted by the actual wire. If the axis was horizontal, and the image of the middle vertical wire was seen at a distance from that wire, the error of collimation was equal to half this distance in angular measure. And when the image was at a distance, and the axis inclined to the horizon, the error of collimation was equal to half the angular distance corrected for the error of inclination. The distance was always measured by moving the micrometer screw of the horizontal wire system until, by estimation, wire 4 and its image, formed with wire D and its image, a square on each side of the stationary horizontal wire. These measures were repeated with wire 4 alternately to the north and south of the stationary horizontal wire until accordance of the results established their accuracy.

All the individual determinations will be found in a table at the end of the introduction, the adopted result for each night at the bottom of the page. In the former it will be seen that on many nights the collimation error obtained prior to the zone observations differed from that found afterwards. Most probably this was occasioned by our mode of observing. A small tin lantern, with glass front and sides, was used for reading the head of the micrometer screw that moved the horizontal system, and was unavoidably brought near to the eye end of the telescope after measuring the differential declination of every star. It stood, when not thus used, upon a box at the left hand of the observer (when looking south), and, as in reading, it was always held on that side of the eyepiece, the two sides of the telescope were, most probably, unequally heated, and slight flexure of the eye end ensued. But with this construction of the instrument a movable lantern is indispensable in zone work, and the difficulty was irremediable. Beyond doubt a portion of these apparent discrepancies are also due to the differently sensitive conditions of the observer's eye before commencing and at the close of a fatiguing series of observations. This appears the more evident, for the changes on different nights, when compared, develop no law, either as to direction or amount, a fact which influenced me to adopt mean values in the reduction of the transit observations from the individual determinations for each night, grouped in periods of greatest accordance, rather than the individual results themselves. To the values thus obtained, corrections are applied for the reduction of wires and diurnal aberration.

When the line of collimation falls to the east of the imaginary mean of wires, the sign of the correction for all stars above the pole is considered positive, and *vice versa;* and if the value of the deviation be represented by c and the declination of a star by δ, the correction for collimation applicable to the observed transit of that star will be found by the formula

$$\text{Correction} = c \text{ Sec. } \delta.$$

All the observations were made with the clamp arm of the instrument in one position.

The *fifteenth* column contains the numerical corrections applied to the observed transits on account of the deviations of the line of collimation from the true meridian. These corrections are regarded positive when the western pivot of the instrument is too far to the north. They have been obtained by the combinations of pairs of stars above and below the pole, differing, as nearly as possible, 12 hours in right ascension; north and south of the zenith; or, wholly to the north of the zenith, according as the Nautical Almanac list of 100 stars furnished suitable ones before and after the zone observations. To facilitate their computation, the following tables, containing factors for the errors of level, collimation, and azimuth, were prepared for all the stars of the Nautical Almanac list which were visible at Santiago, the factor for azimuth of any star being obtained from the expression

$$\text{Sin Z. D. Sec. } \delta.$$

Then, if the tabular difference of right ascension of the selected stars be represented by Δ, the observed interval of their transits corrected for level, collimation, and clock rate by Δ', the azimuth factors by n and n', the meridian error is obtained from the formula

$$\text{Meridian error} = \frac{\Delta - \Delta'}{n - n'}$$

All the computed meridian errors, together with the determining stars, will be found in a table at the end of this introduction.

[Here followed six tables of factors for correcting errors of level, collimation, and azimuth at Santiago; namely, one table for 83 standard stars whose places were given in the English Nautical Almanacs of that time; three tables, respectively, for the circumpolar stars *β Hydri*, *β Chamæleontis*, and *σ Octantis*, at both upper and lower culminations; and two tables, respectively, for each tenth minute of declination from $+0° 04'$ to $+51° 54'$, and from $-0° 04'$ to $-83° 54'$. All these are omitted because they can be reproduced at any time from the above formulæ.]

Three grave obstacles intervene to prevent a satisfactory knowledge of the absolute meridian errors. I shall refer to each separately.

First. Within 30° of the south pole there is but one star—*β Hydri*—so bright as the third magnitude, and as at the inferior culmination its zenith distance exceeds 68°, it could rarely be seen through the haze near the horizon. True, under especially favorable circumstances there were occasional observations of *α Tri. Aust.* at both its transits, the lower being within 12¼° of the horizon. But we could never avail of them, for the stability of the circle adjustments and going of the clock during twelve hours were so liable to be affected by earthquakes that all confidence in that method of determining the azimuth was destroyed. To many, indeed, probably to most, of these earthquakes* we were physically insensible, and suspected them only from otherwise inexplicable changes of our instruments. For our purposes, then, the existence of this subterranean disturbing cause rendered of comparative unimportance the absence of stars of suitable brightness and declination; and, for the same reason, it was impossible to assume the azimuth unchanged even between the commencement and the termination of any night's work, and therefore it became necessary to make, on all possible nights, a double series of observations, as well for the meridional line of the instrument as for the clock error.

Second. Except on one or two occasions, no careful reduction of any observations was ever made in Chile. Only a knowledge of the approximate instrumental errors was cared for, to prevent their accumulation to inconvenient amounts. This proves unfortunate; for early in their investigation, after the return of the expedition to the United States, it became evident that the almanac places of most of the stars south of the zenith of Santiago were erroneous, and the azimuths obtained by using stars above and below the pole differed essentially from the error deduced from the transits of two northern stars. The accuracy of the Nautical Almanac places having been accepted with entire confidence, we availed ourselves of southern stars on all possible occasions, and thus there are many nights when the azimuths are wholly dependent on such pairs. Had they been suspected, whilst using them as azimuth stars, we should have accumulated on other nights a sufficient number of independent observations from which to compute the possible corrections applicable to their right ascensions. In most cases the number of such independent observations afforded by our notebooks is, unfortunately, small. They have all, however, been reduced, and the results are given in subjoined tables. The method was as follows: Adopting the corrections to the Nautical Almanac right ascensions given in the volume of the Greenwich observations for

* We *recorded* more than 160 shocks. (See Vol. I, pp. 515–517.)

1854 (pp. xxxi–xxxiii), a preliminary reduction of every night's observations was made for the purpose of obtaining the clock rate. The correction for rate was then applied to every star of the Nautical Almanac standard list which had been observed within an hour of the one whose right ascension was the subject of investigation, and the azimuthal and clock errors were computed anew. Stars that differed much more than one hour in right ascension could not be availed of without risk of introducing an appreciable error, dependent on the *third* obstacle to satisfactory knowledge of the azimuth, and which will be discussed presently. The names of the stars used in determining the meridional deviations are also given, and, where more than one pair has been employed, the azimuth adopted in the reduction is a mean of the partial results by weights derived from the sums of the azimuth factors for the several pairs.

β Hydri was the first star the accuracy of whose right ascension was investigated. Of this there are forty-three observations on nights when there are suitable pairs of northern stars from which to determine the meridian error. Giving to the resulting Nautical Almanac errors of each night a weight proportionate to the number of wires over which the star was observed, the mean correction to the tabulated place is -0.114^s. But as a portion of these results are very discrepant, on applying the criterion of Professor Peirce,* I find that the observations of December 9, 1850, and June 15, 1852, should be rejected. The correction then becomes -0.158^s, which is applied to the tabulated place, and *β Hydri* is used in finding azimuth errors for other stars; and so, successively, as the correction is ascertained for each star, that corrected place is availed of to determine the tabular errors of any other star.

* Astronomical Journal, No. 45.

Computed Corrections to Nautical Almanac Right Ascensions.

β HYDRI.

Date.		No. of wires.	Determining stars.	N. A.—Obs'n.
				s.
1850.—Apr.	10	6	β Leonis and β Corvi	+0.862
	12	7	β Leonis and β Corvi	.695
	13	7	β Corvi and 12 Canum Venaticûm	+ .955
	14	7	β Leonis, β Corvi, and 12 Canum Venaticûm	— .203
	15	4	β Corvi, 12 Canum Venaticûm, and α Virginis	+ .805
	17	7	12 Canum Venaticûm and α Virginis	— .217
	18	7	β Corvi and 12 Canum Venaticûm	+0.866
	19	7	β Corvi, 12 Canum Venaticûm, and α Virginis	+1.674
	20	7	β Corvi and 12 Canum Venaticûm	—0.304
	21	7	β Corvi and 12 Canum Venaticûm	+ .544
	27	7	α Leonis and β Corvi	+ .507
Oct.	26	7	α Andromedæ and β Ceti	— .199
Dec.	2	7	α Andromedæ and β Ceti	+ .406
	5	7	α Piscis Australis, α Pegasi, α Andromedæ, and β Ceti	0.339
	9	7	α Andromedæ and β Ceti	*1.984
	12	3	α Andromedæ and β Ceti	0.502
	13	7	α Andromedæ and β Ceti	+ .217
1851.—Mar.	19	7	δ Leonis and δ Hydræ et Crateris	—0.947
	25	7	δ Leonis and δ Hydræ et Crateris	.890
June	5	7	δ Leonis, δ Hydræ et Crateris β Leonis, and β Corvi	.493
	15	7	δ Leonis, β Leonis, β Corvi, and α Virginis	.830
	18	7	β Corvi and 12 Canum Venaticûm	— .569
	28	7	β Corvi and 12 Canum Venaticûm	+0.103
Nov.	3	7	α Andromedæ and β Ceti	1.168
	16	7	α Piscis Australis and α Pegasi	+1.105
	25	4	α Piscis Australis and α Pegasi	—0.526
	30	7	α Piscis Australis and α Andromedæ	.539
Dec.	2	7	α Piscis Australis, α Pegasi, α Andromedæ, and γ Pegasi	— .076
	3	7	α Piscis Australis and α Andromedæ	+ .468
	5	7	α Andromedæ and β Ceti	—0.313
	8	7	α Andromedæ and β Ceti	+1.084
	12	7	α Andromedæ and β Ceti	+0.388
	13	7	α Andromedæ and β Ceti	— .040
	15	7	α Andromedæ and β Ceti	— .071
	16	7	α Andromedæ and β Ceti	+ .988
	17	7	γ Pegasi and β Ceti	— .104
1852.—Apr.	26	7	β Corvi and 12 Canum Venaticûm	— .498
May	25	7	β Corvi and 12 Canum Venaticûm	+ .087
	26	7	δ Leonis and δ Hydræ et Crateris	.540
	27	7	δ Hydræ et Crateris, β Leonis, and β Corvi	+0.451
	28	7	δ Leonis, δ Hydræ et Crateris and β Corvi	—1.424
June	3	7	α Virginis and η Ursæ Majoris	0.034
	15	7	β Corvi and η Ursæ Majoris	*—3.485

* Rejected by Peirce's criterion.

Adopted mean *correction* by weight,— $0.158^s \pm 0.107^s$.

Computed corrections to N. A., right ascensions—Continued.

α ERIDINI.

Date.		No. of wires.	Determining stars.	N. A.—Obs'n.
				s.
1850.—Oct.	20	7	β Chamæleontis and β Hydri	+0.096
Dec.	17	7	β Hydri and θ¹ Ceti	— .304
	27	7	θ¹ Ceti and α Arietis	+ .356
1851.—Nov.	16	7	α Andromedæ, β Chamæleontis, and β Hydri	.099
	30	7	α Andromedæ, β Chamæleontis, and β Hydri	+ .024
Dec.	8	7	α Andromedæ and β Hydri	— .371
	16	7	α Andromedæ and β Hydri	.232
	17	7	γ Pegasi and β Hydri	.084
	20	7	β Hydri and θ¹ Ceti	.161
	22	7	β Hydri and θ¹ Ceti	.027
1852.—Sept.	8	7	β Hydri, β Ceti, and α Arietis	—0.161

Adopted mean *correction* by weight, $+0.070^s \pm 0.063^s$.

α ARGUS.

Date.		No. of wires.	Determining stars.	N. A.—Obs'n.
1850.—Mar.	21	7	α Columbæ and μ Geminorum	—0.460
	24	7	β Tauri, α Columbæ, and μ Geminorum	— .271
	29	7	μ Geminorum and α Canis Majoris	+ .129
Dec.	10	5	μ Geminorum, α Canis Majoris, ε Canis Majoris, and δ Geminorum	+ .116
1851.—Feb.	27	7	α Orionis and ε Canis Majoris	— .408
Mar.	1	7	β Tauri and ε Canis Majoris	.318
	27	7	μ Geminorum, α Canis Majoris, ε Canis Majoris, and δ Geminorum	— .248
Dec.	25	7	μ Geminorum and ε Canis Majoris	+ .205
	26	7	μ Geminorum and α Canis Majoris	— .206
	31	7	μ Geminorum and α Canis Majoris	.146
1852.—Jan.	2	7	μ Geminorum and α Canis Majoris	.320
	14	7	μ Geminorum and ε Canis Majoris	.110
Feb.	7	7	μ Geminorum and α Canis Majoris	— .098
	19	7	β Tauri and α Columbæ	+ .096
	20	7	β Tauri and α Canis Majoris	— .405
	21	7	α Canis Majoris and α² Geminorum	.111
	24	7	β Tauri and α Canis Majoris	.089
	25	7	β Tauri and α Canis Majoris	.196
	27	7	α Canis Majoris and β Geminorum	— .248
Mar.	1	7	α Columbæ, μ Geminorum, ε Canis Majoris, and β Geminorum	+ .086
	11	7	α Columbæ, μ Geminorum, ε Canis Majoris, and β Geminorum	+ .213
	15	7	α Columbæ and μ Geminorum	.385
	16	7	μ Geminorum and α Canis Majoris	.048
	25	7	ε Canis Majoris and β Geminorum	+ .299
	29	7	μ Geminorum and α Canis Majoris	— .054
	30	7	μ Geminorum and α Canis Majoris	— .162
	31	7	μ Geminorum and ε Canis Majoris	+0.335

Adopted mean *correction* by weight, $+0.072^s \pm 0.046^s$.

Computed corrections to N. A., right ascensions—Continued.

ι ARGUS.

Date.		No. of wires.	Determining stars.	N. A.—Obs'n.
				s.
1850.—May	7	7	15 Argus and ε Leonis	—0. 109
1851.—Mar.	16	7	α² Geminorum and 15 Argus	— . 179
	23	7	δ Geminorum and 15 Argus	+ . 005
	29	7	α Leonis and η Argus	— . 184
1852.—Jan.	18	7	β Geminorum and 15 Argus	. 299
	19	7	β Geminorum and 15 Argus	. 457
	21	7	α² Geminorum and 15 Argus	. 055
Feb.	16	7	ε Hydræ and ι Ursæ Majoris	. 199
	17	7	ι Ursæ Majoris and α Hydræ	. 189
Mar.	7	7	β Geminorum and 15 Argus	— . 400
	19	7	δ Geminorum, 15 Argus, and ε Hydræ	+ . 409
	20	7	α Hydræ, α Leonis, and η Argus	. 112
	23	7	δ Geminorum, ε Hydræ, and η Argus	. 351
	26	7	ε Leonis and η Argus	+ . 108
June	23	7	ε Canis Majoris and β Geminorum	— . 130
	30	7	β Geminorum and 15 Argus	+0. 343

Adopted mean *correction* by weight, + 0.055ˢ ± 0.063ˢ.

η ARGUS.

Date.		No. of wires.	Determining stars.	N. A.—Obs'n.
1850.—Apr.	27	7	β Chamæleontis and β Hydri	—0. 147
May	24	7	β Chamæleontis and β Hydri	*+ . 901
June	5	7	α Leonis and β Corvi	. 593
1851.—Mar.	3	7	δ Leonis and δ Hydræ et Crateris	+ . 329
	18	7	δ Leonis and δ Hydræ et Crateris	— . 034
	20	7	δ Leonis and δ Hydræ et Crateris	+ . 124
	24	7	δ Leonis and β Chamæleontis	. 559
	29	7	δ Leonis and δ Hydræ et Crateris	+ . 549
Apr.	1	7	δ Leonis and δ Hydræ et Crateris	— . 307
May	17	7	β Chamæleontis and β Hydri	— . 085
	18	7	β Chamæleontis and β Hydri	+ . 087
	26	7	β Chamæleontis and β Hydri	— . 178
June	5	7	β Chamæleontis and β Hydri	— . 195
1852.—Apr.	21	7	δ Leonis and δ Hydræ et Crateris	+ . 160
May	22	7	β Chamæleontis and β Hydri	. 355
	26	7	β Chamæleontis and β Hydri	+0. 151

* Rejected by Peirce's criterion.

Adopted mean *correction* by weight, — 0.131ˢ ± 0.077ˢ.

Computed corrections to N. A., right ascensions—Continued.

β CHAMÆLEONTIS.

Date.		No. of wires.	Determining stars.	N. A.—Obs'n.
				s.
1850.—Apr.	3	7	β Corvi and 12 Canum Venaticûm	—0.514
	10	7	β Hydri and β Corvi	— .416
	12	6	β Hydri and β Corvi	+ .231
	13	7	β Hydri and 12 Canum Venaticûm	.145
	14	7	β Hydri and 12 Canum Venaticûm	+ .190
	15	7	β Hydri, β Corvi, 12 Canum Venaticûm, and α Virginis	— .100
	16	7	β Hydri, β Corvi, and 12 Canum Venaticûm	— .365
	17	7	β Hydri and 12 Canum Venaticûm	+ .333
	18	7	β Hydri and 12 Canum Venaticûm	.638
	19	7	β Hydri and 12 Canum Venaticûm	+ .743
	20	7	β Corvi and 12 Canum Venaticûm	— .595
	21	7	β Hydri, β Corvi, and 12 Canum Venaticûm	+ .307
	25	7	δ Leonis and δ Hydræ et Crateris	— .351
	27	7	β Hydri, β Corvi, and α Virginis	+ .100
May	15	7	δ Hydræ et Crateris and β Hydri	+ .187
	24	7	β Leonis, β Hydri, and 12 Canum Venaticûm	— .142
Sept.	30	7	β Hydri, β Corvi, and α Virginis	— .175
Dec.	5	7	α Andromedæ, β Hydri, and β Ceti	+ .056
	9	7	α Andromedæ and β Hydri	— .849
1851.—Mar.	19	7	β Leonis and β Hydri	.449
	25	7	δ Hydræ et Crateris and β Hydri	— .207
May	16	7	β Corvi and 12 Canum Venaticûm	+ .699
	17	7	β Leonis and β Hydri	—0.571
	18	7	β Leonis and β Hydri	*1.224
	26	7	δ Hydræ et Crateris, β Hydri, and β Corvi	0.163
	31	7	β Leonis, β Hydri, and β Corvi	— .343
June	5	7	δ Hydræ et Crateris, β Hydri, and β Corvi	+ .085
	8	7	β Leonis, β Hydri, β Corvi, and α Virginis	— .185
	15	7	δ Leonis and β Hydri	.381
	18	7	β Hydri, β Corvi, and 12 Canum Venaticûm	.375
	28	7	β Hydri, β Corvi, and 12 Canum Venaticûm	.136
Oct.	1	7	ι Piscium, α Andromedæ, and β Hydri	— .232
Nov.	3	7	α Andromedæ, β Hydri, and β Ceti	+ .239
	16	7	α Andromedæ, β Hydri, and θ¹ Ceti	— .170
	25	4	α Piscis Australis, α Pegasi, ι Piscium, and β Hydri	+ .303
	30	7	α Andromedæ and β Hydri	.155
Dec.	3	7	α Andromedæ and β Hydri	.210
	4	7	ι Piscium, α Andromedæ, and β Hydri	.115
	5	7	β Hydri and β Ceti	.331
	6	4	α Andromedæ and β Hydri	+ .689
1852.—Mar.	5	7	δ Leonis and β Hydri	— .307
	6	4	β Hydri, β Corvi, and 12 Canum Venaticûm	.694
Apr.	3	7	β Hydri and β Corvi	— .575
	5	7	α Leonis, β Hydri, and β Corvi	+ .156
	26	7	β Hydri, β Corvi, and 12 Canum Venaticûm	—0.038

* Rejected by Peirce's criterion.

Computed corrections to N. A., right ascensions—Continued.

β CHAMÆLEONTIS—Continued.

Date.		No. of wires.	Determining stars.	N. A.—Obs'n.
				s.
1852.—May	22	7	δ Leonis, β Leonis, and β Hydri	— 0.094
	25	4	β Hydri and 12 Canum Venaticûm	— .526
	26	7	δ Leonis and β Hydri	+ .478
	27	7	δ Hydræ et Crateris, β Leonis, β Hydri, and β Corvi	— .418
	28	7	δ Leonis, δ Hydræ et Crateris, β Hydri, and β Corvi	+ .160
June	3	7	β Hydri, α Virginis, and η Ursæ Majoris	.026
	15	7	β Hydri, β Corvi, α Virginis, and η Ursæ Majoris	.330
	22	7	12 Canum Venaticûm, α Virginis, and η Ursæ Majoris	+0.304

Adopted mean *correction* by weight, + 0.063ˢ ± 0.053ˢ.

α¹ CRUCIS.

Date.		No. of wires.	Determining stars.	N. A.—Obs'n.
1850.—Mar.	28	7	β Leonis, β Chamæleontis, and β Corvi	—0.420
	30	7	β Chamæleontis and β Corvi	.316
Apr.	1	7	β Leonis and β Chamæleontis	.046
	3	7	β Leonis, β Chamæleontis, β Corvi, and 12 Canum Venaticûm	.367
	5	7	β Leonis, β Chamæleontis, and β Corvi	.239
	6	4	β Leonis and β Chamæleontis	*.714
June	20	7	β Leonis, α Virginis, and β Centauri	.294
July	5	5	β Chamæleontis and β Corvi	—0.235

* Rejected by Peirce's criterion.

Adopted mean *correction* by weight, + 0.275ˢ ± 0.045ˢ.

β CENTAURI.

Date.		No. of wires.	Determining stars.	N. A.—Obs'n.
1850.—Apr.	29	7	α Bootis and α² Centauri	—0.151
May	1	7	α Bootis and α² Centauri	.235
	13	7	α Bootis and α² Centauri	.189
	14	7	α Virginis and η Ursæ Majoris	.272
	16	6	α Virginis, η Ursæ Majoris, α Bootis, and α² Centauri	.111
June	20	7	η Bootis and α² Centauri	.460
July	10	7	α Bootis and α² Centauri	.042
	13	7	α Virginis and α² Centauri	.355
	18	3	α² Centauri and ε Bootis	.512
	19	7	α Bootis, α² Centauri, α² Libræ, and β Libræ	.075
	20	7	α² Centauri, α² Libræ, and α Coronæ Borealis	.066
1851.—Mar.	28	7	α² Centauri and ε Bootis	.091
Apr.	30	7	α² Centauri and ε Bootis	— .077
May	1	7	α² Centauri and ε Bootis	+ .040
	26	7	β Chamæleontis and β Hydri	— .247
June	11	7	η Bootis and α² Centauri	— .213
	26	7	η Bootis, α² Centauri, α Coronæ Borealis, and α Serpentis	+0.130

Computed corrections to N. A., right ascensions—Continued.

β CENTAURI—Continued.

Date.		No. of wires.	Determining stars.	N. A.—Obs'n.
				s.
1852.—June	2	7	β Chamæleontis and β Hydri	+0.262
	3	7	β Chamæleontis and β Hydri	—.083
	15	7	β Corvi and a^2 Centauri	+.048
	23	7	β Chamæleontis and 12 Canum Venaticûm	—.023
	24	7	a Bootis and $β^1$ Scorpii	—.697
July	6	7	η Ursæ Majoris and a Virginis	+0.160

Adopted mean *correction* by weight, + 0.132ˢ ± 0.045ˢ.

a^2 CENTAURI.

Date.		No. of wires.	Determining stars.	N. A.—Obs'n.
1850.—May	16	7	a Virginis and η Ursæ Majoris	+0.151
	21	7	β Corvi and ε Bootis	+.066
	25	7	ε Bootis and a^2 Libræ	*—.480
	31	7	ε Bootis, a^2 Libræ, a Coronæ Borealis, and $β^1$ Scorpii	+.085
June	2	7	ε Bootis and a^2 Libræ	.045
July	6	7	12 Canum Venaticûm and a Scorpii	.087
	12	7	ε Bootis, β Libræ, a Coronæ Borealis, and a Scorpii	.260
	13	7	β Corvi, 12 Canum Venaticûm, a Virginis, and a Bootis	.370
	18	7	ε Bootis and a Scorpii	.218
	19	7	a Coronæ Borealis and a Scorpii	.435
	20	7	a^2 Libræ and a Coronæ Borealis	.274
1851.—Apr.	2	7	ε Bootis and a^2 Libræ	+.042
May	16	7	β Chamæleontis and 12 Canum Venaticûm	—.084
	31	7	β Chamæleontis and β Hydri	+.089
June	12	7	a Bootis and a Scorpii	.371
	14	7	ε Bootis and a^2 Libræ	.561
	16	7	ε Bootis and a^2 Libræ	.408
	26	7	β Corvi, η Bootis, a Coronæ Borealis, and a Scorpii	*.704
1852.—Apr.	19	7	ε Bootis and a^2 Libræ	.289
May	5	7	ε Bootis and a^2 Libræ	.230
June	3	3	β Chamæleontis and β Hydri	.275
	15	7	β Chamæleontis and β Hydri	.001
Aug.	6	7	ε Bootis and a^2 Libræ	+0.243

* Rejected by Peirce's criterion.

Adopted mean *correction* by weight, — 0.202ˢ ± 0.036ˢ.

Computed corrections to N. A., right ascensions—Continued.

α TRIANGULI AUSTRALIS.

Date.		No. of wires.	Determining stars.	N. A.—Obs'n.
				s.
1850.—June	4	7	α Coronæ Borealis and α Scorpii	+0.425
	11	7	α Coronæ Borealis and α Scorpii	— .120
	12	7	α Coronæ Borealis and α Scorpii	+ .486
July	6	7	α Coronæ Borealis and α Scorpii	— .040
	18	7	ε Bootis and α Scorpii	.291
	19	7	α^2 Centauri and α Coronæ Borealis	.228
	20	7	α^2 Centauri and α Coronæ Borealis	.165
	26	7	α Scorpii and β Draconis	— .164
Aug.	17	7	α Scorpii and α Herculis	+ .320
	26	7	α Ophiuchi and σ Octantis	+ .360
	29	7	α Herculis and σ Octantis	— .026
1851.—Feb.	10	7	α Scorpii and α Lyræ	.220
	13	7	γ^1 Eridani and α Aurigæ	.638
	21	5	α Aurigæ, β Tauri, α Leoporis, and α Columbæ	.783
	22	7	α Aurigæ and α Columbæ	.252
June	26	7	α Coronæ Borealis and σ Octantis	— .278
	27	7	α Scorpii and α Herculis	+ .522
	28	7	α Coronæ Borealis and α Scorpii	.487
Aug.	12	7	α Herculis and σ Octantis	.431
	22	7	α Scorpii and α Herculis	+ .543
	26	7	α Scorpii, α Herculis, and σ Octantis	— .105
	27	7	α Ophiuchi and σ Octantis	+ .407
1852.—Feb.	12	7	η Tauri and α Columbæ	.344
June	11	6	α Herculis and γ Draconis	+ .721
July	24	7	μ^1 Sagittarii and α Lyræ	— .363
Aug.	24	7	α Scorpii and α Herculis	—0.237

Adopted mean *correction* by weight, —0.049ˢ ± 0.078ˢ.

α PAVONIS.

Date.		No. of wires.	Determining stars.	N. A.—Obs'n.
1850.—Sept.	19	7	α Cygni and α Piscis Australis	+0.300
Oct.	25	6	α Cygni and β Aquarii	.537
	26	7	α^2 Capricorni and α Cygni	.649
Nov.	28	7	α Pegasi and α Piscis Australis	.473
1851.—Aug.	31	7	μ^1 Sagittarii and α Lyræ	.441
Sept.	4	7	α^2 Capricorni, ζ Pegasi, and α Piscis Australis	.325
Oct.	5	7	61^1 Cygni, β Aquarii, α Pegasi, and α Piscis Australis	.509
	12	7	61^1 Cygni and α Piscis Australis	.774
	21	6	α^2 Capricorni and α Cygni	.488
	23	6	α^2 Capricorni and α Cygni	.911
	24	5	α^2 Capricorni and α Cygni	.641
	27	7	α^2 Capricorni, 61^1 Cygni, ζ Cygni, and β Aquarii	.384
	29	7	α Cygni and β Aquarii	.190
	31	6	α^2 Capricorni and α Cygni	+0.471

Adopted mean *correction* by weight, —0.500ˢ ± 0.052ˢ.

Computed corrections to N. A., right ascensions—Continued.

α GRUIS.

Date.		No. of wires.	Determining stars.	N. A.—Obs'n.
				s.
1850.—Sept.	2	7	ζ Pegasi and α Piscis Australis	—0.111
	10	7	α Pegasi and α Piscis Australis	+ .220
	19	3	ζ Cygni and α Piscis Australis	— .190
Nov.	7	7	ζ Cygni and β Aquarii	— .062
	21	7	α Pegasi and α Piscis Australis	+ .145
1851.—Aug.	14	7	61[1] Cygni and α Piscis Australis	.058
	28	7	ζ Pegasi and α Piscis Australis	+ .110
Sept.	1	7	61[1] Cygni and β Aquarii	— .113
Oct.	12	7	61[1] Cygni and α Piscis Australis	— .023
	17	7	α Cygni and β Aquarii	+ .071
Nov.	12	7	ζ Cygni and β Aquarii	.030
	15	7	ζ Pegasi and α Piscis Australis	.238
	17	7	ζ Pegasi and α Piscis Australis	.020
	18	7	α Pegasi and α Piscis Australis	.231
	21	6	ζ Pegasi, α Pegasi, and α Piscis Australis	+0.141

Adopted mean *correction* by weight, —0.063ˢ ± 0.034ˢ.

Third. The action of the sun caused a lateral diurnal oscillation of the northern extremity, if not the whole hill of Santa Lucia, in the same direction as the apparent motion of the former; so that from the commencement to the close of observations there was a relative motion of the western pier to the north of the eastern one, more or less rapid according to the season of the year, the state of the atmosphere, the temperature which had prevailed during the day, and the hour of commencement and velocity of the cold current of air from the Andes in the evening.

An account of the geological structure of the hill has already been given, but it is proper to repeat here what is said, page XXXII, Vol. III, and subjoin a diagram for readier appreciation. "Santa Lucia is a solid mass of rock. Its horizontal projection is an oval, some 1,300 feet long, from NNE. to SSW., and 500 feet in its greatest transverse diameter. Its highest pinnacles, 200 feet above the city, as well as many others, are columnar, and at a little distance closely resemble basalt. Some of them are vertical; a few are horizontal; most of them, as do also its strata, stand at every inclination toward the east, but not one of them dips to the west. The slope is tolerably regular from the summit to the north and south extremes, though that of the southern portion is the most abrupt and broken. Partially covered with decomposed rock and scanty vegetable mold, its eastern face has an inclination not differing greatly from 45°. The western is precipitous—a bare wall of nearly black porphyry, with occasional injected veins of quartz."

During the forenoon the decomposed rock and vegetable matter of the eastern slope of the hill measurably shields the porphyry from solar action, but from the time that the sun passes the meridian its rays become effective on the extremities of the prisms, and as the dilatation and contraction of crystalline rocks is relatively greater in the direction of the minor than in that of the major axis, it follows that the NW. extremities of the columns must experience a lateral torsion whose amplitudes are dependent on the ranges of temperature, and therefore the maximum oscillation will take place on the day when the extremes of temperature of the rock are greatest.

To determine the law of this oscillation accurately requires a knowledge of the temperatures of the rock strata at the instants of observing the azimuth stars, and these data we do not possess. We have only the corresponding air temperatures, as shown by a thermometer suspended 4 or 5 feet above the surface of the ground and shielded from reflected and radiant heat. If the few observations made at our residence before the black bulb thermometer was broken be taken as approximative, we know from them that at 3 p. m. in the summer months (December, January, and February) the surface rock must have been constantly heated to at least 120° F., and have cooled below 60° before daylight next morning. But the probabilities are that the day temperature was often considerably higher, for we not unfrequently found the rocks on Santa Lucia so hot that we could not venture to rest the hand upon them.

[Here Captain GILLISS's manuscript ends, presumably because he wished to await the result of a further investigation of the azimuth errors, which he had intrusted to Prof. FREDERICK G. HESSE, U. S. N. The paper prepared by that gentleman was completed prior to August, 1862, and is as follows:]

INVESTIGATION OF THE AZIMUTHAL ERRORS OF THE MERIDIAN CIRCLE.

By Prof. FREDERICK W. HESSE, United States Navy.

The meridian errors obtained from two, three, or more pairs of stars before and after zone (during one night's observation) and for the whole period, have established the following undoubted facts, viz:

(1) The azimuthal deviations from the meridian are subjected to a continuous change from sundown, or some time before, to six or more hours after that time.

(2) This change is greatest at or near sundown, reaching a minimum six or more hours later.

(3) The change varies with the different months of the year, being greatest during the summer and least during winter.

In order to trace the law, or to arrive approximately at the curve of change for each month, it becomes necessary not only to find during one night's observations at different periods the deviations from the meridian, but to know also the time corresponding to the computed deviation.

The probable time corresponding to the deviation from the meridian obtained by the formula $\frac{\Delta - \Delta'}{n' - n}$ will in general be the mean of the transits of the observed stars, provided these deviations from the true meridian are due only to accidental causes, such as the imperfection of the instrument or the degree of accuracy of the observed transits; but if the same is subjected to a continuous change under the influence of a well recognized cause, the time which corresponds to a value found by the above formula will be a function dependent on the difference of right ascensions of the selected stars, their azimuth factor and the law of change.

This law being unknown, it was assumed that the variation within a short period of time, not exceeding one hour, was proportional to the time. This assumption is justified from the nature of the projected curves of meridian errors, their variations from a straight line during an interval of less than one hour being very small.

If $\varkappa$ and $\varkappa'$ represent the two meridian errors, it follows that

$$\varkappa' = \varkappa + \varDelta\delta, \text{ provided } \varDelta < 60^{m}$$

where $\varDelta$ denotes the difference of right ascension of the selected stars and δ the increase of the meridian error during one minute of time.

If we represent as before the tabular difference of right ascension of the selected stars by $\varDelta$, the observed interval of their transits corrected for level, collimation, and clock rate by $\varDelta'$, the azimuth factors by n and n', it follows that

$$\varDelta - \varDelta' = \varkappa' n' - \varkappa n$$

and substituting for $\varkappa'$ its equivalent, $\varkappa + \varDelta\delta$

$$\varDelta - \varDelta' = \varkappa(n' - n) + \varDelta\delta n'$$

Therefore

$$\varkappa + \frac{\varDelta n'}{n' - n}\delta = \frac{\varDelta - \varDelta'}{n' - n}$$

but as $\varkappa$ denotes the deviation from the meridian at the time α of the transit of the first star observed, $\alpha + \frac{\varDelta n'}{n' - n}$ will be the time corresponding to the meridian error $\frac{\varDelta - \varDelta'}{n' - n}$.

A table was thus computed for the values of $\alpha + \frac{\varDelta n'}{n' - n}$ for the different combinations of stars used in obtaining the meridian errors.

In order to trace, if possible, a law representing the increase of the meridian errors, it was important to have them all referred to a common origin of time, bearing a relation to the cause of the change itself. The suggestion given in another part of this volume with reference to the cause of the disturbance, and moreover the relation which these changes before and after zone bear to each other, suggest the several times of sundown as a natural origin.

For this purpose a table was prepared giving the sidereal time of sundown at the observatory for each day in the year, and the sidereal time for the known values of the meridian errors was given as their sidereal difference from sundown; — if before, and + if after.

As a first approximation, the mean monthly increase of the deviation from the meridian near sundown or before zone was then computed and applied as a correction for the reduction of the meridian error found nearest to sundown to that at sundown. The meridian errors six hours or more after sundown required no correction on account of the small increase at that time and the sufficient proximity of the determining stars after zone. The next step was to find whether there existed a relation between the air temperatures and the increase of the meridian error from sundown to the after-zone observations. As there were no observations of rock temperatures, which would in all probability have furnished the best argument to be incorporated in the law of change, the air temperatures were used as their best substitute. The temperatures used for this purpose were those taken from the thermometers near the observatory, they being more or less subjected to the radiation of the heated rock.

The mean monthly increase from sundown to six hours after compared with that of the air temperatures exhibited a close dependence. During that period the increase of the meridian errors for 1° of temperature was found to be nearly uniform throughout the year, while the small differences between the monthly means for 1° and the annual mean presented such irregularities as might be attributed rather to accidental causes than to any existing law.

While such a result was well calculated to encourage the idea that the same dependence might exist during the whole interval, yet the observations prove that the increase of the meridian errors was much greater during the first part of the interval than that of the air temperature.

The original idea of adopting the mean monthly curves of air temperatures for the changes in azimuth was therefore abandoned, and in searching for another approximate law the analogy of cooling bodies presented as the simplest function containing all the requisite properties, the logarithmic curve of the form

$$\varkappa = \alpha - \beta\varepsilon^{\tau}$$

where τ denotes the sidereal interval in hours from sundown corresponding to the meridian error $\varkappa$; α, β, and ε are quantities constant for each day.

If $\varkappa$ and $\varkappa'$ as before represent the meridian errors for the intervals τ and τ', we have

$$\varkappa = \alpha - \beta\varepsilon^{\tau}$$

$$\varkappa' = \alpha - \beta\varepsilon^{\tau'}$$

therefore $$\varkappa' - \varkappa = \beta(\varepsilon^{\tau} - \varepsilon^{\tau'})$$

and for $\tau = 0$ $$d = \varkappa' - \varkappa = \beta(1 - \varepsilon^{\tau'})$$

where d denotes the increase of the meridian error during the sidereal interval from sundown to τ'.

If we represent by $\varkappa_0$ the meridian error at sundown, this may be adopted for the value of α, and then

$$\varkappa_0 = \varkappa - d = \varkappa - \beta(1 - \varepsilon^{\tau})$$

The value of ε is independent of β and to be determined only from the ratio of the increase of azimuthal deviation before and after zone.

The following table furnishes the adopted mean monthly values:

Month.	Mean temperature of the air in shade, taken from U. S. Nav. Ast. Exp., Vol. VI, page 382.	I ε	II Adjusted.	I–II Residuals.	Month.	Mean temperature of the air in shade, taken from U. S. Nav. Ast. Exp., Vol. VI, page 382.	I ε	II Adjusted.	I–II Residuals.
	°					°			
January	70.9 F.	0.62	0.62	0.00	August	52.1 F.	0.76	0.77	−0.01
February	70.7	.60	.62	−0.02	September	55.2	.73	.74	−0.01
March	67.0	.68	.65	+0.03	October	58.0	.70	.72	−0.02
April	58.9	.75	.76	−0.01	November	63.1	.67	.68	−0.01
May	54.0	.78	.75	+0.03	December	68.1	0.64	0.65	−0.01
June	48.9	.80	.79	+0.01					
July	47.9	0.78	0.80	−0.02	Mean	59.6	0.71	. . .	. . .

In adjusting the data for ε to the laws of temperature, the following formula was found:

$$\varepsilon = 0.71 - 0.0077\,(T - 59.6^\circ),$$

where T is the mean monthly temperature of the air for the time when ε is required.

In order to facilitate interpolation, the following table furnishes the values of ε adopted for each tenth day of the year:

Month.	ε Day of month. 1	11	21	Month.	ε Day of month. 1	11	21
January	0.63	0.62	.61	July	0.79	0.78	0.78
February	.60	.60	.61	August	.78	.77	.75
March	.63	.66	.69	September	.75	.73	.72
April	.72	.74	.76	October	.71	.70	.69
May	.77	.78	.79	November	.68	.67	.66
June	0.79	0.80	0.80	December	0.66	0.64	0.63

For increasing values of τ the limit of d is β, which represents the maximum increase of the meridian error. From the values of ε it is apparent that this approximate maximum is reached during the summer months, at an interval of six or seven hours after sundown, which gradually increases to eight or nine hours between February and July. It may be stated here that the same law was observed in the mean monthly increase or decrease of air temperatures.

From the equation found above,

$$x' - x = \beta\,(\epsilon^{\tau} - \epsilon^{\tau'})$$

it follows that

$$\beta = \frac{x' - x}{\epsilon^{\tau} - \epsilon^{\tau'}}$$

and by means of the latter expression the values of β were computed for all days whereon meridian errors had been observed both before and after zone, the said values being used individually only for these days.

The following table gives for each monthly mean of the three years the values of β and their weights; the increase, d, of meridian error from 0^h to 6^h, or from sundown to six hours thereafter, which is equal to $d = \beta\,(1 - \epsilon^{n})$; the observed mean difference, $t°$, of the air temperature during the same interval, and finally the increase, b, of the meridian error due to 1° of temperature, or

$$b = \frac{\beta(1 - \epsilon^{n})}{t°} = \frac{d}{t°}$$

Month.	β	Weight.	$t°$	d	b
January	0.50	22	12.6	0.47	0.038
February	.48	24	12.0	.46	.039
March	.62	100	13.4	.56	.042
April	.50	33	10.6	.41	.038
May	.38	8	8.0	.29	.036
June	.43	18	8.1	.32	.040
July	.43	28	9.0	.33	.036
August	.50	12	10.0	.40	.040
September	.49	12	10.0	.42	.042
October	.45	11	10.3	.40	.039
November	.40	13	11.0	.37	.034
December	0.46	48	11.2	0.43	0.039

Adopted mean value of $b = 0.04$.

These results prove that the total increase d is beyond a doubt nearly proportional to the observed difference of air temperature, $t°$; the variations in the observed daily values of d as compared with their corresponding differences in the air temperature go to prove the same, though less obviously, on account of the errors in the observations and the absence of correct temperature differences.

For nights when only one meridian error could be obtained the value of β was computed by the formula

$$\beta = \frac{bt°}{1 - \epsilon^{6}}$$

in which the mean value of b was used, viz., 0.04.

As the daily differences of air temperature from sundown to six hours after were generally not given with the necessary degree of accuracy, the mean was used of the observed daily and the computed mean monthly differences, the latter being taken from the following table:

Month.	t°	Month.	t°
January	12.6	July	9.2
February	12.0	August	9.5
March	13.4	September	9.9
April	12.2	October	10.5
May	10.4	November	11.0
June	8.1	December	11.4

In the application of the foregoing formulæ it was found most convenient to furnish for each day the meridian error at sundown and the coefficient β.

The meridian error $\varkappa$, at the sidereal interval τ from sundown, is therefore found by the formula

$$\varkappa = \varkappa_0 + \beta (1 - \varepsilon^{\tau})$$

The following table gives the values of ε^{τ} for the arguments ε and τ:

Sidereal difference from sundown.	ε										
	0.60	0.62	0.64	0.67	0.68	0.70	0.73	0.75	0.76	0.78	0.80
h. m.											
0 00	1.00	1.00	1.00	1.00	1.00	1.00	1.00	1.00	1.00	1.00	1.00
06	0.95	0.95	0.96	0.96	0.96	0.97	0.97	0.97	0.97	0.98	0.98
12	.90	.91	.91	.92	.93	.93	.94	.94	.95	.95	.95
18	.86	.87	.88	.89	.89	.90	.91	.92	.92	.93	.93
24	.82	.83	.84	.85	.86	.87	.88	.89	.90	.91	.92
30	.77	.79	.80	.82	.82	.84	.85	.87	.87	.88	.89
36	.74	.75	.77	.79	.79	.81	.83	.84	.85	.86	.87
42	.70	.72	.73	.75	.76	.78	.80	.82	.83	.84	.85
48	.66	.68	.70	.72	.73	.75	.78	.79	.80	.82	.84
54	.63	.65	.67	.70	.71	.73	.75	.77	.78	.80	.82
1 00	.60	.62	.64	.67	.68	.70	.73	.75	.76	.78	.80
06	.57	.59	.61	.64	.65	.68	.71	.73	.74	.76	.78
12	.54	.56	.59	.62	.63	.65	.69	.71	.72	.74	.76
18	.52	.54	.56	.59	.61	.63	.66	.69	.70	.72	.75
24	.49	.51	.54	.57	.58	.61	.64	.67	.68	.71	.73
30	.47	.49	.51	.55	.56	.59	.62	.65	.66	.69	.72
36	.44	.47	.49	.53	.54	.57	.60	.63	.65	.67	.70
42	.42	.44	.47	.51	.52	.54	.59	.61	.63	.66	.69
48	.40	.42	.45	.49	.50	.53	.57	.60	.61	.64	.67
54	.38	.40	.43	.47	.48	.51	.55	.58	.59	.62	.65
2 00	.36	.38	.41	.45	.46	.49	.53	.56	.58	.61	.64
20	.31	.33	.36	.40	.41	.44	.48	.51	.52	.56	.59
40	.27	.29	.31	.35	.36	.39	.43	.46	.48	.52	.55
3 00	.22	.24	.26	.30	.31	.34	.39	.42	.44	.47	.51
30	.18	.20	.21	.25	.26	.29	.33	.37	.38	.42	.46
4 00	.13	.15	.17	.20	.21	.24	.28	.32	.33	.37	.41
30	.10	.12	.14	.17	.18	.20	.24	.28	.29	.33	.37
5 00	.08	.10	.11	.14	.15	.17	.21	.24	.25	.29	.33
6 00	0.05	0.06	0.07	0.09	0.10	0.12	0.15	0.18	0.19	0.22	0.26

CLOSURE OF THE OFFICE OF THE EXPEDITION.

After Captain GILLISS's death the office of the Expedition was left with little or no supervision until July 3, 1865, when the chief of the Bureau of Navigation, Captain PERCIVAL DRAYTON, U. S. N., issued an order placing it under the charge of the newly appointed superintendent of the Naval Observatory, Rear-Admiral CHARLES H. DAVIS, U. S. N. At that time there was no reason to anticipate any interruption of the work, and a letter written by Admiral DAVIS on December 28, 1865, shows that he expected to complete and publish the fourth and fifth volumes precisely as planned by Captain GILLISS; but money for the payment of the assistants soon ceased to be available, and orders dispensing with their services after June 30, 1866, were issued by the Secretary of the Navy, Hon. GIDEON WELLES, on June 13, 1866. That necessitated the closure of the office, and in accordance with an order issued August 17, 1866, by the chief of the Bureau of Navigation, Capt. (afterwards Rear-Admiral) THORNTON A. JENKINS, U. S. N., all the papers of the Expedition were finally deposited at the Naval Observatory "for safe-keeping."

HISTORY OF THE SECOND PERIOD.

By Prof. WM. HARKNESS, United States Navy.

When the records of the Expedition were deposited at the Naval Observatory the zone observations had all been copied upon sheets of paper measuring 14 by 17¼ inches, each sheet ruled to contain observations of 52 stars, and each observation extending across a pair of sheets, both of which were given the same number. Each pair of sheets was headed "Zone Observations with the Meridian Circle at Santiago de Chile, 185–;" the words "Zone Observations with the Meridian" being on the left-hand sheet and "Circle at Santiago de Chile, 185–" on the right-hand sheet of the pair.

The left-hand sheets contained eighteen columns, arranged as follows, the numbers referring to the columns: 1. Month and day. 2. Number for reference. 3 to 9. Times of transit over wires, A, B, C, D, E, F, G. 10. Mean of wires. 11 and 12. Micrometer, wire and revolution. 13. Barometer. 14 and 15. Thermometer, attached and external. 16. Transit wire at which declination was observed. 17. Circle reading. 18. Names of stars, so far as they were identified.

The right-hand sheets contained twelve columns, arranged as follows: 19. Number for reference. 20 to 23. Corrections for instrument, clock, micrometer, and refraction. 24 and 25. Precession, etc., in right ascension and in declination. 26 and 27. Mean place for 1850.0, right ascension and declination. 28 Magnitude of star. 29. Initials of observer. 30. Instrumental errors, viz, the constants for level, collimation, azimuth at sundown, β, nadir, coincidence of wires, clock correction and rate.

When the Expedition left the United States Captain GILLISS expected to make at least three observations of every star above the eighth magnitude situated between 30° and 90° of south declination, but that proved impracticable, and the work of the Expedition was confined to stars situated within 25° of the south pole, while at the same time the limit of magnitude was extended from the eighth to what the observers called the tenth magnitude. The method followed in observing the region lying between −65° 00′ and −84° 30′ was to clamp the telescope and allow it to remain fixed throughout the entire duration of each zone, the circles being read frequently from the beginning to the end of the zone and the positions of all stars down to the tenth magnitude being determined as they passed through the field by observing their transits over one or more of the right-ascension wires, and also bisecting them with one of the horizontal wires carried by the zenith-distance micrometer. The width of each zone was thus limited to the range of the micrometer, which was usually about 30 minutes of arc, although an extreme range of 42 minutes was possible, and if there was any sensible change of the circle reading during a zone, it was assumed to have taken place uniformly and was distributed proportionally to the elapsed time. In the region between $-84^\circ\ 30'$ and the south pole a separate circle reading was made for each star. All the zone observations were made with clamp west.

From 371 observations of the stars *σ Octantis*, *β Hydri*, *β Chamæleontis*, and *α Trianguli Australis*, both above and below the pole, 211 Moon culminations, and 24 occultations, Captain GILLISS determined the geographical position of the meridian circle to be*

Latitude, $33^\circ\ 26'\ 25.89''$ South,
Longitude, $4^h\ 42^m\ 33.81^s$ West,

from Greenwich.

The zone observations began February 12, 1850, and ended September 8, 1852; there being for 1850, 152 pairs of sheets containing 7,419 observations; for 1851, 219 pairs of sheets containing 10,964 observations; and for 1852, 176 pairs of sheets containing 8,921 observations, making a total of 547 pairs of sheets containing 27,304 observations of 16,748 stars. In August, 1866, the reduction of these observations had been carried so far upon all the sheets that only columns 20, 24, 25, 26, and 27 remained to be filled, and it was erroneously supposed that the work could be finished by a single computer in about one year.

Although the records of the Expedition were sent to the Naval Observatory merely for safe-keeping, it was manifestly desirable that the zone observations should be reduced and published without unnecessary delay, and as the Observatory staff was too small to permit the employment of any part of it upon them, application was made to Congress for funds to employ a computer. On March 3, 1873, $1,500 was appropriated for that purpose, and with it, first Mr. PARKER PHILLIPS, and later Mr. JOHN T. HEDRICK were employed and placed under the immediate direction of Prof. WILLIAM HARKNESS, U. S. N. In laying out the work for them it became evident that the form of reduction adopted by Captain GILLISS necessitated a good deal of unnecessary labor, but the sheets were then so far advanced that nothing could be gained by a change of plan and it was therefore decided to fill up the vacant columns precisely as Captain GILLISS had intended. That was done as follows:

In the twentieth column, "Corrections for instrument," there was entered for each star the quantity

$$A\ [\varkappa_0 + \beta(1-\varepsilon^\tau)] + Bb + Cc$$

where A, B, and C are respectively the factors for azimuth, level, and collimation, $\varkappa_0$ is the azimuth at sundown, β, ε, and τ are explained on page XLV, and b and c are

* See pages xlix-li of The U. S. Naval Astronomical Expedition to the Southern Hemisphere during the years 1849-1852, vol. 3 (Washington, 1856).

respectively the level and collimation constants. As the zones were about 30 minutes wide it was usually found sufficient to compute the reductions from apparent to mean place at intervals of one hour, for points whose declinations were respectively 10 minutes greater and 10 minutes less than the declination of the center of the zone. Then, having regard to second differences, these reductions were interpolated to every half hour, and from the values so found the reductions for each star were interpolated and entered in columns 24 and 25. Throughout the region lying within $5\frac{1}{2}$ degrees of the pole it was found impracticable to make use of interpolations, and the reductions were computed separately for each star.

After the various columns of the reduction sheets were filled, the mean places of the stars for 1850.0 were derived by means of the following formulæ:

Mean Right Ascension = + Time of transit over mean of wires,
+ Corrections for instrument,
+ Clock correction,
+ Reduction to mean place.

Mean south declination = + 33° 26′ 25.89″ + Circle reading,
+ Correction for Micrometer + Refraction,
— Nadir error, — Coincidence of wires,
— Reduction to mean place.

The several quantities involved in finding the right ascensions have been fully explained on pages XXVI–XL, but it yet remains to explain those involved in the declinations.

The quantity 33° 26′ 25.89″ is the adopted latitude of the instrument, and the quantity called the circle reading is in every case the mean result from both circles, obtained by reading all the microscopes, two of which were applied to the eastern circle and two to the western. It should be remarked, however, that in the region within 5° 30′ of the south pole, where a separate circle reading was necessary for each star, it was sometimes impossible to read more than one or two microscopes, but in all such cases the observed readings were ultimately reduced to the mean of four microscopes by applying proper corrections.

The "correction for micrometer" consists of two parts, either or both of which may vanish. The first part is the distance in arc from the stationary horizontal wire to the point in the field at which the star was bisected by the zenith distance micrometer, and the second part is the correction for curvature of path when the star was bisected elsewhere than at the middle right ascension wire. The readings of the zenith distance micrometer were reckoned both upward and downward from a zero point situated close to the stationary horizontal wire, and were recorded positive or negative according as the system of wires was moved toward or from that part of the field occupied by wire 7. Thus, for about half the stars the observers were obliged to record the complements of the actual readings of the micrometer head, and it is not unlikely that errors may have been introduced in that way. With 27.927″ for the value of one revolution of the micrometer screw, let n be the number of revolutions and parts of a revolution through which wire 4 was moved from the zero point; W, the interval between wire 4 and the horizontal wire by which the star was bisected; I, the equatorial interval between the middle transit wire, D, and the transit wire at which the star was observed; δ, the declination of the star; and M, the total correction for micrometer. Then we shall have for the first part of the correction, $n\,27.927'' + W$; for the second part, $-[6.7367]\ I^2 \tan\delta$; and for the entire correction

$$M = n\,27.927'' + W - [6.7367]\ I^2 \tan\delta$$

where the value of W is to be taken from the table on page XXV, the intervals for wires 1, 2, 3 being regarded as negative, and those for 5, 6, 7 as positive. The quantity in brackets is the logarithm of the coefficient which it represents, and with the values of I from the table on page XXIII, the values of the logarithm of $[6.7367]\ I^2$ for the several right ascension wires were taken as follows:

R. A. wire	A	B	C	E	F	G
$[6.7367]\ I^2$	0.1519	9.7944	9.1976	9.1976	9.7997	0.1501

"Nadir error" is the excess of the circle reading above 180° 00′ 00″ when the telescope was so pointed that the stationary horizontal wire covered its own image, seen reflected from a basin of mercury; and "coincidence of wires" is the excess of the zenith distance micrometer reading above zero when the middle movable horizontal wire was brought into exact coincidence with the stationary one.

The refractions were computed from BESSEL'S constants by means of the tables given in the Appendix to the Washington Observations for 1845, and the reductions from apparent to mean place were computed from the quantities given in the Washington Observations for 1847, Appendix C, Table I.

In November, 1874, work on the zones ceased because the appropriation made in March, 1873, was exhausted. At that time all stars observed in 1850 and 1852 and all stars observed in zones in 1851 had been reduced to their mean places for 1850.0; all the stars in 0 hours had been catalogued ready for publication, and the first page of the catalogue had been put in type to show the proposed arrangement of the pages. Most of the circumpolar stars which had been observed by individual pointings in 1851 were still incompletely reduced. For about 1,500 of them all the columns had been filled up except the mean right ascensions and declinations, and for about 2,250 the reductions from apparent to mean place had not yet been computed.

Nothing further was done until 1881. In that year Ensign (now Professor) STIMSON J. BROWN, U. S. N., was assigned to the work on July 19, and remained upon it until June 10, 1882, and Mr. EMIL WIESSNER was assigned to it on October 19, and remained upon it until June 30, 1883. By March 4, 1882, these gentlemen had completed the determination of the mean places for 1850.0 of all the stars which were left unreduced in 1874. The next step was to collect the various observations from the zones into a catalogue, and that was accomplished by copying each observation upon a separate card, measuring 5 by 8¼ inches, arranged to show the page of the zone sheet; the number of the star upon the page, the date of observation, the magnitude, and the mean right ascension and declination for 1850.0, together with blank spaces in which to enter any systematic corrections in the two coordinates, and the seconds of the finally adopted right ascensions and declinations. The copying of the 27,304 observations upon these cards occupied Messrs. BROWN and WIESSNER 69 working days; that is, the cards were filled in at an average rate of 33 cards per man per hour. When they were sorted out in the order of right ascensions all the observations of the same star fell together, and were copied upon a single card, and thus a complete card catalogue, showing the location of every star upon the zone sheets, was obtained very expeditiously, and without the risk of omitting a single observation.

When all the observations upon each star were thus brought together, numerous discordances appeared, many of which arose from errors of computation or obvious errors of record, and were easily corrected, but enough remained to show that either from the instability of the instrument in azimuth or from other causes the work was affected by serious systematic errors. Meanwhile STONE's Cape Catalogue of 12,441 stars had been published in 1881, and it was thought that a comparison of the GILLISS stars with it would afford the best means of eliminating these systematic errors. The two catalogues were found to have 1,854 stars in common, and as a preliminary step, STONE's places were all reduced to GILLISS's epoch, 1850.0. Then the work of forming the systematic corrections and applying them to all the stars was begun. But before it had progressed far a change in the appropriations for the Naval Observatory cut off the money out of which Mr. WIESSNER was paid, and on June 30, 1883, the work was again suspended. Nevertheless, enough had been accomplished to show that the systematic corrections thus far obtained would not suffice to bring all the observations into harmony. At first the discordances were regarded as accidental, but before the work was suspended enough of them had accumulated to render that hypothesis improbable and to make evident the necessity for a fuller investigation of their source.

No further opportunity of advancing the catalogue occurred until October 1, 1890, when Prof. S. J. BROWN, U. S. N., was again assigned to it, and took up the investigation of the systematic corrections at the point where Mr. WIESSNER had stopped. On January 17, 1891, it was found practicable to assign Prof. EDGAR FRISBY, U. S. N., to the same work, and he and Professor BROWN were constituted a board to complete it.

HISTORY OF THE THIRD PERIOD.

By Prof. EDGAR FRISBY, U. S. N., and Prof. STIMSON J. BROWN, U. S. N.

The third period, in accordance with the instructions of the Superintendent of the Observatory, began October 6, 1890, and extended until the completion of the work for publication. It was begun by Professor BROWN in compliance with the following orders:

"NAVAL OBSERVATORY,
"*Washington, D. C., October* 6, 1890.

"Prof. S. J. BROWN, U. S. N., *Naval Observatory*.

"SIR: You will be pleased to take charge of the books and papers relating to the observations taken by the Expedition in charge of the late Capt. J. M. GILLISS during the years 1850, 1851, and 1852. Under the supervision of Prof. WM. HARKNESS you will reduce these observations and have them ready for the printer at the earliest date practicable.

"Very respectfully,

"F. V. MCNAIR,
"*Captain, U. S. N., Superintendent.*"

It was understood in assuming this task that the systematic corrections to all the zones had been derived, and that there only remained to be done, first, the application of these systematic corrections to nearly all the zones of 1851 and to all those of 1852; second, the derivation of the mean position of those stars which had been observed more than once, and the computation of the general precession in right ascension and declination; third, the preparation of the MS. copy for the printer. It was estimated that this would require about six months. A preliminary examination verified the statement upon which this estimate rested, as the corrections for all the zones were found to have been derived and neatly arranged in detail on 452 catalogue cards, one for each zone, with the exception of a few in which none of the zero stars occurred.

The time estimated for the completion of the zones was, however, soon found to be much too small on account of obstacles the nature of which will be fully explained in the subsequent portions of this introduction. The character of the reductions and discussions to be made was radically changed, and the amount of work to be done was thereby greatly increased. For reasons which do not concern the value of the catalogue in its present shape, and which have been partially detailed in the preceding portion of this introduction, Professors FRISBY and BROWN were designated by the Superintendent of the Observatory as a joint board to complete the work, by the following order:

"NAVAL OBSERVATORY,
"*Washington, January* 17, 1891.

"Prof. EDGAR FRISBY, U. S. N., and Prof. S. J. BROWN, U. S. N.,
"*Naval Observatory.*

"GENTLEMEN: You are appointed a board, until otherwise directed, to complete the work of reducing, discussing, and preparing for publication the zone observations made with the Transit Circle in the city of Santiago, Chile, by the United States Naval Astronomical Expedition in charge of Lieut. J. M. GILLISS, U. S. N., between August 16, 1849, and November 16, 1852, from the data in this Observatory, giving the positions of about 17,000 fixed stars lying within 25° of the south pole.

"The manuscript notes of the Expedition and the card catalogues already prepared under the direction of Prof. W. HARKNESS, U. S. N., must be returned to the fireproof building every evening.

"Respectfully,

"F. V. McNAIR,
"*Captain, U. S. N., Superintendent.*"

Being charged by these orders with the entire responsibility for the character of the results and their completion at the earliest practicable date, it was decided, after a consultation concerning the condition of the work, to make a new and thorough discussion of all the observations. For this purpose the list of 4,000 zero stars, prepared by Professor BROWN, was adopted, and also the plan which he had initiated for the elimination of the errors with which the results were still apparently affected.

The progress of the work was subject to frequent interruptions by duties connected with the removal of the Observatory to the new site, but the manuscript copy of the catalogue was finished and turned over to the officer of the Observatory in charge of the printing in May, 1893. The printing of the Catalogue was begun in January, 1894, and practically completed about the middle of the same year. The proof sheets, first and second revise, and the plate proof were compared with the original positions of the catalogue cards. This introduction was prepared in accordance with the direction of the Superintendent of the Observatory, contained in the subjoined order:

"UNITED STATES NAVAL OBSERVATORY, GEORGETOWN HEIGHTS,
"*Washington, D. C., June* 26, 1894.

"GENTLEMEN: Referring to your letter dated June 25, 1894, regarding the introduction to the GILLISS zones, you are informed that Professor HARKNESS has been directed to prepare the history of all that transpired relative to this work prior to 1890; from which time you will be pleased to write the introduction to date.

"Very respectfully,

"F. V. McNAIR.
"*Captain, U. S. N., Superintendent of the Naval Observatory.*"

The last portion of the introduction, referred to in the above order, obviously could not be completed until the character of the preceding portions was known, together with the amount of discussion of observations which might be contained therein.

On the 3d day of June, 1895, the first portion of the introduction was finished and turned over by the Astronomical Director of the Observatory, and the completion of the remainder was immediately begun in accordance with the orders of the Superintendent.

The subsequent portions of this introduction will be devoted to the conditions of the reductions which necessitated the rediscussion of all the observations; the formation of the new series of systematic corrections to the zones; character of errors which have been eliminated, together with changes which have been made in the original records; corrections to reduce the magnitudes to correspond with those of the Cordoba zones, and a list of the proper motions of stars deduced from the zone observations; and explanation of the printed catalogue.

But little progress had been made in the application of the systematic corrections to the zones when large discrepancies were discovered between the separate positions, particularly in right ascension, of stars which occurred in more than one zone; these were still more apparent in the examination of the corrections derived from the zero stars, the mean of which formed the corrections to a given zone. The one observed on April 10, 1850, is cited in some detail to show in an extreme case the nature of these errors. The corrected positions of the stars in this zone were in some cases more than five seconds greater or smaller in right ascension than the adopted position of the zero stars, or the observed position of the same star in zones unaffected by this error. The separate corrections in right ascension, the mean of which made the adopted systematic correction to the zone, had a range of about eight seconds, while the mean was only $0.^s41$, the mean declination being 66°. In the original record of the zone it was noted by Captain GILLISS: "Extraordinary motion of stars, as though refraction changed 20″ whilst it passed over field." On the catalogue card containing the formation of the zone correction this remark was quoted, with the additional note: "This accounts for variation in $\Delta\alpha$; give observations only one-half weight." The reduction of the individual wires to the mean wire of all the observations of this zone showed that the observations in this respect were of the usual accuracy, and that neither poor observing nor anomalous atmospheric conditions could account for the discordances. Other zones affected with similar discrepancies, but smaller in amount, had also been given one-half weight, while some of the observed positions in right ascension of the zero stars contained in them had been arbitrarily changed one second to bring them into agreement with other observations and the adopted position of the zero stars. On the catalogue cards containing the latter observations was the brief note: "This observation is certainly one second in error, although the source of the error can not be detected It has been changed in the seventh column."

Another source of systematic error lay in the application to several zones differing widely in declination but observed the same evening, of the same mean correction in right ascension, formed from all the zero stars observed in separate zones. A correction of this character a priori would be expected to vary with the declination of the zone, and this was subsequently found to be the case.

The explanation which had been accepted as the cause of these errors did not appear to be quite satisfactory nor to rest upon very extensive investigation. Unless errors of such magnitude could be eliminated the value of the catalogue would be so largely vitiated as to render the results scarcely worth publishing, and it was deemed necessary to make a thorough investigation of their source. This involved the formation of a new and larger list of zero stars and a rediscussion of all the systematic corrections to the zones.

In making the revision of the zones, it was thought desirable to use as zero stars for the systematic corrections the list of 4,000 stars to which allusion has been before made. This would give a much larger number of comparison stars and would make the resulting positions directly comparable with the Cordoba zones, which covered a large portion of the space included in the GILLISS work. In forming this list, all stars of the Cordoba General Catalogue (1875.0) which were contained within the limits of the zones were reduced in duplicate to the epoch 1850, using the precessions and secular variations there given. Those stars lying between 80° south declination and the pole were reduced by the trigonometric formula, using the constants given in the Cordoba Zone Catalogue. As no proper motions were given in these catalogues, they were taken from the Cape Catalogue of 1880, when given there, and the Melbourne General Catalogue of 1870. A few proper motions were revised and a few additional ones derived from the more complete data furnished by the GILLISS zones. In the formation of the preliminary corrections, however, the proper motions depending on the GILLISS places were not used. No attempt was made to reduce these positions to a homogeneous system by the application of corrections to the results for the separate years represented in the Cordoba Catalogue, on the supposition that the weighted mean of the different years would give a system sufficiently accurate for the purpose of correcting the zone observations. A few stars from the Cape Catalogue, 1880, were included, their positions having been reduced to the system of the Cordoba Catalogue by the application of the systematic corrections given in volume 47, Monthly Notices R. A. S., pages 446–454.

The reductions were all made in duplicate and the positions checked by comparison with the position of such of the 1,854 zero stars of the list previously used as were common to both. This comparison showed the presence of about two hundred mistakes in computation of the precession. Although many of these were errors of a second in right ascension or several seconds in declination, they were not sufficient in number or magnitude to materially affect the resulting positions nor account for the discordant results.

For the purpose of further comparison and the detection of errors in the zone positions, all the stars of the Cordoba Zone Catalogues contained within the limits of the GILLISS zones were reduced to the epoch of the latter. These were recorded, as in the case of the zero stars, at the top of the appropriate catalogue card. This list contained 4,659 stars. The reductions were only approximate within the limits of one-tenth of a second in right ascension and one second in declination, but this was nearly enough exact to show any error in either catalogue. This list proved of great service in the detection of errors of record and computation in the GILLISS zones and a few erroneous positions in the Córdoba zones.

The preliminary correction to each zone was then derived from a comparison of the observed position with the adopted position of the corresponding zero star. Using these corrections, an investigation of the most discordant zones showed their peculiar errors to be due to the fact that the transit wires were not vertical. Out of the 452 zones, 113 were found to be affected by this instrumental error, which appeared to be sensibly constant for periods extending over several zones, until accidentally disturbed or altered by the adjustment for collimation. As explained in the preceding portions of the introduction, the telescope was clamped for any given zone and the stars observed as they swept through the field: thus the effect of the inclination of the transit wires on the right ascension of any star would depend directly upon the distance of the star from the fixed horizontal wire. With the increased number of zero stars given by the new list, it was an easy matter to derive the inclination of the wires for any given zone by plotting the corrections as ordinates on paper ruled in squares, using the declination as abscissas. As the inclination was found to remain sensibly constant until disturbed accidentally or by adjustment of the collimation, the mean for such period was determined and used in correcting the positions of all the zones embraced within that period. The paper used for this purpose was ruled in squares of one-tenth of an inch, each of which represented $0^s.1$ in right ascension and $1'$ in declination. From the line drawn through the center of gravity of the plotted corrections was deduced the change in $\Delta\alpha$ for $1'$ of declination, which was reduced to its value at the equator. The mean value thus deduced was applied to the mean systematic correction for any zone by the formula

$$\Delta\alpha = \Delta'\alpha + i \sec.\ \delta_0 (\delta_0 - \delta)$$

where $\Delta'\alpha$ represents the mean correction in right ascension at the mean declination of the zero stars of the zone and $\delta_0 - \delta$ represents the difference, expressed in minutes, between the mean declination of the zone and the declination of the star to be corrected.

The correction for the zone of April 10, quoted in a preceding paragraph, was as follows:

$$\Delta\alpha = + 0.16^s + 0.29^s (\delta_0 - \delta)$$

which for the limits of the zone, 34′, gave a range of 9.86ˢ in the correction. The mean declination of the zone was 66.1°. The application of the correction made its results compare favorably with those zones not affected by this cause. All the other zones to which this correction was applied were brought into satisfactory agreement with the rest of the observations, and all the discrepancies of the character noted in Section I were completely removed. The following table gives the value of *i* in equatorial interval for the period during which it was applied:

Value of i for one minute of declination at the equator.

Period.		*i* for 1′ in δ.	Range of correction applied to the zones.	
From—	To—			
		s.	s.	s.
Apr. 10, 1850	Apr. 10, 1850	0.1180	± 4.86	
Apr. 11, 1850	Apr. 15, 1850	0.0290	From ± 1.05 to	± 1.24
May 25, 1850	July 25, 1850	0.0134	± 0.55	± 2.72
Dec. 20, 1850	Dec. 29, 1850	0.0160	± 0.83	± 2.46
Dec. 30, 1850	Jan. 10, 1851	0.0347	± 2.16	± 8.73
Apr. 3, 1852	Apr. 26, 1852	0.0133	± 0.56	± 0.94
Apr. 27, 1852	May 29, 1852	0.0100	± 0.60	± 1.28
June 21, 1852	July 26, 1852	0.0150	± 0.50	± 1.32

Both the transit wires and the wires of the declination system were equally affected by the inclination; and the declination of stars observed out of the meridian would from this cause contain appreciable errors. The method of reducing the observed declinations of such stars, however, practically eliminated these errors. Investigating their positions, it was found that the reduction to the meridian had been applied to the original micrometer reading by marking over with ink the pencil record in the original observing book. The original record could generally be made out with the aid of a small magnifying glass, and it was found that the reduction applied did not agree with the one computed by the ordinary formula. It was ascertained that it had been obtained from the records of stars which had been observed over the first and last transit wires, and thus included the correction necessary to be applied for inclination of the wire. In cases where the inclination of the wire was large, there would still remain an appreciable error, owing to the fact that bisections in declination would not be made precisely at the recorded wire or time.

Another source of systematic error in the declinations was found to lie in the adoption of sensibly inaccurate wire intervals for the horizontal wires. The source of this inaccuracy is indicated in the description of the method by which they were obtained. A careful search was made for the original records of these measurements, but only a portion of them could be found, which was contained in the original record books of the zone observations. From these no improvement could be made in the adopted intervals. An attempt was made to derive from the zero stars new values, but the material proved insufficient except in the case of those used between July 29 and December 26, 1850. To the observations embraced in this latter period corrections were applied to the declinations, so that their positions would be the same as if their corrected values had been used. The entire system of wire intervals, including the corrected set, is given below:

Adopted intervals of horizontal wires.

Period. From	Period. To—	1.	2.	3.	(4)	5.	6.	7.
		′ ″	′ ″	′ ″		′ ″	′ ″	′ ″
1850, Feb. 4 . .	1850, May 22 . .	. . . . . .	8 30.00	4 14.98		4 14.00	8 27.40	. . . . . .
May 22 . .	July 29 . .	. . . . . .	8 25.32	4 9.64		4 19.24	8 38.08	. . . . . .
July 29 . .	Dec. 27 . .	. . . . . .	8 26.26	4 14.24		4 15.66	8 31.28	. . . . . .
July 29 . .	Dec. 27 . .	. . . . . .	8 31.2	4 16.8		4 13.5	8 31.5	. . . . . .
Dec. 27 . .	1851, July 14 . .	12 26.83	8 29.72	4 16.10		4 17.17	8 28.87	12 35.00
1851, July 14 . .	1852, Jan. 11 . .	12 29.70	8 23.13	4 13.70		4 14.17	8 30.67	12 36.30
1852, Jan. 12 . .	Apr. 23 . .	12 35.46	8 29.48	4 15.81		4 14.23	8 25.44	12 36.90
Apr. 24 . .	Sept. 13 . .	12 33.47	8 29.37	4 15.20		4 13.43	8 25.73	12 37.53

The accuracy of the declinations in a considerable number of zones is noticeably affected by the imperfect performance of the micrometer screw, to which Captain GILLISS refers in his introduction. That the defect was more serious in its nature than the remarks concerning it would indicate, is evident from the positions derived from the observations, as well as from the remarks in the original record books. In order to present exactly the difficulty with the micrometer screw, the notes in the original record are quoted entire:

"*June* 12, 1850.—In measuring No. 68, the micrometer screw became jammed so hard that it could not be moved."

"*June* 14, 1850.—The micrometer screw being immovable, unscrewed eye end and took to pieces. It was found necessary to force the screw."

"*September* 10, 1850.—The micrometer screw, which has since the 12th of June been chafing, finally wore out the female screw, and at star No. 26 ceased to be of service."

"*September* 14, 1850.—The female screw was plugged up and a new one cut, the right-ascension micrometer screw substituted in place of the other one."

From frequent notes in the observing books, it is evident that this screw never gave satisfaction, and on December 27, 1850, it was working so badly that observations with it were suspended. From this date until June 6, 1851, the zones were observed by means of the slow-motion screw and the circle microscopes. The positions derived from zones from December 20 to December 27 were not used except in cases of stars observed but once.

In order to get an expression for the accuracy of the final positions, an approximate probable error of a position resulting from a single observation was derived. The residuals for its determination were obtained by taking the difference between the mean position and each individual observation of a large number of stars observed two or more times. About a thousand stars were taken from zones selected at random, the right ascensions being grouped in zones of 5° each. It is believed that the resulting probable error represents fairly the accuracy of the work. The formula employed in deducing it was

$$r_1 = \pm\ 0.8453\ \frac{\Sigma r}{m}$$

where Σr represents the sum of the residuals independent of their signs, and m their number.

$$r = \pm\ 0^s.042 \text{ sec. } \delta, \text{ in right ascension.}$$
$$r = \pm\ 1''.06 \text{ in declination.}$$

These are somewhat smaller than the probable error of zone observations deduced from results of the independent observations of stars contained in the MS. copy of Volume IV.

In the application to the individual stars of a zone of the systematic correction finally adopted, the observed position of all the zero stars and also those stars on which more than one observation had been made were carefully checked whenever there was any suspicious discrepancy in the final places. Many errors, both of computation and record, were thus detected and removed. In many cases the errors of record were apparent, and showed the star to have been before observed. But in no case was a change made in the original record unless there was found indisputable evidence of error. In the case of zero stars, or Gould zone stars, the card containing the duplicate observation was thrown out and its corrected position transferred to the proper card. In all such cases notes giving explicitly the change made in the observation were made on the catalogue card, and were inserted in the printed catalogue. Over 600 duplicate observations were thus found and thrown out. The changes made were of the following classes:

(*a*) Changes of 5 or 10 seconds in the recorded time on one transit wire where several had been observed.

(*b*) The wire on which the correct time had been recorded was changed so as to throw the whole time of transit backward or forward by one wire interval; very rarely by two.

(*c*) Change in the declination wire recorded, thus changing the star's declination by about 4′ 15″; rarely 8′ 30″.

(*d*) In a few cases the micrometer reading was changed by one revolution, changing the declination by 27″.9, or on account of mistaken numbers the fractional reading of the micrometer head. Changes in the earlier portion of the work appeared to be necessary from a confusion of the German numbers 4 and 7, 3 and 8, and 5 and 8. These latter changes were very rare, and only when indicated beyond question by other observations of the same star, and the position of the same star in the zero list or in the Cordoba zones.

(*e*) Changes in the actual reading of the micrometer, based upon the peculiar method of recording adopted. In all of the stars observed above the middle horizontal wire the micrometer reading was regarded as positive, and the actual reading of the micrometer recorded in the observing book. Below this wire the revolutions were counted negative, the observer apparently subtracting from unity the decimal part of the revolution, and prefixing to the remainder the whole number of revolutions with the negative sign. A number of changes were made on this account: (1) Where the observer had neglected to perform the subtraction indicated; (2) where there was an evident error of several tenths of a revolution in the subtraction.

(*f*) Where the records of two stars had become interchanged, the right ascension of one being recorded with the declination of the other.

The estimates of the magnitude of the zone stars were based upon standards selected from the British Association Catalogue. But few of the stars contained in this are as faint as the ninth magnitude, and the continuance of the estimates below this would be liable to considerable systematic error. In view of the size of the telescope objective, the magnitudes of stars fainter than the ninth appear to be estimated much too faint. In order to render the Gilliss magnitudes directly comparable with those of the Cordoba zones, the magnitudes of all the 4,659 stars common to both catalogues were compared. The differences were arranged according to the order of magnitude given in the Gilliss zones, and the results are given in the following table:

Corrections to reduce Gilliss *magnitudes to* Gould's *for each* 2^h *of R. A.*

[The argument is the nearest *half magnitude* of Gilliss's observation.]

R. A.	Magnitudes.			
	10.0	9.5	9.0	8.5
h. h.				
0 to 2	— 0.60	— 0.29	— 0.07	0.00
2 4	— 0.75	— 0.40	— 0.11	+ 0.07
4 6	— 0.59	— 0.28	— 0.12	+ 0.01
6 8	— 0.79	— 0.34	— 0.15	+ 0.06
8 10	— 0.70	— 0.45	— 0.28	+ 0.01
10 12	— 0.72	— 0.39	— 0.32	0.00
12 14	— 0.65	— 0.46	— 0.16	+ 0.20
14 16	— 0.69	— 0.37	— 0.23	+ 0.11
16 18	— 0.65	— 0.14	— 0.08	+ 0.26
18 20	— 0.67	— 0.49	— 0.16	+ 0.26
20 22	— 0.53	— 0.34	— 0.23	+ 0.27
22 0	— 0.41	— 0.34	— 0.22	+ 0.17
Mean	— 0.65	— 0.36	— 0.18	+ 0.12
	± 0.065	± 0.064	± 0.053	± 0.073

The proper motions used in the reduction of the positions of the zero stars were taken from the Cape Catalogue and the Melbourne General Catalogue. After the preliminary general correction and the correction for inclination of the transit wires had been applied to positions of the zero stars, these were used in computing approximate proper motions. A few of the assumed values used in reducing the positions of the zero stars were thus revised and a few new ones adopted. These, together with a number of other stars where there was a suspicion of considerable proper motion, are given in the following list. The values adopted and used in the reduction of the zero stars are marked with (P). The other proper motions given in the list were not used, as there were generally enough zero stars for deriving the corrections to the zones without including them. In reducing the positions of the various catalogues used, the constants were derived from the values of the precession and secular variation contained in the Cordoba Catalogue, except in the case of positions from the Cape and Melbourne Catalogues. A correction was applied to the positions of the Cape Catalogue to reduce them to the system of the Cordoba Catalogue. The others were not thus corrected, as it was not expected that anything more than approximate values could be determined from the sources available.

The catalogues referred to in the remarks by the initials are as follows:

G., the Cordoba General Catalogue, 1875.0. S., the Cape Catalogue, 1880. M., the Melbourne General Catalogue, 1870.0. Gi., the GILLISS Zone Catalogue, 1850.0. C_{40}, the Cape Catalogue, 1840.0. C_{50}, the Cape Catalogue, 1850.0. B., the BRISBANE Paramatta Catalogue, 1825.0. J., the JOHNSON St. Helena Catalogue, 1830.0. T., the TAYLOR Madras General Catalogue, 1835.0. L., the LACAILLE-BAILY Catalogue, 1750.0.

The positions of BRISBANE and LACAILLE were generally discordant, and no use was made of them in forming the proper motion, but simply to confirm the results derived from other sources.

Proper motions used in the reduction of zero stars and not contained in the list of proper motions.

Name.	μ	μ'	Name.	μ	μ'
	s.	″		s.	″
γ_3 Octant's	− 0.014	− 0.05	5406 Lacaille	− 0.004	− 0.08
ξ Toucani	+ 0.275	+ 1.17	η Muscæ	− 0.012	− 0.04
κ Toucani	+ 0.072	+ 0.053	ι Muscæ	− 0.033	0.23
576 Lacaille	+ 0.070	− 0.06	κ Octantis	− 0.052	− 0.05
η^2 Hydri	+ 0.013	0.10	d Octantis	− 0.053	0.02
μ Hydri	+ 0.032	0.00	γ Trianguli A	− 0.018	0.06
ε Hydri	+ 0.017	0.00	ρ Octantis	+ 0.080	− 0.08
ζ Hydri	+ 0.012	+ 0.08	ε Trianguli A	− 0.004	− 0.09
1092 Lacaille	+ 0.011	0.00	6381 Lacaille	0.000	− 0.05
ι Hydri	+ 0.040	+ 0.05	γ Apodis	− 0.045	− 0.16
γ Hydri	+ 0.012	+ 0.12	β Apodis	− 0.105	− 0.34
1639 Lacaille	0.000	− 0.10	α Trianguli A	0.000	− 0.06
γ Mensæ	0.000	+ 0.24	6947 Lacaille	0.000	− 0.06
δ Dorradus	− 0.002	− 0.02	7088 Lacaille	− 0.008	− 0.06
δ Volantis	− 0.004	0.00	ι Apodis	+ 0.007	− 0.06
α Chamæleontis	+ 0.028	0.00	σ Octantis	+ 0.075	− 0.02
β Volantis	− 0.009	− 0.12	ζ Pavonis	− 0.004	− 0.13
θ Chamæleontis	− 0.050	+ 0.01	8078 Lacaille	− 0.004	0.02
A Octantis	− 0.014	+ 0.02	8156 Lacaille	− 0.002	0.00
3644 Lacaille	− 0.011	− 0.03	e Pavonis	0.000	− 0.13
3709 Lacaille	− 0.006	− 0.01	8229 Lacaille	+ 0.008	− 0.09
E Carinæ	− 0.001	− 0.06	d Pavonis	+ 0.193	− 1.23
β Argus	· 0.032	+ 0.09	8301 Lacaille	+ 0.006	− 0.08
ζ Octantis	− 0.080	+ 0.02	8306 Lacaille	+ 0.007	− 0.07
ι Chamæleontis	− 0.060	+ 0.08	β Octantis	− 0.130	− 0.01
H Carinæ	− 0.007	− 0.04	μ′ Octantis	+ 0.062	0.00
4139 Lacaille	− 0.007	− 0.02	β Pavonis	− 0.009	− 0.06
4367 Lacaille	0.000	− 0.04	8570 Lacaille	− 0.001	− 0.40
γ Chamæleontis	− 0.017	0.00	8671 Lacaille	0.000	0.05
θ Argus	0.000	− 0.02	8672 Lacaille	− 0.035	− 0.10
4724 Lacaille	− 0.011	− 0.05	ε Octantis	+ 0.015	− 0.08
4744 Lacaille	+ 0.004	− 0.03	ν Indi	+ 0.028	− 0.70
4766 Lacaille	− 0.010	0.00	9117 Lacaille	+ 0.025	− 0.03
ε Chamæleontis	0.021	− 0.07	β Octantis	− 0.034	0.00
4975 Lacaille	− 0.018	0.00	τ Octantis	+ 0.022	0.00
β Chamæleontis	− 0.017	+ 0.02	γ^1 Octantis	− 0.030	− 0.03
5235 Lacaille	− 0.088	0.00	γ^2 Octantis	− 0.017	− 0.03
δ Muscæ	+ 0.042	0.00			

Gilliss number.	Gould number.	App. position 1875.0.				μ	μ'	Remarks.
		α		δ				
		h.	m.	°	′	s.	″	
85	182	0	10	80	33	+0.18	−0.03	G. S. Gi (3) L. (P).
125	234	0	14	78	41	(+0.04)	(−0.02)	G. S. Gi (1) L (?).
323	594	0	33	74	39	−0.035	+0.08	G. S. Gi (2). L.
370	646	0	37	83	43	−0.044	. . .	G. S. Gi (3). L.
385	687	0	39	66	18	+0.01	−0.70	G. S. Gi (1). L.
392	678	0	39	86	6	−0.05	−0.11	G. S. Gi (7) L (?) (P).
399	700	0	40	86	23	+0.20	. . .	G. S. Gi (5) L (?).
431	762	0	44	75	36	+0.038	. . .	G. S. M. Gi (1) B. L. (P).
455	782	0	46	86	34	+0.13	. . .	G. S. Gi (6) L (?).
468	833	0	49	75	20	+0.04	. . .	G. S. Gi (2) L.
491	869	0	51	74	59	+0.02	+0.14	G. S. Gi (2) B (?) L (?).
639	1137	1	7	73	41	. . . .	+0.22	G. Gz (3) Gi (1). B.
654	1184	1	10	74	56	+0.04	−0.16	G. Gz (3) Gi (2).
662	1197	1	11	69	29	+0.07	+0.10	G. S. Gi (2) C_{40} B. L (?) (P).
868	1586	1	33	79	8	−0.03	−0.20	G. S. M. Gi (1) C_{40} (P).
1536	2942	2	40	72	0	(−0.05)	. . .	G. S. Gi (1) L (?).
1809	3506	3	9	66	58	(−0.03)	. . .	G. Gi (2).
2070	4040	3	35	78	46	−0.024	−0.04	G. M. Gi (1) B. (P).
2585	5173	4	30	68	9	+0.04	+0.50	G. Gz. S. Gi (2) L. (P).
2770	5564	4	48	76	32	+0.04	+0.15	G. S. Gi (2) B. L. (P).
2786	5600	4	50	76	31	(+0.01)	(+0.20)	G. Gi (2) B (?).
2836	5719	4	55	72	37	−0.01	+0.29	G. S. Gi (2). L.
2968	5980	5	7	76	47	0.00	−0.20	G. S. Gz (2) Gi (2) L. (?).
3028	6091	5	12	77	42	+0.045	−0.36	G. S. Gz Gi (1) L. (P).
3145	6293	5	21	80	19	0.00	−0.20	G. S. Gi (2) L.
3251	6499	5	30	75	47	+0.04	−0.30	G. Gz. Gi (2).
3372	6731	5	39	73	3	+0.04	−0.03	G. Gz (2) Gi (2) B (?).
3511	6907	5	48	80	34	+0.064	+1.10	G. S. M. Gi (2) B. L. (P).
3686	7209	5	59	79	23	0.00	−0.15	G. Gz. S. M. Gi (3) B. L (?) (P).
3839	7579	6	12	65	30	0.00	+0.15	G. S. Gi (3) B.
3840	7581	6	12	65	30	0.00	+0.22	G. S. Gi (3).
3875	7623	6	13	72	59	0.00	(−0.50)	G. S. Gi (1).
3885	7639	6	14	74	43	+0.029	−0.24	G. S. Gi (4) C_{40} B. (P).
4112	8025	6	28	78	20	(−0.04)	. . .	G. S. Gi (3).
4416	8701	6	51	73	9	(−0.04)	. . .	G. S. Gi (2).
4497	8840	6	57	74	34	−0.01	+0.32	G. S. Gi (2) L. (P).
4524	8934	7	0	67	45	. . . .	+0.20	G. S. Gi (1) C_{40} B. L. (P).
4610	9120	7	7	71	5	. . . .	+0.25	G. S. Gi (4) L (?) (P).
4669	9271	7	12	67	34	+0.02	−0.20	G. S. Gi (2) L.
4705	9368	7	16	63	54	. . . .	+0.17	G. S. Gi (5). B. L (?).
4829	9602	7	23	76	11	. . . .	+0.18	G. S. Gi (2). B (?).
5014	Gz. 7^h, 2777	7	28	69	39	(−0.10)	(−0.90)	Gz. (1) Gi (1).
5790	12000	8	45	79	13	+0.06	−0.23	G. S. Gi (2) L (?).
5822	12101	8	48	66	18	+0.01	+0.08	G. S. Gi (3) M. L. (P).
5855	12177	8	52	66	20	−0.045	−0.05	G. Gz. (3) Gi (3).
5954	12378	9	0	65	54	0.00	−0.14	G. S. M. Gi (3) C_{40} B. J. L. (P).
6143	12756	9	17	66	54	0.00	(−0.20)	G. Gz. Gi (2) L (?).

Gilliss number.	Gould number.	App. position 1875·0.				μ	μ′	Remarks.
		α		δ				
		h.	m.	°	′	s.	″	
6200	12853	9	21	77	7	—0.02	+0.17	G. Gz. Gi (2) B. L.
6257	12967	9	25	77	22	—0.07	+0.37	G. S. Gi. L.
6365	9^h, 2661	9	34	71	1	0.00	—0.20	Gz (3) Gi (2).
7365	14817	10	44	79	49	—0.029	—0.08	G. S. M. C_{80} C_{80} Gi. J. B. (P).
7376	14829	10	45	79	53	—0.03	0.00	G. S. M. C_{40} C_{80} C_{80} Gi. J. B. (P).
7415	14897	10	48	76	28	—0.04	0.00	G. S. Gi (2). L.
8582	16878	12	17	66	57	—0.13	+0.16	G. S. Gi (3) B. L.
8624	16970	12	21	68	2	0.00	—0.17	G. S. Gi (2) L.
8685	17104	12	26	68	3	—0.09	—0.20	G. S. Gi (2) L. (P).
8987	17696	12	54	86	53	—0.26	—0.12	G. S. Gi (10) L (?).
9122	17907	13	3	69	17	0.00	+0.18	G. S. Gz. Gi (3) B (?) L (?).
9232	18092	13	11	70	12	(—0.06)	. . .	G. S. Gi (1) B (?) L.
9265	18163	13	14	78	18	—0.05	+0.12	G. S. Gi (3) L.
9370	18352	13	23	76	55	—0.12	. . .	G. S. Gi (2) L.
9474	18517	13	30	67	2	+0.05	—0.28	G. S. Gi (1) B. L.
9794	19078	13	57	74	15	—0.06	+0.30	G. S. Gi (1) L.
9941	19298	14	8	67	28	. . . .	—0.14	G. S. Gi (1) B. L.
10204	19733	14	27	67	22	—0.067	—0.26	G. S. M. Gi (1) C_{40} B. L. (P).
10382	19994	14	39	66	42	—0.01	—0.25	G. M. Gi (2). (P).
10461	20112	14	44	65	54	—0.032	+0.14	G. Gz. Gi (1) C_{40} B. (P)
10563	20305	14	52	75	9	(—0.05)	. . .	G. Gz. Gi (1).
10660	20457	14	58	73	23	(—0.02)	(—0.33)	G. Gz. Gi (1).
11239	21419	15	41	83	52	—0.20	—0.22	G. S. Gi (1). L.
11479	21754	15	56	70	43	—0.044	—0.37	G. S. Gi (2) L. (P).
11661	22027	16	8	69	25	0.00	—0.35	G. Gz. M. Gi (1). (P).
11751	22185	16	15	69	48	+0.040	. . .	G. S. Gi (2) C_{80} B. L. (P).
12293	23081	16	56	68	40	0.00	0.00	G. S. Gi (2) C_{80}. (P).
12358	Gz. 17^h, 107	17	1	75	58	(+0.33)	(+1.05)	Gz (1) Gi (2).
12407	23282	17	5	66	48	+0.01	—0.16	G. Gz. S. Gi (3) C_{40} B. L. (P).
12412	23312	17	6	75	12	—0.238	—0.18	G. S. Gi (2) B (?) L. (P).
12430	Gz. 17^h, 508	17	7	75	51	(—0.36)	(—0.71)	Gz. Gi.
12784	24041	17	36	74	58	. . . .	(—0.25)	G. Gz. Gi (1) B.
12702	24176	17	41	87	39	—0.103	—0.13	G. S. Gi (4) L. (P).
12793	24067	17	37	75	11	(—0.03)	(—0.10)	G. Gi (2) B (2) ?.
12990	24483	17	54	75	53	. . . .	—0.24	G. S. Gi (3) L. (P).
13036	24570	17	57	73	41	. . . .	—0.32	G. S. Gi (1) L.
13559	25845	18	46	70	6		(—0.20)	G. Gz (2) Gi (1).
13956	26680	19	22	69	21	. . . .	—0.20	G. S. Gi (2) B. L.
14080	26897	19	32	66	35	. . . .	(—0.24)	G. Gz. Gi (2).
14161	27074	19	39	65	54	+0.02	—0.23	G. S. Gi (1) C_{80} C_{80} B. L. (P).
14312	27380	19	53	67	39	+0.144	—0.73	G. S. Gi (1) B. L.
14678	28171	20	27	75	47	+0.055	—0.14	G. S. Gi (3) L.
14679	28172	20	27	75	46	+0.03	—0.17	G. S. Gi (3).
14890	28682	20	48	70	3	+0.08	—0.30	G. S. Gi (2) L. (P).
14966	28851	20	56	73	40	+0.08	—0.34	G. S. Gi (3) L. (P).
15035	29062	21	5	80	38	+0.05	—0.12	G. S. Gi (2) L (?).
15139	29274	21	14	68	46	(—0.01)	(—0.25)	G. S. Gi (1) L.

Gilliss number.	Gould number.	App. position 1875·0.				μ	μ′	Remarks.
		α		δ				
		h.	m.	°	′	s.	″	
15141	29300	21	16	82	36	+0.07	+0.13	G. S. Gi (2) L (?).
15163	29309	21	16	65	56	+0.019	-0.83	S. M. Gi (P).
15197	29360	21	19	75	48	+0.05	—0.21	G. Gz. Gi (2).
15266	29533	21	27	77	57	0.00	—0.30	G. S. Gi (3) B. L. (P).
15415	29822	21	41	77	55	+0.09	—0.22	G. S. Gi (2) L.
15423	29847	21	43	80	18	+0.01	—0.20	G. S. Gi (3) L (?).
15437	29865	21	43	74	1	(—0.06)	(—0.46)	G. S. Gi (1) L.
16040	31127	22	46	70	44	—0.018	—0.20	G. S. M. Gi (2) B. L. (P).
16067	31201	22	50	83	22	+0.19	—0.22	G. S. Gi (2) L.
16463	31974	23	32	73	23	+0.03	—0.65	G. S. Gi. C_{40} B. L. (P).
16561	32165	23	43	66	57	+0.06	. . .	G. Gz. Gi (2)
16595	32223	23	46	83	42	(+0.08)	. . .	G. Gi (2).
16646	32310	23	51	66	28	+0.033	0.00	G. S. Gi (2) B.
16692	32376	23	56	77	45	—0.04	—0.22	G. S. Gi. B. L.
16731	Gz. 23^h, 1620	23	59	79	57	(+0.22)	(—0.50)	Gz. Gi (1).

In the comparison between the zone catalogues of GOULD and GILLISS, a large number of GOULD's stars were apparently not contained in the GILLISS zones. In some cases this might be due to error of the record or reduction in either catalogue; in others it may be due to variability, especially as some of the stars of this list are as bright as the eighth magnitude. This might well be the case in regions not included in the Milky Way, where the omissions are few. In the Milky Way, however, there is no doubt that the omissions were due to the fact that the estimates of magnitude of the GILLISS zones are too faint for stars below the eighth magnitude. Accordingly in this region the GOULD stars below the ninth magnitude have not been included in the list. A list is also given of the Cordoba General Catalogue stars, which were not found in the GILLISS zones. This list also omits stars fainter than the ninth and a half magnitude.

Cordoba zone stars not found in the Gilliss zones.

[Such stars fainter than the ninth magnitude in the Milky Way are omitted.]

Number in zone.	Mag.	Number in zone.	Mag.	Number in zone.	Mag.	Number in zone.	Mag.	Number in zone.	Mag.
0^h, 174	9½	6^h, 1561	9	7^h, 3259	9	8^h, 243	8½	8^h, 1947	8½
183	9	2368	9	3772	9	392	9	1977	9
2^h, 870	7½	2903	9	3812	9	541	8½	1999	9
3^h, 178	9	7^h, 687	9	3965	9	734	9	2054	8½
217	8½	1080	9	4174	9	1011	8½	2183	9
350	9	1387	9	4530	8½	1051	9	2536	9
4^h, 1566	8½	1821	8	8^h, 24	9	1374	9	2543	9
5^h, 380	9	1838	9	52	9	1521	9	2608	9
* 444	8½	2258	9	166	9	1659	9	2613	9
6^h, 166	9	2787	9	203	9	1880	8½	2667	8½

* A star same right ascension, but differing 1′ in declination.

Cordoba zone stars not found in the Gilliss zones—Continued.

[Such stars fainter than the ninth magnitude in the Milky Way are omitted.]

Number in zone.	Mag.	Number in zone.	Mag.	Number in zone.	Mag.	Number in zone.	Mag.	Number in zone.	Mag.
8^h, 2670	9	10^h, 590	9	11^h, 668	9	12^h, 399	9	14^h, 690	9
2692	9	728	9	821	9	417	8½	758	9
2904	9	773	9	1043	9	428	9	998	9
3069	8½	845	8½	1062	9	473	9	1592	9
3193	8½	848	9	1111	9	499	8	1609	9
3468	9	851	9	1199	9	666	9	1717	9
3475	9	874	8½	1254	9	759	8	1724	9
3558	9	927	9	1518	9	1113	9	2132	9
3654	9	1019	9	1574	8½	1447	9	2345	9
3842	8½	1271	9	1713	9	1676	9	2346	9
3843	8½	1406	9	1771	9	1954	7	2347	9
4186	9	1484	9	1877	9	1975	9	2365	9
4243	9	1543	8½	2020	9	2090	9	2590	9
4322	9	1583	9	2176	9	2143	8	2822	9
4390	9	1614	9	2191	9	2219	9	2903	9
9^h, 22	9	1618	9	2221	9	2360	9	2930	9
192	9	1622	9	2222	9	2507	9	3228	9
288	9	1624	9	2359	9	2530	9	3675	9
333	9	1683	8¾	2438	9	2566	8½	15^h, 135	9
671	9	1713	9	2517	9	2580	9	201	9
970	9	1788	9	2589	9	2622	9	643	9
1630	8½	1797	9	2722	9	2627	9	938	9
2062	9	1847	8½	2731	9	2652	9	1625	9
2071	9	2254	8¾	2754	9	2834	9	1910	9
2079	9	2261	8½	2882	9	2910	9	1973	8½
2801	8½	2520	9	2909	9	2977	9	1999	9
2940	8½	2531	9	3080	9	3022	9	2154	8½
3329	9	2569	9	3203	9	3099	9	3364	9
3429	9	2818	9	3249	8½	3200	9	16^h, 1350	9
3431	9	3035	8½	3250	9	3318	9	1647	8½
3459	9	3253	9	3261	9	3401	9	1834	9
3616	9	3363	9	3340	9	3411	8½	2337	9
3849	9	3479	9	3381	9	3500	9	3243	9
4104	9	3499	9	3440	8¾	3524	8½	18^h, 979	8½
4124	9	3517	9	3479	9	13^h, 34	9	2068	9
4129	8½	3548	9	3568	9	97	9	2437	9
4266	9	3683	9	3763	8½	316	9	19^h, 403	9
4279	9	3704	9	3781	9	451	9	1466	9
4308	9	3725	9	3864	9	640	9	1851	9
4347	9	3811	9	4050	7	733	9	20^h, 1091	9
4392	9	4127	9	4097	9	888	9	22^h, 806	{ 8 var.
4527	9	4139	9	12^h, 9	9	1059	9		
4547	9	4185	9	53	9	1222	9	23^h, 1042	8½
10^h, 8	9	4209	9	74	8¾	2159	9	1296	8
102	9	11^h, 220	9	218	9	14^h, 492	8½		
432	8½	524	9	321	9	515	9		
526	9	598	8½	350	9	536	9		
543	9	664	9	378	9	688	8		

Cordoba General Catalogue stars not contained in the Gilliss zones.

[None fainter than the ninth and a half magnitude included.]

Gould's No.	Mag.	Gould's No.	Mag.	Gould's No.	Mag.	Gould's No.	Mag.
393	9½	10967	9½	13120	9½	16232	9½
453	9	10988	9	13161	9½	16233	9
596	9	11014	8¾	13231	9	16298	9
946	8	11034	9½	13262	9	16324	8
1171	8½	11072	9½	13334	9½	16326	8¾
1179	9½	11100	10(?)	13337	9½	16418	9½
1795	8½	11114	9	13445	9	16520	9
2782	8½	11166	9	13776	9½	16533	9½
3962	7½	11172	9	13860	9½	16703	7
4886	8½	11186	9	13861	9½	17305	8½
6247	9	11174	9½	13893	9½	17356	9½
8374	9½	11189	9½	14240	8¾	17487	8
9162	9½	11273	9½	14447	9½	18057	7¾
9365	9	11572	9	14457	8½	18176	8½
9670	7¾	11742	9½	14766	8¾	19451	9
9659	9	11745	9½	14768	9½	24605	9½
9676	8½	12069	9	15300	9½	25047	8½
9806	9	12087	9	15492	9½	25212	9½
10234	9½	11091	9½	15521	8½	25319	9
10297	9½	12254	7½	15592	7¾	27133	8½
10511	9½	12401	8¾	15637	8½	28053	9½
10539	9	12439	9½	15685	8½	29457	9½
10770	9	12807	9½	15715	9½	30765	var.
10773	8	12840	9½	15801	8		
10789	9	11888	9	15884	9½		
10892	8¾	12970	9	16152	9½		

As stated in Captain Gilliss's introduction, the lower limit of the zones was intended to be about 60°. The first sweeps in the winter of 1850 extended to 62° 30′. After that it was found impossible with the small force and numerous other works in hand to extend the limit below 65°. The following table gives the northern limits of the zones actually observed:

Northern limits of the zones.

h.	m.		h.	m.	°	′
0	0	to	4	42	65	42
4	42		5	0	65	30
5	0		5	42	65	0
5	42		11	0	62	12
11	0		12	0	62	30
12	0		18	30	65	45
18	30		0	0	66	10

Explanation of the Printed Catalogue.

Column 1 is the catalogue number.

Column 2, the name of the star when contained in previous catalogues, or the constellation in which the star is found, the constellation boundaries being the same as those used in the *Uranometria Argentina.* In naming the stars preference is given according to priority of publication.

Column 3, the observed magnitude.

Column 4, the mean right ascension for 1850.0.

Column 5, the precession in right ascension computed by the formulæ given the star tables of the American Ephemeris. The value of $n \sin \alpha \tan \delta$ was checked by a sliding scale especially designed for that purpose by Professor Frisby.

Column 6, the mean south declination for 1850.0.

Column 7, the precession in declination taken directly from the star tables of the American Ephemeris.

Column 8, the mean year of the observations.

Column 9, the number of observations in each coordinate.

A CATALOGUE

OF

16,748 SOUTHERN STARS

DEDUCED BY THE UNITED STATES NAVAL OBSERVATORY FROM THE ZONE OBSERVATIONS MADE AT SANTIAGO DE CHILE

BY THE

U. S. NAVAL ASTRONOMICAL EXPEDITION

TO

THE SOUTHERN HEMISPHERE

DURING

THE YEARS 1849-'50-'51-'52.

CATALOGUE

OF

16 748 SOUTHERN STARS

FROM THE

ZONE OBSERVATIONS MADE AT SANTIAGO, CHILE,

DURING THE YEARS 1850 TO 1852.

Number.	Constellation, Name of Star, or Synonym.	Magnitude.	Right Ascension, 1850.0.	Annual Precession.	South Declination, 1850.0.	Annual Precession.	Mean year.	No. of obs.
			h. m. s.	s.	° ′ ″	″		
1	Gould, Z. C., 0h 36 . . .	9.0	0 0 0.65	+ 3.071	78 19 46.1	+ 20.06	50.81	2
2	Tucanæ	9.0	3.54	3.071	66 25 15.6	.06	52.28	2
3	Tucanæ	9.8	5.78	3.070	72 38 47.7	.06	51.32	2
4	Gould, 26	9.0	6.04	3.070	67 43 49.6	.06	51.88	3
5	Melbourne, (1), 2	7.5	7.46	+ 3.070	66 45 56.5	+ 20.06	51.90	2
6	Lacaille, 9736	8.8	0 0 36.06	+ 3.054	79 2 44.6	+ 20.06	51.33	2
7	Lacaille, 9734	7.0	39.24	3.052	79 26 1.0	.06	51.30	2
8	Gould, Z. C., 0h 52 . . .	9.0	41.70	3.060	70 53 27.7	.06	51.77	1
9	Lacaille, 9745	7.8	51.30	2.981	86 52 28.9	.06	51.87	6
10	Tucanæ	9.0	58.45	+ 3.058	67 20 55.2	+ 20.06	51.82	1
11	Tucanæ	9.0	0 1 0.52	+ 3.058	67 15 18.5	+ 20.06	51.82	1
12	Hydri	9.8	3.94	3.044	77 25 43.8	.06	51.29	2
13	Lacaille, 9743	7.8	13.53	3.045	75 4 23.4	.06	50.83	3
14	Octantis	9.3	14.38	2.990	84 56 23.5	.06	51.66	5
15	Gould, Z. C., 0h 64 . . .	9.4	16.55	+ 3.054	67 35 40.3	+ 20.06	51.62	4
16	Tucanæ	9.0	0 1 20.96	+ 3.054	66 38 46.2	+ 20.06	51.90	1
17	Hydri	10.0	26.59	3.019	80 49 25.0	.06	51.83	1
18	Hydri	9.8	1 45.39	3.033	75 15 23.2	.05	50.82	2
19	Tucanæ	10.0	2 4.49	3.042	68 2 21.2	.05	50.81	1
20	Tucanæ	9.0	4.79	+ 3.043	66 51 49.2	+ 20.05	51.90	1
21	Octantis	9.2	0 2 13.58	+ 2.942	84 16 2.6	+ 20.05	50.86	2
22	Gould, Z. C., 0h 101 . .	9.8	18.62	3.039	68 17 0.4	.05	51.80	2
23	Hydri	9.5	20.61	·998	79 40 28.4	.05	50.78	1
24	Tucanæ	8.5	28.48	3.039	66 11 30.2	.05	52.28	2
25	Tucanæ	10.0	31.55	- 3.037	66 46 56.3	- 20.05	51.90	1

Number.	Constellation, Name of Star, or Synonym.	Magnitude.	Right Ascension, 1850.0.	Annual Precession.	South Declination, 1850.0.	Annual Precession.	Mean year.	No. of obs.
			h. m. s.	s.	° ′ ″	″		
26	Tucanæ	10.0	0 2 35.79	+ 3.023	72 48 31.4	+ 20.05	50.88	1
27	Gould, Z. C., 0^h 109 . .	9.0	43.01	2.987	79 25 26.2	.05	51.30	2
28	Hydri	9.0	43.08	2.955	82 15 24.6	.05	50.81	1
29	Gould, Z. C., 0^h 110 . .	10.0	45.94	3.032	68 9 19.8	.05	50.81	1
30	Hydri	9.5	47.72	+ 2.993	78 18 56.2	+ 20.05	50.81	2
31	Lacaille, 9750	8.8	0 2 52.86	+ 2.978	79 50 11.3	+ 20.05	50.78	2
32	Gould, Z. C., 0^h 115 . .	10.0	53.16	3.021	71 40 17.9	.05	51.73	1
33	Lacaille, 9752	8.2	2 55.44	2.983	79 8 9.2	.05	51.33	2
34	Gould, Z. C., 0^h 118 . .	10.0	3 0.06	3.011	73 58 1.2	.05	50.87	2
35	γ^3 Octantis	6.5	7.18	+ 2.924	83 3 28.4	+ 20.05	51.36	2
36	Hydri	9.5	0 3 7.35	+ 2.982	78 44 10.7	+ 20.05	50.78	1
37	Tucanæ	10.0	6.87	3.008	74 29 5.2	.05	50.86	1
38	Lacaille, 9755	7.0	14.02	3.006	74 3 38.7	.05	50.87	1
39	Tucanæ	10.0	23.14	3.019	69 22 59.1	.05	51.78	1
40	Gould, Z. C., 0^h 130 . .	9.5	25.62	+ 3.000	74 20 28.1	+ 20.05	50.87	2
41	Hydri	10.0	0 3 29.69	+ 2.929	81 53 4.7	+ 20.05	51.81	1
42	Tucanæ	10.0	4 3.28	2.992	73 26 23.5	.05	50.88	1
43	Gould, Z. C., 0^h 149 . .	9.8	8.60	2.992	73 4 38.0	.05	50.88	2
44	Octantis	9.5	13.58	2.574	87 8 34.2	.05	51.86	3
45	Gould, Z. C., 0^h 151 . .	9.0	18.15	+ 2.989	73 6 34.6	+ 20.05	50.88	2
46	Tucanæ	10.0	0 4 21.31	+ 2.984	73 58 20.3	+ 20.05	50.87	1
47	Lacaille, 9764	7.7	38.81	2.900	81 0 32.4	.05	51.47	3
48	Lacaille, 9765	7.5	4 43.27	2.959	76 18 43.6	.05	50.83	2
49	Octantis	9.6	5 5.55	2.672	85 44 37.4	.05	51.70	7
50	Hydri	9.3	26.25	+ 2.867	81 2 2.9	+ 20.05	51.47	3
51	Hydri	9.5	0 5 34.33	+ 2.914	78 22 25.2	+ 20.05	50.82	2
52	Gould, Z. C., 0^h 189 . .	9.8	54.48	2.966	72 21 27.2	.05	51.78	2
53	Gould, Z. C., 0^h 190 . .	10.0	5 56.05	2.974	70 29 36.5	.05	51.77	1
54	Gould, Z. C., 0^h 198	9.0	6 9.28	2.988	66 46 43.3	.05	51.90	2
55	Gould, Z. C., 0^h 196 . .	9.8	9.37	+ 2.967	71 5 9.2	+ 20.05	51.79	2
56	Hydri	9.8	0 6 9.57	+ 2.860	80 20 30.4	+ 20.05	50.81	2
57	Tucanæ	9.2	18.36	2.976	69 1 54.6	.05	51.79	2
58	Gould, Z. C., 0^h 205 . .	8.5	23.98	2.902	77 34 30.4	.05	51.29	2
59	Tucanæ	10.0	24.26	2.985	66 37 42.7	.05	51.90	1
60	Tucanæ	9.0	27.07	+ 2.987	66 5 13.5	+ 20.05	52.66	1
61	Gould, Z. C., 0^h 209 .	9.5	0 6 33.20	+ 2.915	76 19 8.6	+ 20.05	50.83	2
62	Gould, Z. C., 0^h 213 . .	9.2	43.08	2.920	75 31 14.6	.05	50.88	2
63	Octantis	9.0	46.48	2.943	88 14 58.6	.05	51.86	4
64	Tucanæ	9.8	48.04	2.966	69 27 46.6	.05	51.84	2
65	Tucanæ	9.5	53.13	+ 2.981	66 8 38.7	+ 20.05	52.66	1

Number.	Constellation, Name of Star, or Synonym.	Magnitude.	Right Ascension, 1850.0.	Annual Precession.	South Declination, 1850.0.	Annual Precession.	Mean year.	No. of obs.
			h. m. s.	s.	° ′ ″	″		
66	Lacaille, 15	7.5	0 6 56.45	+ 2.886	75 44 51.0	+ 20.05	50.85	1
67	Gould, Z. C., 0h 223 . .	9.2	6 59.93	2.977	66 43 0.9	.05	51.90	2
68	Gould, Z. C., 0h 222 . .	10.0	7 0.26	2.944	72 18 6.2	.05	51.78	2
69	Gould, Z. C., 0h 224 . .	8.5	1.87	2.976	69 45 11.2	.05	51.89	1
70	Hydri	10.0	5.80	+ 2.891	77 5 36.9	+ 20.05	50.85	2
71	Octantis	9.5	0 7 28.83	+ 2.719	82 56 33.0	+ 20.05	51.88	1
72	Lacaille, 23	6.1	31.14	2.470	85 49 44.9	.05	51.62	9
73	Gould, Z. C., 0h 244 . .	8.0	48.59	2.967	69 22 0.9	.04	51.78	1
74	Gould, 159	9.0	7 57.65	2.946	69 40 57.2	.04	51.89	1
75	Tucanæ	10.0	8 3.32	+ 2.918	73 0 53.9	+ 20.04	50.88	1
76	Hydri	10.0	0 8 5.49	+ 2.824	79 13 58.9	+ 20.04	51.82	1
77	Gould, Z. C., 0h 256 . .	9.5	21.37	2.934	70 35 7.0	.04	51.77	1
78	Gould, Z. C., 0h 259 . .	10.0	31.99	2.913	72 34 28.2	.04	51.75	1
79	Gould, Z. C., 0h 260 . .	9.8	32.30	2.916	72 15 36.5	.04	51.78	2
80	Tucanæ	8.8	33.08	+ 2.950	67 45 27.0	+ 20.04	51.90	2
81	Melbourne, (1), 11 . . .	9.0	0 8 48.28	+ 2.940	68 43 59.9	+ 20.04	51.79	2
82	Hydri	10.0	51.71	2.843	77 15 51.9	.04	50.81	1
83	Gould, Z. C., 0h 271 . .	10.0	53.36	2.912	72 5 2.9	.04	51.81	1
84	Lacaille, 30	6.5	8 59.30	2.849	76 44 45.0	.04	50.86	2
85	Lacaille, 29	7.2	9 10.39	+ 2.746	80 40 59.4	+ 20.04	51.47	3
86	Lacaille, 32	8.0	0 9 15.58	+ 2.913	71 13 35.4	+ 20.04	51.80	1
87	Stone, 88	8.0	16.07	2.941	67 31 31.0	.04	51.87	3
88	Gould, 186	9.5	25.07	2.914	70 47 52.3	.04	51.78	1
89	Gould, Z. C., 0h 293 . .	9.2	33.42	2.771	79 30 37.7	.04	51.32	2
90	Tucanæ	9.5	41.95	+ 2.942	66 28 8.1	+ 20.04	51.91	1
91	Gould, 193	9.5	0 9 57.13	+ 2.905	70 47 53.2	+ 20.04	51.78	1
92	Tucanæ	10.0	9 59.11	2.926	68 17 27.0	.04	51.80	2
93	Lacaille, 33	7.0	10 5.07	2.751	79 36 45.1	.04	50.78	1
94	Melbourne, (1), 13 . . .	8.5	10.05	2.929	67 24 2.6	.04	51.87	3
95	Gould, Z. C., 0h 306 . .	9.5	14.38	+ 2.935	66 29 39.7	+ 20.04	51.91	1
96	Gould, Z. C., 0h 305 . .	8.7	0 10 16.32	+ 2.852	74 47 41.7	+ 20.04	50.86	2
97	Gould, Z. C., 0h 307 . .	10.0	17.63	2.878	72 44 55.2	.04	50.88	1
98	Octantis	10.0	27.20	2.368	85 3 11.7	.04	51.88	2
99	Gould, Z. C., 0h 311 . .	9.0	28.57	2.897	70 42 44.4	.04	51.77	2
100	Tucanæ	9.5	28.86	+ 2.901	70 19 20.7	+ 20.04	51.84	2
101	Stone, 93	8.0	0 10 33.79	+ 2.925	67 11 33.4	+ 20.04	51.86	2
102	Tucanæ	10.0	35.91	2.919	67 58 48.0	.03	51.87	3
103	Gould, Z. C., 0h 314 . . .	9.5	39.15	2.930	66 25 27.4	.03	51.90	1
104	Gould, Z. C., 0h 316 . . .	8.5	42.88	2.846	74 30 39.1	.03	50.86	1
105	Tucanæ	9.8	49.27	+ 2.913	68 19 48.6	+ 20.03	51.80	2

Number.	Constellation, Name of Star, or Synonym.	Magnitude.	Right Ascension, 1850.0.	Annual Precession.	South Declination, 1850.0.	Annual Precession.	Mean year.	No. of obs.
			h. m. s.	s.	° ′ ″	″		
106	Tucanæ	10.0	0 10 53.77	+ 2.891	70 40 30.6	+ 20.03	51.77	2
107	Octantis	10.0	54.39	2.009	86 34 37.2	.03	51.87	2
108	Gould, Z. C., 0^h 323	8.0	56.24	2.898	69 53 45.7	.03	51.90	2
109	Gould, 207	9.0	56.56	2.929	65 53 51.5	.03	52.66	1
110	Gould, Z. C., 0^h 328	10.0	10 59.88	+ 2.866	72 41 16.6	+ 20.03	51.32	2
111	Tucanæ	9.5	0 11 2.03	+ 2.898	69 42 44.3	+ 20.03	51.78	1
112	Lacaille, 39	8.5	18.90	2.695	80 3 53.4	.03	50.82	1
113	Hydri	9.0	25.72	2.803	76 3 9.2	.03	50.84	1
114	Gould, Z. C., 0^h 339	8.0	27.35	2.794	76 29 51.4	.03	50.83	1
115	Tucanæ	9.0	39.87	+ 2.893	69 24 10.1	+ 20.03	51.84	2
116	Tucanæ	10.0	0 11 42.18	+ 2.859	72 12 43.8	+ 20.03	51.78	2
117	Tucanæ	10.0	46.48	2.828	74 18 39.8	.03	50.87	1
118	Tucanæ	9.5	11 56.93	2.896	68 24 35.6	.03	51.80	1
119	ζ Tucanæ	4.0	12 13.97	2.913	65 45 20.8	.03	52.65	1
120	Hydri	9.5	15.31	+ 2.745	77 40 31.4	+ 20.03	51.76	1
121	Gould, 230	9.5	0 12 15.56	+ 2.744	77 42 36.0	+ 20.03	51.76	1
122	Octantis	9.8	18.76	2.400	83 53 39.6	.03	50.85	2
123	Tucanæ	9.2	25.09	2.856	71 29 40.6	.03	51.77	2
124	Tucanæ	9.0	25.32	2.891	68 12 59.2	.03	51.80	1
125	Lacaille, 47	9.0	29.30	+ 2.703	78 49 20.9	+ 20.03	50.83	1
126	Gould, Z. C., 0^h 362	9.0	0 12 33.31	+ 2.871	69 56 6.4	+ 20.03	51.90	2
127	Gould, 240	10.0	50.40	2.713	78 12 51.2	.02	50.81	2
128	Gould, Z. C., 0^h 369	10.0	12 59.35	2.814	73 38 17.9	.02	50.87	1
129	Gould, Z. C., 0^h 376	10.0	13 14.26	2.863	69 44 3.5	.02	51.89	1
130	Tucanæ	10.0	14.30	+ 2.856	70 17 11.8	+ 20.02	51.84	2
131	Tucanæ	10.0	0 13 19.04	+ 2.859	69 57 53.5	+ 20.02	51.90	1
132	Hydri	10.0	29.93	2.743	76 32 5.1	.02	50.83	1
133	Tucanæ	10.0	30.49	2.875	68 14 1.6	.02	51.80	1
134	π Tucanæ	6.0	39.60	2.847	70 27 28.7	.02	51.77	1
135	Gould, 255	10.0	45.27	+ 2.825	72 0 22.3	+ 20.02	51.81	1
136	Lacaille, 57	8.5	0 13 45.90	+ 2.792	78 3 23.2	+ 20.02	50.85	1
137	ο Octantis	7.7	46.34	− 2.658	89 11 50.3	.02	51.86	6
138	Gould, 257	9.0	47.74	+ 2.892	65 56 44.6	.02	52.66	1
139	Brisbane, 30	8.2	48.82	2.811	72 48 40.4	.02	50.89	2
140	Hydri	10.2	53.09	+ 2.668	78 40 8.8	+ 20.02	50.81	2
141	Tucanæ	10.0	0 13 53.76	+ 2.874	67 45 2.8	+ 20.02	51.91	1
142	Gould, Z. C., 0^h 397	10.0	57.57	2.770	74 48 33.6	.02	50.86	1
143	Gould, Z. C., 0^h 398	9.0	13 59.99	2.729	76 35 27.0	.02	50.83	1
144	Hydri	10.0	14 2.73	2.728	76 35 31.4	.02	50.89	1
145	Gould, Z. C., 0^h 402	10.0	6.01	+ 2.827	71 26 33.4	+ 20.02	51.82	2

Number.	Constellation, Name of Star, or Synonym.	Magnitude.	Right Ascension, 1850.0.	Annual Precession.	South Declination, 1850.0.	Annual Precession.	Mean year.	No. of obs.
			h. m. s.	s.	° ′ ″	″		
146	Hydri	10.0	0 14 6.41	+ 2.696	77 39 55.3	+ 20.02	51.76	1
147	Tucanæ	10.0	14.28	2.851	69 22 41.7	.02	51.78	1
148	Gould, Z. C., 0h 406 . .	9.0	14.81	2.874	67 13 43.4	.02	51.86	2
149	Hydri	9.0	20.02	2.744	75 42 39.7	.02	50.85	1
150	Octantis	10.0	22.94	+ 2.210	84 26 42.0	+ 20.02	51.38	2
151	Gould, Z. C., 0h 411 . .	7.5	0 14 24.11	+ 2.878	66 35 4.7	+ 20.02	51.90	1
152	Tucanæ	10.0	25.24	2.871	67 18 56.0	.02	51.82	1
153	Tucanæ	10.0	25.65	2.860	68 12 58.0	.02	51.81	1
154	Gould, 271	8.0	28.40	2.871	67 8 12.8	.02	51.86	2
155	Hydri	10.0	48.14	+ 2.477	81 44 37.3	+ 20.01	51.81	1
156	Octantis	9.3	0 14 53.80	+ 1.854	85 54 55.1	+ 20.01	51.86	6
157	Hydri	10.0	14 57.70	2.414	82 26 43.7	.01	50.83	1
158	Lacaille, 64	6.8	15 1.46	2.650	78 15 34.2	.01	50.81	2
159	Gould, Z. C., 0h 431 . .	9.5	2.54	2.812	71 21 44.5	.01	51.80	1
160	Hydri	9.5	3.54	+ 2.671	77 37 56.8	+ 20.01	51.76	1
161	Gould, 286	8.5	0 15 19.27	+ 2.846	68 27 35.1	+ 20.01	51.79	1
162	Hydri	10.0	22.76	2.526	80 40 10.0	.01	51.83	1
163	Octantis	9.0	39.02	2.378	82 30 37.9	.01	50.83	1
164	Gould, Z. C., 0h 453 . .	9.0	50.64	2.752	73 51 46.4	.01	50.87	2
165	Gould, Z. C., 0h 457 . .	9.8	56.44	+ 2.823	69 29 41.2	+ 20.01	51.84	2
166	Octantis	10.0	0 15 59.32	+ 2.203	83 52 41.3	+ 20.01	50.85	2
167	Gould, 301	8.0	16 3.55	2.862	65 57 4.0	.01	52.66	1
168	Hydri	10.0	38.85	2.513	80 8 41.0	.00	50.82	1
169	Octantis	9.0	41.55	2.047	84 34 32.9	.00	51.18	3
170	Tucanæ	9.0	44.59	+ 2.825	68 32 47.4	+ 20.00	51.80	1
171	Tucanæ	10.0	0 16 45.75	+ 2.772	71 55 33.6	+ 20.00	51.73	1
172	Hydri	9.5	46.07	2.654	76 50 10.7	.00	50.89	1
173	Lacaille, 76	8.5	16 51.21	2.627	77 33 22.0	.00	50.81	1
174	Brisbane, 37	9.2	17 16.76	2.740	73 5 6.8	.00	50.87	3
175	Gould, Z. C., 0h 488 . .	10.0	17.30	+ 2.775	71 16 30.7	+ 20.00	51.80	1
176	Gould, Z. C., 0h 491 . .	10.0	0 17 21.93	+ 2.711	74 20 42.1	+ 20.00	50.86	1
177	Lacaille, 80	Neb.	22.45	2.742	72 54 48.4	.00	50.89	2
178	Hydri	10.0	24.69	2.662	76 4 40.7	.00	50.84	1
179	Brisbane, 41	9.0	30.94	2.742	72 48 40.8	.00	50.89	2
180	Tucanæ	10.0	34.11	+ 2.758	71 51 53.4	+ 20.00	51.73	1
181	Octantis	9.5	0 17 38.75	+ 2.167	83 30 58.8	+ 20.00	50.82	2
182	Gould, 329	9.0	39.66	2.567	78 28 41.0	.00	50.76	1
183	Gould, Z. C., 0h 501 . .	8.5	45.97	2.640	76 30 29.8	.00	50.83	1
184	β Hydri	3.0	47.57	2.579	78 5 56.8	.00	50.85	1
185	Hydri	8.5	51.13	+ 2.600	77 33 28.4	+ 20.00	51.76	1

No. 181. Possibly 83° 30′ 30.9″.

Number.	Constellation, Name of Star, or Synonym.	Magnitude.	Right Ascension, 1850.0.	Annual Precession.	South Declination, 1850.0.	Annual Precession.	Mean year.	No. of obs.
			h. m. s.	s.	° ′ ″	″		
186	Hydri	9.0	0 17 53.99	+ 2.663	75 39 23.1	+ 20.01	50.85	1
187	Tucanæ	8.7	18 6.61	2.832	66 13 46.2	19.99	52.16	3
188	Hydri	10.0	8.91	2.675	75 3 56.4	.99	50.88	2
189	Tucanæ	9.3	10.09	2.832	66 10 56.6	.99	52.16	3
190	Hydri	9.8	10.99	+ 2.330	81 52 {41.8 / 20.1}	+ 19.99	51.31	2
191	Tucanæ	10.0	0 18 12.13	+ 2.754	71 31 37.2	+ 19.99	51.77	2
192	Hydri	9.2	18.73	2.518	79 4 55.8	.99	51.33	2
193	Hydri	10.0	19.93	2.647	75 52 46.2	.99	50.84	1
194	Brisbane, 42	9.5	22.78	2.711	73 26 54.2	.99	50.88	2
195	Hydri	9.5	29.40	+ 2.637	76 5 38.5	+ 19.99	50.84	1
196	Tucanæ	10.0	0 18 33.38	+ 2.822	66 37 8.4	+ 19.99	51.90	1
197	Gould, Z. C., 0h 525 . .	8.0	38.96	2.795	68 34 37.5	.99	51.80	1
198	Octantis	9.9	45.13	1.643	85 37 41.3	.99	51.87	4
199	Gould, Z. C., 0h 528 . .	9.5	45.22	2.766	70 45 53.6	.99	51.77	2
200	Hydri	9.5	18 54.34	+ 2.654	75 14 9.5	+ 19.99	50.89	1
201	Tucanæ	10.0	0 19 25.10	+ 2.740	71 11 11.4	+ 19.98	51.80	1
202	Tucanæ	9.0	30.66	2.663	74 27 46.1	.98	50.86	1
203	Hydri	10.0	30.89	2.574	77 9 27.9	.98	50.81	1
204	Gould, Z. C., 0h 548 . .	9.5	31.23	2.776	69 0 29.4	.98	51.79	1
205	Tucanæ	9.5	35.10	+ 2.741	70 54 18.0	+ 19.98	51.77	1
206	Tucanæ	10.0	0 19 36.33	+ 2.800	67 10 23.1	+ 19.98	51.90	1
207	Lacaille, 93	7.5	54.55	2.750	70 8 15.9	.98	51.91	1
208	Gould, Z. C., 0h 561 . .	8.5	55.59	2.776	68 33 10.8	.98	51.80	1
209	Tucanæ	9.5	19 57.66	2.736	70 54 10.5	.98	51.77	1
210	Tucanæ	10.0	20 4.45	+ 2.738	70 41 46.8	+ 19.98	51.77	2
211	Tucanæ	10.0	0 20 7.19	+ 2.745	70 14 51.0	+ 19.98	51.77	1
212	Tucanæ	10.0	9.22	2.742	70 23 55.6	.98	51.84	2
213	Hydri	11.0	12.10	2.236	81 58 57.5	.98	51.81	1
214	Tucanæ	10.0	12.75	2.728	71 4 47.0	.98	51.77	1
215	Gould, Z. C., 0h 568 . .	10.0	14.96	+ 2.707	72 4 4.3	+ 19.98	51.81	1
216	Tucanæ	10.0	0 20 29.54	+ 2.759	69 9 36.6	+ 19.98	51.79	2
217	Hydri	11.0	36.89	2.222	81 57 24.9	.98	51.81	1
218	Tucanæ	9.5	40.66	2.800	66 7 3.1	.97	52.66	1
219	Hydri	10.0	43.60	2.564	76 36 43.6	.97	50.86	2
220	Hydri	9.2	44.50	+ 2.542	77 8 46.8	+ 19.97	50.85	2
221	Tucanæ	10.0	0 20 45.59	+ 2.717	71 10 52.7	+ 19.97	51.80	1
222	Tucanæ	10.0	20 55.34	2.781	67 16 48.4	.97	51.82	1
223	Brisbane, 48	9.0	21 3.64	2.677	72 42 45.2	.97	51.32	2
224	Gould, 391	9.0	11.42	2.781	67 0 46.3	.97	51.86	2
225	Tucanæ	9.2	15.52	+ 2.711	71 2 58.8	+ 19.97	51.79	2

Number.	Constellation, Name of Star, or Synonym.	Magnitude.	Right Ascension, 1850.0.	Annual Precession.	South Declination, 1850.0.	Annual Precession.	Mean year.	No. of obs.
			h. m. s.	s.	° ′ ″	″		
226	Tucanæ	10.0	0 21 18.24	+ 2.724	70 22 20.0	+ 19.97	51.84	2
227	Gould, Z. C., 0ʰ 599 . . .	8.5	19.51	2.558	76 26 42.4	.97	50.83	1
228	Gould, 395	9.0	23.41	2.782	66 44 30.1	.97	51.91	1
229	Hydri	10.0	27.23	2.298	80 49 37.5	.97	51.83	1
230	Gould, 397	10.0	28.67	+ 2.785	66 44 13.1	+ 19.97	51.91	1
231	Hydri	10.0	0 21 31.39	+ 2.405	79 21 29.3	+ 19.97	51.82	1
232	Gould, Z. C., 0ʰ 613 . . .	10.0	40.48	+ 2.715	70 30 15.2	.97	51.77	1
233	Octantis	8.9	41.65	− 1.761	88 30 4.4	.97	51.87	4
234	Hydri	9.5	21 47.21	+ 2.547	76 24 16.4	.97	50.83	1
235	Lacaille, 113	8.8	22 16.35	+ 1.844	83 58 8.8	+ 19.96	50.86	3
236	Tucanæ	8.8	0 22 22.32	+ 2.778	66 5 11.5	+ 19.96	52.29	2
237	Tucanæ	9.5	40.58	2.705	70 11 9.4	.96	51.90	1
238	Hydri	9.5	22 42.41	2.557	75 36 5.2	.96	50.85	1
239	Gould, Z. C., 0ʰ 642 . . .	9.0	23 8.92	2.754	66 59 38.6	.95	51.86	2
240	Gould, Z. C., 0ʰ 641 . . .	9.0	10.89	+ 2.581	74 36 37.6	+ 19.95	50.86	1
241	Hydri	9.8	0 23 15.65	+ 2.478	77 8 5.8	+ 19.95	50.85	2
242	Octantis	9.3	19.44	1.394	85 21 37.3	.95	51.86	6
243	Hydri	9.5	33.08	2.098	82 3 20.8	.95	50.81	1
244	Tucanæ	10.0	38.81	2.735	67 45 {17.6 / 45.4}	.95	51.90	2
245	Octantis	10.0	23 44.32	+ 1.339	85 26 19.4	+ 19.95	51.88	2
246	Gould, Z. C., 0ʰ 662 . . .	10.0	0 24 4.34	+ 2.586	73 54 46.8	+ 19.95	50.86	2
247	Gould, Z. C., 0ʰ 663 . . .	9.8	6.72	2.678	70 21 55.9	.95	51.84	2
248	Tucanæ	9.2	30.14	2.683	69 51 56.6	.94	51.90	3
249	Octantis	10.0	45.56	2.257	87 4 0.4	.94	51.87	3
250	Tucanæ	9.5	47.78	+ 2.649	71 9 14.6	+ 19.94	51.77	1
251	Gould, Z. C., 0ʰ 680 . .	10.0	0 24 53.18	+ 2.531	74 59 48.4	+ 19.94	50.88	2
252	Gould, Z. C., 0ʰ 682 . .	9.2	55.22	2.666	70 19 25.4	.94	51.84	2
253	Brisbane, 60	10.0	57.47	2.597	72 59 44.3	.94	50.88	1
254	Hydri	9.5	57.56	2.512	75 26 30.7	.94	50.85	1
255	Tucanæ	10.0	24 57.83	+ 2.702	68 31 48.1	+ 19.94	51.80	1
256	Tucanæ	9.5	0 25 0.78	+ 2.744	66 2 47.1	+ 19.94	52.29	2
257	Tucanæ	9.5	5.00	2.675	69 46 34.2	.94	51.90	3
258	Tucanæ	9.0	6.32	2.687	69 11 11.4	.94	51.79	2
259	Gould, 455	9.5	8.00	2.384	77 52 55.4	.94	51.31	2
260	Tucanæ	10.0	14.49	+ 2.687	69 5 37.4	+ 19.93	51.79	1
261	Hydri	9.5	0 25 47.20	+ 2.320	78 42 23.6	+ 19.93	50.81	2
262	Gould, Z. C., 0ʰ 707 . .	9.5	49.50	2.683	68 50 20.3	.93	51.79	1
263	Tucanæ	9.5	55.59	2.717	66 55 25.8	.93	51.90	1
264	Gould, Z. C., 0ʰ 712 . .	10.0	25 57.91	2.650	70 19 32.4	.93	51.77	1
265	Hydri	10.0	26 7.98	+ 2.458	76 5 1.0	+ 19.93	50.84	1

No. 244. One of these declinations is evidently 1 rev. = 27.9″ wrong.

Number.	Constellation, Name of Star, or Synonym.	Magnitude.	Right Ascension, 1850.0.	Annual Precession.	South Declination, 1850.0.	Annual Precession.	Mean year.	No. of obs.
			h. m. s.	s.	° ′ ″	″		
266	Gould, Z. C., 0h 721	9.2	0 26 17.43	+ 2.673	68 59 29.1	+ 19.92	51.79	2
267	Octantis	9.9	27.38	1.118	85 28 21.6	.92	51.87	4
268	Tucanæ	10.0	48.22	2.671	68 44 42.2	.92	51.79	2
269	Tucanæ	10.0	49.02	2.693	67 35 34.6	.92	51.82	1
270	Gould, 489	9.2	57.12	+ 2.408	76 42 45.8	+ 19.92	50.86	2
271	Lacaille, 139	7.2	0 26 59.64	+ 2.585	72 5 37.6	+ 19.92	51.36	2
272	Stone, 208	9.2	27 7.07	2.676	68 14 16.4	.92	51.79	2
273	Gould, Z. C., 0h 745	10.0	8.22	+ 2.686	67 44 42.3	.92	51.90	2
274	Octantis	9.8	24.74	− 5.662	88 57 11.6	.91	51.87	2
275	Gould, Z. C., 0h 753	9.5	48.02	+ 2.200	79 29 32.8	+ 19.91	51.30	2
276	Hydri	9.0	0 27 54.25	+ 2.394	76 31 59.7	+ 19.91	50.83	1
277	Gould, Z. C., 0h 759	10.0	54.60	2.654	68 43 13.2	.91	51.79	2
278	Hydri	9.0	27 58.73	1.837	82 29 3.7	.91	50.83	1
279	Octantis	8.9	28 4.67	1.409	84 23 11.2	.91	51.14	4
280	Gould, Z. C., 0h 767	8.8	13.12	+ 2.401	76 14 52.6	+ 19.90	50.83	2
281	Gould, Z. C., 0h 776	8.7	0 28 23.88	+ 2.632	69 25 29.5	+ 19.90	51.86	3
282	Tucanæ	10.0	27.90	2.596	70 47 5.3	.90	51.77	1
283	Gould, Z. C., 0h 779	10.0	28.18	2.669	67 37 37.4	.90	51.90	2
284	Gould, Z. C., 0h 781	10.0	38.44	2.503	73 40 17.4	.90	50.87	1
285	Hydri	10.0	39.77	+ 2.255	78 27 15.4	+ 19.90	50.78	1
286	Octantis	10.0	0 28 40.34	+ 1.441	84 9 34.1	+ 19.90	50.89	1
287	Hydri	9.0	41.99	2.377	76 25 3.1	.90	50.83	1
288	Gould, Z. C., 0h 790	9.0	28 51.96	2.630	69 10 6.4	.90	51.79	2
289	Tucanæ	10.0	29 25.22	2.638	68 26 32.1	.89	51.80	1
290	Brisbane, 73	9.7	28.59	+ 2.532	72 22 7.3	+ 19.89	51.49	3
291	Gould, Z. C., 0h 811	10.0	0 29 37.46	+ 2.552	71 40 33.3	+ 19.89	51.77	2
292	Tucanæ	10.0	40.36	+ 2.644	68 1 30.7	.89	51.81	1
293	Octantis	9.7	41.36	− 0.284	87 3 34.0	.89	51.86	7
294	Brisbane, 74	8.0	42.63	+ 2.467	74 2 59.7	.89	50.87	1
295	Gould, 543	9.5	45.90	+ 2.439	74 42 17.4	+ 19.89	50.86	2
296	Gould, Z. C., 0h 812	10.0	0 29 49.59	+ 2.151	79 20 7.7	+ 19.89	51.82	1
297	Octantis	9.5	55.15	1.685	82 50 42.4	.89	51.88	1
298	Tucanæ	10.0	57.23	2.617	68 59 9.9	.89	51.79	1
299	Gould, 545	8.0	59.06	2.387	75 41 49.5	.98	50.85	1
300	Lacaille, 151	7.0	29 59.49	+ 2.682	65 57 5.1	+ 19.88	51.92	1
301	Brisbane, 76	9.0	0 30 4.83	+ 2.474	73 40 22.5	+ 19.88	50.87	1
302	Gould, Z. C., 0h 827	9.2	10.24	2.560	71 5 19.4	.88	51.79	2
303	Tucanæ	10.0	11.28	2.657	67 4 6.8	.88	51.90	1
304	Octantis	10.0	13.23	0.435	86 11 44.2	.88	51.88	1
305	Lacaille, 154	8.2	14.06	+ 2.582	70 15 35.0	+ 19.88	51.84	2

Number.	Constellation, Name of Star, or Synonym.	Magnitude.	Right Ascension, 1850.0.	Annual Precession.	South Declination, 1850.0.	Annual Precession.	Mean year.	No. of obs.
			h. m. s.	s.	° ′ ″	″		
306	Tucanæ	10.0	0 30 28.24	+ 2.565	70 43 51.4	+ 19.88	51.77	1
307	Tucanæ	10.0	53.24	2.532	71 36 12.8	.87	51.73	1
308	Hydri	9.5	54.37	1.777	82 5 59.0	.87	50.81	1
309	Octantis	10.0	30 55.29	1.559	83 13 16.3	.87	51.88	1
310	Gould, Z. C., 0h 845	10.0	31 12.32	+ 2.112	79 17 23.6	+ 19.87	51.82	1
311	Lacaille, 161	7.0	0 31 12.68	+ 2.277	77 8 8.4	+ 19.87	50.85	2
312	Tucanæ	10.0	28.36	2.639	67 4 13.6	.87	51.90	1
313	Tucanæ	9.5	29.06	2.602	68 46 19.4	.87	51.79	2
314	Gould, Z. C., 0h 865	9.0	33.72	2.589	69 13 32.7	.87	51.79	2
315	Melbourne, (1), 43	7.5	34.38	+ 2.642	66 52 12.0	+ 19.87	51.90	2
316	Lacaille, 168	8.0	0 31 42.94	+ 2.308	76 26 0.5	+ 19.86	50.83	1
317	Tucanæ	10.0	44.91	2.487	72 28 27.1	.86	51.75	1
318	Gould, Z. C., 0h 873	9.2	50.52	2.262	77 7 7.2	.86	50.85	2
319	Tucanæ	10.0	31 52.10	2.568	69 48 37.9	.86	51.90	1
320	Gould, Z. C., 0h 879	9.2	32 0.28	+ 2.172	78 18 44.6	+ 19.86	50.81	2
321	Hydri	10.0	0 32 6.99	+ 2.025	79 54 33.5	+ 19.86	50.82	1
322	Tucanæ	10.0	27.40	2.473	72 31 28.1	.86	51.75	1
323	Lacaille, 171	8.2	31.90	2.376	74 47 2.0	.85	50.86	2
324	Octantis	9.5	34.98	1.340	83 45 24.0	.85	50.80	1
325	Tucanæ	10.0	40.85	+ 2.591	68 24 38.4	+ 19.85	51.80	1
326	Lacaille, 173	7.0	0 32 45.78	+ 2.409	73 57 46.6	+ 19.85	50.87	2
327	Tucanæ	9.0	59.11	2.440	73 6 28.1	.85	50.88	1
328	Tucanæ	9.0	32 59.36	2.441	73 5 39.0	.85	50.88	1
329	Gould, 605	11.0	33 7.51	2.555	69 33 52.4	.85	51.90	1
330	Gould, Z. C., 0h 909	9.0	7.98	+ 2.300	75 58 34.4	+ 19.85	50.84	2
331	Gould, 606	9.7	0 33 8.08	+ 2.555	69 33 50.4	+ 19.85	51.86	3
332	Gould, Z. C., 0h 913	9.5	9.19	2.585	68 23 57.0	.85	51.80	1
333	Octantis	10.0	10.76	+ 1.541	82 44 57.6	.85	51.88	1
334	Octantis	10.0	16.90	− 0.841	87 10 20.8	.85	51.87	2
335	Tucanæ	10.0	18.89	+ 2.523	70 32 57.9	+ 19.84	51.77	1
336	Tucanæ	9.8	0 33 20.68	+ 2.543	69 52 3.8	+ 19.84	51.90	3
337	Tucanæ	9.5	34.81	2.474	71 55 35.1	.84	51.49	3
338	Octantis	10.5	34.92	0.990	84 38 32.5	.84	51.91	1
339	Hydri	9.0	36.06	2.107	78 33 12.4	.84	50.78	1
340	Tucanæ	10.0	40.34	+ 2.453	72 26 52.6	+ 19.84	51.75	1
341	Hydri	8.8	0 33 50.31	+ 1.672	82 0 5.4	+ 19.84	51.31	2
342	Octantis	10.0	34 2.26	0.087	86 12 31.7	.84	51.88	1
343	Hydri	10.0	9.90	2.127	78 11 26.5	.83	50.85	1
344	Brisbane, 88	9.5	13.14	2.350	74 35 11.6	.83	50.86	2
345	Octantis	9.5	23.26	+ 0.872	84 48 30.6	+ 19.83	51.35	2

Number.	Constellation, Name of Star, or Synonym.	Magnitude.	Right Ascension, 1850.0.	Annual Precession.	South Declination, 1850.0.	Annual Precession.	Mean year.	No. of obs.
			h. m. s.	s.	° ′ ″	″		
346	Gould, 627	9.0	0 34 23.66	+ 2.608	66 40 54.2	+ 19.83	51.90	2
347	Hydri	9.5	25.22	2.147	77 48 34.8	.83	51.76	1
348	Hydri	10.0	28.35	2.156	77 39 25.8	.83	51.76	1
349	Lacaille, 179	9.0	28.50	2.551	69 0 37.8	.83	51.79	2
350	Tucanæ	9.3	29.66	+ 2.523	69 55 15.0	+ 19.83	51.90	3
351	Octantis	9.2	0 34 29.68	+ 1.328	83 26 {44.7 / 14.9}	+ 19.83	50.82	2
352	Tucanæ	10.0	34.88	2.548	69 0 27.6	.83	51.79	2
353	Gould, Z. C., 0h 947 . .	9.0	34 50.27	2.586	67 20 14.5	.83	51.82	1
354	Tucanæ	10.0	35 0.97	2.429	72 26 33.2	.82	51.75	1
355	Gould, Z. C., 0h 951 . .	9.0	1.37	+ 2.534	69 16 47.8	+ 19.82	51.78	1
356	Tucanæ	10.0	0 35 11.07	+ 2.441	72 2 24.9	+ 19.82	50.91	1
357	Hydri	8.5	16.06	2.223	76 25 21.9	.82	50.83	1
358	Hydri	9.0	24.25	2.220	76 25 18.3	.82	50.83	1
359	Lacaille, 197	8.0	31.24	1.691	81 28 52.5	.82	51.30	2
360	Gould, Z. C., 0h 969 . .	9.2	55.90	+ 2.193	76 38 9.2	+ 19.81	50.86	2
361	Gould, 649	10.0	0 35 57.60	+ 2.292	75 0 0.8	+ 19.81	50.88	2
362	ρ Tucanæ	6.2	36 2.58	2.595	66 17 35.3	.81	51.91	2
363	Gould, Z. C., 0h 976 . .	9.5	5.40	2.597	66 12 28.9	.81	51.91	2
364	Tucanæ	10.0	7.92	2.527	68 54 53.7	.81	51.79	1
365	Gould, Z. C., 0h 982 . .	10.0	11.78	+ 2.599	66 1 54.7	+ 19.81	51.92	1
366	Brisbane, 91	9.0	0 36 13.36	+ 2.313	74 30 14.5	+ 19.81	50.86	1
367	Octantis	9.3	13.39	− 1.004	87 2 43.8	.81	51.87	6
368	Tucanæ	10.0	14.01	+ 2.594	66 13 40.4	.81	51.91	1
369	Lacaille, 191	7.0	16.91	2.589	66 25 53.9	.81	51.91	1
370	Lacaille, 212	8.7	21.08	+ 1.110	83 51 25.3	+ 19.80	50.86	3
371	Lacaille, 228	8.1	0 36 22.72	− 0.009	86 4 32.1	+ 19.80	51.86	6
372	Octantis	8.8	33.79	+ 0.685	84 54 46.3	.80	51.53	3
373	Gould, Z. C., 0h 993 . .	9.5	35.92	2.460	70 50 22.5	.80	51.77	1
374	Octantis	10.0	36 44.68	1.064	83 55 56.3	.80	50.87	3
375	Gould, Z. C., 0h 1000 . .	9.5	37 8.00	+ 2.501	69 27 44.2	+ 19.79	51.86	3
376	Gould, Z. C., 0h 1006 . .	9.5	0 37 9.35	+ 2.579	65 52 26.4	+ 19.79	51.92	1
377	Gould, Z. C., 0h 1005 . . .	9.8	12.09	2.440	71 7 7.9	.79	51.79	2
378	Hydri	10.5	21.20	1.994	78 36 54.7	.79	50.83	1
379	Tucanæ	10.0	21.25	2.540	67 48 42.7	.79	51.87	3
380	Tucanæ	10.0	54.40	+ 2.551	67 6 23.9	+ 19.78	51.90	1
381	Brisbane, 98	9.0	0 37 57.86	+ 2.274	74 33 32.5	+ 19.78	50.86	1
382	Octantis	9.0	37 58.24	− 2.040	87 31 49.8	.78	51.87	3
383	Tucanæ	10.0	38 4.62	+ 2.439	70 44 39.7	.78	51.77	1
384	Gould, Z. C., 0h 1031 . .	9.5	5.80	2.479	69 33 8.7	.78	51.86	3
385	Lacaille, 206	7.0	16.05	+ 2.562	66 26 28.3	+ 19.78	51.91	1

No. 351. One of these declinations is evidently 1 rev. = 27.9″ wrong.
No. 385. This Star appears to have a proper motion of about + 0s.01 and − 0.7″.

Number.	Constellation, Name of Star, or Synonym.	Magnitude.	Right Ascension. 1850.0.	Annual Precession.	South Declination. 1850.0.	Annual Precession.	Mean year.	No. of obs.
			h. m. s.	s.	° ′ ″	″		
386	Tucanæ	10.0	0 38 16.67	+ 2.336	73 11 6.6	+ 19.78	50.88	1
387	Lacaille, 209	8.8	34.74	2.231	75 5 8.5	.77	50.88	2
388	Lacaille, 211	9.0	37.55	2.382	71 59 26.4	.77	50.91	1
389	Brisbane, 102	9.5	41.88	2.253	74 39 23.6	.77	50.86	2
390	Tucanæ	10.0	44.77	+ 2.332	73 4 34.9	+ 19.77	50.88	1
391	Octantis	10.5	0 38 57.13	+ 0.704	84 32 43.2	+ 19.77	51.91	1
392	Lacaille, 242	8.0	38 58.85	− 0.368	86 14 3.7	.77	51.86	7
393	Octantis	10.5	39 16.29	+ 0.673	84 34 18.2	.76	51.91	1
394	Hydri	9.8	21.34	1.604	81 9 14.7	.76	51.31	2
395	Tucanæ	10.0	23.83	+ 2.272	74 2 57.9	+ 19.76	50.87	1
396	Tucanæ	10.0	0 39 31.12	+ 2.557	65 59 11.4	+ 19.76	51.92	1
397	Tucanæ	10.0	36.86	2.482	68 41 41.2	.76	51.79	2
398	Lacaille, 221	8.5	39 54.49	+ 1.891	78 54 9.3	.75	50.83	1
399	Lacaille, 248	7.5	40 3.36	− 0.755	86 31 23.8	.75	51.86	5
400	Octantis	10.0	4.00	−12.554	89 8 50.3	+ 19.75	51.87	1
401	Gould, Z. C., 0h 1080	10.0	0 40 5.99	+ 2.322	72 45 12.2	+ 19.75	51.75	1
402	Brisbane, 104	9.5	8.54	2.224	74 38 19.7	.75	50.86	1
403	Hydri	9.0	10.76	+ 2.169	75 32 24.8	.75	50.85	1
404	Octantis	8.9	17.70	− 0.279	86 0 21.7	.75	51.88	1
405	Octantis	9.0	23.26	+ 1.235	82 43 41.8	+ 19.75	51.31	2
406	Tucanæ	10.0	0 40 23.27	+ 2.348	72 3 1.5	+ 19.75	50.91	1
407	Hydri	9.5	23.63	2.072	76 44 20.0	.75	50.86	2
408	Tucanæ	10.0	30.62	2.437	69 40 40.6	.74	51.90	2
409	Tucanæ	10.0	33.85	2.234	74 18 4.6	.74	50.87	1
410	Gould, Z. C., 0h 1100	9.5	39.27	+ 2.319	72 36 10.9	+ 19.74	51.42	2
411	Brisbane, 105	9.8	0 40 50.90	+ 2.276	73 24 57.9	+ 19.74	50.88	2
412	Hydri	10.0	50.98	1.691	80 15 27.2	.74	50.82	1
413	Hydri	9.0	40 58.80	1.656	80 29 9.0	.74	50.81	1
414	Tucanæ	10.0	41 4.38	2.404	70 21 2.5	.74	51.84	2
415	Hydri	9.5	13.38	+ 2.004	77 22 19.9	+ 19.73	50.81	1
416	Hydri	9.5	0 41 26.70	+ 1.756	79 38 29.5	+ 19.73	50.78	1
417	Gould, Z. C., 0h 1116	9.0	31.22	2.459	68 31 49.7	.73	51.80	1
418	Hydri	9.0	31.83	2.133	75 36 20.3	.73	50.85	1
419	Hydri	9.0	40.04	1.959	77 44 31.7	.73	51.76	1
420	Gould, Z. C., 0h 1119	10.0	41 47.83	+ 2.032	76 52 2.5	+ 19.72	50.89	1
421	Tucanæ	10.0	0 42 8.34	+ 2.493	67 5 45.7	+ 19.72	52.14	3
422	Gould, Z. C., 0h 1137	10.0	16.51	2.522	65 54 56.0	.72	51.92	1
423	Octantis	9.8	27.27	1.131	82 46 6.6	.71	51.31	2
424	Gould, Z. C., 0h 1142	8.8	28.48	2.451	68 20 57.2	.71	51.80	2
425	Tucanæ	9.0	28.58	+ 2.508	66 23 17.4	+ 19.71	51.91	2

422. Original record 65° 50′ 49.4″. Corrected by 1 wire interval.

Number.	Constellation, Name of Star, or Synonym.	Magnitude.	Right Ascension, 1850.0.	Annual Precession.	South Declination, 1850.0.	Annual Precession.	Mean year.	No. of obs.
			h. m. s.	s.	° ′ ″	″		
426	Tucanæ	9.8	0 42 35.34	+ 2.381	70 18 52.4	+ 19.71	51.84	2
427	Tucanæ	10.0	43 0.21	2.428	68 50 21.5	.70	51.79	1
428	Tucanæ	10.0	12.42	2.496	66 29 18.9	.70	51.91	1
429	Gould, Z. C., 0h 1164 . .	9.0	13.63	2.513	65 49 18.3	.70	51.92	1
430	Gould, Z. C., 0h 1160 . .	8.5	14.13	+ 2.385	69 56 39.3	+ 19.70	51.90	3
431	λ Hydri	5.0	0 43 21.52	+ 2.082	75 44 24.6	+ 19.70	50.84	2
432	Tucanæ	10.0	24.54	2.415	69 1 41.0	.70	51.79	1
433	Gould, Z. C., 0h 1158 . .	9.8	28.74	1.765	79 4 56.2	.70	51.33	2
434	Hydri	10.0	43 54.70	1.574	80 20 58.4	.69	50.81	2
435	Brisbane, 110	9.0	44 1.42	+ 2.164	74 17 46.0	+ 19.69	50.87	2
436	Tucanæ	10.0	0 44 7.27	+ 2.331	70 56 57.2	+ 19.69	51.77	1
437	Brisbane, 111	9.0	17.76	2.234	72 56 54.1	.68	50.88	1
438	Tucanæ	9.3	18.37	2.466	67 1 19.6	.68	52.14	3
439	Hydri	9.2	24.16	2.054	75 48 42.6	.68	51.84	2
440	Hydri	10.0	27.06	+ 1.831	78 15 50.1	+ 19.68	50.85	1
441	Gould, Z. C., 0h 1192 . .	10.0	0 44 37.65	+ 2.260	72 18 37.1	+ 19.68	51.33	2
442	Hydri	10.0	44 51.75	1.427	81 0 45.2	.67	51.31	2
443	Tucanæ	9.8	45 15.29	2.206	73 8 21.0	.66	50.88	2
444	Lacaille, 244	8.0	17.40	2.265	71 58 11.8	.66	51.36	2
445	Tucanæ	9.5	31.67	+ 2.475	66 9 25.0	+ 19.66	51.91	1
446	Octantis	8.8	0 45 33.33	+ 0.161	84 49 7.2	+ 19.66	51.35	2
447	Octantis	10.0	38.57	0.814	83 19 11.8	.66	50.84	1
448	Tucanæ	9.2	52.35	2.470	66 9 6.4	.66	51.91	2
449	Tucanæ	10.0	45 53.14	2.285	71 19 52.5	.66	51.80	1
450	Gould. 796	10.0	46 4.51	+ 2.171	73 29 41.7	+ 19.65	50.88	1
451	Hydri	9.5	0 46 7.21	+ 1.544	80 1 56.0	+ 19.65	50.82	1
452	Tucanæ	9.8	28.46	+ 2.371	68 58 59.9	.95	51.79	2
453	Octantis	8.7	34.34	− 0.032	85 1 52.5	.64	51.69	6
454	Octantis	9.5	36.30	+ 0.351	84 19 58.7	.64	50.87	2
455	Lacaille, 293	8.3	38.13	− 1.631	86 42 39.5	+ 19.64	51.86	6
456	Lacaille, 250	7.5	0 46 41.02	+ 2.319	70 19 1.6	+ 19.64	51.86	2
457	Tucanæ	10.0	44.20	2.265	71 26 19.6	.64	51.80	1
458	Brisbane, 116	8.0	45.10	2.315	70 18 56.4	.64	51.86	2
459	Tucanæ	9.5	46 56.72	2.408	67 43 1.2	.64	51.90	2
460	Hydri	9.5	47 1.51	+ 1.873	77 12 7.7	+ 19.64	50.81	1
461	Gould, Z. C., 0h 1252 . .	8.5	0 47 3.28	+ 2.004	75 41 3.2	+ 19.64	50.85	1
462	Hydri	10.0	4.57	1.626	79 18 52.2	.63	51.82	1
463	Hydri	10.0	16.03	+ 1.601	79 27 15.2	.63	51.31	2
464	Octantis	9.2	20.47	− 2.336	87 5 49.1	.63	51.86	5
465	Gould, Z. C., 0h 1275 . .	8.8	51.48	− 2.429	66 40 55.9	+ 19.62	52.17	3

Number.	Constellation, Name of Star, or Synonym.	Magnitude.	Right Ascension, 1850.0.	Annual Precession.	South Declination, 1850.0.	Annual Precession.	Mean year.	No. of obs.
			h. m. s.	s.	° ′ ″	″		
466	Tucanæ	10.0	0 47 53.40	+ 2.109	73 55 27.8	+ 19.62	50.87	1
467	Tucanæ	10.0	48 0.89	2.419	66 55 49.7	.62	52.69	1
468	Lacaille, 261	8.5	2.16	1.999	75 28 3.6	.62	50.68	2
469	Gould, Z. C., 0h 1279 . .	10.0	8.18	2.248	71 18 41.3	62	51.80	1
470	Tucanæ	10.0	26.34	+ 2.418	66 47 26.0	+ 19.61	51.90	1
471	Gould, Z. C., 0h 1289 . .	9.0	0 48 35.05	+ 2.303	69 54 3.0	+ 19.61	51.90	3
472	Gould, Z. C., 0h 1288 . .	9.0	36.29	2.258	70 55 52.0	.61	51.77	1
473	Tucanæ	10.0	37.31	2.400	67 16 37.8	.61	51.82	1
474	Brisbane, 120	9.2	38.28	2.044	74 41 1.8	.61	50.86	2
475	Gould, 846	10.2	39.00	+ 2.431	66 16 31.7	+ 19.61	51.92	2
476	Lacaille, 258	8.2	0 48 40.56	+ 2.431	66 16 19.4	+ 19.61	51.92	2
477	Hydri	9.2	41.88	1.257	81 10 23.6	.61	51.31	2
478	Hydri	10.0	44.20	2.008	75 8 51.6	.60	50.90	1
479	Octantis	9.5	45.26	0.506	83 43 22.1	.60	50.80	1
480	Hydri	9.5	52.20	+ 1.805	77 24 39.1	+ 19.60	51.76	1
481	Octantis	10.0	0 48 54.21	+ 0.408	83 55 56.4	+ 19.60	50.90	1
482	Gould, 855	9.8	56.04	2.361	68 15 39.4	.60	51.80	2
483	Gould, 856	10.0	48 59.90	2.361	68 15 34.0	.60	51.80	2
484	Octantis	10.0	49 8.67	0.686	83 11 54.1	.60	51.88	1
485	Melbourne, (1), 62 . . .	8.5	17.44	+ 0.075	84 33 41.8	+ 19.59	51.81	3
486	λ² Tucanæ	6.2	0 49 23.24	+ 2.272	70 20 21.6	+ 19.59	51.86	2
487	Tucanæ	10.0	36.28	2.292	69 47 35.7	.59	51.90	1
488	Lacaille, 263	9.0	39.77	2.305	69 26 28.9	.59	51.86	3
489	Octantis	10.0	41.22	0.224	84 14 15.1	.59	50.87	2
490	Tucanæ	10.0	49 55.30	+ 2.422	66 0 55.8	+ 19.58	51.92	1
491	Lacaille, 267	8.2	0 50 0.97	+ 1.983	75 7 14.6	+ 19.58	50.88	2
492	Gould, Z. C., 0h 1314 . .	10.0	5.69	1.625	78 40 17.8	.58	50.80	2
493	Hydri	9.5	6.42	1.430	79 59 6.7	.58	50.80	2
494	Octantis	10.0	7.86	+ 0.386	83 50 8.8	.58	50.90	1
495	Octantis	10.0	19.43	− 1.163	86 3 57.1	+ 19.57	51.88	1
496	Tucanæ	10.0	0 50 28.33	+ 2.148	72 26 52.2	+ 19.57	51.75	1
497	Tucanæ	9.8	50 46.62	2.230	70 45 31.9	.56	51.79	2
498	Tucanæ	10.0	51 4.02	2.262	69 56 40.9	.56	51.90	3
499	Hydri	10.0	4.64	1.074	81 34 59.0	.56	51.81	1
500	Gould, Z. C., 0h 1343 . .	9.8	5.96	+ 2.046	73 56 6.6	+ 19.56	50.87	2
501	Brisbane, 126	9.5	0 51 16.24	+ 1.992	74 38 12.8	+ 19.56	50.86	2
502	Octantis	9.1	16.99	− 1.011	85 50 38.7	.56	51.87	6
503	Stone, 369	8.7	18.84	− 0.923	85 45 2.2	.56	51.71	7
504	Hydri	9.0	19.11	+ 1.091	81 28 36.6	.56	51.30	2
505	Gould, Z. C., 0h 1354 . .	9.0	25.41	+ 2.367	67 6 19.3	+ 19.55	52.14	3

No. 468. This Star seems to have a proper motion of +0s.04 in right ascension.

Number.	Constellation, Name of Star, or Synonym.	Magnitude.	Right Ascension, 1850.0.	Annual Precession.	South Declination, 1850.0.	Annual Precession.	Mean year.	No. of obs.
			h. m. s.	s.	° ′ ″	″		
506	Tucanæ	10.0	0 51 25.78	+ 2.324	68 18 31.8	+ 19.55	51.80	2
507	Tucanæ	10.0	30.09	2.372	66 56 38.9	.55	51.90	1
508	Gould, Z. C., 0h 1358 . .	9.0	36.85	2.232	70 26 4.7	.55	51.81	1
509	Stone, 377	9.0	44.32	2.036	73 52 49.5	.55	50.87	2
510	Gould, Z. C., 0h 1364 . .	9.0	51.95	+ 1.988	74 31 54.9	+ 19.54	50.86	1
511	Gould, Z. C., 0h 1365 . .	9.0	0 51 57.48	+ 1.846	76 13 8.3	+ 19.54	50.83	2
512	Tucanæ	8.5	51 58.74	2.378	66 34 57.0	.54	52.31	2
513	Lacaille, 272	7.0	52 3.72	2.349	67 22 19.0	.54	51.82	1
514	Hydri	10.0	11.03	1.669	77 51 30.7	.54	51.76	1
515	Gould, 913	9.0	19.92	+ 2.283	69 0 14.6	+ 19.54	51.85	2
516	Octantis	10.0	0 52 21.01	− 0.890	85 37 47.7	+ 19.54	51.88	1
517	Tucanæ	10.0	27.37	+ 2.159	71 37 12.0	.53	51.81	1
518	Gould, Z. C., 0h 1382 . .	10.0	28.55	2.177	71 16 10.9	.53	51.80	1
519	Hydri	9.5	52 34.56	1.869	75 48 39.7	.53	50.85	1
520	Lacaille, 281	7.0	53 8.79	+ 1.701	77 21 57.0	+ 19.52	50.81	1
521	Hydri	9.2	0 53 12.14	+ 1.009	81 31 9.3	+ 19.52	51.30	2
522	Stone, 382	9.5	17.20	0.376	83 28 55.4	.52	50.82	2
523	Hydri	10.0	49.82	1.572	78 16 56.2	.51	50.78	1
524	Tucanæ	9.0	53 58.98	2.268	68 50 10.1	.50	51.91	1
525	Octantis	10.0	0 54 14.14	+ 0.234	83 41 48.6	+ 19.50	50.80	1
526	Melbourne, (1), 65 . . .	9.0	0 54 15.34	+ 2.293	68 3 39.7	+ 19.50	51.81	1
527	Tucanæ	10.0	18.12	2.212	69 57 11.3	.50	51.90	1
528	Gould, Z. C., 0h 1424 . .	10.0	35.68	1.930	74 33 27.5	.49	50.86	1
529	Octantis	9.8	35.88	0.606	82 42 27.5	.49	51.31	2
530	Octantis	10.0	38.13	+ 0.331	83 25 39.2	+ 19.49	50.84	1
531	Gould, Z. C., 0h 1429 . .	10.0	0 54 38.25	+ 2.151	71 4 41.3	+ 19.49	51.79	2
532	Gould, Z. C., 0h 1435 . .	8.5	54 50.69	2.152	70 59 7.8	.48	51.77	1
533	Gould, Z. C., 0h 1441 . .	9.0	55 5.07	+ 2.247	68 54 7.7	.48	51.91	1
534	Octantis	9.8	6.78	− 0.531	84 56 55.3	.48	51.87	3
535	Hydri	10.0	13.06	+ 1.534	78 16 42.6	+ 19.48	50.81	2
536	Lacaille, 295	8.8	0 55 16.05	+ 1.732	76 37 25.4	+ 19.48	50.86	2
537	Gould, Z. C., 0h 1447 . .	9.0	20.85	2.224	69 20 44.2	.47	51.78	1
538	Stone, 396	9.5	23.94	1.726	76 37 21.6	.47	50.86	2
539	Tucanæ	10.0	43.28	2.137	70 59 58.7	.47	51.77	1
540	Hydri	11.0	45.98	+ 1.664	77 6 26.0	+ 19.47	50.81	1
541	Gould, Z. C., 0h 1464 . .	9.0	0 55 51.76	+ 2.214	69 22 44.6	+ 19.46	51.78	1
542	Brisbane, 139	8.2	58.71	1.980	73 30 20.8	.46	50.88	3
543	Gould, Z. C., 0h 1462 . .	9.0	55 58.92	1.762	76 7 34.5	.46	50.84	1
544	Hydri	10.5	56 0.75	1.382	79 9 27.8	.46	50.83	1
545	Hydri	9.5	1.70	+ 1.660	77 5 23.9	+ 19.46	50.84	2

Number.	Constellation, Name of Star, or Synonym.	Magnitude.	Right Ascension, 1850.0.	Annual Precession.	South Declination. 1850.0.	Annual Precession.	Mean year.	No. of obs.
			h. m. s.	s.	° ′ ″	″		
546	Tucanæ	10.0	0 56 3.56	+ 2.316	66 48 52.2	+ 19.46	52.69	1
547	Gould, Z. C., 0h 1471	10.0	5.52	2.054	72 20 49.2	.46	50.91	2
548	Gould, Z. C., 0h 1474	9.0	8.52	2.327	66 27 36.3	.46	51.92	1
549	Tucanæ	10.0	9.85	2.206	69 26 31.8	.46	51.85	2
550	Tucanæ	10.0	16.92	+ 2.181	69 57 31.7	+ 19.45	51.90	1
551	Lacaille, 299	8.8	0 56 28.54	+ 2.046	72 21 20.8	+ 19.45	50.91	2
552	Gould, Z. C., 0h 1481	9.5	33.63	1.702	76 35 9.4	.45	50.83	1
553	Hydri	9.5	38.13	1.798	75 35 41.9	.45	50.85	1
554	Lacaille, 298	7.2	56 57.14	+ 2.324	66 15 47.6	.44	51.92	2
555	Octantis	9.5	57 1.66	− 1.138	85 31 35.8	+ 19.44	51.88	2
556	Hydri	9.0	0 57 30.76	+ 1.561	77 36 21.4	+ 19.43	51.76	1
557	Tucanæ	10.0	38.49	2.067	71 40 39.5	.42	51.81	1
558	Tucanæ	9.5	44.20	2.210	68 50 55.4	.42	51.91	1
559	Tucanæ	10.0	57 55.65	2.229	68 27 35.1	.42	51.80	1
560	Gould, Z. C., 0h 1519	9.2	58 2.28	+ 1.665	76 36 28.8	+ 19.42	50.86	2
561	Brisbane, 143	9.5	0 58 7.44	+ 1.982	72 53 43.4	+ 19.41	50.88	1
562	Lacaille, 306	Neb.	7.80	2.060	71 39 19.4	.41	51.81	2
563	Hydri	10.0	7.93	1.515	77 50 21.3	.41	51.76	1
564	Gould, Z. C., 0h 1527	10.0	9.14	2.321	65 54 30.8	.41	51.91	1
565	Gould, Z. C., 0h 1530	10.0	22.40	+ 1.958	73 10 22.9	+ 19.41	50.88	1
566	Tucanæ	10.0	0 58 30.12	+ 2.146	69 58 25.9	+ 19.41	51.90	1
567	Tucanæ	10.0	58 59.94	2.267	67 4 48.0	.40	51.82	1
568	Lacaille, 313	7.5	59 0.72	+ 1.419	78 21 21.0	.40	50.81	2
569	Octantis	10.0	15.00	− 4.075	87 15 43.2	.39	51.87	2
570	Octantis	9.5	15.16	− 0.153	83 56 53.1	+ 19.39	50.92	2
571	Tucanæ	9.0	0 59 21.76	+ 1.856	74 16 3.9	+ 19.39	50.87	3
572	Gould, Z. C., 1h 7	9.5	24.33	2.057	71 20 43.2	.39	51.80	1
573	Brisbane, 146	9.0	31.06	+ 1.945	73 3 27.4	.38	50.88	2
574	Octantis	9.4	38.47	− 0.847	84 58 59.4	.38	51.84	8
575	Tucanæ	10.0	0 59 45.32	+ 1.870	74 0 17.3	+ 19.38	50.87	1
576	Tucanæ	10.0	1 0 0.34	+ 2.234	67 33 29.2	+ 19.37	51.86	2
577	Tucanæ	9.2	2.12	2.262	66 51 46.4	.37	52.30	2
578	Lacaille, 314	8.5	11.32	1.968	72 32 25.0	.37	50.90	1
579	Lacaille, 315	9.0	27.52	2.015	71 44 15.3	.36	51.81	1
580	Lacaille, 320	9.0	31.96	+ 1.251	79 9 3.6	+ 19.36	51.83	2
581	Brisbane, 150	8.5	1 0 32.73	+ 1.904	73 21 11.8	+ 19.36	50.88	1
582	Tucanæ	9.5	44.24	2.089	70 23 18.2	.36	51.86	2
583	Gould, Z. C., 1h 45	9.5	48.58	2.267	66 27 36.7	.35	51.92	1
584	Gould, Z. C., 1h 46	9.5	51.34	2.168	68 46 58.2	.35	51.91	1
585	Hydri	10.2	52.60	+ 1.242	79 8 20.6	+ 19.35	51.33	2

Number.	Constellation, Name of Star, or Synonym.	Magnitude.	Right Ascension, 1850.0.	Annual Precession.	South Declination, 1850.0.	Annual Precession.	Mean year.	No. of obs.
			h. m. s.	s.	° ′ ″	″		
586	Hydri	10.0	1 1 8.06	+ 1.452	77 43 25.2	+ 19.35	51.76	1
587	Gould, Z. C., 1h 57 . . .	9.5	13.88	2.254	66 39 55.6	.34	52.17	3
588	Lacaille, 322	8.8	40.10	1.226	79 6 5.1	.33	51.33	2
589	Tucanæ	10.0	43.94	1.970	72 6 34.0	.33	50.91	1
590	Gould, Z. C., 1h 76 . . .	9.5	1 57.32	+ 2.124	69 22 6.8	+ 19.33	51.91	1
591	Tucanæ	10.0	1 2 4.05	+ 2.145	68 54 7.5	+ 19.33	51.91	1
592	Octantis	9.8	8.69	— 0.937	84 53 42.7	.32	51.81	4
593	Lacaille, 324	8.2	9.41	— 1.329	78 23 15.7	.32	50.85	3
594	Tucanæ	10.0	32.52	+ 2.020	71 5 38.5	.31	51.80	1
595	Octantis	9.8	2 32.78	+ 0.053	83 11 32.0	+ 19.31	51.36	2
596	Hydri	9.5	1 3 4.72	+ 1.540	76 39 13.3	+ 19.30	50.86	2
597	Hydri	9.5	6.61	1.624	75 53 52.4	.30	50.85	1
598	Hydri	10.0	7.55	0.688	81 19 31.1	.30	50.80	1
599	Tucanæ	9.5	8.86	1.785	74 12 40.4	.30	50.88	1
600	Tucanæ	9.5	9.24	+ 1.740	74 43 28.5	+ 19.30	50.86	3
601	Octantis	9.5	1 3 16.26	+ 0.173	82 50 4.2	+ 19.30	51.88	1
602	Tucanæ	10.0	25.10	2.005	71 5 29.7	.29	51.77	1
603	Hydri	10.0	25.95	0.800	80 51 51.1	.29	51.83	1
604	Hydri	9.0	29.78	0.989	80 2 18.3	.29	50.82	1
605	Brisbane, 161	9.5	36.15	+ 1.810	73 48 51.8	+ 19.29	50.87	2
606	Brisbane, 159	9.5	1 3 39.47	+ 1.978	71 28 4.6	+ 19.29	51.81	2
607	Melbourne, (1), 81 . . .	8.0	41.12	+ 2.178	67 40 51.6	.29	51.90	2
608	Lacaille, 342	8.2	43.48	— 0.455	84 3 30.3	.29	50.92	2
609	Tucanæ	10.0	43.79	+ 1.140	79 14 28.7	.29	51.82	1
610	Lacaille, 330	8.2	51.86	+ 1.416	77 28 31.3	+ 19.28	51.29	2
611	Gould, 1091	9.0	1 3 51.93	+ 2.041	70 22 1.5	+ 19.28	51.86	2
612	Brisbane, 164	9.0	3 55.62	1.894	72 38 10.8	.28	50.89	2
613	Gould, 1095	10.0	4 4.98	2.077	69 39 10.3	.28	51.99	2
614	Gould, 1094	10.0	5.25	1.983	71 16 22.0	.28	51.80	1
615	Gould, 1098	9.0	7.74	+ 2.176	67 35 59.7	+ 19.28	51.87	3
616	Tucanæ	10.0	1 4 8.03	+ 2.060	69 57 8.2	+ 19.28	51.99	2
617	Hydri	10.0	12.60	1.075	79 30 37.5	.27	50.93	1
618	Gould, Z. C., 1h 133 . .	10.0	21.24	2.076	69 35 28.3	.27	51.90	3
619	Hydri	10.0	32.88	1.178	78 53 39.6	.27	50.83	1
620	Brisbane, 165	10.0	43.88	+ 1.962	71 26 38.8	+ 19.26	51.81	2
621	Tucanæ	9.8	1 4 55.21	+ 2.134	68 16 4.5	+ 19.26	51.88	2
622	Tucanæ	10.0	55.48	2.022	70 23 43.6	.26	51.90	1
623	Hydri	10.0	4 56.66	1.323	77 56 7.8	.26	51.76	1
624	Gould, Z. C., 1h 151 . .	8.8	5 5.96	2.201	66 43 18.1	.25	52.17	3
625	Hydri	9.5	10.96	+ 0.597	81 22 50.2	+ 19.25	51.30	2

Number.	Constellation, Name of Star, or Synonym.	Magnitude.	Right Ascension, 1850.0.	Annual Precession.	South Declination, 1850.0.	Annual Precession.	Mean year.	No. of obs.
			h. m. s.	s.	° ′ ″	″		
626	Gould, Z. C., 1h 153 . .	9.0	1 5 14.19	+ 2.014	70 27 30.8	+ 19.25	51.81	1
627	Tucanæ	10.0	17.68	2.144	67 57 26.0	.25	51.85	2
628	Gould, 1117	8.2	18.95	2.187	66 58 54.2	.25	52.30	2
629	Lacaille, 332	6.8	34.98	1.776	73 45 25.8	.24	50.87	2
630	Tucanæ	11.0	35.23	+ 1.776	73 45 20.3	+ 19.24	50.87	1
631	Gould, Z. C., 1h 167 . .	9.0	1 5 50.93	+ 1.713	74 25 18.0	+ 19.23	50.88	1
632	Lacaille, 350	8.0	5 54.98	0.211	82 26 56.7	.23	50.75	1
633	Tucanæ	9.5	6 0.88	1.984	70 45 15.3	.23	51.77	1
634	Gould, 1131	9.5	5.89	2.089	68 50 46.9	.23	51.91	1
635	Gould, Z. C., 1h 172 . .	9.5	14.90	+ 1.070	79 13 10.3	+ 19.22	51.82	1
636	Lacaille, 360	7.9	1 6 28.43	− 0.822	84 23 32.0	+ 19.22	51.35	6
637	Tucanæ	9.6	31.00	+ 1.951	71 8 39.4	.22	51.87	4
638	Hydri	9.0	34.18	0.895	80 1 28.2	.22	50.82	1
639	Brisbane, 170	9.0	41.23	1.750	73 49 2.6	.21	50.87	1
640	Tucanæ	9.5	52.93	+ 1.815	72 59 7.7	+ 19.21	50.88	1
641	Lacaille, 343	9.0	1 6 53.53	+ 1.260	78 0 16.1	+ 19.21	50.85	1
642	Gould, Z. C., 1h 194 . .	10.0	7 2.54	1.073	79 4 52.4	.20	51.82	1
643	Tucanæ	10.0	38.70	+ 1.980	70 23 15.8	.19	51.90	1
644	Octantis	9.5	40.74	0.042	82 52 27.5	.19	51.88	1
645	Octantis	9.0	7 53.42	− 0.466	83 42 14.6	+ 19.18	50.80	1
646	Lacaille, 345	7.8	1 8 0.91	+ 1.642	74 42 19.2	+ 19.18	50.87	2
647	Tucanæ	9.0	12.20	2.153	66 53 8.0	.17	52.30	2
648	Gould, Z. C., 1h 235 . . .	9.0	13.93	2.198	65 49 41.5	.17	51.92	1
649	Hydri	9.2	14.63	0.955	79 30 0.6	.17	51.38	2
650	Octantis	10.0	26.66	+ 0.047	82 35 19.5	+ 19.17	51.31	2
651	Brisbane, 173	9.0	1 8 34.24	+ 1.882	71 40 38.0	+ 19.17	51.81	1
652	Hydri	9.8	9 6.54	1.337	77 6 10.8	.15	50.85	2
653	Hydri	9.5	8.25	1.515	75 41 6.4	.15	50.85	1
654	Gould, 1184	9.5	16.20	1.575	75 4 14.6	.15	50.88	2
655	Hydri	9.5	19.58	+ 1.523	75 34 16.6	+ 19.15	50.85	1
656	Gould, Z. C., 1h 261 . . .	10.0	1 9 22.55	+ 1.897	71 15 17.7	+ 19.14	51.80	1
657	Gould, Z. C., 1h 259 . . .	9.0	30.40	1.482	75 53 56.9	.14	50.84	2
658	Tucanæ	10.0	38.56	2.104	67 32 56.2	.14	51.82	1
659	Hydri	9.5	46.21	0.154	82 10 39.9	.13	50.81	1
660	Gould, Z. C., 1h 271 . . .	10.0	47.01	+ 1.822	71 49 59.6	+ 19.13	51.36	2
661	Octantis	10.5	1 9 48.64	− 0.066	82 43 5.6	+ 19.13	51.31	2
662	Lacaille, 353	7.8	9 53.46	+ 1.991	69 37 3.6	.13	51.89	2
663	Octantis	9.8	10 3.20	− 3.510	86 30 6.8	.13	51.88	3
664	Tucanæ	10.0	3.80	+ 1.857	71 39 53.4	.13	51.81	1
665	Hydri	9.0	6.14	+ 0.555	80 54 33.3	+ 19.13	51.83	1

No. 639. This Star seems to have a proper motion in declination of + 0.22″.
No. 654. This Star seems to have a proper motion of + 0s.04 and − 0.16″.
No. 662. This Star seems to have a proper motion of + 0s.07 and + 0.1″.

Number.	Constellation, Name of Star, or Synonym.	Magnitude.	Right Ascension, 1850.0.	Annual Precession.	South Declination, 1850.0.	Annual Precession.	Mean year.	No. of obs.
			h. m. s.	s.	° ′ ″	″		
666	Gould, 1201	10.0	1 10 9.03	+ 2.155	66 16 23.6	+ 19.12	51.92	1
667	Hydri	9.8	22.63	0.814	79 51 8.3	.12	50.87	2
668	Gould, Z. C., 1^h 281 . . .	10.0	25.19	+ 1.517	75 25 24.2	.12	50.90	1
669	Octantis	10.0	35.60	− 0.841	84 5 3.3	.11	50.94	1
670	κ Tucanæ	5.5	40.44	+ 1.976	69 40 24.4	+ 19.11	51.89	2
671	Tucanæ	9.5	1 10 40.80	+ 1.976	69 40 18.2	+ 19.11	51.89	2
672	Tucanæ	10.0	10 45.70	1.650	74 3 6.5	.10	50.87	1
673	Tucanæ	10.0	11 5.85	1.993	69 16 15.1	.09	51.91	1
674	Lacaille, 359	7.0	20.64	+ 2.046	68 13 27.1	.09	51.88	2
675	Octantis	10.0	25.41	− 3.866	86 37 8.3	+ 19.09	51.88	1
676	Octantis	9.1	1 11 30.56	−38.863	89 26 22.1	+ 19.09	51.86	4
677	Octantis	10.0	31.13	− 0.442	83 20 12.3	.09	50.84	1
678	Gould, Z. C., 1^h 321 . .	8.0	39.03	+ 2.114	66 45 33.0	.08	52.17	3
679	Tucanæ	9.9	41.59	1.875	71 1 50.8	.08	51.87	4
680	Lacaille, 361	6.5	50.73	+ 2.091	67 11 24.0	+ 19.08	52.14	3
681	Tucanæ	10.0	1 11 50.73	+ 2.091	67 11 21.9	+ 19.08	52.69	1
682	Gould, Z. C., 1^h 343 . . .	9.8	12 9.58	2.076	67 25 34.8	.07	51.90	2
683	Lacaille, 363	8.0	15.46	2.140	66 0 11.8	.07	51.92	1
684	Hydri	10.0	27.44	0.268	81 34 3.0	.06	50.79	1
685	Gould, 1239	9.0	40.83	+ 1.300	76 45 23.3	+ 19.06	50.86	2
686	Tucanæ	10.0	1 12 40.83	+ 2.056	67 41 1.7	+ 19.06	51.89	1
687	Tucanæ	9.5	42.22	1.913	70 12 39.4	.06	51.86	2
688	Lacaille, 363	8.0	44.00	2.044	67 54 6.0	.05	51.87	3
689	Tucanæ	9.2	12 59.70	+ 2.086	66 58 36.6	.05	51.86	2
690	Octantis	9.1	13 19.07	− 3.083	86 5 29.2	+ 19.04	51.87	6
691	Gould, Z. C., 1^h 355 . . .	9.5	1 13 19.96	+ 0.872	79 10 31.6	+ 19.04	51.33	2
692	Tucanæ	10.0	25.90	1.956	69 18 54.4	.04	51.91	1
693	Brisbane, 183	8.8	30.48	+ 2.056	67 27 55.0	.03	51.87	3
694	Octantis	10.0	50.16	− 0.849	83 49 59.9	.02	50.94	1
695	Gould, Z. C., 1^h 376 . .	10.0	13 52.68	+ 1.588	74 4 10.7	+ 19.02	50.87	1
696	Octantis	10.0	1 14 4.58	− 1.993	85 12 27.2	+ 19.02	51.88	2
697	Hydri	10.0	6.56	+ 0.476	80 42 18.2	.02	50.81	1
698	Gould, 1269	10.0	12.25	1.490	74 56 35.9	.01	50.86	1
699	Brisbane, 184	10.0	17.90	1.633	73 30 40.5	.01	50.88	2
700	Hydri	9.5	20.64	+ 1.400	75 42 7.7	+ 19.01	50.85	1
701	Gould, Z. C., 1^h 392 . .	9.5	1 14 23.16	+ 2.038	67 35 15.8	+ 19.01	51.90	2
702	Lacaille, 380	9.8	31.07	1.764	71 54 26.3	.01	51.36	2
703	Hydri	10.0	33.90	1.243	76 50 54.8	.00	50.89	1
704	Gould, Z. C., 1^h 402 . .	9.0	44.73	2.063	66 59 50.1	19.00	52.14	3
705	Gould, 1284	9.2	14 56.86	+ 2.022	68 12 54.6	+ 18.99	51.88	2

Number.	Constellation, Name of Star, or Synonym.	Magnitude.	Right Ascension, 1850.0.	Annual Precession.	South Declination, 1850.0.	Annual Precession.	Mean year.	No. of obs.
			h. m. s.	s.	° ′ ″	″		
706	Lacaille, 383	8.0	1 15 0.04	+ 1.913	69 41 28.4	+ 18.99	51.89	2
707	Gould, Z. C., 1h 413 . .	9.5	10.11	1.866	70 20 44.4	.99	51.86	2
708	Hydri	9.0	10.44	+ 0.399	80 50 45.6	.99	51.83	1
709	Octantis	10.0	13.99	— 2.626	85 40 24.6	.99	51.88	1
710	Tucanæ	9.5	26.96	+ 1.890	69 54 21.4	+ 18.98	51.90	1
711	Brisbane, 189	9.8	1 15 29.74	+ 1.692	72 35 30.4	+ 18.98	50.89	2
712	Gould, 1298	9.0	30.48	1.953	68 51 26.8	.98	51.91	1
713	Brisbane, 191	9.5	33.40	1.688	72 37 40.2	.98	50.89	2
714	Hydri	9.0	15 50.15	1.377	75 37 22.2	.97	50.85	1
715	Gould, Z. C., 1h 441 . .	8.8	16 8.35	+ 2.050	66 53 30.4	+ 18.96	52.30	2
716	Hydri	9.5	1 15 12.66	+ 0.522	80 17 8.0	+ 18.96	50.81	2
717	Lacaille, 397	8.3	13.63	1.049	77 49 12.1	.96	51.19	3
718	Lacaille, 393	8.2	15.88	1.594	73 32 11.2	.96	50.88	2
719	Lacaille, 398	8.0	17.32	1.189	76 56 13.9	.96	50.89	1
720	Gould, 1319	9.2	43.83	+ 2.025	67 13 50.2	+ 18.94	51.86	2
721	Hydri	10.0	1 16 45.49	+ 0.105	81 36 9.7	+ 18.94	51.81	1
722	Lacaille, 420	8.2	46.95	— 0.193	82 19 47.0	.94	50.82	2
723	Lacaille, 391	6.8	47.13	+ 2.027	67 10 10.8	.94	52.14	3
724	Octantis	9.1	16 59.57	— 6.713	87 25 15.7	.94	51.86	4
725	Hydri	10.0	17 3.51	+ 1.401	75 12 36.9	+ 18.93	50.90	1
726	Hydri	9.5	1 17 4.20	— 0.075	82 1 4.0	+ 18.93	51.31	2
727	Octantis	9.5	5.51	5.223	86 57 16.4	.93	51.87	2
728	Octantis	10.0	20.02	— 0.892	83 37 40.0	.93	50.80	1
729	Brisbane, 198	9.5	20.94	+ 1.643	72 33 25.6	.92	50.90	1
730	Brisbane, 200	9.5	22.50	+ 1.545	73 49 12.7	+ 18.92	50.87	2
731	Tucanæ	10.0	1 17 24.34	+ 1.788	70 57 49.2	+ 18.92	51.95	1
732	Lacaille, 399	8.0	27.13	1.820	70 30 14.9	.92	51.81	1
733	Gould, Z. C., 1h 471 . .	9.2	30.18	1.549	73 45 43.6	.92	50.87	2
734	Hydri	9.8	37.01	1.244	76 20 28.4	.92	50.83	2
735	Hydri	10.0	37.56	+ 0.059	81 36 46.7	+ 18.92	51.81	1
736	Hydri	10.0	1 17 38.02	+ 0.491	80 13 56.0	+ 18.92	50.82	1
737	Tucanæ	10.0	39.77	1.773	71 6 37.5	.92	51.95	1
738	Tucanæ	10.0	44.08	2.060	66 16 10.6	.91	51.92	1
739	Gould, Z. C., 1h 475 . .	9.5	50.31	1.471	74 27 7.6	.91	50.88	1
740	Gould, Z. C., 1h 470 . .	9.8	54.35	+ 0.579	79 51 42.3	+ 18.91	50.87	2
741	Gould, Z. C., 1h 480 . .	9.5	1 17 56.50	+ 1.474	74 24 23.1	+ 18.91	50.88	1
742	Tucanæ	10.0	18 0.74	+ 1.667	72 22 23.6	.91	50.91	1
743	Octantis	9.3	4.32	— 4.187	86 28 43.8	.90	51.88	3
744	Lacaille, 401	8.3	8.28	+ 1.856	69 52 0.8	.90	51.90	3
745	Tucanæ	9.5	9.72	+ 2.020	66.57 38.7	+ 18.90	51.82	1

Number.	Constellation, Name of Star, or Synonym.	Magnitude.	Right Ascension, 1850.0.	Annual Precession.	South Declination, 1850.0.	Annual Precession.	Mean year.	No. of obs.
			h. m. s.	s.	° ′ ″	″		
746	Hydri	9.0	1 18 14.27	− 0.008	81 43 46.9	+ 18.90	51.81	1
747	Tucanæ	10.0	27.50	+ 2.053	66 13 54.4	.89	51.92	1
748	Gould, Z. C., 1^h 500 . .	9.8	37.98	1.979	67 37 26.3	.89	51.90	2
749	Gould, Z. C., 1^h 502 . .	9.0	39.44	1.977	67 39 13.5	.89	51.90	2
750	Gould, Z. C., 1^h 501 . .	9.0	40.62	+ 1.913	68 46 0.2	+ 18.89	51.93	2
751	Tucanæ	9.7	1 18 52.08	+ 1.833	69 51 28.0	+ 18.88	51.90	3
752	Tucanæ	9.8	56.42	1.889	69 6 41.3	.88	51.91	2
753	Hydri	10.0	56.57	1.124	76 56 50.9	.88	50.89	1
754	Tucanæ	11.0	18 59.73	1.890	69 4 24.8	.88	51.91	1
755	Tucanæ	9.5	19 14.15	+ 2.009	66 54 14.7	+ 18.87	51.90	1
756	Hydri	10.0	1 19 14.48	+ 1.271	75 52 40.4	+ 18.87	50.84	1
757	Gould, Z. C., 1^h 515 . .	9.5	21.19	1.469	74 22 2.3	.87	50.88	1
758	Tucanæ	10.0	30.42	1.565	73 12 42.5	.86	50.88	1
759	Tucanæ	11.0	47.36	1.829	69 51 5.5	.85	51.90	1
760	Tucanæ	9.0	48.45	+ 1.599	72 47 25.9	+ 18.85	50.88	1
761	Lacaille, 417	8.8	1 19 56.34	+ 1.709	71 27 29.4	+ 18.85	51.81	2
762	Brisbane, 206.	9.2	20 1.47	1.639	72 17 57.4	.85	50.91	2
763	Gould, Z. C., 1^h 536 . .	9.0	1.54	1.998	66 55 30.0	.85	52.30	2
764	Lacaille, 421	8.5	4.51	1.565	73 6 22.8	.84	50.88	2
765	Hydri	10.0	7.32	+ 1.956	67 40 58.7	+ 18.84	51.89	1
766	Hydri	9.5	1 20 11.37	+ 1.794	70 15 46.7	+ 18.84	51.90	1
767	Hydri	8.8	32.41	− 0.122	81 47 {40.6 / 66.7}	.83	51.31	2
768	Gould, Z. C., 1^h 549 . .	9.2	36.58	+ 2.020	66 20 50.4	.83	51.92	2
769	Hydri	10.0	43.94	1.239	75 52 15.3	.82	50.84	1
770	Lacaille, 429	9.0	45.04	+ 1.009	77 23 36.4	+ 18.82	50.87	2
771	Gould, Z. C., 1^h 555 . .	9.0	1 20 54.04	+ 1.916	68 11 47.9	+ 18.82	51.88	2
772	Hydri	10.0	20 56.34	1.731	70 58 5.8	.82	51.95	2
773	Hydri	10.0	21 3.48	1.988	66 51 36.3	.81	51.90	1
774	Gould, Z. C., 1^h 559 . .	9.0	10.84	+ 1.732	70 54 35.0	.81	51.95	2
775	Hydri	9.0	13.54	+ 1.570	72 50 6.0	+ 18.81	50.88	1
776	Octantis	9.0	1 21 16.36	−10.279	88 0 35.0	+ 18.81	51.85	2
777	Gould, 1399	9.0	25.18	+ 1.884	68 37 12.5	.80	51.93	2
778	Hydri	10.0	35.55	+ 1.495	73 31 48.3	.80	50.88	1
779	Octantis	10.0	43.50	− 1.527	84 12 17.6	.79	50.94	1
780	Gould, Z. C., 1^h 578 . .	10.0	44.97	+ 1.634	72 0 29.1	+ 18.79	50.91	1
781	Gould, Z. C., 1^h 582 . .	10.0	1 21 48.68	+ 1.917	67 58 8.4	+ 18.79	51.87	3
782	Hydri	10.0	22 5.08	0.694	78 50 56.6	.78	50.83	1
783	Hydri	10.0	12.94	1.214	75 47 11.3	.78	50.84	2
784	Gould, Z. C., 1^h 583 . .	9.5	30.44	0.403	79 59 29.3	.77	50.87	2
785	Brisbane, 211	9.8	22 51.42	+ 1.472	73 31 43.3	+ 18.76	50.88	2

No. 767. One of these declinations is evidently 1 rev. wrong.

Number.	Constellation, Name of Star, or Synonym.	Magnitude.	Right Ascension, 1850.0.	Annual Precession.	South Declination, 1850.0.	Annual Precession.	Mean year.	No. of obs.
			h. m. s.	s.	° ′ ″	″		
786	Hydri	9.5	1 23 41.94	+ 1.633	71 38 8.2	+ 18.73	51.81	1
787	Lacaille, 443	8.0	23 44.66	1.558	72 29 2.6	.73	50.90	1
788	Gould, 1443	9.0	24 9.68	1.824	68 57 1.3	.72	51.91	1
789	Hydri	9.8	15.76	0.146	80 40 20.8	.72	51.32	2
790	Gould, Z. C., 1h 648 . .	8.2	16.50	+ 1.685	70 54 23.4	+ 18.72	51.95	2
791	Gould, Z. C., 1h 645 . . .	8.5	1 24 19.24	+ 1.389	74 2 48.9	+ 18.71	50.87	1
792	Gould, Z. C., 1h 657 . .	8.5	25.63	1.864	68 15 41.1	.71	51.81	1
793	Lacaille, 461	8.5	33.07	0.135	80 40 29.8	.71	51.32	2
794	Gould, Z. C., 1h 663 . . .	10.0	34.80	1.865	68 12 34.6	.71	51.81	1
795	Brisbane, 218	9.5	42.62	+ 1.519	72 43 7 1	+ 18.70	50.89	2
796	Lacaille, 455	7.2	1 24 46.97	+ 1.158	75 49 15.9	+ 18.70	50.84	2
797	Octantis	10.0	52.14	− 2.826	85 18 35.5	.70	51.88	1
798	Hydri	10.0	24 55.98	+ 0.999	76 51 1.0	.69	50.92	2
799	Hydri	10.0	25 21.61	1.483	72 58 15.5	.68	50.88	1
800	Hydri	10.0	33.09	+ 1.257	74 37 26.2	+ 18.68	50.88	1
801	Hydri	8.5	1 25 38.04	− 0.252	81 38 49.9	+ 18.67	51.38	2
802	Gould, 1474	8.5	25 38.91	+ 1.981	65 53 33.0	.67	51.92	1
803	Octantis	9.7	26 10.65	− 6.893	87 10 46.3	.66	51.87	5
804	Hydri	10.0	13.08	+ 0.983	76 45 59.8	.65	50.72	2
805	Gould, 1488	8.0	14.08	+ 1.974	65 53 48.1	+ 18.65	51.92	1
806	Octantis	11.0	1 26 16.82	− 0.829	82 49 3.6	+ 18.65	51.88	1
807	Gould, Z. C., 1h 711 . . .	8.5	18.81	+ 1.898	67 16 11.2	.65	51.82	1
808	Hydri	9.5	31.24	1.901	67 9 53.3	.64	52.69	1
809	Hydri	10.0	37.02	0.736	78 4 23.0	.64	50.85	1
810	Gould, Z. C., 1h 724 . . .	8.8	38.89	+ 1.932	66 34 38.9	+ 18.64	52.17	3
811	Hydri	9.8	1 26 42.06	− 0.112	81 10 53.6	+ 18.64	51.31	2
812	Lacaille, 471	8.0	49.90	+ 0.319	79 48 58.4	.63	50.87	2
813	Brisbane, 223	8.2	26 51.98	1.353	73 56 46.0	.63	50.87	2
814	Gould, Z. C., 1h 734 . . .	9.5	27 3.12	1.600	71 23 16.0	.63	50.88	2
815	Hydri	9.0	15.64	+ 1.708	70 3 16.0	+ 18.62	51.90	1
816	Gould, Z. C., 1h 742 . . .	8.5	1 27 25.66	+ 1.645	70 45 58.4	+ 18.61	51.97	3
817	Hydri	10.0	30.62	0.463	79 11 14.5	.61	51.82	1
818	Hydri	9.5	43.66	1.562	71 42 21.0	.61	51.38	2
819	Gould, Z. C., 1h 746 . . .	9.0	47.45	1.067	76 0 21.0	.60	50.84	1
820	Gould, Z. C., 1h 754 . . .	10.0	27 59.54	+ 1.201	75 0 57.7	+ 18.60	50.88	2
821	Octantis	10.0	1 28 4.29	− 4.969	86 26 0.4	+ 18.59	51.88	2
822	Octantis	10.0	7.43	− 3.268	85 27 44.1	.59	51.87	3
823	Hydri	10.0	11.28	+ 1.509	72 11 45.0	.59	50.91	2
824	Gould, Z. C., 1h 767 . . .	9.8	18.33	1.753	69 8 18.4	.59	51.91	2
825	Hydri	10.0	20.72	+ 1.740	69 19 18.0	+ 18.58	51.91	1

No. 821. Right Ascension may be 1^m greater.

Number.	Constellation, Name of Star, or Synonym.	Magnitude.	Right Ascension, 1850.0.	Annual Precession.	South Declination, 1850.0.	Annual Precession.	Mean year.	No. of obs.
			h. m. s.	s.	° ′ ″	″		
826	Hydri	10.0	1 28 30.47	+ 1.794	68 29 0.1	+ 18.58	51.95	1
827	Gould, Z. C., 1^h 770 . . .	9.5	36.76	1.138	75 23 1.7	.58	50.90	1
828	Hydri	10.0	46.90	+ 0.932	76 43 9.1	.57	50.92	2
829	Octantis	9.1	28 51.41	−33.928	89 13 3.1	.57	51.86	4
830	Gould, Z. C., 1^h 789 . . .	10.0	29 8.11	+ 1.820	67 56 48.1	+ 18.56	51.85	2
831	Hydri	10.0	1 29 9.84	+ 1.784	68 30 14.4	+ 18.56	51.95	1
832	Gould, 1535	9.0	10.14	0.796	77 26 2.3	.56	50.88	2
833	Gould, Z. C., 1^h 783 . . .	10.0	10.27	1.160	75 4 25.7	.56	50.86	1
834	Hydri	9.0	17.06	0.073	80 23 25.9	.55	50.81	2
835	Hydri	10.0	27.28	+ 1.794	68 17 51.2	+ 18.55	51.88	2
836	Gould, Z. C., 1^h 796 . . .	9.0	1 29 38.87	+ 1.599	70 54 17.8	+ 18.54	51.96	1
837	Brisbane, 230	8.8	40.77	1.536	71 38 12.2	.54	51.38	2
838	Hydri	10.0	42.40	1.666	70 3 24.4	.54	51.90	1
839	Hydri	9.7	48.18	1.683	69 48 9.2	.54	51.90	3
840	Gould, Z. C., 1^h 794 . .	9.0	29 51.83	+ 1.731	77 41 4.4	+ 18.53	50.94	1
841	Gould, 1558	8.5	1 30 2.22	+ 1.919	66 3 58.4	+ 18.53	51.92	1
842	Gould, Z. C., 1^h 816 . .	9.8	3.38	1.687	69 42 10.6	.53	51.89	2
843	Hydri	9.5	18.87	1.204	74 37 41.5	.52	50.88	1
844	Hydri	9.5	20.17	+ 1.223	74 28 41.5	.52	50.88	1
845	Hydri	8.5	20.35	− 0.572	81 58 37.4	+ 18.52	51.19	3
846	Gould, Z. C., 1^h 809 . .	8.0	1 30 25.47	+ 0.235	79 43 54.9	+ 18.52	50.93	1
847	Lacaille, 491	8.5	28.40	+ 0.466	78 50 16.5	.51	50.83	1
848	Octantis	9.5	33.21	− 0.887	82 35 35.1	.51	50.95	1
849	Gould, 1568	8.3	37.87	+ 1.803	67 54 12.6	.51	51.87	3
850	Gould, 1570	9.8	43.19	+ 1.661	69 55 18.2	+ 18.50	51.90	2
851	Octantis	9.5	1 30 43.58	− 0.895	82 35 39.3	+ 18.50	51.31	2
852	Gould, Z. C., 1^h 836 . .	10.0	43.90	+ 1.906	66 7 58.0	.50	51.92	1
853	Gould, 1571	11.0	47.18	1.660	69 55 36.0	.50	51.90	1
854	Hydri	10.0	52.44	1.802	67 52 28.8	.50	51.85	2
855	Gould, 1576	9.0	30 53.92	+ 1.909	66 3 29.0	+ 18.50	51.92	1
856	Octantis	9.2	1 31 2.65	− 1.030	82 48 41.1	+ 18.49	51.41	2
857	Octantis	9.5	27.58	− 1.150	82 59 6.2	.48	51.42	2
858	Gould, Z. C., 1^h 855 . .	9.5	29.29	+ 1.818	67 29 30.9	.48	51.87	3
859	Gould, Z. C., 1^h 857 . .	10.0	52.60	1.063	75 26 31.0	.47	50.90	1
860	Gould, Z. C., 1^h 865 . .	9.2	31 58.66	+ 1.710	69 1 1.6	+ 18.46	51.91	2
861	Hydri	9.5	1 32 10.86	+ 1.544	71 5 27.2	+ 18.46	51.92	2
862	Gould, Z. C., 1^h 863 . .	9.0	13.99	0.897	76 27 54.9	.45	50.83	1
863	Gould, Z. C., 1^h 878 . .	9.0	27.75	1.299	73 30 20.0	.45	50.88	2
864	Gould, Z. C., 1^h 871 . .	8.5	28.32	+ 1.014	75 41 19.0	.45	50.85	1
865	Octantis	10.2	31.56	− 0.989	82 37 46.1	+ 18.44	51.19	3

Number.	Constellation, Name of Star, or Synonym.	Magnitude.	Right Ascension, 1850.0.	Annual Precession.	South Declination, 1850.0.	Annual Precession.	Mean year.	No. of obs.
			h. m. s.	s.	° ′ ″	″		
866	Hydri	10.0	1 32 34.10	+ 0.944	76 7 46.8	+ 18 44	50.84	1
867	Gould, Z. C., 1h 883	9.0	41.70	1.288	73 35 51.8	.44	50.88	2
868	Lacaille, 505	6.5	43.62	0.301	79 15 58.7	.44	51.82	1
869	Hydri	9.8	44.74	1.601	70 18 16.8	.44	51.86	2
870	Gould, Z. C., 1h 888	9.0	44.79	+ 1.779	67 50 45.7	+ 18.44	51.87	3
871	Hydri	9.0	1 32 53.36	+ 1.857	66 32 25.6	+ 18.43	51.92	1
872	Hydri	9.8	32 53.66	1.565	70 42 55.6	.43	51.83	2
873	Hydri	10.0	33 5.16	0.252	79 23 20.1	.42	51.38	2
874	Hydri	10.0	6.95	1.608	70 8 40.4	.42	51.90	1
875	Hydri	9.5	10.34	+ 1.563	70 41 17.0	+ 18.42	51.88	2
876	Lacaille, 510	7.2	1 33 14.74	− 0.158	80 41 41.6	+ 18.42	51.32	2
877	Lacaille, 497	8.0	34.91	+ 1.801	67 19 11.5	.41	51.82	1
878	Hydri	9.5	37.99	1.817	67 3 15.5	.41	51.90	1
879	Hydri	9.5	38.93	+ 1.834	66 46 6.5	.40	51.90	1
880	Octantis	10.2	44.40	− 3.222	85 10 25.2	+ 18.40	51.85	3
881	Hydri	9.5	1 33 46.34	+ 0.988	75 40 44.1	+ 18.40	50.85	1
882	Gould, 1621	8.8	51.70	1.731	68 20 1.8	.39	51.88	2
883	Lacaille 499	7.2	33 56.59	+ 1.854	66 22 7.0	.39	51.92	2
884	Hydri	9.8	34 4.64	− 0.601	81 44 5.7	.39	51.38	2
885	Hydri	9.0	8.11	+ 0.231	79 21 25.0	+ 18.38	51.82	1
886	Octantis	9.2	1 34 15.81	− 6.740	86 52 54.3	+ 18.38	51.87	3
887	Gould, Z. C., 1h 927	9.2	20.82	+ 1.691	68 49 16.4	.38	51.93	2
888	Gould, 1628	9.0	34 37.28	0.723	77 8 9.6	.37	50.88	2
889	Hydri	9.2	35 10.87	1.691	68 39 36.2	.36	51.93	2
890	Hydri	10.0	11.42	+ 1.696	68 35 9.8	+ 18.36	51.91	1
891	Gould, 1644	8.2	1 35 21.20	+ 0.873	76 11 30.2	+ 18.35	50.83	2
892	Hydri	10.0	38.27	1.447	71 31 4.3	.34	51.88	1
893	Melbourne (1), 107	8.2	44.92	1.657	69 1 7.5	.33	51.91	2
894	Hydri	10.0	47.93	1.670	68 45 50.7	.33	51.91	1
895	Hydri	9.5	35 54.36	+ 1.585	69 55 23.3	+ 18.32	51.90	1
896	Gould, Z. C., 1h 956	9.2	1 36 0.94	+ 1.205	73 55 21.0	+ 18.32	50.81	2
897	Gould, 1668	9.0	4.46	1.725	67 59 31.1	.32	51.90	3
898	Hydri	10.0	24.73	1.342	72 28 52.1	.31	50.90	1
899	Hydri	10.0	29.12	1.083	74 38 9.3	.31	50.88	1
900	Gould, Z. C., 1h 965	9.0	29.28	+ 1.680	68 33 36.2	+ 18.31	51.93	2
901	Gould, Z. C., 1h 969	9.8	1 36 30.55	+ 1.723	67 56 35.5	+ 18.30	51.87	3
902	Hydri	10.5	32.48	− 1.064	82 28 11.4	.30	50.75	1
903	Octantis	10.2	32.71	− 3.423	85 11 17.2	.30	51.52	3
904	Stone, 679	8.2	39.18	+ 1.505	70 44 46.6	.30	51.88	2
905	Hydri	10.0	41.46	+ 1.101	74 28 28.3	+ 18.30	50.88	1

Number.	Constellation, Name of Star, or Synonym.	Magnitude.	Right Ascension, 1850.0.	Annual Precession.	South Declination, 1850.0.	Annual Precession.	Mean year.	No. of obs.
			h. m. s.	s.	° ′ ″	″		
906	Octantis	10.0	1 36 45.27	−25.474	88 54 1.6	+ 18.30	51.87	1
907	Hydri	9.0	52.14	− 0.925	82 11 16.6	.29	50.81	1
908	Gould, Z. C., 1h 981 . .	9.0	53.77	+ 1.685	68 25 0.6	.29	51.98	1
909	Gould, Z. C., 1h 970 . .	10.0	36 59.24	0.587	77 15 36.5	.29	50.81	1
910	Hydri	10.0	37 0.62	+ 1.729	67 45 21.3	+ 18.29	51.90	2
911	Gould, Z. C., 1h 986 . .	9.0	1 37 3.41	+ 1.806	66 35 18.2	+ 18.28	51.90	1
912	Lacaille, 521	8.8	22.98	0.725	76 46 48.2	.27	50.89	2
913	Lacaille, 517	8.5	31.55	0.900	75 44 37.9	.27	50.85	1
914	Hydri	10.0	34.13	1.775	66 56 17.5	.27	51.90	1
915	Hydri	9.0	41.24	+ 1.783	66 46 32.0	+ 18.26	51.91	2
916	Octantis	10.0	1 37 45.52	−17.743	88 28 39.2	+ 18.26	51.85	2
917	Hydri	9.5	37 49.86	+ 1.649	68 47 18.5	.26	51.91	1
918	Hydri	8.3	38 6.23	1.725	67 35 55.1	.25	51.87	3
919	Gould, Z. C., 1h 1002 . .	10.0	15.77	1.113	74 9 26.5	.24	50.87	1
920	Hydri	9.5	18.74	+ 1.708	67 48 42.6	+ 18.24	51.89	4
921	Octantis	10.0	1 38 29.25	− 3.821	85 22 49.8	+ 18.23	51.88	2
922	Hydri	9.0	29.66	+ 1.497	70 30 50.8	.23	51.81	1
923	Hydri	10.0	33.47	− 0.533	81 13 46.2	.23	50.80	1
924	Hydri	9.7	45.17	+ 1.728	67 25 55.5	.22	51.87	3
925	Hydri	10.0	47.37	+ 1.236	73 4 12.2	+ 18.22	50.88	1
926	Hydri	9.5	1 38 51.47	+ 1.421	71 17 21.2	+ 18.22	51.88	1
927	Hydri	10.0	38 56.76	+ 1.522	70 8 44.3	.22	51.90	1
928	Gould, Z. C., 1h 1011 . .	9.5	39 7.30	− 0.129	80 4 10.6	.21	50.82	1
929	Hydri	10.0	12.13	− 0.900	81 57 41.8	.21	50.81	1
930	Lacaille, 534	8.5	15.66	+ 0.641	77 0 0.5	+ 18.20	50.88	2
931	Lacaille, 533	8.5	1 39 18.11	+ 0.748	76 25 18.1	+ 18.20	50.83	1
932	Gould, Z. C., 1h 1034 . .	9.0	19.12	1.501	70 19 38.8	.20	51.86	2
933	Hydri	10.0	39 21.60	1.298	72 25 54.9	.20	50.90	1
934	Hydri	10.0	40 10.68	1.685	67 47 48.5	.17	51.96	1
935	Hydri	10.0	15.43	+ 1.111	73 53 28.8	+ 18.17	50.87	1
936	Lacaille, 561	9.0	1 40 30.37	− 1.579	83 2 23.2	+ 18.16	50.90	2
937	Gould, 1715	9.8	33.28	− 1.580	83 2 19.6	.16	50.90	2
938	Hydri	10.0	38.43	+ 1.278	72 25 2.7	.15	50.90	1
939	Octantis	7.8	39.52	−14.215	88 6 58.4	.15	51.87	4
940	Gould, Z. C., 1h 1058 . .	9.5	40.09	+ 1.299	72 13 6.6	+ 18.15	50.91	1
941	Gould, Z. C., 1h 1068 . .	9.0	1 40 51.79	+ 1.608	68 43 42.9	+ 18.15	51.93	2
942	Hydri	9.8	41 0.47	− 0.682	81 21 38.4	.14	50.87	2
943	Gould, Z. C., 1h 1073 . .	9.3	3.52	+ 1.517	69 50 34.8	.14	51.90	3
944	Gould, Z. C., 1h 1076 . .	9.5	13.67	+ 1.421	71 18 42.8	.13	51.88	1
945	Lacaille, 551	6.8	21.72	− 0.142	79 54 14.2	+ 18.13	50.87	2

Number.	Constellation, Name of Star, or Synonym.	Magnitude.	Right Ascension, 1850.0.	Annual Precession.	South Declination, 1850.0.	Annual Precession.	Mean year.	No. of obs.
			h. m. s.	s.	° ′ ″	″		
946	Octantis	10.0	1 41 33.50	− 3.259	84 49 31.8	+ 18.12	51.78	3
947	Gould, Z. C., 1h 1087 . .	10.0	36.52	+ 1.775	66 8 7.4	.12	51.92	1
948	Hydri	10.0	42.11	1.210	72 51 36.4	.11	50.88	1
949	Hydri	9.7	42.20	1.505	69 52 49.4	.11	51.90	2
950	Hydri	9.2	41 55.51	+ 0.113	78 59 55.8	+ 18.11	51.33	2
951	Hydri	9.8	1 42 10.60	+ 0.958	74 44 26.6	+ 18.10	50.86	2
952	Gould, Z. C., 1h 1094 . .	9.0	12.33	+ 0.596	76 53 7.7	.09	50.95	1
953	Lacaille, 576	5.6	12.87	− 2.183	83 44 12.1	.09	50.80	1
954	Hydri	9.2	15.04	0.740	81 23 34.6	.09	50.87	2
955	Octantis	10.2	16.56	− 3.584	85 2 43.9	+ 18.09	51.52	3
956	Hydri	10.0	1 42 22.15	+ 1.109	73 35 59.0	+ 18.09	50.88	1
957	Hydri	10.0	24.28	− 1.092	82 6 2.2	.09	50.81	1
958	Hydri	9.0	24.64	+ 1.609	68 26 59.6	.09	51.95	1
959	Lacaille, 558	7.0	29.38	− 0.500	80 48 17.4	.08	51.32	2
960	Hydri	10.0	29.62	+ 1.747	66 24 46.9	+ 18.08	51.92	1
961	Hydri	10.0	1 42 30.34	− 0.656	76 32 17.9	+ 18.08	50.83	1
962	Lacaille, 546	7.0	43.34	1.010	74 18 12.8	.08	50.87	2
963	Gould, Z. C., 1h 1120	9.5	48.18	1.244	72 23 39.2	.07	50.91	2
964	Gould, Z. C., 1h 1123 . .	10.0	48.94	+ 1.687	67 16 43.0	.07	51.82	1
965	Lacaille, 563	7.9	42 55.07	− 0.460	80 40 3.6	+ 18.07	51.32	2
966	Hydri	10.0	1 43 2.47	+ 1.366	71 11 8.5	+ 18.06	51.88	1
967	Hydri	9.8	8.22	− 0.458	80 38 40.6	.06	51.32	2
968	Gould, Z. C., 1h 1132 . .	9.5	9.10	+ 1.575	68 45 34.2	.06	51.93	2
969	Hydri	10.0	22.50	+ 1.592	68 30 9.9	.05	51.95	1
970	Lacaille, 592	7.5	24.40	− 2.478	84 0 10.9	+ 18.05	50.94	1
971	Hydri	10.0	1 43 33.14	+ 1.092	73 34 8.1	+ 18.04	50.88	1
972	Gould, Z. C., 1h 1153 . .	8.2	43 57.61	1.405	70 37 49.2	.03	51.88	2
973	Gould, 1792	9.2	44 10.51	0.600	76 38 18.0	.02	50.89	2
974	Gould, Z. C., 1h 1162 . .	8.5	25.08	1.718	66 30 53.3	.01	51.92	1
975	Lacaille, 564	8.0	30.10	+ 0.276	78 6 23.0	+ 18.01	50.85	1
976	Lacaille, 556	8.0	1 44 30.58	+ 0.797	75 29 28.0	+ 18.01	50.88	2
977	Gould, Z. C., 1h 1163 . .	8.5	38.30	1.409	70 28 26.4	.01	51.81	1
978	Gould, Z. C., 1h 1167 . .	9.2	45.49	1.376	70 48 52.6	.00	51.87	2
979	Hydri	10.0	48.41	+ 1.435	70 10 10.6	18.00	51.90	1
980	Lacaille, 573	8.5	44 56.94	− 0.457	80 29 28.0	+ 17.99	50.81	1
981	Octantis	9.5	1 45 . .	−21.820	88 38 20.8	+ 17.99	51.87	1
982	Hydri	9.5	2.44	+ 1.543	68 50 45.6	.99	51.91	1
983	Octantis	9.8	18.02	− 2.160	83 {31 51.9 / 32 24.8}	.98	50.82	2
984	Gould, Z. C., 1h 1178 . .	9.0	19.85	+ 1.086	73 21 59.6	.98	50.88	1
985	Gould, 1819	10.0	33.30	+ 1.650	67 19 13.9	+ 17.97	51.82	1

No. 981. Right Ascension may be anything between 40m and 50m. No. 983. One rev. wrong.

Number.	Constellation, Name of Star, or Synonym.	Magnitude.	Right Ascension, 1850.0.	Annual Precession.	South Declination, 1850.0.	Annual Precession.	Mean year.	No. of obs.
			h. m. s.	s.	° ′ ″	″		
986	Hydri	10.0	1 45 36.04	+ 0.343	77 42 37.8	+ 17.97	50.94	1
987	Hydri	9.5	46.76	— 0.829	81 19 29.9	.96	50.80	1
988	Hydri	10.0	45 59.11	+ 1.474	69 31 28.6	.96	51.90	3
989	Hydri	9.5	46 11.94	1.550	68 33 47.6	.94	51.95	1
990	Hydri	10.0	13.07	+ 1.558	68 27 19.8	+ 17.94	51.95	1
991	Hydri	9.5	1 46 13.82	— 0.674	80 56 0.5	+ 17.94	51.83	1
992	Gould, Z. C., 1h 1211 . .	10.0	16.04	+ 1.316	71 11 22.1	.94	51.92	2
993	Hydri	10.0	31.08	1.523	68 50 47.1	.93	51.91	1
994	Lacaille, 567	8.0	35.20	1.091	73 9 22.8	.93	50.88	2
995	Hydri	9.0	36.96	+ 0.360	77 31 30.8	+ 17.92	50.88	2
996	Lacaille, 634	6.0	1 46 40.83	— 4.597	85 31 30.1	+ 17.92	51.86	6
997	Hydri	10.0	47 9.56	0.758	81 3 23.8	.90	51.31	2
998	Hydri	9.2	22.64	+ 1.134	72 41 25.6	.89	50.89	2
999	Hydri	10.0	23.62	0.420	77 10 18.2	.89	50.88	2
1 000	Hydri	10.0	29.07	+ 1.725	65 50 26.4	+ 17.89	51.92	1
1 001	Hydri	10.5	1 47 44.71	— 1.527	82 29 47.1	+ 17.88	50.75	1
1 002	Hydri	9.5	44.93	+ 1.212	71 57 40.8	.88	50.93	2
1 003	Gould, 1855	8.0	45.12	0.645	75 59 4.5	.88	50.84	1
1 004	Gould, Z. C., 1h 1250 . .	9.8	54.82	+ 0.713	75 34 8.8	.87	50.88	2
1 005	Octantis	9.3	56.39	— 5.230	85 49 14.1	+ 17.87	51.87	3
1 006	Gould, Z. C., 1h 1258 . .	9.0	1 47 57.45	+ 1.502	68 52 4.4	+ 17.87	51.91	1
1 007	Lacaille, 581	8.0	48 5.38	0.879	74 30 49.2	.87	50.88	1
1 008	Hydri	9.5	9.93	+ 1.231	71 43 36.3	.86	50.95	1
1 009	Octantis	9.2	32.22	— 7.478	86 41 29.2	.85	51.86	4
1 010	Gould, Z. C., 1h 1264 . .	8.5	41.27	+ 0.462	76 50 1.6	+ 17.84	50.95	1
1 011	Octantis	10.0	1 48 43.78	— 8.753	87 2 34.8	+ 17.84	51.87	1
1 012	Gould, Z. C., 1h 1274 . .	9.2	45.86	+ 1.116	72 39 6.2	.84	50.89	2
1 013	η^1 Hydri	6.8	47.41	+ 1.506	68 41 2.4	.84	51.93	2
1 014	Lacaille, 623	7.5	50.47	— 2.859	84 6 51.1	.84	50.94	1
1 015	Octantis	10.5	50.69	— 1.582	82 30 57.8	+ 17.84	50.75	1
1 016	Hydri	10.0	1 48 52.68	+ 0.408	77 4 8.5	+ 17.84	50.88	2
1 017	Gould, Z. C., 1h 1265 . .	9.0	48 56.06	— 0.144	79 13 37.8	.83	51.89	2
1 018	Octantis	10.0	49 6.65	— 2.305	83 29 52.0	.83	50.80	1
1 019	Gould, Z. C., 1h 1290 . .	9.0	11.30	+ 1.630	66 57 15.8	.82	51.90	1
1 020	Hydri	9.5	16.50	+ 0.682	75 36 3.0	+ 17.82	50.85	1
1 021	Gould, Z. C., 1h 1293 . .	9.0	1 49 17.73	+ 1.479	68 55 33.6	+ 17.82	51.91	1
1 022	Lacaille, 606	6.7	18.48	— 0.768	80 55 4.2	.82	51.40	2
1 023	Hydri	10.0	19.82	+ 1.498	68 41 32.2	.82	51.91	1
1 024	Octantis	10.0	22.25	— 8.529	86 58 12.6	.82	51.87	1
1 025	Hydri	9.5	30.97	+ 1.384	69 59 17.7	+ 17.81	51.90	1

Number.	Constellation, Name of Star, or Synonym.	Magnitude.	Right Ascension, 1850.0.	Annual Precession.	South Declination, 1850.0.	Annual Precession.	Mean year.	No. of obs.
			h. m. s.	s.	° ′ ″	″		
1 026	Hydri	10.0	1 49 35.29	+ 1.134	72 23 16.0	+ 17.81	50.91	1
1 027	Hydri	9.5	37.38	+ 0.698	75 27 42.7	.81	50.91	2
1 028	Hydri	9.0	39.53	− 0.581	80 26 0.4	.81	50.81	1
1 029	Octantis	10.0	49 52.01	− 8.345	86 54 24.8	.80	51.87	1
1 030	Gould, 1898	10.0	50 19.91	+ 1.270	71 2 26.3	+ 17.78	51.88	1
1 031	Gould, Z. C., 1h 1318	10.0	1 50 21.48	+ 1.567	67 38 9.3	+ 17.78	51.91	1
1 032	Hydri	9.0	24.71	1.362	70 5 11.9	.77	51.90	1
1 033	Gould, Z. C., 1h 1304	8.8	28.88	0.449	76 43 54.8	.77	50.89	2
1 034	Hydri	9.2	30.54	1.630	66 43 31.8	.77	51.91	2
1 035	Hydri	10.0	31.69	+ 0.215	77 45 5.1	+ 17.77	50.94	1
1 036	Hydri	10.0	1 50 43.80	+ 0.532	76 15 24.1	+ 17.76	50.84	1
1 037	Hydri	8.0	44.81	1.341	70 15 17.0	.76	51.86	2
1 038	Gould, Z. C., 1h 1338	9.5	54.03	1.557	67 42 58.8	.75	51.91	1
1 039	Hydri	9.5	50 54.87	0.665	75 30 17.3	.75	50.85	1
1 040	Gould, Z. C., 1h 1334	8.8	51 0.70	+ 1.037	72 59 21.6	+ 17.75	50.88	2
1 041	η^3 Hydri	5.2	1 51 8.30	+ 1.499	68 23 9.0	+ 17.74	51.95	2
1 042	Gould, 1928	8.5	18.26	1.642	66 25 30.6	.74	51.92	2
1 043	Gould, Z. C., 1h 1350	9.2	21.12	+ 1.060	72 45 39.0	.74	50.89	2
1 044	Gould, Z. C., 1h 1330	9.5	24.21	− 0.224	79 16 7.4	.73	51.82	1
1 045	Hydri	9.0	24.50	+ 0.515	76 16 8.4	+ 17.73	51.96	1
1 046	Hydri	9.8	1 51 25.98	+ 0.680	75 21 47.1	+ 17.73	50.93	2
1 047	Gould, Z. C., 1h 1361	9.5	35.76	1.471	68 38 49.0	.73	51.93	2
1 048	Hydri	10.0	38.07	+ 0.902	73 54 32.8	.72	50.87	1
1 049	Lacaille, 633	9.2	40.70	− 2.352	83 24 55.1	.72	50.82	2
1 050	Hydri	8.2	43.71	− 0.934	81 6 47.7	+ 17.72	51.53	3
1 051	Gould, Z. C., 1h 1377	9.2	1 51 53.34	+ 1.472	68 35 49.8	+ 17.71	51.93	2
1 052	Hydri	11.0	51 56.67	− 0.022	78 32 5.3	.71	50.93	1
1 053	Octantis	10.0	52 7.37	2.024	82 58 14.6	.70	50.95	1
1 054	Octantis	9.5	14.69	− 2.164	83 8 59.5	.70	50.95	1
1 055	Lacaille, 601	7.5	18.70	+ 1.421	69 7 33.4	+ 17.70	51.91	2
1 056	Gould, Z. C., 1h 1384	9.0	1 52 23.23	+ 0.983	73 13 27.1	+ 17.69	50.88	1
1 057	Hydri	10.0	26.66	1.551	67 30 2.6	.69	51.86	2
1 058	Hydri	10.0	31.93	+ 1.629	66 24 4.0	.68	51.92	1
1 059	Lacaille, 628	7.5	33.99	− 1.472	82 5 52.0	.68	50.81	1
1 060	Gould, Z. C., 1h 1400	9.0	52 50.94	+ 0.846	74 8 37.0	+ 17.67	50.87	1
1 061	Gould, Z. C., 1h 1410	9.8	1 53 12.04	+ 1.260	70 43 14.8	+ 17.66	51.88	2
1 062	Gould, Z. C., 1h 1409	9.0	19.34	0.845	74 5 39.0	.65	50.87	1
1 063	Hydri	8.2	19.61	0.555	75 50 52.9	.65	50.84	2
1 064	Hydri	9.5	26.47	+ 1.076	72 20 49.2	.65	50.91	2
1 065	Hydri	10.0	29.00	− 1.526	82 7 52.7	+ 17.65	50.81	1

Number.	Constellation, Name of Star, or Synonym.	Magnitude.	Right Ascension, 1850.0.	Annual Precession.	South Declination, 1850.0.	Annual Precession.	Mean year.	No. of obs.
			h. m. s.	s.	° ′ ″	″		
1066	Gould, Z. C., 1^h 1426 . .	7.8	1 53 33.02	+ 1.596	66 44 45.6	+ 17.64	51.91	2
1067	Hydri	9.0	37.82	0.766	74 34 38.9	.64	50.88	1
1068	Gould, 1972	6.5	48.76	1.630	66 9 20.1	.63	51.92	1
1069	Lacaille, 621	7.2	49.58	0.015	78 13 39.0	.63	50.89	2
1070	Gould, Z. C., 1^h 1420 . .	8.5	53 58.47	+ 0.077	77 58 37.2	+ 17.63	50.90	2
1071	Octantis	9.8	1 54 8.46	− 7.195	86 26 24.7	+ 17.62	51.86	4
1072	Hydri	9.5	14.46	+ 1.126	71 48 36.6	.62	50.93	2
1073	Gould, Z. C., 1^h 1456 . .	9.0	15.31	1.490	67 59 54.7	.62	51.96	1
1074	Gould, Z. C., 1^h 1461 . .	9.0	28.02	1.117	71 51 23.2	.61	50.93	2
1075	Gould, Z. C., 1^h 1455 . .	9.0	30.57	+ 0.838	73 59 49.5	+ 17.60	50.87	1
1076	Gould, 1988	9.0	1 54 35.16	+ 1.460	68 18 53.0	+ 17.60	51.95	2
1077	Hydri	9.0	40.39	− 1.355	81 45 17.4	.60	50.95	1
1078	Hydri	10.0	56.14	+ 1.045	72 24 15.5	.59	50.90	1
1079	Octantis	8.9	54 57.61	− 5.127	85 30 59.4	.59	51.87	5
1080	Octantis	10.0	55 12.37	− 2.117	82 55 26.5	+ 17.58	50.95	1
1081	Gould, Z. C., 1^h 1491 . .	9.0	1 55 21.75	+ 1.284	70 9 34.9	+ 17.57	51.90	1
1082	Gould, Z. C., 1^h 1492 . .	8.5	22.97	+ 1.272	70 17 0.6	.57	51.86	2
1083	Hydri	9.8	24.38	− 1.027	81 3 14.4	.57	51.20	3
1084	Hydri	10.0	26.24	+ 1.236	70 37 48.5	.56	51.81	1
1085	Hydri	10.0	29.12	+ 1.178	71 10 35.9	+ 17.56	51.88	1
1086	Gould, 2006	6.8	1 55 31.03	+ 1.609	66 10 45.0	+ 17.56	51.92	2
1087	Hydri	11.0	32.98	+ 1.190	71 3 30.6	.56	51.88	1
1088	Hydri	9.5	33.43	− 0.571	79 56 {8.4 / 34.9}	.56	50.87	2
1089	Lacaille, 625	7.0	35.39	+ 0.461	76 5 50.0	.56	50.84	1
1090	Lacaille, 616	6.3	45.36	+ 1.563	66 47 41.2	+ 17.55	51.91	2
1091	Hydri	9.0	1 55 45.70	+ 0.466	76 3 15.2	+ 17.55	50.84	1
1092	Gould, 2000	8.8	46.04	− 0.055	78 18 33.4	.55	50.89	2
1093	Gould, Z. C., 1^h 1499 . .	9.0	53.39	+ 0.905	73 21 21.9	.55	50.88	1
1094	Octantis	9.5	55 54.82	− 6.358	86 4 13.5	.55	51.88	3
1095	Hydri	9.8	56 8.82	− 0.458	79 34 52.0	+ 17.54	51.44	2
1096	Lacaille, 637	7.0	1 56 11.50	− 0.294	79 4 54.8	+ 17.53	51.54	3
1097	Gould, 2011	9.0	13.24	+ 0.676	74 50 12.1	.53	50.86	1
1098	Gould, Z. C., 1^h 1512 . .	9.2	18.58	1.208	70 46 55.2	.53	51.88	2
1099	Hydri	9.0	41.02	1.228	70 32 9.4	.51	51.81	1
1100	Gould, 2042	8.5	56 59.00	+ 1.616	65 50 3.4	+ 17.50	51.92	1
1101	Hydri	10.0	1 57 13.40	+ 1.176	70 57 0.9	+ 17.49	51.96	1
1102	Octantis	9.0	15.84	− 7.751	86 32 19.3	.49	51.87	3
1103	Hydri	10.0	19.01	+ 1.607	65 54 59.6	.49	51.92	1
1104	Hydri	9.5	19.49	1.087	71 41 8.4	.46	50.95	1
1105	Gould, Z. C., 1^h 1547 . .	9.7	20.30	+ 1.456	67 55 42.7	+ 17.48	51.92	3

No. 1068. Original record gives 66° 9′ 31.3″. Micrometer reading is recorded — 0.30 rev.; it was probably — 0.70 rev. If corrected by this amount the declination will be 66° 9′ 20.1″.

No. 1088. One rev. wrong.

Number.	Constellation, Name of Star, or Synonym.	Magnitude.	Right Ascension, 1850.0.	Annual Precession.	South Declination, 1850.0.	Annual Precession.	Mean year.	No. of obs.
			h. m. s.	s.	° ′ ″	″		
1106	Gould, 2048	7.5	1 57 22.52	+ 1.513	67 4 26.1	+ 17.48	51.90	3
1107	Hydri	9.2	22.63	— 1.109	81 5 36.2	.48	50.98	3
1108	Gould, Z. C., 1^{h} 1548 . .	10.0	26.35	+ 1.244	70 15 50.1	.48	51.90	1
1109	Hydri	10.0	26.69	1.535	66 53 34.8	.48	51.90	1
1110	Hydri	9.5	31.96	+ 0.006	77 55 19.3	+ 17.47	50.94	1
1111	Gould, Z. C., 1^{h} 1557 . .	9.0	1 57 38.06	+ 1.264	70 2 18.8	+ 17.47	51.90	1
1112	Hydri	10.0	42.19	0.211	77 3 58.2	.47	50.95	1
1113	Hydri	11.0	42.72	1.116	71 26 4.7	.47	51.88	1
1114	Hydri	10.0	44.77	0.630	74 56 6.4	.47	50.86	1
1115	Hydri	10.0	47.61	+ 1.252	70 9 13.4	+ 17.47	51.90	1
1116	Gould, Z. C., 1^{h} 1571 . .	9.5	1 57 51.91	+ 1.585	66 7 10.3	+ 17.46	51.92	1
1117	Hydri	10.0	58 13.63	+ 1.219	70 24 22.7	.45	51.81	1
1118	Hydri	11.0	24.47	— 0.347	79 3 50.0	.44	50.83	1
1119	Hydri	10.0	26.19	+ 1.080	71 39 14.9	.44	50.95	1
1120	Hydri	9.8	36.26	— 0.408	79 14 13.3	+ 17.43	51.89	2
1121	Lacaille, 656	8.2	1 58 41.10	— 0.496	79 29 26.7	+ 17.43	51.57	3
1122	Hydri	9.2	41.69	+ 0.012	77 47 30.4	.43	50.90	2
1123	Hydri	9.8	43.32	0.590	75 3 30.0	.42	50.90	3
1124	Gould, Z. C., 1^{h} 1588 . .	9.5	47.76	1.211	70 24 1.1	.42	51.86	2
1125	Hydri	10.0	49.58	+ 0.372	76 12 40.6	+ 17.42	50.83	2
1126	Brisbane, 296	8.0	1 58 50.39	+ 1.540	66 36 23.2	+ 17.42	51.91	2
1127	Gould, 2070	8.4	57.28	+ 0.536	75 20 29.3	.41	51.63	4
1128	Octantis	9.5	57.68	— 4.039	84 42 26.7	.41	51.79	5
1129	Hydri	10.0	58 58.36	+ 0.890	73 5 14.3	.41	50.88	2
1130	Gould, Z. C., 1^{h} 1591 . .	10.0	59 13.38	+ 0.308	76 28 54.4	+ 17.40	50.83	1
1131	Hydri	9.2	1 59 17.11	+ 0.347	76 17 8.6	+ 17.40	50.83	2
1132	Gould, Z. C., 1^{h} 1596 . .	9.0	19.94	0.677	74 28 40.2	.40	50.88	1
1133	Hydri	9.0	28.74	1.560	66 14 2.3	.39	51.92	1
1134	Hydri	10.0	31.67	+ 1.307	69 19 3.5	.39	51.97	1
1135	Hydri	11.0	43.48	— 0.366	79 1 8.7	+ 17.38	50.83	1
1136	Lacaille, 643	7.0	1 59 47.59	+ 1.118	71 8 30.8	+ 17.38	51.92	2
1137	Hydri	9.5	59 53.69	0.644	74 37 5.5	.37	50.88	1
1138	Hydri	10.0	2 0 1.64	0.282	76 31 22.5	.37	50.83	1
1139	Hydri	10.0	2.48	0.872	73 5 28.8	.37	50.88	2
1140	Hydri	9.0	4.38	+ 1.063	71 35 2.9	+ 17.37	51.42	2
1141	Gould, Z. C., 2^{h} 10 . .	8.8	2 0 5.79	+ 0.703	74 13 20.7	+ 17.37	50.87	2
1142	Lacaille, 652	7.7	23.74	— 0.460	75 9 58.3	.35	51.35	4
1143	Lacaille, 679	7.7	26.00	1.844	82 13 35.8	.35	51.21	3
1144	Hydri	11.0	32.99	— 0.354	78 54 41.1	.35	50.83	1
1145	Gould, Z. C., 2^{h} 32 . . .	9.5	35.03	+ 1.150	70 44 14.8	+ 17.34	51.81	1

Number.	Constellation, Name of Star, or Synonym.	Magnitude.	Right Ascension, 1850.0.	Annual Precession.	South Declination, 1850.0.	Annual Precession.	Mean year.	No. of obs.
			h. m. s.	s.	° ′ ″	″		
1 146	Lacaille, 642	7.5	2 0 35.51	+ 1.573	65 51 33.9	+ 17.34	51.92	1
1 147	Gould, 2121	10.0	1 6.13	1.569	65 50 30.9	.32	51.92	1
1 148	Hydri	9.2	6.21	0.551	75 1 53.3	.32	51.82	2
1 149	Hydri	10.0	6.52	0.538	75 6 2.9	.32	50.86	1
1 150	Hydri	10.0	20.23	+ 1.066	71 23 46.2	+ 17.31	51.88	1
1 151	Hydri	10.0	2 1 22.95	+ 1.228	69 52 38.8	+ 17.31	51.90	1
1 152	Gould, Z. C., 2ʰ 56 . . .	9.5	25.05	1.285	69 17 16.3	.31	51.91	1
1 153	Gould, 2125	10.0	29.52	0.995	71 58 13.5	.30	50.93	2
1 154	Gould, 2126	10.0	30.95	0.994	71 58 18.8	.30	50.93	2
1 155	Hydri	10.0	35.94	+ 0.223	76 38 13.8	+ 17.30	50.89	2
1 156	Gould, 2130	8.2	2 1 43.16	+ 1.306	69 1 5.9	+ 17.29	51.91	2
1 157	Brisbane, 305	9.0	52.14	1.486	66 51 9.6	.28	51.91	2
1 158	Gould, Z. C., 2ʰ 59 . . .	10.0	52.27	0.558	74 54 28.6	.28	50.86	1
1 159	Hydri	10.0	1 53.82	+ 0.963	77 41 42.4	.28	50.94	1
1 160	Lacaille, 675	7.0	2 4.02	— 0.624	79 35 20.6	+ 17.28	51.44	2
1 161	Hydri	10.0	2 2 10.67	+ 0.676	74 9 53.4	+ 17.27	50.87	1
1 162	Hydri	10.0	20.14	1.418	67 38 6.5	.27	51.91	1
1 163	Hydri	10.0	30.46	+ 0.209	76 37 9.1	.26	50.89	2
1 164	Octantis	10.0	51.34	—12.143	87 25 47.8	.24	51.87	1
1 165	Hydri	10.0	56.84	+ 0.221	76 31 3.8	+ 17.24	50.83	1
1 166	Hydri	10.0	2 2 59.32	+ 0.665	74 8 28.8	+ 17.24	50.87	1
1 167	Hydri	9.0	3 1.19	+ 1.150	70 25 3.7	.24	51.81	1
1 168	Octantis	10.0	2.57	— 9.343	86 50 49.6	.23	51.88	1
1 169	Octantis	9.5	4.89	— 2.225	82 38 28.6	.23	50.95	1
1 170	Gould, 2157	9.0	6.93	+ 1.045	71 20 39.4	+ 17.23	51.88	1
1 171	Lacaille, 664	6.8	2 3 15.16	+ 1.485	66 39 32.6	+ 17.23	51.91	2
1 172	Octantis	10.0	15.28	— 4.585	84 53 21.2	.23	52.69	1
1 173	Lacaille, 665	8.7	15.64	+ 0.923	72 19 28.2	.23	50.91	2
1 174	Octantis	10.0	24.58	—10.496	87 6 26.4	.22	51.87	1
1 175	Gould, Z. C., 2ʰ 114 . .	10.0	32.81	+ 1.108	70 44 13.3	+ 17.21	51.96	1
1 176	Octantis	11.0	2 3 34.17	— 3.345	83 53 36.8	+ 17.21	50.94	1
1 177	Hydri	9.5	37.71	+ 0.757	73 28 21.0	.21	50.88	2
1 178	Hydri	10.0	38.76	1.227	69 34 20.8	.21	51.91	1
1 179	Hydri	10.0	45.55	+ 0.779	73 18 23.0	.20	50.88	1
1 180	Hydri	9.0	47.29	— 1.393	81 14 45.4	+ 17.20	50.80	1
1 181	Lacaille, 700	9.2	2 3 49.34	— 2.543	83 1 0.0	+ 17.20	50.90	2
1 182	Hydri	10.0	3 54.39	+ 1.129	70 29 36.2	.20	51.81	1
1 183	Hydri	10.2	4 13.80	— 0.388	78 43 25.2	.18	50.88	2
1 184	Hydri	10.0	31.34	0.247	78 14 8.3	.17	50.93	1
1 185	Octantis	10.0	37.65	— 3.182	83 41 (16.1)	+ 17.16	50.80	1

No. 1 158. Original record 74° 56′ 20.3″. Changed 4 revs.

Number.	Constellation, Name of Star, or Synonym.	Magnitude.	Right Ascension, 1850.0.	Annual Precession.	South Declination, 1850.0.	Annual Precession.	Mean year.	No. of obs.
			h. m. s.	s.	° ′ ″	″		
1186	Hydri	9.8	2 4 41.06	+ 0.244	76 14 49.6	+ 17.16	50.83	2
1187	Octantis	10.0	47.52	− 6.425	85 49 43.8	.16	51.87	4
1188	Lacaille, 686	8.2	54.75	− 0.637	79 24 54.9	.15	51.14	2
1189	Hydri	9.5	4 59.59	+ 0.464	75 6 21.1	.15	51.83	2
1190	Gould, Z. C., 2h 160 . .	10.0	5 7.69	+ 1.428	67 5 55.6	+ 17.14	51.90	1
1191	Hydri	10.0	2 5 9.17	+ 1.036	71 9 47.4	+ 17.14	51.88	1
1192	Gould, Z. C., 2h 167 . .	10.0	30.67	+ 1.191	69 41 14.9	.12	51.89	1
1193	Octantis	10.0	33.09	− 4.774	84 55 41.4	.12	52.69	1
1194	Gould, Z. C., 2h 166 . .	9.0	42.06	+ 0.621	74 7 14.8	.11	50.87	1
1195	Lacaille, 760	8.2	5 56.10	− 6.351	85 45 38.5	+ 17.10	51.87	4
1196	Hydri	10.0	2 6 19.94	− 0.194	77 53 44.1	+ 17.09	50.94	1
1197	Octantis	8.9	25.23	− 5.612	85 23 12.7	.10	51.87	5
1198	Gould, Z. C., 2h 187 . .	9.7	30.34	+ 0.405	75 16 18.4	.08	51.52	3
1199	Hydri	10.0	37.53	+ 0.376	75 24 26.1	.07	51.98	1
1200	Lacaille, 716	10.0	38.78	− 2.836	83 13 29.9	+ 17.07	50.84	1
1201	Hydri	10.0	2 6 44.70	+ 1.262	68 47 27.0	+ 17.07	51.91	1
1202	Gould, Z. C., 2h 199 . .	10.0	6 49.46	+ 0.862	72 21 29.6	.06	50.91	2
1203	Hydri	10.0	7 7.22	− 1.612	81 26 57.2	.05	50.87	2
1204	Gould, 2245	10.0	14.62	+ 1.188	69 29 9.3	.04	51.90	3
1205	Hydri	9.5	25.56	+ 1.234	68 59 20.2	+ 17.04	51.91	2
1206	Octantis	10.0	2 7 40.00	−10.763	87 4 29.7	+ 17.02	51.87	1
1207	Lacaille, 764	7.7	47.70	− 5.863	85 28 19.4	.02	51.87	5
1208	Hydri	9.5	48.25	+ 1.244	68 50 22.2	.02	51.91	1
1209	Octantis	10.0	51.33	− 2.491	82 44 53.0	.02	50.95	1
1210	Gould, 2257	9.2	7 53.02	+ 1.399	67 3 29.6	+ 17.01	51.93	2
1211	Hydri	10.0	2 8 3.85	+ 1.388	67 10 25.1	+ 17.01	51.90	1
1212	Gould, 2255	9.0	4.95	0.723	73 12 3.7	.01	50.88	1
1213	Hydri	10.0	6.91	+ 1.359	67 30 27.8	.00	51.94	2
1214	Octantis	9.8	11.46	− 5.043	85 0 12.5	.00	51.87	3
1215	Hydri	9.5	12.31	+ 0.926	71 41 58.2	+ 17.00	50.95	1
1216	Hydri	9.2	2 8 15.19	− 1.581	81 19 29.1	+ 17.00	50.89	2
1217	Hydri	10.0	15.67	+ 1.295	68 13 34.0	.00	51.95	2
1218	Hydri	9.5	16.25	− 0.458	78 37 46.8	.00	50.88	2
1219	Octantis	9.0	22.20	−49.553	89 13 34.7	17.99	51.86	3
1220	Melbourne (1), 127 . . .	8.5	24.40	+ 1.438	66 28 37.5	+ 16.99	51.92	1
1221	Hydri	10.0	2 8 53.56	+ 1.432	66 29 52.4	+ 16.97	51.92	1
1222	Gould, Z. C., 2h 240 . .	9.5	9 0.58	− 0.102	77 19 39.1	.96	50.81	1
1223	Hydri	9.5	2.36	+ 1.427	66 32 56.4	.96	51.92	1
1224	Lacaille, 691	7.0	11.00	1.400	66 51 29.2	.95	51.90	1
1225	Hydri	10.0	11.24	+ 1.031	70 42 9.6	+ 16.95	51.81	1

Number.	Constellation, Name of Star, or Synonym.	Magnitude.	Right Ascension, 1850.0.	Annual Precession.	South Declination, 1850.0.	Annual Precession.	Mean year.	No. of obs.
			h. m. s.	s.	° ′ ″	″		
1 226	Hydri	9.5	2 9 23.37	− 0.863	79 41 40.9	+ 16.94	50.93	1
1 227	Gould, Z. C., 2h 249 . .	9.5	23.75	− 0.127	77 23 37.9	.94	50.81	1
1 228	Gould, Z. C., 2h 257 . .	9.2	26.64	+ 0.442	74 46 25.6	.94	50.88	2
1 229	Lacaille, 698	8.0	33.96	+ 0.912	71 39 15.5	.94	50.95	1
1 230	Hydri	10.5	55.84	− 0.613	78 58 21.7	+ 16.92	50.83	1
1 231	Octantis	11.0	2 9 59.68	− 3.826	84 3 15.1	+ 16.92	50.94	1
1 232	Gould, 2297	10.0	10 8.24	+ 0.662	73 22 57.8	.91	50.88	1
1 233	Gould, 2300	10.0	9.25	+ 0.662	73 23 3.7	.91	50.88	1
1 234	Octantis	9.5	13.07	−16.793	87 55 32.8	.91	51.87	3
1 235	Lacaille, 702	7.7	16.14	+ 0.346	75 12 16.6	+ 16.90	51.51	3
1 236	Stone, 888	9.8	2 10 24.16	− 3.173	83 26 49.1	+ 16.90	51.41	2
1 237	Lacaille, 709	7.0	25.22	− 0.136	77 19 40.1	.90	50.81	1
1 238	Gould, 2319	9.0	29.81	+ 1.363	67 8 0.3	.89	51.90	1
1 239	Lacaille, 710	7.0	36.69	0.029	76 39 42.2	.89	50.89	2
1 240	Gould, 2322	10.0	37.70	+ 1.330	67 29 43.3	+ 16.89	51.91	1
1 241	Brisbane, 329	9.0	2 10 42.56	+ 0.416	74 47 22.8	+ 16.88	50.88	1
1 242	Hydri	10.0	44.71	+ 1.150	69 24 22.9	.88	51.91	1
1 243	Lacaille, 743	9.0	44.72	− 2.818	83 0 34.6	.88	50.90	2
1 244	Hydri	10.0	10 45.66	− 0.196	77 32 20.2	.88	50.88	2
1 245	π¹ Hydri	6.0	11 6.81	+ 1.230	68 32 34.8	+ 16.86	51.93	2
1 246	Hydri	10.0	2 11 7.12	− 2.085	82 0 31.2	+ 16.86	51.99	1
1 247	Lacaille, 715	8.0	7.39	− 0.063	77 0 1.9	.86	50.95	1
1 248	Hydri	9.5	8.70	+ 1.200	68 51 10.8	.86	51.91	1
1 249	Hydri	10.5	10.13	− 1.080	80 6 17.9	.86	50.82	1
1 250	Hydri	9.0	10.14	+ 0.909	71 29 27.0	+ 16.86	51.42	2
1 251	Hydri	10.5	2 11 18.09	− 0.718	79 10 30.6	+ 16.85	50.83	1
1 252	Gould, 2325	10.0	20.87	− 0.022	76 48 38.8	.85	50.95	1
1 253	Gould, Z. C., 2h 328 . .	9.8	32.69	+ 1.278	67 57 38.5	.84	51.93	2
1 254	Gould, Z. C., 2h 312 . .	9.0	35.36	− 0.176	77 23 49.1	.84	50.81	1
1 255	Hydri	10.0	37.17	+ 1.266	68 5 19.3	+ 16.84	51.96	1
1 256	Hydri	9.0	2 11 39.62	+ 1.055	70 11 9.0	+ 16.84	51.86	2
1 257	Hydri	10.0	45.46	+ 1.205	68 43 27.5	.83	51.91	1
1 258	Stone, 906	10.0	55.82	− 1.426	80 48 30.8	.82	51.32	2
1 259	Hydri	10.0	58.07	+ 1.083	69 53 53.9	.82	51.90	1
1 260	Gould, Z. C., 2h 333 . .	10.0	11 59.54	+ 0.799	72 14 12.0	+ 16.82	50.91	2
1 261	π² Hydri	6.5	2 12 21.76	+ 1.224	68 26 34.5	+ 16.80	51.95	2
1 262	Hydri	9.5	25.42	+ 0.921	71 14 31.1	.80	51.88	1
1 263	Gould, 2344	9.5	26.24	− 0.039	76 47 13.7	.80	50.89	2
1 264	Octantis	10.0	37.25	4.446	84 26 39.4	.79	52.68	1
1 265	Octantis	10.0	12 55.00	− 5.329	85 0 56.4	+ 16.78	51.88	1

Number.	Constellation, Name of Star, or Synonym.	Magnitude.	Right Ascension, 1850.0.	Annual Precession.	South Declination, 1850.0.	Annual Precession.	Mean year.	No. of obs.
			h. m. s.	s.	° ′ ″	″		
1 266	Hydri	10.0	2 13 0.79	+ 1.226	68 20 21.2	+ 16.77	51.95	1
1 267	Hydri	10.0	2.86	1.040	70 9 37.4	.77	51.86	2
1 268	Gould, Z. C., 2^h 357	10.0	3.26	0.785	72 13 8.0	.77	50.91	2
1 269	Lacaille, 714	7.5	7.95	+ 1.041	70 7 48.2	.77	51.90	1
1 270	Hydri	10.0	12.38	− 0.309	77 44 50.1	+ 16.76	50.94	1
1 271	Hydri	10.0	2 13 15.37	+ 1.082	69 44 40.7	+ 16.76	51.89	1
1 272	Octantis	11.0	18.84	− 3.986	84 3 27.6	.76	50.94	1
1 273	Gould, 2373	9.3	29.05	+ 1.109	69 27 22.6	.75	51.90	3
1 274	Hydri	9.5	29.39	0.859	71 36 53.4	.75	50.95	1
1 275	Gould, Z. C., 2^h 369	9.2	39.69	+ 1.141	69 5 8.4	+ 16.74	51.91	2
1 276	Hydri	9.0	2 13 55.27	+ 0.097	76 4 29.5	+ 16.73	50.84	1
1 277	Gould, 2381	9.2	13 58.34	0.928	71 0 33.4	.73	51.92	2
1 278	Gould, 2384	8.0	14 2.74	1.025	70 10 11.8	.72	51.86	2
1 279	Gould, Z. C., 2^h 387	9.0	8.65	+ 1.159	68 53 9.1	.72	51.91	1
1 280	Hydri	10.0	16.24	− 1.124	80 0 22.6	+ 16.71	50.82	1
1 281	Gould, Z. C., 2^h 381	9.8	2 14 23.35	− 0.065	76 43 44.6	+ 16.71	50.89	2
1 282	Lacaille, 734	7.0	26.64	− 0.148	77 3 15.9	.70	50.88	2
1 283	Hydri	10.0	57.07	+ 0.387	74 32 21.0	.68	50.88	1
1 284	Hydri	9.8	14 59.40	+ 0.799	71 54 {33.0 / 5.3}	.68	50.93	2
1 285	Hydri	10.0	15 4.88	− 0.435	78 1 52.7	+ 16.67	50.85	1
1 286	Gould, 2408	9.0	2 15 13.55	+ 1.185	68 28 45.3	+ 16.67	51.95	1
1 287	Hydri	9.3	18.53	+ 0.224	75 21 18.4	.66	51.81	3
1 288	Hydri	10.0	19.62	− 0.314	77 36 5.6	.66	50.94	1
1 289	Octantis	10.0	25.00	−11.646	87 6 10.5	.66	51.87	1
1 290	Gould, Z. C., 2^h 415	9.0	26.11	+ 0.821	71 45 42.4	+ 16.66	50.95	1
1 291	Hydri	10.0	2 15 26.18	− 1.860	81 24 36.6	+ 16.66	50.95	1
1 292	Stone, 932	9.5	35.31	1.040	79 43 22.7	.65	50.93	1
1 293	Gould, Z. C., 2^h 407	10.0	37.73	0.262	77 23 19.5	.65	50.81	1
1 294	Octantis	9.8	15 54.76	− 4.488	84 20 52.2	63	51.82	2
1 295	Gould, Z. C., 2^h 425	9.0	16 2.72	+ 0.748	72 9 29.9	+ 16.63	50.91	1
1 296	Gould, Z. C., 2^h 443	9.0	2 16 28.02	+ 0.945	70 34 36.6	.60	51.90	2
1 297	Gould, 2438	9.0	31.75	+ 1.180	68 22 27.4	.60	51.95	2
1 298	Octantis	10.0	36.56	− 7.138	85 47 44.3	.60	51.87	3
1 299	Gould, Z. C., 2^h 452	9.0	46.34	+ 0.844	71 22 5.0	.59	51.88	1
1 300	Horologii	9.2	50.04	+ 1.296	67 3 53.2	+ 16.59	51.93	2
1 301	Hydri	9.5	2 16 52.46	− 2.497	82 18 37.2	+ 16.58	51.99	1
1 302	Lacaille, 736	8.2	16 54.64	+ 1.135	68 46 24.2	.58	51.93	2
1 303	Gould, Z. C., 2^h 481	9.0	17 31.93	+ 0.984	70 7 9.0	.55	51.90	1
1 304	Octantis	9.0	17 50.29	− 1.858	81 16 34.7	.54	50.99	1
1 305	Gould, Z. C., 2^h 495	9.2	18 3.10	+ 1.118	68 48 23.1	+ 16.53	51.93	2

No. 1 284. Evidently 1 rev. wrong.

Number.	Constellation, Name of Star, or Synonym.	Magnitude.	Right Ascension, 1850.0.	Annual Precession.	South Declination, 1850.0.	Annual Precession.	Mean year.	No. of obs.
			h. m. s.	s.	° ′ ″	″		
1306	Stone, 948	9.0	2 18 3.46	— 1.105	79 43 1.7	+ 16.53	50.93	1
1307	Gould, 2462	8.5	23.68	1.183	79 53 1.6	.51	50.87	2
1308	Gould, Z. C., 2h 483 . .	9.5	25.64	— 0.963	79 20 31.5	.51	51.96	1
1309	Horologii	9.0	27.60	+ 1.317	66 35 47.3	.51	51.91	2
1310	Gould, Z. C., 2h 506 . .	10.0	41.02	+ 0.501	73 31 12.4	+ 16.50	50.88	2
1311	Hydri	10.0	2 18 42.59	— 0.010	76 8 13.0	+ 16.49	50.84	1
1312	Hydri	9.8	55.43	+ 0.927	70 26 39.6	.48	51.90	2
1313	Gould, Z. C., 2h 512 . .	9.5	56.63	0.479	73 37 31.1	.48	50.87	1
1314	Hydri	10.0	18 58.89	0.941	70 19 44.4	.48	51.81	1
1315	δ Hydri	5.5	19 5.72	+ 1.050	69 20 34.6	+ 16.47	51.91	1
1316	Gould, Z. C., 2h 530 . .	9.0	2 19 7.26	+ 1.285	66 53 35.4	+ 16.47	51.90	1
1317	Horologii	9.0	10.36	1.304	66 39 27.6	.47	51.91	2
1318	Hydri	9.5	15.62	+ 0.751	71 47 50.1	.47	50.95	1
1319	Gould, Z. C., 2h 515	9.0	18.58	— 0.092	76 25 54.8	.46	50.83	1
1320	Gould, Z. C., 2h 529 . .	8.8	24.30	+ 0.573	72 59 47.0	+ 16.46	50.88	2
1321	Hydri	9.2	2 19 25.40	+ 0.795	71 26 47.0	+ 16.46	51.42	2
1322	Hydri	9.0	27.63	— 1.880	81 13 34.2	.46	50.99	1
1323	Octantis	10.0	29.44	— 4.188	83 59 21.1	.45	50.94	1
1324	Hydri	10.0	30.66	+ 0.926	70 23 11.2	.45	51.90	1
1325	Gould, 2504	7.8	31.30	+ 0.406	73 59 49.1	+ 16.45	50.87	2
1326	Hydri	9.2	2 19 38.74	+ 0.366	74 12 19.9	+ 16.45	50.87	2
1327	Hydri	10.0	39.95	— 0.495	77 53 27.8	.45	50.90	2
1328	Hydri	10.0	44.58	+ 0.323	74 25 49.2	.44	50.88	1
1329	Gould, Z. C., 2h 527 . .	9.5	53.43	— 0.791	78 46 38.2	.43	50.88	2
1330	Hydri	10.0	55.72	+ 0.418	73 53 16.9	+ 16.43	50.87	1
1331	Hydri	9.0	2 19 56.40	+ 0.673	72 21 14.6	+ 16.43	50.91	2
1332	Hydri	10.0	19 59.11	— 2.670	82 22 45.4	.43	51.98	1
1333	Hydri	10.0	20 2.03	+ 1.105	68 43 18.2	.43	51.91	1
1334	Gould, Z. C., 2h 531 . .	8.0	7.95	0.391	74 1 28.3	.42	50.87	1
1335	Gould, 2536	8.5	30.91	+ 1.342	66 0 59.9	+ 16.40	51.92	1
1336	Octantis	9.5	2 20 34.30	— 4.446	84 9 18.5	+ 16.40	52.68	1
1337	Octantis	9.0	39.15	— 7.362	85 46 44.9	.40	52.01	6
1338	Hydri	10.0	39.43	+ 0.625	72 31 42.5	.40	50.90	1
1339	Gould, 2537	9.0	41.67	0.938	70 9 9.1	.39	51.90	1
1340	Hydri	9.5	42.47	+ 0.785	71 22 54.0	+ 16.39	51.88	1
1341	Hydri	10.0	2 20 50.76	— 0.397	77 28 11.3	+ 16.39	50.94	1
1342	Octantis	9.2	54.35	—11.133	86 53 32.1	.38	51.87	6
1343	Gould, Z. C., 2h 569	10.0	20 57.34	+ 0.985	69 42 54.0	.38	51.89	2
1344	Hydri	9.5	21 2.17	0.481	73 24 38.3	.38	50.88	1
1345	Hydri	10.0	7.02	+ 0.340	74 12 46.5	+ 16.37	50.88	1

Number.	Constellation, Name of Star, or Synonym.	Magnitude.	Right Ascension, 1850.0.	Annual Precession.	South Declination, 1850.0.	Annual Precession.	Mean year.	No. of obs.
			h. m. s.	s.	° ′ ″	″		
1 346	Hydri	10.0	2 21 10.08	+ 0.805	71 10 57.5	+ 16.37	51.88	1
1 347	Gould, Z. C., 2h 582 . .	9.0	23.83	+ 1.074	68 49 51.8	.36	51.93	2
1 348	Octantis	10.0	27.37	− 3.810	83 35 2.2	.36	51.98	1
1 349	Hydri	9.8	39.73	+ 0.708	71 50 58.0	.35	50.93	2
1 350	Gould, Z. C., 2h 594 . .	9.2	55.06	+ 1.068	68 49 34.0	+ 16.33	51.93	2
1 351	Hydri	10.0	2 21 57.05	− 0.611	78 5 50.4	+ 16.33	50.85	1
1 352	κ Hydri	6.5	22 1.18	+ 0.304	74 19 29.2	.33	50.87	2
1 353	Lacaille, 769	6.5	21.27	1.223	67 10 11.8	.31	51.93	2
1 354	Hydri	9.5	21.67	0.693	71 53 0.8	.31	50.95	1
1 355	Hydri	10.0	22.90	+ 0.487	73 14 41.9	+ 16.31	50.88	1
1 356	Hydri	10.0	2 22 25.46	+ 0.834	70 49 0.2	+ 16.31	51.96	1
1 357	Hydri	10.0	28.70	0.380	73 51 54.3	.30	50.87	1
1 358	Horologii	10.0	29.29	1.241	66 57 18.9	.30	51.90	1
1 359	Hydri	10.0	31.46	+ 1.014	69 15 44.6	.30	51.91	1
1 360	Hydri	9.5	31.96	− 2.115	81 27 44.8	+ 16.30	50.97	2
1 361	Hydri	9.0	2 22 33.61	+ 0.704	71 47 7.2	+ 16.30	50.95	1
1 362	Hydri	10.0	36.41	+ 0.845	70 42 46.7	.30	51.97	2
1 363	Octantis	8.8	39.92	− 7.292	85 41 52.8	29	52.00	6
1 364	Gould, Z. C., 2h 618 . .	9.5	41.64	+ 1.070	68 42 54.8	.29	51.93	2
1 365	Gould, 2573	8.0	42.56	+ 1.098	68 26 42.6	+ 16.29	51.95	1
1 366	Hydri	9.2	2 22 42.60	+ 0.165	74 59 4.8	+ 16.29	51.82	2
1 367	Gould, Z. C., 2h 601 . .	9.2	48.45	− 0.705	78 19 40.2	.29	50.89	2
1 368	Octantis	10.0	58.82	− 3.742	83 27 37.4	.28	51.41	2
1 369	Hydri	9.8	22 58.84	+ 0.050	75 30 30.5	.28	52.33	2
1 370	Hydri	10.0	23 1.36	+ 0.531	72 54 36.9	+ 16.28	50.88	1
1 371	Hydri	10.0	2 23 3.15	+ 0.048	75 30 28.7	+ 16.28	50.96	1
1 372	Hydri	9.5	7.85	+ 0.929	69 57 {30.5 / 12.4}	.28	51.90	2
1 373	Hydri	9.5	8.83	− 0.164	76 26 50.7	27	50.83	1
1 374	Lacaille, 777	8.0	21.36	+ 0.826	70 46 39.4	.26	51.97	2
1 375	Octantis	9.2	22.89	− 3.494	83 11 57.8	+ 16.26	50.90	2
1 376	Hydri	10.0	2 23 29.23	+ 1.050	68 48 51.2	+ 16.25	51.91	1
1 377	Hydri	9.5	36.50	0.176	74 50 57.8	.25	50.89	1
1 378	Lacaille, 788	8.3	41.31	0.120	75 6 50.5	.24	51.51	3
1 379	Hydri	9.2	49.36	0.727	71 29 25.8	.24	51.42	2
1 380	Lacaille, 778	8.5	50.17	+ 0.993	69 18 33.2	+ 16.23	51.91	1
1 381	Hydri	9.0	2 23 51.94	+ 0.146	74 58 38.4	+ 16.23	52.69	1
1 382	Hydri	9.5	23 52.28	+ 0.895	70 9 30.1	.23	51.90	1
1 383	Octantis	10.0	24 8.84	− 3.562	83 14 14.9	.22	50.84	1
1 384	Gould, Z. C., 2h 659 . .	9.8	15.39	+ 0.966	69 30 48.7	.21	51.90	3
1 385	Octantis	9.5	2 24 15.78	−11.419	86 53 27.0	+ 16.21	51.87	5

Number.	Constellation, Name of Star, or Synonym.	Magnitude.	Right Ascension, 1850.0.	Annual Precession.	South Declination, 1850.0.	Annual Precession.	Mean year.	No. of obs.
			h. m. s.	s.	° ′ ″	″		
1386	Horologii	10.0	2 24 32.95	+ 1.131	67 25 8.6	+ 16.20	51.97	1
1387	Octantis	9.2	34.54	—18.853	87 56 24.8	.20	51.80	4
1388	Hydri	10.0	34.64	+ 0.772	71 4 23.4	.20	51.96	1
1389	Horologii	10.0	38.09	+ 1.300	65 59 40.8	.19	51.92	1
1390	Octantis	10.0	49.89	— 4.548	84 4 53.9	+ 16.18	50.94	1
1391	Gould, 2624	7.0	2 24 55.36	+ 1.277	66 13 56.0	+ 16.18	51.91	2
1392	Gould, Z. C., 2h 676 . .	9.5	56.34	0.830	70 34 40.4	.18	51.99	1
1393	Gould, Z. C., 2h 672 . .	10.0	24 56.59	0.615	72 9 54.7	.18	50.91	1
1394	Hydri	9.8	25 1.35	1.131	67 50 8.3	.17	51.18	3
1395	Gould, 2629	9.0	10.76	+ 1.243	66 35 54.2	+ 16.17	51.91	2
1396	Gould, 2635	9.2	2 25 28.02	+ 1.227	66 44 39.1	+ 16.15	51.91	2
1397	Lacaille, 790	8.0	32.67	0.869	70 11 43.4	.15	51.95	2
1398	Gould, 2637	10.0	38.74	+ 0.924	69 43 1.9	.14	51.89	2
1399	Gould, Z. C., 2h 685 . .	9.0	48.12	— 0.506	77 28 52.5	.13	51.25	3
1400	Horologii	10.0	25 48.32	+ 1.220	66 46 59.7	+ 16.13	51.90	1
1401	Horologii	8.8	2 26 1.59	+ 1.157	67 26 13.2	+ 16.12	52.18	3
1402	Hydri	9.8	2.99	0.396	73 26 30.8	.12	50.88	2
1403	Hydri	9.5	29.08	+ 1.006	68 53 13.6	.10	51.91	1
1404	Hydri	9.5	29.32	— 2.358	81 38 36.6	.10	50.95	1
1405	Hydri	9.5	34.87	— 0.825	78 25 30.7	+ 16.09	50.93	1
1406	Gould, Z. C., 2h 709 . .	10.0	2 26 35.97	+ 0.146	74 44 33.6	+ 16.09	50.88	2
1407	Gould, Z. C., 2h 722 . .	9.2	36.06	1.226	66 36 59.6	.09	51.91	2
1408	Hydri	10.0	39.88	+ 0.426	73 12 24.8	.09	50.88	1
1409	Octantis	9.8	43.08	— 8.329	85 59 34.3	.08	51.87	4
1410	Gould, Z. C., 2h 706 . .	9.5	26 46.34	— 0.638	77 50 43.4	+ 16.08	50.80	2
1411	Gould, 2669	9.0	2 27 8.68	+ 1.234	66 27 10.6	+ 16.06	51.92	1
1412	Hydri	9.5	9.30	— 0.019	75 28 35.4	.06	51.88	3
1413	Hydri	10.0	17.37	+ 0.833	70 18 4.7	.06	51.90	1
1414	Hydri	9.5	22.06	1.008	68 46 20.6	.05	51.93	2
1415	Lacaille, 800	8.5	27.09	+ 0.951	69 16 47.0	+ 16.05	51.91	1
1416	Gould, 2687	8.0	2 27 48.37	+ 0.701	71 16 20.2	+ 16.03	51.88	1
1417	Gould, Z. C., 2h 745 . .	9.5	48.87	0.410	73 11 53.8	.03	50.88	1
1418	Hydri	10.0	49.86	+ 0.582	72 6 22.0	.03	50.91	1
1419	Lacaille, 817	6.8	53.32	— 0.254	76 24 32.6	.02	50.83	2
1420	Gould, Z. C., 2h 735 . .	10.0	54.91	— 0.332	76 42 25.9	+ 16.02	50.95	1
1421	Gould, Z. C., 2h 754 . .	9.0	2 27 59.27	+ 0.688	71 21 2.1	+ 16.02	51.88	1
1422	Lacaille, 870	8.0	28 3.28	— 4.140	83 37 51.9	.01	51.99	1
1423	Stone, 1014	10.0	4.59	— 1.853	80 43 2.6	.01	51.40	2
1424	Horologii	9.5	9.10	+ 1.194	66 47 29.1	.01	51.92	1
1425	Hydri	9.3	11.88	+ 0.028	75 10 23.2	+ 16.01	51.51	3

Number.	Constellation, Name of Star, or Synonym.	Magnitude.	Right Ascension, 1850.0.	Annual Precession.	South Declination, 1850.0.	Annual Precession.	Mean year.	No. of obs.
			h. m. s.	s.	° ′ ″	″		
1426	Hydri	10.0	2 28 19.42	+ 1.039	68 21 45.4	15.00	51.96	1
1427	Hydri	8.5	19.43	— 2.832	82 13 25.1	.00	51.60	3
1428	Gould, Z. C., 2h 768 . .	8.8	19.50	+ 0.651	71 34 45.5	.00	51.42	2
1429	Lacaille, 807	7.5	36.56	1.008	68 37 50.0	.99	51.93	2
1430	Hydri	10.0	39.04	+ 0.869	69 51 45.5	+ 15.98	51.90	1
1431	Octantis	9.7	2 28 46.06	— 6.769	85 18 16.6	+ 15.98	51.87	3
1432	Octantis	9.2	52.04	— 6.924	85 22 29.2	.97	51.87	4
1433	Gould, 2706	9.2	53.02	+ 0.265	73 55 26.2	.97	50.87	2
1434	Hydri	10.0	28 57.75	+ 0.690	71 14 15.5	.97	51.88	1
1435	Octantis	10.0	29 15.47	5.120	84 20 58.2	+ 15.95	51.82	2
1436	Gould, Z. C., 2h 779 . .	10.0	2 29 26.78	— 0.774	78 5 9.3	+ 15.94	50.85	1
1437	Octantis	10.0	32.35	— 9.498	86 18 16.8	.94	52.15	3
1438	Gould, 2732	9.0	29 34.58	+ 1.155	67 2 43.2	.93	51.93	2
1439	Gould, Z. C., 2h 799 . .	8.0	30 7.84	— 0.316	76 28 46.2	.91	50.83	1
1440	Gould, Z. C., 2h 818 . .	9.0	18.24	+ 0.799	70 15 53.5	+ 15.90	51.95	2
1441	Gould, Z. C., 2h 822 . .	9.2	2 30 20.02	+ 1.214	66 17 56.4	+ 15.89	51.92	2
1442	Horologii	10.0	25.17	1.186	66 36 0.2	.89	51.90	1
1443	Gould, Z. C., 2h 824 . .	9.0	35.92	+ 0.730	70 46 36.2	.88	51.97	2
1444	Lacaille, 835	8.0	36.68	— 0.317	76 26 51.5	.88	50.83	1
1445	Octantis	9.2	53.86	— 3.751	83 8 45.0	+ 15.86	50.90	2
1446	Gould, 2767	9.5	2 30 56.67	+ 1.210	66 16 2.2	+ 15.86	51.92	2
1447	Hydri	9.0	31 7.23	0.419	72 50 28.1	.85	50.88	1
1448	Gould, Z. C., 2h 839 . .	9.2	7.40	+ 0.771	70 24 8.8	.85	51.95	2
1449	Hydri	10.0	10.91	— 1.979	80 47 7.3	.85	51.27	3
1450	Gould, Z. C., 2h 850 . .	9.3	19.94	+ 0.836	69 51 24.3	+ 15.84	51.90	3
1451	Hydri	9.5	2 31 20.26	— 0.961	78 30 31.4	+ 15.84	50.93	1
1452	Hydri	9.0	25.70	— 2.492	81 36 36.6	.84	50.95	1
1453	Gould, Z. C., 2h 866 . .	9.0	36.00	+ 1.188	66 26 48.7	.83	51.92	1
1454	Horologii	10.0	42.21	1.191	66 24 3.2	.82	51.92	1
1455	Hydri	9.5	46.91	+ 0.597	71 37 46.1	+ 15.82	50.95	1
1456	Hydri	10.0	2 31 52.41	— 1.210	79 7 33.8	+ 15.81	51.96	1
1457	Gould, Z. C., 2h 869 . .	9.0	31 58.59	+ 0.319	73 21 8.4	.81	50.88	1
1458	Hydri	10.0	32 5.39	0.236	73 48 17.9	.80	50.87	1
1459	Hydri	10.0	6.18	0.711	70 45 45.6	.80	51.99	1
1460	Lacaille, 836	8.0	7.42	+ 0.360	73 6 11.2	+ 15.80	50.88	2
1461	Hydri	9.8	2 32 14.29	— 1.046	78 40 48.7	+ 15.79	50.88	2
1462	Hydri	10.0	16.20	— 0.039	75 9 14.4	.79	52.69	1
1463	Gould, 2789	8.0	16.92	+ 0.618	71 25 28.9	.79	51.42	2
1464	Hydri	10.0	20.02	— 1.537	79 51 13.6	.79	50.87	2
1465	Octantis	9.3	22.22	— 9.960	86 22 41.0	+ 15.78	52.13	6

Number.	Constellation, Name of Star, or Synonym.	Magnitude.	Right Ascension, 1850.0.	Annual Precession.	South Declination, 1850.0.	Annual Precession.	Mean year.	No. of obs.
			h. m. s.	s.	° ′ ″	″		
1 466	Gould, Z. C., 2h 885 . .	8.5	2 32 26.09	+ 0.316	73 19 39.1	+ 15.78	50.88	1
1 467	Hydri	9.2	29.19	0.945	68 47 22.4	.78	51.93	2
1 468	Lacaille, 839	7.5	37.09	+ 0.369	73 0 18.2	.77	50.88	2
1 469	Gould, Z. C., 2h 878 . .	9.2	53.54	— 1.245	79 9 0.6	.76	51.40	2
1 470	Gould, Z. C., 2h 910 . .	8.8	32 57.56	+ 0.936	68 48 54.4	+ 15.75	51.93	2
1 471	Octantis	10.0	2 33 0.35	— 4.902	84 4 24.6	+ 15.75	50.94	1
1 472	Octantis	9.0	10.95	— 3.289	82 34 39.1	.74	52.00	1
1 473	Hydri	10.0	27.33	+ 0.456	72 23 44.1	.73	50.90	1
1 474	Hydri	9.8	27.36	0.891	69 10 7.3	.73	51.91	2
1 475	Hydri	10.0	34.86	+ 0.600	71 25 {35.2 / 59.3}	+ 15.72	51.42	2
1 476	Hydri	10.0	2 33 44.99	— 0.324	76 14 42.2	+ 15.71	50.84	1
1 477	Lacaille 856	7.5	51.04	+ 0.036	74 40 39.1	.70	50.88	2
1 478	Gould, Z. C., 2h 914 . .	10.0	52.26	— 1.217	79 1 28.3	.70	50.83	1
1 479	Lacaille, 864	7.0	33 59.81	0.409	76 33 0.9	.70	50.83	1
1 480	Hydri	9.0	34 0.29	— 0.202	75 44 12.9	+ 15.70	51.98	1
1 481	Octantis	8.6	2 34 15.57	— 7.004	85 16 16.0	+ 15.68	52.12	7
1 482	Octantis	10.0	33.09	—10.720	86 32 8.5	.66	52.14	3
1 483	Gould, Z. C., 2h 951 . .	8.8	34.20	+ 0.582	71 27 28.4	.66	51.42	2
1 484	Hydri	10.0	46.80	— 1.104	78 40 49.1	.66	50.93	1
1 485	Octantis	8.8	49.11	— 3.438	82 40 53.4	+ 15.66	51.26	3
1 486	Hydri	9.0	2 34 51.23	— 1.240	79 1 28.5	+ 15.65	51.96	1
1 487	Hydri	9.8	34 52.40	0.961	78 17 2.6	.65	50.89	2
1 488	μ Hydri	5.6	35 0.00	1.593	79 45 46.0	.64	50.93	1
1 489	Hydri	9.8	6.02	0.966	78 17 2.4	.64	50.89	2
1 490	Gould, Z. C., 2h 967 . .	8.8	19.53	— 0.090	75 8 54.2	+ 15.62	51.51	3
1 491	Lacaille, 854	7.2	2 35 25.31	+ 1.001	67 56 54.0	+ 15.62	52.18	3
1 492	Hydri	10.0	29.52	0.611	71 10 15.9	.62	51.88	1
1 493	Gould, Z. C., 2h 975 . .	9.0	34.68	+ 0.519	71 47 42.6	.61	50.93	2
1 494	Octantis	10.0	35 36.23	— 5.401	84 20 27.7	.61	52.68	1
1 495	Gould, Z. C., 2h 1003 . .	10.0	36 4.50	+ 1.183	65 58 42.0	+ 15.58	51.92	1
1 496	Horologii	9.5	2 36 4.68	+ 1.151	66 20 23.6	+ 15.58	51.92	2
1 497	Gould, Z. C., 2h 991 . .	10.0	6.53	+ 0.658	70 45 50.6	.58	51.96	1
1 498	Octantis	9.0	10.95	— 3.393	82 34 39.1	.58	52.00	1
1 499	Hydri	10.0	12.25	+ 0.300	73 5 57.6	.58	50.88	2
1 500	Hydri	8.5	17.18	+ 0.258	73 19 34.0	+ 15.57	50.88	1
1 501	Hydri	10.0	2 36 23.02	— 0.024	74 45 44.9	+ 15.57	50.88	1
1 502	Hydri	9.8	23.63	0.194	75 31 18.4	.57	51.87	3
1 503	Lacaille, 894	9.0	28.37	1.565	79 41 18.5	.56	50.93	1
1 504	Gould, Z. C., 2h 1002 . .	9.0	36.66	0.085	75 1 46.6	.55	51.51	3
1 505	Octantis	10.0	38.62	7.022	85 13 6.9	+ 15.55	52.03	6

No. 1 472. This right ascension is wrong; same star as No. 1 498.

No. 1 475. One of these observations is evidently 1 rev. wrong.

Number.	Constellation, Name of Star, or Synonym.	Magnitude.	Right Ascension, 1850.0.	Annual Precession.	South Declination, 1850.0.	Annual Precession.	Mean year.	No. of obs.
			h. m. s.	s.	° ′ ″	″		
1 506	Lacaille, 897	7.3	2 36 57.78	+ 1.020	67 36 0.4	+ 15.53	52.18	3
1 507	Lacaille, 866	7.5	37 1.24	1.102	66 45 34.0	.53	51.91	2
1 508	Hydri	9.5	6.31	0.958	68 11 25.9	.53	51.96	1
1 509	Hydri	11.0	8.14	0.756	69 55 23.3	.52	51.90	1
1 510	ε Hydri	4.5	17.87	+ 0.875	68 54 37.9	+ 15.52	51.91	1
1 511	Lacaille, 877	6.5	2 37 22.64	+ 0.564	71 19 26.8	+ 15.51	51.88	1
1 512	Gould, Z. C., 2h 1044 . .	8.8	37.74	0.204	73 30 42.2	.50	50.88	2
1 513	Hydri	9.5	39.42	0.023	74 26 20.0	.50	50.88	1
1 514	Hydri	9.0	47.27	0.996	67 44 45.9	.49	52.29	2
1 515	Horologii	12.0	37 51.04	+ 0.746	69 55 27.5	+ 15.48	51.90	1
1 516	Hydri	10.0	2 38 12.25	− 0.215	75 28 35.2	+ 15.46	52.33	2
1 517	Hydri	9.5	23.36	+ 0.962	68 0 6.9	.45	51.96	1
1 518	Gould, 2917	8.5	38.44	0.762	69 43 25.6	.44	51.89	2
1 519	Hydri	9.5	39.17	0.888	68 39 28.6	.44	51.95	1
1 520	Gould, Z. C., 2h 1076 . .	9.0	42.20	+ 0.653	70 33 36.7	+ 15.44	51.99	1
1 521	Lacaille, 880	6.7	2 38 42.44	+ 0.743	69 52 24.1	+ 15.44	51.90	3
1 522	Hydri	10.0	46.05	+ 0.877	68 44 27.2	.43	51.95	1
1 523	Gould, Z. C., 2h 1048 . .	8.7	46.13	− 1.901	80 15 35.0	.43	51.61	3
1 524	Octantis	11.0	49.79	4.942	83 54 56.9	.43	50.94	1
1 525	Hydri	9.0	38 56.40	− 0.198	75 21 10.2	+ 15.42	51.87	3
1 526	Hydri	10.0	2 39 15.60	+ 0.081	74 1 38.8	+ 15.41	50.87	1
1 527	Hydri	9.3	28.47	− 0.258	75 33 53.6	.39	51.88	3
1 528	Hydri	9.5	31.41	− 0.054	74 39 47.0	.39	50.88	1
1 529	Gould, 2933	9.0	33.90	+ 1.077	66 43 46.8	.39	51.91	2
1 530	Hydri	9.2	36.25	− 0.238	75 28 15.4	+ 15.39	51.88	3
1 531	Hydri	9.0	2 39 38.49	+ 0.799	69 19 24.9	+ 15.38	51.91	1
1 532	Hydri	10.0	55.66	− 2.205	80 44 53.6	.37	51.48	2
1 533	Hydri	10.0	39 56.84	2.613	81 24 2.0	.37	50.97	2
1 534	Octantis	9.9	40 2.23	− 6.799	85 1 14.9	.36	52.12	7
1 535	Hydri	9.8	2.97	+ 0.992	67 32 46.0	+ 15.36	52.18	3
1 536	Lacaille, 882	9.0	2 40 14.60	+ 0.404	72 5 52.7	+ 15.35	50.91	1
1 537	Hydri	10.0	32.00	0.049	74 4 39.7	.33	50.87	1
1 538	Horologii	10.0	39.62	1.033	67 3 40.2	.33	51.90	1
1 539	Lacaille, 893	7.5	40 49.84	+ 1.003	67 20 46.1	.32	51.97	1
1 540	Lacaille, 904	8.9	41 1.08	− 0.146	74 57 41.3	+ 15.31	50.89	1
1 541	Gould, Z. C., 2h 1140 . .	9.2	2 41 2.04	+ 0.739	69 39 58.9	+ 15.31	51.89	2
1 542	Brisbane, 410	8.8	16.80	+ 0.938	67 55 10.4	.29	52.18	3
1 543	Hydri	9.0	18.86	− 3.201	82 8 30.3	.29	51.99	1
1 544	Lacaille, 928	8.0	28.93	− 2.082	80 27 23.0	.28	51.99	1
1 545	Lacaille, 898	6.8	31.80	+ 0.717	69 47 47.8	+ 15.28	51.90	3

Number.	Constellation, Name of Star, or Synonym.	Magnitude.	Right Ascension, 1850.0.	Annual Precession.	South Declination, 1850.0.	Annual Precession.	Mean year.	No. of obs.
			h. m. s.	s.	° ′ ″	″		
1546	Gould, Z. C., 2ʰ 1153 . .	9.3	2 41 34.60	+ 0.717	69 47 8.1	+ 15.28	51.90	3
1547	Lacaille, 901	8.5	38.91	0.428	71 50 24.4	.27	50.93	2
1548	Gould, Z. C., 2ʰ 1127 . .	9.5	40.16	1.908	80 7 14.8	.27	51.41	2
1549	Brisbane, 412	9.0	40.97	0.685	70 0 54.4	.27	51.90	1
1550	Hydri	10.0	43.41	+ 0.847	68 41 37.9	+ 15.27	51.91	1
1551	Gould, Z. C., 2ʰ 1131 . .	8.5	2 41 45.60	− 1.553	79 22 27.2	+ 15.26	51.44	2
1552	Hydri	9.5	49.62	− 0.028	74 21 25.3	.26	50.88	1
1553	Gould, Z. C., 2ʰ 1155 . .	9.2	52.61	+ 0.154	73 26 37.2	.26	50.88	2
1554	Hydri	8.8	55.67	0.835	68 46 42.5	.26	51.93	2
1555	Hydri	9.5	41 58.58	+ 0.659	70 12 6.4	+ 15.25	51.96	2
1556	Hydri	10.0	2 42 13.15	+ 0.431	71 46 16.9	+ 15.24	50.95	1
1557	Octantis	9.2	30.99	−10.959	86 26 51.0	.22	52.18	8
1558	Gould, 2988	9.0	38.33	0.391	75 52 20.6	.21	51.41	2
1559	Hydri	10.2	40.94	− 2.581	81 14 5.4	.21	51.50	2
1560	Gould, 2997	9.8	46.01	+ 0.953	67 38 2.0	+ 15.21	52.18	3
1561	Gould, 2998	9.5	2 42 46.07	+ 0.952	67 37 59.7	+ 15.21	52.67	1
1562	Hydri	10.0	42 58.47	+ 0.552	70 51 42.7	.20	51.96	1
1563	Hydri	10.0	43 14.08	− 2.574	81 11 54.0	.18	51.50	2
1564	ζ Hydri	5.0	14.98	+ 0.881	68 14 53.1	.18	51.95	2
1565	Horologii	9.0	19.24	+ 1.017	66 56 50.2	+ 15.18	51.93	2
1566	Hydri	10.0	2 43 22.59	− 1.778	79 46 28.1	+ 15.17	50.93	1
1567	Hydri	9.0	32.91	+ 0.905	67 59 49.6	.16	51.96	1
1568	Hydri	9.0	35.05	0.463	71 26 44.5	.16	51.42	2
1569	Brisbane, 422	9.0	36.35	+ 0.690	69 48 45.1	.16	51.90	3
1570	Lacaille, 966	8.0	42.00	− 3.942	82 52 25.6	+ 15.15	51.15	2
1571	Lacaille, 1029	8.3	2 43 54.31	−10.775	86 22 31.2	+ 15.14	52.18	8
1572	Hydri	10.5	43 54.64	1.262	78 33 28.0	.14	50.93	1
1573	Hydri	9.0	44 2.56	− 3.012	81 47 26.4	.13	51.47	2
1574	Hydri	9.5	11.97	+ 0.651	70 0 31.3	.13	51.90	1
1575	Hydri	9.2	13.06	+ 0.439	71 32 38.0	+ 15.12	51.42	2
1576	Hydri	10.0	2 44 14.05	+ 0.122	73 24 51.4	+ 15.12	50.88	1
1577	Gould, Z. C., 2ʰ 1226 . .	8.8	14.38	0.720	69 31 5.7	.12	51.90	3
1578	Lacaille, 916	7.8	30.18	+ 0.388	71 51 50.1	.11	50.93	2
1579	Hydri	10.0	33.47	− 1.774	79 42 22.8	.10	50.93	1
1580	Hydri	10.0	35.15	+ 0.982	67 9 25.6	+ 15.10	51.90	1
1581	Brisbane, 425	8.3	2 44 42.69	+ 0.902	67 54 12.8	+ 15.10	52.18	3
1582	Gould, Z. C., 2ʰ 1214 . .	9.5	47.72	− 1.234	78 26 20.2	.09	50.93	1
1583	Hydri	9.5	47.98	+ 0.629	70 10 17.1	.09	51.90	1
1584	Hydri	10.0	2 44 52.89	0.281	72 28 40.2	.09	50.90	1
1585	Hydri	10.0	44 59.94	+ 0.688	69 42 30.4	+ 15.08	51.89	2

Number.	Constellation, Name of Star, or Synonym.	Magnitude.	Right Ascension, 1850.0.	Annual Precession.	South Declination, 1850.0.	Annual Precession.	Mean year.	No. of obs.
			h. m. s.	s.	° ′ ″	″		
1586	Octantis	10.0	2 45 2.50	− 7.355	85 9 56.8	+ 15.08	51.88	1
1587	Hydri	9.0	17.62	3.295	82 6 16.8	.56	51.99	1
1588	Gould, 3036	9.0	31.30	0.336	75 27 40.6	.05	51.88	3
1589	Gould, Z. C., 2^h 1237 . .	9.2	31.86	− 0.788	77 6 0.6	.05	51.47	2
1590	Hydri	10.0	45 58.16	+ 0.266	72 28 45.8	+ 15.02	50.90	1
1591	Hydri	10.0	2 46 2.60	+ 0.082	74 18 18.7	+ 15.02	50.87	1
1592	Gould, 3060	9.5	7.77	1.095	65 51 41.2	.01	51.92	1
1593	Gould, Z. C., 2^h 1266 . .	9.5	8.66	0.298	72 16 16.4	.01	50.91	2
1594	Gould, 3059	9.5	9.91	0.886	67 54 47.5	.01	52.32	2
1595	Gould, Z. C., 2^h 1270 . .	8.8	22.14	+ 0.064	73 33 24.8	+ 15.00	50.88	2
1596	Hydri	10.0	2 46 39.47	+ 0.700	69 27 33.6	+ 14.98	51.90	3
1597	Hydri	10.2	46 47.80	− 1.400	78 45 9.4	.98	50.88	2
1598	Gould, 3062	10.0	47 19.08	1.513	78 59 39.6	.94	51.40	2
1599	Hydri	9.5	26.58	− 2.212	80 24 52.6	.94	52.01	1
1600	Gould, Z. C., 2^h 1315 . .	9.0	48.21	+ 0.499	70 50 21.5	+ 14.92	51.96	1
1601	Hydri	10.0	2 47 55.06	+ 0.733	69 4 27.3	+ 14.91	51.91	1
1602	Octantis	9.8	48 0.31	− 7.044	84 56 43.4	.90	52.31	4
1603	Hydri	9.5	0.92	2.346	80 38 20.0	.90	52.00	2
1604	Lacaille, 955	7.2	5.47	− 0.750	76 49 4.0	.90	50.89	2
1605	Hydri	9.2	10.62	+ 0.122	73 6 49.2	+ 14.89	50.88	2
1606	Gould, 3079	9.2	2 48 13.66	− 1.535	78 59 55.3	+ 14.89	51.40	2
1607	Gould, Z. C., 2^h 1326 . .	9.2	14.95	+ 0.552	70 25 47.6	.89	51.95	2
1608	Hydri	10.0	17.06	+ 0.501	70 47 6.3	.89	51.96	1
1609	Gould, Z. C., 2^h 1328 . .	8.2	35.80	− 0.189	74 36 49.9	.87	50.88	2
1610	Lacaille, 943	8.2	42.86	+ 0.834	68 8 19.9	+ 14.86	51.95	2
1611	Hydri	10.0	2 48 48.79	+ 0.372	71 35 41.6	+ 14.86	51.88	1
1612	Lacaille, 952	7.0	58.26	− 0.162	74 27 42.6	.85	50.88	1
1613	Hydri	9.5	48 58.48	− 0.401	75 29 5.5	.85	51.88	3
1614	Gould, Z. C., 2^h 1351 . .	8.8	49 5.73	+ 0.057	73 23 21.1	.84	50.88	2
1615	Hydri	9.5	16.39	− 0.538	75 59 53.9	+ 14.83	51.61	3
1616	Gould, Z. C., 2^h 1372 . .	9.8	2 49 21.90	+ 0.818	68 13 0.3	+ 14.82	51.95	2
1617	Gould, Z. C., 2^h 1342 . .	9.5	22.02	− 1.217	78 8 30.1	.82	50.85	1
1618	Hydri	10.0	25.21	− 2.181	80 16 4.8	.82	51.41	2
1619	Gould, 3133	9.3	32.21	+ 0.622	69 48 0.7	.81	51.90	3
1620	Gould, Z. C., 2^h 1367 . .	8.5	36.16	− 0.109	74 11 24.3	+ 14.81	50.87	2
1621	Hydri	10.5	2 49 38.37	− 2.447	80 43 12.0	+ 14.81	52.00	1
1622	Hydri	10.2	49 54.62	1.570	78 59 29.0	.79	51.40	2
1623	Gould. Z. C., 2^h 1365 . .	9.8	50 3.93	− 1.278	78 15 51.0	.78	50.89	2
1624	Gould, 3151	8.5	11.49	+ 0.843	67 55 12.3	.78	52.32	2
1625	Hydri	10.0	12.50	+ 0.359	71 34 20.2	+ 14.77	51.42	2

Number.	Constellation, Name of Star, or Synonym.	Magnitude.	Right Ascension, 1850.0.	Annual Precession.	South Declination, 1850.0.	Annual Precession.	Mean year.	No. of obs.
			h. m. s.	s.	° ′ ″	″		
1 626	Lacaille, 948	7.0	2 50 17.10	+ 1.034	66 3 58.4	+ 14.77	51.92	1
1 627	Hydri	10.0	19.02	0.563	70 10 24.1	.77	51.90	1
1 628	Hydri	10.0	50 32.40	+ 0.141	72 49 38.9	.76	50.88	1
1 629	Hydri	10.0	51 10.56	— 0.840	76 55 35.0	.72	50.95	1
1 630	ν Hydri	4.0	29.59	— 0.492	75 40 46.1	+ 14.70	51.98	1
1 631	Horologii	9.5	2 51 51.25	+ 0.931	66 56 57.2	+ 14.68	51.90	1
1 632	Hydri	9.5	52 1.88	+ 0.123	72 48 57.1	.67	50.88	1
1 633	Gould, Z. C., 2^h 1417 . .	9.0	2.27	— 0.474	75 34 36.0	.67	52.33	2
1 634	Gould, 3190	9.0	8.61	0.338	75 1 1.6	.66	51.51	3
1 635	Hydri	10.0	10.92	— 0.184	74 20 33.1	+ 14.66	50.87	1
1 636	Octantis	10.0	2 52 12.48	— 5.544	83 57 10.4	+ 14.66	51.46	2
1 637	Hydri	10.0	24.15	— 3.514	82 6 7.6	.64	51.99	1
1 638	Horologii	9.5	29.94	+ 0.936	66 48 29.2	.64	51.91	2
1 639	Hydri	9.5	41.25	0.580	69 50 19.8	.63	51.90	1
1 640	Hydri	9.2	49.78	+ 0.488	70 39 25.0	+ 14.62	51.97	2
1 641	Horologii	9.5	2 52 54.16	+ 0.920	66 56 38.5	+ 14.61	51.90	1
1 642	Hydri	9.8	53 2.50	— 3.391	81 55 46.8	.61	51.47	2
1 643	Hydri	9.3	19.47	— 0.448	75 23 26.6	.59	51.88	3
1 644	Hydri	10.0	21.62	+ 0.003	73 21 11.0	.59	50.88	1
1 645	Hydri	9.5	23.77	+ 0.061	73 2 50.2	+ 14.58	50.88	1
1 646	Hydri	10.0	2 53 25.34	— 1.914	79 34 4.5	+ 14.58	51.44	2
1 647	Hydri	10.0	26.72	+ 0.761	68 20 7.8	.58	51.96	1
1 648	Horologii	10.0	44.75	0.872	67 19 11.4	.56	51.97	1
1 649	Horologii	10.0	57.42	0.926	66 47 30.2	.55	51.91	2
1 650	Hydri	10.0	53 59.92	+ 0.245	71 57 59.4	+ 14.55	50.93	2
1 651	Hydri	10.0	2 54 9.86	+ 0.702	68 45 37.9	+ 14.54	51.91	1
1 652	Gould, Z. C., 2^h 1474 . .	9.8	17.94	— 0.705	76 17 24.0	.53	52.00	2
1 653	Hydri	10.0	19.05	— 0.735	76 23 22.8	.53	51.98	1
1 654	Horologii	10.0	26.70	+ 0.928	66 44 10.4	.52	51.90	1
1 655	Octantis	10.0	27.62	— 4.106	82 40 40.3	+ 14.52	50.95	1
1 656	Gould, 3249	8.8	2 54 29.72	+ 0.369	71 9 27.8	+ 14.52	51.92	2
1 657	Gould, 3244	8.0	31.55	— 0.158	74 3 27.5	.52	50.87	1
1 658	Hydri	8.7	36.74	+ 0.805	67 50 54.8	.51	52.32	2
1 659	Gould, Z. C., 2^h 1469 . .	9.0	36.85	— 1.811	79 17 45.6	.51	51.96	1
1 660	Hydri	9.0	44.28	+ 0.684	68 51 32.3	+ 14.50	51.91	1
1 661	Hydri	9.5	2 54 58.74	— 0.647	76 2 21.2	+ 14.49	52.00	2
1 662	Octantis	11.0	54 59.55	4.686	83 12 12.4	.49	52.01	1
1 663	Lacaille, 1146	7.6	55 4.49	— 9.073	85 38 41.5	.48	52.19	5
1 664	Brisbane, 461	9.0	19.25	+ 0.597	69 29 18.1	.47	51.90	3
1 665	Hydri	9.8	34.74	+ 0.641	69 7 30.6	+ 14.45	51.91	2

Number.	Constellation, Name of Star, or Synonym.	Magnitude.	Right Ascension, 1850.0.	Annual Precession.	South Declination, 1850.0.	Annual Precession.	Mean year.	No. of obs.
			h. m. s.	s.	° ′ ″	″		
1 666	Horologii	9.0	2 55 38.86	+ 0.838	67 27 28.2	+ 14.45	52.32	2
1 667	Gould, 3265	9.0	41.48	+ 0.146	72 24 43.9	.45	50.90	1
1 668	Hydri	10.0	49.50	— 1.890	79 24 17.0	.44	51.44	2
1 669	Hydri	9.3	55 56.04	0.520	75 30 31.4	.43	51.87	3
1 670	Hydri	9.2	56 0.27	— 0.777	76 26 4.1	+ 14.43	52.00	2
1 671	Horologii	9.5	2 56 1.18	+ 0.917	66 40 39.6	+ 14.43	51.91	2
1 672	Hydri	8.8	10.46	0.775	67 58 9.3	.42	52.32	2
1 673	Hydri	10.0	10.79	+ 0.352	71 7 52.2	.42	51.88	1
1 674	Hydri	8.5	12.88	+ 0.800	67 44 27.7	.41	52.32	2
1 675	Octantis	10.0	14.72	— 4.726	83 12 0.5	+ 14.41	51.48	2
1 676	Horologii	10.0	2 56 19.88	+ 1.001	65 49 3.0	+ 14.41	51.92	1
1 677	Hydri	10.0	21.55	— 2.697	80 50 20.0	.41	52.00	1
1 678	Lacaille, 995	8.0	25.61	1.489	78 28 5.6	.40	50.93	1
1 679	Octantis	9.2	26.26	34.813	88 35 34.3	.40	52.68	3
1 680	Hydri	10.0	39.46	— 0.197	74 5 40.4	+ 14.39	50.87	1
1 681	Hydri	9.2	2 56 41.67	— 0.562	75 37 10.6	+ 14.39	52.33	2
1 682	Hydri	9.5	48.31	+ 0.226	71 51 46.3	.38	50.95	1
1 683	Horologii	10.0	53.58	0.916	66 36 32.8	.37	51.90	1
1 684	Hydri	9.2	56 58.54	+ 0.674	68 44 24.2	.37	51.93	2
1 685	Hydri	10.0	57 13.88	— 0.542	75 30 30.7	+ 14.35	50.96	1
1 686	Hydri	10.0	2 57 19.74	+ 0.302	71 21 33.0	+ 14.35	51.88	1
1 687	Hydri	10.5	37.18	— 1.653	78 48 0.8	.33	51.43	2
1 688	Hydri	10.0	41.07	+ 0.404	70 40 6.3	.32	51.96	1
1 689	Gould, Z. C., 2ʰ 1558 . .	10.0	46.80	— 1.073	77 16 17.9	.32	52.00	1
1 690	Gould, Z. C., 2ʰ 1572 . .	9.2	48.84	— 0.018	73 8 36.1	+ 14.32	50.88	2
1 691	Hydri	9.5	2 57 58.54	— 0.659	75 54 5.9	+ 14.31	51.98	1
1 692	Hydri	10.0	58 7.32	3.617	82 1 12.6	.30	51.99	1
1 693	Hydri	10.0	9.70	0.716	76 5 35.7	.30	52.02	1
1 694	Hydri	9.5	10.36	— 0.254	74 15 7.8	.29	50.87	2
1 695	Gould, 3308	9.5	15.26	+ 0.453	70 17 28.7	+ 14.29	51.95	2
1 696	Hydri	10.0	2 58 18.13	— 0.236	74 9 41.6	+ 14.29	50.87	1
1 697	Gould, Z. C., 2ʰ 1569 . .	9.7	22.27	1.617	78 40 47.6	.28	51.26	3
1 698	Hydri	10.0	29.10	— 1.519	78 26 19.1	.28	50.93	1
1 699	Hydri	10.0	30.63	+ 0.368	70 50 45.3	.27	51.96	1
1 700	Octantis	10.5	36.24	— 6.312	84 16 53.2	+ 14.27	51.81	2
1 701	Gould, Z. C., 2ʰ 1594 . .	10.0	2 58 40.10	+ 0.836	67 12 4.8	+ 14.26	51.93	2
1 702	Octantis	9.8	41.25	— 5.380	83 39 13.4	.26	52.00	2
1 703	Gould, Z. C., 2ʰ 1578 . .	9.0	42.83	— 1.450	78 15 18.5	.26	51.27	3
1 704	Gould, 3322	8.0	52.59	+ 0.779	67 41 37.5	.25	52.67	1
1 705	Hydri	10.0	52.79	+ 0.360	70 51 55.3	+ 14.25	51.96	1

No. 1 693. Possibly 19ˢ.70. No. 1 698. Possibly 19ˢ.90.

Number.	Constellation, Name of Star, or Synonym.	Magnitude.	Right Ascension, 1850.0.	Annual Precession.	South Declination, 1850.0.	Annual Precession.	Mean year.	No. of obs.
			h. m. s.	s.	° ′ ″	″		
1706	Octantis	9.5	2 59 0.12	− 6.281	84 15 10.1	+ 14.24	51.53	3
1707	Gould, Z. C., 2ʰ 1602 . .	9.2	22.84	− 0.300	74 22 43.6	.22	50.87	2
1708	Hydri	10.0	25.61	+ 0.379	70 41 53.3	.22	51.96	1
1709	Hydri	10.0	46.01	+ 0.043	72 40 44.7	.20	50.88	1
1710	Gould, Z. C., 2ʰ 1609 . .	8.0	59 57.57	− 0.998	76 55 32.7	+ 14.18	50.95	1
1711	Horologii	10.0	3 0 3.90	+ 0.965	65 49 28.3	+ 14.18	51.92	1
1712	Hydri	10.0	10.80	− 3.012	81 9 38.2	.17	52.00	2
1713	Lacaille, 997	8.0	40.20	+ 0.428	70 16 17.6	.14	51.95	2
1714	Hydri	10.0	52.66	0.683	68 19 53.6	.13	51.95	2
1715	Gould, Z. C., 3ʰ 45 . . .	9.0	0 58.32	+ 0.835	67 0 15.6	.12	51.93	2
1716	Gould, Z. C., 3ʰ 37 . . .	9.2	3 1 0.27	+ 0.184	71 48 2.2	+ 14.12	50.93	2
1717	Horologii	10.0	0.60	+ 0.788	67 25 33.0	.12	52.32	2
1718	Hydri	9.2	3.47	− 2.865	80 54 36.4	.12	52.01	2
1719	Hydri	10.0	9.85	− 3.039	81 9 42.1	.11	52.02	1
1720	Gould, Z. C., 3ʰ 44 . . .	8.8	11.37	+ 0.294	71 6 55.5	+ 14.11	51.92	2
1721	Horologii	8.2	3 1 14.57	+ 0.925	66 6 48.0	+ 14.10	51.92	2
1722	Octantis	9.4	24.74	− 6.326	84 13 11.7	.09	51.65	4
1723	Hydri	9.7	28.92	1.504	78 15 6.7	.09	51.27	3
1724	Lacaille, 1036	7.8	37.21	2.434	80 11 12.5	.08	52.00	3
1725	Hydri	10.0	44.57	− 0.688	75 46 59.2	+ 14.07	51.98	1
1726	Hydri	10.0	3 1 51.44	+ 0.492	69 43 34.4	+ 14.07	51.89	2
1727	θ Hydri	6.0	1 59.07	0.050	72 29 18.9	.06	50.90	1
1728	Gould, 3379	9.2	2 15.25	+ 0.292	71 2 44.9	.04	51.92	2
1729	Octantis	8.8	16.19	− 6.875	84 31 1.0	.04	51.82	2
1730	Hydri	10.0	16.84	+ 0.123	72 3 32.9	+ 14.04	50.91	1
1731	Hydri	10.0	3 2 23.03	+ 0.242	71 20 49.3	+ 14.03	51.88	1
1732	Gould, 3384	7.5	25.69	0.949	65 46 3.4	.03	51.92	1
1733	Hydri	9.0	30.41	0.398	70 19 57.8	.03	51.90	1
1734	Gould, 3382	8.5	32.36	+ 0.286	71 3 38.0	.02	51.92	2
1735	Hydri	9.5	38.18	− 1.598	78 25 40.2	+ 14.02	50.93	1
1736	Hydri	10.0	3 2 43.37	− 2.059	79 26 18.2	+ 14.01	51.46	2
1737	Hydri	9.8	43.63	− 0.706	75 47 27.6	.01	52.00	2
1738	Hydri	9.5	45.45	+ 0.204	71 32 58.6	.01	50.95	1
1739	Gould, Z. C., 3ʰ 86 . .	8.2	49.21	− 0.067	73 2 44.2	.01	50.88	2
1740	Hydri	10.0	49.34	+ 0.154	71 50 20.4	+ 14.01	50.95	1
1741	Hydri	10.0	3 2 59.49	− 7.638	84 53 24.6	+ 14.00	52.00	1
1742	Gould, Z. C., 3ʰ 95 . .	9.0	3 3.15	+ 0.426	70 5 58.0	13.99	51.90	1
1743	Gould, Z. C., 3ʰ 101	9.0	4.63	+ 0.646	68 26 59.4	.99	51.95	1
1744	Hydri	9.0	10.70	− 0.737	75 52 24.2	.98	52.00	2
1745	Gould, Z. C., 3ʰ 77 . .	9.5	17.86	− 2.399	80 3 30.8	+ 13.98	52.01	1

Number.	Constellation, Name of Star, or Synonym.	Magnitude.	Right Ascension, 1850.0.	Annual Precession.	South Declination, 1850.0.	Annual Precession.	Mean year.	No. of obs.
			h. m. s.	s.	° ′ ″	″		
1746	Hydri	9.5	3 3 21.00	– 3.889	82 9 14.0	+ 13.97	51.99	2
1747	Gould, Z. C., 3^h 92 . .	9.0	21.96	1.003	76 45 12.2	.97	51.46	2
1748	Gould, Z. C., 3^h 104 . .	10.0	30.35	0.222	73 45 5.9	.96	50.87	1
1749	Octantis	10.0	32.26	6.998	84 33 16.4	.96	52.68	1
1750	Octantis	8.7	3 45.11	– 6.336	84 10 8.2	+ 13.95	51.40	4
1751	Hydri	10.5	3 4 4.60	– 1.027	76 47 8.1	+ 13.93	51.46	2
1752	Octantis	9.2	7.56	–23.988	87 57 47.6	.92	52.68	3
1753	Horologii	10.0	11.27	+ 0.785	67 10 41.8	.92	51.90	1
1754	Hydri	10.0	12.62	– 2.233	79 42 53.2	.92	50.93	1
1755	Hydri	10.0	17.28	– 1.567	78 16 17.2	+ 13.91	50.93	1
1756	Gould, Z. C., 3^h 144 . .	9.0	3 4 36.39	+ 0.615	68 34 20.1	+ 13.89	51.95	1
1757	Gould, Z. C., 3^h 149	9.2	44.43	0.587	68 47 0.0	.89	51.93	2
1758	Gould, Z. C., 3^h 155 . .	9.5	4 44.59	+ 0.902	66 1 27.0	.89	51.92	1
1759	Horologii	9.5	5 9.19	– 0.027	72 40 40.4	.86	50.88	1
1760	Hydri	9.8	17.52	– 0.295	73 58 31.8	+ 13.85	50.87	2
1761	Hydri	10.0	3 5 17.98	+ 0.433	69 52 33.4	+ 13.85	51.90	1
1762	Lacaille, 1031	8.5	18.11	– 0.010	72 34 53.0	.85	50.90	1
1763	Octantis	10.0	18.89	12.129	86 21 34.4	.85	52.69	1
1764	Octantis	10.0	23.48	– 4.629	82 50 22.6	.84	50.95	1
1765	Hydri	10.0	28.26	+ 0.668	68 4 16.6	+ 13.84	51.96	1
1766	Hydri	10.0	3 5 29.36	+ 0.040	72 17 37.9	+ 13.84	50.91	1
1767	Hydri	10.0	33.19	0.738	75 44 31.6	.83	51.98	1
1768	Lacaille, 1022	8.0	34.27	+ 0.846	66 29 37.8	.83	51.92	1
1769	Hydri	10.0	44.35	+ 0.490	69 25 45.2	.82	51.91	1
1770	Lacaille, 1043	7.0	59.01	– 0.321	74 2 50.4	+ 13.81	50.87	1
1771	Gould, Z. C., 3^h 192 . .	8.5	3 5 59.28	+ 0.542	69 1 32.7	+ 13.81	51.91	2
1772	Hydri	10.0	6 2.42	+ 0.672	67 59 28.4	.80	52.32	2
1773	Lacaille, 1047	8.0	3.67	– 0.523	74 53 54.5	.80	50.89	1
1774	Horologii	10.0	5.92	+ 0.804	66 50 20.5	.80	51.90	1
1775	Hydri	10.0	9.28	– 1.081	76 50 51.5	+ 13.80	50.95	1
1776	Hydri	10.0	3 6 9.86	+ 0.481	69 27 43.9	+ 13.80	51.90	3
1777	Horologii	9.5	21.88	+ 0.861	66 16 39.0	.78	51.92	1
1778	Gould, Z. C., 3^h 195 . .	9.0	24.38	– 0.144	73 11 35.0	.78	50.88	1
1779	Stone, 1282	8.2	23.92	–10.031	85 45 45.0	.78	51.85	4
1780	Hydri	9.0	25.32	+ 0.175	71 27 27.6	+ 13.78	51.42	2
1781	Lacaille, 1046	7.8	3 6 40.28	– 0.031	72 35 57.6	+ 13.76	50.89	2
1782	Lacaille, 1035	6.7	41.01	+ 0.423	69 50 14.1	.76	51.90	3
1783	Hydri	9.5	42.68	– 2.105	79 21 28.5	.76	52.00	1
1784	Hydri	9.8	43.12	– 3.427	81 29 11.8	.76	50.97	2
1785	Gould, Z. C., 3^h 218 . .	10.0	6 47.75	+ 0.330	70 27 47.5	+ 13.76	51.99	1

Number.	Constellation, Name of Star, or Synonym.	Magnitude.	Right Ascension, 1850.0.	Annual Precession.	South Declination, 1850.0.	Annual Precession.	Mean year.	No. of obs.
			h. m. s.	s.	° ′ ″	″		
1786	Hydri	10.0	3 7 2.90	+ 0.610	68 24 39.5	+ 13.74	51.95	1
1787	Horologii	10.0	6.48	+ 0.858	66 15 1.7	.74	51.92	1
1788	Hydri	10.0	8.57	− 2.680	80 23 4.6	.73	52.01	3
1789	Hydri	9.5	15.83	− 0.646	75 18 17.5	.73	51.82	2
1790	Hydri	10.0	19.04	+ 0.291	70 40 48.0	+ 13.72	51.99	1
1791	Hydri	8.0	3 7 23.02	− 0.351	74 5 40.6	+ 13.72	50.87	1
1792	Gould, 3467	8.8	25.58	− 0.921	76 16 7.0	.72	52.00	2
1793	Hydri	10.0	26.49	+ 0.612	68 22 11.1	.71	51.95	1
1794	Hydri	10.0	30.88	− 3.277	81 15 38.0	.71	50.99	1
1795	Gould, Z. C., 3h 216 . .	9.5	34.15	− 1.547	78 4 3.3	+ 13.71	52.02	1
1796	Hydri	9.5	3 7 35.08	− 4.193	82 20 47.2	+ 13.71	51.99	2
1797	Hydri	9.5	38.02	0.739	75 37 43.7	.70	51.98	1
1798	Gould, Z. C., 3h 240 . .	9.0	41.94	0.101	72 53 31.6	.70	50.88	1
1799	Lacaille, 1065	8.0	45.98	1.423	77 44 15.0	.69	50.94	1
1800	Hydri	10.0	46.14	− 0.240	73 32 52.9	+ 13.69	50.88	1
1801	Hydri	9.0	3 7 50.59	− 0.755	75 40 26.3	+ 13.69	51.97	1
1802	Gould, Z. C., 3h 258 . .	9.2	7 55.34	+ 0.775	66 56 35.7	.68	51.93	2
1803	Hydri	9.5	8 3.24	0.401	69 53 10.7	.67	51.90	1
1804	Hydri	9.0	7.63	0.684	67 43 15.2	.67	52.67	1
1805	Gould, Z. C., 3h 256 . .	9.8	9.39	+ 0.007	72 17 18.7	+ 13.67	50.91	2
1806	Hydri	9.8	3 8 12.32	+ 0.161	71 24 48.8	+ 13.67	51.42	2
1807	Lacaille, 1090	8.0	20.92	− 3.616	81 40 12.9	.66	50.95	1
1808	Hydri	9.2	24.92	+ 0.160	71 24 28.4	.65	51.42	2
1809	Gould, 3506	9.0	34.91	0.756	67 3 47.5	.64	51.93	2
1810	Hydri	9.0	42.37	+ 0.179	71 16 29.6	+ 13.63	51.88	1
1811	Lacaille, 1054	8.0	3 8 45.53	+ 0.635	68 4 19.1	+ 13.63	51.96	1
1812	Hydri	10.0	47.96	+ 0.066	71 53 9.2	.63	50.95	1
1813	Hydri	9.5	52.35	− 0.610	75 4 29.8	.62	51.51	3
1814	Octantis	10.0	52.94	46.363	88 51 46.2	.62	52.68	1
1815	Lacaille, 1203	8.6	8 56.89	−12.806	86 27 44.4	+ 13.62	51.69	5
1816	Gould, Z. C., 3h 278 . .	10.0	3 9 1.44	− 0.058	72 34 54.0	+ 13.61	50.90	1
1817	Hydri	10.0	32.42	+ 0.312	70 22 28.0	.58	51.90	1
1818	Gould, Z. C., 3h 306 . .	8.8	41.54	+ 0.548	68 46 52.4	.57	51.95	2
1819	Gould, 3450	8.7	9 49.68	− 9.049	85 21 12.9	.56	51.58	3
1820	Hydri	10.0	10 2.22	+ 0.417	69 37 35.7	+ 13.55	51.89	2
1821	Gould, Z. C., 3h 293 . .	9.0	3 10 9.84	− 1.642	78 10 52.0	+ 13.54	51.47	2
1822	Gould, Z. C., 3h 327 . .	9.8	19.56	+ 0.698	67 25 10.1	.53	52.32	2
1823	Hydri	10.0	21.81	− 0.727	75 26 3.0	.53	51.87	3
1824	Horologii	10.0	25.48	+ 0.862	65 55 10.9	.52	51.92	1
1825	Hydri	9.5	36.86	− 4.413	82 28 44.8	+ 13.51	52.00	1

Number.	Constellation, Name of Star, or Synonym.	Magnitude.	Right Ascension, 1850.0.	Annual Precession.	South Declination, 1850.0.	Annual Precession.	Mean year.	No. of obs.
			h. m. s.	s.	° ′ ″	″		
1826	Gould, 3545	9.0	3 10 37.43	+ 0.303	70 21 23.0	+ 13.51	51.94	2
1827	Hydri	9.0	40.21	− 0.653	75 8 18.7	.51	51.81	2
1828	Hydri	9.5	10 59.84	− 3.246	81 5 50.7	.49	52.02	1
1829	Hydri	10.0	11 0.53	+ 0.267	70 33 46.2	.48	51.99	1
1830	Hydri	10.0	4.97	+ 0.138	71 20 42.6	+ 13.48	51.88	1
1831	Lacaille, 1884	8.2	3 11 26.30	−55.694	89 2 0.7	+ 13.46	51.86	6
1832	Gould, Z. C., 3h 360	8.5	30.04	+ 0.505	68 52 43.2	.45	51.91	1
1833	Lacaille, 1075	8.0	34.98	− 0.610	74 52 13.1	.45	50.89	1
1834	Lacaille, 1066	8.8	38.77	+ 0.802	66 23 9.8	.44	51.92	2
1835	Hydri	10.0	39.88	+ 0.137	71 19 4.5	+ 13.44	51.88	1
1836	Horologii	9.5	3 11 47.73	+ 0.637	66 58 3.2	+ 13.43	51.93	2
1837	Hydri	9.5	49.79	− 0.975	76 12 50.1	.43	52.00	2
1838	Hydri	10.0	50.14	+ 0.489	68 58 10.7	.43	51.91	1
1839	Hydri	10.0	56.15	− 3.270	81 5 50.7	.42	52.01	2
1840	Gould, 3564	9.5	11 56.22	− 0.543	74 38 8.4	+ 13.42	50.88	2
1841	Hydri	9.8	3 12 6.46	+ 0.403	69 34 21.5	+ 13.41	51.90	3
1842	Hydri	9.5	7.30	− 3.821	81 47 33.8	.41	51.47	2
1843	Gould, Z. C., 3h 382 . .	9.0	11.64	+ 0.642	67 44 29.4	.41	52.35	2
1844	Gould, Z. C., 3h 375 . .	9.5	20.22	− 0.397	73 59 49.5	.40	50.87	1
1845	Gould, 3583	8.8	21.52	+ 0.703	67 12 42.8	+ 13.40	51.93	2
1846	Gould, 3565	9.0	3 12 44.46	− 2.336	79 33 31.9	+ 13.37	51.47	2
1847	Hydri	10.0	44.51	3.739	81 40 36.2	.37	50.95	1
1848	Hydri	9.5	44.84	0.653	75 1 15.0	.37	51.82	2
1849	Gould, 3579	9.0	47.10	1.008	76 16 17.4	.37	52.00	2
1850	Lacaille, 1105	7.0	48.48	− 2.336	79 33 21.2	+ 13.37	51.47	2
1851	Hydri	10.0	3 12 57.61	− 0.304	73 32 34.6	+ 13.36	50.88	1
1852	Reticuli	9.2	12 58.01	+ 0.671	67 26 20.2	.36	52.32	2
1853	Lacaille, 1086	7.8	13 0.80	− 0.710	75 13 29.7	.35	51.82	2
1854	Hydri	9.2	0.84	0.868	75 47 33.6	.35	52.00	2
1855	Lacaille, 1085	7.8	3.42	− 0.545	74 34 35.0	+ 13.35	50.88	2
1856	Gould, Z. C., 3h 399 . .	9.5	3 13 4.18	− 0.122	72 39 3.0	+ 13.35	50.89	2
1857	Octantis	10.0	5.30	7.467	84 35 28.7	.35	52.00	1
1858	Gould, Z. C., 3h 404 . .	9.0	18.86	0.153	72 47 34.8	.33	50.89	2
1859	Hydri	9.8	24.05	− 3.862	81 48 0.8	.33	51.47	2
1860	Gould, Z. C., 3h 416 . .	10.0	26.60	+ 0.808	66 11 7.0	+ 13.33	51.92	1
1861	Reticuli	10.0	3 13 27.52	+ 0.749	66 43 24.1	+ 13.32	51.92	1
1862	Hydri	8.5	39.31	− 1.053	76 22 10.0	.31	52.00	2
1863	Gould, Z. C., 3h 407 . .	9.0	44.78	0.939	75 59 42.3	.31	52.00	2
1864	Hydri	10.0	13 46.70	− 0.220	73 5 33.6	.30	50.88	2
1865	Lacaille, 1082	7.8	14 11.40	+ 0.268	70 20 6.2	+ 13.28	51.95	2

No. 1827. Some doubt about this Star. The declination may possibly be 75° 16′ 46.6″.

Number.	Constellation, Name of Star, or Synonym.	Magnitude.	Right Ascension, 1850.0.	Annual Precession.	South Declination, 1850.0.	Annual Precession.	Mean year.	No. of obs.
			h. m. s.	s.	° ′ ″	″		
1 866	Hydri	9.5	3 14 34.85	+ 0.478	68 50 50.0	+ 13.25	51.91	1
1 867	Hydri	10.0	40.15	— 0.601	74 42 41.5	.25	50.88	1
1 868	Hydri	10.0	45.03	+ 0.472	68 52 38.2	.24	51.91	1
1 869	Hydri	9.5	49.53	— 3.644	81 29 32.3	.24	50.95	1
1 870	Octantis	10.0	52.58	—16.770	87 6 4.1	+ 13.23	52.67	1
1 871	Hydri	10.0	3 14 53.48	— 0.585	74 38 13.3	+ 13.23	50.88	1
1 872	Octantis	8.8	15 5.22	— 6.352	83 54 32.0	.22	51.46	2
1 873	Hydri	9.5	6.72	+ 0.008	71 49 39.0	.22	50.93	2
1 874	Hydri	10.0	11.52	— 3.394	81 9 22.1	.21	52.01	2
1 875	Reticuli	9.0	12.35	+ 0.692	67 5 13.6	+ 13.21	51.90	1
1 876	Hydri	10.0	3 15 17.14	— 1.835	78 24 32.4	+ 13.21	51.47	2
1 877	Hydri	10.0	20.84	1.815	78 21 43.0	.20	51.47	2
1 878	Hydri	10.1	25.04	2.904	80 26 8.7	.20	52.03	1
1 879	Octantis	9.5	34.65	34.779	88 28 30.4	.19	52.68	1
1 880	Octantis	9.0	39.51	—34.438	88 30 3.5	+ 13.18	52.69	1
1 881	Hydri	10.0	3 15 53.28	+ 0.388	69 24 0.1	+ 13.17	51.91	1
1 882	Octantis	11.0	16 1.91	— 5.902	83 34 59.1	.16	52.01	1
1 883	Hydri	9.7	3.72	— 0.814	75 26 19.8	.15	51.88	3
1 884	Lacaille, 1092	6.5	18.38	+ 0.636	67 28 19.0	.14	52.22	3
1 885	Lacaille, 1094	8.2	27.86	+ 0.478	68 42 21.0	+ 13.13	51.93	2
1 886	Hydri	10.0	3 16 30.72	+ 0.484	68 39 48.9	+ 13.12	51.91	1
1 887	Gould, 3639	9.0	31.12	— 1.520	77 34 58.4	.12	51.47	2
1 888	Hydri	9.5	34.74	3.739	81 33 13.1	.12	50.97	2
1 889	Gould, Z. C., 3h 488 . .	9.0	37.18	0.016	71 51 43.3	.12	50.93	2
1 890	Gould, Z. C., 3h 474 . .	9.0	48.40	— 1.890	78 28 10.2	+ 13.10	50.93	1
1 891	Hydri	9.7	3 16 50.55	+ 0.318	69 48 55.4	+ 13.10	51.93	4
1 892	Lacaille, 1097	6.5	17 4.74	+ 0.750	66 25 40.0	.09	51.92	1
1 893	Lacaille, 1109	7.5	9.89	— 0.485	74 5 44.6	.08	50.87	1
1 894	Lacaille, 1098	7.0	12.36	+ 0.766	66 16 12.8	.08	51.92	2
1 895	Octantis	10.0	24.14	—19.812	87 27 44.2	+ 13.07	52.69	1
1 896	Gould, Z. C., 3h 504 . .	8.8	3 17 24.85	+ 0.122	71 1 22.2	+ 13.06	51.92	2
1 897	Octantis	9.2	32.69	— 6.022	83 37 52.1	.05	52.00	2
1 898	Gould, Z. C., 3h 516 . .	9.5	51.32	+ 0.203	70 29 45.7	.04	51.99	1
1 899	Hydri	9.0	17 59.95	0.255	70 9 21.9	+ 13.03	51.90	1
1 900	Hydri	12.0	18 6.32	+ 0.305	69 49 . .	.02	52.02	1
1 901	Reticuli	10.0	3 18 8.36	+ 0.696	66 49 31.0	+ 13.02	51.90	1
1 902	Hydri	10.0	11.16	0.001	71 40 8.6	.01	50.95	1
1 903	Gould, Z. C., 3h 529 . .	9.0	16.12	+ 0.591	67 41 55.0	.01	52.35	2
1 904	Hydri	10.0	18.89	— 3.732	81 29 27.9	.00	50.95	1
1 905	Gould, Z. C., 3h 499 . .	9.2	22.21	— 2.667	79 56 26.2	+ 13.00	51.47	2

No. 1 879. This Star and No. 1 880 probably the same, but the source of error can not be found. The circle readings differ 1′ 33″.

Number.	Constellation, Name of Star, or Synonym.	Magnitude.	Right Ascension, 1850.0.	Annual Precession.	South Declination, 1850.0.	Annual Precession.	Mean year.	No. of obs.
			h. m. s.	s.	° ′ ″	″		
1906	Hydri	9.5	3 18 22.68	+ 0.240	70 13 28.2	+ 13.00	51.90	1
1907	Hydri	10.0	22.95	0.302	69 48 52.7	.00	51.90	1
1908	Reticuli	10.0	23.97	0.807	65 47 21.3	.00	51.92	1
1909	Reticuli	9.5	26.12	+ 0.674	66 59 17.3	13.00	51.90	1
1910	Octantis	10.0	33.00	— 4.903	82 43 13.0	+ 12.99	51.47	2
1911	Gould, 3701	6.8	3 18 38.64	+ 0.747	66 19 38.4	+ 12.98	51.92	2
1912	Octantis	10.0	46.49	— 6.103	83 39 30.0	.97	51.98	1
1913	Gould, Z. C., 3h 538 . .	10.0	47.58	0.070	72 1 18.3	.97	50.91	1
1914	Octantis	10.0	48.90	8.135	84 48 2.9	.97	52.69	1
1915	Hydri	9.8	54.81	— 1.372	77 4 15.4	+ 12.96	51.47	2
1916	Hydri	9.0	3 18 57.34	+ 0.115	70 57 53.5	+ 12.96	51.88	1
1917	Hydri	9.0	19 5.06	— 0.348	73 22 56.1	.95	50.88	1
1918	Hydri	9.5	5.27	4.550	82 22 16.8	.95	51.99	2
1919	Lacaille, 1133	8.0	15.10	2.372	79 22 26.1	.94	52.00	1
1920	Hydri	10.0	17.02	— 0.907	75 36 3.7	+ 12.94	52.33	2
1921	Hydri	10.0	3 19 19.64	— 0.290	73 5 43.0	+ 12.94	50.88	2
1922	Hydri	9.0	20.83	— 0.401	73 36 16.6	.94	50.87	1
1923	Hydri	10.5	29.64	+ 0.275	69 55 21.0	.93	51.96	2
1924	Gould, 3714	8.2	38.72	— 0.073	71 58 50.5	.92	50.93	2
1925	Octantis	10.0	39.55	— 5.557	83 14 20.5	+ 12.91	52.01	1
1926	Hydri	10.0	3 19 40.21	— 2.495	79 35 13.3	+ 12.91	50.93	1
1927	Hydri	7.0	48.45	— 1.716	77 56 5.2	.90	51.48	2
1928	Lacaille, 1118	7.8	19 55.98	+ 0.395	69 4 20.4	.90	51.91	2
1929	Hydri	10.0	20 3.44	— 2.039	78 39 57.0	.89	51.48	2
1930	Hydri	8.2	18.73	— 1.147	76 20 15.0	+ 12.87	52.00	2
1931	Lacaille, 1126	7.0	3 20 21.43	— 0.631	74 31 11.8	+ 12.87	50.88	1
1932	Gould, Z. C., 3h 594 . .	9.0	24.28	— 0.184	72 30 57.3	.86	50.90	1
1933	Gould, 3738	7.5	32.94	+ 0.733	66 18 53.0	.86	51.92	2
1934	Hydri	9.5	45.94	0.227	70 8 56.1	.84	51.90	1
1935	Gould, 3753	9.0	55.80	+ 0.754	66 5 15.6	+ 12.83	51.92	1
1936	Gould, Z. C., 3h 603 . .	9.5	3 20 56.77	— 0.473	73 49 47.6	+ 12.83	50.87	2
1937	Gould, Z. C., 3h 617 . .	8.2	20 57.14	+ 0.393	69 0 24.6	.83	51.91	2
1938	Hydri	9.5	21 2.07	+ 0.521	68 3 4.8	.82	51.96	1
1939	Hydri	10.0	3.82	— 0.290	72 59 39.6	.82	50.88	2
1940	Gould, Z. C., 3h 634 . .	10.0	27.17	+ 0.560	67 42 28.4	+ 12.79	52.03	1
1941	Gould, Z. C., 3h 631 . .	9.0	3 21 35.03	— 0.107	72 2 36.8	+ 12.79	50.91	1
1942	Hydri	10.0	45.35	0.944	75 36 17.5	.77	51.98	1
1943	Hydri	10.0	51.18	— 0.016	71 32 3.2	.77	50.95	1
1944	Gould, 3765	9.2	21 54.39	+ 0.262	69 50 33.5	.76	51.90	3
1945	Octantis	9.5	22 0.80	—24.844	87 52 59.7	+ 12.76	52.69	2

Number.	Constellation, Name of Star, or Synonym.	Magnitude.	Right Ascension, 1850.0.	Annual Precession.	South Declination, 1850.0.	Annual Precession.	Mean year.	No. of obs.
			h. m. s.	s.	° ′ ″	″		
1946	Hydri	9.5	3 22 15.72	+ 0.341	69 17 21.5	+ 12.74	51.91	1
1947	Hydri	9.5	16.16	0.242	69 57 17.2	.74	51.90	1
1948	Reticuli	9.8	16.65	0.584	67 26 35.4	.74	52.00	2
1949	Hydri	9.2	27.57	0.376	69 1 47.4	.73	51.91	2
1950	Hydri	9.2	36.14	+ 0.365	69 5 42.2	+ 12.72	51.91	2
1951	Gould, Z. C., 3^h 641	9.8	3 22 46.77	− 2.269	79 2 14.8	+ 12.70	52.01	2
1952	Hydri	10.0	48.90	− 1.207	76 24 16.4	.70	51.98	1
1953	Hydri	10.0	49.06	+ 0.012	71 19 2.1	.70	51.88	1
1954	Hydri	10.0	22 51.78	− 0.743	74 49 11.8	.70	50.89	1
1955	Hydri	10.0	23 10.07	− 0.157	72 12 40.5	+ 12.68	50.91	1
1956	Lacaille, 1140	8.3	3 23 14.83	− 0.830	75 7 30.4	+ 12.67	51.51	3
1957	Hydri	9.5	16.52	− 4.151	81 50 2.2	.67	51.47	2
1958	Lacaille, 1132	6.5	25.78	+ 0.199	70 9 6.5	.66	51.95	2
1959	Gould, Z. C., 3^h 662	10.0	32.44	− 2.222	78 54 52.9	.65	52.02	1
1960	Gould, Z. C., 3^h 706	8.5	48.67	+ 0.386	68 51 38.7	+ 12.63	51.91	1
1961	Lacaille, 1182	8.8	3 23 56.40	− 3.821	81 25 37.8	+ 12.63	50.97	2
1962	Octantis	9.8	23 56.99	9.466	85 15 48.1	.63	52.68	2
1963	Octantis	9.0	24 14.85	8.994	85 4 25.5	.61	51.74	6
1964	Gould, Z. C., 3^h 711	9.8	16.76	0.502	73 46 21.9	.60	50.87	2
1965	Gould, 3782	9.0	22.45	− 3.772	81 21 20.3	+ 12.60	50.97	2
1966	Hydri	9.2	3 24 24.56	− 3.517	81 1 20.5	+ 12.59	51.51	2
1967	Gould, Z. C., 3^h 730	9.5	24.56	+ 0.347	69 13 54.6	.59	51.91	1
1968	Gould, Z. C., 3^h 735	9.0	24.60	+ 0.619	67 0 37.1	.59	51.93	2
1969	Hydri	9.0	33.52	− 1.015	75 42 35.4	.58	51.98	1
1970	Hydri	10.0	39.36	+ 0.107	70 38 44.5	+ 12.53	51.96	1
1971	Gould, Z. C., 3^h 734	8.5	3 24 53.38	− 0.532	73 52 7.0	+ 12.56	50.87	2
1972	Lacaille, 1139	6.2	24 56.15	+ 0.229	69 51 41.6	.56	51.90	3
1973	Gould, Z. C., 3^h 753	9.0	25 2.56	0.473	68 8 19.4	.55	51.96	1
1974	Reticuli	10.0	16.18	+ 0.686	66 22 20.2	.54	51.92	1
1975	Octantis	9.0	25.70	−16.863	87 0 3.3	+ 12.52	51.03	1
1976	Gould, Z. C., 3^h 754	9.0	3 25 30.04	− 0.574	74 0 53.1	+ 12.52	50.87	1
1977	Hydri	9.5	35.06	+ 0.503	67 52 2.4	.51	51.96	1
1978	Lacaille, 1236	8.0	36.21	− 7.000	84 4 36.7	.51	50.98	2
1979	Hydri	9.0	48.07	− 1.063	75 48 28.8	.50	52.00	2
1980	Gould, 3849	9.0	25 48.67	+ 0.179	70 7 32.3	+ 12.50	51.90	1
1981	Gould, Z. C., 3^h 765	8.7	3 26 9.10	− 0.983	75 30 18.6	+ 12.48	51.88	3
1982	Hydri	10.0	23.67	− 2.122	78 35 47.0	.46	51.47	2
1983	Reticuli	9.5	33.42	+ 0.580	67 11 4.7	.45	51.90	1
1984	Hydri	10.0	41.15	− 0.458	73 26 57.7	.44	50.88	1
1985	Hydri	10.0	48.71	+ 0.407	68 30 30.2	+ 12.43	51.95	1

Number.	Constellation, Name of Star, or Synonym.	Magnitude.	Right Ascension, 1850.0.	Annual Precession.	South Declination, 1850.0.	Annual Precession.	Mean year.	No. of obs.
			h. m. s.	s.	° ′ ″	″		
1986	Octantis	9.0	3 27 9.00	− 9.066	85 3 12.4	+ 12.41	51.91	3
1987	Hydri	10.0	13.42	+ 0.486	67 53 17.2	.40	51.96	1
1988	Octantis	9.8	18.78	− 5.705	83 11 14.3	.40	51.48	2
1989	Octantis	9.7	23.84	− 7.303	84 12 47.2	.39	51.31	3
1990	Hydri	9.8	24.85	+ 0.375	68 42 0.7	+ 12.39	51.93	2
1991	Gould, Z. C., 3^h 813	9.0	3 27 27.17	− 0.580	73 55 57.8	+ 12.39	50.87	2
1992	Octantis	8.3	30.63	− 8.275	84 42 12.7	.38	51.41	5
1993	Hydri	10.0	37.64	+ 0.490	67 49 26.1	.37	51.96	1
1994	Hydri	9.0	52.24	+ 0.174	70 1 51.0	.36	51.90	1
1995	Hydri	10.0	52.32	− 0.049	71 21 15.2	+ 12.36	51.88	1
1996	Hydri	10.0	3 27 53.85	− 0.469	73 26 12.7	+ 12.35	50.88	1
1997	Hydri	9.0	27 59.42	− 0.520	73 39 4.5	.35	50.87	1
1998	Reticuli	10.0	28 6.61	+ 0.705	65 59 52.4	.34	51.92	1
1999	Lacaille, 1185	7.5	19.21	− 1.594	77 15 43.1	.33	52.00	1
2000	Gould, 3901	9.8	22.92	+ 0.133	70 15 18.8	+ 12.32	51.95	2
2001	Hydri	10.0	3 28 23.87	+ 0.496	67 43 36.3	+ 12.32	52.03	1
2002	Hydri	10.0	26.97	− 0.037	71 15 3.6	.32	51.88	1
2003	Hydri	9.8	35.02	0.170	71 57 51.6	.31	50.93	2
2004	Octantis	9.8	44.75	6.679	83 49 1.2	.30	51.46	2
2005	Octantis	9.5	48.86	− 7.021	84 1 26.5	+ 12.29	50.94	1
2006	Gould, Z. C., 3^h 851	8.2	3 28 54.74	− 0.604	73 57 27.8	+ 12.28	50.87	2
2007	Gould, Z. C., 3^h 874	9.0	29 5.29	+ 0.448	68 3 0.8	.27	51.96	1
2008	Lacaille, 1164	6.0	20.59	+ 0.579	66 59 53.8	.26	51.93	2
2009	Hydri	10.0	28.97	− 0.046	71 14 31.1	.24	51.88	1
2010	Hydri	9.0	32.86	+ 0.354	68 42 21.4	+ 12.24	51.93	2
2011	Gould, 3939	8.0	3 29 34.24	+ 0.601	66 47 56.2	+ 12.24	51.92	2
2012	Gould, Z. C., 3^h 866	9.2	37.98	− 1.566	77 7 54.0	.24	51.47	2
2013	Gould, Z. C., 3^h 877	8.0	40.49	0.638	74 3 43.3	.23	50.87	1
2014	Octantis	9.0	29 45.46	11.423	85 49 3.8	.23	51.69	5
2015	Octantis	9.7	30 12.59	− 9.284	85 5 23.0	+ 12.19	51.35	3
2016	Gould, 3955	8.2	3 30 16.51	+ 0.670	66 9 1.2	+ 12.19	51.92	2
2017	Mensæ	9.5	19.12	− 1.472	76 50 50.3	.19	50.95	1
2018	Mensæ	9.0	20.47	− 4.758	82 16 33.2	.19	51.99	2
2019	Reticuli	9.0	26.26	+ 0.663	66 12 18.9	.18	51.92	2
2020	Gould, 3962	8.0	27.55	+ 0.695	65 55 8.2	+ 12.18	51.92	1
2021	Lacaille, 1210	10.0	3 30 30.27	− 2.880	79 52 39.0	+ 12.17	50.99	3
2022	Hydri	9.0	43.12	0.065	71 13 59.1	.16	51.88	1
2023	Gould, Z. C., 3^h 894	9.2	58.87	2.497	79 10 50.9	.14	52.01	2
2024	Octantis	9.0	30 59.32	12.893	86 11 9.4	.14	51.85	2
2025	Gould, Z. C., 3^h 931	9.2	31 2.68	− 0.016	70 58 40.1	+ 12.14	51.92	2

Number.	Constellation, Name of Star, or Synonym.	Magnitude.	Right Ascension, 1850.0.	Annual Precession.	South Declination, 1850.0.	Annual Precession.	Mean year.	No. of obs.
			h. m. s.	s.	° ′ ″	″		
2026	Hydri	10.0	3 31 17.58	+ 0.098	70 17 35.8	+ 12.12	51.90	1
2027	Gould, Z. C., 3h 934 . .	9.0	21.51	— 0.543	73 34 30.6	.12	50.88	1
2028	Lacaille, 1178	7.8	21.88	+ 0.533	67 13 55.7	.11	51.93	2
2029	Gould, Z. C., 3h 937 . .	9.0	24.11	0.524	73 29 23.7	.11	50.88	1
2030	Octantis	10.0	37.51	— 9.789	85 15 34.3	+ 12.10	52.69	1
2031	Lacaille, 1184	8.0	3 31 50.54	+ 0.441	67 55 6.6	+ 12.08	52.00	2
2032	Mensæ	10.0	51.09	— 1.501	76 51 45.4	.08	50.95	1
2033	Gould, Z. C., 3h 951 . .	9.0	55.95	— 0.393	72 52 30.0	.07	50.88	1
2034	Reticuli	10.0	57.84	+ 0.576	66 50 18.0	.07	51.90	1
2035	Mensæ	9.0	31 57.92	— 3.624	80 56 23.4	+ 12.07	52.02	.
2036	Gould, Z. C., 3h 928 . .	9.0	3 32 2.44	— 2.487	79 7 30.8	+ 12.07	52.01	1
2037	Gould, 3998	9.8	8.64	+ 0.436	67 56 20.9	.06	52.00	2
2038	Mensæ	10.0	12.44	— 1.266	76 9 41.3	.06	51.98	1
2039	Lacaille, 1204	8.0	16.27	0.945	75 6 2.7	.05	51.51	3
2040	Hydri	10.0	25.39	— 0.038	71 1 24.9	+ 12.04	51.96	1
2041	Lacaille, 1263	8.8	3 32 31.96	— 5.375	82 46 56.1	+ 12.03	51.47	2
2042	Lacaille, 1222	8.0	37.24	— 2.015	78 7 18.2	.03	52.02	1
2043	Lacaille, 1188	7.0	44.30	+ 0.638	66 15 44.5	.02	51.92	2
2044	Hydri	10.0	49.73	— 0.897	74 54 9.0	.01	50.89	1
2045	Mensæ	10.0	51.68	— 2.119	78 21 36.6	+ 12.01	50.93	1
2046	Hydri	9.0	3 32 57.14	+ 0.069	70 22 13.2	+ 12.00	51.95	2
2047	Gould, Z. C., 3h 990 . .	9.8	32 57.58	0.204	69 31 9.8	.00	51.91	2
2048	Hydri	9.5	33 5.18	+ 0.338	68 35 55.4	12.99	51.92	2
2049	Mensæ	10.0	7.30	— 3.433	80 38 38.8	11.99	52.03	1
2050	Hydri	10.0	17.82	— 0.799	74 30 51.2	+ 11.98	50.88	1
2051	Hydri	9.5	3 33 19.33	+ 0.085	70 14 58.5	+ 11.98	51.90	1
2052	Octantis	9.8	25.71	—21.098	87 27 30.5	.97	52.68	2
2053	Hydri	10.0	26.27	— 0.884	74 49 27.9	.97	50.89	1
2054	Gould, Z. C., 3h 1008 . .	9.2	32.34	+ 0.247	69 11 44.2	.96	51.91	2
2055	Gould, Z. C., 3h 1014 . .	10.0	36.34	+ 0.363	68 22 53.8	+ 11.96	51.95	2
2056	Hydri	9.5	3 33 38.72	+ 0.255	69 8 22.8	+ 11.95	51.91	2
2057	Gould, Z. C., 3h 1029 . .	9.5	34 3.92	0.367	68 19 43.0	.93	51.95	2
2058	Mensæ	9.0	4.34	4.417	81 49 49.9	.92	51.47	2
2059	Hydri	10.0	11.48	0.432	67 50 4.1	.92	52.03	1
2060	Gould, 4036	9.2	15.78	+ 0.361	68 21 28.2	+ 11.91	51.95	2
2061	Lacaille, 1278	7.8	3 34 17.98	— 5.683	82 59 40.4	+ 11.91	51.48	2
2062	Gould, Z. C., 3h 1042 . .	9.0	30.86	+ 0.285	68 52 32.3	.89	51.91	1
2063	Mensæ	9.0	37.23	— 3.867	81 10 45.0	.89	51.51	2
2064	Hydri	10.0	45.51	— 0.932	74 56 20.2	.88	50.89	1
2065	Hydri	9.5	53.32	+ 0.069	70 15 17.9	+ 11.87	51.95	2

Number.	Constellation, Name of Star, or Synonym.	Magnitude.	Right Ascension, 1850.0.	Annual Precession.	South Declination, 1850.0.	Annual Precession.	Mean year.	No. of obs.
			h. m. s.	s.	° ′ ″	″		
2 066	Mensæ	10.0	3 35 15.67	— 1.123	75 34 25.4	+ 11.84	51.88	3
2 067	Gould, Z. C., 3h 1053 . .	8.5	21.20	— 0.536	73 20 33.7	.83	51.46	2
2 068	Hydri	10.0	23.54	+ 0.158	69 41 9.2	.83	51.90	1
2 069	Mensæ	9.5	25.28	— 3.551	80 44 23.5	.83	52.02	2
2 070	Brisbane, 593	5.3	35.21	— 2.410	78 51 6.3	+ 11.82	52.02	1
2 071	Mensæ	9.0	3 35 36.38	— 1.146	75 37 59.6	+ 11.82	52.33	2
2 072	Gould, Z. C., 3h 1056 . .	9.0	37.57	0.832	74 31 40.5	.82	50.88	1
2 073	Mensæ	10.0	42.46	1.225	75 52 56.8	.81	51.98	1
2 074	Mensæ	9.0	35 47.81	1.203	75 48 27.4	.80	51.98	1
2 075	Mensæ	10.0	36 4.59	— 1.849	77 35 54.1	+ 11.78	50.94	1
2 076	Mensæ	10.0	3 36 4.66	— 1.149	75 37 34.2	+ 11.78	52.33	2
2 077	Gould, Z. C., 3h 1085 . .	9.0	6.66	+ 0.521	67 0 51.6	.78	51.93	2
2 078	Reticuli	10.0	9.45	0.508	67 6 56.6	.78	51.90	1
2 079	Lacaille, 1218	7.5	12.17	+ 0.263	68 55 41.2	.77	51.91	1
2 080	Mensæ	9.0	32.23	— 5.644	82 54 47.2	+ 11.77	50.95	1
2 081	Mensæ	9.5	3 36 57.08	— 3.541	80 40 20.0	+ 11.72	52.02	2
2 082	Mensæ	10.0	37 0.76	— 1.991	77 54 5.5	.72	51.48	2
2 083	Hydri	9.8	3.20	+ 0.235	69 4 1.3	.71	51.91	2
2 084	Hydri	10.0	4.60	0.327	68 25 40.6	.71	51.95	1
2 085	Gould, Z. C., 3h 1125 . .	9.0	22.26	+ 0.570	66 31 28.6	+ 11.69	51.92	1
2 086	Hydri	9.0	3 37 25.75	— 0.088	71 1 26.6	+ 11.69	51.88	1
2 087	Hydri	10.0	27.90	0.641	73 41 7.4	.68	50.04	1
2 088	Mensæ	10.0	28.00	2.868	79 37 53.4	.68	52.00	1
2 089	Gould, Z. C., 3h 1114 . .	10.0	28.66	0.315	72 12 33.7	.68	50.91	2
2 090	Mensæ	10.0	43.01	— 1.804	77 25 29.0	+ 11.67	50.94	1
2 091	Octantis	8.8	3 37 43.86	—19.571	87 15 0.9	+ 11.67	51.30	6
2 092	Lacaille, 1281	8.0	45.36	3.962	81 12 32.8	.66	51.51	2
2 093	Gould, Z. C., 3h 1129 . .	8.5	56.39	0.695	73 53 3.1	.65	51.65	3
2 094	Mensæ	10.0	37 58.44	1.963	77 48 3.3	.65	52.02	1
2 095	Octantis	8.5	38 1.92	—13.993	86 20 59.3	+ 11.64	51.02	1
2 096	Gould, 4111	8.0	3 38 6.35	— 0.845	74 27 50.9	+ 11.64	50.88	1
2 097	Lacaille, 1848	8.5	6.45	46.860	88 45 2.4	.64	51.44	4
2 098	Gould, Z. C., 3h 1119 . .	9.0	17.27	2.078	78 3 20.5	.63	52.02	1
2 099	Gould, Z. C., 3h 1123 . .	9.0	17.75	1.791	77 22 18.1	.63	52.00	1
2 100	Mensæ	10.0	31.00	— 2.559	79 2 29.2	+ 11.61	52.01	2
2 101	Hydri	9.5	3 38 42.30	— 0.334	72 14 37.8	+ 11.60	50.91	2
2 102	Gould, Z. C., 3h 1161 . .	9.0	45.87	+ 0.648	65 45 49.4	.59	51.92	1
2 103	Hydri	9.5	38 51.59	+ 0.409	67 42 53.7	.59	52.03	1
2 104	Hydri	9.2	39 0.83	— 0.163	71 20 45.2	.57	51.96	2
2 105	Mensæ	8.5	9.39	— 6.789	83 40 42.1	+ 11.56	51.99	1

No. 2 083. This observation may be 69° 2′ 37.5″ = 3 revs. wrong.

Number.	Constellation, Name of Star, or Synonym.	Magnitude.	Right Ascension, 1850.0.	Annual Precession.	South Declination, 1850.0.	Annual Precession.	Mean year.	No. of obs.
			h. m. s.	s.	° ′ ″	″		
2 106	Hydri	9.5	3 39 10.99	− 0.132	71 10 12.0	+ 11.56	51.88	1
2 107	Mensæ	10.0	20.06	− 1.136	75 26 21.9	.55	50.96	1
2 108	Gould, 4143	9.2	24.45	+ 0.154	69 27 31.4	.55	51.91	2
2 109	Gould, Z. C., 3^h 1179 . .	9.0	33.38	0.034	70 11 50.2	.54	51.95	2
2 110	Hydri	9.5	37.16	+ 0.237	68 53 53.1	+ 11.53	51.91	1
2 111	Gould, Z. C., 3^h 1187 . .	9.2	3 39 39.63	+ 0.316	68 20 53.6	+ 11.53	51.95	2
2 112	Gould, Z. C., 3^h 1200 . .	9.0	40 5.79	− 0.217	71 34 39.4	.50	51.63	3
2 113	Lacaille, 1245	8.0	11.04	+ 0.496	66 57 13.8	.49	51.93	2
2 114	Mensæ	8.8	11.94	− 5.609	82 48 15.9	.49	51.47	2
2 115	Lacaille, 1261	7.0	15.61	− 1.108	75 18 17.1	+ 11.49	51.88	3
2 116	Gould, Z. C., 3^h 1209 . .	9.0	3 40 16.74	− 0.012	70 25 57.6	+ 11.48	51.90	1
2 117	Mensæ	10.0	19.05	− 1.149	75 26 25.9	.48	51.98	1
2 118	Gould, Z. C., 3^h 1217 . .	9.0	21.16	+ 0.463	67 12 26.3	.48	51.93	2
2 119	Hydri	9.2	35.30	0.211	69 1 3.5	.46	51.91	2
2 120	Hydri	9.2	51.16	+ 0.264	68 38 35.8	+ 11.44	51.93	2
2 121	Hydri	11.0	3 40 53.46	− 0.212	71 30 10.3	+ 11.44	51.88	1
2 122	Gould, Z. C., 3^h 1203 . .	8.8	40 53.78	1.923	77 36 1.3	.44	51.47	2
2 123	Gould, Z. C., 3^h 1234 . .	10.0	41 12.06	− 0.418	72 30 52.2	.42	50.90	1
2 124	Reticuli	9.0	16.08	+ 0.524	66 39 39.8	.41	51.92	1
2 125	Lacaille, 1280	8.2	16.67	− 1.918	77 34 20.9	+ 11.41	51.47	2
2 126	Gould, Z. C., 3^h 1258 . .	9.0	3 41 27.45	+ 0.382	67 45 14.8	+ 11.40	52.03	1
2 127	Hydri	9.0	36.25	− 0.005	70 18 57.8	.39	51.90	1
2 128	Hydri	9.5	39.03	+ 0.211	68 57 7.2	.39	51.91	1
2 129	Hydri	10.0	40.14	− 0.200	71 24 23.9	.38	51.88	1
2 130	Hydri	10.0	45.31	+ 0.150	69 21 8.0	+ 11.38	51.91	1
2 131	Reticuli	10.0	3 41 48.75	+ 0.578	66 10 32.5	+ 11.37	51.92	1
2 132	Lacaille, 1279	8.0	41 53.82	− 1.428	76 14 41.5	.37	52.00	2
2 133	Hydri	10.0	42 0.79	0.363	72 12 45.4	.36	50.91	2
2 134	Lacaille, 1196	7.0	20.14	2.501	78 48 16.3	.34	52.02	1
2 135	Mensæ	9.5	23.27	− 4.503	81 42 49.1	+ 11.33	50.95	1
2 136	Gould, Z. C., 3^h 1281 . .	9.8	3 42 37.05	− 0.629	73 23 23.2	+ 11.32	51.46	2
2 137	Reticuli	10.0	51.59	+ 0.560	66 15 41.2	.30	51.92	1
2 138	Mensæ	10.0	42 57.84	− 3.529	80 29 47.4	.29	52.03	1
2 139	Mensæ	9.5	43 4.70	− 4.805	82 0 40.4	.28	51.47	2
2 140	Gould, 4228	8.8	6.36	+ 0.446	67 9 58.0	+ 11.28	51.93	2
2 141	Gould, 4217	8.0	3 43 13.82	− 0.890	74 24 9.3	11.27	51.26	3
2 142	Lacaille, 1307	7.5	16.94	− 2.942	79 34 37.8	.27	51.47	2
2 143	Gould, Z. C., 3^h 1309 . .	9.0	21.74	+ 0.006	70 9 15.8	.26	51.90	1
2 144	Mensæ	10.0	23.96	− 4.960	82 9 22.1	.26	51.99	1
2 145	Hydri	9.2	32.68	− 0.029	70 21 22.6	+ 11.25	51.95	2

Number.	Constellation, Name of Star, or Synonym.	Magnitude.	Right Ascension, 1850.0.	Annual Precession.	South Declination, 1850.0.	Annual Precession.	Mean year.	No. of obs.
			h. m. s.	s.	° ′ ″	″		
2146	Reticuli	10.0	3 43 35.19	+ 0.446	67 8 22.9	+ 11.25	51.90	1
2147	Reticuli	10.0	41.82	+ 0.573	66 5 53.1	.24	51.92	1
2148	Mensæ	10.2	42.68	− 1.603	76 40 17.6	.24	51.46	2
2149	Gould, Z. C., 3h 1333 . .	9.5	48.97	+ 0.300	68 12 35.3	.23	51.95	2
2150	Mensæ	9.5	51.04	− 1.381	76 1 34.3	+ 11.23	51.98	1
2151	Mensæ	9.8	3 43 51.27	− 2.295	78 20 5.8	+ 11.23	51.47	2
2152	Gould, Z. C., 3h 1331 . .	9.0	44 12.62	0.746	73 47 55.0	.20	51.65	3
2153	Gould, 4247	9.5	22.43	0.239	71 28 10.4	.19	51.63	3
2154	Gould, 4251	9.2	23.40	0.054	70 27 23.6	.19	52.00	2
2155	Gould, 4248	11.0	26.39	− 0.239	71 28 8.1	+ 11.18	50.95	1
2156	Hydri	10.0	3 44 33.62	− 0.072	70 32 59.9	+ 11.17	52.00	2
2157	Hydri	10.0	48.14	+ 0.204	68 49 32.6	.16	51.91	1
2158	Lacaille, 1283	7.8	44 50.22	− 0.064	70 29 13.6	.15	52.00	2
2159	Gould, Z. C., 3h 1359 . .	9.2	45 0.70	− 0.582	73 4 35.5	.14	50.88	2
2160	Lacaille, 1277	8.2	3.60	+ 0.457	66 57 53.6	+ 11.14	51.93	2
2161	Lacaille, 1285	8.0	3 45 5.38	− 0.066	70 29 3.8	+ 11.14	52.00	2
2162	Hydri	9.5	22.27	0.018	70 11 31.4	.12	51.90	1
2163	Mensæ	10.0	27.07	7.300	83 52 25.0	.11	51.46	2
2164	Hydri	9.0	31.01	0.267	71 33 37.1	.11	51.63	3
2165	Lacaille, 1295	7.5	33.65	− 0.936	74 28 28.5	+ 11.10	50.88	1
2166	Gould, Z. C., 3h 1389 . .	9.5	3 45 38.69	− 0.390	72 9 57.6	+ 11.10	50.91	1
2167	Gould, Z. C., 3h 1402 . .	9.0	44.58	+ 0.412	67 16 20.7	.09	51.97	1
2168	Hydri	10.0	45 54.65	+ 0.416	67 13 52.1	.08	51.97	1
2169	Mensæ	10.0	46 12.97	− 7.213	83 48 31.2	.05	51.46	2
2170	Lacaille, 1298	7.0	13.00	− 0.387	72 7 17.7	+ 11.05	50.91	1
2171	Mensæ	9.5	3 46 14.40	− 1.628	76 38 47.4	+ 11.05	50.95	1
2172	Mensæ	10.0	16.62	− 1.517	76 20 2.7	.05	51.98	1
2173	Hydri	9.5	17.93	+ 0.331	67 50 59.2	.05	52.00	2
2174	Octantis	9.5	31.46	−24.368	87 40 10.4	11.03	51.03	2
2175	Gould, Z. C., 3h 1425 . .	9.0	32.18	+ 0.019	69 54 21.9	+ 10.03	51.90	2
2176	Gould, Z. C., 3h 1427 . .	9.0	3 46 32.24	+ 0.422	67 9 3.6	+ 10.03	51.93	2
2177	Hydri	10.0	47 3.99	+ 0.334	67 46 45.2	.99	51.96	1
2178	Gould, Z. C., 3h 1401 . .	9.2	12.95	− 3.239	79 56 48.0	.98	51.48	2
2179	Lacaille, 1340	9.0	16.49	4.263	81 19 30.2	.98	50.99	1
2180	Mensæ	10.0	17.87	− 1.286	75 35 52.8	+ 10.97	51.98	1
2181	Stone, 1624	9.5	3 47 30.84	− 4 287	81 20 48.7	+ 10.96	50.99	1
2182	Gould, 4301	9.2	33.91	2.567	78 45 57.8	.96	51.48	2
2183	Mensæ	10.0	37.07	2.578	78 47 15.9	.95	50.93	1
2184	Mensæ	10.0	38.87	4.416	81 29 27.6	.95	50.95	1
2185	Lacaille, 1301	7.5	47 45.58	− 0.065	70 20 42.7	+ 10.94	51.95	2

Number.	Constellation, Name of Star, or Synonym.	Magnitude.	Right Ascension, 1850.0.	Annual Precession.	South Declination, 1850.0.	Annual Precession.	Mean year.	No. of obs.
			h. m. s.	s.	° ′ ″	″		
2186	Mensæ	10.0	3 48 6.59	− 2.108	77 46 52.7	+ 10.92	50.94	1
2187	Gould, Z. C., 3h 1464	9.2	7.36	0.257	71 22 35.6	.91	51.50	2
2188	Hydri	10.0	14.38	0.970	74 29 7.2	.91	50.88	1
2189	Mensæ	10.0	15.38	− 7.292	83 49 12.6	.90	51.46	2
2190	Hydri	10.0	17.60	+ 0.123	69 9 50.5	+ 10.90	51.91	1
2191	Gould, 4333	11.0	3 48 24.44	− 0.236	71 15 10.7	+ 10.89	52.04	1
2192	Gould, 4334	10.0	25.93	0.237	71 15 26.4	.89	52.04	1
2193	Mensæ	10.0	29.21	2.197	77 58 6.6	.89	51.48	2
2194	Gould, Z. C., 3h 1447	9.5	30.00	− 2.834	79 14 3.1	.89	52.00	1
2195	Hydri	10.0	32.41	+ 0.123	69 9 6.9	+ 10.88	51.91	1
2196	Mensæ	11.0	3 48 52.15	− 2.181	77 55 14.5	+ 10.86	52.02	1
2197	Mensæ	9.5	52.26	3.328	80 2 17.3	.86	52.04	1
2198	Hydri	9.5	52.43	0.057	70 14 16.3	.86	51.90	1
2199	Lacaille, 1311	7.5	48 54.09	0.159	70 48 44.2	.86	51.96	1
2200	Gould, 4349	9.5	49 5.64	− 0.075	70 19 48.4	+ 10.84	51.95	2
2201	Mensæ	9.2	3 49 8.78	− 5.818	82 47 13.9	+ 10.84	51.47	2
2202	Lacaille, 1319	8.8	10.66	1.117	74 58 1.6	.84	51.47	2
2203	Hydri	10.0	13.36	− 0.054	70 12 15.2	.83	51.90	1
2204	Lacaille, 1308	8.2	25.81	+ 0.446	66 47 42.1	.82	51.92	2
2205	γ Hydri	3.2	49 37.38	− 1.045	74 41 49.8	+ 10.80	50.88	2
2206	Gould, Z. C., 3h 1539	9.2	3 50 10.23	+ 0.050	69 31 37.6	+ 10.76	51.91	2
2207	Gould, 4373	8.7	22.36	− 0.097	70 23 36.7	.75	51.95	2
2208	Reticuli	10.0	31.42	+ 0.370	67 18 48.8	.74	51.97	1
2209	Lacaille, 1414	8.0	54.26	−10.393	85 12 2.8	.71	51.51	4
2210	Lacaille, 1328	8.0	55.15	− 1.190	75 8 43.4	+ 10.71	51.30	3
2211	Lacaille, 1334	8.2	3 50 58.86	− 1.893	77 10 9.6	+ 10.70	51.47	2
2212	Hydri	10.0	51 0.01	0.626	72 59 40.2	.70	50.88	1
2213	Hydri	10.0	1.23	0.635	73 2 0.4	.70	50.88	1
2214	Lacaille, 1358	9.0	1.43	3.671	80 28 45.2	.70	52.03	1
2215	Gould, 4360	10.0	3.61	− 3.675	80 29 3.7	+ 10.70	52.03	1
2216	Gould, Z. C., 3h 1573	10.0	3 51 6.52	+ 0.525	66 3 1.2	+ 10.69	51.92	1
2217	Gould, Z. C., 3h 1575	9.5	11.70	+ 0.485	66 22 15.6	.69	51.92	2
2218	Hydri	10.0	13.54	− 0.816	73 46 19.8	.69	50.04	1
2219	Hydri	10.0	18.82	+ 0.130	68 57 40.2	.68	51.91	1
2220	Mensæ	10.0	22.06	− 3.542	80 17 18.2	+ 10.68	52.04	1
2221	Gould, 4385	8.0	3 51 27.67	− 1.784	76 52 29.9	+ 10.67	50.95	1
2222	Reticuli	9.5	36.73	+ 0.368	67 16 4.1	.66	51.97	1
2223	Reticuli	10.0	37.16	+ 0.380	67 10 37.4	.66	51.97	1
2224	Hydri	9.5	37.53	− 0.361	71 44 22.5	.66	50.95	1
2225	Lacaille, 1343	8.0	49.43	− 1.807	76 55 17.0	+ 10.64	50.95	1

Number.	Constellation, Name of Star, or Synonym.	Magnitude.	Right Ascension, 1850.0.	Annual Precession.	South Declination, 1850.0.	Annual Precession.	Mean year.	No. of obs.
			h. m. s.	s.	° ′ ″	″		
2226	Reticuli	10.0	3 51 57.22	+ 0.431	66 45 44.9	+ 10.63	51.90	1
2227	Mensæ	9.5	59.05	− 2.177	77 50 17.5	.63	51.48	2
2228	Mensæ	9.5	51 59.13	2.173	77 48 3.6	.63	51.48	2
2229	Gould, Z. C., 3^h 1598	9.5	52 10.62	0.003	69 44 45.6	.62	51.90	1
2230	Hydri	11.0	14.41	− 0.827	73 46 19.7	+ 10.61	52.04	1
2231	Reticuli	10.0	3 52 19.60	+ 0.405	66 57 0.2	+ 10.60	51.90	1
2232	Octantis	9.2	20.02	−33.020	88 11 56.0	.60	51.28	4
2233	Gould, Z. C., 3^h 1581	10.0	20.84	− 1.531	76 8 53.3	.60	52.02	1
2234	Gould, Z. C., 3^h 1614	9.0	24.56	+ 0.458	66 31 25.6	.60	51.92	1
2235	Octantis	8.3	27.74	−27.115	87 50 46.2	+ 10.59	51.02	3
2236	Hydri	9.5	3 52 39.44	− 0.026	69 51 54.8	+ 10.58	51.90	2
2237	Octantis	9.5	46.40	33.598	88 13 30.5	.57	51.03	1
2238	Octantis	8.8	48.31	12.310	85 46 29.1	.57	51.03	3
2239	Hydri	9.0	51.72	0.049	69 59 27.9	.56	51.90	1
2240	Mensæ	9.0	52 55.47	− 5.289	82 15 27.4	+ 10.56	51.99	2
2241	Hydri	10.0	3 53 6.28	− 0.528	72 28 2.4	+ 10.55	50.90	1
2242	Hydri	9.0	6.30	0.068	70 5 16.7	.55	51.90	1
2243	Mensæ	9.8	7.44	− 4.048	80 55 29.0	.54	51.54	2
2244	Hydri	9.8	24.76	+ 0.274	67 51 37.2	.52	52.00	2
2245	Gould, Z. C., 3^h 1641	10.0	34.57	− 0.164	70 36 44.2	+ 10.51	51.97	2
2246	Mensæ	10.0	3 53 37.85	− 1.190	75 2 17.6	+ 10.51	50.96	1
2247	Reticuli	10.0	46.40	+ 0.410	66 49 47.6	.50	51.90	1
2248	Gould, Z. C., 3^h 1659	9.0	48.48	+ 0.402	66 53 13.9	.49	51.90	1
2249	Gould, Z. C., 3^h 1620	8.8	48.56	− 2.191	77 47 0.6	.49	51.48	2
2250	Hydri	10.0	53 54.09	− 0.600	72 45 27.7	+ 10.49	50.88	1
2251	Lacaille, 1353	7.5	3 54 5.36	− 1.619	76 20 15.6	+ 10.47	52.00	2
2252	Mensæ	9.2	11.77	7.622	83 54 43.6	.46	51.46	2
2253	Gould, Z. C., 3^h 1667	9.2	15.92	0.036	69 50 30.0	.46	51.91	2
2254	Mensæ	10.0	16.93	5.517	82 25 58.8	.46	51.99	2
2255	Octantis	9.0	34.40	−13.588	86 4 45.0	+ 10.44	51.03	2
2256	Hydri	10.0	3 54 50.35	+ 0.065	69 11 46.7	+ 10.42	51.91	1
2257	Hydri	9.0	55 4.72	− 0.095	70 8 46.6	.40	51.90	1
2258	Mensæ	10.0	20.29	1.815	76 49 35.6	.38	50.95	1
2259	Mensæ	8.9	20.52	3.828	80 35 5.1	.38	51.79	4
2260	Hydri	10.0	20.93	− 0.752	73 20 34.7	+ 10.38	50.88	1
2261	Mensæ	9.0	3 55 25.50	− 6.171	82 56 32.2	+ 10.37	51.48	2
2262	Reticuli	9.5	31.66	+ 0.473	66 14 0.1	.37	51.92	2
2263	Gould, Z. C., 3^h 1699	9.0	46.37	− 1.025	74 22 47.6	.35	51.46	2
2264	Gould, Z. C., 3^h 1710	9.5	55 59.72	0.932	74 1 41.7	.33	52.04	1
2265	Mensæ	10.0	56 1.16	− 3.325	79 50 28.0	+ 10.33	51.48	2

Number.	Constellation, Name of Star, or Synonym.	Magnitude.	Right Ascension, 1850.0.	Annual Precession.	South Declination, 1850.0.	Annual Precession.	Mean year.	No. of obs.
			h. m. s.	s.	° ′ ″	″		
2 266	Mensæ	10.0	3 56 6.19	− 5.607	82 28 30.7	+ 10.32	52.00	1
2 267	Hydri	9.5	10.14	− 0.074	69 58 27.8	.32	51.90	1
2 268	Hydri	10.0	10.30	+ 0.101	68 53 42.8	.32	51.91	1
2 269	Mensæ	10.0	17.25	− 1.475	75 50 27.6	.31	51.98	1
2 270	Mensæ	9.4	43.16	− 9.128	84 37 23.6	+ 10.28	51.77	4
2 271	Hydri	10.0	3 56 42.96	− 0.454	71 57 48.1	+ 10.28	50.93	2
2 272	Gould, Z. C., 3^h 1738 . .	9.8	54.39	0.633	72 46 16.4	.26	50.89	2
2 273	Gould, 4482	9.0	56 56.51	− 1.450	75 44 33.7	.26	51.98	1
2 274	Gould, Z. C., 3^h 1755 . .	9.0	57 3.77	+ 0.329	67 16 2.9	.25	51.97	1
2 275	Gould, Z. C., 3^h 1732 . .	9.5	10.75	− 1.871	76 54 26.0	+ 10.24	50.95	1
2 276	Hydri	9.5	3 57 23.50	− 0.064	69 51 16.2	+ 10.23	51.90	1
2 277	Hydri	10.0	26.64	− 0.985	74 10 7.8	.22	52.04	1
2 278	Lacaille, 1352	7.8	41.56	+ 0.449	66 18 17.5	.20	51.92	2
2 279	Gould, Z. C., 3^h 1778 . .	8.7	57 53.72	− 0.084	69 57 5.6	.19	51.90	2
2 280	Mensæ	9.4	58 1.46	− 9.209	84 38 25.2	+ 10.18	51.77	4
2 281	Gould, Z. C., 3^h 1771 . .	9.0	3 58 1.98	− 1.048	74 22 35.8	+ 10.18	51.46	2
2 282	Gould, Z. C., 3^h 1775 . .	9.0	10.52	− 0.943	73 59 3.8	.17	52.04	2
2 283	Lacaille, 1362	8.0	14.76	+ 0.105	68 46 0.7	.16	51.93	2
2 284	Gould, Z. C., 3^h 1783 . .	9.5	29.09	− 1.418	75 35 32.7	.14	51.98	1
2 285	Reticuli	10.0	32.91	+ 0.398	66 39 54.8	+ 10.14	51.90	1
2 286	Hydri	10.0	3 58 35.59	+ 0.096	68 48 25.6	+ 10.13	51.91	1
2 287	Gould, Z. C., 3^h 1769 . .	9.5	36.13	− 2.497	78 17 37.2	.13	51.47	2
2 288	Mensæ	10.0	36.53	− 1.408	75 33 21.1	.13	51.47	2
2 289	Reticuli	10.0	46.01	+ 0.346	67 2 51.9	.12	51.90	1
2 290	Hydri	10.0	58 57.24	− 0.554	72 20 4.3	+ 10.11	50.91	1
2 291	Gould, 4554	8.7	3 59 18.29	+ 0.401	66 35 47.8	+ 10.08	51.96	3
2 292	Octantis	9.0	23.31	−15.399	86 25 12.7	.07	51.03	3
2 293	Hydri	10.0	59 24.21	0.597	72 30 24.8	.07	50.90	1
2 294	Hydri	10.0	4 0 6.09	− 0.497	72 1 17.2	.02	50.91	1
2 295	Gould, Z. C., 4^h 9 .	9.8	6.52	+ 0.267	67 33 53.9	+ 10.02	52.00	2
2 296	Mensæ	9.5	4 0 23.61	− 1.701	76 21 15.1	+ 10.00	51.98	1
2 297	Hydri	10.0	28.31	0.411	71 35 14.6	9.99	50.95	1
2 298	Mensæ	9.5	32.22	3.355	79 46 15.9	.99	51.48	2
2 299	Octantis	8.2	44.87	39.745	88 26 52.2	.97	51.03	2
2 300	Hydri	10.0	49.16	− 0.367	71 21 23.4	+ 9.97	52.04	1
2 301	Stone, 1741	8.8	4 0 49.42	− 0.922	73 48 4.0	+ 9.97	52.04	2
2 302	Mensæ	10.0	1 23.55	2.619	78 27 38.7	.92	50.93	1
2 303	Lacaille, 1380	7.2	38.54	0.420	71 34 57.4	.90	51.49	3
2 304	Mensæ	10.0	40.76	1.318	75 9 53.2	.90	51.52	2
2 305	Lacaille, 1396	7.5	1 45.35	− 1.939	76 55 56.5	+ 9.89	50.95	1

Number.	Constellation, Name of Star, or Synonym.	Magnitude.	Right Ascension, 1850.0.	Annual Precession.	South Declination, 1850.0.	Annual Precession.	Mean year.	No. of obs.
			h. m. s.	s.	° ′ ″	″		
2306	Gould, Z. C., 4^h 20 . . .	9.5	4 2 0.43	− 3.690	80 14 7.0	+ 9.88	52.04	3
2307	Gould, 4594	9.0	22.13	2.894	78 57 10.4	.85	52.02	1
2308	Mensæ	10.0	23.85	2.076	77 15 0.4	.85	52.00	1
2309	Gould, Z. C., 4^h 79 . . .	10.0	27.82	0.244	70 38 29.0	.84	51.96	1
2310	Lacaille, 1383	7.5	33.85	− 0.022	69 21 26.4	+ 9.83	51.91	1
2311	Mensæ	10.0	4 2 35.76	− 3.208	79 29 12.3	+ 9.83	50.93	1
2312	Lacaille, 1471	8.4	52.16	9.101	84 31 38.6	.81	51.63	5
2313	Gould, Z. C., 4^h 95 . . .	9.2	57.08	0.256	70 40 52.8	.80	51.97	2
2314	Hydri	10.0	2 59.21	1.239	74 51 25.4	.80	50.97	1
2315	Mensæ	9.0	3 15.84	− 4.299	81 0 2.4	+ 9.78	51.44	4
2316	Mensæ	10.0	4 3 18.58	− 3.054	79 12 38.5	+ 9.78	52.00	1
2317	Gould, Z. C., 4^h 89 . . .	10.0	22.17	1.637	76 4 28.3	.77	52.02	1
2318	Octantis	9.1	25.94	10.542	85 5 49.0	.77	51.50	4
2319	Gould, Z. C., 4^h 106 . .	9.8	26.07	0.431	71 33 41.4	.77	51.50	2
2320	Gould, Z. C., 4^h 112 . .	9.8	30.66	− 0.430	71 33 13.3	+ 9.76	51.50	2
2321	Hydri	9.5	4 3 46.41	+ 0.240	67 34 27.8	+ 9.74	51.97	1
2322	Gould, Z. C., 4^h 129 . .	9.8	52.90	+ 0.402	66 21 13.9	.73	51.92	2
2323	Mensæ	10.0	3 55.26	− 3.243	79 30 37.6	.73	50.93	1
2324	Gould, Z. C., 4^h 143 . .	9.8	4 17.08	+ 0.420	66 11 28.6	.70	51.92	2
2325	Gould, Z. C., 4^h 127 . .	9.8	4 37.72	− 1.631	76 0 55.4	+ 9.68	52.00	2
2326	Octantis	9.2	4 5 5.74	−16.518	86 34 29.3	+ 9.64	51.02	2
2327	Gould, Z. C., 4^h 153 . .	9.2	8.99	0.495	71 48 10.2	.63	50.93	2
2328	Mensæ	9.8	21.25	3.940	80 30 4.4	.62	52.04	2
2329	Hydri	10.0	29.28	0.537	71 57 47.1	.61	50.93	2
2330	Gould, Z. C., 4^h 174 . .	9.5	45.88	− 0.120	69 48 2.3	+ 9.59	51.90	2
2331	Octantis	9.0	4 5 50.32	−78.581	89 10 33.2	+ 9.58	51.58	3
2332	Hydri	8.5	5 54.06	+ 0.149	68 6 12.8	.58	51.96	1
2333	Lacaille, 1405	7.5	6 28.08	− 0.208	70 16 12.8	.53	51.95	2
2334	Gould, Z. C., 4^h 200 . .	9.2	33.34	+ 0.222	67 34 1.9	.53	52.00	2
2335	Gould, 4689	8.5	36.25	+ 0.321	66 50 25.5	+ 9.52	52.06	1
2336	Gould, Z. C., 4^h 183 . .	9.3	4 6 40.52	− 1.499	75 33 43.8	+ 9.52	51.66	3
2337	Mensæ	10.0	41.06	1.631	75 57 0.2	.52	51.98	1
2338	Hydri	11.0	41.99	0.105	69 40 9.4	.52	51.90	1
2339	Gould, Z. C., 4^h 203 . .	10.0	47.24	0.104	69 39 30.4	.51	51.90	1
2340	Mensæ	9.5	52.24	− 7.242	83 29 14.4	+ 9.50	52.00	2
2341	Mensæ	10.0	4 6 54.23	− 2.447	77 57 20.6	+ 9.50	51.48	2
2342	Hydri	10.0	6 56.18	− 0.280	70 38 31.4	.50	51.96	1
2343	Lacaille, 1401	8.5	7 2.40	+ 0.429	65 58 19.5	.49	51.92	1
2344	Mensæ	9.8	4.76	− 6.197	82 45 5.2	.49	51.47	2
2345	Gould, 4610	9.0	24.02	−12.384	85 38 1.2	+ 9.46	51.02	5

Number.	Constellation, Name of Star, or Synonym.	Magnitude.	Right Ascension, 1850.0.	Annual Precession.	South Declination, 1850.0.	Annual Precession.	Mean year.	No. of obs.
			h. m. s.	s.	° ′ ″	″		
2 346	Hydri	11.0	4 7 50.03	0.075	69 26 43.1	+ 9.43	51.91	1
2 347	Gould, Z. C., 4^h 237 . .	10.0	51.12	− 0.075	69 26 40.0	.43	51.91	2
2 348	Hydri	9.2	51.53	+ 0.036	68 45 9.6	.43	51.93	2
2 349	Hydri	9.5	56.26	− 0.205	70 11 26.2	.42	51.90	1
2 350	Gould, Z. C., 4^h 228 . .	9.5	7 58.12	− 0.961	73 40 39.7	+ 9.42	52.04	1
2 351	Hydri	10.0	4 8 12.77	− 0.206	70 10 46.2	+ 9.40	51.90	1
2 352	Gould, Z. C., 4^h 240 . .	10.0	18.97	− 0.961	73 40 8.8	.39	52.04	1
2 353	Hydri	9.5	26.93	+ 0.008	68 54 17.6	.38	51.91	1
2 354	Melbourne (1), 207 . . .	8.5	27.97	− 0.021	69 5 34.7	.38	51.91	1
2 355	Gould, 4699	8.8	30.30	− 2.985	78 57 8.7	+ 9.38	52.02	2
2 356	Gould, 4739	8.5	4 8 42.89	+ 0.360	66 26 12.4	+ 9.36	51.92	1
2 357	Gould, Z. C., 4^h 221 . .	9.0	43.48	− 3.377	79 36 30.0	.36	51.99	2
2 358	Mensæ	9.2	46.11	4.461	81 4 31.7	.36	51.58	3
2 359	Mensæ	10.0	8 51.74	3.098	79 8 32.5	.35	52.02	1
2 360	Mensæ	9.2	9 4.54	− 7.868	83 49 30.6	+ 9.33	51.46	2
2 361	Mensæ	9.8	4 9 22.56	− 1.652	75 55 32.0	+ 9.31	52.00	2
2 362	Gould, 4751½	10.0	28.59	+ 0.182	67 42 47.8	.30	52.03	·
2 363	Gould, 4753½	9.5	33.16	+ 0.182	67 42 39.3	.30	52.03	1
2 364	Mensæ	8.2	35.28	− 7.239	83 26 39.8	.29	52.00	2
2 365	Hydri	10.0	37.74	− 0.496	71 37 39.8	+ 9.29	50.95	1
2 366	Lacaille, 1444	7.0	4 9 38.86	− 3.043	79 1 53.0	+ 9.29	52.01	2
2 367	Hydri	10.0	40.22	0.594	72 5 12.8	.29	50.91	1
2 368	Lacaille, 1421	9.0	52.26	0.551	72 1 57.0	.27	50.91	1
2 369	Gould, Z. C., 4^h 274 . .	9.0	9 57.63	− 2.660	78 18 40.6	.26	51.47	2
2 370	Reticuli	9.5	10 19.36	+ 0.333	66 34 11.2	+ 9.24	51.99	2
2 371	Reticuli	9.5	4 10 26.45	+ 0.432	65 47 14.5	+ 9.23	51.92	1
2 372	Mensæ	9.2	35.20	− 5.186	81 49 4.0	.22	51.47	2
2 373	Mensæ	10.0	37.98	9.506	84 36 20.0	.21	51.03	2
2 374	Mensæ	10.0	46.80	4.785	81 24 5.5	.20	50.95	1
2 375	Mensæ	10.0	54.40	− 1.604	75 44 22.5	+ 9.19	51.98	1
2 376	Hydri	10.0	4 10 56.24	− 0.843	73 6 47.7	9.19	51.07	1
2 377	Hydri	10.0	11 0.30	0.035	69 3 50.8	.18	51.91	1
2 378	Hydri	9.2	6.67	1.172	74 20 54.4	.17	51.46	2
2 379	Lacaille, 1592	7.1	8.90	−12.713	85 41 31.0	.17	51.02	6
2 380	Stone, 1809	8.5	19.55	+ 0.269	67 0 15.4	+ 9.16	52.02	2
2 381	Gould, Z. C., 4^h 370 . .	9.0	4 11 33.20	+ 0.135	67 56 17.3	+ 9.14	52.00	2
2 382	Hydri	9.5	34.01	− 0.532	71 43 21.4	.14	50.95	1
2 383	Gould, Z. C., 4^h 372 . .	9.0	40.87	+ 0.035	68 35 32.6	.13	51.93	2
2 384	Gould, Z. C., 4^h 354 . .	9.5	11 52.68	− 1.796	76 15 14.0	.11	52.00	2
2 385	Reticuli	9.5	12 0.55	+ 0.397	65 59 29.7	+ 9.10	51.92	1

Number.	Constellation, Name of Star, or Synonym.	Magnitude.	Right Ascension, 1850.0.	Annual Precession.	South Declination, 1850.0.	Annual Precession.	Mean year.	No. of obs.
			h. m. s.	s.	° ′ ″	″		
2 386	Gould, Z. C., 4h 374 .	9.5	4 12 0.99	− 0.611	72 4 25.9	+ 9.10	50.91	1
2 387	Mensæ	10.0	2.16	3.855	80 14 28.4	.10	52.05	2
2 388	Hydri	9.5	15.82	0.181	69 52 47.2	.09	51.90	1
2 389	Gould, Z. C., 4h 391	9.2	23.48	0.414	71 7 5.3	.07	52.00	2
2 390	Octantis	9.2	24.39	−26.420	87 41 4.6	+ 9.07	51.03	3
2 391	Mensæ	10.0	4 12 26.70	− 7.336	83 27 49.7	+ 9.07	52.01	1
2 392	Mensæ	10.0	28.17	4.983	81 34 43.6	.07	50.95	1
2 393	Hydri	10.0	32.46	0.496	71 30 48.3	.06	52.04	1
2 394	Hydri	10.0	50.30	0.349	70 46 1.1	.04	51.97	2
2 395	Gould, Z. C., 4h 400 . .	9.5	52.59	− 1.129	74 8 21.5	+ 9.04	52.04	1
2 396	Mensæ	11.0	4 12 59.44	− 5.000	81 35 11.1	+ 9.03	50.95	1
2 397	Mensæ	9.5	13 3.22	2.521	77 57 11.4	9.02	51.48	2
2 398	Hydri	10.0	26.58	− 0.783	72 46 23.7	8.99	51.06	1
2 399	Gould, Z. C., 4h 436 . .	9.2	30.16	+ 0.240	67 7 1.7	.99	52.02	2
2 400	Gould, Z. C., 4h 433 . .	9.0	31.52	− 0.125	69 29 47.4	+ 8.99	51.91	2
2 401	Hydri	10.0	4 13 31.90	− 0.402	71 0 44.7	+ 8.99	52.04	1
2 402	Mensæ	10.0	37.10	6.284	82 43 2.2	.98	50.95	1
2 403	Gould, Z. C., 4h 411 . .	9.2	42.56	− 2.104	76 59 23.6	.97	51.47	2
2 404	Gould, Z. C., 4h 442 . .	9.5	13 43.96	+ 0.187	67 29 2.0	.97	52.00	2
2 405	Lacaille, 1442	7.5	14 2.16	− 0.364	70 47 48.0	+ 8.95	51.97	2
2 406	Mensæ	10.0	4 14 5.04	− 2.167	77 7 54.8	+ 8.94	50.95	1
2 407	Hydri	10.0	25.87	0.766	72 40 4.4	.92	51.06	1
2 408	Mensæ	10.0	27.76	1.557	75 29 27.1	.91	51.66	3
2 409	Mensæ	10.0	28.54	3.840	80 10 6.1	.91	52.05	2
2 410	Gould, Z. C., 4h 464 . .	9.5	30.72	− 0.387	70 52 47.7	+ 8.91	51.96	1
2 411	Reticuli	9.5	4 14 34.46	+ 0.210	67 17 14.0	+ 8.90	51.97	1
2 412	Gould, Z. C., 4h 439 . .	9.0	41.72	− 2.303	77 25 50.2	.89	51.47	2
2 413	Gould, Z. C., 4h 479 . .	8.8	50.34	0.206	69 54 46.2	.88	51.63	3
2 414	Gould, Z. C., 4h 478 . .	10.0	51.78	0.402	70 57 46.0	.88	51.96	1
2 415	Mensæ	10.0	56.39	− 4.879	81 25 26.8	+ 8.88	50.97	2
2 416	Gould, Z. C., 4h 490 . .	9.0	4 14 58.79	+ 0.302	66 35 15.5	+ 8.87	52.02	3
2 417	Mensæ	10.0	15 0.71	− 7.071	83 15 24.9	.87	52.01	1
2 418	Hydri	10.0	1.55	− 0.561	71 41 30.7	.87	50.95	1
2 419	Reticuli	10.7	2.92	+ 0.303	66 34 37.3	.87	52.02	3
2 420	Mensæ	6.5	4.21	− 4.694	81 16 47.0	+ 8.87	50.99	1
2 421	Gould, Z. C., 4h 488 . .	9.2	4 15 6.10	− 0.063	69 6 15.7	+ 8.86	51.91	2
2 422	Lacaille, 1453	8.0	11.72	0.900	73 11 39.2	.86	51.07	1
2 423	Hydri	9.5	26.96	0.209	69 54 21.4	.84	51.90	1
2 424	Hydri	9.0	31.12	− 0.440	71 7 46.4	.83	52.00	2
2 425	Lacaille, 1445	8.7	39.69	+ 0.339	66 16 38.6	+ 8.82	51.97	3

Number.	Constellation, Name of Star, or Synonym.	Magnitude.	Right Ascension, 1850.0.	Annual Precession.	South Declination, 1850.0.	Annual Precession.	Mean year.	No. of obs.
			h. m. s.	s.	° ′ ″	″		
2426	Hydri	10.0	4 15 43.68	+ 0.104	67 58 12.8	+ 8.81	52.03	1
2427	Gould, Z. C., 4^h 520 . .	9.0	49.72	+ 0.374	65 59 39.6	.81	51.92	1
2428	Hydri	9.2	15 56.72	— 0.961	73 24 35.0	.80	51.55	2
2429	Hydri	9.5	16 1.80	— 0.664	72 9 56.9	.79	50.91	1
2430	Brisbane, 696	8.0	5.66	+ 0.234	67 2 46.6	+ 8.79	52.02	2
2431	Hydri	9.0	4 16 17.04	— 0.522	71 30 8.1	+ 8.77	51.69	3
2432	Gould, Z. C., 4^h 521 . .	9.5	26.37	1.077	73 49 49.5	.76	52.04	2
2433	Mensæ	10.0	26.76	4.617	81 6 31.1	.76	52.69	1
2434	Gould, 4887	8.8	34.50	— 0.088	69 9 4.9	.75	51.91	2
2435	Gould, 4893	9.5	46.17	+ 0.359	66 4 19.1	+ 8.73	51.92	1
2436	Gould, Z. C., 4^h 534 . .	8.5	4 16 49.49	— 0.821	72 48 57.7	+ 8.73	51.06	1
2437	Mensæ	10.0	17 0.79	6.594	82 53 52.0	.71	50.94	1
2438	Hydri	10.0	0.92	— 0.125	69 21 36.7	.71	51.91	1
2439	Hydri	10.0	1.77	+ 0.053	68 15 14.2	.71	51.95	1
2440	Gould, Z. C., 4^h 540 . .	9.2	6.40	— 1.132	74 0 44.2	+ 8.71	52.04	2
2441	Melbourne (1), 218 . . .	9.0	4 17 15.51	+ 0.080	68 3 55.2	+ 8.69	51.96	1
2442	Hydri	9.0	16.86	— 0.951	73 19 35.8	.69	51.07	1
2443	Lacaille, 1460	7.2	18.34	0.256	70 5 28.5	.69	51.49	2
2444	Hydri	9.2	25.13	0.528	71 29 11.2	.68	51.69	3
2445	Hydri	9.5	29.76	— 0.127	69 21 6.0	+ 8.67	51.91	1
2446	Hydri	10.0	4 17 35.23	— 0.115	69 16 48.9	+ 8.67	51.91	1
2447	Mensæ	10.0	37.42	— 1.749	75 57 25.0	.66	51.98	1
2448	Hydri	10.0	40.07	+ 0.115	67 48 43.2	.66	51.96	1
2449	Mensæ	10.0	42.44	— 4.027	80 21 28.6	.66	52.04	1
2450	Lacaille, 1530	8.5	44.09	— 4.637	81 6 24.7	+ 8.66	51.58	3
2451	Lacaille, 1514	9.0	4 17 51.10	— 3.369	79 23 28.8	+ 8.65	51.47	2
2452	Hydri	10.0	56.46	0.290	70 15 24.2	.64	51.07	1
2453	Mensæ	10.0	58.86	4.450	80 53 5.6	.64	51.87	2
2454	Octantis	9.2	17 59.30	17.376	86 37 21.0	.64	51.03	2
2455	Mensæ	10.0	18 2.31	— 1.672	75 43 39.9	+ 8.63	51.98	1
2456	Gould, Z. C., 4^h 533 . .	9.0	4 18 13.94	— 3.583	79 42 59.7	+ 8.62	50.93	1
2457	Mensæ	9.0	18.89	— 7.154	83 15 53.7	.61	52.01	1
2458	Gould, 4927	9.0	25.21	+ 0.317	66 19 24.8	.60	51.97	3
2459	Lacaille, 1481	8.5	30.45	— 0.723	72 21 34.6	.59	50.91	2
2460	Gould, Z. C., 4^h 603 . .	10.0	31.05	+ 0.134	67 38 37.6	+ 8.59	52.03	1
2461	Hydri	9.2	4 18 31.10	— 0.192	69 41 16.8	+ 8.59	51.99	2
2462	Lacaille, 1502	7.5	52.74	1.840	76 10 21.8	.56	52.00	2
2463	Hydri	9.3	18 59.35	0.094	69 5 33.9	.56	51.63	3
2464	Gould, Z. C., 4^h 618 . .	9.0	19 15.53	0.029	68 40 56.4	.54	51.93	2
2465	Hydri	10.0	15.95	— 0.658	72 1 32.9	+ 8.53	50.91	1

Number.	Constellation, Name of Star, or Synonym.	Magnitude.	Right Ascension, 1850.0.	Annual Precession.	South Declination, 1850.0.	Annual Precession.	Mean year.	No. of obs.
			h. m. s.	s.	° ′ ″	″		
2466	Gould, Z. C., 4h 612 . .	9.8	4 19 16.38	− 0.728	72 20 6.5	+ 8.53	50.91	2
2467	Mensæ	9.0	18.11	2.691	78 8 33.2	.53	52.02	1
2468	Mensæ	10.0	28.95	8.004	83 45 45.3	.52	51.98	1
2469	Hydri	9.0	39.22	0.266	70 3 26.5	.50	51.49	2
2470	Hydri	9.8	46.74	− 1.339	74 38 43.7	+ 8.49	51.21	4
2471	Mensæ	9.3	4 19 49.36	− 4.669	81 6 19.6	+ 8.49	51.18	3
2472	Mensæ	9.2	20 0.45	− 1.919	76 21 8.5	.48	52.00	2
2473	Reticuli	10.0	1.63	+ 0.276	66 33 55.8	.47	52.02	3
2474	Lacaille, 1487	8.5	6.98	− 0.166	69 28 25.8	.47	51.99	4
2475	Lacaille, 1482	9.0	8.46	+ 0.272	66 35 33.4	+ 8.47	52.02	3
2476	Gould, Z. C., 4h 648 . .	9.0	4 20 8.59	+ 0.006	68 25 36.5	+ 8.46	51.95	1
2477	Lacaille, 1486	8.5	9.87	− 0.095	69 3 13.5	.46	51.63	3
2478	Gould, Z. C., 4h 627 .	8.0	14.07	1.686	75 42 16.2	.46	51.98	1
2479	Brisbane, 711	8.0	26.34	− 0.216	69 45 2.8	.44	51.99	2
2480	Hydri	9.5	33.10	+ 0.049	68 8 9.6	+ 8.43	51.96	1
2481	Mensæ	10.0	4 20 37.31	− 3.695	79 50 5.7	+ 8.43	51.48	2
2482	Gould, Z. C., 4h 671 . .	9.8	39.06	+ 0.140	67 30 50.2	.42	52.00	2
2483	Gould, Z. C., 4h 676 . .	9.0	52.98	− 0.002	68 27 10.9	.41	51.95	1
2484	Hydri	9.8	54.26	− 0.102	69 4 3.1	.40	51.91	2
2485	Hydri	9.0	55.24	+ 0.010	68 22 15.5	+ 8.40	51.95	1
2486	Reticuli	10.0	4 20 59.18	+ 0.267	66 35 20.9	+ 8.40	52.02	3
2487	Hydri	10.0	20 59.92	− 1.104	73 47 6.0	.40	52.04	1
2488	Hydri	10.0	21 13.82	0.090	68 58 46.3	.38	51.91	1
2489	Hydri	8.5	15.76	− 0.417	70 47 56.7	.38	52.01	3
2490	Reticuli	10.0	20.14	+ 0.160	67 20 53.3	+ 8.37	51.97	1
2491	Mensæ	10.0	4 21 21.18	− 3.237	79 5 53.4	+ 8.37	52.00	1
2492	Reticuli	10.0	25.21	+ 0.162	67 19 52.9	.36	51.97	1
2493	Mensæ	10.0	35.02	− 5.003	81 26 16.4	.35	50.95	1
2494	Gould, 4987	10.0	43.52	+ 0.331	66 4 41.4	.34	51.92	1
2495	Mensæ	10.0	21 44.92	− 1.926	76 19 27 7	+ 8.34	51.98	1
2496	Brisbane, 715	8.8	4 22 0.34	− 0.264	69 57 49.2	+ 8.32	51.68	3
2497	Gould, Z. C., 4h 704 . .	10.0	0.74	0.752	72 21 1.2	.32	50.91	2
2498	Mensæ	10.0	3.53	2.814	78 19 8.9	.31	51.48	2
2499	Hydri	9.5	9.25	0.514	71 15 59.9	.30	52.08	1
2500	Mensæ	9.5	10.48	− 6.373	82 39 25.2	+ 8.30	51.47	2
2501	Gould, Z. C., 4h 724 . .	9.5	4 22 20.02	− 0.313	70 13 4.8	+ 8.29	51.53	2
2502	Gould, Z. C., 4h 742 . .	9.2	27.33	+ 0.311	66 12 0.2	.28	52.00	2
2503	Hydri	10.0	28.71	− 0.262	69 55 56.0	.28	51.68	3
2504	Reticuli	10.0	44.38	+ 0.162	67 16 43.1	.26	51.97	1
2505	Hydri	10.0	46.53	− 0.598	71 37 47.2	+ 8.26	50.95	1

Number.	Constellation, Name of Star, or Synonym.	Magnitude.	Right Ascension, 1850.0.	Annual Precession.	South Declination, 1850.0.	Annual Precession.	Mean year.	No. of obs.
			h. m. s.	s.	° ′ ″	″		
2506	Reticuli	10.0	4 22 51.19	+ 0.305	66 13 47.1	+ 8.25	52.07	1
2507	Reticuli	10.0	53.82	+ 0.145	67 23 38.4	.25	52.00	2
2508	Mensæ	9.2	54.83	— 4.601	80 58 26.1	.24	51.87	2
2509	Mensæ	9.8	22 55.40	— 2.854	78 22 39.0	.24	51.47	2
2510	Hydri	10.0	23 1.18	+ 0.019	68 13 29.2	+ 8.24	51.96	1
2511	Mensæ	8.5	4 23 6.50	— 4.955	81 21 43.0	+ 8.23	50.97	2
2512	Gould, Z. C., 4h 722 . .	10.2	7.92	2.439	77 31 23.2	.23	51.47	2
2513	Hydri	9.5	13.90	0.472	71 0 31.2	.22	52.08	1
2514	Hydri	9.9	16.29	— 1.377	74 40 2.7	.22	51.21	4
2515	Lacaille, 1511	7.8	23.40	+ 0.186	67 4 50.4	+ 8.21	52.02	2
2516	Mensæ	9.0	4 23 24.54	— 5.859	82 13 16.0	+ 8.20	51.99	2
2517	Gould, Z. C., 4h 744 . .	9.0	24.81	— 1.846	76 4 3.3	.20	52.02	1
2518	Reticuli	10.0	31.04	+ 0.159	67 16 18.1	.20	51.97	1
2519	Mensæ	10.0	46.35	— 5.519	81 54 39.3	.18	50.95	1
2520	Gould, Z. C., 4h 776 . .	10.0	48.93	— 0.708	72 5 46.0	+ 8.17	50.91	1
2521	Mensæ	10.0	4 23 51.06	— 2.592	77 50 2.4	+ 8.17	50.94	1
2522	Hydri	10.0	24 2.72	1.004	73 19 0.6	.15	51.07	1
2523	Gould, Z. C., 4h 787 . .	10.0	7.92	— 0.854	72 42 44.2	.15	50.98	2
2524	Gould, Z. C., 4h 814 . .	10.0	17.48	+ 0.340	65 53 56.3	.13	51.92	1
2525	Mensæ	10.0	20.99	— 2.615	77 52 13.4	+ 8.13	50.94	1
2526	Gould, Z. C., 4h 772 . .	9.0	4 24 24.84	— 2.399	77 24 18.1	+ 8.13	51.47	2
2527	Mensæ	9.5	36.63	— 1.462	74 54 21.2	.11	50.97	1
2528	Melbourne (1) 223 . . .	8.5	39.74	+ 0.306	66 8 50.9	.10	51.92	1
2529	Gould, Z. C., 4h 794 . .	9.0	39.89	— 1.742	75 44 32.6	.10	51.98	1
2530	Lacaille, 1524	9.0	24 53.06	— 0.173	69 19 54.7	+ 8.09	51.50	2
2531	Octantis	9.0	4 25 7.66	—32.640	88 2 12.2	+ 8.07	51.02	5
2532	Mensæ	10.0	10.18	2.541	77 41 43.4	.06	50.94	1
2533	Mensæ	9.2	18.63	1.636	75 25 16.0	.05	51.66	3
2534	Mensæ	9.8	20.32	7.651	83 29 5.4	.05	52.00	2
2535	Mensæ	10.0	23.29	— 4.873	81 14 12.3	+ 8.05	50.99	1
2536	Gould, Z. C., 4h 858 . .	9.0	4 25 29.46	+ 0.323	65 59 2.5	+ 8.04	51.92	1
2537	Gould, Z. C., 4h 864 . .	10.0	36.20	+ 0.331	65 54 52.7	.03	51.92	1
2538	Brisbane, 728	8.8	37.39	— 0.245	69 43 30.4	.03	51.99	2
2539	Hydri	9.2	41.12	+ 0.053	67 54 21.3	.02	52.00	2
2540	Gould, Z. C., 4h 831 . .	9.5	42.28	— 1.869	76 4 14.3	+ 8.02	52.02	1
2541	Mensæ	9.5	4 25 49.70	— 6.844	82 57 10.9	+ 8.01	50.95	1
2542	Mensæ	8.0	26 1.12	2.893	78 22 58.1	8.00	51.48	2
2543	Mensæ	10.0	6.23	2.089	76 37 54.2	7.99	50.95	1
2544	Mensæ	9.5	26.04	4.140	80 20 34.4	.96	52.05	2
2545	Octantis	9.5	26.16	—28.168	87 45 1.3	+ 7.96	51.03	3

Number.	Constellation, Name of Star, or Synonym.	Magnitude.	Right Ascension, 1850.0.	Annual Precession.	South Declination, 1850.0.	Annual Precession.	Mean year.	No. of obs.
			h. m. s.	s.	° ′ ″	″		
2 546	Lacaille, 1532	8. 2	4 26 30. 70	— 0. 361	70 19 36. 3	+ 7. 96	51. 53	2
2 547	Octantis	9. 8	26 31. 80	41. 696	88 25 46. 2	. 95	51. 04	2
2 548	Hydri	9. 5	27 5. 36	— 0. 626	71 37 12. 1	. 91	50. 95	1
2 549	Reticuli	10. 0	15. 45	+ 0. 229	66 37 17. 6	. 90	52. 06	1
2 550	Mensæ	10. 0	18. 74	— 6. 483	82 41 8. 8	+ 7. 89	51. 47	2
2 551	Brisbane, 736	9. 0	4 27 31. 85	+ 0. 304	66 2 38. 7	+ 7. 87	51. 92	1
2 552	Gould, 5122	10. 0	34. 73	+ 0. 304	66 2 23. 5	. 87	51. 92	1
2 553	Mensæ	9. 2	38. 76	— 7. 671	83 28 6. 2	. 87	52. 00	2
2 554	Gould, Z. C., 4h 940 . .	9. 0	41. 97	0. 024	68 19 43. 8	. 86	51. 95	2
2 555	Gould, Z. C., 4h 933 . .	9. 5	43. 18	— 0. 443	70 42 37. 1	+ 7. 86	52. 01	3
2 556	Hydri	10. 2	4 27 44. 26	— 1. 068	73 27 3. 8	+ 7. 86	51. 55	2
2 557	Gould, 5119	9. 7	27 51. 64	0. 519	71 4 58. 1	. 85	52. 04	3
2 558	Gould, Z. C., 4h 936 . .	9. 5	28 1. 55	1. 126	73 39 32. 0	. 83	52. 04	1
2 559	Mensæ	9. 5	3. 88	2. 064	76 31 24. 1	. 83	51. 98	1
2 560	Gould, Z. C., 4h 935 . .	9. 5	9. 33	— 1. 472	74 50 26. 0	+ 7. 82	51. 52	2
2 561	Gould, 5126	9. 0	4 28 12. 40	— 0. 326	70 4 58. 2	+ 7. 82	51. 07	1
2 562	Lacaille, 1575	7. 6	13. 05	3. 814	79 51 45. 8	. 82	51. 48	2
2 563	Hydri	10. 0	16. 26	0. 616	71 31 59. 3	. 81	50. 95	1
2 564	Mensæ	6. 2	16. 52	4. 334	80 33 34. 2	. 81	52. 37	2
2 565	Gould, Z. C., 4h 941 . .	8. 5	27. 98	— 1. 846	75 56 16. 0	+ 7. 80	52. 00	2
2 566	Gould, 5111	9. 8	4 28 45. 30	— 3. 858	79 54 55. 0	+ 7. 78	51. 48	2
2 567	Octantis	9. 0	52. 16	—25. 519	87 31 51. 9	. 77	51. 02	1
2 568	Gould, Z. C., 4h 986 . .	9. 0	28 58. 71	+ 0. 167	66 59 50. 9	. 76	52. 06	1
2 569	Mensæ	9. 5	29 0. 48	— 3. 641	79 35 24. 8	. 76	51. 47	2
2 570	Reticuli	9. 0	3. 50	+ 0. 262	66 18 23. 5	+ 7. 75	52. 07	1
2 571	Gould, Z. C., 4h 958 . .	8. 5	4 29 7. 11	— 1. 883	76 1 22. 4	+ 7. 75	52. 00	2
2 572	Reticuli	10. 0	13. 63	+ 0. 158	67 3 17. 4	. 75	52. 06	1
2 573	Lacaille, 1552	8. 0	13. 93	— 0. 953	72 57 35. 9	. 74	51. 06	1
2 574	Mensæ	10. 0	14. 29	2. 111	76 36 40. 0	. 74	51. 46	2
2 575	Mensæ	9. 5	22. 13	— 1. 599	75 12 3. 4	+ 7. 73	51. 70	3
2 576	Gould, Z. C., 4h 983 . .	10. 0	4 29 24. 93	— 0. 898	72 43 54. 0	+ 7. 72	50. 98	2
2 577	Reticuli	10. 0	25. 69	+ 0. 243	66 26 0. 4	. 72	52. 07	1
2 578	Gould, Z. C., 4h 990 . .	10. 0	27. 58	— 0. 800	72 19 18. 8	. 72	50. 91	2
2 579	Mensæ	10. 0	35. 15	2. 065	76 29 18. 2	. 71	51. 98	1
2 580	Gould, Z. C., 4h 989 . .	9. 4	45. 74	— 1. 546	75 1 48. 2	+ 7. 69	51. 51	4
2 581	Lacaille, 1548	8. 8	4 29 45. 77	— 0. 308	69 56 1. 9	+ 7. 69	51. 57	2
2 582	Mensæ	9. 0	54. 38	4. 275	80 27 26. 5	. 68	50. 06	1
2 583	Octantis	9. 5	29 57. 65	—27. 892	87 42 56. 3	. 68	51. 02	4
2 584	Gould, 5177	9. 0	30 3. 60	+ 0. 257	66 18 22. 0	. 67	52. 00	2
2 585	Lacaille, 1545	7. 5	4. 39	— 0. 019	68 12 52. 9	+ 7. 67	51. 95	2

No. 2 567. Reduced correctly; possibly same as 2 583.

No. 2 585. Proper motion in declination + 0.50″.

Number.	Constellation, Name of Star, or Synonym.	Magnitude.	Right Ascension, 1850.0.	Annual Precession.	South Declination, 1850.0.	Annual Precession.	Mean year.	No. of obs.
			h. m. s.	s.	° ′ ″	″		
2 586	Mensæ	10.0	4 30 10.82	— 2.085	76 31 27.5	+ 7.66	51.98	1
2 587	Mensæ	9.5	30 43.17	3.318	79 2 43.3	.62	52.01	1
2 588	Hydri	9.8	31 6.23	0.465	70 42 35.4	.59	52.01	3
2 589	Hydri	10.0	13.62	0.914	72 44 42.3	.58	51.06	1
2 590	Mensæ	10.0	16.04	— 8.302	83 47 14.7	+ 7.57	50.94	1
2 591	Hydri	10.0	4 31 18.32	— 0.282	69 44 14.4	+ 7.57	52.08	1
2 592	Lacaille, 1584	8.5	31.60	2.761	78 0 38.2	.55	51.48	2
2 593	Gould, 5184	8.5	34.38	— 2.761	78 0 33.6	.55	51.48	2
2 594	Reticuli	10.0	38.85	+ 0.304	65 53 14.2	.54	51.92	1
2 595	Hydri	10.0	52.31	— 0.344	70 3 19.1	+ 7.52	51.07	1
2 596	Mensæ	10.0	4 31 56.68	— 2.867	78 12 40.4	+ 7.52	50.93	1
2 597	Brisbane, 751	9.0	32 4.57	0.883	72 35 30.3	.51	50.90	1
2 598	Octantis	9.0	5.76	28.564	87 45 20.0	.51	51.02	5
2 599	Lacaille, 1662	9.0	6.44	8.627	83 57 0.2	.50	51.46	2
2 600	Gould, Z. C., 4h 1046 . .	9.0	6.53	— 2.238	76 51 25.0	+ 7.50	50.95	1
2 601	Hydri	10.0	4 32 10.27	— 1.443	74 38 30.0	+ 7.50	50.98	4
2 602	Mensæ	10.0	36.92	2.306	77 0 18.9	.56	50.95	1
2 603	Hydri	10.0	44.75	0.662	71 36 42.6	.45	50.95	1
2 604	Hydri	9.8	50.49	— 0.649	71 32 46.0	.44	51.02	2
2 605	Gould, 5238	9.2	32 57.14	+ 0.055	67 37 33.7	+ 7.44	52.00	2
2 606	Brisbane, 754	9.0	4 33 3.68	— 0.916	72 42 1.2	+ 7.43	51.48	2
2 607	Hydri	10.0	17.87	0.283	69 40 35.3	.41	52.07	1
2 608	Mensæ	9.8	18.41	2.194	76 43 19.9	.41	51.46	2
2 609	Mensæ	10.0	29.04	— 8.443	83 50 21.8	.39	51.46	2
2 610	Reticuli	9.5	32.63	+ 0.256	66 10 57.1	+ 7.39	52.07	1
2 611	Gould, Z. C., 4h 1136 . .	9.8	4 33 48.78	+ 0.071	67 29 20.7	+ 7.37	52.00	2
2 612	Gould, Z. C., 4h 1139 . .	9.0	51.12	+ 0.212	66 29 46.6	.36	52.07	1
2 613	Gould, Z. C., 4h 1114 . .	9.5	53.50	— 1.895	75 56 23.8	.36	52.01	3
2 614	Hydri	10.0	53.57	0.278	69 37 46.4	.36	52.07	1
2 615	Mensæ	10.0	33 55.04	— 2.829	78 5 51.1	+ 7.36	52.02	1
2 616	Mensæ	9.0	4 34 8.38	— 7.241	83 7 13.0	+ 7.34	51.48	2
2 617	Mensæ	10.0	8.95	— 6.441	82 32 52.3	.34	52.08	1
2 618	Gould, Z. C., 4h 1147 . .	9.8	12.54	+ 0.053	67 35 53.6	.33	52.00	2
2 619	Gould, Z. C., 4h 1143 . .	9.0	20.18	— 0.586	71 12 16.0	.32	52.08	2
2 620	Lacaille, 1639	6.5	28.69	— 5.690	81 54 45.4	+ 7.31	51.47	2
2 621	Mensæ	8.5	4 34 39.23	— 7.588	83 20 4.2	+ 7.30	52.01	1
2 622	Hydri	10.0	34 47.88	1.018	73 3 18.2	.29	51.06	2
2 623	Octantis	9.8	35 10.33	42.976	88 26 55.5	.26	51.05	1
2 624	Brisbane, 759	9.5	24.81	0.244	69 23 14.8	.24	51.58	2
2 625	Mensæ	10.0	25.60	— 1.111	73 24 1.0	+ 7.23	51.07	1

No. 2 623. Declination possibly 88° 26′ 35.5″.

Number.	Constellation, Name of Star, or Synonym.	Magnitude.	Right Ascension, 1850.0.	Annual Precession.	South Declination, 1850.0.	Annual Precession.	Mean year.	No. of obs.
			h. m. s.	s.	° ′ ″	″		
2626	Brisbane, 758	9.0	4 35 25.81	+ 0.104	67 12 39.3	+ 7.23	51.97	1
2627	Doradûs	9.0	32.37	— 0.100	68 31 54.5	.23	51.95	1
2628	Mensæ	10.0	44.10	1.829	75 42 54.8	.21	51.98	1
2629	Mensæ	10.0	46.98	2.943	78 16 46.0	.21	51.47	2
2630	Mensæ	8.5	56.92	— 5.925	82 6 5.6	+ 7.19	51.99	1
2631	Doradûs	9.0	4 35 58.47	— 0.104	68 32 28.0	+ 7.19	51.95	1
2632	Mensæ	9.5	36 7.30	2.914	78 13 8.6	.18	51.47	2
2633	Mensæ	10.0	17.67	— 1.274	73 58 0.1	.16	52.04	2
2634	Gould, 5293	9.2	20.64	+ 0.168	66 43 43.0	.16	52.07	2
2635	Gould, Z. C., 4h 1218 . .	10.0	27.91	+ 0.092	67 15 12.2	+ 7.15	51.97	1
2636	Gould, Z. C., 4h 1219 . .	9.5	4 36 29.46	+ 0.051	67 31 50.4	+ 7.15	52.00	2
2637	Mensæ	9.8	29.62	— 9.618	84 22 38.5	.15	51.50	2
2638	Gould, Z. C., 4h 1199 . .	8.8	36 33.65	1.934	75 59 4.2	.14	52.01	3
2639	Lacaille, 1608	6.8	37 0.97	1.153	73 30 43.4	.10	51.55	2
2640	Lacaille, 1606	7.5	1.05	- 1.027	73 2 9.7	+ 7.10	51.06	2
2641	Gould, Z. C., 4h 1248 .	9.0	4 37 4.75	+ 0.207	66 24 55.7	+ 7.10	52.07	1
2642	Mensæ	10.0	6.06	— 1.447	74 31 46.7	.10	50.92	1
2643	Mensæ	10.0	7.62	— 1.932	75 57 49.2	.10	52.00	2
2644	Gould, Z. C., 4h 1250 . .	9.2	8.82	+ 0.160	66 45 30.9	.09	52.07	2
2645	Mensæ	9.8	23.02	— 1.365	74 15 8.2	+ 7.07	51.48	2
2646	Mensæ	10.0	4 37 23.16	— 0.584	71 6 31.4	+ 7.07	52.08	1
2647	Gould, Z. C., 4h 1208 . .	9.5	23.95	3.105	78 32 55.4	.07	50.93	1
2648	Doradûs	10.0	30.67	0.026	68 0 18.3	.06	51.96	1
2649	Mensæ	10.0	35.61	2.851	78 4 8.6	.06	52.02	1
2650	Mensæ	10.0	36.12	— 4.301	80 21 58.4	+ 7.06	52.05	2
2651	Mensæ	9.5	4 37 56.63	— 0.838	72 14 31.0	+ 7.03	50.91	2
2652	Mensæ	9.8	38 7.16	1.387	74 18 32.2	.01	51.48	2
2653	Mensæ	9.0	10.28	5.723	81 53 34.8	.01	50.47	2
2654	Mensæ	10.0	17.98	3.269	78 49 18.7	.00	52.02	1
2655	Gould, Z. C., 4h 1275 . .	9.5	19.04	— 0.782	71 59 15.5	+ 7.00	50.91	1
2656	Mensæ	10.0	4 38 20.76	— 0.441	70 21 55.5	+ 7.00	51.99	1
2657	Mensæ	9.8	29.52	0.584	71 4 29.2	6.98	52.08	2
2658	Gould, Z. C., 4h 1258 . .	9.5	40.16	2.717	77 46 44.7	.97	50.94	1
2659	Lacaille, 1607	8.1	40.63	0.212	69 6 11.7	.97	51.36	3
2660	Lacaille, 1724	7.2	53.20	—10.760	84 49 8.6	+ 6.95	51.36	6
2661	Doradûs	10.0	4 38 55.26	+ 0.213	66 18 32.2	+ 6.95	52.00	2
2662	Brisbane, 773	9.0	39 1.02	+ 0.260	65 57 21.5	.94	51.92	1
2663	Stone, 2038	9.2	9.84	— 1.504	74 39 58.1	.93	51.32	3
2664	Lacaille, 1645	8.5	11.83	3.346	78 56 10.4	.93	52.02	1
2665	Lacaille, 1718	7.2	15.65	— 9.990	84 30 43.2	+ 6.92	51.59	2

Number.	Constellation, Name of Star, or Synonym.	Magnitude.	Right Ascension, 1850.0.	Annual Precession.	South Declination, 1850.0.	Annual Precession.	Mean year.	No. of obs.
			h. m. s.	s.	° ′ ″	″		
2 666	Mensæ	9.7	4 39 19.44	− 1.952	75 58 20.4	+ 6.92	52.01	3
2 667	Octantis	9.2	19.85	18.803	86 42 58.7	.91	51.03	4
2 668	Mensæ	10.0	22.76	0.801	72 2 25.8	.91	50.91	1
2 669	Mensæ	10.0	26.90	0.653	71 22 27.9	.91	52.08	1
2 670	Mensæ	10.0	31.79	− 0.386	70 2 48.4	+ 6.90	51.07	1
2 671	Brisbane, 778	8.5	4 39 34.43	− 0.330	69 44 37.2	+ 6.89	52.09	2
2 672	Gould, Z. C., 4h 1315	8.5	38.00	0.976	72 46 1.1	.89	50.98	2
2 673	Mensæ	10.0	42.08	3.773	79 36 9.7	.88	52.00	1
2 674	Mensæ	9.7	39 48.14	2.123	76 24 27.0	.88	52.00	2
2 675	Mensæ	10.0	40 12.47	− 1.255	73 48 9.9	+ 6.84	52.04	1
2 676	Gould, Z. C., 4h 1308	9.0	4 40 19.21	− 3.066	78 25 32.6	+ 6.83	50.93	1
2 677	Gould, Z. C., 4h 1369	9.0	23.05	+ 0.301	65 35 38.0	.83	51.10	1
2 678	Gould, Z. C., 4h 1347	7.4	27.13	− 1.352	74 8 4.8	.82	52.04	1
2 679	Mensæ	9.2	29.04	2.127	76 24 16.6	.82	52.00	2
2 680	Mensæ	10.0	32.49	− 1.049	72 39 32.9	+ 6.82	51.06	1
2 681	Lacaille, 1707	6.5	4 40 37.77	− 7.496	83 12 47.8	+ 6.81	52.01	1
2 682	Gould, Z. C., 4h 1320	9.2	45.48	3.205	78 40 0.4	.80	51.48	2
2 683	Mensæ	9.0	45.94	2.441	77 8 47.4	.80	51.47	2
2 684	Octantis	9.5	50.99	18.891	86 43 17.0	.79	51.02	1
2 685	Mensæ	9.8	40 53.43	− 2.440	77 8 32.4	+ 6.79	51.47	2
2 686	Gould, 5387	8.8	4 41 6.47	+ 0.272	65 47 50.0	+ 6.77	51.51	2
2 687	Gould, 5388	9.8	9.42	0.220	66 11 18.8	.76	52.00	2
2 688	Gould, Z. C., 4h 1405	8.2	21.22	+ 0.079	67 10 49.4	.75	52.02	2
2 689	Octantis	11.0	21.75	−11.629	85 6 16.6	.75	52.00	1
2 690	Gould, Z. C., 4h 1348	10.0	33.94	− 3.694	79 27 16.0	+ 6.73	51.96	2
2 691	Mensæ	10.0	4 41 40.20	− 3.064	78 23 50.4	+ 6.72	50.93	1
2 692	Mensæ	10.0	40.75	1.658	75 5 16.3	.72	51.33	3
2 693	Gould, Z. C., 4h 1345	10.0	42.07	4.198	80 10 5.1	.72	52.04	1
2 694	Brisbane, 787	8.2	47.34	0.757	71 47 12.9	.71	50.93	2
2 695	Mensæ	10.0	48.25	− 5.374	81 30 54.3	+ 6.71	50.98	2
2 696	Mensæ	10.0	4 41 58.12	− 2.681	77 38 36.3	+ 6.70	50.94	1
2 697	Brisbane, 786	8.5	41 59.70	0.192	68 53 13.6	.70	51.50	1
2 698	Mensæ	10.0	42 0.42	3.713	79 28 37.0	.69	51.46	2
2 699	Mensæ	10.0	0.88	0.618	71 8 19.6	.69	52.08	2
2 700	Mensæ	10.0	2.25	− 2.183	76 30 42.7	+ 6.69	51.98	1
2 701	Octantis	11.0	4 42 14.24	−11.604	85 5 22.0	+ 6.68	52.10	1
2 702	Mensæ	10.0	30.85	5.340	81 28 19.8	.65	50.97	2
2 703	Gould, Z. C., 4h 1408	9.0	34.45	2.468	77 10 16.9	.65	51.47	2
2 704	Mensæ	10.0	35.96	− 0.686	71 26 29.3	.65	50.95	1
2 705	Gould, Z. C., 4h 1448	9.0	40.52	+ 0.127	66 48 29.8	+ 6.64	52.06	1

No. 2 678. Differs 6s from Gould's Zones.

Number.	Constellation, Name of Star, or Synonym.	Magnitude.	Right Ascension, 1850.0.	Annual Precession.	South Declination, 1850.0.	Annual Precession.	Mean year.	No. of obs.
			h. m. s.	s.	° ′ ″	″		
2706	Mensæ	9.0	4 42 42.92	− 4.538	80 35 9.8	+ 6.64	52.06	1
2707	Octantis	10.0	45.27	11.868	85 11 11.4	.63	52.00	1
2708	Brisbane, 788	9.5	45.81	0.002	67 40 51.2	.63	52.03	1
2709	Mensæ	10.0	42 55.77	− 0.451	70 17 20.2	.62	51.07	1
2710	Doradûs	9.2	43 10.20	+ 0.150	66 38 1.2	+ 6.60	52.07	2
2711	Gould, Z. C., 4h 1460	9.0	4 43 14.76	+ 0.094	67 1 10.6	+ 6.59	52.02	2
2712	Brisbane, 792	9.1	16.50	+ 0.264	65 46 48.2	.59	51.55	4
2713	Mensæ	9.8	17.19	− 1.771	75 23 3.6	.59	51.68	3
2714	Mensæ	9.2	28.68	− 1.664	75 3 50.6	.57	51.51	4
2715	Gould, Z. C., 4h 1469	9.5	29.98	+ 0.085	67 4 26.4	+ 6.57	52.02	2
2716	Mensæ	9.5	4 43 46.44	− 2.193	76 30 5.5	+ 6.55	51.98	1
2717	Lacaille, 1839	7.9	43 46.56	18.141	86 35 25.8	.55	51.02	6
2718	Gould, Z. C., 4h 1456	10.0	44 1.98	2.568	77 21 51.8	.53	52.00	1
2719	Mensæ	10.0	7.49	0.793	71 52 52.4	.52	50.95	1
2720	Brisbane, 800	8.2	14.52	− 0.768	71 46 12.0	+ 6.51	50.93	2
2721	Mensæ	8.0	4 44 17.62	− 5.453	81 33 37.4	+ 6.51	50.97	2
2722	Mensæ	10.0	22.45	2.135	76 20 49.8	.50	51.98	1
2723	Lacaille, 1654	6.4	35.18	0.646	71 12 18.1	.48	52.08	2
2724	Gould, Z. C., 4h 1494	9.0	42.54	1.286	73 48 28.2	.47	52.04	2
2725	Mensæ	9.5	45.17	− 1.792	75 25 46.8	+ 6.47	51.68	3
2726	Mensæ	9.5	4 44 48.32	− 1.729	75 13 54.2	+ 6.46	51.51	2
2727	Doradûs	9.8	49.21	0.119	68 21 43.3	.46	51.97	3
2728	Mensæ	10.0	44 56.53	6.228	82 14 53.6	.45	52.08	1
2729	Lacaille, 1676	7.0	45 15.02	2.854	77 56 8.3	.43	51.48	2
2730	Brisbane, 805	8.0	19.72	− 0.110	68 17 31.3	+ 6.42	51.97	3
2731	Doradûs	9.5	4 45 21.55	− 0.172	68 39 56.4	+ 6.42	51.95	1
2732	Lacaille, 1647	7.3	28.88	+ 0.179	66 20 44.6	.41	52.03	3
2733	Doradûs	9.0	33.55	0.202	66 10 30.0	.40	52.07	1
2734	Doradûs	9.0	33.75	0.116	66 47 33.4	.40	52.06	1
2735	Doradûs	9.8	45 54.20	+ 0.263	65 42 39.5	+ 6.37	51.10	2
2736	Mensæ	10.0	4 46 20.28	− 1.233	73 34 51.4	+ 6.34	52.04	1
2737	Mensæ	10.0	22.83	− 1.141	73 16 35.3	.33	51.06	1
2738	Gould, Z. C., 4h 1602	9.0	36.78	+ 0.185	66 16 8.3	.31	52.03	3
2739	Doradûs	10.0	37.93	+ 0.178	66 19 9.0	.31	52.07	1
2740	Lacaille, 1703	8.5	43.05	− 3.941	79 44 24.7	+ 6.30	50.92	1
2741	Mensæ	10.0	4 46 51.95	− 0.671	71 15 50.9	+ 6.29	52.08	1
2742	Gould, Z. C., 4h 1606	9.5	56.65	0.133	68 23 5.8	.29	51.97	3
2743	Brisbane, 813	9.2	46 58.89	− 0.237	69 0 18.3	.28	51.08	2
2744	Doradûs	10.0	47 1.83	+ 0.168	66 22 50.0	.28	52.03	3
2745	Lacaille, 1660	7.3	4.64	− 0.204	68 48 35.3	+ 6.27	51.68	3

No. 2726. Possibly 75° 18′ 9.9″ one wire interval. Prec. = −1″.754.

Number.	Constellation, Name of Star, or Synonym.	Magnitude.	Right Ascension, 1850.0.	Annual Precession.	South Declination, 1850.0.	Annual Precession.	Mean year.	No. of obs.
			h. m. s.	s.	° ′ ″	″		
2 746	Mensæ	9.8	4 47 16.24	— 1.645	74 55 40.3	+ 6.26	51.54	2
2 747	Doradûs	10.0	19.41	+ 0.016	67 25 30.6	.25	51.97	1
2 748	Doradûs	10.0	24.05	— 0.043	67 48 25.3	.25	51.96	1
2 749	Doradûs	10.0	24.07	+ 0.008	67 28 28.2	.25	51.97	1
2 750	Mensæ	10.0	30.15	— 2.186	76 24 49.1	+ 6.24	51.98	1
2 751	Gould, Z. C., 4h 1599 . .	9.5	4 47 31.09	— 2.232	76 31 33.8	+ 6.24	51.98	1
2 752	Doradûs	10.0	38.21	+ 0.003	67 30 22.5	.23	52.00	2
2 753	Mensæ	9.5	47.02	— 1.108	73 5 2.0	.22	51.06	2
2 754	Mensæ	9.0	49.05	1.026	72 45 58.0	.21	51.06	1
2 755	Mensæ	10.0	47 50.85	— 5.519	81 34 55.8	+ 6.21	50.95	1
2 756	Doradûs	10.0	4 48 0.05	+ 0.013	67 25 30.0	+ 6.20	51.97	1
2 757	Gould, Z. C., 4h 1656 . .	9.2	1.61	+ 0.115	66 43 56.2	.20	52.07	2
2 758	Gould, Z. C., 4h 1593 . .	9.2	4.20	— 3.857	79 35 59.0	.19	51.47	2
2 759	Brisbane, 822	8.8	17.73	0.602	71 22 16.6	.17	51.52	2
2 760	Mensæ	9.8	19.50	— 3.521	79 4 39.4	+ 6.17	52.01	2
2 761	Gould, Z. C., 4h 1646 . .	9.2	4 48 24.68	— 1.543	74 35 11.8	+ 6.16	51.50	2
2 762	Brisbane, 820	9.2	32.04	0.247	69 1 23.7	.15	51.08	2
2 763	Mensæ	10.0	35.18	— 6.241	82 13 5.6	.15	52.08	1
2 764	Doradûs	10.0	40.31	+ 0.234	65 50 27.5	.14	51.55	4
2 765	Mensæ	9.8	41.12	— 2.830	77 49 43.8	+ 6.14	51.48	2
2 766	Gould, Z. C., 4h 1685 . .	9.0	4 48 49.83	+ 0.099	66 48 56.7	+ 6.13	52.06	1
2 767	Lacaille, 1672	7.8	49 6.38	— 0.071	67 58 3.6	.11	52.00	2
2 768	Mensæ	9.8	16.59	— 2.826	77 49 37.4	.09	51.48	2
2 769	Doradûs	10.0	21.49	+ 0.231	65 52 19.8	.08	51.55	4
2 770	Lacaille, 1702	7.8	28.24	— 2.258	76 34 18.9	+ 6.08	51.46	2
2 771	Mensæ	10.0	4 49 35.71	— 3.476	78 59 10.7	+ 6.07	52.00	1
2 772	Mensæ	10.0	47.12	1.023	72 42 23.8	.05	51.06	1
2 773	Mensæ	9.5	49.48	1.822	75 23 52.7	.05	51.68	3
2 774	Octantis	10.0	58.57	12.026	85 10 11.4	.03	51.36	3
2 775	Mensæ	10.0	58.67	— 0.701	71 19 35.1	+ 6.03	52.08	1
2 776	Mensæ	9.2	4 49 59.14	— 1.360	73 56 54.6	+ 6.03	52.04	2
2 777	Gould, Z. C., 4h 1702 . .	9.0	50 8.84	1.428	74 10 29.9	.02	51.48	2
2 778	Brisbane, 827	9.0	10.24	0.532	70 30 40.7	.02	51.99	1
2 779	Mensæ	10.0	16.82	0.691	71 16 25.2	.01	52.08	1
2 780	Mensæ	10.0	16.90	— 0.665	71 9 8.2	+ 6.01	52.08	1
2 781	Mensæ	10.0	4 50 32.34	— 3.169	78 26 32.8	+ 5.99	50.93	1
2 782	Mensæ	9.5	50 51.26	9.915	84 23 8.0	.96	51.50	2
2 783	Doradûs	10.0	51 3.03	0.422	69 55 22.1	.94	51.59	2
2 784	Mensæ	10.0	9.59	2.910	77 56 49.0	.93	50.94	1
2 785	Mensæ	10.0	9.64	- 9.389	84 8 52.1	+ 5.93	50.94	1

Number.	Constellation, Name of Star, or Synonym.	Magnitude.	Right Ascension, 1850.0.	Annual Precession.	South Declination, 1850.0.	Annual Precession.	Mean year.	No. of obs.
			h. m. s.	s.	° ′ ″	″		
2786	Brisbane, 834	8.5	4 51 10.28	− 2.274	76 33 50.8	+ 5.93	51.46	2
2787	Gould, Z. C., 4h 1735	9.0	19.00	1.652	74 51 59.9	.92	51.52	2
2788	Mensæ	9.5	22.40	5.018	81 1 42.5	.92	51.02	2
2789	Mensæ	10.0	22.86	2.856	77 50 14.7	.92	51.48	2
2790	Gould, Z. C., 4h 1725	8.8	28.97	− 3.087	78 16 45.8	+ 5.91	51.49	2
2791	Mensæ	9.5	4 51 32.57	− 4.954	80 57 12.6	+ 5.90	51.06	1
2792	Mensæ	10.0	36.83	1.809	75 19 42.6	.90	51.68	3
2793	Lacaille, 1687	7.0	40.26	− 0.143	68 19 6.5	.89	51.95	2
2794	Gould, Z. C., 4h 1782	9.8	46.20	+ 0.244	65 40 50.8	.89	51.10	2
2795	Octantis	8.6	48.16	−17.900	86 30 40.5	+ 5.88	51.03	6
2796	Lacaille, 1692	8.0	4 51 48.81	− 0.455	70 4 40.1	+ 5.88	51.07	1
2797	Gould, Z. C., 4h 1778	9.0	53.64	0.205	68 41 13.0	.87	51.53	2
2798	Mensæ	8.8	54.43	1.811	75 19 38.7	.87	51.68	3
2799	Gould, 5633	7.8	55.71	0.148	68 20 26.8	.87	51.95	2
2800	Mensæ	10.0	51 57.41	− 1.167	73 12 57.6	+ 5.87	51.07	1
2801	Mensæ	10.0	4 52 1.10	− 4.813	80 47 19.0	+ 5.86	51.06	1
2802	Mensæ	10.0	1.25	3.431	78 52 30.8	.86	52.02	1
2803	Mensæ	9.8	11.42	1.498	74 21 46.9	.85	51.48	2
2804	Gould, Z. C., 4h 1762	9.5	22.60	2.594	77 16 37.1	.83	52.00	1
2805	Mensæ	9.8	22.64	− 1.038	72 42 31.7	+ 5.83	50.98	2
2806	Gould, Z. C., 4h 1745	8.8	4 52 25.00	− 3.755	79 23 8.4	+ 5.83	51.47	2
2807	Gould, Z. C., 4h 1802	9.5	35.86	0.001	67 23 38.7	.82	52.00	2
2808	Mensæ	10.0	52 39.28	5.086	81 5 1.6	.81	51.02	2
2809	Mensæ	10.0	53 0.89	3.024	78 8 15.6	.78	52.02	1
2810	Mensæ	10.0	2.21	− 2.076	76 1 54.4	+ 5.78	52.02	1
2811	Lacaille, 1701	6.0	4 53 17.41	+ 0.067	66 54 53.6	+ 5.76	52.06	1
2812	Mensæ	9.8	17.75	− 1.033	72 40 8.8	.76	50.98	2
2813	Brisbane, 839	8.0	17.77	0.381	69 38 43.5	.76	52.10	1
2814	Mensæ	9.5	19.04	7.151	82 51 29.2	.75	50.95	1
2815	Doradûs	10.0	20.27	− 0.261	68 58 30.9	+ 5.75	51.08	1
2816	Brisbane, 838	10.0	4 53 21.64	− 0.507	70 18 25.7	+ 5.75	51.53	2
2817	Doradûs	10.0	37.20	+ 0.243	65 38 4.3	.73	51.10	2
2818	Mensæ	9.5	42.03	− 2.274	76 31 8.2	.72	51.98	1
2819	Mensæ	10.0	53 55.97	0.821	71 46 20.5	.70	50.95	1
2820	Doradûs	10.0	54 9.95	− 0.208	68 38 49.6	+ 5.68	51.95	1
2821	Gould, Z. C., 4h 1878	9.2	4 54 22.04	− 0.028	67 31 20.2	+ 5.67	52.00	2
2822	Mensæ	9.8	39.70	4.152	79 55 43.6	.64	51.70	3
2823	Gould, Z. C., 4h 1877	9.5	40.96	− 0.819	71 49 0.7	.64	50.93	2
2824	Doradûs	10.0	46.04	+ 0.024	67 10 2.1	.63	52.06	1
2825	Mensæ	8.0	53.01	− 1.190	73 14 21.6	+ 5.62	51.07	1

Number.	Constellation, Name of Star, or Synonym.	Magnitude.	Right Ascension, 1850.0.	Annual Precession.	South Declination, 1850.0.	Annual Precession.	Mean year.	No. of obs.
			h. m. s.	s.	° ′ ″	″		
2 826	Doradûs	9.5	4 54 56.81	− 0.192	68 31 57.2	+ 5.62	51.95	1
2 827	Mensæ	9.0	55 5.00	1.366	73 51 56.5	.61	52.02	2
2 828	Doradûs	9.8	10.51	0.013	67 24 9.2	.60	52.00	2
2 829	Mensæ	9.8	10.88	4.112	79 51 59.6	.60	51.69	3
2 830	Mensæ	9.5	14.93	−10.459	84 34 39.2	+ 5.59	52.00	1
2 831	Mensæ	9.5	5 55 18.62	− 1.394	73 57 24.6	+ 5.59	52.04	2
2 832	Doradûs	10.0	22.14	0.404	69 43 22.0	.58	52.10	1
2 833	Doradûs	10.0	27.08	0.257	68 54 8.6	.57	51.08	1
2 834	Gould, Z. C., 4^h 1903 . .	10.0	29.00	1.049	72 41 20.7	.57	51.06	1
2 835	Mensæ	10.0	31.60	− 0.579	70 36 55.2	+ 5.57	52.08	1
2 836	Lacaille, 1721	7.5	5 55 37.72	− 1.041	72 39 19.0	+ 5.56	50.98	2
2 837	Doradûs	10.0	55 53.22	0.036	67 32 4.7	.54	51.97	1
2 838	Lacaille, 1726	8.0	56 11.61	1.133	72 59 55.6	.51	51.06	2
2 839	Mensæ	10.0	21.06	− 0.933	72 11 55.4	.50	50.91	1
2 840	Lacaille, 1714	7.0	21.64	+ 0.283	65 14 41.4	+ 5.50	52.12	1
2 841	Gould, Z. C., 4^h 1966 . .	8.5	4 56 40.20	− 0.202	68 33 2.7	+ 5.47	52.03	2
2 842	Mensæ	9.5	42.40	4.259	80 2 48.4	.47	52.04	1
2 843	Gould, Z. C., 4^h 1942 . .	10.0	49.96	2.147	76 9 1.3	.46	52.02	1
2 844	Mensæ	9.8	53.70	7.639	83 8 59.0	.45	51.48	2
2 845	Mensæ	10.0	54.79	− 3.175	78 21 38.2	+ 5.45	50.93	1
2 846	Mensæ	10.0	4 56 58.16	− 0.884	71 58 57.7	+ 5.45	50.91	1
2 847	Mensæ	8.0	56 59.82	− 5.808	81 45 15.7	.44	50.95	1
2 848	Doradûs	9.8	57 2.42	+ 0.247	65 30 40.5	.44	51.10	2
2 849	Mensæ	10.0	13.29	− 3.705	79 14 41.9	.43	52.00	1
2 850	Doradûs	10.0	13.69	− 0.291	69 3 22.2	+ 5.42	51.08	1
2 851	Octantis	8.6	4 57 16.71	−11.882	85 4 47.7	+ 5.42	51.17	8
2 852	Mensæ	9.8	16.88	1.350	73 46 3.2	.42	52.04	2
2 853	Gould, 5720–21	7.5	23.86	4.958	80 53 25.4	.41	51.06	1
2 854	Doradûs	10.0	24.05	0.084	67 48 25.3	.41	51.96	1
2 855	Mensæ	9.5	30.36	− 0.784	71 32 0.5	+ 5.40	51.52	2
2 856	Lacaille, 1816	7.7	4 57 32.91	− 8.707	83 45 40.0	+ 5.40	51.31	3
2 857	Mensæ	10.0	32.91	− 1.819	75 14 58.8	.40	50.96	1
2 858	Doradûs	10.0	35.52	+ 0.129	66 22 7.8	.39	52.08	2
2 859	Lacaille, 1768	8.0	54.43	− 4.271	80 2 56.1	.37	52.04	1
2 860	Gould, Z. C., 4^h 2014 . .	9.5	57 57.37	+ 0.111	66 28 57.7	+ 5.36	52.07	1
2 861	Doradûs	9.8	4 58 5.72	− 0.297	69 4 0.6	+ 5.35	51.08	2
2 862	Mensæ	10.0	12.63	10.890	84 43 34.5	.34	52.00	1
2 863	Brisbane, 864	9.0	13.40	0.673	71 0 33.6	.34	52.08	2
2 864	Mensæ	9.5	27.84	0.326	69 13 28.5	.32	51.07	1
2 865	Lacaille, 1733	7.8	32.22	− 0.705	71 8 59.2	+ 5.31	52.08	2

No. 2 853. Gould observes this Star double.

Number.	Constellation, Name of Star, or Synonym.	Magnitude.	Right Ascension, 1850.0.	Annual Precession.	South Declination, 1850.0.	Annual Precession.	Mean year.	No. of obs.
			h. m. s.	s.	° ′ ″	″		
2866	Mensæ	9.0	4 58 36.83	−10.685	84 38 38.4	+ 5.31	52.03	2
2867	Doradûs	9.5	37.18	+ 0.117	66 25 23.3	.31	52.07	1
2868	Doradûs	10.0	38.42	0.191	65 53 16.2	.31	51.59	2
2869	Gould, Z. C., 4h 2041 . .	9.5	40.72	+ 0.041	66 56 58.6	.30	52.02	2
2870	Mensæ	10.0	41.65	− 7.190	82 50 15.9	+ 5.30	50.95	1
2871	Mensæ	9.5	4 58 43.52	− 2.638	77 16 27.0	+ 5.30	52.00	1
2872	Gould, Z. C., 4h 1983 . .	9.2	51.59	3.951	79 35 35.5	.29	51.69	3
2873	Gould, Z. C., 4h 2028 . .	9.0	52.96	1.423	73 59 26.2	.29	52.04	2
2874	Doradûs	9.5	58 57.86	0.015	67 19 24.6	.28	51.97	1
2875	Doradûs	9.5	59 13.54	− 0.342	69 17 41.7	+ 5.26	51.08	1
2876	Gould, Z. C., 4h 2062 . .	9.5	4 59 25.44	− 0.547	70 22 14.6	+ 5.25	51.99	1
2877	Brisbane, 1752	5.9	32.40	1.801	75 9 52.6	.23	51.01	4
2878	Gould, Z. C., 4h 2049 . .	9.5	4 59 51.30	− 2.257	76 22 40.1	.20	52.00	2
2879	Doradûs	9.8	5 0 6.76	+ 0.139	66 13 44.2	.18	52.08	2
2880	Octantis	8.6	17.47	−70.700	88 59 48.1	+ 5.17	51.03	4
2881	Mensæ	8.0	5 0 19.03	− 7.108	82 46 2.0	+ 5.16	51.52	2
2882	Mensæ	9.0	20.23	3.282	78 30 24.6	.16	50.93	1
2883	Mensæ	10.0	22.15	3.548	78 57 23.5	.16	52.02	1
2884	Mensæ	7.8	22.22	5.102	81 1 0.4	.16	51.02	2
2885	Lacaille, 1740	8.2	28.03	− 0.259	68 47 54.4	+ 5.15	51.71	3
2886	Gould, 5827	9.0	5 0 40.75	+ 0.200	65 46 3.9	+ 5.13	51.43	3
2887	Doradûs	9.5	0 42.21	− 0.314	69 6 25.2	.13	51.08	1
2888	Lacaille, 1782	8.5	1 2.36	3.212	78 22 20.0	.10	51.47	2
2889	Mensæ	9.8	6.04	1.690	74 48 20.0	.10	51.52	2
2890	Lacaille, 1784	8.2	11.16	− 3.359	78 37 43.2	+ 5.09	51.48	2
2891	Gould, Z C., 5h 23 . . .	9.8	5 1 14.62	− 0.991	72 20 27.4	+ 5.09	50.91	2
2892	Gould, Z. C., 5h 27 . .	10.0	17.70	0.700	71 4 24.4	.08	52.08	2
2893	Brisbane, 875	9.0	24.91	− 0.400	69 33 48.8	.07	52.10	1
2894	Doradûs	10.0	25.62	+ 0.193	65 47 58.2	.07	52.09	1
2895	Gould, Z. C., 5h 47 . .	9.5	29.42	− 0.076	67 39 39.4	+ 5.06	52.03	1
2896	Mensæ	9.0	5 1 29.63	− 8.011	83 20 30.3	+ 5.06	52.01	1
2897	Doradûs	9.5	31.36	0.163	68 12 4.9	.06	51.95	1
2898	Mensæ	9.5	32.62	1.624	74 35 47.0	.06	51.50	2
2899	Mensæ	10.0	38.47	− 0.616	70 39 53.6	.05	52.08	1
2900	Doradûs	10.0	39.21	+ 0.281	65 7 30.6	+ 5.05	52.12	1
2901	Gould, Z. C., 5h 60 . .	9.0	5 1 40.46	+ 0.282	65 6 52.9	+ 5.05	52.12	1
2902	Mensæ	8.8	43.45	− 8.199	83 27 0.4	.05	52.00	2
2903	Gould, Z. C., 5h 61 . .	8.3	1 55.39	0.171	68 14 22.8	.03	52.01	3
2904	Mensæ	10.0	2 4.80	4.534	80 20 24.4	.01	52.05	2
2905	Gould, Z. C., 5h 65 . .	9.5	9.62	− 0.669	70 54 41.5	+ 5.01	52.08	1

Number.	Constellation, Name of Star, or Synonym.	Magnitude.	Right Ascension, 1850.0.	Annual Precession.	South Declination, 1850.0.	Annual Precession.	Mean year.	No. of obs.
			h. m. s.	s.	° ′ ″	″		
2906	Lacaille, 1758	8.0	5 2 13.91	− 0.654	70 50 11.6	+ 5.00	52.08	1
2907	Gould, Z. C., 5ʰ 79 . .	9.0	17.24	+ 0.032	66 55 28.8	5.00	52.06	1
2908	Doradûs	9.5	34.44	− 0.337	69 11 45.4	4.97	51.08	1
2909	Lacaille, 1755	8.5	36.11	0.431	69 42 12.8	.97	52.10	1
2910	Mensæ	9.8	37.56	− 2.277	76 23 7.0	+ 4.97	52.00	2
2911	Doradûs	10.0	5 2 43.60	+ 0.081	66 34 42.5	+ 4.96	52.06	1
2912	Mensæ	9.5	52.02	− 6.334	82 9 23.9	.95	51.99	1
2913	Mensæ	10.0	2 56.40	5.443	81 20 49.4	.94	50.99	1
2914	Mensæ	8.7	3 0.54	1.936	75 29 28.4	.94	51.68	3
2915	Doradûs	8.5	0.63	− 0.144	68 2 59.2	+ 4.94	51.96	1
2916	Mensæ	10.0	5 3 0.73	− 0.540	70 15 36.6	+ 4.94	51.07	1
2917	Doradûs	9.0	2.99	+ 0.001	67 7 0.2	.93	52.02	2
2918	Doradûs	10.0	7.23	− 0.006	67 9 50.8	.93	52.06	1
2919	Gould, Z. C., 5ʰ 95 . .	9.5	10.14	− 1.081	72 40 2.1	.92	51.06	1
2920	Gould, 5892	8.5	13.68	+ 0.265	65 12 55.1	+ 4.92	52.12	1
2921	Doradûs	10.0	5 3 19.92	+ 0.048	66 47 43.1	+ 4.91	52.06	1
2922	Doradûs	9.5	23.71	+ 0.041	66 50 16.3	.90	52.06	1
2923	Doradûs	9.0	38.73	− 0.257	68 42 53.8	.88	51.08	1
2924	Mensæ	10.0	45.26	3.857	79 23 48.9	.87	52.10	1
2925	Gould, Z. C., 5ʰ 85 . .	9.5	3 54.43	− 3.509	78 50 54.4	+ 4.86	50.02	1
2926	Gould, 5913	9.0	5 4 9.36	+ 0.162	65 57 50.0	+ 4.84	51.43	3
2927	Mensæ	9.8	35.36	− 4.176	79 50 44.5	.80	51.69	3
2928	Gould, Z. C., 5ʰ 145 . .	9.0	39.36	0.935	72 2 45.0	.80	51.49	2
2929	β Mensæ	5.6	40.45	0.812	71 31 4.4	.79	51.52	2
2930	Mensæ	11.0	4 58.23	− 4.001	79 35 41.3	+ 4.77	52.00	1
2931	Mensæ	11.0	5 5 5.35	− 4.002	79 35 47.5	+ 4.76	52.00	1
2932	Lacaille, 1788	6.7	6.76	1.627	74 32 49.8	.76	51.32	3
2933	Gould, Z. C., 5ʰ 118 . .	10.0	17.87	3.836	79 20 57.5	.74	52.00	1
2934	Mensæ	10.0	21.74	4.698	80 30 23.0	.74	52.06	1
2935	Mensæ	10.0	26.62	− 0.838	71 36 57.0	+ 4.73	50.95	1
2936	Octantis	9.0	5 5 30.22	−30.873	87 48 25.4	+ 4.72	51.02	3
2937	Lacaille, 1777	8.5	32.20	0.191	68 17 1.3	.72	52.04	2
2938	Doradûs	8.8	38.70	− 0.331	69 5 42.3	.71	51.08	2
2939	Doradûs	9.5	39.69	+ 0.003	67 2 44.6	.71	52.02	2
2940	Mensæ	10.0	42.74	− 1.710	74 47 34.4	+ 4.71	50.97	1
2941	Gould, Z. C., 5ʰ 189 . .	10.0	5 5 46.37	− 0.677	70 52 44.6	+ 4.70	52.08	1
2942	Doradûs	9.5	47.42	0.380	69 22 0.2	.70	51.59	2
2943	Mensæ	9.5	48.86	2.054	75 46 7.3	.70	52.03	1
2944	Octantis	9.5	5 {53.90 63.90}	−12.197 −12.199	85 8 0.5	{.69 .68}	51.03	1
2945	Gould, 5958	9.0	6 0.84	+ 0.130	66 9 35.8	+ 4.68	52.08	2

No. 2944. 10ˢ discrepancy between wires.

Number.	Constellation, Name of Star, or Synonym.	Magnitude.	Right Ascension, 1850.0.	Annual Precession.	South Declination, 1850.0.	Annual Precession.	Mean year.	No. of obs.
			h. m. s.	s.	° ′ ″	″		
2946	Gould, Z. C., 5^h 199 . .	9.5	5 6 0.94	− 0.481	69 54 7.2	+ 4.68	51.07	2
2947	Doradûs	10.0	4.95	+ 0.151	66 0 17.4	.68	52.09	1
2948	Lacaille, 1776	8.2	5.44	+ 0.209	65 34 17.3	.67	51.44	3
2949	Mensæ	10.0	13.31	− 0.663	70 48 9.2	.66	52.08	1
2950	Octantis	8.8	14.78	−37.581	88 10 3.9	+ 4.66	51.03	3
2951	Doradûs	9.0	5 6 16.74	− 0.160	68 4 49.7	+ 4.66	51.96	1
2952	Mensæ	10.0	19.15	− 4.795	80 36 44.8	.65	51.06	1
2953	Doradûs	10.0	24.67	+ 0.186	65 44 13.3	.65	51.10	1
2954	Lacaille, 1812	8.2	25.63	− 3.262	78 23 41.4	.65	51.48	2
2955	Doradûs	10.0	25.81	+ 0.169	65 51 42.2	+ 4.65	52.09	1
2956	Mensæ	10.0	5 6 26.94	− 6.493	82 15 18.0	+ 4.64	52.04	2
2957	Brisbane, 895	9.0	32.45	0.250	68 36 44.1	.64	52.11	1
2958	Gould, Z. C., 5^h 213 . .	9.5	6 48.22	− 1.751	74 54 8.2	.61	51.52	2
2959	Doradûs	10.0	7 6.14	+ 0.180	65 45 50.3	.59	51.10	1
2960	Mensæ	9.8	26.57	− 4.936	80 45 55.2	+ 4.56	51.56	2
2961	Mensæ	9.8	5 7 28.00	− 0.825	71 31 27.1	+ 4.56	51.52	2
2962	Mensæ	9.5	36.14	1.087	72 36 49.4	.55	50.90	1
2963	Mensæ	10.0	38.65	1.109	72 41 52.5	.54	51.06	1
2964	Brisbane, 897	9.5	40.00	0.684	70 52 31.6	.54	52.08	1
2965	Mensæ	10.0	46.25	− 0.769	71 16 7.6	+ 4.53	52.08	1
2966	Brisbane, 898	10.0	5 7 48.25	− 0.667	70 47 31.5	+ 4.53	52.08	1
2967	Mensæ	10.0	8 0.64	1.010	72 17 49.4	.51	51.49	2
2968	Lacaille, 1814	7.5	1.82	2.492	76 49 16.2	.51	51.46	2
2969	Mensæ	10.0	7.22	0.565	70 17 21.0	.50	51.53	2
2970	Mensæ	10.0	21.69	− 1.834	75 7 19.2	+ 4.48	51.51	2
2971	Lacaille, 1829	7.0	5 8 21.83	− 3.335	78 30 4.3	+ 4.48	50.93	1
2972	Mensæ	10.0	39.34	− 4.020	79 35 0.5	.46	52.10	1
2973	Brisbane, 901	9.0	44.36	+ 0.235	65 18 56.0	.45	52.12	1
2974	Gould, Z. C., 5^h 298 . .	8.8	48.54	− 1.869	75 12 59.7	.44	51.70	3
2975	Gould, Z. C., 5^h 302 . .	8.6	51.22	− 1.833	75 6 48.7	+ 4.44	51.51	4
2976	Doradûs	9.5	5 8 55.32	+ 0.093	66 21 25.7	+ 4.43	52.09	1
2977	Gould, Z. C., 5^h 328 . .	9.5	8 57.02	− 0.054	67 21 16.8	.43	51.97	1
2978	Doradûs	10.0	9 1.88	0.266	68 39 42.6	.42	51.08	1
2979	Lacaille, 1804	8.0	5.54	0.816	71 27 18.9	.42	51.56	2
2980	Gould, Z. C., 5^h 335 . .	10.0	6.94	− 0.055	67 21 38.9	+ 4.42	51.97	1
2981	Lacaille, 1808	7.0	5 9 11.93	− 1.256	73 13 43.3	+ 4.41	51.07	1
2982	Mensæ	9.8	16.29	1.213	73 4 5.4	.40	51.06	2
2983	Mensæ	9.0	19.24	7.105	82 41 44.0	.40	51.52	2
2984	Gould, Z. C., 5^h 321 . .	9.0	25.14	1.415	73 47 18.6	.39	52.04	1
2985	Mensæ	10.0	25.53	− 1.388	73 41 34.6	+ 4.39	52.04	1

Number.	Constellation, Name of Star, or Synonym.	Magnitude.	Right Ascension, 1850.0.	Annual Precession.	South Declination, 1850.0.	Annual Precession.	Mean year.	No. of obs.
			h. m. s.	s.	° ′ ″	″		
2986	Brisbane, 903	9.0	5 9 25.87	+ 0.245	65 13 28.3	+ 4.39	52.12	1
2987	Doradûs	10.0	32.78	+ 0.178	65 43 42.9	.38	51.10	2
2988	Lacaille, 1835	8.0	37.90	− 3.275	78 22 51.8	.37	51.47	2
2989	Mensæ	10.0	43.48	− 4.886	80 41 11.0	.36	51.06	1
2990	Gould, 6035	7.7	9 54.10	+ 0.055	66 36 17.3	+ 4.35	52.07	2
2991	Mensæ	9.0	5 10 7.71	− 7.976	83 15 39.0	+ 4.33	52.01	1
2992	Mensæ	9.0	9.38	2.388	76 33 6.6	.33	51.98	1
2993	Doradûs	10.0	21.99	0.392	69 20 28.8	.31	51.08	1
2994	Brisbane, 912	8.2	36.60	0.497	69 53 50.0	.29	51.59	2
2995	Mensæ	9.2	47.12	− 6.138	81 55 37.2	+ 4.27	51.47	2
2996	Octantis	9.2	5 10 49.26	−12.559	85 13 20.5	+ 4.27	51.36	3
2997	Mensæ	10.0	10 59.66	− 2.550	76 54 57.7	.26	50.95	1
2998	Gould, Z. C., 5^h 398 . .	9.0	11 9.63	+ 0.013	66 51 53.2	.24	52.06	1
2999	Octantis	9.8	18.93	−26.632	87 28 49.9	.23	51.04	2
3000	Mensæ	9.5	19.26	− 1.540	74 10 36.2	+ 4.23	52.04	1
3001	Mensæ	10.0	5 11 23.94	− 0.667	70 43 53.5	+ 4.22	52.08	1
3002	Mensæ	9.2	24.33	− 7.822	83 9 30.0	.22	51.48	2
3003	Lacaille, 1807	7.2	26.89	+ 0.223	65 21 9.6	.22	51.61	2
3004	Gould, Z. C., 5^h 404 . .	9.0	30.40	− 0.424	69 29 49.1	.21	51.14	2
3005	Mensæ	10.0	40.51	− 0.647	70 37 49.2	+ 4.20	52.08	1
3006	Doradûs	10.0	5 11 54.30	+ 0.341	64 24 34.8	+ 4.18	51.11	1
3007	Mensæ	10.0	11 58.98	− 2.064	75 42 52.9	.17	52.03	1
3008	Gould, Z. C., 5^h 422 . .	10.0	12 9.07	1.080	72 30 41.2	.16	50.90	1
3009	Octantis	8.7	12.51	11.986	85 2 5.8	.15	51.69	3
3010	Doradûs	9.5	23.76	− 0.051	67 16 4.5	+ 4.14	51.97	1
3011	Brisbane, 720	8.0	5 12 35.60	− 0.469	69 43 6.2	+ 4.12	52.10	1
3012	Mensæ	10.0	38.91	3.360	78 30 2.0	.11	50.93	1
3013	Lacaille, 1831[1]	9.7	40.52	1.017	72 15 7.5	.11	51.37	3
3014	Lacaille, 1831[2]	9.7	41.89	1.017	72 15 7.3	.11	51.37	3
3015	Gould, Z. C., 5^h 456 . .	9.5	49.15	− 0.187	68 7 9.7	+ 4.10	51.96	1
3016	Mensæ	9.0	5 12 49.23	− 7.939	83 12 56.6	+ 4.10	52.01	1
3017	Lacaille, 1818	8.3	12 51.62	+ 0.212	65 24 23.4	.10	51.44	3
3018	Doradûs	10.0	13 1.89	− 0.197	68 10 37.8	.08	51.96	1
3019	Doradûs	9.5	3.09	+ 0.310	64 38 13.9	.08	51.11	1
3020	Mensæ	10.0	3.83	− 3.319	78 25 26.5	+ 4.08	50.93	1
3021	Mensæ	10.0	5 13 5.55	− 5.906	81 42 11.1	+ 4.08	50.95	1
3022	Gould, Z. C., 5^h 444 . .	8.7	9.48	1.823	75 1 30.2	.07	51.33	3
3023	Mensæ	10.0	10.52	0.638	70 33 43.1	.07	52.04	2
3024	Doradûs	9.8	14.04	0.155	67 54 49.4	.06	52.00	2
3025	Mensæ	9.0	17.60	− 2.408	76 33 42.6	+ 4.06	51.46	2

No. 2 999. Right Ascension possibly 10^s too large.

Number.	Constellation, Name of Star, or Synonym.	Magnitude.	Right Ascension, 1850.0.	Annual Precession.	South Declination, 1850.0.	Annual Precession.	Mean year.	No. of obs.
			h. m. s.	s.	° ′ ″	″		
3026	Octantis	8.0	5 13 24.56	−50.869	88 36 33.5	+ 4.05	51.03	4
3027	Doradûs	10.0	41.49	0.508	69 54 1.9	.03	51.07	1
3028	Lacaille, 1857	7.5	41.84	2.951	77 43 52.4	.02	50.94	1
3029	Gould, Z. C., 5^h 459	9.0	47.20	2.436	76 37 15.0	.02	51.46	2
3030	Mensæ	10.0	50.43	− 4.136	79 41 55.2	+ 4.01	52.10	1
3031	θ Doradûs	5.0	5 13 52.48	− 0.068	67 21 14.7	+ 4.00	51.97	1
3032	Doradûs	10.0	56.62	0.130	67 44 43.3	.00	52.03	1
3033	Octantis	8.2	13 59.02	23.638	87 11 29.8	4.00	51.02	2
3034	Brisbane, 926	9.0	14 4.62	0.202	68 11 15.8	3.99	52.04	2
3035	Mensæ	10.0	6.16	− 1.598	74 19 28.4	+ 3.98	52.04	1
3036	Mensæ	10.0	5 14 7.92	− 4.877	80 38 19.8	+ 3.98	51.06	1
3037	Gould, Z. C., 5^h 441	9.8	10.04	4.278	79 53 27.0	.98	52.07	2
3038	Gould, Z. C., 5^h 486	10.0	36.79	2.701	77 12 16.7	.95	52.00	1
3039	Mensæ	10.0	43.17	0.798	71 17 0.3	.94	52.08	1
3040	Mensæ	9.8	48.10	− 3.424	78 35 16.2	+ 3.93	51.48	2
3041	Brisbane, 927	9.0	5 14 54.14	− 0.786	71 13 32.1	+ 3.92	52.08	1
3042	Gould, Z. C., 5^h 503	9.0	54.98	2.448	76 38 11.2	.92	51.46	2
3043	Mensæ	9.0	56.76	5.205	80 59 52.2	.92	51.02	2
3044	Mensæ	9.0	14 56.95	5.774	81 34 4.3	.92	50.97	2
3045	Mensæ	10.0	15 8.51	− 1.427	73 44 53.3	+ 3.90	52.04	1
3046	Doradûs	10.0	5 15 40.06	+ 0.319	64 30 53.8	+ 3.86	51.11	1
3047	Mensæ	10.0	47.83	− 0.556	70 6 49.2	.84	51.07	1
3048	Gould, Z. C., 5^h 538	9.2	15 48.54	− 2.316	76 18 38.1	.84	52.00	2
3049	Doradûs	10.0	16 0.92	+ 0.021	66 43 27.8	.83	52.07	2
3050	Mensæ	9.5	6.82	− 1.369	73 32 4.3	+ 3.82	51.07	1
3051	Lacaille, 1921	6.5	5 16 6.84	− 7.120	82 39 40.0	+ 3.82	51.52	2
3052	Octantis	8.8	7.30	18.876	86 34 38.0	.82	51.02	3
3053	Brisbane, 936	9.0	7.74	− 0.215	68 13 40.4	.82	52.04	2
3054	Doradûs	10.0	12.72	+ 0.083	66 17 24.7	.81	52.09	1
3055	Gould, Z. C., 5^h 571	9.5	12.86	− 0.687	70 44 48.2	.81	52.04	2
3056	Gould, Z. C., 5^h 577	9.0	5 16 27.65	− 0.890	71 39 54.3	+ 3.79	50.95	1
3057	Gould, Z. C., 5^h 576	9.8	30.70	1.018	72 11 57.6	.78	51.60	2
3058	Brisbane, 939	9.0	46.08	− 0.227	68 17 28.2	.76	52.04	2
3059	Doradûs	10.0	47.22	+ 0.157	65 44 46.5	.76	51.10	1
3060	Mensæ	9.0	51.64	− 1.601	74 18 0.0	+ 3.75	51.48	2
3061	Mensæ	9.0	5 16 52.44	− 9.560	84 3 49.3	+ 3.75	50.94	1
3062	Gould, Z. C., 5^h 579	8.0	16 56.79	2.017	75 31 34.3	.75	51.68	3
3063	Mensæ	9.5	17 0.05	3.649	78 56 26.8	.74	52.02	1
3064	Mensæ	9.2	0.52	7.876	83 9 28.7	.74	51.48	2
3065	Brisbane, 940	9.5	2.57	− 0.131	67 42 2.1	+ 3.74	52.03	1

Number.	Constellation, Name of Star, or Synonym.	Magnitude.	Right Ascension, 1850.0.	Annual Precession.	South Declination, 1850.0.	Annual Precession.	Mean year.	No. of obs.
			h. m. s.	s.	° ′ ″	″		
3066	Mensæ	10.0	5 17 4.53	− 0.578	70 12 7.8	+ 3.73	51.53	2
3067	Mensæ	9.0	12.30	1.939	75 18{39.4 / 21.5}	.72	51.51	2
3068	Brisbane, 941	9.0	17.18	0.237	68 20 36.6	.72	52.04	2
3069	Lacaille, 1881	8.0	18.53	2.799	77 23 3.8	.71	51.47	2
3070	Mensæ	10.0	25.53	− 1.481	73 54 0.3	+ .70	52.04	1
3071	Mensæ	10.0	5 17 27.73	− 1.611	74 19 26.2	+ 3.70	51.48	2
3072	Mensæ	10.0	35.07	7.168	82 41 11.7	.69	50.95	1
3073	Gould, Z. C., 5h 619	8.8	38.89	0.663	70 36 51.0	.69	52.04	2
3074	Doradûs	9.5	42.38	− 0.229	68 17 18.8	.68	52.04	2
3075	Doradûs	9.2	49.26	+ 0.163	65 41 1.6	+ 3.67	51.10	2
3076	Octantis	9.8	5 17 57.55	−16.399	86 8 14.4	+ 3.66	51.04	2
3077	Doradûs	10.0	18 3.67	0.490	69 44 30.3	.65	52.10	1
3078	Brisbane, 947	9.5	3.86	0.287	68 37 32.7	.65	51.77	3
3079	Mensæ	10.0	8.85	1.480	73 53 16.6	.64	52.04	1
3080	Gould, Z. C., 5h 630	9.2	11.82	− 1.112	72 33 17.2	+ 3.64	51.01	2
3081	Lacaille, 1861	9.2	5 18 20.10	− 0.309	68 44 41.8	+ 3.63	51.77	3
3082	Gould, Z. C., 5h 642	9.7	22.64	1.036	72 14 47.2	.62	51.37	3
3083	Mensæ	9.2	22.82	4.345	79 56 44.0	.62	52.07	2
3084	Doradûs	10.0	34.18	0.116	67 34 44.8	.61	52.00	2
3085	Brisbane, 952	9.5	39.79	− 0.269	68 30 24.0	+ 3.60	52.11	1
3086	Brisbane, 953	9.2	5 18 44.26	− 0.285	68 37 1.5	+ 3.59	51.77	3
3087	Mensæ	10.0	45.70	4.093	79 35 48.1	.59	52.00	1
3088	Mensæ	10.0	46.67	− 0.625	70 24 46.3	.59	51.99	1
3089	Doradûs	10.0	47.69	+ 0.274	64 48 53.6	.59	52.13	1
3090	Lacaille, 1878	6.0	18 56.83	− 1.440	73 44 36.2	+ 3.57	52.04	1
3091	Doradûs	10.0	5 19 12.49	− 0.488	69 42 38.6	+ 3.55	52.10	1
3092	Doradûs	9.8	16.36	− 0.429	69 23 58.2	.55	51.59	2
3093	Gould, Z. C., 5h 700	9.5	17.75	+ 0.098	66 7 34.4	.54	52.09	1
3094	Gould, Z. C., 5h 706	9.2	24.69	0.152	65 44 20.1	.53	51.10	2
3095	Doradûs	10.0	26.98	+ 0.270	64 50 22.9	+ 3.53	52.13	1
3096	Gould, 6214	10.0	5 19 28.78	− 3.393	78 29 32.3	+ 3.53	50.93	1
3097	Gould, Z. C., 5h 682	10.0	35.30	1.458	73 47 47.7	.52	52.04	1
3098	Lacaille, 1880	8.0	37.57	0.904	71 40 51.0	.52	50.95	1
3099	Mensæ	11.0	42.03	− 1.896	75 9 29.9	.51	52.06	1
3100	Gould, Z. C., 5h 713	8.5	43.10	+ 0.031	66 35 20.0	+ 3.51	51.60	2
3101	Lacaille, 1869	8.0	5 19 45.84	+ 0.086	66 12 16.5	+ 3.50	52.09	1
3102	Gould, 6255	10.0	47.89	− 0.337	68 53 6.0	.50	51.08	1
3103	Doradûs	9.2	51.90	+ 0.007	66 45 15.2	.49	52.07	2
3104	Gould, 6260	10.0	54.91	− 0.310	68 43 40.4	.49	51.08	1
3105	Gould, 6259	10.0	19 55.20	− 0.342	68 54 32.5	+ 3.49	51.08	1

No. 3085. Original record 49.79s.

Number.	Constellation, Name of Star, or Synonym.	Magnitude.	Right Ascension, 1850.0.	Annual Precession.	South Declination, 1850.0.	Annual Precession.	Mean year.	No. of obs.
			h. m. s.	s.	° ′ ″	″		
3106	Lacaille, 1866	8.5	5 20 5.96	+ 0.345	64 13 10.4	+ 3.48	51.11	1
3107	Gould, Z. C., 5h 711 . .	9.0	13.13	− 1.196	72 51 4.4	.46	51.06	1
3108	Lacaille, 1871	8.0	14.26	+ 0.313	64 28 57.5	.46	51.11	1
3109	Mensæ	10.0	15.35	− 0.999	72 4 22.8	.46	52.08	1
3110	Mensæ	10.0	20.75	− 7.199	82 41 33.0	+ 3.45	51.52	2
3111	Gould, Z. C., 5h 668 . .	9.8	5 20 26.78	− 4.386	79 59 1.8	+ 3.44	52.07	2
3112	Mensæ	10.0	36.33	1.934	75 15 16.5	.43	51.51	2
3113	Mensæ	11.0	38.40	− 1.943	75 16 50.4	.43	52.06	1
3114	Doradûs	9.8	42.28	+ 0.121	65 56 34.9	.42	51.43	3
3115	Octantis	8.7	20 53.73	−14.536	85 43 13.4	+ 3.41	51.02	3
3116	Doradûs	10.0	5 21 11.51	− 1.921	75 12 45.8	+ 3.38	51.70	3
3117	Mensæ	10.0	19.12	− 1.014	72 7 17.7	.37	52.08	1
3118	Doradûs	10.0	21.14	+ 0.142	65 46 53.9	.37	51.10	1
3119	Doradûs	10.0	22.51	+ 0.127	65 53 18.8	.36	51.10	2
3120	Mensæ	10.0	25.28	− 3.334	78 22 21.7	+ 3.36	51.47	2
3121	Gould, Z. C., 5h 764 . .	9.0	5 21 25.64	− 0.211	68 7 30.9	+ 3.36	51.96	1
3122	Gould, Z. C., 5h 722 . .	9.5	33.51	3.802	79 8 38.4	.35	52.01	2
3123	Gould, Z. C., 5h 756 . .	10.0	35.76	1.197	72 50 21.3	.34	51.06	1
3124	Mensæ	9.2	43.26	2.700	77 7 56.4	.33	51.47	2
3125	Mensæ	10.0	52.43	− 1.917	75 11 46.0	+ 3.32	51.70	3
3126	Mensæ	10.0	5 21 55.13	− 0.965	71 54 43.2	+ 3.32	52.08	1
3127	Gould, Z. C., 5h 733 . .	9.0	55.32	3.858	79 13 34.6	.32	52.01	2
3128	Mensæ	10.0	55.53	3.388	78 27 46.1	.32	50.93	1
3129	Doradûs	10.0	56.48	0.436	69 23 56.9	.32	51.59	2
3130	Gould, Z. C., 5h 775 . .	9.5	56.59	− 0.732	70 52 51.2	+ 3.32	52.08	1
3131	Mensæ	10.0	5 21 57.48	− 2.513	76 42 55.0	+ 3.31	51.46	2
3132	Brisbane, 967	9.2	22 7.70	0.781	71 6 18.2	.30	52.08	2
3133	Gould, Z. C., 5h 780 . .	9.5	7.82	0.899	71 37 33.7	.30	50.95	1
3134	Mensæ	9.0	13.18	1.592	74 12 29.8	.29	51.48	2
3135	Lacaille, 1935	7.8	16.53	− 4.869	80 34 9.8	+ 3.29	51.56	2
3136	Mensæ	9.8	5 22 38.02	− 1.969	75 19 58.1	+ 3.26	51.51	2
3137	Mensæ	10.0	38.79	2.643	76 59 57.1	.25	50.95	1
3138	Gould, Z. C., 5h 811 . .	9.0	39.20	0.229	68 12 44.3	.25	52.04	2
3139	Mensæ	10.0	39.31	2.643	77 0 4.5	.25	52.00	1
3140	Lacaille, 1900	7.9	40.47	− 1.798	74 50 31.8	+ 3.25	51.52	2
3141	Gould, Z. C., 5h 797 . .	9.0	5 22 42.42	− 0.912	71 40 28.4	+ 3.25	50.95	1
3142	Doradûs	9.8	51.10	0.113	67 29 34.4	.24	52.00	2
3143	Mensæ	10.0	22 54.72	− 2.422	76 29 33.5	.23	51.98	1
3144	Doradûs	10.0	23 1.64	+ 0.029	66 33 23.1	.22	51.60	2
3145	Lacaille, 1943	8.5	2.39	− 4.684	80 20 35.5	+ 3.22	52.05	2

Number.	Constellation, Name of Star, or Synonym.	Magnitude.	Right Ascension, 1850.0.	Annual Precession.	South Declination, 1850.0.	Annual Precession.	Mean year.	No. of obs.
			h. m. s.	s.	° ′ ″	″		
3146	Mensæ	10.0	5 23 5.64	− 2.417	76 28 45.7	+ 3.22	51.98	1
3147	Mensæ	9.5	7.28	7.271	82 43 42.0	.21	51.52	2
3148	Mensæ	10.0	15.46	9.921	84 11 57.8	.20	51.50	2
3149	Mensæ	10.0	20.43	− 9.039	83 46 46.3	.19	51.98	1
3150	Doradûs	10.0	22.00	+ 0.202	65 18 3.7	+ 3.19	52.12	1
3151	Mensæ	9.2	5 23 23.30	− 0.700	70 42 33.0	+ 3.19	52.04	2
3152	Mensæ	11.0	23.63	1.940	75 14 40.4	.19	52.06	1
3153	Octantis	10.0	25.38	25.309	87 20 13.0	.19	51.04	2
3154	Doradûs	10.0	29.13	0.538	69 54 50.2	.18	51.07	1
3155	Doradûs	9.5	32.31	− 0.228	68 11 52.4	+ 3.18	52.04	2
3156	Gould, Z. C., 5h 833	9.3	5 23 38.88	− 1.048	72 13 52.2	+ 3.17	51.37	3
3157	Octantis	8.7	23 53.36	−13.956	85 33 55.4	.15	51.02	3
3158	Gould, 6377	9.0	24 3.47	+ 0.303	64 29 52.5	.13	51.11	1
3159	Mensæ	9.8	6.30	− 1.610	74 14 52.4	.13	51.48	2
3160	Mensæ	10.0	13.11	−10.703	84 31 22.6	+ 3.12	52.06	1
3161	Gould, Z. C., 5h 840	9.8	5 24 31.92	− 2.673	77 3 0.2	+ 3.09	51.47	2
3162	Lacaille, 1903	7.5	38.58	0.910	71 38 31.1	.08	51.48	2
3163	Doradûs	9.2	50.72	0.488	69 37 59.3	.06	51.62	2
3164	Mensæ	8.5	24 52.62	5.491	81 13 38.7	.06	50.99	1
3165	Gould, Z. C., 5h 898	9.0	25 12.55	− 0.204	68 1 30.8	+ 3.03	51.96	1
3166	Gould, 6398	9.2	5 25 14.67	+ 0.071	66 13 51.6	+ 3.03	51.61	2
3167	Doradûs	10.0	14.99	0.223	65 6 50.5	.03	52.12	1
3168	Doradûs	10.0	22.02	+ 0.253	64 52 24.3	.02	52.13	1
3169	Mensæ	9.8	22.17	− 1.738	74 38 6.2	.02	51.32	3
3170	Doradûs	10.0	23.42	− 0.124	67 31 42.7	+ 3.02	51.97	1
3171	Lacaille, 1910	8.2	5 25 28.52	− 1.113	72 28 16.8	+ 3.01	51.02	2
3172	Gould, 6400	8.0	33.83	+ 0.121	65 51 57.4	.00	51.43	3
3173	Gould, Z. C., 5h 907	8.5	34.63	− 0.221	68 7 38.8	.00	51.96	1
3174	Gould, Z. C., 5h 904	10.0	37.55	− 0.590	70 8 46.9	3.00	51.07	1
3175	Gould, Z. C., 5h 925	9.5	42.04	+ 0.313	64 23 25.1	+ 2.99	51.11	1
3176	Gould, Z. C., 5h 874	9.2	5 25 50.86	− 3.478	78 35 10.8	+ 2.98	51.48	2
3177	Lacaille, 1898	8.0	50.95	+ 0.306	64 26 50.3	.98	51.11	1
3178	Doradûs	9.5	56.01	− 0.184	67 53 48.7	.97	51.96	1
3179	Doradûs	9.2	25 59.96	+ 0.020	66 34 11.6	.96	51.59	2
3180	Gould, 6410	9.0	26 2.48	+ 0.271	64 43 29.1	+ 2.96	52.13	1
3181	Mensæ	9.8	5 26 10.50	− 2.142	75 46 1.2	+ 2.95	52.02	2
3182	Gould, Z. C., 5h 877	10.0	10.86	3.696	78 56 35.3	.95	52.02	1
3183	Lacaille, 1925	6.8	16.88	− 1.520	73 55 52.9	.94	52.04	2
3184	Doradûs	10.0	17.75	+ 0.310	64 24 52.7	.94	51.11	1
3185	Mensæ	9.8	45.17	− 3.348	78 21 17.9	+ 2.90	51.29	3

No. 3.179. Possibly 1′ less in declination.

Number.	Constellation, Name of Star, or Synonym.	Magnitude.	Right Ascension, 1850.0.	Annual Precession.	South Declination, 1850.0.	Annual Precession.	Mean year.	No. of obs.
			h. m. s.	s.	° ′ ″	″		
3 186	Lacaille, 1937	8.5	5 26 46.01	− 2.254	76 2 53.9	+ 2.90	52.02	1
3 187	Mensæ	9.5	52.30	7.493	82 51 43.4	.89	50.95	1
3 188	Doradûs	9.5	53.08	0.509	69 43 13.6	.89	51.62	2
3 189	Doradûs	9.8	26 53.45	− 0.194	67 56 38.6	.89	52.00	2
3 190	Doradûs	11.0	27 0.74	+ 0.274	64 41 23.3	+ 2.88	52.13	1
3 191	Lacaille, 1953	8.4	5 27 3.32	− 1.915	75 8 15.3	+ 2.87	51.01	4
3 192	Lacaille, 1989	6.0	3.80	5.988	81 41 25.4	.87	50.95	1
3 193	Mensæ	10.4	6.60	6.304	81 57 57.3	.87	50.95	1
3 194	Mensæ	10.0	21.26	5.848	81 33 35.1	.85	50.95	1
3 195	Gould, Z. C., 5ʰ 960 . .	9.9	31.68	− 1.914	75 7 50.2	+ 2.83	51.50	4
3 196	Mensæ	9.8	5 27 31.81	− 6.895	82 26 6.1	+ 2.83	52.04	2
3 197	Doradûs	10.0	38.13	0.368	68 57 1.2	.82	51.08	1
3 198	Doradûs	9.5	39.62	0.482	69 34 12.5	.82	51.14	1
3 199	Lacaille, 1920	7.0	47.60	0.331	68 44 23.2	.81	51.60	2
3 200	Octantis	9.5	51.36	−15.166	85 50 53.0	+ 2.80	51.02	2
3 201	Mensæ	10.0	5 27 53.63	− 0.793	71 5 22.5	+ 2.80	52.08	1
3 202	Mensæ	10.0	28 4.70	3.307	78 16 22.6	.78	51.48	2
3 203	Brisbane, 985	8.7	6.70	0.603	70 10 58.8	.78	51.53	2
3 204	Mensæ	9.5	7.15	− 2.177	75 50 29.6	.78	52.03	1
3 205	Doradûs	9.5	7.73	+ 0.004	66 39 2.2	+ 2.78	52.06	1
3 206	Octantis	9.2	5 28 10.88	−16.351	86 5 58.9	+ 2.78	51.03	2
3 207	Gould, Z. C., 5ʰ 1020 . .	10.0	16.58	0.159	67 42 43.1	.77	52.03	1
3 208	Gould, Z. C., 5ʰ 1018 . .	9.0	16.76	− 0.243	68 11 25.3	.77	52.04	2
3 209	Lacaille, 1917	7.2	18.78	+ 0.157	65 34 5.4	.76	51.44	3
3 210	Mensæ	10.0	23.17	− 3.192	78 3 43.2	+ 2.76	52.02	1
3 211	Mensæ	9.0	5 28 28.74	− 1.398	73 29 40.0	+ 2.75	51.13	1
3 212	Mensæ	10.0	30.67	0.744	70 51 19.9	.75	52.08	1
3 213	Mensæ	9.8	36.97	− 1.543	73 59 11.7	.74	52.04	2
3 214	Gould, Z. C., 5ʰ 1040 . .	9.2	37.07	+ 0.217	65 6 49.0	.74	52.13	2
3 215	Lacaille, 1950	8.0	39.86	− 1.789	74 45 19.5	+ 2.73	51.32	3
3 216	Mensæ	10.0	5 28 39.89	− 1.745	74 37 29.8	+ 2.73	52.07	1
3 217	Mensæ	9.5	46.41	3.192	78 3 30.6	.72	52.02	1
3 218	Mensæ	10.0	49.99	8.135	83 15 31.7	.72	52.01	1
3 219	Doradûs	9.5	52.89	0.205	67 59 13.9	.71	51.96	1
3 220	Mensæ	9.5	57.66	− 3.050	77 49 17.0	+ 2.71	51.48	2
3 221	Mensæ	10.0	5 28 58.80	− 1.150	72 35 39.8	+ 2.71	51.13	1
3 222	Gould, 6499	8.8	29 0.27	2.549	76 44 17.6	.70	51.46	2
3 223	Mensæ	9.2	5.76	1.331	73 15 5.4	.70	51.10	2
3 224	Brisbane, 989	8.5	8.12	0.019	66 47 58.4	.69	52.06	1
3 225	Mensæ	10.0	10.58	− 3.349	78 20 23.6	+ 2.69	51.47	2

Number.	Constellation, Name of Star, or Synonym.	Magnitude.	Right Ascension, 1850.0.	Annual Precession.	South Declination, 1850.0.	Annual Precession.	Mean year.	No. of obs.
			h. m. s.	s.	° ′ ″	″		
3 226	Gould, Z. C., 5^h 1065 . .	9.2	5 29 12.66	− 0.008	66 43 38.5	+ 2.69	51.59	2
3 227	Lacaille, 1931	8.5	20.22	+ 0.017	66 33 3.8	.68	51.12	1
3 228	Mensæ	10.0	20.25	5.497	81 12 31.5	.68	50.99	1
3 229	Lacaille, 1922	7.0	22.32	+ 0.350	64 2 23.0	.67	52.13	1
3 230	Mensæ	10.0	23.14	− 0.659	70 38 19.2	+ 2.67	52.08	1
3 231	Gould, Z. C., 5^h 1068 . .	9.0	5 29 35.48	− 0.851	71 19 49.1	+ 2.65	52.08	1
3 232	Gould, Z. C., 5^h 1044 . .	9.0	39.48	2.143	75 44 25.1	.65	52.03	1
3 233	Octantis	9.5	40.06	14.703	85 44 7.9	.65	51.02	3
3 234	Mensæ	9.5	45.69	7.010	82 30 38.3	.64	52.08	1
3 235	Gould, 6485	8.2	29 49.99	− 1.281	73 3 53.5	+ 2.63	51.09	3
3 236	Mensæ	10.0	5 30 3.30	− 1.782	74 43 25.0	+ 2.61	50.95	2
3 237	Mensæ	10.0	4.02	3.320	78 16 55 6	.61	51.49	2
3 238	Doradûs	10.0	4.90	− 0.088	67 14 16.3	.61	51.97	1
3 239	Doradûs	9.5	6.01	+ 0.355	63 59 23.6	.61	52.13	1
3 240	Gould, Z. C., 5^h 1090 . .	9.0	12.72	− 0.825	71 12 32.9	+ 2.60	52.08	1
3 241	Lacaille, 1963	7.0	5 30 16.65	− 1.487	73 46 57.6	+ 2.59	52.04	2
3 242	Doradûs	10.0	20.69	0.379	68 58 45.6	.59	51.08	1
3 243	Gould, Z. C., 5^h 1093 . .	8.5	21.26	− 0.783	71 1 6.2	.59	52.08	2
3 244	Doradûs	9.5	23.05	+ 0.270	64 40 39.4	.58	51.62	2
3 245	Doradûs	10.0	23.71	− 0.004	66 40 57.1	+ 2.58	52.06	1
3 246	Doradûs	9.0	5 30 24.62	− 0.185	67 50 49.6	+ 2.58	51.96	1
3 247	Lacaille, 2066	7.2	32.64	− 9.576	84 0 52.6	.57	51.46	2
3 248	Doradûs	9.2	35.36	+ 0.152	65 34 39.0	.57	51.10	2
3 249	Doradûs	9.5	38.56	0.329	64 12 1.4	.56	51.62	2
3 250	Brisbane, 993	9.0	40.88	+ 0.000	66 39 0.6	+ 2.56	51.59	2
3 251	Gould, 6499	7.2	5 30 43.82	− 2.168	75 47 45.4	+ 2.55	52.02	2
3 252	Mensæ	9.5	31 0.82	2.680	77 0 47.6	.53	51.47	2
3 253	Gould, 6517	8.0	1.25	0.766	70 56 2.6	.53	52.08	1
3 254	Octantis	9.5	2.34	25.579	87 21 19.4	.53	51.03	2
3 255	Gould, Z. C., 5^h 1074 . .	9.0	7.83	− 4.304	79 48 16.8	+ 2.52	52.07	2
3 256	Mensæ	10.0	5 31 15.09	− 0.647	70 22 6.6	+ 2.51	51.99	1
3 257	Mensæ	10.0	20.32	1.227	72 51 1.3	.50	51.06	1
3 258	Mensæ	8.0	23.66	5.394	81 5 37.4	.50	51.02	2
3 259	Doradûs	9.2	24.86	0.552	69 53 26.4	.49	51 10	2
3 260	Brisbane, 1001	8.5	26.80	− 0.590	70 4 58.2	+ 2.49	51.07	1
3 261	Gould, Z. C., 5^h 1151 . .	9.0	5 31 39.55	− 0.021	66 46 48.8	+ 2.47	52.06	1
3 262	Gould, Z. C., 5^h 1131 . .	9.0	40.08	1.231	72 51 41.4	.47	51.06	1
3 263	Mensæ	9.5	40.52	1.834	74 51 59.8	.47	51.52	2
3 264	Mensæ	8.5	31 54.68	8.239	83 18 30.3	.45	52.01	1
3 265	Mensæ	9.7	32 1.50	− 0.634	70 17 52.6	+ 2.44	51.53	2

Number.	Constellation, Name of Star, or Synonym.	Magnitude.	Right Ascension, 1850.0.	Annual Precession.	South Declination, 1850.0.	Annual Precession.	Mean year.	No. of obs.
			h. m. s.	s.	° ′ ″	″		
3 266	Mensæ	10.0	5 32 6.12	− 2.333	76 12 10.0	+ 2.44	51.98	1
3 267	Brisbane, 1004	9.0	8.71	− 0.768	70 56 2.5	.43	52.08	1
3 268	Lacaille, 1949	6.5	10.96	+ 0.310	64 19 37.3	.43	51.62	2
3 269	Gould, Z. C., 5^h 1172 . .	9.2	15.84	+ 0.101	65 55 37.1	.42	51.60	2
3 270	Mensæ	10.0	16.09	− 2.595	76 49 5.3	+ 2.42	50.95	1
3 271	Doradûs	9.8	5 32 26.48	− 0.200	67 54 56.4	+ 2.41	52.00	2
3 272	Doradûs	10.0	30.95	+ 0.121	65 46 55.5	.40	52.09	1
3 273	Mensæ	9.5	33.22	− 2.377	76 18 27.8	.40	52.00	2
3 274	Brisbane, 1009	8.0	38.20	− 1.249	72 55 13.2	.39	51.06	1
3 275	Mensæ	9.0	39.44	+ 0.208	65 7 36.3	+ 2.39	52.13	2
3 276	Brisbane, 1008[1]	12.0	5 32 39.63	− 0.789	71 1 26.1	+ 2.39	52.08	1
3 277	Mensæ	9.5	39.72	5.317	81 0 27.2	.39	51.02	2
3 278	Brisbane, 1008[2]	8.5	42.59	0.789	71 1 22.8	.38	52.08	2
3 279	Doradûs	10.0	42.73	0.085	67 11 28.8	.38	51.97	1
3 280	Mensæ	9.3	42.87	− 2.066	75 30 39.1	+ 2.38	51.68	3
3 281	Doradûs	9.0	5 32 43.75	+ 0.286	64 31 2.4	+ 2.38	51.11	1
3 282	Doradûs	10.0	54.00	0.020	66 29 12.1	.37	51.12	1
3 283	Gould, Z. C., 5^h 1200 . .	8.5	32 55.19	+ 0.089	66 2 27.8	.36	52.09	1
3 284	Doradûs	10.0	33 2.57	− 0.435	69 15 42.1	.35	51.08	1
3 285	Gould, Z. C., 5^h 1147 . .	9.0	3.24	− 3.430	78 27 26.9	+ 2.35	50.93	1
3 286	Mensæ	9.5	5 33 22.16	− 1.970	75 14 43.0	+ 2.33	51.51	2
3 287	Mensæ	8.8	25.58	1.432	73 34 25.6	.32	51.55	2
3 288	Doradûs	9.5	33.70	0.445	69 18 52.2	.31	51.08	1
3 289	Mensæ	10.0	39.78	3.630	78 47 29.1	.30	52.02	1
3 290	Gould, Z. C., 5^h 1215 . .	8.8	33 56.79	− 1.425	73 32 27.5	+ 2.28	51.41	3
3 291	Mensæ	9.8	5 34 4.44	− 3.090	77 49 57.2	+ 2.26	51.48	2
3 292	Doradûs	10.0	9.56	0.078	67 7 59.4	.26	52.02	2
3 293	Mensæ	10.0	14.08	1.807	74 45 55.2	.25	50.97	1
3 294	Gould, Z. C., 5^h 1179 . .	9.5	19.62	4.316	79 48 17.0	.24	52.07	2
3 295	Doradûs	10.0	22.52	− 0.126	67 26 23.6	+ 2.24	51.97	1
3 296	Gould, Z. C., 5^h 1245 . .	9.0	5 34 23.01	− 0.222	68 1 57.6	+ 2.24	51.96	1
3 297	Doradûs	10.0	24.06	0.112	67 20 41.5	.24	51.97	1
3 298	Doradûs	10.0	32.52	0.077	67 6 57.9	.22	52.02	2
3 299	Gould, Z. C., 5^h 1220 . .	9.8	42.28	2.916	77 29 9.8	.21	51.47	2
3 300	Mensæ	10.0	42.32	− 4.358	79 51 36.6	+ 2.21	51.07	2
3 301	Mensæ	11.0	5 34 43.62	− 4.391	79 54 9.2	+ 2.21	52.10	1
3 302	Doradûs	10.0	50.63	− 0.058	66 59 32.9	.20	52.06	1
3 303	Doradûs	10.0	34 57.15	+ 0.345	64 0 57.6	.19	52.13	1
3 304	Doradûs	10.0	35 3.29	− 0.235	68 6 13.0	.18	51.96	1
3 305	Doradûs	10.0	3.49	− 0.058	66 59 18.8	+ 2.18	52.06	1

Number.	Constellation, Name of Star, or Synonym.	Magnitude.	Right Ascension, 1850.0.	Annual Precession.	South Declination, 1850.0.	Annual Precession.	Mean year.	No. of obs.
			h. m. s.	s.	° ′ ″	″		
3306	Mensæ	9.5	5 35 17.82	— 1.482	73 43 35.3	+ 2.16	52.04	1
3307	Brisbane, 1016	8.8	18.37	0.268	68 17 49.4	.16	52.04	2
3308	Doradûs	10.0	21.07	— 0.148	67 34 1.6	.15	52.00	2
3309	Gould, 6647	10.0	23.80	+ 0.214	65 3 15.8	.15	52.13	2
3310	Brisbane, 1018	8.5	25.05	— 1.257	72 55 44.7	+ 2.15	51.06	1
3311	Doradûs	11.0	5 35 37.62	+ 0.239	64 51 24.0	+ 2.13	52.13	1
3312	Lacaille, 2050	7.0	38.46	— 5.469	81 9 6.6	.13	51.02	2
3313	Doradûs	10.0	41.88	+ 0.354	63 55 45.4	.12	52.13	1
3314	Mensæ	8.8	45.63	—10.524	84 24 52.6	.12	51.34	3
3315	Doradûs	10.0	47.97	+ 0.238	64 51 53.0	+ 2.11	52.13	1
3316	Doradûs	9.0	5 35 56.86	— 0.496	69 33 40.9	+ 2.10	51.11	2
3317	Gould, 6630	8.8	35 58.04	— 2.403	76 20 50.4	.10	52.00	2
3318	Gould, 6659	10.0	36 4.12	+ 0.211	65 4 20.1	.09	52.13	2
3319	Gould, Z. C., 5ʰ 1323 . .	9.5	8.47	— 0.213	67 57 30.1	.08	51.96	1
3320	Mensæ	10.0	15.08	— 0.702	70 35 10.8	+ 2.07	52.08	1
3321	Gould, Z. C., 5ʰ 1334 . .	9.2	5 36 16.40	+ 0.031	66 22 38.0	+ 2.07	51.61	2
3322	Doradûs	9.5	18.58	— 0.331	68 39 11.6	.07	51.60	2
3323	Octantis	9.0	19.75	14.343	85 37 58.0	.07	51.03	5
3324	Gould, Z. C., 5ʰ 1325 . .	8.7	20.86	0.634	70 15 33.0	.07	51.53	2
3325	Mensæ	9.8	30.22	— 1.104	72 20 13.6	+ 2.05	51.60	2
3326	Gould, Z. C., 5ʰ 1267 . .	9.0	5 36 35.44	— 4.372	79 52 8.1	+ 2.04	52.07	2
3327	Mensæ	9.2	37.36	0.799	71 2 4.8	.03	52.08	2
3328	Mensæ	10.0	53.63	0.811	71 5 22.5	.02	52.08	1
3329	Lacaille, 1985	6.5	36 55.48	0.009	66 38 40.0	.02	51.59	2
3330	Mensæ	9.8	37 0.40	— 1.642	74 14 52.8	+ 2.01	51.48	2
3331	Octantis	8.8	5 37 1.48	—12.743	85 11 28.9	+ 2.01	51.27	5
3332	Doradûs	9.2	4.26	0.500	69 34 27.0	2.00	51.11	2
3333	Mensæ	9.5	13.82	2.370	76 15 42.8	1.99	51.98	1
3334	Mensæ	10.0	14.69	0.704	70 35 27.3	.99	52.08	1
3335	Doradûs	10.0	15.72	— 0.484	69 29 16.3	+ 1.99	51.14	1
3336	Gould, Z. C., 5ʰ 1352 . .	8.5	5 37 16.02	— 0.983	71 50 3.9	+ 1.99	51.51	2
3337	Brisbane, 1025	7.4	19.05	0.838	71 13 27.4	.98	52.08	1
3338	Octantis	9.8	21.99	—14.532	85 40 29.0	.98	51.04	2
3339	Gould, Z. C., 5ʰ 1386 . .	9.8	28.16	+ 0.220	64 59 23.6	.97	52.13	2
3340	Gould, Z. C., 5ʰ 1375 . .	8.5	41.77	— 0.967	71 45 47.3	+ 1.95	50.95	1
3341	Doradûs	10.0	5 37 45.85	— 0.370	68 51 35.2	+ 1.94	51.08	1
3342	γ Mensæ	5.0	51.04	2.448	76 26 41.1	.94	51.98	1
3343	Lacaille, 2022	8.5	57.46	— 2.011	75 19 37.2	.93	51.51	2
3344	Doradûs	9.7	37 59.34	+ 0.322	64 10 18.0	.92	51.62	2
3345	Gould, Z. C., 5ʰ 1378 . .	9.8	38 11.28	— 1.991	75 16 16.6	+ 1.91	51.51	2

Number.	Constellation, Name of Star, or Synonym.	Magnitude.	Right Ascension, 1850.0.	Annual Precession.	South Declination, 1850.0.	Annual Precession.	Mean year.	No. of obs.
			h. m. s.	s.	° ′ ″	″		
3346	Doradûs	10.0	5 38 12.10	+ 0.150	65 30 36.9	+ 1.90	51.10	2
3347	Doradûs	9.8	17.92	− 0.361	68 48 31.8	.90	51.60	2
3348	Mensæ	10.0	18.77	2.722	77 3 38.9	.90	50.95	1
3349	Mensæ	10.0	19.20	3.481	78 31 15.1	.89	50.93	1
3350	Mensæ	10.0	23.24	− 1.351	73 15 56.2	+ 1.89	51.07	1
3351	Lacaille, 2016	5.2	5 38 29.90	− 1.519	73 49 41.2	+ 1.88	52.04	2
3352	Doradûs	9.0	39.85	− 0.106	67 16 3.3	.86	51.97	1
3353	Gould, Z. C., 5h 1434	9.5	41.72	+ 0.234	64 52 10.0	.86	52.10	2
3354	Gould, 6703	10.0	42.13	− 1.280	72 59 24.8	.86	51.06	1
3355	Mensæ	10.0	43.28	− 8.340	83 20 46.3	+ 1.86	52.01	1
3356	Brisbane, 1034	9.5	5 38 44.76	− 1.281	72 59 35.8	+ 1.86	51.06	1
3357	Mensæ	11.0	45.47	− 6.828	82 20 26.3	.86	52.08	1
3358	Gould, 6728	8.5	56.36	+ 0.458	63 0 40.2	.84	51.16	2
3359	Mensæ	9.0	38 56.86	− 5.280	80 56 31.2	.84	51.06	1
3360	Doradûs	10.0	39 12.65	− 0.430	69 10 55.4	+ 1.82	51.08	1
3361	Gould, Z. C., 5h 1415	9.7	5 39 12.70	− 1.962	75 10 56.8	+ 1.82	51.70	3
3362	Doradûs	9.7	16.69	+ 0.154	65 28 10.9	.81	51.44	3
3363	Gould, Z. C., 5h 1437	10.0	17.44	− 0.744	70 45 46.1	.81	52.04	2
3364	Mensæ	10.0	17.62	1.301	72 52 32.4	.81	51.06	1
3365	Gould, 6723	9.5	18.41	− 0.688	70 29 50.8	+ 1.81	51.99	1
3366	Brisbane, 1037	8.5	5 39 24.13	− 0.595	70 2 29.3	+ 1.80	51.07	1
3367	Mensæ	10.0	24.40	5.949	81 36 8.4	.80	50.95	1
3368	Mensæ	9.5	27.65	1.399	73 24 41.6	.80	51.59	2
3369	Mensæ	9.0	28.60	1.426	73 30 26.0	.79	51.59	2
3370	Mensæ	10.0	33.17	− 2.630	76 51 12.7	+ 1.79	50.94	1
3371	Mensæ	10.5	5 39 40.02	− 7.822	83 1 47.9	+ 1.78	50.95	1
3372	Brisbane, 1039	8.2	43.66	1.301	73 3 43.1	.77	51.10	2
3373	Lacaille, 2007	Neb.	45.00	0.430	69 10 43.9	77	51.08	1
3374	Gould, 6746	8.7	46.84	0.083	67 6 55.0	.77	52.04	3
3375	Gould, Z. C., 5h 1449	9.0	52.40	− 1.676	74 19 53.2	+ 1.76	51.48	2
3376	Doradûs	9.5	5 39 58.04	+ 0.224	64 55 53.5	+ 1.75	52.10	2
3377	Gould, Z. C., 5h 1490	9.8	40 4.87	− 0.066	66 59 45.3	.74	52.04	3
3378	Doradûs	9.0	11.85	0.552	69 48 53.3	.73	51.14	1
3379	Gould, Z. C., 5h 1503	9.8	13.77	0.063	66 58 29.9	.73	52.04	3
3380	Mensæ	19.0	21.36	− 3.405	78 22 37.2	+ 1.72	51.47	2
3381	Gould, Z. C., 5h 1472	9.8	5 40 27.18	− 2.056	75 26 14.7	+ 1.71	51.50	2
3382	Gould, Z. C., 5h 1506	10.0	36.42	0.877	71 21 21.6	.70	52.08	1
3383	Mensæ	9.0	38.00	0.963	71 43 28.5	.69	50.95	1
3384	Lacaille, 2039	6.2	40.64	− 1.506	73 46 17.9	.69	52.04	2
3385	Lacaille, 2009	7.5	47.88	+ 0.459	62 58 53.8	+ 1.68	51.16	2

Number.	Constellation, Name of Star, or Synonym.	Magnitude.	Right Ascension, 1850.0.	Annual Precession.	South Declination, 1850.0.	Annual Precession.	Mean year.	No. of obs.
			h. m. s.	s.	° ′ ″	″		
3386	Gould, Z. C., 5^h 1452 . .	9.2	5 40 48.42	− 3.846	79 5 54.8	+ 1.68	52.01	2
3387	Mensæ	10.0	40 55.40	− 1.340	73 11 47.3	.67	51.13	1
3388	Doradûs	9.3	41 2.13	+ 0.146	65 30 55.1	.66	51.44	3
3389	Mensæ	11.0	11.91	− 6.254	81 52 6.1	.64	51.99	1
3390	Gould, Z. C., 5^h 1511 . .	8.2	12.43	− 1.831	74 47 37.3	+ 1.64	51.30	3
3391	Mensæ	9.0	5 41 16.36	− 0.961	71 42 47.7	+ 1.64	50.95	1
3392	Mensæ	10.0	24.56	3.400	78 21 50.4	.63	51.47	2
3393	Doradûs	9.5	43.27	0.402	69 0 31.6	.60	51.08	2
3394	Mensæ	8.0	45.91	− 6.045	81 40 56.8	.59	50.95	1
3395	Gould, Z. C., 5^h 1561 . .	9.0	47.71	+ 0.056	66 9 28.9	+ 1.59	52.09	1
3396	Mensæ	9.5	5 41 48.61	− 2.505	76 33 28.2	+ 1.59	51.98	1
3397	Mensæ	10.0	50.16	1.986	75 14 14.6	.59	51.51	2
3398	Doradûs	9.8	50.69	− 0.576	69 55 36.6	.59	51.10	2
3399	Doradûs	8.0	41 54.23	+ 0.452	63 2 20.1	.58	51.15	1
3400	Mensæ	10.5	42 0.31	− 7.363	82 43 14.2	+ 1.57	50.95	1
3401	Lacaille, 2032	7.5	5 42 2.98	− 0.141	67 28 8.4	+ 1.57	52.00	2
3402	Doradûs	8.8	7.44	+ 0.188	65 11 42.0	.56	52.10	2
3403	Doradûs	10.0	8.20	+ 0.322	64 8 13.1	.56	52.12	1
3404	Lacaille, 2037	8.5	8.91	− 0.371	68 50 9.0	.56	51.08	1
3405	Mensæ	10.5	10.19	− 7.952	83 6 19.4	+ 1.56	50.95	1
3406	Gould, Z. C., 5^h 1545 . .	9.0	5 42 15.26	− 2.312	76 5 36.6	+ 1.55	52.02	1
3407	Octantis	9.0	23.85	12.699	85 10 5.8	.54	51.27	4
3408	Gould, Z. C., 5^h 1582 . .	10.0	28.56	− 0.391	68 56 39.9	.53	51.08	1
3409	Doradûs	10.0	29.83	+ 0.507	62 32 8.5	.53	52.13	1
3410	Gould, Z. C., 5^h 1557 . .	8.7	37.36	− 2.092	75 31 27.4	+ 1.52	51.68	3
3411	Mensæ	10.0	5 42 47.61	− 0.835	71 9 23.8	+ 1.50	52.08	1
3412	Brisbane, 1049	9.0	47.96	1.168	72 32 32.0	.50	51.13	1
3413	Lacaille, 2038	7.7	54.93	0.100	67 12 0.6	.49	52.04	3
3414	Brisbane, 1050	9.2	55.08	0.361	68 46 29.2	.49	51.60	2
3415	Mensæ	9.5	59.63	− 0.611	70 5 52.5	+ 1.49	51.07	1
3416	Mensæ	10.0	5 42 59.90	− 2.094	75 31 33.3	+ 1.49	52.03	1
3417	Gould, Z. C., 5^h 1613 . .	9.2	43 5.02	+ 0.106	65 47 17.3	.48	51.43	3
3418	Stone, 2449	7.8	5.34	−35.672	88 1 43.6	.48	51.02	6
3419	Doradûs	9.5	28.54	− 0.310	68 28 37.2	.44	52.11	1
3420	Doradûs	9.5	36.20	+ 0.139	65 32 58.1	+ 1.43	51.10	1
3421	Doradûs	9.5	5 43 37.47	+ 0.287	64 24 30.1	+ 1.43	51.11	1
3422	Doradûs	9.2	39.80	+ 0.191	65 9 40.9	.43	52.10	2
3423	Brisbane, 1057	9.0	41.27	− 1.117	72 20 16.0	43	51.60	2
3424	Mensæ	10.0	43.32	6.235	81 50 41.8	.42	51.47	2
3425	Mensæ	9.0	49.28	− 2.590	76 44 42.8	+ 1.41	51.46	2

Number.	Constellation, Name of Star, or Synonym.	Magnitude.	Right Ascension, 1850.0.	Annual Precession.	South Declination, 1850.0.	Annual Precession.	Mean year.	No. of obs.
			h. m. s.	s.	° ′ ″	″		
3426	Gould, Z. C., 5h 1591 . .	9.5	5 43 57.55	− 3.710	78 52 27.2	+ 1.40	52.02	1
3427	Brisbane, 1059	9.0	44 0.98	− 0.341	68 39 11.0	.40	51.60	2
3428	Doradûs	10.0	2.05	+ 0.235	64 48 58.8	.40	52.07	1
3429	Doradûs	8.5	3.39	+ 0.457	62 58 20.3	.39	51.16	2
3430	Gould, Z. C., 5h 1658 . .	10.0	13.51	− 0.683	70 26 35.3	+ 1.38	51.99	1
3431	Lacaille, 2049	8.0	5 44 14.70	− 0.253	68 8 30.2	+ 1.38	51.55	2
3432	Mensæ	9.5	27.10	2.063	75 26 13.1	.36	52.06	1
3433	Mensæ	9.5	29.12	6.033	81 39 50.4	.36	50.95	1
3434	Mensæ	10.0	29.95	4.199	79 36 10.0	.36	52.05	2
3435	Mensæ	10.0	29.99	− 4.750	80 19 16.1	+ 1.36	52.05	2
3436	d Doradûs	5.8	5 44 30.53	+ 0.105	65 47 30.2	+ 1.35	51.43	3
3437	Mensæ	10.0	31.88	− 3.101	77 48 21.9	.35	50.94	1
3438	Doradûs	10.0	44.61	+ 0.136	65 33 38.0	.34	51.10	1
3439	ι Mensæ	6.0	48.83	− 3.724	78 53 37.5	.33	52.02	1
3440	Mensæ	9.8	54.40	− 2.430	76 22 7.5	+ 1.32	52.00	2
3441	Mensæ	10.0	5 44 56.85	− 1.606	74 4 57.7	+ 1.32	52.04	1
3442	Gould, Z. C., 5h 1701 . .	10.0	45 1.87	0.699	70 30 51.2	.31	51.99	1
3443	Gould, Z. C., 5h 1627 . .	9.2	7.62	4.419	79 54 2.8	.30	52.07	2
3444	Mensæ	10.0	8.36	0.620	70 7 39.0	.30	51.07	1
3445	Mensæ	11.0	10.03	− 1.086	72 12 22.2	+ 1.30	51.60	2
3446	Mensæ	9.8	5 45 11.75	−11.542	84 47 0.4	+ 1.29	51.53	4
3447	Mensæ	10.0	14.96	1.084	72 12 1.4	.29	51.60	2
3448	Mensæ	9.5	18.00	0.505	69 32 32.8	.29	51.11	2
3449	Mensæ	10.0	18.43	2.320	76 5 56.5	.28	52.01	1
3450	Gould, Z. C., 5h 1705 . .	10.0	24.24	− 1.428	73 29 2.0	+ 1.28	51.59	2
3451	Mensæ	10.0	5 45 25.21	− 0.894	71 24 8.0	+ 1.27	50.95	1
3452	Mensæ	9.8	39.76	4.739	80 18 16.4	.25	52.05	2
3453	Mensæ	9.5	47.38	− 1.593	74 2 9.6	.24	52.04	1
3454	Doradûs	9.0	48.02	+ 0.416	63 19 22.6	.24	51.16	1
3455	Mensæ	10.0	49.60	− 0.692	70 28 38.0	+ 1.24	51.99	1
3456	Mensæ	9.2	5 45 55.49	− 0.829	71 6 55.6	+ 1.23	52.08	2
3457	Doradûs	10.0	45 59.32	+ 0.455	62 58 32.2	.23	51.16	1
3458	Doradûs	10.0	46 4.01	− 0.244	68 4 23.2	.22	51.14	1
3459	Gould, Z. C., 5h 1754 . .	9.8	14.60	0.218	67 54 55.3	.20	51.59	2
3460	Brisbane, 1070	9.0	16.56	− 0.645	70 14 49.2	+ 1.20	51.53	2
3461	Gould, Z. C., 5h 1762 . .	9.5	5 46 19.36	+ 0.138	65 32 19.3	+ 1.20	51.44	3
3462	Doradûs	9.2	22.70	+ 0.456	62 57 48.2	.19	51.16	2
3463	Mensæ	10.0	22.82	− 2.465	76 26 51.3	.19	51.98	1
3464	Doradûs	9.0	28.68	+ 0.234	64 48 31.3	.18	52.07	1
3465	Lacaille, 2071	9.0	34.15	− 0.192	67 45 32.0	+ 1.17	52.03	1

Number.	Constellation, Name of Star, or Synonym.	Magnitude.	Right Ascension, 1850.0.	Annual Precession.	South Declination, 1850.0.	Annual Precession.	Mean year.	No. of obs.
			h. m. s.	s.	° ′ ″	″		
3466	Mensæ	9.5	5 46 36.85	− 1.782	74 37 26.8	+ 1.17	51.50	2
3467	Gould, Z. C., 5h 1743 . .	9.5	39.91	− 1.755	74 32 30.6	.17	50.92	1
3468	Gould, Z. C., 5h 1780 . .	9.5	40.18	+ 0.142	65 30 18.6	.17	51.44	3
3469	Mensæ	10.0	41.93	− 3.614	78 42 35.6	.16	50.93	1
3470	Mensæ	10.0	42.60	− 1.195	72 37 38.0	+ 1.16	51.09	2
3471	Doradûs	9.2	5 46 46.12	+ 0.258	64 37 17.8	+ 1.16	51.59	2
3472	Gould, Z. C., 5h 1782 . .	10.0	46 49.33	− 0.191	67 44 56.0	.15	52.03	1
3473	Mensæ	11.0	47 0.35	2.018	75 18 9.6	.14	52.06	1
3474	Gould, Z. C., 5h 1798 . .	9.5	7.60	0.147	67 28 34.8	.13	52.00	2
3475	Mensæ	9.5	8.68	− 7.125	82 32 28.1	+ 1.12	52.08	1
3476	Doradûs	10.0	5 47 16.61	+ 0.481	62 44 17.6	+ 1.11	52.13	1
3477	Doradûs	9.5	17.16	+ 0.400	63 27 7.1	.11	51.64	2
3478	Doradûs	9.2	17.50	− 0.035	66 44 43.3	.11	51.59	2
3479	Gould, Z. C., 5h 1766[1] . .	9.8	18.20	1.881	74 54 55.0	.11	51.53	2
3480	Gould, Z. C., 5h 1766[2] . .	10.0	18.41	− 1.882	74 54 58.5	+ 1.11	51.52	2
3481	Gould, Z. C., 5h 1781 . .	8.5	5 47 21.14	− 1.462	73 35 32.0	+ 1.11	52.04	1
3482	Mensæ	10.0	22.95	4.425	79 54 11.0	.10	52.10	1
3483	Mensæ	10.0	27.20	2.265	75 57 17.1	.10	52.03	1
3484	Lacaille, 2125	9.0	27.73	4.485	79 58 49.2	.10	52.07	2
3485	Octantis	9.5	28.85	−14.041	85 32 20.7	+ 1.10	51.02	1
3486	Gould, Z. C., 5h 1788 . .	9.2	5 47 39.18	− 1.893	74 56 56.2	+ 1.08	51.53	2
3487	Mensæ	9.5	39.22	− 3.615	78 42 33.1	.08	51.48	2
3488	Doradûs	10.0	39.35	+ 0.532	62 16 18.5	.08	52.13	1
3489	Mensæ	9.5	41.43	− 1.586	74 0 15.1	.08	52.04	1
3490	Mensæ	9.8	42.12	− 2.019	75 18 16.8	+ 1.08	51.51	2
3491	Doradûs	10.0	5 47 44.03	+ 0.330	64 1 57.7	+ 1.07	52.12	1
3492	Brisbane, 1099	8.5	45.42	− 1.217	72 42 32.4	.07	51.09	2
3493	Lacaille, 2103	8.8	48.44	2.239	75 53 18.6	.07	52.02	2
3494	Gould, Z. C., 5h 1827 . .	9.0	51.38	0.200	67 48 11.5	.06	51.59	2
3495	Doradûs	9.2	47 59.30	− 0.030	66 42 44.3	+ 1.05	51.59	2
3496	Mensæ	9.5	5 48 1.11	− 4.844	80 25 31.0	+ 1.05	52.06	1
3497	Gould, 6879	9.0	5.74	− 5.282	80 55 8.5	.04	51.06	1
3498	Lacaille, 2079	7.5	7.26	+ 0.187	65 12 35.6	.04	52.10	2
3499	Doradûs	9.5	9.10	+ 0.404	63 24 32.5	.04	51.64	2
3500	Doradûs	8.0	22.14	− 0.025	66 40 27.4	+ 1.02	51.59	2
3501	Doradûs	9.0	5 48 22.52	+ 0.444	63 3 41.9	+ 1.02	51.16	2
3502	Mensæ	9.5	25.00	− 1.772	74 35 12.3	.01	50.92	1
3503	Mensæ	10.0	25.06	2.122	75 34 55.0	.01	52.03	1
3504	Doradûs	10.0	25.48	0.014	66 36 4.4	.01	52.06	1
3505	Mensæ	9.5	27.67	−10.778	84 29 33.1	+ 1.01	52.06	1

Number.	Constellation, Name of Star, or Synonym.	Magnitude.	Right Ascension, 1850.0.	Annual Precession.	South Declination, 1850.0.	Annual Precession.	Mean year.	No. of obs.
			h. m. s.	s.	° ′ ″	″		
3506	Mensæ	10.0	5 48 31.59	− 1.592	74 1 15.7	+ 1.00	52.04	1
3507	Octantis	9.0	48 54.52	15.077	85 47 28.9	0.97	51.02	4
3508	Mensæ	10.0	49 2.75	0.818	71 3 7.9	.96	52.08	1
3509	Mensæ	9.5	5.73	1.479	73 38 38.4	.95	52.04	1
3510	Mensæ	9.5	9.31	− 0.954	71 38 52.8	+ 0.95	50.95	1
3511	Lacaille, 2138	6.0	5 49 11.26	− 4.974	80 34 29.6	+ 0.95	51.56	2
3512	Mensæ	9.5	21.30	3.151	77 45 6.8	.93	50.94	1
3513	Gould, 6945	9.0	22.33	0.556	69 47 26.9	.93	51.14	1
3514	Mensæ	9.5	23.27	0.935	71 33 59.7	.93	50.95	1
3515	Mensæ	11.0	23.30	− 2.025	75 18 54.1	+ 0.93	52.06	1
3516	Mensæ	10.0	5 49 26.33	− 1.658	74 13 54.1	+ 0.92	51.48	2
3517	Doradûs	10.0	30.64	0.307	68 25 59.8	.92	52.11	1
3518	Gould, Z. C., 5ʰ 1893 . .	9.2	35.20	0.814	71 1 52.7	.91	52.08	2
3519	Mensæ	9.8	35.90	− 1.998	75 14 26.8	.91	51.46	2
3520	Gould, Z. C., 5ʰ 1919 . .	9.0	38.89	+ 0.379	63 37 0.4	+ 0.91	52.13	1
3521	Doradûs	10.0	5 49 40.96	+ 0.059	66 5 19.2	+ 0.90	52.09	1
3522	Mensæ	10.0	48.75	− 2.224	75 50 35.9	.89	52.03	1
3523	Gould, Z. C., 5ʰ 1931 . .	9.0	55.40	+ 0.408	63 21 52.4	.88	51.16	1
3524	Doradûs	10.0	57.32	0.002	66 30 38.9	.88	51.12	1
3525	Doradûs	9.5	58.02	+ 0.462	62 53 39.2	+ 0.88	51.16	1
3526	Gould, 6966	9.0	5 49 59.17	− 0.676	70 23 1.4	+ 0.88	51.99	1
3527	Gould, Z. C., 5ʰ 1870 . .	9.5	50 0.12	2.954	77 30 5.8	.87	51.47	2
3528	Gould, Z. C., 5ʰ 1879 . .	9.0	0.82	2.635	76 49 30.8	.87	51.96	2
3529	ε Doradûs	5.5	2.94	0.066	66 56 19.2	.87	52.02	2
3530	Gould, Z. C., 5ʰ 1918 . .	9.5	8.48	− 0.808	71 0 10.6	+ 0.86	52.08	2
3531	Mensæ	9.0	5 50 8.84	− 1.615	74 5 33.7	+ 0.86	52.04	1
3532	Gould, Z. C., 5ʰ 1865 . .	9.2	11.94	− 3.604	78 41 1.8	.86	51.48	2
3533	Lacaille, 2089	8.0	13.03	+ 0.168	65 17 43.8	.86	52.12	1
3534	Mensæ	10.0	18.14	− 1.568	73 56 21.0	.85	52.04	1
3535	Mensæ	9.5	26.64	−11.772	84 51 27.2	+ 0.84	52.00	1
3536	Doradûs	8.8	5 50 32.28	− 0.089	67 5 15.0	+ 0.83	52.02	2
3537	Gould, 6993	8.5	33.32	+ 0.385	63 33 53.2	.83	52.13	1
3538	Lacaille, 2091	7.0	34.19	+ 0.325	64 3 59.6	.83	52.12	1
3539	Mensæ	10.0	40.45	−10.253	84 16 29.6	.82	51.50	2
3540	Lacaille, 2171	7.0	42.84	− 6.058	81 40 28.0	+ 0.81	50.95	1
3541	Doradûs	9.0	5 50 44.28	+ 0.500	62 32 54.4	+ 0.81	52.13	1
3542	Mensæ	10.0	48.30	− 3.485	78 29 2.8	.80	50.93	1
3543	Lacaille, 2111	7.0	51.08	1.229	72 44 32.2	.80	51.09	2
3544	Gould, 6989	9.0	52.57	0.564	69 49 26.6	.80	51.11	2
3545	Mensæ	10.0	53.69	− 1.044	72 0 52.1	+ 0.80	52.07	1

Number.	Constellation, Name of Star, or Synonym.	Magnitude.	Right Ascension, 1850.0.	Annual Precession.	South Declination, 1850.0.	Annual Precession.	Mean year.	No. of obs.
			h. m. s.	s.	° ′ ″	″		
3546	Mensæ	9.8	5 50 54.21	− 2.031	75 19 41.0	+ 0.80	51.51	2
3547	Mensæ	10.0	50 55.88	2.212	75 48 37.2	.79	52.03	1
3548	Mensæ	12.0	51 2.19	1.444	73 31 3.2	.78	51.13	1
3549	Mensæ	11.0	4.06	− 1.443	73 30 51.2	.78	51.13	1
3550	Doradûs	9.5	13.77	+ 0.497	62 34 45.9	+ 0.77	52.13	1
3551	Mensæ	9.2	5 51 19.78	− 2.046	75 22 7.1	+ 0.76	51.51	2
3552	Gould, 7007	9.0	19.86	0.544	69 43 9.9	.76	51.14	1
3553	Doradûs	9.0	26.71	0.314	68 27 49.2	.75	52.11	1
3554	Doradûs	10.0	34.16	0.460	69 16 33.1	.74	51.08	1
3555	Gould, Z. C., 5h 1946 . .	10.0	39.03	− 2.903	77 23 40.0	+ 0.73	52.00	1
3556	Mensæ	9.8	5 51 39.66	− 2.624	76 47 58.6	+ 0.73	51.46	2
3557	Mensæ	9.8	44.00	− 1.849	74 48 31.0	.72	51.52	2
3558	Doradûs	9.2	48.65	+ 0.126	65 36 12.6	.72	51.10	2
3559	Mensæ	10.0	49.22	− 5.112	80 43 35.2	.72	51.06	1
3560	Doradûs	10.0	51.16	− 0.091	67 5 57.4	+ 0.71	52.02	2
3561	Mensæ	9.8	5 51 52.04	− 6.296	81 52 55.0	+ 0.71	51.97	2
3562	Doradûs	10.0	51 54.55	+ 0.328	64 1 53.7	.71	52.12	1
3563	Doradûs	10.0	52 2.82	+ 0.221	64 53 19.6	.70	52.07	1
3564	Mensæ	10.0	9.04	− 3.058	77 42 4.5	.69	50.94	1
3565	Mensæ	10.0	11.51	− 1.703	74 21 50.7	+ 0.68	50.92	1
3566	Doradûs	9.8	5 52 13.59	+ 0.468	62 49 53.3	+ 0.68	51.16	2
3567	Mensæ	9.8	17.58	− 2.021	75 17 50.7	.67	51.51	2
3568	Gould, Z. C., 5h 2036 . .	8.0	28.15	+ 0.410	63 20 31.7	.66	51.16	1
3569	Lacaille, 2116	7.8	28.34	− 0.586	69 55 44.6	.66	51.10	2
3570	Gould, Z. C., 5h 1984 . .	9.2	30.96	− 2.373	76 12 37.3	+ 0.65	52.00	2
3571	Stone, 2708	9.0	5 52 31.59	− 0.896	71 23 8.6	+ 0.65	52.08	1
3572	Mensæ	10.0	38.40	− 1.155	72 27 23.2	.64	51.13	1
3573	Gould, Z. C., 5h 2041 . .	8.2	48.12	+ 0.245	64 41 39.4	.63	51.59	2
3574	Mensæ	9.0	51.59	− 1.483	73 38 54.5	.62	52.04	1
3575	Doradûs	9.3	52.02	+ 0.127	65 35 30.9	+ 0.62	51.44	3
3576	Doradûs	9.2	5 52 52.59	+ 0.151	65 24 50.7	+ 0.62	51.44	3
3577	Doradûs	10.0	52.97	− 0.170	67 35 54.9	.62	52.03	1
3578	Lacaille, 2106	6.0	52 57.30	+ 0.433	63 8 9.4	.62	51.16	3
3579	Mensæ	9.5	53 8.77	− 1.018	71 54 7.6	.60	51.51	2
3580	Mensæ	10.0	9.03	− 1.876	74 53 5.1	+ 0.60	50.97	1
3581	Gould, Z. C., 5h 2039 . .	9.0	5 53 10.92	− 0.704	70 30 26.4	+ 0.60	51.99	1
3582	Gould, 7061	9.0	12.42	0.566	69 49 44.6	.59	51.10	2
3583	Gould, Z. C., 5h 2065 . .	9.5	29.06	0.409	68 59 41.1	.57	51.08	1
3584	Mensæ	10.0	29.48	− 1.040	71 59 42.1	.57	52.08	1
3585	Lacaille, 2113	7.0	30.97	+ 0.269	64 30 24.5	+ 0.57	51.11	1

Number.	Constellation, Name of Star, or Synonym.	Magnitude.	Right Ascension, 1850.0.	Annual Precession.	South Declination, 1850.0.	Annual Precession.	Mean year.	No. of obs.
			h. m. s.	s.	° ′ ″	″		
3586	Mensæ	9.5	5 53 32.99	— 9.50[illegible]	83 56 2.6	+ 0.56	51.46	2
3587	Gould, Z. C., 5h 1997 . .	9.5	34.80	4.196	79 34 45.8	.56	52.05	2
3588	Gould, Z. C., 5h 2042 . .	9.1	42.88	1.934	75 3 0.5	.55	51.51	4
3589	Doradûs	10.0	53 46.56	0.315	68 27 58.7	.54	52.11	1
3590	Gould, Z. C., 5h 2049 . .	8.8	54 1.56	— 2.343	76 8 7.6	+ 0.52	52.00	2
3591	Doradûs	10.0	5 54 12.70	— 0.432	69 7 12.3	+ 0.51	51.08	2
3592	Mensæ	10.0	16.46	3.451	78 25 10.3	.50	51.47	2
3593	Mensæ	10.0	18.85	1.034	71 57 57.4	.50	50.95	1
3594	Mensæ	10.0	21.88	7.770	82 58 16.4	.49	51.48	2
3595	Gould, Z. C., 5h 2044 . .	9.5	37.92	4.034	79 20 48.0	+ 0.47	52.00	1
3596	Doradûs	10.0	5 54 40.84	+ 0.074	65 58 17.9	+ 0.47	51.43	3
3597	Doradûs	9.8	41.10	+ 0.437	63 6 7.7	.46	51.16	2
3598	Doradûs	10.0	41.68	— 0.418	69 2 35.6	.46	51.08	2
3599	Doradûs	9.5	42.72	— 0.004	66 30 31.7	.46	51.12	1
3600	Gould, 7107	8.0	44.20	+ 0.083	65 54 10.8	+ 0.46	51.43	3
3601	Mensæ	10.0	5 54 45.46	— 1.609	74 3 46.6	+ 0.46	52.04	1
3602	Gould, Z. C., 5h 2120 . .	8.7	46.59	+ 0.072	65 58 49.9	.46	51.43	3
3603	Gould, Z. C., 5h 2087 . .	9.5	47.58	— 2.628	76 34 26.0	.46	51.98	1
3604	Octantis	9.5	48.15	14.061	85 32 19.1	.45	51.04	1
3605	Mensæ	10.0	57.15	— 0.933	71 32 24.4	+ 0.44	50.95	1
3606	Gould, Z. C., 5h 2116 . .	9.5	5 54 57.77	— 0.778	70 51 9.2	+ 0.44	52.08	1
3607	Doradûs	9.0	55 2.14	0.459	69 15 39.0	.43	51.08	1
3608	Mensæ	10.0	5.43	3.012	77 36 25.7	.43	50.94	1
3609	Mensæ	9.2	10.66	1.134	72 21 58.9	.42	51.60	2
3610	Gould, Z. C., 5h 2097 . .	8.5	21.41	— 2.912	77 24 26.3	+ 0.41	51.47	2
3611	Gould, Z. C., 5h 2142 . .	10.0	5 55 21.44	— 0.193	67 43 51.6	+ 0.41	52.03	1
3612	Mensæ	10.0	21.93	1.959	75 7 6.8	.41	52.06	1
3613	Gould, Z. C., 5h 2141 . .	9.5	24.67	0.110	67 12 35.5	.40	51.97	1
3614	Gould, Z. C., 5h 2111 . .	8.5	35.92	— 2.337	76 7 0.4	.38	52.02	1
3615	Doradûs	10.0	41.00	+ 0.202	65 1 32.3	+ 0.38	52.07	1
3616	Mensæ	10.0	5 55 43.23	— 1.134	72 22 4.3	+ 0.37	51.13	1
3617	Doradûs	10.0	54.13	0.352	68 40 10.9	.36	51.08	1
3618	Gould, 7131	8.2	55 54.66	— 0.169	67 34 53.3	.36	52.00	2
3619	Pictoris	9.5	56 7.05	+ 0.512	62 25 14.2	.34	52.13	1
3620	Mensæ	8.5	7.07	— 8.323	83 18 35.6	+ 0.34	52.01	1
3621	Gould, Z. C., 5h 2176 . .	9.0	5 56 12.87	— 0.091	67 5 10.7	+ 0.33	52.06	1
3622	Lacaille, 2134	8.0	14.67	0.248	68 3 53.3	.33	51.15	1
3623	Gould, 7150	8.2	27.22	0.164	67 33 12.8	.31	52.00	2
3624	Mensæ	10.0	31.92	1.109	72 16 4.9	.30	52.08	1
3625	Doradûs	10.0	34.97	— 0.284	68 16 37.0	+ 0.30	52.11	1

No. 3611. This Right Ascension is 8.75s less than Gould's.

Number.	Constellation, Name of Star, or Synonym.	Magnitude.	Right Ascension, 1850.0.	Annual Precession.	South Declination, 1850.0.	Annual Precession.	Mean year.	No. of obs.
			h. m. s.	s.	° ′ ″	″		
3 626	Mensæ	9.0	5 56 36.62	− 1.835	74 45 31.0	+ 0.30	51.32	3
3 627	Mensæ	9.0	37.93	0.710	70 31 51.9	.29	51.99	1
3 628	Doradus	10.0	45.30	0.488	69 24 48.1	.28	51.14	1
3 629	Doradûs	9.5	46.82	− 0.502	69 30 50.3	.28	51.08	1
3 630	Doradûs	9.0	49.12	+ 0.023	66 19 18.4	+ 0.28	51.60	2
3 631	Mensæ	8.5	5 56 54.96	− 5.526	81 9 41.8	+ 0.27	51.02	2
3 632	Doradûs	9.0	56 56.90	+ 0.166	65 17 48.0	.27	52.12	1
3 633	Gould, Z. C., 5h 2209 . .	9.0	57 2.74	+ 0.475	62 45 44.8	.26	51.16	2
3 634	Mensæ	10.0	6.93	− 0.803	70 57 44.5	.25	52.08	1
3 635	Doradûs	10.0	14.27	+ 0.293	64 18 32.4	+ 0.24	51.11	1
3 636	Mensæ	9.2	5 57 23.48	− 4.787	80 20 42.2	+ 0.23	52.05	2
3 637	Gould, Z. C., 5h 2222 . .	9.0	28.46	+ 0.152	65 23 51.6	.22	51.61	2
3 638	Doradûs	8.5	34.29	+ 0.310	64 10 19.3	.21	52.12	1
3 639	Gould, Z. C., 5h 2225 . .	9.2	42.78	− 0.146	67 26 15.3	.20	52.00	2
3 640	Mensæ	10.0	46.07	− 1.137	72 22 33.2	+ 0.18	51.60	2
3 641	Mensæ	8.5	5 57 50.92	− 8.367	83 20 0.5	+ 0.19	52.01	1
3 642	Pictoris	10.0	58 0.25	+ 0.362	63 44 17.9	.17	52.13	1
3 643	Mensæ	11.0	1.73	−10.326	84 18 3.0	.17	50.94	1
3 644	Pictoris	9.8	7.08	+ 0.403	63 23 12.7	.16	51.64	2
3 645	Doradûs	10.0	8.97	+ 0.017	66 20 46.7	+ 0.16	52.09	1
3 646	Gould, Z. C., 5h 2236 . .	9.5	5 58 11.96	− 0.663	70 17 56.4	+ 0.16	51.53	2
3 647	Gould, Z. C., 5h 2228 . .	8.5	13.20	− 1.254	72 49 28.3	.16	51.06	1
3 648	Doradûs	9.7	15.38	+ 0.304	64 12 53.4	.15	51.62	2
3 649	Doradûs	9.5	16.22	+ 0.283	64 23 9.2	.15	51.11	1
3 650	Lacaille, 2157	8.5	28.08	− 0.502	69 29 21.4	+ 0.13	51.10	2
3 651	Mensæ	10.0	5 58 30.57	− 0.923	71 29 49.2	+ 0.13	52.08	1
3 652	Mensæ	10.0	31.62	− 0.916	71 27 58.9	.13	50.95	1
3 653	Pictoris	10.0	32.38	+ 0.402	63 24 9.2	.13	51.16	1
3 654	Mensæ	10.0	32.67	− 5.465	81 5 56.7	.13	51.06	1
3 655	Mensæ	9.5	33.13	− 1.518	73 45 35.9	+ 0.13	52.04	1
3 656	Mensæ	10.0	5 58 34.37	− 1.148	72 25 3.9	+ 0.12	51.13	1
3 657	Mensæ	8.2	34.42	1.346	73 9 41.4	.12	51.10	2
3 658	Mensæ	8.7	40.21	− 2.040	75 21 33.0	.12	51.68	3
3 659	Gould, 7216	8.2	43.94	+ 0.469	62 48 43.6	.11	51.16	2
3 660	Gould, Z. C., 5h 2272 . .	9.5	45.96	− 0.153	67 28 42.2	+ 0.11	52.00	2
3 661	Doradûs	9.5	5 58 48.09	+ 0.272	64 28 27.4	+ 0.10	51.11	1
3 662	Mensæ	9.5	54.25	1.616	74 4 55.2	.10	52.04	1
3 663	Doradûs	9.8	55.24	+ 0.152	65 23 41.6	.09	51.61	2
3 664	Mensæ	9.5	55.27	− 2.145	75 37 30.8	.09	52.03	1
3 665	Mensæ	9.5	57.87	− 1.596	74 0 59.9	+ 0.09	52.04	1

No. 3 651 and No. 3 652. Probably same Star. Differ 4 revolutions = 1′ 51.7″.

Number.	Constellation, Name of Star, or Synonym.	Magnitude.	Right Ascension, 1850.0.	Annual Precession.	South Declination, 1850.0.	Annual Precession.	Mean year.	No. of obs.
			h. m. s.	s.	° ′ ″	″		
3666	Gould, Z. C., 5h 2264 . .	8.5	5 59 11.23	— 1.484	73 38 34.8	+ 0.07	52.04	1
3667	Pictoris	9.0	15.02	+ 0.529	62 16 10.9	.07	52.13	1
3668	Gould, 7221	9.3	17.12	0.145	65 26 57.1	.06	51.44	3
3669	Gould, 7225	9.5	19.65	+ 0.149	65 25 14.5	.06	51.44	3
3670	Lacaille, 2296	6.8	20.72	—11.776	84 50 33.7	+ 0.06	51.31	7
3671	Lacaille, 2147	7.5	5 59 23.58	+ 0.292	64 18 42.4	+ 0.05	51.62	2
3672	Gould, Z. C., 5h 2307 . .	8.8	25.82	+ 0.469	62 48 43.6	.05	51.16	2
3673	Mensæ	10.0	29.34	— 0.912	71 26 46.2	.04	51.47	2
3674	Gould, Z. C., 5h 2295 . .	9.5	34.82	— 0.646	70 13 19.2	.04	51.53	2
3675	Gould, Z. C., 5h 2319 . .	9.5	36.65	+ 0.463	62 52 0.8	+ 0.03	51.16	2
3676	Gould, Z. C., 5h 2262 . .	9.0	5 59 37.97	— 2.655	76 51 30.5	+ 0.03	50.95	1
3677	Gould, Z. C., 5h 2322 . .	10.0	41.22	+ 0.469	62 48 37.9	.03	51.16	2
3678	Mensæ	9.2	45.59	-10.464	84 21 32.2	.02	51.50	2
3679	Gould, 7235	8.5	5 59 57.39	0.574	69 51 33.4	.00	51.10	2
3680	Gould, Z. C., 5h 2291 . .	8.6	6 0 6.11	— 2.088	75 28 18.9	— 0.01	51.33	4
3681	Doradûs	9.8	6 0 9.30	— 0.212	67 50 40.6	— 0.01	51.59	2
3682	Gould, Z. C., 6h 17 . . .	9.0	10.49	+ 0.363	63 43 39.4	.02	52.13	1
3683	Doradûs	9.8	12.55	+ 0.037	66 13 18.4	.02	51.61	2
3684	Mensæ	11.0	18.11	— 9.671	84 0 36.4	.03	50.94	1
3685	Mensæ	9.5	24.44	— 6.650	82 10 9.2	— 0.04	51.99	1
3686	Lacaille, 2210	6.3	6 0 24.93	— 4.060	79 22 53.6	— 0.04	51.42	3
3687	Mensæ	9.0	25.52	12.091	84 57 37.8	.04	51.27	4
3688	Stone, 2774	8.5	32.48	0.885	71 19 42.2	.05	52.08	1
3689	Gould, Z. C., 5h 2283 . .	9.3	39.16	4.017	79 19 6.5	.06	51.42	3
3690	Mensæ	9.0	41.82	— 0.895	71 22 21.3	— 0.06	52.08	1
3691	Gould, Z. C., 6h 10 . . .	9.0	6 0 51.08	— 1.470	73 35 52.0	— 0.07	51.59	2
3692	Lacaille, 2209	8.2	1 1.04	3.560	78 36 2.6	.09	51.08	5
3693	Gould, Z. C., 5h 2286 . .	9.0	2.16	— 4.631	80 9 11.3	.09	52.04	1
3694	Gould, Z. C., 6h 62 . . .	9.0	12.46	+ 0.386	63 32 20.6	.11	51.64	2
3695	Gould, Z. C., 6h 18 . . .	9.5	18.24	— 2.301	76 1 33.5	— 0.11	52.02	1
3696	Mensæ	9.5	6 1 18.34	— 1.520	73 45 58.6	— 0.11	52.04	2
3697	Gould, Z. C., 6h 9 . . .	8.2	19.94	— 2.635	76 48 51.1	.12	51.46	2
3698	Lacaille, 2175	7.0	20.80	+ 0.224	64 51 9.6	.12	52.07	1
3699	Gould, Z. C., 6h 53 . . .	10.0	23.80	— 0.596	69 58 7.1	.12	51.07	1
3700	Gould, Z. C., 6h 51 . . .	8.5	25.61	— 0.854	71 11 27.0	— 0.12	52.08	2
3701	Pictoris	9.2	6 1 33.86	+ 0.485	62 40 1.3	— 0.14	51.48	3
3702	Mensæ	9.8	34.79	— 1.462	73 34 5.0	.14	51.59	2
3703	Doradûs	11.0	36.83	+ 0.184	65 9 12.9	.14	52.07	1
3704	Doradûs	9.5	39.59	— 0.015	66 35 0.0	.15	52.06	1
3705	Doradûs	10.0	42.28	+ 0.186	65 8 25.3	— 0.15	52.07	1

No. 3699. Original record 0 minutes.

Number.	Constellation, Name of Star, or Synonym.	Magnitude.	Right Ascension, 1850.0.	Annual Precession.	South Declination, 1850.0.	Annual Precession.	Mean year.	No. of obs.
			h. m. s.	s.	° ′ ″	″		
3706	Doradûs	9.0	6 1 43.17	+ 0.160	65 20 9.0	− 0.15	52.12	1
3707	Gould, Z. C., 6h 39 . . .	9.5	54.00	− 2.331	76 6 0.7	.17	52.02	1
3708	Mensæ	10.0	1 58.55	9.724	84 2 4.7	.17	50.94	1
3709	Mensæ	9.8	2 1.86	5.762	81 23 46.1	.18	50.97	2
3710	Mensæ	10.0	13.81	− 1.938	75 3 30.2	− 0.20	52.06	1
3711	Mensæ	10.0	6 2 14.73	− 1.949	75 5 15.5	− 0.20	52.06	1
3712	Gould, Z. C., 6h 99 . . .	9.8	15.82	− 0.214	67 51 33.1	.20	51.59	2
3713	Gould, Z. C., 6h 105 . .	9.0	16.07	+ 0.362	63 44 33.4	.20	52.13	1
3714	Mensæ	9.0	16.96	− 2.178	75 42 38.5	.20	52.03	1
3715	Gould, 7319	8.0	18.23	+ 0.408	63 20 44.2	− 0.20	51.16	1
3716	Doradûs	9.7	6 2 25.22	+ 0.095	65 48 40.5	− 0.21	51.43	3
3717	Mensæ	9.2	30.70	− 7.350	82 41 23.2	.22	51.52	2
3718	Doradûs	10.0	33.04	+ 0.214	64 55 38.6	.22	52.07	1
3719	Doradûs	9.8	35.98	0.086	65 52 47.9	.23	51.43	3
3720	Brisbane, 1154	9.5	38.64	+ 0.523	62 19 10.3	− 0.23	52.13	1
3721	Doradûs	10.0	6 2 41.74	+ 0.102	65 45 51.9	− 0.24	51.43	3
3722	Doradûs	9.5	43.49	+ 0.172	65 14 46.4	.24	52.12	1
3723	Mensæ	10.0	45.49	− 0.934	71 32 36.1	.24	51.52	2
3724	Gould, 7322	8.5	48.54	0.603	70 0 34.2	.25	51.07	1
3725	Mensæ	9.5	50.20	− 6.405	81 58 11.1	− 0.25	51.04	3
3726	Mensæ	8.5	6 3 17.44	− 0.622	70 6 13.9	− 0.29	51.07	1
3727	Mensæ	9.2	20.71	3.160	77 53 30.1	.29	51.48	2
3728	Lacaille, 2193	7.2	27.08	0.124	67 17 54.2	.30	52.02	2
3729	Mensæ	10.0	29.95	1.024	71 55 22.5	.31	50.95	1
3730	Mensæ	10.0	30.65	− 3.408	78 20 33.6	− 0.31	51.47	2
3731	Doradûs	9.8	6 3 33 17	− 0.280	68 15 22.0	− 0.31	51.62	2
3732	Mensæ	10.0	36.66	3.260	78 4 38.9	.32	52.02	1
3733	Mensæ	9.5	36.77	2.802	77 10 36.7	.32	51.47	2
3734	Mensæ	9.5	40.19	2.824	77 13 29.5	.32	51.47	2
3735	Mensæ	9.2	40.65	− 3.136	77 50 50.1	− 0.32	51.48	2
3736	Gould, Z. C., 6h 128 . .	9.0	6 3 49.88	− 1.974	75 9 36.9	− 0.34	51.78	4
3737	Doradûs	10.0	51.04	+ 0.001	66 28 24.2	.34	51.12	1
3738	Mensæ	10.0	52.12	− 2.058	75 23 31.0	.34	51.51	2
3739	Mensæ	9.0	55.62	0.738	70 39 51.7	.34	52.36	2
3740	Mensæ	9.5	55.91	− 6.629	82 9 13.0	− 0.34	51.99	1
3741	Brisbane, 1161	9.0	6 3 56.14	+ 0.527	62 16 46.5	− 0.34	52.13	1
3742	Gould, Z. C., 6h 169 . .	9.5	4 6.44	− 0.488	69 25 9.7	.36	51.11	2
3743	Mensæ	9.2	13.07	0.818	71 2 5.9	.37	52.08	2
3744	Mensæ	9.0	21.96	7.006	82 26 38.4	.38	51.11	2
3745	Mensæ	10.0	23.52	− 1.031	71 57 9.8	− 0.38	51.69	3

Number.	Constellation, Name of Star, or Synonym.	Magnitude.	Right Ascension, 1850.0.	Annual Precession.	South Declination, 1850.0.	Annual Precession.	Mean year.	No. of obs.
			h. m. s.	s.	° ′ ″	″		
3746	Mensæ	8.5	6 4 25.98	− 1.076	72 8 11.6	− 0.39	52.08	1
3747	Doradûs	9.5	33.43	0.324	68 30 46.4	.40	52.11	1
3748	Mensæ	10.0	49.51	− 5.816	81 26 48.4	.42	50.95	1
3749	Gould, Z. C., 6^h 206 . .	9.2	4 55.64	+ 0.093	65 49 44.4	.43	51.10	2
3750	Gould, Z. C., 6^h 202 . .	9.0	5 7.18	− 0.541	69 41 43.1	− 0.45	51.14	1
3751	Gould, Z. C., 6^h 215 . .	9.2	6 5 8.70	+ 0.090	65 51 2.4	− 0.45	51.10	2
3752	Gould, Z. C., 6^h 235 . .	9.0	19.66	+ 0.444	63 2 24.0	.47	51.15	3
3753	Doradûs	10.0	23.99	− 0.317	68 28 18.4	.47	52.11	1
3754	Pictoris	9.5	30.09	+ 0.486	62 40 2.2	.48	51.16	2
3755	Mensæ	10.0	33.73	− 3.665	78 46 39.5	− 0.49	52.02	1
3756	Gould, Z. C., 6^h 243 . .	9.0	6 5 34.00	+ 0.168	65 17 10.8	− 0.49	52.12	1
3757	Octantis	9.5	41.06	−14.386	85 37 17.8	.50	51.03	2
3758	Brisbane, 1172	6.0	41.48	+ 0.544	62 7 42.8	.50	52.13	1
3759	Stone, 2825	8.0	44.51	− 1.627	74 7 11.3	.50	52.04	1
3760	Doradûs	10.0	47.78	− 0.300	68 22 45.5	− 0.51	52.11	1
3761	Doradûs	9.5	6 5 54.32	+ 0.135	65 31 32.7	− 0.52	51.10	1
3762	Mensæ	10.0	56.00	−12.143	84 58 46.2	.52	52.00	1
3763	η^1 Doradûs	6.0	5 58.63	+ 0.067	66 1 5.8	.52	52.09	1
3764	Doradûs	10.0	6 16.54	+ 0.061	66 3 38.2	.55	52.09	1
3765	Gould, 7391	9.5	18.20	− 2.840	77 15 40.3	− 0.55	52.00	1
3766	Doradûs	10.0	6 6 18.45	− 0.462	69 16 48.0	− 0.55	51.08	1
3767	Mensæ	10.0	21.06	4.153	79 31 9.2	.56	52.00	1
3768	Gould, 7392	9.8	22.99	3.003	77 35 29.0	.56	51.47	2
3769	Doradûs	10.0	27.10	− 0.412	69 0 40.5	.56	51.08	1
3770	Gould, Z. C., 6^h 296 . .	8.2	28.11	+ 0.287	64 21 48.6	− 0.57	51.62	2
3771	Doradûs	10.0	6 6 34.28	+ 0.265	64 32 8.5	− 0.57	51.11	1
3772	Doradûs	9.8	34.50	+ 0.100	65 46 56.0	.58	51.10	2
3773	Mensæ	10.0	36.14	− 1.232	72 44 19.6	.58	51.06	1
3774	Mensæ	9.0	38.44	11.075	84 36 11.2	.58	51.54	2
3775	Gould, 7430	8.3	6 44.40	− 0.138	67 23 53.5	− 0.59	52.02	3
3776	Gould, Z. C., 6^h 259 . .	9.0	6 7 1.78	+ 2.462	76 25 21.0	− 0.62	51.10	2
3777	Doradûs	9.5	3.08	− 0.579	69 53 30.8	.62	51.10	2
3778	Gould, Z. C., 6^h 313 . .	9.0	5.78	+ 0.281	64 24 57.1	.62	51.11	1
3779	Mensæ	10.0	5.99	− 4.056	79 22 48.4	.62	52.00	1
3780	Lacaille, 2212	7.2	7.96	− 0.117	67 15 31.6	− 0.62	52.02	2
3781	Gould, Z. C., 6^h 317 . .	9.5	6 7 8.17	+ 0.417	63 16 48.3	− 0.62	51.16	1
3782	Doradûs	9.5	9.35	− 0.491	69 26 14.6	.63	51.08	1
3783	Mensæ	11.0	10.32	6.343	81 55 15.9	.63	51.17	1
3784	Doradûs	10.0	13.25	− 0.251	68 5 18.4	.63	51.15	1
3785	Gould, Z. C., 6^h 326 . .	9.5	16.57	+ 0.427	63 11 30.2	− 0.64	51.16	1

Number.	Constellation, Name of Star, or Synonym.	Magnitude.	Right Ascension, 1850.0.	Annual Precession.	South Declination, 1850.0.	Annual Precession.	Mean year.	No. of obs.
			h. m. s.	s.	° ′ ″	″		
3786	Pictoris	9.5	6 7 24.17	+ 0.454	62 57 11.8	− 0.65	51.15	1
3787	Gould, Z. C., 6^h 327 . .	10.0	25.27	+ 0.173	65 15 0.1	.65	52.12	1
3788	Mensæ	10.0	27.16	− 3.364	78 16 13.0	.65	52.16	1
3789	Mensæ	10.0	28.70	1.522	73 46 44.7	.65	52.04	1
3790	Doradûs	9.0	29.04	− 0.452	69 13 57.4	− 0.65	51.08	1
3791	Gould, Z. C., 6^h 241 . .	9.8	6 7 32.24	− 4.589	80 6 18.5	− 0.66	51.09	2
3792	Gould, Z. C., 6^h 340 . .	9.5	33.43	+ 0.383	63 34 10.8	.66	52.13	1
3793	Octantis	9.5	41.09	−21.177	86 50 44.2	.67	51.04	2
3794	Mensæ	10.0	43.56	1.224	72 43 21.5	.68	51.06	1
3795	Octantis	9.3	44.88	−23.590	87 7 50.2	− 0.68	51.02	3
3796	Doradûs	10.0	6 7 55.31	− 0.072	66 58 25.4	− 0.69	52.06	1
3797	Pictoris	10.0	8 0.20	+ 0.521	62 20 56.7	.70	52.13	1
3798	Pictoris	10.0	2.34	0.451	62 59 18.0	.70	51.15	1
3799	Pictoris	10.0	17.18	+ 0.341	63 55 50.8	.72	52.13	2
3800	Octantis	9.5	32.03	−27.325	87 28 58.8	− 0.75	51.03	2
3801	Stone, 2863	8.5	6 8 32.12	+ 0.199	65 3 17.0	− 0.75	52.10	2
3802	Mensæ	9.5	9 3.98	− 3.055	77 41 54.9	.79	50.94	1
3803	Doradûs	9.8	4.41	+ 0.120	65 38 44.1	.79	51.10	2
3804	Doradûs	10.0	7.57	+ 0.119	65 39 11.0	.80	51.10	1
3805	Gould, 7501	9.0	10.58	− 0.531	69 39 21.7	− 0.80	51.14	1
3806	Gould, Z. C., 6^h 390 . .	9.5	6 9 23.45	− 1.157	72 28 2.8	− 0.82	51.13	1
3807	Mensæ	9.5	24.17	6.370	81 56 43.9	.82	51.45	3
3808	Mensæ	10.0	24.71	− 3.333	78 13 1.9	.82	52.16	1
3809	Gould, Z. C., 6^h 419 . .	9.5	28.09	+ 0.295	64 18 18.5	.83	51.62	2
3810	Gould, Z. C., 6^h 389 . .	9.5	31.42	− 1.489	73 40 21.5	− 0.83	52.04	1
3811	Brisbane, 1185	8.0	6 9 41.32	+ 0.171	65 16 35.7	− 0.85	52.12	1
3812	ν Doradûs	6.0	41.76	− 0.374	68 48 37.7	.85	51.59	2
3813	Doradûs	9.5	44.56	− 0.568	69 50 51.5	.85	51.14	1
3814	Gould, Z. C., 6^h 439 . .	9.0	9 48.86	+ 0.090	65 52 5.6	.86	51.59	2
3815	Pictoris	9.0	10 0.11	+ 0.423	63 14 9.1	− 0.87	51.16	1
3816	Doradûs	10.0	6 10 2.80	− 0.001	66 30 14.1	− 0.88	51.12	1
3817	Mensæ	9.7	8.69	− 5.925	81 33 18.4	.89	50.73	3
3818	Gould, Z. C., 6^h 459 . .	9.5	14.81	+ 0.120	65 39 8.4	.90	51.10	1
3819	Gould, Z. C., 6^h 384 . .	9.2	25.28	− 3.931	79 12 4.8	.91	51.08	2
3820	Gould, Z. C., 6^h 402 . .	9.2	25.42	− 2.833	77 15 12.2	− 0.91	51.09	2
3821	Mensæ	9.0	6 10 28.24	− 7.582	82 51 12.8	− 0.92	50.95	1
3822	Gould, Z. C., 6^h 456 . .	9.0	33.60	− 0.826	71 5 7.6	.92	52.08	2
3823	Gould, Z. C., 6^h 469 . .	9.5	35.15	+ 0.159	65 22 5.9	.93	52.12	1
3824	Mensæ	11.0	50.68	− 3.371	78 17 17.6	.95	52.16	1
3825	Doradûs	9.8	51.93	− 0.141	67 25 34.6	− 0.95	52.05	2

No. 3806. Declination 10″ less than Gould.

Number.	Constellation, Name of Star, or Synonym.	Magnitude.	Right Ascension, 1850.0.	Annual Precession.	South Declination, 1850.0.	Annual Precession.	Mean year.	No. of obs.
			h. m. s.	s.	° ′ ″	″		
3 826	η^2 Doradûs	6.2	6 10 55.38	+ 0.134	65 33 13.2	− 0.96	51.44	3
3 827	Pictoris	9.0	10 57.39	+ 0.347	63 53 18.7	.96	52.13	2
3 828	Mensæ	10.0	11 7.86	− 5.729	81 22 16.6	.97	51.06	1
3 829	Doradûs	10.0	9.95	− 0.221	67 55 16.9	.98	51.15	1
3 830	Gould, Z. C., 6h 488 . .	9.2	12.90	− 0.650	70 15 47.8	− 0.98	51.53	2
3 831	Octantis	9.2	6 11 16.51	−14.402	85 21 56.2	− 0.99	51.03	3
3 832	Doradûs	10.0	19.33	+ 0.207	65 0 26.2	.99	52.07	1
3 833	Gould, Z. C., 6h 489 . .	9.2	21.16	− 0.912	71 28 11.2	0.99	52.06	2
3 834	Mensæ	9.2	27.57	2.227	75 51 14.0	1.00	51.10	2
3 835	Doradûs	10.0	28.13	− 0.060	66 54 27.6	− 1.00	52.06	1
3 836	Stone, 2883	8.0	6 11 30.74	+ 0.141	65 30 21.9	− 1.01	51.61	3
3 837	Doradûs	10.0	31.70	+ 0.288	64 22 14.4	.01	51.11	1
3 838	Gould, Z. C., 6h 452 . .	11.0	38.91	− 3.489	78 29 43.8	.02	52.16	1
3 839	Brisbane, 1201	8.0	49.24	+ 0.143	65 29 36.7	.03	51.44	3
3 840	Stone, 2888	8.3	52.57	+ 0.142	65 29 46.8	− 1.04	51.44	3
3 841	Gould, Z. C., 6h 514 . .	9.0	6 11 52.77	− 0.966	71 42 1.4	− 1.04	52.03	1
3 842	Mensæ	9.5	11 54.29	2.631	76 49 23.8	.04	50.95	1
3 843	Mensæ	10.0	12 1.13	1.225	72 44 12.7	.05	51.06	1
3 844	Gould, 7576	9.0	5.56	0.663	70 19 33.4	.06	51.53	2
3 845	Gould, Z. C., 6h 463 . .	9.2	16.74	− 4.385	79 50 53.6	− 1.07	50.15	2
3 846	Pictoris	9.2	6 12 23.42	+ 0.477	62 46 12.6	− 1.08	51.64	2
3 847	Gould, Z. C., 6h 510	9.2	25.43	− 2.416	76 19 35.1	.09	51.10	2
3 848	Gould, Z. C., 6h 559 . .	9.5	29.54	− 0.447	69 13 16.2	.09	51.08	1
3 849	Doradûs	9.8	40.64	+ 0.106	65 46 4.2	.11	51.60	2
3 850	Lacaille, 2248	7.5	42.76	− 0.541	69 43 8.9	− 1.11	51.14	1
3 851	Mensæ	10.0	6 12 45.64	− 4.970	80 34 27.5	− 1.12	51.10	2
3 852	Stone, 2846	9.0	48.86	11.910	84 54 29.0	.12	51.03	2
3 853	Mensæ	9.5	49.36	1.451	73 33 20.2	.12	51.59	2
3 854	Mensæ	9.5	49.55	− 1.695	74 21 17.1	.12	51.38	3
3 855	Gould, Z. C., 6h 583 . .	9.5	12 57.16	+ 0.074	65 59 57.5	− 1.13	52.09	1
3 856	Gould, Z. C., 6h 598 . .	9.2	6 13 7.66	+ 0.151	65 26 6.0	− 1.15	51.61	2
3 857	Mensæ	10.0	11.16	− 1.257	72 51 42.6	.15	51.06	1
3 858	Doradûs	9.5	16.36	+ 0.257	64 37 56.4	.16	51.09	2
3 859	Doradûs	10.0	17.97	+ 0.051	66 9 44.2	.16	52.09	1
3 860	Mensæ	10.0	20.00	− 1.844	74 48 27.6	− 1.17	51.52	2
3 861	Gould, Z. C., 6h 536 . .	9.2	6 13 26.32	− 3.504	78 31 33.8	− 1.18	51.16	2
3 862	Brisbane, 1204	8.5	30.10	+ 0.537	62 14 5.6	.18	52.13	1
3 863	Gould, Z. C., 6h 613 . .	9.0	34.37	+ 0.146	65 28 38.7	.19	52.12	1
3 864	Doradûs	9.8	37.54	− 0.155	67 31 39.8	.19	52.05	2
3 865	Pictoris	9.5	39.34	+ 0.410	63 17 2.0	− 1.19	51.16	1

No. 3 856. 15″ greater than Gould.
No. 3 860. The two observations differ by 1 rev. Declination may be 74° 47′ 59.7″.

Number.	Constellation, Name of Star, or Synonym.	Magnitude.	Right Ascension, 1850.0.	Annual Precession.	South Declination, 1850.0.	Annual Precession.	Mean year.	No. of obs.
			h. m. s.	s.	° ′ ″	″		
3866	Lacaille, 2249	7.5	6 13 43.84	− 0.019	66 37 56.6	− 1.20	51.59	2
3867	Gould, Z. C., 6h 620 . .	9.5	44.43	+ 0.536	62 14 34.4	.20	52.13	1
3868	Gould, 7624	8.5	45.70	− 0.440	69 11 32.2	.20	51.08	1
3869	Gould, Z. C., 6h 623 . .	7.5	46.58	+ 0.057	66 7 27.5	.20	52.09	1
3870	Gould, 7633	9.2	48.23	− 0.035	66 45 0.6	− 1.21	51.59	2
3871	Doradûs	9.5	6 13 50.02	+ 0.117	65 41 40.5	− 1.21	51.10	2
3872	Brisbane, 1206	9.0	52.26	+ 0.528	62 19 4.5	.21	52.13	1
3873	Mensæ	9.0	54.33	− 1.065	72 7 12.4	.22	52.08	1
3874	Gould, Z. C., 6h 583 . .	9.0	55.04	2.554	76 39 17.4	.22	50.56	2
3875	Lacaille, 2272	8.0	13 56.87	− 1.284	72 57 55.7	− 1.22	51.06	1
3876	Mensæ	9.0	6 14 5.80	− 2.659	76 53 28.3	− 1.23	50.95	1
3877	Doradûs	10.0	9.14	0.150	67 30 0.0	.24	52.06	1
3878	Gould, Z. C., 6h 628 . .	9.0	9.65	0.543	69 44 6.5	.24	51.14	1
3879	Doradûs	9.0	13.39	0.296	64 19 29.8	.24	51.62	2
3880	Gould, Z. C., 6h 636 . .	10.0	14.12	− 0.464	69 19 22.7	− 1.25	51.08	1
3881	Mensæ	9.0	6 14 15.11	− 5.458	81 6 29.5	− 1.25	51.06	1
3882	Gould, Z. C., 6h 587 . .	9.0	17.79	− 3.163	77 55 5.6	.25	50.17	2
3883	Pictoris	9.5	27.32	+ 0.409	63 22 49.0	.26	51.64	2
3884	Mensæ	10.0	27.34	− 0.982	71 46 51.8	.26	52.05	2
3885	α Mensæ	5.7	42.05	− 1.805	74 41 56.0	− 1.29	51.28	4
3886	Mensæ	9.5	6 14 52.21	− 1.448	73 33 7.2	− 1.30	52.04	1
3887	Doradûs	10.0	15 10.83	0.258	68 9 57.3	.33	52.11	1
3888	Doradûs	9.5	12.07	0.400	68 58 43.8	.33	51.08	1
3889	Doradûs	10.0	13.45	0.218	67 55 27.2	.33	51.15	1
3890	Gould, Z. C., 6h 673 . .	9.0	25.46	− 1.227	72 45 31.3	1.35	51.06	1
3891	Mensæ	10.0	6 15 31.47	− 1.141	72 25 42.6	− 1.36	51.13	1
3892	Lacaille, 2266	7.5	33.30	+ 0.042	66 14 9.6	.36	51.61	2
3893	Doradûs	9.8	34.02	− 0.419	69 5 10.0	.36	51.08	2
3894	Mensæ	10.0	34.02	1.640	74 11 26.9	.36	52.04	1
3895	Gould, Z. C., 6h 652 . .	9.0	35.70	− 2.699	76 59 5.1	− 1.36	50.95	1
3896	Gould, Z. C., 6h 715	9.2	6 15 47.24	− 0.026	66 42 3.5	− 1.38	51.59	2
3897	Mensæ	10.0	15 54.54	2.362	76 12 29.9	.39	52.02	1
3898	Doradûs	9.5	16 1.12	0.417	69 4 39.6	.40	51.08	1
3899	Mensæ	10.0	5.22	5.384	81 2 11.6	.41	51.06	1
3900	Mensæ	10.0	14.55	− 0.614	70 6 19.0	− 1.42	51.07	1
3901	Doradûs	10.0	6 16 16.31	− 0.441	69 12 35.2	− 1.42	51.08	1
3902	Herschel, 3855[1]	10.0	19.04	1.725	74 27 40.8	.43	51.16	1
3903	Pictoris	9.0	20.88	0.538	62 14 36.9	.43	52.13	1
3904	Herschel 3855[2]	10.5	21.32	1.724	74 27 39.8	.43	51.16	1
3905	Gould, Z. C., 6h 750 . .	10.0	34.75	− 0.702	70 32 9.1	− 1.45	51.99	1

Number.	Constellation, Name of Star, or Synonym.	Magnitude.	Right Ascension, 1850.0.	Annual Precession.	South Declination, 1850.0.	Annual Precession.	Mean year.	No. of obs.
			h. m. s.	s.	° ′ ″	″		
3906	Doradûs	10.0	6 16 37.63	+ 0.329	64 4 17.6	− 1.45	52.12	1
3907	Gould, Z. C., 6^h 756	10.0	43.24	− 0.674	70 24 14.1	.46	51.99	1
3908	Lacaille, 2304	8.0	47.37	− 1.916	75 1 56.7	.47	51.51	4
3909	Gould, Z. C., 6^h 783	9.0	53.73	+ 0.271	64 32 27.8	.48	51.11	1
3910	Lacaille, 2273	8.0	16 58.68	+ 0.342	63 58 2.3	− 1.48	52.13	2
3911	Pictoris	9.0	6 17 4.29	+ 0.535	62 16 42.2	− 1.49	52.13	1
3912	Mensæ	10.0	5.10	− 1.037	72 1 20.5	.49	52.08	1
3913	Doradûs	10.0	8.23	0.035	66 46 29.2	.50	52.06	1
3914	Lacaille, 2294	7.0	16.16	0.936	71 35 51.5	.51	52.05	2
3915	Lacaille, 2316	7.5	18.68	− 2.729	77 3 21.4	− 1.51	50.56	2
3916	Mensæ	9.8	6 17 24.06	− 8.098	83 11 37.0	− 1.52	52.02	2
3917	Gould, Z. C., 6^h 813	9.2	26.46	+ 0.447	63 4 22.9	.52	51.15	2
3918	Lacaille, 2298	7.0	29.22	− 0.946	71 38 44.5	.53	52.06	2
3919	Gould, 7732	8.0	42.80	− 0.403	69 0 46.9	.55	51.08	1
3920	Pictoris	9.5	45.82	+ 0.466	62 54 23.9	− 1.55	51.15	1
3921	Gould, 7727	9.5	6 17 49.25	− 1.140	72 26 9.5	− 1.56	51.13	1
3922	Mensæ	9.5	17 50.36	− 1.651	74 14 20.1	.56	52.04	1
3923	Doradûs	9.5	18 4.44	+ 0.279	64 29 7.6	.58	51.11	1
3924	Gould, Z. C., 6^h 791	8.5	6.05	− 2.169	75 43 51.8	.58	50.18	1
3925	Gould, Z. C., 6^h 834	9.0	8.30	+ 0.249	64 43 26.6	− 1.59	51.59	2
3926	Mensæ	9.5	6 18 13.73	− 1.643	74 12 49.2	− 1.59	52.04	1
3927	Mensæ	9.0	15.45	1.329	73 9 6.0	.60	51.06	1
3928	Mensæ	10.0	16.07	− 7.446	82 46 40.4	.60	50.13	1
3929	Pictoris	10.0	17.07	+ 0.526	62 22 11.1	.60	52.13	1
3930	Gould, Z. C., 6^h 803	9.0	18.60	− 2.223	75 52 12.3	− 1.60	50.79	3
3931	Doradûs	10.0	6 18 19.52	+ 0.234	64 50 44.4	− 1.60	52.07	1
3932	Doradûs	10.0	19.76	+ 0.145	65 30 53.0	.60	52.12	1
3933	Lacaille, 2308	7.2	20.40	− 1.447	73 34 2.0	.60	51.59	2
3934	Gould, Z. C., 6^h 844	9.0	20.42	+ 0.208	65 2 45.6	.60	52.07	1
3935	Gould, Z. C., 6^h 806	8.8	28.56	− 2.234	75 53 57.1	− 1.61	50.79	3
3936	Mensæ	10.0	6 18 34.42	− 2.916	77 27 6.2	− 1.62	50.17	2
3937	Mensæ	9.8	41.70	2.918	77 27 28.4	.63	50.17	2
3938	Gould, Z. C., 6^h 842	10.0	42.92	1.752	70 47 19.5	.64	52.08	1
3939	Gould, Z. C., 6^h 830	9.0	43.12	− 1.396	73 23 32.0	.64	51.59	2
3940	Lacaille, 2286	7.2	43.19	+ 0.369	63 45 19.6	− 1.64	52.13	2
3941	Mensæ	9.5	6 18 45.28	− 9.679	84 2 5.0	− 1.64	50.12	1
3942	Gould, Z. C., 6^h 816	9.0	48.27	2.667	76 55 41.9	.64	50.95	1
3943	Mensæ	10.0	51.33	1.385	73 21 18.7	.65	51.13	1
3944	Gould, Z. C., 6^h 817	9.0	53.66	2.842	77 18 9.4	.65	50.17	1
3945	Mensæ	10.0	54.31	− 4.244	79 40 38.3	− 1.65	50.16	1

Number.	Constellation, Name of Star, or Synonym.	Magnitude.	Right Ascension, 1850.0.	Annual Precession.	South Declination, 1850.0.	Annual Precession.	Mean year.	No. of obs.
			h. m. s.	s.	° ′ ″	″		
3946	Pictoris	9.2	6 18 57.82	+ 0.391	63 2 18.0	− 1.66	51.15	2
3947	Gould, Z. C., 6h 886	9.0	19 12.25	+ 0.082	65 59 2.3	.68	52.10	1
3948	Lacaille, 2512	7.3	13.93	− 15.587	85 54 56.4	.68	51.02	5
3949	Gould, Z. C., 6h 863	9.0	14.96	1.328	73 9 9.7	.68	51.13	1
3950	Lacaille, 2305	8.2	23.78	− 0.416	69 5 28.8	− 1.69	51.08	2
3951	Gould, Z. C., 6h 881	10.0	6 19 28.27	− 0.739	70 43 56.2	− 1.70	52.08	1
3952	Mensæ	9.5	34.56	+ 0.180	65 15 51.3	.71	52.12	1
3953	Melbourne (1) 303	8.0	34.99	+ 0.537	62 16 45.4	.71	52.13	1
3954	Doradûs	9.5	36.26	− 0.149	67 31 36.8	.71	52.05	2
3955	Doradûs	9.8	42.83	− 0.348	68 43 1.5	− 1.72	51.60	2
3956	Mensæ	10.0	6 19 43.83	− 1.393	73 23 6.5	− 1.72	52.04	1
3957	Gould, Z. C., 6h 926	8.8	52.14	+ 0.358	63 51 22.6	.74	52.13	2
3958	Mensæ	9.4	53.22	− 1.938	75 6 34.4	.74	51.51	4
3959	Gould, Z. C., 6h 837	9.2	57.02	− 3.849	79 9 15.2	.74	50.16	2
3960	Doradûs	10.0	19 57.76	+ 0.030	66 20 55.5	− 1.74	52.09	1
3961	Doradûs	11.0	6 20 4.09	+ 0.030	66 21 1.5	− 1.75	52.09	1
3962	Gould, Z. C., 6h 931	9.5	7.83	0.012	66 28 48.3	.76	51.12	1
3963	Gould, Z. C., 6h 939	8.8	9.04	+ 0.343	63 58 51.8	.76	52.13	2
3964	Doradûs	9.8	10.26	− 0.355	68 45 45.6	.76	51.60	2
3965	Doradûs	9.5	14.93	+ 0.261	64 38 41.1	− 1.77	51.11	1
3966	Doradûs	10.0	6 20 15.62	− 0.571	69 55 9.6	− 1.77	51.10	2
3967	Gould, Z. C., 6h 942	9.0	18.74	+ 0.034	66 19 39.6	.78	51.61	2
3968	Mensæ	10.0	24.03	− 1.967	75 11 39.2	.78	50.96	1
3969	Gould, Z. C., 6h 955	9.0	25.74	+ 0.535	62 18 18.9	.78	52.13	1
3970	Mensæ	9.5	26.60	− 1.185	72 37 41.4	− 1.79	51.06	1
3971	Lacaille, 2385	8.0	6 20 33.62	− 5.350	81 0 52.8	− 1.80	50.60	2
3972	Doradûs	9.8	34.82	0.293	68 21 38.6	.80	51.63	2
3973	Lacaille, 2314	8.0	36.14	0.531	69 42 50.5	.80	51.14	1
3974	Lacaille, 2322	8.0	51.58	1.041	72 3 44.6	.82	52.07	1
3975	Doradûs	9.5	55.45	− 0.203	67 52 16.8	− 1.83	51.59	2
3976	Mensæ	10.0	6 20 58.78	− 1.321	73 8 15.0	− 1.83	51.06	1
3977	Mensæ	9.5	21 3.14	1.020	71 58 31.8	.84	52.05	2
3978	Doradûs	9.8	5.94	− 0.327	68 36 34.6	.84	51.60	2
3979	Gould, Z. C., 6h 987	10.0	12.80	+ 0.512	62 31 23.9	.85	52.13	1
3980	Mensæ	8.8	14.00	− 2.321	76 7 59.6	− 1.86	51.10	2
3981	Gould, Z. C., 6h 995	9.0	6 21 21.23	+ 0.508	62 33 26.8	− 1.87	52.13	1
3982	Lacaille, 2403	8.7	21.47	− 6.333	81 56 34.4	.87	50.48	3
3983	Doradûs	10.0	23.19	0.050	66 54 23.0	.87	52.06	1
3984	Doradûs	10.0	31.99	0.119	67 21 23.6	.88	52.06	1
3985	Gould, Z. C., 6h 993	9.5	36.49	− 0.143	67 30 20.7	− 1.89	52.05	2

Number.	Constellation, Name of Star, or Synonym.	Magnitude.	Right Ascension, 1850.0.	Annual Precession.	South Declination, 1850.0.	Annual Precession.	Mean year.	No. of obs.
			h. m. s.	s.	° ′ ″	″		
3986	Doradûs	10.0	6 21 51.59	− 0.089	67 9 49.3	− 1.91	52.06	1
3987	Doradûs	10.0	53.64	+ 0.128	65 40 6.3	.91	51.10	1
3988	Lacaille, 2363	8.5	21 59.45	− 2.686	76 59 7.0	.92	50.43	3
3989	Doradûs	9.5	22 3.61	+ 0.300	64 21 0.1	.93	51.11	1
3990	Lacaille, 2312	6.0	3.75	+ 0.391	63 36 4.6	− 1.93	52.13	1
3991	Gould, 7835	9.0	6 22 7.24	− 5.326	80 59 43.3	− 1.93	51.10	2
3992	Gould, Z. C., 6h 1026	8.0	10.21	+ 0.286	64 27 55.0	.94	51.11	1
3993	Gould, Z. C., 6h 1003	8.0	16.85	− 1.360	73 17 12.0	.95	51.13	1
3994	Mensæ	10.0	22.56	− 0.680	70 28 21.0	.96	51.99	1
3995	Gould, Z. C., 6h 1040	9.5	23.17	+ 0.286	64 28 5.0	− 1.96	51.11	1
3996	Mensæ	9.5	6 22 33.06	− 2.498	76 33 51.1	− 1.97	50.18	1
3997	Gould, Z. C., 6h 1046	9.0	37.06	+ 0.051	66 13 41.0	.98	51.61	2
3998	Mensæ	9.0	40.47	− 0.936	71 38 4.5	1.98	52.03	1
3999	Gould, Z. C., 6h 1027	9.0	50.70	− 1.283	73 0 36.5	2.00	51.10	2
4000	Doradûs	9.8	22 56.14	+ 0.154	65 29 17.4	− 2.00	51.10	2
4001	Mensæ	9.8	6 23 0.06	− 1.232	72 40 21.2	− 2.01	51.09	2
4002	Doradûs	10.0	4.68	0.082	67 7 37.5	.02	52.06	1
4003	Mensæ	10.0	7.39	2.084	82 32 15.1	.02	50.13	1
4004	Mensæ	9.5	15.24	− 9.357	83 53 28.6	.03	50.12	1
4005	Gould, 7903	10.0	17.30	+ 0.380	63 42 8.8	− 2.03	52.13	1
4006	Doradûs	9.5	6 23 17.46	+ 0.226	64 57 8.4	− 2.03	52.07	1
4007	Gould, Z. C., 6h 1083	9.0	18.35	− 0.050	66 55 20.3	.04	52.06	1
4008	Mensæ	10.0	19.41	0.883	71 24 34.4	.04	52.03	1
4009	Doradûs	9.0	23.13	− 0.002	66 35 53.5	.04	52.06	1
4010	Lacaille, 2321	7.0	23.65	+ 0.423	63 20 16.4	− 2.04	51.16	1
4011	Doradûs	9.5	6 23 29.10	− 0.114	67 20 20.6	− 2.05	52.06	1
4012	Gould, Z. C., 6h 1098	8.0	33.88	+ 0.489	62 45 0.6	.06	51.15	1
4013	Gould, 7881	8.8	50.20	− 2.702	77 1 46.6	.08	50.43	3
4014	Brisbane, 1258	8.2	57.39	+ 0.142	65 35 22.2	.09	51.44	3
4015	Mensæ	10.0	23 59.08	− 0.945	71 40 52.3	− 2.10	52.03	1
4016	Lacaille, 2329	6.0	6 24 0.08	+ 0.377	63 44 21.5	− 2.10	52.13	1
4017	Brisbane, 1261	8.8	1.09	− 1.942	75 8 47.2	.10	51.51	4
4018	Doradûs	10.0	1.27	1.525	73 51 55.0	.10	52.04	1
4019	π^1 Doradûs	6.5	2.94	− 0.562	69 53 59.7	.10	51.10	2
4020	Gould, Z. C., 6h 1123	9.0	4.96	+ 0.535	62 20 6.3	− 2.10	51.13	1
4021	Lacaille, 2426	8.0	6 24 6.95	− 6.373	81 59 10.8	− 2.11	50.48	3
4022	Doradûs	9.5	7.92	+ 0.261	64 40 52.3	.11	52.07	1
4023	Doradûs	9.5	16.09	0.244	64 49 16.9	.12	52.07	1
4024	Pictoris	9.0	19.42	+ 0.437	63 13 34.5	.12	51.16	1
4025	Gould, 7900	8.5	22.11	− 2.495	76 34 0.2	− 2.13	50.18	1

No. 3991. Gould's declination 1′ greater.

Number.	Constellation, Name of Star, or Synonym.	Magnitude.	Right Ascension, 1850.0.	Annual Precession.	South Declination, 1850.0.	Annual Precession.	Mean year.	No. of obs.
			h. m. s.	s.	° ′ ″	″		
4026	Doradûs	10.0	6 24 25.56	— 0.043	66 53 18.9	— 2.13	52.06	1
4027	Gould, 7928	8.5	30.38	+ 0.166	65 25 5.0	.14	51.44	3
4028	Gould, Z. C., 6h 1137 . .	8.2	32.67	+ 0.130	65 41 7.1	.14	51.10	2
4029	Gould, 7908	8.5	39.51	— 2.113	75 37 10.4	.15	50.18	1
4030	Doradûs	11.0	40.07	— 3.351	78 18 24.0	— 2.15	52.16	1
4031	Pictoris	10.0	6 24 40.87	+ 0.421	63 21 47.4	— 2.16	51.16	1
4032	Doradûs	10.0	42.12	— 0.458	69 21 51.9	.16	51.08	1
4033	Doradûs	9.0	44.72	0.305	68 30 44.6	.16	52.11	1
4034	Octantis	9.3	47.56	508.550	89 51 4.1	.17	51.03	6
4035	Mensæ	9.8	51.30	— 4.657	80 14 31.4	- 2.17	51.10	2
4036	Doradûs	9.5	6 24 57.66	— 0.207	67 55 50.4	- 2.18	51.15	1
4037	Mensæ	9.0	25 2.18	9.977	84 11 2.8	.19	50.12	1
4038	Lacaille, 2358	8.0	2.95	0.761	70 51 27.9	.19	52.08	1
4039	Mensæ	10.0	2.98	— 3.572	78 41 12.7	.19	50.16	1
4040	Gould, Z. C., 6h 1175 . .	8.5	16.98	+ 0.318	64 14 11.4	— 2.21	51.62	2
4041	Mensæ	9.2	6 25 21.38	— 2.530	76 39 7.6	— 2.21	50.56	2
4042	Doradûs	9.0	26.04	+ 0.291	64 27 29.9	.22	51.11	1
4043	Lacaille, 2439	8.0	27.04	— 6.500	82 5 45.0	.22	50.15	2
4044	Lacaille, 2357	8.0	33.06	0.598	70 5 51.0	.23	51.07	1
4045	Doradûs	9.2	37.08	— 0.120	67 24 3.2	— 2.24	52.05	2
4046	Mensæ	11.0	6 25 38.81	— 3.379	78 21 38.3	— 2.24	52.16	1
4047	Octantis	9.2	41.66	13.419	85 23 36.6	.24	51.03	2
4048	Doradûs	9.5	43.12	— 0.338	68 42 38.8	.25	51.60	2
4049	Pictoris	10.0	56.25	+ 0.479	62 52 25.4	.27	51.15	1
4050	Gould, Z. C., 6h 1178 . .	9.0	57.78	— 1.086	72 16 44.6	— 2.27	51.60	2
4051	Mensæ	9.5	6 25 59.47	— 1.626	74 12 30.9	— 2.27	52.04	1
4052	Gould, Z. C., 6h 1176 . .	9.0	26 1.42	1.344	73 15 22.3	.27	51.13	1
4053	Mensæ	10.0	2.20	3.568	78 41 13.1	.27	52.16	1
4054	Gould, Z. C., 6h 1147 . .	10.0	4.51	2.798	77 14 52.7	.28	50.17	1
4055	Mensæ	9.5	4.97	— 5.018	80 39 37.3	— 2.28	52.06	1
4056	Doradûs	9.2	6 26 4.98	+ 0.304	64 21 53.5	— 2.28	51.62	2
4057	Brisbane 1274	8.2	6.69	— 1.900	75 2 30.8	.28	51.51	4
4058	Mensæ	10.0	8.52	— 1.660	74 19 6.7	.28	52.04	1
4059	Doradûs	9.0	14.26	+ 0.132	65 41 6.8	.29	51.10	1
4060	Doradûs	9.5	16.35	— 0.325	68 38 39.9	— 2.29	51.60	2
4061	Mensæ	9.0	6 26 28.14	— 8.601	83 31 30.4	2.31	51.42	3
4062	Mensæ	9.5	32.50	1.036	72 5 2.3	.32	52.08	1
4063	Mensæ	9.0	33.71	1.587	74 5 23.5	.32	52.04	1
4064	Lacaille, 2440	8.0	41.20	5.791	81 28 39.2	.33	51.20	2
4065	π[1] Doradus	7.0	44.81	— 0.500	69 36 15.1	— 2.34	51.14	1

Number.	Constellation, Name of Star, or Synonym.	Magnitude.	Right Ascension, 1850.0.	Annual Precession.	South Declination, 1850.0.	Annual Precession.	Mean year.	No. of obs.
			h. m. s.	s.	° ′ ″	″		
4 066	Mensæ	10.0	6 26 46.04	— 8.061	83 11 53.6	2.34	52.02	2
4 067	Gould, Z. C., 6h 1244 . .	10.0	48.19	+ 0.090	65 59 37.2	.34	52.09	1
4 068	Pictoris	10.0	48.25	+ 0.482	62 51 19.2	.34	51.15	1
4 069	Mensæ	10.0	55.38	— 1.251	72 55 26.0	.35	51.06	1
4 070	Pictoris	10.0	56.00	+ 0.529	62 25 52.9	— 2.35	52.13	1
4 071	Gould, Z. C., 6h 1180 . .	9.8	6 26 57.75	— 3.423	78 26 45.9	— 2.35	51.16	2
4 072	Gould, Z. C., 6h 1232 . .	9.5	26 59.64	— 0.764	70 54 18.3	.36	52.08	1
4 073	Doradûs	9.5	27 4.63	+ 0.348	64 0 50.9	.36	52.12	1
4 074	Doradûs	10.0	14.75	— 0.320	68 37 21.2	.38	51.08	1
4 075	Doradûs	10.0	14.77	— 0.143	67 33 31.3	— 2.38	52.05	2
4 076	Octantis	10.5	6 27 18.25	—14.992	85 47 48.1	— 2.38	51.04	2
4 077	Gould, Z. C., 6h 1278 . .	8.8	23.09	+ 0.527	62 27 15.0	.39	51.66	2
4 078	Doradûs	9.5	25.45	+ 0.144	65 36 23.9	.39	51.10	2
4 079	Mensæ	9.5	29.07	— 0.880	71 25 56.2	.40	52.08	1
4 080	Doradûs	9.2	29.74	+ 0.273	64 37 17.4	— 2.40	51.59	2
4 081	Doradûs	9.5	6 27 31.80	+ 0.007	66 34 43.0	— 2.40	51.59	2
4 082	Doradûs	9.5	32.33	0.002	66 37 0.4	.40	52.06	1
4 083	Pictoris	10.0	32.70	0.526	62 27 55.1	.40	51.66	2
4 084	Doradûs	9.5	39.03	+ 0.142	65 37 29.9	.41	51.10	2
4 085	Mensæ	9.7	44.63	— 4.998	80 39 34.6	— 2.42	50.78	3
4 086	Doradûs	9.0	6 27 47.58	— 0.099	67 17 19.8	— 2.43	52.06	1
4 087	Mensæ	10.0	48.16	5.516	81 12 50.2	.43	51.06	1
4 088	Mensæ	9.5	50.23	5.660	81 21 26.8	.43	51.06	1
4 089	Doradûs	10.0	50.78	0.023	66 47 10.0	.43	52.06	1
4 090	Mensæ	10.0	50.83	— 1.277	73 1 40.8	— 2.43	51.06	1
4 091	Mensæ	9.5	6 27 54.85	— 0.884	71 27 11.2	— 2.44	52.08	1
4 092	Gould, Z. C., 6h 1289 . .	8.5	59.63	0.241	68 10 17.2	.44	51.63	2
4 093	Brisbane, 1282	9.2	27 59.75	0.003	66 39 14.3	.44	51.59	2
4 094	Mensæ	9.5	28 0.62	1.400	73 28 13.3	.44	52.04	1
4 095	Mensæ	9.0	4.68	— 5.244	80 55 58.1	— 2.45	50.14	1
4 096	Doradûs	9.5	6 28 8.64	— 0.419	69 11 8.7	— 2.46	51.08	1
4 097	Doradûs	9.5	12.56	0.091	67 14 11.9	.46	52.06	1
4 098	Mensæ	9.2	13.01	0.999	71 56 41.6	.46	52.05	2
4 099	Doradûs	10.0	15.45	0.040	66 54 11.5	.47	52.06	1
4 100	Gould, Z. C., 6h 1306 . .	9.5	17.79	— 0.215	68 0 47.6	— 2.47	51.15	1
4 101	Doradûs	9.0	6 28 20.94	+ 0.335	64 7 55.5	— 2.47	52.13	1
4 102	Doradûs	9.5	27.44	0.168	65 27 36.1	.48	51.10	1
4 103	Gould, Z. C., 6h 1315 . .	9.0	31.83	+ 0.155	65 32 28.5	.49	51.44	3
4 104	Doradûs	10.0	40.88	— 0.026	66 49 15.2	.50	52.06	1
4 105	Doradûs	9.5	46.09	+ 0.344	64 4 5.2	— 2.51	52.12	1

Number.	Constellation, Name of Star, or Synonym.	Magnitude.	Right Ascension, 1850.0.	Annual Precession.	South Declination, 1850.0.	Annual Precession.	Mean year.	No. of obs.
			h. m. s.	s.	° ′ ″	″		
4106	Pictoris	9.2	6 28 47.14	+ 0.504	62 40 32.2	— 2.51	51.64	2
4107	Mensæ	9.8	28 53.28	— 7.745	83 1 31.4	.52	52.02	2
4108	Pictoris	9.0	29 0.69	+ 0.459	63 5 7.6	.53	51.15	2
4109	Gould, Z. C., 6h 1335 . .	9.5	1.31	— 0.198	67 55 18.8	.53	51.15	1
4110	Gould, Z. C., 6h 1359 . .	9.5	23.72	— 0.215	68 1 30.3	— 2.57	51.15	1
4111	Gould, Z. C., 6h 1329 . .	9.0	6 29 26.42	— 1.483	73 46 4.0	— 2.57	52.04	2
4112	Lacaille, 2431	9.0	29.29	— 3.342	78 19 9.9	.57	50.82	3
4113	Lacaille, 2381	7.0	38.50	+ 0.168	65 27 42.1	.59	51.44	3
4114	Mensæ	10.0	41.25	— 1.964	75 14 58.2	.59	52.02	1
4115	Mensæ	9.8	43.32	— 1.377	73 24 16.5	— 2.59	51.59	2
4116	Mensæ	10.0	6 29 44.33	— 1.680	74 24 37.1	— 2.60	51.16	1
4117	Mensæ	12.0	47.58	1.597	74 8 44.2	.60	52.04	1
4118	Mensæ	10.0	29 50.66	0.601	70 9 12.3	.60	51.07	1
4119	Octantis	9.2	30 0.34	—15.331	85 52 47.0	.62	51.26	5
4120	Brisbane, 1295	8.8	1.62	+ 0.555	62 13 34.8	— 2.62	51.66	2
4121	Gould, Z. C., 6h 1366 . .	9.5	6 30 3.60	— 1.594	74 8 25.9	— 2.62	52.04	1
4122	Gould, 8076	8.3	8.18	+ 0.106	65 54 55.6	.63	51.43	3
4123	Gould, Z. C., 6h 1376 . .	8.8	9.30	— 0.988	71 55 12.3	.63	52.05	2
4124	Doradûs	10.0	10.94	0.547	69 53 9.4	.63	51.10	2
4125	Mensæ	9.0	16.10	— 0.945	71 44 20.7	2.64	52.03	1
4126	Doradûs	9.5	6 30 21.33	+ 0.298	64 27 50.9	— 2.65	51.11	1
4127	Lacaille, 2442	8.0	21.76	— 3.731	78 58 29.3	.65	50.16	2
4128	Pictoris	10.0	25.75	+ 0.423	63 25 30.6	.66	52.13	1
4129	Doradûs	9.0	30.47	— 0.260	68 18 41.7	.66	52.11	1
4130	Gould, Z. C., 6h 1384 . .	9.5	30.80	— 1.297	73 7 26.2	— 2.66	51.10	2
4131	Mensæ	8.0	6 30 34.31	— 8.917	83 41 32.8	— 2.67	50.12	1
4132	Doradûs	9.5	34.86	+ 0.257	64 47 12.1	.67	52.08	2
4133	Mensæ	9.5	39.80	— 1.359	73 21 0.4	.68	51.13	1
4134	Mensæ	9.2	51.66	3.515	78 37 36.6	.69	51.82	3
4135	Gould, Z. C., 6h 1425 . .	9.0	54.56	— 0.126	67 29 31.0	— 2.70	52.05	2
4136	Gould, Z. C., 6h 1433 . .	9.0	6 30 56.33	— 0.051	67 0 41.2	— 2.70	52.06	2
4137	Doradûs	9.3	31 0.50	+ 0.282	64 35 46.4	.71	51.75	3
4138	Brisbane, 1300	8.8	3.80	— 0.013	66 45 27.8	.71	51.59	2
4139	Doradûs	9.5	7.03	+ 0.303	64 25 48.7	.71	51.11	1
4140	Mensæ	9.5	10.63	— 5.841	81 32 46.0	— 2.72	51.06	1
4141	Mensæ	9.1	6 31 17.13	— 8.463	83 26 55.9	2.73	51.07	4
4142	Gould, Z. C., 6h 1468 . .	10.0	20.02	+ 0.516	62 36 18.1	.73	52.13	1
4143	Gould, 8099	8.2	21.59	— 0.350	68 50 9.7	.73	51.60	2
4144	Gould, Z. C., 6h 1451 . .	8.0	34.50	0.863	71 23 47.2	.75	52.08	1
4145	Gould, 8114	8.2	34.80	— 0.120	67 28 2.0	— 2.75	52.05	2

Number.	Constellation, Name of Star, or Synonym.	Magnitude.	Right Ascension, 1850.0.	Annual Precession.	South Declination, 1850.0.	Annual Precession.	Mean year.	No. of obs.
			h. m. s.	s.	° ′ ″	″		
4 146	Mensæ	10.0	6 31 40.88	− 0.641	70 22 9.1	− 2.76	51.99	1
4 147	Gould, 8117	7.3	41.08	+ 0.205	65 12 10.6	.76	52.09	3
4 148	Mensæ	10.0	41.50	8.548	83 29 53.7	.76	52.02	1
4 149	Gould, Z. C., 6h 1479 . .	9.5	45.58	+ 0.130	65 45 53.5	.77	51.60	2
4 150	Gould, 8124	7.3	31 50.65	+ 0.199	65 15 11.9	− 2.78	52.09	3
4 151	Doradûs	10.0	6 32 1.24	− 0.009	66 44 30.9	− 2.79	52.06	1
4 152	Gould, Z. C., 6h 1442 . .	10.0	9.39	2.615	76 53 43.2	.81	50.17	1
4 153	Mensæ	10.0	16.79	0.866	71 24 59.6	.82	52.08	1
4 154	Brisbane, 1311	7.5	23.36	0.205	68 0 1.4	.83	51.15	1
4 155	Mensæ	10.0	38.64	− 8.575	83 31 0.8	− 2.85	52.02	2
4 156	Mensæ	10.0	6 32 41.26	− 0.805	71 9 0.2	− 2.85	52.08	2
4 157	Gould, Z. C., 6h 1537 . .	9.0	41.38	+ 0.507	62 42 28.9	.85	51.48	3
4 158	Gould, Z. C., 6h 1520 . .	8.8	45.24	− 0.301	68 34 24.2	.86	51.60	2
4 159	Mensæ	9.5	46.81	1.389	73 28 36.3	.86	52.04	1
4 160	Lacaille, 2419	7.8	51.80	0.645	70 24 2.2	− 2.87	51.53	2
4 161	Octantis	8.2	6 32 52.70	−16.003	86 1 52.6	− 2.87	51.08	7
4 162	Doradûs	10.0	32 53.49	+ 0.324	64 17 18.9	.87	51.11	1
4 163	Gould, Z. C., 6h 1525 . .	10.0	33 6.49	− 1.018	72 4 15.2	.89	52.08	1
4 164	Gould, Z. C., 6h 1563 . .	9.0	18.04	+ 0.384	63 47 45.4	.90	52.13	2
4 165	Pictoris	9.0	25.95	+ 0.537	62 26 20.8	− 2.92	51.18	1
4 166	Doradûs	10.0	6 33 29.94	+ 0.320	64 19 46.6	− 2.92	51.62	2
4 167	Mensæ	10.0	35.53	− 1.180	72 43 5.8	.93	51.06	1
4 168	Lacaille, 2410	7.8	38.80	+ 0.461	63 8 12.9	.93	51.15	2
4 169	Pictoris	10.0	39.01	0.541	62 24 32.0	.93	51.18	1
4 170	Pictoris	9.0	52.85	+ 0.542	62 22 41.2	− 2.95	51.18	1
4 171	Doradûs	9.0	6 33 56.25	+ 0.105	65 58 30.5	− 2.96	52.09	1
4 172	Doradûs	9.0	34 0.68	0.268	64 44 56.8	.97	52.08	2
4 173	Doradûs	10.0	6.22	+ 0.256	64 50 48.0	.97	52.08	2
4 174	Mensæ	10.0	10.80	− 1.164	72 39 47.7	.98	51.06	1
4 175	Pictoris	10.0	17.18	+ 0.500	62 47 35.8	− 2.99	51.15	1
4 176	Mensæ	10.0	6 34 19.43	− 0.598	70 11 14.2	− 2.99	51.07	1
4 177	Gould, Z. C., 6h 1617 . .	8.2	30.38	+ 0.252	64 53 9.3	3.01	52.08	2
4 178	Mensæ	10.0	32.00	− 1.144	72 34 30.8	.01	51.13	1
4 179	Mensæ	10.0	38.74	0.784	71 4 45.0	.02	52.08	1
4 180	Gould, Z. C., 6h 1575 . .	10.0	46.42	2.235	76 0 48.6	− 3.03	50.18	2
4 181	Doradûs	10.0	6 34 50.55	− 0.258	68 21 7.0	3.04	52.11	1
4 182	Mensæ	10.0	53.38	4.136	79 36 36.7	.04	50.16	1
4 183	Mensæ	9.5	34 56.66	6.474	82 7 5.0	.05	51.15	2
4 184	Stone, 2901	7.6	35 6.66	42.502	88 20 21.0	.06	51.09	8
4 185	Gould, Z. C., 6h 1625 . .	9.5	8.70	− 1.008	72 3 20.8	− 3.06	52.08	1

No. 4 158. Gould's Declination is 15″ greater than this.

No. 4 171. The Right Ascension may be 10 or 20s greater.

Number.	Constellation, Name of Star, or Synonym.	Magnitude.	Right Ascension, 1850.0.	Annual Precession.	South Declination, 1850.0.	Annual Precession.	Mean year.	No. of obs.
			h. m. s.	s.	° ′ ″	″		
4186	Lacaille, 2466	8.0	6 35 10.30	− 2.733	77 10 34.5	− 3.07	50.17	1
4187	Melbourne (1), 315 . . .	8.2	14.93	+ 0.270	64 45 13.1	.07	51.76	3
4188	Volantis	10.0	19.97	+ 0.005	66 41 26.3	.08	51.12	1
4189	Volantis	8.0	25.86	− 0.883	71 31 35.7	.09	52.06	2
4190	Pictoris	9.5	30.42	+ 0.506	62 13 7.7	− 3.10	51.18	1
4191	Gould, 8237	8.0	6 35 30.71	− 0.055	67 5 48.8	− 3.10	52.06	2
4192	Volantis	10.0	31.37	+ 0.358	64 2 34.7	.10	52.12	1
4193	Gould, Z. C., 6h 1675 . .	9.0	42.60	− 0.270	68 25 55.3	.11	52.11	1
4194	Gould, Z. C., 6h 1696 . .	8.5	50.18	+ 0.294	64 34 3.1	.12	51.11	1
4195	Volantis	10.0	51.12	− 1.091	72 23 57.2	− 3.12	51.13	1
4196	Volantis	9.0	6 35 51.87	− 0.740	70 53 21.1	− 3.13	52.08	1
4197	Pictoris	9.5	57.91	+ 0.503	62 47 21.8	.13	51.15	1
4198	Gould, Z. C., 6h 1686 . .	9.5	35 58.90	− 0.509	69 45 30.9	.14	51.14	1
4199	Gould, Z. C., 6h 1699 . .	9.5	36 2.17	0.140	67 38 59.5	.14	52.03	1
4200	Volantis	9.5	4.40	− 0.489	69 39 7.0	− 3.14	51.14	1
4201	Mensæ	10.0	6 36 6.04	− 2.354	76 19 21.6	− 3.16	50.18	1
4202	Gould, Z. C., 6h 1684 . .	10.0	10.56	− 1.040	72 11 50.1	.15	52.08	1
4203	Pictoris	10.0	20.02	+ 0.397	63 43 45.7	.17	52.13	1
4204	Gould, 8232	8.5	20.22	− 2.255	76 4 47.0	.17	50.18	2
4205	Volantis	10.0	34.92	− 0.261	68 23 28.0	− 3.19	52.11	1
4206	Lacaille, 2457	7.4	6 36 38.76	− 1.420	73 37 29.6	− 3.19	52.04	1
4207	Gould, Z. C., 6h 1739 . .	8.8	39.28	+ 0.253	64 54 29.8	.19	52.08	2
4208	Gould, Z. C., 6h 1716 . .	8.5	40.99	− 1.025	72 8 26.8	.20	52.08	1
4209	Volantis	10.0	44.66	0.850	71 24 0.0	.20	52.03	1
4210	Octantis	9.0	47.18	14.130	85 36 43.9	− 3.21	51.20	1
4211	Volantis	9.8	6 36 53.48	− 0.667	70 33 31.9	3.21	52.04	2
4212	Mensæ	9.0	55.70	6.556	82 11 41.3	.22	51.17	1
4213	Volantis	10.0	36 58.16	1.782	74 47 26.6	.22	50.97	1
4214	Gould, Z. C., 6h 1751 . .	9.0	37 8.32	0.547	69 57 54.9	.24	51.10	2
4215	Octantis	8.8	18.47	−17.548	86 20 19.7	− 3.25	51.08	6
4216	Volantis	9.0	6 37 19.84	− 0.088	67 20 9.3	− 3.25	52.06	1
4217	Pictoris	9.8	21.08	+ 0.378	63 54 19.2	.25	52.13	2
4218	Volantis	9.8	24.08	− 0.237	68 15 49.4	.26	51.63	2
4219	Mensæ	7.5	26.72	7.473	82 52 7.5	.26	50.13	1
4220	Volantis	9.0	28.34	− 0.950	71 50 25.3	− 3.26	51.05	2
4221	Gould, Z. C., 6h 1785 . .	9.2	6 37 31.08	+ 0.436	63 24 42.2	− 3.27	51.64	2
4222	Volantis	9.8	37.04	− 1.138	72 36 15.0	.28	51.09	2
4223	Gould, Z. C., 6h 1786 . .	9.5	42.30	+ 0.076	66 14 22.9	.28	51.61	2
4224	Lacaille, 2502	8.2	43.73	− 3.581	78 47 11.0	.29	51.16	2
4225	Gould, Z. C., 6h 1799 . .	9.0	46.95	+ 0.445	63 19 50.4	− 3.29	51.16	1

Number.	Constellation, Name of Star, or Synonym.	Magnitude.	Right Ascension, 1850.0.	Annual Precession.	South Declination, 1850.0.	Annual Precession.	Mean year.	No. of obs.
			h. m. s.	s.	° ′ ″	″		
4226	Octantis	9.1	6 37 59.11	−20.980	86 51 42.1	− 3.31	51.11	5
4227	Volantis	10.0	38 3.01	+ 0.139	65 47 17.7	.31	51.43	3
4228	Volantis	9.5	3.61	− 0.419	69 18 30.3	.32	51.08	1
4229	Volantis	10.0	3.66	− 1.706	74 34 10.0	.32	51.16	1
4230	Lacaille, 2451	7.5	9.35	+ 0.117	65 56 54.6	− 3.32	51.43	3
4231	Mensæ	10.0	6 38 21.48	− 5.522	81 16 42.6	− 3.34	51.06	1
4232	Volantis	10.9	24.09	1.670	74 27 47.1	.34	51.16	1
4233	Gould, Z. C., 6^h 1755	9.5	27.50	3.521	78 41 29.4	.35	52.16	1
4234	Mensæ	9.0	30.44	7.687	83 0 52.0	.35	50.13	1
4235	Gould, Z. C., 6^h 1741	10.0	30.73	− 4.167	79 40 43.4	− 3.35	50.16	1
4236	Volantis	10.0	6 38 34.25	+ 0.181	65 29 14.0	− 3.36	51.10	2
4237	Volantis	10.0	34.40	0.134	65 50 11.1	.36	52.09	1
4238	Gould, Z. C., 6^h 1862	9.0	47.78	+ 0.332	64 18 39.6	.38	51.62	2
4239	Octantis	10.0	57.75	−13.530	85 26 49.8	.39	52.20	1
4240	Gould, 8323	8.2	38 58.61	− 0.708	70 46 46.2	− 3.39	52.03	2
4241	Gould, Z. C., 6^h 1836	9.8	6 39 2.68	− 1.146	72 38 54.9	− 3.40	51.09	2
4242	Volantis	12.0	4.44	+ 0.228	65 8 8.8	.40	52.10	2
4243	Gould, Z. C., 6^h 1874	8.5	8.22	+ 0.229	65 7 56.7	.41	52.10	2
4244	Volantis	9.2	11.84	− 1.360	73 26 36.2	.43	51.59	2
4245	Lacaille, 2472	8.0	17.30	− 0.641	70 27 42.7	− 3.44	51.99	1
4246	Volantis	10.0	6 39 34.72	− 1.259	73 5 3.2	− 3.45	51.06	1
4247	Mensæ	10.0	34.82	1.846	75 0 19.0	.45	51.32	3
4248	Volantis	9.5	35.60	0.538	69 57 8.6	.45	51.10	2
4249	Gould, Z. C., 6^h 1853	9.1	35.94	− 1.873	75 5 1.8	.45	51.24	5
4250	Pictoris	9.0	40.10	+ 0.490	62 58 5.3	− 3.45	51.15	1
4251	Octantis	8.8	6 39 41.48	−16.894	86 13 31.7	− 3.46	51.08	6
4252	Gould, Z. C., 6^h 1885	8.5	44.17	0.711	70 48 13.4	.46	52.08	1
4253	Volantis	10.0	47.20	− 0.988	72 1 42.3	.46	52.08	1
4254	Volantis	9.5	49.30	+ 0.143	65 47 29.4	.47	51.10	2
4255	Volantis	10.0	39 56.01	− 0.018	66 54 51.7	− 3.48	52.06	1
4256	Volantis	9.0	6 40 21.89	− 1.444	73 44 48.3	− 3.51	52.04	1
4257	Brisbane, 1347	8.5	24.22	+ 0.348	64 12 38.8	.52	51.62	2
4258	Gould, Z. C., 6^h 1951	9.2	47.76	− 0.665	70 35 58.6	.55	52.04	2
4259	Gould, Z. C., 6^h 1915	9.0	48.05	2.085	75 41 9.7	.55	50.18	1
4260	Octantis	10.0	48.99	−16.429	86 8 21.4	− 3.55	51.20	1
4261	Brisbane, 1356	8.0	6 40 52.41	+ 0.317	64 28 10.3	− 3.56	51.11	1
4262	Volantis	10.0	57.08	0.010	66 44 43.6	.56	52.06	1
4263	Gould, Z. C., 6^h 1983	9.5	59.31	0.140	65 49 55.0	.57	51.43	3
4264	Brisbane, 1357	8.5	40 59.65	+ 0.209	65 18 57.1	.57	52.12	1
4265	Volantis	10.0	41 0.10	− 0.335	68 53 9.7	− 3.57	51.08	1

Number.	Constellation, Name of Star, or Synonym.	Magnitude.	Right Ascension, 1850.0.	Annual Precession.	South Declination, 185 .0.	Annual Precession.	Mean year.	No. of obs.
			h. m. s.	s.	° ′ ″	″		
4 266	Volantis	10.0	6 41 0.40	− 1.248	73 3 43.1	− 3.57	51.06	1
4 267	Lacaille, 2551	9.5	2.21	4.718	80 24 47.0	.57	50.15	1
4 268	Gould, Z. C., 6^h 1937 . .	9.0	6.69	1.072	75 39 14.3	.58	50.18	1
4 269	Gould, Z. C., 6^h 2010 . .	9.0	17.65	+ 0.472	62 37 23.0	.59	51.17	2
4 270	Brisbane, 1358	8.5	19.00	+ 0.330	64 22 21.5	− 3.60	51.62	2
4 271	Mensæ	10.0	6 41 19.75	− 2.562	76 51 31.4	− 3.60	50.17	1
4 272	Lacaille, 2527	7.5	23.45	2.885	77 32 54.0	.60	50.17	2
4 273	Volantis	9.0	27.99	1.721	74 39 16.3	.61	51.40	3
4 274	Volantis	9.0	28.34	− 1.565	74 9 47.5	.61	51.60	2
4 275	Volantis	10.0	31.96	+ 0.267	64 52 43.9	− 3.61	52.07	1
4 276	Gould, Z. C., 6^h 2027 . .	9.5	6 41 32.19	+ 0.460	62 44 32.7	− 3.62	51.15	1
4 277	Volantis	9.5	41.68	+ 0.254	64 58 56.4	.63	52.07	1
4 278	Gould, Z. C., 6^h 1971 . .	9.5	43.19	− 2.021	75 31 25.0	.63	51.49	2
4 279	Gould, 8350	9.5	43.51	− 4.756	80 27 52.2	.63	50.15	1
4 280	Volantis	9.0	45.85	+ 0.004	66 47 44.7	− 3.63	52.06	1
4 281	Gould, Z. C., 6^h 2034 . .	9.2	6 41 51.46	+ 0.127	65 56 35.8	− 3.64	51.59	2
4 282	Volantis	9.0	51.61	− 0.573	70 9 47.5	.64	51.07	1
4 283	Volantis	9.0	52.29	+ 0.136	65 52 16.9	.64	51.10	1
4 284	Volantis	10.0	55.84	− 1.593	74 15 38.0	.65	51.60	2
4 285	Volantis	9.0	59.49	+ 0.175	65 35 18.9	− 3.65	51.10	1
4 286	Gould, Z. C., 6^h 2051 . .	9.0	6 41 59.74	+ 0.550	62 27 44.1	− 3.65	51.18	1
4 287	Volantis	9.5	42 0.13	0.165	65 39 48.6	.66	51.10	1
4 288	Melbourne, (1), 317 . . .	8.2	3.72	+ 0.494	62 58 56.5	.66	51.15	2
4 289	Volantis	10.0	7.50	− 0.333	68 53 13.9	.67	51.08	1
4 290	Volantis	10.0	12.59	+ 0.258	64 57 39.2	− 3.67	52.07	1
4 291	Pictoris	9.8	6 42 19.29	+ 0.446	63 24 24.8	− 3.68	51.64	2
4 292	Pictoris	6.0	22.58	+ 0.407	63 44 38.4	.69	52.13	1
4 293	Volantis	9.5	23.30	− 0.529	69 56 51.4	.69	51.10	2
4 294	Volantis	9.5	25.65	1.188	72 51 14.9	.69	51.06	1
4 295	Octantis	9.2	31.16	−19.526	86 40 18.6	− 3.70	51.09	3
4 296	Lacaille, 2504	9.8	6 42 41.33	− 0.639	70 30 7.7	− 3.71	51.99	1
4 297	Lacaille, 2495	8.0	47.14	0.130	67 41 25.8	.72	52.03	1
4 298	Volantis	10.0	54.22	0.226	68 16 50.6	.73	51.63	2
4 299	Volantis	8.5	57.20	0.198	68 6 51.5	.74	51.15	1
4 300	Mensæ	10.0	57.74	− 4.942	80 42 25.7	− 3.74	50.14	1
4 301	Gould, 8440	7.5	6 42 58.30	− 0.737	70 58 23.6	3.74	52.08	2
4 302	Volantis	9.8	42 59.52	1.595	74 16 46.1	.74	51.60	2
4 303	Gould, Z. C., 6^h 2040 . .	9.5	43 2.70	2.467	76 39 32.9	.74	50.18	2
4 304	Mensæ	8.5	4.43	− 7.762	83 5 13.6	.75	50.13	2
4 305	Gould, Z. C., 6^h 2110 . .	8.8	7.66	+ 0.073	66 20 41.2	− 3.75	51.61	2

Number.	Constellation, Name of Star, or Synonym.	Magnitude.	Right Ascension, 1850.0.	Annual Precession.	South Declination, 1850.0.	Annual Precession.	Mean year.	No. of obs.
			h. m. s.	s.	° ′ ″	″		
4 306	Volantis	9.5	6 43 8.79	− 0.196	68 6 9.0	− 3.75	51.15	1
4 307	Lacaille, 2515	6.0	11.81	0.881	71 37 13.9	.76	52.03	1
4 308	Lacaille, 2508	8.0	14.53	0.451	69 33 8.2	.76	51.14	1
4 309	Volantis	9.2	18.25	0.378	69 9 32.3	.77	51.08	2
4 310	Volantis	10.0	22.54	− 0.778	71 9 53.3	− 3.77	52.08	1
4 311	Mensæ	9.0	6 43 24.13	− 5.274	81 3 29.6	− 3.78	51.06	1
4 312	Mensæ	10.0	33.93	− 3.495	78 41 32.4	.79	52.16	1
4 313	Volantis	11.0	36.01	+ 0.321	64 29 18.4	.79	51.11	1
4 314	Volantis	11.0	37.01	+ 0.321	64 29 11.4	.79	51.11	1
4 315	Gould, 8463	7.8	38.14	− 0.755	71 3 43.8	− 3.80	52.08	2
4 316	Mensæ	9.0	6 43 49.04	− 1.972	75 24 44.8	− 3.81	50.19	1
4 317	Gould, Z. C., 6h 2169 . .	9.0	43 54.17	+ 0.321	64 29 34.4	.82	51.11	1
4 318	Volantis	9.5	44 0.75	+ 0.203	65 24 52.9	.83	51.10	1
4 319	Volantis	9.2	4.95	− 0.372	69 8 24.5	.83	51.08	2
4 320	Gould, Z. C., 6h 2187 . .	9.5	8.54	+ 0.477	63 10 32.8	− 3.84	51.16	1
4 321	Gould, Z. C., 6h 2105 . .	9.5	6 44 8.98	− 2.515	76 46 52.5	− 3.84	50.18	2
4 322	Gould, Z. C., 6h 2125 . .	9.2	11.69	1.951	75 21 31.3	.84	51.06	3
4 323	Gould, Z. C., 6h 2108 . .	9.2	21.12	2.863	77 32 1.1	.86	50.17	2
4 324	Volantis	10.0	22.08	1.629	74 24 19.3	.86	51.16	1
4 325	Volantis	10.0	24.08	− 0.936	71 52 27.3	− 3.86	52.03	1
4 326	Gould, 8486	10.0	6 44 24.19	− 0.641	70 32 10.6	− 3.86	51.99	1
4 327	Volantis	10.0	25.17	0.005	66 54 28.4	.86	52.06	1
4 328	Gould, 8482	7.0	25.78	0.903	71 44 3.0	.86	52.03	1
4 329	Volantis	9.2	31.73	− 0.102	67 32 32.6	.87	52.05	2
4 330	Volantis	9.5	44.18	+ 0.062	66 27 14.4	− 3.89	51.12	1
4 331	Lacaille, 3228	7.0	6 44 45.39	+ 0.274	64 52 49.7	− 3.89	52.07	1
4 332	Volantis	9.8	47.20	− 0.571	70 11 48.2	.89	51.53	2
4 333	Brisbane, 1387	9.0	48.61	− 2.791	77 23 21.8	.90	50.17	1
4 334	Gould, Z. C., 6h 2218 . .	9.0	44 51.76	+ 0.026	66 42 1.4	.90	51.59	2
4 335	Gould, Z. C., 6h 2195 . .	9.0	45 0.15	− 1.181	72 51 49.9	− 3.91	51.06	1
4 336	Pictoris	10.0	6 45 3.32	+ 0.518	62 49 19.2	− 3.92	51.15	1
4 337	Mensæ	10.0	10.35	− 4.605	80 18 27.3	.93	50.15	1
4 338	Volantis	9.5	18.94	− 0.119	67 39 45.0	.94	52.03	1
4 339	Gould, Z. C., 6h 2246 . .	9.0	28.46	+ 0.103	66 10 39.0	.95	52.08	1
4 340	Volantis	9.5	33.85	− 0.134	67 45 36.7	− 3.96	52.03	1
4 341	Volantis	9.0	6 45 42.96	− 1.660	74 31 5.6	− 3.97	51.16	1
4 342	Volantis	9.5	45 47.91	+ 0.209	65 24 6.5	3.98	51.10	1
4 343	Gould, Z. C., 6h 2257 . .	9.2	46 7.13	− 0.917	71 48 59.2	4.01	52.05	2
4 344	Volantis	10.0	15.01	1.932	75 19 48.7	.02	51.49	2
4 345	Volantis	10.0	20.16	− 1.583	74 17 4.3	− 4.03	51.16	1

No. 4 332. Declination may be 70° 7′ 40.6″.

Number.	Constellation, Name of Star, or Synonym.	Magnitude.	Right Ascension, 1850.0.	Annual Precession.	South Declination, 1850.0.	Annual Precession.	Mean year.	No. of obs.
			h. m. s.	s.	° ′ ″	″		
4 346	Pictoris	9.5	6 46 23.92	+ 0.470	63 16 41.7	— 4.03	51.16	1
4 347	Volantis	10.0	26.06	— 0.690	70 48 15.1	.04	52.08	1
4 348	Lacaille, 2536	6.7	26.10	— 0.580	70 16 5.4	.04	51.53	2
4 349	Gould, 8565	8.5	35.41	+ 0.274	64 55 4.8	.05	52.07	1
4 350	Lacaille, 2547	8.0	39.61	— 1.198	72 57 0.3	— 4.06	51.06	1
4 351	Volantis	10.0	6 46 41.25	— 0.052	67 15 22.3	— 4.06	52.06	1
4 352	Volantis	10.0	45.89	0.039	67 10 20.2	.06	52.06	1
4 353	Gould, 8553	10.0	47.53	— 1.080	72 29 45.8	.07	51.13	1
4 354	Volantis	10.0	48.04	+ 0.212	65 24 7.5	.07	52.12	1
4 355	Mensæ	9.5	49.93	— 9.103	83 51 43.6	— 4.07	50.12	1
4 356	Gould, Z. C., 6^h 2368 . .	10.0	6 46 54.63	— 0.116	67 40 21.6	— 4.08	52.03	1
4 357	Pictoris	9.5	47 0.16	+ 0.472	63 16 24.3	.08	51.16	1
4 358	Volantis	10.0	3.99	+ 0.191	65 33 57.4	.09	51.10	1
4 359	Volantis	10.0	12.33	— 0.653	70 38 27.7	.10	52.08	1
4 360	Gould, 8571	7.7	21.01	— 0.594	70 21 13.1	— 4.11	51.53	2
4 361	Mensæ	10.0	6 47 32.95	— 8.786	83 42 11.7	— 4.13	51.20	1
4 362	Volantis	9.0	36.38	0.775	71 13 3.5	.14	52.08	1
4 363	Volantis	9.8	47.72	0.904	71 47 25.2	.15	52.05	2
4 364	Lacaille, 2592	9.0	53.54	3.235	78 16 57.3	.16	50.79	3
4 365	Gould, Z. C., 6^h 2332 . .	10.0	58.54	— 2.787	77 25 1.0	— 4.17	50.17	2
4 366	Volantis	10.0	6 47 59.56	— 0.697	70 51 40.4	— 4.17	52.08	1
4 367	Volantis	10.0	48 7.49	+ 0.151	65 52 41.6	.18	51.43	3
4 368	Gould, Z. C., 6^h 2422 . .	9.0	32.26	— 0.577	70 17 45.4	.22	51.53	2
4 369	Gould, 8618	9.0	32.77	+ 0.139	65 58 30.5	.22	52.09	1
4 370	Gould, Z. C., 6^h 2405 . .	9.2	39.14	— 1.563	74 15 0.4	— 4.23	51.60	2
4 371	Volantis	10.0	6 48 40.17	+ 0.145	65 56 9.7	— 4.23	52.09	1
4 372	Volantis	8.8	41.40	— 0.204	68 14 52.0	.23	51.63	2
4 373	Volantis	10.0	44.15	0.355	69 7 26.8	.23	51.08	1
4 374	Gould, Z. C., 6^h 2407 . .	9.5	48.96	— 1.950	75 24 46.9	.24	51.06	3
4 375	Volantis	10.0	48 58.76	+ 0.269	65 0 21.9	— 4.25	52.12	1
4 376	Volantis	9.8	6 49 4.40	— 1.226	73 5 24.9	— 4.26	51.10	2
4 377	Gould, Z. C., 6^h 2477 . .	8.2	14.38	0.032	67 10 22.8	.28	52.06	2
4 378	Volantis	9.5	25.34	0.135	67 51 45.6	.29	51.15	1
4 379	Volantis	10.0	29.26	0.033	67 11 12.2	.30	52.06	1
4 380	Volantis	10.0	33.00	— 1.041	72 23 6.5	4.30	51.13	1
4 381	Gould, Z. C., 6^h 2479 . .	9.8	6 49 34.10	— 0.933	71 56 24.4	— 4.30	52.05	2
4 382	Volantis	10.0	44.59	— 1.501	74 3 47.4	.32	52.04	1
4 383	Lacaille, 2556	7.0	45.61	+ 0.124	66 6 49.2	.32	52.09	1
4 384	Gould, Z. C., 6^h 2519 . .	9.5	48.05	+ 0.524	62 51 52.7	.32	51.15	1
4 385	Mensæ	9.5	49.87	— 5.052	80 52 12.6	— 4.33	50.14	1

No. 4 353. Declination 1′ greater than Gould.

No. 4 365. Declination 1′ greater than Gould.

Number.	Constellation, Name of Star, or Synonym.	Magnitude.	Right Ascension, 1850.0.	Annual Precession.	South Declination, 1850.0.	Annual Precession.	Mean year.	No. of obs.
			h. m. s.	s.	° ′ ″	″		
4 386	Mensæ	9.5	6 49 52.20	— 2.375	76 31 16.3	— 4.33	50.18	1
4 387	Gould, Z. C., 6h 2526 . .	9.0	55.03	+ 0.521	62 53 36.8	.33	51.15	1
4 388	Gould, Z. C., 6h 2512 . .	10.0	55.18	— 0.018	67 5 50.0	.33	52.06	1
4 389	Pictoris	9.5	56.79	+ 0.573	62 25 14.7	.34	51.18	1
4 390	Gould, Z. C., 6h 2514 . .	9.2	49 58.78	— 0.080	67 30 10.4	— 4.34	52.05	2
4 391	Volantis	9.0	6 50 15.02	— 0.214	68 20 10.4	— 4.36	51.63	2
4 392	Carinæ	10.0	18.98	+ 0.439	63 37 59.8	.37	52.13	1
4 393	Volantis	9.5	42.28	— 0.353	69 8 49.3	.40	51.08	2
4 394	Carinæ	10.0	42.42	+ 0.416	63 50 43.2	.40	52.13	2
4 395	Volantis	9.5	49.98	— 0.474	69 48 7.3	— 4.41	51.14	1
4 396	Mensæ	10.0	6 50 51.71	— 2.254	76 14 17.2	— 4.41	50.18	1
4 397	Volantis	10.0	50 51.71	1.597	74 23 28.6	.41	51.16	1
4 398	Gould, Z. C., 6h 2562 . .	10.0	51 4.43	0.951	72 2 29.9	.43	52.08	1
4 399	Gould, Z. C., 6h 2518 . .	9.0	6.21	2.611	77 4 34.9	.44	50.17	1
4 400	Volantis	10.0	6.70	— 0.831	71 31 39.2	— 4.44	52.06	2
4 401	Gould, Z. C., 6h 2596 . .	9.0	6 51 14.05	— 0.242	68 31 25.6	— 4.45	52.11	1
4 402	Gould, Z. C., 6h 2603 . .	9.0	17.57	+ 0.220	65 26 1.9	.45	51.44	3
4 403	Volantis	9.3	18.10	— 1.008	72 16 50.5	.45	51.14	3
4 404	Mensæ	8.8	22.00	— 5.671	81 31 10.3	.46	51.10	2
4 405	Volantis	10.0	31.31	+ 0.394	64 2 54.4	— 4.47	52.12	1
4 406	Volantis	9.8	6 51 34.79	— 1.581	74 21 13.7	— 4.48	51.60	2
4 407	Volantis	10.0	36.95	— 1.069	72 31 47.5	.48	51.13	1
4 408	Volantis	9.0	38.10	+ 0.070	66 31 43.7	.48	51.12	1
4 409	Lacaille, 2586	7.5	38.73	— 0.471	69 48 3.0	.48	51.10	2
4 410	Gould, Z. C., 6h 2632 . .	9.0	40.37	+ 0.268	65 4 13.8	— 4.48	52.10	2
4 411	Gould, Z. C., 6h 2634 . .	9.0	6 51 41.58	+ 0.289	64 54 23.0	— 4.49	52.07	1
4 412	Gould, Z. C., 6h 2626 . .	9.0	44.29	— 0.241	68 31 43.8	.49	52.11	1
4 413	Volantis	9.5	53.20	+ 0.338	64 31 5.3	.50	51.11	1
4 414	Volantis	9.5	56.47	— 1.615	74 27 50.3	.51	51.16	1
4 415	Volantis	10.0	51 56.82	— 0.661	70 45 39.4	— 4.51	52.04	2
4 416	Lacaille, 2596	8.5	6 52 6.68	— 1.222	73 7 18.4	— 4.52	51.10	2
4 417	Carinæ	9.5	10.21	+ 0.435	63 42 35.1	.53	52.13	1
4 418	Volantis	9.5	10.52	0.121	66 11 6.9	.53	52.09	1
4 419	Carinæ	9.2	11.87	+ 0.418	63 51 28.4	.53	52.13	2
4 420	Gould, Z. C., 6h 2651 . .	9.5	12.19	— 0.291	68 49 48.8	— 4.53	51.09	1
4 421	Gould, Z. C., 6h 2550 . .	9.8	6 52 16.60	— 4.282	79 57 23.8	— 4.54	50.15	2
4 422	Volantis	9.0	18.39	+ 0.057	66 38 2.8	.54	51.94	2
4 423	Gould, Z. C., 6h 2671 . .	9.0	18.69	+ 0.249	65 14 0.7	.54	52.12	1
4 424	Volantis	9.8	18.96	— 1.448	73 55 28.0	.54	52.04	2
4 425	ζ Mensæ	6.5	26.81	— 4.835	80 38 53.4	— 4.55	50.14	2

No. 4 404. Declination may be 81° 31′ 39.″2.

Number.	Constellation, Name of Star, or Synonym.	Magnitude.	Right Ascension, 1850.0.	Annual Precession.	South Declination, 1850.0.	Annual Precession.	Mean year.	No. of obs.
			h. m. s.	s.	° ′ ″	″		
4 426	Mensæ	8. 5	6 52 27. 43	—10. 549	84 32 20. 2	— 4. 55	51. 10	6
4 427	Volantis	10. 0	36. 34	+ 0. 057	66 38 18. 8	. 56	52. 06	1
4 428	Mensæ	9. 0	36. 53	— 5. 432	81 17 43. 9	. 56	51. 06	1
4 429	Gould, Z. C., 6h 2699 . .	9. 0	40. 36	+ 0. 529	62 53 14. 7	. 57	51. 15	1
4 430	Carinæ	10. 0	42. 37	+ 0. 460	63 10 36. 0	— 4. 57	51. 16	1
4 431	Carinæ	8. 5	6 52 43. 36	+ 0. 595	62 16 13. 3	— 4. 57	51. 18	1
4 432	Carinæ	9. 5	44. 58	+ 0. 423	63 49 29. 4	. 58	52. 13	2
4 433	Mensæ	10. 5	44. 65	—11. 303	84 49 30. 8	. 58	51. 05	1
4 434	Gould, Z. C., 6h 2686 . .	9. 5	46. 08	0. 360	69 13 37. 5	. 58	51. 08	1
4 435	Gould, 8729	9. 2	48. 44	— 0. 399	69 26 26. 5	— 4. 58	51. 11	2
4 436	Gould, Z. C., 6h 2701 . .	9. 0	6 52 49. 29	+ 0. 193	65 40 5. 5	— 4. 58	51. 10	2
4 437	Volantis	10. 0	55. 32	— 0. 472	69 49 55. 3	. 59	51. 08	1
4 438	Volantis	10. 0	52 58. 55	+ 0. 128	66 9 3. 4	. 59	52. 09	1
4 439	Volantis	10. 0	53 3. 24	— 1. 286	73 22 21. 0	. 60	51. 59	2
4 440	Volantis	9. 3	6. 50	+ 0. 353	64 25 37. 1	— 4. 61	51. 11	1
4 441	ι Volantis	5. 5	6 53 9. 19	— 0. 660	70 46 31. 1	4. 61	52. 04	2
4 442	Gould, Z. C., 6h 2712 . .	10. 0	11. 98	0. 296	68 52 36. 5	. 61	51. 08	1
4 443	Volantis	10. 0	12. 57	0. 288	68 49 55. 3	. 61	51. 08	1
4 444	Gould, Z. C., 6h 2616 . .	10. 0	17. 38	— 4. 221	79 53 6. 7	. 62	50. 15	1
4 445	Gould, Z. C., 6h 2726 . .	8. 8	17. 81	+ 0. 188	65 43 14. 7	— 4. 62	51. 10	2
4 446	Gould, Z. C., 6h 2738 . .	9. 0	6 53 20. 98	+ 0. 521	62 58 22. 5	— 4. 63	51. 15	1
4 447	Lacaille, 2630	7. 8	40. 16	— 2. 417	76 40 6. 0	. 65	50. 18	2
4 448	Gould, Z. C., 6h 2711 . .	9. 2	43. 60	1. 578	74 22 22. 0	. 66	51. 60	2
4 449	Gould, Z. C., 6h 2747 . .	9. 5	49. 78	0. 301	68 55 6. 1	. 67	51. 08	1
4 450	Octantis	9. 3	53. 25	—19. 354	86 40 54. 3	— 4. 67	51. 12	8
4 451	Carinæ	9. 5	6 53 59. 28	+ 0. 514	63 3 26. 5	— 4. 68	51. 15	2
4 452	Volantis	10. 0	54 1. 26	+ 0. 040	66 47 15. 0	. 68	52. 06	1
4 453	Lacaille, 2604	8. 0	4. 29	— 0. 563	70 19 8. 8	. 69	51. 53	2
4 454	Volantis	10. 0	10. 95	+ 0. 125	66 11 35. 1	. 70	52. 09	1
4 455	Volantis	9. 5	14. 40	— 0. 475	69 52 18. 1	4. 70	51. 14	1
4 456	Volantis	9. 0	6 54 23. 72	+ 0. 265	65 9 8. 5	— 4. 72	52. 07	1
4 457	Volantis	9. 5	34. 03	+ 0. 390	64 9 0. 3	. 73	52. 12	1
4 458	Volantis	9. 2	34. 08	— 0. 138	67 57 46. 3	. 73	51. 59	2
4 459	Carinæ	9. 2	35. 71	+ 0. 509	63 7 0. 6	. 73	51. 15	2
4 460	Volantis	9. 5	45. 25	— 0. 781	71 22 3. 0	— 4. 75	52. 03	1
4 461	Volantis	10. 0	6 54 45. 38	— 0. 927	72 0 4. 3	— 4. 75	52. 07	1
4 462	Gould, Z. C., 6h 2815 . .	9. 2	54. 64	0. 321	69 3 0. 1	. 76	51. 08	2
4 463	Gould, Z. C., 6h 2824 .	9. 2	54 58. 84	— 0. 270	68 45 31 0	. 77	51. 60	2
4 464	Gould, Z. C., 6h 2842 . .	9. 2	55 6. 79	+ 0. 189	65 44 58. 0	. 78	51. 10	2
4 465	Volantis	10. 0	7. 78	+ 0. 313	64 47 48. 5	— 4. 78	52. 07	1

Nos. 4 451 and 4 459. Possibly the declinations of these two stars are interchanged.

Number.	Constellation, Name of Star, or Synonym.	Magnitude.	Right Ascension, 1850.0.	Annual Precession.	South Declination, 1850.0.	Annual Precession.	Mean year.	No. of obs.
			h. m. s.	s.	° ′ ″	″		
4466	Volantis	9.5	6 55 9.52	— 0.095	67 42 4.5	— 4.78	50.03	1
4467	Volantis	10.0	29.47	+ 0.215	65 33 36.8	.81	51.10	1
4468	Mensæ	9.0	30.72	— 7.537	83 1 26.0	.81	50.13	2
4469	Gould, Z. C., 6h 2885 . .	9.0	41.50	+ 0.372	64 22 23.6	.82	51.62	2
4470	Volantis	10.0	45.70	— 0.016	67 12 9.2	— 4.82	52.06	1
4471	Gould, Z. C., 6h 2860 . .	8.8	6 55 50.90	— 0.784	71 23 54.8	— 4.84	52.06	2
4472	Volantis	8.8	55 53.20	0.462	69 50 15.8	.84	50.82	3
4473	Carinæ	9.5	56 4.72	0.519	63 5 49.0	.86	51.15	2
4474	Lacaille, 2614	7.0	11.55	0.196	68 18 18.6	.87	51.17	3
4475	Mensæ	10.0	16.15	— 3.386	78 38 34.2	4.87	52.16	1
4476	Gould, Z. C., 6h 2937 . .	9.5	6 56 24.73	+ 0.154	66 2 22.1	— 4.89	52.09	1
4477	Volantis	9.5	24.91	0.323	64 44 42.2	.89	51.59	2
4478	Volantis	9.0	25.46	+ 0.300	64 55 41.7	.89	52.07	1
4479	Mensæ	10.0	28.38	— 3.603	79 0 21.3	.89	50.16	1
4480	Volantis	10.0	34.35	+ 0.147	66 5 38.1	— 4.90	52.09	1
4481	Volantis	10.0	6 56 35.68	— 1.136	72 52 29.6	— 4.90	51.06	1
4482	Lacaille, 2627	7.7	41.09	0.546	70 17 15.3	.91	51.10	3
4483	Volantis	10.0	41.88	0.392	69 28 50.7	.91	51.14	1
4484	Gould, Z. C., 6h 2953 . .	9.5	45.83	0.273	68 49 6.7	.92	51.08	1
4485	Volantis	10.0	46.82	— 5.572	81 28 14.3	— 4.92	51.06	1
4486	Gould, Z. C., 6h 2966 . .	9.2	6 56 50.04	+ 0.015	67 1 25.2	— 4.92	52.06	2
4487	Mensæ	9.5	56 53.78	—10.504	84 53 47.2	.93	51.03	1
4488	Volantis	9.2	57 1.08	+ 0.337	64 38 37.4	.94	51.59	2
4489	Mensæ	9.5	5.04	—11.716	84 59 31.6	.94	51.03	1
4490	Carinæ	10.0	5.28	+ 0.043	66 50 4.1	— 4.94	52.06	1
4491	Carinæ	10.0	6 57 8.87	+ 0.491	63 20 16.9	— 4.95	51.16	1
4492	Volantis	10.0	12.67	— 0.534	70 14 15.2	.95	51.07	1
4493	Volantis	9.2	16.16	— 1.319	73 33 40.0	.96	51.59	2
4494	Volantis	10.0	17.47	+ 0.024	66 58 8.9	.96	52.06	1
4495	Lacaille, 2658	9.0	17.50	— 2.971	77 53 55.2	— 4.96	50.17	2
4496	Volantis	10.0	6 57 22.63	+ 0.400	64 8 9.7	— 4.97	52.12	1
4497	Lacaille, 2644	9.0	26.38	— 1.610	74 32 1.4	.97	50.68	2
4498	Volantis	9.5	37.66	0.328	69 8 56.2	.98	51.08	1
4499	Volantis	10.0	37.73	0.185	68 18 37.9	.98	51.15	1
4500	Volantis	10.0	43.57	— 0.689	70 59 55.0	— 4.99	52.08	1
4501	Gould, Z. C., 6h 3019 . .	8.8	6 57 43.74	+ 0.053	66 46 53.6	— 4.99	51.59	2
4502	Volantis	9.0	45.00	+ 0.034	66 54 52.4	5.00	52.06	1
4503	Volantis	9.5	54.81	— 0.155	68 8 23.9	.01	51.15	1
4504	Carinæ	9.8	57 55.24	+ 0.564	62 41 45.0	.01	51.17	2
4505	Mensæ	9.0	58 2.51	— 8.887	83 49 20.1	— 5.03	50.12	1

Number.	Constellation, Name of Star, or Synonym.	Magnitude.	Right Ascension, 1850.0.	Annual Precession.	South Declination, 1850.0.	Annual Precession.	Mean year.	No. of obs.
			h. m. s.	s.	° ′ ″	″		
4506	Lacaille, 2363	8.2	6 58 12.33	− 2.756	77 28 48.5	− 5.04	50.17	2
4507	Gould, Z. C., 6h 3038 . .	9.8	14.72	− 0.326	69 8 57.6	.04	51.08	2
4508	Gould, Z. C., 6h 3058 . .	9.0	19.89	+ 0.361	64 28 51.6	.05	51.11	1
4509	Volantis	9.2	24.26	+ 0.297	65 0 8.9	.06	52.10	2
4510	Volantis	10.0	29.98	− 0.005	67 11 49.4	− 5.06	52.06	1
4511	Volantis	9.5	6 58 32.89	− 1.464	74 4 47.5	− 5.07	52.04	1
4512	Volantis	[illegible]	34.05	− 1.462	74 4 31.0	.07	52.04	1
4513	Volantis	[illegible]	42.46	+ 0.397	64 11 22.2	.08	52.12	1
4514	Carinæ	9.5	58 57.50	+ 0.574	62 37 40.0	.10	51.17	2
4515	Gould, Z. C., 6h 3086 . .	9.2	59 3.18	− 0.326	69 10 11.6	− 5.11	51.08	2
4516	Volantis	9.5	6 59 5.48	− 1.014	72 26 22.9	− 5.11	50.52	3
4517	Volantis	13.0	17.81	0.866	71 49 33.1	.13	52.08	1
4518	Gould, 8902	9.0	18.98	0.865	71 49 26.8	.13	52.05	2
4519	Mensæ	11.0	19.48	− 6.141	82 0 48.0	.13	51.17	1
4520	Gould, Z. C., 6h 3126 . .	9.0	23.10	+ 0.216	65 39 7.2	− 5.14	51.10	1
4521	Volantis	10.0	6 59 27.11	+ 0.044	66 53 2.7	− 5.14	52.06	1
4522	Carinæ	10.0	56.76	0.590	62 30 22.1	.19	51.18	1
4523	Gould, Z. C., 7h 5 . . .	8.8	6 59 58.30	+ 0.212	65 41 46.1	.19	51.10	2
4524	Lacaille, 2646	5.3	7 0 5.31	− 0.071	67 42 34.8	.20	52.03	1
4525	Volantis	9.8	13.00	− 0.386	69 31 14.6	− 5.21	51.11	2
4526	Volantis	9.5	7 0 13.87	+ 0.039	66 56 16.7	− 5.21	52.06	1
4527	Volantis	9.5	24.50	+ 0.022	67 3 23.8	.23	52.06	2
4528	Lacaille, 2724	8.5	25.81	− 4.963	80 52 20.8	.23	50.14	1
4529	Volantis	9.5	28.70	− 0.394	69 34 28.3	.23	51.14	1
4530	Gould, Z. C., 7h 53 . . .	8.3	31.74	+ 0.494	63 23 59.4	− 5.23	51.47	3
4531	Volantis	9.5	7 0 33.08	− 1.284	73 29 39.4	− 5.24	51.13	1
4532	Lacaille, 2645	7.8	33.57	+ 0.241	65 29 23.1	.24	51.44	3
4533	Volantis	9.5	34.49	1.284	73 29 44.1	.24	52.04	1
4534	Volantis	9.5	46.22	+ 0.093	66 34 53.4	.26	51.12	1
4535	Volantis	9.5	46.32	− 0.867	71 51 44.8	− 5.26	52.05	1
4536	Gould, 8953	8.8	7 0 51.74	− 0.396	69 35 34.8	− 5.26	51.19	2
4537	Gould, Z. C., 7h 18 . . .	9.6	53.90	− 1.516	74 17 29.0	.27	51.90	4
4538	Volantis	9.8	55.73	+ 0.183	65 56 5.7	.27	51.07	3
4539	Gould, Z. C., 7h 52 . . .	9.2	0 59.99	− 0.683	71 2 23.8	.28	51.46	3
4540	Carinæ	10.0	1 7.69	+ 0.610	62 20 55.5	− 5.29	51.18	1
4541	Gould, Z. C., 7h 47 . . .	10.0	7 1 8.39	− 0 975	72 19 19.7	− 5.29	52.08	1
4542	Mensæ	9.2	10.44	9.325	84 3 38.6	.29	50.66	2
4543	Volantis	9.8	11.38	0.777	71 28 18.8	.29	50.04	2
4544	Lacaille, 2689	8.2	12.00	2.782	77 34 29.2	.29	50.17	2
4545	Lacaille, 2719	9.5	17.95	− 4.192	79 56 1.2	− 5.30	50.15	2

Number.	Constellation, Name of Star, or Synonym.	Magnitude.	Right Ascension, 1850.0.	Annual Precession.	South Declination, 1850.0.	Annual Precession.	Mean year.	No. of obs.
			h. m. s.	s.	° ′ ″	″		
4546	Gould, Z. C., 7^h, 64 . . .	9.5	7 1 19.50	− 1.097	72 48 43.5	− 5.30	51.06	1
4547	Carinæ	10.0	26.97	+ 0.439	63 54 40.5	.31	52.12	1
4548	Gould, Z. C., 7^h 112 . .	9.0	29.46	0.264	65 20 12.1	.32	52.12	1
4549	Carinæ	10.0	1 55.41	0.465	63 41 56.3	.35	51.63	2
4550	Gould, Z. C., 7^h 156 . .	9.0	2 5.70	+ 0.264	65 21 25.6	− 5.37	52.12	1
4551	Volantis	10.0	7 2 5.85	+ 0.078	66 42 51.2	− 5.37	51.12	1
4552	Gould, Z. C., 7^h 127 . .	9.3	8.58	− 0.685	71 4 21.9	.37	51.46	3
4553	Mensæ	8.5	10.47	− 6.080	81 59 16.8	.37	50.65	2
4554	Carinæ	9.5	22.22	+ 0.511	63 17 52.6	.39	51.16	1
4555	Mensæ	9.2	22.98	− 5.973	81 53 47.8	5.39	50.65	2
4556	Volantis	10.0	7 2 29.41	− 2.005	75 46 . .	− 5.40	50.18	1
4557	Gould, Z. C., 7^h 163 . .	8.7	36.16	0.705	71 10 32.7	.41	51.46	3
4558	Mensæ	9.0	47.26	− 2.261	76 25 42.6	.43	50.18	1
4559	Carinæ	10.0	2 53.96	+ 0.606	62 26 22.2	.44	51.18	1
4560	Lacaille, 2664	7.2	3 0.89	+ 0.089	66 39 55.4	− 5.45	51.59	2
4561	Volantis	10.0	7 3 3.22	+ 0.016	67 10 4.5	− 5.45	52.06	1
4562	Gould, 9010	8.8	9.38	− 0.530	70 20 31.0	.46	51.09	3
4563	Lacaille, 2662	7.8	9.83	+ 0.403	64 15 39.4	.46	51.62	2
4564	Volantis	9.2	9.93	− 0.079	67 47 16.8	.46	51.59	2
4565	Mensæ	8.5	11.48	− 1.797	75 11 47.5	− 5.46	50.19	1
4566	Lacaille, 2788	7.5	7 3 13.62	− 6.971	82 41 59.7	− 5.46	50.13	2
4567	Mensæ	10.0	15.17	1.885	75 27 . .	.47	50.18	1
4568	Gould, Z. C., 7^h 218 . .	8.8	27.71	− 0.968	72 20 16.2	.48	50.83	3
4569	Carinæ	10.0	41.20	+ 0.554	62 56 43.7	.50	51.15	1
4570	Carinæ	9.8	44.30	+ 0.543	63 2 52.0	− 5.51	51.15	2
4571	Octantis	9.7	7 3 45.92	−23.909	87 16 20.9	− 5.51	51.12	6
4572	Volantis	10.0	48.32	1.034	72 36 44.1	.51	50.21	1
4573	Gould, 9025	8.0	3 51.56	1.092	72 50 29.1	.52	51.06	1
4574	Volantis	9.5	4 0.24	− 0.622	70 48 53.8	.53	52.08	1
4575	Volantis	9.8	1.38	+ 0.105	66 34 50.6	− 5.53	51.59	2
4576	Gould, 9038	8.5	7 4 20.29	− 1.218	73 19 36.6	− 5.56	50.67	2
4577	Volantis	9.5	20.38	+ 0.436	64 0 37.1	.56	52.12	1
4578	Volantis	10.0	22.20	− 1.716	74 58 50.8	.56	50.20	1
4579	Volantis	9.5	29.56	+ 0.433	64 2 34.2	.57	52.12	1
4580	Gould, Z. C., 7^h 280	9.0	30.14	− 1.347	73 47 31.8	− 5.57	50.82	3
4581	Lacaille, 2686	8.2	7 4 31.60	− 0.518	70 18 41.4	− 5.57	50.65	2
4582	Carinæ	9.0	35.60	+ 0.564	62 52 54.4	.58	51.15	1
4583	Volantis	9.0	51.44	+ 0.431	64 4 12.9	.60	52.12	1
4584	Gould, Z. C., 7^h 347 . .	8.0	53.85	− 0.108	68 0 57.4	.60	51.15	1
4585	Gould, Z. C., 7^h 360 . .	9.0	55.96	+ 0.457	63 50 48.1	− 5.61	52.13	2

Number.	Constellation, Name of Star, or Synonym.	Magnitude.	Right Ascension, 1850.0.	Annual Precession.	South Declination, 1850.0.	Annual Precession.	Mean year.	No. of obs.
			h. m. s.	s.	° ′ ″	″		
4586	Volantis	9.8	7 4 58.86	− 0.354	69 27 18.0	− 5.61	50.82	3
4587	Volantis	9.5	5 5.72	1.134	73 1 38.4	.62	51.10	2
4588	Gould, Z. C., 7^h 299	10.0	9.96	2.268	76 29 5.7	.63	50.18	1
4589	Volantis	9.5	28.35	− 0.751	71 26 59.2	.65	52.03	1
4590	Carinæ	10.0	31.28	+ 0.594	62 37 47.9	− 5.66	51.18	1
4591	Gould, Z. C., 7^h 404	9.2	7 5 44.89	+ 0.469	63 46 2.0	− 5.67	51.63	2
4592	Volantis	9.0	47.12	0.822	71 46 7.4	.68	52.05	2
4593	Gould, Z. C., 7^h 302	10.0	50.04	3.782	79 24 12.1	.68	50.16	1
4594	Gould, Z. C., 7^h 402	9.5	54.92	0.127	68 9 19.6	.69	51.15	1
4595	θ Mensæ	6.0	5 57.96	− 3.649	79 11 56.6	− 5.69	50.16	1
4596	Mensæ	10.0	7 6 6.79	− 5.478	81 28 21.6	− 5.71	51.06	1
4597	Gould, Z. C., 7^h 370	9.0	8.28	1.846	75 23 32.1	.71	50.18	2
4598	Octantis	9.5	23.70	29.547	87 45 1.2	.73	51.09	3
4599	Volantis	9.5	26.68	− 0.741	71 24 38.6	.73	52.03	1
4600	Carinæ	9.5	36.34	+ 0.602	62 34 47.2	− 5.75	51.18	1
4601	Lacaille, 2694	7.0	7 6 36.49	+ 0.086	66 46 28.6	5.75	52.06	1
4602	Carinæ	9.5	38.62	+ 0.572	62 51 57.4	.75	51.15	1
4603	Gould, 9108	9.0	42.79	− 0.226	68 46 30.4	.76	51.09	1
4604	Lacaille, 2704	7.2	46.58	− 0.196	68 35 55.7	.76	50.67	2
4605	Gould, Z. C., 7^h 488	9.5	52.95	+ 0.533	63 14 8.8	− 5.77	51.16	1
4606	Gould, Z. C., 7^h 472	9.5	7 6 54.00	+ 0.047	67 3 10.7	− 5.77	52.06	2
4607	Volantis	10.0	56.59	0.288	65 17 44.4	.78	52.12	1
4608	Carinæ	9.0	6 58.51	+ 0.607	62 32 39.0	.78	51.18	1
4609	Gould, Z. C., 7^h 456	10.0	7 16.16	− 1.468	74 15 13.3	.80	51.16	1
4610	Lacaille, 2723	8.2	19.56	− 0.654	71 2 39.5	− 5.81	51.15	4
4611	Mensæ	9.5	7 7 21.52	− 4.900	80 52 48.5	− 5.81	50.14	1
4612	Octantis	9.8	37.12	18.648	86 37 44.0	.83	51.16	5
4613	Lacaille, 2794	8.8	43.56	− 4.668	80 36 57.8	.84	50.14	2
4614	Gould, Z. C., 7^h 545	8.5	47.88	+ 0.505	63 30 52.2	.85	51.64	2
4615	Lacaille, 2716	9.0	7 56.52	− 0.045	67 41 35.2	− 5.86	52.03	1
4616	Volantis	8.2	7 8 0.88	− 1.047	72 45 7.4	− 5.86	50.64	2
4617	Volantis	9.8	6.00	+ 0.424	64 13 17.5	.87	51.46	3
4618	Volantis	9.5	15.38	− 0.950	72 21 56.9	.88	52.08	1
4619	Volantis	8.0	19.22	− 1.050	72 46 25.5	.89	51.06	1
4620	Gould, Z. C., 7^h 583	9.5	25.75	+ 0.501	63 33 50.6	− 5.90	52.13	1
4621	Volantis	9.9	7 8 30.22	+ 0.428	64 12 29.0	− 5.91	51.38	4
4622	Volantis	9.8	33.36	0.259	65 47 42.6	.91	51.59	2
4623	Carinæ	9.5	43.99	0.589	62 46 17.9	.93	51.15	1
4624	Volantis	9.0	45.60	+ 0.354	64 49 45.0	.93	52.07	1
4625	Octantis	9.2	45.88	−31.809	87 54 9.5	− 5.93	51.12	8

No. 4595. Micrometer reading recorded — $1^r.26$. Assumed to be — $1^r.74$.
No. 4616. Blue and nebulous. No. 4619. Red and sharply defined.

Number.	Constellation, Name of Star, or Synonym.	Magnitude.	Right Ascension, 1850.0.	Annual Precession.	South Declination, 1850.0.	Annual Precession.	Mean year.	No. of obs.
			h. m. s.	s.	° ′ ″	″		
4626	Lacaille, 2845	6.8	7 8 46.20	− 8.662	83 47 14.2	− 5.93	50.12	2
4627	Gould, Z. C., 7^h 550 . .	8.8	49.67	1.692	74 59 24.6	.93	50.19	2
4628	Mensæ	9.5	8 58.01	− 5.882	81 53 1.2	.94	51.17	1
4629	Lacaille, 2722 . . .	8.1	9 7.90	+ 0.430	64 12 31.9	.96	51.38	4
4630	Mensæ	10.0	10.16	− 3.260	78 36 0.0	− 5.96	52.16	1
4631	Gould, Z. C., 7^h 639 . .	9.5	7 9 12.13	+ 0.525	63 22 51.0	− 5.96	51.64	2
4632	Mensæ	10.0	29.70	− 5.488	81 31 15.9	5.99	51.05	2
4633	Volantis	10.0	35.25	− 0.880	72 6 19.6	6.00	52.08	1
4634	Carinæ	10.0	40.22	+ 0.621	62 30 14.1	.00	51.18	1
4635	Gould, Z. C., 7^h 633 . .	9.2	40.76	− 0.847	71 58 1.8	− 6.00	52.05	2
4636	Volantis	9.0	7 9 46.60	+ 0.293	65 20 27.4	− 6.01	52.12	1
4637	γ^2 Volantis	7.5	57.65	− 0.480	70 15 5.8	.03	50.65	2
4638	Volantis	10.0	58.14	0.080	67 58 6.2	.03	51.15	1
4639	γ^1 Volantis	4.8	9 59.92	0.481	70 15 12.8	.03	50.65	2
4640	Gould, Z. C., 7^h 652 . .	10.0	10 1.39	− 0.853	72 0 8.2	− 6.03	52.05	2
4641	Volantis	10.0	7 10 4.17	+ 0.105	66 44 46.9	− 6.04	52.06	1
4642	Volantis	9.5	11.77	+ 0.273	65 30 25.6	.05	51.44	3
4643	Volantis	10.0	20.77	− 0.085	68 0 53.8	.06	51.15	1
4644	Volantis	9.5	25.87	− 1.132	73 8 7.8	.07	51.06	1
4645	Volantis	9.0	26.55	+ 0.200	66 4 9.3	− 6.07	52.09	1
4646	Lacaille, 2735	6.5	7 10 36.93	+ 0.581	62 56 4.6	− 6.08	51.15	1
4647	Volantis	9.5	37.09	0.377	64 41 30.5	.08	52.07	1
4648	Volantis	9.5	39.67	0.087	66 53 0.4	.09	52.06	1
4649	Volantis	10.0	39.87	+ 0.371	64 44 35.4	.09	52.07	1
4650	Gould, Z. C., 7^h 648 . .	9.0	42.39	− 2.773	77 42 8.5	− 6.09	50.17	1
4651	Octantis	9.8	7 10 43.73	−19.077	86 42 29.3	− 6.09	51.14	8
4652	Lacaille, 2743	7.0	45.37	+ 0.098	66 48 48.6	.09	52.06	1
4653	Gould, Z. C., 7^h 746 . .	9.5	52.59	0.093	66 50 49.6	.10	52.06	1
4654	Gould, Z. C., 7^h 760 . .	8.0	52.75	+ 0.590	62 49 44.1	.10	51.15	1
4655	Gould, Z. C., 7^h 725 . .	10.2	10 58.62	− 0.862	72 3 42.8	− 6.11	51.15	2
4656	Lacaille, 2751	8.5	7 11 7.84	+ 0.034	67 42 24.0	− 6.13	52.03	1
4657	Volantis	10.0	9.80	− 0.853	72 1 34.2	.13	52.08	1
4658	Mensæ	10.0	14.72	5.268	81 19 14.4	.13	51.06	1
4659	Gould, Z. C., 7^h 772 . .	12.0	22.98	0.216	68 50 9.5	.15	51.08	1
4660	Gould, Z. C., 7^h 776 . .	10.0	26.86	− 0.213	68 49 20.5	− 6.15	51.08	1
4661	Gould, Z. C., 7^h 803 . .	9.0	7 11 28.16	+ 0.415	64 24 19.0	− 6.15	51.62	2
4662	Gould, Z. C., 7^h 784 . .	9.5	36.41	− 0.442	70 5 44.4	.16	51.07	1
4663	Gould. Z. C., 7^h 706 . .	9.0	48.30	− 3.203	78 32 23.6	.18	51.16	2
4664	Carinæ	9.2	50.24	+ 0.564	63 6 32.0	.18	51.15	2
4665	Gould, Z. C., 7^h 810 . .	9.5	51.31	− 0.170	68 34 41.3	− 6.18	50.25	1

Number.	Constellation, Name of Star, or Synonym.	Magnitude.	Right Ascension, 1850.0.	Annual Precession.	South Declination, 1850.0.	Annual Precession.	Mean year.	No. of obs.
			h. m. s.	s.	° ′ ″	″		
4666	Gould, Z. C., 7^h 835 . .	9.0	7 11 53.66	+ 0.580	62 57 24.6	— 6.19	51.15	1
4667	Volantis	9.5	12 0.33	— 0.485	70 19 33.7	.20	51.07	1
4668	Carinæ	10.0	13.77	+ 0.501	63 41 1.3	.22	52.13	1
4669	Lacaille, 2760	8.0	15.20	0.001	67 30 55.3	.22	51.04	2
4670	Lacaille, 2755	7.2	18.72	+ 0.393	64 36 45.2	— 6.22	51.59	2
4671	Lacaille, 2765	7.5	7 12 20.46	— 0.395	69 51 57.1	— 6.23	50.66	2
4672	Volantis	9.5	23.09	+ 0.329	65 8 21.5	.23	52.12	1
4673	Gould, Z. C., 7^h 876 . . .	8.0	28.44	+ 0.602	62 46 26.3	.24	51.15	1
4674	Volantis	10.0	39.02	— 0.730	71 31 12.4	.25	52.03	1
4675	Volantis	9.5	42.32	+ 0.323	65 11 36.1	— 6.26	52.12	1
4676	Lacaille, 2800	8.2	7 12 43.96	— 1.947	75 47 47.9	— 6.26	50.18	3
4677	Lacaille, 2775	8.8	43.97	— 0.858	72 5 9.4	.26	51.15	2
4678	Carinæ	9.5	47.05	+ 0.652	62 18 10.6	.26	51.18	1
4679	Lacaille, 2799	8.2	49.93	— 1.685	75 2 59.1	.27	50.19	2
4680	Gould, Z. C., 7^h 802 . .	10.0	12 52.56	— 2.994	78 10 25.4	— 6.27	50.17	1
4681	Carinæ	10.0	7 13 11.88	+ 0.557	63 12 50.2	— 6.30	51.16	1
4682	Gould, Z. C., 7^h 943 . . .	8.0	14.68	0.602	62 48 3.3	.30	51.15	1
4683	Carinæ	10.0	36.85	+ 0.244	65 50 18.2	.33	51.10	2
4684	Mensæ	9.8	45.65	— 5.447	81 31 50.6	.34	50.60	2
4685	Mensæ	10.0	47.79	— 3.250	78 39 21.8	— 6.35	52.16	1
4686	Lacaille, 2795	9.2	7 13 48.06	— 0.950	72 29 44.4	— 6.35	50.21	2
4687	Gould, Z. C., 7^h 983 . .	9.2	48.91	+ 0.391	64 40 53.8	.35	51.59	2
4688	Volantis	10.0	58.08	0.468	64 1 48.7	.36	51.46	3
4689	Gould, Z. C., 7^h 991 . .	9.2	13 58.28	0.383	64 45 2.6	.36	51.59	2
4690	Gould, Z. C., 7^h 1007 . .	8.5	14 11.47	+ 0.593	62 54 42.1	— 6.38	51.15	1
4691	Mensæ	10.0	7 14 15.12	— 3.207	78 35 10.9	— 6.38	52.16	1
4692	Gould, Z. C., 7^h 901 . .	10.0	16.11	— 3.046	78 17 43.3	.39	52.16	1
4693	Gould, Z. C., 7^h 1005 . .	9.2	18.72	+ 0.025	67 24 47.8	.39	52.05	2
4694	Gould, Z. C., 7^h 973 . .	10.0	24.52	— 1.320	73 54 19.9	.40	50.20	1
4695	Mensæ	10.0	37.31	—11.250	84 56 41.8	— 6.42	51.20	2
4696	Volantis	10.0	7 14 39.44	+ 0.368	64 53 44.2	— 6.42	52.07	1
4697	Volantis	10.0	42.84	+ 0.189	66 16 35.8	.42	52.09	1
4698	Volantis	9.2	44.06	— 0.460	70 16 10.0	.42	50.65	2
4699	Gould, Z. C., 7^h 1048 . .	9.0	55.32	+ 0.432	64 22 23.3	.44	51.11	1
4700	Mensæ	10.0	14 59.02	— 1.797	75 25 24.2	— 6.45	50.19	1
4701	Brisbane, 1568	9.0	7 15 3.81	— 0.030	67 47 53.8	— 6.45	52.03	1
4702	Volantis	9.5	4.55	+ 0.342	65 7 5.4	.45	52.07	1
4703	Herschel, 3955[1]	9.5	11.52	0.239	65 55 34.4	.46	51.43	2
4704	Herschel, 3955[2]	11.0	13.94	0.239	65 55 17.2	.47	51.10	1
4705	Lacaille, 2782	7.9	24.35	+ 0.499	63 51 37.8	— 6.48	51.53	5

Number.	Constellation, Name of Star, or Synonym.	Magnitude.	Right Ascension, 1850.0.	Annual Precession.	South Declination, 1850.0.	Annual Precession.	Mean year.	No. of obs.
			h. m. s.	s.	° ′ ″	″		
4706	Mensæ	9.2	7 15 32.88	− 8.754	83 53 33.4	− 6.49	50.12	2
4707	Volantis	10.0	41.20	1.236	73 27 36.8	.50	50.21	1
4708	Volantis	10.0	41.69	− 0.735	71 44 20.6	.50	52.03	1
4709	Lacaille, 2787	9.0	43.05	+ 0.567	63 12 21.1	.51	51.16	1
4710	Carinæ	10.0	44.37	+ 0.630	62 36 59.6	− 6.51	51.18	1
4711	Volantis	10.0	7 15 45.58	− 1.444	74 21 13.5	− 6.51	51.16	1
4712	Gould, 9387	8.8	52.34	+ 0.623	62 44 47.8	.52	51.16	2
4713	Gould, Z. C., 7h 1113	9.0	52.84	+ 0.159	66 32 3.6	.52	51.12	1
4714	Octantis	9.2	57.28	−13.652	85 40 37.7	.53	51.03	2
4715	Gould, Z. C., 7h 1061	10.0	15 59.88	− 1.692	75 8 15.9	− 6.53	50.20	1
4716	Gould, Z. C., 7h 1137	10.0	7 16 4.23	+ 0.346	65 7 2.4	− 6.54	52.12	1
4717	Carinæ	10.0	18.06	+ 0.493	63 53 41.6	.56	51.14	2
4718	Volantis	9.5	25.13	− 0.696	71 27 42.2	.56	52.03	1
4719	Brisbane, 1587	9.0	27.04	− 1.296	73 52 2.8	.57	50.20	2
4720	Gould, Z. C., 7h 1164	9.5	28.24	+ 0.576	63 9 21.4	− 6.57	51.15	2
4721	Carinæ	8.0	7 16 29.88	+ 0.527	63 35 42.7	− 6.57	51.13	1
4722	Volantis	10.0	40.44	0.261	65 48 13.1	.59	51.10	1
4723	Gould, Z. C., 7h 1166	9.0	40.62	0.134	66 44 5.1	.59	51.15	3
4724	Volantis	10.0	43.78	+ 0.328	65 17 17.9	.59	52.12	1
4725	Brisbane, 1593	9.0	48.34	− 1.299	73 53 6.6	− 6.60	50.20	2
4726	δ Volantis	4.2	7 16 53.38	− 0.004	67 40 57.6	− 6.60	51.14	2
4727	Volantis	9.5	16 55.53	+ 0.372	64 56 29.3	.60	52.07	1
4728	Volantis	9.5	17 1.99	− 0.762	71 46 40.3	.62	52.03	1
4729	Volantis	9.5	6.94	+ 0.197	66 17 58.7	.62	51.61	2
4730	Carinæ	9.0	7.46	+ 0.678	62 12 7.8	− 6.62	51.18	1
4731	Carinæ	9.0	7 17 13.16	+ 0.584	63 6 32.3	− 6.63	51.15	1
4732	Volantis	9.5	13.27	0.106	66 57 0.1	.63	52.06	1
4733	Volantis	10.0	13.30	0.419	64 33 35.0	.63	51.11	1
4734	Lacaille, 2805	7.8	13.42	0.584	63 6 31.8	.63	51.15	2
4735	Volantis	9.3	17.71	+ 0.509	63 47 11.6	− 6.64	51.46	3
4736	Gould, Z. C., 7h 1210	9.5	7 17 18.38	+ 0.007	67 37 13.8	− 6.64	51.14	2
4737	Mensæ	9.5	25.60	− 4.816	80 54 47.4	.65	50.14	1
4738	Volantis	10.0	32.73	+ 0.320	65 22 28.0	.66	52.12	1
4739	Volantis	10.0	38.56	− 2.455	77 8 39.4	.67	50.17	1
4740	Volantis	9.5	50.18	− 0.022	67 49 42.8	− 6.68	51.15	1
4741	Mensæ	9.0	7 17 56.97	− 5.112	81 14 38.4	− 6.69	51.06	1
4742	Gould, 9399	9.0	17 58.38	− 3.224	78 40 44.8	.69	50.82	3
4743	Volantis	10.0	18 2.28	+ 0.213	66 12 42.9	.70	52.09	1
4744	Volantis	10.0	4.62	+ 0.355	64 36 31.5	.70	51.11	1
4745	Gould, Z. C., 7h 1169	9.0	5.06	− 3.213	78 38 10.2	− 6.70	51.16	2

Number.	Constellation, Name of Star, or Synonym.	Magnitude.	Right Ascension, 1850.0.	Annual Precession.	South Declination, 1850.0.	Annual Precession.	Mean year.	No. of obs.
			h. m. s.	s.	° ′ ″	″		
4746	Lacaille, 2825	9.5	7 18 5.71	— 0.917	72 27 59.6	— 6.70	50.21	2
4747	Gould, Z. C., 7h 1286 . .	9.5	7.67	+ 0.416	64 56 41.2	.71	52.07	1
4748	Volantis	10.0	7.92	0.275	65 44 34.8	.71	51.10	1
4749	Volantis	10.0	9.44	0.291	65 37 28.0	.71	51.10	1
4750	Carinæ	9.0	14.01	+ 0.641	62 36 17.9	— 6.71	51.18	1
4751	Volantis	10.0	7 18 19.81	+ 0.131	66 48 46.1	— 6.72	52.06	1
4752	Volantis	10.0	21.66	+ 0.413	64 38 54.8	.72	51.11	1
4753	Lacaille, 2856	8.8	36.68	— 3.509	79 10 3.2	.75	50.19	3
4754	Lacaille, 2828	8.5	46.30	0.685	71 28 27.1	.76	50.78	4
4755	Mensæ	9.5	56.44	— 5.723	81 51 23.5	— 6.77	50.65	2
4756	Brisbane, 1600	9.0	7 18 58.90	+ 0.120	66 54 24.2	— 6.78	51.16	2
4757	Volantis	10.0	19 5.04	+ 0.083	67 10 9.7	.78	52.06	1
4758	Gould, Z. C., 7h 1330 . .	9.5	10.63	— 0.593	71 3 17.6	.79	50.22	2
4759	Gould, Z. C., 7h 1346 . .	8.8	12.65	— 0.213	69 2 40.3	.79	51.08	2
4760	Volantis	10.0	17.10	+ 0.072	67 15 6.8	— 6.80	52.06	1
4761	Volantis	10.0	7 19 23.99	+ 0.385	64 54 55.2	— 6.81	52.07	1
4762	Carinæ	10.0	28.82	0.656	62 29 59.3	.82	51.18	1
4763	Gould, Z. C., 7h 1400 . .	9.2	31.48	0.553	63 28 9.4	.82	51.14	2
4764	Carinæ	9.5	40.70	+ 0.650	62 34 0.6	.83	51.18	1
4765	Gould, Z. C., 7h 1294 . .	10.0	43.28	— 3.229	78 43 2.9	— 6.84	52.16	1
4766	Volantis	10.0	7 19 45.28	+ 0.277	65 47 15.2	— 6.84	51.10	1
4767	Gould, Z. C., 7h 1399 . .	9.5	50.21	— 0.192	68 56 14.1	.85	51.08	1
4768	Volantis	10.0	19 52.09	+ 0.330	65 22 19.4	.85	52.12	1
4769	Volantis	9.5	20 5.56	— 0.996	72 50 15.0	.87	51.06	1
4770	Volantis	10.0	5.92	— 0.087	68 18 51.8	— 6.87	51.15	1
4771	Lacaille, 2846	8.2	7 20 12.88	— 1.816	75 35 20.9	— 6.88	50.18	2
4772	Volantis	10.0	23.33	+ 0.202	66 22 9.0	.89	51.12	1
4773	Lacaille, 2838	8.5	28.00	— 0.363	69 55 14.4	.90	50.66	2
4774	Gould, Z. C., 7h 1480 . .	9.0	33.39	+ 0.542	63 36 32.8	.91	51.13	1
4775	Lacaille, 2936	7.0	38.96	— 7.952	83 30 14.4	— 6.91	50.12	2
4776	Gould, Z. C., 7h 1482 . .	9.0	7 20 42.15	+ 0.335	65 21 59.5	— 6.92	52.12	1
4777	Gould, 9531	9.2	48.76	— 0.768	71 54 8.8	.92	51.13	2
4778	Gould, Z. C., 7h 1458 . .	9.5	50.07	0.957	72 41 58.5	.93	51.06	1
4779	Octantis	9.5	53.55	—12.928	85 30 58.2	.94	51.13	5
4780	Carinæ	10.0	20 58.37	+ 0.667	62 26 51.7	— 6.94	51.18	1
4781	Volantis	10.0	7 21 2.81	— 0.278	69 28 7.5	— 6.94	51.08	1
4782	Gould, Z. C., 7h 1431 . .	9.0	7.82	— 2.253	76 44 45.4	.95	50.18	2
4783	Gould, Z. C., 7h 1503 . .	9.0	9.69	+ 0.115	67 0 12.0	.96	52.06	1
4784	Gould, 9544	8.8	14.03	— 0.351	69 52 52.4	.96	50.66	2
4785	Volantis	10.0	19.36	+ 0.281	65 46 36.6	— 6.97	51.10	2

Number.	Constellation, Name of Star, or Synonym.	Magnitude.	Right Ascension, 1850.0.	Annual Precession.	South Declination, 1850.0.	Annual Precession.	Mean year.	No. of obs.
			h. m. s.	s.	° ′ ″	″		
4 786	Volantis	10.0	7 21 20.93	− 0.077	68 17 14.2	− 6.97	51.15	1
4 787	Gould, Z. C., 7^h 1513	9.5	26.10	0.222	69 9 37.5	.98	51.08	1
4 788	Mensæ	10.0	33.88	− 3.833	79 42 45.4	.99	50.16	1
4 789	Volantis	10.0	34.41	+ 0.496	64 3 24.7	.99	51.14	1
4 790	Volantis	9.5	37.34	+ 0.156	66 44 32.3	− 6.99	51.12	1
4 791	Volantis	9.0	7 21 38.58	+ 0.256	66 0 24.1	− 6.99	52.09	1
4 792	Volantis	10.0	41.36	− 0.375	70 1 16.3	7.00	51.07	1
4 793	Volantis	10.0	45.11	0.008	67 51 56.4	.00	51.15	1
4 794	Gould, Z. C., 7^h 1551 . .	9.2	47.32	+ 0.463	64 20 5.1	.01	51.13	3
4 795	Gould, Z. C., 7^h 1558 . .	9.5	53.30	+ 0.446	64 30 0.2	− 7.02	51.11	1
4 796	Gould, Z. C., 7^h 1505 . .	9.5	7 21 57.59	− 1.729	75 23 46.0	− 7.02	50.18	2
4 797	Carinæ	9.0	22 4.31	+ 0.561	63 29 47.9	.03	51.13	1
4 798	Brisbane, 1621	8.0	13.20	0.366	65 10 21.4	.04	52.10	2
4 799	Gould, Z. C., 7^h 1569 . .	9.0	13.30	+ 0.122	67 0 11.3	.04	50.86	3
4 800	Stone, 3658	8.0	17.98	+ 0.228	66 14 41.6	− 7.05	51.61	2
4 801	Gould, Z. C., 7^h 1590 . .	8.5	7 22 19.79	+ 0.460	64 23 46.7	− 7.05	51.11	1
4 802	Volantis	9.5	20.28	+ 0.416	64 45 51.4	.05	52.07	1
4 803	Volantis	9.0	25.18	− 1.189	73 37 53.1	.06	50.20	1
4 804	Lacaille, 2853	8.8	32.10	0.612	71 14 24.8	.07	50.43	2
4 805	Volantis	9.5	33.74	− 0.368	70 0 41.6	− 7.07	51.07	1
4 806	Volantis	9.5	7 22 36.73	− 0.360	69 58 8.0	− 7.07	51.07	1
4 807	Gould, Z. C., 7^h 1605 . .	8.7	38.63	+ 0.114	67 4 15.6	.08	50.86	3
4 808	Brisbane, 1626	9.0	40.57	− 0.500	70 41 45.1	.08	50.23	2
4 809	Volantis	9.8	40.93	+ 0.424	64 42 58.3	.08	51.59	2
4 810	Volantis	10.0	52.08	+ 0.270	65 56 39.2	− 7.10	51.10	1
4 811	Mensæ	10.0	7 22 52.80	− 2.370	77 3 34.6	− 7.10	50.17	1
4 812	Volantis	10.0	22 55.93	0.057	68 12 49.0	.10	51.15	1
4 813	Gould, Z. C., 7^h 1593 . .	9.2	23 3.92	1.246	73 50 59.4	.11	50.20	2
4 814	Gould, Z. C., 7^h 1542 . .	9.8	4.65	2.886	78 8 53.8	.11	51.16	2
4 815	Gould, Z. C., 7^h 1596 . .	9.0	15.06	− 1.575	74 56 46.1	− 7.13	50.20	1
4 816	Mensæ	9.0	7 23 22.69	− 8.147	83 38 39.2	− 7.14	50.12	1
4 817	Lacaille, 2891	9.0	27.59	− 2.824	78 2 3.8	.14	50.17	1
4 818	Volantis	9.2	27.82	+ 0.022	67 43 21.8	.14	51.14	2
4 819	Octantis	9.8	28.29	−18.386	86 40 3.0	.14	51.13	4
4 820	Carinæ	10.0	28.95	+ 0.581	63 21 55.0	− 7.15	51.16	1
4 821	Volantis	10.0	7 23 30.42	− 0.200	69 6 3.2	− 7.15	51.08	1
4 822	Volantis	9.8	32.36	+ 0.490	64 10 43.8	.15	51.14	2
4 823	Gould, Z. C., 7^h 1665 . .	9.0	33.50	+ 0.234	66 14 31.2	.15	51.61	2
4 824	Volantis	10.0	38.28	− 0.322	69 47 43.3	.16	50.24	1
4 825	Volantis	10.0	49.08	− 0.872	72 25 49.3	− 7.17	50.21	1

Number.	Constellation, Name of Star, or Synonym.	Magnitude.	Right Ascension, 1850.0.	Annual Precession.	South Declination, 1850.0.	Annual Precession.	Mean year.	No. of obs.
			h. m. s.	s.	° ′ ″	″		
4 826	Volantis	10.0	7 23 52.77	+ 0.218	66 22 18.8	− 7.18	51.12	1
4 827	Gould, Z. C., 7ʰ 1706 . .	9.2	24 0.96	0.647	62 45 29.9	.19	51.17	2
4 828	Volantis	9.0	11.66	+ 0.074	67 24 6.2	.20	50.76	4
4 829	Stone, 3664	8.8	15.56	− 1.982	76 7 55.4	.21	50.18	2
4 830	Lacaille, 2862	7.2	15.76	− 0.420	70 20 19.0	− 7.21	50.65	2
4 831	Volantis	10.0	7 24 16.53	− 0.052	68 13 26.1	− 7.21	51.15	1
4 832	Volantis	9.8	22.18	+ 0.491	64 11 55.6	.22	51.14	2
4 833	Volantis	9.2	32.74	+ 0.504	64 5 37.4	.23	51.14	2
4 834	Gould, Z. C., 7ʰ 1719 . .	9.0	37.09	− 0.404	70 15 49.7	.24	51.07	1
4 835	Volantis	10.0	37.15	+ 0.319	65 37 55.2	− 7.24	51.10	1
4 836	Volantis	9.0	7 24 39.33	− 0.965	72 50 . .	− 7.24	50.21	1
4 837	Volantis	9.2	39.64	− 0.690	71 39 40.5	.24	51.13	2
4 838	Gould, Z. C., 7ʰ 1762 . .	8.2	46.50	+ 0.650	62 45 54.0	.25	51.17	2
4 839	Gould, 9646	9.0	50.52	0.398	65 0 48.1	.26	52.07	1
4 840	Gould, Z. C., 7ʰ 1767 . .	9.0	24 57.58	+ 0.369	65 14 56.0	− 7.27	52.12	1
4 841	Gould, Z. C., 7ʰ 1750 . .	9.0	7 25 5.17	− 0.441	70 28 19.7	− 7.28	50.23	1
4 842	Volantis	10.0	27.64	+ 0.368	65 16 8.5	.31	52.12	1
4 843	Volantis	9.5	30.23	− 0.762	72 0 20.7	.31	50.22	1
4 844	Volantis	8.5	31.82	+ 0.253	66 9 9.0	.31	52.09	1
4 845	Carinæ	10.0	35.04	+ 0.583	63 25 37.5	− 7.32	51.16	1
4 846	Volantis	9.5	7 25 35.42	− 0.931	72 43 18.7	− 7.32	51.06	1
4 847	Volantis	9.0	37.83	0.366	70 5 59.9	.32	51.07	1
4 848	Gould, 9655	9.5	43.04	0.762	72 0 55.1	.33	51.04	1
4 849	Volantis	10.0	51.21	− 0.007	67 59 43.8	.34	52.02	1
4 850	Volantis	10.0	51.36	+ 0.285	65 56 25.6	− 7.34	51.60	2
4 851	Gould, Z. C., 7ʰ 1778 . .	9.0	7 25 54.87	− 1.590	75 3 18.2	− 7.34	50.19	2
4 852	Gould, Z. C., 7ʰ 1831 . .	8.7	55.34	+ 0.528	63 55 53.9	.34	51.13	3
4 853	Gould, Z. C., 7ʰ 1788 . .	9.5	25 56.54	− 1.163	73 37 41.1	.35	50.20	1
4 854	Volantis	9.8	26 1.52	0.180	69 3 54.9	.35	51.08	2
4 855	Volantis	10.0	2.21	− 0.006	67 59 42.6	− 7.35	52.02	1
4 856	Gould, 9677	9.0	7 26 3.66	+ 0.368	65 17 22.8	− 7.36	52.12	1
4 857	Gould, Z. C., 7ʰ 1799 . .	10.0	7.56	− 1.216	73 49 17.4	.36	50.20	1
4 858	Lacaille, 2885	8.2	8.64	0.575	71 10 4.3	.36	50.22	2
4 859	Gould, Z. C., 7ʰ 1830 . .	9.5	8.90	0.074	68 25 43.4	.36	50.25	1
4 860	Mensæ	9.0	11.26	− 9.933	84 32 14.3	− 7.37	51.03	2
4 861	Gould, Z. C., 7ʰ 1855 . .	9.0	7 26 18.73	+ 0.204	66 33 41.8	− 7.38	51.12	1
4 862	Gould, Z. C., 7ʰ 1870 . .	9.0	20.20	+ 0.662	63 14 37.1	.38	51.16	1
4 863	Octantis	10.0	24.75	− 30.445	87 52 32.9	.38	51.21	1
4 864	Volantis	9.5	29.85	− 0.100	68 35 52.0	.39	50.53	3
4 865	Volantis	8.8	35.72	+ 0.518	64 2 54.4	− 7.40	51.14	2

No. 4 857. Differs 8.8″ from Gould. No. 4 862. Differs 10″ from Gould.

Number.	Constellation, Name of Star, or Synonym.	Magnitude.	Right Ascension, 1850.0.	Annual Precession.	South Declination, 1850.0.	Annual Precession.	Mean year.	No. of obs.
			h. m. s.	s.	° ′ ″	″		
4866	Gould, Z. C., 7^{h} 1883 . .	9.0	7 26 41.97	+ 0.288	65 56 43.4	− 7.41	51.60	2
4867	Volantis	9.5	50.16	0.341	65 32 8.7	.42	51.44	3
4868	Volantis	10.0	26 53.40	0.187	66 42 20.8	.42	51.12	1
4869	Gould, Z. C., 7^{h} 1925 . .	8.8	27 4.88	+ 0.619	63 8 56.0	.44	51.15	2
4870	Octantis	9.1	8.51	−23.773	87 21 7.0	− 7.44	51.13	7
4871	Mensæ	10.0	7 27 10.56	− 5.054	81 18 44.8	− 7.45	51.06	1
4872	Lacaille, 2887	8.0	16.44	0.208	69 16 3.8	.45	51.08	1
4873	Lacaille, 2886	7.5	17.01	0.049	68 18 29.7	.46	51.14	3
4874	Lacaille, 2942	9.5	17.06	− 4.707	80 56 1.6	.46	50.14	1
4875	Gould, Z. C., 7^{h} 1951 . .	9.2	25.46	+ 0.490	64 19 21.6	− 7.47	51.12	2
4876	Gould, 9708	8.5	7 27 27.30	+ 0.194	66 40 16.4	− 7.47	50.69	2
4877	Gould, Z. C., 7^{h} 1936 . .	8.5	34.02	− 0.195	69 12 14.2	.48	51.08	1
4878	Lacaille, 2877	7.0	37.54	+ 0.506	64 11 43.4	.48	51.14	2
4879	Carinæ	8.8	37.76	+ 0.634	63 1 47.8	.48	51.15	2
4880	Mensæ	9.5	38.42	− 5.545	81 48 29.1	− 7.48	51.17	1
4881	Gould, Z. C., 7^{h} 1973 . .	9.0	7 27 40.22	+ 0.476	64 27 39.2	− 7.49	51.11	1
4882	Volantis	9.0	42.64	0.524	64 2 43.9	.49	51.14	2
4883	Volantis	10.0	47.32	0.274	66 5 23.5	.50	52.09	1
4884	Volantis	9.2	55.26	+ 0.027	67 50 37.2	.51	51.14	3
4885	Octantis	9.2	27 57.83	− 11.526	85 9 23.2	− 7.51	51.16	5
4886	Volantis	9.8	7 28 2.58	+ 0.489	64 21 29.0	− 7.52	51.12	2
4887	Gould, Z. C., 7^{h} 1960 . .	9.5	3.00	− 0.727	71 55 32.1	.52	50.22	1
4888	Gould, Z. C., 7^{h} 1985 . .	9.0	4.56	0.066	68 26 43.0	.52	50.25	1
4889	Volantis	10.0	10.69	− 1.093	73 25 37.7	.53	50.21	1
4890	Carinæ	8.5	13.25	+ 0.572	63 37 57.3	− 7.53	51.13	1
4891	Carinæ	9.5	7 28 33.12	+ 0.691	62 31 10.4	− 7.56	51.18	1
4892	Volantis	10.0	41.96	+ 0.268	66 10 28.8	.57	52.09	1
4893	Gould, Z. C., 7^{h} 1954 . .	10.0	44.52	− 2.569	77 37 13.1	.57	50.17	1
4894	Volantis	9.5	49.96	+ 0.523	64 5 40.6	.58	51.14	1
4895	Gould, Z. C., 7^{h} 2035 . .	9.0	28 57.02	− 0.412	70 26 22.4	− 7.59	50.23	1
4896	Volantis	10.0	7 29 2.02	+ 0.380	65 18 49.7	− 7.60	52.12	1
4897	Gould, Z. C., 7^{h} 2055 . .	9.0	2.96	− 0.032	68 15 51.5	.60	51.14	3
4898	Volantis	9.0	3.10	− 0.872	72 34 45.6	.60	51.06	1
4899	Carinæ	10.0	18.46	+ 0.596	63 27 7.9	.62	51.13	1
4900	Volantis	10.0	20.19	+ 0.262	66 14 37.4	− 7.62	52.09	1
4901	Volantis	9.7	7 29 22.02	+ 0.509	64 14 13.3	− 7.62	51.13	3
4902	Gould, Z. C., 7^{h} 2086 . .	8.2	31.49	+ 0.037	67 50 8.8	.64	51.14	3
4903	Gould, 9749	8.5	38.80	− 0.916	72 46 44.2	.65	50.64	2
4904	Mensæ	9.5	43.33	− 5.704	81 59 1.2	.65	51.17	1
4905	Gould, Z. C., 7^{h} 2121 . .	8.8	47.28	+ 0.635	63 6 26.2	− 7.66	51.15	2

Number.	Constellation, Name of Star, or Synonym.	Magnitude.	Right Ascension, 1850.0.	Annual Precession.	South Declination, 1850.0.	Annual Precession.	Mean year.	No. of obs.
			h. m. s.	s.	° ′ ″	″		
4906	Volantis	9.8	7 29 48.58	+ 0.003	68 3 54.0	− 7.66	51.58	2
4907	Carinæ	10.0	55.68	+ 0.558	63 49 1.0	.67	51.13	1
4908	Volantis	10.0	55.97	− 0.250	69 35 57.9	.67	50.24	1
4909	Volantis	10.0	56.18	0.170	69 8 13.5	.67	51.08	1
4910	Lacaille, 2948	8.0	29 59.10	− 3.203	78 51 49.3	− 7.68	50.16	1
4911	Lacaille, 2901	7.5	7 30 6.45	+ 0.524	64 3 8.6	− 7.68	51.14	2
4912	Gould, Z. C., 7^h 2103 . .	9.5	9.05	− 0.880	72 38 47.8	.69	51.06	1
4913	Gould, Z. C., 7^h 2132 . .	8.8	11.16	+ 0.295	66 1 35.1	.69	51.54	3
4914	Volantis	10.0	15.65	0.310	65 54 40.8	.70	52.09	1
4915	Gould, Z. C., 7^h 2156 . .	9.0	22.50	+ 0.388	65 17 31.1	− 7.71	52.12	1
4916	Lacaille, 2927	8.0	7 30 24.92	− 1.626	75 16 36.9	− 7.71	50.19	1
4917	Volantis	10.5	36.35	0.322	70 1 11.2	.72	50.65	2
4918	Volantis	10.0	44.09	0.088	68 40 33.1	.73	50.24	2
4919	Gould, Z. C., 7^h 2152 . .	9.5	44.42	− 0.655	71 41 3.7	.73	50.22	1
4920	Volantis	9.0	46.74	+ 0.134	67 13 20.0	− 7.74	50.26	2
4921	Brisbane, 1669	8.5	7 30 56.38	+ 0.186	66 51 36.6	− 7.75	50.26	1
4922	Gould, Z. C., 7^h 2194 . .	10.0	30 56.95	0.095	67 29 46.5	.75	50.26	1
4923	Volantis	10.0	31 0.93	+ 0.350	65 37 34.3	.76	51.10	1
4924	Gould, Z. C., 7^h 2089 . .	10.0	7.15	− 3.683	79 39 29.1	.77	50.16	1
4925	Gould, Z. C., 7^h 2230 . .	8.8	8.06	+ 0.652	63 0 2.1	− 7.77	51.15	2
4926	Carinæ	9.0	7 31 10.23	+ 0.664	62 53 27.1	− 7.77	51.15	1
4927	Mensæ	10.0	16.28	− 2.213	76 52 20.7	.78	50.17	1
4928	Volantis	10.0	17.30	+ 0.369	65 29 12.9	.78	51.10	1
4929	Mensæ	10.0	19.43	− 2.324	77 8 8.5	.78	50.17	1
4930	Mensæ	9.8	25.02	− 2.003	76 21 13.1	− 7.79	50.18	3
4931	Gould, Z. C., 7^h 2211 . .	9.8	7 31 29.14	− 0.857	72 35 33.4	7.80	50.64	2
4932	Gould, Z. C., 7^h 2185 . .	9.5	30.43	− 1.453	74 46 13.0	.80	50.20	1
4933	Gould, Z. C., 7^h 2259 . .	9.0	50.72	+ 0.160	67 4 52.0	.82	50.26	2
4934	Volantis	9.8	31 52.45	− 1.346	74 25 50.8	.83	50.68	2
4935	Gould, 9820	9.5	32 5.13	− 0.967	73 3 18.2	− 7.85	51.06	1
4936	Volantis	9.8	7 32 30.96	− 0.857	72 37 19.9	− 7.88	50.49	3
4937	Volantis	10.0	33.47	1.285	74 14 25.5	.88	51.16	1
4938	Gould, 9832	9.0	45.20	− 1.543	75 5 9.4	.90	50.19	2
4939	Gould, Z. C., 7^h 2351 . .	8.8	32 56.86	+ 0.644	63 9 45.1	.91	51.15	2
4940	Gould, Z. C., 7^h 2377 . .	9.5	33 20.96	+ 0.396	65 21 26.2	− 7.94	52.12	1
4941	Gould, Z. C., 7^h (2375) .	10.0	7 33 21.84	+ 0.331	65 52 24.7	− 7.95	51.10	1
4942	Gould, Z. C., 7^h 2393 . .	9.0	23.76	+ 0.693	62 42 24.6	.95	51.17	2
4943	Volantis	9.5	26.09	− 0.635	71 40 38.8	.95	50.22	1
4944	Volantis	10.0	28.25	+ 0.335	65 50 44.2	.95	51.10	1
4915	Carinæ	10.0	28.91	+ 0.706	62 34 39.9	− 7.96	51.18	1

No. 4941. Differs 5′ from Gould.

Number.	Constellation, Name of Star, or Synonym.	Magnitude.	Right Ascension, 1850.0.	Annual Precession.	South Declination, 1850.0.	Annual Precession.	Mean year.	No. of obs.
			h. m. s.	s.	° ′ ″	″		
4946	Lacaille, 2935	8.0	7 33 30.59	− 0.034	68 26 15.2	− 7.96	50.25	1
4947	Volantis	9.2	32.24	+ 0.423	65 8 30.1	.96	52.10	2
4948	Volantis	9.0	33.83	0.312	66 1 25.5	.96	50.28	1
4949	Gould, Z. C., 7^h 2409	9.0	39.68	+ 0.703	62 37 19.4	.97	51.17	2
4950	ε Mensæ	6.0	46.29	− 3.110	78 46 30.7	− 7.98	51.17	4
4951	Volantis	10.0	7 33 46.68	+ 0.503	64 28 19.8	− 7.98	51.11	1
4952	Gould, Z. C., 7^h 2413	9.5	50.56	+ 0.414	65 13 55.6	.98	52.12	1
4953	Octantis	9.3	33 57.79	−25.065	87 30 15.9	7.99	51.13	7
4954	Carinæ	10.0	34 0.86	+ 0.610	63 31 36.1	8.00	51.14	2
4955	Gould, Z. C., 7^h 2400	10.0	1.50	− 0.481	70 57 38.7	− 8.00	50.23	1
4956	Gould, Z. C., 7^h 2432	9.0	7 34 6.78	+ 0.277	66 19 6.2	− 8.01	51.69	2
4957	Mensæ	8.0	15.15	− 7.898	83 37 29.8	.02	50.12	2
4958	Volantis	9.0	15.47	+ 0.456	64 54 20.0	.02	52.07	1
4959	Gould, 9878	9.0	19.59	− 0.216	69 33 26.5	.02	50.24	2
4960	Gould, Z. C., 7^h 2404	9.3	27.65	− 1.376	74 36 2.2	− 8.03	50.52	3
4961	Lacaille, 2961	8.0	7 34 31.44	− 1.185	73 57 42.8	− 8.04	50.20	2
4962	Lacaille, 2967	8.2	36.81	1.303	74 21 42.9	.05	50.52	3
4963	Mensæ	10.0	45.52	1.687	75 34 11.3	.06	50.18	1
4964	Mensæ	9.5	34 48.18	5.951	82 16 21.8	.06	50.65	2
4965	Gould, Z. C., 7^h 2470	10.0	35 1.28	0.752	72 15 13.9	− 8.08	50.22	2
4966	Gould, 9936	9.7	7 35 10.51	+ 0.577	63 52 32.3	− 8.09	51.13	3
4967	Gould, Z. C., 7^h 2533	9.5	11.28	+ 0.675	62 57 42.0	.09	51.15	1
4968	Gould, Z. C., 7^h 2412	9.8	12.36	− 2.868	78 21 43.9	.10	51.17	4
4969	Volantis	9.5	21.08	− 0.862	72 43 41.0	.11	50.64	2
4970	Volantis	10.0	22.29	+ 0.519	64 24 23.6	− 8.11	51.11	1
4971	Gould, Z. C., 7^h 2514	9.0	7 35 23.71	− 0.297	70 2 47.6	− 8.11	50.22	1
4972	Lacaille, 3096	7.5	23.80	− 8.908	84 10 16.6	.11	50.12	1
4973	Gould, Z. C., 7^h 2539	9.0	26.00	+ 0.197	66 57 27.4	.11	50.26	2
4974	Gould, Z. C., 7^h 2523	9.5	28.64	− 0.253	69 48 38.3	.12	50.24	1
4975	Gould, Z. C., 7^h 2452	9.0	29.28	− 2.445	77 30 16.3	− 8.12	50.17	2
4976	Volantis	10.0	7 35 33.44	− 0.858	72 43 8.6	− 8.12	51.06	1
4977	Volantis	10.0	39.50	+ 0.167	67 10 35.3	.13	50.26	1
4978	Gould, Z. C., 7^h 2575	9.5	40.62	+ 0.681	62 55 34.3	.13	51.15	1
4979	Gould, Z. C., 7^h 2548	9.0	47.58	− 0.269	69 54 24.0	.14	50.24	2
4980	Gould, Z. C., 7^h 2510	9.0	49.42	− 1.551	75 11 36.8	− 8.14	50.19	1
4981	Volantis	10.0	7 35 51.10	+ 0.114	67 33 28.3	− 8.15	50.25	1
4982	Lacaille, 2953	8.5	35 52.08	0.188	67 2 17.2	.15	50.26	2
4983	Carinæ	9.5	36 3.39	0.577	63 54 46.1	.16	51.13	3
4984	Gould, Z. C., 7^h 2610	9.2	12.01	0.419	65 16 59.1	.17	51.51	3
4985	Volantis	9.0	15.06	+ 0.158	67 16 12.4	− 8.18	50.26	2

Number.	Constellation, Name of Star, or Synonym.	Magnitude.	Right Ascension, 1850.0.	Annual Precession.	South Declination, 1850.0.	Annual Precession.	Mean year.	No. of obs.
			h. m. s.	s.	° ′ ″	″		
4 986	Gould, Z. C., 7h 2619 . .	9.0	7 36 15.88	+ 0.601	63 42 29.0	— 8.18	51.13	1
4 987	Volantis	9.9	25.81	+ 0.394	65 30 7.4	.19	51.61	4
4 988	Lacaille, 2968	8.5	32.52	— 0.314	70 10 59.3	.20	50.23	2
4 989	Lacaille, 2966	7.8	32.60	— 0.060	68 42 57.2	.20	50.24	3
4 990	Brisbane, 1714	8.0	35.85	+ 0.614	63 36 8.9	— 8.21	51.13	1
4 991	Lacaille, 2955	7.6	7 36 41.32	+ 0.386	65 34 23.8	— 8.21	51.61	4
4 992	Carinæ	8.5	53.00	0.670	63 5 6.0	.23	51.15	2
4 993	Gould, Z. C., 7h 2653 . .	8.8	54.71	0.028	68 10 3.2	.23	50.70	1
4 994	Volantis	9.5	36 58.45	0.256	66 35 11.1	.24	51.12	1
4 995	Gould, Z. C., 7h 2678 . .	9.0	37 4.88	+ 0.444	65 6 52.7	— 8.24	51.49	3
4 996	Carinæ	10.0	7 37 10.21	+ 0.737	62 26 8.3	— 8.25	51.18	1
4 997	Carinæ	9.5	14.28	0.749	62 19 6.5	.26	51.18	1
4 998	Carinæ	9.0	15.88	0.657	63 13 35.9	.26	51.16	1
4 999	Gould, Z. C., 7h 2710 . .	8.5	26.64	+ 0.673	63 4 41.8	.27	51.15	2
5 000	Mensæ	9.5	30.22	— 5.935	82 17 56.0	— 8.28	50.65	2
5 001	Gould, Z. C., 7h 2712 . .	9.2	7 37 30.24	+ 0.579	63 57 44.5	— 8.28	51.14	2
5 002	Octantis	8.8	42.04	—13.098	85 41 39.9	.29	51.21	4
5 003	Lacaille, 2977	8.2	48.34	+ 0.202	67 6 1.5	.30	50.26	2
5 004	Mensæ	10.0	51.01	— 1.760	75 51 45.2	.31	50.18	3
5 005	Gould, Z. C., 7h 2684 . .	9.5	57.25	— 1.386	74 43 57.1	— 8.31	51.16	1
5 006	Lacaille, 3274	7.1	7 37 58.08	—18.444	86 45 46.7	— 8.31	51.15	7
5 007	Volantis	9.0	38 3.61	+ 0.240	66 45 18.9	.32	51.12	1
5 008	Mensæ	10.0	5.44	— 4.193	80 29 39.7	.32	50.15	1
5 009	Volantis	9.5	16.95	— 0.311	70 13 43.1	.34	50.23	1
5 010	Gould, 10026	9.0	22.96	+ 0.293	66 22 25.8	— 8.35	51.12	1
5 011	Lacaille, 3010	7.8	7 38 23.02	— 1.149	73 55 57.2	— 8.35	50.20	2
5 012	Gould, Z. C., 7h 2779 . .	8.8	28.18	+ 0.075	67 55 16.6	.35	50.70	2
5 013	Volantis	10.0	30.47	— 0.268	70 0 0.9	.36	50.23	1
5 014	Gould, Z. C., 7h 2777 . .	9.0	36.31	0.216	69 42 53.9	.37	50.24	1
5 015	Octantis	9.4	37.46	—20.932	87 6 7.6	— 8.37	51.16	5
5 016	Gould, Z. C., 7h 2794 . .	9.8	7 38 46.17	— 0.089	68 58 39.5	— 8.38	50.24	3
5 017	Octantis	9.2	53.02	17.758	86 39 46.4	.39	51.14	7
5 018	Lacaille, 3029	7.3	54.12	— 2.083	76 44 34.8	.39	50.18	2
5 019	Lacaille, 2983	8.8	38 59.35	+ 0.567	64 8 2.2	.40	51.14	2
5 020	Lacaille, 3040	9.0	39 0.45	— 2.520	77 44 53.6	— 8.40	50.17	1
5 021	Gould, Z. C., 7h 2758 . .	10.0	7 39 6.58	— 2.052	76 40 13.3	— 8.41	50.18	1
5 022	Gould, 10057	8.8	7.34	+ 0.586	63 58 12.2	.41	51.13	2
5 023	Gould, Z. C., 7h 2790 . .	9.5	14.36	— 1.385	74 46 3.7	.42	51.16	1
5 024	Gould, 10071	9.0	23.98	+ 0.690	63 0 26.4	.43	51.15	2
5 025	Carinæ	9.0	35.30	+ 0.754	62 22 34.2	— 8.44	51.18	1

No. 5 014. Differs from Gould 2s in right ascension and 24″ in declination.

Number.	Constellation, Name of Star, or Synonym.	Magnitude.	Right Ascension, 1850.0.	Annual Precession.	South Declination, 1850.0.	Annual Precession.	Mean year.	No. of obs.
			h. m. s.	s.	° ′ ″	″		
5026	Gould, Z. C., 7^h 2879 . .	9.0	7 39 36.32	− 0.608	63 47 16.0	− 8.44	51.13	1
5027	Lacaille, 3027	8.0	41.10	− 1.517	75 11 55.5	.45	50.19	1
5028	Volantis	10.0	50.07	+ 0.419	65 26 19.9	.46	51.58	4
5029	Volantis	10.0	50.16	− 0.498	71 14 36.9	.46	50.22	1
5030	Gould, Z. C., 7^h 2901 . .	8.8	50.44	+ 0.599	63 53 0.7	− 8.46	51.13	3
5031	Gould, Z. C., 7^h 2856 . .	10.0	7 39 54.03	− 0.985	73 22 3.1	− 8.47	50.21	1
5032	Gould, Z. C., 7^h 2793 . .	9.8	40 0.00	3.010	78 45 23.3	.48	51.18	2
5033	Chamæleontis	10.0	11.18	− 4.700	81 4 58.0	.49	51.60	2
5034	Volantis	9.5	16.26	+ 0.087	67 54 47.5	.50	51.15	1
5035	Gould, Z. C., 7^h 2905 . .	10.0	17.77	− 0.272	70 5 35.0	− 8.50	50.23	1
5036	Gould, Z. C., 7^h 2940 . .	9.0	7 40 17.90	+ 0.735	62 36 27.5	− 8.50	51.18	1
5037	Carinæ	9.3	27.42	+ 0.607	63 50 21.0	.51	51.13	3
5038	Gould, Z. C., 7^h 2929 . .	9.5	31.26	− 0.100	69 6 44.2	.52	50.24	2
5039	Gould, Z. C., 7^h 2964 . .	8.0	34.91	+ 0.709	62 52 49.9	.52	51.15	1
5040	Gould, Z. C., 7^h 2914 . .	9.5	38.93	− 0.843	72 49 16.8	− 8.53	50.21	1
5041	Gould, Z. C., 7^h 2969 . .	8.8	7 40 44.39	+ 0.425	65 25 46.8	− 8.53	51.69	5
5042	Gould, Z. C., 7^h 2994 . .	8.5	46.66	0.684	63 7 37.0	.54	51.15	2
5043	Volantis	10.0	51.43	0.583	64 4 57.7	.54	51.14	1
5044	Gould, Z. C., 7^h 2997 . .	8.9	54.11	+ 0.459	65 9 36.4	.55	51.65	4
5045	Lacaille, 3018	8.7	40 59.62	− 0.161	69 29 31.9	− 8.55	50.24	2
5046	Octantis	9.8	7 41 9.68	−18.159	86 44 27.3	− 8.57	52.00	5
5047	Melbourne (1), 371 . . .	9.5	10.36	+ 0.179	67 19 13.2	.57	50.26	2
5048	Carinæ	9.5	19.61	0.664	63 20 30.9	.58	51.16	1
5049	Volantis	10.0	25.00	0.178	67 20 22.1	.59	50.26	2
5050	Carinæ	10.0	32.24	+ 0.660	63 23 45.8	− 8.60	51.13	1
5051	Brisbane, 1757	7.5	7 41 38.25	+ 0.288	66 32 36.7	− 8.61	51.12	1
5052	Chamæleontis	9.5	39.51	− 5.528	82 1 23.3	.61	51.17	1
5053	Gould, Z. C., 7^h 3055 . .	9.0	41.86	+ 0.292	66 31 19.1	.61	51.12	1
5054	Gould, Z. C., 7^h 3024 . .	9.0	41.91	− 0.696	72 13 44.8	.61	50.21	3
5055	Chamæleontis	10.0	43.17	− 1.456	75 4 1.4	− 8.61	50.20	1
5056	Volantis	10.0	7 41 49.14	+ 0.363	65 58 56.9	− 8.62	52.25	1
5057	Volantis	9.0	51.72	0.345	66 7 26.2	.62	50.28	1
5058	Gould, 10168	8.5	54.40	0.699	63 2 19.1	.63	51.15	2
5059	Volantis	9.5	41 58.46	0.123	67 44 37.6	.63	50.25	1
5060	Gould, Z. C., 7^h 3089 . .	9.5	42 1.03	+ 0.710	62 55 54.3	− 8.64	51.15	1
5061	Lacaille, 3066	7.5	7 42 1.76	− 2.276	77 17 7.5	− 8.64	50.17	1
5062	Volantis	10.0	9.47	+ 0.513	64 45 31.9	.65	52.07	1
5063	Gould, Z. C., 7^h 3073 . .	9.5	10.35	− 0.403	70 51 8.1	.65	50.23	1
5064	Lacaille, 3038	8.8	10.70	0.512	71 24 18.6	.65	50.22	2
5065	Gould, Z. C., 7^h 3075 . .	10.0	13.34	− 0.383	70 45 7.5	− 8.65	50.23	1

Number.	Constellation, Name of Star, or Synonym.	Magnitude.	Right Ascension, 1850.0.	Annual Precession.	South Declination, 1850.0.	Annual Precession.	Mean year.	No. of obs.
			h. m. s.	s.	° ′ ″	″		
5 066	Gould, Z. C., 7h 3088 . .	10.0	7 42 25.19	− 0.321	70 26 4.9	− 8.67	50.23	1
5 067	Volantis	9.5	28.01	0.044	68 50 48.8	.67	50.24	1
5 068	Chamæleontis	10.0	30.55	− 4.218	80 36 34.6	.67	50.14	1
5 069	Gould, Z. C., 7h 3121 . .	9.3	31.20	+ 0.571	64 15 40.5	.68	51.13	3
5 070	Chamæleontis	9.5	38.11	− 2.940	78 39 50.4	− 8.68	52.21	1
5 071	Chamæleontis	9.5	7 42 41.92	− 5.087	81 35 24.4	− 8.69	50.60	2
5 072	Volantis	9.2	43 4.18	1.274	74 30 48.7	.72	51.18	2
5 073	Octantis	9.0	22.47	− 6.558	82 52 36.7	.74	50.13	1
5 074	Lacaille, 3037	8.5	25.07	+ 0.259	66 50 19.7	.75	50.26	1
5 075	ζ1 Volantis	5.0	37.59	− 0.685	72 14 41.5	− 8.76	50.21	3
5 076	ζ2 Volantis	10.2	7 43 41.26	− 0.684	72 14 47.9	− 8.77	50.21	3
5 077	Chamæleontis	11.0	42.44	− 5.551	82 3 39.0	.77	51.17	1
5 078	Gould, Z. C., 7h 3219 . .	9.0	43.36	+ 0.480	65 6 47.1	.77	52.11	3
5 079	Lacaille, 3054	7.2	45.76	0.408	65 42 22.8	.77	51.41	3
5 080	Carinæ	10.0	46.78	+ 0.713	62 59 29.0	− 8.77	51.15	2
5 081	Carinæ	10.0	7 43 48.97	+ 0.720	62 55 48.2	− 8.78	51.15	1
5 082	Lacaille, 3055	7.0	50.49	− 0.177	69 41 41.6	.78	50.24	1
5 083	Carinæ	10.0	50.68	+ 0.712	63 1 11.5	.78	51.15	1
5 084	Chamæleontis	10.0	52.11	− 4.776	81 17 27.4	.78	51.06	1
5 085	Volantis	9.2	53.43	+ 0.487	65 3 49.0	− 8.78	52.11	3
5 086	Carinæ	9.5	7 43 53.53	+ 0.738	62 44 42.7	− 8.78	51.18	1
5 087	Volantis	9.8	54.17	0.582	64 13 39.4	.78	51.12	2
5 088	Lacaille, 3034	8.5	43 55.85	+ 0.628	63 49 33.5	.79	51.13	3
5 089	Lacaille, 3058	9.0	44 6.28	− 0.453	71 10 41.0	.80	50.22	1
5 090	Gould, Z. C., 7h 3164 . .	9.0	18.68	− 2.572	78 0 32.3	− 8.82	50.17	1
5 091	Volantis	9.7	7 44 19.88	+ 0.408	65 45 2.7	− 8.82	51.48	3
5 092	Gould, Z. C., 7h 3278 . .	9.8	20.24	0.633	63 47 18.8	.82	51.13	2
5 093	Gould, Z. C., 7h 3282 . .	9.0	25.58	0.455	65 21 15.9	.83	51.51	3
5 094	Gould, Z. C., 7h 3288 . .	9.2	25.96	0.606	64 2 4.8	.83	51.14	2
5 095	Volantis	10.0	36.65	+ 0.091	68 4 7.4	− 8.84	51.15	1
5 096	Gould, Z. C., 7h 3274 . .	9.6	7 44 37.40	− 0.079	69 8 50.4	− 8.84	50.24	2
5 097	Lacaille, 3057	7.5	38.83	0.130	69 27 14.2	.84	50.24	2
5 098	Octantis	8.8	41.56	−52.786	88 46 9.5	.85	51.15	7
5 099	Gould, Z. C., 7h 3320 . .	9.0	43.97	+ 0.752	62 39 16.6	.85	51.17	2
5 100	Carinæ	9.8	50.97	+ 0.628	63 51 13.8	− 8.86	51.13	3
5 101	Volantis	9.0	7 44 50.91	+ 0.044	68 23 17.0	− 8.86	50.25	1
5 102	Carinæ	10.0	44 52.80	0.646	63 41 30.7	.86	51.13	1
5 103	Brisbane, 1786	8.5	45 2.06	+ 0.288	66 41 56.8	.87	50.69	2
5 104	Chamæleontis	8.7	9.75	5.348	81 54 14.3	.88	50.48	3
5 105	Gould, Z. C., 7h 3349 . .	10.0	16.74	+ 0.225	67 10 9.6	− 8.89	50.26	2

No. 5 071. Declination possibly 81° 35′ 53.4″.

Number.	Constellation, Name of Star, or Synonym.	Magnitude.	Right Ascension, 1850.0.	Annual Precession.	South Declination, 1850.0.	Annual Precession.	Mean year.	No. of obs.
			h. m. s.	s.	° ′ ″	″		
5 106	Volantis	10.0	7 45 26.19	+ 0.430	65 36 18.6	— 8.90	51.10	1
5 107	Gould, 10255	9.5	26.19	— 1.272	74 34 56.1	.90	51.18	2
5 108	Lacaille, 3062	8.0	30.27	+ 0.244	67 3 38.0	.91	50.26	1
5 109	Chamæleontis	9.5	38.50	— 1.527	75 24 11.4	.92	50.18	2
5 110	Carinæ	9.8	44.92	+ 0.761	62 36 40.0	— 8.93	51.67	2
5 111	Gould, Z. C., 7h 3411 . .	9.5	7 45 49.15	+ 0.710	63 7 19.2	— 8.94	51.16	1
5 112	Gould, Z. C., 7h 3400 . .	9.0	50.33	0.340	66 20 11.2	.94	51.21	3
5 113	Volantis	10.0	51.66	+ 0.326	66 27 1.7	.94	51.12	1
5 114	Gould, Z. C., 7h 3315 . .	10.0	45 58.86	— 2.517	77 54 50.0	.95	50.17	2
5 115	Gould, Z. C., 7h 3421 . .	9.0	46 3.04	+ 0.688	63 20 40.3	— 8.95	51.16	1
5 116	Gould, Z. C., 7h 3415 . .	9.2	7 46 15.01	— 0.064	69 7 30.9	— 8.97	50.24	3
5 117	Gould, Z. C., 7h 3355 . .	9.0	16.88	— 2.071	76 54 35.7	.97	50.17	1
5 118	Herschel, 4011[1]	10.0	17.10	+ 0.295	66 41 52.5	.97	51.12	1
5 119	Herschel, 4011[2]	10.0	17.85	+ 0.296	66 41 43.0	.97	51.12	1
5 120	Chamæleontis	10.0	18.29	— 2.812	78 31 40.4	— 8.97	52.21	1
5 121	Chamæleontis	8.5	7 46 19.75	— 5.781	82 18 27.1	— 8.97	50.48	3
5 122	Octantis	9.5	31.14	8.739	84 13 14.7	.99	50.12	1
5 123	Gould, 10292	9.0	32.73	1.065	73 53 6.6	.99	50.20	2
5 124	Gould, Z. C., 7h 3371 . .	9.0	33.13	2.359	77 35 22.5	8.99	50.17	1
5 125	Octantis	9.5	38.60	—51.671	88 44 58.3	— 9.00	51.16	7
5 126	Chamæleontis	10.0	7 46 49.40	— 3.338	79 26 40.7	— 9.01	50.21	2
5 127	Octantis	9.8	55.10	—11.341	85 15 44.0	.02	51.20	2
5 128	Gould, 10322	8.5	56.01	+ 0.697	63 18 36.1	.02	51.16	1
5 129	Gould, 10323	9.5	56.92	0.696	63 18 45.6	.02	51.16	1
5 130	Volantis	11.0	46 58.39	+ 0.560	64 34 42.2	— 9.03	52.07	1
5 131	Gould, Z. C., 7h 3382 . .	9.7	7 47 0.18	— 3.104	79 3 27.4	— 9.03	50.19	3
5 132	Volantis	10.0	0.65	+ 0.551	64 39 8.1	.03	52.07	1
5 133	Lacaille, 3107	6.5	4.47	— 2.557	78 1 29.9	.03	50.17	1
5 134	Volantis	10.0	8.19	+ 0.561	64 34 28.2	.04	52.07	1
5 135	Gould, 10329	9.0	10.64	+ 0.576	64 26 40.8	— 9.04	51.11	1
5 136	Volantis	9.0	7 47 18.54	+ 0.548	64 41 48.0	— 9.05	51.59	2
5 137	Gould, Z. C., 7h 3529 . .	9.0	28.65	+ 0.397	65 58 8.4	.06	51.26	2
5 138	Octantis	10.0	30.43	—11.324	85 15 51.2	.07	51.19	1
5 139	Gould, Z. C., 7h 3484 . .	9.5	31.00	0.938	73 28 15.0	.07	50.21	2
5 140	Lacaille, 3085	8.8	33.42	— 0.461	71 20 44.5	— 9.07	50.22	2
5 141	Octantis	8.8	7 47 36.04	— 8.270	83 59 54.4	— 9.07	50.12	2
5 142	Volantis	9.5	37.63	0.088	69 19 31.1	.08	50.24	1
5 143	Chamæleontis	9.2	47 54.78	— 5.534	82 7 0.8	.10	50.65	2
5 144	Volantis	9.2	48 7.36	+ 0.118	68 2 36.0	.11	50.70	2
5 145	Gould, Z. C., 7h 3575 . .	9.5	11.21	+ 0.258	67 3 44.4	— 9.12	50.26	2

No. 5 144. Possibly 67° 58′ 20.0″.

Number.	Constellation, Name of Star, or Synonym.	Magnitude.	Right Ascension, 1850.0.	Annual Precession.	South Declination, 1850.0.	Annual Precession.	Mean year.	No. of obs.
			h. m. s.	s.	° ′ ″	″		
5 146	Volantis	10.0	7 48 16.17	+ 0.427	65 45 59.0	− 9.13	51.10	2
5 147	Gould, Z. C., 7^h 3590	9.2	16.41	+ 0.466	65 26 32.5	.13	50.83	3
5 148	Octantis	. .	20.34	−33.978	88 9 34.8	.13	51.03	1
5 149	Volantis	9.5	33.18	+ 0.122	68 1 51.1	.15	50.70	2
5 150	Gould, Z. C., 7^h 3631	9.5	39.10	+ 0.659	63 45 36.9	− 9.16	51.13	1
5 151	Lacaille, 3083	6.2	7 48 40.26	+ 0.423	65 48 44.2	− 9.16	51.18	4
5 152	Volantis	10.0	44.46	0.113	68 5 58.6	.16	51.15	1
5 153	Carinæ	10.0	44.56	0.668	63 29 17.6	.16	51.13	1
5 154	Gould, Z. C., 7^h 3634	9.0	52.54	0.169	67 43 25.6	.17	50.25	1
5 155	Carinæ	10.0	53.76	+ 0.798	62 23 48.2	− 9.18	51.18	1
5 156	Gould, 10383	8.0	7 48 59.85	+ 0.794	62 26 37.3	− 9.18	51.18	1
5 157	Volantis	10.0	49 12.87	− 0.295	70 34 8.2	.20	50.23	1
5 158	Gould, 10393	7.0	18.68	+ 0.786	62 32 46.5	.21	51.18	1
5 159	Brisbane, 1820	9.0	28.46	− 0.992	73 42 46.0	.22	50.20	1
5 160	Volantis	9.5	35.63	+ 0.467	65 30 7.5	− 9.23	51.10	1
5 161	Gould, 10412	8.0	7 49 54.11	+ 0.791	62 31 16.5	− 9.25	51.18	1
5 162	Volantis	9.0	58.88	+ 0.369	66 18 40.6	.26	51.19	2
5 163	Gould, Z. C., 7^h 3710	9.7	49 59.70	− 0.029	69 4 20.4	.26	50.24	3
5 164	Chamæleontis	10.0	50 14.28	1.672	75 58 17.2	.28	50.18	2
5 165	Octantis	8.8	23.30	− 6.101	82 38 22.0	− 9.29	50.13	2
5 166	Brisbane, 1824	9.0	7 50 29.56	+ 0.242	67 17 5.6	− 9.30	50.26	2
5 167	Lacaille, 3092	7.8	31.53	0.653	63 54 13.6	.30	51.13	3
5 168	Carinæ	10.0	48.60	0.790	62 35 2.3	.32	51.18	1
5 169	Volantis	9.0	50 51.34	+ 0.545	64 54 7.3	.33	52.07	1
5 170	Volantis	10.0	51 6.99	− 1.119	74 14 13.1	− 9.35	51.16	1
5 171	Gould, Z. C., 7^h 3781	10.0	7 51 13.85	− 0.562	71 58 12.3	− 9.36	50.21	1
5 172	Volantis	10.0	15.12	− 0.197	70 6 40.0	.36	50.23	1
5 173	Volantis	9.5	16.62	+ 0.389	66 12 41.6	.36	51.12	1
5 174	Carinæ	9.8	20.60	+ 0.652	63 57 32.3	.37	51.14	2
5 175	Brisbane, 1830	9.0	21.16	− 0.145	69 42 5.6	− 9.37	50.24	1
5 176	Volantis	9.8	7 51 21.98	+ 0.639	64 4 44.2	− 9.37	51.14	2
5 177	Volantis	9.8	26.30	0.627	64 11 40.8	.37	51.14	2
5 178	Gould, Z. C., 7^h 3847	9.0	29.16	+ 0.578	64 38 17.7	.38	51.60	2
5 179	Lacaille, 3111	9.0	30.75	− 0.245	70 23 22.6	.38	50.23	1
5 180	Carinæ	10.0	34.56	+ 0.795	62 34 31.3	− 9.38	51.18	1
5 181	Volantis	9.8	7 51 36.54	+ 0.648	64 0 42.7	− 9.39	51.14	2
5 182	Octantis	9.2	37.93	−13.504	85 55 21.7	.39	51.20	4
5 183	Volantis	9.0	40.32	+ 0.551	64 53 10.0	.39	52.07	1
5 184	Volantis	9.3	45.04	0.617	64 18 17.7	.40	51.13	3
5 185	Gould, Z. C., 7^h 3872	9.5	47.95	+ 0.719	63 22 33.4	− 9.40	51.16	1

No. 5 146. Possibly 65° 46′ 59.0″. No. 5 149. Possibly 67° 57′ 35.0″.

Number.	Constellation, Name of Star, or Synonym.	Magnitude.	Right Ascension, 1850.0.	Annual Precession.	South Declination, 1850.0.	Annual Precession.	Mean year.	No. of obs.
			h. m. s.	s.	° ′ ″	″		
5186	Chamæleontis	9.0	7 51 54.13	− 2.405	77 50 . .	− 9.41	50.17	1
5187	Volantis	9.5	56.00	+ 0.491	65 25 11.7	.41	51.10	1
5188	Gould, Z. C., 7h 3888 . .	8.5	51 56.87	+ 0.734	63 15 27.0	.41	51.16	1
5189	Volantis	9.5	52 23.42	− 0.297	70 42 12.0	.45	50.23	1
5190	Volantis	9.8	32.12	+ 0.057	68 38 23.4	− 9.46	50.25	2
5191	Gould, Z. C., 7h 3875 . .	9.5	7 52 35 24	− 1.108	74 14 52.6	− 9.46	50.52	3
5192	Carinæ	9.5	37.32	+ 0.807	62 30 2.3	.46	51.18	1
5193	Gould, Z. C., 7h 3849 . .	9.0	38.63	− 2.138	77 15 2.4	.47	50.17	1
5194	Gould, Z. C., 7h 3910 . .	9.0	38.91	− 0.466	71 37 6.7	.47	50.22	1
5195	Carinæ	9.2	39.66	+ 0.792	62 39 44.4	9.47	51.17	2
5196	Gould, Z. C., 7h 3965 . .	9.5	7 52 47.26	+ 0.596	64 32 37.3	− 9.48	51.11	1
5197	Gould, Z. C., 7h 3993 . .	9.5	53 13.50	0.291	67 3 28.2	.51	50.26	2
5198	Gould, Z. C., 7h 3995 . .	9.2	18.32	0.214	67 36 48.4	.52	50.25	2
5199	Gould, Z. C., 7h 4019 . .	9.8	20.28	0.588	64 38 52.2	.52	51.59	2
5200	Volantis	9.0	27.34	+ 0.344	66 40 8.6	9.53	51.12	1
5201	Volantis	9.8	7 53 27.94	+ 0.585	64 40 47.9	− 9.53	51.59	2
5202	Gould, Z. C., 7h 3944 . .	8.5	32.52	− 1.784	76 22 59.3	.53	50.18	1
5203	Chamæleontis	9.8	33.32	3.788	80 16 9.6	.54	50.15	2
5204	Gould, Z. C., 7h 3946 . .	9.0	38.42	− 1.927	76 45 43.4	.54	50.18	2
5205	Gould, Z. C., 7h 4046 . .	9.5	42.98	+ 0.453	65 49 6.6	− 9.55	51.18	4
5206	Octantis	8.8	7 53 43.03	−19.322	86 59 39.3	− 9.55	51.17	7
5207	Volantis	10.0	43.23	+ 0.576	64 36 42.4	.55	52.07	1
5208	Octantis	9.7	43.54	−10.978	85 12 58.3	.55	51.14	3
5209	Octantis	10.0	44.49	7.030	83 21 40.5	.55	50.12	1
5210	Chamæleontis	9.5	45.23	− 1.378	75 12 2.0	− 9.55	50.19	1
5211	Chamæleontis	11.0	7 53 50.02	− 5.424	82 7 20.6	− 9.56	51.17	1
5212	Volantis	9.5	50.29	+ 0.087	68 30 30.8	.56	50.25	1
5213	Gould, Z. C., 7h 3919 . .	9.5	53.66	− 3.141	79 17 11.4	.56	50.21	2
5214	Carinæ	10.0	53.69	+ 0.837	62 15 48.2	.56	51.18	1
5215	Gould, Z. C., 7h 4075 . .	10.0	54.96	+ 0.681	63 48 48.7	− 9.56	51.13	1
5216	Carinæ	10.0	7 53 55.53	+ 0.808	62 34 10.1	− 9.56	51.18	1
5217	Gould, Z. C., 7h 4063 . .	9.5	53 59.38	0.200	67 44 55.1	.57	50.25	1
5218	Gould, Z. C., 7h 4090 . .	10.0	54 6.44	+ 0.686	63 47 27.2	.58	51.13	1
5219	Gould, Z. C., 7h 4065 . .	10.0	15.31	− 0.385	71 13 58.2	.59	50.22	1
5220	Gould, Z. C., 7h 4099 . .	9.5	17.95	+ 0.631	64 18 35.0	− 9.59	51.13	3
5221	Gould, Z. C., 7h 4094 . .	9.2	7 54 24.68	+ 0.076	68 36 8.8	− 9.60	50.24	2
5222	Volantis	9.0	28.00	+ 0.569	64 52 47.3	.61	50.28	1
5223	Chamæleontis	10.0	36.78	− 4.664	81 22 33.2	.62	51.06	1
5224	Gould, Z. C., 7h 4131 . .	10.0	37.97	+ 0.708	63 36 22.5	.62	51.13	1
5225	Volantis	9.5	41.34	+ 0.057	68 44 23.7	− 9.62	50.24	2

Number.	Constellation, Name of Star, or Synonym.	Magnitude.	Right Ascension, 1850.0.	Annual Precession.	South Declination, 1850.0.	Annual Precession.	Mean year.	No. of obs.
			h. m. s.	s.	° ′ ″	″		
5 226	Lacaille, 3204	7. 5	7 54 44. 80	— 5. 497	82 12 19. 6	— 9. 63	50. 65	2
5 227	Gould, Z. C., 7h 4153	10. 0	51. 81	+ 0. 692	63 46 11. 5	. 64	51. 13	1
5 228	Carinæ	10. 0	54 59. 50	+ 0. 703	63 40 35. 2	. 65	51. 13	1
5 229	Gould, Z. C., 7h 4109 . .	10. 0	55 4. 02	— 0. 967	73 49 3. 2	. 65	50. 20	1
5 230	Chamæleontis	10. 0	6. 41	— 3. 780	80 17 37. 6	— 9. 66	50. 15	1
5 231	Volantis	10. 0	7 55 9. 21	+ 0. 510	65 25 3. 3	— 9. 66	51. 10	1
5 232	Chamæleontis	8. 5	13. 99	— 4. 562	81 16 27. 8	. 66	51. 06	1
5 233	Brisbane, 1855	7. 0	17. 42	+ 0. 783	62 53 30. 4	. 67	50. 73	2
5 234	Gould, Z. C., 7h 4186 . .	9. 5	20. 74	0. 536	65 12 32. 2	. 67	50. 28	1
5 235	Gould, Z. C., 7h 4200 . .	9. 2	24. 54	+ 0. 715	63 34 33. 2	— 9. 68	50. 71	2
4 236	Gould, Z. C., 7h 4216 . .	8. 8	7 55 32. 46	+ 0. 743	63 18 24. 2	— 9. 69	50. 73	2
5 237	Gould, Z. C., 7h 4172 . .	10. 0	32. 46	— 0. 309	70 54 4. 1	. 69	50. 23	1
5 238	Lacaille, 3214	7. 8	46. 44	— 5. 212	81 57 28. 4	. 71	50. 48	3
5 239	Gould, Z. C., 7h 4237 . .	9. 3	48. 37	+ 0. 639	64 19 13. 8	. 71	51. 13	3
5 240	Carinæ	10. 0	49. 90	+ 0. 796	62 47 33. 9	9. 71	51. 15	1
5 241	Gould, Z. C., 7h 4220 . .	9. 5	7 55 52. 50	+ 0. 095	68 32 59. 5	— 9. 71	50. 25	1
5 242	Chamæleontis	10. 5	55. 73	— 4. 578	81 18 23. 9	. 71	51. 06	1
5 243	Carinæ	9. 2	57. 27	+ 0. 683	63 54 51. 8	. 72	51. 14	2
5 244	Volantis	9. 2	57. 30	+ 0. 266	67 22 12. 6	. 72	50. 26	2
5 245	Volantis	9. 5	55 57. 81	— 1. 291	74 59 32. 3	— 9. 72	50. 20	1
5 246	Chamæleontis	10. 5	7 56 3. 24	— 4. 583	81 18 55. 0	— 9. 73	51. 06	1
5 247	Gould, Z. C., 7h 4281 . .	9. 0	8. 42	+ 0. 795	62 49 17. 9	. 73	51. 15	1
5 248	Carinæ	10. 0	10. 20	0. 740	63 22 33. 1	. 74	51. 16	1
5 249	Volantis	9. 2	15. 62	0. 402	66 21 40. 2	. 74	51. 21	3
5 250	Volantis	10. 0	18. 84	+ 0. 346	66 47 47. 9	— 9. 75	50. 26	1
5 251	Brisbane, 1864	8. 5	7 56 19. 14	+ 0. 806	62 43 21. 4	— 9. 75	50. 88	3
5 252	Gould, Z. C., 7h 4292 . .	9. 5	22. 96	+ 0. 516	65 25 51. 9	. 75	51. 10	2
5 253	Gould, Z. C., 7h 4273 . .	10. 0	25. 79	— 0. 011	69 14 50. 4	. 76	50. 24	1
5 254	Lacaille, 3142	7 7	27. 99	+ 0. 509	65 29 55. 7	. 76	50. 83	3
5 255	Gould, Z. C., 7h 4305 . .	9. 0	34. 33	+ 0. 283	67 17 3. 2	— 9. 78	50. 26	1
5 256	Volantis	10. 0	7 56 54. 38	+ 0. 499	65 36 20. 6	— 9. 79	51. 10	1
5 257	Carinæ	9. 0	56. 46	0. 685	63 57 4. 2	. 80	51. 14	1
5 258	Gould, Z. C., 7h 4351 . .	8. 8	56 57. 47	+ 0. 615	64 36 3. 8	. 80	50. 56	3
5 259	Octantis	9. 6	57 0. 03	—19. 338	87 1 12. 2	. 80	51. 20	6
5 260	Octantis	10. 0	5. 27	— 6. 804	83 15 51. 9	— 9. 81	50. 12	1
5 261	Gould, Z. C., 7h 4376 . .	10. 0	7 57 8. 29	+ 0. 760	63 13 42. 0	— 9. 81	51. 16	1
5 262	Gould, 10620	9. 0	9 44	— 0. 673	72 42 13. 7	. 81	50. 21	3
5 263	Volantis	8. 8	10. 74	+ 0. 667	64 8 1. 0	. 81	50. 85	3
5 264	Gould, Z. C., 7h 4363 . .	9. 2	12. 20	0. 289	67 15 57. 6	. 82	50. 26	1
5 265	Volantis	11. 5	12. 30	+ 0. 289	67 15 53. 1	— 9. 82	50. 26	1

Number.	Constellation, Name of Star, or Synonym.	Magnitude.	Right Ascension, 1850.0.	Annual Precession.	South Declination, 1850.0.	Annual Precession.	Mean year.	No. of obs.
			h. m. s.	s.	° ′ ″	″		
5 266	Gould, Z. C., 7h 4288 . .	9.0	7 57 16.91	- 1.678	76 12 39.2	— 9.82	50.18	3
5 267	Gould, Z. C., 7h 4356 . .	9.2	20.49	0.185	70 18 31.4	.83	50.23	2
5 268	Gould, Z. C., 7h 4346 . .	9.0	21.20	— 0.429	71 35 3.3	.83	50.22	1
5 269	Gould, Z. C., 7h 4400 . .	8.8	34.32	+ 0.410	66 21 48.8	.84	51.21	3
5 270	Gould, Z. C., 7h 4365 . .	9.5	42.97	— 0.919	73 44 3.2	9.85	50.20	1
5 271	Gould, Z. C., 7h 4362 . .	9.8	7 57 49.06	— 1.191	74 43 23.3	— 9.86	50.52	3
5 272	Lacaille, 3226	8.5	57 59.12	— 4.787	81 34 34.0	.88	50.14	1
5 273	Volantis	10.0	58 1.74	+ 0.517	65 30 50.0	.88	51.10	1
5 274	Volantis	10.0	8.18	0.512	65 33 35.8	.89	51.10	2
5 275	Gould, Z. C., 7h 4482 . .	9.8	10.72	+ 0.782	63 4 25.2	— 9.89	52.15	2
5 276	Gould, Z. C., 7h 4467 . .	8.8	7 58 12.44	+ 0.490	65 45 6.9	— 9.89	51.48	3
5 277	Gould, Z. C., 7h 4465 . .	9.0	16.06	0.296	67 16 10.2	.90	50.26	2
5 278	D[1] Carinæ	4.9	25.94	+ 0.775	63 9 6.9	.91	50.84	5
5 279	Gould, Z. C., 7h 4455 . .	9.0	29.18	— 0.419	71 35 9.8	.91	50.22	1
5 280	Gould, Z. C., 7h 4411 . .	10.0	30.66	— 1.647	76 9 46.4	— 9.92	50.18	2
5 281	Volantis	9.8	7 58 36.88	+ 0.101	68 38 19.8	9.92	50.24	2
5 282	Volantis	9.3	41.79	0.665	64 13 54.8	.93	51.13	3
5 283	Gould, Z. C., 7h 4524 . .	9.0	50.23	0.645	64 25 34.9	.94	50.70	2
5 284	Gould, Z. C., 7h 4550 . .	9.5	58 59.84	+ 0.832	62 36 17.2	.95	51.18	1
5 285	Gould, Z. C., 7h 4401 . .	9.0	59 9.09	— 3.357	79 46 6.4	— 9.96	50.16	1
5 286	Gould, Z. C., 7h 4526 . .	9.5	7 59 14.92	— 0.317	71 6 11.5	— 9.97	50.22	2
5 287	Lacaille, 3238	7.0	17.82	— 4.421	81 11 58.9	.97	51.06	1
5 288	Gould, Z. C., 7h 4552 . .	9.8	18.97	+ 0.091	68 44 8.8	.98	50.24	2
5 289	Gould, Z. C., 7h 4578 . .	9.0	21.45	0.838	62 33 57.7	.98	51.18	1
5 290	Gould, Z. C., 7h 4590 . .	9.0	32.02	+ 0.850	62 26 15.1	— 9.99	51.18	1
5 291	Gould, Z. C., 7h 4511 . .	9.0	7 59 37.46	— 1.608	76 5 26.6	— 10.00	50.18	2
5 292	Carinæ	9.0	47.34	+ 0.841	62 33 10.5	.01	51.18	1
5 293	Melbourne, (1), 147 . . .	9.0	53.38	— 0.731	73 3 28.3	.02	50.21	1
5 294	Chamæleontis	9.5	7 59 57.63	— 3.863	80 31 9.4	.03	50.15	1
5 295	Gould, Z. C., 8h 37 . . .	9.0	8 0 3.87	+ 0.791	63 5 15.7	— 10.03	51.15	1
5 296	Gould, Z. C., 8h 3 . . .	9.0	8 0 8.84	— 0.293	71 1 3.9	— 10.04	50.22	2
5 297	Volantis	10.0	11.36	+ 0.019	69 14 3.9	.04	50.24	1
5 298	Carinæ	9.5	16.19	+ 0.810	62 54 22.6	.05	51.15	1
5 299	Octantis	9.5	20.78	—22.503	87 25 35.7	.05	51.20	2
5 300	Gould, Z. C., 8h 45	9.5	33.17	— 0.065	69 45 49.8	— 10.07	50.24	1
5 301	Lacaille, 3245	9.0	8 0 36.09	— 4.070	80 48 7.9	— 10.07	50.14	1
5 302	Gould, 10725	9.5	0 55.81	0.677	72 52 28.6	.10	50.21	1
5 303	Lacaille, 3188	6.5	1 4.12	— 0.675	72 49 30.0	.11	50.21	2
5 304	Volantis	9.5	12.05	+ 0.577	65 10 10.9	.12	50.28	1
5 305	Lacaille, 3174	8.0	18.51	+ 0.529	65 35 28.6	— 10.13	51.10	2

No. 5 281. Possibly 68° 42′ 31″.5.

Number.	Constellation, Name of Star, or Synonym.	Magnitude.	Right Ascension, 1850.0.	Annual Precession.	South Declination, 1850.0.	Annual Precession.	Mean year.	No. of obs.
			h. m. s.	s.	° ′ ″	″		
5 306	Gould, Z. C., 8^h 133 . .	9.5	8 1 23.67	+ 0.764	63 25 56.1	— 10.13	50.30	1
5 307	Gould, Z. C., 8^h 84 . . .	9.2	24.41	— 0.668	72 51 13.7	.13	50.21	2
5 308	Gould, Z. C., 8^h 132 . .	9.5	27.64	+ 0.550	65 24 52.4	.14	51.10	2
5 309	Gould, Z. C., 8^h 142 . .	9.0	34.72	0.610	64 53 55.8	.15	50.28	1
5 310	Gould, Z. C., 8^h 156 . .	9.5	36.46	+ 0.747	63 37 14.6	— 10.15	50.29	1
5 311	Volantis	9.5	8 1 37.38	+ 0.315	67 17 56.0	— 10.15	50.26	2
5 312	Volantis	9.2	40.69	+ 0.306	67 22 17.4	.15	50.26	3
5 313	Lacaille, 3182	8.5	42.28	· 0.029	69 36 20.3	.16	50.24	1
5 314	Gould, Z. C., 8^h 157 . .	8.8	43.83	+ 0.508	65 47 12.9	.16	51.48	3
5 315	Volantis	10.0	1 47.34	+ 0.341	67 7 0.2	— 10.16	50.26	1
5 316	Gould, 10759	9.0	8 2 0.95	— 0.026	69 36 14.4	— 10.18	50.24	1
5 317	Carinæ	9.5	2.65	+ 0.878	62 17 59.2	.18	51.18	1
5 318	Octantis	11.0	2.86	—15.684	86 29 7.6	.18	51.05	1
5 319	Gould, Z. C., 8^h 174 . .	9.2	4.84	+ 0.463	66 10 45.8	.19	50.27	2
5 320	Gould, Z. C., 8^h 181 . .	9.0	8.99	+ 0.689	64 12 9.5	— 10.19	50.70	2
5 321	Brisbane, 1905	9.0	8 2 11.30	+ 0.016	69 21 10.0	— 10.19	50.24	1
5 322	Brisbane, 1904	9.0	19.44	0.436	66 24 43.5	.20	50.26	1
5 323	Brisbane, 1903	8.0	21.61	+ 0.870	62 24 40.0	.21	51.18	1
5 324	Chamæleontis	10.0	28.72	— 2.565	78 27 46.0	.22	52.21	1
5 325	D^2 Carinæ	7.0	34.26	+ 0.871	62 24 28.0	— 10.22	51.18	1
5 326	Volantis	9.5	8 2 34.38	+ 0.477	66 5 31.8	— 10.22	50.28	1
5 327	Gould, Z. C., 8^h 211 . .	9.2	36.88	+ 0.498	65 55 1.2	.23	51.14	3
5 328	Octantis	9.8	2 58.96	—15.732	86 30 8.4	.25	51.21	2
5 329	Gould, 10786	9.5	3 5.27	— 0.017	69 36 8.9	.26	50.24	1
5 230	Volantis	10.0	13.71	+ 0.227	68 1 14.6	— 10.27	50.25	1
5 331	Volantis	10.0	8 3 16.94	— 0.140	70 19 41.6	— 10.28	50.23	1
5 332	Lacaille, 3203	8.2	27.44	— 0.005	69 32 36.4	.29	50.24	2
5 333	Volantis	9.5	32.49	+ 0.173	68 24 29.6	.30	50.25	1
5 334	Lacaille, 3202	7.8	34.30	0.158	68 30 26.1	.30	50.25	1
5 335	Lacaille, 3194	8.5	35.62	+ 0.398	66 46 42.2	— 10.30	50.26	2
5 336	Volantis	9.0	8 3 36.98	+ 0.506	65 54 52.8	— 10.30	50.28	1
5 337	Gould, Z. C., 8^h 248 . .	9.0	37.46	— 0.873	73 46 59.5	.30	50.20	1
5 338	Gould, Z. C., 8^h 238 . .	9.2	3 39.28	— 1.233	75 4 42.4	.30	50.19	2
5 339	Gould, Z. C., 8^h 328 . .	10.0	4 0.26	+ 0.776	63 28 3.6	.33	50.29	1
5 340	Lacaille, 3215	9.0	5.39	— 0.061	69 54 29.4	— 10.34	50.23	1
5 341	Lacaille, 3196	8.0	8 4 16.89	+ 0.730	63 56 38.1	— 10.35	50.29	1
5 342	Octantis	9.1	21.67	—10 145	85 2 50.8	.36	51.20	5
5 343	Volantis	10.0	21.91	+ 0.517	65 51 20.4	.36	51.10	1
5 344	Melbourne (1), 400 . . .	9.0	23.57	0.657	64 38 4.4	.36	50.56	3
5 345	Gould, Z. C., 8^h 384 . .	8.8	25.50	+ 0.791	63 21 3.8	— 10.36	50.30	2

Number.	Constellation, Name of Star, or Synonym.	Magnitude.	Right Ascension, 1850.0.	Annual Precession.	South Declination, 1850.0.	Annual Precession.	Mean year.	No. of obs.
			h. m. s.	s.	° ′ ″	″		
5 346	Gould, Z. C., 8h 288	9.5	8 4 27.29	− 1.448	75 47 20.0	− 10.36	50.18	1
5 347	Gould, Z. C., 8h 389 . .	10.0	28.78	+ 0.779	63 28 21.5	.37	50.29	1
5 348	Volantis	10.0	29.32	0.519	65 50 45.9	.37	51.10	1
5 349	Lacaille, 3200	7.0	36.06	0.818	63 4 56.3	.37	50.30	2
5 350	Volantis	9.8	40.78	+ 0.544	65 39 10.0	− 10.38	51.10	2
5 351	Brisbane, 1916	5.4	8 4 51.96	+ 1.850	46 54 19.7	− 10.39	51.15	5
5 352	γ Argus	3.0	54.66	1.851	46 53 48.8	.40	51.15	5
5 353	Brisbane, 1918	9.3	57.84	+ 1.850	46 54 41.7	.40	51.15	5
5 354	Volantis	11.0	4 59.78	− 1.011	74 21 36.9	.40	51.16	1
5 355	Velorum	9.8	5 0.65	+ 1.849	46 55 59.2	− 10.41	51.15	5
5 356	Gould, Z. C., 8h 390 . .	11.5	8 5 15.85	− 1.006	74 21 0.9	− 10.42	51.16	1
5 357	Gould, Z. C., 8h 332 . .	9.8	27.38	− 2.486	78 23 27.1	.44	50.84	3
5 358	Gould, Z. C., 8h 436 . .	9.2	31.06	+ 0.030	69 25 44.4	.44	50.24	2
5 359	Gould, Z. C., 8h 438 . .	9.0	31.15	+ 0.030	69 25 51.2	.44	50.24	1
5 360	Lacaille, 3270	8.5	36.92	− 2.916	79 12 34.3	− 10.45	50.17	3
5 361	Volantis	10.0	8 5 56.54	− 0.193	70 45 0.6	− 10.48	50.23	1
5 362	Volantis	9.0	5 59.54	+ 0.082	69 7 28.7	.48	50.24	2
5 363	Octantis	9.8	6 0.06	−12.828	85 53 55.6	.48	50.22	2
5 364	Gould, Z. C., 8h 499 . .	9.0	5.93	+ 0.225	68 10 51.4	.49	50.25	1
5 365	Lacaille, 3225	8.0	10.00	+ 0.567	65 32 13.6	− 10.49	50.83	3
5 366	D° Carinæ	6.5	8 6 26.97	+ 0.802	63 21 38.4	− 10.51	50.30	1
5 367	Volantis	9.0	30.03	+ 0.218	68 15 5.9	.52	50.25	1
5 368	Volantis	10.0	6 33.30	− 0.074	70 6 16.6	.52	50.23	1
5 369	Gould, Z. C., 8h 576 . .	10.0	7 11.08	− 0.300	71 22 29.7	.57	50.22	1
5 370	Stone, 4195	8.5	21.10	+ 0.194	68 27 33.2	− 10.58	50.25	1
5 371	ε¹ Volantis	6.5	8 7 25.45	+ 0.236	68 10 35.8	− 10.59	50.26	2
5 372	ε² Volantis	10.0	25.94	0.236	68 10 31.6	.59	50.26	2
5 373	Lacaille 3235	8.5	28.38	0.710	64 19 56.5	.59	50.29	2
5 374	Gould, Z. C., 8h 643 . .	9.5	28.64	+ 0.757	63 52 45.6	.59	50.29	1
5 375	Octantis	9.2	31.77	− 9.975	85 1 35.2	− 10.59	51.20	5
5 376	Chamæleontis	10.0	8 7 38.96	− 5.008	82 0 15.4	− 10.60	51.17	1
5 377	Octantis	9.8	40.71	19.079	87 4 1.6	.60	51.20	4
5 378	Lacaille, 3282	8.0	41.90	− 2.621	78 43 37.6	.61	50.81	3
5 379	Volantis	9.5	42.54	+ 0.298	67 44 59.6	.61	50.25	1
5 380	Volantis	9.5	48.62	− 0.878	73 58 30.1	− 10.61	50.20	1
5 381	Volantis	9.0	8 7 49.95	+ 0.044	69 27 28.6	− 10.62	50.24	2
5 382	Volantis	10.0	8 23.10	+ 0.585	65 30 49.8	.66	51.10	1
5 383	Volantis	9.5	23.97	− 0.355	71 42 41.6	.66	50.22	1
5 384	Gould, Z. C., 8h 686 . .	9.0	28.80	0.269	71 16 37.1	.66	50.22	1
5 385	Octantis	9.2	32.57	−10.720	85 18 24.3	− 10.67	51.53	3

No. 5 374. Gould's declination is 10″ greater.

Number.	Constellation, Name of Star, or Synonym.	Magnitude.	Right Ascension, 1850.0.	Annual Precession.	South Declination, 1850.0.	Annual Precession.	Mean year.	No. of obs.
			h. m. s.	s.	° ′ ″	″		
5386	Lacaille, 3254	8.0	8 8 32.83	+ 0.191	68 32 49.2	− 10.67	50.24	2
5387	Gould, Z. C., 8h 731 . .	8.0	34.76	· 0.901	62 27 16.1	.67	51.18	1
5388	Gould, Z. C., 8h 663 . .	9.5	44.21	− 1.557	76 15 41.9	.68	50.15	3
5389	Carinæ	9.0	44.54	+ 0.908	62 23 16.9	.68	51.18	1
5390	Volantis	9.0	54.64	− 0.348	71 41 53.7	− 10.70	50.22	1
5391	Volantis	10.0	8 8 55.88	+ 0.583	65 33 34.4	− 10.70	51.10	1
5392	Gould, Z. C., 8h 765 . .	10.0	9 9.57	+ 0.584	65 34 11.4	.71	51.10	1
5393	Volantis	9.2	13.22	− 0.840	73 53 8.2	.72	50.20	2
5394	Gould, Z. C., 8h 664 . .	9.8	20.07	2.877	79 14 49.6	.73	50.17	2
5395	Chamæleontis	9.5	25.02	− 2.563	78 39 56.7	− 10.73	52.21	1
5396	Chamæleontis	10.0	8 9 25.32	− 5.011	82 2 42.6	− 10.73	51.17	1
5397	Volantis	10.0	27.85	0.799	73 44 13.5	.74	50.20	1
5398	Lacaille, 3268	8.5	42.25	0.258	71 16 29.3	.75	50.22	1
5399	Gould, Z. C., 8h 773 . .	9.0	9 44.06	0.627	73 2 5.8	.76	50.21	2
5400	Chamæleontis	10.0	10 15.86	− 1.676	76 38 50.2	− 10.80	50.17	1
5401	Volantis	9.0	8 10 16.44	+ 0.389	67 12 57.6	− 10.80	50.26	2
5402	Gould, Z. C., 8h 766 . .	10.0	20.70	− 2.253	78 3 9.0	.80	50.17	1
5403	Volantis	9.7	27.64	+ 0.569	65 46 27.8	.81	51.48	3
5404	Gould, Z. C., 8h 837 . .	10.0	27.80	− 0.436	72 11 57.2	.81	50.21	1
5405	Gould, Z. C., 8h 872 . .	9.0	32.93	+ 0.426	66 54 15.0	− 10.82	50.26	1
5406	Chamæleontis	10.0	8 10 46.43	− 3.604	80 25 56.5	− 10.83	50.15	1
5407	Gould, 11030	9.0	51.02	+ 0.813	63 31 21.9	.84	50.29	2
5408	Chamæleontis	9.5	51.31	− 5.104	82 9 56.0	.84	50.65	2
5409	Gould, Z. C., 8h 857 . .	9.0	51.45	0.871	74 4 30.4	.84	50.20	1
5410	Gould, Z. C., 8h 886 . .	9.0	51.68	− 0.076	70 19 57.9	− 10.84	50.23	2
5411	Volantis	10.0	8 10 52.92	+ 0.648	65 6 0.1	− 10.84	50.28	1
5412	Lacaille, 3371	9.0	11 6.64	− 6.494	83 17 50.1	.86	50.12	1
5413	Gould, Z. C., 8h 924 . .	10.0	9.19	+ 0.417	67 3 22.0	.86	50.26	1
5414	Gould, Z. C., 8h 944 . .	9.1	21.14	+ 0.556	65 56 8.8	.88	51.18	4
5415	Gould, Z. C., 8h 920 . .	10.0	21.93	− 0.168	70 52 33.7	− 10.88	50.23	1
5416	Gould, Z. C., 8h 940 . .	10.0	8 11 31.24	+ 0.107	69 15 23.8	− 10.89	50.24	1
5417	Gould, 11043	9.5	41.14	− 0.022	70 3 32.0	.90	50.23	1
5418	Gould, 11048	9.0	51.22	+ 0.104	69 17 46.4	.91	50.24	1
5419	Gould, Z. C., 8h 923 . .	9.0	11 51.68	− 1.301	75 36 46.7	.91	50.18	1
5420	Gould, 11061	10.0	12 1.79	+ 0.253	68 18 35.9	− 10.93	50.25	1
5421	Gould, 11069	9.8	8 12 16.64	+ 0.423	67 4 9.8	− 10.94	50.26	2
5422	Volantis	10.0	22.84	− 1.808	77 4 14.6	.95	50.17	2
5423	Volantis	10.0	23.43	− 1.255	75 29 14.6	.95	50.18	1
5424	Gould, Z. C., 8h 1047 . .	9.8	30.70	+ 0.540	66 8 38.6	.96	51.26	2
5425	Gould, Z. C., 8h 1064 . .	8.7	40.53	+ 0.621	65 27 19.5	− 10.97	50.83	3

No. 5401. Declination possibly 67° 17′ 12.4″

Number.	Constellation, Name of Star, or Synonym.	Magnitude.	Right Ascension, 1850.0.	Annual Precession.	South Declination, 1850.0.	Annual Precession.	Mean year.	No. of obs.
			h. m. s.	s.	° ′ ″	″		
5426	Gould, 11085	9.5	8 12 41.04	+ 0.608	65 34 1.0	10.97	51.10	2
5427	Carinæ	10.0	41.85	+ 0.800	63 46 19.4	.97	50.29	1
5428	Gould, Z. C., 8h 992 . .	9.2	41.90	− 1.478	76 10 52.5	.97	50.18	2
5429	Gould, 11079	8.5	42.12	0.143	70 48 21.7	.97	50.22	1
5430	Brisbane, 2007	9.0	43.58	−11.169	85 30 25.1	− 10.98	51.20	5
5431	Gould, Z. C., 8h 1070 . .	9.5	8 12 45.96	+ 0.581	65 48 27.8	− 10.98	51.10	1
5432	Gould, Z. C., 8h 1039 . .	10.0	58.38	− 0.839	74 2 36.7	10.99	50.20	1
5433	ζ Carinæ	5.5	12 59.15	+ 0.928	62 27 13.2	11.00	51.18	1
5434	Gould, 11103	9.8	13 16.03	+ 0.585	65 48 31.7	.02	51.10	2
5435	Octantis	9.6	27.99	− 11.134	85 30 19.2	− 11.03	51.20	5
5436	Gould, Z. C., 8h 1155 . .	12.0	8 13 34.39	+ 0.937	62 23 47.0	− 11.04	51.18	1
5437	Gould, Z. C., 8h 1156 . .	9.0	36.64	+ 0.937	62 23 56.0	.04	51.18	1
5438	Gould, Z. C., 8h 1119 . .	9.3	48.46	− 0.515	72 43 26.8	.06	50.21	3
5439	Gould, Z. C., 8h 1181 . .	9.5	13 56.83	+ 0.783	64 1 27.5	.07	50.29	1
5440	Gould, Z. C., 8h 1132 . .	8.5	14 8.40	− 0.958	74 32 25.0	11.08	51.18	2
5441	Lacaille, 3332	7.8	8 14 12.72	− 2.281	78 14 28.1	− 11.09	50.71	4
5442	Gould, Z. C., 8h 1203 . .	10.0	13.96	+ 0.862	63 14 41.5	.09	50.30	1
5443	Gould, Z. C., 8h 1208 . .	9.5	19.73	0.784	64 2 27.1	.09	50.29	1
5444	Volantis	8.8	33.68	+ 0.148	69 9 32.8	.11	50.24	2
5445	Carinæ	9.5	38.34	+ 0.956	62 15 23.1	− 11.12	51.18	1
5446	Lacaille, 3297	9.0	8 14 50.12	+ 0.798	63 56 12.9	− 11.13	50.29	1
5447	Lacaille, 3296	7.0	14 52.65	+ 0.820	63 43 13.1	.13	50.29	1
5448	Gould, Z. C., 8h 1238 . .	9.5	15 0.46	− 0.043	70 21 40.2	.14	50.23	1
5449	Carinæ	9.0	9.00	+ 0.947	62 23 50.1	.15	51.18	1
5450	Gould, Z. C., 8h 1201 . .	9.5	11.09	− 1.481	76 17 1.5	− 11.16	50.18	1
5451	Chamæleontis	9.0	8 15 12.46	− 3.530	80 26 47.9	− 11.16	50.15	1
5452	Gould, 11161	10.0	16.45	+ 0.322	68 0 22.1	.16	50.25	1
5453	Gould, Z. C., 8h 1231 . .	8.0	31.62	− 1.602	76 38 19.2	.18	50.18	1
5454	Octantis	9.2	35.71	14.417	86 22 7.3	.19	51.20	5
5455	Gould, Z. C., 8h 1227 . .	9.5	39.05	− 1.960	77 33 51.6	11.19	50.17	2
5456	Volantis	10.0	8 15 41.79	+ 0.355	67 47 24.1	− 11.19	50.25	1
5457	Gould, 11180	9.5	52.82	+ 0.475	66 52 50.0	.21	50.26	1
5458	Gould, Z. C., 8h 1251 . .	9.2	15 59.25	− 2.104	77 54 43.8	.21	50.17	2
5459	Gould, 11192	10.0	16 11.68	+ 0.681	65 8 7.2	.23	50.28	1
5460	Gould, Z. C., 8h 1366 . .	9.0	11.74	+ 0.712	64 51 17.4	− 11.23	50.28	1
5461	Volantis	9.0	8 16 13.87	+ 0.681	65 8 17.9	− 11.23	50.28	1
5462	Volantis	8.5	14.26	+ 0.039	69 56 3.2	.23	50.23	1
5463	Lacaille, 3329	7.0	27.59	− 0.628	73 20 32.2	.25	50.21	1
5464	Chamæleontis	9.8	30.12	− 4.367	81 32 41.4	.25	50.14	2
5465	Volantis	9.5	35.11	+ 0.069	69 46 14.8	− 11.26	50.24	1

Number.	Constellation, Name of Star, or Synonym.	Magnitude.	Right Ascension, 1850.0.	Annual Precession.	South Declination, 1850.0.	Annual Precession.	Mean year.	No. of obs.
			h. m. s.	s.	° ′ ″	″		
5466	Lacaille, 3313	5.7	8 16 38.58	+ 0.684	65 8 33.2	− 11.26	50.28	2
5467	Octantis	9.8	39.65	−20.136	87 16 14.8	.26	51.20	6
5468	Gould, Z. C., 8^h 1403	8.0	42.94	+ 0.583	66 2 27.4	.27	51.26	2
5469	Gould, Z. C., 8^h 1307	9.0	44.28	1.975	77 38 20.2	.27	50.17	1
5470	Gould, Z. C., 8^h 1388 . .	9.8	46.10	+ 0.241	68 39 32.2	− 11.27	50.24	2
5471	Gould, Z. C., 8^h 1406 . .	9.0	8 16 55.40	+ 0.193	68 59 35.2	− 11.28	50.24	3
5472	Gould, Z. C., 8^h 1435 . .	9.5	17 10.87	0.245	68 39 23.2	.30	50.24	2
5473	Volantis	10.0	11.57	0.439	67 14 37.2	.30	50.26	1
5474	Gould, 12220	9.0	14.12	+ 0.724	64 48 13.1	.30	50.28	1
5475	Chamæleontis	10.0	14.26	− 2.211	78 11 29.0	11.31	52.21	1
5476	Gould, Z. C., 8^h 1466 . .	9.5	8 17 15.65	+ 0.959	62 24 33.1	− 11.31	51.18	1
5477	Volantis	10.0	18.61	0.983	74 46 5.6	.31	51.16	1
5478	Gould, Z. C., 8^h 1434 .	9.3	28.41	0.467	72 40 44.4	.32	50.21	3
5479	Gould, Z. C., 8^h 1381 . .	10.0	32.42	1.885	77 27 0.6	.33	50.17	1
5480	Gould, Z. C., 8^h 1436 . .	9.0	35.70	− 0.716	73 45 38.4	− 11.33	51.23	2
5481	Carinæ	9.5	8 17 40.97	+ 0.954	62 29 0.3	− 11.34	51.18	1
5482	Gould, Z. C., 8^h 1495 . .	9.5	43.97	− 0.970	62 18 38.4	.34	51.18	1
5483	Gould, Z. C., 8^h 1460 . .	9.8	46.97	− 0.435	72 32 52.1	.34	50.21	2
5484	Volantis	10.0	47.69	+ 0.101	69 43 24.1	.35	50.24	1
5485	Volantis	10.0	48.02	+ 0.778	64 19 57.5	− 11.35	50.29	1
5486	Volantis	10.0	8 17 51.06	+ 0.447	67 13 8.6	− 11.35	50.26	1
5487	Gould, Z. C., 8^h 1463 . .	9.8	57.56	− 0.767	73 59 1.2	.36	51.23	2
5488	Gould, Z. C., 8^h 1523 . .	8.2	17 58.30	+ 0.908	63 1 35.3	.36	50.30	2
5489	Gould, Z. C., 8^h 1516 . .	9.2	18 4.30	+ 0.581	66 8 40.8	.37	51.26	2
5490	Volantis	9.2	21.75	− 0.280	71 48 47.5	− 11.39	50.22	2
5491	Carinæ	8.5	8 18 25.41	+ 0.862	63 31 50.9	− 11.39	50.29	1
5492	Lacaille, 3404	8.5	27.48	3.996	81 9 2.2	.39	50.14	2
5493	Volantis	9.5	34.57	+ 0.607	65 56 56.9	.40	52.25	1
5494	Gould, 11255	9.0	34.76	0.599	66 1 8.4	.40	51.26	2
5495	Gould, Z. C., 8^h 1549 . .	9.0	34.96	+ 0.368	67 51 41.4	− 11.40	50.25	2
5496	Gould, Z. C., 8^h 1570 . .	9.0	8 18 36.85	+ 0.857	63 35 34.4	− 11.40	50.29	1
5497	Gould, Z. C., 8^h 1464 . .	10.0	48.29	− 2.742	79 17 34.2	.42	50.16	1
5498	Gould, Z. C., 8^h 1573 . .	9.0	52.58	+ 0.278	68 31 26.8	.42	50.25	1
5499	Octantis	9.2	18 56.82	−18.938	87 8 31.3	.43	51.20	6
5500	Lacaille, 3330	7.0	19 1.57	+ 0.857	63 37 30.7	− 11.43	50.29	1
5501	Gould, Z. C., 8^h 1554 . .	10.0	8 19 2.92	− 0.601	73 21 0.8	− 11.44	50.21	1
5502	Chamæleontis	9.5	11.47	1.048	75 4 35.5	.45	50.19	2
5503	Volantis	9.0	17.39	0.209	71 30 1.0	.45	50.22	2
5504	Chamæleontis	9.5	20.38	− 3.230	80 7 16.8	.46	50.15	1
5505	Volantis	10.0	21.30	+ 0.396	67 41 51.8	− 11.46	50.25	2

Number.	Constellation, Name of Star, or Synonym.	Magnitude.	Right Ascension, 1850.0.	Annual Precession.	South Declination, 1850.0.	Annual Precession.	Mean year.	No. of obs.
			h. m. s.	s.	° ′ ″	″		
5506	Gould, Z. C., 8h 1578	9.3	8 19 26.12	— 0.939	74 42 1.6	11.46	50.52	3
5507	Volantis	10.0	32.24	+ 0.400	67 40 55.2	.47	50.25	2
5508	Gould, 11283	9.8	40.23	— 0.580	66 15 13.9	.48	51.27	2
5509	Gould, Z. C., 1587 . . .	9.0	42.31	— 1.339	76 2 33.2	.48	50.18	1
5510	Lacaille, 3346	8.0	42.38	+ 0.277	68 34 51.4	— 11.48	50.24	2
5511	Gould, Z. C., 8h 1524 . .	10.0	8 19 42.44	3.147	80 0 10.5	— 11.48	50.15	1
5512	Gould, Z. C., 8h 1630 . .	9.8	49.05	— 0.365	72 18 53.6	.49	50.22	2
5513	Carinæ	9.5	19 49.59	+ 0.968	62 29 25.8	.49	51.18	1
5514	Gould, 11296	8.5	20 0.83	+ 0.575	66 18 55.2	.50	51.26	2
5515	Gould, Z. C., 8h 1562 . .	10.0	3.70	— 2.850	79 31 25.8	— 11.51	50.16	2
5516	Gould, Z. C., 8h 1692 . .	9.0	8 20 4.38	+ 0.703	65 11 23.7	11.51	50.28	1
5517	Volantis	9.0	9.40	— 0.433	72 39 5.9	.52	50.21	1
5518	Lacaille, 3355	7.5	13.12	— 0.114	71 2 13.8	.52	50.24	3
5519	Gould, Z. C., 8h 1716 . .	9.0	23.78	0.726	64 59 53.8	.53	50.28	1
5520	Lacaille, 3357	8.0	24.09	— 0.110	71 1 36.7	— 11.53	50.24	3
5521	Gould, Z. C., 8h 1647 . .	9.0	8 20 25.31	— 1.300	75 57 13.6	— 11.53	50.18	3
5522	Brisbane, 2023	9.3	27.94	0.108	71 1 3.8	.54	50.24	3
5523	Gould, Z. C., 8h 1726 . .	9.0	28.71	+ 0.722	65 2 26.4	.54	50.28	2
5524	Carinæ	9.5	34.23	0.980	62 24 17.6	.54	51.18	1
5525	Volantis	10.0	35.36	+ 0.783	64 28 22.0	— 11.55	50.29	1
5526	Chamæleontis	9.5	8 20 39.39	3.818	80 59 6.8	— 11.55	50.14	1
5527	Lacaille, 3351	7.5	42.16	+ 0.515	66 51 10.1	.55	50.26	1
5528	Volantis	9.5	42.44	+ 0.660	65 37 14.4	.55	50.28	1
5529	Volantis	9.5	44.48	— 0.805	74 15 36.9	.56	51.16	1
5530	Gould, Z. C., 8h 1684 . .	10.0	48.44	— 1.271	75 52 33.8	— 11.56	50.18	2
5531	Gould, 11318	8.5	8 20 51.33	+ 0.904	63 15 21.4	— 11.57	50.30	1
5532	Volantis	9.2	20 53.74	— 0.898	74 37 7.6	.57	50.68	2
5533	Carinæ	10.0	21 1.58	+ 0.971	62 32 3.1	.58	51.18	1
5534	Volantis	9.0	4.71	— 0.471	72 52 22.9	.58	50.21	1
5535	Lacaille, 3379	8.0	7.38	— 0.874	74 32 8.6	— 11.58	51.18	2
5536	Gould, 11322	9.0	8 21 17.49	+ 0.231	68 59 24.7	— 11.60	50.24	3
5537	Octantis	9.8	18.54	—16.355	86 47 6.8	.60	51.22	2
5538	Gould, Z. C., 8h 1811 . .	9.0	26.08	+ 0.909	63 14 36.0	.61	50.30	1
5539	Volantis	9.5	32.26	+ 0.766	64 42 . .	.61	50.29	1
5540	Lacaille, 3415	7.5	37.64	— 2.455	78 50 44.3	— 11.62	50.16	1
5541	Gould, Z. C., 8h 1796 . .	8.5	8 21 44.90	— 0.181	71 29 5.1	— 11.63	50.22	2
5542	Octantis	9.5	52.35	18.224	87 4 21.5	.64	51.20	6
5543	Gould, Z. C., 8h 1710 . .	10.0	21 57.47	— 3.153	80 4 46.1	.64	50.15	1
5544	Volantis	10.0	22 2.28	+ 0.089	69 57 18.8	.65	50.23	1
5545	α Chamæleontis	4.0	18.54	1.438	76 26 33.6	— 11.67	50.18	1

No. 5511. Gould's declination is 10″ greater.

Number.	Constellation, Name of Star, or Synonym.	Magnitude.	Right Ascension, 1850.0.	Annual Precession.	South Declination, 1850.0.	Annual Precession.	Mean year.	No. of obs.
			h. m. s.	s.	° ′ ″	″		
5 546	Chamæleontis	10.0	8 22 29.09	— 1.414	76 22 46.7	— 11.68	50.18	1
5 547	Carinæ	9.0	31.24	+ 0.965	62 42 41.1	.68	51.18	1
5 548	Gould, Z. C., 4h 1917 . .	9.0	31.65	+ 0.973	62 37 18.0	.68	51.24	2
5 549	Volantis	10.0	22 43.51	— 0.018	70 37 23.8	.70	50.23	1
5 550	Gould, Z. C., 8h 1919 . .	9.2	23 6.04	— 0.363	72 28 23.2	— 11.73	50.21	2
5 551	Volantis	9.0	8 23 7.03	0.000	70 32 55.1	— 11.73	50.23	1
5 552	Gould, Z. C., 8h 1965 . .	9.0	7.63	+ 0.997	62 24 21.1	.73	51.18	1
5 553	Lacaille, 3383	8.0	11.15	+ 0.131	69 45 35.5	.73	50.24	1
5 554	Volantis	9.7	13.71	— 0.774	74 15 17.4	.73	51.21	3
5 555	η Volantis	7.0	22.83	— 0.454	72 54 51.1	— 11.74	50.21	1
5 556	Volantis	9.5	8 23 25.06	+ 0.604	66 17 47.8	— 11.75	51.26	2
5 557	Volantis	10.0	40.21	— 0.762	74 13 37.9	.77	51.16	1
5 558	Gould, Z. C., 8h 2016 . .	9.0	43.50	+ 0.856	63 57 26.5	.77	50.29	2
5 559	Lacaille, 3378	7.0	48.03	+ 0.843	64 6 29.3	.77	50.29	1
5 560	Chamæleontis	10.0	53.25	— 2.214	78 25 50.3	— 11.78	52.21	1
5 561	Gould, Z. C., 8h 1984 . .	9.0	8 23 54.70	— 0.392	72 39 12.2	— 11.78	50.21	2
5 562	β Volantis	5.5	24 5.39	+ 0.684	65 38 12.5	.80	50.28	1
5 563	Volantis	10.0	16.09	0.558	66 43 56.3	.81	50.26	1
5 564	Volantis	10.0	18.36	0.156	69 39 41.0	.81	50.24	1
5 565	Volantis	9.5	23.34	+ 0.122	69 53 14.4	— 11.82	50.23	1
5 566	Gould, Z. C., 8h 2089 . .	9.5	8 24 53.00	+ 0.209	69 21 25.4	— 11.85	50.24	1
5 567	θ Chamæleontis	4.5	25 2.26	— 1.598	76 59 50.5	.86	50.17	2
5 568	Volantis	10.0	3.84	— 0.545	73 24 11.2	.86	50.21	1
5 569	Volantis	10.0	9.26	+ 0.487	67 22 7.2	.87	50.26	2
5 570	Lacaille, 3440	8.7	11.71	— 1.930	77 50 31.7	— 11.87	50.24	2
5 571	Gould, Z. C., 8h 2117 . .	9.0	8 25 16.86	— 0.098	71 13 40.5	— 11.88	50.22	1
5 572	Lacaille, 3420	9.0	20.01	— 0.606	73 40 30.5	.88	50.20	1
5 573	Gould, Z. C., 8h 2171 . .	9.2	30.80	+ 0.839	64 15 40.8	.90	50.29	2
5 574	Chamæleontis	9.5	32.25	— 3.277	80 22 37.8	.90	50.15	1
5 575	Volantis	9.0	32.54	+ 0.023	70 33 17.3	— 11.90	50.23	1
5 576	Chamæleontis	10.0	8 25 45.81	— 2.400	78 52 41.6	— 11.91	50.16	1
5 577	Gould, Z. C., 8h 2203 . .	10.0	25 55.06	+ 0.950	63 7 46.8	.92	50.30	1
5 578	Chamæleontis	10.0	26 1.24	— 5.068	82 28 40.8	.93	50.13	1
5 579	Lacaille, 3437	8.5	6.62	1.218	75 56 20.9	.94	50.18	3
5 580	Volantis	9.0	6.81	— 0.700	74 5 57.3	— 11.94	51.23	2
5 581	Gould, 11439	9.2	8 26 12.24	— 1.218	75 56 36.4	— 11.94	50.18	3
5 582	A Octantis	7.6	22.98	35.949	88 25 27.0	.96	51.20	6
5 583	Volantis	10.0	24.88	— 0.769	74 23 6.5	.96	51.16	1
5 584	Gould, Z. C., 8h 2221 . .	9.5	27.77	+ 0.240	69 14 46.0	.96	50.24	1
5 585	Gould, Z. C., 8h 2240 . .	9.8	29.91	+ 0.630	66 16 29.9	— 11.97	50.28	2

Number.	Constellation, Name of Star, or Synonym.	Magnitude.	Right Ascension, 1850.0.	Annual Precession.	South Declination, 1850.0.	Annual Precession.	Mean year.	No. of obs.
			h. m. s.	s.	° ′ ″	″		
5 586	Chamæleontis	9.0	8 26 31.23	— 1.715	77 21 51.5	— 11.97	50.17	1
5 587	Lacaille, 3424	7.0	54.10	+ 0.191	69 35 43.6	11.99	50.24	1
5 588	Volantis	10.0	57.60	— 0.514	73 21 47.9	12.00	50.21	1
5 589	Gould, 11484	9.0	26 58.40	+ 0.989	62 46 27.0	.00	50.30	1
5 590	Lacaille, 3442	7.8	27 6.20	— 0.768	74 24 59.8	— 12.01	51.20	3
5 591	Chamæleontis	10.0	8 27 7.39	— 2.249	78 37 13.7	— 12.01	52.21	1
5 592	Volantis	10.0	10.65	+ 0.589	66 40 17.0	.01	50.26	1
5 593	Gould, 11480	9.0	14.50	— 0.044	71 2 28.4	.02	50.22	2
5 594	Gould, Z. C., 8h 2259 . .	9.0	17.58	0.745	74 20 3.4	.02	50.95	4
5 595	Chamæleontis	10.0	39.89	— 2.650	79 25 . .	— 12.05	50.16	1
5 596	Gould, Z. C., 8h 2327 . .	9.0	8 27 40.83	+ 0.411	68 6 48.8	— 12.05	50.25	1
5 597	Octantis	9.0	27 53.18	—12.820	86 9 15.9	.06	51.19	4
5 598	Gould, 11512	7.8	28 0.03	+ 1.014	62 34 29.6	.07	51.24	2
5 599	Volantis	10.0	2.69	+ 0.365	68 28 29.2	.07	50.24	1
5 600	Gould, Z. C., 8h 2276 . .	9.5	7.85	— 2.030	78 11 0.8	— 12.08	51.18	2
5 601	Lacaille, 3432	8.2	8 28 16.15	+ 0.602	66 38 4.7	— 12.09	50.27	3
5 602	Gould, Z. C., 8h 2403 . .	9.0	22.57	0.799	64 51 47.5	.10	50.28	1
5 603	Gould, Z. C., 8h 2432 . .	9.5	42.93	+ 1.027	62 28 57.8	.12	51.18	1
5 604	Volantis	10.0	28 53.19	— 0.728	74 20 45.3	.13	51.16	1
5 605	Gould, Z. C., 8h 2445 . .	10.0	29 6.00	+ 0.323	68 7 3.1	— 12.15	50.25	1
5 606	Gould, Z. C., 8h 2470 . .	10.0	8 29 12.72	+ 0.774	65 9 51.9	— 12.15	50.28	1
5 607	Chamæleontis	9.2	17.19	— 2.027	78 13 6.8	.16	50.67	3
5 608	Lacaille, 3453	8.0	19.66	— 0.120	71 34 39.3	.16	50.22	1
5 609	Lacaille, 3436	8.0	25.95	+ 0.791	65 0 47.7	.17	50.28	1
5 610	Octantis	7.2	26.61	— 6.981	83 57 56.5	— 12.17	50.12	2
5 611	Gould, Z. C., 8h 2458 . .	9.3	8 29 29.47	— 0.329	72 38 53.2	— 12.17	50.21	3
5 612	Gould, Z. C., 8h 2405 . .	9.0	30.05	1.637	77 17 7.8	.17	50.17	1
5 613	Gould, Z. C., 8h 2419 . .	9.2	30.21	— 1.222	76 5 59.7	.18	50.18	2
5 614	Lacaille, 3444	8.6	32.49	+ 0.645	66 21 26.3	.18	50.77	4
5 615	Gould, Z. C., 8h 2500 . .	10.0	33.16	+ 0.774	65 11 27.8	— 12.18	50.28	1
5 616	Octantis	9.5	8 29 34.44	— 9.640	85 13 22.6	— 12.18	51.20	2
5 617	Gould, Z. C., 8h 2519 . .	9.0	44.60	+ 0.971	63 11 16.0	.19	50.30	1
5 618	Chamæleontis	10.0	47.12	— 3.082	80 12 36.7	.19	50.15	1
5 619	Chamæleontis	10.0	49.36	— 4.098	81 34 45.4	.20	50.14	2
5 620	Volantis	9.0	50.54	+ 0.860	64 21 47.6	— 12.20	50.29	1
5 621	Lacaille, 3464	8.0	8 29 54.62	— 0.849	74 51 20.9	— 12.20	50.20	1
5 622	Volantis	9.5	30 3.75	— 0.739	74 26 54.4	.21	50.20	1
5 623	Carinæ	10.0	6.24	+ 1.041	62 25 32.7	.22	51.18	1
5 624	Volantis	9.8	7.26	+ 0.510	67 30 52.2	.22	50.25	2
5 625	Octantis	9.5	11.39	- 6.639	83 46 11.2	— 12.22	50.12	1

No. 5 625. Declination may be 83° 45′ 13.2″.

Number.	Constellation, Name of Star, or Synonym.	Magnitude.	Right Ascension, 1850.0.	Annual Precession.	South Declination, 1850.0.	Annual Precession.	Mean year.	No. of obs.
			h. m. s.	s.	° ′ ″	″		
5 626	Octantis	9.1	8 30 20.78	−30.211	88 10 35.8	− 12.23	51.20	6
5 627	Gould, Z. C., 8h 2549 . .	9.0	33.34	+ 0.110	70 19 55.0	.25	50.23	2
5 628	Gould, Z. C., 8h 2525 . .	9.2	33.41	− 0.733	74 26 55.6	.25	50.68	2
5 629	Gould, Z. C., 8h 2585 . .	9.0	38.76	+ 0.732	65 34 20.7	.25	50.28	1
5 630	Chamæleontis	9.0	40.92	− 2.045	78 18 48.0	− 12.26	50.21	1
5 631	Volantis	9.5	8 30 44.79	+ 0.468	67 52 55.6	− 12.26	50.25	1
5 632	Gould, Z. C., 8h 2586 . .	9.5	45.15	+ 0.504	67 36 10.2	.26	50.25	2
5 633	Chamæleontis	10.0	45.52	− 2.943	80 1 24.3	.26	50.15	1
5 634	Gould, Z. C., 8h 2529 . .	9.5	47.09	1.247	76 14 3.9	.26	50.18	1
5 635	Gould, Z. C., 8h 2553 . .	9.8	52.90	− 0.580	73 51 49.5	12.27	51.23	2
5 636	Gould, Z. C., 8h 2609 . .	8.0	8 30 55.72	+ 0.967	63 19 43.2	− 12.27	50.30	1
5 637	Volantis	9.0	30 57.86	0.037	71 13 2.5	.28	50.22	1
5 638	Carinæ	10.0	31 14.88	+ 1.057	62 19 39.2	.30	51.18	1
5 639	Carinæ	9.5	33.07	+ 1.028	62 41 22.6	.32	51.18	1
5 640	Gould, Z. C., 8h 2581 . .	9.0	35.78	− 1.581	77 13 30.2	− 12.32	50.17	1
5 641	Gould, Z. C., 8h 2647 . .	9.0	8 31 40.58	+ 0.415	68 20 55.7	− 12.33	50.25	2
5 642	Gould, Z. C., 8h 2685 . .	9.2	32 3.00	0.705	66 0 24.5	.35	51.26	2
5 643	Gould, Z. C., 8h 2712 . .	9.2	20.89	0.708	66 0 31.1	.37	51.26	2
5 644	Carinæ	9.0	40.56	1.057	62 26 34.7	.39	51.18	1
5 645	Gould, 11651	8.0	46.32	+ 1.016	62 55 16.6	− 12.40	50.30	1
5 646	Carinæ	10.0	8 32 47.06	+ 1.044	62 36 15.9	− 12.40	51.18	1
5 647	Volantis	10.0	49.34	− 0.665	74 17 51.7	.40	51.16	1
5 648	Gould, 11643	10.0	52.70	+ 0.013	71 2 39.6	.41	50.23	1
5 649	Gould, Z. C., 8h 2741 . .	9.0	56.18	− 0.027	71 16 45.3	.41	50.22	1
5 650	Gould, Z. C., 7h 2752 . .	9.0	57.21	+ 0.249	69 35 42.2	− 12.41	50.24	1
5 651	Lacaille, 3537	5.0	8 32 59.83	− 3.150	80 25 3.8	− 12.42	50.15	1
5 652	Gould, Z. C., 8h 2772 . .	9.0	33 8.76	+ 0.362	68 49 54.9	.43	50.24	2
5 653	Octantis	9.0	8.80	− 6.759	83 54 21.8	.43	50.12	2
5 654	Gould, Z. C., 8h 2814 . .	10.0	32.98	+ 0.626	66 48 22.6	.45	50.26	2
5 655	Lacaille, 3499	7.5	36.11	− 0.666	74 20 18.4	− 12.46	50.95	4
5 656	Lacaille, 3489	8.0	8 33 40.31	− 0.321	72 50 31.2	− 12.46	50.21	1
5 657	Gould, Z. C., 8h 2833 . .	9.0	44.79	+ 0.853	64 43 56.2	.47	50.28	1
5 658	Gould, Z. C., 8h 2832 . .	9.0	47.09	0.737	65 50 22.4	.47	50.93	3
5 659	Gould, Z. C., 8h 2848 . .	8.8	33 55.93	+ 0.903	64 17 28.5	.48	50.29	2
5 660	Gould, Z. C., 8h 2777 . .	9.0	34 0.66	− 1.620	77 25 35.3	− 12.49	50.17	1
5 661	Gould, Z. C., 8h 2870 . .	9.0	8 34 2.03	+ 1.074	62 21 41.6	− 12.49	51.18	1
5 662	Gould, 11691	8.5	3.34	0.914	64 8 6.4	.49	50.29	1
5 663	Volantis	9.0	16.50	0.495	67 55 12.5	.50	50.25	2
5 664	Gould, Z. C., 8h 2881 . .	9.0	16.52	+ 0.952	63 45 1.6	.50	50.29	1
5 665	Lacaille, 3523	8.2	18.16	− 2.025	78 24 44.7	− 12.51	50.75	3

Number.	Constellation, Name of Star, or Synonym.	Magnitude.	Right Ascension, 1850.0.	Annual Precession.	South Declination, 1850.0.	Annual Precession.	Mean year.	No. of obs.
			h. m. s.	s.	° ′ ″	″		
5 666	Carinæ	9.0	8 34 23.01	+ 1.012	63 6 5.9	12.51	50.30	2
5 667	Volantis	10.0	24.81	0.299	69 21 3.8	.51	50.24	1
5 668	Volantis	10.0	32.92	0.634	66 48 23.9	.52	50.27	1
5 669	Lacaille, 3475	5.5	34 38.21	1.081	62 19 37.0	.53	51.18	1
5 670	Gould, Z. C., 8h 2926 . .	9.0	35 1.67	+ 0.647	66 43 50.6	— 12.56	50.26	3
5 671	Gould, Z. C., 8h 2879 . .	10.0	8 35 12.74	— 1.406	76 53 58.3	— 12.57	50.17	1
5 672	Octantis	9.8	20.18	— 6.152	83 33 26.8	.58	50.12	2
5 673	Carinæ	10.0	40.14	+ 0.964	63 44 7.7	.60	50.29	1
5 674	Volantis	10.0	42.21	+ 0.403	68 42 29.0	.60	50.25	1
5 675	Lacaille, 3533	8.0	45.16	— 1.232	76 24 51.3	— 12.61	50.18	1
5 676	Lacaille, 3491	7.0	8 35 45.82	+ 1.090	62 18 52.0	— 12.61	51.18	1
5 677	Brisbane, 3167	9.0	56.74	0.811	65 18 49.7	.62	50.28	1
5 678	Lacaille, 3510	8.5	35 57.89	+ 0.193	70 9 4.6	.62	50.23	1
5 679	Gould, Z. C., 8h 2921 . .	8.8	36 1.68	— 1.896	78 11 16.6	.62	51.04	3
5 680	Carinæ	9.5	2.72	+ 1.062	62 39 32.7	— 12.62	51.18	1
5 681	Gould, 11752	9.0	8 36 23.36	+ 0.035	71 8 1.2	— 12.65	50.22	2
5 682	Volantis	10.0	31.71	0.602	67 13 6.3	.66	50.26	1
5 683	Volantis	9.0	37.16	+ 0.571	67 28 50.0	.66	50.25	2
5 684	Gould, Z. C., 8h 3017 . .	10.0	42.25	— 0.287	72 51 7.1	.67	50.21	1
5 685	Gould, Z. C., 8h 3065 . .	9.5	36 59.48	+ 0.628	67 2 19.1	— 12.69	50.26	1
5 686	Gould, Z. C., 8h 3052 . .	9.0	8 37 0.65	+ 0.356	69 8 3.4	— 12.69	50.24	2
5 687	Volantis	9.0	20.04	+ 0.559	67 37 47.2	.71	50.25	2
5 688	Volantis	10.0	43.41	— 0.674	74 35 14.3	.74	51.16	1
5 689	Gould, Z. C., 8h 3086 . .	9.0	48.26	— 0.735	74 49 37.1	.74	50.20	1
5 690	Brisbane, 2171	8.8	37 51.49	+ 1.076	62 39 4.6	— 12.75	50.74	2
5 691	Gould, Z. C., 8h 3110 . .	10.0	8 38 0.71	— 0.278	72 52 59.9	— 12.76	50.21	1
5 692	Gould, 11800	9.2	1.32	+ 0.928	64 18 12.7	.76	50.29	2
5 693	Volantis	10.0	10.04	— 0.463	73 44 24.6	.77	50.20	1
5 694	Lacaille, 3527	8.3	15.68	+ 0.423	68 44 24.9	.78	50.24	3
5 695	θ Volantis	7.0	30.22	+ 0.264	69 51 4.8	— 12.79	50.24	2
5 696	Volantis	10.0	8 38 34.83	+ 0.850	65 8 18.7	— 12.80	50.28	1
5 697	Gould, Z. C., 8h 3202 . .	10.0	36.06	0.732	66 15 8.0	.80	50.26	1
5 698	Gould, Z. C., 8h 3197 . .	8.8	36.08	0.647	66 59 42.4	.80	50.26	3
5 699	Lacaille, 3522	8.7	39.32	0.980	63 47 54.5	.80	50.29	2
5 700	Gould, Z. C., 8h 3224 . .	8.5	46.29	+ 0.956	64 3 56.8	— 12.81	50.29	1
5 701	Gould, Z. C., 8h 3211 . .	10.0	8 38 50.60	+ 0.424	68 46 25.4	— 12.81	50.24	1
5 702	Gould, Z. C., 8h 3227 . .	9.0	53.05	+ 0.809	65 33 39.7	.82	52.27	1
5 703	Volantis	9.0	53.28	— 0.137	72 13 51.5	.82	50.22	1
5 704	Gould, Z. C., 8h 3230 . .	10.0	56.93	+ 0.641	67 4 20.7	.82	50.26	1
5 705	Gould, Z. C., 8h 3233 . .	9.0	38 59.88	+ 0.616	67 17 0.2	— 12.82	50.26	2

No. 5 701. Declination may be 21.5″.

Number.	Constellation, Name of Star, or Synonym.	Magnitude.	Right Ascension, 1850.0.	Annual Precession.	South Declination, 1850.0.	Annual Precession.	Mean year.	No. of obs.
			h. m. s.	s.	° ′ ″	″		
5706	Gould, Z. C., 8h 3238 . .	9.5	8 39 1.64	+ 0.726	66 20 25.4	− 12.83	51.26	2
5707	Volantis	9.0	3.58	− 0.057	71 48 46.2	.83	50.22	2
5708	Lacaille, 3563	7.5	5.18	− 1.199	76 28 17.3	.83	50.18	1
5709	Gould, Z. C., 8h 3253 . .	10.0	17.74	+ 0.380	69 7 17.3	.84	50.24	1
5710	Chamæleontis	10.0	17.85	− 1.681	77 48 26.6	− 12.84	50.17	1
5711	Lacaille, 3535	7.5	8 39 18.35	+ 0.735	66 16 42.7	− 12.85	50.93	3
5712	Volantis	9.0	22.67	− 0.013	71 35 29.9	.85	50.22	1
5713	Lacaille, 3550	7.2	25.08	− 0.517	74 2 23.0	.85	51.23	2
5714	Gould, Z. C., 8h 3282 . .	9.2	26.22	+ 0.796	65 43 39.2	.85	51.27	2
5715	Lacaille, 3559	8.8	37.99	− 0.635	74 32 3.8	12.87	50.68	2
5716	Gould, Z. C., 8h 3239 . .	9.0	8 39 43.31	− 0.930	75 37 52.8	− 12.87	50.18	1
5717	Carinæ	9.5	45.06	+ 1.091	62 38 4.5	.88	51.18	1
5718	Lacaille, 3555	7.5	46.14	− 0.399	73 32 43.4	.88	50.21	2
5719	Gould, Z. C., 8h 3312 . .	9.8	49.62	+ 0.625	67 16 19.4	.88	50.26	2
5720	Gould, Z. C., 8h 3271 . .	9.5	52.14	− 0.593	74 22 47.8	− 12.88	51.70	2
5721	Gould, Z. C., 8h 3212 . .	9.0	8 39 52.78	− 2.049	78 41 5.4	− 12.88	50.83	3
5722	Volantis	9.5	39 59.87	+ 0.147	70 42 . .	.89	50.23	1
5723	Chamæleontis	10.0	40 5.08	− 2.013	78 36 49.5	.90	52.21	1
5724	Volantis	9.0	7.45	− 0.441	73 45 12.6	.90	50.20	1
5725	Gould, Z. C., 8h 3337 . .	8.4	10.75	+ 0.907	64 41 56.0	− 12.90	50.28	1
5726	Gould, Z. C., 8h 3293 . .	9.0	8 40 14.61	− 0.635	74 34 2.3	− 12.91	51.16	1
5727	Octantis	8.3	15.49	7.695	84 34 16.2	.91	51.20	3
5728	Volantis	9.0	16.18	− 0.032	71 45 16.5	.91	50.22	1
5729	Gould, Z. C., 8h 3350 . .	9.0	26.61	+ 0.843	65 21 1.7	.92	50.28	1
5730	Lacaille, 3586	8.0	28.42	− 2.515	79 37 46.2	− 12.92	50.16	1
5731	Carinæ	9.0	8 40 31.67	+ 1.093	62 41 37.2	− 12.93	51.18	1
5732	Carinæ	9.0	44.38	+ 1.065	63 1 19.4	.94	50.30	1
5733	Octantis	10.0	44.62	− 7.446	84 27 10.7	.94	51.20	1
5734	Lacaille, 3576	8.3	45.50	1.088	76 12 11.8	.94	50.18	3
5735	Octantis	10.0	46.72	−19.890	87 27 11.3	− 12.94	51.18	1
5736	Carinæ	10.0	8 40 47.48	+ 1.039	63 19 17.1	− 12.94	50.30	1
5737	Lacaille, 3581	9.0	41 4.86	− 1.708	77 57 55.6	.96	50.24	2
5738	Brisbane, 2205	8.5	7.62	+ 0.943	64 23 39.6	.97	50.28	1
5739	Volantis	9.0	14.59	+ 0.052	71 20 37.0	.98	50.22	1
5740	Gould, Z. C., 8h 3351 . .	9.8	23.55	− 1.369	77 4 35.6	− 12.99	50.17	2
5741	Carinæ	10.0	8 41 24.91	+ 1.112	62 31 55.7	− 12.99	51.18	1
5742	Carinæ	9.5	31.68	+ 1.102	62 39 40.7	12.99	51.18	1
5743	Gould, Z. C., 8h 3396 . .	8.5	35.44	− 0.778	75 10 59.3	13.00	50.19	1
5744	Lacaille, 3571	8.5	43.52	0.070	72 3 3.5	.01	50.22	1
5745	Gould, Z. C., 8h 3410 . .	9.0	41 46.67	− 0.750	75 5 33.8	− 13.01	50.19	1

No. 5719. Gould's declination is 4′ greater.

Number.	Constellation, Name of Star, or Synonym.	Magnitude.	Right Ascension, 1850.0	Annual Precession.	South Declination, 1850.0.	Annual Precession.	Mean year.	No. of obs.
			h. m. s.	s.	° ′ ″	″		
5746	Volantis	10.0	8 42 11.32	+ 0.519	68 17 41.6	− 13.04	50.25	1
5747	Gould, Z. C., 8^h 3502 . .	9.0	14.31	+ 1.122	62 28 30.5	.04	51.18	1
5748	Volantis	10.0	17.04	− 0.180	72 39 22.3	.04	50.21	2
5749	Lacaille, 3578	8.0	21.27	− 0.264	73 4 21.5	.06	50.24	3
5750	Lacaille, 3562	7.5	23.96	+ 0.866	65 16 56.8	− 13.06	50.28	1
5751	Volantis	9.5	8 42 24.80	+ 0.781	66 6 3.2	− 13.06	50.28	1
5752	Octantis	9.3	24.95	− 59.992	89 4 40.0	.06	51.20	5
5753	Lacaille, 3568	7.2	27.94	+ 0.601	67 39 59.5	.07	50.25	2
5754	Gould, 11937	8.8	30.39	0.881	65 8 58.8	.07	50.28	2
5755	Carinæ	9.5	40.19	+ 1.119	62 33 21.8	− 13.07	51.18	1
5756	Volantis	9.0	8 42 41.06	− 0.444	73 54 28.8	− 13.07	50.20	1
5757	Gould, Z. C., 8^h 3456 . .	9.0	42 41.62	1.345	77 4 13.0	.07	50.17	2
5758	Carinæ	9.5	43 0.62	+ 1.129	62 28 7.5	.09	51.18	1
5759	Gould, 11955	9.0	2.03	+ 1.054	63 20 49.6	.09	50.29	2
5760	Octantis	9.0	10.54	− 6.310	83 50 31.6	− 13.10	50.12	2
5761	Gould, Z. C., 8^h 3583 . .	8.8	8 43 20.50	+ 0.845	65 34 6.5	− 13.11	51.27	2
5762	Lacaille, 3583	8.2	25.89	− 0.470	74 3 53.4	.12	51.23	2
5763	Volantis	9.0	26.78	− 0.633	74 43 57.0	.12	50.68	2
5764	Gould, Z. C., 8^h 3587 . .	9.0	29.85	+ 0.602	67 44 25.9	.12	50.25	1
5765	Gould, Z. C., 8^h 3585 . .	9.0	39.51	+ 0.090	71 16 41.2	− 13.14	50.22	1
5766	Gould, Z. C., 8^h 3591 . .	10.0	8 43 44.69	+ 0.153	70 54 46.7	− 13.14	50.23	1
5767	Gould, Z. C., 8^h 3617 . .	9.2	49.43	0.823	65 49 25.6	.15	51.60	3
5768	Gould, Z. C., 8^h 3599 . .	10.0	51.52	0.105	71 12 25.2	.15	50.22	1
5769	Volantis	10.0	51.57	0.083	71 20 9.0	.15	50.22	1
5770	Volantis	10.0	57.04	+ 0.214	70 32 50.4	− 13.15	50.23	1
5771	Lacaille, 3573	7.5	8 43 57.29	+ 1.122	62 38 16.7	− 13.16	50.74	2
5772	Volantis	10.0	43 59.85	+ 0.213	70 33 39.4	.16	50.23	1
5773	Chamæleontis	9.0	44 16.40	− 4.178	82 5 3.8	.18	50.15	2
5774	Chamæleontis	10.0	26.07	2.870	80 22 39.5	.19	50.15	1
5775	Octantis	9.5	28.78	− 5.854	83 33 40.4	− 13.19	50.12	2
5776	Volantis	9.0	8 44 31.11	+ 0.220	70 32 56.1	− 13.19	50.23	1
5777	Gould, Z. C., 8^h 3683 . .	8.5	31.69	1.073	63 15 26.7	.19	50.30	1
5778	Gould, Z. C., 8^h 3676 . .	9.0	33.16	+ 0.812	65 59 5.4	.19	51.60	3
5779	Volantis	9.5	44.43	− 0.059	72 10 53.4	.21	50.22	1
5780	Gould, Z. C., 8^h 3705 . .	9.0	49.92	+ 0.974	64 23 12.7	− 13.21	50.29	1
5781	Gould, Z. C., 8^h 3647 . .	9.2	8 44 51.54	− 0.716	75 7 44.0	− 13.21	50.19	2
5782	Carinæ	10.0	44 58.21	+ 1.020	63 54 . .	.22	50.29	1
5783	Lacaille, 3599	8.5	45 7.56	− 0.376	73 45 12.6	.23	52.26	1
5784	Volantis	9.5	15.73	+ 0.472	68 52 37.6	.24	50.24	1
5785	Volantis	10.0	25.82	+ 0.626	67 38 31.5	− 13.25	50.25	1

No. 5780 Probably 0.5 rev. wrong. Gould's declination is 17″ greater.

Number.	Constellation, Name of Star, or Synonym.	Magnitude.	Right Ascension, 1850.0.	Annual Precession.	South Declination, 1850.0.	Annual Precession.	Mean year.	No. of obs.
			h. m. s.	s.	° ′ ″	″		
5786	Gould, Z. C., 8h 3724 . .	9.0	8 45 32.69	— 0.242	73 9 31.6	— 13.26	50.21	1
5787	Volantis	9.5	37.23	+ 0.548	68 19 38.2	.26	50.25	1
5788	Lacaille, 3653	7.0	41.06	— 4.090	82 1 54.3	.27	50.65	2
5789	Gould, Z. C., 8h 3762 . .	9.0	45 48.54	+ 0.096	71 23 20.1	.28	50.22	1
5790	Lacaille, 3627	8.5	46 1.56	— 2.140	79 7 28.3	— 13.29	50.16	2
5791	Lacaille, 3611	8.5	8 46 1.85	— 1.121	76 34 9.8	— 13.29	50.18	1
5792	Lacaille, 3588	8.5	4.50	+ 0.937	64 52 23.8	.29	50.28	1
5793	Lacaille, 3608	7.4	5.06	— 0.475	74 14 10.6	.30	50.95	4
5794	Gould, Z. C., 8h 3802 . .	9.2	7.01	+ 0.709	67 2 22.7	.30	50.26	3
5795	η Chamæleontis	6.0	16.97	— 1.809	78 25 0.4	— 13.31	51.18	2
5796	Gould, Z. C., 8h 3796 . .	9.5	8 46 21.38	— 0.081	72 24 12.8	— 13.31	50.21	2
5797	Lacaille, 3616	9.0	37.24	1.089	76 29 51.5	.32	50.18	1
5798	Lacaille, 3630	7.2	46.70	— 1.836	78 30 2.0	.33	51.18	2
5799	Gould, Z. C., 8h 3859 . .	8.9	47.78	+ 1.003	64 14 21.9	.34	50.61	4
5800	Brisbane, 2251	7.8	49.68	+ 1.144	62 37 18.1	— 13.34	50.74	2
5801	Gould, 12053	9.0	8 46 55.13	+ 0.589	68 6 27.7	— 13.35	50.25	1
5802	Carinæ	9.2	46 58.04	1.035	63 54 27.7	.35	51.21	2
5803	Gould, Z. C., 8h 3874 . .	9.0	47 9.70	0.498	68 48 50.9	.37	50.24	2
5804	Gould, Z. C., 8h 3895 . .	8.8	14.20	1.012	64 11 14.9	.37	50.92	3
5805	Gould, 12067	10.0	16.80	+ 0.579	68 12 43.2	— 13.37	50.25	1
5806	Volantis	9.5	8 47 17.44	— 1.001	76 15 13.2	— 13.37	50.18	2
5807	Gould, Z. C., 8h 3840 . .	9.0	17.66	— 0.678	75 7 15.4	.37	50.19	2
5808	Volantis	9.5	21.55	+ 0.882	65 32 42.4	.38	52.27	1
5809	Volantis	10.0	23.44	— 0.453	74 13 18.6	.38	51.16	1
5810	Gould, Z. C., 8h 3893 . .	9.0	29.46	+ 0.357	69 51 31.3	— 13.39	50.24	2
5811	Octantis	9.8	8 47 34.54	-12.237	86 16 48.9	— 13.39	51.20	5
5812	Lacaille, 3610	9.0	36.46	+ 0.221	70 45 31.8	.39	50.23	1
5813	Gould, Z. C., 8h 3885 . .	9.5	40.94	0 377	73 54 30.0	.40	52.26	1
5814	Carinæ	9.5	49.73	+ 1.15	62 33 43.4	.41	51.18	1
5815	Volantis	9.5	54.80	+ 0.859	65 48 33.9	— 13.41	52.27	1
5816	Chamæleontis	8.8	8 47 57.86	— 0.728	75 20 29.4	13.42	50.18	2
5817	Chamæleontis	9.8	48 2.77	0.995	76 16 18.2	.42	50.18	2
5818	Lacaille, 3644	6.0	11.77	— 1.816	78 31 6.0	.43	51.18	2
5819	Gould, Z. C., 8h 3988 . .	10.0	26.98	+ 0.993	64 29 40.0	.45	50.29	1
5820	Gould, Z. C., 8h 3991 . .	9.5	30.25	+ 0.924	65 13 0.1	— 13.45	50.28	1
5821	Lacaille, 3609	6.0	8 48 32.73	+ 0.820	66 14 0.8	— 13.46	50.93	3
5822	Lacaille, 3632	8.0	34.29	— 1.063	76 31 5.7	.46	50.18	1
5823	Gould, Z. C., 8h 3999 . .	9.0	39.61	+ 0.763	66 46 13.6	.46	50.26	2
5824	Gould, 12092	9.0	46.80	— 0.596	74 52 53.0	.47	50.20	1
5825	Chamæleontis	10.0	51.56	1.740	78 22 15.4	— 13.48	52.21	1

Number.	Constellation, Name of Star, or Synonym.	Magnitude.	Right Ascension, 1850.0.	Annual Precession.	South Declination, 1850.0.	Annual Precession.	Mean year.	No. of obs.
			h. m. s.	s.	° ′ ″	″		
5826	Chamæleontis	9.2	8 48 53.06	− 1.713	78 18 31.4	− 13.48	51.25	2
5827	Volantis	10.0	48 55.75	+ 0.415	69 34 1.4	.48	50.24	1
5828	Volantis	10.0	49 3.62	0.544	68 37 15.6	.49	50.25	1
5829	Carinæ	9.5	8.21	+ 1.121	63 6 42.9	.49	50.30	1
5830	Lacaille, 3759	7.5	20.39	−11.206	86 2 24.5	− 13.51	51.20	6
5831	Gould, Z. C., 8h 4049 . .	9.0	8 49 23.59	+ 0.808	66 25 6.9	− 13.51	50.26	1
5832	Gould, Z. C., 8h 4082 . .	8.5	35.64	1.192	62 17 34.7	.52	51.18	1
5833	Gould, Z. C., 8h 4080 . .	9.5	36.25	1.135	62 59 25.1	.52	50.30	1
5834	Gould, 12129	10.0	43.70	0.642	67 53 58.6	.53	50.25	1
5835	Gould, Z. C., 8h 4092 . .	8.5	46.62	+ 1.132	63 2 13.5	− 13.54	50.30	2
5836	Octantis	9.0	8 49 47.14	− 6.552	84 8 46.7	− 13.54	50.12	1
5837	Lacaille, 3629	8.0	51.91	+ 0.039	71 59 15.7	.54	50.22	1
5838	Chamæleontis	10.0	55.03	− 3.161	81 0 34.3	.54	50.14	2
5839	Gould, Z. C., 8h 4104 . .	9.0	49 56.42	+ 1.033	64 11 16.0	.55	50.31	2
5840	Gould, Z. C., 8h 4119 . .	9.5	50 9.44	+ 0.975	64 49 49.4	− 13.56	50.28	1
5841	Gould, Z. C., 8h 4086 . .	8.5	8 50 22.60	− 0.476	74 29 42.7	13.57	50.68	2
5842	Volantis	9.5	23.65	+ 0.577	68 28 29.7	.58	50.25	1
5843	Chamæleontis	10.0	30.89	− 2.594	80 8 58.6	.58	50.15	1
5844	Gould, Z. C., 8h 4151 . .	9.0	31.38	+ 0.972	64 54 4.5	.58	50.28	1
5845	Chamæleontis	9.5	37.67	− 1.677	78 18 4.2	− 13.59	51.26	2
5846	Chamæleontis	10.0	8 50 49.77	− 0.687	75 21 . .	− 13.60	50.18	1
5847	Gould, 12153	9.0	50 49.93	+ 0.620	68 9 55.8	.60	50.25	1
5848	Volantis	10.0	51 3.96	+ 0.725	67 18 0.4	.62	50.26	2
5849	Lacaille, 3669	6.5	16.83	1.960	78 56 48.8	.63	50.16	1
5850	Gould, Z. C., 8h 4125 . .	9.2	26.93	− 2.031	79 7 43.0	− 13.64	50.16	2
5851	Chamæleontis	10.0	8 51 39.77	− 2.547	80 6 50.6	− 13.66	50.15	1
5852	Lacaille, 3640	8.5	40.85	+ 0.486	69 15 37.8	.66	50.24	1
5853	Carinæ	9.5	42.43	+ 1.210	62 15 23.6	.66	51.18	1
5854	Gould, Z. C., 8h 4198 . .	9.0	46.54	− 0.236	73 31 6.4	.66	50.21	2
5855	Gould, 12177	9.0	57.93	+ 0.851	66 13 58.6	− 13.68	50.93	3
5856	Gould, Z. C., 8h 4200 . .	9.0	8 51 59.48	− 0.601	75 5 27.6	− 13.68	50.19	2
5857	Octantis	8.7	52 1.10	−10.404	85 50 58.9	.68	51.20	6
5858	Volantis	9.2	5.32	+ 0.657	67 57 51.0	.68	50.25	2
5859	Carinæ	9.0	8.94	1.208	62 19 37.3	.69	51.18	1
5860	Volantis	10.0	24.56	+ 0.890	65 53 33.0	− 13.70	52.27	1
5861	Gould, Z. C., 8h 4213 . .	9.8	8 52 25.07	− 1.167	77 1 55.8	− 13.71	50.17	2
5862	Gould, Z. C., 8h 4269 . .	10.0	37.42	+ 0.159	71 29 5.3	.72	50.22	1
5863	Volantis	9.5	42.42	+ 0.670	67 54 18.5	.72	50.25	2
5864	Carinæ	10.0	43.36	− 3.944	82 5 19.0	.72	51.17	1
5865	Volantis	9.0	45.18	− 0.410	74 22 56.6	− 13.73	51.16	1

Number.	Constellation, Name of Star, or Synonym.	Magnitude.	Right Ascension, 1850.0.	Annual Precession.	South Declination, 1850.0.	Annual Precession.	Mean year.	No. of obs.
			h. m. s.	s.	° ′ ″	″		
5 866	Volantis	10.0	8 52 48.16	+ 0.176	71 23 45.8	— 13.73	50.22	1
5 867	Volantis	9.5	52 49.70	0.664	67 57 54.6	.73	50.25	1
5 868	Gould, 12212	9.0	53 10.67	0.866	66 11 45.2	.75	50.93	3
5 869	Gould, 12214	8.5	16.89	0.814	66 41 25.1	.76	50.26	3
5 870	Carinæ	9.5	22.01	+ 1.194	62 37 47.8	— 13.77	51.18	1
5 871	Gould, Z. C., 8ʰ 4359 . .	9.0	8 53 25.33	+ 1.173	62 52 59.6	— 13.77	50.30	1
5 872	Gould, Z. C., 8ʰ 4297 . .	9.2	29.50	— 1.040	76 42 2.4	.77	50.18	2
5 873	Octantis	10.0	37.97	— 9.804	85 41 9.0	.78	51.21	2
5 874	Brisbane, 2297	8.5	43.04	+ 1.209	62 28 10.6	.79	51.18	1
5 875	Chamæleontis	10.0	44.63	— 0.642	75 21 . .	— 13.79	50.18	1
5 876	Lacaille, 3649	8.0	8 53 47.53	+ 1.169	62 58 3.9	— 13.79	50.30	1
5 877	Volantis	9.6	53 52.29	— 0.359	74 12 33.3	.80	50.95	4
5 878	Volantis	9.2	54 3.36	+ 0.601	68 34 12.0	.81	50.25	2
5 879	Chamæleontis	8.5	10.45	— 2.756	80 33 26.9	.82	50.15	1
5 880	Volantis	10.0	10.93	+ 1.045	64 26 29.8	— 13.82	50.29	1
5 881	Chamæleontis	9.8	8 54 15.11	— 1.289	77 28 28.2	— 13.82	50.17	2
5 882	Volantis	9.2	23.40	+ 0.479	69 31 20.5	.83	50.24	2
5 883	Volantis	10.0	25.48	0.002	72 29 37.3	.83	50.21	1
5 884	Gould, Z. C., 8ʰ 4450 . .	9.5	27.34	1.227	62 19 0.3	.83	51.18	1
5 885	Gould, Z. C., 8ʰ 4447 . .	9.0	28.01	+ 1.279	62 54 55.6	— 13.84	50.30	1
5 886	Gould, Z. C., 8ʰ 4429 . .	9.0	8 54 43.57	+ 0.148	71 42 7.1	— 13.85	50.22	1
5 887	Gould, Z. C., 8ʰ 4464 . .	9.0	44.40	1.109	63 46 27.0	.85	50.29	1
5 888	Gould, Z. C., 8ʰ 4469 . .	9.5	45.27	+ 1.146	63 20 36.0	.85	50.30	1
5 889	Gould, Z. C., 8ʰ 4392 . .	9.0	47.64	— 1.045	76 47 13.4	.86	50.18	2
5 890	Carinæ	9.0	54 48.72	+ 1.165	63 7 15.4	— 13.86	50.30	1
5 891	Gould, Z. C., 8ʰ 4479 . .	9.5	8 55 0.78	+ 0.829	66 42 16.7	— 13.87	50.26	2
5 892	Lacaille, 3666	8.5	3.53	0.941	65 37 29.4	.87	51.27	2
5 893	Gould, 12257	10.0	4.70	+ 0.941	65 37 35.0	.87	51.27	2
5 894	Lacaille, 3680	7.8	7.06	— 0.365	74 18 33.8	.88	50.95	4
5 895	Gould, Z. C., 8ʰ 4466 . .	10.0	10.95	+ 0.060	72 10 . .	— 13.88	50.22	1
5 896	Brisbane, 2310	8.7	8 55 15.28	— 0.178	73 27 59.2	— 13.89	50.21	2
5 897	Volantis	9.5	19.87	+ 0.634	68 25 4.0	.89	50.25	1
5 898	Chamæleontis	9.5	46.50	— 1.947	79 8 27.8	.92	51.11	1
5 899	Gould, Z. C., 8ʰ 4543 . .	9.5	48.06	+ 0.636	68 26 24.8	.92	50.25	1
5 900	Octantis	9.7	55 56.35	—13.162	86 36 30.3	— 13.93	51.24	3
5 901	Lacaille, 3668	7.7	8 56 2.14	+ 1.105	63 56 45.0	— 13.93	50.29	2
5 902	Lacaille, 3674	8.3	3.56	0.559	69 3 37.7	.94	50.24	3
5 903	Gould, Z. C., 8ʰ 4587 . .	9.5	11.94	0.905	66 5 1.1	.94	51.26	2
5 904	Gould, Z. C., 8ʰ 4564 . .	9.5	18.35	0.174	71 39 51.1	.95	50.22	1
5 905	Lacaille, 3679	7.5	18.60	+ 0.369	70 26 5.3	— 13.95	50.23	1

No. 5 871. Gould's declination is 10″ less.

No. 5 873. Declination possibly 1′ greater.

Number.	Constellation, Name of Star, or Synonym.	Magnitude.	Right Ascension, 1850.0.	Annual Precession.	South Declination, 1850.0.	Annual Precession.	Mean year.	No. of obs.
			h. m. s.	s.	° ′ ″	″		
5906	Lacaille, 3670	8.5	8 56 23.48	+ 1.197	62 52 53.9	— 13.96	50.30	1
5907	Carinæ	10.0	24.17	1.231	62 27 5.1	.96	51.18	1
5908	Gould, Z. C., 8h 4596 . .	10.0	26.72	0.625	68 34 31.9	.96	50. 5	1
5909	Gould, Z. C., 8h 4600 . .	9.0	30.46	0.150	71 49 31.0	.97	50.22	2
5910	Gould, Z. C., 8h 4609 . .	9.0	44.45	+ 0.358	70 33 13.5	— 13.97	50.23	1
5911	Gould, Z. C., 8h 4532 . .	9.8	8 56 50.18	— 2.101	79 31 38.2	— 13.98	50.10	2
5912	Chamæleontis	10.0	56 52.75	— 3.787	82 2 49.0	13.99	51.17	1
5913	Gould, Z. C., 8h 4668 . .	8.5	57 2.71	+ 1.240	62 24 28.2	14.00	50.75	2
5914	Lacaille, 3682	8.2	6.16	+ 0.755	67 33 16.0	.00	50.25	2
5915	Chamæleontis	10.0	7.69	— 2.127	79 34 6.7	— 14.00	51.10	1
5916	Chamæleontis	9.0	8 57 12.26	— 2.875	80 51 31.1	— 14.01	51.14	1
5917	Gould, 12324	9.5	14.31	+ 0.739	67 41 48.9	.01	50.26	1
5918	Carinæ	10.0	16.07	+ 1.241	62 24 28.3	.01	51.18	1
5919	Lacaille, 3688	8.0	23.72	— 0.123	73 20 18.4	.02	50.21	2
5920	Volantis	10.0	30.35	— 2.139	79 36 25.9	— 14.03	51.10	1
5921	Gould, Z. C., 8h 4605 . .	9.8	8 57 47.86	— 2.198	79 43 57.8	— 14.05	51.13	2
5922	Gould, Z. C., 8h 4665 . .	8.5	50.29	0.662	75 39 47.8	.05	51.17	2
5923	Chamæleontis	9.5	50.52	— 2.355	80 1 {26.3 / 36.9}	.05	50.62	2
5924	Volantis	9.0	50.64	+ 0.692	68 9 6.2	.05	50.25	1
5925	Gould, Z. C., 8h 4656 . .	8.5	51.60	— 1.070	77 1 29.2	— 14.05	50.65	2
5926	Gould, Z. C., 8h 4684 . .	9.8	8 57 52.34	— 0.321	74 17 44.0	— 14.05	51.71	2
5927	Gould, Z. C., 8h 4626 . .	9.5	57 59.40	— 2.138	79 37 31.4	.06	50.63	2
5928	Gould, Z. C., 8h 4736 . .	9.0	58 2.19	+ 0.954	65 45 38.8	.06	51.60	3
5929	Lacaille, 3683	8.5	3.10	0.769	67 30 19.0	.06	50.25	2
5930	Carinæ	10.0	7.64	+ 1.244	62 27 41.7	— 14.07	51.18	1
5931	Volantis	10.0	8 58 8.42	— 0.358	74 28 20.0	— 14.07	51.16	1
5932	Chamæleontis	9.0	9.12	1.191	77 23 52.2	.07	51.13	1
5933	Gould, Z. C., 8h 4709 . .	8.2	21.61	0.693	75 48 18.3	.08	50.42	4
5934	Chamæleontis	9.5	37.20	— 3.278	81 28 57.6	.10	51.13	1
5935	Gould, Z. C., 8h 4780 . .	8.5	37.52	+ 1.093	64 20 2.6	— 14.10	50.29	2
5936	Gould, 12356	9.0	8 58 39.96	+ 0.793	67 21 0.8	— 14.10	50.26	2
5937	Chamæleontis	10.0	48.49	— 3.722	82 2 12.1	.11	51.17	1
5938	Gould, Z. C., 8h 4774 . .	9.5	51.62	+ 0.416	70 19 4.5	.11	50.23	2
5939	Chamæleontis	9.0	58 55.97	— 1.180	77 24 30.7	.12	51.13	1
5940	Volantis	9.3	59 2.01	+ 0.036	72 37 52.1	— 14.12	50.21	3
5941	Octantis	10.0	8 59 9.73	—57.861	89 6 28.0	— 14.13	51.20	3
5942	Gould, Z. C., 8h 4824 . .	9.0	14.74	+ 0.921	66 12 2.2	.14	50.93	3
5943	Volantis	9.0	15.57	0.524	69 34 39.9	.14	50.24	1
5944	Volantis	9.5	19.68	0.930	66 7 6.7	.14	51.26	2
5945	Volantis	9.5	24.15	+ 0.579	69 10 47.8	— 14.14	50.24	2

Number.	Constellation, Name of Star, or Synonym.	Magnitude.	Right Ascension, 1850.0.	Annual Precession.	South Declination. 1850.0.	Annual Precession.	Mean year.	No. of obs.
			h. m. s.	s.	° ′ ″	″		
5 946	Lacaille, 3694	7.0	8 59 24.73	+ 0.715	68 5 31.0	− 14.15	50.25	1
5 947	Gould, Z. C., 8^h 4822	10.0	25.40	0.433	70 14 44.7	.15	50.23	1
5 948	Gould, Z. C., 8^h 4847	9.2	35.43	+ 0.955	65 53 57.5	.16	51.60	3
5 949	Gould, 12359	9.2	43.46	− 1.446	78 9 8.7	.16	51.26	2
5 950	Carinæ	9.0	54.76	+ 1.206	63 7 2.6	− 14.18	50.30	2
5 951	Gould, Z. C., 9^h 6	9.5	8 59 56.75	+ 0.235	71 33 46.2	− 14.18	50.22	1
5 952	Gould, Z. C., 9^h 36	9.5	8 59 58.26	1.208	63 6 11.1	.18	50.30	1
5 953	Volantis	10.0	9 0 2.72	1.025	65 15 37.7	.18	50.28	1
5 954	α Volantis	4.8	4.00	0.969	65 47 51.3	.19	51.60	3
5 955	Volantis	9.5	7.93	+ 0.524	69 39 0.3	− 14.19	50.24	1
5 956	Gould, Z. C., 9^h 45	9.0	9 0 29.03	+ 0.193	71 51 21.0	− 14.21	50.22	2
5 957	Octantis	9.8	29.12	− 5.969	84 2 33.9	.21	50.76	2
5 958	Gould, Z. C., 9^h 52	9.0	41.20	0.057	73 14 4.9	.22	50.21	1
5 959	Lacaille, 3709	7.5	42.90	0.174	73 48 31.7	.23	50.74	4
5 960	Gould, 12376	9.3	45.22	− 1.443	78 12 38.8	− 14.23	50.95	3
5 961	Chamæleontis	9.5	9 0 52.07	− 3.711	82 5 40.8	− 14.24	51.17	1
5 962	Volantis	9.0	0 52.79	+ 0.874	66 48 36.9	.24	50.26	1
5 963	Chamæleontis	9.0	1 1.85	− 2.379	80 12 12.8	.25	50.62	2
5 964	Brisbane, 2339	8.5	1.64	+ 1.135	64 5 16.0	.25	50.29	1
5 965	Gould, Z. C., 9^h 131	10.0	21.43	+ 0.378	70 45 54.7	− 14.27	50.23	1
5 966	Lacaille, 3724	8.2	9 1 27.66	− 0.467	75 7 56.4	− 14.27	50.19	2
5 967	Volantis	9.0	32.44	+ 0.264	71 30 33.6	.28	50.22	1
5 968	Melbourne (1), 456	9.1	1 48.76	0.861	67 1 26.5	.29	50.27	4
5 969	Gould, Z. C., 9^h 213	8.5	2 7.35	1.178	63 41 21.0	.31	50.29	1
5 970	Lacaille, 3714	8.0	9.48	+ 0.498	70 0 31.5	− 14.31	50.23	1
5 971	Gould, Z. C., 9^h 232	9.5	9 2 17.45	+ 1.171	63 45 28.3	− 14.32	50.29	1
5 972	Gould, Z. C., 9^h 125	8.5	17.92	− 2.002	79 32 58.4	.32	51.13	2
5 973	Carinæ	10.0	20.14	+ 0.689	68 33 47.7	.33	50.25	1
5 974	Carinæ	10.0	20.92	1.294	62 14 33.3	.33	51.18	1
5 975	Carinæ	9.2	21.54	+ 0.680	68 38 2.5	− 14.33	50.25	2
5 976	Gould, Z. C., 9^h 197	9.0	9 2 24.96	− 0.104	73 35 29.2	− 14.33	50.20	1
5 977	Octantis	9.3	25.30	11.287	86 16 21.2	.33	51.20	5
5 978	Gould, Z. C., 9^h 174	8.8	36.23	1.144	77 30 2.5	.34	50.49	3
5 979	Chamæleontis	9.0	45.58	− 3.268	81 37 8.2	.35	50.47	3
5 980	Gould, Z. C., 9^h 272	9.5	48.06	+ 1.127	64 21 11.1	− 14.35	50.29	1
5 981	Gould, Z. C., 9^h 285	9.5	9 2 50.08	+ 1.298	62 14 43.8	− 14.36	51.18	1
5 982	Gould, Z. C., 9^h 241	9.3	2 52.31	+ 0.075	72 41 50.1	.36	50.21	3
5 983	Chamæleontis	10.0	3 9.26	− 2.016	79 36 55.6	.38	51.10	1
5 984	Gould, Z. C., 9^h 259	9.5	11.10	− 0.073	73 29 11.8	.38	50.21	1
5 985	Gould, Z. C., 9^h 302	10.0	12.14	0.977	66 0 50.3	− 14.38	50.28	1

Number.	Constellation, Name of Star, or Synonym.	Magnitude.	Right Ascension, 1850.0.	Annual Precession.	South Declination, 1850.0.	Annual Precession.	Mean year.	No. of obs.
			h. m. s.	s.	° ′ ″	″		
5 986	Lacaille, 3712	7.5	9 3 13.48	+ 1.169	63 53 53.0	— 14.38	50.29	2
5 987	Chamæleontis	9.0	22.38	— 1.710	78 58 42.1	.39	51.11	1
5 988	Gould, 12445	9.0	22.52	+ 0.822	67 30 36.8	.39	50.25	2
5 989	Gould, Z. C., 9^h 294 . .	9.5	23.35	+ 0.241	71 47 13.9	.39	50.22	1
5 990	Carinæ	9.5	24.18	— 0.254	74 21 24.8	— 14.39	51.16	1
5 991	Carinæ	9.5	9 3 27.93	+ 1.112	64 35 14.9	— 14.39	50.28	1
5 992	Carinæ	10.0	28.18	0.647	68 59 46.0	.40	50.24	2
5 993	Lacaille, 3708	8.2	31.04	+ 1.303	62 14 44.8	.40	50.75	2
5 994	Chamæleontis	9.5	31.25	— 0.583	75 42 40.7	.50	51.15	1
5 995	Gould, Z. C., 9^h 341 . .	10.0	39.15	+ 0.842	67 21 8.2	— 14.41	50.26	1
5 996	Chamæleontis	10.0	9 3 57.98	— 0.559	75 39 1.9	— 14.43	51.15	1
5 997	Gould, Z. C., 9^h 346 . .	9.0	4 5.76	0.064	73 30 31.6	.43	50.21	2
5 998	Gould, Z. C., 9^h 327 . .	8.6	7.01	— 0.736	76 17 53.1	.43	50.42	4
5 999	Gould, Z. C., 9^h 399 . .	9.5	9.49	+ 1.297	62 23 28.2	.44	51.20	1
6 000	Gould, Z. C., 9^h 396	9.8	15.87	+ 1.029	65 34 26.5	— 14.44	50.28	2
6 001	Carinæ	9.0	9 4 18.38	+ 0.659	68 58 17.5	— 14.45	50.24	3
6 002	Gould, Z. C., 9^h 405 . .	9.0	21.73	1.032	65 33 2.1	.45	50.28	2
6 003	Carinæ	9.5	23.11	1.094	64 53 20.7	.45	50.28	1
6 004	*E* Carinæ	6.0	23.34	0.534	69 56 7.3	.45	50.23	1
6 005	Gould, 12476	9.5	28.60	+ 1.308	62 17 16.4	— 14.46	51.18	1
6 006	Carinæ	10.0	9 4 34.45	— 0.281	74 33 15.7	— 14.46	51.16	1
6 007	*G* Carinæ	5.5	42.43	+ 0.223	71 59 56.0	.47	50.22	1
6 008	Carinæ	9.0	43.35	+ 0.579	69 37 41.4	.47	50.24	1
6 009	Chamæleontis	9.5	54.06	— 1.653	78 55 20.2	.48	51.11	1
6 010	Gould, 12486	10.0	57.39	+ 1.222	63 25 42.2	— 14.49	50.30	1
6 011	Gould, Z. C., 9^h 446 . .	9.2	9 4 57.81	+ 0.796	67 53 5.2	— 14.49	50.25	2
6 012	Gould, 12488	10.5	4 58.79	+ 1.223	63 25 40.8	.49	50.30	1
6 013	Chamæleontis	9.0	5 29.98	— 3.559	82 4 48.4	.52	51.00	4
6 014	Carinæ	9.0	41 15	+ 0.489	70 21 43.0	.53	50.23	1
6 015	Gould, Z. C., 9^h 433 . .	9.0	46.73	— 1.605	78 51 25.1	— 14.54	51.11	1
6 016	Brisbane, 2381	9.0	9 5 46.91	+ 0.775	68 8 52.9	— 14.54	50.25	1
6 017	Gould, 12498	9.3	6 3.19	— 0.163	74 7 14.0	.55	50.92	3
6 018	Carinæ	9.2	4.26	+ 0.804	67 55 11.6	.55	50.25	2
6 019	Chamæleontis	9.0	7.54	— 3.964	82 33 7.1	.56	50.13	1
6 020	Octantis	9.5	26.70	—22.544	87 56 47.9	— 14.58	51.20	6
6 021	Carinæ	10.0	9 6 34.68	+ 0.861	67 27 13.4	— 14.58	50.26	1
6 022	Carinæ	9.5	50.56	+ 1.326	62 18 7.9	.60	51.18	1
6 023	Gould, Z. C., 9^h 533 . .	9.2	6 56.04	— 0.465	75 28 30.6	.60	50.67	2
6 024	Lacaille, 3767	8.2	7 4.60	— 0.153	74 8 46.8	.61	50.74	4
6 025	Gould, Z. C., 9^h 597 . .	9.5	9.56	+ 1.023	65 55 29.3	14.62	51.26	2

No. 5 995. Gould's declination is 25.8″ less.

Number.	Constellation, Name of Star, or Synonym.	Magnitude.	Right Ascension, 1850.0.	Annual Precession.	South Declination, 1850.0.	Annual Precession.	Mean year.	No. of obs.
			h. m. s.	s.	° ′ ″	″		
6 026	Lacaille, 3752	8.0	9 7 22.17	+ 1.121	64 52 45.9	— 14.63	50.28	1
6 027	Chamæleontis	10.0	48.85	— 3.239	81 46 17.2	.66	51.13	1
6 028	Chamæleontis	9.5	50.08	3.672	82 17 34.6	.66	52.12	1
6 029	Lacaille, 3778	7.7	51.66	— 0.577	75 57 29.2	.66	50.18	2
6 030	Gould, Z. C., 9h 670 . .	9.5	7 59.52	+ 1.333	62 19 50.1	— 14.67	51.18	1
6 031	Gould, Z. C., 9h 651 . .	9.5	9 8 4.50	+ 0.498	70 29 58.0	— 14.67	50.23	1
6 032	Carinæ	9.2	13.36	+ 0.850	67 42 19.0	.68	50.25	2
6 033	Gould, Z. C., 9h 633 . .	9.0	21.60	— 0.628	76 11 40.2	.69	50.18	2
6 034	Chamæleontis	10.0	30.79	+ 1.291	62 56 28.6	.70	50.30	1
6 035	Gould, Z. C., 9h 700 . .	9.2	34.14	+ 1.005	66 14 46.8	— 14.70	51.26	2
6 036	Chamæleontis	10.0	9 8 40.51	— 0.638	76 13 57.2	— 14.71	51.15	1
6 037	Gould, Z. C., 9h 714 . .	9.0	42.21	+ 1.103	65 13 23.5	.71	50.28	1
6 038	Chamæleontis	9.3	57.70	— 3.421	82 2 26.8	.73	51.21	3
6 039	Gould, Z. C., 9h 741 . .	9.0	58.19	+ 1.163	64 34 33.7	.73	50.29	1
6 040	Gould, Z. C., 9h 734 . .	10.0	8 58.36	+ 1.023	66 6 19.8	— 14.73	50.28	1
6 041	Carinæ	10.0	9 9 1.15	+ 1.323	62 34 23.9	— 14.73	51.18	1
6 042	Gould, Z. C., 9h 754 . .	8.8	2.21	1.344	62 17 55.8	.73	50.75	2
6 043	Carinæ	10.0	3.23	0.989	66 27 55.6	.73	50.26	1
6 044	Carinæ	11.0	6.67	+ 0.989	66 27 27.7	.73	50.26	1
6 045	Chamæleontis	9.0	14.38	— 2.284	80 23 35.8	— 14.74	50.62	2
6 046	Carinæ	9.5	9 9 15.20	+ 1.077	65 34 2.4	— 14.74	50.28	1
6 047	Lacaille, 3911	7.9	17.62	—15.310	87 10 37.8	.75	51.20	6
6 048	Carinæ	10.0	19.51	+ 1.101	65 17 56.6	.75	50.28	1
6 049	Gould, Z. C., 9h 743 . .	9.0	20.03	0.385	71 21 41.3	.75	50.22	1
6 050	Carinæ	9.0	23.36	+ 1.034	66 1 51.8	14.75	50.28	1
6 051	Chamæleontis	10.0	9 9 23.87	— 1.647	79 7 56.2	— 14.75	51.11	1
6 052	Gould, Z. C., 9h 763 . .	10.0	24.05	+ 0.699	69 6 11.4	.75	50.24	1
6 053	Gould, Z. C., 9h 711 . .	9.0	25.15	— 0.664	76 22 1.1	.75	51.15	1
6 054	Lacaille, 3775	7.7	32.30	+ 0.757	68 38 21.6	.76	50.24	3
6 055	Chamæleontis	9.0	33.84	— 2.993	81 30 41.7	— 14.76	50.64	2
6 056	Gould, Z. C., 9h 770 . .	9.5	9 9 35.66	+ 0.356	71 34 9.8	— 14.76	50.22	1
6 057	Gould, Z. C., 9h 769 . .	9.2	37.28	0.272	72 5 29.9	.76	50.22	2
6 058	Lacaille, 3774	8.4	41.05	0.932	67 6 15.5	.77	50.27	4
6 059	Carinæ	10.0	43.40	0.583	70 1 38.3	.77	50.23	1
6 060	Gould, Z. C., 9h 815 . .	10.0	46.08	+ 1.031	66 6 23.0	— 14.77	51.26	2
6 061	Gould, Z. C., 9h 827 . .	8.5	9 9 54.30	+ 1.124	65 6 30.2	— 14.78	50.28	2
6 062	Lacaille, 3779	8.2	9 57.08	0.768	68 35 9.6	.78	50.25	2
6 063	Carinæ	9.0	10 7.40	0.823	68 7 34.7	.79	50.25	1
6 064	Gould, Z. C., 9h 868 . .	8.5	22.98	+ 1.257	63 34 46.9	.81	50.29	1
6 065	Gould, Z. C., 9h 829 . .	9.0	29.70	— 0.223	74 42 16.1	— 14.82	51.18	2

No. 6 040. Declination uncertain by several seconds.
No. 6 049. Gould makes this star double, components 3″ apart, 8¾ magnitude.

Number.	Constellation, Name of Star, or Synonym.	Magnitude.	Right Ascension, 1850.0.	Annual Precession.	South Declination, 1850.0.	Annual Precession.	Mean year.	No. of obs.
			h. m. s.	s.	° ′ ″	″		
6066	Gould, Z. C., 9h 882 . .	9.8	9 10 36.45	+ 1.298	63 3 43.2	14.82	50.30	2
6067	Carinæ	9.5	39.04	0.162	72 48 26.3	.83	50.27	1
6068	Carinæ	9.5	42.14	1.336	62 35 8.9	.83	51.18	1
6069	Gould, Z. C., 9h 901 . .	9.5	51.87	+ 1.262	63 34 22.6	.84	50.29	1
6070	Octantis	8.9	56.32	− 28.468	88 22 0.4	− 14.84	51.20	4
6071	Gould, Z. C., 9h 860 . .	10.0	9 10 57.73	− 0.509	75 54 12.8	14.84	51.15	1
6072	Gould, Z. C., 9h 913 . .	9.0	58.80	+ 1.302	63 4 29.5	.84	50.30	2
6073	Octantis	8.7	57.92	16.473	88 21 59.4	.84	51.19	5
6074	Octantis	9.8	10 59.72	− 12.263	86 38 41.0	.85	51.20	2
6075	Gould, Z. C., 9h 893	9.2	11 6.80	+ 0.380	71 32 54.2	14.85	50.22	2
6076	Carinæ	9.5	9 11 14.15	+ 1.343	62 33 4.4	14.86	51.18	1
6077	Lacaille, 3783	8.0	25.57	1.197	64 25 53.2	.87	50.29	1
6078	Gould, Z. C., 9h 949 . .	9.5	27.01	1.100	65 32 36.5	.87	50.28	1
6079	Gould, Z. C., 9h (953) . .	9.0	27.83	+ 1.248	63 48 48.2	.87	50.29	1
6080	Gould, Z. C., 9h 902 . .	9.0	30.52	− 0.179	74 34 47.9	− 14.88	51.16	1
6081	β Argus	2.3	9 11 32.00	+ 0.724	69 5 59.0	− 14.88	50.24	3
6082	Chamæleontis	10.0	35.81	− 0.443	75 41 38.9	.88	51.15	1
6083	Gould, Z. C., 9h 923	8.8	36.67	+ 0.121	73 6 10.0	.88	50.21	2
6084	Gould, Z. C., 9h 964	9.2	11 56.41	+ 0.289	72 10 28.6	.90	50.22	2
6085	Carinæ	9.2	12 4.12	− 0.046	73 59 35.4	− 14.91	51.23	2
6086	Carinæ	9.2	9 12 4.30	+ 1.111	65 28 53.2	− 14.91	50.28	2
6087	Gould, 12650	8.5	8.06	0.826	68 17 35.5	.91	50.25	2
6088	Carinæ	10.0	16.99	+ 1.348	62 36 7.2	.92	51.18	1
6089	Gould, Z. C., 9h 960 . .	10.0	21.57	− 0.765	76 53 45.8	.93	50.17	1
6090	Gould, Z. C., 9h 1029 . .	9.0	30.47	+ 0.900	67 39 33.2	− 14.93	50.25	2
6091	Gould, Z. C., 9h 1016 . .	9.0	9 12 32.86	+ 0.481	70 59 40.8	− 14.94	50.22	2
6092	Gould, Z. C., 9h 999 . .	10.0	38.05	0.269	75 3 15.0	.94	50.19	2
6093	Gould, Z. C., 9h 1004 . .	8.8	38.36	− 0.088	74 14 14.9	.94	50.82	5
6094	Carinæ	10.0	42.81	+ 0.805	68 32 1.2	.95	50.25	1
6095	Carinæ	9.5	45.27	+ 1.378	62 15 3.6	− 14.95	51.18	1
6096	Gould, Z. C., 9h 1022 . .	9.2	9 12 46.84	+ 0.087	73 22 30.6	− 14.95	50.21	2
6097	Carinæ	10.0	50.69	1.060	66 6 39.4	.95	51.26	2
6098	Gould, Z. C., 9h 1057 . .	10.0	58.25	0.924	67 28 50.8	.96	50.26	1
6099	Carinæ	9.5	12 59.90	0.693	69 29 3.8	.96	50.24	1
6100	Carinæ	9.0	13 28.55	+ 0.697	69 29 28.5	− 14.99	50.24	1
6101	Brisbane, 2432	8.0	9 13 39.04	+ 1.319	63 8 46.4	− 15.00	50.30	1
6102	F Carinæ	7.0	39.24	0.498	76 2 17.4	.00	50.18	2
6103	Gould, Z. C., 9h 1146 . .	9.0	13 51.58	+ 1.384	62 17 14.1	.01	50.75	2
6104	Gould, Z. C., 9h 1147 . .	9.0	14 1.60	1.034	66 30 8.0	.02	50.26	1
6105	Gould, Z. C., 9h 1178 . .	8.8	13.10	+ 1.353	62 45 27.8	− 15.03	50.74	2

No. 6079. Gould's declination is 16′ less.

Number.	Constellation, Name of Star, or Synonym.	Magnitude.	Right Ascension, 1850.0.	Annual Precession.	South Declination, 1850.0.	Annual Precession.	Mean year.	No. of obs.
			h. m. s.	s.	° ′ ″	″		
6106	Carinæ	10.0	9 14 17.06	+ 0.765	69 0 54.2	− 15.04	50.24	2
6107	Lacaille, 3822	8.5	21.69	− 0.220	74 57 48.8	.04	50.19	2
6108	Gould, Z. C., 9h 1143	9.0	23.26	0.057	74 13 13.2	.04	50.82	5
6109	Chamæleontis	9.5	36.58	− 2.936	81 38 7.5	.06	51.13	1
6110	Gould, Z. C., 9h 1181	9.0	36.97	+ 0.423	71 33 38.6	− 15.06	50.22	1
6111	Lacaille, 3840	8.5	9 14 39.19	− 1.362	78 44 10.8	− 15.06	51.12	2
6112	Gould, Z. C., 9h 1192	9.0	43.69	+ 0.426	71 32 52.3	.06	50.22	1
6113	Gould, Z. C., 9h 1224	9.5	46.07	1.096	65 55 35.8	.07	52.25	1
6114	Lacaille, 3806	7.7	47.18	+ 0.988	67 2 43.4	.07	50.28	3
6115	Gould, Z. C., 9h 1195	9.0	59.20	− 0.105	74 29 24.8	− 15.08	51.16	1
6116	Octantis	9.2	9 14 59.84	− 5.591	84 11 20.8	− 15.08	50.33	2
6117	Carinæ	10.0	15 0.57	+ 1.391	62 19 29.8	.08	51.18	1
6118	Carinæ	10.0	4.49	0.896	67 56 59.6	.08	50.25	1
6119	Gould, Z. C., 9h 1243	9.0	4.99	+ 1.087	66 3 33.0	.08	51.26	2
6120	Gould, Z. C., 9h 1206	9.0	8.36	− 0.141	74 39 57.1	− 15.09	51.16	1
6121	Lacaille, 3811	7.0	9 15 10.54	+ 0.885	68 3 26.9	− 15.09	50.25	1
6122	Carinæ	10.0	13.76	1.137	65 31 30.4	.09	50.28	1
6123	Lacaille, 3809	6.5	15.90	1.055	66 25 8.8	.09	50.26	1
6124	Carinæ	10.0	16.69	+ 0.779	68 59 36.3	.10	50.24	1
6125	Chamæleontis	10.0	18.09	1.254	78 29 44.5	− 15.10	51.12	1
6126	Brisbane, 2439	9.0	9 15 19.57	+ 1.202	64 47 55.2	− 15.10	50.28	1
6127	Chamæleontis	10.0	33.64	− 1.957	80 4 59.1	.11	51.10	1
6128	Chamæleontis	9.5	35.48	1.645	79 26 36.8	.11	51.10	1
6129	Octantis	9.5	54.36	6.721	84 52 43.8	.13	50.32	1
6130	Chamæleontis	9.5	56.62	− 0.318	75 29 20.3	− 15.13	51.15	1
6131	Lacaille, 3882	8.5	9 15 59.43	− 2.944	81 42 11.2	− 15.14	51.13	1
6132	Gould, Z. C., 9h 1312	9.0	15 59.46	+ 0.949	67 32 50.0	.14	50.25	2
6133	Gould, Z. C., 9h 1251	9.0	16 0.88	− 0.918	77 36 0.5	.14	50.19	1
6134	Brisbane, 2452[1]	9.5	3.76	+ 0.767	69 10 21.6	.14	50.24	1
6135	Brisbane, 2452[2]	9.5	4.59	+ 0.767	69 10 8.4	− 15.14	50.24	1
6136	Gould, Z. C., 9h 1285	9.5	9 16 12.36	− 0.306	75 27 17.4	− 15.15	51.15	1
6137	Octantis	10.0	18.88	7.161	85 6 28.4	.15	51.20	2
6138	Chamæleontis	8.5	29.60	2.848	81 35 31.0	.17	51.14	2
6139	Chamæleontis	9.5	35.56	− 2.459	81 0 56.9	.17	51.14	1
6140	Gould, Z. C., 9h 1343	9.0	46.95	+ 0.317	72 24 12.0	− 15.18	50.21	2
6141	Gould, Z. C., 9h 1283	9.0	9 16 51.63	− 1.901	80 2 21.0	− 15.19	51.10	1
6142	Gould, Z. C., 9h 1360	10.0	53.27	+ 0.602	70 31 48.8	.19	50.23	1
6143	Lacaille, 3826	8.5	16 56.29	1.035	66 47 37.2	.19	50.28	2
6144	Carinæ	10.0	17 7.59	+ 1.000	67 10 20.0	.20	50.26	1
6145	Gould, Z. C., 9h 1363	9.5	16.41	− 0.333	75 38 37.5	− 15.21	51.15	1

Number.	Constellation, Name of Star, or Synonym.	Magnitude.	Right Ascension, 1850.0.	Annual Precession.	South Declination, 1850.0.	Annual Precession.	Mean year.	No. of obs.
			h. m. s.	s.	° ′ ″	″		
6146	Carinæ	10.0	9 17 23.27	− 1.403	62 26 9.2	− 15.22	51.18	1
6147	Gould, Z. C., 9^h 1391 . .	9.5	24.52	+ 0.187	73 12 6.0	.22	50.21	1
6148	ξ Octantis	6.1	26.71	− 6.991	85 3 15.5	.22	50.98	8
6149	Lacaille, 3845	6.5	35.96	+ 0.017	74 6 6.5	.23	50.92	3
6150	Carinæ	11.5	37.38	− 0.015	74 15 34.2	− 15.23	51.61	2
6151	Lacaille, 3846	6.6	9 17 38.16	− 0.015	74 15 40.1	− 15.23	50.82	5
6152	Chamæleontis	10.0	38.62	− 0.847	77 29 5.4	.23	51.13	1
6153	Gould, Z. C., 9^h 1452 . .	8.7	42.55	+ 1.132	65 51 7.4	.23	50.93	3
6154	Octantis	10.0	43.89	−13.074	86 55 2.2	.24	51.21	2
6155	Chamæleontis	10.0	50.81	− 1.077	78 10 15.3	− 15.24	50.30	1
6156	Carinæ	9.0	9 17 50.96	+ 1.417	62 17 35.8	− 15.24	50.32	1
6157	Gould, Z. C., 9^h 1482 . .	9.5	17 55.54	1.410	62 24 32.5	.25	51.18	1
6158	Gould, 12792	10.0	18 3.40	0.952	67 43 58.3	.25	50.25	1
6159	Carinæ	9.5	7.57	+ 1.303	63 52 14.7	.26	50.29	1
6160	Chamæleontis	10.0	15.34	− 0.548	76 31 44.1	− 15.27	51.15	1
6161	Chamæleontis	8.6	9 18 15.71	− 3.285	82 13 53.0	− 15.27	50.50	4
6162	Gould, Z. C., 9^h 1490 .	9.0	18.54	+ 0.642	70 22 8.9	.27	50.23	2
6163	Melbourne (1), 470 . . .	9.0	19.32	0.956	67 43 11.6	.27	50.25	2
6164	Gould, Z. C., 9^h 1517 . .	9.0	20.23	+ 1.411	62 25 59.2	.27	51.18	1
6165	Octantis	9.0	20.37	− 6.020	84 33 11.9	− 15.27	50.33	1
6166	Lacaille, 3850	9.0	9 18 26.70	+ 0.293	72 41 5.1	− 15.28	50.21	1
6167	Carinæ	9.0	26.73	0.982	67 29 10.4	.28	50.25	2
6168	Carinæ	9.8	26.80	1.372	62 59 41.9	.28	50.30	2
6169	Gould, Z. C., 9^h 1482 . .	10.0	31.70	1.425	62 15 52.2	.28	51.18	1
6170	Carinæ	10.0	35.00	+ 0.777	69 19 53.8	− 15.28	50.24	1
6171	Gould, Z. C., 9^h 1467 . .	9.5	9 18 35.90	− 0.551	76 33 40.2	− 15.28	51.15	1
6172	Gould, Z. C., 9^h 1505 . .	9.5	44.10	+ 0.205	73 12 37.3	.29	50.21	1
6173	Gould, Z. C., 9^h 1464 . .	9.0	49.97	− 1.255	78 41 54.2	.30	51.12	2
6174	Gould, Z. C., 9^h 1550 . .	9.8	50.88	+ 0.846	68 46 35.4	.30	50.24	2
6175	Octantis	9.2	51.51	− 7.095	85 8 14.4	− 15.30	50.98	8
6176	Gould, Z. C., 9^h 1565 . .	9.0	9 18 59.41	+ 0.860	68 39 53.0	− 15.31	50.25	2
6177	Gould, Z. C., 9^h 1566 . .	10.0	19 8.23	+ 0.641	70 26 40.1	.32	50.23	1
6178	Lacaille, 3906	7.0	24.28	− 2.458	81 8 19.6	.33	50.59	2
6179	Chamæleontis	9.5	34.59	0.662	77 0 31.2	.33	50.65	2
6180	Gould, Z. C., 9^h 1533½ .	9.0	36.71	− 1.318	78 54 5.7	15.33	51.11	1
6181	Carinæ	10.0	9 19 40.77	+ 0.782	69 23 41.7	− 15.35	50.24	1
6182	Lacaille, 3868	9.0	46.85	0.264	72 57 51.1	.35	50.21	1
6183	Carinæ	9.0	19 57.06	1.238	64 52 58.4	.36	50.28	1
6184	Carinæ	10.0	20 3.12	0.619	70 41 34.9	.36	50.23	1
6185	Carinæ	10.0	15.80	+ 1.282	64 23 16.1	− 15.38	50.29	1

Number.	Constellation, Name of Star, or Synonym.	Magnitude.	Right Ascension, 1850.0.	Annual Precession.	South Declination, 1850.0.	Annual Precession.	Mean year.	No. of obs.
			h. m. s.	s.	° ′ ″	″		
6 186	Carinæ	9.5	9 20 23.96	+ 0.625	70 40 54.8	15.39	50.23	1
6 187	Gould, Z. C., 9h 1613 . .	9.0	25.98	— 0.891	77 47 21.3	.39	50.17	1
6 188	Octantis	9.0	32.67	— 9.038	85 57 7.4	.39	51.20	6
6 189	Gould, Z. C., 9h 1680 . .	9.0	38.74	+ 0.888	68 34 35.5	.40	50.25	1
6 190	Octantis	9.8	46.36	—22.183	88 3 31.4	— 15.41	51.20	4
6 191	Gould, Z. C., 9h 1718 . .	9.2	9 20 53.87	+ 1.388	63 4 5.2	— 15.41	50.30	2
6 192	Lacaille, 3888	8.0	56.83	— 0.137	75 4 19.9	.42	50.19	2
6 193	Gould, 12861	10.0	20 57.32	+ 1.072	66 50 46.9	.42	50.30	1
6 194	Gould, Z. C., 9h 1693 . .	9.0	21 10.20	— 0.041	74 39 19.6	.43	50.52	3
6 195	Gould, Z. C., 9h 1695 . .	9.2	11.46	— 0.034	74 37 18.4	— 15.43	50.68	2
6 196	Gould, Z. C., 9h 1716 . .	9.5	9 21 12.17	+ 0.586	71 2 7.2	— 15.43	50.22	1
6 197	Gould, 12871	9.5	17.02	1.151	66 2 38.7	.44	51.26	2
6 198	Carinæ	10.0	17.50	+ 1.035	67 15 46.9	.44	50.26	1
6 199	Gould, 12852	11.0	19.60	— 0.627	77 0 27.4	.44	50.65	2
6 200	Lacaille, 3898	9.0	20.46	0.627	77 0 20.4	— 15.44	50.65	2
6 201	Gould, 12849	9.0	9 21 28.48	1.118	78 29 41.7	15.45	51.12	1
6 202	Octantis	9.0	29.12	— 4.688	83 43 41.1	.45	50.12	1
6 203	Gould, Z. C., 9h 1732 . .	9.0	29.21	+ 0.301	72 53 40.4	.45	50.21	1
6 204	Chamæleontis	9.5	29.49	— 2.178	80 46 26.9	.45	51.14	1
6 205	Gould, Z. C., 9h 1752 . .	9.0	31.08	+ 0.963	67 59 17.7	— 15.45	50.25	1
6 206	Chamæleontis	9.5	9 21 35.30	— 2.750	81 40 17.5	— 15.45	51.13	1
6 207	Chamæleontis	9.0	37.93	2.688	81 35 6.0	.46	50.64	2
6 208	Gould, Z. C., 9h 1731 . .	10.0	39.66	— 0.045	74 42 37.4	.46	51.16	1
6 209	Lacaille, 3869	9.2	39.68	+ 1.320	64 3 43.1	.46	50.29	2
6 210	Octantis	10.0	45.42	— 4.662	83 43 0.3	— 15.46	50.12	1
6 211	Gould, Z. C., 9h 1739 . .	9.8	9 21 49.18	0.118	75 3 34.0	— 15.47	50.19	2
6 212	Gould, Z. C., 9h 1709 . .	9.0	55.17	1.346	79 5 52.9	.47	51.11	1
6 213	Chamæleontis	10.0	55.29	1.466	79 22 50.8	.47	50.16	1
6 214	Lacaille, 3955	8.0	21 58.06	— 3.968	83 6 35.0	.47	50.23	2
6 215	Gould, Z. C., 9h 1793 . .	9.0	22 1.82	+ 0.579	71 9 48.7	— 15.48	50.22	1
6 216	Gould, Z. C., 9h 1776 . .	9.0	9 22 4.11	+ 0.243	73 16 22.5	— 15.48	50.21	1
6 217	Lacaille, 3878	7.7	6.30	+ 1.280	64 37 22.8	.48	50.28	2
6 218	Gould, Z. C., 9h 1728 . .	8.2	15.22	— 1.647	79 47 46.0	.49	51.10	2
6 219	Carinæ	8.5	24.42	+ 1.045	67 16 38.6	.50	50.26	2
6 220	Gould, Z. C., 9h 1850 . .	9.0	33.81	+ 1.401	63 5 18.0	— 15.51	50.30	2
6 221	Gould, Z. C., 9h 1856 . .	10.0	9 22 36.41	+ 1.403	63 4 26.1	— 15.51	50.30	1
6 222	Lacaille, 3891	8.0	39.96	+ 0.814	69 25 35.7	.51	50.24	1
6 223	Octantis	10.0	42.54	— 6.175	84 45 53.8	.52	50.33	1
6 224	Lacaille, 3893	9.0	46.30	+ 0.775	69 45 42.9	.52	50.21	1
6 225	Carinæ	8.8	22 48.11	+ 1.006	67 42 48.0	— 15.52	50.25	2

No. 6 193. Gould observes this star double.

Number.	Constellation, Name of Star, or Synonym.	Magnitude.	Right Ascension, 1850.0.	Annual Precession.	South Declination, 1850.0.	Annual Precession.	Mean year.	No. of obs.
			h. m. s.	s.	° ′ ″	″		
6 226	Gould, Z. C., 5h 1868 . .	9.5	9 23 6.19	+ 0.601	71 6 31.5	− 15.54	50.22	1
6 227	Gould, Z. C., 9h 1916 . .	9.0	22.11	1.416	62 58 40.1	.55	50.30	1
6 228	Gould, Z. C., 9h 1887 . .	9.5	25.27	0.412	72 23 31.2	.55	50.21	2
6 229	Gould, 12918	9.5	26.05	0.783	69 45 47.2	.56	50.24	1
6 230	Gould, Z. C., 9h 1905 . .	9.0	26.06	+ 0.875	68 59 9.4	− 15.56	50.24	3
6 231	Stone, 5087	7.0	9 23 28.82	+ 1.320	64 16 51.8	− 15.56	50.29	2
6 232	Chamæleontis	10.0	31.46	− 2.343	81 8 43.6	.56	51.14	1
6 233	Carinæ	9.5	23 31.81	+ 1.009	67 45 21.2	.56	50.25	1
6 234	Carinæ	9.5	24 1.24	1.473	62 14 57.4	.59	51.18	1
6 235	Gould, 12952	9.0	2.93	+ 1.377	63 35 53.8	− 15.59	50.29	1
6 236	Carinæ	9.5	9 24 5.26	+ 1.217	65 36 22.5	− 15.59	50.28	1
6 237	Gould, Z. C., 9h 1975 . .	9.0	7.43	0.924	68 37 2.6	.59	50.25	3
6 238	Carinæ	9.8	11.48	0.788	69 47 27.6	.60	50.24	2
6 239	Carinæ	10.0	19.08	1.449	62 38 3.1	.60	51.18	1
6 240	Gould, Z. C., 9h 2006 . .	8.5	21.24	+ 1.192	65 55 12.4	− 15.61	50.93	3
6 241	Gould, Z. C., 9h 1976 . .	9.8	9 24 25.61	+ 0.187	73 46 32.5	− 15.61	51.23	2
6 242	Carinæ	9.5	33.85	+ 1.375	63 41 17.3	.62	50.29	1
6 243	Chamæleontis	10.0	34.04	− 2.648	81 39 25.1	.62	51.13	1
6 244	Carinæ	9.5	34.58	+ 0.443	72 18 24.2	.62	50.21	1
6 245	Octantis	9.5	43.78	− 3.491	82 43 19.2	− 15.63	50.23	2
6 246	Chamæleontis	10.0	9 24 45.50	− 1.226	78 58 6.1	− 15.63	51.11	1
6 247	Gould, 12976	9.0	52.59	+ 1.239	65 26 38.1	.63	50.28	1
6 248	Carinæ	10.0	52.86	0.428	72 25 30.4	.64	50.21	1
6 249	Gould, Z. C., 9h 2065 . .	9.0	24 59.58	1.410	63 16 8.3	.64	50.30	1
6 250	Carinæ	10.0	25 0.07	+ 0.800	69 46 27.2	− 15.64	50.24	1
6 251	Carinæ	9.5	9 25 3.78	− 1.767	80 11 29.7	− 15.65	51.10	1
6 252	Gould, Z. C., 9h 2044 . .	10.0	4.07	− 0.575	71 28 35.4	.65	50.22	1
6 253	Gould, Z. C., 9h 2082 . .	9.2	8.77	1.461	62 33 33.1	.65	50.75	2
6 254	Gould, Z. C., 9h 2059 . .	10.0	10.47	+ 0.858	69 18 32.0	.65	50.24	1
6 255	Lacaille, 3951	7.0	11.00	− 1.552	79 44 58.9	15.65	51.10	1
6 256	Carinæ	9.0	9 25 20.78	+ 1.069	67 21 {34.7 / 23.9}	− 15.66	50.26	2
6 257	Lacaille, 3941	7.5	26.91	− 0.619	77 15 17.4	.67	51.13	1
6 258	Carinæ	9.5	27.03	+ 1.449	62 46 58.4	.67	50.30	1
6 259	Lacaille, 3931	9.0	30.87	− 0.294	76 4 35.6	.67	50.18	2
6 260	Lacaille, 3909	5.8	34.35	+ 1.194	66 2 50.4	− 15.67	51.26	2
6 261	Lacaille, 3914	7.2	9 25 36.42	+ 0.656	70 57 1.1	15.67	50.27	2
6 262	Gould, Z. C., 9h 2099 . .	9.2	46.90	+ 0.330	73 5 40.1	.68	50.21	2
6 263	Octantis	9.8	48.78	−22.511	88 8 4.7	.69	51.22	2
6 264	Lacaille, 3933	8.8	52.37	− 0.063	75 7 11.8	.69	50.19	2
6 265	Chamæleontis	9.5	25 58.75	+ 1.298	64 51 55.0	− 15.70	50.28	1

No. 6 256. One observation is probably 0.3 of a revolution of the micrometer wrong.

Number.	Constellation, Name of Star, or Synonym.	Magnitude.	Right Ascension, 1850.0.	Annual Precession.	South Declination, 1850.0.	Annual Precession.	Mean year.	No. of obs.
			h. m. s.	s.	° ′ ″	″		
6 266	Carinæ	10.0	9 26 1.20	+ 0.097	74 22 11.6	− 15.70	51.16	1
6 267	Gould, 12961	10.0	1.33	− 2.235	81 5 12.6	.70	51.14	1
6 268	Carinæ	9.0	14.36	+ 1.063	67 31 15.4	.71	50.26	1
6 269	Lacaille, 3947	8.0	19.34	− 0.416	76 35 57.6	.71	50.66	2
6 270	Carinæ	10.0	19.44	+ 1.490	62 17 44.0	− 15.71	51.18	1
6 271	Lacaille, 3922	7.7	9 26 20.00	+ 0.642	71 7 34.3	− 15.71	50.27	2
6 272	Lacaille, 3934	9.0	37.15	0.386	72 50 23.0	.73	50.21	1
6 273	Gould, Z. C., 9ʰ 2186 . .	9.5	42.59	0.664	70 59 53.6	.73	50.23	1
6 274	Gould, Z. C., 9ʰ 2210 . .	9.5	43.10	1.455	62 50 48.8	.74	50.30	1
6 275	Gould, Z. C., 9ʰ 2230 . .	9.0	53.69	+ 1.416	63 25 29.0	− 15.74	50.29	2
6 276	Carinæ	9.0	9 26 54.98	+ 1.055	67 40 50.0	− 15.75	50.26	1
6 277	Carinæ	10.0	27 0.56	1.015	68 4 36.5	.75	50.25	1
6 278	Gould, Z. C., 9ʰ 2217 . .	8.8	7.44	0.561	71 45 56.2	.76	50.26	2
6 279	Gould, 13037	10.0	14.99	1.162	66 35 30.2	.76	50.30	1
6 280	Gould, Z. C., 9ʰ 2246 . .	9.0	16.46	+ 1.167	66 32 38.7	− 15.77	50.28	2
6 281	Lacaille, 3962	8.5	9 27 27.50	− 0.958	78 25 16.3	− 15.78	51.12	1
6 282	Carinæ	9.0	32.32	+ 1.258	65 31 44.2	.78	51.27	2
6 283	Chamæleontis	10.0	33.82	− 1.483	79 43 55.9	.78	51.10	1
6 284	Gould, 13012	9.5	33.90	− 2.177	81 3 56.0	.78	51.14	1
6 285	Gould, Z. C., 9ʰ 2248 . .	9.2	34.90	+ 0.487	72 17 52.8	− 15.78	50.22	2
6 286	Gould, 13045	9.2	9 27 37.18	+ 1.163	66 37 52.8	− 15.78	50.28	2
6 287	Carinæ	9.0	41.20	0.997	68 19 36.3	.79	50.25	2
6 288	Carinæ	9.0	52.12	+ 1.276	65 22 24.0	.80	50.28	1
6 289	Gould, Z. C., 9ʰ 2245 . .	9.5	27 54.56	− 0.381	76 35 2.4	.80	51.15	1
6 290	Gould, 13056	9.5	28 6.71	+ 1.162	66 41 29.0	− 15.81	50.28	2
6 291	Gould, Z. C., 9ʰ 2294 . .	9.0	9 28 11.04	+ 0.574	71 46 41.8	− 15.81	50.26	2
6 292	Gould, Z. C., 9ʰ 2331 . .	8.0	11.69	1.417	63 34 34.2	.82	50.29	1
6 293	Gould, Z. C., 9ʰ 2319 . .	9.0	13.90	+ 1.085	67 31 12.9	.82	50.25	1
6 294	Gould, 13046	9.0	22.20	− 0.495	77 1 45.2	.82	50.65	2
6 295	Lacaille, 4009	7.7	23.35	− 4.719	83 58 51.9	− 15.83	50.55	3
6 296	Gould, Z. C., 9ʰ 2323 . .	9.0	9 28 27.68	+ 0.601	71 37 19.1	− 15.83	50.22	1
6 297	Lacaille, 3940	7.0	31.80	+ 1.224	66 3 22.1	.83	51.26	2
6 298	Chamæleontis	9.8	32.78	− 0.095	75 28 16.8	.83	50.67	2
6 299	Chamæleontis	9.8	33.22	− 1.543	79 55 2.6	.83	51.10	2
6 300	Gould, Z. C., 9ʰ 2365 . .	9.5	46.09	+ 1.058	67 49 54.1	− 15.84	50.25	1
6 301	Gould, Z. C., 9ʰ 2380 . .	9.0	9 28 55.20	+ 1.077	67 40 50.1	− 15.85	50.25	1
6 302	Chamæleontis	6.5	28 57.24	− 1.637	80 8 11.2	.86	51.10	1
6 303	Lacaille, 3957	8.5	29 1.57	+ 0.397	72 59 16.9	.86	50.21	1
6 304	Gould, 13085	9.0	5.73	1.331	64 49 46.3	.86	50.28	1
6 305	Gould, 13083	9.8	5.92	+ 1.176	66 39 49.2	− 15.86	50.28	2

No. 6 281. Declination has been increased 10″. Microscope probably read 10″ wrong.

Number.	Constellation, Name of Star, or Synonym.	Magnitude.	Right Ascension, 1850.0.	Annual Precession.	South Declination, 1850.0.	Annual Precession.	Mean year.	No. of obs.
			h. m. s.	s.	° ′ ″	″.		
6 306	Gould, Z. C., 9^h 2385 . .	9.2	9 29 8.47	+ 0.708	70 55 0.7	15.87	51.25	2
6 307	Gould, 13084	10.0	9.80	+ 1.177	66 39 1.8	.87	50.26	1
6 308	Chamæleontis	10.0	11.03	− 0.476	77 1 17.4	.87	51.13	1
6 309	Lacaille, 3944	8.5	12.44	+ 1.341	64 42 47.8	.87	50.28	2
6 310	Gould, Z. C., 9^h 2393 . .	8.8	15.19	+ 0.684	71 6 21.5	− 15.87	50.91	3
6 311	Carinæ	9.5	9 29 18.69	+ 0.128	74 28 57.4	− 15.88	51.16	1
6 312	Chamæleontis	9.8	25.27	− 0.166	75 50 38.6	.88	50.67	2
6 313	Chamæleontis	10.0	29.27	1.434	79 43 50.0	.88	51.10	1
6 314	Gould, 13063	9.0	32.19	1.784	80 27 33.9	.89	50.15	1
6 315	Chamæleontis	10.0	32.62	0.179	75 54 28.6	− 15.89	51.15	1
6 316	Brisbane, 2580	8.0	9 29 34.30	− 4.693	84 0 4.2	15.89	50.22	2
6 317	Gould, Z. C., 9^h 2422 . .	9.0	38.79	+ 0.917	69 16 0.3	.89	50.24	1
6 318	Lacaille, 3954	8.0	41.00	0.855	69 48 17.2	.89	50.24	2
6 319	Carinæ	9.5	41.68	+ 1.271	65 38 40.8	.90	51.27	2
6 320	Chamæleontis	9.5	51.89	− 1.579	80 3 54.8	− 15.90	51.10	1
6 321	Gould, 13104	9.0	9 29 59.68	+ 1.488	62 48 21.3	− 15.91	50.30	1
6 322	Gould, 13105	8.8	59.68	1.504	62 33 52.9	.91	50.75	2
6 323	Brisbane, 2578	10.0	59.69	1.524	62 15 35.6	.91	51.18	1
6 324	Carinæ	9.0	29 59.97	1.120	67 22 6.8	.91	50.26	1
6 325	Gould, 13100	10.0	30 0.38	+ 1.156	66 58 34.9	− 15.91	50.30	1
6 326	Carinæ	9.2	9 30 7.09	+ 0.750	70 41 22.1	− 15.92	51.26	2
6 327	Gould, Z. C., 9^h 2411 . .	9.2	7.72	− 0.626	77 35 43.3	.92	50.65	2
6 328	Chamæleontis	9.2	13.98	− 0.625	77 35 55.0	.92	50.65	2
6 329	Gould, Z. C., 9^h 2476 . .	9.0	17.15	+ 1.071	67 53 57.8	.93	50.25	1
6 330	Carinæ	8.8	18.63	+ 0.331	73 29 24.4	− 15.93	50.21	2
6 331	Lacaille, 4027	7.0	9 30 23.32	− 4.671	84 0 42.6	− 15.93	50.22	2
6 332	Gould, Z. C., 9^h 2471 . .	8.8	25.81	+ 0.602	71 48 9.6	.93	50.26	2
6 333	// Carinæ	5.0	26.70	0.510	72 24 55.4	.94	50.21	2
6 334	Gould, Z. C., 9^h 2479 . .	9.2	27.77	0.747	70 44 40.4	.94	51.26	1
6 335	Gould, Z. C., 9^h 2474 . .	10.0	28.57	+ 0.579	71 57 48.3	− 15.94	50.21	1
6 336	Carinæ	8.8	9 30 30.40	+ 0.222	74 6 19.3	− 15.94	51.23	2
6 337	Lacaille, 3970	7.7	32.04	0.404	73 4 58.8	.94	50.22	2
6 338	Carinæ	9.8	35.67	0.122	74 37 17.0	.94	51.18	2
6 339	Carinæ	9.0	41.64	0.336	73 29 30.1	.95	50.21	2
6 340	Gould, Z. C., 9^h 2512 . .	9.3	53.97	+ 0.955	69 3 55.2	− 15.96	50.24	3
6 341	Gould, Z. C., 9^h 2523 . .	9.2	9 30 56.21	+ 0.970	68 55 47.2	− 15.96	50.24	2
6 342	Brisbane, 2575	8.0	30 57.19	1.458	63 21 48.5	.96	50.29	2
6 343	Carinæ	10.0	31 5.57	1.506	62 40 26.9	.97	51.18	1
6 344	Gould, Z. C., 9^h 2533 . .	9.2	7.48	0.963	69 1 22.4	.97	50.24	3
6 345	Lacaille, 3963	8.3	16.00	+ 1.303	65 27 11.5	− 15.98	50.94	3

No. 6 314. Gould's declination is 13.3″ greater. No. 6 326. Mean of two declinations differing 9.6″.
Nos. 6 327 and 6 328 are Herschel, 4226.

Number.	Constellation, Name of Star, or Synonym.	Magnitude.	Right Ascension, 1850.0.	Annual Precession.	South Declination, 1850.0.	Annual Precession.	Mean year.	No. of obs.
			h. m. s.	s.	° ′ ″	″		
6346	Chamæleontis	10.0	9 31 23.08	− 1.518	80 2 41.4	− 15.99	51.10	1
6347	Gould, Z. C., 9h 2542 . .	9.0	26.58	+ 0.586	72 0 28.2	.99	50.22	1
6348	Lacaille, 3965	9.0	29.41	1.393	64 19 49.2	.99	50.29	2
6349	Gould, Z. C., 9h 2536 . .	10.0	31.93	0.051	75 2 50.1	.99	50.19	1
6350	Gould, Z. C., 9h 2557 . .	9.0	32.76	+ 0.665	71 28 16.1	− 15.99	50.25	3
6351	Chamæleontis	10.0	9 31 51.62	+ 1.438	63 46 3.2	− 16.01	50.29	1
6352	Gould, Z. C., 9h 2546 . .	9.5	31 56.31	− 0.397	76 56 4.0	.01	50.17	1
6353	Carinæ	9.5	32 8.06	+ 1.018	68 37 30.0	.02	50.25	1
6354	Carinæ	9.8	15.96	0.833	70 15 42.8	.03	50.23	2
6355	Carinæ	9.5	19.65	+ 1.544	62 17 35.7	− 16.03	51.18	1
6356	Lacaille, 3972	8.5	9 32 19.87	+ 1.430	63 52 58.0	− 16.04	50.29	2
6357	Gould, Z. C., 9h 2639 . .	10.0	26.03	1.481	63 13 45.2	.04	50.30	1
6358	Carinæ	10.0	29.12	1.545	62 16 31.2	.04	51.18	1
6359	Gould, Z. C., 9h 2622 . .	10.0	34.86	0.479	72 48 53.7	.05	50.21	1
6360	Lacaille, 3973	7.5	35.84	+ 1.407	64 16 44.4	− 16.05	50.29	2
6361	Octantis	9.5	9 32 42.96	−16.328	87 38 22.3	− 16.06	51.20	6
6362	Lacaille, 3977	8.2	53.95	+ 1.137	67 32 19.8	.06	50.25	2
6363	Gould, Z. C., 9h 2642 . .	9.0	54.20	0.481	72 50 4.4	.07	50.21	1
6364	Carinæ	9.2	32 54.23	1.222	66 36 23.9	.07	50.28	2
6365	Gould, Z. C., 9h 2661 . .	9.8	33 13.47	+ 0.762	70 55 4.9	− 16.08	51.25	2
6366	Lacaille, 4013	8.0	9 33 16.89	− 1.993	81 2 18.5	− 16.09	51.14	2
6367	Carinæ	10.0	19.11	+ 1.300	65 44 29.6	.09	52.27	1
6368	Gould, Z. C., 9h 2631 . .	10.0	24.07	− 1.150	79 17 39.0	.09	50.16	1
6369	Lacaille, 4019	8.2	36.48	− 2.063	81 10 34.2	.10	51.14	2
6370	Carinæ	8.5	40.94	+ 1.304	65 44 9.6	− 16.11	51.27	2
6371	Chamæleontis	10.0	9 33 58.06	− 0.007	75 30 44.9	− 16.12	51.15	1
6372	Carinæ	10.0	34 10.18	+ 0.990	69 6 42.4	.13	50.24	1
6373	Gould, Z. C., 9h 2750 . .	10.0	15.52	+ 1.174	67 18 2.4	.14	50.26	2
6374	Octantis	9.8	17.97	−11.513	86 53 3.0	.14	51.20	2
6375	Lacaille, 4017	8.0	22.75	− 1.697	80 33 16.6	− 16.14	50.15	1
6376	Gould, Z. C., 9h 2678 . .	9.0	9 34 22.89	− 1.301	79 42 49.6	− 16.14	51.11	1
6377	Carinæ	9.5	23.88	+ 1.537	62 39 25.0	.14	51.18	1
6378	Carinæ	9.0	25.00	1.290	66 0 5.7	.14	51.26	2
6379	Brisbane, 2600	9.0	29.48	1.508	63 6 36.2	.15	50.30	2
6380	Gould, Z. C., 9h 2762 . .	9.5	30.43	+ 0.831	70 30 56.8	− 16.15	50.23	1
6381	Carinæ	9.0	9 34 40.24	+ 1.311	65 46 38.2	− 16.16	51.60	3
6382	Lacaille, 3986	7.0	42.65	1.468	63 43 26.7	.16	50.29	1
6383	Gould, 13210	9.0	49.75	1.489	63 26 2.4	.17	50.30	1
6384	Gould, Z. C., 9h 2764 . .	8.8	50.00	+ 0.119	74 59 54.3	.17	50.19	2
6385	Chamæleontis	9.5	50 61	− 0.717	78 12 9.0	− 16.17	50.30	1

Number.	Constellation, Name of Star, or Synonym.	Magnitude.	Right Ascension, 1850.0.	Annual Precession.	South Declination, 1850.0.	Annual Precession.	Mean year.	No. of obs.
			h. m. s.	s.	° ′ ″	″		
6 386	Carinæ	9.5	9 34 58.57	+ 1.556	62 27 35.2	− 16.17	51.18	1
6 387	Carinæ	9.5	35 7.20	1.553	62 31 21.9	.18	51.18	1
6 388	Lacaille, 3989	7.5	10.22	1.466	63 48 41.0	.18	50.29	2
6 389	Gould, Z. C., 9ʰ 2808 . .	9.5	10.65	0.777	71 0 46.7	.18	51.25	2
6 390	Carinæ	10.0	14.38	+ 1.297	66 1 13.6	− 16.19	51.26	2
6 391	Carinæ	9.5	9 35 14.87	+ 1.557	62 28 11.3	− 16.19	51.18	1
6 392	Carinæ	10.0	20.99	1.431	64 18 52.6	.19	50.29	1
6 393	Gould, Z. C., 9ʰ 2804 . .	9.8	22.02	0.135	74 57 54.7	.19	50.19	2
6 394	Carinæ	9.0	28.31	1.074	68 27 43.3	.20	50.25	1
6 395	Carinæ	9.5	33.90	+ 1.303	65 59 25.1	− 16.20	51.26	2
6 396	Stone, 5241	8.2	9 35 34.82	+ 1.287	66 10 54.6	− 16.20	50.93	3
6 397	Carinæ	9.5	34.92	0.208	74 36 58.8	.20	51.16	1
6 398	Gould, Z. C., 9ʰ 2856 . .	8.8	35.32	1.553	62 34 36.3	.20	50.75	2
6 399	Lacaille, 3993	7.5	35 49.66	1.576	62 15 51.3	.21	50.18	1
6 400	Gould, Z. C., 9ʰ 2855 . .	9.2	36 4.07	+ 0.312	74 6 40.8	− 16.23	51.23	2
6 401	Gould, Z. C., 9ʰ 2879 . .	9.5	9 36 5.86	+ 1.135	67 55 33.4	− 16.23	50.25	2
6 402	Gould, Z. C., 9ʰ 2881 . .	8.8	7.35	+ 1.231	66 53 37.5	.23	50.28	2
6 403	Lacaille, 4041	8.0	7.60	− 2.243	81 35 39.5	.23	51.13	1
6 404	Carinæ	9.5	9.49	+ 1.290	66 13 31.2	.23	50.26	1
6 405	Carinæ	10.0	20.16	+ 1.292	66 13 3.7	− 16.24	50.26	1
6 406	Gould, Z. C., 9ʰ 2852 . .	10.0	9 36 20.73	− 0.517	77 40 25.1	− 16.24	50.17	1
6 407	Gould, 13239	9.5	27.47	+ 1.579	62 19 27.7	.25	51.18	1
6 408	Chamæleontis	10.0	34.66	− 3.167	82 50 41.2	.26	50.33	1
6 409	Lacaille, 4042	8.0	54.97	− 1.874	81 0 11.1	.27	51.14	1
6 410	Gould, Z. C., 9ʰ 2904 . .	9.5	36 55.60	+ 0.325	74 7 8.9	− 16.27	50.20	1
6 411	Gould, 13250	9.5	9 37 0.65	+ 1.440	64 25 2.7	− 16.28	50.29	1
6 412	Gould, Z. C., 6ʰ 2932 . .	9.5	8.24	0.669	72 0 14.4	.28	50.22	1
6 413	Gould, 13256	9.8	15.60	1.369	65 23 8.0	.29	51.27	2
6 414	Gould, Z. C., 9ʰ 2944 . .	9.5	15.92	+ 0.669	72 1 13.0	.29	50.22	1
6 415	Chamæleontis	10.0	17.76	− 2.652	82 14 32.7	− 16.29	51.17	1
6 416	Octantis	10.0	9 37 16.85	−10.836	86 47 28.6	− 16.29	51.22	2
6 417	Chamæleontis	9.5	24.49	− 1.814	80 56 14.4	.30	51.14	1
6 418	Carinæ	9.5	29.10	+ 1.548	62 55 32.3	.30	50.30	1
6 419	Chamæleontis	9.0	35.97	− 1.598	80 32 12.5	.31	50.15	1
6 420	Gould, Z. C., 9ʰ 2917 . .	10.0	38.94	− 1.005	79 11 43.1	− 16.31	50.16	1
6 421	Chamæleontis	9.5	9 37 39.05	− 0.891	78 53 23.9	− 16.31	51.11	1
6 422	Lacaille, 4005	7.8	43.18	+ 1.373	65 23 55.6	.31	50.94	3
6 423	Gould, 13261	9.3	48.38	+ 1.336	65 52 56.4	.32	51.60	3
6 424	Chamæleontis	9.8	50.17	− 1.294	79 54 23.8	.32	51.10	2
6 425	Gould, Z. C., 9ʰ 2951 . .	9.0	37 50.64	− 0.483	77 40 3.2	− 16.32	50.17	1

No. 6 416. Mean of two observations differing 5ˢ.

Number.	Constellation, Name of Star, or Synonym.	Magnitude.	Right Ascension, 1850.0.	Annual Precession.	South Declination, 1850.0.	Annual Precession.	Mean year.	No. of obs.
			h. m. s.	s.	° ′ ″	″		
6426	ζ Chamæleontis	5.5	9 38 5.94	− 1.452	80 15 57.0	− 16.33	50.62	2
6427	Gould, 13271	9.5	7.51	+ 1.598	62 14 13.4	.33	51.18	1
6428	Chamæleontis	8.0	8.70	− 2.270	81 44 11.3	.34	50.61	3
6429	Carinæ	9.5	9.82	+ 0.986	69 36 20.7	.34	50.24	1
6430	Carinæ	10.0	10.77	− 1.120	79 30 59.7	− 16.34	51.10	1
6431	Gould, 13269	9.0	9 38 11.56	+ 1.326	66 3 6.3	− 16.34	51.26	2
6432	Gould, Z. C., 9h 3010	9.8	13.26	1.059	68 56 3.0	.34	50.24	2
6433	Gould, Z. C., 9h 3023	9.5	34.04	0.809	71 7 16.8	.36	51.25	2
6434	Gould, Z. C., 9h 3045	9.0	38.76	1.227	67 15 17.1	.36	50.26	2
6435	Lacaille, 4020	8.0	43.24	+ 1.176	67 49 11.8	− 16.36	50.25	2
6436	Carinæ	10.0	9 38 43.56	+ 0.952	69 57 47.3	− 16.37	50.24	1
6437	Carinæ	9.5	44.00	1.340	65 56 48.6	.37	52.25	1
6438	Lacaille, 4018	7.5	38 57.00	1.319	66 13 45.4	.38	50.93	3
6439	Carinæ	9.5	39 7.35	1.432	64 49 22.2	.39	50.28	1
6440	Carinæ	9.5	28.98	+ 1.078	68 53 51.6	− 16.40	50.24	1
6441	Gould, Z. C., 9h 3133	9.5	9 39 38.65	+ 1.550	63 11 57.7	− 16.41	50.30	1
6442	Gould, Z. C., 9h 3123	8.5	39.17	+ 1.193	67 45 5.3	.41	50.25	1
6443	Chamæleontis	10.0	42.02	− 0.294	77 8 44.7	.41	51.13	1
6444	Chamæleontis	10.0	42.96	+ 0.092	75 32 29.3	.41	51.15	1
6445	Gould, Z. C., 9h 3064	8.8	44.91	− 0.939	79 9 28.5	− 16.42	50.64	2
6446	Gould, Z. C., 9h 3121	9.8	9 39 47.69	+ 0.777	71 30 4.3	− 16.42	50.25	3
6447	Chamæleontis	10.0	40 0.68	− 1.085	79 32 48.9	.43	51.10	1
6448	Chamæleontis	10.0	14.61	+ 0.111	75 29 50.6	.44	51.15	1
6449	Gould, Z. C., 9h 3174	9.5	14.63	1.112	68 39 22.0	.44	50.25	1
6450	Gould, Z. C., 9h 3165	9.5	21.27	+ 0.776	71 34 2.6	− 16.45	50.22	1
6451	Gould, Z. C., 9h 3159	9.0	9 40 22.26	+ 0.588	72 52 48.5	− 16.45	50.21	1
6452	Gould, 13320	10.0	30.48	1.534	63 34 12.3	.45	50.29	1
6453	Carinæ	10.0	39.10	1.612	62 22 47.4	.46	51.18	1
6454	Gould, Z. C., 9h 3213	9.5	39.28	1.329	66 20 5.3	.46	50.26	1
6455	Carinæ	9.5	43.17	+ 1.151	68 19 27.6	− 16.47	50.25	1
6456	Lacaille, 4040	7.2	9 40 45.64	+ 0.790	71 30 9.6	− 16.47	50.24	3
6457	Gould, 13327	9.5	40 58.57	1.084	69 1 29.5	.48	50.24	1
6458	Gould, Z. C., 9h 3228	9.0	41 0.29	1.195	67 53 52.8	.48	50.25	2
6459	Chamæleontis	10.0	3.74	+ 0.028	75 56 47.3	.48	51.15	1
6460	Gould, Z. C., 9h 3167	9.0	9.00	− 1.104	79 39 58.3	− 16.49	51.10	1
6461	Carinæ	8.0	9 41 10.03	+ 1.514	63 57 43.2	− 16.49	50.29	1
6462	Carinæ	9.0	12.67	1.503	64 7 15.0	.49	50.29	1
6463	Carinæ	9.2	26.42	0.894	70 45 54.4	.50	51.25	2
6464	Gould, 13340	9.2	30.40	1.082	69 6 7.8	.50	50.24	2
6465	Lacaille, 4044	10.0	31.96	+ 1.082	69 6 16.0	− 16.51	50.24	2

No. 6426. One of the declinations has been increased 10″. No. 6457. Probably variable.

Number.	Constellation, Name of Star, or Synonym.	Magnitude.	Right Ascension, 1850.0.	Annual Precession.	South Declination, 1850.0.	Annual Precession.	Mean year.	No. of obs.
			h. m. s.	s.	° ′ ″	″		
6466	Gould, Z. C., 9h 3286 . .	10.0	9 41 35.63	+ 1.621	62 22 17.8	— 16.51	51.18	1
6467	Gould, 13353	8.8	37.88	+ 1.529	63 48 28.3	.51	50.29	2
6468	Octantis	9.7	44.32	— 7.813	86 0 47.7	.52	51.21	3
6469	Lacaille, 4043	8.5	52.44	+ 1.362	66 5 59.9	.52	51.26	2
6470	Gould, Z. C., 9h 3235 . .	8.8	41 55.78	— 0.850	79 3 42.6	— 16.53	50.64	2
6471	Carinæ	9.0	9 41 59.32	+ 1.416	65 25 25.9	— 16.53	50.94	3
6472	Carinæ	9.5	42 0.12	1.625	62 22 40.4	.53	51.18	1
6473	Carinæ	10.0	13.64	+ 1.410	65 32 17.0	.54	50.28	1
6474	Brisbane, 2685	8.2	14.07	— 1.737	81 3 48.2	.54	50.64	2
6475	Chamæleontis	9.0	14.73	+ 0.202	75 14 10.8	— 16.54	50.19	1
6476	Gould, Z. C., 9h 3250 . .	10.0	9 42 19.20	— 1.089	79 42 7.0	— 16.54	51.10	1
6477	Gould, Z. C., 9h 3309 . .	9.5	20.64	+ 0.658	72 36 53.8	.55	50.21	2
6478	Carinæ	9.5	21.95	0.990	70 3 0.6	.55	50.23	1
6479	Lacaille, 4050	8.2	22.82	1.096	69 4 15.5	.55	50.24	2
6480	Carinæ	10.0	36.98	+ 1.632	62 20 36.3	— 16.56	51.18	1
6481	Gould, 13366	9.0	9 42 38.78	+ 0.468	73 50 46.8	— 16.56	50.89	3
6482	Gould, Z. C., 9h 3291 . .	8.5	41.87	— 0.804	78 59 29.5	.56	51.11	1
6483	Chamæleontis	10.0	42.80	+ 0.159	75 29 22.7	.56	51.15	1
6484	Gould, Z. C., 9h 3300 . .	9.5	54.53	— 1.073	79 42 5.5	.57	51.10	1
6485	Gould, 13384	8.5	42 56.87	+ 1.559	63 32 48.9	— 16.58	50.29	1
6486	Gould, Z. C., 9h 3351 . .	9.2	9 43 2.14	+ 0.698	72 24 51.8	— 16.58	50.21	2
6487	Carinæ	9.5	2.61	1.394	65 51 21.1	.58	51.26	3
6488	Lacaille, 4054	8.0	2.99	1.068	69 25 31.8	.58	50.24	2
6489	Gould, Z. C., 9h 3377 . .	9.0	8.18	1.303	66 57 58.6	.59	50.28	2
6490	Gould, Z. C., 9h 3357 . .	9.0	10.97	+ 0.581	73 12 30.7	— 16.59	50.21	1
6491	Lacaille, 4080	6.2	9 43 17.50	— 1.685	81 1 25.3	— 16.59	50.64	2
6492	υ Argûs	3.0	20.88	+ 1.506	64 22 39.5	.60	50.29	1
6493	Carinæ	9.5	21.54	+ 1.506	64 22 43.7	.60	50.29	1
6494	Octantis	8.3	22.94	— 3.890	83 50 44.9	.60	50.27	3
6495	Gould, 13393	8.5	27.41	+ 1.605	62 54 25.4	— 16.60	50.30	1
6496	Chamæleontis	9.5	9 43 33.56	— 0.552	78 18 35.1	— 16.61	50.30	1
6497	Chamæleontis	10.0	35.62	— 0.005	76 18 28.0	.61	51.15	1
6498	Carinæ	10.0	38.48	+ 1.172	68 28 36.5	.61	50.25	1
6499	Gould, Z. C., 9h 3394 . .	9.0	46.07	+ 0.433	74 9 21.3	.62	51.23	2
6500	Chamæleontis	11.0	47.89	— 0.107	76 44 54.6	— 16.62	51.15	1
6501	Lacaille, 4064	9.0	9 43 49.62	+ 0.053	76 4 30.6	— 16.62	50.18	2
6502	Gould, 13395	8.8	50.79	0.887	71 5 34.3	.62	51.25	3
6503	Gould, 13392	9.2	43 53.31	0.484	73 52 17.4	.62	50.89	3
6504	Gould, 13406	11.0	44 2.67	1.510	64 25 27.5	.63	50.29	1
6505	Gould, 13407	11.0	4.16	+ 1.510	64 25 34.0	16.63	50.29	1

Number.	Constellation, Name of Star, or Synonym.	Magnitude.	Right Ascension, 1850.0.	Annual Precession.	South Declination, 1850.0.	Annual Precession.	Mean year.	No. of obs.
			h. m. s.	s.	° ′ ″	″		
6 506	Gould, Z. C., 9^h 3451 . .	9.5	9 44 4.73	+ 1.144	68 48 28.0	— 16.63	50.24	1
6 507	Gould, Z. C., 9^h 3463 . .	9.0	9.88	+ 1.262	67 34 38.0	.64	50.25	2
6 508	Octantis	10.0	17.32	—42 152	89 1 57.3	.64	51.33	1
6 509	Brisbane, 2697	8.0	18.87	+ 1.649	62 19 24.9	.64	51.18	1
6 510	Gould, Z. C., 9^h 3442 . .	9.5	21.16	+ 0.247	75 12 21.5	— 16.64	50.19	1
6 511	Gould, Z. C., 9^h 3483 . .	9.0	9 44 24.32	+ 1.284	67 21 29.4	— 16.65	50.26	2
6 512	Gould, Z. C., 9^h 3498 . .	10.0	26.75	1.639	62 30 48.3	.65	51.18	1
6 513	Stone, 5323	8.8	27.10	1.384	66 9 56.1	.65	51.26	2
6 514	Gould, Z. C., 9^h 3502 . .	9.0	29.78	1.591	63 17 18.2	.65	50.30	1
6 515	Gould, Z. C., 9^h 3484 . .	10.0	31.39	+ 1.033	69 55 23.3	— 16.65	50.23	1
6 516	Gould, Z. C., 9^h 3492 . .	9.5	9 44 35.38	+ 1.144	68 52 37.0	— 16.66	50.24	1
6 517	Carinæ	10.0	35.93	1.022	70 1 40.7	.66	50.23	1
6 518	Brisbane, 2700	8.5	41.57	0.572	73 25 13.5	.66	50.21	2
6 519	Gould, Z. C., 9^h 3490 . .	9.0	41.72	0.876	71 19 22.4	.66	50.22	1
6 520	Gould, Z. C., 9^h 3513 . .	9.5	48.19	+ 1.223	68 5 19.6	— 16.67	50.25	1
6 521	Brisbane, 2697	9.5	9 44 55.25	+ 1.655	62 19 23.6	— 16.67	50.32	1
6 522	Gould, Z. C., 9^h 3516 . .	9.0	45 4.25	+ 0.661	72 53 0.9	.68	50.21	1
6 523	Octantis	9.5	9.05	— 3.989	83 59 59.3	.68	50.33	1
6 524	Brisbane, 2701	8.8	9.11	+ 1.649	62 27 13.0	.68	50.75	2
6 525	Chamæleontis	10.0	11.02	— 2.388	82 15 39.7	— 16.69	51.17	1
6 526	Lacaille, 4083	7.5	9 45 11.28	— 0.710	78 53 59.3	- 16.69	51.11	1
6 527	Carinæ	9.5	11.66	+ 0.423	74 21 16.4	.69	50.68	2
6 528	Carinæ	9.5	15.97	1.115	69 14 32.8	.69	50.24	1
6 529	Carinæ	9.5	18.87	1.115	69 15 5.7	.69	50.24	1
6 530	Carinæ	9.0	23.63	+ 1.570	63 44 16.9	— 16.70	50.29	1
6 531	Gould, Z. C., 9^h 3569 . .	9.5	9 45 26.84	+ 1.642	62 36 23.8	— 16.70	51.18	1
6 532	Octantis	9.8	40.71	—10.205	86 48 27.5	.71	51.22	6
6 533	Gould, Z. C., 9^h 3573 . .	8.8	47.52	+ 1.015	70 13 14.6	.71	50.23	2
6 534	Carinæ	9.5	45 53.75	1.455	65 26 49.4	.72	51.61	3
6 535	Gould, Z. C., 9^h 3609 . .	10.0	46 12.56	+ 1.039	70 4 28.3	— 16.73	50.23	1
6 536	ν Chamæleontis	6.0	9 46 13.14	+ 0.099	76 4 37.6	— 16.74	50.18	2
6 537	Chamæleontis	10.0	27.71	— 1.183	80 11 22.0	.75	51.10	1
6 538	Gould, 13472	8.8	28.55	+ 1.117	69 22 28.3	.75	50.24	1
6 539	Gould, Z. C., 9^h 3657 . .	8.8	39.67	1.466	65 24 42.9	.76	50.94	3
6 540	Gould, Z. C., 9^h 3656 . .	9.0	41.65	+ 1.419	66 2 23.0	— 16.76	51.26	2
6 541	Carinæ	10.0	9 46 45.33	+ 1.486	65 9 0.4	— 16.76	50.28	1
6 542	Lacaille, 4071	7.7	46.93	1.280	67 43 6.9	.76	50.25	2
6 543	Carinæ	9.5	48.53	0.458	74 18 42.2	.76	50.68	2
6 544	Gould, Z. C., 9^h 3693 . .	9.0	46 58.23	1.665	62 28 30.3	.77	51.18	1
6 545	Gould, 13472	9.0	47 1.79	+ 1.518	64 44 39.9	— 16.77	50.28	1

No. 6 506. Recorded wire 4. Reduced as wire 2.

No. 6 541. The right ascension may be 10^s too large.

Number.	Constellation, Name of Star, or Synonym.	Magnitude.	Right Ascension, 1850.0.	Annual Precession.	South Declination, 1850.0.	Annual Precession.	Mean year.	No. of obs.
			h. m. s.	s.	° ′ ″	″		
6546	Gould, Z. C., 9^h 3700 . .	8.5	9 47 4.08	+ 1.599	63 32 56.0	− 16.78	50.29	1
6547	Gould, 13471	9.0	4.15	+ 1.392	66 26 17.3	.78	50.26	1
6548	Chamæleontis	10.0	18.59	− 2.339	82 18 0.7	.79	51.17	1
6549	Gould, Z. C., 9^h 3717 . .	10.0	21.80	+ 1.678	62 18 48.0	.79	51.18	1
6550	Gould, Z. C., 9^h 3697 . .	9.5	26.22	+ 0.651	73 11 50.5	− 16.79	50.21	1
6551	Gould, Z. C., 9^h 3735 . .	10.0	9 47 32.16	+ 1.681	62 17 14.3	− 16.80	51.18	1
6552	Chamæleontis	10.8	45.76	− 0.988	79 48 50.6	.81	51.10	2
6553	Lacaille, 4086	7.7	45.76	+ 0.334	75 4 50.1	.81	50.25	3
6554	Chamæleontis	9.5	47 48.92	− 0.985	79 48 41.8	.81	51.10	2
6555	Gould, Z. C., 9^h 3732 . .	9.0	48 5.78	+ 0.266	75 27 39.5	− 16.83	51.15	1
6556	Gould, 13505	9.0	9 48 19.04	+ 1.433	66 5 20.2	− 16.84	51.26	2
6557	Chamæleontis	8.8	38.34	− 0.300	77 51 53.3	.85	50.24	2
6558	Octantis	9.4	38.62	− 8.758	86 29 33.6	.85	51.23	7
6559	Carinæ	10.0	39.69	+ 1.225	68 33 47.8	.85	50.25	1
6560	Gould, Z. C., 9^h 3786 . .	8.5	41.62	+ 0.625	73 30 3.4	− 16.85	50.55	3
6561	Gould, 13516	9.0	9 48 44.62	+ 1.412	66 25 19.6	− 16.86	50.26	1
6562	Octantis	8.8	46.36	−14.350	87 37 8.2	.86	51.22	6
6563	Gould, Z. C., 9^h 3798 . .	9.2	47.42	+ 0.914	71 26 27.1	.86	50.22	2
6564	Octantis	8.9	48.22	−22.803	88 23 47.7	.86	51.24	9
6565	Carinæ	9.5	51.54	+ 1.663	62 48 20.0	− 16.86	50.30	1
6566	Chamæleontis	10.0	9 48 53.69	− 0.196	77 30 30.6	− 16.86	51.13	1
6567	Lacaille, 4125	8.3	54.81	− 2.139	82 5 34.5	.86	50.63	3
6568	Gould, Z. C., 9^h 3824 . .	9.0	. 48 55.89	+ 1.376	66 53 22.1	.86	50.30	1
6569	Carinæ	9.5	49 3.58	1.180	69 5 0.2	.87	50.24	2
6570	Gould, 13509	8.7	5.57	+ 0.104	76 18 39.9	− 16.87	50.50	3
6571	Chamæleontis	8.5	9 49 22.08	− 0.346	78 4 47.4	− 16.89	50.30	1
6572	Gould, Z. C., 9^h 3863 . .	10.0	28.10	+ 1.320	67 37 55.8	.89	50.25	2
6573	Gould, Z. C., 9^h 3865 . .	9.5	28.58	1.314	67 41 44.1	.89	50.25	2
6574	Gould, 13536	9.5	32.22	1.627	63 29 5.4	.89	50.29	2
6575	Carinæ	10.0	39.98	+ 1.241	68 31 32.3	16.90	50.25	1
6576	Gould, Z. C., 9^h 3883 . .	9.8	9 49 42.06	+ 1.372	67 2 33.0	− 16.90	50.28	2
6577	Gould, Z. C., 9^h 3906 . .	9.2	52.19	+ 1.578	64 17 49.3	.91	50.29	2
6578	Lacaille, 4103	8.0	53.64	− 0.192	77 34 41.1	.91	50.65	2
6579	Carinæ	10.5	55.38	+ 1.680	62 41 36.7	.91	51.18	1
6580	Carinæ	12.0	49 56.63	+ 1.680	62 41 33.7	− 16.91	51.18	1
6581	Gould, Z. C., 9^h 3875 . .	10.0	9 50 3.64	+ 0.144	76 13 2.3	− 16.92	51.15	1
6582	Lacaille, 4097	8.5	6.40	0.958	71 14 13.4	.92	50.22	1
6583	Lacaille, 4096	8.2	16.38	1.019	70 44 45.2	.93	51.25	2
6584	Gould, Z. C., 9^h 3941 . .	9.2	20.45	+ 1.681	62 44 28.9	.93	50.60	3
6585	Octantis	10.0	21.54	−11.872	87 15 16.3	− 16.93	51.18	1

Number.	Constellation, Name of Star, or Synonym.	Magnitude.	Right Ascension, 1850.0.	Annual Precession.	South Declination, 1850.0.	Annual Precession.	Mean year.	No. of obs.
			h. m. s.	s.	° ′ ″	″		
6586	Melbourne (1), 495 . . .	9.0	9 50 21.72	+ 1.415	66 36 48.7	− 16.93	50.26	1
6587	Gould, 13480	8.5	27.89	− 5.675	85 19 11.1	.94	51.26	7
6588	Carinæ	10.0	32.61	+ 0.755	72 50 . .	.94	50.21	1
6589	Gould, 13570	9.8	39.82	+ 1.695	62 33 17.7	.95	50.75	2
6590	Lacaille, 4169	8.3	41.34	− 5.772	85 19 10.7	− 16.95	51.26	7
6591	Lacaille, 4122	8.5	9 50 45.06	− 1.136	80 21 37.9	− 16.95	50.57	2
6592	Gould, Z. C., 9^h 3945 . .	9.0	48.88	+ 0.679	73 22 26.2	.95	50.21	1
6593	Lacaille, 4099	9.0	51.21	1.193	69 11 11.9	.96	50.24	1
6594	Gould, Z. C., 9^h 3974 . .	9.2	53.41	1.389	67 0 40.3	.96	50.28	2
6595	Gould, Z. C., 9^h 3935 . .	9.0	50 55.62	+ 0.072	76 36 59.0	− 16.96	50.66	2
6596	Gould, Z. C., 9^h 3992 . .	9.0	9 51 2.12	+ 1.585	64 21 50.2	− 16.96	50.29	2
6597	Gould, 13579	9.0	6.22	1.223	68 54 41.0	.97	50.24	1
6598	Gould, Z. C., 9^h 4002 . .	9.0	14.47	1.281	68 18 28.4	.97	50.25	1
6599	Chamæleontis	10.0	29.74	+ 0.247	75 52 19.1	.99	51.15	1
6600	Chamæleontis	9.0	34.03	− 2.116	82 11 57.7	− 16.99	50.33	1
6601	Gould, 13597	9.5	9 51 48.52	+ 1.715	62 23 55.3	− 17.00	51.18	1
6602	Carinæ	10.0	56.85	1.274	68 28 48.6	.01	50.25	1
6603	Lacaille, 4102	8.5	57.70	1.275	68 28 41.8	.01	50.25	1
6604	Brisbane, 2761	9.0	51 59.07	1.309	68 6 1.2	.01	50.25	1
6605	Gould, Z. C., 9^h 4070 . .	10.0	52 21.27	+ 1.106	70 13 41.5	− 17.03	50.23	1
6606	Carinæ	10.0	9 52 24.38	+ 0.878	72 8 53.1	− 17.03	50.22	1
6607	Gould, 13615	8.5	40.43	1.707	62 40 19.6	.04	50.74	2
6608	Carinæ	9.5	40.56	0.741	73 9 52.3	.04	50.21	1
6609	Gould, Z. C., 9^h 4080 . .	8.7	43.70	0.500	74 40 8.7	.04	50.52	3
6610	Gould, Z. C., 9^h 4112 . .	9.0	48.44	+ 1.318	68 6 58.4	− 17.05	50.25	1
6611	Gould, Z. C., 6^h 4103 . .	9.0	9 52 49.28	+ 0.924	71 50 32.8	− 17.05	50.22	2
6612	Gould, Z. C., 9^h 4097 . .	9.0	53 0.62	0.038	76 56 38.7	.06	50.17	1
6613	Gould, Z. C., 9^h 4122 . .	9.0	3.40	+ 1.096	70 24 6.4	.06	50.23	1
6614	Lacaille, 4139	7.0	4.59	− 0.668	79 21 7.9	.06	50.63	2
6615	Lacaille, 4105	7.0	11.49	+ 1.645	63 46 39.0	− 17.06	50.29	1
6616	Carinæ	10.0	9 53 19.89	+ 1.685	63 9 34.4	− 17.07	50.30	1
6617	Gould, Z. C., 9^h 4238 . .	8.5	28.18	1.405	67 11 1.0	.08	50.26	2
6618	Lacaille, 4116	7.2	32.07	1.073	70 40 30.0	.08	50.91	3
6619	Gould, Z. C., 9^h 4149 . .	9.5	37.38	0.794	72 54 22.0	.08	50.21	1
6620	Gould, Z. C., 9^h 4134 . .	10.0	40.72	+ 0.058	76 55 12.0	− 17.09	50.17	1
6621	Gould, Z. C., 9^h 4191 . .	9.0	9 53 42.94	+ 1.684	63 13 52.4	− 17.09	50.30	1
6622	Lacaille, 4113	8.0	46.01	1.306	68 23 13.7	.09	50.25	1
6623	Gould, Z. C., 9^h 4174 . .	9.5	47.05	1.035	71 2 10.4	.09	52.28	1
6624	Gould, Z. C., 9^h 4207 . .	10.0	48.61	1.720	62 38 37.6	.09	51.18	1
6625	Carinæ	10.0	49.20	+ 1.728	62 30 27.6	− 17.09	51.18	1

No. 6617. Gould's right ascension is 1^m too great.

Number.	Constellation, Name of Star, or Synonym.	Magnitude.	Right Ascension, 1850.0.	Annual Precession.	South Declination, 1850.0.	Annual Precession.	Mean year.	No. of obs.
			h. m. s.	s.	° ′ ″	″		
6 626	Gould, Z. C., 9^h 4179	9.5	9 53 51.07	+ 1.027	71 7 16.7	− 17.09	52.28	1
6 627	Gould, Z. C., 9^h 4177	10.0	51.85	+ 0.951	71 45 13.1	.09	50.22	1
6 628	Chamæleontis	10.0	53 56.68	− 1.285	80 53 13.7	.10	51.14	1
6 629	Gould, 13645	9.0	54 7.17	+ 1.664	63 37 6.0	.11	50.29	1
6 630	Carinæ	9.0	14.18	+ 1.517	65 50 47.7	− 17.11	50.28	1
6 631	Chamæleontis	10.0	9 54 15.72	− 1.283	80 54 8.5	− 17.11	51.14	1
6 632	Gould, Z. C., 9^h 4228	10.0	17.89	+ 1.563	65 12 28.0	.11	50.28	1
6 633	Gould, Z. C., 9^h 4186	9.0	27.40	− 0.197	77 58 25.2	.12	50.17	1
6 634	Carinæ	9.0	33.12	+ 1.487	66 17 46.1	.13	50.27	2
6 935	Gould, 13657	9.0	37.84	+ 1.600	64 42 40.7	− 17.13	50.28	1
6 636	Chamæleontis	10.0	9 54 43.13	− 0.453	78 50 41.8	− 17.13	51.11	1
6 637	Gould, Z. C., 9^h 4264	11.0	44.25	+ 1.730	62 36 48.7	.13	51.18	1
6 638	Lacaille, 4117	8.0	45.40	1.730	62 37 32.3	.14	50.60	3
6 639	Carinæ	10.0	56.36	0.737	73 27 1.2	.14	50.73	2
6 640	Chamæleontis	9.0	54 58.07	+ 0.021	77 11 21.8	− 17.15	51.13	1
6 641	Lacaille, 4121	8.0	9 55 10.37	+ 1.568	65 15 54.5	− 17.15	50.28	1
6 642	Gould, 13658	8.0	20.20	0.173	76 34 28.1	.16	51.15	1
6 643	Gould, 13676	9.0	28.89	1.670	63 44 37.6	.17	50.29	1
6 644	Gould, Z. C., 9^h 4329	9.5	38.33	1.548	65 37 40.5	.18	50.28	1
6 645	Gould, Z. C., 9^h 4314	9.0	41.10	+ 1.027	71 20 47.1	− 17.18	50.22	1
6 646	Gould, 13688	9.0	9 55 54.20	+ 1.573	65 18 18.6	− 17.19	50.28	1
6 647	Gould, Z. C., 9^h 4337	9.0	57.72	1.075	70 58 15.8	.19	51.25	2
6 648	Carinæ	9.5	55 58.23	0.892	72 27 59.6	.19	50.56	3
6 649	Gould, Z. C., 9^h 4365	9.0	56 10.36	+ 1.559	65 33 23.5	.20	50.28	1
6 650	Octantis	8.7	16.84	− 8.312	86 32 42.5	− 17.20	51.25	7
6 651	Gould, Z. C., 9^h 4375	9.5	9 56 19.75	+ 1.457	66 56 50.8	− 17.21	50.30	1
6 652	Brisbane, 2811	8.8	32.74	− 1.924	82 11 55.5	.22	50.64	3
6 653	Carinæ	10.0	33.90	+ 1.295	68 59 52.9	.22	50.24	1
6 654	Herschel, 4281[1]	11.0	37.42	− 0.703	79 42 18.9	.22	51.10	1
6 655	Herschel, 4281[2]	9.8	40.68	− 0.701	79 42 21.9	− 17.22	51.18	2
6 656	Gould, Z. C., 9^h 4410	8.8	9 56 42.80	+ 1.424	67 25 41.0	− 17.22	50.26	2
6 657	Octantis	9.7	45.50	− 6.281	85 48 42.5	.23	51.24	3
6 658	Brisbane, 2804	8.8	47.59	+ 0.956	72 4 14.4	.23	50.22	2
6 659	Carinæ	10.0	56 58.22	+ 1.390	67 52 48.9	.24	51.34	1
6 660	Chamæleontis	10.0	57 18.85	− 0.752	79 53 5.2	− 17.25	51.10	2
6 661	Octantis	11.0	9 57 25.77	− 2.309	82 46 54.5	− 17.26	50.33	1
6 662	Gould, Z. C., 9^h 4474	9.5	39.98	+ 1.776	62 18 19.7	.27	51.18	1
6 663	Carinæ	9.5	42.66	1.078	71 10 15.1	.27	52.28	1
6 664	Gould, Z. C., 9^h 4468	9.0	43.08	1.292	69 5 2.3	.27	50.24	2
6 665	Gould, Z. C., 9^h 4461	8.8	43.42	+ 0.999	71 50 3.2	− 17.27	50.22	2

Nos. 6 654, 6 655. Record confused.

Number.	Constellation, Name of Star, or Synonym.	Magnitude.	Right Ascension, 1850.0.	Annual Precession.	South Declination, 1850.0.	Annual Precession.	Mean year.	No. of obs.
			h. m. s.	s.	° ′ ″	″		
6 666	Gould, Z. C., 9^h 4492 . .	9.5	9 57 53.99	+ 1.609	65 6 1.8	− 17.28	50.28	1
6 667	Carinæ	10.0	55.33	1.533	66 11 36.5	.28	50.28	1
6 668	Chamæleontis	10.0	57 58.56	0.076	77 14 7.6	.28	51.13	1
6 669	Lacaille, 4149	8.5	58 15.65	1.606	65 11 37.8	.29	50.28	1
6 670	Lacaille, 4154	9.0	18.82	+ 1.211	70 0 22.6	− 17.30	50.23	1
6 671	Gould, Z. C., 9^h 4495 . .	9.0	9 58 28.35	+ 0.352	76 2 10.0	− 17.30	50.18	2
6 672	Chamæleontis	9.0	32.32	− 1.004	80 35 0.5	.31	50.15	1
6 673	Carinæ	8.5	37.31	+ 1.041	71 35 56.6	.31	50.22	1
6 674	Gould, Z. C., 9^h 4553 . .	9.5	39.77	1.500	66 44 50.7	.31	50.26	1
6 675	Carinæ	10.0	40.49	+ 1.255	69 36 27.2	− 17.31	50.24	1
6 676	Carinæ	9.0	9 58 42.81	+ 1.448	67 25 41.3	− 17.31	50.26	1
6 677	Carinæ	9.5	44.79	0.938	72 27 6.6	.31	51.28	1
6 678	Gould, Z. C., 9^h 4576 . .	9.0	53.90	1.734	63 14 59.0	.32	50.30	1
6 679	Carinæ	9.5	58 58.05	1.780	62 27 32.7	.32	51.18	1
6 680	Gould, Z. C., 9^h 4586 . .	9.0	59 3.67	+ 1.630	64 57 52.1	− 17.33	50.29	2
6 681	Gould, Z. C., 9^h 4578 . .	9.5	9 59 4.30	+ 1.354	68 36 7.1	− 17.33	50.25	1
6 682	Gould, Z. C., 9^h 4593 . .	9.0	4.94	1.771	62 37 52.0	.33	50.74	2
6 683	Gould, Z. C., 9^h 4602 . .	9.5	23.17	1.332	68 53 42.2	.34	50.24	1
6 684	Gould, Z. C., 10^h 28 . .	9.5	40.15	1.696	64 1 12.2	.35	50.29	1
6 685	Chamæleontis	10.0	48.40	+ 0.388	75 59 31.4	− 17.36	50.18	2
6 686	Carinæ	10.0	9 59 48.48	+ 1.717	63 42 16.5	− 17.36	50.29	1
6 687	Gould, Z. C., 10^h 56 . .	9.5	10 0 4.35	+ 1.786	62 32 2.2	.37	51.18	1
6 688	Brisbane, 2846	8.0	17.39	− 1.074	80 51 30.9	.38	51.14	1
6 689	Lacaille, 4163	7.2	17.82	+ 1.709	63 55 13.0	.38	50.29	2
6 690	Gould, Z. C., 10^h 70 . .	9.5	18.86	+ 1.791	62 29 38.4	− 17.38	51.18	1
6 691	Lacaille, 4195	8.5	10 0 29.34	− 1.189	81 6 55.0	− 17.39	50.64	2
6 692	Gould, Z. C., 10^h 74 . .	9.8	32.92	+ 1.433	67 53 25.0	.39	50.80	2
6 693	Octantis	9.5	39.03	− 5.852	85 44 8.7	.40	51.25	3
6 694	Gould, Z. C., 10^h 60 . .	10.0	56.50	− 0.180	78 27 47.9	.41	51.12	1
6 695	Gould, Z. C., 10^h 81 . .	9.5	57.33	+ 0.686	74 27 19.8	− 17.41	50.68	2
6 696	Octantis	10.0	10 0 59.98	− 4.584	85 2 48.7	− 17.41	50.32	1
6 697	Chamæleontis	11.0	1 4.12	− 0.174	78 27 14.5	.42	51.12	1
6 698	Carinæ	9.5	5.38	+ 1.372	68 41 46.5	.42	50.24	1
6 699	Octantis	9.0	10.53	− 2.934	83 42 14.4	.42	50.84	2
6 700	Octantis	10.5	16.03	− 2.928	83 42 9.3	− 17.42	50.35	1
6 701	Gould, Z. C., 10^h 129 . .	9.0	10 1 17.39	+ 1.586	65 59 18.8	− 17.43	50.28	1
6 702	Carinæ	9.5	18.52	1.547	66 31 45.7	.43	50.26	1
6 703	Gould, Z. C., 10^h 126 . .	10.0	20.33	1.231	70 13 23.5	.43	50.23	1
6 704	Gould, Z. C., 10^h 134 . .	9.2	23.10	+ 1.625	65 25 24.7	.43	50.28	2
6 705	Octantis	10.0	26.80	−10.859	87 16 56.9	− 17.43	51.23	4

Number.	Constellation, Name of Star, or Synonym.	Magnitude.	Right Ascension, 1850.0.	Annual Precession.	South Declination, 1850.0.	Annual Precession.	Mean year.	No. of obs.
			h. m. s.	s.	° ′ ″	″		
6706	Lacaille, 4166 . . .	7.5	10 1 27.34	+ 1.768	63 6 15.6	− 17.43	50.30	2
6707	Gould, Z. C., 10^{h} 137 .	10.0	33.77	1.201	70 32 59.5	.44	50.23	1
6708	Gould, Z. C., 10^{h} 165 . .	9.0	42.53	1.688	64 30 23.4	.44	50.29	1
6709	Gould, Z. C., 10^{h} 170 . .	9.0	44.73	1.766	63 11 55.7	.45	50.30	1
6710	Carinæ	9.5	1 53.47	+ 1.796	62 41 5.7	− 17.45	51.17	1
6711	Gould, Z. C., 10^{h} 192 . .	10.0	10 2 3.48	+ 1.795	62 43 46.8	17.46	51.18	1
6712	Lacaille, 4175	8.5	11.83	1.479	67 34 3.4	.46	50.25	2
6713	Gould, Z. C., 10^{h} 198 . .	10.0	13.17	1.539	66 47 35.9	.47	50.30	1
6714	Gould, Z. C., 10^{h} 210 . .	9.5	17.66	1.531	66 54 32.2	.47	50.28	2
6715	Gould, Z. C., 10^{h} 196 . .	9.0	21.93	+ 1.128	71 20 40.7	− 17.47	50.22	1
6716	Gould, Z. C., 10^{h} 200 . .	9.5	10 2 23.60	+ 1.150	71 8 31.4	− 17.47	51.25	1
6717	Gould, Z. C., 10^{h} 216 . .	9.5	27.48	+ 1.283	69 51 26.1	.48	50.24	1
6718	Octantis	9.7	38.87	− 4.178	84 49 47.2	.48	50.95	3
6719	Gould, Z. C., 10^{h} 204 . .	9.2	40.18	+ 0.568	75 20 18.6	.48	50.29	2
6720	Gould, Z. C., 10^{h} 244 . .	9.5	43.50	+ 1.738	63 51 12.1	− 17.49	50.29	1
6721	Gould, Z. C., 10^{h} 221 . .	9.2	10 2 47.82	+ 0.570	75 20 27.3	− 17.49	50.29	2
6722	Octantis	10.0	52.78	− 2.208	82 56 21 7	.49	50.33	1
6723	Carinæ	9.5	2 54.76	+ 1.086	71 46 31.6	.50	50.22	1
6724	Chamæleontis	9.2	3 3.26	− 1.238	81 23 2.0	.50	50.64	2
6725	Gould, Z. C., 10^{h} (193) .	10.0	8.26	+ 1.405	68 36 54.7	− 17.51	50.25	1
6726	Chamæleontis	9.0	10 3 8.50	− 1.287	81 29 7.7	− 17.51	51.13	1
6727	Gould, Z. C., 10^{h} (271) .	10.0	8.67	+ 1.429	68 19 52.3	.51	51.34	1
6728	Carinæ	10.0	9.60	0.748	74 19 0.8	.51	51.16	1
6729	Carinæ	10.0	12.91	1.407	68 36 16.2	.51	50.25	1
6730	Carinæ	9.5	17.10	+ 1.034	72 15 37.1	− 17.51	51.28	1
6731	Gould, Z. C., 10^{h} 295 . .	9.5	10 3 19.15	+ 1.765	63 29 49.4	− 17.51	50.29	1
6732	Carinæ	10.0	20.20	1.833	62 15 44.5	.51	51.18	1
6733	Gould, Z. C., 10^{h} 307 . .	9.5	29.98	1.625	65 47 0.8	.52	50.28	1
6734	Gould, Z. C., 10^{h} 293 . .	10.0	31.56	+ 1.283	70 0 56.9	.52	50.23	1
6735	Chamæleontis	10.0	35.12	− 1.221	81 23 8.5	− 17.52	51.13	1
6736	Gould, Z. C., 10^{h} 299 . .	8.8	10 3 36.57	+ 1.179	71 2 29.9	− 17.53	50.90	3
6737	Octantis	10.0	40.53	− 2.084	82 48 56.4	.53	50.33	1
6738	Carinæ	10.0	41.41	+ 1.141	71 23 52.0	.53	50.22	1
6739	Gould, Z. C., 10^{h} 327 . .	9.5	47.43	1.495	67 36 45.9	.53	50.25	2
6740	Carinæ	10.0	3 59.18	+ 0.899	73 23 4.9	− 17.54	50.21	1
6741	Lacaille, 4194	7.5	10 4 0.94	+ 1.216	70 44 38.0	− 17.54	51.25	2
6742	Octantis	9.5	1.46	− 6.147	85 58 43.0	.54	51.20	2
6743	Gould, Z. C., 10^{h} 346 . .	10.0	4.56	+ 1.426	68 30 53.1	.54	50.25	1
6744	Gould, Z. C., 10^{h} 298 . .	9.0	8.76	− 0.088	78 25 16.1	.55	51.12	1
6745	Carinæ	10.0	13.95	+ 0.924	73 14 6.7	− 17.55	50.21	1

No. 6744. Gould's declination is 19.2″ greater.

Number.	Constellation, Name of Star, or Synonym.	Magnitude.	Right Ascension, 1850.0.	Annual Precession.	South Declination, 1850.0.	Annual Precession.	Mean year.	No. of obs.
			h. m. s.	s.	° ′ ″	″		
6746	Gould, Z. C., 10ʰ 365 . .	9.0	10 4 17.32	+ 1.504	67 35 19.7	− 17.55	50.25	2
6747	Gould, Z. C., 10ʰ 323 . .	9.0	21.82	− 0.095	78 27 47.7	.56	51.12	1
6748	Lacaille, 4191	7.9	26.69	+ 1.701	64 46 34.3	.56	50.31	4
6749	Gould, Z. C., 10ʰ 388 . .	9.5	30.16	+ 1.802	63 2 22.4	.56	50.30	2
6750	μ^1 Chamæleontis	5.5	31.04	− 1.242	81 29 14.8	− 17.56	50.64	2
6751	Brisbane, 2871	5.6	10 4 31.64	+ 1.682	65 4 55.2	− 17.56	50.31	5
6752	Carinæ	9.5	40.68	0.839	73 53 22.0	.57	50.73	2
6753	Gould, Z. C., 10ʰ 394 . .	8.5	46.70	+ 1.248	70 32 53.7	.57	51.26	2
6754	Chamæleontis	9.0	52.75	− 0.980	80 58 17.6	.58	51.14	1
6755	Gould, Z. C., 10ʰ 418 . .	9.5	4 55.52	+ 1.688	65 3 45.2	− 17.58	50.31	3
6756	Gould, Z. C., 10ʰ 359 . .	9.0	10 5 0.81	− 0.364	79 23 55.2	− 17.58	50.16	1
6757	Gould, Z. C., 10ʰ 411 . .	9.0	12.51	+ 0.818	74 6 1.2	.59	51.23	2
6758	Gould, Z. C., 10ʰ 433 . .	9.0	12.86	1.606	66 20 46.1	.59	50.26	1
6759	Gould, Z. C., 10ʰ 437 . .	9.5	14.60	1.645	65 46 29.8	.59	50.28	2
6760	Carinæ	10.0	15.07	+ 1.134	71 40 41.3	− 17.59	50.22	1
6761	Carinæ	9.0	10 5 33.37	+ 1.735	64 24 43.5	− 17.61	50.29	1
6762	Gould, Z. C., 10ʰ 441 . .	9.8	35.82	+ 0.826	74 5 46.9	.61	51.23	2
6763	Lacaille, 4226	7.8	37.74	− 0.027	78 19 52.4	.61	50.71	2
6764	Gould, Z. C., 10ʰ 469 . .	9.0	39.64	+ 1.522	67 34 3.2	.61	50.25	2
6765	Octantis	9.8	40.10	− 10.205	87 14 29.5	− 17.61	51.27	7
6766	Lacaille, 4203	7.0	10 5 46.90	+ 1.494	67 56 46.8	− 17.62	50.66	3
6767	Gould, Z. C., 10ʰ 480 . .	9.0	48.95	1.351	69 37 42.6	.62	50.24	1
6768	Gould, Z. C., 10ʰ 497 . .	9.0	52.58	1.840	62 36 8.7	.62	51.18	1
6769	Gould, Z. C., 10ʰ 492 . .	8.2	55.32	1.647	65 51 50.6	.62	50.28	2
6770	Lacaille, 4199	8.0	5 56.62	+ 1.783	63 38 50.1	− 17.62	50.29	1
6771	Gould, Z. C., 10ʰ 445 . .	9.5	10 6 5.76	− 0.187	78 55 43.9	− 17.63	51.11	1
6772	Carinæ	9.8	7.52	+ 1.489	68 4 14.7	.63	50.80	2
6773	Chamæleontis	10.0	10.53	− 1.480	82 1 54.4	.63	51.17	1
6774	Brisbane, 2883	9.0	14.23	+ 1.808	63 15 3.4	.64	50.30	1
6775	Chamæleontis	9.0	19.70	+ 0.413	76 32 23.6	− 17.64	51.15	2
6776	Lacaille, 4205	8.0	10 6 20.39	+ 1.607	66 31 33.0	− 17.64	50.26	1
6777	Chamæleontis	10.0	23.34	− 0.531	79 59 20.9	.64	51.16	1
6778	Carinæ	9.5	27.17	+ 1.815	63 10 13.2	.64	50.30	1
6779	Carinæ	9.5	29.29	1.692	65 16 15.8	.65	50.28	1
6780	Gould, Z. C., 10ʰ 523 . .	9.5	30.37	+ 1.373	69 29 19.3	− 17.65	50.24	2
6781	Chamæleontis	10.0	10 6 33.96	+ 0.363	76 48 24.7	− 17.65	50.17	1
6782	Octantis	9.5	34.52	− 2.273	83 13 28.0	.65	50.39	1
6783	Gould, Z. C., 10ʰ 544 . .	9.0	34.69	+ 1.789	63 39 57.8	.65	50.29	1
6784	Carinæ	10.0	34.92	1.037	72 40 11.7	.65	51.28	1
6785	Lacaille, 4209	7.7	35.46	+ 1.568	67 6 11.2	− 17.65	50.26	2

Number.	Constellation, Name of Star, or Synonym.	Magnitude.	Right Ascension, 1850.0.	Annual Precession.	South Declination, 1850.0.	Annual Precession.	Mean year.	No. of obs.
			h. m. s.	s.	° ′ ″	″		
6786	Carinæ	10.0	10 6 35.71	+ 1.308	70 12 21.1	− 17.65	50.23	1
6787	Chamæleontis	10.5	40.22	− 0.526	79 59 53.9	.65	51.16	1
6788	Gould, Z. C., 10^h 545 . .	9.5	40.75	+ 1.628	66 16 17.7	.65	50.26	1
6789	μ^2 Chamæleontis	6.5	49.94	− 0.856	80 50 2.8	.66	51.14	1
6790	Carinæ	10.0	6 58.12	+ 1.301	70 19 48.6	− 17.67	50.23	1
6791	Gould, Z. C., 10^h 577 . .	9.5	10 7 1.03	+ 1.857	62 29 4.8	− 17.67	51.18	1
6792	Carinæ	9.0	1.50	1.759	64 15 52.8	.67	50.29	2
6793	Gould, Z. C., 10^h 593 . .	9.0	13.42	1.828	63 4 55.5	.68	50.30	2
6794	Carinæ	10.0	14.23	1.454	68 40 11.0	.68	50.25	1
6795	Gould, Z. C., 10^h 594 . .	9.0	10 7 16.21	+ 1.796	63 40 11.1	− 17.68	50.29	1
6796	Lacaille, 4219	7.5	10 7 17.82	+ 1.574	67 9 8.4	− 17.68	50.28	2
6797	Lacaille, 4254	7.2	23.56	− 1.294	81 46 37.7	.68	50.75	2
6798	Chamæleontis	9.0	35.00	+ 0.897	73 51 5.6	.68	51.25	1
6799	Chamæleontis	9.5	35.06	0.558	75 55 56.5	.68	50.51	3
6800	Carinæ	9.5	35.86	+ 0.752	74 48 50.0	− 17.69	50.20	1
6801	Gould, Z. C., 10^h 604 . .	8.5	10 7 36.90	+ 1.438	68 55 27.9	− 17.69	50.24	1
6802	Carinæ	10.0	55.68	1.514	68 2 47.6	.71	51.34	1
6803	Brisbane, 2902	9.0	7 56.51	1.697	65 26 29.4	.71	50.28	2
6804	Carinæ	9.5	8 4.97	1.864	62 32 58.1	.71	51.18	1
6805	Gould, Z. C., 10^h 662 . .	9.0	10.04	+ 1.816	63 28 30.1	− 17.72	50.30	1
6806	Carinæ	10.0	10 8 10.49	+ 1.324	70 16 10.5	− 17.72	50.23	1
6807	Gould, Z. C., 10^h 642 . .	9.0	10.84	1.442	68 57 41.6	.72	50.24	1
6808	Brisbane, 2905	9.0	11.59	1.694	65 32 42.1	.72	50.28	1
6809	Chamæleontis	9.0	12.58	0.168	77 49 18.4	.72	50.17	1
6810	Gould, 13956	9.0	15.57	+ 1.807	63 39 14.1	− 17.72	50.29	1
6811	Gould, Z. C., 10^h 667 . .	8.5	10 8 15.66	+ 1.668	65 57 2.0	− 17.72	50.28	1
6812	Carinæ	9.8	18.34	− 9.492	87 8 48.9	.72	51.25	5
6813	Gould, Z. C., 10^h 675 . .	9.0	24.69	+ 1.574	67 19 34.3	.73	50.26	1
6814	Lacaille, 4225	8.0	8 32.30	1.766	64 25 32.0	.73	50.29	1
6815	Carinæ	9.0	9 1.84	+ 0.783	74 47 5.2	− 17.75	50.20	1
6816	Carinæ	9.0	10 9 8.48	+ 1.583	67 19 41.0	− 17.76	50.26	1
6817	Gould, Z. C., 10^h 669 . .	9.8	9.02	− 0.309	79 34 40.4	.76	51.18	2
6818	Gould, Z. C., 10^h 727 . .	9.0	9.03	+ 1.843	63 9 15.6	.76	50.30	1
6819	Gould, Z. C., 10^h 717 . .	9.0	9.40	1.445	69 4 51.4	.76	50.24	1
6820	M Carinæ	6.5	16.15	+ 1.700	65 37 49.0	− 17.76	50.28	1
6821	Gould, Z. C., 10^h 683 . .	9.5	10 9 18.64	− 0.317	79 36 55.4	− 17.76	51.18	2
6822	Gould, Z. C., 10^h 714 . .	8.0	18.90	+ 0.945	73 43 20.1	.76	51.25	1
6823	Chamæleontis	10.0	21.38	0.142	78 2 13.4	.76	50.30	1
6824	Carinæ	9.0	23.36	1.677	66 1 51.8	.77	50.27	1
6825	Gould, Z. C., 10^h 751 . .	9.5	26.37	+ 1.871	62 41 10.1	− 17.77	51.18	1

Number.	Constellation, Name of Star, or Synonym.	Magnitude.	Right Ascension, 1850.0.	Annual Precession.	South Declination, 1850.0.	Annual Precession.	Mean year.	No. of obs.
			h. m. s.	s.	° ′ ″	″		
6826	Gould, Z. C., 10h 745 . .	9.2	10 9 26.42	+ 1.744	64 57 47.1	− 17.77	50.30	3
6827	Carinæ	9.2	26.98	+ 1.744	64 57 36.1	.77	50.30	3
6828	Octantis	10.0	43.11	− 2.870	84 3 12.2	.78	50.33	1
6829	Gould, Z. C., 10h 764 . .	9.0	46.69	+ 1.513	68 21 12.7	.78	50.25	1
6830	Gould, Z. C., 10h 780 . .	9.0	48.40	+ 1.815	63 47 44.0	− 17.78	50.29	1
6831	Carinæ	9.5	10 9 48.88	+ 1.889	62 23 49.7	− 17.78	51.18	1
6832	Gould, Z. C., 10h 779 . .	9.0	48.99	1.762	64 43 53.6	.78	50.29	2
6833	Lacaille, 4238	7.7	50.63	1.593	67 18 54.8	.78	50.26	2
6834	Carinæ	10.0	9 59.03	1.458	69 3 22.6	.79	50.24	1
6835	Carinæ	10.0	10 2.19	+ 1.209	71 39 59.3	− 17.79	50.22	1
6836	Carinæ	9.0	10 10 2.64	+ 1.347	70 18 45.6	− 17.79	50.23	2
6837	Chamæleontis	10.0	5.39	0.700	75 25 28.8	.79	51.15	1
6838	ω Argus	4.0	9.69	1.441	69 17 35.3	.80	50.24	1
6839	Carinæ	10.0	15.46	1.464	69 1 57.5	.80	50.24	1
6840	Gould, Z. C., 10h 813 . .	9.5	17.65	+ 1.580	67 33 53.0	− 17.80	50.25	1
6841	Gould, 14011	8.0	10 10 19.04	+ 1.291	70 55 30.8	− 17.80	51.25	1
6842	Carinæ	10.0	20.74	1.381	69 59 56.6	.80	50.23	1
6843	Carinæ	10.0	27.37	1.377	70 3 11.8	.81	50.23	1
6844	Gould, 14020	9.0	27.76	1.664	66 23 54.4	.81	50.26	1
6845	Gould, Z. C., 10h 810 . .	9.2	32.14	+ 0.914	74 6 12.8	− 17.81	51.23	2
6846	Carinæ	9.5	10 10 32.32	+ 1.304	70 49 52.8	− 17.81	51.25	2
6847	Gould, 14022	9.3	37.68	1.542	68 7 43.5	.82	50.66	3
6848	Gould, 14025	9.0	40.14	1.543	68 7 41.5	.82	50.66	3
6849	Lacaille, 4255	8.0	45.54	0.728	75 19 58.0	.82	50.40	1
6850	Carinæ	9.0	48.20	+ 1.214	71 43 47.5	− 17.82	50.22	1
6851	Gould, Z. C., 10h 838 . .	8.5	10 10 50.85	+ 1.117	72 35 23.6	− 17.82	51.28	1
6852	Lacaille, 4248²	9.0	52.22	1.659	66 33 6.4	.83	50.28	2
6853	Lacaille, 4248¹	8.2	10 56.21	1.660	66 32 22.6	.83	50.28	2
6854	Gould, Z. C., 10h 862 . .	8.5	11 2.26	1.196	71 55 59.2	.83	51.24	2
6855	Gould, Z. C., 10h 871 . .	9.0	14.54	+ 1.173	72 9 53.5	− 17.84	51.24	2
6856	Carinæ	9.0	10 11 14.63	+ 1.872	63 1 1.6	− 17.84	50.30	1
6857	Gould, Z. C., 10h 894 . .	10.0	15.32	+ 1.745	65 16 44.9	.84	50.28	1
6858	Chamæleontis	9.2	24.25	− 0.024	78 50 59.4	.85	51.20	2
6859	Gould, 14042	10.0	30.59	+ 1.493	68 53 29.7	.85	50.24	1
6860	Gould, Z. C., 10h 899 . .	9.0	36.41	+ 0.998	73 38 4.4	− 17.85	51.25	1
6861	Chamæleontis	9.5	10 11 37.87	− 0.500	80 19 33.5	− 17.86	51.16	1
6862	Carinæ	10.0	42.83	+ 1.136	72 33 6.4	.86	51.27	1
6863	Gould, Z. C., 10h 918 . .	9.5	50.25	1.464	69 17 37.1	.86	50.24	1
6864	Chamæleontis	10.0	11 57.77	0.721	75 31 10.0	.87	51.15	1
6865	Gould, Z. C., 10h 934 . .	9.0	12 2.01	+ 1.447	69 30 55.6	− 17.87	50.24	2

Number.	Constellation, Name of Star, or Synonym.	Magnitude.	Right Ascension, 1850.0.	Annual Precession.	South Declination, 1850.0.	Annual Precession.	Mean year.	No. of obs.
			h. m. s.	s.	° ′ ″	″		
6 866	Gould, Z. C., 10^h 940	9.5	10 12 3.03	+ 1.780	64 47 30.8	− 17.87	50.28	1
6 867	Gould, Z. C., 10^h 896	8.7	9.79	− 0.335	79 54 39.8	.88	51.17	3
6 868	Lacaille, 4297[1]	9.5	12.76	2.125	83 20 56.4	.88	50.37	2
6 869	Lacaille, 4297[2]	8.5	14.98	− 2.124	83 20 58.2	.88	50.37	2
6 870	Gould, Z. C., 10^h 952	10.0	16.82	+ 1.561	68 9 47.9	17.88	51.34	1
6 871	Octantis	9.4	10 12 25.77	− 4.159	85 13 7.3	− 17.89	51.29	6
6 872	Carinæ	9.5	50.40	+ 1.728	65 50 31.0	.90	50.28	1
6 873	Carinæ	11.0	51.03	1.728	65 50 31.0	.90	50.28	1
6 874	Chamæleontis	10.0	53.37	0.369	77 26 2.4	.91	51.13	1
6 875	Chamæleontis	10.0	55.29	+ 0.548	76 34 42.9	− 17.91	51.15	1
6 876	Carinæ	10.0	10 12 57.15	+ 1.594	67 49 42.0	− 17.91	50.25	1
6 877	Gould, Z. C., 10^h 1008	9.0	12 58.08	1.908	62 39 28.7	.91	50.74	1
6 878	Lacaille, 4259	7.7	13 0.81	1.800	64 40 43.4	.91	50.29	4
6 879	Gould, Z. C., 10^h 1026	9.0	10.18	1.913	62 34 50.1	.92	51.18	1
6 880	Carinæ	9.0	13.48	+ 1.629	67 23 36.9	− 17.92	50.26	2
6 881	Gould, Z. C., 10^h 987	9.0	10 13 16.82	+ 0.091	78 36 47.8	− 17.92	51.20	2
6 882	Carinæ	10.0	22.43	1.760	65 25 20.0	.92	50.28	1
6 883	Gould, Z. C., 10^h 1028	9.0	24.54	1.445	69 45 51.9	.93	50.24	1
6 884	Carinæ	9.5	29.30	1.313	71 10 56.3	.93	52.28	1
6 885	Gould, Z. C., 10^h 1029	10.0	30.10	+ 1.279	71 30 49.3	17.93	50.22	1
6 886	Carinæ	10.0	10 13 38.04	+ 1.768	65 20 26.3	− 17.93	50.28	1
6 887	Lacaille, 4284	8.2	42.23	0.194	78 15 24.8	.94	50.71	2
6 888	Gould, Z. C., 10^h 1056	9.5	42.32	1.685	66 39 45.8	.94	50.30	1
6 889	Gould, Z. C., 10^h 1042	9.2	43.44	1.387	70 27 22.2	.94	51.26	2
6 890	Gould, Z. C., 10^h 1047	9.2	44.56	+ 1.392	70 24 11.8	− 17.94	51.26	2
6 891	Gould, Z. C., 10^h 1067	9.0	10 13 46.93	+ 1.918	62 36 36.1	− 17.94	51.18	1
6 892	Gould, Z. C., 10^h 1050	9.3	47.13	1.375	70 35 49.4	.94	51.58	3
6 893	Gould, Z. C., 10^h 1075	9.0	52.04	1.922	62 32 57.4	.94	51.18	1
6 894	Carinæ	10.0	13 59.89	1.585	68 8 16.4	.95	51.34	1
6 895	Gould, Z. C., 10^h 1041	9.3	14 0.87	+ 0.617	76 20 20.6	− 17.95	50.50	3
6 896	Lacaille, 4342	7.7	10 14 1.12	− 5.857	86 10 40.8	17.95	51.27	7
6 897	Octantis	10.0	1.66	− 3.084	84 27 56.2	.95	50.33	1
6 898	Brisbane, 2969	8.8	14.96	+ 0.380	77 32 1.3	.96	50.65	2
6 899	Gould, Z. C., 10^h 1081	8.5	18.38	1.040	73 40 29.8	.96	51.25	1
6 900	Gould, Z. C., 10^h 1102	9.5	18.55	+ 1.816	64 38 24.4	− 17.96	50.29	2
6 901	Carinæ	10.0	10 14 20.33	+ 1.386	70 33 41.7	− 17.96	52.28	1
6 902	Lacaille, 4268[1]	9.0	26.09	1.857	63 55 24.5	.97	50.29	1
6 903	Lacaille, 4268 mean	4.5	26.31	1.857	63 55 29.3	.97	50.29	1
6 904	Lacaille, 4268[2]	9.0	26.33	1.857	63 55 26.8	.97	50.29	1
6 905	Gould, Z. C., 10^h 1117	9.5	26.93	+ 1.868	63 43 55.3	− 17.97	50.29	1

No. 6 877. Gould's declination is 1′ less.

Number.	Constellation, Name of Star, or Synonym.	Magnitude.	Right Ascension, 1850.0.	Annual Precession.	South Declination, 1850.0.	Annual Precession.	Mean year.	No. of obs.
			h. m. s.	s.	° ′ ″	″		
6906	Gould, Z. C., 10h 1108 .	9.5	10 14 27.92	+ 1.735	66 2 28.4	— 17.97	50.28	1
6907	Herschel, 4308[1]	11.0	29.30	1.315	71 18 45.0	.97	51.16	1
6908	Gould, Z. C., 10h 1119 .	9.0	31.30	1.725	66 12 8.6	.97	50.27	2
6909	Herschel, 4308[2]	11.0	32.34	1.317	71 18 37.0	.97	51.16	1
6910	Gould, Z. C., 10h 1124 .	9.2	38.10	+ 1.552	68 40 39.9	— 17.97	50.25	2
6911	Chamæleontis	9.0	10 14 40.10	— 0.485	80 31 54.9	— 17.98	50.15	1
6912	Chamæleontis	9.5	47.73	+ 0.438	77 19 19.9	.98	51.13	1
6913	Chamæleontis	10.0	52.51	— 0.849	81 24 48.8	.98	51.13	1
6914	Carinæ	9.5	54.37	+ 1.048	73 41 53.3	.98	51.25	1
6915	Gould, Z. C., 10h 1104 .	9.0	54.45	+ 0.485	77 6 48.4	— 17.98	50.65	2
6916	Gould, 14127	8.8	10 14 58.56	+ 1.705	66 35 35.8	— 17.99	50.29	2
6917	Gould, Z. C., 10h 1138 .	9.0	15 4.07	1.121	73 8 48.0	.99	50.21	2
6918	Lacaille, 4274	8.0	5.29	1.839	64 23 1.9	17.99	50.29	1
6919	Gould, Z. C., 10h 1176 .	10.0	16.07	1.706	66 38 9.3	18.00	50.27	1
6920	Gould, Z. C., 10h 1203 .	8.5	33.39	+ 1.937	62 34 57.0	— 18.01	51.18	1
6921	Gould, Z. C., 10h 1199 .	9.0	10 15 36.76	+ 1.655	67 26 57.6	— 18.01	50.25	2
6922	Lacaille, 4280	8.0	44.15	1.726	66 24 48.7	.02	50.26	1
6923	Carinæ	9.5	48.89	1.432	70 18 20.1	.02	52.28	1
6924	Lacaille, 4279	8.5	49.01	1.795	65 18 37.2	.02	50.28	1
6925	Carinæ	9.5	53.45	+ 1.367	71 0 28.0	— 18.02	52.28	1
6926	Carinæ	10.0	10 15 53.61	+ 1.357	71 6 25.3	— 18.02	50.23	1
6927	Gould, 14147	9.0	56.48	0.875	75 3 17.6	.02	50.30	2
6928	Gould, Z. C., 10h 1237 .	9.0	15 57.60	1.942	62 34 3.5	.03	51.18	1
6929	Gould, Z. C., 10h 1241 .	9.0	16 3.00	1.934	62 41 47.2	.03	50.74	2
6930	Gould, Z. C., 10h 1251 .	9.0	10.87	+ 1.958	62 16 41.2	— 18.03	51.18	1
6931	Carinæ	9.5	10 16 11.52	+ 1.123	73 17 10.3	— 18.03	50.21	1
6932	Octantis	9.2	14.16	— 3.159	84 38 30.8	.04	50.33	2
6933	Chamæleontis	10.0	21.91	0.319	80 13 22.6	.04	51.16	1
6934	Chamæleontis	9.0	26.41	0.401	80 27 11.3	.04	50.15	1
6935	Melbourne (1), 518 . . .	9.5	31.89	+ 1.713	66 46 14.8	— 18.05	50.30	1
6936	Gould, 14168	9.5	10 16 32.91	+ 1.713	66 45 56.8	— 18.05	50.30	1
6937	Gould, 14167	8.8	33.58	1.677	67 18 35.3	.05	50.26	2
6938	Lacaille, 4285	9.0	38.95	1.853	64 26 24.0	.05	50.29	1
6939	Gould, Z. C., 10h 1284 .	9.2	16 44.30	1.663	67 32 59.8	.05	50.25	2
6940	Lacaille, 4290[1]	9.3	17 3.04	+ 1.441	70 24 1.0	— 18.07	51.26	2
6941	Lacaille, 4290[2]	9.0	10 17 5.55	+ 1.442	70 23 36.2	— 18.07	50.91	2
6942	Gould, Z. C., 10h 1299 .	8.5	6.08	1.350	71 22 34.5	.07	50.22	2
6943	Gould, 14188	9.0	9.31	+ 1.807	65 21 30.7	.07	50.28	1
6944	Octantis	9.2	14.27	— 2.800	84 21 44.7	.07	50.33	2
6945	Carinæ	9.2	22.35	+ 1.444	70 24 25.0	— 18.08	51.26	2

Number.	Constellation, Name of Star, or Synonym.	Magnitude.	Right Ascension, 1850.0.	Annual Precession.	South Declination, 1850.0.	Annual Precession.	Mean year.	No. of obs.
			h. m. s.	s.	° ′ ″	″		
6946	Carinæ	10.0	10 17 22.65	+ 1.151	73 13 38.6	− 18.08	50.21	1
6947	Gould, Z. C., 10^h 1340 .	9.2	32.01	1.595	68 37 20.9	.08	50.30	3
6948	Gould, Z. C., 10^h 1342 .	9.2	33.18	1.600	68 33 43.4	.09	50.33	2
6949	Gould, 14197	8.6	36.16	1.837	64 56 2.9	.09	50.30	4
6950	Carinæ	9.3	39.37	+ 1.629	68 11 22.8	− 18.09	50.66	3
6951	Octantis	9.0	10 17 40.22	− 1.866	83 19 49.9	− 18.09	50.37	1
6952	Gould, Z. C., 10^h 1350 .	9.0	41.78	+ 1.601	68 34 38.9	.09	50.33	2
6953	Carinæ	10.0	43.67	1.338	71 35 39.4	.09	50.22	1
6954	Carinæ	9.0	46.73	1.965	62 29 0.8	.09	51.18	1
6955	Gould, Z. C., 10^h 1378 .	9.2	17 54.80	+ 1.976	62 16 9.3	− 18.10	51.74	2
6956	Gould, Z. C., 10^h 1386 .	9.2	10 18 1.78	+ 1.985	62 38 35.7	− 18.10	51.74	2
6957	Chamæleontis	8.5	13.69	− 0.244	80 10 13.9	.11	51.16	1
6958	Gould, Z. C., 10^h 1392 .	9.5	21.76	+ 1.356	71 30 29.4	.12	50.22	1
6959	Gould, Z. C., 10^h 1417 .	9.5	27.10	1.959	62 44 29.7	.12	51.18	1
6960	Carinæ	9.0	27.39	+ 1.939	63 9 44.2	− 18.12	50.30	1
6961	*l* Carinæ	5.8	10 18 31.56	+ 1.776	66 8 36.6	− 18.12	50.27	2
6962	Gould, Z. C., 10^h 1428 .	9.5	34.04	1.963	62 41 19.1	.12	51.18	1
6963	Carinæ	10.0	42.22	1.520	69 46 14.7	.13	52.29	1
6964	Gould, Z. C., 10^h 1404 .	8.7	48.53	+ 0.743	76 12 43.9	.13	50.50	3
6965	Carinæ	9.5	18 56.81	+ 0.979	74 44 41.6	− 18.14	51.16	1
6966	Gould, Z. C., 10^h 1462 .	9.2	10 19 7.42	+ 1.799	65 53 9.6	− 18.14	50.28	2
6967	Brisbane, 3005	9.5	8.94	1.687	67 38 56.2	.15	50.93	3
6968	Brisbane, 3006	9.0	15.40	+ 1.697	67 31 9.1	.15	50.93	3
6969	Chamæleontis	9.8	15.56	− 0.885	81 48 58.0	.15	51.15	2
6970	Gould, Z. C., 10^h 1458 .	8.5	24.59	+ 1.054	74 16 25.0	− 18.15	51.01	4
6971	Carinæ	10.0	10 19 26.63	+ 1.294	72 16 53.2	− 18.16	51.28	1
6972	Chamæleontis	10.0	30.29	0.578	77 10 36.7	.16	51.13	1
6973	Gould, Z. C., 10^h 1500 .	10.0	39.48	1.958	64 28 6.3	.16	50.29	1
6974	Gould, Z. C., 10^h 1494 .	9.5	43.46	1.483	70 22 34.0	.17	50.28	1
6975	Gould, Z. C., 10^h 1485 .	9.0	49.26	+ 0.957	75 1 9.5	− 18.17	50.20	1
6976	Gould, Z. C., 10^h 1503 .	8.7	10 19 50.67	+ 1.520	69 58 9.2	− 18.17	50.92	3
6977	Brisbane, 3012	8.8	19 57.01	1.716	67 22 11.3	.18	50.94	3
6978	Carinæ	9.5	20 12.19	1.984	62 35 19.9	.18	52.30	1
6979	Gould, Z. C., 10^h 1521 .	8.8	13.86	1.438	70 57 20.0	.19	51.25	2
6980	Gould, 14253	9.0	16.00	+ 1.591	69 9 25.5	− 18.19	50.24	1
6981	Carinæ	9.4	10 20 18.98	+ 1.072	74 15 56.5	18.19	51.01	4
6982	Octantis	11.0	27.54	− 2.608	84 20 35.1	.19	50.33	1
6983	Chamæleontis	9.5	29.23	− 0.510	81 4 31.7	.20	51.14	1
6984	Gould, Z. C., 10^h 1551 .	9.0	33.38	+ 1.952	63 21 12.3	.20	50.30	1
6985	Octantis	9.5	38.51	− 4.641	85 50 5.0	− 18.20	51.23	6

Number.	Constellation, Name of Star, or Synonym.	Magnitude.	Right Ascension, 1850.0.	Annual Precession.	South Declination, 1850.0.	Annual Precession.	Mean year.	No. of obs.
			h. m. s.	s.	° ′ ″	″		
6986	Gould, Z. C., 10h 1541	9.5	10 20 48.52	+ 0.770	76 18 16.8	− 18.21	50.67	2
6987	Chamæleontis	10.0	49.72	1.478	70 36 50.1	.21	52.28	1
6988	Gould, Z. C., 10h 1575	9.3	53.35	2.000	62 23 50.1	.21	51.93	3
6989	Gould, 14268	8.2	20 56.52	1.459	70 51 0.2	.21	51.25	2
6990	Gould, 14275	8.0	21 4.22	+ 2.015	62 5 51.1	− 18.22	51.38	1
6991	Gould, Z. C., 10h 1597	9.8	10 21 12.70	+ 1.999	62 29 58.0	18.22	51.75	2
6992	Lacaille, 4311	7.4	13.32	+ 1.997	62 32 8.9	.22	51.80	4
6993	Lacaille, 4346	7.5	14.66	− 0.980	82 9 9.3	.22	50.64	3
6994	Gould, 14281	9.5	22.74	+ 1.826	65 52 24.0	.23	50.28	2
6995	/ Carinæ	4.5	24.05	+ 1.217	73 16 8.4	− 18.23	50.21	1
6996	Carinæ	9.0	10 21 27.37	+ 1.441	71 8 0.8	− 18.23	51.25	2
6997	Gould, Z. C., 10h 1621	9.0	38.39	1.723	67 34 39.3	.24	50.94	3
6998	Lacaille, 4322	6.5	43.06	+ 1.229	73 12 40.1	.24	50.21	1
6999	Gould, Z. C., 10h 1573	8.8	43.31	− 0.073	79 58 31.6	.24	51.32	3
7000	Lacaille, 4341	7.5	43.46	− 0.396	80 53 28.9	− 18.24	51.14	1
7001	Chamæleontis	9.5	10 21 44.45	− 0.121	80 7 20.8	− 18.24	51.16	1
7002	Lacaille, 4312	8.5	50.32	+ 1.954	63 35 16.8	.24	50.31	2
7003	Gould, 14294	9.8	56.74	1.881	65 0 49.6	.25	50.28	2
7004	Carinæ	9.5	21 59.04	1.290	72 42 52.3	.25	51.28	1
7005	Gould, Z. C., 10h 1629	9.0	22 7.70	+ 0.974	75 12 38.6	− 18.26	50.40	1
7006	Chamæleontis	10.0	10 22 9.29	+ 0.941	75 26 5.1	− 18.26	51.15	1
7007	Octantis	10.0	11.61	−12.126	87 54 52.0	.26	51.27	2
7008	Chamæleontis	9.0	21.03	− 0.339	80 47 47.7	.26	51.14	1
7009	Gould, 14303	9.0	21.96	+ 1.628	64 16 28.2	.26	50.29	2
7010	Gould, Z. C., 10h 1662	9.2	25.18	+ 1.393	71 47 42.2	.27	51.24	2
7011	Lacaille, 4311	8.5	10 22 29.49	+ 2.000	62 45 33.6	− 18.27	50.74	2
7012	Lacaille, 4316	7.2	30.20	2.023	62 15 31.3	.27	51.82	4
7013	Carinæ	10.0	30.22	1.198	73 35 41.4	.27	51.25	1
7014	Chamæleontis	9.5	32.71	+ 0.809	76 17 55.6	− 18.27	51.15	1
7015	Chamæleontis	10.0	32.74	+ 0.264	78 53 7.1	.27	51.28	1
7016	Lacaille, 4321	6.1	10 22 41.89	+ 1.894	64 56 26.5	− 18.28	50.30	5
7017	Gould, Z. C., 10h 1705	9.0	47.38	1.728	67 44 0.4	.28	51.28	2
7018	Gould, Z. C., 10h 1709	9.0	22 51.75	1.631	69 7 5.4	.28	50.24	2
7019	Gould, Z. C., 10h 1728	9.2	23 0.42	1.745	67 31 25.6	.29	51.28	2
7020	Gould, Z. C., 10h 1736	9.0	5.03	+ 1.940	64 8 35.7	− 18.29	50.29	1
7021	Lacaille, 4318	7.1	10 23 13.52	+ 2.024	62 23 56.6	− 18.29	51.82	4
7022	Gould, Z. C., 10h 1743	9.0	14.30	1.805	66 36 25.5	.30	50.28	3
7023	Gould, Z. C., 10h 1746	9.0	23.89	1.573	69 56 51.2	.30	51.26	2
7024	Lacaille, 4335	8.0	28.70	+ 1.348	72 24 14.8	.30	51.28	1
7025	Chamæleontis	10.0	32.46	− 0.376	80 59 44.5	− 18.31	51.14	1

Number.	Constellation, Name of Star, or Synonym.	Magnitude.	Right Ascension, 1850.0.	Annual Precession.	South Declination, 1850.0.	Annual Precession.	Mean year.	No. of obs.
			h. m. s.	s.	° ′ ″	″		
7026	Carinæ	9.0	10 23 32.52	+ 1.698	68 19 9.8	− 18.31	50.25	1
7027	Gould, Z. C., 10h 1773	10.2	35.80	1.905	64 55 40.6	.31	50.30	2
7028	Gould, Z. C., 10h 1771	9.5	36.02	1.845	66 0 23.0	.31	50.28	1
7029	Gould, Z. C., 10h 1756	9.5	38.88	1.341	72 29 50.1	.31	51.28	1
7030	Chamæleontis	10.0	47.70	+ 0.293	78 54 14.2	− 18.32	51.28	1
7031	Lacaille, 4330	5.2	10 23 53.66	+ 1.984	63 24 18.8	− 18.32	50.58	4
7032	Carinæ	9.0	56.79	0.939	75 41 26.4	.32	51.15	1
7033	Carinæ	8.0	23 57.17	1.127	74 23 7.8	.32	51.28	3
7034	Chamæleontis	9.0	24 1.06	0.431	78 21 55.2	.32	51.12	1
7035	Chamæleontis	10.0	4.76	+ 0.959	75 34 31.7	− 18.33	51.15	1
7036	Carinæ	10.0	10 24 4.88	+ 1.609	69 37 35.8	− 18.33	51.26	2
7037	Gould, Z. C., 10h 1808	9.2	6.51	1.713	68 12 20.2	.33	50.60	4
7038	Lacaille, 4331	7.0	10.20	1.938	64 24 38.8	.33	50.29	1
7039	Gould, Z. C., 10h 1811	9.0	10.87	1.609	69 38 22.5	.33	51.26	2
7040	Gould, 14341-2	9.2	10.90	+ 1.857	65 55 48.1	− 18.33	50.28	2
7041	Gould, Z. C., 10h 1812	9.2	10 24 13.02	+ 1.584	69 57 36.4	− 18.33	51.26	2
7042	Chamæleontis	10.0	22.37	+ 0.954	75 38 44.9	.34	51.15	1
7043	Octantis	9.7	29.68	− 9.956	87 37 17.2	.34	51.25	5
7044	Lacaille, 4354	8.0	34.56	+ 0.564	77 49 59.4	.34	50.39	2
7045	Chamæleontis	10.0	34.58	+ 1.046	75 3 22.8	− 18.34	51.29	1
7046	Gould, Z. C., 10h 1846	9.8	10 24 38.86	+ 1.907	65 6 31.8	− 18.35	50.28	2
7047	Gould, 14350	9.5	51.04	+ 1.878	65 41 31.1	.35	50.28	1
7048	Octantis	10.0	57.92	− 2.749	84 42 43.8	.36	50.83	2
7049	Carinæ	10.0	58.22	+ 1.660	69 6 45.7	.36	50.24	2
7050	Gould, Z. C., 10h 1851	9.2	24 59.82	+ 1.426	71 53 4.6	− 18.36	51.24	4
7051	Chamæleontis	9.0	10 25 7.34	− 0.241	80 46 38.4	− 18.36	51.14	1
7052	Carinæ	9.2	10.08	+ 1.551	70 32 7.5	.36	51.26	2
7053	Gould, Z. C., 10h 1871	9.2	12.38	1.508	71 1 49.3	.37	51.25	2
7054	Gould, Z. C., 10h 1878	9.0	12.88	1.683	68 50 28.9	.37	50.24	1
7055	Gould, Z. C., 10h 1898	10.0	19.86	+ 2.028	62 47 18.3	− 18.37	50.30	1
7056	Gould, Z. C., 10h 1856	9.0	10 25 24.00	+ 0.555	77 58 10.4	− 18.37	50.35	2
7057	Lacaille, 4343	6.8	29.76	1.856	66 12 55.0	.38	50.27	2
7058	Chamæleontis	9.0	32.32	1.006	75 27 57.0	.38	50.78	2
7059	Chamæleontis	9.0	35.32	0.070	79 53 12.6	.38	51.20	2
7060	Gould, Z. C., 10h 1902	9.0	36.34	+ 1.559	70 31 12.2	− 18.38	51.30	2
7061	Gould, Z. C., 10h 1916	8.8	10 25 41.71	+ 1.762	67 46 58.4	− 18.38	50.90	3
7062	Gould, Z. C., 10h 1924	9.8	51.38	1.563	70 29 57.6	.39	51.26	2
7063	Chamæleontis	10.0	52.38	0.458	78 27 27.1	.39	51.12	1
7064	Carinæ	9.2	52.78	1.647	69 27 41.1	.39	51.26	2
7065	Gould, Z. C., 10h 1937	10.0	25 57.68	+ 1.765	67 47 39.6	− 18.39	50.25	1

Number.	Constellation, Name of Star, or Synonym.	Magnitude.	Right Ascension, 1850.0.	Annual Precession.	South Declination, 1850.0.	Annual Precession.	Mean year.	No. of obs.
			h. m. s.	s.	° ′ ″	″		
7 066	Gould, 14374	9.0	10 26 0.58	+ 1.872	66 2 8.9	− 18.39	50.28	1
7 067	Gould, Z. C., 10^h 1949 .	10.2	1.31	2.008	63 24 32.2	.39	50.84	2
7 068	Gould, Z. C., 10^h 1952 .	9.0	1.51	2.054	62 21 24.0	.39	51.74	2
7 069	Carinæ	9.2	2.88	1.677	69 5 3.0	.39	50.24	2
7 070	Gould, Z. C., 10^h 1935 .	8.5	3.31	+ 1.445	71 51 36.9	− 18.39	51.24	4
7 071	Gould, Z. C., 10^h 1957 .	10.2	10 26 5.39	+ 2.007	63 24 41.8	− 18.40	51.34	2
7 072	Brisbane, 3068	9.2	8.73	1.600	70 7 11.6	.40	50.23	1
7 073	Gould, Z. C., 10^h 1969 .	8.2	14.10	2.063	62 12 9.0	.40	51.74	2
7 074	Gould, Z. C., 10^h 1956 .	9.5	16.93	1.546	70 47 36.5	.40	51.60	3
7 075	Chamæleontis	9.0	17.94	+ 1.016	75 30 19.3	− 18.40	51.15	1
7 076	Gould, Z. C., 10^h 1985 .	9.0	10 26 30.10	+ 2.054	62 28 37.3	− 18.41	51.18	1
7 077	Lacaille, 4351	8.0	32.60	1.827	66 55 57.0	.41	50.28	2
7 078	*K* Carinæ	6.2	32.66	1.512	71 13 20.2	.41	51.25	2
7 079	Carinæ	10.0	43.76	1.258	73 43 28.0	.42	51.25	1
7 080	Melbourne (1) 526 . . .	8.0	44.06	+ 1.754	68 7 37.7	− 18.42	50.86	2
7 081	Lacaille, 4353	8.8	10 26 45.08	+ 2.026	63 10 11.0	− 18.42	50.84	2
7 082	Gould, Z. C., 10^h 1995 .	9.5	48.00	1.578	70 29 57.6	.42	51.26	2
7 083	Gould, Z. C., 10^h 2012 .	9.5	52.14	1.902	65 40 51.2	.42	50.28	1
7 084	Gould, Z. C., 10^h 2004	8.7	52.23	1.751	68 11 53.6	.42	50.71	3
7 085	Carinæ	10.0	26 58.22	+ 1.258	73 46 28.6	− 18.43	51.25	1
7 086	Carinæ	9.5	10 27 6.55	+ 1.669	69 24 7.1	− 18.43	52.28	1
7 087	Gould, 14387	9.8	11.73	+ 0.366	78 59 1.6	.43	51.28	2
7 088	Octantis	10.0	11.76	− 1.024	82 40 20.0	.43	50.35	2
7 089	Gould, Z. C., 10^h 2047 .	9.5	18.76	+ 1.901	65 47 52.7	.44	50.28	1
7 090	Gould, Z. C., 10^h 2041 .	9.0	27.72	+ 1.397	72 35 21.0	− 18.44	51.28	1
7 091	Chamæleontis	9.5	10 27 30.80	+ 0.480	78 32 59.7	− 18.45	51.12	1
7 092	Lacaille, 4367	6.0	31.54	+ 1.412	72 27 3.5	.45	51.28	1
7 093	Chamæleontis	10.0	36.27	− 0.501	81 39 48.6	.45	51.13	1
7 094	Carinæ	9.5	45.21	+ 1.571	70 45 51.5	.45	52.28	1
7 095	Gould, 14414	8.7	48.43	+ 1.775	68 1 14.2	− 18.46	50.66	3
7 096	Gould, Z. C., 10^h 2075 .	9.5	10 27 48.78	+ 1.928	65 24 41.0	− 18.46	50.28	1
7 097	Herschel, 4335[1]	10.0	49.64	1.681	69 23 6.5	.46	50.24	1
7 098	Herschel, 4335[2] . . .	10.0	50.99	+ 1.681	69 23 0.6	.46	50.24	1
7 099	Octantis	9.2	51.72	− 1.292	83 9 53.7	.46	50.36	2
7 100	Carinæ	9.5	27 55.42	+ 1.788	67 50 37.2	− 18.46	50.25	1
7 101	Brisbane, 3086[1]	9.3	10 28 6.79	+ 1.537	71 20 20.0	− 18.47	51.58	3
7 102	Brisbane, 3086[2]	9.3	8.80	1.527	71 20 15.3	.47	51.58	3
7 103	Gould, Z. C., 10^h 2101 .	9.5	18.16	1.591	70 37 38.3	.47	52.29	1
7 104	Carinæ	9.5	20.02	1.563	70 57 52.9	.47	52.28	1
7 105	Chamæleontis	9.5	24.39	+ 0.031	80 17 27.1	− 18.48	50.98	3

No 7 090. Gould's declination is 9.2″ less.

No. 7 099. The declination may be 83° 9′ 25.8″.

Number.	Constellation, Name of Star, or Synonym.	Magnitude.	Right Ascension, 1850.0.	Annual Precession.	South Declination, 1850.0.	Annual Precession.	Mean year.	No. of obs.
			h. m. s.	s.	° ′ ″	″		
7 106	Octantis	10.0	10 28 27.57	− 3.937	85 45 26.1	− 18.48	51.29	3
7 107	Brisbane, 3088	8.7	30.34	+ 1.978	64 35 42.0	.48	50.28	2
7 108	Carinæ	9.5	31.60	0.963	76 9 19.2	.48	51.30	1
7 109	Gould, Z. C., 10h 2127 .	9.5	34.50	1.994	64 16 4.0	.48	50.29	1
7 110	Carinæ	11.0	38.68	+ 1.200	74 30 7.7	− 18.48	51.16	1
7 111	Carinæ	9.7	10 28 41.20	+ 1.528	71 25 35.6	− 18.49	51.55	3
7 112	Carinæ	9.5	43.90	1.207	74 27 47.9	.49	51.16	1
7 113	Carinæ	10.0	44.53	1.535	71 21 20.1	.49	52.27	1
7 114	Lacaille, 4392	7.9	46.76	0.044	80 17 14.4	.49	50.77	4
7 115	Gould, Z. C., 10h 2128 .	9.0	48.14	+ 1.509	71 39 12.5	− 18.49	52.26	1
7 116	Gould, Z. C., 10h 2131 .	9.0	10 28 51.13	+ 1.509	71 40 0.9	− 18.49	52.26	1
7 117	Gould, Z. C., 10h 2142 .	9.5	28 55 36	1.802	67 49 35.7	.49	52.30	1
7 118	Gould, Z. C., 10h 2161 .	9.0	29 6.96	+ 1.983	64 37 37.1	.50	50.28	1
7 119	Octantis	9.8	9.06	− 1.815	83 58 10.6	.50	50.34	2
7 120	Chamæleontis	9.0	15.01	+ 0.222	79 44 33.7	− 18.50	51.24	1
7 121	Gould, 14436	8.8	10 29 21.02	+ 0.422	78 59 51.8	− 18.51	50.85	2
7 122	Gould, Z. C., 10h 2148 .	8.8	23.62	0.768	77 23 52.4	.51	50.76	2
7 123	Brisbane, 3098	9.5	29.24	1.991	64 33 24.2	.51	50.29	1
7 124	Gould, 14465	9.0	36.86	+ 1.820	67 41 3.8	.52	50.93	3
7 125	Octantis	8.6	39.66	− 3.142	85 16 32.5	− 18.52	51.28	7
7 126	Lacaille, 4376[1]	8.5	10 29 47.27	+ 2.050	63 21 34.8	− 18.52	50.84	2
7 127	Octantis	9.2	49.11	− 2.360	84 36 30.3	.52	50.33	2
7 128	Lacaille, 4376[2]	9.0	49.61	+ 2.050	63 21 15.3	.52	50.84	2
7 129	Gould, 14471	10.0	50.66	2.001	64 25 37.4	.52	50.29	1
7 130	Carinæ	10.0	29 52.32	+ 1.618	70 35 25.8	− 18.53	52.28	1
7 131	Gould, Z. C., 10h 2234 .	9.0	10 30 2.12	+ 1.902	66 23 24.5	− 18.53	50.26	1
7 132	Gould, Z. C., 10h 2241 .	9.0	5.65	2.049	63 26 42.5	.53	50.58	4
7 133	Gould, Z. C., 10h 2246 .	9.5	7.63	2.059	63 12 56.4	.53	51.38	1
7 134	Gould, Z. C., 10h 2239 .	9.5	13.53	1.639	70 23 27.7	.54	52.28	1
7 135	Gould, Z. C., 10h 2212 .	9.8	14.74	+ 1.004	76 8 0.4	− 18.54	50.74	2
7 136	Lacaille, 4391	8.0	10 30 18.16	+ 1.133	75 15 48.3	− 18.54	50.40	1
7 137	Gould, Z. C., 10h 2249 .	9.0	18.82	1.702	69 34 39.2	.54	52.28	1
7 138	Gould, Z. C., 10h 2268 .	9.2	25.82	2.050	63 30 52.2	.54	50.84	2
7 139	Carinæ	10.0	27.42	2.075	62 56 55.1	.54	50.30	1
7 140	Gould, Z. C., 10h 2264 .	9.0	31.14	+ 1.533	71 41 53.2	− 18.55	52.26	1
7 141	Gould, 14495 (Red) . .	7.8	10 30 38.58	+ 1.777	68 34 27.6	− 18.55	50.33	2
7 142	Octantis	10.5	42.23	− 1.819	84 4 18.7	.55	50.33	1
7 143	Brisbane, 3113	8.2	49.31	+ 2.056	63 28 1.1	.56	50.58	4
7 144	Carinæ	10.0	51.85	1.645	70 26 7.1	.56	52.28	1
7 145	Gould, Z. C., 10h 2303 .	9.2	53.95	+ 1.918	66 16 45.8	− 18.56	50.27	2

Number.	Constellation, Name of Star, or Synonym.	Magnitude.	Right Ascension, 1850.0.	Annual Precession.	South Declination, 1850.0.	Annual Precession.	Mean year.	No. of obs.
			h. m. s.	s.	° ′ ″	″		
7 146	Carinæ	9.0	10 30 54.28	+ 1.692	69 50 15.9	— 18.56	51.26	2
7 147	Carinæ	9.5	30 59.12	1.700	69 44 20.5	.56	52.29	1
7 148	Gould, Z. C., 10h 2322 .	9.5	31 3.72	2.103	62 25 58.4	.57	51.18	1
7 149	Gould, 14509	9.0	7.72	1.752	69 2 28.6	.57	50.24	2
7 150	Chamæleontis	9.5	9.73	+ 0.646	78 13 36.4	— 18.57	51.12	1
7 151	Gould, Z. C., 10h 2311 .	9.2	10 31 17.24	+ 1.302	74 4 45.4	— 18.57	51.23	2
7 152	Chamæleontis	10.0	26.32	0.647	78 15 44.0	.58	51.12	1
7 153	Gould, Z. C., 10h 2345 .	9.5	31.44	2.112	62 18 51.6	.58	51.18	1
7 154	Carinæ	9.5	41.55	2.027	64 18 18.4	.59	50.29	1
7 155	Carinæ	10.0	51.49	+ 1.576	71 27 52.6	— 18.59	52.27	1
7 156	Carinæ	9.5	10 31 51.49	+ 1.861	67 29 53.4	— 18.59	52.30	1
7 157	Carinæ	10.0	31 52.47	1.502	72 17 1.8	.59	51.28	1
7 158	Gould, Z. C., 10h 2365 .	8.8	32 2.83	1.700	69 57 1.6	.60	51.26	2
7 159	Carinæ	10.0	4.00	1.405	73 38 39.3	.60	51.25	1
7 160	Carinæ	9.8	6.38	+ 1.870	67 23 25.6	— 18.60	51.28	2
7 161	Gould, Z. C., 10h 2358 .	8.9	10 32 7.07	+ 1.293	74 17 35.9	— 18.60	51.01	4
7 162	Carinæ	9.5	7.11	2.110	62 30 34.6	.60	51.18	1
7 163	Gould, Z. C., 10h 2364 .	8.8	10.64	+ 1.349	73 49 24.6	.60	51.75	2
7 164	Gould, Z. C., 10h 2375 .	9.0	15.04	1.715	69 47 50.0	.60	52.29	1
7 165	Gould, Z. C., 10h 2386 .	9.5	16.49	+ 2.118	62 21 29.8	— 18.60	51.18	1
7 166	Chamæleontis	10.8	10 32 19.66	— 0.134	81 9 8.3	— 18.61	50.64	2
7 167	Gould, 14507	9.5	20.78	— 1.218	83 21 56.6	.61	50.37	2
7 168	Carinæ	9.0	26.58	+ 1.555	71 48 29.8	.61	51.24	2
7 169	Carinæ	8.5	27.72	1.562	71 43 48.1	.61	52.26	1
7 170	Chamæleontis	10.0	29.67	+ 0.852	77 20 54.0	— 18.61	51.13	1
7 171	Cape (1840), 1327 . . .	8.0	10 32 30.71	— 3.695	85 47 24.6	— 18.61	51.25	6
7 172	Chamæleontis	10.0	30.93	+ 1.048	76 10 39.0	.61	51.20	1
7 173	Octantis	8.6	33.50	—13.544	88 17 1.6	.61	51.27	8
7 174	Lacaille, 4430	7.5	35.06	— 0.123	81 8 44.0	.62	50.64	2
7 175	Carinæ	10.0	35.07	+ 1.870	67 27 50.8	— 18.62	50.26	1
7 176	Gould, Z. C., 10h 2385 .	9.2	10 32 36.70	+ 1.309	74 14 18.4	— 18.62	50.79	2
7 177	Lacaille, 4411	7.2	37.28	1.144	75 31 55.2	.62	50.78	2
7 178	Chamæleontis	9.0	42.19	0.893	77 8 59.9	.62	51.25	1
7 179	Brisbane, 3125	7.5	45.03	2.127	62 16 34.6	.62	51.18	1
7 180	Gould, 14549	9.0	49.32	+ 1.806	68 35 1.4	— 18.62	50.33	2
7 181	Lacaille, 4423	8.1	10 32 50.60	+ 0.142	80 23 13.3	— 18.62	50.82	4
7 182	Chamæleontis	10.0	51.55	— 0.445	81 58 8.3	.62	51.17	1
7 183	Lacaille, 4432	8.0	52.93	— 0.005	80 50 27.1	.62	51.14	1
7 184	Lacaille, 4414	8.0	56.56	+ 1.220	75 0 57.6	.63	51.29	1
7 185	Gould, 14551	7.4	56.82	+ 1.639	70 54 49.0	— 18.63	52.28	1

Number.	Constellation, Name of Star, or Synonym.	Magnitude.	Right Ascension, 1850.0.	Annual Precession.	South Declination, 1850.0.	Annual Precession.	Mean year.	No. of obs.
			h. m. s.	s.	° ′ ″	″		
7 186	Carinæ	10.0	10 32 58.80	+ 2.125	62 22 44.4	− 18.63	51.18	1
7 187	Brisbane, 3131	9.0	59.05	1.736	69 39 12.4	.63	52.29	1
7 188	Gould, Z. C., 10h 2449 .	9.0	32 59.36	1.706	70 3 41.0	.63	50.23	1
7 189	Carinæ	9.2	33 2.11	1.586	71 33 59.2	.63	52.27	2
7 190	Lacaille, 4405	7.7	9.02	+ 2.046	64 15 43.2	− 18.63	50.29	2
7 191	Carinæ	10.0	10 33 17.82	+ 2.127	62 23 53.0	18.64	51.18	1
7 192	Lacaille, 4431	7.0	19.79	0.344	79 44 33.9	.64	51.24	1
7 193	Carinæ	9.8	23.16	1.887	67 23 17.0	.64	51.28	2
7 194	Carinæ	9.5	24.20	1.727	69 51 23.2	.64	52.29	1
7 195	Carinæ	10.0	37.77	+ 1.692	70 22 43.3	18.65	52.28	1
7 196	Lacaille, 3659	7.4	10 33 38.46	+ 2.076	63 43 5.3	− 18.66	50.67	6
7 197	γ Chamæleontis	4.7	38.80	0.791	77 49 49.0	.66	50.39	2
7 198	Chamæleontis	9.2	48.76	0.513	79 8 21.4	.66	50.85	2
7 199	Gould, Z. C., 10h 2513 .	8.5	50.72	1.826	68 29 35.1	.66	50.42	1
7 200	Gould, Z. C., 10h 2524 .	9.0	52.72	+ 2.057	64 11 43.4	+ 18.66	50.99	1
7 201	Carinæ	9.5	10 33 58.42	+ 1.488	72 48 4.9	− 18.66	50.21	1
7 202	Chamæleontis	10.0	34 0.55	0.227	80 13 43.8	.66	51.16	1
7 203	Carinæ	10.0	2.95	1.489	72 48 10.2	.66	50.21	1
7 204	Lacaille, 4416	8.8	5.86	1.761	69 31 45.3	.66	51.26	2
7 205	Gould, Z. C., 10h 2550 .	9.0	6.45	+ 2.111	62 59 48.6	− 18.66	50.82	2
7 206	Lacaille, 4410	7.7	10 34 6.48	+ 2.097	63 19 53.5	− 18.66	51.02	3
7 207	Gould, Z. C., 10h 2559 .	9.0	15.54	2.117	62 54 36.8	.67	50.30	1
7 208	Carinæ	9.0	16.55	1.624	71 21 0.2	.67	51.25	2
7 209	Gould, Z. C., 10h 2571 .	9.0	22.88	1.982	65 52 44.9	.67	50.28	3
7 210	Gould, Z. C., 10h 2573 .	9.0	27.14	+ 1.961	66 18 7.9	− 18.68	50.26	1
7 211	Gould, Z. C., 10h 2577 .	9.0	10 34 29.53	+ 1.957	66 23 2.1	− 18.68	50.26	1
7 212	Chamæleontis	9.5	40.58	+ 0.951	77 6 23.2	.68	51.19	2
7 213	Gould, 14570	9.5	52.36	− 1.151	83 26 40.5	.69	50.37	2
7 214	Gould, Z. C., 10h 2612 .	10.0	34 58.70	+ 2.147	62 20 1.5	.69	51.18	1
7 215	Lacaille, 4418	6.7	35 0.07	+ 2.064	64 19 5.0	− 18.69	50.29	2
7 216	Gould, 14610	8.7	10 35 2.67	+ 2.112	63 13 35.2	− 18.69	51.02	3
7 217	Carinæ	10.0	4.56	2.112	63 14 18.0	.70	51.38	1
7 218	Gould, Z. C., 10h 2620 .	9.5	6.15	2.154	62 12 13.2	.70	51.18	1
7 219	Gould, Z. C., 10h 2576 .	8.2	9.00	0.461	79 31 7.6	.70	50.83	2
7 220	Chamæleontis	10.0	15.91	+ 0.757	78 12 42.7	− 18.70	51.12	1
7 221	Chamæleontis	8.8	10 35 22.04	− 0.163	81 30 53.2	− 18.70	50.64	2
7 222	Gould, Z. C., 10h 2621 .	9.0	25.59	+ 1.516	72 46 42.1	.71	50.21	1
7 223	Lacaille, 4424	7.2	27.22	2.089	63 52 11.3	.71	50.74	5
7 224	Chamæleontis	9.0	30.86	0.332	80 1 32.4	.71	51.16	1
7 225	Lacaille, 4422	7.0	34.26	+ 2.274	58 53 37.3	− 18.71	52.36	1

Number.	Constellation, Name of Star, or Synonym.	Magnitude.	Right Ascension, 1850.0.	Annual Precession.	South Declination, 1850.0.	Annual Precession.	Mean year.	No. of obs.
			h. m. s.	s.	° ′ ″	″		
7226	Carinæ	10.0	10 35 34.64	+ 1.879	68 1 29.9	— 18.71	50.25	1
7227	Octantis	9.0	36.90	— 0.626	82 35 5.1	.71	50.36	2
7228	Lacaille, 4441	7.3	41.80	+ 1.353	74 22 32.9	.71	50.98	3
7229	Lacaille, 4439	6.5	43.27	+ 1.427	73 42 40.6	.72	51.25	1
7230	Lacaille, 4427	7.5	49.58	— 2.101	63 41 29.4	— 18.72	50.62	7
7231	Gould, Z. C., 10h 2676 .	10.0	10 35 52.83	+ 1.950	66 50 13.0	— 18.72	50.30	1
7232	Gould, Z. C., 10h 2661 .	8.5	53.60	1.559	72 24 27.3	.72	51.28	1
7233	Brisbane, 3167	9.5	56.86	2.029	65 18 48.8	.72	50.28	1
7234	Chamæleontis	9.5	35 57.75	1.179	75 47 12.9	.72	51.15	1
7235	Gould, Z. C., 10h 2695 .	9.0	36 4.11	+ 2.132	62 59 48.6	— 18.73	51.37	1
7236	Brisbane, 3171	8.3	10 36 15.27	+ 2.114	63 19 22.8	— 18.73	51.02	3
7237	Carinæ	9.5	17.67	1.308	74 50 56.1	.73	51.29	1
7238	Gould, Z. C, 10h 2724 .	9.5	31.46	2.132	63 6 53.8	.74	51.37	1
7239	Lacaille, 4434	7.5	35.73	2.150	62 43 22.0	.74	50.74	2
7240	Gould, Z. C., 10h 2708 .	9.5	38.43	+ 1.216	75 37 30.6	— 18.74	51.15	1
7241	Chamæleontis	9.8	10 36 38.96	— 0.263	81 52 55.6	— 18.74	50.75	2
7242	Gould, 14646	9.0	42.89	+ 1.956	66 55 31.8	.75	50.30	1
7243	Chamæleontis	9.5	47.04	0.986	77 10 24.7	.75	51.13	1
7244	Gould, Z. C., 10h 2738 .	9.2	48.34	1.947	67 6 55.5	.75	50.28	2
7245	Lacaille, 4435	6.5	53.41	2.300	58 25 50.2	— 18.75	52.36	1
7246	Lacaille, 4460	8.0	10 36 53.80	+ 0.692	78 43 44.3	— 18.75	51.20	2
7247	Lacaille, 4440	5.5	55.30	2.113	63 40 55.7	.75	50.62	7
7248	Gould, Z. C., 10h 2743 .	9.5	36 59.90	+ 1.707	70 51 13.4	.76	52.28	1
7249	Chamæleontis	9.0	37 5.28	— 0.236	81 51 32.6	.76	50.75	2
7250	Brisbane, 3178	8.4	5.78	+ 2.118	63 36 53.4	— 18.76	50.50	6
7251	Gould, Z. C., 10h 2752 .	8.5	10 37 10.15	+ 1.573	72 29 54.2	— 18.76	51.28	1
7252	Carinæ	9.5	13.34	1.856	68 45 56.1	.76	50.24	1
7253	Gould, Z. C., 10h 2754 .	9.0	13.73	1.504	73 14 9.5	.76	50.21	1
7254	Carinæ	9.5	14.37	1.564	72 36 14.8	.76	51.28	1
7255	Gould, Z. C., 10h 2772 .	9.2	16.88	+ 1.953	67 6 40.8	— 18.76	50.28	2
7256	Chamæleontis	10.0	10 37 19.61	+ 0.990	77 13 32.8	— 18.77	51.13	1
7257	Carinæ	8.5	20.33	2.109	63 53 32.0	.77	50.83	4
7258	Gould, Z. C., 10h 2781 .	9.0	33.62	1.664	71 30 37.8	.77	52.27	2
7259	Carinæ	9.0	34.85	2.123	63 36 51.0	.77	51.36	1
7260	θ Argus	2.8	36.85	+ 2.124	63 36 32.7	— 18.77	50.62	7
7261	Chamæleontis	9.5	10 37 42.72	— 0.072	81 30 1.7	— 18.78	51.13	1
7262	Chamæleontis	9.0	46.04	— 0.330	82 8 26.6	.78	50.37	1
7263	Gould, Z. C., 10h 2816 .	9.0	50.03	+ 2.181	62 13 25.9	.78	51.18	1
7264	Carinæ	9.4	51.22	1.892	68 19 10.8	.78	50.60	4
7265	Gould, Z. C., 10h 2821 .	10.0	37 54.39	+ 2.137	63 23 53.3	— 18.78	51.38	1

Number.	Constellation, Name of Star, or Synonym.	Magnitude.	Right Ascension, 1850.0.	Annual Precession.	South Declination, 1850.0.	Annual Precession.	Mean year.	No. of obs.
			h. m. s.	s.	° ′ ″	″		
7 266	Octantis	8.4	10 38 3.84	− 5.737	86 57 33.9	18.79	51.25	4
7 267	Brisbane, 3189	6.8	6.34	+ 2.143	63 16 56.3	.79	51.02	3
7 268	Lacaille, 4449	8.0	10.07	2.301	58 45 39.3	.79	52.36	1
7 269	Gould, Z. C., 10h 2841	9.0	10.55	2.160	62 51 37.0	.79	50.30	1
7 270	Gould, Z. C., 10h 2840	10.0	13.77	+ 1.988	66 41 39.3	− 18.79	50.26	1
7 271	Gould, Z. C., 10h 2845	9.0	10 38 15.16	+ 1.998	66 30 15.5	− 18.79	50.26	1
7 272	Stone, 5777	7.3	15.84	−18.313	88 44 58.1	.79	51.30	5
7 273	Carinæ	9.5	18.56	+ 1.662	71 41 40.4	.80	52.26	1
7 274	Carinæ	10.0	21.54	2.178	62 27 23.4	.80	51.18	1
7 275	Carinæ	9.5	24.32	+ 2.086	64 41 37.2	− 18.80	50.28	2
7 276	Chamæleontis	9.5	10 38 25.00	− 0.024	81 26 27.1	− 18.80	51.13	1
7 277	Lacaille, 4452	6.7	28.17	+ 2.139	63 27 45.6	.80	50.79	9
7 278	Gould, Z. C., 10h 2868	10.0	28.84	1.992	66 41 8.2	.80	50.26	1
7 279	Gould, Z. C., 10h 2861	9.2	33.24	1.718	71 2 38.2	.80	52.28	2
7 280	Gould, Z. C., 10h 2876	9.5	34.36	+ 2.183	62 23 9.7	− 18.80	51.18	1
7 281	Gould, Z. C., 10h 2869	9.0	10 38 36.13	+ 1.724	70 57 49.2	− 18.80	52.28	2
7 282	Lacaille, 4455	6.5	41.93	2.154	63 10 29.3	.81	51.02	3
7 283	Gould, Z. C., 10h 2886	9.0	45.83	1.987	66 49 55.4	.81	50.30	1
7 284	Carinæ	9.5	51.05	1.958	67 24 30.6	.81	52.30	1
7 285	Gould, Z. C., 10h 2882	7.7	51.62	+ 1.694	71 24 40.2	− 18.81	52.27	2
7 286	Gould, Z. C., 10h 2897	9.0	10 38 54.35	+ 1.885	68 39 41.3	− 18.81	50.33	2
7 287	Octantis	9.0	39 4.38	− 1.831	84 36 42.4	.82	50.33	2
7 288	Lacaille, 4466	7.5	9.49	+ 1.386	74 40 41.4	.82	50.85	2
7 289	Gould, Z. C., 10h 2923	9.0	14.24	+ 2.166	63 1 19.3	.82	50.84	2
7 290	Lacaille, 4510	6.9	14.35	− 2.555	85 18 41.0	− 18.82	51.29	5
7 291	η Argus	1.0	10 39 15.50	+ 2.307	58 53 50.1	− 18.82	52.36	1
7 292	Gould, Z. C., 10h 2935	9.0	24.83	2.019	65 53 11.0	.83	50.28	3
7 293	Gould, Z. C., 10h 2941	9.0	29.18	1.944	67 47 58.1	.83	52.30	1
7 294	Chamæleontis	10.0	33.14	1.015	77 23 58.3	.83	51.13	1
7 295	Carinæ	9.8	41.48	+ 1.966	67 27 36.3	− 18.84	51.28	2
7 296	Octantis	8.2	10 39 42.85	− 1.695	84 30 7.0	− 18.84	50.32	1
7 297	Carinæ	10.0	45.08	+ 1.666	71 55 50.7	.84	52.26	1
7 298	Lacaille, 4467	7.0	48.73	1.808	70 4 20.6	.84	50.23	1
7 299	Gould, Z. C., 10h 2964	9.2	51.30	1.731	71 8 7.2	.84	52.28	2
7 300	Gould, Z. C., 10h 2965	8.0	51.61	+ 1.720	71 16 39.1	− 18.84	52.27	1
7 301	Gould, Z. C., 10h 2970	11.0	10 39 53.14	+ 1.809	70 3 50.7	− 18.84	50.23	1
7 302	Brisbane, 3202	8.0	53.44	2.321	58 37 10.2	.84	52.36	1
7 303	Gould, Z. C., 10h 2977	10.0	39 56.50	2.046	65 57 2.4	.85	50.27	2
7 304	Stone, 5946	7.0	40 0.63	1.811	70 4 4.5	.85	50.23	1
7 305	Brisbane, 3204	7.5	0.73	+ 2.320	58 42 1.0	− 18.85	52.36	1

Number.	Constellation, Name of Star, or Synonym.	Magnitude.	Right Ascension, 1850.0.	Annual Precession.	South Declination, 1850.0.	Annual Precession.	Mean year.	No. of obs.
			h. m. s.	s.	° ′ ″	″		
7 306	Gould, Z. C., 10h 3000 .	8.8	10 40 11.22	+ 2.140	63 54 54.0	— 18.85	50.73	5
7 307	Lacaille, 4489	7.0	13.01	0.726	78 59 52.2	.85	50.85	2
7 308	Chamæleontis	9.0	16.29	1.271	75 47 35.5	.85	51.15	1
7 309	Lacaille, 4474	7.0	16.48	+ 1.697	71 39 26.6	.86	52.28	2
7 310	Chamæleontis	9.0	23.87	— 0.307	82 19 46.2	— 18.86	50.38	2
7 311	Carinæ	9.5	10 40 40.24	+ 2.164	63 26 27.2	— 18.87	51.38	2
7 312	Carinæ	9.0	44.78	2.110	64 49 36.0	.87	50.28	1
7 313	Brisbane, 3210	8.7	45.32	+ 2.150	63 48 21.8	.87	50.62	7
7 314	Octantis	9.5	46.06	—12.404	88 19 22.7	.87	51.20	4
7 315	Gould, Z. C., 10h 3041 .	9.2	50.56	+ 1.988	67 20 26.6	18.87	50.26	2
7 316	Lacaille, 4471	6.9	10 40 51.32	+ 2.155	63 43 31.4	— 18.87	50.59	8
7 317	Carinæ	10.0	53.07	1.811	70 15 35.9	.87	52.28	1
7 318	Lacaille, 4470	7.0	40 53.64	2.202	62 30 28.7	.87	51.18	1
7 319	Lacaille, 4473	6.9	41 2.38	2.167	63 28 28.1	.88	50.85	8
7 320	Chamæleontis	10.0	7.61	+ 1.051	77 25 8.9	— 18.88	51.13	1
7 321	Gould, Z. C., 10h 3062 .	9.0	10 41 12.50	+ 1.857	69 38 21.6	— 18.88	52.29	1
7 322	Carinæ	9.2	16.67	+ 2.155	63 49 35.2	.88	51.00	3
7 323	Octantis	10.5	17.85	— 1.590	84 29 0.9	.89	50.33	1
7 324	Carinæ	9.5	22.34	+ 1.985	67 30 58.8	.89	51.28	2
7 325	Lacaille, 4475	6.2	24.10	+ 2.166	63 35 37.5	— 18.89	50.68	6
7 326	Gould, Z. C., 10h 3083 .	10.0	10 41 28.06	+ 2.061	66 1 49.4	— 18.89	50.28	1
7 327	Gould, Z. C., 10h 3079 .	8.2	35.74	1.668	72 16 38.4	.89	51.77	2
7 328	Brisbane, 3222	8.7	41.03	2.174	63 28 28.3	.90	51.15	5
7 329	Gould, Z. C., 10h 3092 .	9.5	44.02	+ 1.993	67 28 0.9	.90	51.28	2
7 330	Octantis	9.5	44.69	— 1.578	84 30 38.9	— 18.90	50.32	1
7 331	Carinæ	8.5	10 41 47.06	+ 1.508	74 1 56.3	— 18.90	52.26	1
7 332	Lacaille, 4490	8.0	49.72	1.489	74 13 3.5	.90	51.33	2
7 333	Lacaille, 4486	7.5	55.09	1.942	68 25 11.4	.90	50.42	1
7 334	Gould, Z. C., 10h 3101 .	8.0	41 56.82	1.747	71 22 33.0	.90	52.27	2
7 335	Gould, 14789	8.8	42 2.00	+ 2.110	65 7 35.4	— 18.91	50.28	2
7 336	Chamæleontis	9.0	10 42 3.65	+ 1.220	76 27 24.5	— 18.91	51.15	1
7 337	Carinæ	10.0	12.30	1.733	71 36 37.8	.91	52.26	1
7 338	Lacaille, 4485	7.2	12.72	2.169	63 45 24.3	.91	50.68	6
7 339	Lacaille, 4504	9.0	17.19	0.184	81 14 47.7	.91	50.14	1
7 340	Gould, Z. C., 10h 3116 .	9.8	18.08	+ 2.113	65 7 46.5	— 18.91	50.28	2
7 341	Lacaille, 4487	7.6	10 42 24.00	+ 2.182	63 28 18.9	— 18.92	50.85	8
7 342	Carinæ	9.0	30.41	1.740	71 34 46.1	.92	52.26	1
7 343	Carinæ	9.0	34.47	2.200	63 2 28.7	.92	51.37	1
7 344	Carinæ	10.0	35.14	+ 1.887	69 29 49.5	.92	50.24	1
7 345	Octantis	8.5	35.46	— 1.455	84 24 35.6	— 18.92	50.33	1

No. 7 311. The right ascension may be $10^h\,40^m\,2.74^s$.

Number.	Constellation, Name of Star, or Synonym.	Magnitude.	Right Ascension, 1850.0.	Annual Precession.	South Declination, 1850.0.	Annual Precession.	Mean year.	No. of obs.
			h. m. s.	s.	° ′ ″	″		
7 346	Gould, 14805	9.0	10 42 35.71	+ 2.199	63 5 17.8	− 18.92	50.84	2
7 347	Lacaille, 4491	7.5	37.94	+ 1.939	68 38 46.4	.92	50.33	2
7 348	Octantis	9.1	50.91	− 9.017	87 54 27.6	.93	51.28	6
7 349	Carinæ	9.5	53.60	+ 1.739	71 40 26.1	93	52.26	1
7 350	Octantis	8.0	42 56.14	− 2.644	85 35 11.1	− 18.93	51.29	5
7 351	Carinæ	9.0	10 43 3.05	+ 1.551	73 38 49.1	− 18.94	51.25	1
7 352	Carinæ	9.5	4.39	1.850	70 11 4.6	.94	52.28	1
7 353	Gould, Z. C., 10h 3206	9.5	7.23	1.881	69 43 43.1	.94	52.29	1
7 354	Gould, 14819	9.0	8.62	2.207	63 2 28.2	.94	50.84	2
7 355	Gould, Z. C., 10h 3228	9.5	15.98	+ 2.200	63 15 32.0	− 18.94	51.38	2
7 356	Octantis	10.0	10 43 18.16	− 2.528	85 30 57.8	− 18.94	51.23	1
7 357	Chamæleontis	9.0	20.60	+ 0.125	81 31 39.8	.94	50.64	2
7 358	Brisbane, 3240	9.0	30.20	2.064	66 30 45.4	.95	50.26	1
7 359	Gould, Z. C., 10h 3245	9.5	34.00	1.877	69 52 36.7	.95	52.28	1
7 360	Gould, Z. C., 10h 3243	10.0	37.17	+ 1.754	71 38 41.4	− 18.95	52.26	1
7 361	Chamæleontis	9.5	10 43 41.15	+ 1.207	76 48 34.8	− 18.95	51.25	1
7 362	Gould, Z. C., 10h 3255	9.5	42.05	2.054	66 46 46.5	.96	50.30	1
7 363	Carinæ	9.0	43.16	1.851	70 19 1.6	.96	52.28	1
7 364	Carinæ	10.0	44.26	1.757	71 37 42.9	.96	52.26	1
7 365	δ^1 Chamæleontis	5.5	47.26	+ 0.675	79 40 39.4	− 18.96	51.20	2
7 366	Gould, Z. C., 10h 3267	9.3	10 43 55.36	+ 1.995	67 58 28.7	− 18.96	50.66	3
7 367	Gould, Z. C., 10h 3236	8.8	43 59.86	0.651	79 48 23.6	.96	51.20	2
7 368	Gould, Z. C., 10h 3275	8.8	44 8.04	2.019	67 34 7.9	.97	51.28	2
7 369	Carinæ	9.5	13.92	1.687	72 35 43.6	.97	51.28	1
7 370	Gould, Z. C., 10h 3281	9.5	15.21	+ 1.891	69 49 26.3	− 18.97	51.26	2
7 371	Carinæ	10.7	10 44 16.96	+ 2.001	67 57 18.4	− 18.97	50.66	3
7 372	Gould, Z. C., 10h 3288	10.0	17.09	2.039	67 13 41.6	.97	50.26	1
7 373	Chamæleontis	10.0	17.17	0.401	80 46 37.9	.97	51.14	1
7 374	Gould, Z. C., 10h 3286	10.5	18.65	2.001	67 57 13.6	.97	50.66	3
7 375	Chamæleontis	10.0	18.72	+ 0.838	79 1 5.3	− 27.97	51.28	1
7 376	δ^2 Chamæleontis	4.8	10 44 19.47	+ 0.675	79 44 58.6	− 18.97	51.20	2
7 377	Gould, Z. C., 10h 3294	10.0	20.39	2.215	63 10 2.6	.97	51.38	1
7 378	Brisbane, 3246	8.8	28.96	2.195	63 44 51.7	.98	50.62	7
7 379	Gould, Z. C., 10h 3302	9.0	45.43	1.465	74 58 46.9	.99	51.29	1
7 380	Gould, Z. C., 10h 3334	9.5	49.12	+ 2.219	63 11 14.3	− 18.99	50.84	2
7 381	Gould, Z. C., 10h 3336	9.2	10 44 56.67	+ 2.042	67 19 37.4	− 18.99	50.26	2
7 382	Gould, Z. C., 10h 3348	9.5	45 0.49	2.243	62 35 3.8	.99	51.18	1
7 383	Chamæleontis	10.0	0.86	0.186	81 31 55.9	.99	51.13	1
7 384	Gould, Z. C., 10h 3333	9.5	5.31	1.572	74 1 38.0	18.99	52.26	1
7 385	Gould, 14848	8.0	7.03	+ 1.836	70 51 39.2	− 19.00	52.28	1

Number.	Constellation, Name of Star, or Synonym.	Magnitude.	Right Ascension, 1850.0.	Annual Precession.	South Declination, 1850.0.	Annual Precession.	Mean year.	No. of obs.
			h. m. s.	s.	° ′ ″	″		
7386	Carinæ	9.5	10 45 8.15	+ 1.518	74 33 46.7	− 19.00	50.41	1
7387	Carinæ	9.5	10.10	2.251	62 25 25.8	.00	51.18	1
7388	Gould, Z. C., 10^h 3356	10.0	10.62	2.046	67 18 31.0	.00	50.26	1
7389	Gould, Z. C., 10^h 3313	8.5	12.04	0.673	79 52 10.2	.00	51.20	2
7390	Brisbane, 3250	8.8	12.94	+ 2.205	63 41 6.2	− 19.00	50.62	7
7391	Gould, Z. C., 10^h 3355	8.8	10 45 15.84	+ 1.826	71 1 44.2	− 19.00	52.28	2
7392	Gould, Z. C., 10^h 3379	9.5	33.72	1.874	70 23 14.9	.01	52.28	1
7393	Lacaille, 4512	8.5	37.54	1.430	75 26 38.5	.01	50.78	2
7394	Chamæleontis	9.5	38.68	1.410	75 37 3.3	.01	51.15	1
7395	Chamæleontis	10.0	41.25	+ 1.310	76 24 42.5	− 19.01	51.15	1
7396	Gould, Z. C., 10^h 3401	9.2	10 45 49.40	+ 2.042	67 34 6.4	− 19.02	51.28	2
7397	Brisbane, 3261	8.0	46 1.20	2.118	66 1 12.1	.02	50.28	1
7398	Carinæ	10.0	1.75	1.614	73 46 48.4	.02	51.24	1
7399	Gould, Z. C., 10^h 3413	9.0	10.79	1.539	74 33 40.4	.02	50.41	1
7400	Octantis	10.0	12.22	− 1.632	84 51 34.1	− 19.03	51.33	1
7401	Chamæleontis	9.0	10 46 16.96	+ 1.375	76 18 53.0	19.03	51.22	2
7402	Octantis	10.0	27.56	− 3.017	86 2 16.8	.03	51.20	2
7403	Gould, Z. C., 10^h 3439	9.0	52.56	+ 0.837	79 22 9.7	.04	50.42	1
7404	Gould, Z. C., 10^h 3473	9.0	55.26	2.047	67 44 54.9	.05	52.30	1
7405	Brisbane, 3269	9.0	46 57.41	+ 2.227	63 37 3.2	− 19.05	50.32	5
7406	Chamæleontis	10.0	10 47 10.52	+ 1.426	75 45 15.9	− 19.05	51.15	1
7407	Gould, 14901	7.5	16.23	2.277	62 17 25.1	.05	51.18	1
7408	Gould, Z. C., 10^h 3494	9.0	17.18	1.980	69 5 56.2	.06	50.24	2
7409	Carinæ	9.5	19.22	1.637	73 48 14.4	.06	51.25	1
7410	Gould, Z. C., 10^h 3500	9.0	29.00	+ 1.654	73 38 55.3	− 19.06	51.25	1
7411	Gould, Z. C., 10^h 3509	9.0	10 47 31.65	+ 1.765	72 20 54.4	− 19.06	51.77	2
7412	Gould, Z. C., 10^h 3516	9.8	31.78	2.030	68 15 6.3	.06	50.80	2
7413	Chamæleontis	10.0	33.27	1.247	77 10 20.5	.06	51.13	1
7414	Lacaille, 4521	7.5	33.92	2.073	67 24 51.2	.06	50.94	3
7415	Lacaille, 4529	8.0	34.97	+ 1.364	76 19 43.8	− 19.06	51.22	2
7416	Lacaille, 4528	7.0	10 47 44.10	+ 1.514	75 5 12.2	− 19.07	50.84	2
7417	Octantis	10.0	46.40	− 2.909	86 2 8.6	.07	51.20	2
7418	Gould, 14918	8.6	48.94	+ 2.030	68 19 28.2	.07	50.60	4
7419	Octantis	8.0	53.73	− 13.788	88 35 38.9	.07	51.61	6
7420	Gould, Z. C., 10^h 3550	9.5	47 59.38	+ 1.770	72 23 25.0	− 19.07	51.77	2
7421	Gould, Z. C., 10^h 3559	9.0	10 48 8.68	+ 1.728	72 56 14.4	− 19.08	50.21	1
7422	Carinæ	9.5	12.14	1.852	71 19 47.8	.08	52.27	1
7423	Carinæ	10.0	12.37	1.553	74 49 31.5	.08	51.29	1
7424	Carinæ	9.0	21.50	1.811	71 55 58.7	.08	52.26	2
7425	Carinæ	9.5	25.74	+ 1.845	71 29 34.0	− 19.09	52.27	2

Number.	Constellation, Name of Star, or Synonym.	Magnitude.	Right Ascension, 1850.0.	Annual Precession.	South Declination, 1850.0.	Annual Precession.	Mean year.	No. of obs.
			h. m. s.	s.	° ′ ″	″		
7426	Brisbane, 3290	8.5	10 48 27.67	+ 0.710	80 8 19.1	— 19.09	50.82	2
7427	Gould, Z. C., 10ʰ 3581	9.5	27.80	1.905	70 36 48.9	.09	52.28	1
7428	Gould, Z. C., 10ʰ 3589 .	10.0	27.88	2.149	66 1 12.9	.09	50.28	1
7429	Carinæ	9.0	37.25	1.899	70 44 45.8	.09	52.28	2
7430	Carinæ	9.5	38.65	+ 1.829	71 45 53.8	— 19.09	52.26	1
7431	Lacaille, 4578	7.8	10 48 40.01	— 2.948	86 6 30.8	— 19.09	51.77	5
7432	Gould, Z. C., 10ʰ 3614 .	9.5	45.31	+ 2.129	66 33 23.9	.10	50.26	1
7433	Gould, Z. C., 10ʰ 3617 .	9.0	46.33	2.263	63 10 22.8	.10	50.30	1
7434	Lacaille, 4531	6.8	47.35	1.954	69 55 18.6	.10	51.26	2
7435	Carinæ	9.0	51.93	+ 1.846	71 34 16.8	— 19.10	52.26	1
7436	Gould, Z. C., 10ʰ 3620 .	9.5	10 48 54.03	+ 2.134	66 29 11.8	— 19.10	50.26	1
7437	Gould, Z. C., 10ʰ 3621 .	9.5	48 54.64	2.287	62 32 8.5	.10	51.18	1
7438	Chamæleontis	10.0	49 14.72	0.358	81 29 13.6	.11	51.13	1
7439	Gould, Z. C., 10ʰ 3638 .	8.2	18.43	+ 1.869	71 21 25.2	.11	52.27	2
7440	Octantis	9.8	21.63	— 5.748	87 22 0.7	— 19.11	51.33	3
7441	Carinæ	9.5	10 49 24.51	+ 1.609	74 30 29.9	— 19.11	50.41	1
7442	Carinæ	10.0	27.14	1.786	72 31 9.1	.11	51.18	1
7443	Gould, Z. C., 10ʰ 3646 .	9.5	27.36	1.962	69 56 37.3	.11	51.26	2
7444	Gould, 14946	8.8	29.30	0.956	79 9 59.9	.11	50.85	2
7445	Gould, Z. C., 10ʰ 3649 .	9.0	29.87	+ 2.003	69 15 19.9	— 19.11	50.24	1
7446	Gould, Z. C., 10ʰ 3667 .	9.0	10 49 44.92	+ 2.096	67 31 55.8	— 19.12	51.28	2
7447	Octantis	8.8	50.84	— 8.030	87 55 17.5	.12	51.29	6
7448	Lacaille, 4544	6.8	49 58.06	+ 1.047	78 45 39.7	.13	51.25	3
7449	Octantis	9.5	50 5.24	— 2.504	85 52 49.4	.13	51.18	1
7450	Gould, Z. C., 10ʰ 3732 .	9.0	29.54	+ 2.084	67 59 26.2	— 19.14	50.80	2
7451	Chamæleontis	9.2	10 50 31.60	+ 1.062	78 46 9.0	— 19.14	51.20	2
7452	Gould, Z. C., 10ʰ 3737 .	9.5	33.32	2.040	68 51 52.7	.14	50.24	1
7453	Gould, Z. C., 10ʰ 3740 .	9.5	34.17	2.097	67 45 37.7	.14	52.30	1
7454	Chamæleontis	9.0	37.58	0.076	82 25 16.0	.14	50.39	1
7455	Carinæ	10.0	40.07	+ 1.807	72 30 47.6	— 19.15	51.27	1
7456	Carinæ	9.2	10 50 46.60	+ 1.797	72 40 37.4	— 19.15	50.74	2
7457	Gould, Z. C., 10ʰ 3752 .	9.5	50.70	1.940	70 39 9.4	.15	52.28	1
7458	Chamæleontis	9.5	50.82	1.162	78 14 55.5	.15	51.24	2
7459	Carinæ	10.0	51.84	1.388	76 43 25.4	.15	51.25	1
7460	Gould, Z. C., 10ʰ 3769 .	10.0	50 54.94	+ 2.239	64 30 16.4	— 19.15	50.29	1
7461	Brisbane, 3304	8.8	10 51 13.06	+ 1.657	74 23 18.4	— 19.16	51.33	2
7462	Gould, Z. C., 10ʰ (3754) .	9.5	14.67	1.075	78 48 32.5	.16	51.28	1
7463	Carinæ	9.0	19.16	2.008	69 38 52.9	.16	52.29	1
7464	Gould, Z. C., 10ʰ 3809 .	9.5	28.60	2.252	64 19 56.0	.17	50.29	2
7465	Carinæ	9.5	38.96	+ 2.131	67 19 30.2	— 19.17	50.26	2

No. 7462. Gould's declination is 78° 51′ 28.1″.

Number.	Constellation, Name of Star, or Synonym.	Magnitude.	Right Ascension, 1850.0.	Annual Precession.	South Declination, 1850.0.	Annual Precession.	Mean year.	No. of obs.
			h. m. s.	s.	° ′ ″	″		
7 466	Lacaille, 4545	8.2	10 51 45.06	+ 2.107	67 52 0.2	− 19.17	51.30	3
7 467	Gould, Z. C., 10^h 3840	10.0	52 0.50	2.044	69 10 41.5	.18	50.24	1
7 468	Lacaille, 4548[1]	7.8	3.97	2.094	68 14 10.1	.18	50.60	4
7 469	Gould, 15010	8.5	4.36	1.966	70 33 8.4	.18	52.28	2
7 470	Lacaille, 4548[2]	9.6	5.51	+ 2.094	68 14 14.8	− 19.18	50.60	4
7 471	Gould, 15011	8.8	10 52 6.10	+ 1.966	70 33 31.5	− 19.18	52.28	2
7 472	Gould, Z. C., 10^h 3848	. .	7.67	2.081	68 30 43.5	.18	50.43	1
7 473	Carinæ	9.5	11.34	1.729	73 48 37.2	.18	51.25	1
7 474	Gould, Z. C., 10^h 3857	. .	13.17	2.086	68 25 36.2	.19	50.43	1
7 475	Carinæ	9.0	18.59	+ 2.283	63 43 35.6	− 19.19	50.29	1
7 476	Carinæ	9.8	10 52 19.61	+ 1.889	71 47 {46.0, 18.1}	− 19.19	52.26	2
7 477	Gould, Z. C., 10^h 3862	9.5	20.14	2.005	69 57 50.9	.19	52.29	1
7 478	Carinæ	10.0	43.29	2.144	67 21 47.5	.19	50.26	1
7 479	Brisbane, 3320	8.8	46.14	1.993	70 18 17.2	.20	51.26	2
7 480	Carinæ	9.8	52.14	+ 1.898	71 47 {35.8, 7.9}	− 19.20	52.26	2
7 481	Gould, Z. C., 10^h 3914	9.5	10 52 59.21	+ 2.328	62 36 23.1	− 19.21	51.18	1
7 482	Lacaille, 4564	7.0	53 0.04	1.701	74 17 50.6	.21	51.33	2
7 483	Gould, Z. C., 10^h 3911	9.0	1.75	2.062	69 7 37.1	.21	50.24	2
7 484	Carinæ	10.0	12.40	1.735	73 58 13.0	.21	51.25	1
7 485	Carinæ	9.8	18.38	+ 1.871	72 17 26.4	− 19.21	51.77	2
7 486	Gould, Z. C., 10^h 3942	9.2	10 53 23.18	+ 2.318	63 2 49.9	− 19.22	50.30	2
7 487	Carinæ	11.0	28.38	2.319	63 3 2.4	.22	50.30	1
7 488	Carinæ	9.0	39.56	1.854	72 36 1.0	.22	50.74	2
7 489	Gould, Z. C., 10^h 3964	9.5	43.32	2.333	62 42 14.8	.22	50.74	2
7 490	Gould, Z. C., 10^h 3967	10.0	47.36	+ 2.222	65 50 55.3	+ 19 23	50.28	1
7 491	Gould, Z. C., 10^h 3954	9.0	10 53 49.86	+ 1.671	74 47 53.2	− 19.23	51.29	1
7 492	Carinæ	9.0	54 2.81	1.753	73 56 58.8	.23	51.75	2
7 493	Carinæ	9.5	28.40	1.943	71 31 42.4	.24	52.27	2
7 494	Gould, Z. C., 10^h 3982	9.0	31.80	1.001	79 41 11.0	.24	51.24	1
7 495	Chamæleontis	9.5	38.74	+ 0.730	80 52 35.9	− 19.25	51.14	1
7 496	Gould, Z. C., 10^h 4025	9.0	10 54 39.12	+ 2.237	65 44 14.7	− 19.25	50.28	1
7 497	Brisbane, 3339	9.2	44.14	2.140	68 2 10.5	.25	50.86	2
7 498	Gould, Z. C., 10^h 4035	10.0	50.49	2.100	68 53 55.2	.25	50.24	1
7 499	Gould, Z. C., 10^h 4037	9.0	51.58	2.096	68 59 36.7	.25	50.24	1
7 500	Gould, Z. C., 10^h 4040	9.0	52.52	+ 2.270	64 57 10.2	− 19.25	50.28	1
7 501	Gould, 15083	8.5	10 54 56.28	+ 2.260	65 13 35.7	− 19.25	50.28	1
7 502	Carinæ	10.0	55 7.36	1.962	71 23 51.1	.26	52.27	2
7 503	Gould, Z. C., 10^h 4061	9.0	12.91	2.270	65 2 43.7	.26	50.28	1
7 504	Gould, Z. C., 10^h 4056	9.5	14.42	1.980	71 8 26.4	.26	52.27	1
7 505	Chamæleontis	9.5	16.82	+ 1.544	76 17 45.4	− 19.26	51.22	2

No. 7 494. Gould's declination is 4′ greater.

Number.	Constellation, Name of Star, or Synonym.	Magnitude.	Right Ascension, 1850.0.	Annual Precession.	South Declination, 1850.0.	Annual Precession.	Mean year.	No. of obs.
			h. m. s.	s.	° ′ ″	″		
7 506	Carinæ	9.8	10 55 22.56	+ 1.911	72 13 50.7	− 19.26	51.77	2
7 507	Octantis	9.4	31.20	− 3.219	86 37 20.8	.27	51.27	6
7 508	Carinæ	10.0	31.65	+ 1.887	72 36 40.7	.27	51.28	1
7 509	Gould, Z. C., 10h 4078 .	9.2	42.14	2.019	70 39 8.3	.27	52.28	2
7 510	Carinæ	10.0	45.24	+ 1.922	72 10 29.3	− 19.27	52.26	1
7 511	Gould, Z. C., 10h 4090 .	8.5	10 55 55.11	+ 2.210	66 48 39.9	− 19.28	50.30	1
7 512	Lacaille, 4573	7.0	55 56.75	2.332	63 29 47.9	.28	50.29	1
7 513	Chamæleontis	10.0	56 1.62	0.782	80 51 44.1	.28	51.14	1
7 514	Brisbane, 3355	8.0	6.40	0.979	80 1 44.7	.28	51.16	1
7 515	Gould, Z. C., 10h 4125 .	8.5	10.44	+ 2.276	65 12 24.3	− 19.28	50.28	1
7 516	Gould, Z. C., 10h 4137 .	9.0	10 56 18.73	+ 2.256	65 47 34.7	− 19.29	50.28	1
7 517	Gould, Z. C., 10h 4150 .	10.0	22.17	2.337	63 30 . .	.29	50.30	1
7 518	Gould, Z. C., 10h 4096 .	8.5	22.97	0.997	79 59 15.5	.29	51.16	1
7 519	Gould, Z. C., 10h 4153 .	9.2	29.63	2.183	67 38 0.2	.29	51.28	2
7 520	Gould, Z. C., 10h 4164 .	10.0	35.93	+ 2.366	62 40 44.0	− 19.29	51.18	1
7 521	Gould, Z. C., 10h 4166 .	9.5	10 56 40.77	+ 2.195	67 24 49.3	− 19.30	52.30	1
7 522	Gould, Z. C., 10h 4184 .	9.5	47.40	2.367	62 41 13.2	.30	50.74	2
7 523	Gould, Z. C., 10h 4175 .	9.0	50.10	2.032	70 42 26.2	.30	52.28	2
7 524	Chamæleontis	9.0	56.59	2.237	66 29 9.4	.30	50.26	1
7 525	Carinæ	9.0	56.96	+ 2.035	70 41 29.2	.30	52.28	1
7 526	Gould, Z. C., 10h 4186 .	8.0	10 56 58.16	+ 2.021	70 56 23.5	− 19.30	52.28	1
7 527	Chamæleontis	10.0	56 59.98	0.830	80 48 8.9	.30	51.14	1
7 528	Lacaille, 4589	7.5	57 8.19	1.734	74 51 33.1	.31	51.29	1
7 529	Lacaille, 4608	7.0	8.43	0.566	81 46 34.6	.31	50.95	3
7 530	Chamæleontis	10.0	10.61	+ 1.111	79 32 41.2	− 19.31	51.24	1
7 531	Gould, Z. C., 10h 4207 .	8.3	10 57 11.50	+ 1.911	72 40 52.6	− 19.31	50.64	3
7 532	Gould, Z. C., 10h 4217 .	8.0	12.05	2.384	62 15 36.2	.31	51.18	1
7 533	Brisbane, 3366	9.0	15.15	2.334	63 53 15.5	.31	50.29	1
7 534	Chamæleontis	10.0	20.94	1.630	75 56 49.4	.31	51.23	2
7 535	Lacaille, 4618	9.0	21.68	+ 2.217	67 6 9.5	− 19.31	50.28	2
7 536	Gould, 15141	8.0	10 57 22.91	+ 2.249	66 19 8.2	− 19.31	50.26	1
7 537	Lacaille, 4605	7.5	23.72	0.861	80 45 12.6	.31	50.78	2
7 538	Gould, Z. C., 10h 4228 .	9.2	26.16	2.043	70 41 23.9	.31	52.28	2
7 539	Chamæleontis	9.5	31.61	1.537	76 47 55.0	.32	51.25	1
7 540	Carinæ	10.0	32.88	+ 1.947	72 15 20.0	− 19.32	51.28	1
7 541	Gould, Z. C., 10h 4235 .	9.0	10 57 34.53	+ 1.920	72 39 15.0	− 19.32	50.64	3
7 542	Chamæleontis	9.5	36.33	1.196	79 9 6.2	.32	50.85	2
7 543	Gould, Z. C., 10h 4238 . .	9.0	41.32	1.709	75 14 43.7	.32	50.40	1
7 544	Chamæleontis	10.0	41.74	+ 1.140	79 28 18.9	.32	51.24	1
7 545	Octantis	9.0	42.09	− 0.167	83 40 27.4	− 19.32	50.35	1

No. 7 514. Observed declination 80° 2′ 44.7″; changed to agree with Brisbane and Gould.

Number.	Constellation, Name of Star, or Synonym.	Magnitude.	Right Ascension, 1850.0.	Annual Precession.	South Declination, 1850.0.	Annual Precession.	Mean year.	No. of obs.
			h. m. s.	s.	° ′ ″	″		
7 546	Octantis	10.0	10 57 48.38	− 0.737	84 37 27.6	− 19.32	50.32	1
7 547	Lacaille, 4594	7.0	51.94	+ 2.006	71 26 6.0	.32	52.27	2
7 548	Octantis	9.5	55.12	− 0.386	84 5 34.3	.33	50.83	2
7 549	Carinæ	9.0	57 55.55	+ 1.877	73 20 30.2	.33	50.21	1
7 550	Carinæ	9.0	58 1.06	+ 2.018	71 16 53.4	− 19.33	52.27	1
7 551	Gould, Z. C., 10h 4292 .	9.0	10 58 8.78	+ 2.281	65 43 27.6	− 19.33	50.28	1
7 552	Brisbane, 3380	8.0	16.60	2.049	70 49 3.5	.33	52.28	1
7 553	Gould, Z. C., 10h 4298 .	9.5	18.44	2.062	70 37 40.1	.33	52.28	1
7 554	Brisbane, 3383	8.0	20.39	1.765	74 47 12.9	.33	51.29	1
7 555	Chamæleontis	10.0	25.80	+ 0.493	82 9 46.6	− 19.34	50.37	1
7 556	Gould, Z. C., 10h 4322 .	9.5	10 58 29.86	+ 2.385	62 43 2.7	− 19.34	51.18	1
7 557	Gould, Z. C., 10h 4320 .	10.0	32.06	2.276	65 59 5.9	.34	50.28	1
7 558	Carinæ	9.0	41.08	2.092	70 11 32.7	.34	50.23	1
7 559	Gould, 15177	9.0	47.15	2.323	64 45 54.9	.35	50.28	2
7 560	Carinæ	9.2	47.76	+ 1.852	73 52 19.4	− 19.35	51.75	2
7 561	Carinæ	9.0	10 58 52.04	+ 2.238	67 5 46.8	− 19.35	50.26	1
7 562	Lacaille, 4597	8.2	53.64	2.356	63 48 27.0	.35	50.29	2
7 563	Gould, 15181	9.0	54.83	2.278	66 4 41.3	.35	50.28	1
7 564	Gould, Z. C., 10h 4358 .	9.0	56.04	2.234	67 12 23.5	.35	50.26	2
7 565	Gould, Z. C., 10h 4357 .	9.0	58 57.64	+ 2.149	69 7 13.2	− 19.35	50.24	2
7 566	Gould, Z. C., 10h 4370 .	9.5	10 59 4.29	+ 2.182	68 27 32.9	− 19.35	50.41	1
7 567	Gould, Z. C., 10h 4365 .	8.5	4.34	2.047	71 4 58.3	.35	52.28	2
7 568	Octantis	10.0	4.48	0.159	83 6 25.7	.35	50.39	1
7 569	Gould, Z. C., 11h 7 . .	10.0	4.62	2.332	64 36 27.7	.35	50.29	1
7 570	Chamæleontis	9.5	15.42	+ 1.278	78 57 9.3	− 19.36	51.28	1
7 571	Gould, Z. C., 11h 15 . .	9.2	10 59 20.51	+ 2.112	69 58 45.2	− 19.36	51.26	2
7 572	Gould, Z. C., 11h 11 . .	8.2	20.66	1.896	73 26 24.5	.36	50.63	3
7 573	Gould, Z. C., 11h 28 . .	9.8	28.24	2.243	67 10 17.4	.36	50.26	2
7 574	Carinæ	9.2	30.40	1.869	73 49 56.5	.36	51.75	2
7 575	Chamæleontis	8.9	39.58	+ 0.747	81 29 8.7	− 19.37	50.75	4
7 576	Gould, Z. C., 11h 52 . .	9.5	10 59 52.21	+ 2.251	67 5 33.1	− 19.37	50.30	1
7 577	Carinæ	10.0	58.15	1.921	73 15 30.1	.37	50.21	1
7 578	Carinæ	10.0	10 59 58.72	1.908	73 26 11.7	.37	50.73	2
7 579	Carinæ	10.0	11 0 2.61	2.026	71 41 48.5	.37	52.26	1
7 580	Gould, Z. C., 11h 71 . .	9.0	6.55	+ 2.120	70 3 45.8	− 19.38	50.23	1
7 581	Octantis	10.0	11 0 7.52	− 2.541	86 28 45.3	− 19.38	51.33	2
7 582	η Octantis	6.2	11.42	− 0.098	83 47 17.1	.38	50.34	2
7 583	Lacaille, 4620	8.0	12.64	+ 1.841	74 20 43.6	.38	51.33	2
7 584	Lacaille, 4612	8.5	21.90	2.344	64 42 0.2	.38	50.28	2
7 585	Carinæ	9.5	22.01	+ 2.112	70 16 28.4	− 19.38	52.28	1

No. 7 559. Gould's declination is 64° 44′ 50.8″.

Number.	Constellation, Name of Star, or Synonym.	Magnitude.	Right Ascension, 1850.0.	Annual Precession.	South Declination, 1850.0.	Annual Precession.	Mean year.	No. of obs.
			h. m. s.	s.	° ′ ″	″		
7586	Gould, Z. C., 11^h 91 . .	10.0	11 0 22.35	+ 2.253	67 13 29.6	− 19.38	50.26	1
7587	Lacaille, 4613	8.0	26.77	2.367	64 1 46.1	.38	50.29	1
7588	Stone, 6168	8.2	34.55	2.225	67 58 17.2	.39	50.74	5
7589	Gould, Z. C., 11^h 108 . .	9.5	36.88	2.091	70 44 50.1	.39	52.28	2
7590	Gould, Z. C., 11^h 118 . .	8.0	47.54	+ 2.029	71 51 45.4	− 19.39	52.26	2
7591	Lacaille, 4635	8.3	11 0 49.60	+ 0.961	80 48 50.5	− 19.39	50.95	3
7592	Gould, Z. C., 11^h 127 . .	10.0	50.09	2.333	65 11 45.1	.39	50.28	1
7593	Carinæ	11.0	50.47	2.422	62 17 13.5	.39	51.18	1
7594	Gould, Z. C., 11^h 106 . .	9.0	53.89	1.459	78 3 41.9	.39	50.30	1
7595	Gould, Z. C., 11^h 140 . .	8.0	54.87	2.422	62 18 26.0	− 19.39	51.18	1
7596	Lacaille, 4618	8.2	11 0 54.90	+ 2.264	67 6 50.6	− 19.39	50.28	2
7597	Gould, Z. C., 11^h 111 . .	9.5	0 55.32	1.419	78 21 6.1	.39	51.35	1
7598	Gould, Z. C., 11^h 137 . .	9.0	1 2.33	+ 2.023	72 1 29.9	.40	52.26	1
7599	Octantis	9.4	16.79	− 4.795	87 32 1.5	.40	51.29	4
7600	Lacaille, 4622	8.0	23.76	+ 2.142	70 1 9.4	− 19.40	50.23	1
7601	Lacaille, 4625	7.0	11 1 26.10	+ 2.140	70 3 59.0	− 19.41	50.23	1
7602	Lacaille, 4632	7.0	29.26	1.783	75 19 8.4	.41	50.40	1
7603	Gould, Z. C., 11^h 191 . .	10.0	1 42.87	2.108	70 45 17.0	.41	52.28	1
7604	Gould, Z. C., 11^h 228 . .	9.0	2 8.18	2.397	63 40 35.7	.42	50.29	1
7605	Gould, Z. C., 11^h 232 . .	9.0	13.30	+ 2.255	67 47 24.4	− 19.42	50.39	1
7606	Gould, Z. C., 11^h 239 . .	9.5	11 2 15.80	+ 2.269	67 26 38.4	− 19.42	51.28	2
7607	Gould, 15267	9.5	17.21	2.395	63 47 28.3	.42	50.29	1
7608	Carinæ	10.0	19.54	2.414	63 8 46.7	.42	50.30	1
7609	Gould, Z. C., 11^h 252 . .	9.5	22.30	+ 2.301	66 40 0.9	.43	50.30	1
7610	Octantis	10.0	23.45	− 0.451	84 36 26.2	− 19.43	50.32	1
7611	Gould, Z. C., 11^h 254 . .	9.5	11 2 25.49	+ 2.219	68 42 42.8	− 19.43	50.41	1
7612	Lacaille, 4631	8.5	27.76	2.372	64 37 2.0	.43	50.28	2
7613	Gould, Z. C., 11^h 265 . .	9.2	33.60	2.276	67 23 3.0	.43	51.28	2
7614	Gould, Z. C., 11^h 273 . .	9.0	37.49	2.444	62 17 13.7	.43	51.18	1
7615	Carinæ	10.0	43.95	+ 2.145	70 22 3.4	− 19.43	52.28	1
7616	Gould, Z. C., 11^h 280 . .	10.0	11 2 44.94	+ 2.428	62 50 25.9	− 19.43	50.30	1
7617	Carinæ	9.5	46.23	+ 2.066	71 49 38.8	.43	52.26	2
7618	Octantis	10.0	54.52	− 2.285	86 27 48.2	.44	51.33	1
7619	Carinæ	9.5	2 59.18	+ 2.442	62 24 29.0	.44	51.18	1
7620	Octantis	9.0	3 9.60	− 7.480	88 13 5.7	− 19.44	51.29	5
7621	Gould, Z. C., 11^h 307 . .	8.7	11 3 23.48	+ 1.992	73 9 48.8	− 19.45	50.36	3
7622	Gould, Z. C., 11^h 330 . .	9.5	32.90	2.362	65 19 23.0	.45	50.28	1
7623	Carinæ	9.5	37.90	2.187	69 48 21.2	.45	52.29	1
7624	Chamæleontis	10.0	40.72	1.777	75 53 40.8	.45	51.23	2
7625	Gould, Z. C., 11^h 340 . .	9.5	43.62	+ 2.329	66 21 42.1	− 19.45	50.26	1

Number.	Constellation, Name of Star, or Synonym.	Magnitude.	Right Ascension, 1850.0.	Annual Precession.	South Declination, 1850.0.	Annual Precession.	Mean year.	No. of obs.
			h. m. s.	s.	° ′ ″	″		
7626	Gould, Z. C., 11ʰ 346 . .	9.2	11 3 48.96	+ 2.336	66 12 19.1	− 19.46	50.27	2
7627	Chamæleontis	9.0	49.02	1.684	76 50 34.6	.46	51.25	1
7628	Gould, Z. C., 11ʰ 319 . .	9.0	50.67	1.296	79 39 6.6	.46	51.24	1
7629	Gould, Z. C., 11ʰ 333 . .	9.5	3 59.35	1.417	78 56 41.3	.46	51.28	1
7630	Gould, Z. C., 11ʰ 356 . .	9.2	4 0.76	+ 2.167	70 19 32.2	− 19.46	51.26	2
7631	Gould, Z. C., 11ʰ 372 . .	9.5	11 4 7.76	+ 2.302	67 16 12.0	− 19.46	50.26	2
7632	Gould, Z. C., 11ʰ 373 . .	9.2	10.02	2.207	69 33 14.8	.46	51.26	2
7633	Carinæ	9.0	14.11	2.110	71 28.49.9	.47	52.27	2
7634	Gould, Z. C., 11ʰ 388 . .	10.0	21.80	2.440	63 3 21.6	.47	50.30	1
7635	Carinæ	10.0	22.57	+ 2.093	71 50 1.2	− 19.47	52.26	1
7636	Gould, 15311	9.0	11 4 49.16	+ 1.063	80 58 50.7	− 19.48	50.78	2
7637	Carinæ	11.0	51.26	2.401	64 35 29.8	.48	50.28	1
7638	Gould, 15326	9.0	52.94	2.401	64 36 13.2	.48	50.28	2
7639	Chamæleontis	9.0	4 59.96	1.698	76 58 45.0	.48	51.25	1
7640	Gould, Z. C., 11ʰ 416 . .	9.5	5 9.24	+ 1.350	79 34 26.1	− 19.48	51.24	1
7641	Chamæleontis	9.0	11 5 14.17	+ 1.821	75 47 59.2	− 19.49	51.14	2
7642	Carinæ	9.0	19.12	2.048	72 50 32.8	.49	50.42	1
7643	Chamæleontis	9.5	22.82	1.268	80 4 27.5	.49	51.16	1
7644	Carinæ	10.0	29.19	2.130	71 25 8.6	.49	52.27	1
7645	Chamæleontis	10.0	29.88	+ 1.247	80 12 28.6	− 19.49	51.16	1
7646	Carinæ	10.0	11 5 40.66	+ 2.246	69 10 47.8	− 19.50	50.24	2
7647	Gould, Z. C., 11ʰ 477 . .	9.0	44.93	2.188	70 28 21.0	.50	52.28	1
7648	Gould, Z. C., 11ʰ 484 . .	9.5	51.46	2.181	70 38 33.6	.50	52.28	2
7649	Carinæ	10.0	53.70	2.128	71 39 49.3	.50	52.26	1
7650	Lacaille, 4654	6.3	5 58.48	+ 2.184	70 37 19.2	− 19.50	52.28	2
7651	Carinæ	10.0	11 6 1.00	+ 2.201	70 17 48.2	− 19.50	52.28	1
7652	Brisbane, 3463	9.5	9.12	2.422	64 24 14.2	.51	50.29	1
7653	Gould, Z. C., 11ʰ 498 . .	9.0	9.25	1.886	75 17 49.4	.51	50.40	1
7654	Gould, Z. C., 11ʰ 503 . .	8.0	12.97	2.012	73 38 53.6	.51	51.25	1
7655	Octantis	10.0	22.82	+ 0.299	83 37.20.8	− 19.51	50.35	1
7656	Lacaille, 4664	7.5	11 6 31.15	+ 2.203	70 24 8.5	− 19.51	52.29	1
7657	Lacaille, 4657	6.5	33.82	2.456	63 21 19.4	.51	50.29	2
7658	Carinæ	9.0	36.58	2.023	73 35 45.5	.51	51.25	1
7659	Carinæ	9.5	36.55	2.131	71 50 31.8	.51	52.26	2
7660	Lacaille, 4663	9.0	40.52	+ 2.339	67 11 28.2	− 19.52	50.26	2
7661	Carinæ	9.5	11 6 47.05	+ 2.170	71 9 53.6	− 19.52	52.27	1
7662	Gould, Z. C., 11ʰ 555 . .	8.5	49.58	2.372	66 16 59.2	.52	50.27	2
7663	Gould, Z. C., 11ʰ 560 . .	9.0	50.39	2.477	62 39 37.8	.52	50.30	1
7664	Brisbane, 3475	7.7	6 58.70	2.185	70 56 2.0	.52	52.28	2
7665	Carinæ	9.5	7 8.22	+ 2.253	69 32 9.4	− 19.53	52.29	1

Number.	Constellation, Name of Star, or Synonym.	Magnitude.	Right Ascension, 1850.0.	Annual Precession.	South Declination, 1850.0.	Annual Precession.	Mean year.	No. of obs.
			h. m. s.	s.	° ′ ″	″		
7 666	Carinæ	10.0	11 7 12.74	+ 2.171	71 17 2.0	− 19.53	52.27	1
7 667	Gould, Z. C., 11h 572	9.0	12.92	2.049	73 23 37.4	.53	50.44	1
7 668	Gould, 15384	8.5	15.51	2.386	66 1 59.7	.53	50.28	1
7 669	Lacaille, 4698	9.0	21.32	0.347	83 37 25.9	.53	50.35	1
7 670	Gould, Z. C., 11h 611	9.5	32.83	+ 2.205	70 42 46.7	− 19.53	52.28	1
7 671	Gould, Z. C., 11h 632	8.8	11 7 51.01	+ 2.251	69 49 17.2	− 19.54	51.26	2
7 672	Lacaille, 4708	7.7	7 57.01	− 0 326	84 56 10.1	.54	51.21	5
7 673	Gould, Z. C., 11h 630	9.5	8 0.69	+ 1.945	75 3 12.0	.54	51.29	1
7 674	Gould, Z. C., 11h 659	9.0	9.98	2.427	65 4 30.6	.55	50.28	2
7 675	Carinæ	9.5	12.69	+ 2.279	69 17 49.6	− 19.55	50.24	1
7 676	Chamæleontis	10.0	11 8 13.22	+ 1.389	79 54 21.3	− 19.55	51.16	1
7 677	Carinæ	9.2	25.87	2.139	72 16 24.6	.55	51.77	2
7 678	Gould, Z. C., 11h 686	8.7	34.66	+ 2.308	68 42 49.4	.55	50.37	3
7 679	Octantis	10.0	34.84	− 0.797	85 36 10.0	.55	51.28	3
7 680	Octantis	10.0	44.78	+ 0.600	83 9 32.6	− 19.56	50.36	2
7 681	Lacaille, 4689	8.0	11 8 50.15	+ 1.969	74 58 25.1	− 19.56	51.29	1
7 682	Gould, Z. C., 11h 721	9.5	54.15	2.245	70 19 48.9	.56	52.28	1
7 683	Lacaille, 4681	8.0	8 57.40	2.358	67 30 39.3	.56	51.28	2
7 684	Lacaille, 4684	8.0	9 3.26	2.282	69 32 38.2	.56	52.29	1
7 685	Gould, Z. C., 11h 715	9.5	9.45	+ 1.618	78 33 46.3	− 19.56	51.35	1
7 686	Lacaille, 4682	7.5	11 9 14.82	+ 2.422	65 41 39.5	− 19.57	50.28	1
7 687	Carinæ	9.0	14.90	2.204	71 17 35.5	.57	52.27	1
7 688	Carinæ	10.0	16.97	2.148	72 23 1.8	.57	51.27	1
7 689	Chamæleontis	9.8	29.38	1.754	77 29 44.4	.57	51.88	2
7 690	Gould, Z. C., 11h 773	9.8	36.51	+ 2.351	67 58 35.4	− 19.57	50.89	2
7 691	Octantis	9.1	11 9 44.23	− 0.323	85 6 7.8	− 19.58	51.17	7
7 692	Gould, Z. C., 11h 788	9.0	48.03	+ 2.449	65 0 32.2	.58	50.28	1
7 693	Chamæleontis	9.5	9 49.32	1.358	79 35 1.3	.58	51.24	1
7 694	Lacaille, 4692	8.0	10 7.93	2.379	67 22 50.2	.58	50.94	3
7 695	Carinæ	9.0	21.51	+ 2.226	71 14 16.4	− 19.59	52.27	1
7 696	Chamæleontis	10.0	11 10 22.42	+ 1.901	76 13 14.6	− 19.59	51.30	1
7 697	Carinæ	9.8	30.76	2.361	68 2 23.6	.59	50.89	2
7 698	Lacaille, 4704	8.0	34.85	1.862	76 42 0.6	.59	51.20	2
7 699	Gould, Z. C., 11h 837	8.5	38.15	2.324	69 5 17.8	.59	50.31	3
7 700	Gould, Z. C., 11h 848	9.5	42.97	+ 2.428	66 6 19.8	− 19.59	50.28	1
7 701	Octantis	10.0	11 10 45.86	− 0.609	85 34 18.8	− 19.60	51.26	2
7 702	Gould, Z. C., 11h 853	9.0	51.04	+ 2.225	71 25 19.4	.60	52.27	2
7 703	Gould, 15462	9.5	52.95	2.432	66 2 28.0	.60	50.28	1
7 704	Gould, Z. C., 11h 862	8.8	10 57.69	+ 2.344	68 41 18.6	.60	50.43	2
7 705	Lacaille, 4731	7.3	11 0.37	− 0.465	85 24 54.7	− 19.60	51.30	5

No. 7 673. Gould's declination is 6′ greater.

Number.	Constellation, Name of Star, or Synonym.	Magnitude.	Right Ascension, 1850.0.	Annual Precession.	South Declination, 1850.0.	Annual Precession.	Mean year.	No. of obs.
			h. m. s.	s.	° ′ ″	″		
7706	Gould, Z. C., 11^h 870	9.0	11 11 6.14	+ 2.233	71 20 59.6	19.60	52.27	2
7707	Lacaille, 4700	7.5	8.04	2.525	62 39 27.8	.60	50.30	1
7708	Gould, Z. C., 11^h 885 . .	10.0	9.45	2.511	63 13 25.4	.60	50.30	1
7709	Gould, 15472	10.0	11.56	2.399	67 12 49.9	.60	50.26	1
7710	Lacaille, 4701	7.0	13.27	+ 2.406	67 0 14.8	− 19.60	50.30	1
7711	Gould, Z. C., 11^h 893 . .	9.0	11 11 25.57	+ 2.064	74 24 11.6	− 19.61	51.33	2
7712	Gould, Z. C., 11^h 900 . .	9.0	26.28	2.351	68 41 12.2	.61	50.43	2
7713	Lacaille, 4706	8.2	32.36	2.401	67 17 56.2	.61	50.26	2
7714	Carinæ	9.5	32.77	2.373	68 6 59.2	.61	50.89	2
7715	Chamæleontis	10.0	34.53	+ 1.508	79 50 8.7	− 19.61	51.16	1
7716	Gould, Z. C., 11^h 907 . .	8.0	11 11 35.78	+ 2.516	63 13 25.4	− 19.61	50.30	1
7717	Carinæ	10.0	42.44	2.290	70 19 4.3	.61	52.28	1
7718	Gould, Z. C., 11^h 912 . .	9.5	48.90	2.238	71 29 34.0	.62	52.27	2
7719	Gould, 15483	10.0	49.06	2.410	67 8 13.4	.62	50.30	1
7720	Carinæ	9.5	51.85	+ 2.231	71 39 45.6	− 19.62	52.26	1
7721	Gould, Z. C., 11^h 920 . .	9.0	11 11 53.30	+ 2.300	70 9 28.0	− 19.62	50.23	1
7722	Carinæ	9.5	55.23	2.313	69 50 50.6	.62	52.29	1
7723	Gould, Z. C., 11^h 930 . .	9.0	11 59.16	2.344	69 4 44.9	.62	50.31	3
7724	Carinæ	9.5	12 0.04	2.397	67 36 59.4	.62	52.30	1
7725	Chamæleontis	10.0	0.37	+ 1.733	78 16 5.4	− 19.62	51.35	1
7726	Carinæ	9.5	11 12 5.89	+ 2.347	69 2 34.8	− 19.62	50.24	1
7727	Gould, Z. C., 11^h 944 . .	9.5	12.70	2.486	64 43 19.0	.62	50.28	1
7728	Gould, Z. C., 11^h 933 . .	10.0	13.67	1.956	76 4 7.7	.62	51.30	1
7729	Chamæleontis	10.0	14.31	1.957	76 3 58.0	.62	51.30	1
7730	Chamæleontis	9.5	24.07	+ 1.723	78 26 46.4	− 19.63	51.35	1
7731	Gould, Z. C., 11^h 969 . .	9.0	11 12 30.10	+ 2.521	63 26 40.2	19.63	50.29	2
7732	Chamæleontis	9.0	30.19	1.382	80 45 11.6	.63	50.90	2
7733	Carinæ	10.0	41.86	2.399	67 50 5.6	.63	51.34	1
7734	Gould, Z. C., 11^h 974 . .	9.5	42.05	2.308	70 16 40.6	.63	52.28	1
7735	Gould, Z. C., 11^h 983 . .	9.3	47.41	+ 2.415	67 24 21.3	− 19.63	51.61	3
7736	Gould, 15501	9.0	11 12 48.05	+ 2.157	73 21 14.3	− 19.63	50.44	1
7737	Carinæ	9.2	12 52.60	2.506	64 13 47.7	.63	50.29	2
7738	Lacaille, 4712	7.2	13 1.11	2.520	63 45 50.2	.64	50.29	2
7739	Lacaille, 4716	8.0	11.52	2.164	73 22 12.3	.64	50.44	1
7740	Chamæleontis	9.5	14.33	+ 1.735	78 32 29.4	− 19.64	51.35	1
7741	Gould, Z. C., 11^h 1018 .	9.2	11 13 19.93	+ 2.421	67 27 18.4	− 19.64	51.28	2
7742	Carinæ	10.0	21.21	2.265	71 28 48.0	.64	52.26	1
7743	Chamæleontis	10.0	23.23	1.964	76 18 8.6	.64	51.22	2
7744	Gould, Z. C., 11^h 1028	9.0	26.14	2.488	65 12 44.0	.64	50.28	1
7745	Gould, Z. C., 11^h 1016 .	9.0	26.70	+ 2.093	74 35 45.5	− 19.64	50.41	1

Number.	Constellation, Name of Star, or Synonym.	Magnitude.	Right Ascension, 1850.0.	Annual Precession.	South Declination, 1850.0.	Annual Precession.	Mean year.	No. of obs.
			h. m. s.	s.	° ′ ″	″		
7 746	Carinæ	10.0	11 13 28.92	+ 2.268	71 27 59.4	− 19.64	52.26	1
7 747	Gould, Z. C., 11h 1030 .	9.0	29.13	2.545	62 55 47.0	.64	50.30	1
7 748	Carinæ	9.5	42.08	2.289	71 4 55.2	.65	52.27	1
7 749	Gould, 15518	9.0	42.78	2.230	72 20 4.4	.65	51.77	2
7 750	Gould, Z. C., 11h 1049 .	10.0	45.16	+ 2.512	64 25 51.6	− 19.65	50.29	1
7 751	Lacaille, 4720	8.0	11 13 48.60	+ 2.073	75 0 15.3	− 19.65	50.84	2
7 752	Gould, Z. C., 11h 1046 .	9.5	48.74	2.319	70 25 37.4	.65	52.28	1
7 753	Gould, 15526	8.5	52.11	2.465	66 13 40.7	.65	50.27	2
7 754	Gould, Z. C., 11h 1058 .	8.8	13 56.96	2.347	69 47 52.7	.65	51.26	2
7 755	Carinæ	9.0	14 1.74	+ 2.190	73 11 9.4	− 19.65	50.43	2
7 756	Brisbane, 3545	8.0	11 14 3.91	+ 2.446	66 57 24.3	− 19.65	52.29	1
7 757	Lacaille, 4722	7.0	9.72	2.248	72 8 13.4	.66	52.26	1
7 758	Lacaille, 4729	7.0	10.97	1.725	78 50 48.8	.66	51.28	1
7 759	Lacaille, 4724	7.0	13.94	2.127	74 19 16.8	.66	51.33	2
7 760	Lacaille, 4721	6.8	15.82	+ 2.294	71 10 24.0	− 19.66	52.28	2
7 761	Gould, Z. C., 11h 1095 .	10.0	11 14 27.14	+ 2.547	63 19 8.0	− 19.66	50.30	1
7 762	Gould, Z. C., 11h 1089 .	10.0	35.55	1.949	76 50 0.5	.66	51.25	1
7 763	Octantis	9.2	37.26	1.054	82 34 21.0	.66	50.36	2
7 764	Chamæleontis	10.0	45.45	1.550	80 13 34.0	.67	51.16	1
7 765	Carinæ	10.0	49.37	+ 2.451	67 8 12.0	− 19.67	50.26	1
7 766	Octantis	9.8	11 14 54.60	+ 0.638	83 52 14.6	− 19.67	50.34	2
7 767	Octantis	9.5	14 59.04	− 0.001	85 8 49.8	.67	51.14	6
7 768	Gould, Z. C., 11h 1129 .	9.5	15 4.00	+ 2.446	67 23 16.5	.67	51.28	4
7 769	Gould, 15552	8.8	21.68	2.362	69 58 50.3	.68	51.26	2
7 770	Gould, Z. C., 11h 1151 .	9.0	29.31	+ 2.193	73 38 38.3	− 19.68	51.25	1
7 771	Gould, Z. C., 11h 1160 .	9.5	11 15 33.05	+ 2.232	72 56 30.3	− 19.68	50.42	1
7 772	Gould, Z. C., 11h 1162 .	8.5	34.66	2.326	70 57 50.7	.68	52.28	1
7 773	Carinæ	10.0	39.52	2.280	72 1 21.5	.68	52.26	1
7 774	Carinæ	10.0	42.70	2.300	71 35 14.5	.68	52.26	1
7 775	Carinæ	9.0	45.94	+ 2.318	71 12 45.7	− 19.68	52.27	1
7 776	Chamæleontis	10.0	11 15 56.40	+ 1.640	79 53 2.2	− 19.69	51.16	1
7 777	Gould, Z. C., 11h 1191 .	9.5	16 2.89	2.145	74 37 39.1	.69	51.29	1
7 778	Chamæleontis	10.0	6.00	2.092	75 26 23.2	.69	51.15	1
7 779	Gould, Z. C., 11h 1210 .	9.5	6.97	2.578	62 43 33.5	.69	50.30	1
7 780	Gould, Z. C., 11h 1205 .	9.0	13.27	+ 2.257	72 41 48.4	− 19.69	51.28	1
7 781	Gould, Z. C., 11h 1228 .	9.5	11 16 30.78	+ 2.557	63 53 17.6	− 19.70	50.29	1
7 782	Chamæleontis	10.0	49.66	2.029	76 30 18.2	.70	51.15	1
7 783	Muscæ	8.5	53.49	2.556	64 7 52.5	.70	50.29	1
7 784	Lacaille, 4737	7.0	53.59	2.556	64 7 56.7	.70	50.29	1
7 785	Lacaille, 4742	7.0	54.18	+ 2.007	76 47 8.4	− 19.70	51.25	1

No. 7 751. Original record gives 25.3″.

No. 7 777. Gould's declination is 74° 37′ 24.6″.

Number.	Constellation, Name of Star, or Synonym.	Magnitude.	Right Ascension, 1850.0.	Annual Precession.	South Declination, 1850.0.	Annual Precession.	Mean year.	No. of obs.
			h. m. s.	s.	° ′ ″	″		
7 786	Gould, Z. C., 11h 1257 .	9.8	11 16 55.18	+ 2.470	67 26 52.8	− 19.70	51.27	2
7 787	Brisbane, 3567	9.0	17 0.23	1.808	78 50 22.0	.70	51.28	1
7 788	Muscæ	10.0	0.80	2.319	71 40 46.2	.70	52.26	1
7 789	Chamæleontis	10.0	3.45	1.683	79 49 59.3	.71	51.16	1
7 790	Gould, Z. C., 11h 1268 .	10.0	4.89	+ 2.487	66 56 15.8	− 19.71	50.30	1
7 791	Gould, Z. C., 11h 1267 .	10.0	11 17 5.57	+ 2.444	68 23 33.7	− 19.71	50.44	1
7 792	Lacaille, 4738	9.0	11.66	2.551	64 31 10.2	.71	50.29	1
7 793	Muscæ	10.0	13.05	2.229	73 35 53.3	.71	51.25	1
7 794	Gould, 15594	9.0	13.81	2.540	65 0 40.2	.71	50.28	2
7 795	Gould, Z. C., 11h 1276 .	9.5	14.17	+ 2.475	67 25 45.0	− 19.71	51.29	2
7 796	Muscæ	9.5	11 17 18.30	+ 2.238	73 27 15.9	− 19.71	51.25	1
7 797	Gould, Z. C., 11h 1275 .	10.0	19.09	2.289	72 27 34.5	.71	51.28	1
7 798	Octantis	9.5	22.55	0.898	83 30 43.0	.71	50.39	1
7 799	Gould, Z. C., 11h 1300 .	9.5	39.00	2.086	76 0 37.7	.71	51.30	1
7 800	Chamæleontis	9.0	40.60	+ 2.144	75 11 1.6	− 19.72	50.40	1
7 801	Octantis	9.0	11 17 41.61	+ 1.085	82 57 27.2	− 19.72	50.33	1
7 802	Chamæleontis	10.0	42.20	1.963	77 31 9.5	.72	51.36	1
7 803	Gould, Z. C., 11h 1305 .	9.2	43.29	2.096	75 53 44.4	.72	51.28	3
7 804	Chamæleontis	9.5	53.78	1.674	80 5 11.6	.72	51.16	1
7 805	Gould, Z. C., 11h 1331 .	9.0	54.52	+ 2.519	66 10 42.6	− 19.72	50.27	2
7 806	Gould, Z. C., 11h 1332 .	10.0	11 17 57.27	+ 2.468	68 0 43.5	− 19.72	50.74	3
7 807	Gould, Z. C., 11h 1328 .	10.0	18 5.21	2.031	76 51 12.8	.72	51.25	1
7 808	Gould, Z. C., 11h 1350 .	9.5	12.46	2.580	63 46 17.6	.72	50.29	1
7 809	Lacaille, 4744	6.0	14.07	2.350	71 25 59.2	.72	52.27	2
7 810	Gould, Z. C., 11h 1349 .	9.0	14.11	+ 2.508	66 44 47.3	− 19.72	50.30	1
7 811	Gould, Z. C., 11h 1353 .	9.0	11 18 15.00	+ 2.591	63 16 25.8	− 19.72	50.30	1
7 812	Gould, Z. C., 11h 1360 .	8.5	19.91	2.604	62 36 . .	.73	50.30	1
7 813	Octantis	9.2	25.42	0.949	83 31 5.8	.73	50.37	2
7 814	Gould, 15621	8.2	25.78	2.522	66 18 21.4	.73	50.29	3
7 815	Muscæ	10.0	48.60	+ 2.251	73 46 23.6	− 19.73	51.75	2
7 816	Gould, Z. C., 11h 1389 .	9.5	11 18 49.20	+ 2.448	69 3 6.3	− 19.73	50.24	1
7 817	Gould, 15632	8.2	18 55.68	2.380	70 59 12.2	.74	52.28	2
7 818	Chamæleontis	9.5	19 7.59	1.506	81 23 11.8	.74	50.38	1
7 819	Gould, Z. C., 11h 1409 .	9.8	8.50	2.507	67 12 45.6	.74	51.27	2
7 820	Chamæleontis	10.0	8.82	+ 1.550	81 8 38.8	− 19.74	51.37	1
7 821	Gould, 15635	9.0	11 19 9.37	+ 2.341	72 2 27.0	− 19.74	52.26	1
7 822	Gould, 15638	9.0	9.64	2.607	62 59 6.5	.74	50.30	1
7 823	Lacaille, 4747	6.5	14.87	2.605	63 8 44.6	.74	50.30	1
7 824	Lacaille, 4752	8.0	22.32	2.310	72 48 36.7	.74	50.42	1
7 825	Gould, Z. C., 11h 1417 .	9.2	24.02	+ 2.142	75 47 15.7	− 19.74	51.27	2

No. 7 787. Gould's declination is 8.7″ greater.

Number.	Constellation, Name of Star, or Synonym.	Magnitude.	Right Ascension, 1850.0.	Annual Precession.	South Declination, 1850.0.	Annual Precession.	Mean year.	No. of obs.
			h. m. s.	s.	° ′ ″	″		
7826	Octantis	8.7	11 19 25.96	− 2.390	87 31 53.8	− 19.74	51.31	6
7827	Octantis	9.0	38.32	+ 0.627	84 31 38.5	.75	50.33	1
7828	Gould, Z. C., 11ʰ 1446 .	9.2	47.38	2.301	73 9 48.8	.75	50.43	2
7829	Chamæleontis	10.0	53.95	1.956	78 13 17.1	.75	50.30	1
7830	Gould, Z. C., 11ʰ 1454 .	9.0	54.50	2.163	75 37 46.6	− 19.75	51.27	2
7831	Chamæleontis	9.5	11 19 55.02	+ 2.200	75 3 57.9	− 19.75	51.29	1
7832	Muscæ	10.0	19 59.49	2.519	67 12 26.8	.75	52.29	1
7833	Chamæleontis	10.0	20 2.30	2.077	76 52 34.4	.75	51.25	1
7834	Octantis	10.0	2.65	1.282	82 37 6.8	.75	50.39	1
7835	Gould, Z. C., 11ʰ 1477 .	9.2	5.61	+ 2.545	66 15 17.2	− 19.75	50.27	2
7836	Gould, Z. C., 11ʰ 1471 .	9.5	11 20 7.74	+ 2.315	72 59 50.4	− 19.75	50.42	1
7837	Octantis	10.0	7.99	0.977	83 41 42.6	.75	50.35	1
7838	Chamæleontis	10.0	12.09	1.726	80 16 2.2	.75	51.16	1
7839	Gould, Z. C., 11ʰ 1480 .	9.0	16.12	2.168	75 40 59.8	.76	51.27	2
7840	Gould, Z. C., 11ʰ 1498 .	9.5	24.95	+ 2.509	67 46 57.1	− 19.76	52.30	1
7841	Gould, Z. C., 11ʰ 1490 .	9.5	11 20 25.82	+ 2.220	74 54 11.8	− 19.76	51.29	1
7842	Chamæleontis	10.0	32.82	2.103	76 42 11.4	.76	51.20	2
7843	Gould, 15665	9.0	39.25	2.500	68 13 19.2	.76	50.60	5
7844	Chamæleontis	9.0	42.85	1.818	79 41 38.9	.76	51.24	1
7845	Chamæleontis	9.0	46.26	+ 1.842	79 30 21.7	− 19.76	51.24	1
7846	Brisbane, 3589	9.0	11 20 48.79	+ 2.628	62 47 34.0	− 19.76	50.30	1
7847	Chamæleontis	10.0	50.01	1.751	80 13 52.5	.76	51.16	1
7848	Muscæ	9.5	53.10	2.407	71 7 57.4	.76	52.27	1
7849	Octantis	10.0	20 55.34	1.326	82 35 46.5	.77	50.39	1
7850	Chamæleontis	10.0	21 3.16	+ 2.069	77 17 30.9	− 19.77	51.36	1
7851	Chamæleontis	10.0	11 21 9.80	+ 1.688	80 44 40.5	− 19.77	51.37	1
7852	Muscæ	9.5	12.59	2.397	71 32 50.8	.77	52.26	1
7853	Gould, Z. C., 11ʰ 1554 .	10.0	18.83	2.477	69 18 2.5	.77	50.24	1
7854	Gould, Z. C., 11ʰ 1565 .	10.0	24.45	2.625	63 21 57.9	.77	50.30	1
7855	Gould, Z. C., 11ʰ 1566 .	8.8	26.20	+ 2.230	75 6 30.8	− 19.77	50.84	2
7856	Gould, 15680	8.5	11 21 28.11	+ 2.584	65 21 54.3	− 19.77	50.28	1
7857	Gould, Z. C., 11ʰ 1579 .	9.0	38.97	2.416	71 14 39.5	.78	52.27	1
7858	Gould, Z. C., 11ʰ 1602 .	9.5	53.10	2.617	64 3 36.7	.78	50.29	1
7859	Melbourne (1), 563 . . .	8.0	21 54.75	2.583	65 39 49.6	.78	50.28	1
7860	Lacaille, 4767	7.0	22 3.56	+ 2.063	77 41 47.6	− 19.78	50.40	1
7861	Brisbane, 3599	9.0	11 22 4.31	+ 2.644	62 45 1.0	− 19.78	50.30	1
7862	Lacaille, 4763	8.5	8.65	2.458	70 17 30.5	.78	51.26	2
7863	Lacaille, 4765	7.0	11.68	2.410	71 38 52.6	.78	52.26	1
7864	Lacaille, 4762	7.0	32.70	2.650	62 43 43.1	.79	50.30	1
7865	Chamæleontis	10.0	22 34.46	+ 2.147	76 46 30.8	− 19.79	51.25	1

Number.	Constellation, Name of Star, or Synonym.	Magnitude.	Right Ascension, 1850.0.	Annual Precession.	South Declination, 1850.0.	Annual Precession.	Mean year.	No. of obs.
			h. m. s.	s.	° ′ ″	″		
7866	Gould, 15723	10.0	11 23 8.39	+ 2.552	67 37 2.0	— 19.80	52.30	1
7867	Lacaille, 4784	8.0	9.28	0.990	84 7 45.4	.80	50.33	1
7868	Chamæleontis	9.0	13.17	1.608	81 41 30.4	.80	51.34	1
7869	Gould, Z. C., 11^h 1700 .	9.2	14.18	2.587	66 14 59.7	.80	50.29	3
7870	Gould, Z. C., 11^h 1690 .	9.5	20.33	+ 2.008	78 40 38.1	— 19.80	51.32	2
7871	Gould, 15724	9.0	11 23 23.63	+ 2.345	73 41 50.6	— 19.80	51.25	1
7872	Gould, Z. C., 11^h 1726 .	9.5	43.70	2.652	63 21 7.4	.81	50.30	1
7873	Gould, Z. C., 11^h 1729 .	9.0	51.66	2.404	72 32 53.4	.81	51.28	1
7874	Gould, Z. C., 11^h 1733 .	9.0	23 54.51	2.445	71 30 24.0	.81	52.27	2
7875	Gould, Z. C., 11^h 1744 .	9.2	24 0.58	+ 2.576	67 8 54.5	— 19.81	51.27	2
7876	Chamæleontis	10.0	11 24 2.26	+ 2.171	76 56 32.0	— 19.81	51.25	1
7877	Octantis	8.3	16.66	— 3.316	88 8 19.6	.81	51.61	6
7878	Gould, Z. C., 11^h 1768 .	10.0	20.33	+ 2.661	63 12 57.8	.81	50.30	1
7879	Gould, Z. C., 11^h 1770 .	9.5	24.64	2.464	71 12 58.5	.82	52.27	1
7880	Gould, 15745	9.5	27.61	+ 2.362	73 46 53.9	— 19.82	51.75	2
7881	Gould, Z. C., 11^h 1786 .	9.0	11 24 31.00	+ 2.631	64 54 54.5	— 19.82	50.28	1
7882	Gould, Z. C., 11^h 1789 .	9.2	37.55	2.597	66 35 22.2	.82	50.29	2
7883	Gould, Z. C., 11^h 1799 .	10.0	48.33	2.530	69 20 2.6	.82	50.24	1
7884	Chamæleontis	9.5	50.11	1.873	80 19 25.0	.82	50.79	2
7885	Gould, Z. C., 11^h 1798 .	9.5	51.13	+ 2.418	72 38 25.2	— 19.82	51.28	1
7886	Gould, Z. C., 11^h 1790 .	10.0	11 24 51.74	+ 2.102	78 6 54.4	— 19.82	50.30	1
7887	Chamæleontis	10.0	55.27	2.245	76 8 49.1	.82	51.30	1
7888	Muscæ	9.5	24 57.50	2.643	64 34 31.9	.82	50.29	1
7889	Gould, Z. C., 11^h 1817 .	9.0	25 0.95	2.595	66 54 15.7	.82	50.30	1
7890	Muscæ	9.5	2.24	+ 2.624	65 34 45.2	— 19.82	50.28	1
7891	Lacaille, 4780	9.2	11 25 16.50	+ 2.591	67 13 49.7	— 19.83	51.28	2
7892	Gould, Z. C., 11^h 1835 .	9.0	19.17	2.675	63 4 53.8	.83	50.30	1
7893	Gould, Z. C., 11^h 1833 .	10.0	22.03	2.515	70 6 36.7	.83	50.23	1
7894	Gould, Z. C., 11^h 1838 .	8.2	25.42	2.485	71 6 1.9	.83	52.28	2
7895	Gould, Z. C., 11^h 1844 .	9.2	27.83	+ 2.617	66 12 36.6	— 19.83	50.27	2
7896	Gould, Z. C., 11^h 1848 .	9.5	11 25 29.66	+ 2.662	63 54 38.5	— 19.83	50.29	2
7897	Chamæleontis	9.2	31.76	2.253	76 14 49.8	.83	51.22	2
7898	Lacaille, 4786	7.2	38.86	2.416	73 4 30.6	.83	50.43	2
7899	Gould, Z. C., 11^h 1854 .	9.0	39.05	2.470	71 39 14.2	.83	52.26	1
7900	Lacaille 4782	6.2	41.32	+ 2.621	66 8 1.5	— 19.83	51.32	2
7901	Muscæ	9.5	11 25 44.75	+ 2.460	71 57 55.9	— 19.83	52.26	1
7902	Gould, Z. C., 11^h 1872 .	8.2	48.08	2.489	71 9 20.4	.83	52.28	2
7903	Muscæ	9.8	50.66	2.466	71 50 36.7	.83	52.26	2
7904	Gould, Z. C., 11^h 1881 .	9.5	52.45	2.666	63 58 32.1	.83	50.29	1
7905	Gould, Z. C., 11^h 1893 .	9.0	25 59.98	+ 2.647	65 2 5.3	— 19.84	50.99	3

No. 7872. Observed place 63° 25′ 20.8″; changed 1 wire interval

No. 7903. Declination may be 71° 50′ 8.8″.

Number.	Constellation, Name of Star, or Synonym.	Magnitude.	Right Ascension, 1850.0.	Annual Precession.	South Declination, 1850.0.	Annual Precession.	Mean year.	No. of obs.
			h. m. s.	s.	° ′ ″	″		
7906	Chamæleontis	9.0	11 26 0.23	+ 1.966	79 52 7.8	− 19.84	51.20	2
7907	Chamæleontis	9.0	1.48	2.310	75 28 31.0	.84	50.89	2
7908	Gould, Z. C., 11h 1903 .	9.0	5.14	2.688	62 49 5.7	.84	50.30	1
7909	Octantis	10.0	30.44	1.177	84 8 4.4	.84	50.33	1
7910	Velorum	. .	31.04	+ 2.736	59 56 27.9	− 19.84	52.39	1
7911	Gould, Z. C., 11h 1930 .	9.0	11 26 34.42	+ 2.500	71 14 4.8	− 19.84	52.27	1
7912	Gould, Z. C., 11h 1932 .	10.0	34.46	2.535	70 5 5.1	.84	50.23	1
7913	Lacaille, 4811	9.0	38.47	1.429	83 16 7.6	.84	50.39	1
7914	Gould, 15801	8.8	39.75	2.632	66 13 32.0	.84	50.99	4
7915	Gould, Z. C., 11h 1946 .	9.0	42.79	+ 2.593	68 0 38.3	− 19.85	50.37	1
7916	Gould, Z. C., 11h 1928 .	9.0	11 26 43.64	+ 2.102	78 43 27.1	− 19.85	51.32	2
7917	Octantis	8.8	52.68	− 1.210	87 25 30.5	.85	51.30	5
7918	Stone, 6433	8.0	53.15	+ 2.738	60 3 25.6	.85	52.39	1
7919	Lacaille, 4791	7.5	56.06	2.561	69 21 56.4	.85	51.26	2
7920	Chamæleontis	9.5	58.46	+ 2.265	76 36 40.2	− 19.85	51.20	2
7921	Stone, 6429	9.0	11 26 59.64	+ 1.913	80 36 8.7	− 19.85	50.43	1
7922	Gould, Z. C., 11h 1969 .	9.5	27 3.18	+ 2.552	69 45 23.3	.85	52.28	1
7923	Octantis	9.0	5.37	− 8.344	89 2 22.8	.85	51.33	5
7924	Octantis	9.2	6.76	− 4.826	88 36 47.0	.85	51.32	3
7925	Chamæleontis	9.0	10.35	+ 2.197	77 41 26.8	19.85	50.40	1
7926	Muscæ	9.0	11 27 14.93	+ 2.531	70 35 1.7	− 19.85	52.28	2
7927	Muscæ	9.0	24.16	2.665	65 0 5.3	.85	50.28	1
7928	Lacaille, 4793	8.2	26.74	2.674	64 34 53.8	.85	51.37	2
7929	Octantis	9.5	28.01	1.273	83 59 45.2	.85	50.34	2
7930	Gould, Z. C., 11h 1998 .	9.5	31.44	+ 2.381	74 42 23.7	− 19.86	50.41	1
7931	Gould, Z. C., 11h 1997 .	9.5	11 27 37.30	+ 2.134	78 38 51.8	− 19.86	51.32	2
7932	Gould, Z. C., 11h 2017 .	9.0	42.30	+ 2.701	63 7 29.6	.86	50.30	1
7933	Stone, 6404	7.8	51.57	− 3.695	88 25 5.0	.86	51.33	5
7934	Brisbane, 3659	7.5	54.34	+ 2.749	59 57 28.1	.86	52.40	2
7935	Gould, Z. C., 11h 2033 .	9.5	57.89	+ 2.616	67 46 13.3	− 19.86	52.30	1
7936	Muscæ	10.0	11 27 58.55	+ 2.437	73 39 13.2	− 19.86	51.25	1
7937	Chamæleontis	9.5	28 0.76	2.316	76 10 29.5	.86	51.30	1
7938	Gould, Z. C., 11h 2046 .	9.0	16.17	2.624	67 37 24.1	.86	52.30	1
7939	Lacaille, 4798	7.0	19.66	2.752	60 3 51.9	.87	52.38	3
7940	Muscæ	9.0	22.58	+ 2.632	67 19 15.4	− 19.87	52.29	1
7941	Gould, Z. C., 11h 2041 .	8.0	11 28 23.38	+ 2.090	79 25 16.6	− 19.87	50.83	2
7942	Velorum	9.0	35.84	2.757	59 55 5.4	.87	52.40	2
7943	Velorum	9.0	40.19	2.704	63 39 20.2	.87	50.33	1
7944	Octantis	8.5	41.22	1.659	82 39 19.4	.87	50.36	2
7945	Gould, 15844	9.5	51.24	+ 2.643	67 6 15.4	− 19.87	51.28	2

Number.	Constellation, Name of Star, or Synonym.	Magnitude.	Right Ascension, 1850.0.	Annual Precession.	South Declination, 1850.0.	Annual Precession.	Mean year.	No. of obs.
			h. m. s.	s.	° ′ ″	″		
7946	Gould, Z. C., 11^h 2084	9.5	11 28 57.69	+ 2.419	74 32 45.2	− 19.87	50.41	1
7947	Gould, Z. C., 11^h 2100	8.5	29 9.98	2.595	69 22 40.7	.88	51.32	3
7948	Gould, Z. C., 11^h 2104	9.0	11.04	2.683	65 15 39.6	.88	51.34	2
7949	Gould, Z. C., 11^h 2105	9.8	15.84	2.616	68 33 52.0	.88	50.43	2
7950	Gould, Z. C., 11^h 2108	9.8	18.14	+ 2.523	71 58 5.7	− 19.88	52.26	2
7951	Octantis	9.5	11 29 19.60	+ 0.406	86 9 47.6	− 19.88	51.28	3
7952	Lacaille, 4810	7.0	23.47	2.761	60 13 25.0	.88	52.39	3
7953	Lacaille, 4813	7.0	24.82	2.524	72 0 51.0	.88	52.26	2
7954	Chamæleontis	8.8	26.22	2.293	77 8 25.2	.88	51.31	2
7955	Stone, 6464	8.2	34.84	+ 2.765	60 3 51.6	− 19.88	52.38	4
7956	Velorum	10.0	11 29 39.96	+ 2.719	63 26 42.0	19.88	50.30	1
7957	Chamæleontis	9.5	43.97	2.671	66 17 45.9	.88	52.38	1
7958	Gould, Z. C., 11^h 2155	9.2	49.06	2.638	67 58 48.0	.88	51.34	2
7959	Gould, 15872	8.3	50.83	2.586	70 10 0.9	.88	51.32	3
7960	Gould, Z. C., 11^h 2163	10.0	51.45	+ 2.689	65 22 5.0	− 19.88	52.41	1
7961	Chamæleontis	9.5	11 29 55.40	+ 2.369	76 0 53.9	− 19.88	51.30	1
7962	Gould, Z. C., 11^h 2180	10.0	30 1.35	2.631	68 25 19.8	.89	50.41	1
7963	Chamæleontis	9.2	6.48	2.311	77 8 3.4	.89	51.31	2
7964	Gould, Z. C., 11^h 2193	8.5	14.00	2.522	72 31 24.6	.89	51.28	1
7965	Gould, Z. C., 11^h 2196	10.0	14.38	+ 2.639	68 12 20.9	19.89	50.41	1
7966	Velorum	8.5	11 30 19.45	+ 2.774	59 53 54.8	− 19.89	52.39	3
7967	Gould, Z. C., 11^h 2198	8.8	22.70	2.315	77 10 36.6	.89	51.31	2
7968	Gould, Z. C., 11^h 2210	10.0	24.79	2.715	64 13 14.4	.89	52.45	1
7969	Gould, Z. C., 11^h 2212	9.5	25.93	2.675	66 34 49.6	.89	50.26	1
7970	Gould, Z. C., 11^h 2218	9.2	32.05	+ 2.686	66 4 6.1	− 19.89	51.33	2
7971	Gould, Z. C., 11^h 2220	9.5	11 30 32.29	+ 2.710	64 39 10.2	− 19.89	52.45	1
7972	Gould, Z. C., 11^h 2224	9.0	32.31	2.738	62 51 12.0	.89	50.30	1
7973	Lacaille, 4820	7.0	34.53	2.775	60 4 24.0	.89	52.36	4
7974	Lacaille, 4822	7.0	35.17	2.630	68 50 41.9	.89	50.45	1
7975	Chamæleontis	10.0	37.07	+ 2.350	76 40 24.2	− 19.89	51.15	1
7976	Brisbane, 3687	7.0	11 30 37.33	+ 2.774	60 9 14.4	− 19.89	52.38	4
7977	Velorum	9.5	39.32	2.724	63 52 23.9	.89	50.29	1
7978	Gould, Z. C., 11^h 2226	9.2	40.19	2.394	75 51 41.6	.89	51.34	2
7979	Chamæleontis	10.0	41.04	+ 2.612	69 38 54.8	.89	52.29	1
7980	Octantis	7.9	41.25	− 6.445	88 58 24.1	− 19.89	51.31	5
7981	Gould, Z. C., 11^h 2227	9.0	11 30 46.98	+ 2.154	79 30 22.1	− 19.89	50.83	2
7982	Lacaille, 4826	6.2	31 2.68	2.679	66 47 19.4	.90	50.31	2
7983	Octantis	10.0	5.49	1.389	84 17 32.5	.90	50.33	2
7984	Muscæ	9.5	6.02	2.670	67 17 8.6	.90	52.29	1
7985	π Chamæleontis	7.2	6.06	+ 2.441	75 3 59.5	− 19.90	50.71	3

Number.	Constellation, Name of Star, or Synonym.	Magnitude.	Right Ascension, 1850.0.	Annual Precession.	South Declination, 1850.0.	Annual Precession.	Mean year.	No. of obs.
			h. m. s.	s.	° ′ ″	″		
7986	Lacaille, 4828	8.5	11 31 11.04	+ 2.715	64 49 30.6	19.90	50.28	1
7987	Velorum	8.0	18.44	2.785	59 49 29.4	.90	52.39	2
7988	Chamæleontis	10.0	19.30	2.374	76 32 38.3	.90	51.15	1
7989	Gould, Z. C., 11^h 2282	9.2	25.90	2.343	77 9 2.1	.90	51.31	2
7990	Muscæ	9.5	33.56	+ 2.670	67 37 26.3	− 19.90	52.30	1
7991	Muscæ	9.0	11 31 41.43	+ 2.603	70 39 15.1	− 19.90	52.28	1
7992	Gould, Z. C., 11^h 2313	10.0	43.25	2.700	66 9 12.2	.90	52.38	1
7993	Chamæleontis	10.0	31 56.48	2.382	76 41 17.4	.91	51.15	1
7994	Gould, Z. C., 11^h 2318	9.5	32 0.59	2.277	78 24 44.9	.91	51.35	1
7995	Lacaille, 4840	7.0	3.24	+ 2.585	71 32 11.8	− 19.91	52.27	2
7996	Gould, 15924	8.5	11 32 12.43	+ 2.626	70 3 9.5	− 19.91	50.84	2
7997	Gould, Z. C., 11^h 2348	9.0	17.53	2.671	68 5 15.1	.91	50.37	1
7998	Chamæleontis	9.2	18.07	2.168	79 53 5.1	.91	51.25	3
7999	Herschel, 4462[1]	10.0	22.26	1.892	82 14 26.0	.91	50.39	1
8000	Brisbane, 3700	9.0	25.27	+ 2.790	60 19 58.3	− 19.91	52.40	1
8001	Herschel, 4462[2]	9.0	11 32 25.29	+ 1.894	82 14 26.2	19.91	50.38	2
8002	Lacaille, 4843	6.2	34.30	2.736	64 33 59.4	.91	51.37	2
8003	Gould, Z. C., 11^h 2367	9.5	35.42	2.616	70 43 46.0	.91	52.28	1
8004	Muscæ	10.0	37.11	2.692	67 16 18.1	.91	52.29	1
8005	Chamæleontis	9.5	38.37	+ 2.239	79 10 24.0	− 19.91	50.42	1
8006	Gould, 15947	9.0	11 32 45.52	+ 2.767	62 38 48.6	− 19.92	50.30	1
8007	Chamæleontis	10.0	47.48	2.357	77 28 14.0	.92	50.88	2
8008	Gould, Z. C., 11^h 2379	9.0	49.06	2.555	72 58 55.3	.92	50.42	1
8009	Gould, 15946	9.0	49.26	2.593	71 43 42.7	.92	52.26	1
8010	Brisbane, 3706	8.5	49.52	+ 2.766	62 38 55.3	− 19.92	50.30	1
8011	Chamæleontis	9.5	11 32 49.60	+ 2.287	78 35 53.2	− 19.92	51.35	1
8012	Gould, Z. C., 11^h 2386	9.2	32 50.93	2.750	63 52 9.0	.92	50.29	2
8013	Gould, Z. C., 11^h 2400	7.5	33 2.71	2.796	60 21 18.3	.92	52.38	2
8014	Gould, Z. C., 11^h 2403	9.5	5.97	2.744	64 26 59.4	.92	52.45	1
8015	Chamæleontis	10.0	24.57	+ 2.400	77 1 12.6	− 19.92	51.36	1
8016	Gould, Z. C., 11^h 2424	9.0	11 33 27.86	+ 2.760	63 37 52.4	− 19.92	50.29	1
8017	Lacaille, 4855	7.5	33.06	2.674	68 50 25.9	.92	50.45	1
8018	Chamæleontis	9.8	34.81	2.424	76 38 10.3	.92	51.25	3
8019	Chamæleontis	9.5	39.38	2.046	81 29 56.2	.92	50.86	2
8020	Brisbane, 3713	7.3	43.87	+ 2.808	60 8 59.5	− 19.92	52.39	4
8021	Gould, 15971	9.2	11 33 50.36	+ 2.636	70 44 17.6	− 19.93	52.28	2
8022	Muscæ	9.0	52.30	2.708	67 19 34.6	.93	52.29	1
8023	Lacaille, 4865	7.8	56.53	1.446	84 39 22.1	.93	50.60	4
8024	Stone, 6525	7.5	33 57.45	2.808	60 5 29.6	.93	52.38	4
8025	Velorum	7.5	34 5.44	+ 2.807	60 18 31.6	− 19.93	52.39	3

Number.	Constellation, Name of Star, or Synonym.	Magnitude.	Right Ascension, 1850.0.	Annual Precession.	South Declination, 1850.0.	Annual Precession.	Mean year.	No. of obs.
			h. m. s.	s.	° ′ ″	″		
8026	Chamæleontis	10.0	11 34 6.74	+ 2.391	77 30 45.0	− 19.93	50.88	2
8027	Gould, Z. C., 11h 2469	9.0	7.07	2.639	70 49 27.4	.93	52.28	1
8028	Muscæ	9.5	16.58	2.712	67 23 39.0	.93	52.29	1
8029	Gould, Z. C., 11h 2481	8.5	21.18	2.654	70 20 7.4	.93	51.86	2
8030	Chamæleontis	10.0	22.18	+ 2.252	79 41 26.6	− 19.93	51.24	1
8031	Chamæleontis	9.8	11 34 25.58	+ 2.236	79 53 42.1	− 19.93	51.26	3
8032	Gould, Z. C., 11h 2482	10.0	25.92	2.540	74 12 18.8	.93	52.27	2
8033	Muscæ	10.0	29.53	2.643	70 30 29.8	.93	52.28	1
8034	Brisbane, 3723	8.5	35.34	2.781	62 59 51.4	.93	50.30	1
8035	Chamæleontis	10.0	43.10	+ 2.284	79 25 14.2	− 19.93	51.24	1
8036	Gould, 15997	9.2	11 34 43.23	+ 2.724	67 4 58.8	− 19.93	51.30	2
8037	Gould, Z. C., 11h 2504	9.5	44.59	2.666	70 3 57.4	.93	51.44	1
8038	Gould, Z. C., 11h 2509	9.5	46.04	2.703	68 16 33.6	.94	50.37	1
8039	Brisbane, 3725	9.0	52.19	2.785	62 57 52.6	.94	50.30	1
8040	Gould, 16000	8.5	57.22	+ 2.656	70 40 31.1	− 19.94	52.28	2
8041	Gould, Z. C., 11h 2518	9.2	11 34 58.60	+ 2.749	65 44 27.6	− 19.94	51.35	2
8042	Muscæ	10.0	35 7.41	2.643	71 19 49.2	.94	52.27	1
8043	Lacaille, 4864	7.0	10.72	2.482	76 13 19.5	.94	51.26	3
8044	Gould, 16006	7.8	10.89	2.820	60 7 48.8	.94	52.39	4
8045	Octantis	9.2	21.27	+ 0.766	86 26 16.1	− 19.94	51.33	5
8046	Gould, Z. C., 11h 2540	9.0	11 35 22.26	+ 2.731	67 10 55.0	− 19.94	52.29	1
8047	Brisbane, 3732	8.8	22.37	2.562	74 17 10.2	.94	51.65	3
8048	Octantis	10.0	23.36	1.550	84 37 6.6	.94	50.32	1
8049	Gould, Z. C., 11h 2559	10.0	30.59	2.790	63 11 47.3	.94	50.30	1
8050	Gould, 16014	9.5	31.13	+ 2.725	67 33 58.7	− 19.94	52.30	1
8051	Muscæ	10.0	11 35 35.82	+ 2.666	70 42 21.0	− 19.94	52.28	1
8052	Gould, Z. C., 11h 2568	9.5	37.64	2.762	65 22 56.7	.94	51.34	2
8053	Muscæ	10.0	39.04	2.762	65 21 15.5	.94	52.41	1
8054	Lacaille, 4866	7.2	40.10	2.564	74 23 43.4	.94	51.34	2
8055	Gould, Z. C., 11h 2562	9.0	41.12	+ 2.443	77 19 11.0	− 19.94	51.36	1
8056	Gould, 16020	9.0	11 35 45.36	+ 2.792	63 15 3.2	− 19.94	50.30	1
8057	Lacaille, 4873	6.5	51.56	2.036	82 16 6.4	.95	50.38	2
8058	Gould, Z. C., 11h 2586	9.5	35 59.10	2.731	67 40 32.5	.95	52.30	1
8059	Muscæ	9.5	36 0.56	2.657	71 21 57.1	.95	52.27	1
8060	Gould, Z. C., 11h 2593	9.5	5.42	+ 2.725	68 8 10.9	− 19.95	50.37	1
8061	Muscæ	10.0	11 36 5.95	+ 2.645	71 55 25.9	− 19.95	52.26	1
8062	Gould, Z. C., 11h 2599	8.5	8.13	2.780	64 32 47.4	.95	52.45	1
8063	Muscæ	9.0	9.31	2.629	72 35 16.5	.95	50.42	1
8064	Gould, Z. C., 11h 2598	10.0	9.70	2.694	69 49 44.7	.95	52.29	1
8065	Gould, 16031	9.0	17.92	+ 2.770	65 25 33.6	− 19.95	51.34	2

Number.	Constellation, Name of Star, or Synonym.	Magnitude.	Right Ascension, 1850.0.	Annual Precession.	South Declination, 1850.0.	Annual Precession.	Mean year.	No. of obs.
			h. m. s.	s.	° ′ ″	″		
8 066	Muscæ	10.0	11 36 21.12	+ 2.759	66 12 54.5	− 19.95	52.38	1
8 067	Lacaille, 4870	7.0	21.73	2.658	71 36 45.2	.95	52.26	1
8 068	Muscæ	10.0	28.33	2.669	71 13 9.1	.95	52.27	1
8 069	Muscæ	10.0	35.58	2.672	71 10 57.9	.95	52.27	1
8 070	Muscæ	10.0	35.71	+ 2.683	70 40 51.7	− 19.95	52.28	1
8 071	Lacaille, 4874	6.5	11 36 37.51	+ 2.404	78 28 25.0	− 19.95	51.35	1
8 072	Lacaille, 4871	9.5	38.47	2.741	67 38 46.8	.95	52.30	1
8 073	Muscæ	10.0	45.04	2.671	71 19 57.7	.95	52.27	1
8 074	Chamæleontis	8.0	47.16	2.469	77 22 {16.4 / 63.1}	.95	50.88	2
8 075	Gould, Z. C., 11ʰ 2660 .	9.0	36 55.89	+ 2.737	68 8 11.7	− 19.95	50.37	1
8 076	Gould, Z. C., 11ʰ 2665 .	9.0	11 37 0.75	+ 2.752	67 18 30.8	− 19.96	52.29	1
8 077	Gould, Z. C., 11ʰ 2667 .	9.8	9.07	2.544	75 51 33.8	.96	51.34	2
8 078	Gould, Z. C., 11ʰ 2679 .	8.0	26.92	2.844	60 1 4.6	.96	52.38	2
8 079	Muscæ	10.0	29.84	2.698	70 40 51.7	.96	52.28	1
8 080	Chamæleontis	10.0	30.44	+ 2.311	80 14 5.5	− 19.96	51.16	1
8 081	Muscæ	10.0	11 37 30.48	+ 2.726	69 14 46.8	− 19.96	51.45	1
8 082	Gould, Z. C., 11ʰ 2681 .	9.0	30.85	2.771	66 28 14.5	.96	50.32	1
8 083	Muscæ	9.0	32.66	2.658	72 28 21.6	.96	51.28	1
8 084	Lacaille, 4880	6.5	33.59	2.356	79 39 2.0	.96	51.24	1
8 085	Gould, Z. C., 11ʰ 2707 .	10.0	53.08	+ 2.731	69 15 53.5	− 19.96	51.45	1
8 086	Gould, Z. C., 11ʰ 2714 .	9.5	11 37 56.20	+ 2.823	62 42 30.9	− 19.96	50.30	1
8 087	Gould, 16067	8.2	38 2.56	2.651	73 6 40.0	.96	50.43	2
8 088	Gould, Z. C., 11ʰ 2726 .	9.5	2.84	2.824	62 44 38.5	.96	50.30	1
8 089	Lacaille, 4879	7.5	4.00	2.671	72 19 5.0	.96	51.77	2
8 090	Chamæleontis	8.5	8.87	+ 2.521	76 59 4.5	− 19.97	51.25	1
8 091	Gould, Z. C., 11ʰ 2736 .	9.5	11 38 13.43	+ 2.775	66 51 57.8	− 19.97	50.30	1
8 092	Melbourne (1), 575 . . .	8.5	13.76	2.798	65 9 33.1	.97	51.35	2
8 093	Gould, Z. C., 11ʰ 2738 .	9.5	13.94	2.739	69 7 9.4	.97	50.95	2
8 094	Gould, Z. C., 11ʰ 2745 .	9.5	15.20	2.782	66 22 42.6	.97	50.32	1
8 095	Lacaille, 4883	3.5	33.56	+ 2.793	65 53 49.8	+ 19.97	51.35	2
8 096	Gould, Z. C., 11ʰ (2757) .	9.0	11 38 34.37	+ 2.653	73 24 33.8	− 19.97	51.25	1
8 097	Gould, 16084	9.0	34.39	2.714	70 46 14.0	.97	52.28	2
8 098	Gould, 16083	9.0	35.43	2.636	74 2 27.2	.97	52.28	1
8 099	Gould, 16088	8.5	36.89	2.722	70 24 37.4	.97	51.86	2
8 100	Chamæleontis	9.8	48.13	+ 2.491	77 59 18.0	− 19.97	50.35	2
8 101	Chamæleontis	8.5	11 38 48.95	+ 2.502	77 47 6.3	− 19.97	50.35	2
8 102	Gould, Z. C., 11ʰ 2779 .	9.5	38 57.54	2.582	75 56 53.8	.97	51.34	2
8 103	Gould, Z. C., 11ʰ 2793 .	9.5	39 4.77	2.771	67 57 41.0	.97	52.30	1
8 104	Brisbane, 3761	8.0	8.30	2.583	76 1 37.7	.97	51.30	1
8 105	Muscæ	9.5	12.30	+ 2.812	64 58 57.1	− 19.97	52.41	1

No. 8 096. Gould's declination is 12′ 0.1″ less.

No. 8 103. Gould's declination is 10″ greater.

Number.	Constellation, Name of Star, or Synonym.	Magnitude.	Right Ascension, 1850.0.	Annual Precession.	South Declination, 1850.0.	Annual Precession.	Mean year.	No. of obs.
			h. m. s.	s.	° ′ ″	″		
8 106	Lacaille, 4885	6.0	11 39 16.96	+ 2.860	60 20 41.7	− 19.97	52.40	1
8 107	Gould, Z. C., 11h 2806 .	9.0	18.82	2.683	72 46 13.0	.97	50.42	1
8 108	Gould, 16101	9.0	21.12	2.740	70 4 17.2	.97	51.44	1
8 109	Gould, Z. C., 11h 2814 .	9.5	21.98	2.817	64 45 25.0	.97	51.37	2
8 110	Gould, Z. C., 11h 2820 .	9.0	25.70	+ 2.802	66 1 26.0	− 19.97	52.38	1
8 111	Gould, Z. C., 11h 2821 .	9.2	11 39 26.05	+ 2.805	65 49 30.6	− 19.97	51.35	2
8 112	Lacaille, 4889	7.7	26.81	2.827	63 56 0.2	.97	50.29	2
8 113	Muscæ	9.5	29.33	2.790	67 3 5.2	.97	52.29	1
8 114	Lacaille, 4891	9.0	29.62	2.742	70 4 38.2	.97	51.44	1
8 115	Muscæ	9.8	31.22	+ 2.754	69 26 22.6	− 19.97	51.87	2
8 116	Gould, Z. C., 11h 2826 .	9.8	11 39 37.24	+ 2.601	75 49 52.8	− 19.97	51.34	2
8 117	Muscæ	9.0	38.78	2.758	69 17 29.0	.97	51.45	1
8 118	Gould, 16111	9.0	45.36	2.753	69 40 56.3	.98	52.29	1
8 119	Chamæleontis	10.0	45.37	2.531	77 41 24.1	.98	50.40	1
8 120	Muscæ	9.5	46.53	+ 2.755	69 35 53.7	− 19.98	52.29	1
8 121	Gould, Z. C., 11h 2837 .	9.0	11 39 48.55	+ 2.654	74 16 32.8	− 19.98	51.64	2
8 122	Muscæ	9.5	52.61	2.717	71 41 30.6	.98	52.26	1
8 123	Gould, Z. C., 11h 2849 .	9.5	54.72	2.756	69 38 26.3	.98	52.29	1
8 124	Muscæ	9.5	39 58.10	2.826	64 37 26.0	.98	51.37	2
8 125	Gould, Z. C., 11h 2865 .	10.0	40 7.79	+ 2.663	74 12 34.4	− 19.98	52.28	1
8 126	Gould, Z. C., 11h 2876 .	10.0	11 40 10.04	+ 2.802	66 47 36.5	− 19.98	50.30	1
8 127	Gould, Z. C., 11h 2883 .	9.0	18.96	2.764	69 34 58.0	.98	52.29	1
8 128	Velorum	10.0	24.51	2.844	63 25 24.0	.99	50.30	1
8 129	Gould, Z. C., 11h 2896 .	9.0	27.30	2.826	65 10 26.4	.99	52.41	1
8 130	Velorum	10.0	28.93	+ 2.844	63 28 4.4	− 19.99	50.30	1
8 131	Chamæleontis	10.0	11 40 29.67	+ 2.554	77 36 42.8	19.99	50.38	1
8 132	Gould, Z. C., 11h 2898 .	9.5	30.12	2.743	70 56 4.9	.99	52.28	1
8 133	Brisbane, 3774	8.0	30.44	2.623	75 47 2.7	.99	51.38	1
8 134	Gould, Z. C., 11h 2891 .	9.0	32.04	2.580	77 4 11.2	.99	51.31	2
8 135	Lacaille, 4896	7.0	33.56	+ 2.807	66 51 31.2	− 19.99	50.30	1
8 136	Muscæ	9.5	11 40 36.50	+ 2.790	68 9 10.6	− 19.99	50.37	1
8 137	Gould, Z. C., 11h 2913 .	9.0	44.20	2.847	64 29 23.4	.99	52.45	1
8 138	Muscæ	10.0	57.01	2.831	65 17 9.8	.99	52.41	1
8 139	Muscæ	9.0	40 57.68	2.831	65 17 18.2	.99	52.41	1
8 140	Lacaille, 4899	6.0	41 4.26	+ 2.824	65 58 49.6	− 19.99	52.38	1
8 141	Gould, Z. C., 11h 2926 .	8.5	11 41 4.73	+ 2.803	67 42 8.1	− 19.99	52.30	1
8 142	Gould, Z. C., 11h 2921 .	8.5	4.87	2.592	77 3 14.8	.99	51.31	2
8 143	Gould, Z. C., 11h 2939 .	9.0	15.53	2.812	67 11 42.2	.99	52.29	1
8 144	Muscæ	9.0	16.17	2.825	66 9 22.1	.99	52.38	1
8 145	Brisbane, 3781	8.0	16.42	+ 2.632	76 3 50.1	− 19.99	51.30	1

No. 8 143. Gould's declination is 10″ less.

Number.	Constellation, Name of Star, or Synonym.	Magnitude.	Right Ascension, 1850.0.	Annual Precession.	South Declination, 1850.0.	Annual Precession.	Mean year	No. of obs.
			h. m. s.	s.	° ′ ″	″		
8146	Gould, Z. C., 11h 2940	9.0	11 41 23.54	+ 2.725	72 38 43.0	− 19.99	50.85	2
8147	Gould, Z. C., 11h 2943	9.0	29.32	2.610	76 51 43.2	.99	51.25	1
8148	Muscæ	10.0	35.21	2.746	71 46 27.0	.99	52.26	1
8149	Chamæleontis	10.0	37.53	2.463	80 1 52.8	.99	51.37	1
8150	Gould, 16142	9.5	41.30	+ 2.820	67 1 27.4	− 19.99	52.29	1
8151	Octantis	8.8	11 41 48.47	+ 2.009	84 16 30.1	− 19.49	50.36	3
8152	Gould, Z. C., 11h 2969	9.0	52.38	2.787	69 37 11.3	.99	52.29	1
8153	Gould, Z. C., 11h 2975	9.5	41 59.24	2.802	68 44 53.1	19.99	50.43	2
8154	Gould, Z. C., 11h 2982	9.0	42 4.26	2.760	71 28 14.6	20.00	52.27	2
8155	Gould, Z. C., 11h 2986	9.0	5.74	+ 2.815	67 51 52.2	− 20.00	52.30	1
8156	Chamæleontis	9.0	11 42 5.87	+ 2.365	81 36 17.7	− 20.00	51.34	1
8157	Gould, Z. C., 11h 2997	9.8	16.55	2.833	66 38 25.1	.00	50.31	2
8158	Chamæleontis	10.0	22.77	2.426	80 57 57.9	.00	51.37	1
8159	Lacaille, 4903	4.5	24.83	2.871	62 57 16.7	.00	50.30	1
8160	Chamæleontis	10.0	37.30	+ 2.686	75 19 31.6	− 20.00	50.44	1
8161	Gould, Z. C., 11h 3025	9.8	11 42 39.45	+ 2.851	65 24 38.0	− 20.00	51.34	2
8162	Velorum	10.0	41.77	2.870	63 25 48.7	.00	50.30	1
8163	Lacaille, 4907	6.0	48.30	2.805	69 23 31.0	.00	51.87	2
8164	Chamæleontis	10.0	48.94	2.648	76 41 17.6	.00	51.30	2
8165	Octantis	10.0	42 54.61	+ 1.995	84 43 0.6	− 20.00	50.32	1
8166	Gould, Z. C., 11h 3056	9.5	11 43 13.17	+ 2.712	74 48 16.1	− 20.00	51.29	1
8167	Gould, Z. C., 11h 3068	9.2	19.69	2.839	67 21 12.4	.00	52.30	2
8168	Gould, Z. C., 11h 3075	9.0	26.09	2.884	62 48 57.6	.00	50.30	1
8169	Muscæ	10.0	34.69	2.818	69 21 0.9	.00	51.45	1
8170	Lacaille, 4915	8.8	35.53	+ 2.765	72 42 2.8	− 20.00	50.85	2
8171	Gould, Z. C., 11h 3078	10.0	11 43 36.21	+ 2.638	77 34 25.6	− 20.01	50.40	1
8172	Lacaille, 4912	8.0	42.23	2.845	67 14 45.6	.01	52.29	1
8173	Gould, Z. C., 11h 3098	9.0	51.43	2.804	70 39 31.4	.01	52.28	2
8174	Gould, Z. C., 11h 3101	9.8	43 53.04	2.815	69 58 43.9	.01	51.87	2
8175	Muscæ	10.0	44 2.61	+ 2.807	70 35 47.6	− 20.01	52.28	1
8176	Melbourne (1), 583	8.5	11 44 7.40	+ 2.842	68 1 40.0	− 20.01	50.37	1
8177	Gould, Z. C., 11h 3117	9.5	9.76	2.657	77 27 5.4	.01	50.88	2
8178	Gould, Z. C., 11h 3138	10.0	22.30	2.854	67 20 24.9	.01	52.29	1
8179	Gould, 16197	11.0	29.26	2.888	63 45 46.7	.01	50.29	1
8180	Lacaille, 4919	8.5	29.66	+ 2.888	63 45 44.2	− 20.01	50.29	1
8181	Lacaille, 4920	6.0	11 44 32.88	+ 2.884	64 22 17.0	− 20.01	51.37	2
8182	Muscæ	9.0	33.78	2.856	67 19 35.7	.01	52.29	1
8183	Gould, Z. C., 11h 3157	9.5	42.97	2.860	67 13 11.5	.01	52.29	1
8184	Gould, Z. C., 11h 3159	9.5	43.93	2.872	65 57 34.7	.01	52.38	1
8185	Muscæ	9.5	44.34	+ 2.852	67 56 15.3	− 21.01	50.37	1

Number.	Constellation, Name of Star, or Synonym.	Magnitude.	Right Ascension, 1850.0.	Annual Precession.	South Declination, 1850.0.	Annual Precession.	Mean year.	No. of obs.
			h. m. s.	s.	° ′ ″	″		
8 186	Chamæleontis	10.0	11 44 59.38	+ 2.671	77 40 30.3	— 20.01	50.40	1
8 187	Chamæleontis	9.5	44 59.60	2.807	71 43 54.0	.01	52.26	1
8 188	Gould, 16210	9.0	45 1.14	2.855	68 4 50.1	.01	50.37	1
8 189	Chamæleontis	9.8	13.63	2.706	76 45 1.9	.01	51.30	2
8 190	Muscæ	10.0	15.84	+ 2.808	71 56 56.6	— 20.01	52.26	1
8 191	Chamæleontis	10.0	11 45 16.91	+ 2.535	80 55 53.5	— 20.02	51.37	1
8 192	Gould, Z. C., 11^{h} 3196 .	9.5	18.24	2.880	65 55 47.7	.02	52.38	1
8 193	Muscæ	9.5	18.28	2.813	71 39 16.3	.02	52.26	1
8 194	Muscæ	9.5	21.12	2.812	71 46 34.1	.02	52.26	1
8 195	Gould, Z. C., 11^{h} 3201 .	9.5	24.56	+ 2.849	69 7 10.8	— 20.02	50.45	1
8 196	Muscæ	9.5	11 45 26.30	+ 2.819	71 25 55.6	— 20.02	52.27	1
8 197	Gould, Z. C., 11^{h} 3205 .	9.5	27.18	2.849	69 8 33.2	.02	50.45	1
8 198	Muscæ	9.0	27.40	2.822	71 14 21.1	.02	52.27	1
8 199	Brisbane, 3812	8.5	34.81	2.896	64 24 25.8	.02	52.45	1
8 200	Gould, Z. C., 11^{h} 3215 .	8.8	36.86	+ 2.853	69 3 0.6	— 20.02	50.95	2
8 201	Gould, Z. C., 11^{h} 3235 .	9.5	11 45 57.14	+ 2.743	76 0 12.7	— 20.02	51.30	1
8 202	Lacaille, 4927	8.2	45 58.86	2.900	64 34 9.2	.02	51.37	2
8 203	Gould, Z. C., 11^{h} 3239 .	9.5	46 3.17	2.849	69 59 34.7	.02	51.86	2
8 204	Gould, Z. C., 11^{h} 3256 .	9.0	15.17	2.797	73 42 12.6	.02	51.25	1
8 205	Lacaille, 4929	8.0	21.69	+ 2.905	64 27 20.1	— 20.02	52.45	1
8 206	Gould, Z. C., 11^{h} 3273 .	9.0	11 46 24.03	+ 2.887	66 48 43.2	— 20.02	50.30	1
8 207	Chamæleontis	10.0	30.96	2.548	81 27 34.5	.02	51.34	1
8 208	Gould, Z. C., 11^{h} 3283 .	9.2	33.31	2.844	70 58 50.3	.02	52.28	2
8 209	Gould, Z. C., 11^{h} 3284 .	9.2	34.08	2.822	72 35 35.2	.02	50.85	2
8 210	Muscæ	9.5	39.25	+ 2.853	70 27 44.1	— 20.02	52.28	1
8 211	Gould, Z. C., 11^{h} 3294 .	9.2	11 46 42.56	+ 2.869	69 6 7.0	— 20.02	50.95	2
8 212	Muscæ	9.0	45.66	2.897	66 11 32.0	.02	51.35	2
8 213	Gould, Z. C., 11^{h} 3299 .	9.0	47.68	2.892	66 50 16.5	.02	50.30	1
8 214	Muscæ	10.0	46 57.33	2.856	70 37 9.3	.02	52.28	1
8 215	Gould, Z. C., 11^{h} 3312 .	9.8	47 4.61	+ 2.873	69 10 4.6	— 20.02	50.95	2
8 216	Gould, Z. C., 11^{h} 3317 .	9.5	11 47 6.33	+ 2.898	66 37 13.8	— 20.02	50.32	1
8 217	Brisbane, 3823	9.0	7.97	2.927	62 51 56.5	.02	50.30	1
8 218	Gould, Z. C., 11^{h} 3320 .	9.8	8.80	2.875	69 8 18.0	.02	50.95	2
8 219	Gould, Z. C., 11^{h} 3330 .	9.5	16.26	2.893	67 29 29.4	.03	52.30	2
8 220	Brisbane, 3824	9.0	17.59	+ 2.917	64 23 4.5	— 20.03	52.45	1
8 221	Chamæleontis	9.0	11 47 19.61	+ 2.637	80 21 23.4	— 20.03	51.37	1
8 222	Lacaille, 4935	7.5	20.90	2.910	65 32 20.0	.03	51.34	2
8 223	Gould, Z. C., 11^{h} 3339 .	9.0	29.16	2.911	65 35 49.1	.03	51.34	2
8 224	Gould, Z. C., 11^{h} 3353 .	8.5	36.02	2.873	70 2 34.4	.03	51.44	1
8 225	Brisbane, 3839	9.0	46.16	+ 2.535	82 26 13.6	— 20.03	50.38	1

Number.	Constellation, Name of Star, or Synonym.	Magnitude.	Right Ascension, 1850.0.	Annual Precession.	South Declination, 1850.0.	Annual Precession.	Mean year.	No. of obs.
			h. m. s.	s.	° ′ ″	″		
8 226	Gould, 16267	9.2	11 47 50.96	+ 2.916	65 34 17.2	— 20.03	51.38	2
8 227	Gould, Z. C., 11h 3372 .	10.0	54.47	2.933	63 12 41.3	.03	50.30	1
8 228	Brisbane, 3830	9.0	55.37	2.934	62 52 6.9	.03	50.30	1
8 229	Chamæleontis	9.0	47 58.54	2.667	80 10 22.7	.03	50.90	2
8 230	Gould, Z. C., 11h 3388 .	8.8	48 2.37	+ 2.869	70 59 34.3	— 20.03	52.28	2
8 231	Chamæleontis	9.0	11 48 5.67	+ 2.656	80 30 55.4	— 20.03	50.43	1
8 232	Gould, Z. C., 11h 3395 .	9.2	6.99	2.904	67 32 13.2	.03	52.30	2
8 233	Gould, Z. C., 11h 3399	9.5	11.15	2.923	65 5 39.6	.03	52.41	1
8 234	Gould, 16281	8.0	19.87	2.906	67 42 51.6	.03	52.30	1
8 235	Gould, Z. C., 11h 3407 .	10.0	22.90	+ 2.727	78 52 53.8	— 20.03	51.28	1
8 236	Brisbane, 3835	8.8	11 48 32.44	+ 2.936	63 47 55.6	— 20.03	50.29	2
8 237	Gould, Z. C., 11h 3435 .	8.5	46.82	2.805	76 10 26.0	.03	51.30	1
8 238	Gould, 16291	9.0	48 56.36	2.941	63 42 44.8	.03	50.29	1
8 239	Brisbane, 3837	9.0	49 1.80	2.946	62 57 26.8	.03	50.30	1
8 240	Gould, 16296	9.5	5.90	+ 2.889	70 49 0.8	— 20.03	52.28	1
8 241	Chamæleontis	10.0	11 49 10.16	+ 2.772	78 6 8.9	— 20.03	50.30	1
8 242	Muscæ	9.5	18.40	2.921	67 28 46.0	.03	52.30	2
8 243	Gould, Z. C., 11h 3473 .	9.0	21.48	2.907	69 19 18.9	.03	51.45	1
8 244	Muscæ	9.5	37.55	2.901	70 29 59.4	.04	52.28	1
8 245	Brisbane, 3842	9.0	37.81	+ 2.954	62 50 33.6	— 20.04	50.30	1
8 246	Velorum	10.0	11 49 38.84	+ 2.951	63 27 0.8	— 20.04	50.30	1
8 247	Muscæ	10.0	44.25	2.938	65 50 22.2	.04	52.38	1
8 248	Gould, Z. C., 11h 3501 .	9.5	49 46.70	2.939	65 45 28.0	.04	52.38	1
8 249	Gould, 16322	9.5	50 1.54	2.828	76 36 56.1	.04	51.34	1
8 250	Gould, Z. C., 11h 3518 .	7.5	6.46	+ 2.756	79 39 30.5	— 20.04	51.24	1
8 251	Octantis	9.2	11 50 15.53	+ 2.125	86 34 3.0	— 20.04	51.33	5
8 252	Gould, Z. C., 11h 3534 .	9.0	18.15	2.949	65 18 25.3	.04	52.41	1
8 253	Herschel, 4483[1]	12.0	21.00	2.913	70 31 48.8	.04	52.29	1
8 254	Herschel, 4483[2]	13.0	22.99	2.913	70 31 54.9	.04	52.29	1
8 255	Gould, Z. C., 11h 3541 .	9.0	26.89	+ 2.870	74 32 30.3	— 20.04	50.41	1
8 256	Lacaille, 4956	8.5	11 50 28.44	+ 2.955	64 29 12.5	— 20.04	52.45	1
8 257	Gould, Z. C., 11h 3547 .	8.8	34.18	2.913	70 56 49.4	.04	52.28	2
8 258	Muscæ	9.5	39.40	2.948	66 15 25.8	.04	52.38	1
8 259	Gould, Z. C., 11h 3552 .	9.5	39.52	2.908	71 33 45.5	.04	52.26	1
8 260	Muscæ	10.0	46.71	+ 2.920	70 32 35.8	— 20.04	52.28	1
8 261	Muscæ	9.0	11 50 52.00	+ 2.707	81 41 17.8	— 20.04	51.34	1
8 262	Lacaille, 4962	8.0	51 2.28	2.929	69 54 17.1	.04	51.86	2
8 263	Octantis	10.0	11.33	2.635	83 17 30.2	.04	50.39	1
8 264	Brisbane, 3855	7.0	15.49	2.851	76 59 22.2	.04	51.25	1
8 265	Lacaille, 4963	6.0	15.68	+ 2.967	63 30 17.0	— 20.04	50.29	2

Number.	Constellation, Name of Star, or Synonym.	Magnitude.	Right Ascension, 1850.0.	Annual Precession.	South Declination, 1850.0.	Annual Precession.	Mean year.	No. of obs.
			h. m. s.	s.	° ′ ″	″		
8 266	Gould, Z. C., 11h 3604	9.2	11 51 21.56	+ 2.899	73 46 34.4	− 20.04	51.76	2
8 267	Chamæleontis	10.0	23.40	2.877	75 33 35.9	.04	51.38	1
8 268	Chamæleontis	9.5	30.77	2.840	77 57 14.2	.04	51.01	3
8 269	Gould, 16364	8.5	35.40	2.910	73 10 24.5	.04	50.43	2
8 270	Chamæleontis	10.0	36.44	+ 2.878	75 49 20.5	− 20.04	51.30	1
8 271	Gould, 16367	9.0	11 51 39.13	+ 2.970	64 27 48.3	− 20.04	52.45	1
8 272	Crucis	9.5	39.51	2.973	63 55 26.4	.04	50.29	1
8 273	Muscæ	9.5	45.06	2.957	67 19 8.1	.04	52.29	1
8 274	Octantis	9.5	49.75	2.397	85 57 30.2	.04	51.31	4
8 275	Muscæ	9.8	50.28	+ 2.864	77 8 58.4	− 20.04	51.31	2
8 276	Brisbane, 3863	9.5	11 51 57.90	+ 2.972	64 33 34.8	− 20.04	52.45	1
8 277	Gould, Z. C., 11h 3644	10.0	52 6.92	2.872	77 3 3.1	.04	51.31	2
8 278	Chamæleontis	9.0	9.54	2.733	82 18 16.8	.04	50.38	2
8 279	ε Chamæleontis	5.0	15.50	2.870	77 23 13.1	.04	50.88	2
8 280	Brisbane, 3866	8.8	16.66	+ 2.963	67 29 57.8	− 20.05	52.30	2
8 281	Gould, Z. C., 11h 3664	9.2	11 52 21.84	+ 2.955	69 2 58.4	− 20.05	50.95	2
8 282	Gould, Z. C., 11h 3666	9.2	24.60	2.875	77 21 37.8	.05	50.83	2
8 283	Crucis	10.0	33.10	2.987	62 58 59.3	.05	50.30	1
8 284	Gould, Z. C., 11h 3682	9.0	37.73	2.952	70 12 1.6	.05	51.44	1
8 285	Gould, Z. C., 11h 3683	10.0	38.09	+ 2.974	65 25 7.8	− 20.05	52.41	1
8 286	Gould, Z. C., 11h 3688	9.5	11 52 38.36	+ 2.979	65 14 32.8	− 20.05	52.41	1
8 287	Gould, Z. C., 11h 3691	9.5	40.96	2.975	66 13 35.8	.05	51.35	2
8 288	Lacaille, 4975	7.0	41.02	2.882	77 21 29.1	.05	50.88	2
8 289	Muscæ	10.0	41.93	2.980	65 3 58.9	.05	52.41	1
8 290	Muscæ	9.5	49.15	+ 2.949	71 12 59.9	− 20.05	52.27	1
8 291	Gould, 16397	9.5	11 52 49.74	+ 2.971	67 31 5.7	− 20.05	52.30	2
8 292	Muscæ	9.8	51.72	2.972	67 24 10.9	.05	52.30	2
8 293	Gould, Z. C., 11h 3708	10.0	56.40	2.963	69 19 18.0	.05	51.45	1
8 294	Gould, Z. C., 11h 3709	10.0	52 57.43	2.962	69 29 38.4	.05	52.29	1
8 295	Chamæleontis	8.5	53 3.35	+ 2.908	76 8 28.9	− 20.05	51.30	1
8 296	Muscæ	10.0	11 53 7.60	+ 2.954	71 10 19.4	− 20.05	52.27	1
8 297	Crucis	9.0	10.49	2.994	63 0 49.1	.05	50.30	1
8 298	Muscæ	9.8	16.46	2.916	75 49 33.4	.05	51.34	2
8 299	Gould, Z. C., 11h 3732	9.0	23.18	2.985	66 3 31.0	.05	52.38	1
8 300	Gould, Z. C., 11h 3728	9.5	24.64	+ 2.901	77 19 48.4	− 20.05	51.36	1
8 301	Gould, 16410	10.0	11 53 27.91	+ 2.978	68 21 8.2	− 20.05	50.41	1
8 302	Gould, Z. C., 11h 3746	9.5	35.51	2.997	63 25 5.8	.05	50.30	1
8 303	Muscæ	10.0	41.38	2.968	70 29 31.0	.05	52.28	1
8 304	Gould, Z. C., 11h 3749	10.0	41.79	2.898	78 2 32.2	.05	52.35	1
8 305	Gould, Z. C., 11h 3753	9.5	43.55	+ 2.999	63 14 41.2	− 20.05	50.30	1

No. 8 276. Brisbane's declination agrees with 8 271.

No. 8 302. Gould's declination is 20″ less.

Number.	Constellation, Name of Star, or Synonym.	Magnitude.	Right Ascension, 1850.0.	Annual Precession.	South Declination, 1850.0.	Annual Precession.	Mean year.	No. of obs.
			h. m. s.	s.	° ′ ″	″		
8306	Gould, Z. C., 11h 3757	9.0	11 53 48.54	+ 2.971	70 23 31.5	− 20.05	52.28	1
8307	Muscæ	10.0	53 53.69	2.963	71 54 16.9	.05	52.26	1
8308	Gould, Z. C., 11h 3770	9.0	54 0.30	3.000	63 53 16.8	.05	50.29	2
8309	Chamæleontis	9.0	3.07	2.833	81 43 32.2	.05	51.34	1
8310	Gould, Z. C., 11h 3782	9.0	7.96	+ 2.980	69 34 18.5	− 20.05	52.29	1
8311	Muscæ	11.0	11 54 8.19	+ 2.964	72 26 55.9	− 20.05	51.28	1
8312	Lacaille, 4980	7.5	8.30	2.986	68 21 42.4	.05	50.42	1
8313	Gould, Z. C., 11h 3783	9.0	9.01	2.909	78 9 7.5	.05	51.32	2
8314	Gould, Z. C., 11h 3789	9.5	11.18	2.998	65 26 4.2	.05	51.34	2
8315	Lacaille, 4981	7.5	16.36	+ 2.968	72 27 42.9	− 20.05	51.28	1
8316	Chamæleontis	9.5	11 54 26.27	+ 2.940	76 8 50.0	− 20.05	51.30	1
8317	Gould, Z. C., 11h 3806	9.0	28.35	2.969	72 39 13.5	.05	50.85	2
8318	Gould, 16427	8.5	31.68	3.007	63 57 45.1	.05	50.29	2
8319	Gould, 16430	9.2	36.85	2.998	67 0 58.5	.05	51.11	3
8320	Chamæleontis	10.0	40.84	+ 2.889	80 21 25.7	− 20.05	51.37	1
8321	Muscæ	9.0	11 54 42.07	+ 2.994	67 58 42.1	− 20.05	50.37	1
8322	Octantis	10.0	46.88	2.752	84 25 46.8	.05	50.33	1
8323	Gould, Z. C., 11h 3833	9.5	45.93	2.903	66 3 13.7	.05	52.38	1
8324	Brisbane, 3885	9.0	51.98	2.996	68 33 29.7	.05	50.41	1
8325	Lacaille, 4984	7.0	52.56	+ 2.987	70 39 16.4	− 20.05	52.28	2
8326	Muscæ	10.0	11 54 54.44	+ 2.970	73 45 46.1	− 20.05	52.28	1
8327	Gould, Z. C., 11h 3845	9.0	58.76	2.971	73 48 19.8	.05	51.76	2
8328	Gould, Z. C., 11h 3847	8.8	59.31	2.970	73 56 33.8	.05	51.77	2
8329	Lacaille, 4991	7.1	59.34	2.736	84 47 48.9	.05	50.75	5
8330	Gould, Z. C., 11h 3848	9.0	54 59.77	+ 2.992	70 1 26.2	− 20.05	51.44	1
8331	Lacaille, 4985	6.5	11 55 0.76	+ 2.999	68 21 23.3	− 20.05	50.42	1
8332	Gould, Z. C., 11h 3855	9.5	9.11	3.004	67 15 44.6	.05	52.29	1
8333	Octantis	9.7	9.93	2.762	84 48 7.6	.05	50.83	4
8334	Muscæ	9.0	11.36	3.012	64 48 7.4	.05	50.28	1
8335	Gould, Z. C., 11h 3859	9.5	12.19	+ 3.012	65 5 42.4	− 20.05	52.41	1
8336	Muscæ	8.5	11 55 13.01	+ 3.004	67 44 57.6	− 20.05	52.30	1
8337	Muscæ	8.5	21.00	3.014	64 52 49.2	.05	50.28	1
8338	Gould, Z. C., 11h 3879	9.5	25.04	3.013	65 22 26.3	.05	51.34	2
8339	Chamæleontis	9.0	30.18	2.886	81 57 10.3	.05	51.34	1
8340	Gould, Z. C., 11h 3887	8.8	30.84	+ 3.016	64 59 54.4	− 20.05	51.35	2
8341	Octantis	9.7	11 55 31.43	+ 2.770	85 3 39.2	− 20.05	51.13	5
8342	Gould, Z. C., 11h 3888	9.5	32.47	2.980	74 6 26.0	.05	52.28	1
8343	Melbourne (1), 591	9.0	33.83	3.020	63 30 53.0	.05	50.29	1
8344	Octantis	10.0	42.09	2.780	84 57 59.2	.05	51.33	4
8345	Chamæleontis	10.0	44.35	+ 2.969	76 26 40.3	− 20.00	51.34	1

No. 8330. Recorded 55m 59s.

Number.	Constellation, Name of Star, or Synonym.	Magnitude.	Right Ascension, 1850.0.	Annual Precession.	South Declination, 1850.0.	Annual Precession.	Mean year.	No. of obs.
			h. m. s.	s.	° ′ ″	″		
8 346	Gould, Z. C., 11^h 3902	9.0	11 55 45.84	+ 2.998	71 30 25.8	− 20.05	52.27	2
8 347	Gould, Z. C., 11^h 3907	7.7	52.10	3.000	71 21 28.0	.05	52.27	2
8 348	Chamæleontis	10.0	54.53	2.921	81 0 56.4	.05	50.88	2
8 349	Chamæleontis	10.0	55 55.60	2.969	76 55 58.2	.05	51.25	1
8 350	Gould, 16461	9.0	56 0.40	+ 3.016	67 28 3.5	− 20.05	52.30	2
8 351	Lacaille, 4996	7.0	11 56 4.65	+ 2.995	73 22 43.4	− 20.05	50.44	1
8 352	Muscæ	9.0	15.13	3.009	70 59 53.7	.05	52.27	1
8 353	Gould, Z. C., 11^h 3941	7.5	23.38	3.030	63 42 16.2	.05	50.29	2
8 354	Gould, Z. C., 11^h 3945	10.0	26.69	3.031	64 22 11.7	.05	52.45	1
8 355	Brisbane, 3898	7.2	27.76	+ 3.019	68 41 9.8	− 20.05	50.43	2
8 356	Brisbane, 3899	9.0	11 56 32.34	+ 3.029	64 32 58.8	− 20.05	52.45	1
8 357	Gould, Z. C., 11^h 3949	9.0	36.38	3.022	68 53 28.9	.05	50.45	1
8 358	Gould, 16478	9.0	38.26	3.009	72 46 19.8	.05	50.42	1
8 359	Octantis	9.8	46.68	2.850	85 8 1.9	.05	51.33	5
8 360	Gould, Z. C., 11^h 3958	8.5	47.43	+ 3.030	65 58 9.8	− 20.05	52.38	1
8 361	Gould, Z. C., 11^h 3959	9.0	11 56 50.64	+ 3.036	62 44 24.0	− 20.05	50.30	1
8 362	Gould, Z. C., 11^h 3960	9.5	51.85	3.030	66 12 52.6	.05	51.69	3
8 363	Muscæ	10.0	53.34	3.009	73 50 48.8	.05	51.77	2
8 364	Gould, Z. C., 11^h 3963	8.0	55.05	3.021	70 23 3.3	.05	52.28	1
8 365	Lacaille, 5000	6.5	56 56.97	+ 3.030	67 29 35.6	− 20.05	52.30	2
8 366	κ Chamæleontis	5.6	11 57 4.82	+ 3.005	75 41 6.4	− 20.05	51.38	1
8 367	Chamæleontis	9.0	5.03	2.959	81 25 14.9	.05	50.38	1
8 368	Gould, Z. C., 11^h 3984	9.0	7.81	3.021	71 41 59.9	.05	52.26	1
8 369	Gould, Z. C., 11^h 3985	8.5	8.71	3.022	71 37 16.6	.05	52.26	1
8 370	Lacaille, 5012	7.2	9.19	+ 3.039	64 42 39.8	− 20.05	51.37	2
8 371	Brisbane, 3909	8.0	11 57 15.83	+ 3.030	69 9 59.9	− 20.05	51.45	1
8 372	Brisbane, 3908	9.0	15.97	3.030	68 58 33.8	.05	50.95	2
8 373	Gould, 16492	8.5	17.24	3.013	74 51 7.3	.05	51.29	1
8 374	Muscæ	9.5	27.02	3.036	66 17 7.4	.06	52.37	2
8 375	Gould, Z. C., 11^h 4009	9.5	28.15	+ 3.036	67 34 34.0	− 20.06	52.30	1
8 376	Chamæleontis	10.0	11 57 28.64	+ 2.997	79 15 34.5	− 20.06	50.42	1
8 377	Gould, Z. C., 11^h 4021	9.5	39.34	3.040	66 24 3.0	.06	52.36	1
8 378	Gould, Z. C., 11^h 4024	9.5	42.65	3.039	67 43 59.5	.06	52.30	1
8 379	Muscæ	10.0	44.17	3.038	69 33 32.9	.06	52.29	1
8 380	Gould, Z. C., 11^h 4028	9.0	44.96	+ 3.045	64 2 45.3	− 20.06	50.29	1
8 381	Brisbane, 3913	8.8	11 57 45.57	+ 3.041	67 0 50.4	− 20.06	51.38	4
8 382	Lacaille, 5010	8.5	47.95	3.046	63 26 16.0	.06	50.30	1
8 383	Gould, Z. C., 11^h 4034	9.5	48.46	3.038	69 4 41.6	.06	50.95	2
8 384	Gould, Z. C., 11^h 4037	9.5	50.48	3.011	78 12 19.4	.06	51.33	3
8 385	Muscæ	9.5	57 58.48	+ 3.038	70 49 24.4	− 20.06	52.28	1

Number.	Constellation, Name of Star, or Synonym.	Magnitude.	Right Ascension, 1850.0.	Annual Precession.	South Declination, 1850.0.	Annual Precession.	Mean year.	No. of obs.
			h. m. s.	s.	° ′ ″	″		
8386	Brisbane, 3914	7.7	11 58 4.70	+ 3.043	68 44 14.0	− 20.06	50.43	2
8387	Gould, Z. C., 11h 4054	9.5	12.70	3.050	64 39 25.5	.06	52.45	1
8388	Octantis	10.0	17.18	2.903	86 36 49.1	.06	51.33	1
8389	Lacaille, 5018	9.0	23.80	3.048	68 17 33.4	.06	50.37	1
8390	Lacaille, 5017	7.5	23.98	+ 3.042	72 43 10.0	− 20.06	50.85	2
8391	Brisbane, 3920	8.0	11 58 32.49	+ 3.050	68 57 44.8	− 20.06	50.45	1
8392	Lacaille, 5019	7.0	33.64	3.052	67 48 58.4	.06	51.34	2
8393	Brisbane, 3919	8.4	35.74	2.979	84 56 41.6	.06	51.19	6
8394	Lacaille, 5020	7.5	38.22	3.055	64 52 29.0	.06	50.28	1
8395	Gould, Z. C., 11h 4094	9.5	39.54	+ 3.055	64 42 25.4	− 20.06	50.28	1
8396	Octantis	9.3	11 58 41.32	+ 2.962	85 55 28.9	− 20.06	51.33	3
8397	Muscæ	9.5	44.58	3.050	71 2 46.9	.06	52.28	1
8398	Gould, Z. C., 12h 4	9.0	48.40	3.057	64 39 29.4	.06	52.45	1
8399	Gould, 16538	9.8	58 54.40	3.059	63 6 16.2	.06	50.34	2
8400	Chamæleontis	10.0	59 0.15	+ 3.042	78 53 10.2	− 20.06	51.28	1
8401	Gould, Z. C., 12h 21	8.8	11 59 0.29	+ 3.054	72 14 11.9	− 20.06	51.95	3
8402	Muscæ	10.0	3.37	3.055	71 22 3.4	.06	52.27	1
8403	η Crucis	4.8	5.86	3.061	63 46 37.1	.06	50.29	3
8404	Gould, Z. C., 12h 26	9.0	6.29	3.054	73 29 2.4	.06	51.25	1
8405	Gould, Z. C., 12h 27	9.2	6.70	+ 3.047	78 18 3.8	− 20.06	51.84	2
8406	Chamæleontis	9.5	11 59 12.48	+ 3.047	79 25 16.9	− 20.06	50.83	2
8407	Muscæ	9.0	12.86	3.059	70 2 22.3	.06	51.44	1
8408	Chamæleontis	9.2	12.99	3.050	77 56 15.8	.06	51.87	2
8409	Lacaille, 5024	7.8	13.09	3.051	77 54 19.5	.06	51.01	3
8410	Muscæ	9.0	18.49	+ 3.057	74 57 4.4	− 20.06	51.29	1
8411	Gould, Z. C., 12h 43	9.5	11 59 18.66	+ 3.063	66 22 34.2	− 20.06	52.37	2
8412	Gould, Z. C., 12h 48	9.5	24.16	3.065	65 6 47.8	.06	52.41	1
8413	Gould, Z. C., 12h 50	9.5	26.32	3.063	69 33 23.8	.06	52.29	1
8414	Gould, Z. C., 12h 55	10.0	28.60	3.064	69 36 14.1	.06	52.29	1
8415	Lacaille, 5026	8.0	32.60	+ 3.066	63 50 47.8	− 20.06	50.29	3
8416	Gould, Z. C., 12h 64	9.0	11 59 38.11	+ 3.067	65 3 51.5	− 20.06	52.41	1
8417	λ Chamæleontis	7.0	11 59 59.32	3.072	74 31 57.9	.06	50.41	1
8418	Muscæ	10.0	12 0 2.73	3.073	71 41 48.3	.06	52.26	1
8419	Gould, Z. C., 12h 84	9.5	4.91	3.073	67 2 52.0	.06	51.67	3
8420	Gould, Z. C., 12h 102	9.0	20.19	+ 3.076	66 8 54.3	− 20.06	52.38	1
8421	Chamæleontis	10.0	12 0 30.58	+ 3.084	76 33 38.6	− 20.06	51.34	1
8422	Muscæ	10.0	32.87	3.081	71 48 42.3	.06	52.26	1
8423	Chamæleontis	10.0	42.44	3.093	79 9 22.7	.06	50.42	1
8424	Gould, Z. C., 12h 130	11.0	50.01	3.091	75 39 7.3	.06	51.38	1
8425	Chamæleontis	13.0	50.07	+ 3.091	75 38 49.3	− 20.06	51.38	1

Number.	Constellation, Name of Star, or Synonym.	Magnitude.	Right Ascension, 1850.0.	Annual Precession.	South Declination, 1850.0.	Annual Precession.	Mean year.	No. of obs.
			h. m. s.	s.	° ′ ″	″		
8426	Gould, Z. C., 12^h 128 . .	9.8	12 0 50.73	+ 3.083	66 58 21.4	− 20.06	51.40	3
8427	Chamæleontis	9.0	53.12	3.095	77 15 42.5	.06	51.36	1
8428	Gould, 16582	8.5	54.56	3.084	67 48 15.1	.06	52.30	1
8429	Chamæleontis	10.0	56.40	3.095	76 51 8.3	.06	51.25	1
8430	Gould, Z. C., 12^h 142 . .	9.8	0 57.93	+ 3.085	66 57 58.5	− 20.06	51.73	3
8431	Chamæleontis	10.0	12 1 2.59	+ 3.098	77 8 54.7	− 20.06	51.36	1
8432	Gould, 16586	8.5	3.47	3.089	70 12 1.7	.06	51.44	1
8433	Lacaille, 5038	7.5	7.08	3.102	77 56 43.8	.06	50.87	4
8434	Gould, Z. C., 12^h 160 . .	9.5	13.34	3.100	75 48 7.2	.06	51.38	1
8435	Muscæ	9.5	19.94	+ 3.088	64 46 46.4	− 20.06	50.28	1
8436	Gould, Z. C., 12^h 166 . .	8.0	12 1 25.82	+ 3.089	65 16 20.8	− 20.06	52.41	1
8437	Gould, Z. C., 12^h 171 . .	9.5	31.70	3.093	69 22 3.7	.06	51.87	2
8438	Gould, Z. C., 12^h 180 . .	9.5	35.84	3.094	66 58 38.7	.06	51.37	3
8439	Gould, Z. C., 12^h 185 . .	9.0	39.15	3.098	70 0 57.6	.06	51.44	1
8440	Muscæ	10.0	43.13	+ 3.093	64 47 39.8	− 20.06	50.28	1
8441	Lacaille, 5040	7.7	12 1 46.19	+ 3.096	66 37 6.4	− 20.06	50.99	3
8442	Gould, Z. C., 12^h 197 . .	9.5	52.56	3.104	70 56 34.4	.06	52.28	2
8443	Muscæ	9.5	56.03	3.097	65 52 57.0	.06	51.32	2
8444	Gould, Z. C., 12^h 204 . .	9.8	1 58.96	3.098	66 35 33.9	.06	51.34	2
8445	Muscæ	10.0	2 4.58	+ 3.099	65 52 44.6	− 20.06	51.32	2
8446	Gould, Z. C., 12^h 219 . .	9.5	12 2 15.87	+ 3.111	70 54 44.3	− 20.06	52.28	1
8447	Chamæleontis	10.0	19.30	3.149	80 3 15.8	.06	51.37	1
8448	Chamæleontis	10.0	27.63	3.145	78 53 54.2	.06	51.28	1
8449	Gould, Z. C., 12^h 238 . .	8.8	35.78	3.106	66 20 32.2	.06	52.37	2
8450	Brisbane, 3944	8.0	37.85	+ 3.196	82 56 35.3	− 20.06	50.33	1
8451	Chamæleontis	9.8	12 2 43.76	+ 3.183	81 53 13.1	− 20.05	50.85	2
8452	Mensæ	10.0	53.05	3.120	71 28 27.8	.05	52.26	1
8453	Gould, Z. C., 12^h 258 . .	9.5	2 55.08	3.112	66 54 2.3	.05	51.46	2
8454	Lacaille, 5047	7.2	3 5.09	3.115	67 25 31.6	.05	52.30	2
8455	Mensæ	10.0	7.56	+ 3.180	80 13 15.7	− 20.05	51.37	1
8456	Muscæ	9.8	12 3 21.40	+ 3.130	71 26 11.4	− 20.05	52.27	2
8457	Gould, Z. C., 12^h 286 . .	10.0	22.40	3.121	68 16 52.9	.05	50.37	1
8458	Gould, Z. C., 12^h 296 . .	9.8	30.32	3.123	68 19 9.0	.05	50.39	2
8459	Gould, Z. C., 12^h 297 . .	9.8	30.76	3.120	66 57 16.2	.05	51.46	2
8460	Gould, Z. C., 12^h 305 . .	9.0	39.16	+ 3.130	70 20 30.6	− 20.05	52.28	1
8461	Chamæleontis	8.0	12 3 40.94	+ 3.196	80 13 51.3	− 20.05	50.90	2
8462	Octantis	8.0	42.67	3.237	82 31 13.6	.05	50.39	1
8463	Gould, Z. C., 12^h 323 . .	8.5	52.02	3.150	73 55 21.6	.05	51.76	2
8464	Gould, Z. C., 12^h 325 . .	9.5	53.91	3.131	68 53 30.9	.05	50.45	1
8465	Gould, Z. C., 12^h 337 . .	9.0	3 58.02	+ 3.124	65 51 41.7	− 20.05	52.38	1

Number.	Constellation, Name of Star, or Synonym.	Magnitude.	Right Ascension, 1850.0.	Annual Precession.	South Declination, 1850.0.	Annual Precession.	Mean year.	No. of obs.
			h. m. s.	s.	° ′ ″	″		
8466	Gould, Z. C., 12^h 341 . .	9.0	12 4 1.76	+ 3.155	74 6 15.6	— 20.05	52.28	1
8467	Muscæ	10.0	22.78	3.152	72 24 10.8	.05	51.80	2
8468	Gould, Z. C., 12^h 365 . .	9.5	23.60	3.138	68 43 25.6	.05	50.43	2
8469	Muscæ	10.0	23.92	3.153	72 23 24.2	.05	51.80	2
8470	Muscæ	10.0	30.89	+ 3.132	66 46 37.2	— 20.05	51.46	1
8471	Lacaille, 5060	6.2	12 4 45.65	+ 3.145	69 19 0.4	— 20.05	51.89	2
8472	Chamæleontis	7.8	47.40	3.230	79 59 46.8	.05	51.31	2
8473	Chamæleontis	8.0	4 57.01	3.275	81 54 6.8	.65	50.85	2
8474	Gould, Z. C., 12^h 401 . .	8.8	5 4.01	3.153	69 56 55.9	.05	52.02	3
8475	Gould, Z. C., 12^h 415 . .	9.2	14.73	+ 3.156	69 57 41.2	— 20.05	52.02	3
8476	Gould, Z. C., 12^h 421 . .	9.5	12 5 17.36	+ 3.173	73 4 37.4	+ 20.05	50.42	1
8477	Gould, Z. C., 12^h 422 . .	9.8	20.31	3.145	66 50 38.4	.05	51.46	2
8478	Gould, Z. C., 12^h 431 . .	9.5	26.64	3.151	68 12 15.2	.05	50.39	2
8479	Gould, Z. C., 12^h 439 . .	9.5	32.48	3.165	70 51 51.1	.05	52.28	1
8480	Lacaille, 5064	7.0	35.06	+ 3.222	77 44 18.1	— 20.05	51.06	3
8481	Gould, Z. C., 12^h 447 . .	9.0	12 5 35.71	+ 3.179	73 7 31.2	— 20.05	50.42	1
8482	Muscæ	10.0	41.84	3.188	74 2 4.2	.05	52.28	1
8483	Gould, Z. C., 12^h 453 . .	9.0	41.84	3.197	75 11 18.6	.05	50.40	1
8484	Muscæ	9.5	42.90	3.170	71 15 27.3	.05	52.27	1
8485	Octantis	11.5	44.02	+ 3.402	84 13 25.3	— 20.05	50.33	1
8486	Gould, Z. C., 12^h 456 . .	9.0	12 5 46.76	+ 3.148	66 3 1.8	— 20.05	52.38	1
8487	Gould, Z. C., 12^h 462 . .	9.5	53.40	3.149	66 0 20.4	.05	52.38	1
8488	Gould, Z. C., 12^h 464 . .	9.5	5 54.60	3.158	68 13 53.6	.05	50.40	2
8489	Lacaille, 5096	6.9	6 2.75	3.907	87 34 51.9	.05	51.33	5
8490	Gould, 16696	10.0	4.89	+ 3.165	69 12 21.3	— 20.05	51.48	1
8491	Muscæ	9.0	12 6 8.67	+ 3.178	71 17 38.1	— 20.05	52.27	1
8492	Chamæleontis	7.8	13.92	3.286	80 21 33.0	.05	50.90	2
8493	Gould, Z. C., 12^h 485 . .	9.5	14.34	3.157	66 54 24.2	.05	51.45	2
8494	Gould, Z. C., 12^h 497 . .	9.3	21.11	3.159	66 58 36.7	.05	51.73	3
8495	Chamæleontis	9.5	24.00	+ 3.309	81 3 51.5	— 20.05	50.88	2
8496	Muscæ	10.0	12 6 27.53	+ 3.183	71 21 14.5	— 20.05	52.27	1
8497	Muscæ	10.0	28.14	3.204	74 2 17.0	.05	52.28	1
8498	Gould, Z. C., 12^h 512 . .	9.0	36.02	3.162	66 49 10.8	.05	51.46	2
8499	Gould, Z. C., 12^h 517 . .	9.0	41.17	3.183	70 41 59.8	.05	52.28	2
8500	Muscæ	9.5	6 56.20	+ 3.163	66 5 40.9	— 20.05	52.38	1
8501	Chamæleontis	10.0	12 7 0.21	+ 3.295	79 36 56.0	— 20.05	51.24	1
8502	Gould, 16721	8.8	2.52	3.200	72 13 33.9	.05	51.80	2
8503	Gould, Z. C., 12^h 549 . .	7.5	6.44	3.306	79 56 58.3	.05	51.30	2
8504	Muscæ	10.0	11.25	3.217	73 54 6.9	.05	52.28	1
8505	Gould, Z. C., 12^h 573 . .	8.8	31.83	+ 3.176	67 3 37.0	— 20.05	51.73	3

No. 8490. Probably variable.

Number.	Constellation, Name of Star, or Synonym.	Magnitude.	Right Ascension, 1850.0.	Annual Precession.	South Declination, 1850.0.	Annual Precession.	Mean year.	No. of obs.
			h. m. s.	s.	° ′ ″	″		
8506	Gould, Z. C., 12h 577 . .	9.5	12 7 39.46	+ 3.175	66 32 30.0	— 20.05	52.36	1
8507	Gould, Z. C., 12h 583 . .	9.0	42.43	3.197	70 14 34.5	.05	52.28	1
8508	Chamæleontis	10.0	7 45.96	3.279	77 41 32.6	.04	50.40	1
8509	Gould, 16743	8.7	8 1.91	3.200	69 59 35.1	.04	52.03	3
8510	Brisbane, 5079	6.2	10.43	+ 3.213	71 46 43.8	— 20.04	52.30	2
8511	Gould, Z. C., 12h 618 . .	9.8	12 8 14.05	+ 3.183	66 40 17.5	— 20.04	51.91	2
8512	Gould, Z. C., 12h 625 . .	9.8	14.78	3.204	69 58 0.6	.04	51.89	2
8513	Octantis	9.0	15.05	3.491	83 27 21.2	.04	50.27	2
8514	Chamæleontis	10.0	26.08	3.266	75 47 0.4	.04	51.38	1
8515	Chamæleontis	10.0	39.01	+ 3.300	77 42 22.0	— 20.04	50.40	1
8516	Chamæleontis	10.0	12 8 39.37	+ 3.280	76 32 6.8	— 20.04	51.34	1
8517	Gould, Z. C., 12h 649 . .	8.0	41.15	3.352	79 45 18.9	.04	51.24	1
8518	Chamæleontis	10.0	41.84	3.291	77 0 24.8	.04	51.25	1
8519	Lacaille, 5081	9.0	41.86	3.241	73 18 36.8	.04	50.44	1
8520	Chamæleontis	9.5	45.74	+ 3.290	76 48 49.0	— 20.04	51.25	1
8521	Gould, Z. C., 12h 654 . .	9.5	12 8 48.90	+ 3.237	72 41 56.7	— 20.04	50.85	2
8522	Octantis	10.0	9 2.72	3.636	84 39 24.4	.04	50.82	2
8523	Gould, Z. C., 12h 674 . .	9.0	8.84	3.219	70 3 32.0	.04	51.44	1
8524	Gould, Z. C., 12h 677 . .	9.2	12.41	3.206	68 15 57.7	.04	50.39	2
8525	Muscæ	10.0	22.65	+ 3.197	66 28 46.4	— 20.04	52.36	1
8526	Brisbane, 3984	8.5	12 9 29.09	+ 3.197	66 8 3.0	— 20.04	52.38	1
8527	ε Muscæ	5.8	30.65	3.203	67 7 35.2	.04	51.73	3
8528	Gould, Z. C., 12h 696 . .	9.2	35.46	3.222	69 31 59.9	.04	52.31	2
8529	Chamæleontis	10.0	35.99	3.307	76 35 11.4	.04	51.34	1
8530	Gould, Z. C., 12h 701 . .	9.5	39.51	+ 3.214	68 23 6.2	— 20.04	50.42	1
8531	β Chamæleontis	5.0	12 9 39.65	+ 3.348	78 28 43.7	— 20.04	51.35	1
8532	Gould, Z. C., 12h 705 . .	10.0	43.75	3.204	66 50 29.0	.04	51.46	2
8533	Gould, Z. C., 12h 707 . .	8.5	44.91	3.221	69 6 36.4	.04	50.95	2
8534	Chamæleontis	10.0	9 45.67	3.320	77 5 38.1	.04	51.31	2
8535	Gould, Z. C., 12h 736 . .	9.5	10 7.70	+ 3.207	66 26 0.2	— 20.04	52.36	1
8536	Chamæleontis	10.0	12 10 11.90	+ 3.353	78 2 10.8	— 20.04	52.35	1
8537	Muscæ	9.0	13.58	3.285	74 24 6.9	.04	50.41	1
8538	Gould, 16784	9.0	26.89	3.249	71 3 1.5	.04	52.28	2
8539	Brisbane, 3984	8.0	54.92	3.214	65 49 30.8	.03	52.38	1
8540	Gould, Z. C., 12h 777 . .	9.0	55.28	+ 3.258	71 5 3.4	— 20.03	52.28	2
8541	Muscæ	10.0	12 10 57.18	+ 3.253	70 34 48.7	— 20.03	52.28	1
8542	Gould, Z. C., 12h 793 . .	9.5	11 8.65	3.250	70 2 6.4	.03	51.44	1
8543	Chamæleontis	9.2	13.33	3.526	81 47 32.1	.03	50.85	2
8544	Lacaille, 5093	7.2	15.24	3.334	75 57 54.7	.03	51.34	2
8545	Octantis	9.0	17.65	+ 3.784	84 43 28.6	— 20.03	50.33	2

No. 8517. Gould's declination is 18.6″ less. No. 8528. Gould's declination is 10.9″ less.

Number.	Constellation, Name of Star, or Synonym.	Magnitude.	Right Ascension, 1850.0.	Annual Precession.	South Declination, 1850.0.	Annual Precession.	Mean year.	No. of obs.
			h. m. s.	s.	° ′ ″	″		
8546	Gould, Z. C., 12h 808	10.0	12 11 19.82	+ 3.225	66 41 20.5	— 20.03	51.76	3
8547	Gould, Z. C., 12h 811	10.0	29.97	3.243	68 36 27.2	.03	50.45	1
8548	Octantis	9.5	33.78	3.733	84 10 44.5	.03	50.33	1
8549	Muscæ	9.0	41.54	3.226	66 10 31.2	.03	52.40	1
8550	Muscæ	9.0	54.24	+ 3.229	66 11 50.9	— 20.03	52.38	1
8551	Chamæleontis	10.0	12 11 55.38	+ 3.379	77 13 18.6	— 20.03	51.36	1
8552	Gould, Z. C., 12h 853	9.5	12 17.21	3.300	72 33 58.6	.03	50.85	2
8553	Lacaille, 5105	9.0	22.86	3.451	79 13 41.1	.03	50.43	2
8554	Chamæleontis	9.5	24.97	3.394	77 19 56.1	.03	51.36	1
8555	Chamæleontis	9.5	28.24	+ 3.623	82 29 12.2	— 20.03	50.39	1
8556	Muscæ	9.8	12 12 30.66	+ 3.241	66 39 17.4	— 20.03	51.76	3
8557	Muscæ	9.2	35.61	3.242	66 42 17.1	.03	51.76	3
8558	Muscæ	9.2	38.47	3.298	71 58 47.2	.03	52.30	2
8559	Muscæ	10.0	40.75	3.243	66 40 32.8	.03	52.36	1
8560	Lacaille, 5104	7.2	46.90	+ 3.443	78 38 0.2	— 20.03	51.32	2
8561	Gould, Z. C., 12h 893	9.8	12 12 54.66	+ 3.448	78 41 32.5	— 20.02	51.32	2
8562	Gould, Z. C., 12h 886	9.5	55.28	3.248	66 47 25.5	.02	51.46	2
8563	Gould, Z. C., 12h 892	10.0	12 58.46	3.261	68 10 50.7	.02	50.37	1
8564	Octantis	10.2	13 8.34	3.710	83 9 9.0	.02	50.36	2
8565	Gould, 16834	9.2	15.58	+ 3.267	68 16 40.0	— 20.02	50.39	2
8566	Gould, Z. C., 12h 915	9.0	12 13 16.35	+ 3.343	74 4 12.4	— 20.02	52.28	1
8567	Octantis	9.9	19.97	4.012	85 16 21.2	.02	51.33	5
8568	Lacaille, 5111	7.5	43.36	3.328	72 40 9.5	.02	50.85	2
8569	ζ² Muscæ	6.8	49.66	3.259	66 41 20.6	.02	51.76	3
8570	ζ¹ Muscæ	6.8	52.17	+ 3.267	67 28 20.2	— 20.02	52.30	2
8571	Gould, Z. C., 12h 956	9.0	12 13 59.47	+ 3.259	66 26 15.2	— 20.02	52.36	1
8572	Lacaille, 5107	6.7	14 5.13	4.074	85 19 2.5	.02	51.33	5
8573	Gould, Z. C., 12h 977	9.2	15.48	3.279	68 16 42.9	.02	50.39	2
8574	Octantis	10.2	17.58	3.721	83 11 17.4	.02	50.36	2
8575	Chamæleontis	9.0	27.97	+ 3.565	80 31 16.4	— 20.02	50.43	1
8576	Gould, 16860	8.8	12 14 28.10	+ 3.266	66 58 52.6	— 20.02	51.87	2
8577	Gould, 16865	9.0	36.26	3.383	74 40 21.4	.02	50.41	1
8578	Chamæleontis	8.2	37.36	3.437	76 50 56.4	.02	51.35	2
8579	Muscæ	9.5	14 39.41	3.378	74 23 54.3	.02	50.41	1
8580	Chamæleontis	8.8	15 7.96	+ 3.471	77 32 7.2	— 20.01	50.88	2
8581	Muscæ	9.5	12 15 11.43	+ 3.269	65 50 35.3	— 20.01	52.38	1
8582	Lacaille, 5123	7.7	12.01	3.279	66 48 33.9	.01	51.76	3
8583	Muscæ	9.5	27.19	3.354	72 17 41.2	.01	51.80	2
8584	Gould, Z. C., 12h 1050	9.5	30.10	3.274	65 56 16.9	.01	52.38	1
8585	Lacaille, 5124	8.0	33.40	+ 3.446	76 22 57.8	— 20.01	51.34	1

No. 8553. Probably variable.

Number.	Constellation, Name of Star, or Synonym.	Magnitude.	Right Ascension, 1850.0.	Annual Precession.	South Declination, 1850.0.	Annual Precession.	Mean year.	No. of obs.
			h. m. s.	s.	° ′ ″	″		
8 586	Gould, Z. C., 12^h 1069 .	9.5	12 15 49.25	+ 3.329	70 17 24.6	— 20.01	52.28	1
8 587	Muscæ	9.8	15 54.98	3.291	67 6 6.8	.01	51.87	2
8 588	Gould, Z. C., 12^h 1077 .	8.8	16 0.43	3.284	66 13 42.8	.01	52.37	2
8 589	Muscæ	9.5	4.56	3.399	74 0 21.0	.01	52.28	1
8 590	Chamæleontis	10.0	6.52	+ 3.684	81 16 23.7	— 20.01	50.38	1
8 591	Gould, Z. C., 12^h 1086 .	9.8	12 16 10.66	+ 3.396	73 47 52.4	— 20.01	51.76	2
8 592	Lacaille, 5132	7.0	16.40	3.360	71 46 14.9	.01	52.30	2
8 593	Gould, Z. C., 12^h 1100 .	9.5	23.01	3.304	67 38 54.5	.01	52.30	1
8 594	Lacaille, 5133	8.2	25.77	3.295	66 48 3.6	.00	51.76	3
8 595	Gould, Z. C., 12^h 1111 .	8.8	30.05	+ 3.335	69 53 45.1	— 20.00	52.03	3
8 596	Gould, Z. C., 12^h 1117 .	9.0	12 16 33.89	+ 3.306	67 36 11.3	— 20.00	52.30	1
8 597	Chamæleontis	9.0	42.58	3.441	75 12 34.7	.00	50.40	1
8 598	Gould, Z. C., 12^h 1129 .	9.5	42.87	3.293	66 14 20.4	.00	52.38	1
8 599	Gould, Z. C., 12^h 1148 .	8.5	54.05	3.468	76 3 5.8	.00	51.30	1
8 600	Chamæleontis	10.0	16 55.33	+ 3.709	81 12 17.4	— 20.00	50.38	1
8 601	Lacaille, 5139	8.8	12 17 1.48	+ 3.326	68 38 41.5	— 20.00	50.43	2
8 602	Muscæ	9.5	1.54	3.307	67 8 46.7	.00	52.29	1
8 603	Gould, 16921	9.2	3.68	3.326	68 38 35.7	.00	50.43	2
8 604	Gould, 16924	9.0	13.25	3.315	67 35 23.0	.00	52.30	1
8 605	Muscæ	10.0	24.75	+ 3.386	72 6 48.2	-- 20.00	52.33	1
8 606	Octantis	8.8	12 17 27.83	+ 4.179	84 44 55.4	— 20.00	50.32	2
8 607	Chamæleontis	9.0	30.43	3.488	76 13 44.0	.00	51.32	2
8 608	Chamæleontis	10.0	17 49.26	3.501	76 22 57.1	20.00	51.34	1
8 609	Gould, Z. C., 12^h 1212 .	10.0	18 6.83	3.464	74 55 27.3	19.99	51.29	1
8 610	Gould, Z. C., 12^h 1217 .	9.5	13.68	3.370	70 22 41.8	— 19.99	51.86	2
8 611	Chamæleontis	9.5	12 18 14.17	+ 3.503	76 9 49.8	— 19.99	51.30	1
8 612	Gould, 16948	8.2	21.68	3.499	75 55 26.2	.99	51.34	2
8 613	Lacaille, 5145	7.0	22.19	3.940	82 58 20.2	.99	50.36	2
8 614	Chamæleontis	9.5	26.63	3.511	76 15 0.4	.99	51.30	1
8 615	Gould, Z. C., 12^h 1236 .	8.5	26.94	+ 3.498	75 50 1.0	— 19.99	51.34	2
8 616	Gould, Z. C., 12^h 1240 .	9.5	12 18 38.91	+ 3.385	70 50 10.1	— 19.99	52.28	1
8 617	Lacaille, 5149	9.0	41.75	3.468	74 38 29.2	.99	50.85	2
8 618	Gould, Z. C., 12^h 1254 .	9.0	50.80	3.402	71 37 20.5	.99	52.26	1
8 619	Gould, Z. C., 12^h 1259 .	9.2	55.11	3.385	70 35 45.8	.99	52.28	2
8 620	Chamæleontis	9.5	57.12	+ 3.844	81 51 37.2	— 19.99	51.34	1
8 621	Chamæleontis	9.5	12 18 59.36	+ 3.497	75 24 7.6	— 19.99	50.40	1
8 622	Muscæ	9.7	18 59.97	3.403	71 30 31.4	.99	52.29	3
8 623	Gould, Z. C., 12^h 1269 .	9.0	19 3.62	3.323	66 6 58.6	.99	52.38	1
8 624	Lacaille, 5158	8.0	29.40	3.351	67 53 42.3	.98	51.34	2
8 625	Gould, 16972	8.0	37.00	+ 3.381	69 42 51.0	-- 19.98	52.31	2

No. 8 587. Possibly 67° 5′ 6.8″.

Number.	Constellation, Name of Star, or Synonym.	Magnitude.	Right Ascension, 1850.0.	Annual Precession.	South Declination, 1850.0.	Annual Precession.	Mean year.	No. of obs.
			h. m. s.	s.	° ′ ″	″		
8 626	Gould, Z. C., 12h 1305	10.0	12 19 41.50	+ 3.340	66 48 18.4	− 19.98	51.46	2
8 627	Muscæ	10.0	55.60	3.433	72 11 34.1	.98	52.03	3
8 628	Muscæ	9.5	56.00	3.345	66 58 48.5	.98	52.29	1
8 629	Muscæ	10.0	19 57.82	3.340	66 31 20.7	.98	52.36	1
8 630	Muscæ	9.8	20 2.21	+ 3.390	69 52 21.0	− 19.98	51.89	2
8 631	Brisbane, 4063	9.0	12 20 4.80	+ 3.502	74 47 17.0	− 19.98	51.29	1
8 632	Chamæleontis	10.0	12.02	3.874	81 39 0.5	.98	51.34	1
8 633	Gould, Z. C., 12h 1342	9.5	15.28	3.380	69 4 35.0	.98	51.45	1
8 634	Octantis	9.5	19.25	4.138	83 39 51.6	.98	50.35	1
8 635	Gould, Z. C., 12h 1350	9.2	20.88	+ 3.441	72 12 17.3	− 19.98	52.41	2
8 636	Gould, Z. C., 12h 1351	9.5	12 20 25.70	+ 3.350	66 49 46.8	− 19.98	51.46	2
8 637	Octantis	10.0	27.30	6.079	87 43 46.8	.98	51.33	1
8 638	Gould, Z. C., 12h 1357	10.0	27.77	3.394	69 42 0.6	.98	52.34	1
8 639	Gould, Z. C., 12h 1374	8.8	44.74	3.448	72 11 6.1	.97	52.03	3
8 640	Muscæ	10.0	47.48	+ 3.351	66 33 57.5	− 19.97	52.36	1
8 641	Gould, 17001	7.0	12 20 59.73	+ 3.347	66 1 48.2	− 19.97	52.38	1
8 642	Gould, Z. C., 12h 1390	9.0	21 8.84	3.346	65 46 3.3	.97	52.38	1
8 643	Gould, Z. C., 12h 1401	9.2	10.65	3.611	77 6 13.2	.97	51.35	3
8 644	Lacaille, 5166	7.3	26.20	3.460	72 9 17.4	.97	52.42	3
8 645	Gould, Z. C., 12h 1415	10.0	34.25	+ 3.381	67 55 6.1	− 19.97	52.30	1
8 646	Gould, 17015	8.2	12 21 42.10	+ 3.443	71 11 34.3	− 19.97	52.30	3
8 647	Muscæ	9.5	45.46	3.449	71 25 56.8	.97	52.31	2
8 648	Chamæleontis	10.0	47.83	3.561	75 26 47.9	.97	51.38	1
8 649	Lacaille, 5170	7.3	48.61	3.469	72 15 31.4	.97	52.14	4
8 650	Lacaille, 5168	9.0	21 49.49	+ 3.532	74 34 14.2	− 19.97	50.41	1
8 651	Muscæ	9.0	12 22 1.44	+ 3.471	72 12 18.5	− 19.96	51.28	1
8 652	Chamæleontis	9.5	2.24	3.755	79 21 48.4	.96	51.42	1
8 653	Muscæ	9.5	4.36	3.381	67 22 43.2	.96	52.29	1
8 654	Muscæ	10.0	14.80	3.446	70 54 12.6	.96	52.28	1
8 655	Lacaille, 5171	7.5	31.52	+ 3.744	78 57 18.7	− 19.96	50.85	2
8 656	Chamæleontis	9.8	12 22 44.32	+ 3.647	76 52 18.6	− 19.96	51.35	2
8 657	Lacaille, 5177	8.2	48.71	3.665	77 21 52.4	.96	50.88	2
8 658	Chamæleontis	9.2	22 49.45	3.584	75 25 54.4	.96	50.89	2
8 659	Gould, Z. C., 12h 1499	9.2	23 4.78	3.438	69 48 58.9	.95	52.31	2
8 660	Muscæ	10.0	5.40	+ 3.536	73 48 26.4	− 19.95	52.28	1
8 661	Lacaille, 5181	6.5	12 23 10.64	+ 3.492	72 10 16.6	− 19.95	52.42	3
8 662	Gould, Z. C., 12h 1514	9.0	17.67	3.400	67 31 49.1	.95	52.29	1
8 663	Muscæ	9.5	23.94	3.555	74 13 38.7	.95	52.28	1
8 664	Muscæ	10.0	34.38	3.563	74 21 36.1	.95	52.28	1
8 665	γ Muscæ	4.5	34.44	+ 3.478	71 18 13.3	− 19.95	52.30	3

Number.	Constellation, Name of Star, or Synonym.	Magnitude.	Right Ascension, 1850.0.	Annual Precession.	South Declination, 1850.0.	Annual Precession.	Mean year.	No. of obs.
			h. m. s.	s.	° ′ ″	″		
8 666	Muscæ	9.5	12 23 35.60	+ 3.546	73 49 48.6	− 19.95	51.76	2
8 667	Lacaille, 5183	9.0	37.44	3.517	72 49 35.4	.95	50.42	1
8 668	Lacaille, 5182	9.0	39.40	3.572	74 36 15.0	.95	50.41	1
8 669	Gould, Z. C., 12h 1536 .	10.0	41.24	3.412	67 55 18.6	.95	50.37	1
8 670	Gould, 17083	9.0	44.89	+ 3.779	78 55 29.8	− 19.95	50.76	2
8 671	Octantis	8.5	12 23 49.82	+ 4.419	84 2 44.7	− 19.95	50.83	2
8 672	Gould, 17082	8.8	51.66	3.488	71 32 37.2	.95	52.29	3
8 673	Muscæ	9.5	52.42	3.400	67 2 17.2	.95	52.29	1
8 674	Gould, Z. C., 12h 1558 .	9.0	23 58.23	3.406	67 17 1.0	.95	52.29	1
8 675	Chamæleontis	9.5	24 1.45	+ 4.087	82 9 8.9	− 19.95	50.37	1
8 676	Gould, Z. C., 12h 1561 .	9.5	12 24 2.88	+ 3.439	69 5 42.9	− 19.95	51.45	1
8 677	Gould, Z. C., 12h 1578 .	10.0	18.75	3.443	69 6 38.8	.94	51.45	1
8 678	Muscæ	10.0	18.75	3.585	74 33 24.0	.94	50.41	1
8 679	Octantis	9.0	19.62	4.307	83 27 13.8	.94	50.37	2
8 680	Gould, Z. C., 12h 1586 .	9.5	24.90	+ 3.389	65 51 35.5	− 19.94	52.38	1
8 681	Gould, Z. C., 12h 1605 .	9.5	12 24 40.53	+ 3.590	74 29 37.8	− 19.94	50.41	1
8 682	Gould, Z. C., 12h 1596 .	9.5	41.82	3.428	68 0 23.8	.94	50.37	1
8 683	Gould, Z. C., 12h 1602 .	9.5	47.25	3.430	68 0 30.8	.94	50.37	1
8 684	Muscæ	9.5	24 57.80	3.538	72 40 14.0	.94	50.85	2
8 685	Lacaille, 5194	7.5	25 0.70	+ 3.431	67 54 54.6	− 19.94	51.34	2
8 686	Muscæ	10.0	12 25 2.88	+ 3.498	71 4 59.3	− 19.94	52.28	1
8 687	Muscæ	9.5	4.83	3.463	69 29 28.9	.94	52.34	2
8 688	Gould, Z. C., 12h 1631 .	9.0	9.82	3.731	77 28 8.0	.94	50.88	2
8 689	Muscæ	10.0	21.70	3.545	72 39 18.9	.93	51.28	1
8 690	Muscæ	9.2	40.53	+ 3.489	70 17 17.2	− 19.93	51.86	2
8 691	Muscæ	10.0	12 25 54.72	+ 3.555	72 39 59.9	− 19.93	51.28	1
8 692	Gould, Z. C., 12h 1672 .	10.0	26 0.25	3.459	69 3 49.9	.93	51.45	1
8 693	Octantis	9.5	6.45	3.667	75 40 23.0	.93	51.38	1
8 694	Lacaille, 5203	8.0	10.07	3.430	66 55 42.2	.93	51.45	1
8 695	Gould, 17128	8.0	11.82	+ 3.538	71 53 19.8	− 19.93	52.30	2
8 696	Gould, Z. C., 12h 1708 .	9.2	12 26 21.22	+ 3.804	78 9 30.3	− 19.92	51.38	2
8 697	Gould, Z. C., 12h 1705 .	9.2	29.11	3.484	69 29 37.2	.92	51.89	2
8 698	Muscæ	10.0	31.49	3.597	73 36 18.8	.92	51.25	1
8 699	Gould, Z. C., 12h 1721 .	9.5	42.50	3.457	68 0 21.5	.92	50.37	1
8 700	Lacaille, 5206	8.5	44.65	+ 3.448	67 31 24.2	− 19.92	52.30	2
8 701	Gould, Z. C., 12h 1735 .	9.2	12 26 47.37	+ 3.741	76 52 29.8	− 19.92	51.35	2
8 702	Muscæ	10.0	48.04	3.432	66 35 57.0	.92	51.45	1
8 703	Gould, Z. C., 12h 1742 .	9.0	51.21	3.425	66 6 12.9	.92	52.38	1
8 704	Muscæ	9.0	26 59.98	3.445	67 8 50.7	.92	51.87	2
8 705	Gould, Z. C., 12h 1742 .	9.0	27 1.61	+ 3.423	65 51 27.6	− 19.92	52.38	1

No. 8 671. The declination may be 84° 6′ 59.4″.

No. 8 679. The declination may be 83° 26′ 55.9″.

Number.	Constellation, Name of Star, or Synonym.	Magnitude.	Right Ascension, 1850.0.	Annual Precession.	South Declination, 1850.0.	Annual Precession.	Mean year.	No. of obs.
			h. m. s.	s.	° ′ ″	″		
8706	Gould, Z. C., 12h 1749	9.5	12 27 7.96	+ 3.439	66 42 39.9	— 19.92	51.96	2
8707	Muscæ	10.0	15.19	3.579	72 38 26.9	.91	51.28	1
8708	Muscæ	9.5	18.56	3.617	73 44 11.3	.91	51.25	1
8709	Muscæ	9.5	25.06	3.455	67 23 21.6	.91	52.30	2
8710	Muscæ	9.5	26.48	+ 3.518	70 18 38.6	— 19.91	51.86	2
8711	Muscæ	10.0	12 27 30.50	+ 3.640	74 15 55.3	— 19.91	52.28	1
8712	Gould, Z. C., 12h 1769	9.0	40.49	3.442	66 28 30.1	.91	52.36	1
8713	Lacaille, 5210	7.2	41.44	3.440	66 20 38.2	.91	52.37	2
8714	Lacaille, 5235	7.1	43.13	12.085	88 58 28.7	.91	51.50	7
8715	Gould, Z. C., 12h 1785	8.5	55.90	+ 3.444	66 23 13.6	— 19.91	52.37	2
8716	Muscæ	10.0	12 27 56.89	+ 3.647	74 11 57.9	— 19.91	52.28	1
8717	Muscæ	10.0	58.43	3.646	74 10 5.0	.91	52.28	1
8718	Muscæ	9.0	27 59.55	3.531	70 27 21.0	.91	52.28	1
8719	Gould, Z. C., 12h 1796	9.5	28 6.73	3.444	66 17 33.8	.91	52.37	2
8720	Gould, Z. C., 12h 1807	9.5	16.24	+ 3.546	70 52 26.9	— 19.90	52.28	1
8721	α Muscæ	4.0	12 28 18.46	+ 3.486	68 18 29.2	— 19.90	50.39	2
8722	Octantis	9.3	27.33	6.579	87 17 48.4	.90	51.33	5
8723	Muscæ	10.0	37.27	3.601	72 32 14.2	.90	51.28	1
8724	Gould, Z. C., 12h 1827	9.0	47.60	3.467	67 2 16.6	.90	51.87	2
8725	Octantis	9.3	56.76	+ 5.052	85 8 15.6	— 19.90	51.00	3
8726	Gould, Z. C., 12h 1836	8.5	12 28 58.00	+ 3.531	69 50 13.1	— 19.90	51.89	2
8727	Octantis	9.5	59.49	4.431	82 55 27.7	.90	50.33	1
8728	Gould, Z. C., 12h 1837	9.5	8 59.92	3.455	66 14 43.5	.90	52.37	2
8729	Lacaille, 5219	7.5	29 4.18	3.507	68 46 33.0	.90	50.47	2
8730	Muscæ	10.0	7.28	+ 3.658	73 52 16.1	— 19.89	51.25	1
8731	Gould, Z. C., 12h 1860	9.0	12 29 8.48	+ 4.040	80 3 51.0	— 19.89	51.37	1
8732	Gould, Z. C., 12h 1849	9.5	8.57	3.454	66 5 18.0	.89	52.38	1
8733	Muscæ	9.5	10.94	3.456	66 8 7.3	.89	52.38	1
8734	Gould, Z. C., 12h 1852	9.5	13.06	3.491	67 54 32.6	.89	52.30	1
8735	Lacaille, 5217	7.8	18.51	+ 3.719	75 14 19.6	— 19.89	50.93	2
8736	Muscæ	10.0	12 29 26.40	+ 3.543	70 0 24.4	— 19.89	51.44	1
8737	Chamæleontis	9.2	27.22	3.815	77 1 10.2	.89	51.35	2
8738	Muscæ	10.0	32.19	3.606	72 10 47.6	.89	52.33	1
8739	Muscæ	9.2	39.09	3.672	73 56 56.2	.89	51.76	2
8740	Gould, Z. C., 12h 1885	9.2	40.53	+ 3.664	73 44 50.8	— 19.89	51.76	2
8741	Muscæ	10.0	12 29 42.53	+ 3.588	71 28 35.9	— 19.89	52.26	1
8742	Lacaille, 5221	9.0	42.62	3.697	74 32 35.8	.89	50.41	1
8743	Gould, Z. C., 12h 1886	8.8	47.44	3.499	67 56 12.4	.89	51.34	2
8744	Muscæ	9.0	58.61	3.592	71 27 18.7	.89	52.31	2
8745	Gould, 17194	8.5	29 59.28	+ 3.461	65 52 14.9	— 19.88	52.38	1

Number.	Constellation, Name of Star, or Synonym.	Magnitude.	Right Ascension, 1850.0.	Annual Precession.	South Declination, 1850.0.	Annual Precession.	Mean year.	No. of obs.
			h. m. s.	s.	° ′ ″	″		
8 746	Gould, Z. C., 12h 1895	8.5	12 30 0.36	+ 3.675	73 51 17.6	− 19.88	51.76	2
8 747	Lacaille, 5224	7.0	1.70	3.471	66 22 2.2	.88	52.36	1
8 748	Gould, Z. C., 12h 1901	9.2	9.71	3.576	70 47 44.0	.88	52.28	2
8 749	Chamæleontis	10.0	14.13	4.157	80 47 35.1	.88	51.37	1
8 750	Gould, Z. C., 12h 1904	9.2	14.76	+ 3.523	68 42 57.0	− 19.88	50.46	2
8 751	Chamæleontis	9.0	12 30 14.77	+ 3.875	77 38 9.8	− 19.88	50.40	1
8 752	Muscæ	8.8	19.00	3.601	71 34 33.2	.88	52.31	2
8 753	Muscæ	10.0	20.07	3.633	72 32 58.5	.88	51.28	1
8 754	Muscæ	9.5	26.17	3.667	73 26 6.2	.88	50.44	1
8 755	Gould, Z. C., 12h 1921	9.2	32.18	+ 3.843	77 1 56.5	− 19.88	51.35	2
8 756	Chamæleontis	9.8	12 30 42.12	+ 3.934	78 17 34.0	− 19.88	51.84	2
8 757	Gould, Z. C., 12h 1942	9.0	42.28	3.853	77 7 21.1	.88	51.35	2
8 758	Gould, Z. C., 12h 1940	9.8	49.46	3.532	68 43 45.2	.88	50.46	2
8 759	Muscæ	9.5	57.30	3.601	71 11 31.9	.87	52.35	1
8 760	Gould, 17217	8.2	30 58.86	+ 3.481	66 12 55.0	− 19.87	52.37	2
8 761	Muscæ	9.0	12 31 2.96	+ 3.606	71 19 7.9	− 19.87	52.35	1
8 762	Gould, Z. C., 12h 1952	9.8	3.46	3.500	67 7 36.4	.87	51.87	2
8 763	Gould, 17220	8.2	5.38	3.592	70 49 9.4	.87	52.28	2
8 764	Gould, Z. C., 12h 1959	9.5	8.40	3.498	66 59 29.6	.87	51.87	2
8 765	Gould, Z. C., 12h 1964	9.5	12.85	3.542	68 54 11.3	− 19.87	50.50	1
8 766	Gould, Z. C., 12h 1970	10.0	12 31 17.27	+ 3.538	68 40 38.8	− 19.87	50.42	1
8 767	Lacaille, 5228	7.5	25.62	3.567	69 43 18.1	.87	52.34	1
8 768	Muscæ	9.2	25.80	3.615	71 24 3.8	.87	52.31	2
8 769	Chamæleontis	9.0	31.40	4.356	81 52 23.8	.87	50.85	2
8 770	Muscæ	9.0	39.79	+ 3.583	70 11 49.9	− 19.87	51.44	1
8 771	Gould, Z. C., 12h 1993	8.0	12 31 46.25	+ 3.569	69 35 45.7	− 19.86	52.34	1
8 772	Chamæleontis	10.0	49.04	4.270	81 13 9.6	.86	50.38	1
8 773	Gould, Z. C., 12h 2003	9.0	53.84	3.698	73 29 32.0	.86	51.25	1
8 774	Gould, Z. C., 12h 1998	9.5	31 54.29	3.507	66 54 49.7	.86	51.45	1
8 775	Muscæ	9.5	32 1.40	+ 4.206	80 40 26.3	− 19.86	50.90	2
8 776	Gould, Z. C., 12h 2020	9.0	12 32 5.30	+ 3.724	74 2 34.1	− 19.86	52.28	1
8 777	Chamæleontis	10.0	7.64	3.819	75 57 35.7	.86	51.30	1
8 778	Muscæ	9.5	9.58	3.624	71 16 25.0	.86	52.35	1
8 779	Chamæleontis	9.2	12.37	4.390	81 54 37.8	.86	50.85	2
8 780	Gould, Z. C., 12h 2031	9.0	19.42	+ 3.531	67 44 58.0	− 19.86	52.30	1
8 781	Gould, Z. C., 12h 2035	9.5	12 32 22.63	+ 3.703	73 23 52.1	− 19.86	50.44	1
8 782	Chamæleontis	10.0	35.94	4.295	81 11 36.2	.85	50.38	1
8 783	Gould, 17252	8.5	38.38	3.501	66 8 42.2	.85	52.38	1
8 784	Muscæ	10.0	39.07	3.617	70 47 16.7	.85	52.28	1
8 785	Chamæleontis	9.5	40.02	+ 3.804	75 27 35.0	− 19.85	50.89	2

No. 8 755. Gould's declination is 8.4″ less.

Number.	Constellation, Name of Star, or Synonym.	Magnitude.	Right Ascension, 1850.0.	Annual Precession.	South Declination, 1850.0.	Annual Precession.	Mean year.	No. of obs.
			h. m. s.	s.	° ′ ″	″		
8786	Octantis	10.0	12 32 53.66	+ 5.358	85 13 2.1	− 19.85	51.33	1
8787	Gould, 17260	9.5	55.16	3.522	66 58 5.3	.85	51.45	1
8788	Lacaille, 5236	7.7	32 59.04	3.561	68 34 56.6	.85	50.46	2
8789	Gould, Z. C., 12h 2069	9.0	33 2.71	3.874	76 31 35.5	.85	51.34	1
8790	Octantis	12.0	4.52	+ 4.853	83 50 19.1	− 19.85	50.33	1
8791	Octantis	10.0	12 33 7.27	+ 4.605	82 50 26.2	− 19.85	50.33	1
8792	Chamæleontis	9.0	10.59	4.444	81 59 42.8	.85	50.37	1
8793	Gould, Z. C., 12h 2081	9.5	22.59	3.587	69 22 13.4	.84	51.89	2
8794	Chamæleontis	10.0	27.09	4.198	80 12 14.4	.84	50.90	2
8795	Gould, Z. C., 12h 2096	9.5	40.29	+ 3.586	69 9 53.8	− 19.84	51.45	1
8796	Herschel, 4540[1]	9.0	12 33 42.18	+ 3.674	71 57 45.6	− 19.84	52.30	2
8797	Herschel, 4540[2]	9.2	42.70	3.674	71 57 57.2	.84	52.30	2
8798	Gould, Z. C., 12h 2108	9.2	54.69	3.545	67 21 41.6	.84	52.30	2
8799	Gould, Z. C., 12h 2112	10.0	33 58.50	3.579	68 43 14.8	.84	50.46	2
8800	Muscæ	9.0	34 10.50	+ 3.670	71 37 25.2	− 19.83	52.26	1
8801	Muscæ	10.0	12 34 12.52	+ 3.695	72 18 10.6	− 19.83	51.80	2
8802	Muscæ	9.0	12.68	3.755	73 46 33.0	.83	51.76	2
8803	Gould, Z. C., 12h 2138	9.5	23.58	3.518	65 51 50.8	.83	52.38	1
8804	Gould, Z. C., 12h 2141	9.0	29.06	3.518	65 47 55.3	.83	52.38	1
8805	Gould, 17306	8.2	30.41	+ 3.629	70 12 50.0	− 19.83	51.86	2
8806	Gould, Z. C., 12h 2158	9.0	12 34 49.45	+ 3.543	66 45 24.0	− 19.83	51.90	2
8807	Muscæ	8.5	50.04	3.694	71 58 38.2	.83	52.30	2
8808	Muscæ	9.5	50.13	3.818	74 48 56.5	.83	51.29	1
8809	Chamæleontis	9.2	52.74	3.831	75 2 45.4	.82	50.93	2
8810	Muscæ	9.0	34 55.46	+ 3.765	73 40 50.0	− 19.82	51.25	1
8811	Octantis	10.0	12 35 6.89	+ 4.692	82 49 25.1	− 19.82	50.33	1
8812	Gould, Z. C., 12h 2183	9.2	8.11	3.795	74 14 0.6	.82	51.34	2
8813	Gould, Z. C., 12h 2195	9.0	13.08	4.090	78 38 21.2	.82	51.32	2
8814	Chamæleontis	10.0	23.81	4.248	80 4 54.8	.82	51.37	1
8815	Gould, Z. C., 12h 2193	10.0	26.05	+ 3.575	67 43 48.9	− 19.82	52.30	1
8816	Gould, 17341	9.0	12 35 45.44	+ 3.750	72 58 28.4	− 19.81	50.43	2
8817	Muscæ	9.0	50.54	3.758	73 7 25.2	.81	50.42	1
8818	Brisbane, 4166	9.0	50.69	3.593	68 12 29.2	.81	50.39	2
8819	Muscæ	9.0	51.52	3.696	71 33 6.9	.81	52.31	2
8820	Lacaille, 5255	7.5	35 52.21	+ 3.588	68 0 31.9	− 19.81	50.37	1
8821	Brisbane, 4168	8.0	12 36 1.82	+ 3.676	70 53 18.2	− 19.81	52.28	1
8822	Herschel, 4545[1]	9.8	2.36	3.820	74 21 54.5	.81	51.34	2
8823	Herschel, 4545[2]	9.8	3.06	3.820	74 21 43.0	.81	51.34	2
8824	Gould, 17351	10.0	6.24	3.677	70 53 37.6	.81	52.28	1
8825	Chamæleontis	10.0	13.15	+ 4.174	79 11 24.9	− 19.81	50.42	1

Number.	Constellation, Name of Star, or Synonym.	Magnitude.	Right Ascension, 1850.0.	Annual Precession.	South Declination, 1850.0.	Annual Precession.	Mean year.	No. of obs.
			h. m. s.	s.	° ′ ″	″		
8826	Gould, Z. C., 12h 2257	9.5	12 36 13.78	+ 3.541	65 49 24.3	− 19.81	52.38	1
8827	Gould, Z. C., 12h 2273	9.8	15.56	4.042	77 45 16.0	.81	51.06	3
8828	Chamæleontis	9.0	17.97	4.544	81 50 45.0	.81	51.34	1
8829	Melbourne (1), 632	9.5	29.38	3.566	66 47 16.8	.80	51.90	2
8830	Muscæ	9.5	40.66	+ 3.811	73 55 36.1	− 19.80	51.25	1
8831	Octantis	9.5	12 36 48.76	+ 5.530	85 1 37.1	− 19.80	51.08	4
8832	β Muscæ	4.5	37 8.03	3.587	67 17 9.8	.79	52.29	1
8833	Muscæ	9.5	12.86	3.737	72 0 13.4	.79	52.33	1
8834	Gould, Z. C., 12h 2314	9.5	13.26	3.772	72 49 55.4	.79	50.42	1
8835	Muscæ	10.0	14.07	+ 3.860	74 38 50.5	− 19.79	51.46	1
8836	Gould, Z. C., 12h 2322	8.5	12 37 24.20	+ 3.560	65 59 46.4	− 19.79	52.38	1
8837	Muscæ	10.0	25.02	3.663	69 49 34.9	.79	51.44	1
8838	Muscæ	10.0	31.00	3.744	72 1 59.5	.79	52.33	1
8839	Gould, 17386	9.0	32.14	3.595	67 23 39.4	.79	52.30	2
8840	Muscæ	9.2	41.16	+ 3.568	66 12 55.6	− 19.79	52.37	2
8841	Gould, 17395	8.2	12 37 53.14	+ 3.791	72 59 31.0	− 19.78	50.43	2
8842	Muscæ	10.0	38 1.37	3.661	69 27 10.3	.78	52.34	1
8843	Gould, 17396	9.0	5.75	3.709	70 51 44.3	.78	52.28	1
8844	Lacaille, 5266	7.0	8.22	4.313	79 52 54.4	.78	51.30	2
8845	Gould, Z. C., 12h 2373	9.0	8.88	+ 3.733	71 28 13.4	− 19.78	52.31	2
8846	Muscæ	9.0	12 38 22.90	+ 3.743	71 38 21.2	− 19.78	52.26	1
8847	Brisbane, 4184	9.0	24.59	3.711	70 46 30.2	.78	52.28	2
8848	Muscæ	9.5	43.72	3.760	71 53 49.6	.77	52.33	1
8849	Muscæ	10.0	45.22	3.674	69 30 15.7	.77	52.34	1
8850	Gould, Z. C., 12h 2407	9.5	47.00	+ 3.626	67 53 58.8	− 19.77	52.30	1
8851	Gould, Z. C., 12h 2411	9.0	12 38 49.49	+ 3.615	67 27 19.9	− 19.77	52.30	2
8852	Muscæ	10.0	49.82	3.885	74 30 9.7	.77	50.41	1
8853	Gould, Z. C., 12h 2424	9.2	58.22	3.702	70 15 19.3	.77	51.86	2
8854	Muscæ	10.0	38 59.66	3.743	71 21 41.2	.77	52.35	1
8855	Octantis	10.0	39 0.54	+ 5.125	83 42 18.2	− 19.77	50.35	1
8856	Octantis	9.4	12 39 6.44	+ 7.198	86 51 1.4	− 19.77	51.33	5
8857	Octantis	9.8	8.24	5.054	83 27 31.9	.76	50.37	2
8858	Lacaille, 5275	8.5	8.30	3.590	66 18 37.6	.76	52.37	2
8859	Octantis	9.6	9.94	8.307	87 30 46.6	.76	51.33	4
8860	Gould, Z. C., 12h 2438	9.0	10.54	+ 3.590	66 18 40.2	− 19.76	52.37	2
8861	Muscæ	9.5	12 39 12.67	+ 3.711	70 23 13.3	− 19.76	52.28	1
8862	Muscæ	9.0	18.49	3.878	74 11 40.5	.76	52.28	1
8863	Gould, Z. C., 12h 2446	9.0	18.68	3.598	66 32 11.4	.76	52.36	1
8864	Muscæ	9.5	31.86	3.803	72 34 8.5	.76	51.28	1
8865	Gould, 17423	9.0	34.87	+ 3.630	67 36 33.6	− 15.76	52.30	1

Number.	Constellation, Name of Star, or Synonym.	Magnitude.	Right Ascension, 1850.0.	Annual Precession.	South Declination, 1850.0.	Annual Precession.	Mean year.	No. of obs.
			h. m. s.	s.	° ′ ″	″		
8 866	Gould, 17428	8.5	12 39 45.82	+ 3.660	68 34 21.2	− 19.76	50.46	2
8 867	ι Octantis	5.8	46.61	5.388	84 18 23.6	.76	50.33	2
8 868	Brisbane, 4194	8.5	39 54.87	3.741	70 54 41.6	.75	52.28	1
8 869	Lacaille, 5279	6.2	40 6.64	3.754	71 10 1 0	.75	52.31	2
8 870	Muscæ	9.2	11.49	+ 3.900	74 16 30.3	− 19.75	51.34	2
8 871	Muscæ	9.5	12 40 20.53	+ 3.819	72 35 17.8	− 19.75	51.28	1
8 872	Gould, 17439	8.5	22.50	3.593	65 46 4.0	.75	52.38	1
8 873	Gould, Z. C., 12ʰ 2509	9.0	25.82	3.628	67 8 11.5	.75	52.29	1
8 874	Octantis	11.0	30.69	5.425	84 17 39.5	.74	50.33	1
8 875	Muscæ	9.5	39.04	+ 3.822	72 32 0.5	− 19.74	51.28	1
8 876	Octantis	9.5	12 40 41.65	+ 7.099	86 38 36.0	− 19.74	51.93	1
8 877	Brisbane, 4199	9.0	40 55.10	3.754	70 48 13.0	.74	52.28	1
8 878	Gould, 17447	9.0	41 3.02	3.684	68 43 12.5	.74	50.46	2
8 879	Muscæ	10.0	13.20	3.785	71 27 27.7	.73	52.26	1
8 880	Gould, Z. C., 12ʰ 2570	9.5	18.07	+ 3.644	67 16 19.5	− 19.73	52.29	1
8 881	Gould, Z. C., 12ʰ 2573	9.5	12 41 19.68	+ 3.652	67 33 14.2	− 19.73	52.30	1
8 882	Gould, Z. C., 12ʰ 2574	9.0	21.52	3.646	67 18 36.8	.73	52.29	1
8 883	Chamæleontis	9.5	22.25	4.120	77 5 44.1	.73	51.25	1
8 884	Muscæ	9.5	25.15	3.758	70 42 25.5	.73	52.28	1
8 885	Muscæ	10.0	25.25	+ 3.891	73 39 10.1	− 19.73	51.25	1
8 886	Chamæleontis	10.0	12 41 25.66	+ 4.457	80 9 20.6	− 19.73	51.37	1
8 887	Chamæleontis	9.0	27.20	4.864	82 21 16.0	.73	50.39	1
8 888	Chamæleontis	10.0	30.90	4.406	79 45 47.1	.73	51.24	1
8 889	Gould, Z. C., 12ʰ 2585	9.0	31.30	3.629	66 36 9.7	.73	51.90	2
8 890	Gould, Z. C., 12ʰ 2590	9.0	39.88	+ 3.671	68 2 10.6	− 29.73	50.37	1
8 891	Gould, 17459	8.0	12 41 44.84	+ 3.697	68 49 37.0	− 19.72	50.50	1
8 892	Gould, Z. C., 12ʰ 2621	8.8	42 3.62	3.875	73 6 10.7	.72	50.43	2
8 893	Muscæ	9.5	10.08	3.622	66 2 9.8	.72	52.38	1
8 894	Gould, Z. C., 12ʰ 2622	8.0	15.51	3.903	73 34 10.3	.22	51.25	1
8 895	Muscæ	9.2	19.20	+ 3.924	73 56 4.3	− 19.72	52.36	3
8 896	Chamæleontis	8.8	12 42 21.83	+ 4.775	81 47 27.0	− 19.71	50.85	2
8 897	Gould, Z. C., 12ʰ 2640	9.0	28.92	3.734	69 35 43.0	.71	52.34	1
8 898	Gould, Z. C., 12ʰ 2641	8.5	42 32.80	3.627	66 2 17.1	.71	52.38	1
8 899	Brisbane, 4214	9.5	43 1.01	3.790	70 50 6.7	.70	52.28	1
8 900	Chamæleontis	10.0	1.44	+ 4.387	79 15 23.4	− 19.70	50.42	1
8 901	Muscæ	10.0	12 43 4.64	+ 3.916	73 30 32.8	− 19.70	51.25	1
8 902	Muscæ	10.0	7.41	3.748	69 42 25.8	.70	52.34	1
8 903	Gould, Z. C., 12ʰ 2675	9.2	9.32	3.781	70 33 32.1	.70	52.28	2
8 904	Muscæ	9.5	24.88	3.923	73 30 44.5	.70	51.25	1
8 905	Muscæ	9.5	29.40	+ 3.956	74 4 46.9	− 19.70	52.28	1

No. 8 891. Declination wire not specified.

Number.	Constellation, Name of Star, or Synonym.	Magnitude.	Right Ascension, 1850.0.	Annual Precession.	South Declination, 1850.0.	Annual Precession.	Mean year.	No. of obs.
			h. m. s.	s.	° ′ ″	″		
8906	Gould, Z. C., 12^{h} 2702	9.0	12 43 35.67	+ 3.733	69 3 44.7	− 19.69	51.45	1
8907	Octantis	10.0	39.62	6.804	86 7 7.3	.69	51.33	2
8908	Octantis	10.0	45.75	5.188	83 9 46.4	.69	50.39	1
8909	Gould, Z. C., 12^{h} 2710	9.5	46.39	3.655	66 29 22.1	.69	52.36	1
8910	Chamæleontis	10.0	57.35	+ 4.047	75 21 30.2	− 19.69	50.40	1
8911	Lacaille, 5297	7.0	12 43 57.79	+ 3.790	70 27 22.4	− 19.69	52.28	1
8912	Muscæ	9.5	44 7.30	3.649	66 5 39.4	.69	52.38	1
8913	Gould, Z. C., 12^{h} 2734	9.5	20.53	3.666	66 35 19.6	.68	52.36	1
8914	Muscæ	9.5	29.58	3.653	66 4 12.8	.68	52.38	1
8915	Octantis	8.4	34.94	+ 7.462	86 37 50.6	− 19.68	51.63	10
8916	Gould, Z. C., 12^{h} 2749	9.5	12 44 37.65	+ 3.668	66 31 38.9	− 19.68	52.36	1
8917	Gould, Z. C., 12^{h} 2756	9.0	40.06	3.768	69 35 59.1	.68	52.34	1
8918	Gould, 17511	7.5	41.23	3.755	69 13 12.8	.68	51.45	1
8919	Gould, Z. C., 12^{h} 2761	9.2	44.14	3.806	70 33 5.5	.68	52.28	2
8920	Gould, Z. C., 12^{h} 2772	9.5	44 53.37	+ 3.715	67 57 27.3	− 19.67	52.30	1
8921	Chamæleontis	9.5	12 45 9.46	+ 4.844	81 35 47.8	− 19.67	50.86	2
8922	Gould, Z. C., 12^{h} 2791	8.5	10.35	3.978	73 52 59.9	.67	51.98	4
8923	Muscæ	9.2	10.54	3.955	73 28 58.8	.67	50.85	2
8924	Brisbane, 4230	9.0	16.71	3.939	73 9 8.4	.67	50.43	2
8925	Gould, 17539	8.5	38.36	+ 3.730	68 5 28.1	− 19.66	50.37	1
8926	Muscæ	9.5	12 45 42.48	+ 3.663	65 51 14.0	− 19.66	52.38	1
8927	Muscæ	9.5	45.60	3.789	69 42 14.2	.66	52.34	1
8928	Lacaille, 5310	7.2	48.75	3.900	72 13 4.2	.66	51.80	2
8929	Octantis	8.2	45 49.57	7.308	86 24 45.5	.66	51.60	9
8930	Gould, 17552	9.0	46 0.40	+ 3.738	68 10 56.4	− 19.65	50.39	2
8931	Gould, Z. C., 12^{h} 2830	8.8	12 46 1.50	+ 3.714	67 26 57.3	− 19.65	52.30	2
8932	Gould, Z. C., 12^{h} 2839	9.0	6.51	3.822	70 12 53.9	.65	52.28	1
8933	Muscæ	10.0	6.88	3.806	70 0 7.9	.65	51.44	1
8934	Muscæ	10.0	8.48	3.693	66 41 40.9	.65	52.36	1
8935	Muscæ	8.5	9.86	+ 3.985	73 40 38.2	− 19.65	51.89	3
8936	Muscæ	9.5	12 46 18.49	+ 3.855	71 4 51.0	− 19.65	52.35	1
8937	Muscæ	10.0	26.02	4.047	74 34 52.8	.65	50.41	1
8938	Muscæ	10.0	30.32	3.842	70 42 26.7	.64	52.28	1
8939	Lacaille, 5318	6.5	35.40	3.873	71 22 12.6	.64	52.31	2
8940	Muscæ	10.0	37.87	+ 4.009	73 54 47.9	− 19.64	52.28	1
8941	Gould, Z. C., 12^{h} 2879	9.5	12 46 45.91	+ 4.070	74 48 32.7	− 19.64	51.46	1
8942	Gould, Z. C., 12^{h} 2869	9.0	45.97	3.685	66 8 34.1	.64	52.38	1
8943	Muscæ	9.5	47.14	3.680	65 58 31.7	.64	52.38	1
8944	Muscæ	10.0	47.45	3.680	65 58 7.3	.64	52.38	1
8945	Gould, Z. C., 12^{h} 2874	0.2	46 49.13	+ 3.703	66 43 45.2	− 19.64	51.90	2

Number.	Constellation, Name of Star, or Synonym.	Magnitude.	Right Ascension, 1850.0.	Annual Precession.	South Declination, 1850.0.	Annual Precession.	Mean year.	No. of obs.
			h. m. s.	s.	° ′ ″	″		
8 946	Gould, Z. C., 12h 2892	9.5	12 47 9.52	+ 3.710	66 48 34.2	− 19.63	51.45	1
8 947	Gould, Z. C., 12h 2894	9.2	10.66	3.707	66 43 44.8	.63	51.91	2
8 948	Gould, Z. C., 12h 2906	8.8	21.86	3.956	72 45 13.7	.63	50.85	2
8 949	Muscæ	9.5	23.80	4.063	74 30 34.7	.63	50.41	1
8 950	Chamæleontis	9.5	24.07	+ 5.062	82 8 42.9	− 19.63	50.37	1
8 951	Octantis	10.0	12 47 24.52	+ 8.804	87 15 24.9	− 19.63	51.93	2
8 952	Muscæ	10.0	30.52	3.891	71 25 9.6	.63	52.31	2
8 953	Gould, Z. C., 12h 2920	9.0	38.32	3.978	73 3 50.6	.62	50.43	2
8 954	Gould, 17586	9.5	48.70	3.770	68 21 41.7	.62	50.42	1
8 955	Gould, Z. C., 12h 2929	9.5	49.84	+ 3.776	68 31 47.2	− 19.62	50.42	1
8 956	Gould, Z. C., 12h 2950	8.8	12 47 51.28	+ 4.516	79 7 58.8	− 19.62	50.85	2
8 957	Gould, Z. C., 12h 2931	8.5	54.71	3.707	66 23 58.1	.62	52.36	1
8 958	Lacaille, 5323	7.5	55.79	3.731	67 8 48.8	.62	52.06	4
8 959	Chamæleontis	9.5	47 56.16	4.735	80 31 23.2	.62	50.43	1
8 960	Gould, Z. C., 12h 2946	9.5	48 7.20	+ 3.731	67 4 54.1	− 19.62	52.29	1
8 961	Gould, Z. C., 12h 2945	10.0	12 48 8.22	+ 3.708	66 20 31.5	− 19.62	52.38	1
8 962	Gould, Z. C., 12h 2954	9.0	9.44	3.896	71 18 34.2	.62	52.35	1
8 963	Gould, Z. C., 12h 2961	9.5	13.88	3.893	71 12 29.7	.61	52.35	1
8 964	Octantis	9.0	15.46	6.007	84 33 37.6	.61	50.33	1
8 965	Muscæ	9.0	17.17	+ 3.911	71 34 41.3	− 19.61	52.26	1
8 966	Gould, Z. C., 12h 2965	9.0	12 48 18.36	+ 3.829	69 44 0.4	− 19.61	52.34	1
8 967	Muscæ	9.5	20.60	3.903	71 23 13.2	.61	52.31	2
8 968	Gould, Z. C., 12h 2974	9.0	20.87	3.734	67 0 38.8	.61	52.29	1
8 969	Muscæ	9.0	31.72	4.019	73 28 25.4	.61	51.84	2
8 970	Muscæ	9.5	31.80	+ 3.732	66 56 25.3	− 19.61	51.45	1
8 971	Muscæ	10.0	12 48 50.73	+ 3.859	70 14 3.1	− 19.60	52.28	1
8 972	Muscæ	9.0	52.43	4.023	73 25 54.4	.60	51.25	1
8 973	Muscæ	9.5	54.64	3.950	72 7 3.1	.60	52.33	1
8 974	Gould, Z. C., 12h 2995	9.0	55.25	3.728	66 39 28.4	.60	51.45	1
8 975	Muscæ	9.5	57.96	+ 3.754	67 25 17.6	− 19.60	52.29	2
8 976	Gould, Z. C., 12h 3000	9.2	12 48 59.76	+ 4.0[illegible]	73 46 49.5	− 19.60	51.76	2
8 977	Chamæleontis	9.5	49 1.60	5.04	81 49 38.4	.60	51.34	1
8 978	Muscæ	10.0	18.42	3.820	69 21 9.8	.59	51.45	1
8 979	Gould, Z. C., 12h 3023	9.0	22.42	3.881	70 32 49.7	.59	52.28	1
8 980	Gould, Z. C., 12h 3048	9.5	27.96	4.574	79 12 24.5	− 19.59	50.42	1
8 981	Gould, 17639	8.3	12 49 33.97	+ 3.853	69 49 23.0	− 19.59	52.08	3
8 982	Chamæleontis	9.0	34.34	4.443	78 11 35.8	.59	51.40	2
8 983	Lacaille, 5335	9.0	41.92	3.863	70 1 20.5	.59	51.95	2
8 984	Gould, Z. C., 12h 3052	9.0	51.22	3.735	66 29 5.3	.58	52.36	1
8 985	Gould, Z. C., 12h 3055	9.5	54.16	+ 3.746	66 47 55.9	− 19.58	51.45	1

Nos. 8 962 and 3. Right ascensions have been interchanged to agree with Gould.

Number.	Constellation, Name of Star, or Synonym.	Magnitude.	Right Ascension, 1850.0.	Annual Precession.	South Declination, 1850.0.	Annual Precession.	Mean year.	No. of obs.
			h. m. s.	s.	° ′ ″	″		
8 986	Gould, Z. C., 12h 3059	9.8	12 49 55.26	+ 3.860	69 52 27.6	— 19.58	51.89	2
8 987	Lacaille, 5325	7.5	50 17.44	8.198	86 44 59.4	.58	51.63	10
8 988	Octantis	10.0	18.81	6.010	84 20 24.4	.57	50.33	1
8 989	Octantis	11.0	27.17	6.031	84 21 53.6	.57	50.33	1
8 990	Chamæleontis	9.0	27.86	+ 5.208	82 12 55.8	— 19.57	50.38	2
8 991	Gould, Z. C., 12h 3088	9.0	12 50 34.53	+ 3.934	71 15 11.0	— 19.57	52.35	1
8 992	Muscæ	10.0	34.73	3.954	71 39 5.6	.57	52.26	1
8 993	Muscæ	9.2	38.62	3.964	71 48 50.8	.57	52.30	2
8 994	Muscæ	8.0	39.95	3.954	71 36 59.2	.57	52.26	1
8 995	Chamæleontis	9.3	41.91	+ 4.317	76 44 27.7	— 19.57	51.35	3
8 996	Gould, Z. C., 12h 3091	10.0	12 50 42.68	+ 3.853	69 23 56.3	— 19.57	51.45	1
8 997	Gould, Z. C., 12h 3118	9.5	45.29	4.523	78 33 18.9	.57	51.35	1
8 998	Chamæleontis	10.0	45.56	4.588	79 2 10.2	.57	51.28	1
8 999	Stone, 7146	8.5	46.37	3.733	66 2 10.3	.57	52.38	1
9 000	Gould, Z. C., 12h 3097	10.0	50 46.70	+ 3.792	67 47 30.2	— 19.57	52.30	1
9 001	Muscæ	9.2	12 51 1.74	+ 3.971	71 49 10.2	— 19.56	52.30	2
9 002	Muscæ	9.5	7.76	4.083	73 41 41.5	.56	51.25	1
9 003	Chamæleontis	10.0	19.37	4.297	76 22 37.4	.56	51.34	1
9 004	Gould, Z. C., 12h 3141	8.5	23.86	3.963	71 32 21.4	.55	52.31	2
9 005	Chamæleontis	10.0	26.10	+ 4.210	75 20 43.9	— 19.55	50.40	1
9 006	Muscæ	10.0	12 51 34.52	+ 3.893	70 2 10.2	— 19.55	51.44	1
9 007	Gould, Z. C., 12h 3146	10.0	35.30	3.768	66 47 41.8	.55	51.45	1
9 008	Lacaille, 5338	7.0	40.67	4.829	80 20 29.2	.55	51.40	2
9 009	Gould, Z. C., 12h 3158	9.0	48.32	3.787	67 16 32.6	.55	52.29	1
9 010	Muscæ	9.5	54.94	+ 3.746	65 59 34.7	— 19.54	52.38	1
9 011	Gould, Z. C., 12h 3166	8.8	12 51 57.05	+ 3.781	67 2 46.4	— 19.54	51.87	2
9 012	Lacaille, 5343	8.2	51 57.14	4.679	79 24 33.5	.54	50.83	2
9 013	d Muscæ	4.5	52 2.37	3.933	70 44 18.1	.54	52.28	2
9 014	Octantis	10.0	7.04	7.433	86 2 46.0	.54	51.63	2
9 015	Stone, 7162	9.0	10.62	+ 3.835	68 25 7.3	— 19.54	50.42	1
9 016	Octantis	10.0	12 52 10.76	+11.771	88 0 55.1	— 19.54	51.92	2
9 017	Lacaille, 5339	7.5	19.16	5.265	82 8 36.2	.54	50.38	2
9 018	Chamæleontis	9.2	20.04	4.307	76 13 22.5	.54	51.32	1
9 019	Muscæ	9.5	28.76	4.046	72 42 9.2	.53	50.85	2
9 020	Gould, Z. C., 12h 3209	9.0	28.86	+ 3.985	71 37 18.2	— 19.53	52.26	1
9 021	Chamæleontis	10.0	12 52 29.77	+ 5.050	81 16 34.6	— 19.53	50.38	1
9 022	Gould, Z. C., 12h 3221	9.5	42.48	3.753	65 53 49.5	.53	52.38	1
9 023	Gould, Z. C., 12h 3226	9.0	45.74	3.778	66 37 37.4	.53	51.91	2
9 024	Gould, Z. C., 12h 3228	10.0	48.78	3.757	65 58 20.5	.53	52.38	1
9 025	Muscæ	10.0	56.16	+ 3.827	67 55 50.6	— 19.52	50.37	1

No. 9 016. Right ascension may be 5m greater.

No. 9 018. Declination may be 1′ greater.

Number.	Constellation, Name of Star, or Synonym.	Magnitude.	Right Ascension, 1850.0.	Annual Precession.	South Declination, 1850.0.	Annual Precession.	Mean year.	No. of obs.
			h. m. s.	s.	° ′ ″	″		
9026	Lacaille, 5356	6.8	12 52 58.27	+ 3.945	70 40 1.6	− 19.52	52.28	2
9027	Octantis	11.0	53 8.51	5.447	82 37 46.1	.52	50.33	1
9028	Gould, Z. C., 12h 3263	8.8	10.43	4.612	78 42 28.1	.52	51.32	2
9029	Gould, Z. C., 12h 3244	9.8	10.64	3.811	67 25 14.4	.52	52.30	2
9030	Gould, 17724	9.0	17.42	+ 3.834	67 59 14.8	− 19.52	50.37	1
9031	Chamæleontis	10.0	12 53 21.96	+ 5.321	82 11 18.6	− 19.52	50.37	1
9032	Gould, Z. C., 12h 3261	9.5	28.41	3.812	67 20 5.3	.51	52.29	1
9033	Chamæleontis	8.5	44.34	4.302	75 50 27.2	.51	51.34	2
9034	Gould, 17759	9.0	48.62	4.013	71 42 15.3	.51	52.26	1
9035	Gould, Z. C., 12h 3283	9.0	53 54.54	+ 3.904	69 27 35.8	− 19.50	51.89	2
9036	Gould, Z. C., 12h 3292	9.8	12 54 0.13	+ 3.887	69 2 7.8	− 19.50	50.97	2
9037	Octantis	9.8	18.75	5.619	82 58 28.9	.50	50.33	1
9038	Lacaille, 5353	8.7	23.13	5.284	81 54 51.4	.49	50.74	3
9039	Muscæ	9.5	33.45	4.072	72 30 33.5	.49	51.28	1
9040	Gould, Z. C., 12h 3324	10.0	35.90	+ 3.827	67 19 30.5	− 19.49	52.29	1
9041	Muscæ	9.0	12 54 43.84	+ 4.211	74 28 36.7	− 19.49	50.41	1
9042	Gould, Z. C., 12h 3333	9.2	45.52	3.904	69 10 48.2	.49	50.97	2
9043	Muscæ	9.0	45.71	4.191	74 12 31.1	.49	52.28	1
9044	Muscæ	9.2	50.72	3.956	70 17 16.6	.48	51.86	2
9045	Muscæ	9.0	54 55.79	+ 4.185	74 5 5.5	− 19.48	52.28	1
9046	Gould, Z. C., 12h 3346	9.0	12 55 4.23	+ 3.788	66 2 46.0	− 19.48	52.38	1
9047	Chamæleontis	10.0	8.68	4.668	78 42 43.0	.48	51.32	2
9048	Chamæleontis	9.5	11.98	4.713	79 0 13.5	.48	51.28	1
9049	Gould, Z. C., 12h 3364	9.2	14.04	4.044	71 49 40.8	.48	52.30	2
9050	Gould, 17759	9.0	16.91	+ 3.895	68 47 21.2	− 19.48	50.50	1
9051	Chamæleontis	8.8	12 55 18.15	+ 4.459	77 3 52.8	− 19.48	51.35	2
9052	Chamæleontis	11.0	23.20	4.348	75 55 22.9	.47	51.38	1
9053	Gould, Z. C., 12h 3370	9.5	33.27	4.059	71 59 37.3	.47	52.33	1
9054	Muscæ	9.5	46.28	4.027	71 21 20.9	.47	52.35	1
9055	Muscæ	10.0	49.40	+ 4.064	71 59 38.7	− 19.46	52.33	1
9056	Muscæ	9.8	12 55 50.02	+ 4.192	73 55 52.7	− 19.46	51.76	2
9057	Muscæ	10.0	56 1.57	3.855	67 32 44.1	.46	52.30	1
9058	Gould, Z. C., 12h 3393	9.5	2.80	3.842	67 11 42.7	.46	52.29	1
9059	Gould, Z. C., 12h 3398	9.0	2.91	4.068	71 59 41.8	.46	52.33	1
9060	Chamæleontis	10.0	5.54	+ 4.422	76 30 8.7	− 19.46	51.34	1
9061	Gould, Z. C., 12h 3400	9.5	12 56 7.50	+ 3.912	68 54 30.0	− 19.46	50.50	1
9062	Muscæ	9.5	14.51	3.943	69 33 3.0	.46	52.34	1
9063	Gould, Z. C., 12h 3419	9.0	21.22	3.852	67 21 9.8	.45	52.30	2
9064	Gould, Z. C., 12h 3424	9.5	26.68	3.812	66 14 23.4	.45	52.37	2
9065	Lacaille, 5369	7.0	26.79	+ 4.560	77 38 23.4	− 19.45	50.42	2

No. 9058. Gould's declination is 13.2″ greater. No. 9059. Gould's declination is 9.1″ less.
Nos. 9060 and 9070. The declinations of these two stars may have been interchanged.

Number.	Constellation, Name of Star, or Synonym.	Magnitude.	Right Ascension, 1850.0.	Annual Precession.	South Declination, 1850.0.	Annual Precession.	Mean year.	No. of obs.
			h. m. s.	s.	° ′ ″	″		
9 066	Muscæ	9.5	12 56 27.50	+ 4.219	74 8 10.6	— 19.45	52.28	1
9 067	Gould, Z. C., 12^h 3453 .	9.0	28.09	4.799	79 18 29.0	.45	50.42	1
9 068	Muscæ	9.5	28.40	3.802	65 56 33.2	.45	52.38	1
9 069	Chamæleontis	10.0	28.45	5.332	81 47 20.0	.45	51.49	1
9 070	Gould, Z. C., 12^h 3436 .	9.0	28.76	+ 4.422	76 24 55.9	— 19.45	51.34	1
9 071	Gould, Z. C., 12^h 3428 .	9.5	12 56 30.63	+ 3.838	66 56 23.5	— 19.45	51.45	1
9 072	Muscæ	10.0	33.62	4.224	74 10 39.9	.45	52.28	1
9 073	Muscæ	10.0	40.36	3.818	66 18 50.5	.45	52.36	1
9 074	Gould, Z. C., 12^h 3440 .	9.0	42.27	3.855	67 17 52.3	.45	52.29	1
9 075	Octantis	8.8	51.26	+ 6.651	84 45 35.1	— 19.44	50.77	5
9 076	Muscæ	10.0	12 56 51.42	+ 5.260	81 27 58.6	— 19.44	50.86	2
9 077	Gould, Z. C., 12^h 3454 .	9.0	55.37	3.812	66 2 44.7	.44	52.38	1
9 078	Lacaille, 5378	8.5	56.84	4.133	72 46 50.8	.44	50.85	2
9 079	Muscæ	9.0	56 58.29	4.282	74 47 12.3	.44	51.46	1
9 080	Muscæ	10.0	57 7.91	+ 3.989	70 12 59.2	— 19.44	51.44	1
9 081	Muscæ	9.5	12 57 17.57	+ 3.963	69 38 22.2	— 19.43	52.34	1
9 082	Muscæ	9.5	18.04	3.967	69 42 37.5	.43	52.34	1
9 083	Gould, Z. C., 12^h 3476 .	9.0	21.58	3.821	66 9 13.2	.43	52.37	2
9 084	Gould, Z. C., 12^h 3489 .	9.5	27.88	3.939	69 3 34.7	.43	51.45	1
9 085	Gould, Z. C., 12^h 3505 .	10.0	40.47	+ 4.113	72 16 11.6	— 19.42	54.33	1
9 086	Muscæ	10.0	12 57 56.74	+ 4.233	73 55 52.6	— 19.42	51.25	1
9 087	Gould, Z. C., 12^h 3516 .	9.2	57 58.58	3.839	66 25 6.0	.42	52.38	2
9 088	Gould, 17830	7.8	58 4.56	4.105	72 1 8.5	.42	52.30	2
9 089	Gould, Z. C., 13^h 18 . .	9.5	28.16	4.471	76 26 37.9	.41	51.34	1
9 090	Lacaille, 5393	7.5	35.39	+ 3.868	66 59 35.1	— 19.40	51.87	2
9 091	Gould, Z. C., 13^h 14 . .	9.5	12 58 40.60	+ 3.873	67 4 59.2	— 19.40	51.87	2
9 092	Gould, 17846	9.5	41.04	3.950	68 55 9.4	.40	50.50	1
9 093	Muscæ	10.0	58 55.40	3.843	66 11 53.6	.40	52.38	1
9 094	Octantis	10.2	59 4.92	5.865	83 2 28.2	.39	50.36	2
9 095	Gould, Z. C., 13^h 52 . .	9.0	6.25	+ 4.036	70 31 52.2	— 19.39	52.28	1
9 096	Octantis	9.2	12 59 6.38	+ 11.557	87 41 48.7	— 19.39	51.63	8
9 097	Lacaille, 5373	9.0	10.27	6.145	83 39 39.5	.39	50.35	1
9 098	Gould, 17863	9.0	31.81	4.107	71 38 57.0	.38	52.26	1
9 099	Muscæ	9.2	53.08	4.268	73 53 38.5	.38	51.76	2
9 100	Muscæ	10.0	12 59 56.44	+ 3.855	66 10 54.2	— 19.37	52.38	1
9 101	Octantis	9.2	13 0 7.69	+ 7.097	85 4 33.2	— 19.37	50.21	5
9 102	Gould, 17876	9.5	9.27	3.930	67 58 50.9	.37	50.37	1
9 103	Muscæ	10.0	10.66	4.123	71 43 40.2	.37	52.26	1
9 104	Gould, Z. C., 13^h 104 . .	8.5	12.64	3.849	65 55 36.0	.37	52.38	1
9 105	Muscæ	9.5	14.83	+ 4.262	73 42 20.5	— 19.37	51.25	1

Nos. 9 060 and 9 070. The declinations of these two stars may have been interchanged.

Number.	Constellation, Name of Star, or Synonym.	Magnitude.	Right Ascension, 1850.0.	Annual Precession.	South Declination, 1850.0.	Annual Precession.	Mean year.	No. of obs.
			h. m. s.	s.	° ′ ″	″		
9 106	Gould, Z. C., 13h 109 . .	9.2	13 0 19.42	+ 3.882	66 44 55.9	− 19.37	52.07	3
9 107	Muscæ	9.5	22.02	4.169	72 23 57.4	.36	51.28	1
9 108	Muscæ	10.0	26.18	4.027	69 57 40.3	.36	52.34	1
9 109	Gould, Z. C., 13h 122 . .	9.0	30.41	4.093	71 7 49.2	.36	52.31	2
9 110	Gould, 17887	8.5	31.81	+ 4.031	70 0 23.5	− 19.36	51.89	2
9 111	Muscæ	10.0	13 0 35.96	+ 4.050	70 20 12.2	− 19.36	52.28	1
9 112	Gould, Z. C., 13h 124 . .	9.0	37.32	3.860	66 5 11.8	.36	52.38	1
9 113	Gould, 17890	8.2	38.78	4.088	71 0 47.3	.36	52.32	2
9 114	Muscæ	10.0	48.33	4.344	74 35 28.8	.35	50.94	2
9 115	Muscæ	10.0	0 54.49	+ 4.072	70 39 28.5	− 19.35	52.28	1
9 116	Muscæ	10.0	13 1 5.86	+ 4.272	73 38 50.9	− 19.35	51.25	1
9 117	Gould, Z. C., 13h 154 . .	9.3	10.88	3.893	66 45 51.1	.35	52.07	3
9 118	Muscæ	10.0	13.25	4.001	69 11 59.1	.35	51.45	1
9 119	Muscæ	9.5	19.57	4.078	70 37 54.5	.34	52.28	1
9 120	Gould, Z. C., 13h 167 . .	9.5	25.39	+ 3.875	66 12 24.3	− 19.34	52.39	2
9 121	Gould, Z. C., 13h 179 . .	9.0	13 1 30.66	+ 4.219	72 49 6.2	− 19.34	50.42	1
9 122	Lacaille, 5409	6.8	35.22	4.004	69 8 33.8	.34	51.12	3
9 123	Brisbane, 4342	9.0	37.87	3.956	68 7 1.2	.34	50.92	2
9 124	Gould, Z. C., 13h 182 . .	9.2	41.12	3.896	66 39 57.3	.33	51.93	2
9 125	Gould, Z. C., 13h 185 . .	10.0	41.36	+ 3.986	68 44 15.3	− 19.33	50.42	1
9 126	Muscæ	10.0	13 1 44.70	+ 4.047	69 57 34.9	− 19.33	51.89	2
9 127	Gould, Z. C., 13h 199 . .	9.5	48.78	4.150	71 42 41.1	.33	52.26	1
9 128	Chamæleontis	9.5	51.72	4.673	77 26 42.2	.33	51.36	1
9 129	Gould, Z. C., 13h 198 . .	9.8	55.52	3.898	66 38 56.2	.33	51.93	2
9 130	Gould, Z. C., 13h 229 . .	8.2	57.74	+ 4.811	78 23 39.7	− 19.33	51.84	2
9 131	Chamæleontis	8.7	13 1 58.74	+ 4.582	76 39 36.2	− 19.33	51.35	3
9 132	Lacaille, 5406	6.5	1 59.34	4.703	77 38 54.4	.33	50.42	2
9 133	Gould, Z. C., 13h 213 . .	9.3	2 4.62	4.030	69 31 21.7	.33	52.07	3
9 134	Gould, Z. C., 13h 240 . .	9.0	5.20	4.825	78 29 24.6	.32	51.35	1
9 135	Octantis	9.6	5.41	+ 8.783	86 24 52.4	− 19.33	51.47	5
9 136	Gould, Z. C., 13h 246 . .	9.8	13 2 12.92	+ 4.897	78 19 18.6	− 19.32	51.84	2
9 137	Muscæ	10.0	18.73	4.131	71 16 22.6	.32	52.35	1
9 138	Gould, Z. C., 13h 228 . .	9.0	21.95	3.874	65 51 56.3	.32	52.38	1
9 139	Muscæ	9.8	24.65	4.351	74 17 44.9	.32	51.34	2
9 140	Gould, 17930	8.5	31.55	+ 4.164	71 44 38.1	− 19.31	52.28	2
9 141	Muscæ	9.5	13 2 33.78	+ 4.183	72 1 51.8	− 19.31	52.33	1
9 142	Muscæ	9.5	36.24	4.268	73 13 23.2	.31	50.44	1
9 143	Gould, Z. C., 13h 259 . .	9.5	2 48.82	4.032	69 21 13.2	.31	51.94	2
9 144	Muscæ	8.2	3 7.29	4.372	74 22 21.4	.30	51.34	2
9 145	Muscæ .	9.5	17.11	+ 4.114	70 43 37.0	− 19.30	52.28	2

No. 9 131. One observation gives 38′.

Number.	Constellation, Name of Star, or Synonym.	Magnitude.	Right Ascension, 1850.0.	Annual Precession.	South Declination, 1850.0.	Annual Precession.	Mean year.	No. of obs.
			h. m. s.	s.	° ′ ″	″		
9 146	Gould, Z. C., 13^h 291 . .	9.2	13 3 17.50	+ 4.271	73 5 5.0	− 19.30	50.43	2
9 147	Muscæ	10.2	18.32	4.087	70 13 56.4	.30	51.86	2
9 148	Gould, Z. C., 13^h 288 . .	9.2	19.05	4.134	71 2 44.5	.30	52.33	3
9 149	Gould, 17947	8.0	20.33	3.961	67 42 0.8	.30	52.30	1
9 150	Gould, Z. C., 13^h 289 . .	9.5	20.99	+ 4.089	70 13 40.3	− 19.30	51.86	2
9 151	Chamæleontis	10.0	13 3 22.97	+ 4.635	76 51 13.5	− 19.29	51.25	1
9 152	Muscæ	10.0	30.50	4.216	72 15 51.4	.29	51.28	1
9 153	Muscæ	10.0	39.39	4.052	69 30 3.6	.29	52.43	1
9 154	Gould, Z. C., 13^h 320 . .	9.0	41.35	4.293	73 16 33.5	.29	50.44	1
9 155	Gould, Z. C., 13^h 312 . .	9.5	41.28	+ 3.942	67 7 39.9	− 19.29	51.87	2
9 156	Muscæ	9.0	13 3 44.11	+ 4.338	73 49 41.3	− 19.29	51.76	2
9 157	Gould, Z. C., 13^h 326 . .	9.0	47.40	4.250	72 40 48.4	.28	50.85	2
9 158	Muscæ	9.5	48.87	3.889	65 47 47.2	.28	52.38	1
9 159	Gould, Z. C., 13^h 349 . .	10.0	51.31	4.868	78 25 43.5	.28	51.35	1
9 160	Muscæ	9.8	3 51.93	+ 4.054	69 28 43.4	− 19.28	52.07	3
9 161	Muscæ	9.0	13 4 17.68	+ 4.363	73 59 48.1	− 19.27	52.28	1
9 162	Muscæ	10.0	23.03	4.063	69 29 36.1	.27	52.43	1
9 163	Octantis	10.3	29.30	9.378	86 37 48.3	.27	51.73	3
9 164	Muscæ	9.0	30.01	4.957	78 51 18.5	.27	51.28	1
9 165	Gould, 17975	9.5	35.64	+ 4.325	73 27 56.4	− 19.26	51.25	1
9 166	Muscæ	9.5	13 4 36.23	+ 4.165	71 12 34.1	− 19.26	52.35	1
9 167	Chamæleontis	9.8	40.74	4.647	76 41 58.2	.26	51.30	2
9 168	Lacaille, 5430	7.5	4 51.07	4.080	69 40 50.1	.26	52.34	1
9 169	Brisbane, 4367	8.2	5 4.18	3.958	67 4 58.4	.25	51.87	2
9 170	Lacaille, 5432	8.0	7.56	+ 4.042	68 52 52.9	− 19.25	50.50	1
9 171	η Muscæ	6.0	13 5 9.03	+ 3.959	67 5 49.8	− 19.25	51.87	2
9 172	Muscæ	9.2	24.70	3.926	66 12 54.7	.25	52.39	2
9 173	Octantis	9.3	25.97	8.033	85 39 32.3	.24	51.46	8
9 174	Lacaille, 5424	7.0	30.47	5.044	79 10 48.9	.24	50.42	1
9 175	Gould, 18003	9.0	42.06	+ 4.044	68 45 16.0	− 19.24	50.50	1
9 176	Lacaille, 5427	7.3	13 5 44.04	+ 4.853	78 1 40.7	− 19.24	50.91	4
9 177	Muscæ	9.5	45.72	4.226	71 50 52.4	.24	52.26	1
9 178	Gould, Z. C., 13^h 441 . .	9.0	48.09	3.998	67 45 39.0	.24	52.30	1
9 179	Muscæ	9.5	49.59	4.174	71 1 53.3	.23	52.36	1
9 180	Octantis	9.8	54.21	+ 6.570	83 48 48.3	− 19.23	50.34	2
9 181	Gould, Z. C., 13^h 454 . .	9.2	13 5 55.03	+ 4.105	69 50 7.9	− 19.23	51.89	2
9 182	Chamæleontis	10.0	56.56	4.581	75 53 26.4	.23	51.34	2
9 183	Chamæleontis	9.5	5 58.65	5.016	78 57 5.7	.23	51.28	1
9 184	Chamæleontis	9.5	6 4.18	4.217	71 37 46.5	.23	52.29	2
9 185	Gould, Z. C., 13^h 482 . .	10.0	21.48	+ 4.131	70 10 48.8	− 19.22	51.44	1

Number.	Constellation, Name of Star, or Synonym.	Magnitude.	Right Ascension, 1850.0.	Annual Precession.	South Declination, 1850.0.	Annual Precession.	Mean year.	No. of obs.
			h. m. s.	s.	° ′ ″	″		
9 186	Muscæ	9.5	13 6 27.80	+ 4.278	72 24 51.1	19.22	51.28	1
9 187	Chamæleontis	10.0	32.05	4.950	78 28 53.6	.22	51.35	1
9 188	Muscæ	9.5	32.50	4.428	74 14 11.0	.22	51.34	2
9 189	Chamæleontis	10.0	33.26	4.869	77 58 18.8	.22	51.39	2
9 190	Gould, Z. C., 13ʰ 497 . .	9.0	34.86	+ 4.132	70 7 56.1	− 19.22	51.44	1
9 191	Muscæ	10.0	13 6 39.24	+ 4.397	73 51 28.3	− 19.21	51.77	2
9 192	Octantis	9.0	49.16	6.411	83 26 3.8	.21	50.37	2
9 193	Gould, Z. C., 13ʰ 517 . .	9.0	50.93	4.132	70 3 34.1	.21	51.44	1
9 194	Muscæ	10.0	57.08	4.404	73 52 58.6	.21	51.76	2
9 195	Lacaille, 5445	7.8	6 57.09	+ 4.297	72 32 56.2	− 19.21	50.85	2
9 196	Gould, Z. C., 13ʰ 541 . .	9.0	13 7 0.88	+ 4.269	72 9 21.1	− 19.21	52.33	1
9 197	Muscæ	9.0	3.53	4.156	70 26 1.2	.20	52.28	1
9 198	Gould, Z. C., 13ʰ 556 . .	9.0	4.20	4.928	78 15 46.4	.20	51.84	2
9 199	Gould, Z. C., 13ʰ 537 . .	10.0	4.62	4.061	68 41 37.6	.20	50.50	1
9 200	Chamæleontis	9.0	7.23	+ 4.533	75 11 56.9	− 19.20	50.40	1
9 201	Octantis	8.4	13 7 8.99	+ 9.852	86 44 25.4	− 19.20	51.51	7
9 202	Lacaille, 5451	5.2	10.25	3.939	65 59 19.9	.20	52.41	2
9 203	Gould, Z. C., 13ʰ 546 . .	9.5	13.16	3.974	66 47 57.4	.20	51.45	1
9 204	Muscæ	9.5	22.90	4.205	71 7 15.3	.20	52.36	1
9 205	Chamæleontis	9.5	23.00	+ 4.544	75 15 7.9	− 19.20	50.40	1
9 206	Chamæleontis	10.0	13 7 32.61	+ 5.145	79 23 24.5	− 19.19	50.42	1
9 207	Gould, Z. C., 13ʰ 558 . .	8.5	32.66	3.951	66 9 33.7	.19	52.41	2
9 208	Lacaille, 5455	9.0	35.79	4.001	67 14 42.1	.19	52.29	1
9 209	Gould, Z. C., 13ʰ 565 . .	9.5	35.80	4.020	67 43 11.7	.19	52.30	1
9 210	Lacaille, 5456	7.0	39.05	+ 4.020	67 42 4.3	− 19.19	52.30	1
9 211	Muscæ	8.5	13 7 43.64	+ 4.434	74 2 40.6	− 19.19	52.28	1
9 212	Octantis	10.0	46.01	7.724	85 12 46.3	.19	51.73	3
9 213	Gould, Z. C., 13ʰ 574 . .	9.0	48.77	3.975	66 38 35.4	.18	52.41	1
9 214	Gould, 18057	7.8	7 51.16	4.784	77 10 1.9	.18	51.31	2
9 215	Gould, Z. C., 13ʰ 616 . .	8.5	8 8.21	+ 4.600	75 37 38.0	− 19.18	51.38	1
9 216	Muscæ	9.2	13 8 10.38	+ 4.356	73 1 33.4	− 19.18	50.43	2
9 217	Gould, Z. C., 13ʰ 610 . .	9.8	17.93	4.039	67 54 7.0	.17	51.88	2
9 218	Octantis	10.0	22.82	6.408	83 16 48.7	.17	50.39	1
9 219	Muscæ	10.0	25.88	4.310	72 22 12.3	.17	51.28	1
9 220	Chamæleontis	10.0	26.08	+ 4.624	75 47 0.4	− 19.17	51.38	1
9 221	Lacaille, 5459	6.8	13 8 26.98	+ 4.230	71 14 28.0	− 19.17	52.35	2
9 222	Muscæ	9.5	29.24	3.958	66 2 57.8	.17	52.38	1
9 223	Chamæleontis	10.0	38.26	5.196	79 28 41.1	.16	51.24	1
9 224	Gould, Z. C., 13ʰ 655 . .	9.0	43.30	4.830	77 20 19.9	.16	51.36	1
9 225	Gould, Z. C., 13ʰ 643 . .	9.0	50.80	+ 3.986	66 35 17.1	− 19.16	52.41	1

No. 9 193. Gould's declination is 12.3″ greater.

Number.	Constellation, Name of Star, or Synonym.	Magnitude.	Right Ascension, 1850.0.	Annual Precession.	South Declination, 1850.0.	Annual Precession.	Mean year.	No. of obs.
			h. m. s.	s.	° ′ ″	″		
9226	Brisbane, 4383	8.0	13 8 53.54	+ 5.441	80 30 53.0	− 19.16	50.43	1
9227	Gould, Z. C., 13h 652 . .	10.0	8 59.86	3.998	66 49 48.2	.15	51.45	1
9228	Gould, Z. C., 13h 659 . .	10.0	9 8.40	4.055	68 0 11.2	.15	51.46	1
9229	Muscæ	9.5	13.76	4.294	71 57 51.0	.15	52.30	2
9230	Chamæleontis	10.0	16.12	+ 4.808	77 5 11.1	− 19.15	51.25	1
9231	Gould, Z. C., 13h 692 . .	8.8	13 9 16.92	+ 4.851	77 23 17.2	− 19.15	50.88	2
9232	Lacaille, 5463	8.5	20.61	4.171	70 3 43.4	.15	51.44	1
9233	Muscæ	10.0	30.15	4.450	73 50 9.8	.14	51.25	1
9234	Chamæleontis	9.2	32.00	5.598	81 0 48.6	.14	50.88	2
9235	Melbourne (1), 666 . . .	8.0	32.17	+ 3.994	66 34 20.0	− 19.14	52.41	1
9236	Chamæleontis	10.0	13 9 39.62	+ 5.258	79 37 30.8	− 19.14	51.24	1
9237	Gould, Z. C., 13h 717 . .	8.0	45.42	4.438	73 39 3.5	.13	51.25	1
9238	Muscæ	9.5	55.27	3.985	66 15 50.8	.13	52.38	1
9239	Gould, Z. C., 13h 724 . .	8.0	55.34	4.344	72 29 0.3	.13	51.28	1
9240	Gould, Z. C., 13h 738 . .	9.3	56.23	+ 4.978	78 5 48.6	− 19.13	51.08	3
9241	Gould, Z. C., 13h 722 . .	9.2	13 9 57.24	+ 4.147	69 29 57.2	− 19.13	51.89	2
9242	Chamæleontis	10.0	10 1.73	5.593	80 56 3.8	.13	51.37	1
9243	Gould, 18116	8.0	5.02	4.483	74 4 50.0	.13	52.28	1
9244	Lacaille, 5470	8.0	7.41	4.115	68 53 23.0	.12	50.50	1
9245	Muscæ	9.0	8.74	+ 4.193	70 14 3.2	− 19.12	52.28	1
9246	Octantis	9.5	13 10 10.80	+ 7.780	85 6 24.6	− 19.12	51.56	5
9247	Muscæ	10.0	31.84	4.526	74 26 14.2	.11	50.41	1
9248	Gould, Z. C., 13h 756 . .	9.2	33.51	4.155	69 29 11.7	.11	51.89	2
9249	Muscæ	10.0	40.00	4.500	74 8 0.9	.11	52.28	1
9250	Gould, Z. C., 13h 760 . .	9.5	10 45.35	+ 3.975	65.47 9.6	− 19.11	52.38	1
9251	Gould, 18131	8.8	13 11 2.92	+ 4.143	69 8 59.4	− 19.10	50.97	2
9252	Gould, Z. C., 13h 781 . .	9.0	2.92	4.252	70 56 21.8	.10	52.36	1
9253	Muscæ	10.0	5.51	4.206	70 12 37.3	.10	52.28	1
9254	Chamæleontis	10.0	7.97	5.516	80 17 47.4	.10	50.90	2
9255	Gould, Z. C., 13h 783 . .	9.0	9.26	+ 4.104	68 24 12.5	− 19.10	50.42	1
9256	Chamæleontis	9.5	13 11 9.46	+ 4.734	76 11 36.4	− 19.10	51.30	1
9257	Brisbane, 4409	7.8	12.53	4.023	66 44 48.0	.10	51.93	2
9258	Chamæleontis	11.0	16.63	5.467	80 18 12.3	.09	50.90	2
9259	Lacaille, 5481	7.8	23.21	4.183	69 45 27.9	.09	52.37	2
9260	Muscæ	10.0	38.20	+ 4.216	70 13 47.2	−. 19.08	52.28	1
9261	Octantis	9.2	13 11 38.35	+ 6.659	83 27 21.4	− 19.08	50.37	2
9262	Lacaille, 5480	6.3	50.88	4.294	71 21 27.4	.08	52.31	3
9263	Gould, 18156	8.5	53.78	4.627	75 8 32.4	.08	50.93	2
9264	Gould, 18154	10.0	11 55.46	4.124	68 34 19.5	.08	50.42	1
9265	Lacaille, 5477; (5473)	8.2	12 2.65	+ 5.045	78 9 49.6	− 19.07	51.08	3

Number.	Constellation, Name of Star, or Synonym.	Magnitude.	Right Ascension, 1850.0.	Annual Precession.	South Declination, 1850.0.	Annual Precession.	Mean year.	No. of obs.
			h. m. s.	s.	° ′ ″	″		
9 266	Gould, Z. C., 13^h 844 . .	9.0	13 12 7.62	+ 4.426	73 0 25.2	− 19.07	50.43	2
9 267	Gould, Z. C., 13^h 840 . .	9.5	8.63	4.104	68 8 22.6	.07	51.46	1
9 268	Muscæ	10.0	10.55	4.060	67 15 58.9	.07	52.29	1
9 269	Chamæleontis	10.0	15.90	5.162	78 46 49.4	.07	51.28	1
9 270	Muscæ	9.5	16.50	4.057	67 10 43.7	19.07	52.29	1
9 271	Chamæleontis	8.5	13 12 25.24	+ 4.911	77 16 2.5	− 19.06	51.36	1
9 272	Muscæ	10.0	50.17	4.631	74 59 42.3	.05	54.46	1
9 273	Lacaille, 5452	7.6	50.71	7.890	85 2 37.7	.05	51.47	7
9 274	Muscæ	10.0	53.62	4.361	72 1 38.6	.05	52.33	1
9 275	Gould, Z. C., 13^h 879 . .	9.0	12 54.83	+ 4.107	68 0 16.6	− 19.05	51.46	1
9 276	Gould, 18180	9.0	13 13 0.67	+ 4.198	69 36 4.5	− 19.05	52.34	1
9 277	Octantis	10.0	17.50	6.621	83 14 47.8	.04	50.39	1
9 278	Muscæ	10.0	17.91	4.389	72 17 46.7	.04	51.28	1
9 279	Gould, Z. C., 13^h 910 . .	8.0	18.11	4.227	69 59 46.8	.04	51.44	2
9 280	ι^{1} Muscæ	5.8	25.44	+ 4.550	74 5 46.8	− 19.04	52.38	3
9 281	Chamæleontis	10.0	13 13 27.41	+ 5.154	78 33 42.7	− 19.03	51.35	1
9 282	Lacaille, 5444	8.1	28.61	9.025	85 56 53.6	.03	51.53	6
9 283	Gould, 18197	10 0	33.85	4.841	76 35 5.8	.03	51.34	1
9 284	Octantis	9.5	34.64	6.497	82 58 35.4	.03	51.43	2
9 285	Muscæ	9.5	41.01	+ 4.340	71 33 59.2	− 19.03	52.32	1
9 286	Gould, Z. C., 13^h 929 . .	10.0	13 13 43.32	+ 4.024	66 3 45.3	− 19.03	52.45	1
9 287	Chamæleontis	9.5	44.89	4.713	75 32 58.7	.03	51.38	1
9 288	Muscæ	9.2	46.31	4.247	70 12 25.0	.03	51.86	2
9 289	Brisbane, 4427	9.0	53.81	4.048	66 31 40.8	.02	52.41	1
9 290	Chamæleontis	9.2	54.94	+ 5.937	81 35 9.0	− 19.02	50.86	2
9 291	Gould, 18200	10.0	13 13 55.49	+ 4.047	66 29 51.7	− 19.02	52.41	1
9 292	Lacaille, 5497	7.0	13 58.37	4.215	69 38 21.9	.02	52.34	1
9 293	Muscæ	10.0	14 0.87	4.645	74 54 20.5	.02	51.46	1
9 294	Chamæleontis	10.0	5.62	4.826	76 23 6.9	.02	51.34	1
9 295	Muscæ	9.5	9.08	+ 4.406	72 19 43.9	− 19.02	51.28	1
9 296	Melbourne (1), 670 . . .	9.2	13 14 9.94	+ 4.121	67 55 58.8	− 19.02	51.88	2
9 297	Octantis	8.9	12.82	8.972	85 52 32.0	.01	51.53	6
9 298	Chamæleontis	9.5	23.29	4.832	76 22 53.4	.01	51.34	1
9 299	Gould, Z. C., 13^h 984 . .	9.0	29.54	4.517	73 32 31.2	.01	51.25	1
9 300	Octantis	9.6	29.58	+15.399	88 0 58.1	− 19.01	51.63	4
9 301	Octantis	9.5	13 14 29.74	+ 7.829	84 52 35.0	19.01	51.22	2
9 302	Chamæleontis	10.5	30.96	6.202	82 13 43.2	.01	50.94	2
9 303	Muscæ	10.0	33.45	4.054	66 29 42.2	.00	52.41	1
9 304	Gould, Z. C., 13^h 977 . .	9.5	37.78	4.057	66 31 47.7	.00	52.41	1
9 305	Muscæ	9.5	43.59	+ 4.270	70 19 50.3	− 19.00	52.28	1

Number.	Constellation, Name of Star, or Synonym.	Magnitude.	Right Ascension, 1850.0.	Annual Precession.	South Declination, 1850.0.	Annual Precession.	Mean year.	No. of obs.
			h. m. s.	s.	° ′ ″	″		
9306	Gould, Z. C., 13h 1031	8.5	13 14 49.40	+ 5.379	79 28 27.9	− 19.00	50.83	2
9307	Gould, 18231	9.0	49.85	4.870	76 35 7.3	.00	51.34	1
9308	Octantis	9.5	52.61	15.335	87 59 40.8	19.00	51.63	4
9309	Gould, Z. C., 13h 1003	9.0	55.47	4.170	68 39 12.3	18.99	50.50	1
9310	Gould, Z. C., 13h 1006	9.0	57.51	+ 4.103	67 23 33.9	− 18.99	52.30	2
9311	Muscæ	10.0	13 14 58.15	+ 4.533	73 36 57.5	− 18.99	51.25	1
9312	Lacaille, 5506	5.7	58.60	4.242	69 50 34.4	.99	51.78	3
9313	Lacaille, 5508	8.0	14 58.94	4.122	67 44 34.5	.99	52.30	1
9314	Gould, Z. C., 13h 1019	10.0	15 4.77	4.146	68 10 7.0	.99	51.46	1
9315	Gould, Z. C., 13h 1027	9.0	7.31	+ 4.516	73 24 17.1	− 18.99	50.44	1
9316	Gould, Z. C., 13h 1040	9.0	13 15 24.60	+ 4.371	71 35 45.2	− 18.98	52.32	1
9317	ι¹ Muscæ	6.8	30.85	4.571	73 54 29.2	.98	52.09	4
9318	Gould, Z. C., 13h 1050	9.0	35.92	4.246	69 45 17.8	.98	52.34	1
9319	Chamæleontis	10.0	15 52.85	5.177	78 20 7.1	97	51.35	1
9320	Gould, Z. C., 13h 1067	9.5	16 0.19	+ 4.040	65 47 23.0	− 18.96	52.45	1
9321	Chamæleontis	10.0	13 16 5.95	+ 4.751	75 26 48.8	18.96	51.38	1
9322	Chamæleontis	9.0	16.01	5.762	80 46 42.0	.96	51.37	1
9323	Muscæ	9.0	19.67	4.367	71 20 39.1	.95	52.35	1
9324	Muscæ	9.5	28.26	4.294	70 17 17.6	.95	51.86	2
9325	Muscæ	10.0	30.61	+ 4.568	73 40 52.1	− 18.95	51.25	1
9326	Gould, Z. C., 13h 1111	9.5	13 16 34.37	+ 4.346	71 0 33.2	− 18.95	52.36	1
9327	Gould, Z. C., 13h 1121	8.5	41.24	4.444	72 15 33.2	.94	51.80	2
9328	Gould, 18261	8.2	46.12	4.070	66 14 21.0	.94	52.43	2
9329	Gould, Z. C., 13h 1141	8.8	16 51.47	5.155	78 4 27.9	.94	51.08	3
9330	Gould, 18263	9.0	17 3.04	+ 4.061	65 57 52.6	− 18.93	52.45	1
9331	Octantis	10.0	13 17 7.26	+ 6.738	83 8 2.4	− 18.93	50.39	1
9332	Chamæleontis	9.0	9.64	4.768	75 24 22.6	.93	50.40	1
9333	Chamæleontis	10.0	20.38	6.033	81 29 49.8	.93	50.86	2
9334	Octantis	8.6	20.62	10.470	86 34 32.0	.93	51.63	6
9335	Gould, Z. C., 13h 1146	9.0	22.93	+ 4.132	67 19 13.5	− 18.92	52.29	1
9336	Lacaille, 5518	8.5	13 17 29.64	+ 4.824	75 47 37.4	− 18.92	51.38	1
9337	Chamæleontis	9.5	35.60	4.825	75 47 15.9	.92	51.38	1
9338	κ Octantis	7.1	38.78	8.173	85 0 43.8	.92	51.53	8
9339	Gould, Z. C., 13h 1181	9.0	44.24	4.636	74 7 30.0	.91	52.34	2
9340	Chamæleontis	10.0	45.73	+ 5.043	77 16 42.5	− 18.91	51.36	1
9341	Chamæleontis	9.0	13 17 48.41	+ 4.826	75 45 21.4	− 18.91	51.38	1
9342	Lacaille, 5528	7.8	52.28	4.126	67 5 16.4	.91	51.87	2
9343	Lacaille, 5529	7.5	58.42	4.225	68 50 41.6	.91	50.50	1
9344	Gould, Z. C., 13h 1188	9.2	17 59.67	4.345	70 41 17.3	.91	52.32	2
9345	Gould, Z. C., 13h 1190	9.0	18 1.16	+ 4.403	71 27 24.1	− 18.91	52.23	2

Number.	Constellation, Name of Star, or Synonym.	Magnitude.	Right Ascension, 1850.0.	Annual Precession.	South Declination, 1850.0.	Annual Precession.	Mean year.	No. of obs.
			h. m. s.	s.	° ′ ″	″		
9 346	Gould, Z. C., 13ʰ 1187	9.5	13 18 3.45	+ 4.160	67 40 59.3	− 18.90	52.30	1
9 347	Lacaille, 5516	6.5	6.65	5.346	78 52 52.0	.90	51.28	1
9 348	Gould, Z. C., 13ʰ 1215	9.2	7.42	5.206	78 10 4.1	.90	51.39	2
9 349	Muscæ	10.0	24.72	4.728	74 50 46.7	.89	51.46	1
9 350	Gould, Z. C., 13ʰ 1205	9.2	29.83	+ 4.101	66 11 13.6	− 18.89	52.43	2
9 351	Muscæ	9.5	13 18 35.60	+ 4.346	70 33 56.6	− 18.89	52.28	1
9 352	Gould, Z. C., 13ʰ 1244	10.0	51.59	4.547	73 0 0.6	.88	50.42	1
9 353	Lacaille, 5520	7.8	18 53.96	5.250	78 17 51.0	.88	51.84	2
9 354	Gould, Z. C., 13ʰ 1281	9.5	19 7.30	5.398	78 59 34.2	.87	50.85	2
9 355	Chamæleontis	9.5	11.76	+ 5.232	78 9 40.9	− 18.87	51.39	2
9 356	Muscæ	9.0	13 19 22.15	+ 4.667	74 1 16.5	− 18.87	52.34	2
9 357	Muscæ	9.5	28.70	4.363	70 36 13.1	.86	52.28	1
9 358	Gould, Z. C., 13ʰ 1278	10.0	31.21	4.272	69 15 15.3	.86	51.45	1
9 359	Gould, Z. C., 13ʰ 1282	10.0	36.15	4.272	69 14 5.8	.86	51.45	1
9 360	Gould, 18333	9.5	36.16	+ 5.311	78 30 44.7	− 18.86	51.35	1
9 361	Gould, Z. C., 13ʰ 1288	9.5	13 19 45.01	+ 4.084	66 14 43.0	− 18.85	52.41	1
9 362	Gould, Z. C., 13ʰ 1292	9.5	49.03	4.250	68 50 13.1	.85	50.48	2
9 363	Muscæ	9.5	19 57.41	4.365	70 32 8.1	.85	52.28	1
9 364	Gould, 18334	8.5	20 0.96	4.333	70 3 52.2	.85	51.44	1
9 365	Chamæleontis	9.5	5.35	+ 5.079	77 8 58.7	− 18.84	51.25	1
9 366	Chamæleontis	9.0	13 20 16.61	+ 5.793	80 25 40.7	− 18.84	50.43	1
9 367	Gould, 18346	9.0	16.88	4.577	73 2 44.6	.84	50.43	2
9 368	Chamæleontis	10.0	19.87	4.994	76 34 0.4	.84	51.34	1
9 369	Gould, Z. C., 13ʰ 1337	8.5	23.10	4.657	73 50 0.5	.84	51.77	2
9 370	Lacaille, 5541	6.5	23.56	+ 5.029	76 47 16.4	− 18.84	51.30	2
9 371	Gould, 18353	9.5	13 20 28.64	+ 5.031	76 47 32.6	− 18.83	51.30	2
9 372	Gould, Z. C., 13ʰ 1342	9.0	36.42	4.232	68 20 46.8	.83	50.94	2
9 373	Gould, 18373	10.0	40.98	7.236	83 40 57.8	.83	50.35	1
9 374	Muscæ	10.0	47.55	4.217	68 3 4.7	.82	51.46	1
9 375	Lacaille, 5519	7.9	48.72	+ 6.908	83 8 12.7	− 18.82	50.96	2
9 376	Chamæleontis	9.2	13 20 50.38	+ 5.086	77 5 7.6	− 18.82	51.31	2
9 377	Gould, Z. C., 13ʰ 1368	9.0	20 53.86	4.727	74 23 42.6	.82	50.41	1
9 378	Muscæ	10.0	21 1.70	4.400	70 47 7.1	.82	52.36	1
9 379	Chamæleontis	9.5	2.91	5.089	77 4 25.8	.82	51.25	1
9 380	Gould, Z. C., 13ʰ 1362	9.5	4.29	+ 4.102	65 48 0.9	− 18.81	52.45	1
9 381	Muscæ	9.5	13 21 7.58	+ 4.381	70 30 22.6	− 18.81	52.28	1
9 382	Gould, 18389	9.0	7.86	5.691	79 57 52.8	.81	51.30	2
9 383	Gould, Z. C., 13ʰ 1373	9.2	11.39	4.146	66 38 41.8	.81	51.93	2
9 384	Gould, Z. C., 13ʰ 1401	9.5	14.11	4.728	74 21 1.0	.81	50.41	1
9 385	Gould, Z. C., 13ʰ 1397	9.0	25.27	+ 4.157	66 47 39.8	− 18.80	51.93	2

No. 9 360. Gould's declination is 1′ greater.

Number.	Constellation, Name of Star, or Synonym.	Magnitude.	Right Ascension, 1850.0.	Annual Precession.	South Declination, 1850.0.	Annual Precession.	Mean year.	No. of obs.
			h. m. s.	s.	° ′ ″	″		
9386	Gould, Z. C., 13h 1407 .	9.0	13 21 29.23	+ 4.418	70 55 38.4	− 18.80	52.36	1
9387	Gould, 18367	9.0	38.54	4.249	68 23 10.6	.80	50.42	1
9388	Chamæleontis	9.5	38.94	5.272	78 2 6.0	.80	51.39	2
9389	Brisbane, 4490	8.0	39.58	4.253	68 27 18.6	.80	50.42	1
9390	Gould, Z. C., 13h 1417 .	9.5	46.33	+ 4.162	66 48 12.8	− 18.79	51.95	2
9391	Octantis	10.0	13 21 48.96	+ 8.404	84 59 31.2	− 18.79	51.93	1
9392	Octantis	9.4	49.46	9.538	85 51 58.6	.79	51.59	7
9393	Chamæleontis	10.0	53.01	5.764	80 8 52.5	.79	51.37	1
9394	Chamæleontis	10.0	54.22	5.518	79 10 30.8	.79	50.42	1
9395	Gould, Z. C., 13h 1429 .	8.0	57.08	+ 4.334	69 39 30.1	− 18.79	52.34	1
9396	Muscæ	10.0	13 21 57.36	+ 4.334	69 39 5.7	− 18.79	52.34	1
9397	Gould, Z. C., 13h 1449 .	7.2	22 4.78	4.684	73 47 21.0	.78	51.77	2
9398	Muscæ	9.5	11.93	4.322	69 24 51.7	.78	52.34	1
9399	Muscæ	10.0	17.58	4.392	70 24 2.5	.78	52.28	1
9400	Muscæ	9.0	19.74	+ 4.445	71 5 39.4	− 18.78	52.35	2
9401	Gould, Z. C., 13h 1472 .	10.0	13 22 30.96	+ 4.803	74 46 33.6	− 18.77	51.46	1
9402	Chamæleontis	9.5	48.46	6.103	81 8 10.3	.76	50.88	2
9403	Chamæleontis	10.0	22 52.32	5.769	80 3 9.5	.76	51.37	1
9404	Muscæ	9.5	23 0.64	6.727	74 1 5.7	.76	52.28	1
9405	Gould, Z. C., 13h 1487 .	9.5	1.41	+ 4.227	67 44 38.9	− 18.75	52.30	1
9406	Chamæleontis	8.3	13 23 1.62	+ 4.434	81 58 36.4	− 18.75	51.12	3
9407	Gould, Z. C., 13h 1501 .	9.2	2.87	4.706	73 49 31.2	.75	51.77	2
9408	Chamæleontis	10.0	10.32	5.924	80 33 0.2	.75	50.43	1
9409	Brisbane, 4505	9.0	10.58	4.136	65 57 30.0	.75	52.45	1
9410	Gould, Z. C., 13h 1508 .	9.2	11.44	+ 4.713	73 51 49.8	− 18.75	51.76	2
9411	Octantis	9.5	13 23 12.54	+ 6.970	83 3 21.0	− 18.75	50.96	2
9412	Octantis	10.0	23.61	6.742	82 36 44.5	.74	50.39	1
9413	Muscæ	9.5	46.54	4.423	70 31 4.8	.73	52.28	1
9414	Octantis	10.0	23 58.17	6.736	82 33 8.0	.73	50.39	1
9415	Octantis	9.9	24 2.43	+ 9.738	85 53 12.9	− 18.72	51.59	7
9416	Muscæ	10.0	13 24 6.47	+ 4.695	73 32 3.1	− 18.72	51.25	1
9417	Muscæ	10.0	8.26	4.532	71 48 16.3	.72	52.32	1
9418	Lacaille, 5568	7.5	9.02	4.599	72 33 19.5	.72	51.28	1
9419	Lacaille, 5565	8.0	9.40	4.960	75 44 20.8	.72	51.38	1
9420	Muscæ	9.5	12.82	+ 4.522	71 40 30.0	− 18.72	52.32	1
9421	Gould, 18452	10.0	13 24 18.34	+ 5.613	79 17 11.4	− 18.71	50.42	1
9422	Muscæ	9.5	21.80	4.180	66 31 22.2	.71	52.41	1
9423	Stone, 7461	8.8	21.97	11.848	86 51 43.2	.71	51.45	5
9424	Chamæleontis	10.0	26.34	5.863	80 12 38.9	.71	51.37	1
9425	Gould, Z. C., 13h 1572 .	10.0	27.67	+ 4.279	68 14 38.5	− 18.71	51.46	1

Number.	Constellation, Name of Star, or Synonym.	Magnitude.	Right Ascension, 1850.0.	Annual Precession.	South Declination, 1850.0.	Annual Precession.	Mean year.	No. of obs.
			h. m. s.	s.	° ′ ″	″		
9426	Gould, Z. C., 13h 1588	10.0	13 24 29.18	+ 4.842	74 46 33.6	− 18.71	51.46	1
9427	Muscæ	10.0	33.03	4.431	70 28 13.6	.71	52.34	2
9428	Gould, Z. C., 13h 1607	9.0	24 54.91	4.503	71 18 46.6	.70	52.35	1
9429	Gould, Z. C., 13h 1610	9.5	25 0.66	4.321	68 47 36.1	.69	50.48	2
9430	Gould, Z. C., 13h 1612	9.8	5.51	+ 4.323	68 48 16.0	− 18.69	50.48	2
9431	Gould, Z. C., 13h 1615 .	9.5	13 25 5.70	+ 4.369	69 29 57.9	− 18.69	51.78	2
9432	Chamæleontis	11.0	6.00	5.824	80 0 3.5	.69	51.37	1
9433	Gould, Z. C., 13h 1628 .	10.0	17.95	4.297	68 21 32.2	.68	51.46	1
9434	Gould, 18467	8.5	30.13	4.536	71 35 26.5	.68	52.32	2
9435	Gould, Z. C., 13h 1672 .	9.2	34.08	+ 5.596	79 3 54.0	− 18.67	50.85	2
9436	Chamæleontis	9.5	13 25 46.11	+ 5.186	76 58 58.4	− 18.67	51.25	1
9437	Chamæleontis	9.5	48.24	4.923	75 12 18.4	.67	50.40	1
9438	Gould, Z. C., 13h 1673 .	9.5	48.49	4.961	75 29 21.4	.67	51.38	1
9439	Chamæleontis	10.0	48.62	5.144	76 43 11.6	.67	51.30	2
9440	Chamæleontis	9.0	52.69	+ 4.923	75 11 22.0	− 18.66	50.40	1
9441	Gould, Z. C., 13h 1681 .	9.0	13 25 53.14	+ 4.961	75 28 23.4	− 18.66	51.38	1
9442	Chamæleontis	10.0	26 1.70	5.400	78 6 41.4	.66	51.40	2
9443	Gould, Z. C., 13h 1679 .	9.0	2.83	4.584	72 2 20.7	.66	52.33	1
9444	Herschel, 4594[1]	10.0	16.59	5.807	79 48 39.7	.65	51.24	1
9445	Herschel, 4594[2]	9.5	19.24	+ 5.808	79 48 38.7	− 18.65	51.24	1
9446	Gould, Z. C., 13h 1703 .	10.0	13 26 27.20	+ 4.229	66 56 34.0	− 18.65	51.45	1
9447	Chamæleontis	9.0	29.04	5.063	76 6 19.9	.65	51.30	1
9448	Lacaille, 5577	8.0	30.21	4.900	74 54 58.9	.64	51.46	1
9449	Gould, Z. C., 13h 1729	9.2	34.74	4.819	74 14 19.4	.64	51.34	2
9450	Gould, Z. C., 13h 1715 .	9.2	41.09	+ 4.307	68 12 18.8	− 18.64	50.94	2
9451	Gould, Z. C., 13h 1722 .	8.5	13 26 41.57	+ 4.403	69 38 34.8	− 18.64	52.34	1
9452	Chamæleontis	9.5	44.44	5.204	76 57 14.7	.64	51.25	1
9453	Gould, Z. C., 13h 1730 .	9.8	47.94	4.308	68 12 17.8	.64	50.94	2
9454	Gould, Z. C., 13h 1737 .	9.0	48.16	4.606	72 8 13.9	.63	52.33	1
9455	Gould, Z. C., 13h 1739	10.0	26 58.97	+ 4.236	66 56 31.2	− 18.63	51.45	1
9456	Chamæleontis	9.8	13 27 2.14	+ 5.158	76 38 3.4	− 18.63	51.30	2
9457	Muscæ	10.0	26.92	4.890	74 40 24.0	.61	51.46	1
9458	Lacaille, 5587	7.0	28.35	4.416	69 40 31.4	.61	52.34	1
9459	Gould, Z. C., 13h 1764 .	9.5	29.50	4.398	69 25 6.4	.61	51.79	2
9460	Muscæ	9.5	29.64	+ 4.338	68 31 36.5	− 18.61	50.42	1
9461	Gould, Z. C., 13h 1766	10.0	13 27 34.39	+ 4.395	69 20 51.5	− 18.61	51.45	1
9462	Gould, Z. C., 13h 1768	9.0	44.00	4.290	67 42 42.3	.60	52.30	1
9463	Gould, 18503	8.0	44.38	4.585	71 44 13.4	.60	52.32	1
9464	Octantis	8.5	48.81	11.257	86 30 20.0	.60	51.64	8
9465	Gould, Z. C., 13h 1771	9.2	50.56	+ 4.255	67 5 54.0	− 18.60	51.87	2

No. 9435. Gould's declination is 20.8″ greater.
No. 9452. Record ambiguous; 25m and 26m recorded; 26m probably correct.

Number.	Constellation, Name of Star, or Synonym.	Magnitude.	Right Ascension, 1850.0.	Annual Precession.	South Declination, 1850.0.	Annual Precession.	Mean year.	No. of obs.
			h. m. s.	s.	° ′ ″	″		
9466	Chamæleontis	10.0	13 27 54.95	+ 6.687	82 7 5.5	— 18.60	51.49	1
9467	Gould, Z. C., 13h 1798	9.2	55.34	4.843	74 13 6.3	.60	51.34	2
9468	Muscæ	9.5	59.07	4.208	66 13 18.0	.60	52.41	1
9469	Gould, Z. C., 13h 1800	9.5	27 59.17	4.742	73 18 43.4	.60	50.44	1
9470	Octantis	8.7	28 5.91	+ 8.840	85 1 54.6	— 18.59	51.41	5
9471	Muscæ	10.0	13 28 10.20	+ 4.211	66 13 37.0	— 18.59	52.41	1
9472	Gould, Z. C., 13h 1827	8.2	10.86	5.203	76 44 52.3	.59	51.30	2
9473	Octantis	8.7	15.59	9.056	85 12 8.5	.59	51.78	6
9474	Lacaille, 5594	8.0	21.41	4.250	66 53 44.4	.58	51.45	1
9475	Muscæ	10.0	23.46	+ 4.483	70 22 53.8	— 18.58	52.40	1
9476	κ Muscæ	7.0	13 28 34.96	+ 4.458	70 1 18.4	— 18.58	51.44	1
9477	Muscæ	9.0	35.92	4.831	74 0 43.6	.58	52.28	1
9478	Gould, 18538	10.0	41.70	5.694	79 6 21.9	.57	50.42	1
9479	Gould, Z. C., 13h 1854	10.0	28 58.15	4.200	65 50 23.9	.56	52.45	1
9480	Muscæ	10.0	29 1.61	+ 4.653	72 14 39.1	— 18.56	51.28	1
9481	Lacaille, 5592	7.8	13 29 2.14	+ 4.880	74 21 4.5	— 18.56	51.34	2
9482	Muscæ	9.2	2.79	4.662	72 19 46.5	.56	51.80	2
9483	Gould, Z. C., 13h 1867	9.0	3.18	4.545	71 1 27.4	.56	52.35	2
9484	Chamæleontis	9.5	3.36	6.275	81 0 50.2	.56	50.88	2
9485	Gould, Z. C., 13h 1863	10.0	7.64	+ 4.279	67 13 20.7	— 18.56	52.29	1
9486	Gould, Z. C., 13h 1866	9.0	13 29 9.15	+ 4.278	67 11 28.7	— 18.56	52.29	1
9487	Muscæ	9.5	16.23	4.629	71 56 35.0	.55	52.32	1
9488	Muscæ	10.0	18.84	4.330	68 0 35.5	.55	51.46	1
9489	Chamæleontis	9.5	23.90	5.291	77 5 52.5	.55	51.37	2
9490	Chamæleontis	8.5	28.22	+ 5.117	76 1 41.1	— 18.55	51.30	1
9491	Muscæ	9.0	13 29 29.71	+ 4.224	66 9 47.2	— 18.55	52.41	1
9492	Gould, 18552	8.5	30.48	4.848	74 0 43.2	.55	52.28	1
9493	Lacaille, 5597	7.2	31.50	4.559	71 6 20.5	.55	52.35	2
9494	Muscæ	10.0	32.59	4.414	69 13 35.7	.54	51.45	1
9495	Gould, Z. C., 13h 1911	9.2	29 50.88	+ 4.559	71 2 45.6	18.53	52.35	2
9496	Muscæ	10.0	13 30 1.82	+ 4.854	73 58 43.7	— 18.53	52.28	1
9497	Chamæleontis	10.0	2.20	5.033	75 22 26.7	.53	50.40	1
9498	Muscæ	10.0	3.11	4.506	70 21 32.6	.53	52.40	1
9499	Gould, 18557	9.0	6.95	4.214	65 51 6.9	.53	52.45	1
9500	Gould, Z. C., 13h 1966	8.8	12.86	+ 5.712	79 0 19.6	— 18.52	50.85	2
9501	Gould, Z. C., 13h 1946	9.0	13 30 27.63	+ 4.410	68 59 7.3	— 18.51	50.45	1
9502	Gould, Z. C., 13h 1942	8.8	27.82	4.287	67 7 6.0	.51	51.87	2
9503	Gould, Z. C., 13h 1960	9.0	42.93	4.306	67 19 44.1	.51	52.29	1
9504	Chamæleontis	9.2	45.10	5.637	78 38 4.0	.50	51.32	2
9505	Octantis	9.8	57.29	+10.274	85 53 43.9	— 18.50	51.73	3

Number.	Constellation, Name of Star, or Synonym.	Magnitude.	Right Ascension, 1850.0.	Annual Precession.	South Declination, 1850.0.	Annual Precession.	Mean year.	No. of obs.
			h. m. s.	s.	° ′ ″	″		
9 506	Muscæ	10.0	13 30 58.81	+ 4.826	73 34 8.4	− 18.50	51.25	1
9 507	Chamæleontis	10.0	31 3.78	5.893	79 36 23.4	.49	51.24	1
9 508	Gould, Z. C., 13h 1987 .	8.0	5.50	6.468	69 39 15.4	.49	52.34	1
9 509	Octantis	10.0	11.05	8.432	84 28 58.2	.49	51.49	1
9 510	Gould, Z. C., 13h 1988 .	9.0	14.51	+ 4.387	68 29 6.0	− 18.49	50.42	1
9 511	Gould, Z. C., 13h 1992 .	9.0	13 31 16.42	+ 4.494	69 57 53.7	− 18.49	51.79	2
9 512	Gould, Z. C., 13h 2010 .	9.5	17.96	5.325	77 2 8.3	.49	51.37	2
9 513	Chamæleontis	10.0	22.81	5.496	77 54 56.3	.48	50.40	1
9 514	Gould, Z. C., 13h 2004 .	9.0	28.99	4.609	71 19 12.7	.48	52.35	1
9 515	Gould, Z. C., 13h 2009 .	8.2	33.32	+ 4.695	72 13 44.6	− 18.48	51.80	2
9 516	Gould, Z. C., 13h 2035 .	9.8	13 31 45.82	+ 4.707	72 19 36.3	− 18.47	51.80	2
9 517	Muscæ	9.2	49.29	4.281	66 39 55.2	.47	51.93	2
9 518	Muscæ	10.0	53.32	4.872	73 49 55.0	.47	51.25	1
9 519	Gould, Z. C., 13h 2031 .	10.0	55.85	4.473	69 34 6.2	.46	52.34	1
9 520	Gould, Z. C., 13h 2028 .	9.0	31 55.87	+ 4.388	68 21 59.9	− 18.46	50.42	1
9 521	Gould, Z. C., 13h 2053 .	9.2	13 32 1.06	+ 5.218	76 18 55.8	− 18.46	51.32	2
9 522	Gould, Z. C., 13h 2032 .	9.5	2.29	4.236	65 49 2.7	.46	52.45	1
9 523	Lacaille, 5619	8.5	3.99	4.772	72 52 30.0	.46	50.42	1
9 524	Muscæ	9.2	10.92	4.288	66 42 23.0	.46	51.93	2
9 525	Muscæ	9.5	19.78	+ 4.286	66 38 26.6	− 18.45	51.93	2
9 526	Gould, Z. C., 13h 2055 .	10.0	13 32 21.84	+ 4.474	69 30 0.1	− 18.45	51.45	1
9 527	Octantis	10.0	32 38.06	7.124	82 36 22.2	.44	50.39	1
9 528	Gould, Z. C., 13h 2107 .	9.5	33 11.12	4.591	70 48 25.4	.42	52.36	1
9 529	Octantis	9.8	15.88	9.966	85 36 36.3	.42	51.78	4
9 530	Gould, 18625	7.2	18.85	+ 5.323	76 45 58.2	− 18.42	51.30	2
9 531	Gould, Z. C., 13h 2124 .	9.0	13 33 20.10	+ 4.667	71 37 56.4	− 18.42	52.32	1
9 532	Lacaille, 5630	7.8	22.40	4.591	70 46 27.4	.41	52.38	2
9 533	Chamæleontis	9.2	38.24	6.437	81 1 44.1	.41	50.88	2
9 534	Gould, Z. C., 13h 2144 .	9.5	44.35	4.269	66 3 10.7	.40	52.45	1
9 535	Gould, Z. C., 13h 2148 .	10.0	44.82	+ 4.359	67 42 46.8	− 18.40	52.30	1
9 536	Chamæleontis	9.8	13 33 46.92	+ 5.634	78 16 8.1	− 18.40	51.40	2
9 537	Gould, Z. C., 13h 2160 .	8.5	52.64	4.638	71 13 37.8	.40	52.35	1
9 538	Muscæ	9.5	54.62	4.639	71 13 36.4	.40	52.35	1
9 539	Muscæ	10.0	33 57.88	4.342	67 14 34.2	.39	52.29	1
9 540	Chamæleontis	9.8	34 1.18	+ 5.649	78 18 34.0	− 18.39	51.40	2
9 541	Gould, Z. C., 13h 2183 .	9.0	13 34 1.54	+ 5.414	77 10 21.3	− 18.39	51.37	2
9 542	Gould, Z. C., 13h 2176 .	9.5	5.12	4.984	74 24 17.1	.39	50.41	1
9 543	Chamæleontis	10.0	10.48	5.146	75 33 45.3	.39	51.38	1
9 544	Gould, Z. C., 13h 2200 .	9.0	11.18	5.367	76 54 0.1	.39	51.25	1
9 545	Octantis	9.5	12.35	+ 7.304	82 48 17.0	− 18.39	50.95	2

No. 9 516. Gould's right ascension is 10s greater.

Number.	Constellation, Name of Star, or Synonym.	Magnitude.	Right Ascension, 1850.0.	Annual Precession.	South Declination, 1850.0.	Annual Precession.	Mean year.	No. of obs.
			h. m. s.	s.	° ′ ″	″		
9546	Muscæ	10.0	13 34 13.94	+ 4.934	73 59 27.5	— 18.38	52.28	1
9547	Muscæ	9.0	22.46	4.542	69 59 59.1	.38	51.44	1
9548	Lacaille, 5642	9.0	24.13	4.274	65 59 49.4	.38	52.45	1
9549	Muscæ	10.0	27.58	4.941	74 0 34.6	.38	52.28	1
9550	Gould, Z. C., 13^h 2209 .	9.0	42.98	+ 4.384	67 44 3.2	— 18.37	52.30	1
9551	Muscæ	9.5	13 34 46.95	+ 4.981	74 16 39.4	— 18.37	52.28	1
9552	Muscæ	9.5	34 48.83	4.579	70 22 4.5	.36	52.40	1
9953	Muscæ	10.0	35 10.01	4.363	67 19 45.0	.35	52.29	1
9954	Gould, Z. C., 13^h 2248 .	9.5	11.78	5.091	75 2 [illegible]	.35	50.93	2
9555	Octantis	10.0	12.02	+ 7.668	83 18 17.3	— 18.35	50.39	1
9556	Muscæ	9.2	13 35 14.40	+ 4.367	67 22 55.0	— 18.35	52.30	2
9557	Muscæ	10.0	23.52	4.892	73 27 36.8	.34	51.25	1
9558	Muscæ	10.5	26.68	5.076	74 53 41.4	.34	51.46	1
9559	Gould, Z. C., 13^h 2249 .	9.5	27.66	4.509	69 22 45.8	.34	51.79	2
9560	Muscæ	10.0	36.43	+ 5.079	74 53 41.4	— 18.34	51.46	1
9561	Gould, Z. C., 13^h 2293 .	9.8	13 35 38.59	+ 5.445	77 8 8.0	— 18.34	51.37	2
9562	Gould, Z. C., 13^h 2262 .	9.5	41.26	4.458	68 38 54.4	.33	50.45	1
9563	Gould, Z. C., 13^h 2260 .	9.5	41.51	4.408	67 55 20.6	.33	51.88	2
9564	Lacaille, 5651	7.0	46.29	4.602	70 28 34.8	.33	52.40	1
9565	Chamæleontis	10.0	35 59.88	+ 5.112	75 4 30.9	— 18.32	50.40	1
9566	Gould, Z. C., 13^h 2320 .	10.0	13 36 19.30	+ 5.086	74 50 43.3	— 18.31	51.46	1
9567	Chamæleontis	10.0	22.03	5.180	75 29 8.9	.31	51.38	1
9568	Octantis	9.5	25.03	7.713	83 17 25.4	.31	50.39	1
9569	Gould, Z. C., 13^h 2359 .	9.0	34.85	4.999	74 9 24.7	.30	52.28	1
9570	Octantis	9.9	36.98	+10.448	85 45 28.3	— 18.30	51.67	7
9571	Chamæleontis	9.0	13 36 38.67	+ 7.007	82 4 23.4	— 18.30	50.53	1
9572	Gould, Z. C., 13^h 2377 .	8.8	52.58	5.470	77 6 51.4	.29	51.37	2
9573	Chamæleontis	10.0	56.24	5.420	76 50 31.7	.29	51.25	1
9574	Lacaille, 5633	5.3	56.42	6.937	81 55 2.5	.29	51.07	3
9575	Gould, 18701	8.2	36 58.04	+ 4.390	67 23 32.6	— 18.29	52.30	2
9576	Gould, Z. C., 13^h 2363 .	9.5	13 37 2.42	+ 4.346	66 40 57.4	— 18.29	51.93	2
9577	Octantis	10.0	16.01	8.092	83 44 30.5	.28	50.35	1
9578	Gould, Z. C., 13^h 2379 .	9.5	21.34	4.351	66 42 4.4	.27	51.93	2
9579	Gould, Z. C., 13^h 2390 .	9.0	25.67	4.584	69 57 30.3	.27	51.40	2
9580	Muscæ	9.2	26.86	+ 4.612	70 17 54.0	— 18.27	51.92	2
9581	Gould, Z. C., 13^h 2400 .	8.5	13 37 29.98	+ 4.928	73 26 45.8	— 18.27	50.84	2
9582	Gould, Z. C., 13^h 2429 .	8.2	30.17	6.056	79 31 27.0	.27	51.83	2
9583	Chamæleontis	9.0	41.29	5.313	76 8 44.7	.26	51.30	1
9584	Gould, Z. C., 13^h 2424 .	9.5	46.02	5.114	74 49 52.9	.26	51.46	1
9585	Muscæ	9.5	46.49	+ 4.623	70 22 11.5	— 18.26	52.40	1

Number.	Constellation, Name of Star, or Synonym.	Magnitude.	Right Ascension, 1850.0.	Annual Precession.	South Declination, 1850.0.	Annual Precession.	Mean year.	No. of obs.
			h. m. s.	s.	° ′ ″	″		
9586	Gould, Z. C., 13h 2410	8.8	13 37 51.39	+ 4.583	69 52 52.0	− 18.26	51.79	2
9587	Muscæ	10.0	53.76	4.349	66 33 32.8	.25	52.41	1
9588	Octantis	10.0	37 59.36	8.618	84 17 29.8	.25	51.32	1
9589	Octantis	8.5	38 0.29	8.127	83 44 24.5	.25	50.55	1
9590	Gould, Z. C., 13h 2425	9.5	3.67	+ 4.576	69 45 26.1	− 18.25	52.34	1
9591	Muscæ	10.0	13 38 25.98	+ 4.731	71 27 21.9	− 18.24	52.32	1
9592	Muscæ	9.5	29.42	4.337	66 13 55.3	.23	52.43	2
9593	Gould, 18747	7.8	35.31	5.668	77 52 37.4	.23	51.06	3
9594	Chamæleontis	9.8	40.44	6.507	80 46 24.1	.22	50.90	2
9595	Muscæ	9.5	45.28	+ 4.344	66 17 49.7	− 18.22	52.43	2
9596	Muscæ	9.2	13 38 46.70	+ 5.114	74 42 17.9	− 18.22	50.94	2
9597	Lacaille, 5666	7.2	38 57.83	4.835	72 23 39.5	.22	51.80	2
9598	Muscæ	9.8	39 1.99	4.666	70 38 34.8	.21	52.38	2
9599	Gould, Z. C., 13h 2503	9.2	3.83	5.633	77 39 43.7	.21	50.40	2
9600	Gould, Z. C., 13h 2484	9.5	14.21	+ 4.336	66 4 22.5	− 18.21	52.45	1
9601	Muscæ	9.8	13 39 17.47	+ 5.067	74 17 9.6	− 18.20	51.34	2
9602	Gould, Z. C., 13h 2505	9.5	17.62	5.097	74 30 34.1	.20	50.41	1
9603	Muscæ	9.0	37.91	4.642	70 16 29.8	.19	51.92	2
9604	Gould, Z. C., 13h 2536	9.5	49.90	4.794	71 51 35.8	.18	52.32	2
9605	Apodis	10.0	55.23	+ 5.040	73 59 26.4	− 18.18	52.28	1
9606	Gould, 18775	9.2	13 39 59.74	+ 5.089	74 21 14.0	− 18.18	51.34	2
9607	Gould, Z. C., 13h 2533	10.0	40 0.48	4.357	66 4 27.1	.18	52.45	1
9608	Gould, 18802	9.0	0.61	8.496	84 3 8.6	.18	51.49	1
9609	Gould, Z. C., 13h 2557	9.5	2.94	4.796	71 49 36.0	.18	52.34	2
9610	Gould, Z. C., 13h 2550	9.0	3.98	+ 4.626	70 0 30.7	− 18.17	51.44	1
9611	Gould, Z. C., 13h 2562	9.2	13 40 6.46	+ 4.850	72 21 18.8	− 18.17	51.80	2
9612	Lacaille, 5678	6.5	12.91	4.521	68 39 12.9	.17	50.43	2
9613	Apodis	9.0	21.76	4.771	71 33 5.1	.16	52.33	2
9614	Apodis	9.5	23.96	4.991	73 32 9.7	.16	51.25	1
9615	Lacaille, 5677	7.5	34.06	+ 4.876	72 31 19.0	− 18.16	51.28	1
9616	Apodis	10.0	13 40 36.52	+ 5.011	73 40 2.5	− 18.15	51.25	1
9617	Octantis	10.0	39.10	9.125	84 38 47.9	.15	51.00	2
9618	Apodis	9.5	39.30	5.614	77 23 43.6	.15	50.94	2
9619	Gould, Z. C., 13h 2601	9.5	39.71	4.671	70 25 46.0	.15	52.40	1
9620	Gould, Z. C., 13h 2604	9.5	47.66	+ 4.505	68 20 9.8	− 18.15	50.94	2
9621	Gould, Z. C., 13h 2605	9.0	13 40 51.56	+ 4.415	67 1 19.2	− 18.15	51.87	2
9622	Apodis	10.0	40 52.79	5.841	78 22 33.7	.14	52.35	1
9623	Circini	10.0	41 2.87	4.537	68 43 4.1	.14	50.45	1
9624	Apodis	9.5	3.03	4.764	71 22 8.1	.14	52.35	1
9625	Gould, Z. C., 13h 2621	9.5	7.71	+ 4.501	68 13 26.6	− 18.14	50.94	2

Number.	Constellation, Name of Star, or Synonym.	Magnitude.	Right Ascension, 1850.0.	Annual Precession.	South Declination, 1850.0.	Annual Precession.	Mean year.	No. of obs.
			h. m. s.	s.	° ′ ″	″		
9 626	Gould, Z. C., 13^h 2637	9.5	13 41 12.51	+ 4.971	73 15 19.5	− 18.13	50.44	1
9 627	Lacaille, 5679	8.0	15.08	4.585	69 18 4.3	.13	51.45	1
9 628	Apodis	9.5	24.37	5.563	77 3 29.0	.12	51.25	1
9 629	Circini	10.0	25.74	4.475	67 48 6.1	.12	51.46	1
9 630	Apodis	9.0	42.80	+ 5.296	75 31 33.4	− 18.11	50.89	2
9 631	Apodis	9.5	13 41 43.52	+ 6.032	79 1 23.3	− 18.11	50.42	1
9 632	Apodis	10.0	43.56	5.294	75 30 41.4	.11	51.38	1
9 633	Gould, Z. C., 13^h 2711	9.2	41 58.73	5.715	77 43 5.0	.10	50.40	2
9 634	Apodis	9.5	42 10.26	5.102	74 8 49.1	.10	52.28	1
9 635	Lacaille, 5687	7.5	11.92	+ 4.353	65 46 1.3	− 18.10	52.45	1
9 636	Apodis	9.0	13 42 14.18	+ 5.565	76 58 26.5	− 18.09	51.25	1
9 637	Gould, Z. C., 13^h 2723	9.0	16.08	5.475	76 29 44.3	.09	51.34	1
9 638	Gould, Z. C., 13^h 2702	10.0	17.30	4.516	68 12 47.5	.09	51.46	1
9 639	Gould, Z. C., 13^h 2704	10.0	23.43	4.435	67 2 23.3	.09	51.45	1
9 640	Apodis	9.8	26.99	+ 5.375	75 54 49.8	− 18.09	51.34	2
9 641	Apodis	10.0	13 42 28.98	+ 4.711	70 34 21.0	− 18.08	52.40	1
9 642	Octantis	9.2	31.87	9.826	85 6 19.5	.08	51.29	5
9 643	Lacaille, 5689	8.5	35.80	4.487	67 45 39.4	.08	52.30	1
9 644	Gould, Z. C., 13^h 2740	9.0	44.75	4.854	71 59 25.6	.07	52.33	1
9 645	Apodis	10.0	45.25	+ 6.605	80 40 59.9	− 18.07	51.37	1
9 646	Gould, Z. C., 13^h 2757	9.5	13 42 48.87	+ 5.109	74 6 50.0	− 18.07	52.28	1
9 647	Circini	10.0	53.46	4.398	66 21 52.9	.07	52.41	1
9 648	Gould, Z. C., 13^h 2761	9.5	42 56.31	4.964	72 56 21.0	.07	50.42	1
9 649	Lacaille, 5693	8.2	43 8.63	4.476	67 29 34.6	.06	52.30	2
9 650	Apodis	10.0	14.48	+ 5.433	76 8 53.5	− 18.06	51.30	1
9 651	Lacaille, 5672	8.0	13 43 22.10	+ 7.318	82 11 5.6	− 18.05	50.53	1
9 652	Lacaille, 5698	9.0	25.46	4.391	66 9 28.2	.05	52.45	1
9 653	Apodis	9.0	28.42	4.812	71 27 47.1	.05	52.33	2
9 654	Gould, 18846	9.5	28.46	4.392	66 9 14.8	.05	52.45	1
9 655	Lacaille, 5696	7.0	28.97	+ 4.440	66 54 29.8	− 18.05	51.45	1
9 656	Apodis	10.0	13 43 31.02	+ 5.988	78 41 4.1	− 18.04	51.35	1
9 657	Apodis	8.5	32.65	6.843	81 12 2.8	.04	50.38	1
9 658	Gould, Z. C., 13^h 2800	10.0	35.17	5.216	74 45 56.4	.04	51.46	1
9 659	Circini	9.0	35.54	4.527	68 8 24.1	.04	51.46	1
9 660	Gould, Z. C., 13^h 2802	8.8	40.10	+ 5.098	73 50 34.4	− 18.04	51.76	2
9 661	Gould, Z. C., 13^h 2796	10.0	13 43 45.27	+ 4.692	70 8 56.2	− 18.04	51.44	1
9 662	Gould, Z. C., 13^h 2806	9.0	45.30	5.080	73 45 46.0	.04	51.76	2
9 663	Apodis	9.8	48.54	6.329	79 48 55.6	.03	51.31	2
9 664	Octantis	8.8	43 57.95	18.864	87 52 30.3	.03	51.81	5
9 665	Apodis	9.0	44 1.62	+ 5.436	76 4 33.3	− 18.03	51.30	1

Number.	Constellation, Name of Star, or Synonym.	Magnitude.	Right Ascension, 1850.0.	Annual Precession.	South Declination, 1850.0.	Annual Precession.	Mean year.	No. of obs.
			h. m. s.	s.	° ′ ″	″		
9 666	Gould, Z. C., 13h 2842	8.0	13 44 20.51	+ 4.693	70 3 50.6	− 18.01	51.44	1
9 667	Apodis	9.5	28.41	4.802	71 12 26.1	.01	52.35	1
9 668	Circini	10.0	29.10	4.548	68 15 50.9	.01	51.46	1
9 669	Apodis	8.5	31.36	6.839	81 6 53.1	.01	50.88	2
9 670	Gould, Z. C., 13h 2852	10.0	35.24	+ 4.650	69 31 58.8	− 18.00	51.79	2
9 671	Apodis	9.5	13 44 47.64	+ 4.949	72 32 27.1	− 18.00	51.28	1
9 672	Circini	9.0	47.74	4.425	66 25 50.2	18.00	52.41	1
9 673	Apodis	9.2	50.72	5.553	76 36 40.8	17.99	51.30	2
9 674	Brisbane, 4614	7.4	53.78	28.686	88 40 42.7	.99	51.81	5
9 675	Gould, Z. C., 13h 2906	9.0	44 57.93	+ 5.451	76 2 46.7	− 17.99	51.30	1
9 676	Circini	9.5	13 45 10.05	+ 4.573	68 28 7.5	− 17.98	50.42	1
9 677	Lacaille, 5694	6.8	17.52	5.827	77 51 12.1	.98	51.06	3
9 678	Apodis	10.0	22.09	6.084	78 51 30.1	.97	51.28	1
9 679	Apodis	10.0	23.88	6.452	80 2 30.9	.97	51.37	1
9 680	Gould, Z. C., 13h 2910	10.0	24.23	+ 4.649	69 22 48.2	− 17.97	51.45	1
9 681	Gould, Z. C., 13h 2920	9.0	13 45 25.28	+ 5.001	72 53 59.2	− 17.97	50.42	1
9 682	Apodis	9.8	31.10	5.818	77 47 33.6	.97	51.42	2
9 683	Circini	9.5	37.11	4.430	66 20 57.2	.97	52.41	1
9 684	Gould, Z. C., 13h 2927	9.0	40.14	4.796	70 58 2.3	.96	52.36	1
9 685	Apodis	10.0	41.73	+ 4.815	71 9 8.2	− 17.96	52.35	1
9 686	Gould, Z. C., 13h 2951	10.0	13 45 43.79	+ 5.284	75 16 14.9	− 17.96	50.40	1
9 687	Gould, Z. C., 13h 2925	9.5	46.43	4.560	68 11 29.1	.96	51.46	1
9 688	Gould, Z. C., 13h 2938	9.5	49.85	4.672	69 34 58.7	.96	52.34	1
9 689	Gould, Z. C., 13h 2946	8.0	51.43	4.861	71 35 5.9	.95	52.32	1
9 690	Apodis	10.0	54.46	+ 7.130	81 38 39.5	− 17.95	51.34	1
9 691	Apodis	9.5	13 45 56.76	+ 4.827	71 14 6.4	− 17.95	52.35	1
9 692	Gould, Z. C., 13h 2943	9.2	45 57.70	4.511	67 29 45.1	.95	52.30	2
9 693	Lacaille, 5721	7.0	46 0.69	4.496	67 16 1.7	.95	52.29	1
9 694	Apodis	9.0	7.00	5.761	77 28 50.4	.94	50.94	2
9 695	Circini	9.5	8.07	+ 4.592	68 33 21.4	− 17.94	50.42	1
9 696	Apodis	9.5	13 46 9.04	+ 4.816	71 5 20.2	− 17.94	52.35	2
9 697	Octantis	9.8	9.64	13.281	86 39 4.4	.94	51.78	4
9 698	Apodis	10.0	14.92	7.203	81 45 55.2	.94	51.34	1
9 699	Gould, Z. C., 13h 2969	10.0	22.75	4.557	68 3 0.0	.93	51.46	1
9 700	Gould, Z. C., 13h 2972	10.0	23.79	+ 4.556	68 2 9.8	− 17.93	51.46	1
9 701	Circini	10.0	13 46 28.62	+ 4.505	67 18 43.5	− 17.93	52.29	1
9 702	Apodis	10.0	47 12.10	5.977	78 16 41.0	.90	51.39	2
9 703	Circini	10.0	15.07	4.617	68 40 54.1	.90	50.45	1
9 704	Circini	9.0	18.67	4.625	68 46 12.5	.90	50.45	1
9 705	Gould, Z. C., 13h 3047	9.0	20.69	+ 5.483	75 56 35.0	− 17.90	51.34	2

No. 9 689. Gould's declination is 11.5″ less.

Number.	Constellation, Name of Star, or Synonym.	Magnitude.	Right Ascension, 1850.0.	Annual Precession.	South Declination, 1850.0.	Annual Precession.	Mean year.	No. of obs.
			h. m. s.	s.	° ′ ″	″		
9706	Gould, Z. C., 13h 3067	9.5	13 47 29.60	+ 5.837	77 40 15.6	− 17.89	50.40	1
9707	Apodis	9.5	29.62	5.837	77 40 10.5	.89	50.40	1
9708	Circini	9.0	35.58	4.666	69 13 21.5	.89	51.45	1
9709	Gould, Z. C., 13h 3065 .	9.5	49.44	5.022	72 44 17.6	.88	50.85	2
9710	Lacaille, 5745	8.0	50.66	+ 4.507	67 6 20.0	− 17.88	51.87	2
9711	Apodis	10.0	13 47 59.22	+ 6.089	78 37 29.6	− 17.87	51.32	2
9712	Apodis	10.0	47 59.23	6.561	80 7 54.2	.87	51.37	1
9713	Gould, Z. C., 13h 3075 .	9.0	48 9.33	4.435	65 58 33.6	.86	52.45	1
9714	Brisbane, 4696	8.8	9.34	4.512	67 6 47.2	.86	51.87	2
9715	Octantis	10.0	11.15	+10.519	85 19 59.0	− 17.86	51.93	2
9716	Gould, Z. C., 13h 3090 .	9.0	13 48 12.72	+ 4.933	71 54 23.6	− 17.86	52.34	2
9717	Apodis	10.0	16.98	7.683	82 28 57.7	.86	50.39	1
9718	Circini	9.5	23.25	4.501	66 55 8.6	.85	51.45	1
9719	Circini	10.0	36.07	4.504	66 55 26.5	.85	51.45	1
9720	Gould, Z. C., 13h 3138 .	7.8	40.19	+ 5.722	77 1 36.3	− 17.84	51.37	2
9721	Octantis	9.5	13 48 41.09	+17.877	87 38 18.2	− 17.84	51.67	7
9722	Gould, 18956	7.0	41.75	4.479	66 32 49.3	.84	52.41	1
9723	Gould, 18961	7.5	48.90	4.496	66 46 18.4	.84	51.93	2
9724	Apodis	9.5	50.25	6.323	79 21 0.8	.84	50.42	1
9725	Gould, Z. C., 13h 3127 .	10.0	52.97	+ 4.848	71 0 10.2	− 17.84	52.36	1
9726	Gould, Z. C., 13h 3146 .	9.8	13 48 56.88	+ 5.350	74 57 58.4	− 17.83	51.93	2
9727	Apodis	10.0	57.30	6.504	79 53 24.2	.83	51.37	1
9728	Apodis	10.0	48 57.70	4.975	72 10 37.1	.83	52.33	1
9729	Apodis	10.0	49 6.78	5.577	76 15 13.7	.83	51.30	1
9730	Apodis	10.0	18.67	+ 5.839	77 29 33.0	− 17.82	51.49	1
9731	Circini	10.0	13 49 18.85	+ 4.565	67 39 16.8	17.82	52.30	1
9732	Lacaille, 5691	7.5	22.29	8.747	83 49 26.3	.82	50.80	3
9733	Apodis	10.0	23.45	5.358	74 57 54.4	.81	51.73	2
9734	Gould, Z. C., 13h 3176 .	9.5	38.65	4.975	72 4 48.4	.80	52.33	1
9735	Apodis	9.5	45.12	+ 5.038	72 36 13.3	− 17.80	51.28	1
9736	Apodis	10.0	13 49 48.06	+ 5.333	74 45 16.6	− 17.80	51.73	2
9737	Gould, Z. C., 13h 3191 .	8.0	49 51.14	4.866	71 2 2.2	.80	52.35	2
9738	Gould, Z. C., 13h 3192 .	9.0	50 3.52	4.465	66 4 59.2	.79	52.49	2
9739	Apodis	10.0	9.49	5.280	79 5 39.4	.78	50.42	1
9740	Gould, Z. C., 13h 3202 .	9.0	14.22	+ 4.627	68 18 33.4	− 17.78	50.94	2
9741	Gould, Z. C., 13h 3214 .	9.0	13 50 23.00	+ 4.942	71 40 40.9	− 17.77	52.32	1
9742	Gould, Z. C., 13h 3217 .	9.0	33.78	4.676	68 51 32.2	.77	50.45	1
9743	Lacaille, 5736	7.5	34.95	6.604	80 2 21.0	.77	51.37	1
9744	Gould, 19012	10.5	37.77	6.023	78 7 27.4	.76	50.51	1
9745	Gould, Z. C., 13h 3221 .	9.0	41.00	+ 4.496	66 26 53.7	− 17.76	52.41	1

No 9711. Mean of two observations which differ 10″.

Number.	Constellation, Name of Star, or Synonym.	Magnitude.	Right Ascension, 1850.0.	Annual Precession.	South Declination, 1850.0.	Annual Precession.	Mean year.	No. of obs.
			h. m. s.	s.	° ′ ″	″		
9746	Gould, Z. C., 13h 3236	8.5	13 50 45.70	+ 4.818	70 24 46.5	− 17.76	52.40	1
9747	Circini	10.0	46.10	4.487	66 17 20.4	.76	52.52	1
9748	Gould, Z. C., 13h 3227	9.0	46.30	4.487	66 17 27.1	.76	52.47	2
9749	Apodis	9.5	48.70	4.871	70 56 54.3	.76	52.36	1
9750	Apodis	9.5	49.88	+ 4.891	71 8 13.7	− 17.76	52.35	1
9751	Brisbane, 4714	7.8	13 50 51.12	+ 4.523	66 48 19.7	17.76	51.93	2
9752	θ Apodis	5.0	52.08	5.579	76 4 6.6	.76	51.30	1
9753	Circini	9.5	58.06	4.771	69 52 55.5	.75	51.44	1
9754	Gould, Z. C., 13h 3240	9.2	50 58.74	4.476	66 3 35.2	.75	52.48	2
9755	Gould, Z. C., 13h 3259	9.0	51 9.59	+ 4.653	68 29 11.5	− 17.74	50.42	1
9756	Apodis	10.0	13 51 14.18	+ 6.592	79 57 9.0	− 17.74	51.30	2
9757	Apodis	9.5	15.23	7.259	81 32 45.2	.74	50.86	2
9758	Circini	10.0	34.50	4.762	69 41 46.5	.73	52.34	1
9759	Gould, Z. C., 13h 3328	8.5	38.76	6.156	78 31 39.2	.72	51.35	1
9760	Gould, Z. C., 13h 3316	10.0	44.16	+ 5.582	75 59 18.5	− 17.72	51.30	1
9761	Circini	8.0	13 51 44.70	+ 4.605	67 46 31.2	− 17.72	52.30	1
9762	Circini	10.0	47.77	4.767	69 42 41.9	.72	52.34	1
9763	Circini	9.0	55.14	4.574	67 19 41.5	.71	52.29	1
9764	Gould, 19028	7.5	51 59.20	4.931	71 20 31.4	.71	52.33	2
9765	Apodis	9.5	52 3.83	+ 4.804	70 4 2.2	− 17.71	51.44	1
9766	Gould, Z. C., 13h 3354	8.8	13 52 9.70	+ 5.818	77 6 29.3	− 17.70	51.37	2
9767	Gould, Z. C., 13h 3314	8.5	12.34	4.484	66 0 24.4	.70	52.48	2
9768	Circini	9.2	16.76	4.627	67 58 44.4	.70	51.88	2
9769	Gould, Z. C., 13h 3351	9.2	17.48	5.421	75 0 17.6	.70	51.73	2
9770	Circini	9.0	21.50	+ 4.622	67 53 26.2	− 17.69	51.88	2
9771	Apodis	9.0	13 52 27.81	+ 5.271	74 0 47.7	− 17.69	52.28	1
9772	Apodis	9.5	32.33	5.871	77 18 17.0	.69	51.49	1
9773	Gould, Z. C., 13h 3348	10.0	33.40	5.652	68 14 26.7	.69	51.46	1
9774	Brisbane, 4723	8.0	39.12	4.728	69 8 44.8	.68	50.95	2
9775	Apodis	9.0	40.37	+ 5.486	75 20 55.8	− 17.68	50.40	1
9776	Apodis	8.0	13 52 43.94	+ 5.439	75 3 38.5	− 17.68	50.40	1
9777	Apodis	10.0	52.07	4.882	70 44 43.6	.67	52.40	1
9778	Gould, Z. C., 13h 3373	9.2	52.74	4.728	69 6 20.6	.67	50.94	2
9779	Brisbane, 4731	9.0	52 57.74	4.556	66 54 46.1	.67	51.45	1
9780	Apodis	11.0	53 0.27	+ 5.050	72 15 27.7	− 17.67	51.28	1
9781	Circini	9.5	13 53 3.28	+ 4.545	66 44 37.8	− 17.67	52.41	1
9782	Apodis	10.0	4.13	5.050	72 14 59.9	.66	51.28	1
9783	Apodis	9.5	20.74	5.304	78 53 26.3	.65	51.28	1
9784	Gould, Z. C., 13h 3398	10.0	24.86	4.749	69 15 45.8	.65	51.45	1
9785	Octantis	8.8	39.68	+18.919	87 42 3.3	− 17.64	51.93	4

Number.	Constellation, Name of Star, or Synonym.	Magnitude.	Right Ascension, 1850.0.	Annual Precession.	South Declination, 1850.0.	Annual Precession.	Mean year.	No. of obs.
			h. m. s.	s.	° ′ ″	″		
9786	Gould, Z. C., 13h 3422 .	8.5	13 53 40.09	+ 5.257	73 45 58.8	— 17.64	51.77	2
9787	Gould, 19058	8.5	43.98	5.458	75 3 36.6	.64	51.93	2
9788	Apodis	10.0	53 56.47	5.296	73 58 4.4	.63	52.28	1
9789	Apodis	8.2	54 8.63	6.026	80 49 23.5	.62	50.79	3
9790	Apodis	8.5	15.85	+ 7.142	81 4 21.8	— 17.62	50.88	2
9791	Apodis	9.5	13 54 30.02	+ 5.053	72 4 45.8	— 17.61	52.33	1
9792	Apodis	10.0	44.53	5.340	74 12 6.6	.60	52.28	1
9793	Apodis	9.5	44.89	4.927	70 54 55.2	.60	52.36	1
9794	Lacaille, 5781	7.0	54 49.47	5.332	74 8 14.9	.59	52.28	1
9795	Apodis	9.5	55 2.60	+ 5.891	77 8 45.0	— 17.58	51.37	2
9796	Gould, Z. C., 13h 3494 .	9.0	13 55 6.47	+ 4.809	69 39 58.7	— 17.58	52.34	1
9797	Circini	10.0	7.06	4.653	67 50 48.7	.58	51.88	2
9798	Octantis	9.5	15.53	8.198	82 50 4.3	.57	51.52	1
9799	Circini	10.0	17.94	4.594	67 2 43.4	.57	52.29	1
9800	Apodis	9.5	19.49	+ 4.854	70 6 28.7	— 17.57	51.44	1
9801	Gould, Z. C., 13h 3516 .	8.5	13 55 22.70	+ 4.891	70 28 52.0	— 17.57	52.40	1
9802	Apodis	9.0	24.89	5.141	72 41 8.9	.57	51.28	1
9803	Gould, Z. C., 13h 3536 .	9.0	29.21	5.322	73 59 32.5	.56	52.28	1
9804	Apodis	10.0	31.89	6.219	78 24 16.1	.56	51.35	1
9805	Brisbane, 4745	8.5	55 55.10	+ 4.630	67 25 29.3	— 17.55	52.30	2
9806	Gould, Z. C., 13h 3564 .	9.0	13 56 4.98	+ 4.827	69 42 54.2	— 17.54	52.34	1
9807	Apodis	10.0	7.47	5.047	71 48 49.9	.54	52.33	1
9808	Gould, Z. C., 13h 3600 .	9.0	26.18	5.300	73 43 25.5	.52	51.25	1
9809	Gould, Z. C., 13h 3580 .	9.5	28.04	4.557	66 20 26.0	.52	52.47	2
9810	Gould, Z. C., 13h 3581 .	9.5	28.05	+ 4.652	67 37 25.4	— 17.52	52.30	1
9811	Apodis	10.0	13 56 46.64	+ 5.141	72 25 32.1	— 17.51	51.28	1
9812	Gould, Z. C., 13h 3627 .	9.0	56 59.98	4.844	69 46 2.9	.50	52.34	1
9813	Circini	9.2	57 3.32	4.673	67 47 50.0	.50	52.39	2
9814	Apodis	10.0	4.20	5.338	73 54 35.3	.50	51.76	2
9815	Gould, Z. C., 13h 3638 .	9.0	5.23	+ 4.961	70 54 49.4	— 17.50	52.36	1
9816	Brisbane, 4755	8.3	13 57 5.82	+ 4.644	67 25 18.3	— 17.49	52.36	3
9817	Circini	10.0	8.03	4.839	69 41 45.2	.49	52.34	1
9818	Gould, Z. C., 13h 3641 .	8.5	8.31	4.917	70 29 10.2	.49	52.40	1
9819	Apodis	10.0	9.74	5.771	76 22 33.8	.49	51.34	1
9820	Apodis	10.0	22.37	+ 5.136	72 23 25.3	— 17.48	51.28	1
9821	Apodis	10.0	13 57 26.73	+ 7.185	80 56 39.6	— 17.48	51.37	1
9822	Circini	11.0	29.58	4.797	69 11 21.1	.48	51.45	1
9823	Octantis	9.5	31.72	9.196	83 53 8.3	.48	51.02	2
9824	Apodis	11.0	34.17	4.798	69 11 3.5	.47	51.45	1
9825	Gould, Z. C., 13h 3704 .	9.8	37.62	+ 5.530	75 2 51.9	— 17.47	51.42	3

No. 9790. Declination may be 81° 4′ 32.4″.

Number.	Constellation, Name of Star, or Synonym.	Magnitude.	Right Ascension, 1850.0.	Annual Precession.	South Declination, 1850.0.	Annual Precession.	Mean year.	No. of obs.
			h. m. s.	s.	° ′ ″	″		
9 826	Circini	10.0	13 57 40.16	+ 4.547	65 59 52.3	− 17.47	52.52	1
9 827	Apodis	9.0	41.39	5.277	73 24 57.2	.47	50.44	1
9 828	Apodis	10.0	42.12	5.317	73 41 31.5	.47	51.25	1
9 829	Circini	9.0	43.68	4.575	66 23 23.3	.47	52.41	1
9 830	Apodis	9.5	45.38	+ 6.213	78 11 5.4	− 17.47	51.40	2
9 831	Apodis	7.0	13 57 51.66	+ 5.998	77 19 40.0	− 17.46	51.49	1
9 832	Gould, Z. C., 14h 31	9.5	52.07	6.049	77 32 15.5	.46	50.40	1
9 833	Gould, Z. C., 14h 49	9.2	56.24	6.753	79 51 52.8	.46	51.30	2
9 834	Gould, 19151	8.0	57.16	5.039	71 30 6.8	.46	52.33	2
9 835	Gould, Z. C., 13h 3695	10.0	57 57.48	+ 4.540	65 51 23.1	− 17.46	52.52	1
9 836	Apodis	10.0	13 58 1.79	+ 5.628	75 32 15.4	− 17.45	51.38	1
9 837	Lacaille, 5804	6.0	6.38	4.843	69 35 24.1	.45	52.34	1
9 838	Apodis	9.3	6.89	5.088	71 54 4.4	.45	52.35	3
9 839	Apodis	10.0	9.89	5.048	71 32 55.6	.45	52.32	1
9 840	Gould, Z. C., 14h 11	9.5	11.82	+ 4.648	67 17 38.0	− 17.45	52.29	1
9 841	Apodis	9.5	13 58 18.44	+ 5.888	76 48 39.8	− 17.44	51.25	1
9 842	Gould, Z. C., 14h 40	8.3	30.87	4.944	70 32 57.5	.43	52.41	3
9 843	Gould, 19163	9.0	36.86	5.103	71 57 46.4	.43	52.35	3
9 844	Lacaille, 5811	7.2	45.82	4.798	69 0 15.9	.42	51.44	3
9 845	Gould, Z. C., 14h 57	9.0	55.71	+ 4.555	65 53 15.6	− 17.42	52.52	1
9 846	Gould, Z. C., 14h 61	9.0	13 58 56.27	+ 4.660	67 20 26.0	− 17.42	52.29	1
9 847	Apodis	10.0	58.09	7.198	80 52 7.2	.41	51.37	1
9 848	Circini	9.5	58.94	4.559	65 57 52.2	.41	52.52	1
9 849	Octantis	10.0	58 59.19	8.723	83 18 17.3	.41	50.50	1
9 850	Gould, Z. C., 14h 90	9.0	59 2.75	+ 5.464	74 29 34.7	− 17.41	50.41	1
9 851	Octantis	8.5	13 59 7.48	+ 8.167	82 34 25.3	− 17.41	50.39	1
9 852	Circini	10.0	8.63	4.723	68 5 7.7	.41	51.46	1
9 853	Gould, Z. C., 14h 146	9.0	17.01	6.864	80 3 15.8	.40	51.37	1
9 854	Apodis	10.0	18.45	6.413	78 44 34.1	.40	51.28	1
9 855	Apodis	10.0	22.12	+ 6.145	77 47 4.1	− 17.40	50.44	1
9 856	Apodis	10.0	13 59 25.31	+ 5.012	71 3 41.8	− 17.39	52.35	1
9 857	Circini	9.5	30.67	4.889	69 52 27.6	.39	51.44	1
9 858	Octantis	10.0	32.58	8.187	82 34 46.5	.39	50.39	1
9 859	Apodis	10.0	34.64	5.459	74 24 14.7	.39	50.41	1
9 860	Gould, Z. C., 14h 109	8.8	36.72	+ 4.851	69 27 40.4	− 17.39	52.07	3
9 861	Gould, Z. C., 14h 171	9.5	13 59 40.67	+ 4.707	67 48 32.5	− 17.38	52.48	1
9 862	Lacaille, 5805	8.0	43.24	5.698	75 44 51.7	.38	51.38	1
9 863	η Apodis	4.8	44.50	6.974	80 17 53.4	.38	50.90	2
9 864	Gould, Z. C., 14h 129	8.8	45.96	5.216	72 42 53.0	.38	50.85	2
9 865	Circini	10.0	48.70	+ 4.716	67 53 49.1	− 17.38	52.48	1

No. 9 861. Gould's right ascension is 1m greater.

Number.	Constellation, Name of Star, or Synonym.	Magnitude.	Right Ascension, 1850.0.	Annual Precession.	South Declination, 1850.0.	Annual Precession.	Mean year.	No. of obs.
			h. m. s.	s.	° ′ ″	″		
9 866	Gould, Z. C., 14^h 128 . .	8.2	13 59 50.33	+ 4.972	70 38 12.0	− 17.38	52.40	3
9 867	Apodis	10.0	59 52.39	6.149	77 45 21.2	.37	50.42	2
9 868	Gould, Z. C., 14^h 132 . .	9.2	14 0 0.64	4.642	66 55 48.7	.37	51.94	2
9 869	Apodis	9.0	2.92	7.384	81 11 5.8	.37	50.88	2
9 870	Gould, Z. C., 14^h 157 . .	8.4	3.00	+ 5.576	75 2 58.3	− 17.37	51.20	4
9 871	Octantis	9.2	14 0 6.06	+18.793	87 33 47.1	− 17.36	51.93	4
9 872	Apodis	9.5	13.18	4.922	70 5 59.3	.36	51.44	1
9 873	Octantis	9.0	15.15	8.301	82 42 8.1	.36	50.39	1
9 874	Apodis	10.0	17.67	5.393	73 53 58.4	.36	51.76	2
9 875	Gould, Z. C., 14^h 163 . .	8.8	25.80	+ 5.066	71 24 46.4	− 17.35	52.33	2
9 876	Gould, Z. C., 14^h 160 . .	9.2	14 0 28.52	+ 4.730	67 58 32.2	− 17.35	51.88	2
9 877	Gould, Z. C., 14^h 180 . .	9.0	31.66	5.267	72 59 52.8	.35	50.43	2
9 878	Apodis	9.0	45.72	5.834	76 19 9.0	.34	51.34	1
9 879	Apodis	9.5	51.40	5.508	74 33 34.6	.33	50.41	1
9 880	Apodis	8.5	0 55.04	+ 5.878	76 30 37.2	− 17.33	51.34	1
9 881	Apodis	10.0	14 1 1.93	+ 5.634	75 16 0.7	− 17.32	50.56	1
9 882	Apodis	9.5	2.26	7.250	80 50 23.3	.32	51.37	1
9 883	Gould, Z. C., 14^h 219 . .	9.0	5.27	5.248	72 47 25.1	.32	50.42	1
9 884	Gould, Z. C., 14^h 207 . .	8.5	11.10	4.569	65 44 30.1	.32	52.52	1
9 885	Lacaille, 5816	7.0	19.16	+ 5.987	76 57 28.5	− 17.31	51.37	2
9 886	Apodis	10.0	14 1 20.69	+ 5.485	74 22 0.4	− 17.31	50.41	1
9 887	Apodis	9.5	25.34	5.214	72 29 56.6	.31	51.28	1
9 888	Gould, Z. C., 14^h 260 . .	9.0	27.90	5.994	76 58 26.6	.31	51.37	2
9 889	Apodis	10.0	30.53	6.846	79 50 28.2	.30	51.24	1
9 890	Circini	9.8	33.22	+ 4.630	66 32 16.2	− 17.30	52.46	2
9 891	Gould, Z. C., 14^h 243 . .	9.5	14 1 40.10	+ 4.832	68 58 8.1	− 17.30	52.42	1
9 892	Circini	9.8	46.22	4.611	66 14 30.3	.29	52.46	2
9 893	Gould, Z. C., 14^h 277 . .	9.5	56.03	5.615	75 4 3.5	.28	52.40	1
9 894	Gould, Z. C., 14^h 271 . .	9.8	1 59.35	5.134	71 47 8.3	.28	52.35	3
9 895	Lacaille, 5831	8.0	2 4.30	+ 4.710	67 28 54.2	− 17.28	52.36	3
9 896	Apodis	9.5	14 2 11.78	+ 5.497	78 46 31.0	− 17.27	51.85	2
9 897	Gould, Z. C., 14^h 285 . .	9.8	16.77	5.142	71 49 10.8	.27	52.35	3
9 898	Gould, Z. C., 14^h 290 . .	8.0	23.65	4.997	70 31 35.7	.26	52.40	1
9 899	Apodis	10.0	24.68	5.301	73 0 56.0	.26	50.44	1
9 900	Apodis	10.0	25.04	+ 6.852	79 47 26.0	− 17.26	51.24	1
9 901	Circini	10.0	14 2 30.27	+ 4.819	68 41 38.9	− 17.26	50.42	1
9 902	Gould, Z. C., 14^h 292 . .	8.8	31.75	4.742	67 48 58.9	.26	52.08	3
9 903	Gould, Z. C., 14^h 311 . .	8.8	33.34	5.206	72 17 27.2	.26	52.01	3
9 904	Circini	10.0	35.36	4.883	69 22 50.8	.25	51.94	2
9 905	Gould, Z. C., 14^h 300 . .	9.0	37.41	+ 4.761	68 1 24.5	− 17.25	51.46	1

No. 9 874. Mean of two observations which differ 12.4″.

Number.	Constellation, Name of Star, or Synonym.	Magnitude.	Right Ascension, 1850.0.	Annual Precession.	South Declination, 1850.0.	Annual Precession.	Mean year.	No. of obs.
			h. m. s.	s.	° ′ ″	″		
9906	Lacaille, 5801	8.0	14 2 43.78	+ 8.020	82 8 57.9	− 17.25	50.53	1
9907	Circini	9.0	45.07	4.918	69 43 10.0	.25	52.34	1
9908	Octantis	9.3	45.72	11.106	85 8 41.6	.25	51.46	3
9909	Gould, Z. C., 14ʰ 310	9.2	45.90	4.687	67 5 37.6	.25	52.40	2
9910	Octantis	10.0	47.16	+23.550	88 5 27.2	− 17.25	51.92	1
9911	Circini	10.0	14 2 54.03	+ 4.684	67 2 5.7	− 17.24	52.29	1
9912	Gould, Z. C., 14ʰ 333	9.2	55.70	4.980	70 20 46.1	.24	51.92	2
9913	Lacaille, 5830	8.5	2 57.57	5.693	75 23 4.4	.24	50.48	2
9914	Apodis	9.5	3 0.22	5.164	71 54 30.0	.24	52.37	2
9915	Gould, Z. C., 14ʰ 343	9.2	1.88	+ 4.981	70 17 26.0	− 17.24	51.92	2
9916	Gould, Z. C., 14ʰ 346	8.0	14 3 12.26	+ 4.685	67 0 6.7	− 17.23	52.40	2
9917	Gould, Z. C., 14ʰ 351	7.2	17.58	4.682	66 57 4.6	.22	52.40	2
9918	d Octantis	5.5	30.02	8.638	82 58 23.3	.21	51.01	2
9919	Gould, Z. C., 14ʰ 374	9.5	32.00	5.204	72 9 16.0	.21	52.37	2
9920	Apodis	10.0	41.40	+ 6.445	78 29 16.6	− 17.21	51.35	1
9921	Circini	10.0	14 3 43.21	+ 4.892	69 18 39.1	− 17.20	52.42	1
9922	Apodis	10.0	44.48	5.911	76 23 33.4	.20	51.34	1
9923	Circini	8.3	3 45.46	4.759	67 50 3.7	.20	52.08	3
9924	Apodis	8.8	4 17.14	7.704	81 31 33.8	.18	50.47	2
9925	Octantis	10.0	22.97	+ 8.561	82 49 47.4	− 17.17	51.52	1
9926	Circini	9.2	14 4 23.01	+ 4.637	66 11 52.6	− 17.17	52.52	2
9927	ε Apodis	6.2	31.95	6.771	79 24 40.5	.17	50.83	2
9928	Apodis	10.0	41.40	6.124	77 13 19.8	.16	51.49	1
9929	Gould, Z. C., 14ʰ 436	9.0	44.42	4.989	70 8 32.8	.16	51.44	1
9930	Apodis	8.7	45.37	+ 7.271	80 38 9.9	− 17.16	50.79	3
9931	Lacaille, 5846	7.0	14 4 51.99	+ 4.620	65 53 3.7	− 17.15	52.52	1
9932	Circini	9.5	54.64	4.664	66 27 57.6	.15	52.52	1
9933	Circini	9.0	4 59.73	4.764	67 42 13.4	.15	52.39	2
9934	Circini	10.0	5 1.29	4.936	69 35 9.4	.15	52.34	1
9935	Gould, Z. C., 14ʰ 473	9.0	2.18	+ 5.569	74 28 32.0	− 17.14	50.41	1
9936	Gould, Z. C., 14ʰ 518	8.2	14 5 7.54	+ 6.650	79 1 6.9	− 17.14	51.38	3
9937	Gould, Z. C., 14ʰ 489	8.0	14.98	5.465	73 48 26.6	.14	51.76	2
9938	Circini	9.5	15.46	4.665	66 26 18.5	.13	52.52	1
9939	Circini	10.0	39.13	4.670	66 26 18.5	.12	52.52	1
9940	Apodis	9.0	44.25	+ 5.869	76 0 6.5	− 17.11	51.30	1
9941	Lacaille, 5853	8.5	14 5 47.36	+ 4.743	67 20 27.4	− 17.11	52.29	1
9942	Circini	9.5	48.81	4.660	66 13 50.4	.11	52.52	1
9943	Gould, Z. C., 14ʰ 535	9.2	51.05	5.612	74 38 20.3	.11	51.42	3
9944	Gould, Z. C., 14ʰ 504	10.0	52.00	4.624	65 46 55.9	.11	52.52	1
9945	Gould, 19304	8.2	56.18	+ 5.065	70 41 48.2	− 17.10	52.42	2

No. 9935. Gould's declination is 9.3″ greater.

No. 9936. Gould's declination is 9.8″ greater.

Number.	Constellation, Name of Star, or Synonym	Magnitude.	Right Ascension, 1850.0.	Annual Precession.	South Declination, 1850.0.	Annual Precession.	Mean year.	No. of obs.
			h. m. s.	s.	° ′ ″	″		
9 946	Gould, Z. C., 14^h 531	9.0	14 5 58.21	+ 5.131	71 15 46.4	− 17.10	52.35	1
9 947	Lacaille, 5847	6.5	6 2.71	5.397	73 16 12 1	.10	50.44	1
9 948	Gould, Z. C., 14^h 532	9.2	4.46	4.910	69 10 48.1	.10	51.44	2
9 949	Gould, 19305	10.0	5.50	4.755	67 21 0.0	.10	52.29	1
9 950	Circini	10.0	8.18	+ 4.840	68 25 28.9	− 17.09	50.42	1
9 951	Apodis	10.0	14 6 15.75	+ 5.133	71 14 50.2	− 17.09	52.35	1
9 952	Gould, Z. C., 14^h 541	9.0	16.04	4.905	69 6 0.6	.09	51.44	2
9 953	Gould, Z. C., 14^h 546	8.2	17.41	5.110	71 2 37.4	.09	52.40	2
9 954	Gould, 19308	9.5	18.52	4.748	67 19 57.9	.09	52.29	1
9 955	Gould, Z. C., 14^h 570	9.5	24.31	+ 5.584	74 25 3.7	− 17.08	50.41	1
9 956	Apodis	9.0	14 6 25.42	+ 5.324	72 43 3.7	− 17.08	51.28	1
9 957	Gould, Z. C., 14^h 587	9.5	25.78	5.287	77 42 13.0	.08	50.42	2
9 958	Gould, Z. C., 14^h 586	9.5	33.43	5.857	75 51 51.7	.08	51.34	2
9 959	Apodis	12.0	6 39.31	5.857	75 51 24 2	.07	51.38	1
9 960	Apodis	10.0	7 0.64	+ 6.943	79 41 49.5	− 17.05	51.24	1
9 961	Gould, Z. C., 14^h 621	9.5	14 7 4.97	+ 5.897	76 0 23.8	− 17.05	51.30	1
9 962	Apodis	10.0	6.16	5.227	71 54 17.8	.05	52.32	1
9 963	Gould, Z. C., 14^h 596	9.5	7.45	4.927	69 12 52.9	.05	52.42	1
9 964	Gould, Z. C., 14^h 602	9.5	16.14	4.811	67 56 9.5	.04	51.97	2
9 965	Gould, Z. C., 14^h 607	9.0	20.15	+ 4.772	67 27 59.2	− 17.04	52.38	2
9 966	Gould, Z. C., 14^h 637	9.0	14 7 38.52	+ 5.419	73 14 21.0	− 17.03	50.44	1
9 967	Gould, Z. C., 14^h 625	8.8	42.54	4.668	66 6 11.0	.02	52.55	2
9 968	Apodis	9.5	48.38	5.466	73 32 0.4	.02	51.25	1
9 969	Apodis	9.5	52.63	5.099	70 44 45.0	.01	52.40	1
9 970	Gould, Z. C., 14^h 634	9.0	54.31	+ 4.737	66 57 59.4	− 17.01	52.52	1
9 971	Octantis	10.0	14 7 59.84	+ 9.860	84 2 26.8	− 17.01	51.49	1
9 972	Gould, Z. C., 14^h 659	9.0	8 0.12	5.335	72 37 5.4	.01	50.85	2
9 973	Gould, Z. C., 14^h 649	9.8	2.68	4.812	67 50 25.0	.01	51.97	2
9 974	Gould, 19335	7.0	3.89	5.205	71 37 19.6	.01	52.32	1
9 975	Gould, Z. C., 14^h 655	9.0	4.66	+ 4.929	69 6 27.8	− 17.01	51.44	2
9 976	Gould, Z. C., 14^h 668	9.0	14 8 10.48	+ 5.344	72 39 51.0	− 17.00	50.85	2
9 977	Circini	8.5	11.57	5.009	69 53 4.8	.00	51.44	1
9 978	Apodis	10.0	14.28	7.010	79 46 59.1	17.00	51.24	1
9 979	Octantis	9.2	23.36	10.609	84 36 49.9	16.99	51.00	2
9 980	Circini	9.5	29.38	+ 4.790	67 31 7.4	− 16.99	52.38	2
9 981	Apodis	9.5	14 8 34.38	+ 5.105	70 43 12.7	− 16.98	52.40	1
9 982	Circini	9.5	37.05	4.704	66 26 33.4	.98	52.52	1
9 983	Gould, 19346	8.0	37.16	5.222	71 41 23.9	.98	52.32	1
9 984	Apodis	9.8.	41.89	5.547	73 57 1.6	.98	51.76	2
9 985	Apodis	9.5	47.26	+ 6.180	77 5 29.4	− 16.97	51.25	1

No. 9 984. Mean of two observations which differ 14.4″.

Number.	Constellation, Name of Star, or Synonym.	Magnitude.	Right Ascension, 1850.0.	Annual Precession.	South Declination, 1850.0.	Annual Precession.	Mean year.	No. of obs.
			h. m. s.	s.	° ′ ″	″		
9986	Lacaille, 5865[1]	8.5	14 8 47.51	+ 5.382	72 51 46.6	− 16.97	50.42	1
9987	Lacaille, 5865[2]	9.0	8 48.01	5.382	72 51 48.6	.97	50.42	1
9988	Gould, Z. C., 14h 721	9.0	9 11.75	5.022	69 53 2.7	.95	52.34	1
9989	Apodis	8.8	15.89	6.369	77 46 4.5	.95	51.06	3
9990	Gould, 19360	8.8	24.54	+ 5.484	73 28 48.9	− 16.94	50.85	2
9991	Apodis	9.5	14 9 25.08	+ 5.069	70 17 27.0	− 16.94	52.40	1
9992	Apodis	9.5	45.36	5.214	71 29 36.2	.93	51.35	1
9993	Circini	9.0	47.84	4.685	66 1 40.2	.93	52.55	2
9994	Gould, Z. C., 14h 757	8.0	50.58	4.883	68 23 26.0	.92	50.42	1
9995	Gould, Z. C., 14h 780	9.0	9 57.14	+ 5.479	73 23 11.6	− 16.92	50.44	1
9996	Gould, Z. C., 14h 781	9.5	14 10 2.58	+ 5.299	72 6 55.2	− 16.91	52.41	1
9997	Gould, Z. C., 14h 775	9.5	4.77	4.836	67 49 49.1	.91	51.46	1
9998	Stone, 7818	9.2	12.64	5.639	74 20 42.6	.91	51.34	2
9999	Circini	8.5	13.11	4.721	66 26 5.0	.91	52.52	1
10000	Lacaille, 5835	8.0	22.85	+ 9.365	83 28 14.2	− 16.90	50.53	2
10001	Lacaille, 5866	7.2	14 10 30.40	+ 6.444	77 55 54.3	− 16.89	51.06	3
10002	Brisbane, 4843	9.0	31.96	5.799	75 11 34.9	.89	50.56	1
10003	Apodis	10.0	32.14	5.718	74 45 31.9	.89	52.40	1
10004	Circini	8.5	36.75	4.727	66 26 47.6	.89	52.52	1
10005	Apodis	8.0	52.52	+ 7.013	79 39 36.8	− 16.87	51.24	1
10006	Apodis	9.5	14 10 53.02	+ 7.013	79 36 26.1	− 16.87	51.24	1
10007	Gould, Z. C., 14h 842	9.5	11 5.57	4.985	69 16 43.4	.86	52.42	1
10008	Lacaille, 5864[1]	8.5	8.82	6.948	79 25 14.4	.86	50.83	2
10009	Lacaille, 5864[2]	9.2	11.06	6.950	79 25 18.4	.86	50.83	2
10010	Gould, Z. C., 14h 851	8.5	11.82	+ 5.098	70 19 41.0	− 16.86	52.40	1
10011	Apodis	9.0	14 11 12.83	+ 5.092	70 16 30.4	− 16.86	51.44	1
10012	Apodis	9.5	16.20	6.822	79 3 49.5	.86	51.28	1
10013	Apodis	10.0	18.94	6.089	76 29 19.9	.85	51.34	1
10014	Gould, 19400	9.0	21.86	4.696	65 56 27.9	.85	52.57	1
10015	Apodis	9.2	23.88	+ 6.118	76 36 20.2	− 16.85	51.30	2
10016	Lacaille, 5876	7.2	14 11 27.68	+ 5.946	75 49 58.6	− 16.85	51.34	2
10017	Apodis	9.5	28.40	5.080	70 8 8.3	.85	51.44	1
10018	Octantis	7.6	32.09	24.323	88 2 36.7	.84	51.93	4
10019	Gould, Z. C., 14h 898	9.0	32.29	6.103	76 31 54.0	.84	51.34	1
10020	Stone, 7826	6.5	33.29	+ 4.699	65 57 19.0	− 16.84	52.57	1
10021	Gould, 19410	7.2	14 11 42.76	+ 5.300	71 55 57.0	− 16.83	52.36	2
10022	Apodis	11.0	48.76	8.099	81 46 1.8	.83	50.56	1
10023	Apodis	10.0	11 55.74	5.665	74 19 12.5	.82	52.28	1
10024	Apodis	9.8	12 3.82	6.049	76 15 2.2	.82	51.32	2
10025	Gould, Z. C., 14h 906	9.5	4.08	+ 5.174	70 52 54.2	− 16.82	52.49	2

Number.	Constellation, Name of Star, or Synonym.	Magnitude.	Right Ascension, 1850.0.	Annual Precession.	South Declination, 1850.0.	Annual Precession.	Mean year.	No. of obs.
			h. m. s.	s.	° ′ ″	″		
10 026	Gould, 19425	8.0	14 12 8.21	+ 5.297	71 51 47.2	− 16.81	52.37	2
10 027	Apodis	10.0	17.95	5.836	75 12 45.1	.81	50.56	1
10 028	Gould, Z. C., 14h 912 . .	8.5	18.94	5.007	69 20 29.5	.81	52.42	1
10 029	Apodis	9.8	25.33	5.255	71 30 28.3	.80	52.39	3
10 030	Gould, Z. C., 14h 926 . .	9.0	28.17	+ 5.415	72 40 52.5	− 16.80	51.28	1
10 031	Circini	8.8	14 12 32.56	+ 5.064	69 51 6.6	− 16.79	52.79	2
10 032	Apodis	9.0	34.33	8.582	82 26 36.0	.79	50.39	1
10 033	Apodis	9.5	34.99	6.450	77 47 25.8	.79	52.35	1
10 034	Apodis	10.0	37.94	5.629	74 2 24.8	.79	52.28	1
10 035	Apodis	9.5	40.39	+ 6.039	76 9 1.0	− 16.79	51.30	1
10 036	Gould, Z. C., 14h 953 . .	8.0	14 12 40.50	+ 5.578	73 43 51.5	− 16.79	51.25	1
10 037	Gould, Z. C., 14h 936 . .	8.5	40.50	5.108	70 14 15.0	.79	51.92	2
10 038	Lacaille, 5890	6.7	45.02	4.839	67 30 30.9	.78	52.42	4
10 039	Apodis	9.5	53.08	6.525	78 1 23.0	.78	52.35	1
10 040	Gould, 19436	8.8	12 53.50	+ 4.856	67 40 42.2	− 16.78	52.50	2
10 041	Apodis	10.0	14 13 0.12	+ 6.669	78 28 48.5	− 16.77	51.35	1
10 042	Apodis	10.0	3.26	5.179	70 49 31.3	.77	52.49	2
10 043	Octantis	9.8	3.92	9.937	83 54 8.4	.77	51.02	2
10 044	Gould, Z. C., 14h 958 . .	9.8	6.32	4.856	67 38 58.7	.77	52.50	2
10 045	Gould, Z. C., 14h 996 . .	9.0	12.26	+ 6.126	76 29 11.4	− 16.76	51.34	1
10 046	Gould, Z. C., 14h 1016 .	8.8	14 13 14.76	+ 6.377	77 28 32.0	− 16.76	50.94	2
10 047	Gould, Z. C., 14h 981 . .	9.0	15.71	5.283	71 37 37.6	.76	52.32	1
10 048	Apodis	10.0	21.23	5.992	75 52 24.2	.76	51.38	1
10 049	Melbourne (1), 728 . . .	7.3	24.10	12.930	85 44 8.4	.75	51.93	3
10 050	Apodis	9.5	28.26	+ 6.937	79 15 9.9	− 16.75	50.42	1
10 051	Apodis	9.0	14 13 34.16	+ 7.877	81 17 34.7	− 16.75	50.38	1
10 052	Apodis	9.0	40.90	7.777	81 6 18.5	.74	50.56	1
10 053	Circini	9.5	47.49	4.912	68 10 13.7	.73	51.46	1
10 054	Gould, 19460	9.0	50.55	5.301	71 41 50.4	.73	52.32	1
10 055	Gould, Z. C., 14h 1013 .	9.5	53.74	+ 4.793	66 48 11.3	− 16.73	52.53	2
10 056	Gould, Z. C., 14h 1017 .	9.0	14 13 55.10	+ 4.793	66 48 21.8	− 16.73	52.53	2
10 057	Lacaille, 5885	6.0	13 55.79	6.041	76 2 47.8	.73	51.30	1
10 058	Gould, Z. C., 14h 1077 .	8.8	14 0.56	6.580	78 7 9.8	.72	51.40	2
10 059	Gould, Z. C., 14h 1056 .	8.5	12.69	5.412	72 30 11.6	.71	51.28	1
10 060	Gould, 19470	9.1	33.19	+ 4.865	67 33 33.6	− 16.70	52.42	4
10 061	Gould, Z. C., 14h 1063 .	8.6	14 14 34.94	+ 4.794	66 43 48.0	− 16.70	52.54	5
10 062	Gould, Z. C., 14h 1070 .	9.5	42.83	4.723	65 49 3.4	.69	52.57	1
10 063	Circini	9.5	47.44	4.790	66 39 12.0	.69	52.54	5
10 064	Gould, 19481	9.0	50.79	4.724	65 49 2.0	.68	52.57	1
10 065	Gould, Z. C., 14h 1095 .	9.5	56.31	+ 5.164	70 27 21.7	− 16.68	52.40	1

Number.	Constellation, Name of Star, or Synonym.	Magnitude.	Right Ascension, 1850.0.	Annual Precession.	South Declination, 1850.0.	Annual Precession.	Mean year.	No. of obs.
			h. m. s.	s.	° ′ ″	″		
10 066	Gould, Z. C., 14h 1090	9.0	14 14 56.33	+ 4.974	68 40 19.1	− 16.68	50.45	1
10 067	Apodis	11.0	57.46	5.770	74 36 43.5	.68	52.48	1
10 068	Apodis	11.0	59.26	5.770	74 36 37.4	.68	52.48	1
10 069	Apodis	9.5	14 59.58	6.456	77 37 7.6	.68	50.40	1
10 070	Gould, Z. C., 14h 1141	9.5	15 1.16	+ 6.825	78 48 9.0	− 16.68	51.28	1
10 071	Lacaille, 5900	7.0	14 15 7.65	+ 4.733	65 53 8.8	− 16.67	52.57	2
10 072	Apodis	9.8	18.72	5.640	73 50 13.7 ·	.66	51.76	2
10 073	Circini	9.2	19.74	5.105	69 53 17.2	.66	52.07	3
10 074	Circini	9.5	20.00	4.835	67 6 16.4	.66	52.38	1
10 075	Lacaille, 5884	7.0	20.58	+ 7.006	79 16 48.8	− 16.66	51.42	2
10 076	Gould, Z. C., 14h 1114	10.0	14 15 21.32	+ 4.938	68 15 0.5	− 16.66	50.94	2
10 077	Brisbane, 4888	8.2	25.19	5.224	70 54 37.8	.66	52.48	3
10 078	Apodis	8.5	30.33	6.022	75 49 15.8	.65	51.38	1
10 079	Circini	9.8	42.46	5.111	69 53 57.8	.64	52.38	2
10 080	Gould, Z. C., 14h 1168	10.0	52.14	+ 5.632	73 44 1.7	− 16.63	51.25	1
10 081	Apodis	10.0	14 15 53.40	+ 6.977	79 10 3.6	− 16.63	52.41	1
10 082	Apodis	10.0	16 3.84	5.592	73 28 2.9	.62	51.25	1
10 083	Apodis	10.0	14.90	8.209	81 42 19.9	.62	50.56	1
10 084	Lacaille, 5902	7.0	18.46	5.989	75 35 52.8	.61	50.92	2
10 085	Apodis	9.0	20.49	+ 6.190	76 29 14.1	− 16.61	51.34	1
10 086	Gould, Z. C., 14h 1185	9.5	14 16 30.21	+ 4.789	66 24 31.7	− 16.60	52.56	1
10 087	Gould, 19528	9.0	32.79	5.085	69 33 37.8	.60	52.38	2
10 088	Gould, 19526	9.0	34.60	4.884	67 30 8.0	.60	52.34	2
10 089	Apodis	10.0	34.96	5.598	73 27 11.9	.60	51.25	1
10 090	Gould, Z. C., 14h 1190	8.8	38.26	+ 4.761	66 2 24.6	− 16.60	52.57	2
10 091	Octantis	9.8	14 16 39.42	+20.097	87 28 29.6	− 16.60	51.93	3
10 092	Gould, Z. C., 14h 1195	9.5	39.49	4.836	66 56 43.5	.60	52.52	1
10 093	Circini	10.0	39.82	5.074	69 26 40.2	.59	52.34	1
10 094	Circini	9.5	40.00	4.858	67 11 40.1	.59	52.38	1
10 095	Apodis	9.8	50.69	+ 5.420	72 14 41.6	− 16.59	51.84	2
10 096	Apodis	10.0	14 16 51.07	+ 5.611	73 30 21.0	− 16.59	51.25	1
10 097	Apodis	10.0	17 3.29	5.526	72 56 55.0	.58	50.42	1
10 098	Brisbane, 4901	7.8	4.23	5.649	73 42 56.2	.57	51.76	2
10 099	Apodis	10.0	9.02	6.266	76 43 46.0	.57	51.25	1
10 100	Gould, Z. C., 14h 1232	9.2	13.52	+ 5.045	69 5 57.6	− 19.57	51.43	2
10 101	Apodis	8.5	14 17 19.93	+ 6.280	76 46 15.4	− 16.56	51.86	2
10 102	Apodis	10.0	44.04	5.410	72 4 58.2	.54	52.41	1
10 103	Apodis	9.8	44.44	5.228	70 40 44.5	.54	52.45	2
10 104	Octantis	10.0	17 51.66	10.582	84 14 48.2	.54	51.16	3
10 105	Apodis	9.8	18 5.04	+ 5.439	72 14 55.4	− 16.52	51.84	2

· No. 10 072. Mean of two observations which differ 10.2″.

Number.	Constellation, Name of Star, or Synonym.	Magnitude.	Right Ascension, 1850.0.	Annual Precession.	South Declination, 1850.0.	Annual Precession.	Mean year.	No. of obs.
			h. m. s.	s.	° ′ ″	″		
10 106	Apodis	10.0	14 18 6.48	+ 5.832	74 38 51.0	— 16.52	51.40	3
10 107	Gould, Z. C., 14h 1308 .	9.5	7.32	5.722	74 2 33.9	.52	52.40	2
10 108	Octantis	9.2	9.14	12.723	85 30 31.7	.52	51.93	2
10 109	Circini	9.5	9.72	4.811	66 26 54.8	.52	52.56	1
10 110	Octantis	8.8	13.87	+15.844	86 36 14.8	— 16.52	51.93	4
10 111	Gould, Z. C., 14h 1310 .	10.0	14 18 21.62	+ 5.242	70 43 20.6	— 16.51	52.45	2
10 112	Apodis	9.5	31.17	5.429	72 7 57.4	.50	52.41	1
10 113	Gould, 19568 . .	9.3	33.21	4.833	66 39 30.8	.50	52.55	3
10 114	Octantis	8.8	42.96	15.651	86 32 20.5	.49	51.93	4
10 115	Gould, Z. C., 14h 1339 .	9.0	49.75	+ 5.260	70 48 54.0	— 16.49	52.49	2
10 116	Brisbane, 4918	6.3	14 18 49.93	+ 4.869	67 2 27.2	— 16.49	52.49	3
10 117	Apodis	9.5	55.20	6.257	76 32 53.2	.48	51.34	1
10 118	Apodis	9.5	55.50	6.065	75 43 8.1	.48	51.38	1
10 119	Octantis	9.0	57.81	16.624	86 46 55.2	.48	51.93	4
10 120	Gould, Z. C., 14h 1405 .	9.0	18 58.84	+ 7.425	80 4 11.6	— 16.48	51.37	1
10 121	Gould, 19582	9.0	14 19 2.96	+ 4.932	67 43 9.2	— 16.48	52.52	1
10 122	Apodis	10.0	4.58	5.396	71 50 21.0	.48	52.32	1
10 123	Apodis	9.0	4.73	5.337	71 23 44.1	.48	52.48	1
10 124	Gould, Z. C., 14h 1362 .	9.2	17.40	4.974	68 7 47.4	.46	52.00	2
10 125	Lacaille, 5913	8.5	18.44	+ 7.125	79 19 46.8	— 16.46	51.42	2
10 126	Apodis	9.5	14 19 22.64	+ 5.223	70 26 51.4	— 16.46	52.40	1
10 127	Gould, Z. C., 14h 1385 .	9.8	24.42	5.915	74 57 22.6	.46	51.83	2
10 128	Apodis	9.5	24.84	6.056	75 38 18.9	.46	51.87	2
10 129	Apodis	9.2	26.82	6.058	75 38 36.0	.46	51.87	2
10 130	Gould, Z. C., 14h 1391 .	9.5	27.70	+ 5.927	75 0 40.1	— 16.46	51.47	3
10 131	Apodis	9.2	14 19 27.86	+ 8.230	81 34 14.6	— 16.46	50.47	2
10 132	Brisbane, 4920	8.5	28.64	5.590	73 6 58.0	.46	50.43	2
10 133	Gould, Z. C., 14h 1374 .	9.5	28.89	5.303	71 5 9.5	.46	52.48	1
10 134	Gould, 19589	9.0	30.22	5.108	69 25 0.3	.45	52.38	2
10 135	Lacaille, 5924	7.0	41.72	+ 6.246	76 26 35.4	— 16.44	51.34	1
10 136	Circini	9.5	14 19 46.59	+ 4.882	67 4 21.5	— 16.44	52.38	1
10 137	Gould, Z. C., 14h 1393 .	8.8	48.38	5.170	69 56 53.5	.44	52.07	3
10 138	Apodis	9.5	19 54.10	5.416	71 53 31.6	.43	52.38	2
10 139	Apodis	9.8	20 14.16	5.480	72 18 45.3	.43	51.84	2
10 140	Lacaille, 5823	7.0	14.56	+20.805	87 31 11.5	— 16.43	51.93	3
10 141	Gould, 19610	8.5	14 20 14.80	+ 5.064	68 54 58.6	— 16.43	50.45	1
10 142	Gould, Z. C., 14h 1428 .	9.5	17.10	5.278	70 47 48.0	.42	52.52	1
10 143	Octantis	8.8	19.06	17.437	86 56 16.4	.41	51.93	4
10 144	Gould, Z. C., 14h 1469 .	9.5	19.97	6.655	77 53 37.7	.41	51.09	3
10 145	Apodis	10.0	21.76	+ 5.482	72 18 50.9	+ 16.41	51.84	2

Number.	Constellation, Name of Star, or Synonym.	Magnitude.	Right Ascension, 1850.0.	Annual Precession.	South Declination, 1850.0.	Annual Precession.	Mean year.	No. of obs.
			h. m. s.	s.	° ′ ″	″		
10 146	Circini	9.0	14 20 25.32	+ 4.944	67 40 9.2	− 16.41	52.52	1
10 147	Apodis	10.0	29.38	6.836	78 26 56.3	.40	51.35	1
10 148	Gould, Z. C., 14ʰ 1458	9.0	34.24	5.692	73 37 55.2	.40	51.89	2
10 149	Apodis	9.8	36.38	6.123	75 50 18.6	.40	51.87	2
10 150	Gould, Z. C., 14ʰ 1471	9.8	41.97	+ 5.842	74 27 45.6	− 16.39	51.45	2
10 151	Gould, Z. C., 14ʰ 1452	9.0	14 20 42.04	+ 5.025	68 28 51.2	− 16.39	51.47	2
10 152	Apodis	10.0	47.70	7.698	80 32 32.0	.39	50.43	1
10 153	Gould, 19623	8.0	53.47	4.847	66 31 0.3	.38	52.56	1
10 154	Circini	9.0	20 53.75	4.931	67 28 10.5	.38	52.45	2
10 155	Apodis	10.0	21 0.09	+ 7.028	78 57 42.1	− 16.38	51.28	1
10 156	Apodis	9.5	14 21 4.51	+ 6.064	75 31 52.4	− 16.38	51.87	2
10 157	Lacaille, 5944	7.3	4.55	5.601	73 1 36.4	.38	51.07	3
10 158	Apodis	9.2	19.01	5.507	72 23 5.4	.36	50.84	2
10 159	Apodis	9.0	19.29	7.847	80 48 9.3	.36	50.50	2
10 160	Lacaille, 5948	7.5	21.50	+ 4.889	66 56 56.6	− 16.36	52.55	2
10 161	Gould, Z. C., 14ʰ 1487	10.0	14 21 22.80	+ 4.846	66 26 42.7	− 16.36	52.56	1
10 162	Gould, Z. C., 14ʰ 1495	10.0	27.22	4.844	66 24 56.9	.35	52.56	1
10 163	Gould, Z. C., 14ʰ 1500	10.0	28.68	5.044	68 34 17.5	.35	51.47	2
10 164	Apodis	10.0	28.82	6.519	77 20 50.3	.35	51.49	1
10 165	Apodis	9.5	31.96	+ 5.791	74 6 20.3	− 16.35	52.40	2
10 166	Octantis	10.0	14 21 49.86	+10.599	84 7 0.2	− 16.34	51.49	1
10 167	Gould, Z. C., 14ʰ 1529	9.5	54.04	5.202	69 59 6.8	.33	51.44	1
10 168	Apodis	8.8	21 56.44	6.421	76 57 8.5	.33	51.86	2
10 169	Gould, 19659	8.0	22 1.95	5.073	68 47 9.4	.33	50.45	1
10 170	Apodis	10.0	5.78	+ 5.840	74 19 21.1	− 16.32	52.28	1
10 171	Gould, Z. C., 14ʰ 1570	8.8	14 22 16.23	+ 5.952	74 53 12.4	− 16.31	51.92	2
10 172	Brisbane, 4939	8.5	21.52	4.897	66 54 36.2	.31	52.55	1
10 173	Octantis	9.2	22.28	11.859	84 56 19.8	.31	51.22	2
10 174	Gould, Z. C., 14ʰ 1582	9.5	22.78	5.893	74 34 49.9	.31	52.48	1
10 175	Gould, Z. C., 14ʰ 1564	9.5	35.55	+ 5.002	68 1 2.1	− 16.30	51.46	1
10 176	Apodis	9.2	14 22 46.13	+ 8.233	81 24 17.7	− 16.29	50.50	3
10 177	Gould, 19680	8.0	46.17	5.137	69 18 20.0	.29	52.42	1
10 178	Gould, Z. C., 14ʰ 1584	9.0	47.49	5.009	68 4 30.1	.29	51.46	1
10 179	Apodis	10.0	49.29	5.757	73 47 53.7	.29	51.25	1
10 180	Brisbane, 4947	9.0	58.30	+ 4.891	66 45 45.4	− 16.28	52.56	2
10 181	Gould, Z. C., 14ʰ 1608	9.0	14 22 58.46	+ 5.175	69 37 49.4	− 16.28	52.34	1
10 182	Gould, Z. C., 14ʰ 1603	8.5	23 0.92	5.041	68 21 21.2	.28	51.50	2
10 183	Circini	10.0	6.17	5.177	69 37 30.5	.27	52.34	1
10 184	Gould, 19690	8.5	11.94	5.271	70 25 10.1	.27	52.40	1
10 185	Gould, 19742	9.2	15.79	+10.396	83 54 7.0	− 16.26	51.02	2

Number.	Constellation, Name of Star, or Synonym.	Magnitude.	Right Ascension, 1850.0.	Annual Precession.	South Declination, 1850.0.	Annual Precession.	Mean year.	No. of obs.
			h. m. s.	s.	° ′ ″	″		
10 186	Circini	9.5	14 23 21.67	+ 4.834	66 2 24.8	− 16.26	52.57	1
10 187	Gould, Z. C., 14h 1659 .	9.0	25.28	5.900	74 31 8.7	.26	52.48	1
10 188	Gould, Z. C., 14h 1630 .	9.5	25.56	4.931	67 9 38.9	.26	52.38	1
10 189	Apodis	10.0	37.26	5.258	70 15 56.9	.25	52.40	1
10 190	Apodis	10.0	45.17	+ 5.362	71 5 12.3	− 16.24	52.48	1
10 191	Gould, Z. C., 14h 1653 .	9.0	14 23 45.52	+ 4.864	66 21 18.8	− 16.24	52.57	2
10 192	Apodis	10.0	23 50.58	5.853	74 14 11.9	.23	52.28	1
10 193	Apodis	9.7	24 1.81	6.393	76 41 3.0	.22	51.69	3
10 194	Apodis	9.5	6.54	5.881	74 21 31.8	.22	52.48	1
10 195	Circini	8.5	18.95	+ 4.842	66 1 9.5	− 16.21	52.57	1
10 196	Lacaille, 5882	7.8	14 24 25.54	+13.909	85 50 29.4	− 16.20	50.93	4
10 197	Gould, Z. C., 14h 1718 .	10.0	34.64	5.202	69 41 17.7	.20	52.34	1
10 198	Gould, Z. C., 14h 1740 .	9.8	35.53	5.955	74 42 7.6	.20	52.10	3
10 199	Octantis	9.2	41.44	{11.379 / 11.391}	84 34 {23.2 / 50.4}	.19	51.00	2
10 200	Gould, Z. C., 14h 1727 .	10.0	42.40	+ 5.212	69 45 37.7	− 16.19	52.34	1
10 201	Gould, Z. C., 14h 1746 .	9.8	14 24 43.12	+ 5.982	74 49 27.4	− 16.19	51.93	2
10 202	Gould, 19730	9.0	25 10.30	5.060	68 17 29.2	.17	52.00	2
10 203	Circini	9.2	12.20	4.978	67 26 57.8	.16	52.45	2
10 204	Lacaille, 5976	7.0	15.93	4.961	67 15 39.7	.16	52.38	1
10 205	Circini	10.0	16.16	+ 4.977	67 26 3.8	− 16.16	52.52	1
10 206	Apodis	9.5	14 25 17.76	+ 5.594	72 34 3.3	− 16.16	52.52	1
10 207	Brisbane, 4963	9.0	29.83	4.910	66 40 7.1	.15	52.57	3
10 208	Gould, Z. C., 14h 1773 .	9.5	32.21	4.891	66 26 30.7	.15	52.56	1
10 209	Lacaille, 5957	6.5	32.79	6.340	76 21 23.4	.15	51.32	2
10 210	Octantis	7.0	37.49	+18.754	87 6 10.9	− 16.14	51.93	3
10 211	Apodis	10.0	14 25 44.85	+ 5.401	71 10 27.3	− 16.14	52.48	1
10 212	Apodis	9.0	49.45	5.670	72 59 43.9	.13	52.35	2
10 213	Apodis	8.7	55.76	6.428	76 40 21.0	.13	51.69	3
10 214	Apodis	10.0	25 59.75	8.340	81 24 56.1	.12	50.38	1
10 215	Circini	10.0	26 . .	4.856	65 58 12.3	.12	52.57	1
10 216	Gould, Z. C., 14h 1800 .	9.5	14 26 1.37	+ 4.986	67 25 17.8	− 16.12	52.45	2
10 217	Gould, Z. C., 14h 1811 .	8.5	5.69	5.215	69 37 58.2	.12	52.34	1
10 218	Gould, Z. C., 14h 1854 .	8.8	17.29	6.536	77 2 50.0	.11	51.98	2
10 219	Gould, 19763	9.0	21.10	5.707	73 10 11.8	.10	51.40	2
10 220	Apodis	10.0	25.76	+ 6.454	76 44 10.4	− 16.10	51.25	1
10 221	Lacaille, 5972	7.0	14 26 26.75	+ 5.944	74 29 5.6	− 16.10	52.48	1
10 222	Apodis	9.5	27.32	5.578	72 20 48.1	.10	52.46	2
10 223	Apodis	9.5	27.69	5.529	72 1 11.6	.10	52.41	1
10 224	Apodis	9.5	30.08	5.387	70 59 25.0	.10	52.50	2
10 225	Circini	10.0	41.88	+ 5.189	69 20 10.4	− 16.09	52.42	1

Number.	Constellation, Name of Star, or Synonym.	Magnitude.	Right Ascension, 1850.0.	Annual Precession.	South Declination, 1850.0.	Annual Precession.	Mean year.	No. of obs.
			h. m. s.	s.	° ′ ″	″		
10 226	Lacaille, 5986	7.8	14 26 45.30	+ 5.005	67 32 51.2	− 16.08	52.45	2
10 227	Apodis	9.8	52.17	5.389	70 58 2.7	.08	52.50	2
10 228	Apodis	10.0	52.33	5.711	73 8 50.1	.08	52.35	1
10 229	Apodis	9.8	26 52.85	8.762	82 0 21.2	.08	50.54	2
10 230	Gould, Z. C., 14h 1876	9.0	27 0.12	+ 5.486	71 39 53.1	− 16.07	52.32	1
10 231	Gould, Z. C., 14h 1911	9.0	14 27 2.78	+ 6.379	76 23 47.0	− 16.07	51.34	1
10 232	Gould, 19793	8.5	7.86	6.096	75 10 20.1	.06	51.48	2
10 233	Apodis	9.0	9.63	7.497	79 44 31.8	.06	51.24	1
10 234	Apodis	10.0	19.53	6.736	77 39 41.7	.05	52.36	1
10 235	Apodis	9.8	20.44	+ 5.571	72 13 7.6	− 16.05	52.46	2
10 236	Apodis	10.0	14 27 21.06	+ 5.907	74 12 44.9	− 16.05	52.50	2
10 237	Gould, Z. C., 14h 1889	9.5	28.55	4.953	66 54 15.8	.05	52.57	1
10 238	Apodis	10.0	29.88	5.765	73 24 42.3	.04	52.52	1
10 239	Gould, 18798	8.5	30.39	6.571	77 5 9.4	.04	51.98	2
10 240	Octantis	9.5	38.66	+12.660	85 12 45.6	− 16.04	51.93	2
10 241	Gould, Z. C., 14h 1937	9.0	14 27 40.19	+ 5.811	73 39 34.1	− 16.04	52.52	1
10 242	Circini	10.0	40.90	5.091	68 18 34.1	.03	52.53	1
10 243	Circini	9.5	41.98	5.073	68 7 40.6	.03	52.53	1
10 244	Apodis	9.5	44.62	6.190	75 33 4.7	.03	52.36	1
10 245	Gould, Z. C., 14h 1930	9.5	45.66	+ 5.328	70 23 45.9	− 16.03	52.40	1
10 246	Gould, Z. C., 14h 1978	10.0	14 27 48.32	+ 6.943	78 16 17.0	− 16.03	51.35	1
10 247	Apodis	10.0	54.04	6.404	76 25 50.5	.02	51.34	1
10 248	Gould, Z. C., 14h 1957	9.0	56.16	5.769	73 23 39.4	.02	51.48	2
10 249	Gould, Z. C., 14h 1929	9.0	57.04	4.893	66 9 47.1	.02	52.58	2
10 250	Gould, 19821	9.0	27 57.54	+ 7.016	78 28 22.6	− 16.02	51.35	1
10 251	Gould, Z. C., 14h 1952	9.5	14 28 1.44	+ 5.435	71 11 30.3	− 16.02	52.48	1
10 252	Gould, 19811	9.0	5.85	5.784	73 28 6.4	.01	51.48	2
10 253	Gould, Z. C., 14h 1954	9.5	7.73	5.252	69 43 33.8	.01	52.34	1
10 254	Gould, Z. C., 14h 1959	7.8	19.96	4.908	66 17 32.2	.00	52.57	3
10 255	Apodis	10.0	23.47	+ 6.913	78 8 36.1	− 16.00	50.44	1
10 256	Apodis	9.8	14 28 27.11	+ 5.596	72 16 31.6	− 15.99	52.46	2
10 257	Octantis	9.5	27.38	12.838	85 16 38.4	.99	51.93	3
10 258	Gould, Z. C., 14h 2020	10.0	28.33	6.967	78 17 49.7	.99	51.35	1
10 259	Circini	9.5	28.34	4.866	65 46 50.1	.99	52.57	1
10 260	Circini	10.0	33.12	+ 4.960	66 50 52.2	− 15.99	52.57	1
10 261	Apodis	9.0	14 28 34.35	+ 6.603	77 7 14.6	− 15.99	51.98	2
10 262	Gould, Z. C., 14h 2017	10.0	38.39	6.489	76 42 6.7	.98	51.34	1
10 263	Apodis	10.0	39.88	6.118	75 8 48.2	.98	52.40	1
10 264	Gould, Z. C., 14h 2012	8.5	47.62	5.953	74 20 21.4	.97	52.50	2
10 265	Gould, Z. C., 14h 1996	9.5	49.91	+ 5.118	68 26 16.3	− 15.97	52.53	1

Number.	Constellation, Name of Star, or Synonym.	Magnitude.	Right Ascension, 1850.0.	Annual Precession.	South Declination, 1850.0.	Annual Precession.	Mean year.	No. of obs.
			h. m. s.	s.	° ′ ″	″		
10 266	Apodis	10.0	14 28 53.35	+ 6.153	75 17 26.6	− 15.97	50.56	1
10 267	Gould, Z. C., 14^h 2001	9.0	29 0.67	5.038	67 37 14.7	.96	52.52	1
10 268	Apodis	9.0	4.49	5.638	72 29 19.5	.96	52.52	1
10 269	Gould, Z. C., 14^h 2007	10.0	5.65	5.039	67 37 28.4	.96	52.52	1
10 270	Gould, Z. C., 14^h 2026	9.5	7.51	+ 5.807	73 30 30.7	− 15.96	52.52	1
10 271	Octantis	10.0	14 29 10.76	+12.959	85 18 56.0	− 15.96	51.93	1
10 272	Octantis	10.0	17.18	10.207	83 31 5.3	.95	50.53	2
10 273	Apodis	9.0	19.26	8.874	82 2 42.2	.95	50.53	1
10 274	Gould, 19834	10.0	24.20	5.008	66 47 44.6	.94	52.57	1
10 275	Gould, Z. C., 14^h 2036	9.0	24.68	+ 5.779	73 19 8.1	− 15.94	50.44	1
10 276	α Apodis	4.5	14 29 28.95	+ 7.026	78 24 1.2	− 15.94	51.35	1
10 277	Gould, Z. C., 14^h 2072	9.5	42.53	6.439	76 25 59.7	.93	51.34	1
10 278	Gould, Z. C., 14^h 2040	9.0	45.23	5.112	68 16 58.1	.93	52.00	2
10 279	Apodis	10.0	46.06	6.888	77 58 26.4	.92	50.44	1
10 280	Octantis	9.2	46.74	+13.942	85 43 21.8	− 15.92	51.93	4
10 281	Gould, Z. C., 14^h 2066	9.0	14 29 50.62	+ 6.021	74 34 52.3	− 15.92	52.44	2
10 282	Gould, Z. C., 14^h 2059	9.0	29 58.68	5.413	70 49 57.4	.91	52.52	1
10 283	Gould, Z. C., 14^h 2092	8.5	30 0.67	6.316	75 54 54.6	.91	51.83	2
10 284	Apodis	9.8	9.64	7.499	79 34 21.8	.90	51.83	2
10 285	Apodis	8.5	13.19	+ 5.520	71 33 37.2	− 15.90	52.40	2
10 286	Apodis	9.8	14 30 17.04	+ 5.955	74 12 30.0	− 15.90	52.50	2
10 287	Gould, Z. C., 14^h 2098	9.5	20.24	5.808	73 24 15.3	.89	51.48	2
10 288	Apodis	9.5	21.89	5.695	72 43 47.9	.89	52.43	2
10 289	Gould, Z. C., 14^h 2085	9.0	25.70	5.122	68 17 55.8	.89	52.00	2
10 290	Apodis	10.0	29.28	+ 5.808	73 23 17.6	− 15.89	52.52	1
10 291	Gould, Z. C., 14^h 2089	10.0	14 30 29.32	+ 5.119	68 15 55.0	− 15.89	51.46	1
10 292	Apodis	9.5	33.76	5.641	72 22 6.8	.88	52.52	1
10 293	Apodis	9.5	34.73	6.238	75 32 7.1	.88	52.36	1
10 294	Apodis	9.0	38.20	8.342	81 11 19.7	.88	50.57	1
10 295	Apodis	9.0	47.69	+ 9.109	82 17 12.1	− 17.87	50.47	3
10 296	Gould, Z. C., 14^h 2105	9.0	14 30 51.48	+ 4.920	66 7 54.4	− 15.87	52.57	1
10 297	Lacaille, 6011	7.3	30 57.89	5.207	69 1 28.8	.86	51.12	3
10 298	Apodis	9.2	31 0.94	5.346	70 12 22.4	.86	51.92	2
10 299	Gould, Z. C., 14^h 2120	9.5	2.03	5.277	69 38 4.1	.86	52.34	1
10 300	Apodis	9.0	2.41	+ 8.370	81 12 52.6	− 15.86	50.57	1
10 301	Apodis	9.5	14 31 23.05	+ 7.151	78 37 44.5	− 15.84	51.35	1
10 302	Gould, 19879	8.5	24.40	5.516	71 26 36.7	.84	52.40	2
10 303	Gould, Z. C., 14^h 2145	8.8	26.24	5.257	69 25 8.4	.83	52.38	2
10 304	Circini	8.6	29.81	4.983	66 45 30.6	.83	52.57	4
10 305	Gould, 19887	10.0	30.23	+ 6.210	75 20 23.3	− 15.83	50.56	1

Number.	Constellation, Name of Star, or Synonym.	Magnitude.	Right Ascension, 1850.0.	Annual Precession.	South Declination, 1850.0.	Annual Precession.	Mean year.	No. of obs.
			h. m. s.	s.	° ′ ″	″		
10 306	Circini	9.8	14 31 30.42	+ 5.006	66 59 52.2	− 15.83	52.50	3
10 307	Octantis	9.1	38.14	13.950	85 40 51.7	.82	51.93	4
10 308	Octantis	9.8	41.56	13.389	85 26 45.1	.82	51.93	4
10 309	Gould, Z. C., 14^h 2157	8.5	44.20	4.902	65 48 56.2	.82	52.57	1
10 310	Gould, 19883	9.2	45.14	+ 5.270	69 29 58.0	− 15.82	52.38	2
10 311	Apodis	9.5	14 31 51.23	+ 8.643	81 36 2.8	− 15.81	50.57	1
10 312	Apodis	9.2	31 57.80	7.715	79 56 55.6	.81	51.31	2
10 313	Gould, Z. C., 14^h 2190	8.2	32 12.24	5.428	70 43 14.1	.79	52.45	2
10 314	Octantis	10.0	13.37	40.087	88 43 27.9	.79	51.93	1
10 315	Gould, Z. C., 14^h 2241	9.5	22.56	+ 7.127	78 30 9.0	− 15.78	51.35	1
10 316	Gould, 19898	7.7	14 32 23.54	+ 5.218	68 58 35.7	− 15.78	51.12	3
10 317	Apodis	9.5	26.94	7.610	79 41 39.4	.78	51.25	1
10 318	Apodis	10.0	28.60	6.471	76 21 6.5	.78	51.34	1
10 319	Gould, Z. C., 14^h 2214	9.8	49.86	4.954	66 16 34.3	.75	52.58	2
10 320	Gould, Z. C., 14^h 2272	9.5	32 54.64	+ 7.089	78 21 45.6	− 15.76	51.35	1
10 321	Circini	9.5	14 33 2.12	+ 5.133	68 7 9.7	− 15.75	52.01	2
10 322	Apodis	9.0	4.65	5.707	72 33 19.3	.75	52.52	1
10 323	Apodis	10.0	5.60	6.483	76 21 15.8	.75	51.34	1
10 324	Apodis	9.8	11.54	7.750	79 57 22.8	.74	51.31	2
10 325	Gould, Z. C., 14^h 2270	9.5	13.96	+ 6.262	75 26 7.3	− 15.74	51.46	2
10 326	Apodis	9.2	14 33 22.93	+ 7.852	80 9 18.6	− 15.73	50.90	2
10 327	Circini	10.0	24.28	5.180	68 31 21.0	.73	52.53	1
10 328	Apodis	10.0	28.03	5.618	71 56 36.5	.73	52.41	1
10 329	Octantis	10.2	30.92	42.443	88 47 31.6	.72	51.93	2
10 330	Gould, Z. C., 14^h 2254	9.0	31.72	+ 5.040	67 7 45.9	− 15.72	52.48	2
10 331	Gould, Z. C., 14^h 2276	8.0	14 33 36.18	+ 5.545	71 26 1.4	− 15.72	52.40	2
10 332	Gould, Z. C., 14^h 2266	9.0	38.22	5.075	67 28 52.9	.72	52.49	3
10 333	Apodis	9.5	43.08	5.648	72 7 19.7	.71	52.41	1
10 334	Apodis	10.0	48.12	5.679	72 18 46.5	.71	52.41	1
10 335	Apodis	9.8	49.16	+ 5.383	70 13 22.4	− 15.71	51.92	2
10 336	Gould, Z. C., 14^h 2287	9.2	14 33 53.73	+ 5.224	68 52 11.9	− 15.70	50.48	2
10 337	Gould, Z. C., 14^h 2321	9.5	34 8.93	5.706	72 27 8.3	.69	52.52	1
10 338	Gould, Z. C., 14^h 2314	9.5	9.42	5.535	71 18 56.4	.69	52.48	1
10 339	Gould, Z. C., 14^h 2309	9.5	13.47	5.233	68 54 58.9	.68	50.45	1
10 340	Gould, Z. C., 14^h 2349	10.0	13.54	+ 6.525	76 25 59.5	− 15.68	51.34	1
10 341	Circini	9.5	14 34 15.05	+ 5.034	66 58 56.7	− 15.68	52.48	2
10 342	Octantis	9.0	17.16	26.722	87 58 52.3	.68	51.93	2
10 343	Gould, Z. C., 14^h 2378	9.0	20.36	7.069	78 12 59.2	.68	50.44	1
10 344	Apodis	9.2	24.47	7.360	78 59 23.5	.67	51.84	2
10 345	Apodis	9.0	27.58	+ 5.765	72 46 56.5	− 15.67	52.35	1

Number.	Constellation, Name of Star, or Synonym.	Magnitude.	Right Ascension, 1850.0.	Annual Precession.	South Declination, 1850.0.	Annual Precession.	Mean year.	No. of obs.
			h. m. s.	s.	° ′ ″	″		
10 346	Lacaille, 6022	6.7	14 34 28.87	+6.564	76 32 34.5	−15.67	51.34	1
10 347	Circini	9.5	47.16	5.188	68 26 37.9	.65	52.53	1
10 348	Gould, Z. C., 14ʰ 2352	10.0	50.94	5.148	68 4 2.4	.65	51.46	1
10 349	Gould, Z. C., 14ʰ 2368 .	9.5	51.60	5.481	70 51 35.9	.65	52.52	1
10 350	Lacaille, 6046	7.8	51.74	+5.129	67 52 56.6	−15.65	52.18	3
10 351	Lacaille, 6045	7.5	14 34 58.75	+5.366	69 57 44.8	−15.64	51.89	2
10 352	Gould, Z. C., 14ʰ 2367 .	9.0	35 0.01	5.132	67 53 38.7	.64	52.18	3
10 353	Circini	9.2	3.88	4.975	66 15 13.2	.64	52.58	2
10 354	Octantis	9.5	4.01	12.503	84 55 36.9	.64	50.51	1
10 355	Gould, Z. C., 14ʰ 2371 .	8.8	6.04	+5.064	67 12 1.8	−15.64	52.47	2
10 356	Apodis	10.0	14 35 10.53	+6.546	76 26 36.2	−15.63	51.34	1
10 357	Gould, 19971	8.5	13.44	6.176	74 53 45.8	.63	52.40	1
10 358	Gould, Z. C., 14ʰ 2393 .	9.0	15.13	5.443	70 31 56.9	.63	52.40	1
10 359	Circini	9.5	15.96	5.466	70 42 23.5	.63	52.45	2
10 360	Gould, Z. C., 14ʰ 2410 .	10.0	37.58	+5.319	69 30 43.4	−15.61	52.42	1
10 361	Brisbane, 5033	8.5	14 35 37.92	+4.941	65 48 10.4	−15.61	52.58	2
10 362	Octantis	10.0	49.00	11.838	84 31 21.6	.60	51.00	2
10 363	Lacaille, 6044	7.5	52.16	6.064	74 18 6.8	.59	52.50	2
10 364	Brisbane, 5036	8.2	54.52	4.951	65 53 26.8	.59	52.58	2
10 365	Apodis	10.0	58.44	+5.469	70 39 49.3	.59	52.46	2
10 366	Brisbane, 5037	8.0	14 35 59.28	+4.956	65 56 20.8	−15.59	52.58	2
10 367	Gould, Z. C., 14ʰ 2486 .	9.0	36 2.87	7.476	79 10 48.2	.58	52.41	1
10 368	Brisbane, 5022	6.2	3.50	9.563	82 36 40.2	.58	51.48	3
10 369	Octantis	9.8	4.20	14.420	85 45 27.0	.58	51.93	4
10 370	Apodis	8.8	4.68	+6.404	75 49 17.5	−15.58	51.87	2
10 371	Apodis	9.5	14 36 6.01	+5.407	70 10 31.4	+15.58	52.40	1
10 372	Lacaille, 6059	6.8	6.90	4.959	65 57 32.2	.58	52.58	2
10 373	Apodis	8.2	8.32	19.411	87 2 58.6	.58	51.93	3
10 374	Circini	9.0	11.24	4.982	66 12 36.4	.58	52.58	2
10 375	Gould, 19983	9.0	11.42	+4.959	65 56 53.6	−15.58	52.58	2
10 376	Apodis	9.5	14 36 12.16	+6.925	77 40 7.5	−15.58	52.36	1
10 377	Apodis	10.0	12.91	8.153	80 35 13.2	.57	50.43	1
10 378	Circini	9.5	13.89	5.079	67 13 45.7	.57	52.38	1
10 379	Apodis	9.5	21.91	8.154	80 34 53.6	.57	50.43	1
10 380	Lacaille, 6058	7.7	22.19	+5.191	68 18 9.3	−15.57	52.18	3
10 381	Apodis	9.8	14 36 25.80	+6.639	76 41 55.8	−15.56	51.91	2
10 382	Melbourne (1), 749 . . .	7.5	38.56	5.022	66 35 22.2	.55	52.58	2
10 383	Apodis	9.8	40.17	9.360	82 21 3.4	.55	50.51	2
10 384	Apodis	9.8	40.84	5.597	71 30 19.9	.55	52.40	2
10 385	Apodis	9.5	46.17	+6.157	74 41 11.0	−15.54	52.48	1

No. 10 370. Mean of two observations differing 8.0″.

Number.	Constellation, Name of Star, or Synonym.	Magnitude.	Right Ascension, 1850.0.	Annual Precession.	South Declination, 1850.0.	Annual Precession.	Mean year.	No. of obs.
			h. m. s.	s.	° ′ ″	″		
10386	Apodis	10.0	14 36 46.45	+ 6.594	76 31 31.8	− 15.54	51.34	1
10387	Lacaille, 6061	8.5	47.27	5.381	69 54 10.4	.54	51.89	2
10388	Herschel 4693[1]	10.0	47.87	5.807	72 49 59.6	.54	52.40	1
10389	Gould, 20007	9.5	49.06	5.382	69 54 24.8	.54	51.89	2
10390	Herschel 4693[2]	12.0	36 49.54	+ 5.808	72 50 1.8	− 15.54	52.35	1
10391	Lacaille, 6036	8.2	14 37 0.13	+ 7.140	78 15 25.4	− 15.53	50.90	2
10392	Lacaille, 6009	9.2	11.26	9.745	82 46 4.7	.52	51.97	2
10393	Apodis	10.0	16.40	5.486	70 39 47.1	.52	52.52	1
10394	Circini	10.0	24.69	5.154	67 50 41.5	.51	52.52	1
10395	Apodis	9.8	25.44	+ 9.425	82 23 54.6	− 15.51	50.51	2
10396	Apodis	9.0	14 37 30.78	+ 5.952	73 35 27.6	− 15.50	52.52	1
10397	Gould, 20020 . . .	7.5	36.56	5.058	66 52 0.9	.50	52.57	1
10398	Octantis	8.8	50.09	10.027	83 2 3.4	.49	51.48	3
10399	Gould, Z. C., 14h 2545 .	8.8	52.31	5.159	67 50 46.6	.48	52.54	2
10400	Gould, Z. C., 14h 2557 .	9.0	53.09	+ 5.454	70 21 47.9	− 15.48	52.40	1
10401	Gould, Z. C., 14h 2563 .	9.0	14 37 58.81	+ 5.470	70 28 33.0	− 15.48	52.40	1
10402	Gould, Z. C., 14h 2551 .	9.5	38 0.65	4.969	65 51 23.5	.48	52.57	1
10403	Apodis	9.2	19.97	5.671	71 50 59.2	.46	52.40	2
10404	Apodis	10.0	20.82	6.743	76 56 4.8	.46	52.48	1
10405	Lacaille, 6066	5.8	21.22	+ 5.785	72 33 52.3	− 15.46	52.43	2
10406	Gould, Z. C., 14h 2630 .	9.5	14 38 32.21	+ 6.768	77 1 23.1	− 15.45	51.99	2
10407	Gould, 20061	8.2	32.60	6.250	74 58 36.9	.45	51.48	2
10408	Gould, Z. C., 14h 2605 .	9.0	38.26	5.551	71 0 15.6	.44	52.50	2
10409	Apodis	9.2	43.52	6.029	73 53 45.3	.44	52.52	2
10410	Octantis	10.0	43.92	+15.163	85 57 38.8	− 15.44	51.93	1
10411	Octantis	9.0	14 38 44.14	+12.036	84 33 34.4	− 15.43	51.00	2
10412	Circini	9.5	44.31	5.121	67 22 41.8	.43	52.38	1
10413	Apodis	10.0	38 51.58	6.579	76 18 39.4	.43	51.34	1
10414	Apodis	9.8	39 13.24	5.560	71 0 55.2	.41	52.50	2
10415	Gould, 20066	9.0	14.19	+ 5.036	66 26 46.4	− 15.41	52.58	1
10416	Brisbane, 5046	6.7	14 39 14.21	+ 9.510	82 25 34.6	− 15.41	51.15	3
10417	Gould, 20071	8.8	14.86	6.092	74 10 34.5	.41	52.50	2
10418	Apodis	8.5	16.13	5.710	72 1 23.7	.41	52.41	1
10419	Lacaille, 6006	7.0	28.31	11.487	84 11 4.4	.39	51.00	2
10420	Gould, Z. C., 14h 2639 .	9.5	28.92	+ 5.068	66 45 54.8	− 15.39	52.58	2
10421	Apodis	10.0	14 39 35.44	+ 7.315	78 34 24.6	− 15.39	52.33	1
10422	Circini	10.0	39.17	5.326	69 9 49.9	.38	52.42	1
10423	Gould, Z. C., 14h 2660 .	9.5	41.03	5.297	68 54 47.9	.38	50.45	1
10424	Circini	10.0	42.03	5.297	68 54 47.9	.38	50.45	1
10425	Circini	9.5	42.53	+ 5.135	67 24 39.8	− 15.38	52.45	2

No. 10423. Gould's declination is 68° 54′ 34.4″ (1850).

Number.	Constellation, Name of Star, or Synonym.	Magnitude.	Right Ascension, 1850.0.	Annual Precession.	South Declination, 1850.0.	Annual Precession.	Mean year.	No. of obs.
			h. m. s.	s.	° ′ ″	″		
10426	Gould, Z. C., 14h 2678	9.8	14 39 46.16	+ 5.573	71 3 34.8	− 15.38	52.50	2
10427	Apodis	9.0	46.33	5.738	72 8 59.2	.38	52.41	1
10428	Apodis	10.0	39 52.06	5.573	71 2 48.8	.37	52.52	1
10429	Gould, Z. C., 14h 2737	9.5	40 6.27	7.276	78 26 26.9	.36	51.35	1
10430	Gould, 20086	8.2	13.24	+ 5.177	67 45 58.6	− 15.35	52.54	2
10431	Apodis	10.0	14 40 15.50	+ 5.956	73 23 9.1	− 14.35	52.52	1
10432	Octantis	8.5	15.58	20.753	87 12 48.2	.35	51.93	2
10433	Apodis	8.8	17.43	5.705	71 53 49.0	.35	52.40	2
10434	Octantis	8.6	18.10	16.098	86 12 37.8	.35	51.93	4
10435	Gould, Z. C., 14h 2723	9.0	24.78	+ 6.037	73 48 4.0	− 14.34	52.52	2
10436	Gould, Z. C., 14h 2753	10.0	14 40 26.17	+ 7.232	78 18 0.5	− 15.34	51.35	1
10437	Apodis	9.5	28.14	6.074	73 59 10.2	.34	52.52	1
10438	Gould, Z. C., 14h 2705	8.5	30.44	5.002	65 56 19.4	.34	52.57	1
10439	Gould, Z. C., 14h 2725	8.5	32.74	5.914	73 7 51.3	.33	51.39	2
10440	Lacaille, 6089	8.0	35.20	+ 5.289	68 45 1.5	− 15.33	51.49	2
10441	Octantis	10.0	14 40 41.04	+ 18.781	86 51 26.3	− 15.33	51.93	2
10442	Gould, Z. C., 14h 2760	9.0	42.83	6.666	76 30 19.0	.32	51.34	1
10443	Lacaille, 6085	8.2	52.80	5.818	72 32 39.6	.31	52.43	2
10444	Gould, Z. C., 14h 2770	9.0	53.92	6.655	76 27 2.2	.31	51.34	1
10445	Gould, Z. C., 14h 2740	9.2	55.39	+ 5.379	69 28 39.1	− 15.31	52.38	2
10446	Apodis	10.0	14 40 56.39	+ 5.491	70 21 9.6	− 15.31	52.40	1
10447	Lacaille, 6077	6.0	40 58.25	6.548	76 2 38.7	.31	51.30	1
10448	Circini	9.0	41 2.52	5.385	69 30 51.2	.31	52.34	1
10449	Circini	9.0	2.98	5.410	69 43 13.4	.30	52.34	1
10450	Apodis	9.5	4.90	+ 5.882	72 54 21.5	− 15.30	52.35	1
10451	Apodis	9.5	14 41 8.35	+ 5.831	72 36 10.1	− 15.30	52.35	1
10452	Gould, Z. C., 14h 2802	9.0	10.23	7.104	77 53 47.9	.30	52.36	1
10453	Lacaille, 6093	7.0	13.30	5.353	69 14 24.3	.30	52.42	1
10454	Lacaille, 6030	9.0	14.96	10.442	83 18 22.1	.29	50.50	1
10455	Apodis	10.0	18.31	+ 5.498	70 22 5.1	− 15.29	52.40	1
10456	Apodis	10.0	14 41 19.20	+ 6.337	75 9 19.8	− 15.29	52.40	1
10457	Gould, Z. C., 14h 2771	9.5	21.34	5.615	71 12 13.3	.29	52.48	1
10458	Apodis	9.5	25.41	5.807	72 26 4.7	.28	52.52	1
10459	Circini	10.0	26.18	5.320	68 56 6.2	.28	50.45	1
10460	Circini	9.2	28.64	+ 5.114	67 1 26.2	− 15.28	52.57	2
10461	Brisbane, 5071	7.0	14 41 33.94	+ 4.999	65 47 28.2	− 15.28	52.57	1
10462	Apodis	9.5	33.96	5.957	73 17 5.8	.28	50.44	1
10463	Gould, Z. C., 14h 2782	10.0	37.78	5.314	68 51 46.9	.27	50.45	1
10464	Gould, Z. C., 14h 2776	9.5	38.18	5.140	67 15 45.2	.27	52.56	1
10465	Gould, Z. C., 14h 2864	9.0	44.04	+ 7.827	79 39 57.3	− 15.27	51.24	1

Number.	Constellation, Name of Star, or Synonym.	Magnitude.	Right Ascension, 1850.0.	Annual Precession.	South Declination, 1850.0.	Annual Precession.	Mean year.	No. of obs.
			h. m. s.	s.	° ′ ″	″		
10 466	Circini	9.5	14 41 51.37	+ 5.394	69 30 52.3	− 15.26	52.42	1
10 467	Circini	8.2	51.70	5.443	69 54 1.0	.26	51.38	2
10 468	Apodis	10.0	41 53.28	6.074	73 52 34.7	.26	52.52	1
10 469	Gould, Z. C., 14ʰ 2821 .	8.2	42 5.38	5.558	70 44 14.2	.25	52.46	2
10 470	Gould, Z. C., 14ʰ 2824 .	9.0	10.51	+ 5.380	69 22 8.3	− 15.24	52.42	1
10 471	Circini	9.5	14 42 19.07	+ 5.287	68 33 30.7	− 15.23	52.53	1
10 472	Gould, Z. C., 14ʰ 2883	9.5	19.63	6.939	77 19 21.5	.23	51.49	1
10 473	Apodis	10.0	25.47	5.656	71 23 12.9	.23	52.32	1
10 474	Apodis	9.0	26.47	6.304	74 55 46.0	.23	52.40	1
10 475	Apodis	8.5	31.91	+ 6.189	74 23 37.2	− 15.22	52.48	1
10 476	Gould, Z. C., 14ʰ 2839 .	8.0	14 42 34.11	+ 5.032	66 2 24.6	15.22	52.57	1
10 477	Octantis	8.5	42.74	13.220	85 5 29.9	.21	51.46	3
10 478	Gould, Z. C., 14ʰ 2863 .	9.8	46.54	5.259	68 15 49.7	.21	52.54	2
10 479	Lacaille, 6088	6.8	46.83	6.717	76 32 51.3	.21	52.09	3
10 480	Apodis	10.0	42 54.12	+ 6.694	76 27 36.2	− 15.20	51.91	2
10 481	Gould, 20146	8.2	14 43 4.91	+ 5.569	70 43 33.4	− 15.19	52.46	2
10 482	Apodis	8.2	15.04	5.563	70 40 2.0	.18	52.46	2
10 483	Apodis	9.5	15.15	5.478	70 2 9.9	.18	51.44	1
10 484	Gould, Z. C., 14ʰ 2907 .	9.5	16.04	5.634	71 9 47.9	.18	52.48	1
10 485	Apodis	9.5	20.09	+ 6.376	75 10 39.8	− 15.17	52.40	1
10 486	Apodis	9.5	14 43 33.06	+ 6.255	74 37 29.3	− 15.16	52.44	2
10 487	Gould, 20156	8.0	33.61	5.582	70 46 45.8	.16	52.46	2
10 488	Gould, 20162	10.0	34.37	6.699	76 25 58.9	.16	51.34	1
10 489	Gould, Z. C., 14ʰ 2917 .	9.0	34.99	5.263	68 13 29.8	.16	52.54	2
10 490	Apodis	9.5	37.18	+ 5.830	72 23 17.8	− 15.16	52.46	2
10 491	Apodis	10.0	14 43 47.95	+ 6.470	75 32 19.5	− 15.15	52.36	1
10 492	Circini	10.0	49.14	5.108	66 42 42.3	.15	52.57	1
10 493	Gould, Z. C., 14ʰ 2940 .	10.0	52.10	5.637	71 8 8.7	.14	52.48	1
10 494	Lacaille, 6105	7.0	54.55	6.217	74 25 14.3	.14	52.48	1
10 495	Apodis	10.0	43 59.01	+ 5.666	71 19 16.0	− 15.14	52.48	1
10 496	Apodis	9.0	14 44 4.23	+ 5.864	72 33 6.1	− 15.13	52.52	1
10 497	Apodis	9.5	6.24	6.829	76 51 3.7	.13	52.48	1
10 498	Apodis	9.0	8.32	5.526	70 18 59.0	.13	51.92	2
10 499	Apodis	10.0	9.56	6.047	73 33 38.6	.13	52.52	1
10 500	Gould, Z. C., 14ʰ 2960 .	10.0	10.73	+ 5.649	71 11 2.1	− 15.13	52.48	1
10 501	Octantis	10.0	14 44 14.47	+ 14.647	85 39 40.0	− 15.12	51.93	2
10 502	Gould, 20164	7.8	18.21	5.069	66 15 27.3	.12	52.58	3
10 503	Apodis	9.5	21.98	7.771	79 24 28.8	.12	52.41	1
10 504	Apodis	9.5	26.55	5.522	70 15 45.0	.11	52.40	1
10 505	Gould, 20182	8.5	27.72	+ 5.980	73 10 40.6	− 15.11	51.39	2

No. 10 493. Gould's declination agrees with No. 10 495.

Number.	Constellation, Name of Star, or Synonym.	Magnitude.	Right Ascension, 1850.0.	Annual Precession.	South Declination, 1850.0.	Annual Precession.	Mean year.	No. of obs.
			h. m. s.	s.	° ′ ″	″		
10 506	Apodis	9.5	14 44 32.75	+ 5.752	71 49 53.1	− 15.11	52.41	1
10 507	Circini	10.0	35.62	5.144	66 59 43.8	.10	52.56	1
10 508	Circini	8.8	37.39	5.427	69 31 1.0	.10	52.38	2
10 509	Gould, Z. C., 14^h 2999 .	9.0	38.89	5.768	71 55 16.7	.10	52.41	1
10 510	Gould, Z. C., 14^h 3024 .	9.0	47.00	+ 6.794	76 41 22.8	− 15.09	52.19	4
10 511	Gould, Z. C., 14^h 3013 .	7.8	14 44 49.54	+ 5.940	72 55 34.3	− 15.09	52.44	3
10 512	Circini	9.0	50.80	5.250	67 58 37.9	.09	52.56	1
10 513	Apodis	9.0	53.74	6.072	73 38 3.3	.09	52.52	1
10 514	Apodis	9.0	44 56.85	5.700	71 27 42.6	.08	52.40	2
10 515	Apodis	10.0	45 7.77	+ 6.125	73 52 58.5	− 15.07	52.52	1
10 516	Octantis	9.8	14 45 9.76	+ 10.356	83 5 35.5	− 15.07	51.01	2
10 517	Gould, 20194	8.3	11.70	5.970	73 3 44.2	.07	51.94	4
10 518	Gould, Z. C., 14^h 3016 .	9.8	12.67	5.237	67 49 41.9	.07	52.54	2
10 519	Gould, Z. C., 14^h 3034 .	8.5	18.68	5.900	72 39 40.6	.06	52.46	4
10 520	Gould, Z. C., 14^h 3038 .	9.0	27.80	+ 5.613	70 49 31.5	− 15.05	52.52	1
10 521	Apodis	9.0	14 45 30.03	+ 5.789	71 58 46.1	− 15.05	52.41	1
10 522	Gould, Z. C., 14^h 3052 .	8.5	30.36	5.996	73 11 7.4	.05	51.40	2
10 523	Gould, Z. C., 14^h 3045 .	8.2	37.44	5.387	69 5 47.2	.04	51.44	2
10 524	Octantis	10.0	37.47	14.787	85 40 59.6	.04	51.93	2
10 525	Gould, Z. C., 14^h 3058 .	10.0	45 54.43	+ 5.152	66 56 39.2	− 15.03	52.57	1
10 526	Gould, 20213	8.0	14 46 7.94	+ 5.305	68 20 36.0	− 15.01	52.54	2
10 527	Apodis	8.2	10.97	5.923	72 43 29.2	.01	52.46	4
10 528	Apodis	7.5	11.73	5.819	72 6 43.7	.01	52.41	1
10 529	Gould, Z. C., 14^h 3085 .	8.5	16.66	5.335	68 35 52.6	.01	51.49	2
10 530	Gould, 20227	7.0	18.52	+ 6.191	74 7 13.2	− 15.00	52.52	1
10 531	Gould, Z. C., 14^h 3098 .	9.5	14 46 29.76	+ 5.403	69 8 35.7	− 14.99	51.44	2
10 532	Apodis	9.5	30.68	5.830	72 8 50.2	.99	52.41	1
10 533	Apodis	10.0	34.71	5.696	71 17 44.2	.99	52.48	1
10 534	Apodis	9.5	47.41	6.197	74 6 44.4	.98	52.52	1
10 535	Trianguli Australis . . .	9.0	51.87	+ 5.485	69 45 36.4	− 14.97	52.34	1
10 536	Gould, 20241	9.0	14 46 52.66	+ 6.488	75 23 55.4	− 14.97	52.36	1
10 537	Gould, Z. C., 14^h 3160 .	8.0	53.76	7.108	77 33 57.4	.97	51.92	2
10 538	Apodis	9.8	55.14	5.559	70 18 52.3	.97	51.92	2
10 539	Lacaille, 6128	7.0	56.36	5.640	70 53 5.6	.97	52.51	1
10 540	Gould, Z. C., 14^h 3133 .	9.5	46 58.36	+ 5.660	71 1 3.9	− 14.96	52.52	1
10 541	Apodis	10.0	14 47 16.32	+ 5.562	70 18 2.9	− 14.95	51.92	2
10 542	Apodis	9.2	30.28	5.736	71 28 36.6	.93	52.40	2
10 543	Lacaille, 6108	6.8	31.81	7.994	79 43 4.6	.93	51.28	3
10 544	Gould, Z. C., 14^h 3187 .	9.0	33.33	6.467	75 15 56.4	.93	50.56	1
10 545	Gould, Z. C., 14^h 3156 .	8.5	43.37	+ 5.351	68 36 47.7	− 14.93	51.49	2

No. 10 528. Possibly variable.

Number.	Constellation, Name of Star, or Synonym.	Magnitude.	Right Ascension, 1850.0.	Annual Precession.	South Declination, 1850.0.	Annual Precession.	Mean year.	No. of obs.
			h. m. s.	s.	° ′ ″	″		
10546	Gould, Z. C., 14h 3189	9.0	14 47 37.34	+ 6.384	74 54 34.5	− 14.93	52.40	1
10547	Gould, Z. C., 14h 3198	8.5	46.77	6.350	74 44 55.9	.92	52.44	2
10548	Apodis	10.0	48.10	7.445	78 26 52.6	.92	51.35	1
10549	Trianguli Australis	10.5	56.99	5.218	67 22 35.4	.91	52.54	2
10550	Lacaille, 6136	8.0	47 57.29	+ 5.218	67 22 42.6	− 14.91	52.54	2
10551	Apodis	9.0	14 48 1.53	+ 5.710	71 16 4.8	− 14.90	52.48	1
10552	Octantis	9.0	14.43	29.994	88 5 39.8	.89	51.93	3
10553	Apodis	8.0	17.86	5.696	71 9 2.7	.89	52.48	1
10554	Gould, 20285	9.5	21.67	6.273	74 21 37.0	.88	52.50	2
10555	Apodis	10.0	25.77	+ 7.806	79 16 39.7	− 14.88	52.41	1
10556	Gould, 20302	9.0	14 48 27.53	+ 7.198	77 44 24.7	− 14.88	51.40	2
10557	Trianguli Australis	9.5	27.83	5.273	67 50 11.2	.88	52.56	1
10558	Apodis	9.5	28.90	6.090	73 27 10.8	.88	52.52	1
10559	Gould, Z. C., 14h 3220	9.0	30.40	5.135	66 30 11.4	.88	52.58	1
10560	Octantis	9.2	39.00	+10.947	83 29 52.3	− 14.87	50.53	2
10561	Circini	9.0	14 48 45.09	+ 5.159	66 43 7.1	− 14.86	52.58	1
10562	Gould, Z. C., 14h 3237	9.2	45.66	5.166	66 47 42.2	.86	52.58	2
10563	Gould, 20305	8.5	50.02	6.434	75 2 29.6	.86	52.40	1
10564	Apodis	8.5	52.38	5.911	72 26 27.1	.85	52.52	1
10565	Lacaille, 6144	8.2	58.23	+ 5.566	70 11 10.7	− 14.85	52.13	3
10566	Apodis	9.8	14 48 59.62	+ 6.872	76 41 25.6	− 14.85	51.81	2
10567	Apodis	9.5	49 14.20	6.873	76 40 55.5	.83	51.61	2
10568	Apodis	9.5	19.36	6.263	74 14 30.1	.83	52.50	2
10569	Gould, 20309	9.2	20.06	5.524	69 50 11.2	.83	51.89	2
10570	Gould, Z. C., 14h 3305	9.0	20.48	+ 6.403	74 52 23.0	− 14.83	52.40	1
10571	Gould, Z. C., 14h 3364	9.5	14 49 29.50	+ 8.074	79 47 17.4	− 14.82	51.27	3
10572	Gould, Z. C., 14h 3311	9.5	30.40	6.200	73 55 48.8	.82	52.52	1
10573	Circini	9.3	32.28	5.166	66 42 34.3	.81	52.58	3
10574	Gould, 20317	9.5	33.99	6.071	73 16 13.3	.81	50.44	1
10575	Gould, Z. C., 14h 3348	9.5	37.16	+ 7.285	77 55 11.1	− 14.81	51.40	2
10576	Trianguli Australis	9.5	14 49 39.61	+ 5.178	66 49 2.9	− 14.81	52.57	1
10577	Gould, Z. C., 14h 3323	9.5	48.71	6.077	73 17 26.3	.80	50.44	1
10578	Gould, Z. C., 14h 3309	8.5	49.11	5.477	69 26 21.0	.80	52.38	2
10579	Gould, 20327	8.5	50.57	6.220	73 59 58.0	.80	52.52	1
10580	Apodis	10.0	51.60	+ 7.296	77 56 18.0	− 14.80	51.40	2
10581	Octantis	9.5	14 49 52.38	+14.666	85 32 52.9	− 14.79	51.93	2
10582	Gould, Z. C., 14h 3327	9.5	52.52	5.948	72 34 31.3	.79	52.49	3
10583	Apodis	9.5	55.87	5.956	72 37 2.6	.79	52.50	3
10584	Apodis	9.5	49 56.81	6.727	76 7 19.4	.79	51.30	1
10585	Lacaille, 6126	7.8	50 5.26	+ 7.731	79 1 21.4	− 14.78	52.37	2

No. 10567. The declination may be 1′ larger.

Number.	Constellation, Name of Star, or Synonym.	Magnitude.	Right Ascension, 1850.0.	Annual Precession.	South Declination, 1850.0.	Annual Precession.	Mean year.	No. of obs.
			h. m. s.	s.	° ′ ″	″		
10 586	Apodis	9.5	14 50 12.08	+ 6.329	74 29 12.3	− 14.78	52.48	1
10 587	Apodis	10.0	14.43	8.253	80 5 54.7	.77	51.37	1
10 588	Apodis	9.0	19.67	6.687	75 56 56.5	.77	51.30	1
10 589	Lacaille, 6150	7.0	21.47	5.791	71 35 31.8	.77	52.40	2
10 590	Gould, Z. C., 14h 3351 .	9.2	22.73	+ 5.500	69 33 25.1	− 14.76	52.38	2
10 591	Gould, 20332	8.8	14 50 27.68	+ 5.591	70 14 2.7	− 14.76	52.03	3
10 592	Gould, Z. C., 14h 3395 .	9.5	27.93	6.841	76 29 40.0	.76	52.47	2
10 593	Apodis	9.5	35.04	6.046	73 3 38.8	.75	52.47	2
10 594	Apodis	9.8	41.59	6.999	77 1 19.0	.75	51.98	2
10 595	Octantis	12.0	43.72	+18.921	86 43 36.3	− 14.74	51.93	1
10 596	Trianguli Australis . . .	10.0	14 50 45.36	+ 5.139	66 12 59.6	− 14.74	52.59	1
10 597	Apodis	9.5	47.34	6.315	74 22 51.7	.74	52.48	1
10 598	Apodis	9.5	50 49.26	6.289	74 15 24.4	.74	52.52	1
10 599	Apodis	9.0	51 1.82	6.161	73 37 31.7	.73	52.52	1
10 600	Apodis	10.0	8.64	+ 8.583	80 38 34.4	− 14.72	50.56	1
10 601	Apodis	11.0	14 51 12.12	+ 7.716	78 55 58.4	− 14.72	52.33	1
10 602	Gould, Z. C., 14h 3445 .	9.5	19.78	6.857	76 28 44.8	.71	52.47	2
10 603	Gould, Z. C., 14h 3413 .	9.5	22.95	5.270	67 31 47.0	.71	52.56	1
10 604	Gould, Z. C., 14h 3422 .	10.0	31.02	5.274	67 33 10.8	.70	52.56	1
10 605	Lacaille, 6158	7.0	31.40	+ 5.803	71 34 33.0	− 14.70	52.40	2
10 606	Octantis	9.8	14 51 31.81	+12.164	84 17 6.4	− 14.70	51.49	2
10 607	Gould, Z. C., 14h 3438 .	8.5	34.05	5.953	72 28 19.8	.69	52.52	1
10 608	Apodis	9.8	35.36	8.083	79 42 24.2	.69	51.24	2
10 609	Trianguli Australis . . .	10.0	43.38	5.182	66 39 24.6	.68	52.58	2
10 610	Lacaille, 6163	7.5	44.52	+ 5.649	70 32 16.2	− 14.68	52.40	1
10 611	Trianguli Australis . . .	9.8	14 51 47.78	+ 5.225	67 4 16.5	− 14.68	52.57	2
10 612	Circini	9.0	50.33	5.111	65 55 19.6	.68	52.59	1
10 613	Octantis	8.2	54.25	18.615	86 38 37.7	.67	51.93	4
10 614	Gould, Z. C., 14h 3454 .	9.5	57.40	5.707	70 55 11.6	.67	52.52	1
10 615	Gould, Z. C., 14h 3488 .	10.0	51 58.08	+ 6.850	76 25 51.9	− 14.67	52.48	1
10 616	Apodis	9.0	14 52 0.43	+ 6.636	75 38 49.7	− 14.67	51.87	2
10 617	Gould, Z. C., 14h 3450 .	9.5	0.97	5.349	68 10 17.5	.67	52.54	2
10 618	Apodis	9.5	8.79	5.871	71 56 47.8	.66	52.41	1
10 619	Octantis	9.8	9.49	12.152	84 15 39.5	.66	51.49	2
10 620	Apodis	9.5	12.96	+ 6.634	75 37 33.9	− 14.66	51.87	2
10 621	Gould, Z. C., 14h 3473 .	9.0	14 52 16.34	+ 5.535	69 39 36.7	− 14.65	52.34	1
10 622	Trianguli Australis . . .	10.0	17.18	5.230	67 3 56.3	.65	52.56	1
10 623	Lacaille, 6167	7.0	21.01	5.772	71 18 47.8	.65	52.48	1
10 624	Octantis	7.5	24.83	19.048	86 43 36.9	.64	51.93	4
10 625	Apodis	9.0	26.06	+ 5.997	72 39 18.0	− 14.64	52.50	3

Number.	Constellation, Name of Star, or Synonym.	Magnitude.	Right Ascension, 1850.0.	Annual Precession.	South Declination, 1850.0.	Annual Precession.	Mean year.	No. of obs.
			h. m. s.	s.	° ′ ″	″		
10 626	Gould, Z. C., 14h 3544	8.2	14 52 26.77	+ 7.833	79 8 5.8	− 14.64	52.37	2
10 627	Gould, Z. C., 14h 3474	9.0	27.56	5.148	66 14 31.2	.64	52.59	2
10 628	Gould, Z. C., 14h 3482	9.5	28.99	5.309	67 46 43.2	.64	52.55	2
10 629	Gould, Z. C., 14h 3519	9.0	31.77	6.949	76 43 59.8	.64	52.48	1
10 630	Gould, Z. C., 14h 3568	8.8	52 53.85	+ 7.544	78 25 41.4	− 14.61	51.81	2
10 631	Brisbane, 5143	8.8	14 53 0.59	+ 8.366	80 10 55.8	− 14.61	50.90	2
10 632	Apodis	9.5	2.94	5.592	70 0 55.2	.61	52.56	1
10 633	Apodis	9.5	12.58	6.824	76 15 55.7	.60	52.48	1
10 634	Gould, Z. C., 14h 3533	9.5	13.47	5.603	70 5 4.1	.60	52.56	1
10 635	Gould, Z. C., 14h 3539	9.2	16.92	+ 5.582	69 55 28.0	− 14.59	52.45	2
10 636	Gould, Z. C., 14h 3579	9.5	14 53 23.50	+ 6.855	76 21 52.8	− 14.58	52.47	2
10 637	Gould, Z. C., 14h 3543	9.2	23.62	5.532	69 32 27.5	.58	52.38	2
10 638	Apodis	9.5	23.76	6.407	74 36 54.8	.58	52.44	2
10 639	Lacaille, 6169	7.8	29.40	6.469	74 52 31.5	.58	52.40	1
10 640	Apodis	10.0	35.22	+ 5.957	72 20 18.5	− 14.57	52.41	1
10 641	Apodis	10.0	14 53 38.07	+ 5.955	72 19 28.5	− 14.57	52.41	1
10 642	Trianguli Australis	9.8	38.52	5.303	67 37 7.4	.57	52.55	2
10 643	Trianguli Australis	9.5	43.20	5.587	69 55 18.0	.57	52.34	1
10 644	Circini	9.5	53 43.43	5.118	65 48 47.7	.56	52.59	1
10 645	Apodis	9.2	54 0.11	+ 7.822	79 2 6.6	− 14.55	52.37	2
10 646	Apodis	10.0	14 54 0.27	+ 6.664	75 37 39.5	− 14.55	51.38	1
10 647	Lacaille, 6184	7.0	0.30	5.858	71 43 12.5	.55	52.32	1
10 648	Lacaille, 6185	8.0	27.42	5.623	70 7 20.3	.52	52.00	2
10 649	Gould, Z. C., 14h 3609	9.0	30.00	5.403	68 24 35.1	.52	52.53	1
10 650	Gould, Z. C., 14h 3639	9.5	30.57	+ 5.256	73 50 28.3	− 14.52	52.52	1
10 651	Gould, Z. C., 14h 3680	9.0	14 54 33.20	+ 7.762	78 52 13.7	− 14.52	52.33	1
10 652	Gould, Z. C., 14h 3653	9.0	40.54	6.254	73 49 17.6	.51	52.52	2
10 653	Apodis	9.0	49.04	5.952	72 13 0.8	.50	52.46	2
10 654	Gould, Z. C., 14h 3662	8.5	54 54.72	5.331	74 10 5.4	.49	52.50	2
10 655	Apodis	9.5	55 3.91	+ 6.958	76 36 40.6	− 14.48	52.48	1
10 656	Trianguli Australis	9.5	14 55 5.54	+ 5.180	66 18 20.8	− 14.48	52.59	2
10 657	Trianguli Australis	9.0	10.12	5.592	69 50 10.7	.48	52.34	1
10 658	Gould, Z. C., 14h 3734	9.2	10.82	7.807	78 56 33.4	.48	52.37	2
10 659	Gould, Z. C., 14h 3672	8.8	16.45	5.710	70 39 47.8	.47	52.46	2
10 660	Gould, 20457	9.0	17.12	+ 6.154	73 16 46.4	− 14.47	50.44	1
10 661	Apodis	9.5	14 55 17.14	+ 6.928	76 30 10.8	− 14.47	52.48	1
10 662	Octantis	9.5	19.02	17.179	86 14 40.9	.47	51.93	2
10 663	Gould, Z. C., 14h 3660	9.5	19.19	5.240	66 52 22.0	.47	52.59	1
10 664	Gould, Z. C., 14h 3673	9.5	25.70	5.392	68 14 14.4	.46	52.54	2
10 665	Gould, 20452	7.5	25.80	+ 5.142	65 53 23.0	− 14.46	52.59	1

Number.	Constellation, Name of Star, or Synonym.	Magnitude.	Right Ascension, 1850.0.	Annual Precession.	South Declination, 1850.0.	Annual Precession.	Mean year.	No. of obs.
			h. m. s.	s.	° ′ ″	″		
10 666	Herschel, 4729[1]	10.0	14 55 31.60	+ 5.563	69 35 22.4	− 14.46	52.34	2
10 667	Lacaille, 6174	7.2	31.99	7.651	78 33 31.0	.46	52.01	3
10 668	Herschel, 4729[4]	10.0	32.38	5.563	69 35 28.7	.45	52.34	1
10 669	Gould, Z. C., 14^h 3718 .	9.2	40.80	6.023	72 33 21.0	.45	52.43	2
10 670	Apodis	9.5	42.77	+ 6.415	74 29 32.3	− 14.44	52.48	1
10 671	Gould, Z. C., 14^h 3686 .	8.0	14 55 42.87	+ 5.157	66 0 51.3	− 14.44	52.59	1
10 672	Apodis	8.5	45.34	6.714	75 42 30.8	.44	51.87	2
10 673	Lacaille, 6189	7.2	45.38	5.779	71 4 50.4	.44	52.50	2
10 674	Octantis	9.8	45.87	10.281	82 40 0.2	.44	51.00	2
10 675	Apodis	9.5	48.10	+ 6.457	74 40 17.8	− 14.44	52.44	2
10 676	Gould, Z. C., 14^h 3728 .	7.0	14 55 52.17	+ 6.079	72 50 44.7	− 14.44	53.35	1
10 677	Lacaille, 6193	7.0	53.44	5.385	68 8 10.5	.43	52.54	2
10 678	Apodis	10.0	55 58.37	8.299	79 55 32.0	.43	51.37	1
10 679	Apodis	9.0	56 3.77	8.728	80 40 18.8	.42	50.50	2
10 680	Octantis	9.2	4.38	+16.920	86 9 44.5	− 14.42	51.94	2
10 681	Gould, Z. C., 14^h 3746 .	8.2	14 56 6.80	+ 6.032	72 34 7.3	− 14.42	52.43	2
10 682	Gould, Z. C., 14^h 3740 .	8.8	8.53	5.729	70 43 17.8	.42	52.46	2
10 683	Gould, Z. C., 14^h 3724 .	9.5	11.09	5.148	65 53 3.2	.42	52.59	1
10 684	Gould, Z. C., 14^h 3754 .	9.0	11.66	6.010	72 26 43.0	.42	52.52	1
10 685	Lacaille, 6197	6.5	14.26	+ 5.210	66 30 2.4	− 14.41	52.58	1
10 686	Apodis	9.5	14 56 16.56	+ 7.062	76 52 50.8	− 14.41	52.48	1
10 687	Apodis	10.0	32.78	6.676	75 30 47.8	.39	50.97	2
10 688	Apodis	9.5	34.22	5.786	71 3 50.8	.39	52.50	2
10 689	Apodis	9.5	43.72	5.692	70 25 13.0	.38	52.40	1
10 690	Apodis	10.0	45.39	+ 9.628	81 54 34.7	− 14.38	50.53	1
10 691	Apodis	10.0	14 56 47.34	+ 7.334	77 39 54.2	− 14.38	52.36	1
10 692	Circini	9.0	49.47	5.334	67 36 7.4	.38	52.58	1
10 693	Gould, Z. C., 14^h 3834	9.8	56 55.27	7.152	77 7 24.1	.37	51.98	2
10 694	Apodis	10.0	57 1.16	6.275	73 45 38.5	.37	52.52	1
10 695	Gould, Z. C., 14^h 3797 .	9.8	5.05	+ 5.467	68 42 25.8	− 14.36	52.55	2
10 696	Octantis	9.0	14 57 5.66	+24.557	87 30 44.6	− 14.36	51.93	4
10 697	Gould, Z. C., 14^h 3800 .	9.0	11.72	5.271	66 59 43.5	.35	52.58	2
10 698	Lacaille, 6196	6.2	14.76	5.981	72 11 30.7	.35	52.46	2
10 699	Apodis	10.0	17.53	6.441	74 30 13.9	.35	52.48	1
10 700	Gould, 20508	7.5	22.17	+ 6.404	74 20 2.8	− 14.34	52.51	3
10 701	Gould, Z. C., 14^h 3835 .	8.0	14 57 23.60	+ 6.000	72 17 59.7	− 14.34	52.46	2
10 702	Trianguli Australis . . .	10.0	25.61	5.510	69 1 22.0	.34	52.57	1
10 703	Gould, Z. C., 14^h 3823 .	9.2	33.29	5.202	66 17 42.1	.33	52.59	2
10 704	Gould, 20505	7.8	41.12	5.321	67 24 24.6	.32	52.57	2
10 705	Gould, Z. C., 15^h 5 . . .	9.5	42.48	+ 5.619	70 2 25.3	− 14.32	52.56	1

Number.	Constellation, Name of Star, or Synonym.	Magnitude.	Right Ascension, 1850.0.	Annual Precession.	South Declination, 1850.0.	Annual Precession.	Mean year.
			h. m. s.	s.	° ′ ″	″	
10706	Gould, Z. C., 15h 37 . .	8.2	14 57 49.08	+ 6.560	74 58 29.0	− 14.32	51.48
10707	Trianguli Australis . . .	10.0	57 59.89	5.506	68 56 43.6	.31	52.57
10708	Lacaille, 6191	8.2	58 6.77	7.064	76 46 56.1	.30	52.47
10709	Gould, Z. C., 15h 31 . .	9.5	7.03	5.734	70 35 46.3	.30	52.40
10710	Gould, 20520	8.5	9.76	+ 6.010	72 17 46.6	− 14.30	52.46
10711	Lacaille, 6213	7.8	14 58 10.02	+ 5.296	67 7 52.4	− 14.30	52.57
10712	Apodis	9.5	12.55	6.480	74 36 45.8	.29	52.44
10713	Apodis	9.5	13.97	8.249	79 43 45.6	.29	51.34
10714	Trianguli Australis . . .	9.5	19.70	5.567	69 23 8.5	.29	52.42
10715	Trianguli Australis . . .	10.0	30.58	+ 5.427	68 15 21.6	− 14.27	52.53
10716	Octantis	10.0	14 58 33.87	+11.488	83 38 41.8	− 14.27	50.55
10717	Gould, Z. C., 15h 80 . .	9.0	50.36	5.754	70 40 30.4	.25	52.46
10718	Octantis	9.8	55.56	25.087	87 33 8.8	.25	51.93
10719	Gould, 20533	9.0	58 56.90	5.671	70 5 51.6	.25	52.56
10720	Lacaille, 6194	7.8	59 9.65	+ 7.468	77 54 16.9	− 14.23	51.72
10721	Apodis	10.0	14 59 11.46	+ 7.253	77 18 6.5	− 14.23	51.49
10722	Gould, Z. C., 15h 169 . .	9.2	14.22	7.681	78 26 46.5	.23	51.87
10723	Apodis	9.0	14.86	9.248	81 19 28.5	.23	50.57
10724	Trianguli Australis . . .	9.2	19.26	5.511	68 52 27.6	.22	52.57
10725	Apodis	10.0	20.05	+ 7.237	77 14 47.0	− 14.22	51.49
10726	Apodis	10.0	14 59 40.68	+ 8.346	79 51 1.9	− 14.20	51.37
10727	Octantis	10.0	42.89	13.058	84 35 53.9	.20	51.49
10728	Lacaille, 6220	7.8	44.40	5.665	69 59 12.8	.20	52.45
10729	Gould, Z. C., 15h 134 . .	9.5	48.81	5.276	66 48 15.9	.19	52.59
10730	Trianguli Australis . . .	10.0	14 59 50.50	+ 5.245	66 29 54.9	− 14.19	52.58
10731	Lacaille, 6222	6.0	15 0 1.15	+ 5.602	69 30 25.8	− 14.18	52.38
10732	Gould, Z. C., 15h 226 . .	9.3	14.00	7.496	77 55 36.4	.17	51.72
10733	Apodis	10.0	17.31	7.822	78 43 41.0	.16	51.84
10734	Trianguli Australis . . .	9.5	17.44	5.348	67 24 51.0	.16	52.57
10735	Octantis	10.0	20.77	+14.683	85 20 16.2	− 14.16	51.93
10736	Trianguli Australis . . .	9.7	15 0 21.83	+ 5.599	69 27 26.0	− 14.16	52.38
10737	Apodis	8.2	23.04	5.957	71 49 41.4	.16	52.37
10738	Apodis	9.0	27.62	{8.850 8.849}	80{41 17.9 40 59.9}	.15	50.50
10739	Octantis	10.0	33.38	12.176	84 3 24.1	.15	51.48
10740	Trianguli Australis . . .	10.0	34.87	+ 5.347	67 23 5.4	− 14.15	52.58
10741	Gould, Z. C., 15h 192 . .	9.5	15 0 39.88	+ 5.309	67 1 52.3	− 14.14	52.59
10742	Apodis	9.5	45.23	6.959	76 17 14.7	.14	52.01
10743	Lacaille, 6227	7.2	0 50.70	5.671	69 56 36.4	.13	52.45
10744	Trianguli Australis . . .	10.0	1 7.02	5.446	68 11 39.2	.11	52.54
10745	Gould, Z. C., 15h 231 . .	9.0	11.29	+ 5.368	67 31 7.2	− 14.11	52.57

Number.	Constellation, Name of Star, or Synonym.	Magnitude.	Right Ascension, 1850.0.	Annual Precession.	South Declination, 1850.0.	Annual Precession.	Mean year.	No. of obs.
			h. m. s.	s.	° ′ ″	″		
10 746	Apodis	10.0	15 1 11.42	+ 6.327	73 43 25.6	− 14.11	52.52	1
10 747	Trianguli Australis . . .	10.0	12.55	5.350	67 21 1.5	.11	52.56	1
10 748	Gould, Z. C., 15h 256 . .	8.5	15.17	6.203	73 6 39.4	.10	51.39	2
10 749	Apodis	10.0	15.90	7.266	77 13 44.9	.10	51.49	1
10 750	Gould, Z. C., 15h 285 . .	8.5	28.95	+ 6.393	74 0 37.6	− 14.09	52.52	1
10 751	Apodis	9.5	15 1 29.41	+ 5.764	70 32 13.4	− 14.09	52.40	1
10 752	Lacaille, 6234	7.2	30.96	5.665	69 50 40.8	.09	52.45	2
10 753	Apodis	10.0	34.20	6.289	73 31 3.2	.08	52.52	1
10 754	Trianguli Australis . . .	9.5	37.45	5.426	67 58 48.9	.08	52.56	1
10 755	Trianguli Australis . . .	8.2	41.73	+ 5.244	66 19 18.4	− 14.08	52.59	2
10 756	Gould, Z. C., 15h 300 . .	9.5	15 1 44.15	+ 6.217	73 9 1.7	− 14.07	52.35	1
10 757	Trianguli Australis . . .	9.5	45.69	5.678	69 55 23.7	.07	52.34	1
10 758	Apodis	10.0	47.36	5.826	70 55 10.7	.07	52.52	1
10 759	Apodis	10.0	48.67	6.847	75 49 55.5	.07	51.38	1
10 760	Apodis	9.5	56.29	+ 5.980	71 50 54.7	− 14.06	52.37	2
10 761	Gould, Z. C., 15h 295 . .	8.8	15 1 58.73	+ 5.580	69 10 59.6	− 14.06	52.52	3
10 762	Gould, Z. C., 15h 325 . .	10.0	2 1.51	6.626	74 58 49.7	.06	52.40	1
10 763	Octantis	9.0	1.90	11.030	83 9 57.7	.06	51.01	2
10 764	Apodis	9.5	2.65	5.934	71 34 17.0	.05	52.32	1
10 765	Octantis	9.8	16.35	+24.158	87 24 26.4	− 14.04	51.93	4
10 766	Gould, Z. C., 15h 338 . .	9.0	15 2 19.63	+ 6.315	73 35 27.5	− 14.04	52.52	1
10 767	Gould, Z. C., 15h 324 . .	8.5	27.27	5.602	69 18 41.9	.03	52.42	1
10 768	Brisbane, 5195	8.2	27.66	8.428	79 53 3.7	.03	51.35	2
10 769	Gould, Z. C., 15h 378 . .	9.5	39.45	7.200	76 57 27.1	.02	52.48	1
10 770	Octantis	10.0	41.44	+12.088	83 56 40.2	− 14.01	51.02	2
10 771	Gould, Z. C., 15h 337 . .	8.5	15 2 44.10	+ 5.316	66 54 34.8	− 14.01	52.59	1
10 772	Trianguli Australis . . .	9.0	44.90	5.628	69 28 47.9	.01	52.38	2
10 773	Apodis	10.0	47.35	6.015	71 59 19.2	.01	52.41	1
10 774	Apodis	10.0	47.74	6.836	75 44 15.8	.01	51.38	1
10 775	Apodis	15.0	49.69	+ 6.837	75 44 15.8	− 14.01	51.38	1
10 776	Gould, Z. C., 15h 349 . .	9.0	15 2 50.63	+ 5.410	67 44 12.0	− 14.00	52.58	1
10 777	Trianguli Australis . . .	9.0	53.40	5.362	67 18 42.2	.00	52.56	1
10 778	Gould, 20609	8.5	54.02	6.039	72 6 57.0	.00	52.41	1
10 779	Gould, Z. C., 15h 382 . .	8.0	2 57.32	6.588	74 45 52.7	14.00	52.47	3
10 780	Gould, Z. C., 15h 357 . .	9.2	3 2.62	+ 5.339	67 5 41.8	− 13.99	52.58	2
10 781	Apodis	10.0	15 3 2.83	+ 7.995	78 59 18.8	− 13.99	52.33	1
10 782	Trianguli Australis . . .	9.0	9.56	5.378	67 26 14.4	.99	52.57	2
10 783	Gould, 20625	8.0	14.24	6.651	75 0 14.9	.98	51.48	2
10 784	Apodis	9.0	14.62	6.511	74 25 17.0	.98	52.48	1
10 785	Apodis	10.0	18.94	+ 6.220	73 3 19.8	− 13.98	52.35	1

No. 10 771. Declination decreased 1 rev. to agree with Gould. No. 10 782. Mean of two observations which differ 9.8″.

Number.	Constellation, Name of Star, or Synonym.	Magnitude.	Right Ascension, 1850.0.	Annual Precession.	South Declination, 1850.0.	Annual Precession.	Mean year.	No. of obs.
			h. m. s.	s.	° ′ ″	″		
10 786	Gould, Z. C., 15^h 417 . .	9.5	15 3 20.97	+ 6.418	74 0 24.6	− 13.97	52.52	1
10 787	Gould, Z. C., 15^h 416 . .	8.5	39.54	5.678	69 43 42.4	.95	52.34	1
10 788	Gould, Z. C., 15^h 485 . .	9.5	3 53.61	7.286	77 8 57.2	.94	51.93	2
10 789	Gould, Z. C., 15^h 443 . .	7.5	4 6.66	5.454	68 1 10.5	.93	52.56	1
10 790	Apodis	10.0	7.34	5.886	71 7 26.5	− 13.92	52.48	1
10 791	Apodis	10.0	15 4 9.84	+ 6.989	76 11 42.1	− 13.92	52.56	1
10 792	Apodis	9.0	13.66	6.312	73 27 1.5	.92	50.44	1
10 793	Gould, Z. C., 15^h 458 . .	9.0	13.88	5.567	68 54 2.9	.92	52.57	1
10 794	Gould, Z. C., 15^h 508 . .	9.0	15.35	7.234	76 58 27.6	.92	52.48	1
10 795	Gould, 20653	8.8	20.44	+ 6.467	74 9 48.7	− 13.91	52.24	2
10 796	Lacaille, 6247	7.0	15 4 23.01	+ 5.907	71 13 56.2	− 13.91	52.48	1
10 797	Trianguli Australis . . .	9.5	55.77	5.398	67 27 38.3	.87	52.58	1
10 798	Gould, Z. C., 15^h 499 . .	9.0	58.55	5.231	65 54 0.8	.87	52.59	1
10 799	Trianguli Australis . . .	9.5	58.75	5.391	67 23 19.2	.87	52.58	1
10 800	γ Trianguli Australis . . .	4.0	4 59.04	+ 5.477	68 7 7.7	− 13.87	52.54	2
10 801	Gould, 20671	9.8	15 5 2.90	+ 6.683	75 1 19.5	− 13.87	51.48	2
10 802	Gould, Z. C., 15^h 510 . .	8.5	9.02	5.223	65 48 43.3	.86	52.59	1
10 803	Trianguli Australis . . .	9.0	28.19	5.293	66 27 22.3	.84	52.58	1
10 804	Apodis	9.5	30.31	5.952	71 25 23.5	.84	52.48	1
10 805	Apodis	9.0	38.81	+ 6.072	72 6 27.9	− 13.83	52.41	1
10 806	Apodis	10.0	15 5 45.71	+ 5.826	70 37 4.0	− 13.82	52.46	2
10 807	Apodis	10.0	46.62	7.591	77 53 57.2	.82	52.36	1
10 808	Apodis	10.0	47.10	6.090	72 11 50.6	.82	52.41	1
10 809	Trianguli Australis . . .	9.5	50.15	5.305	66 32 34.2	.82	52.58	1
10 810	Gould, Z. C., 15^h 572 . .	9.0	51.59	+ 5.600	69 1 24.4	− 13.81	52.50	2
10 811	Trianguli Australis . . .	9.5	15 5 55.94	+ 5.401	67 24 11.7	− 13.81	52.58	1
10 812	Apodis	9.5	5 56.92	6.218	72 52 9.8	.81	52.35	1
10 813	Lacaille, 6252	7.0	6 0.28	6.420	73 50 45.0	.81	52.52	2
10 814	Trianguli Australis . . .	10.0	6.88	5.381	67 12 32.5	.80	52.56	1
10 815	Apodis	9.5	8.26	+ 8.220	79 19 30.2	− 13.80	52.41	1
10 816	Apodis	10.0	15 6 13.58	+ 6.423	73 50 35.9	− 13.79	52.52	2
10 817	Apodis	9.0	13.94	6.258	73 3 18.1	.79	52.35	1
10 818	Gould, Z. C., 15^h 595 . .	9.0	14.20	5.319	66 38 3.8	.79	52.58	2
10 819	Gould, Z. C., 15^h 613 . .	10.0	14.40	6.190	72 41 30.8	.79	52.52	1
10 820	Octantis	9.7	16.56	+12.677	84 13 39.1	− 13.79	51.16	2
10 821	Trianguli Australis . . .	9.0	15 6 20.38	+ 5.473	67 58 35.7	− 13.78	52.56	1
10 822	Octantis	9.5	20.95	13.341	84 35 48.2	.78	51.00	2
10 823	Lacaille, 6264	7.5	25.38	5.248	65 56 22.2	.78	52.59	1
10 824	Gould, Z. C., 15^h 609 . .	9.5	34.88	5.317	66 35 18.4	.77	52.58	2
10 825	Apodis	9.0	37.05	+ 5.959	71 23 23.9	− 13.77	52.48	1

Number.	Constellation, Name of Star, or Synonym.	Magnitude.	Right Ascension, 1850.0.	Annual Precession.	South Declination, 1850.0.	Annual Precession.	Mean year.	No. of obs.
			h. m. s.	s.	° ′ ″	″		
10826	Gould, Z. C., 15h 617 . .	9.5	15 6 40.84	+ 5.256	65 59 51.8	− 13.76	52.59	1
10827	Gould, 20699	8.0	41.96	5.664	69 26 6.2	.76	52.38	2
10828	Gould, Z. C., 15h 625 . .	9.0	42.98	5.617	69 4 57.0	.76	52.50	2
10829	Octantis	9.0	45.31	15.682	85 34 18.1	.76	51.93	3
10830	Gould, Z. C., 15h 674 . .	9.0	58.41	+ 6.622	74 39 42.1	− 13.74	52.47	3
10831	Apodis	9.2	15 6 58.52	+ 5.835	70 35 27.2	− 13.74	52.46	2
10832	Apodis	9.0	6 59.54	6.623	74 39 47.0	.74	52.47	3
10833	Lacaille, 6254	7.0	7 1.00	6.971	75 58 31.4	.74	52.56	1
10834	Lacaille, 6242	8.0	1.80	7.974	78 45 46.2	.74	52.42	2
10835	Trianguli Australis . . .	9.5	2.32	+ 5.329	66 39 27.5	− 13.74	52.58	1
10836	Lacaille, 6268	6.5	15 7 2.42	+ 5.358	66 55 35.4	− 13.74	52.59	1
10837	Gould, Z. C., 15h 660 . .	9.0	7.42	5.720	69 47 59.0	.73	52.34	1
10838	Apodis	8.8	7.56	6.589	74 30 49.8	.73	52.50	2
10839	Gould, Z. C., 15h 651 . .	8.0	11.82	5.359	66 55 19.6	.73	52.59	1
10840	Gould, Z. C., 15h 687 . .	9.2	14.60	+ 6.371	73 32 23.0	− 13.73	51.48	2
10841	Apodis	9.2	15 7 21.99	+ 6.632	74 40 39.5	− 13.72	52.47	3
10842	Gould, Z. C., 15h 707 . .	8.8	24.10	6.593	74 30 48.4	.72	52.50	2
10843	Apodis	9.2	27.17	7.793	78 19 23.0	.71	52.35	2
10844	Gould, Z. C., 15h 703 . .	9.0	27.31	6.395	73 38 13.3	.71	52.52	1
10845	Apodis	9.5	29.58	+ 8.167	79 9 26.9	− 13.71	52.41	1
10846	Trianguli Australis . . .	9.5	15 7 30.45	+ 5.511	68 11 28.3	− 13.71	52.53	1
10847	Octantis	9.8	38.98	40.263	88 29 45.6	.70	51.93	4
10848	Trianguli Australis . . .	9.5	44.76	5.502	68 6 8.5	.69	52.56	1
10849	Apodis	9.0	7 49.30	9.081	80 45 47.8	.69	50.50	2
10850	Apodis	9.2	8 11.81	+ 6.071	71 55 41.8	− 13.67	52.37	2
10851	Gould, 20742	9.0	15 8 28.43	+ 6.834	75 24 5.7	− 13.65	50.56	1
10852	Gould, Z. C., 15h 785 . .	9.0	35.22	7.046	76 8 46.2	.64	52.56	1
10853	Apodis	9.8	36.78	7.645	77 54 4.4	.64	51.72	3
10854	Octantis	9.3	39.16	13.755	84 45 23.1	.64	51.31	3
10855	Apodis	9.5	49.79	+ 6.508	74 3 53.8	− 13.63	52.52	1
10856	Apodis	10.0	15 8 55.11	+ 9.804	81 42 17.5	− 13.62	50.56	1
10857	Gould, 20738	8.5	9 1.48	5.282	66 2 56.0	.61	52.59	1
10858	Trianguli Australis . . .	9.5	2.75	5.347	66 39 26.6	.61	52.59	1
10859	Gould, Z. C., 15h 798 . .	9.5	3.76	6.391	73 31 2.0	.61	52.52	1
10860	Octantis	10.0	6.68	+12.157	83 49 43.7	13.61	51.02	2
10861	Apodis	10.0	15 9 8.02	+ 9.908	81 49 23.1	− 13.61	50.56	1
10862	Trianguli Australis . . .	9.0	10.86	5.599	68 45 16.9	.60	52.57	1
10863	Gould, Z. C., 15h 819 . .	8.7	19.09	6.672	74 43 39.4	.59	50.47	3
10864	Trianguli Australis . . .	9.0	19.45	5.513	68 4 2.9	.59	52.56	1
10865	Gould, Z. C., 15h 824 . .	10.0	20.71	+ 6.787	75 10 33.9	− 13.59	52.40	1

No. 10859. Gould's declination is 8.4″ less.

Number.	Constellation, Name of Star, or Synonym.	Magnitude.	Right Ascension, 1850.0.	Annual Precession.	South Declination, 1850.0.	Annual Precession.	Mean year.	No. of obs.
			h. m. s.	s.	° ′ ″	″		
10 866	Trianguli Australis . . .	9.5	15 9 28.45	+ 5.543	68 17 24.5	— 13.58	52.53	1
10 867	Gould, Z. C., 15^h 851 . .	9.5	31.35	7.060	76 8 30.1	.58	52.56	1
10 868	ρ Octantis	6.8	32.82	12.354	83 56 55.2	.58	51.02	2
10 869	Lacaille, 6281	7.2	34.57	— 6.013	71 30 3.4	.58	52.40	2
10 870	Trianguli Australis . . .	9.5	35.37	+ 5.547	68 18 51.2	— 13.58	52.53	1
10 871	Apodis	10.0	15 9 38.40	+ 7.827	78 18 15.3	— 13.57	52.36	1
10 872	Lacaille, 6285	7.2	46.88	5.896	70 46 35.9	.56	52.46	2
10 873	Trianguli Australis . . .	9.8	51.98	5.440	67 24 44.7	.56	52.57	2
10 874	Apodis	10.0	9 55.32	8.652	79 59 1.5	.55	51.37	1
10 875	Trianguli Australis . . .	9.5	10 0.75	+ 5.564	68 24 55.6	— 13.55	52.53	1
10 876	Apodis	10.0	15 10 14.04	+ 5.881	70 38 46.1	— 13.53	52.52	1
10 877	Apodis	9.5	15.18	6.017	71 28 38.4	.53	52.40	2
10 878	Apodis	10.0	17.06	6.541	74 7 9.4	.53	52.52	1
10 879	Octantis	9.2	17.38	10.785	82 42 29.4	.53	51.00	2
10 880	Gould, 20780	9.5	21.41	+ 6.859	75 23 24.7	— 13.53	50.56	1
10 881	Apodis	9.5	15 10 21.70	+ 6.293	72 57 44.7	— 13.53	52.35	1
10 882	Trianguli Australis . . .	10.0	24.35	5.414	67 8 19.1	.52	52.56	1
10 883	Gould, Z. C., 15^h 865 . .	10.0	27.49	5.279	65 53 54.9	.52	52.59	1
10 884	Gould, Z. C., 15^h 930 . .	9.0	40.42	7.032	75 59 6.3	.51	52.65	1
10 885	Gould, Z. C., 15^h 881 . .	9.5	47.00	+ 5.302	66 4 58.1	— 13.50	52.59	1
10 886	Gould, Z. C., 15^h 885 . .	9.0	15 10 50.69	+ 5.279	65 51 57.3	— 13.50	52.59	1
10 887	Lacaille, 6269	8.0	53.17	7.814	78 12 58.8	.49	51.68	4
10 888	Gould, Z. C., 15^h 895 . .	9.0	10 54.50	5.489	67 44 13.4	.49	52.58	1
10 889	Trianguli Australis . . .	9.5	11 7.65	5.558	68 16 46.6	.48	52.53	1
10 890	Apodis	10.0	10.23	+ 7.122	76 15 33.6	— 13.47	52.48	1
10 891	Apodis	9.0	15 11 11.37	+ 9.472	81 12 11.0	— 13.47	50.57	1
10 892	Gould, Z. C., 15^h 927 . .	9.2	12.84	5.762	69 47 0.9	.47	52.45	2
10 893	Apodis	10.0	16.44	5.812	70 7 12.5	.47	52.56	1
10 894	Trianguli Australis . . .	8.5	17.50	5.346	66 27 15.7	.47	52.58	1
10 895	Gould, Z. C., 15^h 929 . .	8.2	18.88	+ 5.543	68 9 4.2	— 13.46	52.54	2
10 896	Apodis	9.5	15 11 23.32	+ 6.250	72 40 30.4	— 13.46	52.52	1
10 897	Gould, Z. C., 15^h 950 . .	9.2	29.24	6.032	71 28 44.5	.45	52.45	2
10 898	Apodis	9.5	45.77	9.818	81 37 46.6	.44	50.56	1
10 899	Gould, Z. C., 15^h 997 . .	10.0	48.96	6.889	75 25 16.8	.43	51.38	1
10 900	Trianguli Australis . . .	9.0	49.04	+ 5.589	68 28 17.5	— 13.43	52.53	1
10 901	Gould, Z. C., 15^h 967 . .	9.0	15 11 50.83	+ 5.743	69 36 27.9	— 15.43	52.34	1
10 902	Gould, Z. C., 15^h 981 . .	8.8	52.52	6.182	72 17 30.6	.43	52.47	2
10 903	Octantis	8.8	11 59.98	22.631	87 5 28.3	.42	51.93	3
10 904	Apodis	10.0	12 4.80	8.107	78 49 59.9	.42	52.33	1
10 905	Lacaille, 6308	6.8	10.64	+ 5.505	67 46 11.9	— 13.41	52.57	2

Number.	Constellation, Name of Star, or Synonym.	Magnitude.	Right Ascension, 1850.0.	Annual Precession.	South Declination, 1850.0.	Annual Precession.	Mean year.	No. of obs.
			h. m. s.	s.	° ′ ″	″		
10906	Trianguli Australis . . .	10.0	15 12 20.73	+ 5.341	66 19 39.6	− 13.40	52.58	1
10907	Apodis	9.5	20.99	8.297	79 13 8.0	.40	52.41	1
10908	Trianguli Australis . . .	9.5	26.68	5.553	68 8 20.8	.39	52.53	1
10909	Apodis	10.0	27.21	5.945	70 53 38.8	.39	52.52	1
10910	Trianguli Australis . . .	10.0	30.52	+ 5.678	69 5 18.8	− 13.39	52.42	1
10911	Gould, Z. C., 15^h 998 . .	9.0	15 12 30.68	+ 5.295	65 52 34.6	− 13.39	52.59	1
10912	Gould, Z. C., 15^h 1017 .	9.0	30.70	5.943	70 52 54.3	.39	52.52	1
10913	Gould, Z. C., 15^h 1051 .	8.5	31.38	7.006	75 47 52.2	.39	51.97	2
10914	Trianguli Australis . . .	9.2	39.00	5.799	69 55 59.1	.38	52.45	2
10915	Apodis	10.0	43.87	+ 6.987	75 43 15.9	− 13.37	51.38	1
10916	Apodis	9.0	15 12 54.50	+ 5.319	72 55 37.5	− 13.36	52.35	1
10917	Trianguli Australis . . .	10.0	12 55.86	5.720	69 21 39.3	.36	52.42	1
10918	Gould, Z. C., 15^h 1070 .	8.0	13 1.66	6.462	73 36 6.0	.35	51.48	2
10919	Gould, Z. C., 15^h 1076 .	9.5	6.44	6.809	75 3 . .	.35	52.40	1
10920	Apodis	9.5	11.76	+ 6.265	72 38 19.5	− 13.34	52.52	1
10921	Gould, Z. C., 15^h 1067 .	9.8	15 13 15.92	+ 5.845	70 11 47.3	− 13.34	52.46	2
10922	Apodis	9.0	20.66	8.559	79 40 58.9	.33	51.34	1
10923	Herschel, 4760[1]	9.5	22.78	7.394	76 59 12.4	.33	51.93	2
10924	Gould, Z. C., 15^h 1071 .	8.2	24.45	5.639	68 43 56.2	.33	52.55	2
10925	Herschel, 4760[2]	10.2	24.86	+ 7.393	76 58 54.0	− 13.33	51.98	2
10926	Trianguli Australis . . .	9.5	15 13 28.83	+ 5.775	69 42 29.2	− 13.32	52.34	1
10927	Gould, Z. C., 15^h 1100 .	8.5	30.22	6.571	74 3 23.8	.32	52.52	1
10928	Apodis	9.5	34.12	6.100	71 43 51.3	.32	52.32	1
10929	Apodis	9.0	41.14	7.225	76 27 29.9	.31	52.48	1
10930	Trianguli Australis . . .	10.0	50.39	+ 5.720	69 17 42.7	− 13.30	52.42	1
10931	Gould, Z. C., 15^h 1131 .	9.5	15 13 50.52	+ 6.613	74 13 8.2	− 13.30	52.52	2
10932	Trianguli Australis . . .	9.5	53.69	5.692	69 5 19.6	.30	52.50	2
10933	Gould, Z. C., 15^h 1177 .	9.5	13 54.86	7.971	78 27 8.2	.30	52.36	1
10934	Apodis	9.0	14 2.96	7.237	76 28 45.2	.29	52.48	1
10935	Gould, 20853	8.8	6.26	+ 6.449	73 28 35.9	− 13.28	51.48	2
10936	Gould, Z. C., 15^h 1123 .	8.5	15 14 12.33	+ 5.556	68 1 32.8	− 13.28	52.56	1
10937	Gould, Z. C., 15^h 1140 .	9.0	14.97	5.965	70 53 18.0	.27	52.52	1
10938	Apodis	10.0	18.71	7.756	77 55 8.3	.27	52.36	1
10939	Gould, Z. C., 15^h 1163 .	10.0	24.76	6.479	73 35 40.4	.26	52.52	1
10940	Gould, Z. C., 15^h 1173 .	9.5	26.46	+ 6.631	74 15 39.6	− 13.26	52.52	2
10941	Apodis	10.0	15 14 34.90	+ 6.470	73 32 44.7	− 13.25	52.52	1
10942	Apodis	9.5	37.88	6.295	72 42 0.5	.25	52.52	1
10943	Gould, Z. C., 15^h 1162 .	9.5	49.12	5.514	67 38 38.4	.24	52.58	1
10944	Apodis	10.0	53.94	6.477	73 33 28.0	.23	52.52	1
10945	Gould, Z. C., 15^h 1179 .	9.5	57.61	+ 5.537	67 49 7.2	− 13.23	52.57	2

Number.	Constellation, Name of Star, or Synonym.	Magnitude.	Right Ascension, 1850.0.	Annual Precession.	South Declination, 1850.0.	Annual Precession.	Mean year.	No. of obs.
			h. m. s.	s.	° ′ ″	″		
10 946	Trianguli Australis . . .	10.0	15 14 59.54	+ 5.544	67 52 35.2	− 13.22	52.57	2
10 947	Lacaille, 6300	7.5	15 3.63	8.043	78 33 50.4	.22	52.35	2
10 948	Gould, Z. C., 15h 1239 .	8.5	6.77	7.184	76 15 23.1	.22	52.52	2
10 949	Trianguli Australis . . .	9.5	9.55	5.540	67 49 42.9	.21	52.58	1
10 950	Apodis	9.5	12.06	+ 7.094	75 57 29.8	− 13.21	52.56	1
10 951	κ¹ Apodis	6.0	15 15 17.10	+ 6.335	72 51 42.0	− 13.21	52.35	1
10 952	Apodis	9.5	21.69	7.820	78 1 50.6	.20	52.35	1
10 953	Melbourne (1), 792 . . .	8.0	21.99	54.509	88 52 45.1	.21	51.93	3
10 954	Gould, Z. C., 15h 1215 .	9.0	22.46	5.818	69 51 53.9	.20	52.45	2
10 955	Trianguli Aus ralis . . .	9.5	25.26	+ 5.477	67 16 55.6	− 13.20	52.56	1
10 956	Lacaille, 6331	7.0	15 15 27.05	+ 6.017	71 7 31.4	− 13.19	52.50	2
10 957	Apodis	9.0	28.56	10.377	82 9 2.0	.19	50.51	2
10 958	Apodis	10.0	42.89	8.261	79 1 27.5	.18	52.41	1
10 959	Octantis	11.0	43.77	16.702	85 46 12.5	.18	51.93	3
10 960	Apodis	9.0	49.93	+ 8.303	79 5 21.0	− 13.17	52.37	2
10 961	Trianguli Australis . . .	9.5	15 15 51.15	+ 5.800	69 42 33.1	− 13.17	52.35	1
10 962	Gould, Z. C., 15h 1262 .	9.5	15 57.39	5.942	70 38 2.1	.16	52.40	1
10 963	Lacaille, 6346	7.7	16 1.31	5.444	66 57 20.3	.16	52.59	3
10 964	Apodis	10.0	16.82	6.178	71 59 10.8	.14	52.41	1
10 965	Apodis	9.2	17.20	+ 6.962	75 26 39.0	− 13.14	50.97	2
10 966	Apodis	8.5	15 16 17.26	+ 7.037	75 42 24.2	− 13.14	51.38	1
10 967	Lacaille, 6340	7.0	19.50	5.931	70 32 7.0	.14	52.40	1
10 968	Trianguli Australis . . .	10.0	22.13	5.614	68 19 23.5	.13	52.56	1
10 969	Gould, Z. C., 15h 1307 .	9.0	23.49	6.554	73 49 3.1	.13	52.52	2
10 970	Gould, Z. C., 15h 1321 .	9.0	29.05	+ 6.838	74 58 22.8	− 13.13	51.48	2
10 971	Gould, Z. C., 15h 1289 .	9.0	15 16 29.84	+ 5.664	68 41 23.4	− 13.13	52.55	2
10 972	Lacaille, 6311	8.0	30.82	7.979	78 21 29.7	.13	52.10	3
10 973	Octantis	9.5	34.57	16.715	85 45 38.1	.12	51.93	3
10 974	Gould, 20905	8.5	40.47	5.750	69 18 31.4	.11	52.42	1
10 975	Trianguli Australis . . .	9	41.10	+ 5.543	67 44 11.6	− 13.11	52.58	1
10 976	Gould, Z. C., 15h 1302 .	9.0	15 16 44.52	+ 5.666	68 41 33.3	− 13.11	52.55	2
10 977	Octantis	9.0	45.89	24.650	87 18 53.2	.11	51.93	3
10 978	Apodis	9.0	50.86	10.446	82 10 58.8	.10	50.51	2
10 979	Apodis	9.0	50.95	6.202	72 4 37.6	.10	52.41	1
10 980	Trianguli Australis . .	9.0	52.32	+ 5.470	67 6 44.8	− 13.10	52.59	4
10 981	Trianguli Australis . . .	9.2	15 16 53.42	+ 5.485	67 14 16.8	− 13.10	52.59	3
10 982	Gould, Z. C., 15h 1332 .	8.0	53.54	6.388	73 1 14.2	.10	51.39	2
10 983	Gould, Z. C., 15h 1311 .	9.4	54.32	5.462	67 2 36.4	.10	52.59	4
10 984	Apodis	10.0	16 57.02	8.822	80 0 41.3	.10	51.37	1
10 985	Apodis	9.5	17 6.04	+ 8.516	79 27 16.5	− 13.09	52.41	1

Number.	Constellation, Name of Star, or Synonym.	Magnitude.	Right Ascension, 1850.0.	Annual Precession.	South Declination, 1850.0.	Annual Precession.	Mean year.	No. of obs.
			h. m. s.	s.	° ′ ″	″		
10 986	Gould, Z. C., 15^h 1319 .	9.5	15 17 6.42	+ 5.333	65 51 45.7	− 13.08	52.59	1
10 987	Gould, Z. C., 15^h 1331 .	9.0	10.38	5.707	68 57 30.7	.08	52.57	1
10 988	Gould, Z. C., 15^h 1364 .	9.5	11.38	6.895	75 8 55.3	.08	52.40	1
10 989	Gould, Z. C., 15^h 1328 .	9.0	13.56	5.461	67 0 16.6	.08	52.58	3
10 990	Gould, Z. C., 15^h 1378 .	9.0	18.78	+ 6.885	75 6 22.9	− 13.07	51.48	2
10 991	Gould, Z. C., 15^h 1375 .	9.0	15 17 23.43	+ 6.518	73 35 56.4	− 13.07	52.52	1
10 992	Apodis	10.0	30.89	5.885	70 9 39.4	.06	52.56	1
10 993	Brisbane, 5326	8.0	40.08	5.670	68 38 58.5	.05	52.55	2
10 994	Gould, Z. C., 15^h 1373 .	9.5	41.79	5.777	69 25 30.7	.05	52.38	2
10 995	Apodis	9.5	45.29	+10.780	82 29 35.5	− 13.04	50.49	1
10 996	Gould, Z. C., 15^h 1395 .	9.0	15 17 59.26	+ 5.680	68 42 17.6	− 13.03	52.55	2
10 997	Octantis	9.5	18 1.86	13.080	84 11 54.3	.02	51.49	1
10 998	Gould, Z. C., 15^h 1428 .	8.0	4.63	6.817	74 48 29.6	.02	52.40	1
10 999	Gould, Z. C., 15^h 1405 .	9.5	9.34	5.781	69 25 3.4	.02	52.38	2
11 000	Apodis	10.0	16.84	+ 6.084	71 19 53.0	− 13.01	52.48	1
11 001	Apodis	9.8	15 18 17.22	+ 6.352	72 45 45.6	− 13.01	52.43	2
11 002	Apodis	9.0	17.22	7.097	75 48 28.8	.01	51.97	2
11 003	Apodis	11.0	18.54	6.713	74 22 50.9	.01	52.52	1
11 004	Apodis	10.0	19.48	6.713	74 22 57.4	.00	52.52	1
11 005	Trianguli Australis . . .	9.5	20.26	+ 5.425	66 36 36.9	− 13.00	52.59	3
11 006	Apodis	8.5	15 18 20.66	+ 6.254	72 15 37.0	− 13.00	52.46	2
11 007	Gould, Z. C., 15^h 1416 .	9.0	21.43	5.543	67 36 28.4	13.00	52.58	1
11 008	Apodis	9.0	36.75	7.051	75 38 9.0	12.98	51.38	1
11 009	Trianguli Australis . . .	9.0	37.28	5.372	66 6 14.2	.98	52.59	1
11 010	Gould, Z. C., 15^h 1459 .	9.0	51.84	+ 5.814	69 35 22.6	− 12.97	52.38	2
11 011	Lacaille, 6339	7.8	15 18 57.64	+ 7.760	77 43 18.0	− 12.96	51.40	2
11 012	Apodis	8.8	19 12.26	7.226	76 11 13.0	.95	52.52	2
11 013	Apodis	10.0	12.31	10.610	82 17 3.5	.95	50.53	1
11 014	Trianguli Australis . . .	9.5	14.72	5.402	66 20 0.6	.94	52.59	2
11 015	Trianguli Australis . . .	9.0	20.19	+ 5.360	65 55 59.2	− 12.94	52.59	1
11 016	Apodis	10.0	15 19 24.74	+ 6.114	71 25 52.3	− 12.93	52.32	1
11 017	Octantis	9.8	31.83	34.349	88 7 39.2	.92	51.93	3
11 018	Apodis	9.5	32.79	7.223	76 9 43.9	.92	52.56	1
11 019	Apodis	9.5	33.46	6.133	71 31 42.2	.92	52.40	2
11 020	Lacaille, 6348	7.0	35.07	+ 7.748	77 24 9.1	− 12.92	52.11	3
11 021	Brisbane, 5335[1]	9.0	15 19 37.78	+ 6.602	73 50 29.6	− 12.92	52.52	2
11 022	Gould, Z. C., 15^h 1565 .	10.0	38.15	7.766	77 42 24.3	.92	52.36	1
11 023	Brisbane, 5335[2]	9.0	39.84	6.602	73 50 28.0	.91	52.52	2
11 024	Trianguli Australis . . .	9.5	40.21	5.352	65 49 58.8	.91	52.59	1
11 025	Lacaille, 6371	8.2	41.34	+ 5.461	66 49 13.1	+ 12.91	55.59	3

No. 11 022. Gould's declination is 9.4″ less.

Number.	Constellation, Name of Star, or Synonym.	Magnitude.	Right Ascension, 1850.0.	Annual Precession.	South Declination, 1850.0.	Annual Precession.	Mean year.	No. of obs.
			h. m. s.	s.	° ′ ″	″		
11 026	Gould, Z. C., 15ʰ 1506	8.5	15 19 45.23	+ 5.637	68 15 12.4	− 12.91	52.54	2
11 027	Octantis	9.1	44.95	43.249	88 32 27.3	.91	51.93	5
11 028	Apodis	10.0	52.70	6.133	71 30 35.2	.90	52.40	2
11 029	Octantis	9.8	55.28	12.725	83 56 40.7	.90	51.02	2
11 030	Gould, Z. C., 15ʰ 1518	9.5	19 56.92	+ 5.468	66 51 21.6	− 12.90	52.59	2
11 031	Gould, 20983	8.0	15 20 2.95	+ 5.619	68 5 36.5	− 12.89	52.56	1
11 032	Apodis	9.0	7.50	8.437	79 11 9.6	.88	52.41	1
11 033	Lacaille, 6369	7.0	8.92	6.288	72 19 31.[illegible]	.88	52.46	2
11 034	Apodis	9.8	11.42	9.031	80 14 28.2	.88	51.37	2
11 035	Apodis	9.8	15.05	+ 6.727	73 54 56.2	− 12.88	52.52	2
11 036	Trianguli Australis	9.0	15 20 17.54	+ 5.421	66 25 3.2	− 12.87	52.59	2
11 037	Trianguli Australis	9.5	17.88	5.863	69 49 50.6	.87	52.45	2
11 038	Octantis	9.4	25.50	16.683	85 41 23.3	.87	51.93	5
11 039	Stone, 8429	8.0	26.10	5.659	68 22 25.4	.86	52.54	2
11 040	Gould, Z. C., 15ʰ 1597	8.5	47.16	+ 6.005	70 42 8.8	− 12.84	52.46	2
11 041	Gould, Z. C., 15ʰ 1643	9.0	15 20 48.65	+ 7.764	77 39 3.6	− 12.84	52.36	1
11 042	Gould, Z. C., 15ʰ 1582	9.0	49.61	5.573	67 40 16.3	.84	52.58	1
11 043	Apodis	10.0	54.20	7.120	75 45 19.8	.83	51.38	1
11 044	Gould, Z. C., 15ʰ 1599	9.5	59.05	5.627	68 5 12.5	.83	52.56	1
11 045	Stone, 8438	8.2	20 59.09	+ 6.636	73 54 38.0	− 12.83	52.52	2
11 046	Apodis	9.0	15 21 1.02	+ 6.140	71 28 7.2	− 12.82	52.40	2
11 047	Gould, Z. C., 15ʰ 1635	9.5	2.63	6.925	75 3 29.2	.82	52.40	1
11 048	Trianguli Australis	9.0	8.48	5.600	67 51 37.8	.82	52.58	1
11 049	Apodis	10.5	10.76	8.813	79 50 29.2	.81	51.35	2
11 050	Apodis	10.0	31.11	+ 6.133	71 24 13.8	− 12.79	52.48	1
11 051	Trianguli Australis	9.5	15 21 32.70	+ 5.641	68 10 23.9	− 12.79	52.53	1
11 052	Gould, 21032	8.5	55.63	6.189	71 41 32.7	.76	52.32	1
11 053	Apodis	8.8	56.10	6.368	72 37 23.0	.76	52.43	2
11 054	Apodis	9.3	21 57.65	9.168	80 23 43.0	.76	50.80	3
11 055	Trianguli Australis	9.5	22 23.37	+ 5.731	68 46 6.0	− 12.73	52.57	1
11 056	Trianguli Australis	9.0	15 22 24.93	+ 5.600	67 46 0.8	− 12.73	52.58	1
11 057	Apodis	10.0	30.82	10.019	81 32 13.0	.72	50.56	2
11 058	Apodis	9.0	36.88	6.386	72 40 20.8	.72	52.43	2
11 059	Apodis	9.5	44.79	8.030	78 12 50.1	.71	51.98	2
11 060	Lacaille, 6394	8.0	45.14	+ 5.399	66 2 11.0	− 12.71	52.59	1
11 061	Brisbane, 5364	8.8	15 22 46.12	+ 6.662	73 55 21.9	− 12.71	52.52	2
11 062	Gould, Z. C., 15ʰ 1707	8.0	53.09	5.507	66 58 29.3	.70	52.58	2
11 063	Gould, Z. C., 15ʰ 1712	9.0	53.53	5.588	67 38 26.4	.70	52.58	1
11 064	Gould, Z. C., 15ʰ 1705	10.0	55.81	5.378	65 49 26.9	.69	52.59	1
11 065	Gould, Z. C., 15ʰ 1714	9.0	22 59.86	+ 5.401	66 2 14.2	− 12.69	52.59	1

Number.	Constellation, Name of Star, or Synonym.	Magnitude.	Right Ascension, 1850.0.	Annual Precession.	South Declination, 1850.0.	Annual Precession.	Mean year.	No. of obs.
			h. m. s.	s.	° ′ ″	″		
11 066	Trianguli Australis	10.0	15 23 0.63	+ 5.486	66 47 6.1	− 12.69	52.60	1
11 067	Gould, Z. C., 15^h 1723	9.5	2.29	5.694	68 26 53.9	.69	52.53	1
11 068	Apodis	10.0	2.40	10.068	81 34 49.7	.69	50.56	2
11 069	ε Trianguli Australis	4.8	3.52	5.377	65 48 20.3	.69	52.60	2
11 070	Apodis	10.0	3.77	+ 7.940	77 59 27.9	− 12.69	52.36	1
11 071	Gould, 21061	8.8	15 23 5.72	+ 6.329	72 21 38.2	− 12.68	52.46	2
11 072	Gould, Z. C., 15^h 1727	9.0	6.79	5.634	67 59 21.5	.68	52.56	1
11 073	Lacaille, 6386	7.5	13.36	6.222	71 47 28.4	.67	52.37	2
11 074	Brisbane, 5367	8.2	21.16	6.673	73 56 22.0	.67	52.52	2
11 075	Gould, Z. C., 15^h 1851	8.5	21.41	+ 8.521	79 13 38.4	− 12.67	52.41	1
11 076	Gould, Z. C., 15^h 1764	9.2	15 23 31.59	+ 5.778	69 1 31.2	− 12.65	52.50	2
11 077	Octantis	10.0	36.61	13.055	84 3 57.5	.65	51.49	1
11 078	Lacaille, 6381	6.2	38.22	7.108	75 34 48.6	.65	50.97	2
11 079	Gould, Z. C., 15^h 1778	9.5	39.10	5.857	69 33 38.3	.65	52.38	2
11 080	Apodis	10.0	41.64	+ 6.334	72 20 54.8	− 12.64	52.46	2
11 081	κ^2 Apodis	6.0	15 23 50.54	+ 6.457	72 56 35.5	− 12.63	51.35	2
11 082	Gould, Z. C., 15^h 1799	10.0	50.55	5.796	69 8 0.7	.63	52.42	1
11 083	Gould, 21079	9.0	52.80	6.462	72 58 1.6	.63	51.40	2
11 084	Gould, Z. C., 15^h 1790	9.5	54.55	5.471	66 35 19.6	.63	52.60	2
11 085	Apodis	9.5	58.07	+ 6.007	70 30 16.3	− 12.62	52.40	1
11 086	Apodis	10.0	15 23 59.69	+ 6.884	74 45 10.7	− 12.62	52.40	1
11 087	Trianguli Australis	10.2	24 2.45	5.472	66 35 16.4	.62	52.60	2
11 088	Trianguli Australis	9.8	13.66	5.737	68 41 1.9	.61	52.55	2
11 089	Apodis	10.0	21.22	6.904	74 48 43.9	.60	52.40	1
11 090	Gould, Z. C., 15^h 1838	9.0	21.82	+ 5.741	68 42 13.8	− 12.60	52.55	2
11 091	Apodis	8.8	15 24 22.23	+ 6.327	72 16 27.3	− 12.60	52.46	2
11 092	Gould, Z. C., 15^h 1844	9.8	25.08	5.865	69 33 56.0	.59	52.38	2
11 093	Apodis	9.8	28.42	7.791	77 33 45.8	.59	52.42	2
11 094	Apodis	10.0	29.09	6.130	71 12 16.3	.59	52.48	1
11 095	Apodis	9.5	33.52	+ 8.450	79 2 38.6	− 12.58	52.41	1
11 096	Gould, Z. C., 15^h 1868	9.2	15 24 46.81	+ 6.109	71 3 47.7	− 12.57	52.50	2
11 097	Trianguli Australis	9.5	24 54.55	5.877	69 37 0.3	.56	52.34	1
11 098	Apodis	9.2	25 3.02	6.252	71 50 20.1	.55	52.37	2
11 099	Gould, Z. C., 15^h 1888	9.8	9.42	5.869	69 32 46.6	.54	52.38	2
11 100	Gould, Z. C., 15^h 1915	8.5	10.67	+ 6.368	72 26 6.0	− 12.54	52.52	1
11 101	Lacaille, 6403	7.0	15 25 11.25	+ 5.896	69 43 28.1	− 12.54	52.34	1
11 102	Lacaille, 6401	8.0	18.32	6.146	71 14 41.2	.53	52.48	1
11 103	Apodis	10.0	24.01	7.191	75 46 23.7	.53	51.38	1
11 104	Gould, Z. C., 15^h 1907	8.8	24.20	5.626	67 45 56.8	.53	52.57	2
11 105	Apodis	10.0	26.53	+ 7.028	75 12 48.9	− 12.52	50.56	1

Number.	Constellation, Name of Star, or Synonym.	Magnitude.	Right Ascension, 1850.0.	Annual Precession.	South Declination, 1850.0.	Annual Precession.	Mean year.	No. of obs.
			h. m. s.	s.	° ′ ″	″		
11 106	Gould, Z. C., 15h 1935	9.0	15 25 37.87	+ 6.094	70 55 37.3	− 12.51	52.52	1
11 107	Gould, Z. C., 15h 1921	9.2	40.02	5.420	66 0 16.6	.51	52.60	2
11 108	Apodis	9.5	47.69	6.281	71 57 10.3	.50	52.36	2
11 109	Apodis	9.5	49.27	6.066	70 44 53.6	.50	52.40	1
11 110	Gould, Z. C., 15h 1966	9.0	51.23	+ 6.386	72 29 9.3	− 12.50	52.52	1
11 111	Trianguli Australis	9.5	15 25 55.99	+ 5.728	68 30 0.2	− 12.49	52.53	1
11 112	Gould, Z. C., 15h (1999)	10.0	26 0.30	7.036	75 12 43.4	.49	50.56	1
11 113	Octantis	9.2	3.32	76.717	89 11 0.8	.48	51.93	2
11 114	Apodis	9.8	5.40	7.228	75 51 47.0	.48	51.97	2
11 115	Apodis	9.8	7.90	+ 7.255	75 56 57.3	− 12.48	51.97	2
11 116	Apodis	9.5	15 26 34.04	+ 7.339	76 11 42.1	− 12.45	52.47	1
11 117	Apodis	9.5	34.15	5.432	66 2 57.6	.45	52.60	1
11 118	Trianguli Australis	9.0	38.59	5.974	70 8 1.2	.44	52.56	1
11 119	Apodis	9.5	42.86	5.582	67 19 9.4	.44	52.56	1
11 120	Gould, Z. C., 15h 2015	9.5	45.48	+ 6.081	70 46 48.9	− 12.43	52.40	1
11 121	Apodis	10.0	15 26 46.30	+ 7.258	75 55 43.4	− 12.43	51.97	2
11 122	Gould, Z. C., 15h 2067	8.8	46.64	7.597	76 56 45.8	.43	52.48	2
11 123	Gould, Z. C., 15h 2006	9.2	26 50.54	5.463	66 18 27.6	.43	52.60	2
11 124	Apodis	10.0	27 5.72	7.801	77 28 34.0	.41	52.36	1
11 125	Herschel, 4787[1]	9.5	7.99	+ 8.545	79 8 8.6	− 12.41	52.37	2
11 126	Gould, Z. C., 15h 2032	9.5	15 27 9.35	+ 5.653	67 51 15.9	− 12.41	52.56	1
11 127	Herschel, 4787[2]	9.8	11.14	8.547	79 8 15.4	.40	52.37	2
11 128	Gould, Z. C., 15h 2053	8.8	14.09	6.051	70 34 5.4	.40	52.46	2
11 129	Gould, Z. C., 15h 2082	9.5	16.18	7.115	75 25 40.6	.40	50.97	2
11 130	Gould, Z. C., 15h 2095	9.0	19.50	+ 7.469	76 33 19.8	− 12.39	52.48	1
11 131	Trianguli Australis	9.8	15 27 21.00	+ 5.694	68 9 10.2	− 12.39	52.54	2
11 132	Gould, Z. C., 15h 2045	9.2	21.77	5.560	67 5 36.5	.39	52.58	2
11 133	Trianguli Australis	10.0	30.19	5.614	67 31 19.6	.38	52.56	1
11 134	Apodis	10.0	32.89	7.821	77 30 34.2	.38	52.36	1
11 135	Gould, Z. C., 15h 2075	10.0	38.14	+ 5.972	70 3 29.4	− 12.37	52.56	1
11 136	Gould, Z. C., 15h 2094	9.0	15 27 40.43	+ 6.495	72 54 30.4	− 12.37	52.35	1
11 137	Trianguli Australis	10.0	41.42	5.446	66 5 39.4	.37	52.60	1
11 138	Gould, Z. C., 15h 2134	8.2	51.80	7.624	76 58 27.6	.36	52.48	2
11 139	Gould, 21182	10.0	27 59.79	6.847	74 25 14.9	.35	52.52	1
11 140	Gould, 21185	8.0	28 6.58	+ 6.838	74 22 0.2	− 12.34	52.52	2
11 141	Trianguli Australis	10.0	15 28 13.45	+ 5.564	67 4 14.7	− 12.33	52.60	1
11 142	Octantis	10.0	15.16	17.356	85 46 40.5	.33	51.93	2
11 143	Apodis	10.0	23.73	6.450	72 39 28.8	.32	52.35	1
11 144	Lacaille, 6423	7.8	24.38	5.935	69 46 26.9	.32	52.45	2
11 145	Gould, Z. C., 15h 2111	9.2	26.75	+ 5.482	66 21 15.8	− 12.32	52.60	2

No. 11 139. Gould's declination is 1′ 4.2″ greater. No. 11 142. Declination may be 2′ less.

Number.	Constellation, Name of Star, or Synonym.	Magnitude.	Right Ascension, 1850.0.	Annual Precession.	South Declination, 1850.0.	Annual Precession.	Mean year.	No. of obs.
			h. m. s.	s.	° ′ ″	″		
11 146	Apodis	9.5	15 28 38.55	+ 6.217	71 26 35.6	− 12.30	52.48	1
11 147	Gould, 21203	9.5	28 51.30	6.863	74 25 37.1	.29	52.52	1
11 148	Apodis	9.5	29 1.78	8.138	78 12 41.5	.28	52.36	1
11 149	Trianguli Australis . . .	9.5	6.08	5.779	68 39 46.2	.27	52.55	2
11 150	Apodis	9.5	11.91	6.642	73 29 44.0	.27	52.52	1
11 151	Stone, 8505	8.2	15 29 12.38	+ 5.948	69 48 22.0	− 12.26	52.50	3
11 152	Gould, Z. C., 15h 2180 .	8.8	22.51	5.778	68 38 14.2	.25	52.56	3
11 153	Gould, Z. C., 15h 2185 .	9.5	24.16	5.802	68 48 23.5	.25	52.57	1
11 154	Trianguli Australis . . .	9.5	27.42	5.614	67 23 20.8	.25	52.58	1
11 155	Gould, Z. C., 15h 2199 .	9.5	34.78	+ 5.818	68 54 31.5	− 12.24	52.58	2
11 156	Gould, 21219	8.8	15 29 37.49	+ 6.298	71 49 8.6	− 12.24	52.37	2
11 157	Gould, Z. C., 15h 2273 .	8.0	45.94	7.560	76 42 43.5	.23	52.48	2
11 158	Octantis	10.0	47.68	20.089	86 26 7.6	.22	51.93	2
11 159	Gould, Z. C., 15h 2223 .	9.0	55.06	5.644	67 35 35.4	.22	52.58	1
11 160	Stone, 8510	9.0	29 59.26	+ 5.458	66 1 53.5	− 12.21	52.60	1
11 161	Gould, Z. C., 15h 2231 .	9.0	15 30 0.98	+ 5.569	66 58 46.2	− 12.21	52.60	2
11 162	Trianguli Australis . . .	9.3	1.00	5.682	67 52 45.9	.21	52.58	3
11 163	Gould, Z. C., 15h 2268 .	8.0	3.91	6.668	73 33 36.1	.21	51.48	2
11 164	Trianguli Australis . . .	10.0	10.55	5.871	69 14 6.0	.20	52.42	1
11 165	Gould, Z. C., 15h 2395 .	9.0	33.91	+ 9.133	80 3 16.8	− 12.17	50.60	1
11 166	Gould, Z. C., 15h 2274 .	9.2	15 30 34.64	+ 5.640	67 30 57.0	− 12.17	52.57	2
11 167	Gould, Z. C., 15h 2280 .	9.5	35.94	5.832	68 56 29.5	.17	52.58	2
11 168	Lacaille, 6411	8.0	37.26	8.505	78 55 44.6	.17	51.47	2
11 169	Gould, Z. C., 15h 2277 .	8.8	38.83	5.527	66 35 4.4	.16	52.60	2
11 170	Trianguli Australis . . .	9.2	39.56	+ 5.501	66 21 54.3	− 12.16	52.60	2
11 171	Gould, Z. C., 15h 2305 .	8.5	15 30 42.82	+ 6.269	71 36 14.5	− 12.16	52.32	1
11 172	Gould, Z. C., 15h 2297 .	9.0	44.95	6.068	70 27 31.5	.16	52.40	1
11 173	Apodis	10.0	49.36	6.312	71 49 25.7	.15	52.41	1
11 174	Apodis	9.5	49.78	8.181	78 14 23.1	.15	52.11	3
11 175	Gould, Z. C., 15h 2318 .	8.0	54.07	+ 6.273	71 36 58.2	− 12.15	52.32	1
11 176	Gould, Z. C., 15h 2324 .	9.0	15 30 55.92	+ 6.339	71 57 41.0	− 12.14	52.37	2
11 177	Gould, Z. C., 15h 2302 .	9.0	31 1.91	5.506	66 22 57.4	.14	52.60	2
11 178	Gould, 21247	8.5	16.64	6.040	70 15 22.0	.12	52.46	2
11 179	Gould, Z. C., 15h 2341 .	8.8	17.29	6.033	70 12 57.0	.12	52.46	2
11 180	Apodis	9.5	26.07	+ 6.757	73 52 7.6	12.11	52.52	2
11 181	Gould, Z. C., 15h 2348 .	9.0	15 31 27.44	+ 5.904	69 22 40.1	− 12.11	52.44	3
11 182	Apodis	10.0	29.27	6.140	70 50 23.8	.11	52.52	1
11 183	Apodis	9.0	43.49	6.308	71 45 7.2	.09	52.32	1
11 184	Apodis	9.0	44.33	6.441	72 25 48.3	.09	52.52	1
11 185	Trianguli Australis . . .	9.5	51.94	+ 5.734	68 9 12.1	12.08	52.56	2

No. 11 158. Mean of two observations differing 11.0″.

Number.	Constellation, Name of Star, or Synonym.	Magnitude.	Right Ascension, 1850.0.	Annual Precession.	South Declination, 1850.0.	Annual Precession.	Mean year.	No. of obs.
			h. m. s.	s.	° ′ ″	″		
11 186	Gould, Z. C., 15h 2427 .	9.0	15 31 55.21	+ 6.923	74 30 28.6	— 12.08	52.52	1
11 187	Apodis	9.0	31 59.08	9.860	81 3 36.1	.07	50.57	2
11 188	Gould, Z. C., 15h 2392 .	9.5	32 2.41	5.583	66 57 36.2	.07	52.61	1
11 189	Apodis	9.5	4.58	6.746	73 47 27.4	.06	52.52	1
11 190	Gould, Z. C., 15h 2410 .	9.5	4.88	+ 5.896	69 17 11.1	— 12.06	52.42	1
11 191	Apodis	9.8	15 32 10.88	+ 8.828	79 29 6.4	— 12.06	50.98	2
11 192	Gould, Z. C., 15h 2411 .	9.0	13.40	5.580	66 55 10.5	.05	52.61	1
11 193	Trianguli Australis . . .	10.0	14.10	5.889	69 13 25.2	.05	52.42	1
11 194	Apodis	10.0	23.05	6.137	70 46 6.1	.04	52.52	1
11 195	Octantis	10.0	23.54	+55.961	88 50 30.6	— 12.04	51.93	2
11 196	Octantis	9.0	15 32 30.80	+12.210	83 19 25.6	— 12.03	50.50	1
11 197	Trianguli Australis . . .	9.8	47.03	5.942	69 32 50.6	.02	52.50	2
11 198	Apodis	10.0	32 55.85	6.468	72 29 35.3	.01	52.52	1
11 199	Apodis	9.5	33 2.18	7.376	76 1 17.9	12.00	52.59	2
11 200	Gould, Z. C., 15h 2470 .	9.5	6.25	+ 5.797	68 31 49.5	— 11.99	52.53	1
11 201	Trianguli Australis . . .	9.5	15 33 10.05	+ 5.932	69 27 25.0	— 11.99	52.50	2
11 202	Gould, Z. C., 15h 2510 .	8.5	11.45	6.423	72 15 36.4	.99	52.46	2
11 203	Apodis	9.5	11.54	9.829	80 59 7.4	.99	50.57	2
11 204	Lacaille, 6435	8.0	16.39	7.986	77 41 22.7	.98	52.36	1
11 205	Gould, Z. C., 15h 2516 .	9.0	21.35	+ 6.189	71 1 38.9	— 11.98	52.50	2
11 206	Gould, Z. C., 15h 2534 .	9.0	15 33 23.22	+ 6.698	73 31 15.8	— 11.97	51.48	2
11 207	Gould, Z. C., 15h 2526 .	9.0	32.74	6.087	70 24 32.4	.96	52.40	1
11 208	Apodis	10.0	33.91	7.628	76 44 36.4	.96	52.48	1
11 209	Apodis	8.5	35.78	6.226	71 12 14.2	.96	52.48	1
11 210	Trianguli Australis . . .	9.5	38.33	+ 5.748	68 8 18.3	— 11.96	52.59	1
11 211	Octantis	9.7	15 33 59.79	+17.049	85 36 10.0	— 11.93	51.93	3
11 212	Apodis	9.5	34 1.71	6.788	73 52 1.6	.93	52.52	2
11 213	Gould, Z. C., 15h 2546 .	10.0	2.26	5.807	68 32 54.9	.93	52.53	1
11 214	Gould, Z. C., 15h 2550 .	9.5	8.60	5.805	68 31 21.3	.92	52.53	1
11 215	Apodis	10.0	10.63	+ 5.805	68 31 17.4	11.92	52.52	1
11 216	Gould, Z. C., 15h 2594 .	8.5	15 34 11.39	+ 7.285	75 40 46.4	11.92	52.12	3
11 217	Apodis	8.0	11.60	9.360	80 17 37.6	.92	50.99	2
11 218	Gould, Z. C., 15h 2588 .	9.5	42.18	5.855	68 50 15.1	.88	52.59	1
11 219	Gould, Z. C., 15h 2581 .	9.0	42.73	5.551	66 30 57.3	.88	52.60	1
11 220	Apodis	9.0	44.28	+ 7.407	76 2 34.4	— 11.88	52.59	2
11 221	Trianguli Australis . . .	9.0	15 34 45.91	+ 5.574	66 42 10.8	— 11.88	52.60	2
11 222	Apodis	9.5	34 46.35	6.284	71 27 24.2	.88	52.48	1
11 223	Gould, Z. C., 15h 2710 .	9.5	35 3.46	8.696	79 8 41.8	.86	52.41	1
11 224	Trianguli Australis . . .	9.5	9.88	5.719	67 49 43.0	.85	52.59	2
11 225	Lacaille, 6449	7.5	15.44	+ 7.848	77 16 5.6	11.84	52.52	2

Number.	Constellation, Name of Star, or Synonym.	Magnitude.	Right Ascension, 1850.0.	Annual Precession.	South Declination, 1850.0.	Annual Precession.	Mean year.	No. of obs.
			h. m. s.	s.	° ′ ″	″		
11 226	Gould, Z. C., 15^h 2628	9.5	15 35 24.70	+ 5.510	66 6 47.9	− 11.83	52.60	1
11 227	Trianguli Australis . .	9.5	28.65	5.698	67 38 46.7	.83	52.58	1
11 228	Apodis	9.0	29.68	6.634	73 8 1.0	.82	52.35	1
11 229	Gould, Z. C., 15^h 2659 .	9.0	30.39	6.192	70 54 55.7	.82	52.52	1
11 230	Gould, Z. C., 15^h 2645 .	8.8	36.30	+ 5.581	66 42 27.8	− 11.82	52.60	2
11 231	Trianguli Australis . . .	10.0	15 35 36.97	+ 5.882	68 58 23.0	− 11.82	52.42	1
11 232	Apodis	10.0	37.00	7.509	76 18 49.8	.82	52.56	1
11 233	Trianguli Australis . . .	10.0	37.42	5.650	67 15 50.5	.82	52.56	1
11 234	Gould, Z. C., 15^h 2671 .	9.0	38.66	6.330	71 39 3.2	.81	52.32	1
11 235	Apodis	9.2	39.86	+ 7.491	76 15 33.8	− 11.81	52.52	2
11 236	Apodis	9.0	15 35 49.60	+ 7.097	74 57 58.3	− 11.80	52.40	1
11 237	Gould, Z. C., 15^h 2668 .	9.5	54.98	5.503	66 1 8.2	.79	52.60	1
11 238	Gould, Z. C., 15^h 2707 .	8.0	55.62	6.580	72 52 9.8	.79	52.35	1
11 239	Lacaille, 6404	7.5	58.98	13.025	83 47 28.0	.79	50.55	1
11 240	Apodis	9.0	35 59.66	+ 6.159	70 41 29.6	− 11.79	52.46	2
11 241	Apodis	9.5	15 36 5.00	+ 6.860	74 3 32.6	− 11.78	52.52	1
11 242	Gould, Z. C., 15^h 2697 .	9.5	11.14	5.745	67 57 21.0	.78	52.59	2
11 243	Apodis	10.0	11.32	5.666	67 21 7.3	.78	52.56	1
11 244	Gould, Z. C., 15^h 2774 .	9.0	11.77	8.028	77 40 50.4	.77	52.36	1
11 245	Gould, Z. C., 15^h 2705 .	9.0	12.69	+ 5.813	68 27 12.4	− 11.77	52.53	1
11 246	Trianguli Australis . . .	9.0	15 36 14.82	+ 5.706	67 39 40.0	− 11.77	52.58	1
11 247	Apodis	10.0	15.57	9.597	80 34 43.1	.77	51.37	1
11 248	Gould, Z. C., 15^h 2818 .	9.0	20.55	9.045	79 43 25.3	.76	50.61	1
11 249	Gould, Z. C., 15^h 2725 .	8.5	26.38	5.920	69 10 43.8	.76	52.51	2
11 250	Apodis	8.2	30.12	+ 6.107	70 21 16.4	− 11.75	52.38	2
11 251	Gould, Z. C., 15^h 2737 .	9.0	15 36 35.66	+ 5.817	68 27 22.1	− 11.75	52.53	1
11 252	Gould, Z. C., 15^h 2785 .	8.8	44.09	7.126	75 1 34.2	.74	51.48	2
11 253	Trianguli Australis . . .	9.2	46.39	5.977	69 31 47.4	.73	52.51	2
11 254	Gould, Z. C., 15^h 2752 .	9.0	51.46	5.663	67 17 18.1	.73	52.56	1
11 255	Gould, Z. C., 15^h 2776 .	8.0	51.62	+ 6.592	72 52 22.0	− 11.73	52.35	1
11 256	Gould, Z. C., 15^h 2757 .	8.2	15 36 52.21	+ 5.782	68 11 2.8	− 11.73	52.56	2
11 257	Trianguli Australis . . .	9.5	55.03	6.052	69 59 51.4	.72	52.56	1
11 258	Apodis	10.0	55.10	7.684	76 45 49.8	.72	52.48	2
11 259	Gould, Z. C., 15^h 2763 .	9.0	36 57.14	5.776	68 8 31.5	.72	52.56	2
11 260	Apodis	9.5	37 0.40	+ 6.054	70 0 6.1	− 11.72	52.56	1
11 261	Gould, Z. C., 15^h 2837 .	8.5	15 37 10.84	+ 7.910	77 21 5.2	− 11.71	52.52	2
11 262	Gould, Z. C., 15^h 2815 .	9.0	19.06	6.622	72 59 19.0	.70	51.39	2
11 263	Trianguli Australis . . .	9.0	27.10	5.771	68 4 12.4	.69	52.59	1
11 264	Apodis	9.5	32.92	10.006	81 5 27.2	.68	50.56	1
11 265	Octantis	10.0	39.56	+11.614	82 44 41.5	− 11.67	50.49	1

No. 11 237. Gould's declination is 2′ greater.

Number.	Constellation, Name of Star, or Synonym.	Magnitude.	Right Ascension, 1850.0.	Annual Precession.	South Declination, 1850.0.	Annual Precession.	Mean year.	No. of obs.
			h. m. s.	s.	° ′ ″	″		
11 266	Apodis	9.8	15 37 46.00	+ 8.505	78 41 2.0	− 11.66	51.48	2
11 267	Apodis	8.0	38 0.91	6.272	71 12 54.3	.65	52.48	1
11 268	Apodis	10.0	4.97	11.116	82 17 31.3	.64	50.53	1
11 269	Gould, Z. C., 15h 2881	9.0	7.05	7.009	74 32 25.0	.64	52.52	1
11 270	Lacaille, 6501	7.2	13.73	+ 5.660	67 10 34.6	− 11.63	52.58	2
11 271	Lacaille, 6494	8.0	15 38 16.64	+ 6.459	72 10 7.0	− 11.63	52.46	2
11 272	Octantis	8.9	22.21	24.946	87 8 52.3	.62	51.93	4
11 273	Apodis	10.0	41.07	6.941	74 15 22.8	.60	52.52	1
11 274	Trianguli Australis	10.0	42.75	5.988	69 29 24.9	.60	52.42	1
11 275	Gould, Z. C., 15h 2898	8.8	54.69	+ 5.543	66 10 5.6	− 11.58	52.60	3
11 276	Trianguli Australis	9.5	15 38 57.59	+ 5.667	67 11 10.1	− 11.58	52.56	1
11 277	Gould, Z. C., 15h 2909	9.0	39 0.32	5.736	67 43 34.5	.58	52.58	1
11 278	Gould, Z. C., 15h 2930	9.8	0.34	6.197	70 44 20.7	.58	52.46	2
11 279	Apodis	10.0	3.75	7.312	75 33 26.8	.57	52.53	1
11 280	Octantis	10.0	6.44	+20.901	86 29 39.4	− 11.57	51.93	2
11 281	Lacaille, 6484	6.5	15 39 12.95	+ 8.031	77 34 28.5	− 11.56	52.47	3
11 282	Apodis	9.5	16.56	6.895	74 2 56.5	.56	52.52	1
11 283	Apodis	10.0	17.10	7.755	76 51 51.6	.56	52.48	1
11 284	Trianguli Australis	10.0	17.19	5.804	68 11 59.1	.55	52.52	1
11 285	Trianguli Australis	10.0	20.14	+ 5.936	69 6 31.0	− 11.55	52.42	1
11 286	Gould, Z. C., 15h 2954	10.0	15 39 20.32	+ 6.187	70 39 45.2	− 11.55	52.52	1
11 287	Gould, Z. C., 15h 2932	9.0	21.19	5.534	66 3 50.0	.55	52.60	1
11 288	Apodis	10.0	30.94	9.498	80 20 31.7	.54	51.37	1
11 289	Apodis	9.9	38.58	8.184	77 55 16.5	.53	52.23	4
11 290	Octantis	9.0	40.74	+12.496	83 22 38.4	− 11.53	50.50	1
11 291	Apodis	9.0	15 39 46.26	+ 7.517	76 10 2.0	− 11.52	52.52	2
11 292	Apodis	9.0	39 50.81	6.512	72 20 44.5	.51	52.46	2
11 293	Apodis	10.0	40 0.62	8.777	79 7 54.2	.50	52.41	1
11 294	Gould, Z. C., 15h 3066	8.1	6.36	8.344	78 15 41.8	.50	52.23	4
11 295	Apodis	10.0	15.52	+ 8.883	79 19 3.7	− 11.49	52.41	1
11 296	Gould, Z. C., 15h 3024	9.0	15 40 18.82	+ 6.349	71 30 9.0	− 11.48	52.40	2
11 297	Apodis	10.0	27.17	7.425	75 51 43.2	.47	52.54	1
11 298	Gould, Z. C., 15h 3031	9.5	30.72	6.206	70 42 42.4	.47	52.52	1
11 299	Trianguli Australis	9.0	31.32	5.821	68 14 56.0	.47	52.56	2
11 300	Octantis	9.5	32.36	+12.588	83 25 25.7	− 11.47	50.50	1
11 301	Octantis	8.6	15 40 38.70	+15.564	84 58 45.0	− 11.46	51.46	3
11 302	Gould, Z. C., 15h 3041	8.5	38.82	6.144	70 20 29.6	.46	52.47	2
11 303	Trianguli Australis	9.2	38.96	6.016	69 33 31.8	.46	52.51	2
11 304	Octantis	8.5	43.76	13.016	83 42 3.1	.45	50.55	1
11 305	κ Trianguli Australis	5.2	44.49	+ 5.809	68 8 52.7	− 11.45	52.56	2

Number.	Constellation, Name of Star, or Synonym.	Magnitude.	Right Ascension, 1850.0.	Annual Precession.	South Declination, 1850.0.	Annual Precession.	Mean year.	No. of obs.
			h. m. s.	s.	° ′ ″	″		
11 306	Apodis	10.0	15 40 45.80	+ 6.904	74 0 48.9	− 11.45	52.52	1
11 307	Apodis	10.0	46.00	8.182	77 52 32.9	.45	52.41	3
11 308	Apodis	10.0	46.31	6.886	73 56 34.7	.45	52.52	1
11 309	Trianguli Australis	10.0	50.41	6.057	69 48 29.0	.44	52.60	1
11 310	Apodis	9.0	52.18	+ 9.390	80 8 22.3	− 11.44	50.60	1
11 311	Apodis	10.0	15 40 53.57	+ 8.207	77 55 37.8	− 11.44	52.44	3
11 312	Gould, Z. C., 15^h 3057	8.0	54.02	6.168	70 28 3.9	.44	52.40	1
11 313	Apodis	10.0	40 54.42	9.223	79 52 34.4	.44	50.60	1
11 314	Trianguli Australis	9.0	41 4.53	5.926	68 56 35.3	.43	52.59	1
11 315	Trianguli Australis	9.5	12.67	5.998	69 24 42.4	− 11.42	52.60	1
11 316	Gould, Z. C., 15^h 3087	8.0	15 41 16.78	+ 6.181	70 31 29.7	− 11.41	52.40	1
11 317	Gould, Z. C., 15^h 3075	9.0	18.49	5.656	66 56 58.9	.41	52.61	1
11 318	Gould, Z. C., 15^h 3093	9.5	18.90	6.223	70 45 40.7	.41	52.52	1
11 319	Apodis	9.2	20.22	10.807	81 54 31.6	.41	50.54	2
11 320	Apodis	9.2	26.30	+ 6.155	70 21 42.0	− 11.40	52.47	2
11 321	Apodis	9.5	15 41 28.58	+ 9.346	80 3 15.3	− 11.40	50.60	1
11 322	Lacaille, 6512	8.2	35.15	6.605	72 41 55.3	.39	52.43	2
11 323	Gould, Z. C., 15^h 3103	9.2	39.36	5.664	66 59 15.7	.38	52.58	2
11 324	Octantis	10.0	40.96	17.130	85 31 57.2	.38	51.93	2
11 325	Gould, Z. C., 15^h 3144	9.2	44.20	+ 6.969	74 13 28.4	− 11.38	52.52	2
11 326	Trianguli Australis	8.8	15 41 44.58	+ 5.770	67 47 47.6	− 11.38	52.59	2
11 327	Apodis	10.0	47.49	6.628	72 47 31.7	.38	52.35	1
11 328	Octantis	9.3	41 58.09	26.900	87 21 9.0	.36	51.93	3
11 329	Gould, 21512	8.0	42 9.43	6.957	74 9 33.4	.35	52.52	2
11 330	Gould, Z. C., 15^h 3195	8.8	10.34	+ 7.861	77 2 9.7	− 11.35	52.51	3
11 331	Gould, Z. C., 15^h 3132	9.0	15 42 11.22	+ 5.595	66 23 57.5	− 11.35	52.60	2
11 332	Gould, Z. C., 15^h 3136	9.5	15.64	5.545	65 58 19.1	.34	52.60	1
11 333	Gould, Z. C., 15^h 3160	9.5	31.72	5.679	67 3 5.4	.32	52.58	2
11 334	Lacaille, 6513	8.0	38.00	7.243	75 10 32.6	.31	51.48	2
11 335	Apodis	9.2	46.76	+ 6.464	71 57 46.7	− 11.30	52.37	2
11 336	Trianguli Australis	9.5	15 42 47.51	+ 5.618	66 33 4.5	− 11.30	52.61	1
11 337	Trianguli Australis	9.5	50.87	5.856	68 21 15.8	.30	52.53	1
11 338	Apodis	10.0	56.36	6.603	72 37 17.4	.29	52.52	1
11 339	Trianguli Australis	9.2	42 56.98	5.725	67 23 25.6	.29	52.57	2
11 340	Apodis	9.0	43 3.24	+ 6.625	72 43 51.0	− 11.28	52.43	2
11 341	Trianguli Australis	10.0	15 43 3.64	+ 5.990	69 15 2.8	− 11.28	52.42	1
11 342	Gould, Z. C., 15^h 3198	8.2	5.80	5.793	67 53 29.0	.28	52.59	2
11 343	Apodis	9.5	7.76	8.015	77 23 31.4	.28	52.42	2
11 344	Gould, 21518	9.5	9.00	6.938	74 2 20.0	.28	52.52	1
11 345	Apodis	9.5	22.98	+ 6.615	72 39 35.0	− 11.26	52.52	1

Number.	Constellation, Name of Star, or Synonym.	Magnitude.	Right Ascension, 1859.0.	Annual Precession.	South Declination, 1850.0.	Annual Precession.	Mean year.	No. of obs.
			h. m. s.	s.	° ′ ″	″		
11 346	Gould, Z. C., 15h 3216	8.3	15 43 24.45	+ 5.640	66 41 18.7	− 11.26	52.60	3
11 347	Gould, Z. C., 15h 3240	8.5	27.90	6.730	73 9 55.3	.25	51.40	2
11 348	Apodis	9.5	28.18	6.838	73 37 20.9	.25	52.52	1
11 349	Apodis	10.0	28.96	8.796	79 3 21.6	.25	51.51	2
11 350	Trianguli Australis	9.0	34.01	+ 5.607	66 24 44.5	− 11.25	52.61	1
11 351	Apodis	9.0	15 43 37.03	+ 6.227	70 39 37.5	− 11.24	52.46	2
11 352	Lacaille, 6534	8.2	42.04	6.229	70 40 2.6	.24	52.46	2
11 353	Apodis	10.0	42.68	8.849	79 8 50.3	.24	52.41	1
11 354	Gould, Z. C., 15h 3264	9.2	48.36	6.735	73 10 20.6	.23	51.40	2
11 355	Gould, Z. C., 15h 3242	8.8	51.52	+ 5.836	68 9 29.2	− 11.23	52.56	2
11 356	Gould, Z. C., 15h 3328	10.0	15 43 56.76	+ 8.681	78 49 22.9	− 11.22	50.62	1
11 357	Apodis	9.2	43 57.88	6.168	70 18 15.8	.22	52.48	2
11 358	Apodis	9.0	44 3.34	7.479	75 52 49.6	.21	52.52	3
11 359	Trianguli Australis	9.5	6.30	6.013	69 20 27.8	.21	52.42	1
11 360	Gould, Z. C., 15h 3344	9.3	15.09	+ 8.421	78 17 1.5	− 11.20	52.44	3
11 361	Lacaille, 6536	6.8	15 44 15.16	+ 6.487	72 1 13.2	− 11.20	52.37	2
11 362	Gould, Z. C., 15h 3285	9.0	28.69	5.827	68 3 29.5	.18	52.59	1
11 363	Apodis	9.8	41.04	6.403	71 33 33.5	.17	52.40	2
11 364	Apodis	10.0	42.41	8.868	79 9 5.7	.16	52.41	1
11 365	Apodis	9.3	46.82	+ 8.042	77 23 56.0	− 11.16	52.47	3
11 366	Apodis	8.5	15 44 51.63	+ 6.389	71 28 50.5	− 11.15	52.40	2
11 367	Gould, Z. C., 15h 3362	8.5	45 0.82	6.980	74 7 8.2	.14	53.52	2
11 368	Lacaille, 6542	6.2	6.36	6.726	73 4 29.1	.14	51.40	2
11 369	Octantis	9.5	12.82	16.785	85 21 45.8	.13	51.93	2
11 370	Trianguli Australis	9.5	16.95	+ 5.644	66 36 25.3	− 11.12	52.61	1
11 371	Lacaille, 6527	7.8	15 45 17.34	+ 7.817	76 48 15.4	− 11.12	52.48	2
11 372	Gould, Z. C., 15h 3431	9.3	21.16	8.439	78 17 11.2	.12	52.44	3
11 373	Apodis	9.0	26.22	6.298	70 57 48.2	.11	52.52	1
11 374	Apodis	9.8	29.02	6.196	70 23 0.8	.11	52.48	2
11 375	Trianguli Australis	9.5	29.19	+ 5.600	66 13 59.6	− 11.11	52.60	2
11 376	Apodis	10.0	15 45 39.82	+ 6.393	71 27 33.8	− 11.09	52.48	1
11 377	Octantis	9.5	44.82	12.984	83 35 12.1	.09	50.55	1
11 378	Gould, Z. C., 15h 3407	9.0	45.46	6.751	73 9 10.4	.09	51.40	2
11 379	Apodis	9.5	46.30	7.837	76 50 19.7	.09	52.48	1
11 380	Gould, 21578	8.2	45 58.43	+ 5.865	68 14 22.6	− 11.07	52.56	2
11 381	Trianguli Australis	10.0	46 1.88	+ 5.799	67 45 42.5	− 11.07	52.58	1
11 382	Apodis	8.8	8.94	6.574	72 19 55.6	.06	52.46	2
11 383	Apodis	9.5	9.42	6.886	73 41 46.3	.06	52.52	1
11 384	Trianguli Australis	9.0	9.45	5.789	67 40 41.2	.06	52.58	1
11 385	Octantis	9.2	27.61	+22.901	86 46 38.6	− 11.04	51.71	4

No. 11 368. Mean of two observations differing 0.2″.
No. 11 369. Right ascension may be 15h 48m 42.5s. Differing 1 wire interval.

Number.	Constellation, Name of Star, or Synonym.	Magnitude.	Right Ascension, 1850.0.	Annual Precession.	South Declination, 1850.0.	Annual Precession.	Mean year.	No. of obs.
			h. m. s.	s.	° ′ ″	″		
11 386	Gould, Z. C., 15h 3444 .	10.0	15 46 32.14	+ 5.983	69 0 47.8	− 11.03	52.59	1
11 387	Apodis	10.0	32.95	9.247	79 44 56.3	.03	50.61	1
11 388	Apodis	9.0	35.05	6.379	71 20 25.4	.03	52.48	1
11 389	Octantis	8.5	36.89	13.177	83 41 32.2	.02	50.55	1
11 390	Gould, Z. C., 15h 3467 .	9.5	42.20	+ 6.293	70 52 8.1	− 11.02	52.52	1
11 391	Octantis	8.5	15 46 55.88	+16.165	85 7 10.8	− 11.00	51.46	3
11 392	Octantis	10.0	46 56.97	15.932	85 1 51.9	11.00	51.92	1
11 393	Melbourne (1), 801 . . .	7.0	47 8.94	5.634	66 24 45.0	10.99	52.61	1
11 394	Apodis	9.5	9.08	7.358	75 22 24.8	.99	51.55	2
11 395	Apodis	9.0	10.81	+ 6.379	71 18 39.4	− 10.98	52.48	1
11 396	Gould, Z. C., 15h 3512 .	9.2	15 47 16.44	+ 6.432	71 34 41.2	− 10.98	52.40	2
11 397	Apodis	9.5	16.68	6.659	72 40 18.6	.98	52.60	1
11 398	Apodis	9.2	25.86	10.584	81 31 24.2	.97	50.56	2
11 399	Gould, Z. C., 15h 3524 .	9.0	29.33	6.437	71 35 38.2	.96	51.40	2
11 400	Apodis	9.0	30.88	+ 6.465	71 44 1.4	− 10.96	52.32	1
11 401	Octantis	9.5	15 47 32.84	+13.822	84 3 8.9	− 10.96	51.49	1
11 402	Gould, Z. C., 15h 3600 .	8.5	43.46	8.634	78 36 46.4	.94	50.49	2
11 403	Gould, Z. C., 15h 3520 .	9.5	48.05	5.587	65 58 44.5	.94	52.60	1
11 404	Gould, Z. C., 15h 3607 .	9.2	55.94	8.424	78 10 5.3	.93	52.48	2
11 405	Lacaille, 6554	8.5	57.84	+ 7.513	75 49 48.8	− 10.93	52.52	3
11 406	Lacaille, 6549	7.5	15 47 58.11	+ 7.877	76 51 51.7	− 10.93	52.48	1
11 407	Apodis	9.2	59.42	7.028	74 10 36.8	.92	52.52	2
11 408	Apodis	9.0	47 59.66	7.557	75 57 46.4	.92	52.55	2
11 409	Apodis	10.0	48 1.80	10.816	81 45 36.3	.92	50.56	1
11 410	Apodis	9.5	2.75	+ 6.555	72 9 0.7	− 10.92	52.41	1
11 411	Apodis	9.2	15 48 4.86	+10.586	81 30 35.6	− 10.92	50.56	2
11 412	Gould, Z. C., 15h 3623 .	9.2	9.12	8.402	78 6 49.7	.91	52.48	2
11 413	Octantis	9.3	11.04	29.148	87 32 10.4	.91	51.42	7
11 414	Trianguli Australis . . .	9.5	14.97	5.677	66 41 41.9	.91	52.61	2
11 415	Lacaille, 6552	7.5	15.91	+ 7.862	76 48 56.2	− 10.90	52.48	2
11 416	Apodis	10.0	15 48 17.24	+ 7.562	75 57 56.2	− 10.90	52.56	1
11 417	Gould, Z. C., 15h 3581 .	9.0	19.49	6.419	71 28 12.8	.90	52.48	1
11 418	Apodis	9.3	25.65	6.859	73 29 15.0	.89	51.84	3
11 419	Gould, Z. C., 15h 3563 .	9.0	26.64	5.587	65 56 39.5	.89	52.60	1
11 420	Apodis	10.0	28.13	+ 6.742	72 59 20.1	− 10.89	52.60	1
11 421	Trianguli Australis . . .	9.5	15 48 32.19	+ 5.609	66 7 34.2	− 10.88	52.60	1
11 422	Trianguli Australis . . .	9.0	33.84	6.062	69 24 48.1	.88	51.52	2
11 423	Gould, Z. C., 15h 3602 .	8.5	35.53	6.619	72 25 51.5	.88	52.52	1
11 424	Apodis	9.0	35.81	7.479	75 42 9.7	.88	52.54	1
11 425	Gould, 21668	8.5	48 48.52	+ 7.958	77 2 47.2	− 10.86	52.52	2

No. 11 411. Possibly 81° 31′ 3.5″; 1 revolution different.

Number.	Constellation, Name of Star, or Synonym.	Magnitude.	Right Ascension, 1850.0.	Annual Precession.	South Declination, 1850.0.	Annual Precession.	Mean year.	No. of obs.
			h. m. s.	s.	° ′ ″	″		
11 426	Apodis	10.0	15 49 2.60	+ 9.743	80 25 52.0	− 10.85	51.37	1
11 427	Trianguli Australis	10.0	2.73	5.729	67 3 24.7	.85	52.56	1
11 428	Apodis	10.0	8.57	6.593	72 16 54.4	.84	52.52	1
11 429	Gould, Z. C., 15^h 3613	9.5	8.86	5.669	66 34 57.6	.84	52.61	1
11 430	Gould, Z. C., 15^h 3611	7.5	8.87	+ 5.621	66 11 24.1	− 10.84	52.60	2
11 431	Apodis	9.2	15 49 17.08	+ 11.023	81 56 39.0	− 10.83	50.54	2
11 432	Lacaille, 6573	6.0	18.70	6.531	71 58 42.6	.83	52.37	2
11 433	Apodis	9.5	25.94	6.206	70 14 26.4	.82	52.48	2
11 434	Apodis	10.0	35.81	7.179	74 39 54.1	.81	52.40	1
11 435	Octantis	9.5	35.90	+ 12.043	82 50 35.3	− 10.81	51.52	1
11 436	Apodis	9.0	15 49 36.59	+ 6.284	70 40 38.7	− 10.80	52.40	1
11 437	Apodis	9.5	39.16	7.111	74 24 52.4	.80	52.52	1
11 438	Trianguli Australis	9.0	46.44	6.108	69 37 57.9	.79	52.60	1
11 439	Apodis	9.2	49 56.27	6.603	72 17 33.8	.78	52.46	2
11 440	Gould, Z. C., 15^h 3705	9.5	50 7.76	+ 6.037	69 10 14.3	− 10.77	52.42	1
11 441	Octantis	9.4	15 50 19.44	+ 28.507	87 27 33.0	− 10.75	51.49	6
11 442	Gould, Z. C., 15^h 3722	8.8	21.44	6.438	71 27 53.0	.75	52.40	2
11 443	Apodis	9.0	23.25	7.772	76 29 39.4	.75	52.48	1
11 444	Gould, Z. C., 15^h 3725	10.0	26.57	6.325	70 51 36.4	.74	52.52	1
11 445	Gould, Z. C., 15^h 3757	9.0	37.66	+ 6.983	73 53 20.6	− 10.73	52.52	2
11 446	Gould, Z. C., 15^h 3721	9.0	15 50 38.98	+ 5.611	66 0 57.4	− 10.73	52.60	1
11 447	Gould, Z. C., 15^h 3739	9.5	39.83	6.339	70 55 34.8	.73	52.52	1
11 448	Octantis	7.9	42.66	54.478	88 44 27.0	.72	51.42	7
11 449	Gould, Z. C., 15^h 3765	9.0	45.18	6.949	73 45 6.1	.72	52.52	1
11 450	Lacaille, 6597	8.0	50.64	+ 5.871	68 0 44.0	− 10.71	52.59	1
11 451	Apodis	10.0	15 50 51.05	+ 7.330	75 8 5.2	− 10.71	52.40	1
11 452	Apodis	9.5	50 54.82	6.200	70 7 51.5	.71	52.56	1
11 453	Apodis	9.0	51 5.97	8.505	78 14 22.8	.70	52.43	1
11 454	Apodis	10.0	10.59	6.867	73 23 56.9	.69	52.52	1
11 455	Trianguli Australis	9.5	15.94	+ 6.144	69 46 52.2	− 10.68	52.58	2
11 456	Octantis	10.0	15 51 20.58	+ 28.776	87 29 14.8	− 10.68	51.93	2
11 457	Gould, Z. C., 15^h 3792	9.5	29.54	6.243	70 21 1.7	.67	52.47	2
11 458	Lacaille, 6591	7.2	31.35	6.461	71 31 28.8	.66	52.40	2
11 459	Apodis	8.5	34.04	6.658	72 28 19.8	.66	52.52	1
11 460	Apodis	7.5	37.61	+ 9.766	80 23 42.3	− 10.66	51.37	1
11 461	Apodis	9.5	15 51 41.79	+ 7.674	76 10 11.8	− 10.65	52.56	1
11 462	Trianguli Australis	8.8	44.49	5.763	67 9 58.2	.65	51.93	3
11 463	Apodis	9.7	46.95	6.472	71 34 6.2	.64	52.35	2
11 464	Apodis	9.0	48.11	6.912	73 33 26.0	.64	51.84	3
11 465	Octantis	9.5	51.90	+ 22.717	86 41 51.9	− 10.64	51.93	3

Number.	Constellation, Name of Star, or Synonym.	Magnitude.	Right Ascension, 1850.0.	Annual Precession.	South Declination, 1850.0.	Annual Precession.	Mean year.	No. of obs.
			h. m. s.	s.	° ′ ″	″		
11 466	Octantis	10.0	15 51 57.23	+ 12.235	82 56 46.5	− 10.63	51.52	1
11 467	Apodis	9.5	52 1.69	6.972	73 47 11.2	.63	52.52	2
11 468	Trianguli Australis	9.5	11.20	6.131	69 39 17.5	.61	52.60	1
11 469	Gould, Z. C., 15^h 3840	9.5	16.61	6.258	70 23 50.6	.61	52.40	1
11 470	Apodis	9.0	21.75	+ 6.972	73 46 19.2	− 10.60	52.52	2
11 471	Gould, Z. C., 15^h 3881	9.0	15 52 22.89	+ 7.196	74 36 46.4	− 10.60	52.46	2
11 472	Apodis	10.0	24.38	7.193	74 35 55.3	.60	52.52	1
11 473	Trianguli Australis	9.8	24.54	5.799	67 23 58.7	.60	52.57	2
11 474	Apodis	9.5	34.91	7.821	76 33 2.6	.60	52.48	1
11 475	Apodis	9.0	40.13	+ 6.576	72 2 8.8	− 10.59	52.41	1
11 476	Apodis	10.0	15 52 55.84	+ 6.723	72 42 20.9	− 10.56	52.52	1
11 477	Lacaille, 6606	8.2	53 0.72	6.703	72 36 46.2	.55	52.56	2
11 478	Gould, Z. C., 15^h 3889	8.5	12.36	5.603	65 47 54.8	.54	52.60	1
11 479	Lacaille, 6613	7.8	17 10	6.310	70 38 23.6	.53	52.46	2
11 480	Gould, Z. C., 15^h 3906	8.5	20.89	+ 5.734	66 51 5.9	− 10.53	52.61	1
11 481	Apodis	10.0	15 53 22.11	+ 6.331	70 44 53.3	− 10.53	52.52	1
11 482	Apodis	9.0	22.98	6.920	73 31 20.1	.53	52.53	1
11 483	Trianguli Australis	10.0	23.87	5.884	67 58 3.2	.52	52.58	1
11 484	Apodis	8.8	25.76	7.134	74 20 39.9	.52	52.52	2
11 485	Apodis	9.0	30.41	+ 6.678	72 28 37.7	− 10.52	52.52	1
11 486	Apodis	9.0	15 53 32.54	+ 6.608	72 9 46.0	− 10.51	52.41	1
11 487	Gould, 21774	8.5	36.05	7.221	74 39 8.2	.51	52.46	2
11 488	Gould, Z. C., 15^h 4011	9.0	36.96	8.699	78 33 28.5	.51	51.36	2
11 489	Gould, 21779	9.0	46.24	7.208	74 36 9.0	.50	52.46	2
11 490	Apodis	9.5	46.74	+ 6.628	72 14 26.0	− 10.50	52.46	2
11 491	Lacaille, 6604	8.0	15 53 50.18	+ 7.291	74 53 3.0	− 10.49	52.40	1
11 492	Gould, Z. C., 15^h 3969	9.0	50.24	6.360	70 53 10.8	.49	52.52	1
11 493	Gould, Z. C., 15^h 3962	8.2	50.98	6.068	69 10 37.6	.49	52.51	2
11 494	Octantis	9.2	55.70	23.890	86 51 57.8	.48	51.64	6
11 495	Gould, Z. C., 15^h 3959	9.0	53 58.70	+ 5.746	66 54 30.2	− 10.48	52.61	1
11 496	Apodis	10.0	15 54 0.03	+ 7.012	73 51 46.7	− 10.48	52.52	1
11 497	Trianguli Australis	9.5	4.48	5.782	67 10 57.8	.47	50.62	1
11 498	Gould, Z. C., 15^h 4019	9.2	10.24	7.601	75 51 54.7	.47	52.58	3
11 499	Gould, Z. C., 15^h 3999	9.0	31.74	5.981	68 34 24.5	.44	52.56	2
11 500	Trianguli Australis	10.0	34.90	+ 5.629	65 56 21.4	− 10.44	52.60	1
11 501	Apodis	9.5	15 54 42.26	+ 6.412	71 7 25.8	− 10.43	52.48	1
11 502	Trianguli Australis	9.8	48.14	5.803	67 17 57.0	.42	51.59	2
11 503	Octantis	9.0	49.34	12.494	83 5 10.6	.42	51.01	2
11 504	Trianguli Australis	9.8	50.42	5.806	67 19 10.8	.42	51.59	2
11 505	Trianguli Australis	10.0	52.44	+ 5.746	66 51 53.1	− 10.41	52.61	1

Number.	Constellation, Name of Star, or Synonym.	Magnitude.	Right Ascension, 1850.0.	Annual Precession.	South Declination, 1850.0.	Annual Precession.	Mean year.	No. of obs.
			h. m. s.	s.	° ′ ″	″		
11 506	Apodis	10.0	15 54 57.35	+ 9.008	79 6 8.4	— 10.41	52.41	1
11 507	Gould, Z. C., 15h 4075	8.5	54 59.96	7.757	76 17 14.6	.40	52.52	2
11 508	Apodis	9.0	55 7.61	8.331	77 44 2.4	.40	52.36	1
11 509	Apodis	9.5	10.15	7.464	75 24 19.6	.39	52.58	2
11 510	Gould, Z. C., 15h 4126	9.0	15.00	+ 9.038	79 8 46.0	— 10.39	51.51	2
11 511	Apodis	9.8	15 55 16.27	+ 7.466	75 24 26.2	— 10.38	52.58	2
11 512	Apodis	9.5	32.10	6.346	70 43 43.6	.36	52.40	1
11 513	Trianguli Australis	9.0	45.37	5.917	68 4 23.0	.35	52.59	1
11 514	Apodis	10.0	45.60	6.666	72 19 20.1	.35	52.46	2
11 515	Apodis	10.0	45.87	+ 9.419	79 46 14.6	— 10.35	50.61	1
11 516	Gould, Z. C., 15h 4084	9.5	15 55 47.26	+ 6.183	69 47 15.6	— 10.35	52.58	2
11 517	Apodis	9.0	48.98	6.390	70 57 14.0	.34	52.52	1
11 518	Gould, Z. C., 15h 4090	9.0	49.12	6.250	70 10 48.6	.34	52.47	2
11 519	Gould, Z. C., 15h 4088	8.8	55 52.13	6.069	69 4 38.4	.34	52.51	2
11 520	Trianguli Australis	9.0	56 3.30	+ 5.696	66 24 7.9	— 10.33	52.61	1
11 521	Apodis	8.0	15 56 21.22	+ 6.436	71 10 16.7	— 10.30	52.48	1
11 522	Trianguli Australis	9.5	26.08	5.755	66 50 46.2	.30	52.61	1
11 523	Gould, Z. C., 15h 4137	9.0	30.15	6.379	70 51 40.8	.29	52.52	1
11 524	Octantis	8.5	32.16	15.171	84 34 53.1	.29	51.00	2
11 525	Gould, Z. C., 15h 4118	8.0	34.23	+ 5.644	65 56 59.8	— 10.29	52.60	1
11 526	Gould, Z. C., 15h 4150	8.0	15 56 49.33	+ 6.098	69 12 56.1	— 10.27	52.42	1
11 527	Apodis	7.5	54.78	9.835	80 21 29.4	.26	50.99	2
11 528	Trianguli Australis	9.5	56 58.59	5.922	68 2 38.2	.26	52.59	1
11 529	Octantis	10.0	57 3.22	12.034	82 41 34.1	.25	51.52	1
11 530	Lacaille, 6575	7.5	6.76	+11.441	82 10 46.4	— 10.25	50.51	2
11 531	Gould, Z. C., 16h 21	8.8	15 57 6.94	+ 7.567	75 39 26.1	— 10.25	52.68	2
11 532	Gould, Z. C., 16h 2	8.8	15.27	6.595	71 55 31.4	.24	52.37	2
11 533	Octantis	8.4	15.58	24.897	86 59 2.8	.24	51.42	7
11 534	Apodis	10.0	16.24	8.640	78 19 57.4	.23	52.36	1
11 535	Gould, Z. C., 15h 4189	9.0	19.08	+ 6.027	68 43 51.8	— 10.23	52.56	2
11 536	Apodis	9.5	15 57 25.56	+ 6.871	73 9 11.4	— 10.22	52.56	2
11 537	Trianguli Australis	10.0	26.23	5.685	66 14 19.1	.22	52.61	1
11 538	Octantis	9.5	36.97	13.217	83 31 39.7	.21	50.50	1
11 539	Apodis	10.0	43.17	7.334	74 53 11.2	.20	52.40	1
11 540	Gould, Z. C., 16h 10	9.5	43.35	+ 5.926	68 2 2.9	— 10.20	52.59	1
11 541	Trianguli Australis	8.8	15 57 43.76	+ 5.852	67 30 8.7	— 10.20	51.92	3
11 542	Apodis	8.8	44.70	7.526	75 30 29.5	.20	51.91	2
11 543	Gould, Z. C., 16h 138	8.0	52.04	9.252	79 26 39.2	.19	51.51	2
11 544	Apodis	9.8	55.89	7.759	76 11 37.4	.18	52.52	2
11 545	Apodis	9.5	59.86	+ 6.855	73 3 53.2	— 10.18	52.58	2

No. 11 514. Right ascension may be 15^h 56^m 42.15^s, differing 1 wire interval.

Number.	Constellation, Name of Star, or Synonym.	Magnitude.	Right Ascension, 1850.0.	Annual Precession.	South Declination, 1850.0.	Annual Precession.	Mean year.	No. of obs.
			h. m. s.	s.	° ′ ″	″		
11 546	Gould, Z. C., 16^h 33 . . .	10.0	15 57 59.98	+ 5.933	68 3 59.3	− 10.18	52.59	1
11 547	Gould, Z. C., 16^h 57 . . .	9.0	58 3.74	6.463	71 14 1.5	.17	52.48	1
11 548	δ^1 Apodis	5.0	7.65	8.640	78 18 22.3	.17	52.44	3
11 549	Gould, Z. C., 16^h 100 . .	10.0	11.64	7.345	74 54 10.3	.16	52.40	1
11 550	δ^2 Apodis	5.3	15.30	+ 8.628	78 16 43.3	− 10.16	55.44	3
11 551	Apodis	9.5	15 58 20.92	+ 8.013	76 51 51.3	− 10.15	52.48	1
11 552	Apodis	9.5	25.58	6.573	71 46 2.3	.15	52.32	1
11 553	Gould, Z. C., 16^h 144 . .	9.0	26.81	7.972	76 45 17.6	.15	52.48	2
11 554	Gould, Z. C., 16^h 93 . .	9.5	41.80	5.962	68 13 44.4	.13	52.56	2
11 555	Apodis	9.8	52.13	+ 7.922	76 36 32.7	− 10.11	52.48	2
11 556	Lacaille, 6572	8.5	15 58 53.41	+12.788	83 13 22.2	− 10.11	50.50	1
11 557	Gould, Z. C., 16^h 114 . .	9.0	53.90	5.951	68 8 43.3	.11	52.56	2
11 558	Apodis	9.0	54.57	7.035	73 45 16.2	.11	52.52	1
11 559	Trianguli Australis . . .	9.8	58 56.49	6.090	69 3 50.7	.11	52.51	2
11 560	Octantis	9.0	59 6.27	+30.692	87 36 17.3	− 10.10	51.81	5
11 561	Trianguli Australis . . .	9.5	15 59 11.48	+ 6.168	69 31 51.5	− 10.09	52.60	1
11 562	Brisbane, 5580	9.0	14.93	12.768	83 12 7.8	.09	50.50	1
11 563	Lacaille, 6441	6.5	17.07	27.256	87 15 49.1	.08	51.55	7
11 564	Melbourne (1), 815 . . .	7.8	21.14	5.754	66 40 45.6	.08	52.61	2
11 565	Trianguli Australis . . .	9.5	23.00	+ 5.862	67 29 30.9	− 10.07	52.58	1
11 566	Apodis	10.0	15 59 25.60	+ 8.610	78 12 23.7	− 10.07	52.57	1
11 567	Apodis	9.0	27.42	8.900	78 45 32.1	.07	51.48	2
11 568	Lacaille, 6603	7.3	27.67	10.889	81 35 9.6	.07	50.58	3
11 569	Trianguli Australis . . .	9.5	31.33	6.093	69 3 0.4	.06	52.51	2
11 570	Apodis	9.8	32.36	+ 7.979	76 44 11.6	− 10.06	52.48	2
11 571	Apodis	10.0	15 59 37.20	+10.207	80 47 24.7	− 10.06	50.56	1
11 572	Trianguli Australis . . .	9.5	44.88	5.706	66 17 1.6	.05	52.61	2
11 573	Gould, Z. C., 16^h 183 . .	10.0	15 59 46.23	5.897	67 43 22.8	.05	52.58	1
11 574	Apodis	9.5	16 0 0.10	8.457	77 52 42.1	.03	52.43	1
11 575	Apodis	10.0	0.53	+ 8.603	78 10 30.6	− 10.03	52.57	1
11 576	Gould, Z. C., 16^h 219 . .	9.5	16 0 7.96	+ 6.281	70 9 2.2	− 10.02	52.56	1
11 577	Gould, 21909	9.5	16.75	6.510	71 20 31.2	.01	52.48	2
11 578	Gould, 21910	8.5	18.15	6.538	71 30 50.7	.01	52.40	2
11 579	Gould, 21926	8.0	18.20	8.220	77 18 51.5	.01	52.56	1
11 580	Trianguli Australis . . .	9.8	19.66	+ 5.749	66 35 35.6	− 10.00	52.61	2
11 581	Octantis	10.0	16 0 21.66	+28.123	87 21 3.8	− 10.00	51.93	4
11 582	Apodis	9.0	26.20	10.526	81 9 39.3	10.00	50.58	3
11 583	Trianguli Australis . . .	9.5	36.52	5.834	67 13 26.6	9.98	50.62	1
11 584	Apodis	10.0	37.98	6.444	71 1 20.3	.98	52.52	1
11 585	Gould, Z. C., 16^h 286 .	9.5	41.09	+ 7.304	74 40 26.1	− 9.98	52.50	3

No. 11 560. Mean of two declinations differing 10.4″.

Number.	Constellation, Name of Star, or Synonym.	Magnitude.	Right Ascension, 1850.0.	Annual Precession.	South Declination, 1850.0.	Annual Precession.	Mean year.	No. of obs.
			h. m. s.	s.	° ′ ″	″		
11 586	Apodis	9.5	16 0 47.51	+ 7.361	74 51 46.3	− 9.97	52.40	1
11 587	Gould, Z. C., 16^{h} 259 . .	9.0	49.68	6.276	70 5 33.8	.97	52.56	1
11 588	Gould, Z. C., 16^{h} 277 . .	9.5	0 59.66	6.327	70 22 29.0	.95	52.47	2
11 589	Lacaille, 6681	7.5	1 2.92	6.368	70 36 10.7	.95	52.46	2
11 590	Lacaille, 6675	6.8	6.91	+ 6.790	72 39 22.8	− 9.94	52.56	2
11 591	Apodis	9.5	16 1 9.65	+ 6.309	70 15 58.4	− 9.94	52.47	2
11 592	Gould, 21922	8.0	9.85	6.500	71 17 5.7	.94	52.48	1
11 593	Trianguli Australis . . .	8.5	12.32	5.999	68 21 36.2	.94	52.53	1
11 594	Trianguli Australis . . .	10.0	20.49	5.842	67 14 59.2	.93	50.62	1
11 595	Gould, Z. C., 16^{h} 315 . .	7.2	21.48	+ 6.772	72 34 7.8	− 9.93	52.56	2
11 596	Apodis	10.0	16 1 21.59	+ 9.964	80 25 51.4	− 9.92	51.37	1
11 597	Apodis	10.0	23.49	8.528	77 58 40.1	.92	52.36	1
11 598	Trianguli Australis . . .	9.8	26.48	5.777	66 45 18.6	.92	52.61	2
11 599	Octantis	8.8	39.42	15.474	84 38 38.4	.90	51.00	2
11 600	Gould, Z. C., 16^{h} 329 . .	9.5	41.17	+ 6.334	70 22 45.2	− 9.90	52.47	2
11 601	Trianguli Australis . . .	9.0	16 1 43.59	+ 5.664	65 50 14.6	− 9.90	52.60	1
11 602	Octantis	9.8	47.20	15.661	84 43 17.2	.89	51.00	2
11 603	Gould, Z. C., 16^{h} 331 . .	9.0	1 54.12	5.908	67 41 38.0	.88	52.58	1
11 604	Trianguli Australis . . .	9.0	2 0.22	6.015	68 25 37.4	.88	52.53	1
11 605	Gould, Z. C., 16^{h} 352 . .	9.5	2.60	+ 6.328	70 19 42.9	− 9.87	52.47	2
11 606	Lacaille, 6681	7.0	16 2 8.86	+ 7.112	73 55 28.1	− 9.87	52.52	2
11 607	Lacaille, 6698	7.0	11.60	5.890	67 33 3.8	.86	52.58	1
11 608	Trianguli Australis . . .	8.0	15.33	5.848	67 14 43.9	.86	50.62	1
11 609	Trianguli Australis . . .	10.0	17.35	5.676	65 54 13.4	.85	52.60	1
11 610	Apodis	9.0	27.08	+ 7.779	76 5 55.4	− 9.84	52.56	1
11 611	Gould, Z. C., 16^{h} 372 . .	9.0	16 2 33.85	+ 5.798	66 51 23.7	− 9.83	52.61	1
11 612	Gould, Z. C., 16^{h} 414 . .	9.0	34.85	7.337	74 42 59.5	.83	52.50	3
11 613	Gould, Z. C., 16^{h} 385 . .	9.5	41.07	5.822	67 1 46.1	.82	52.61	1
11 614	Apodis	10.0	42.98	6.530	71 22 22.1	.82	52.48	1
11 615	Apodis	10.0	50.37	+ 6.507	71 14 53.4	− 9.81	52.48	1
11 616	Trianguli Australis . . .	10.0	16 2 56.15	+ 5.989	68 12 24.8	− 9.80	52.53	1
11 617	Trianguli Australis . . .	9.0	2 58.33	5.686	65 57 6.2	.80	52.60	1
11 618	Octantis	9.8	3 1.29	49.358	88 33 22.0	.80	51.93	2
11 619	Gould, Z. C., 16^{h} 427 . .	9.2	2.32	6.880	72 58 3.2	.80	52.58	2
11 620	Apodis	9.0	9.62	+ 8.458	77 46 30.0	− 9.79	52.47	2
11 621	Apodis	9.5	16 3 15.71	+ 8.122	76 59 6.0	− 9.78	52.48	1
11 622	Lacaille, 6687	8.0	15.79	7.523	75 18 15.9	.78	50.56	1
11 623	Gould, Z. C., 16^{h} 464 . .	8.5	17.76	7.596	75 31 43.9	.78	51.59	2
11 624	Trianguli Australis . . .	9.8	19.02	6.127	69 5 19.0	.78	52.51	2
11 625	Gould, Z. C., 16^{h} 443 . .	9.0	24.32	+ 6.476	71 4 3.2	− 9.77	52.50	2

No. 11 599. The declination may be 84° 38′ 10.5″.

Number.	Constellation, Name of Star, or Synonym.	Magnitude.	Right Ascension, 1850.0.	Annual Precession.	South Declination, 1850.0.	Annual Precession.	Mean year.	No. of obs.
			h. m. s.	s.	° ′ ″	″		
11 626	Brisbane, 5606	9.0	16 3 24.41	+12.861	83 11 49.2	− 9.77	50.50	1
11 627	Apodis	9.0	25.34	8.166	77 5 19.4	.77	52.52	2
11 628	Lacaille, 6688	7.0	26.75	7.620	75 35 55.6	.77	51.59	2
11 629	Apodis	9.5	29.41	6.734	72 18 45.8	.76	52.46	2
11 630	Apodis	9.5	30.25	+ 6.585	71 36 40.3	− 9.76	52.32	1
11 631	Octantis	8.4	16 3 34.07	+16.964	85 11 35.2	− 9.76	51.13	4
11 632	Apodis	9.5	35.60	6.424	70 47 1.1	.76	52.52	1
11 633	Apodis	9.5	48.44	7.305	74 33 48.8	.74	52.56	1
11 634	Gould, Z. C., 16h 512 . .	9.0	3 48.71	8.048	76 46 51.3	.74	52.48	2
11 635	Gould, Z. C., 16h 529 . .	9.0	4 0.38	+ 8.057	76 47 52.2	− 9.72	52.48	2
11 636	Apodis	9.0	16 4 1.00	+ 8.667	78 11 39.8	− 9.72	52.57	1
11 637	Apodis	9.5	6.28	6.732	72 16 31.6	.72	52.46	2
11 638	Gould, Z. C., 16h 520 . .	8.8	8.78	7.480	75 8 13.2	.71	51.48	2
11 639	Gould, Z. C., 16h 476 . .	8.8	9.36	5.792	66 43 48.2	.71	52.61	2
11 640	Trianguli Australis . . .	10.0	15.08	+ 6.272	69 54 53.7	− 9.70	52.60	1
11 641	Trianguli Australis . . .	8.8	16 4 19.22	+ 6.216	69 35 1.0	− 9.70	52.51	2
11 642	Octantis	9.5	22.48	12.002	82 31 57.5	.69	50.49	1
11 643	Gould, Z. C., 16h 494 . .	9.2	26.56	5.885	67 24 14.2	.69	51.60	2
11 644	Trianguli Australis . . .	10.0	29.59	6.053	68 33 13.0	.69	52.64	1
11 645	Trianguli Australis . . .	9.5	30.30	+ 6.076	68 42 13.2	− 9.68	52.59	3
11 646	Apodis	9.5	16 4 38.23	+ 8.453	77 43 16.3	− 9.67	52.36	1
11 647	Apodis	9.5	40.06	9.537	79 43 49.1	.67	50.61	1
11 648	Gould, Z. C., 16h 519 . .	9.0	47.78	5.886	67 23 31.8	.66	51.60	2
11 649	Gould, Z. C., 16h 515 . .	8.5	48.28	5.761	66 27 28.5	.66	52.61	1
11 650	Lacaille, 6714	7.0	56.42	+ 6.574	71 29 51.5	− 9.65	52.44	3
11 651	Gould, Z. C., 16h 534 . .	9.0	16 4 57.00	+ 5.920	67 38 5.7	− 9.65	52.58	1
11 652	Apodis	9.0	5 7.28	6.420	70 41 53.8	.64	52.46	2
11 653	Apodis	9.5	15.91	7.119	73 50 16.8	.63	52.52	2
11 654	Gould, Z. C., 16h 553 . .	9.0	18.16	5.920	67 37 2.5	.62	52.58	1
11 655	Lacaille, 6723	7.2	22.55	+ 5.999	68 9 35.6	− 9.62	52.59	3
11 656	Apodis	9.5	16 5 30.71	+10.468	80 58 58.3	− 9.61	50.56	1
11 657	Apodis	9.5	30.99	7.947	73 8 59.8	.61	52.58	2
11 658	Gould, Z. C., 16h 573 . .	9.5	36.91	6.174	69 16 23.3	.60	52.42	1
11 659	Gould, Z. C., 16h 695 . .	8.8	41.78	9.705	79 57 39.7	.59	50.60	2
11 660	Trianguli Australis . . .	9.2	44.50	+ 5.962	67 56 53.0	− 9.59	52.59	2
11 661	Melbourne (1), 822 . . .	9.0	16 5 49.54	+ 6.187	69 20 38.7	− 9.58	52.42	1
11 662	Gould, Z. C., 16h 632 . .	8.5	54.15	7.308	74 29 54.9	.58	52.56	1
11 663	Trianguli Australis . . .	10.0	5 54.54	5.732	66 10 24.2	.58	52.62	1
11 664	Lacaille, 6545	6.6	6 17.25	20.123	86 3 18.1	.55	51.42	7
11 665	Apodis	10.0	17.26	+ 7.500	75 7 49.9	− 9.55	52.40	1

No. 11 654. Gould's declination 12.4″ greater.

Number.	Constellation, Name of Star, or Synonym.	Magnitude.	Right Ascension, 1850.0.	Annual Precession.	South Declination, 1850.0.	Annual Precession.	Mean year.	No. of obs.
			h. m. s.	s.	° ′ ″	″		
11 666	Brisbane, 5650	7.7	16 6 17.30	+ 6.595	71 32 44.8	− 9.55	52.43	4
11 667	Gould, Z. C., 16h 615 . .	9.2	18.09	5.974	67 56 31.0	.55	52.59	2
11 668	Apodis	9.2	21.68	8.953	78 41 30.6	.54	51.49	2
11 669	Octantis	9.5	23.39	14.125	83 55 34.4	.54	50.55	1
11 670	Apodis	10.0	27.58	+ 7.365	74 40 42.3	− 9.53	52.40	1
11 671	Gould, Z. C., 16h 687 . .	9.0	16 6 29.52	+ 7.654	75 36 3.4	− 9.53	51.59	2
11 672	Gould, Z. C., 16h 633 . .	8.5	32.52	5.846	67 0 55.8	.53	51.61	2
11 673	Apodis	9.0	33.95	6.415	70 36 43.5	.53	52.46	2
11 674	Gould, Z. C., 16h 649 . .	8.0	46.71	5.770	66 25 33.8	.51	52.61	1
11 675	Gould, Z. C., 16h 672 . .	9.0	46.86	+ 6.486	70 58 34.5	− 9.51	52.52	1
11 676	Gould, Z. C., 16h 698 . .	8.8	16 6 49.28	+ 7.123	73 47 48.3	− 9.51	52.52	2
11 677	Gould, Z. C., 16h 655 . .	8.0	49.88	5.839	66 57 10.7	.51	52.61	1
11 678	Apodis	9.0	50.11	6.866	72 45 39.8	.51	52.56	2
11 679	Trianguli Australis . . .	10.0	57.82	5.708	65 55 44.9	.50	52.61	2
11 680	Trianguli Australis . . .	10.0	6 58.10	+ 6.116	68 50 49.7	− 9.50	52.59	1
11 681	Gould, Z. C., 16h 676 . .	10.0	16 7 8.35	+ 5.841	66 57 15.1	− 9.48	52.61	1
11 682	Gould, Z. C., 16h 675 . .	9.2	8.96	5.741	66 10 44.3	.48	52.62	2
11 683	Trianguli Australis . . .	10.0	13.47	6.131	68 56 6.7	.48	52.59	1
11 684	Gould, Z. C., 16h 744 . .	9.5	14.32	7.674	75 38 17.1	.47	52.62	1
11 685	Gould, Z. C., 16h 734 . .	10.0	25.62	+ 6.991	73 15 26.1	− 9.46	52.58	1
11 686	Trianguli Australis . . .	9.8	16 7 28.44	+ 6.142	68 59 29.0	− 9.46	52.51	2
11 687	Trianguli Australis . . .	10.0	29.78	6.141	68 58 57.6	.45	52.51	2
11 688	Trianguli Australis . . .	9.2	32.65	5.739	66 8 53.4	.45	52.61	2
11 689	Gould, Z. C., 16h 740 . .	8.0	36.48	6.750	72 13 12.0	.45	52.46	2
11 690	Gould, Z. C., 16h 776 . .	8.8	37.21	+ 7.648	75 32 44.9	− 9.45	50.59	2
11 691	Apodis	9.5	16 7 39.07	+ 8.295	77 16 22.5	− 9.44	52.56	1
11 692	Apodis	9.0	43.31	10.182	80 34 45.1	.44	51.37	1
11 693	Trianguli Australis . . .	8.8	43.72	6.055	68 25 15.1	.44	52.58	2
11 694	Gould, Z. C., 16h 758 . .	9.2	45.46	6.970	73 9 41.4	.43	52.58	3
11 695	Apodis	9.5	51.08	+ 7.734	75 47 25.8	− 9.43	52.59	2
11 696	Apodis	10.0	16 7 52.28	+ 7.409	74 46 43.4	− 9.43	52.40	1
11 697	Gould, Z. C., 16h 780 . .	9.0	8 6.16	6.590	71 26 51.8	.41	52.50	2
11 698	Lacaille, 6737	7.0	9.90	6.278	69 46 36.0	.40	52.58	2
11 699	Trianguli Australis . . .	9.5	11.14	5.746	66 10 12.8	.40	52.62	2
11 700	Trianguli Australis . . .	9.3	11.39	+ 6.108	68 44 38.1	− 9.40	52.59	3
11 701	Lacaille, 6749	6.5	16 8 16.00	+ 5.791	66 31 8.9	− 9.40	52.61	1
11 702	Apodis	9.0	25.04	7.188	73 58 59.4	.38	52.52	1
11 703	Gould, Z. C., 16h 783 . .	8.5	27.55	5.793	66 31 21.7	.38	52.61	1
11 704	Octantis	9.2	37.65	13.640	83 36 58.7	.37	50.52	2
11 705	Gould, Z. C., 16h 822 . .	9.5	49.30	+ 6.430	70 35 48.3	− 9.35	52.40	1

No. 11 667. Gould's declination is 19.4″ greater.

No. 11 694. Gould's declination is 1′ greater.

Number.	Constellation, Name of Star, or Synonym.	Magnitude.	Right Ascension, 1850.0.	Annual Precession.	South Declination, 1850.0.	Annual Precession.	Mean year.	No. of obs.
			h. m. s.	s.	° ′ ″	″		
11 706	Trianguli Australis . . .	10.0	16 8 54.74	+ 6.254	69 36 24.8	− 9.35	52.60	1
11 707	Apodis	9.5	54.91	8.498	77 42 3.2	.35	52.36	1
11 708	Trianguli Australis . . .	10.0	8 55.03	6.269	69 41 41.2	.34	52.60	1
11 709	Gould, Z. C., 16ʰ 868 . .	9.5	9 6.37	7.474	74 57 3.6	.33	52.40	1
11 710	Gould, Z. C., 16ʰ 848 . .	8.0	27.84	+ 5.709	65 48 32.5	− 9.30	52.60	1
11 711	Octantis	9.2	16 9 27.90	+25.463	86 58 18.0	− 9.30	51.64	6
11 712	Trianguli Australis . . .	9.2	30.30	6.042	68 15 27.0	.30	52.58	2
11 713	Trianguli Australis . . .	9.0	31.03	5.991	67 54 28.2	.30	52.59	1
11 714	Gould, Z. C., 16ʰ 874 . .	8.5	34.25	6.645	71 39 13.2	.29	52.52	1
11 715	Apodis	8.8	38.81	+ 9.191	79 2 22.2	− 9.29	51.51	2
11 716	Gould, Z. C., 16ʰ 873 . .	9.0	16 9 42.72	+ 6.282	69 44 17.4	− 9.28	52.60	1
11 717	Octantis	9.0	43.14	50.536	88 39 16.3	.28	51.22	5
11 718	Apodis	9.5	46.35	7.686	75 35 34.8	.28	52.62	1
11 719	Gould, 22131	9.0	9 50.28	6.765	72 12 17.2	.27	52.46	2
11 720	Gould, 22142	7.2	10 3.48	+ 6.598	71 24 35.0	− 9.26	52.50	2
11 721	Gould, Z. C., 16ʰ 888 . .	9.0	16 10 5.24	+ 5.974	67 45 54.2	− 9.25	52.59	2
11 722	Trianguli Australis . . .	9.8	12.48	5.877	67 4 19.4	.24	51.61	2
11 723	Lacaille, 6753	7.5	16.67	6.889	72 43 48.3	.24	52.56	2
11 724	Trianguli Australis . . .	10.0	18.95	6.167	69 1 15.0	.24	52.42	1
11 725	Lacaille, 6754	7.2	29.98	+ 6.876	72 40 0.0	− 9.22	52.56	2
11 726	Apodis	9.5	16 10 32.80	+ 7.823	75 58 0.2	− 9.22	52.59	2
11 727	Apodis	9.5	36.40	7.118	73 38 30.1	.21	52.52	1
11 728	γ Apodis	4.8	37.26	8.935	78 32 53.0	.21	51.49	2
11 729	Gould, Z. C., 16ʰ 934 . .	9.0	44.18	5.754	66 6 27.9	.20	52.60	1
11 730	Trianguli Australis . . .	9.5	45.72	+ 5.828	66 40 39.0	− 9.20	52.61	2
11 731	Lacaille, 6750	8.0	16 10 55.45	+ 7.462	74 51 11.6	− 9.19	52.46	2
11 732	Gould, Z. C., 16ʰ 954 . .	9.0	11 0.57	5.904	67 13 42.1	.18	50.62	1
11 733	Apodis	9.0	6.71	6.580	71 16 37.0	.17	52.48	1
11 734	Trianguli Australis . . .	9.2	7.16	5.785	66 19 59.0	.17	52.60	2
11 735	Apodis	9.5	15.09	+ 7.658	75 27 46.1	− 9.16	52.62	1
11 736	Apodis	9.5	16 11 20.21	+ 8.893	78 27 0.2	− 9.16	51.49	2
11 737	Trianguli Australis . . .	10.0	31.28	6.112	68 37 8.5	.14	52.64	1
11 738	Trianguli Australis . . .	9.0	33.91	6.038	68 8 1.8	.14	52.59	1
11 739	Gould, Z. C., 16ʰ 1043 .	8.5	37.65	7.849	76 0 28.3	.13	52.56	1
11 740	Apodis	10.0	39.20	+11.504	81 57 52.5	− 9.13	50.56	1
11 741	Trianguli Australis . . .	10.0	16 11 50.93	+ 6.110	68 35 41.4	− 9.12	52.64	1
11 742	Lacaille, 6769	6.8	51.94	6.170	68 58 14.0	.12	52.51	2
11 743	Apodis	9.0	11 57.38	9.878	80 4 22.4	.11	50.60	1
11 744	Apodis	9.5	12 1.76	9.251	79 5 12.2	.10	52.41	1
11 745	Lacaille, 6768	7.2	1.90	+ 6.260	69 30 28.9	− 9.10	52.51	2

Number.	Constellation, Name of Star, or Synonym.	Magnitude.	Right Ascension, 1850.0.	Annual Precession.	South Declination, 1850.0.	Annual Precession.	Mean year.	No. of obs.
			h. m. s.	s.	° ′ ″	″		
11 746	Gould, Z. C., 16h 1060	9.0	16 12 9.43	+ 7.152	73 43 2.3	— 9.09	52.52	1
11 747	Gould, Z. C., 16h 1044	9.2	9.74	6.638	71 31 7.4	.09	52.50	2
11 748	Gould, Z. C., 16h 1081	8.2	10.88	7.909	76 9 31.8	.09	52.52	2
11 749	Apodis	9.0	13.40	7.222	73 58 35.7	.09	52.52	1
11 750	Trianguli Australis	9.0	22.24	+ 5.955	67 32 0.4	— 9.08	51.60	2
11 751	ζ Trianguli Australis	5.8	16 12 24.02	+ 6.302	69 44 11.7	— 9.07	52.58	2
11 752	Apodis	10.0	29.75	8.633	77 53 47.6	.07	52.36	1
11 753	Apodis	10.0	30.78	6.751	72 2 13.2	.07	52.41	1
11 754	Apodis	9.5	37.03	7.417	74 38 54.0	.06	52.40	1
11 755	Gould, Z. C., 16h 1092	9.2	42.18	+ 7.000	73 6 5.0	— 9.05	52.58	3
11 756	Apodis	10.0	16 12 42.36	+ 6.779	72 9 31.1	— 9.05	52.41	1
11 757	Octantis	9.0	46.05	15.936	84 41 59.8	.05	51.00	2
11 758	Lacaille, 6696	7.0	47.42	12.676	82 54 59.1	.04	51.52	1
11 759	Lacaille, 6775	7.5	47.96	6.494	70 46 21.8	.04	52.46	2
11 760	Gould, Z. C., 16h 1088	9.0	53.68	+ 6.378	70 9 0.6	— 9.04	52.56	1
11 761	Apodis	10.0	16 12 55.38	+ 6.718	71 52 16.0	— 9.03	52.41	1
11 762	Gould, Z. C., 16h 1068	9.0	12 57.39	5.751	65 59 5.4	.03	52.60	1
11 763	Trianguli Australis	8.8	13 9.92	6.055	68 10 57.0	.02	52.62	2
11 764	Trianguli Australis	10.0	10.05	5.783	66 13 33.2	.01	52.61	1
11 765	Trianguli Australis	10.0	20.01	+ 5.758	66 0 58.4	— 9.00	52.60	1
11 766	Lacaille, 6762	8.0	16 13 24.72	+ 7.729	75 36 32.2	— 9.00	51.58	2
11 767	Gould, Z. C., 16h 1190	9.0	32.62	8.204	76 53 17.1	8.99	52.48	1
11 768	Trianguli Australis	9.0	39.92	6.042	68 4 6.9	.98	52.59	1
11 769	Gould, Z. C., 16h 1174	9.0	47.41	7.156	73 40 50.1	.97	52.52	1
11 770	Trianguli Australis	9.0	49.01	+ 6.048	68 6 15.6	— 8.96	52.59	1
11 771	Apodis	9.5	16 13 54.23	+ 6.362	70 2 4.3	— 8.96	52.56	1
11 772	Gould, Z. C., 16h 1146	8.8	56.00	5.909	67 8 6.2	.95	52.61	2
11 773	Gould, Z. C., 16h 1188	9.2	13 58.76	7.042	73 13 44.8	.95	52.58	2
11 774	Trianguli Australis	9.5	14 11.61	6.046	68 4 19.7	.93	52.59	1
11 775	Gould, Z. C., 16h 1175	9.0	17.62	+ 5.951	67 25 11.4	— 8.93	51.60	2
11 776	Apodis	9.5	16 14 29.06	+ 7.983	76 17 34.0	— 8.91	52.52	2
11 777	Gould, Z. C., 16h 1193	9.5	33.05	5.873	66 50 57.4	.91	52.61	2
11 778	Octantis	10.0	36.70	14.478	84 0 12.1	.90	51.49	1
11 779	Gould, Z. C., 16h 1224	9.8	42.62	6.498	70 43 10.2	.89	52.46	2
11 780	Trianguli Australis	9.5	43.68	+ 6.101	68 24 56.1	— 8.89	52.64	1
11 781	Gould, 22269	11.0	16 14 47.64	+11.048	81 27 13.2	— 8.89	50.56	1
11 782	Apodis	9.0	48.19	8.438	77 24 29.6	.89	52.46	2
11 783	Gould, Z. C., 16h 1265	10.0	14 50.54	7.372	74 25 24.7	.88	52.56	1
11 784	Trianguli Australis	9.5	15 9.79	5.834	66 31 32.9	.86	52.61	1
11 785	Gould, Z. C., 16h 1289	9.5	19.34	+ 6.779	72 3 45.4	— 8.85	52.41	1

No. 11 760. Gould's declination is 9.6″ greater.

Number.	Constellation, Name of Star, or Synonym.	Magnitude.	Right Ascension, 1850.0.	Annual Precession.	South Declination, 1850.0.	Annual Precession.	Mean year.	No. of obs.
			h. m. s.	s.	° ′ ″	″		
11 786	Gould, Z. C., 16h 1254 .	10.0	16 15 19.83	+ 5.875	66 49 32.1	— 8.84	52.62	1
11 787	Trianguli Australis . . .	9.5	20.49	5.836	66 31 53.3	.84	52.61	1
11 788	Gould, Z. C., 16h 1255 .	9.2	21.42	5.895	66 58 24.4	.84	52.61	2
11 789	Gould, Z. C., 16h 1261 .	9.0	22.48	6.033	67 56 23.4	.84	52.59	2
11 790	Gould, Z. C., 16h 1273 .	9.0	30.32	+ 6.003	67 43 39.5	— 8.83	52.58	1
11 791	Lacaille, 6773	7.8	16 15 31.64	+ 8.276	77 0 36.6	— 8.83	52.52	2
11 792	Gould, Z. C., 16h 1353 .	9.5	38.35	8.941	78 26 9.9	.82	50.64	1
11 793	Gould, Z. C., 16h 1310 .	8.5	15 47.63	6.684	71 36 12.0	.81	52.52	1
11 794	Apodis	10.0	16 2.29	10.438	80 44 1.3	.79	50.56	1
11 795	Apodis	10.0	4.44	+ 6.683	71 35 27.6	— 8.79	52.52	1
11 796	Trianguli Australis . . .	10.0	16 16 14.18	+ 6.199	68 58 13.5	— 8.77	52.59	1
11 797	Trianguli Australis . . .	8.8	16.79	5.821	66 22 32.2	.77	52.60	2
11 798	Trianguli Australis . .	9.0	18.32	6.276	69 25 33.5	.77	52.51	2
11 799	Octantis	9.0	23.26	13.500	83 25 8.2	.76	50.53	2
11 800	Gould, Z. C., 16h 1366 .	9.3	24.21	+ 7.615	75 10 17.7	— 8.76	51.83	3
11 801	Octantis	9.0	16 16 28.44	+19.972	85 55 42.1	— 8.75	51.26	4
11 802	Apodis	9.5	29.83	8.851	78 14 26.1	.75	52.57	1
11 803	Gould, Z. C., 16h 1346 .	9.0	34.76	6.488	70 35 55.6	.75	52.46	2
11 804	Apodis	9.5	35.99	6.700	71 39 7.9	.74	52.52	1
11 805	Apodis	9.5	37.78	+ 7.781	75 39 52.3	— 8.74	52.62	1
11 806	Apodis	9.5	16 16 40.82	+ 6.966	72 49 36.5	— 8.74	52.60	1
11 807	Trianguli Australis . . .	9.0	48.70	5.868	66 41 45.0	.73	52.61	3
11 808	Apodis	9.0	50.77	8.860	78 14 58.2	.73	52.57	1
11 809	Gould, Z. C., 16h 1354 .	9.0	54.99	5.892	66 52 35.0	.72	52.62	2
11 810	Gould, Z. C., 16h 1381 .	8.0	16 56.52	+ 6.910	72 34 43.5	— 8.72	52.56	2
11 811	Gould, 22278	9.0	16 17 0.27	+ 6.530	70 48 6.4	— 8.71	52.52	1
11 812	Gould, Z. C., 16h 1363 .	9.0	7.04	5.904	66 57 20.7	.70	52.61	2
11 813	Apodis	9.8	9.28	8.497	77 29 0.0	.70	52.46	2
11 814	Gould, Z. C., 16h 1384 .	9.0	27.18	5.869	66 40 59.3	.68	52.61	3
11 815	Apodis	10.0	38.98	+ 8.819	78 9 1.2	— 8.66	52.57	1
11 816	Apodis	10.0	16 17 44.07	+ 8.741	77 59 23.5	— 8.66	52.36	1
11 817	Lacaille, 6809	6.5	50.28	6.508	70 39 18.4	.65	52.46	2
11 818	Apodis	9.5	17 56.58	6.569	70 57 56.5	.64	52.52	1
11 819	Trianguli Australis . . .	9.5	18 3.97	5.929	67 5 53.0	.63	50.62	1
11 820	Trianguli Australis . . .	10.0	4.76	+ 6.043	67 53 17.2	— 8.63	52.59	1
11 821	Gould, Z. C., 16h 1450 .	9.0	16 18 6.32	+ 7.059	73 9 20.6	— 8.63	52.58	2
11 822	Trianguli Australis . . .	9.0	10.64	5.853	66 31 54.7	.62	52.61	1
11 823	Trianguli Australis . . .	10.0	17.76	6.043	67 52 48.7	.61	52.59	1
11 824	Trianguli Australis . . .	9.0	17.89	6.187	68 48 41.4	.61	52.59	1
11 825	Apodis	9.5	20.28	+ 7.417	74 27 48.1	— 8.61	52.56	1

Number.	Constellation, Name of Star, or Synonym.	Magnitude.	Right Ascension, 1850.0.	Annual Precession.	South Declination, 1850.0.	Annual Precession.	Mean year.	No. of obs.
			h. m. s.	s.	° ′ ″	″		
11 826	Apodis	10.0	16 18 21.32	+10.200	80 22 59.4	− 8.61	51.20	3
11 827	Gould, 22303	9.0	25.32	6.764	71 53 3.1	.60	52.47	2
11 828	Apodis	9.0	38.23	10.281	80 29 6.3	.58	51.50	2
11 829	Trianguli Australis . . .	10.0	38.49	−6.053	67 56 0.1	.58	52.59	1
11 830	Lacaille, 6814	7.0	18 50.52	+ 6.699	71 33 51.2	− 8.57	52.50	2
11 831	Octantis	10.0	16 19 2.15	+16.908	85 0 16.3	− 8.55	50.51	1
11 832	Apodis	9.0	2.56	6.737	71 44 21.0	.55	52.52	1
11 833	Gould, Z. C., 16h 1487 .	9.0	3.43	6.338	69 40 51.3	.55	52.60	1
11 834	Gould, Z. C., 16h 1523 .	9.5	13.80	7.086	73 13 37.5	.54	52.58	1
11 835	Apodis	9.5	26.73	+ 6.613	71 7 41.5	8.52	52.48	1
11 836	Lacaille, 6791	8.0	16 19 28.32	+ 9.134	78 42 28.2	8.52	50.63	2
11 837	Apodis	10.0	29.52	10.592	80 51 20.5	.52	50.56	1
11 838	Trianguli Australis . . .	10.5	36.69	6.043	67 49 49.4	.51	52.59	2
11 839	Octantis	9.0	37.15	17.519	85 12 32.0	.51	51.04	4
11 840	Gould, Z. C., 16h 1516	9.8	37.83	+ 6.044	67 49 54.5	− 8.51	52.59	2
11 841	Lacaille, 6808	8.7	16 19 42.32	+ 7.645	75 9 59.9	− 8.50	51.24	3
11 842	Gould, Z. C., 16h 1533 .	8.8	56.72	5.924	66 58 53.7	.48	52.61	2
11 843	Octantis	9.5	19 57.10	12.305	82 31 24.0	.48	50.49	1
11 844	Trianguli Australis . . .	10.0	20 1.15	6.287	69 20 42.8	.47	52.42	1
11 845	Apodis	9.2	3.22	+ 7.713	75 21 59.8	− 8.47	50.60	2
11 846	Gould, Z. C., 16h 1550 .	9.5	16 20 6.88	+ 6.276	69 16 43.3	− 8.47	52.42	1
11 847	Trianguli Australis . . .	9.5	16.05	5.784	65 54 58.6	.45	52.60	1
11 848	Trianguli Australis . . .	9.0	27.80	5.950	67 8 56.7	.44	52.61	1
11 849	Apodis	9.5	28.58	6.624	71 8 37.7	.44	52.48	1
11 850	Trianguli Australis . . .	9.0	29.88	+ 5.932	67 0 58.4	− 8.44	52.61	2
11 851	Trianguli Australis . . .	9.5	16 20 30.64	+ 5.876	66 36 37.4	− 8.44	52.61	1
11 852	Apodis	9.2	31.19	7.364	74 13 13.6	.43	52.54	2
11 853	Gould, Z. C., 16h 1581 .	9.2	36.70	5.951	67 8 56.6	.43	51.62	2
11 854	Gould, Z. C., 16h 1591 .	8.8	50.76	5.904	66 48 4.1	.41	52.61	2
11 855	Gould, Z. C., 16h 1597 .	9.2	55.06	+ 5.949	67 7 28.4	− 8.40	51.61	2
11 856	Apodis	9.5	16 20 57.52	+ 6.643	71 13 27.8	− 8.40	52.48	1
11 857	Gould, Z. C., 16h 1679 .	8.5	21 0.43	8.431	77 14 3.6	.40	52.56	1
11 858	Apodis	9.5	0.44	9.239	78 51 38.9	.40	50.61	1
11 859	Apodis	10.0	8.16	6.925	72 30 9.8	.39	52.52	1
11 860	Trianguli Australis . . .	9.5	8.29	+ 6.110	68 12 44.6	− 8.39	52.64	1
11 861	Gould, Z. C., 16h 1617	9.0	16 21 13.52	+ 6.052	67 49 27.6	− 8.38	52.59	2
11 862	Trianguli Australis . . .	9.5	19.18	6.126	68 18 14.1	.37	52.64	1
11 863	Apodis	10.0	24.05	8.400	77 9 2.5	.36	52.48	1
11 864	Apodis	10.0	24.47	9.336	79 1 13.3	.36	52.41	1
11 865	Apodis	10.0	30.58	+ 6.516	70 33 48.1	− 8.36	52.40	1

No. 11 861. Gould's declination probably 4′ too small.

Number.	Constellation, Name of Star, or Synonym.	Magnitude.	Right Ascension, 1850.0.	Annual Precession.	South Declination, 1850.0.	Annual Precession.	Mean year.	No. of obs.
			h. m. s.	s.	° ′ ″	″		
11 866	Lacaille, 6846	8.8	16 21 39.39	+ 5.893	66 40 52.4	8.34	52.61	3
11 867	Gould, Z. C., 16h 1654 .	9.2	44.00	5.951	67 6 11.5	.34	51.95	3
11 868	Octantis	9.1	45.19	18.011	85 20 46.1	.34	51.04	4
11 869	β Apodis	4.8	47.70	8.422	77 11 30.6	.33	52.48	3
11 870	Gould, 22366	8.2	51.70	+ 5.891	66 39 42.7	− 8.33	52.61	3
11 871	Trianguli Australis . . .	10.0	16 21 53.05	+ 5.832	66 12 48.6	− 8.33	52.60	1
11 872	Apodis	10.0	21 53.93	6.434	70 6 27.2	.33	52.56	1
11 873	Gould, Z. C., 16h 1809 .	9.0	22 6.27	10.015	80 3 31.2	.31	50.60	1
11 874	Gould, 22372	8.5	10.64	5.892	66 39 14.9	.30	52.61	2
11 875	Trianguli Australis . . .	10.0	13.34	+ 6.214	68 49 31.5	− 8.30	52.59	1
11 876	Apodis	9.5	16 22 21.89	+11.982	82 13 6.1	− 8.29	50.54	3
11 877	Apodis	10.0	24.92	6.425	70 2 26.1	.28	52.56	1
11 878	Octantis	9.8	25.14	12.882	82 55 27.0	.28	51.04	2
11 879	Apodis	9.5	34.49	9.411	79 7 20.8	.27	52.41	1
11 880	Apodis	9.0	34.64	+ 6.427	70 2 42.3	− 8.27	52.56	1
11 881	Apodis	9.5	16 22 36.50	+ 8.308	76 54 9.6	− 8.27	52.44	2
11 882	Apodis	9.0	37.08	6.442	70 7 42.1	.27	52.56	1
11 883	Gould, Z. C., 16h 1768 .	9.0	39.03	7.355	74 7 22.9	.27	52.52	1
11 884	Gould, Z. C., 16h 1777 .	9.0	42.16	7.559	74 48 32.1	.26	52.53	1
11 885	Gould, Z. C., 16h 1728 .	9.0	47.77	+ 6.044	67 42 23.3	− 8.25	52.58	1
11 886	Gould, Z. C., 16h 1785 .	8.5	16 22 51.09	+ 7.480	74 32 47.4	− 8.25	52.56	1
11 887	Gould, Z. C., 16h 1769	8.2	22 53.90	6.802	71 54 16.8	.25	52.47	2
11 888	Apodis	10.0	23 3.36	10.011	80 2 3.9	.23	50.60	1
11 889	Apodis	10.0	4.84	8.036	76 11 56.5	.23	52.52	2
11 890	Apodis	9.5	8.13	+ 6.609	70 58 35.8	− 8.23	52.52	1
11 891	Trianguli Australis . .	10.0	16 23 8.43	+ 6.140	68 19 35.4	− 8.23	52.59	1
11 892	Gould, 22410	7.5	9.02	6.152	68 24 11.8	.23	52.62	2
11 893	Gould, Z. C., 16h 1783 .	9.0	9.74	6.645	71 9 21.5	.22	52.48	1
11 894	Trianguli Australis . . .	9.0	10.12	6.396	69 51 10.6	.22	52.60	1
11 895	Apodis	10.0	17.52	+ 6.816	71 57 18.5	− 8.21	52.52	1
11 896	Trianguli Australis . . .	10.0	16 23 21.43	+ 6.060	67 47 38.8	− 8.21	52.58	1
11 897	Apodis	9.5	26.62	7.050	72 56 47.0	.20	52.60	1
11 898	Lacaille, 6847	7.0	28.65	6.825	71 59 17.6	.20	52.47	2
11 899	Gould, Z. C., 16h 1788 .	8.5	35.66	5.906	66 42 14.6	.19	52.61	1
11 900	Gould, Z. C., 16h 1896 .	9.0	38.34	+ 9.572	79 21 54.0	− 8.19	51.51	2
11 901	Octantis	9.2	16 23 38.55	+51.280	88 32 58.3	− 8.19	51.04	3
11 902	Gould, Z. C., 16h 1842 .	9.0	43.96	7.328	73 59 36.9	.18	52.52	2
11 903	Octantis	8.0	44.74	14.742	84 1 38.5	.18	51.49	1
11 904	Trianguli Australis . . .	9.2	46.94	5.844	66 13 38.6	.18	52.60	2
11 905	Apodis	9.0	48.14	+ 7.096	73 7 18.0	− 8.17	52.59	2

Number.	Constellation, Name of Star, or Synonym.	Magnitude.	Right Ascension, 1850.0.	Annual Precession.	South Declination, 1850.0.	Annual Precession.	Mean year.	No. of obs.
			h. m. s.	s.	° ′ ″	″		
11 906	Gould, Z. C., 16^h 1807	8.8	16 23 49.54	+ 5.895	66 36 46.9	− 8.17	52.61	2
11 907	Gould, Z. C., 16^h 1824	9.2	23 56.80	6.129	68 13 28.0	.16	52.62	2
11 908	Apodis	9.5	24 22.98	8.956	78 15 48.2	.13	52.57	1
11 909	Apodis	8.5	33.15	10.914	81 8 15.4	.11	50.62	1
11 910	Apodis	10.0	37.54	+ 8.927	78 12 5.4	− 8.11	52.57	1
11 911	Apodis	9.2	16 24 40.79	+ 9.646	79 27 36.2	− 8.10	51.51	2
11 912	Octantis	8.0	44.74	14.741	84 0 55.3	.10	51.49	1
11 913	Trianguli Australis	8.8	45.30	6.356	69 34 23.4	.10	52.51	2
11 914	Gould, Z. C., 16^h 1925	9.0	49.10	7.365	74 5 42.1	.09	52.52	1
11 915	Trianguli Australis	9.5	51.74	+ 6.369	69 38 37.9	− 8.09	52.60	1
11 916	Apodis	9.0	16 24 57.02	+ 9.405	79 3 52.0	− 8.08	51.51	2
11 917	Gould, Z. C., 16^h 1913	9.5	25 4.41	6.221	68 45 42.0	.07	52.62	2
11 918	Trianguli Australis	9.8	8.99	6.207	68 40 11.4	.07	52.62	2
11 919	Trianguli Australis	9.5	10.06	6.396	69 46 50.8	.06	52.60	1
11 920	Gould, Z. C., 16^h 1936	10.0	12.00	+ 6.876	72 9 32.0	− 8.06	52.41	1
11 921	Gould, Z. C., 16^h 1919	8.8	16 25 12.48	+ 6.278	69 6 1.2	− 8.06	52.51	2
11 922	Apodis	9.5	21.94	6.679	71 14 50.3	.05	52.48	1
11 923	Apodis	10.0	24.87	11.436	81 40 7.0	.04	50.62	1
11 924	Trianguli Australis	9.8	30.46	6.360	69 33 51.4	.04	52.51	2
11 925	Apodis	9.2	40.44	+ 7.702	75 10 33.3	− 8.02	51.54	2
11 926	Gould, Z. C., 16^h 1957	8.5	16 25 46.43	+ 6.176	68 27 7.8	− 8.02	52.64	1
11 927	Gould, Z. C., 16^h 1952	9.0	50.20	5.867	66 19 21.4	.01	52.60	2
11 928	Gould, Z. C., 16^h 1958	9.0	55.45	5.822	65 58 9.8	.00	52.60	1
11 929	η^1 Trianguli Australis	6.0	25 57.68	6.105	67 59 18.0	8.00	52.59	2
11 930	Apodis	10.0	26 2.99	+ 7.065	72 55 38.7	− 7.99	52.60	1
11 931	Trianguli Australis	9.5	16 26 5.31	+ 5.926	66 44 57.6	− 7.99	52.61	1
11 932	Octantis	9.2	6.50	15.581	84 24 1.3	.99	51.19	3
11 933	Lacaille, 6861	8.0	14.46	6.493	70 16 28.4	.98	52.47	2
11 934	Gould, 22465	9.5	16.07	5.841	66 6 8.6	.98	52.60	1
11 935	Gould, Z. C., 16^h 2011	9.5	19.27	+ 6.328	69 21 8.4	− 7.97	52.42	1
11 936	Gould, Z. C., 16^h 2036	9.2	16 26 27.58	+ 6.748	71 32 24.0	− 7.96	52.50	2
11 937	Apodis	9.5	28.32	10.207	80 14 26.0	.96	51.63	1
11 938	Apodis	9.5	37.40	6.657	71 5 50.7	.95	52.50	2
11 939	Apodis	10.0	39.42	7.619	74 53 25.1	.94	52.53	1
11 940	Octantis	9.0	39.88	+13.882	83 31 13.8	− 7.94	50.53	2
11 941	Octantis	10.0	16 26 50.67	+12.813	82 48 50.6	− 7.93	51.04	2
11 942	Apodis	10.0	53.08	8.385	76 59 2.0	.93	52.40	1
11 943	Trianguli Australis	10.0	26 55.88	6.086	67 49 51.8	.92	52.59	2
11 944	Gould, Z. C., 16^h 2075	9.0	27 0.29	6.887	72 9 12.1	.92	52.41	1
11 945	Could, Z. G., 16^h 2052	8.0	3.76	+ 5.992	67 10 49.3	− 7.91	51.61	2

Number.	Constellation, Name of Star, or Synonym.	Magnitude.	Right Ascension, 1850.0.	Annual Precession.	South Declination, 1850.0.	Annual Precession.	Mean year.	No. of obs.
			h. m. s.	s.	° ′ ″	″		
11 946	Gould, Z. C., 16h 2103 .	9.0	16 27 23.83	+ 6.777	71 38 32.2	- 7.89	52.52	1
11 947	Octantis	9.5	30.36	42.497	88 12 48.6	.88	51.04	2
11 948	Apodis	9.5	44.84	7.795	75 24 3.9	.86	52.62	1
11 949	Gould, 22495	9.5	27 56.26	5.990	67 8 2.6	.84	51.61	2
11 950	Apodis	10.0	28 3.80	+ 6.912	72 13 39.3	- 7.83	52.52	1
11 951	Octantis	10.0	16 28 7.09	+14.865	84 2 25.5	— 7.83	51.49	1
11 952	Lacaille, 6881	7.0	13.68	5.991	67 7 51.7	.82	51.95	3
11 953	Lacaille, 6877	8.0	18.15	5.762	71 32 44.3	.81	52.50	2
11 954	Apodis	9.5	23.68	6.586	70 41 14.5	.80	52.40	1
11 955	Apodis	10.0	24.93	+ 8.925	78 6 58.9	— 7.80	52.57	1
11 956	Apodis	10.0	16 28 26.25	+ 9.121	78 29 25.0	— 7.80	50.64	1
11 957	Gould, Z. C., 16h 2176 .	9.5	33.10	6.322	69 14 26.6	.79	52.42	1
11 958	Apodis	10.0	41.28	8.968	78 11 39.2	.78	51.61	2
11 959	Trianguli Australis . . .	10.0	28 57.80	5.845	66 1 58.3	.76	52.60	1
11 960	Trianguli Australis . . .	9.0	29 0.27	+ 6.203	68 30 28.3	— 7.76	52.64	1
11 961	Trianguli Australis . . .	10.0	16 29 2.02	+ 5.841	65 59 35.3	— 7.75	52.60	1
11 962	Gould, Z. C., 16h 2224 .	9.0	17.59	6.331	69 15 50.4	.73	52.42	1
11 963	Apodis	9.5	25.99	6.870	72 1 10.0	.72	52.41	1
11 964	Gould, Z. C., 16h 2265 .	10.0	39.68	6.661	71 1 1.5	.70	52.52	1
11 965	Apodis	9.5	49.81	+ 6.540	70 23 58.8	— 7.69	52.40	1
11 966	Octantis	9.5	16 29 53.06	+14.310	83 43 42.0	— 7.68	50.55	1
11 967	Apodis	10.0	30 9.35	8.946	78 7 12.5	.66	52.57	1
11 968	Gould, Z. C., 16h 2346 .	10.0	13.68	8.569	77 19 47.4	.66	52.56	1
11 969	Apodis	10.0	29.70	7.908	75 39 32.2	.64	52.62	1
11 970	Gould, Z. C., 16h 2301 .	8.2	29.94	+ 5.876	66 12 10.0	— 7.63	52.61	2
11 971	Gould, Z. C., 16h 2336 .	9.0	16 30 31.94	+ 7.533	74 30 27.3	— 7.63	52.56	1
11 972	Lacaille, 6892	7.5	39.54	6.051	67 27 25.0	.62	51.60	2
11 973	Apodis	9.5	41.58	8.725	77 39 35.2	.62	52.36	1
11 974	Apodis	10.0	43.66	6.911	72 8 27.5	.62	52.41	1
11 975	Gould, Z. C., 16h 2394 .	9.2	48.02	+ 8.642	77 28 44.4	— 7.61	52.46	2
11 976	Apodis	10.0	16 30 49.26	+11.707	81 50 50.2	7.61	50.53	1
11 977	Gould, Z. C., 16h 2317 .	8.8	49.50	5.949	66 44 6.9	.61	52.61	2
11 978	Apodis	10.0	30 52.89	7.008	72 33 7.8	.60	52.52	1
11 979	Apodis	9.2	31 0.09	12.474	82 30 5.4	.59	50.52	2
11 980	Octantis	10.0	1.76	+13.881	83 28 4.6	— 7.59	50.53	2
11 981	Apodis	9.8	16 31 3.28	+ 6.601	70 40 22.6	— 7.59	52.46	2
11 982	Trianguli Australis . . .	9.0	5.78	5.839	65 54 17.2	.59	52.60	1
11 983	Gould, Z. C., 16h 2424 .	8.5	6.66	8.770	77 44 39.6	.59	52.36	1
11 984	Apodis	9.0	8.40	6.603	70 41 1.5	.58	52.46	2
11 985	Apodis	9.2	13.79	+10.644	80 42 51.1	— 7.58	51.10	2

Number.	Constellation, Name of Star, or Synonym.	Magnitude.	Right Ascension, 1850.0.	Annual Precession.	South Declination, 1850.0.	Annual Precession.	Mean year.	No. of obs.
			h. m. s.	s.	° ′ ″	″		
11 986	Trianguli Australis . . .	9.5	16 31 14.02	+ 6.408	69 38 53.2	7.58	52.60	1
11 987	Gould, Z. C., 16ʰ 2415 .	9.0	22.68	7.809	75 20 55.7	.56	50.56	1
11 988	Gould, Z. C., 16ʰ 2364 .	9.0	25.09	6.154	68 6 30.0	.56	52.59	1
11 989	Brisbane, 5796	8.0	28.23	6.276	68 51 48.9	.56	52.59	1
11 990	η² Trianguli Australis . . .	7.0	29.56	+ 6.109	67 48 53 6	7.55	52.59	2
11 991	Octantis	8.8	16 31 34.61	+13.880	83 27 20.2	7.55	50.53	2
11 992	Apodis	10.0	36.20	8.109	76 11 0.2	.55	52.51	3
11 993	Lacaille, 6906	6.0	36.84	5.965	66 49 13.0	.54	52.61	2
11 994	Apodis	9.2	36.92	7 271	73 33 47.4	.54	52.55	2
11 995	Apodis	9.5	40.63	+10.659	80 43 31.1	— 7.54	51.10	2
11 996	Apodis	10.0	16 31 42.24	+10.236	80 11 17.1	— 7.54	50.60	1
11 997	Apodis	10.0	52.88	7.469	74 15 25.0	.52	52.52	1
11 998	Trianguli Australis . .	9.0	31 59.59	5.859	66 1 10.7	.51	52.60	1
11 999	Gould, Z. C., 16ʰ 2404 .	8.5	32 0.86	5.927	66 31 35.5	.51	52.61	1
12 000	Apodis	10.0	9.85	+ 6.735	71 17 46.4	— 7.50	52.48	1
12 001	Gould, Z. C., 16ʰ 2428 .	8.5	16 32 19.27	+ 5.937	66 35 33.8	— 7.49	52.61	1
12 002	Gould, Z. C., 16ʰ 2437 .	9.0	23.20	6.183	68 15 32.2	.48	52.62	2
12 003	Gould, Z. C., 16ʰ 2439 .	9.5	28.68	6.058	67 26 2.2	.47	51.60	2
12 004	Apodis	10.0	37.14	9.179	78 30 44.6	.46	50.64	1
12 005	Lacaille, 6869	7.5	39.58	+ 9.973	79 48 16.4	— 7.46	50.60	2
12 006	Apodis	9.5	16 32 39.69	+11.436	81 33 34.3	— 7.46	50.62	1
12 007	Gould, Z. C., 16ʰ 2451 .	8.0	47.17	5.832	65 47 13.0	.45	52.60	1
12 008	α Trianguli Australis . . .	2.0	50.06	6.263	68 44 35.8	.45	52.63	3
12 009	Gould, Z. C., 16ʰ 2472 .	10.0	50.27	6.195	68 19 20.7	.45	52.59	1
12 010	Gould, Z. C., 16ʰ 2478 .	9.0	50.34	+ 6.376	69 24 31.2	— 7.45	52.51	2
12 011	Gould, Z. C., 16ʰ 2521 .	9.0	16 32 58.10	+ 7.617	74 43 11.7	— 7.43	52.54	2
12 012	Gould, Z. C., 16ʰ 2503 .	9.5	33 11.61	6.351	69 15 2.8	.42	52.42	1
12 013	Gould, Z. C., 16ʰ 2533 .	7.5	11.66	7.568	74 33 12.3	.42	52.56	1
12 014	Apodis	10.0	16.96	7.043	72 37 26.8	.41	52.56	2
12 015	Gould, Z. C., 16ʰ 2612 .	8.5	21.11	+ 9.745	79 27 4.0	— 7.40	51.51	2
12 016	Lacaille, 6901	7.2	16 33 22.59	+ 7.252	73 26 42.4	— 7.40	52.55	2
12 017	Apodis	10.0	23.10	8.364	76 47 4.8	.40	52.44	2
12 018	Gould, Z. C., 16ʰ 2553 .	10.0	31.04	7.257	73 27 25.8	.39	52.55	2
12 019	Apodis	9.0	35.03	11.080	81 10 33.9	.38	50.62	1
12 020	Octantis	8.0	38.89	+24.572	86 41 26.0	— 7.38	51.04	3
12 021	Brisbane, 5810	8.0	16 33 38.94	+ 5.973	66 48 6.5	— 7.38	52.61	1
12 022	Gould, Z. C., 16ʰ 2522 .	8.0	39.25	5.999	66 59 9.1	.38	52.61	1
12 023	Apodis	10.0	40.77	8.265	76 32 9.7	.38	52.48	1
12 024	Gould, Z. C., 16ʰ 2542 .	9.5	42.72	6.660	70 53 18.4	.37	52.52	1
12 025	Trianguli Australis . . .	9.0	51.42	+ 6.108	67 43 37.2	— 7.36	52.58	1

Number.	Constellation, Name of Star, or Synonym.	Magnitude.	Right Ascension, 1850.0.	Annual Precession.	South Declination, 1850.0.	Annual Precession.	Mean year.	No. of obs.
			h. m. s.	s.	° ′ ″	″		
12 026	Apodis	9.0	16 33 57.10	+ 6.882	71 55 12.9	− 7.35	52.47	2
12 027	Trianguli Australis . . .	9.0	34 6.88	6.121	67 48 12.8	.34	52.59	2
12 028	Trianguli Australis . . .	9.5	7.05	6.058	67 22 32.4	.34	50.62	1
12 029	Gould, Z. C., 16h 2575 .	8.8	8.00	6.554	70 20 13.6	.34	52.47	2
12 030	Trianguli Australis . . .	9.5	13.32	+ 6.078	67 30 47.6	− 7.33	52.58	1
12 031	Apodis	9.5	16 34 14.77	+11.058	81 8 31.8	− 7.33	50.62	1
12 032	Apodis	10.0	20.24	7.395	73 56 13.4	.32	52.52	1
12 033	Gould, Z. C., 16h 2689 .	9.2	21.14	10.007	79 49 25.8	.32	50.60	2
12 034	Apodis	10.0	21.74	8.107	76 6 53.5	.32	52.48	1
12 035	Gould, Z. C., 16h 2602 .	9.0	22.58	+ 6.907	72 0 53.4	− 7.32	52.41	1
12 036	Apodis	9.5	16 34 45.34	+ 6.966	72 15 47.3	− 7.29	52.52	1
12 037	Trianguli Australis . . .	8.8	34 45.78	6.080	67 30 13.9	.29	51.60	2
12 038	Trianguli Australis . . .	9.0	35 4.13	6.085	67 31 40.6	.26	52.58	1
12 039	Brisbane, 5816	8.5	4.96	5.973	66 44 56.0	.26	52.61	2
12 040	Gould, Z. C., 16h 2652 .	9.0	5.80	+ 6.910	72 0 32.6	− 7.26	52.41	1
12 041	Gould, Z. C., 16h 2672 .	8.8	16 35 7.42	+ 7.485	74 13 47.4	− 7.26	52.54	2
12 042	Apodis	9.0	16.95	7.130	72 55 22.5	.25	52.60	1
12 043	Gould, Z. C., 16h 2649 .	9.0	21.81	6.020	67 4 30.1	.24	52.61	1
12 044	Apodis	10.0	24.32	6.998	72 22 39.4	.24	52.52	1
12 045	Trianguli Australis . . .	9.5	28.12	+ 5.975	66 45 16.4	− 7.23	52.61	1
12 046	Apodis	9.5	16 35 29.19	+ 6.744	71 14 21.2	− 7.23	52.48	1
12 047	Gould, Z. C., 16h 2666 .	9.0	29.81	6.375	69 18 58.3	.23	52.42	1
12 048	Gould, Z. C., 16h 2691 .	8.5	43.16	6.815	71 34 1.6	.21	52.50	2
12 049	Apodis	9.5	47.58	6.759	71 18 9.3	.20	52.48	1
12 050	Octantis	9.1	53.87	+24.046	86 34 42.2	− 7.20	51.04	4
12 051	Apodis	9.5	16 35 54.59	+ 6.663	70 50 17.0	− 7.20	52.52	1
12 052	Trianguli Australis . . .	9.5	36 7.07	5.944	66 30 14.1	.18	52.61	1
12 053	Trianguli Australis . . .	9.5	10.28	5.941	66 29 1.9	.17	52.61	1
12 054	Trianguli Australis . . .	9.5	10.40	5.934	66 25 24.0	.17	52.61	1
12 055	Trianguli Australis . . .	9.5	13.05	+ 6.130	67 47 22.7	− 7.17	52.59	1
12 056	Trianguli Australis . . .	9.5	16 36 15.82	+ 6.056	67 17 23.7	− 7.17	50.62	1
12 057	Gould, Z. C., 16h 2705 .	9.5	16.82	6.056	67 17 17.0	.16	50.62	1
12 058	Apodis	8.5	20.06	6.703	71 1 8.0	.16	52.52	1
12 059	Trianguli Australis . . .	9.5	24.09	6.133	67 48 8.3	.15	52.59	1
12 060	Gould, Z. C., 16h 2743 .	8.0	25.06	+ 6.879	71 50 4.6	− 7.15	52.47	2
12 061	Apodis	9.5	16 36 25.68	+ 6.755	71 15 51.7	− 7.15	52.48	1
12 062	Gould, 22682	8.0	28.14	7.123	72 50 42.4	.15	52.60	1
12 063	Octantis	9.2	31.97	13.154	82 56 7.8	.14	51.04	2
12 064	Apodis	10.0	36.57	7.666	74 47 4.0	.14	52.53	1
12 065	Trianguli Australis . . .	9.2	38.29	+ 6.207	68 16 4.8	− 7.14	52.62	2

Number.	Constellation, Name of Star, or Synonym.	Magnitude.	Right Ascension, 1850.0.	Annual Precession.	South Declination, 1850.0.	Annual Precession.	Mean year.	No. of obs.
			h. m. s.	s.	° ′ ″	″		
12 066	Gould, Z. C., 16h 2750 .	8.5	16 36 39.42	+ 6.544	70 12 27.6	− 7.13	52.48	2
12 067	Apodis	10.0	40.11	6.688	70 56 7.2	.13	52.52	1
12 068	Apodis	10.0	42.65	6.677	70 52 44.8	.13	52.52	1
12 069	Gould, Z. C., 16h 2817 .	8.5	42.74	8.578	77 12 42.4	.13	52.48	2
12 070	Apodis	8.5	46.56	+ 8.143	76 9 18.7	− 7.12	52.48	2
12 071	Gould, Z. C., 16h 2774	8.2	16 36 56.30	+ 6.569	70 19 53.5	− 7.11	52.48	2
12 072	Lacaille, 6947	7.5	36 59.38	6.077	67 24 31.8	.11	51.60	2
12 073	Apodis	9.5	37 5.25	8.551	77 8 25.9	.10	52.40	1
12 074	Apodis	8.5	6.24	10.827	80 50 21.4	.10	50.56	1
12 075	Gould, 22690	8.0	8.39	+ 7.125	72 51 4.9	− 7.09	52.60	1
12 076	Trianguli Australis . .	9.0	16 37 16.16	+ 6.424	69 32 1.0	− 7.08	52.51	2
12 077	Gould, Z. C., 16h 2794 .	9.5	22.72	6.080	67 25 7.3	.07	52.58	1
12 078	Apodis	9.0	22.89	10.889	80 54 22.8	.07	50.56	1
12 079	Lacaille, 6945	7.2	25.64	6.492	69 54 16.0	.07	52.58	2
12 086	Apodis	8.8	37.35	+ 7.550	74 23 3.2	− 7.05	52.54	2
12 081	Apodis	9.5	16 37 40.73	+ 6.846	71 39 11.9	− 7.05	52.52	1
12 082	Gould, Z. C., 16h 2841 .	8.5	43.01	7.057	72 33 43.6	.05	52.56	2
12 083	Brisbane, 5835	8.2	47.38	6.298	68 47 16.8	.04	52.63	3
12 084	Gould, Z. C., 16h 2898 .	9.2	59.76	8.315	76 34 0.6	.02	52.44	2
12 085	Trianguli Australis . . .	9.5	37 59.83	+ 6.080	67 23 44.7	− 7.02	52.58	1
12 080	Apodis	9.5	16 38 0.20	+ 8.146	76 8 11.8	− 7.02	52.48	1
12 087	Trianguli Australis . . .	9.0	4.92	5.935	66 22 11.6	.02	52.60	2
12 088	Apodis	9.2	11.86	7.650	74 41 49.2	.01	52.54	2
12 089	Gould, Z. C., 16h 2863 .	9.2	20.44	6.099	67 30 51.2	7.00	51.60	2
12 090	Trianguli Australis . . .	10.0	21.14	− 5.930	66 19 22.3	− 6.99	52.60	1
12 091	Gould, Z. C., 16h 2929 .	8.5	16 38 21.48	+ 8.596	77 13 11.9	− 6.99	52.48	2
12 092	Apodis	10.0	45.31	6.570	70 16 38.0	.96	52.40	1
12 093	Lacaille, 6828	8.2	45.74	12.087	82 4 53.1	.96	50.57	2
12 094	Stone, 9149	9.8	48.66	12.078	82 4 21.0	.96	50.57	2
12 095	Apodis	9.2	54.80	+ 9.661	79 13 19.2	− 6.95	52.03	2
12 096	Apodis	9.5	16 38 54.96	+ 6.766	71 14 41.2	− 6.95	52.48	1
12 097	Gould, Z. C., 16h 2914 .	8.8	39 0.04	6.722	71 2 0.6	.94	52.50	2
12 098	Trianguli Australis . . .	9.0	5.97	6.379	69 13 22.1	.93	52.42	1
12 099	Gould, Z. C., 16h 2909 .	9.5	9.18	6.040	67 4 58.2	.93	52.61	1
12 100	Gould, Z. C., 16h 2974 .	8.8	9.20	+ 7.942	75 33 9.7	− 6.93	51.59	2
12 101	Trianguli Australis . . .	9.0	16 39 20.53	+ 5.955	66 28 39.8	− 6.91	52.61	1
12 102	Brisbane, 5842	7.5	21 84	6.345	69 1 9.2	.91	52.51	2
12 103	Gould, Z. C., 16h 2993 .	9.2	23.38	7.917	75 28 35.8	.91	51.59	2
12 104	Trianguli Australis . . .	9.0	35.14	6.958	66 29 32.6	.89	52.61	1
12 105	Apodis	9.2	37.08	+12.212	82 10 38.7	− 6.89	50.54	3

Number.	Constellation, Name of Star, or Synonym.	Magnitude.	Right Ascension, 1850.0.	Annual Precession.	South Declination, 1850.0.	Annual Precession.	Mean year.	No. of obs.
			h. m. s.	s.	° ′ ″	″		
12 106	Trianguli Australis . . .	9.0	16 39 37.58	+ 5.923	66 13 51.6	— 6.89	52.60	2
12 107	Gould, Z. C., 16h 3023 .	10.0	45.10	8.090	75 56 51.8	.88	52.55	2
12 108	Apodis	9.2	45.38	7.007	72 18 2.2	.88	52.46	2
12 109	Apodis	10.2	46.87	8.091	75 56 59.4	.88	52.55	2
12 110	Trianguli Australis . . .	9.5	47.51	+ 6.456	69 38 6.3	— 6.88	52.60	1
12 111	Trianguli Australis . . .	9.5	16 39 51.10	+ 5.969	66 33 28.6	— 6.87	52.61	1
12 112	Apodis	9.5	39 57.49	7.268	73 19 56.7	.86	52.58	1
12 113	Apodis	9.5	40 1.51	7.024	72 21 42.6	.86	52.46	2
12 114	Trianguli Australis . . .	10.0	4.82	6.454	69 37 3.9	.85	52.60	1
12 115	Trianguli Australis . . .	10.0	5.88	+ 6.424	69 27 7.7	— 6.85	52.51	2
12 116	Lacaille, 6961	8.0	16 40 6.61	+ 6.691	70 51 13.8	— 6.85	52.51	1
12 117	Apodis	8.5	11.93	8.639	77 16 43.2	.84	52.56	1
12 118	Trianguli Australis . . .	9.0	12.91	5.901	66 2 48.5	.84	52.60	1
12 119	Apodis	12.0	13.09	8.098	75 57 33.9	.84	52.48	1
12 120	Lacaille, 6948	8.2	22.40	+ 8.101	75 57 49.9	— 6.83	52.55	2
12 121	Apodis	9.0	16 40 23.58	+ 8.168	76 8 25.8	— 6.83	52.48	1
12 122	Gould, Z. C., 16h 3010 .	9.0	25.22	6.011	66 50 26.0	.83	52.61	1
12 123	Gould, Z. C., 16h 3030 .	9.0	25.94	6.764	71 11 4.6	.82	52.48	1
12 124	Apodis	9.5	36.04	9.168	78 20 35.9	.81	51.61	2
12 125	Apodis	9.0	43.59	+ 8.688	77 22 29.5	— 6.80	52.46	2
12 126	Apodis	9.0	16 40 45.76	+11.095	81 5 20.0	— 6.80	50.59	2
12 127	Gould, Z. C., 16h 3060 .	9.5	51.76	6.696	70 51 25.7	.79	52.52	1
12 128	Lacaille, 6939	7.8	40 53.11	8.909	77 50 7.1	.79	52.47	2
12 129	Trianguli Australis . . .	9.0	41 1.16	6.202	68 5 45.8	.78	52.59	1
12 130	Apodis	9.2	9.26	+ 8.697	77 23 11.0	— 6.76	52.44	2
12 131	Stone, 9273	7.8	16 41 13.29	+28.964	87 12 59.2	— 6.76	51.04	4
12 132	Apodis	9.0	13.54	9.114	78 13 51.1	.76	51.61	2
12 133	Gould, Z. C., 16h 3153 .	9.0	24.86	8.028	75 44 37.3	.74	52.62	1
12 134	Lacaille, 6905	8.2	28.66	12.219	82 9 40.2	.74	50.54	3
12 135	Gould, 22783	8.0	33.89	+ 6.026	66 54 31.8	— 6.73	52.61	1
12 136	Stone, 9270	8.0	16 41 35.52	+27.841	87 5 19.8	— 6.73	51.04	4
12 137	Lacaille, 6969	7.0	35.84	7.237	73 10 25.0	.73	52.58	2
12 138	Trianguli Australis . . .	9.0	36.84	6.159	67 48 6.6	.73	52.59	2
12 139	Gould, Z. C., 16h 3111 .	8.0	43.79	5.882	65 51 15.0	.72	52.60	1
12 140	Apodis	10.0	54.20	+ 7.027	72 19 39.4	— 6.70	52.52	1
12 141	Apodis	9.5	16 41 55.94	+ 7.018	72 17 12.8	— 6.70	52.46	2
12 142	Trianguli Australis . . .	9.8	42 7.22	6.290	68 36 35.9	.69	52.63	3
12 143	Trianguli Australis . . .	9.5	8.18	6.262	68 26 4.0	.68	52.64	2
12 144	Trianguli Australis . . .	10.0	12.58	6.414	69 20 0.5	.68	52.42	1
12 145	Octantis	9.8	13.08	+15.706	84 18 5.0	— 6.68	51.49	2

No. 12 108. Declination possibly 72° 26′ 20.5″.

Number.	Constellation, Name of Star, or Synonym.	Magnitude.	Right Ascension, 1850.0.	Annual Precession.	South Declination, 1850.0.	Annual Precession.	Mean year.	No. of obs.
			h. m. s.	s.	° ′ ″	″		
12 146	Trianguli Australis . . .	10.0	16 42 15.64	+ 6.305	68 41 35.8	− 6.67	52.64	2
12 147	Gould, 22806	8.2	19.61	6.226	68 12 32.8	.67	52.62	2
12 148	Gould, Z. C., 16h 3215 .	9.2	22.56	7.676	74 41 49.4	.66	52.54	2
12 149	Trianguli Australis . . .	9.5	24.25	6.002	66 42 52.7	.66	52.61	1
12 150	Gould, Z. C., 16h 3229 .	10.0	39.25	+ 7.270	73 16 22.4	− 6.64	52.58	1
12 151	Apodis	10.0	16 42 54.83	+ 7.303	73 23 20.1	− 6.62	52.52	1
12 152	Trianguli Australis . . .	9.5	58.43	5.989	66 36 1.7	.61	52.61	2
12 153	Gould, Z. C., 16h 3248	9.0	58.93	6.887	71 41 41.0	.61	52.52	1
12 154	Gould, Z. C., 16h 3300 .	8.5	42 59.13	8.404	76 40 53.4	.61	52.44	2
12 155	Gould, Z. C., 16h 3216 .	9.0	43 2.80	+ 6.085	67 16 8.6	− 6.61	50.62	1
12 156	Gould, Z. C., 16h 3289 .	9.0	16 43 4.02	+ 8.052	75 46 28.7	− 6.61	52.62	1
12 157	Gould, Z. C., 16h 3212 .	9.5	13.67	6.033	66 54 10.0	.59	52.61	1
12 158	Trianguli Australis . . .	9.0	18.31	6.044	66 58 11.6	.59	52.61	1
12 159	Trianguli Australis . . .	9.0	20.94	5.944	66 15 45.0	.58	52.60	2
12 160	Trianguli Australis . . .	8.5	24.35	+ 6.395	69 11 30.0	− 6.58	52.42	1
12 161	Trianguli Australis . . .	9.5	16 43 25.87	+ 5.963	66 23 51.4	− 6.58	52.61	1
12 162	Octantis	9.0	26.75	14.110	83 28 18.4	.58	50.53	2
12 163	Octantis	10.0	27.10	17.428	84 58 16.7	.58	50.51	1
12 164	Gould, Z. C., 16h 3258 .	9.0	28.47	6.128	67 32 23.8	.57	52.58	1
12 165	Lacaille, 6989	6.2	30.15	+ 6.367	69 1 19.4	− 6.57	52.51	2
12 166	Apodis	10.0	16 43 41.06	+ 7.331	73 28 23.9	− 6.56	52.52	1
12 167	Apodis	9.5	48.11	9.182	78 18 45.3	.55	51.61	2
12 168	Apodis	9.0	43 48.71	9.560	78 58 37.8	.55	50.61	1
12 169	Apodis	9.0	44 1.02	8.188	76 7 16.0	.53	52.48	1
12 170	Apodis	10.0	1.89	+ 6.584	70 9 32.7	− 6.53	52.56	1
12 171	Apodis	9.5	16 44 10.26	+ 8.033	75 42 5.0	− 6.52	52.62	1
12 172	Apodis	9.8	10.84	9.744	79 16 6.0	.51	52.03	2
12 173	Trianguli Australis . . .	10.0	19.74	6.287	68 31 35.9	.50	52.64	1
12 174	Gould, Z. C., 16h 3329	9.0	22.11	6.045	66 57 15.7	.50	52.61	1
12 175	Gould, Z. C., 16h 3343	9.0	30.24	+ 6.221	68 6 41.9	− 6.49	52.59	1
12 176	Gould, Z. C., 16h 3371 .	8.8	16 44 32.10	+ 6.929	71 50 23.2	− 6.49	52.47	2
12 177	Trianguli Australis . . .	10.0	36.30	6.219	68 5 46.3	.48	52.59	1
12 178	Lacaille, 6988	7.5	37.62	7.053	72 22 7.6	.48	52.46	2
12 179	Gould, Z. C., 16h 3429 .	9.0	42.23	8.405	76 38 59.2	.47	52.44	2
12 180	Apodis	10.0	45.03	+ 6.805	71 16 18.5	− 6.47	52.48	1
12 181	Apodis	9.5	16 44 46.22	+ 7.315	73 23 29.9	− 6.47	52.52	1
12 182	Trianguli Australis . . .	10.0	51.36	5.911	65 58 9.4	.46	52.60	1
12 183	Gould, Z. C., 16h 3413	9.5	52.78	7.540	74 10 56.6	.46	52.54	2
12 184	Trianguli Australis . . .	10.0	53.21	6.237	68 11 56.7	.46	52.66	3
12 185	Trianguli Australis . . .	10.0	53.51	+ 5.924	66 3 57.4	− 6.46	52.60	1

No. 12 159. Possibly 66° 15′ 17.1″.

Number.	Constellation, Name of Star, or Synonym.	Magnitude.	Right Ascension, 1850.0.	Annual Precession.	South Declination, 1850.0.	Annual Precession.	Mean year.	No. of obs.
			h. m. s.	s.	° ′ ″	″		
12 186	Trianguli Australis . . .	9.8	16 44 55.84	+ 6.245	68 14 48.0	− 6.45	52.66	3
12 187	Apodis	10.0	45 4.48	7.417	73 45 14.2	.44	52.52	1
12 188	Apodis	9.8	27.95	7.717	74 44 42.0	.41	52.54	2
12 189	Apodis	9.5	28.11	7.972	72 0 18.3	.41	52.41	1
12 190	Apodis	10.0	30.92	+ 7.599	74 21 51.4	− 6.40	52.52	1
12 191	Gould, 22917	8.2	16 45 40.43	+ 6.403	69 10 18.4	− 6.39	52.46	2
12 192	Apodis	10.0	41.66	7.238	73 4 49.4	.39	52.60	1
12 193	Gould, Z. C., 16h 3521 .	9.2	50.42	8.728	77 22 9.0	.38	52.46	2
12 194	Apodis	9.5	45 59.54	6.933	71 49 7.6	.36	52.52	1
12 195	Trianguli Australis . . .	8.5	46 0.24	+ 5.957	66 16 23.3	− 6.36	52.60	2
12 196	Apodis	10.0	16 46 0.46	+ 8.171	76 2 6.5	6.36	52.48	1
12 197	Gould, Z. C., 16h 3488 .	9.5	8.68	6.980	72 1 22.1	.35	52.41	1
12 198	Octantis	8.6	16.61	70.300	88 55 8.8	.34	51.04	4
12 199	Gould, Z. C., 16h 3489 .	9.0	24.29	6.401	69 8 18.1	.33	52.51	2
12 200	Trianguli Australis . .	9.5	32.22	+ 6.036	66 49 17.1	− 6.32	52.61	1
12 201	Gould, Z. C., 16h 3644 .	9.2	16 46 37.20	+10.180	79 52 36.7	− 6.31	50.60	2
12 202	Lacaille, 6992	7.5	38.14	8.160	75 59 39.6	.31	52.50	2
12 203	Apodis	9.8	40.32	7.367	73 32 13.5	.31	52.55	2
12 204	Apodis	9.5	49.42	6.561	70 0 25.5	.30	52.56	1
12 205	Gould, Z. C., 16h 3539 .	9.5	49.77	+ 7.015	72 9 14.0	− 6.30	52.41	1
12 206	Octantis	9.2	16 46 54.08	+21.087	85 58 6.5	− 6.29	51.05	2
12 207	Gould, Z. C., 16h 36[illegible] .	9.0	47 0.32	8.709	77 18 21.6	.28	52.56	1
12 208	Trianguli Australis . .	9.0	0.38	5.948	66 10 49.0	.28	52.60	2
12 209	Trianguli Australis . . .	9.5	1.24	6.489	69 36 56.0	.28	52.60	1
12 210	Gould, Z. C., 16h 3541 .	9.0	2.06	+ 6.573	70 3 48.3	− 6.28	52.56	1
12 211	Brisbane, 5893	8.5	16 47 11.99	+ 6.279	68 23 45.5	− 6.26	52.64	2
12 212	Apodis	10.0	16.72	7.586	74 17 5.5	.26	52.52	1
12 213	Gould, Z. C., 16h 3547 .	9.0	20.72	6.020	66 41 12.3	.25	52.61	2
12 214	Brisbane, 5898	8.3	21.32	6.316	68 36 39.6	.25	52.66	3
12 215	Trianguli Australis . . .	10.0	24.59	+ 5.931	66 2 32.2	− 6.25	52.60	1
12 216	Apodis	10.0	16 47 31.06	+11.498	81 25 26.2	− 6.24	50.57	3
12 217	Trianguli Australis . . .	9.0	37.36	6.146	67 32 24.7	.23	52.58	1
12 218	Gould, Z. C., 16h 3629 .	9.5	43.14	7.198	72 52 44.2	.22	52.60	1
12 219	Gould, Z. C., 16h 3604 .	9.0	44.96	6.601	70 11 24.3	.22	52.48	2
12 220	Trianguli Australis . . .	9.5	47 59.67	+ 6.495	69 37 21.7	6.20	52.60	1
12 221	Trianguli Australis . . .	9.5	16 48 2.58	+ 6.007	66 34 15.7	− 6.19	52.61	1
12 222	Trianguli Australis . . .	9.0	2.85	6.166	67 39 18.6	.19	52.58	1
12 223	Gould, Z. C., 16h 3641 .	9.2	13.35	6.283	68 23 37.7	.18	52.64	2
12 224	Trianguli Australis . . .	9.2	13.92	6.134	67 26 43.1	.18	51.60	2
12 225	Gould, Z. C., 16h 3652 .	9.2	18.68	+ 6.291	68 26 14.2	− 6.17	52.64	2

Number.	Constellation, Name of Star, or Synonym.	Magnitude.	Right Ascension, 1850.0.	Annual Precession.	South Declination, 1850.0.	Annual Precession.	Mean year.	No. of obs.
			h. m. s.	s.	° ′ ″	″		
12 226	Trianguli Australis	9.0	16 48 28.27	+ 6.210	67 55 40.5	− 6.16	52.59	1
12 227	Gould, Z. C., 16h 3732	9.0	28.43	8.890	77 39 49.6	.16	52.36	1
12 228	Trianguli Australis	9.5	38.71	6.158	67 35 15.4	.14	52.58	1
12 229	Apodis	9.2	38.82	8.603	77 2 25.8	.14	52.48	2
12 230	Apodis	9.5	42.62	+ 8.092	75 46 22.2	− 6.14	52.55	2
12 231	Apodis	10.0	16 48 43.66	+ 7.417	73 40 9.2	− 6.14	52.52	1
12 232	Apodis	9.5	47.12	11.438	81 20 53.8	.13	50.62	1
12 233	Lacaille, 7039	7.0	51.37	6.231	68 2 59.3	.13	52.59	1
12 234	Apodis	9.0	48 52.67	11.311	81 12 55.9	.12	50.59	2
12 235	Gould, Z. C., 16h 3775	8.5	49 0.28	+ 9.195	78 15 9.6	− 6.11	51.61	2
12 236	Apodis	9.5	16 49 1.56	+ 7.469	73 50 58.4	− 6.11	52.52	2
12 237	Gould, Z. C., 16h 3710	9.0	7.47	6.604	70 10 17.4	.11	52.48	2
12 238	Apodis	10.0	9.05	10.163	79 49 4.4	.10	50.60	1
12 239	Apodis	9.0	18.33	11.091	80 58 22.4	.09	50.59	2
12 240	Octantis	9.5	23.32	+37.009	87 50 58.6	− 6.08	51.04	2
12 241	Apodis	8.2	16 49 30.04	+ 7.077	72 21 3.0	− 6.07	52.47	2
12 242	Apodis	8.2	33.56	8.031	75 35 15.8	.07	51.59	2
12 243	Trianguli Australis	9.8	43.81	6.483	69 30 31.9	.05	52.53	3
12 244	Lacaille, 7020	8.0	44.23	7.779	74 50 49.9	.05	52.53	1
12 245	Trianguli Australis	10.0	45.57	+ 6.291	68 23 55.0	− 6.05	52.65	1
12 246	Gould, Z. C., 16h 3743	8.5	16 49 48.55	+ 5.997	66 27 10.4	− 6.05	52.61	1
12 247	Trianguli Australis	9.5	49.45	6.369	68 51 38.5	.05	52.59	1
12 248	Octantis	9.5	49 50.07	14.450	83 36 25.3	.04	50.53	2
12 249	Trianguli Australis	9.0	50 3.18	6.470	69 25 54.3	.03	52.53	3
12 250	Apodis	9.5	18.42	+ 7.888	75 9 54.6	− 6.01	51.54	2
12 251	Lacaille, 7028	8.2	16 50 18.52	+ 7.887	75 9 44.2	− 6.01	51.54	2
12 252	Apodis	9.5	19.53	6.664	70 26 54.8	6.00	52.40	1
12 253	Gould, Z. C., 16h 3834	9.0	25.01	7.694	74 34 3.4	5.99	52.56	1
12 254	Apodis	10.0	35.39	6.709	70 39 44.7	.98	52.52	1
12 255	Trianguli Australis	9.5	37.74	+ 5.967	66 12 50.4	− 5.98	52.61	1
12 256	Trianguli Australis	9.5	16 50 42.07	+ 6.567	69 56 14.0	− 5.97	52.59	2
12 257	Apodis	8.8	43.96	7.963	75 22 25.2	.97	51.59	2
12 258	Apodis	9.5	48.46	6.679	70 30 49.4	.96	52.40	1
12 259	Trianguli Australis	9.5	48.97	6.374	68 51 45.0	.96	52.59	1
12 260	Apodis	10.0	49.80	+ 6.865	71 23 58.7	− 5.96	52.48	1
12 261	Gould, Z. C., 16h 3817	9.0	16 50 56.18	+ 6.007	66 29 33.1	− 5.95	52.61	1
12 262	Apodis	10.0	51 0.98	6.735	70 46 53.5	.95	52.47	2
12 263	Trianguli Australis	8.8	10.52	6.410	69 2 28.5	.93	52.53	3
12 264	Apodis	10.0	13.68	7.455	73 45 15.2	.93	52.52	1
12 265	Gould, 23068	9.0	20.55	+ 9.312	78 25 45.9	− 5.92	50.64	1

No. 12 253. Gould's declination is 12.1″ greater.

No. 12 262. Possibly 70° 46′ 29.4″.

Number.	Constellation, Name of Star, or Synonym.	Magnitude.	Right Ascension, 1850.0.	Annual Precession.	South Declination, 1850.0.	Annual Precession.	Mean year.	No. of obs.
			h. m. s.	s.	° ′ ″	″		
12 266	Trianguli Australis . . .	10.0	16 51 21.59	+ 6.227	67 57 15.7	− 5.92	52.58	1
12 267	Trianguli Australis . . .	10.0	24.19	6.218	67 53 47.2	.91	52.58	1
12 268	Trianguli Australis . . .	9.0	25.91	5.919	65 49 53.9	.91	52.60	1
12 269	Trianguli Australis . . .	9.0	34.38	5.968	66 11 14.0	.90	52.61	2
12 270	Gould, Z. C., 16h 3909 .	8.8	36.75	+ 7.585	74 11 26.5	− 5.90	52.54	2
12 271	Apodis	9.5	16 51 40.66	+ 8.158	75 53 40.0	− 5.89	52.55	2
12 272	Gould, Z. C., 16h 3914 .	9.5	46.42	7.352	73 22 14.9	.88	52.58	1
12 273	Gould, Z. C., 16h 3880 .	9.2	49.65	6.097	67 5 34.8	.88	51.61	2
12 274	Lacaille, 7018	7.8	51 57.60	9.190	78 11 43.4	.87	51.61	2
12 275	Apodis	8.8	52 5.59	+ 7.468	73 46 48.7	− 5.86	52.52	2
12 276	Apodis	9.8	16 52 5.63	+ 7.302	73 10 50.4	− 5.86	52.59	2
12 277	Apodis	10.0	6.04	7.313	73 13 7.9	.86	52.58	1
12 278	Gould, Z. C., 16h 3944 .	9.0	11.48	7.381	73 28 9.8	.85	52.55	2
12 279	Apodis	9.5	14.44	7.262	73 1 29.4	.84	52.58	1
12 280	Gould, 23066	8.2	16.72	+ 6.887	71 28 0.8	− 5.84	52.50	2
12 281	Apodis	9.8	16 52 30.60	+ 6.872	71 23 25.6	− 5.82	52.50	2
12 282	Gould, 23062	8.0	31.35	6.082	66 58 18.4	.82	52.61	1
12 283	Gould, Z. C., 16h 3929 .	8.8	36.04	6.027	66 34 26.0	.81	52.61	2
12 284	Trianguli Australis . . .	9.4	44.32	6.366	68 45 49.9	.80	52.62	2
12 285	Apodis	9.0	48.53	+ 9.031	77 52 37.6	5.80	52.57	1
12 286	Gould, Z. C., 16h 3955 .	9.5	16 52 51.74	+ 6.143	67 22 17.7	− 5.79	50.62	1
12 287	Octantis	9.5	53.62	16.440	84 31 48.2	.79	51.49	1
12 288	Apodis	9.0	55.20	9.492	78 43 22.6	.79	50.61	1
12 289	Lacaille, 7061	8.0	52 56.26	7.167	72 38 27.8	.79	52.56	2
12 290	Gould, Z. C., 16h 4012 .	9.0	53 6.32	+ 7.443	73 40 14.1	− 5.77	52.52	1
12 291	Gould, 23082	8.5	16 53 7.51	+ 6.616	70 8 9.0	− 5.77	52.59	2
12 292	Apodis	9.5	9.81	7.030	72 4 19.8	.77	52.42	1
12 293	Lacaille, 7069	6.5	10.98	6.345	68 38 1.1	.77	52.62	2
12 294	Gould, Z. C., 16h 3997 .	8.8	12.28	6.631	70 12 43.1	.76	52.52	3
12 295	Apodis	9.0	14.14	+ 7.267	73 1 30.3	− 5.76	52.60	1
12 296	Trianguli Australis . . .	10.0	16 53 19.72	+ 6.549	69 38 54.6	− 5.75	52.60	1
12 297	Gould, Z. C., 16h 3998	8.3	24.31	6.163	67 29 26.8	.75	51.29	3
12 298	Gould, Z. C., 16h 4061 .	10.0	33.85	7.924	75 12 24.3	.73	50.56	1
12 299	Gould, Z. C., 16h 4010 .	9.0	38.52	6.071	66 51 43.6	.73	52.61	1
12 300	Trianguli Australis . . .	9.5	41.08	+ 6.247	68 1 22.8	− 5.72	52.59	1
12 301	Gould, Z. C., 16h 4055 .	9.0	16 53 52.16	+ 6.907	71 31 3.1	− 5.71	52.50	2
12 302	Apodis	9.5	52.75	7.465	73 42 6.0	.71	52.52	1
12 303	Trianguli Australis . . .	9.5	53 57.97	6.017	66 28 30.4	.70	52.61	1
12 304	Trianguli Australis . . .	8.8	54 2.62	6.034	66 35 46.2	.69	52.61	2
12 305	Trianguli Australis . . .	10.0	11.03	+ 6.586	69 57 1.5	− 5.68	52.60	1

Number.	Constellation, Name of Star, or Synonym.	Magnitude.	Right Ascension, 1850.0.	Annual Precession.	South Declination, 1850.0.	Annual Precession.	Mean year.	No. of obs.
			h. m. s.	s.	° ′ ″	″		
12 306	Apodis	9.5	16 54 13.74	+ 7.353	73 19 34.9	− 5.68	52.58	1
12 307	Apodis	9.5	16.38	7.889	75 4 29.5	.67	51.54	2
12 308	Trianguli Australis . . .	9.5	19.46	5.976	66 10 28.7	.67	52.60	1
12 309	Trianguli Australis . . .	9.0	19.58	5.941	65 54 56.1	.67	52.60	1
12 310	Gould, Z. C., 16h 4097 .	8.5	26.21	+ 7.036	72 4 7.8	− 5.66	52.41	1
12 311	Trianguli Australis . . .	9.5	16 54 26.70	+ 6.256	68 3 23.8	5.66	52.63	2
12 312	Apodis	10.0	30.84	5.999	71 54 32.2	.65	52.47	2
12 313	Gould, Z. C., 16h 4116 .	8.5	42.24	7.130	72 27 37.9	.64	52.52	1
12 314	Apodis	10.0	49.64	10.027	79 32 44.3	.63	51.64	1
12 315	Octantis	9.7	51.62	+29.179	87 11 8.0	− 5.62	51.04	3
12 316	Apodis	10.0	16 54 53.49	+ 8.917	77 35 56.1	− 5.62	52.36	1
12 317	Gould, Z. C., 16h 4106 .	9.2	54.68	6.357	68 36 57.8	.62	52.62	2
12 318	Aræ	9.5	54 59.35	6.057	66 43 48.2	.61	52.61	1
12 319	Trianguli Australis . . .	8.7	55 7.07	6.443	69 9 1.9	.60	52.53	3
12 320	Gould, Z. C., 16h 4183 .	9.0	7.40	+ 8.281	76 9 12.4	− 5.60	52.48	2
12 321	Aræ	9.0	16 55 13.19	+ 6.081	66 53 30.0	− 5.59	52.61	1
12 322	Gould, Z. C., 16h 4161 .	9.0	14.61	7.146	72 30 17.3	.59	52.52	1
12 323	Apodis	9.5	15.35	7.146	72 30 15.6	.59	52.52	1
12 324	Octantis	9.5	18.65	35.953	87 45 48.2	.59	51.04	2
12 325	Gould, Z. C., 16h 4164 .	10.0	19.01	+ 7.042	72 4 35.6	− 5.59	52.41	1
12 326	Lacaille, 7079	6.5	16 55 24.70	+ 6.702	70 30 53.5	− 5.58	52.40	1
12 327	Gould, Z. C., 16h 4193 .	10.0	24.89	7.962	75 17 5.7	.58	50.56	1
12 328	Apodis	10.0	31.29	7.007	71 55 18.6	.57	52.52	1
12 329	Apodis	10.0	31.74	8.175	75 52 18.0	.57	52.55	2
12 330	Gould, Z. C., 16h 4195 .	9.0	33.08	+ 7.664	74 22 15.6	− 5.57	52.54	2
12 331	Lacaille, 7002	6.8	16 55 33.41	+12.971	82 36 24.2	− 5.57	51.00	2
12 332	Brisbane, 5954	8.5	34.48	6.323	68 26 17.7	.56	52.65	1
12 333	Apodis	9.2	41.64	7.734	74 35 30.2	.55	52.54	2
12 334	Apodis	10.0	41.72	11.792	81 37 8.7	.55	50.55	2
12 335	Gould, Z. C., 16h 4258 .	8.5	43.01	+ 9.494	78 41 15.2	5.55	50.61	1
12 336	Apodis	10.0	16 55 46.43	+10.299	79 55 10.0	− 5.55	50.61	1
12 337	Apodis	9.5	55.47	8.043	75 30 22.5	.54	52.62	1
12 338	Gould, Z. C., 16h 4191 .	9.2	57.75	6.581	69 53 1.0	.54	52.58	2
12 339	Apodis	10.0	55 57.91	8.228	76 0 11.6	.54	52.48	1
12 340	Aræ	9.0	56 9.06	+ 5.934	65 49 4.8	− 5.52	52.60	1
12 341	Apodis	10.0	16 56 17.96	+ 8.208	75 56 42.9	− 5.50	52.48	1
12 342	Gould, Z. C., 16h 4248 .	8.8	35.30	7.093	72 15 43.0	.48	52.47	2
12 343	Gould, 23158	8.2	39.50	6.682	70 23 23.4	.47	52.47	2
12 344	Gould, Z. C., 16h 4235 .	10.0	41.10	6.364	68 39 33.2	.47	52.62	2
12 345	Apodis	8.2	48.23	+ 9.062	77 52 39.7	− 5.46	52.39	2

Number.	Constellation, Name of Star, or Synonym.	Magnitude.	Right Ascension, 1850.0.	Annual Precession.	South Declination, 1850.0.	Annual Precession.	Mean year.	No. of obs.
			h. m. s.	s.	° ′ ″	″		
12 346	Gould, Z. C., 16h 4250	9.8	16 56 51.61	+ 6.511	69 29 15.2	− 5.46	52.58	2
12 347	Aræ	10.0	56 53.57	5.948	65 54 3.8	.45	52.60	1
12 348	Lacaille, 7062	8.0	57 8.93	9.651	78 52 54.6	.43	50.61	1
12 349	Lacaille, 7081	6.2	14.96	7.663	74 20 20.5	.42	52.54	2
12 350	Lacaille, 7103	8.0	16.12	+ 6.020	66 24 35.7	− 5.42	52.61	1
12 351	Octantis	10.0	16 57 16.64	+15.290	83 59 7.7	− 5.42	50.55	1
12 352	Trianguli Australis	10.0	24.48	6.362	68 37 40.9	.41	52.65	1
12 353	Trianguli Australis	9.0	25.03	6.302	68 16 8.7	.41	52.63	3
12 354	Gould, Z. C., 17h 46	9.2	29.88	7.409	73 27 50.0	.40	52.55	2
12 355	Gould, 23168	8.5	36.51	+ 6.021	66 24 35.7	− 5.39	52.61	1
12 356	Apodis	10.0	16 57 37.00	+ 8.301	76 9 50.1	− 5.39	52.48	1
12 357	Octantis	10.0	46.18	16.221	84 24 6.4	.38	51.49	1
12 358	Gould, Z. C., 17h 107	9.5	56.10	8.217	75 56 31.0	.36	52.55	2
12 359	Aræ	9.5	57.07	6.039	66 31 35.6	.36	52.61	1
12 360	Gould, 23179	9.2	57.62	+ 6.106	66 59 23.5	− 5.36	52.61	2
12 361	Lacaille, 7107	6.0	16 57 59.62	+ 6.107	66 59 50.2	− 5.36	52.61	2
12 362	Apodis	9.0	58 3.96	7.255	72 52 52.2	.35	52.60	1
12 363	Gould, Z. C., 17h 60	8.5	14.52	6.090	66 52 42.3	.34	52.61	2
12 364	Lacaille, 7094	8.2	18.92	7.127	72 21 55.2	.33	52.46	2
12 365	Gould, Z. C., 17h 87	8.8	20.02	+ 6.829	71 3 57.8	− 5.33	52.50	2
12 366	Octantis	9.7	16 58 22.59	+20.164	85 41 13.8	− 5.33	51.04	3
12 367	Apodis	9.0	24.10	7.589	74 4 17.0	.33	52.52	1
12 368	Gould, Z. C., 17h 76	8.8	25.38	6.135	67 10 27.0	.32	50.64	2
12 369	Gould, Z. C., 17h 93	9.0	31.77	6.513	69 27 38.0	.32	52.58	2
12 370	Apodis	10.0	36.70	+10.764	80 29 3.4	− 5.31	51.63	1
12 371	Apodis	9.0	16 58 41.36	+ 6.944	71 34 46.6	− 5.30	52.50	2
12 372	Trianguli Australis	9.1	49.93	6.234	67 48 44.3	.29	52.62	4
12 373	Gould, Z. C., 17h 108	9.0	58 53.05	6.208	67 38 40.9	.29	52.60	2
12 374	Apodis	9.5	59 7.76	6.849	71 8 19.8	.26	52.50	2
12 375	Apodis	10.0	8.95	+10.665	80 21 25.7	− 5.26	51.63	1
12 376	Aræ	9.5	16 59 12.43	+ 6.032	66 26 37.9	− 5.26	52.61	1
12 377	Gould, Z. C., 17h 219	9.0	18.40	9.440	78 32 41.3	.25	50.63	2
12 378	Apodis	8.5	24.28	6.991	71 46 25.3	.24	52.41	1
12 379	Aræ	10.0	24.82	6.110	66 59 8.4	.24	52.61	1
12 380	Aræ	9.5	36.11	+ 6.219	67 2 33.2	− 5.22	51.61	2
12 381	Gould, Z. C., 17h 193	9.0	16 59 42.22	+ 7.492	73 43 11.1	− 5.22	52.52	1
12 382	Apodis	10.0	43.69	7.850	74 52 49.0	.21	52.53	1
12 383	Trianguli Australis	10.0	47.53	6.454	69 6 20.6	.21	52.57	1
12 384	Apodis	9.5	50.58	7.793	74 42 14.3	.20	52.56	1
12 385	Gould, 23256	8.0	55.22	+ 8.671	77 0 55.2	− 5.20	51.10	2

No. 12 358. Gould's place is 7s greater in right ascension and 22″ less in declination.

Number.	Constellation, Name of Star, or Synonym.	Magnitude.	Right Ascension, 1850.0.	Annual Precession.	South Declination, 1850.0.	Annual Precession.	Mean year.	No. of obs.
			h. m. s.	s.	° ′ ″	″		
12 386	Apodis	9.2	16 59 55.98	+ 7.280	72 58 29.4	— 5.20	52.59	2
12 387	Apodis	10.0	17 0 12.22	7.361	73 14 19.2	.17	52.58	1
12 388	Aræ	9.5	12.56	6.022	66 21 1.2	.17	52.60	2
12 389	Apodis	9.0	13.28	6.291	68 7 59.5	.17	52.63	2
12 390	Lacaille, 7104	7.5	19.37	+ 7.707	74 25 26.5	— 5.16	52.56	1
12 391	Gould, Z. C., 17ʰ 234 . .	9.0	17 0 20.66	+ 7.389	73 20 15.0	— 5.16	52.58	1
12 392	Gould, Z. C., 17ʰ 225 . .	9.5	20.87	7.160	72 29 54.6	.16	52.52	1
12 393	Apodis	10.0	29.37	10.303	79 52 4.2	.15	50.61	1
12 394	Apodis	10.0	38.17	6.821	70 58 48.1	.14	52.52	1
12 395	Aræ	9.5	44.24	+ 6.031	66 24 9.0	— 5.13	52.61	1
12 396	Apodis	9.5	17 0 46.76	+ 6.590	69 49 28.8	— 5.13	52.58	2
12 397	Gould, 23265	8.5	0 47.68	7.626	74 9 12.3	.12	52.52	1
12 398	Brisbane, 5974	8.5	1 1.40	8.593	76 49 7.7	.10	51.64	1
12 399	Apodis	9.5	1.44	6.306	68 14 33.0	.10	52.62	2
12 400	Apodis	9.0	11.81	+11.678	81 27 11.9	— 5.09	50.57	3
12 401	Apodis	9.2	17 1 20.84	+ 6.662	70 11 1.4	— 5.08	52.48	2
12 402	Apodis	9.5	25.72	7.562	73 55 38.7	.07	52.52	2
12 403	Apodis	9.5	30.93	8.210	75 51 56.2	.06	52.48	1
12 404	Gould, Z. C., 17ʰ 338 . .	9.0	33.50	8.383	76 18 33.0	.06	52.48	1
12 405	Gould, Z. C., 17ʰ 320 . .	9.0	1 36.64	+ 7.628	74 10 46.9	— 5.05	52.52	1
12 406	Brisbane, 5979	8.5	17 2 2.12	+ 8.334	76 10 41.0	— 5.02	52.48	2
12 407	Lacaille, 7142	7.2	7.21	6.087	66 45 48.7	.01	52.61	3
12 408	Apodis	9.5	8.31	7.252	72 47 28.1	.01	52.60	1
12 409	Gould, Z. C., 17ʰ 312 . .	8.5	15.07	6.003	66 9 57.2	5.00	52.60	2
12 410	Gould, Z. C., 17ʰ 336 . .	8.5	25.86	+ 6.218	67 37 46.6	— 4.99	52.61	2
12 411	Apodis	10.0	17 2 26.95	+ 6.452	69 2 10.2	— 4.98	52.57	1
12 412	Lacaille, 7127	7.2	43.26	7.964	75 10 1.1	.96	51.54	2
12 413	Apodis	9.5	46.65	11.006	80 43 32.4	.96	51.28	3
12 414	Octantis	8.5	49.22	20.371	85 42 59.8	.95	51.04	4
12 415	Apodis	9.8	49.96	+ 6.611	69 53 37.1	— 4.95	52.58	2
12 416	Apodis	10.0	17 2 57.48	+ 8.329	76 9 12.5	— 4.94	52.48	1
12 417	Gould, Z. C., 17ʰ 382 . .	9.8	3 2.24	6.492	69 14 47.3	.93	52.61	2
12 418	Apodis	9.0	11.47	10.193	79 40 59.4	.92	50.61	1
12 419	Apodis	10.0	32.36	9.861	79 11 14.6	.89	51.64	1
12 420	Lacaille, 7088	7.5	32.38	+10.993	80 42 9.8	— 4.89	51.28	3
12 421	Gould, Z. C., 17ʰ 415, 418	9.5	17 3 34.38	+ 6.616	69 54 14.8	— 4.89	52.58	2
12 422	Apodis	9.5	35.34	8.279	76 0 52.5	.89	52.48	1
12 423	Lacaille, 7146	7.0	35.51	6.608	69 51 50.1	.89	52.58	2
12 424	Aræ	9.5	37.63	6.058	66 31 41.6	.88	52.61	1
12 425	Apodis	9.5	37.96	+ 9.675	78 53 21.9	— 4.88	50.62	1

No. 12 401. Possibly 70° 11′ 57.4″. There is an error of two revolutions in one of these observations.
No. 12 421. Gould gives two observations differing 1.3ˢ in right ascension, 9.3″ in declension, but does not note it as double.

Number.	Constellation, Name of Star, or Synonym.	Magnitude.	Right Ascension, 1850.0.	Annual Precession.	South Declination, 1850.0.	Annual Precession.	Mean year.	No. of obs.
			h. m. s.	s.	° ′ ″	″		
12426	Gould, Z. C., 17h 430 . .	9.8	17 3 49.55	+ 6.470	69 6 35.4	− 4.87	52.61	2
12427	Apodis	9.5	4 2.27	6.400	68 42 17.4	.85	52.65	1
12428	Apodis	9.5	13.66	6.873	71 8 55.6	.83	52.50	2
12429	Apodis	9.8	24.34	7.332	73 3 26.8	.82	52.59	2
12430	Gould, Z. C., 17h 508 . .	9.5	25.11	+ 8.205	75 48 32.7	− 4.82	52.48	1
12431	Gould, Z. C., 17h 515 . .	9.0	17 4 27.10	+ 8.215	75 50 5.2	− 4.81	52.55	2
12432	Gould, Z. C., 17h 471 . .	9.5	30.12	8.622	69 55 4.2	.81	52.58	2
12433	Apodis	8.0	30.90	6 440	68 55 20.9	.81	52.59	1
12434	Apodis	10.0	38.39	8.087	75 29 11.9	.80	52.62	1
12435	Octantis	9.4	40.19	+21.681	86 0 32.9	− 4.80	51.04	4
12436	Apodis	10.0	17 4 46.70	+11.640	81 22 59.2	− 4.79	50.59	2
12437	Apodis	8.7	47.51	12.661	82 17 17.7	.78	50.55	3
12438	Gould, Z. C., 17h 533 . .	8.5	4 51.84	7.877	74 52 41.9	.78	52.53	1
12439	Aræ	10.0	5 0.03	6.071	66 35 2.6	.77	52.62	1
12440	Apodis	9.0	4.37	+ 6.370	68 30 16.4	− 4.76	52.65	1
12441	Apodis	9.0	17 5 11.02	+11.399	81 8 2.0	− 4.75	50.58	2
12442	Apodis	9.5	13.00	6.877	71 9 7.7	.75	52.50	2
12443	Apodis	10.0	18.23	7.146	72 18 45.4	.74	52.52	1
12444	ι Apodis	6.5	24.11	6.633	69 57 19.2	.73	52.58	2
12445	Apodis	10.0	39.47	+ 8.471	76 27 51.4	− 4.71	52.48	1
12446	Gould, 23366	9.5	17 5 40.18	+ 6.630	69 55 56.1	− 4.71	52.58	2
12447	Lacaille, 7157	8.0	5 41.48	6.696	70 16 24.9	.71	52.47	2
12448	Apodis	9.0	6 4.46	8.872	77 21 52.5	.68	52.46	2
12449	Lacaille, 7105	7.8	14.74	11.234	80 55 35.5	.66	50.59	2
12450	Gould, Z. C., 17h 639 . .	9.0	15.78	+ 7.861	74 48 22.8	− 4.66	52.53	1
12451	Apodis	10.0	17 6 19.96	+10.415	79 57 23.5	− 4.65	50.60	2
12452	ζ Apodis	4.0	20.64	6.228	67 36 17.6	.65	52.63	1
12453	Gould, Z. C., 17h 627 . .	9.0	22.85	7.376	73 11 4.2	.65	52.59	2
12454	Gould, Z. C., 17h 600 . .	9.0	22.90	6.542	69 27 18.1	.65	52.64	3
12455	Aræ	9.2	26.48	+ 6.021	66 12 5.2	− 4.64	52.60	2
12456	Gould, Z. C., 17h 681 . .	9.5	17 6 32.77	+ 8.209	75 47 27.4	− 4.63	52.55	2
12457	Apodis	9.5	44.19	7.712	74 20 12.3	.62	52.54	2
12458	Apodis	10.0	49.16	7.462	73 29 27.8	.61	52.52	1
12459	Apodis	10.0	50.89	8.820	77 14 38.1	.61	52.56	1
12460	Gould, Z. C., 17h 680 . .	9.5	17 6 54.72	+ 7.305	72 54 44.4	− 4.60	52.60	1
12461	Apodis	9.2	17 6 55.52	+10.762	80 23 42.5	− 4.60	51.13	4
12462	Gould, Z. C., 17h 682 . .	9.5	6 58.02	7.372	73 9 46.2	.60	52.59	2
12463	Gould, Z. C., 17h 668 . .	9.0	7 8.15	6.509	69 15 26.4	.59	52.61	2
12464	Apodis	10.0	16.74	10.776	80 24 30.1	.57	51.12	2
12465	Aræ	9.5	20.70	+ 6.027	66 13 33.6	− 4.57	52.61	1

No. 12430. Gould's place is 8.2s less in right ascension, and 16.2″ greater in declination.

Number.	Constellation, Name of Star, or Synonym.	Magnitude.	Right Ascension, 1850.0.	Annual Precession.	South Declination, 1850.0.	Annual Precession.	Mean year.	No. of obs.
			h. m. s.	s.	° ′ ″	″		
12 466	Brisbane, 6014	8.0	17 7 22.11	+ 6.011	66 6 18.2	— 4.57	52.60	1
12 467	Gould, Z. C., 17ʰ 711 . .	9.5	32.29	6.744	70 28 30.1	.55	52.40	1
12 468	Gould, Z. C., 17ʰ 700 . .	10.0	33.92	6.483	69 6 15.2	.55	52.59	1
12 469	Apodis	9.5	34.28	8.674	76 54 49.9	.55	51.64	1
12 470	Gould, Z. C., 17ʰ 708 . .	9.0	46.00	+ 6.258	67 46 3.9	— 4.53	52.65	2
12 471	Gould, Z. C., 17ʰ 744 . .	9.0	17 7 46.18	+ 7.331	72 59 44.1	— 4.53	52.59	2
12 472	Gould, Z. C., 17ʰ 734 . .	8.2	8 8.46	6.023	66 10 42.6	.50	52.60	2
12 473	Apodis	10.0	14.57	6.147	67 1 57.2	.49	52.62	1
12 474	Apodis	9.5	26.52	7.484	73 32 41.5	.47	52.52	1
12 475	Apodis	10.0	28.91	+ 6.872	71 4 11.2	— 4.47	52.48	1
12 476	Apodis	9.5	17 8 35.23	+ 8.184	75 41 40.5	— 4.46	52.62	1
12 477	Apodis	10.0	40.28	7.064	71 55 0.8	.45	52.52	1
12 478	Gould, Z. C., 17ʰ 765 . .	9.0	40.46	6.016	66 7 4.8	.45	52.60	1
12 479	Octantis	9.7	43.27	17.364	84 47 16.4	.45	50.91	3
12 480	Aræ	10.0	8 58.98	+ 6.155	67 4 18.4	— 4.43	52.62	1
12 481	Apodis	9.0	17 9 3.69	+ 6.963	71 28 14.0	— 4.42	52.50	2
12 482	Apodis	9.0	17.10	7.684	74 12 27.0	.40	52.54	2
12 483	Apodis	9.0	17.30	6.917	71 15 40.8	.40	52.48	1
12 484	Apodis	9.0	21.82	6.935	71 20 19.6	.40	52.48	1
12 485	Apodis	8.5	24.63	+ 7.034	71 46 28.3	— 4.39	52.52	1
12 486	Apodis	10.0	17 9 27.54	+ 7.168	72 19 53.0	— 4.39	52.52	1
12 487	Aræ	10.0	30.62	6.209	67 25 0.6	.38	52.63	1
12 488	Aræ	9.5	39.00	6.163	67 6 40.8	.37	52.62	1
12 489	Apodis	9.0	39.35	7.458	73 26 53.7	.37	52.52	1
12 490	Apodis	9.5	50.44	+ 6.829	70 50 29.9	— 4.35	52.52	1
12 491	Apodis	9.0	17 9 53.33	+ 6.319	68 5 58.6	— 4.35	52.67	1
12 492	Gould, Z. C., 17ʰ 851 . .	9.5	9 54.92	6.177	67 12 5.6	.35	50.67	1
12 493	Gould, Z. C., 17ʰ 885 . .	9.0	10 3.25	6.893	71 8 15.4	.34	52.50	2
12 494	Apodis	10.0	4.07	7.077	71 57 47.4	.34	52.41	1
12 495	Apodis	10.0	7.26	+ 9.251	78 4 12.0	— 4.33	52.43	1
12 496	Lacaille, 7197	7.5	17 10 17.88	+ 6.037	66 14 6.0	— 4.32	52.60	2
12 497	Gould, Z. C., 17ʰ 932 . .	8.2	26.79	7.867	74 45 49.4	.30	52.54	2
12 498	Gould, Z. C., 17ʰ 893 . .	9.5	29.18	6.163	67 5 58.7	.30	51.65	2
12 499	Gould, Z. C., 17ʰ 909 . .	10.0	29.20	6.752	70 27 37.1	.30	52.40	1
12 500	Apodis	10.0	31.15	+ 7.637	74 2 4.3	— 4.30	52.52	1
12 501	Apodis	9.5	17 10 38.34	+ 6.994	71 34 55.3	— 4.29	52.52	1
12 502	Apodis	10.0	38.61	7.868	74 45 54.7	.29	52.52	1
12 503	Apodis	9.5	39.19	6.790	70 38 32.7	.29	52.52	1
12 504	Gould, Z. C., 17ʰ 904 . .	10.0	39.42	6.018	66 5 33.0	.28	52.60	1
12 505	Apodis	9.0	39.83	+ 6.978	71 30 40.1	— 4.28	52.48	1

No. 12 470. Gould's declination is 10.1″ greater.

Number.	Constellation, Name of Star, or Synonym.	Magnitude.	Right Ascension, 1850.0.	Annual Precession.	South Declination, 1850.0.	Annual Precession.	Mean year.	No. of obs.
			h. m. s.	s.	° ′ ″	″		
12 506	Apodis	9.5	17 10 43.36	+ 7.202	72 27 0.9	− 4.28	52.52	1
12 507	Gould, Z. C., 17^h 1001 .	8.2	43.42	9.217	77 59 54.5	.28	52.40	2
12 508	Octantis	9.1	45.89	17.488	84 49 20.3	.28	50.94	4
12 509	Apodis	9.2	48.12	6.874	71 2 20.6	.27	52.50	2
12 510	Aræ	9.5	49.72	+ 6.155	67 2 8.7	− 4.27	52.62	1
12 511	Apodis	10.0	17 10 50.12	+ 8.888	77 20 26.5	− 4.27	52.56	1
12 512	Gould, Z. C., 17^h 915 . .	9.0	10 50.99	6.179	67 11 39.5	.27	50.67	1
12 513	Gould, Z. C., 17^h 920 . .	10.0	11 1.45	6.004	65 59 1.1	.25	52.60	1
12 514	Gould, Z. C., 17^h 925 . .	10.0	6.84	6.014	66 3 13.4	.25	52.60	1
12 515	Gould, Z. C., 17^h 962 . .	8.8	9.20	+ 7.347	73 0 10.2	− 4.24	52.59	2
12 516	Apodis	10.0	17 11 9.32	+ 7.477	73 28 31.3	− 4.24	52.52	1
12 517	Apodis	10.0	11.40	9.111	77 47 25.7	.24	52.43	[illegible]
12 518	Apodis	9.5	25.73	6.283	67 50 59.1	.22	52.67	1
12 519	Gould, Z. C., 17^h 965 . .	8.5	29.61	6.648	69 55 9.6	.21	52.58	2
12 520	Apodis	9.8	32.86	+ 6.574	69 31 35.8	4.21	52.62	2
12 521	Apodis	9.5	17 11 38.20	+ 9.075	77 42 51.2	− 4.20	52.36	1
12 522	Apodis	9.5	43.26	7.002	71 35 57.2	.19	52.09	2
12 523	Lacaille, 7198	7.2	50.79	6.762	70 29 17.3	.18	52.48	2
12 524	Apodis	9.0	54.68	6.552	69 24 12.9	.18	52.62	2
12 525	Aræ	9.2	55.82	+ 6.111	66 42 56.0	− 4.18	52.62	2
12 526	Apodis	9.0	17 11 59.23	+12.627	82 12 17.4	− 4.17	50.57	2
12 527	Apodis	10.0	12 9.47	8.618	76 43 45.4	.16	51.64	1
12 528	Apodis	9.0	11.28	9.070	77 41 48.4	.15	52.36	1
12 529	Lacaille, 7090	8.5	16.58	15.535	84 0 28.5	.15	51.49	1
12 530	Apodis	10.0	36.27	+ 6.835	70 49 30.4	− 4.12	52.52	1
12 531	Apodis	10.0	17 12 39.43	+ 6.867	70 58 31.3	− 4.11	52.52	1
12 532	Octantis	10.2	40.51	19.613	85 28 35.2	.11	51.05	2
12 533	Apodis	10.0	41.53	6.865	70 57 55.3	.11	52.52	1
12 534	Gould, Z. C., 17^h 1072 .	8.8	42.52	7.359	73 1 29.2	.11	52.59	2
12 535	Apodis	10.0	47.06	+ 7.150	72 12 32.2	− 4.10	52.46	2
12 536	Apodis	9.5	17 12 50.19	+ 6.446	68 47 51.2	− 4.10	52.59	1
12 537	Brisbane, 6047	8.5	51.42	6.031	66 8 14.3	.10	52.60	1
12 538	Apodis	10.0	12 51.65	7.129	72 7 12.0	.10	52.41	1
12 539	Apodis	10.0	13 0.01	6.869	70 58 45.0	.08	52.52	1
12 540	Apodis	9.0	5.30	+ 6.383	68 25 36.3	− 4.08	52.05	1
12 541	Apodis	10.0	17 13 13.42	+ 7.833	74 37 24.4	− 4.06	52.53	1
12 542	Gould, Z. C., 17^h 1123 .	9.2	21.58	7.473	73 25 54.6	.05	52.55	2
12 543	Gould, Z. C., 17^h 1120 .	8.5	24.66	7.304	72 48 25.5	.05	52.60	1
12 544	Apodis	10.0	27.58	7.649	74 1 56.2	.04	52.52	1
12 545	Gould, Z. C., 17^h 1107 .	9.0	34.13	+ 6.623	69 45 17.8	4.04	52.60	1

No. 12 509. Mean of two observations differing 7.1″.

No. 12 522. Possibly 71° 31′ 43.6″. One wire interval wrong.

Number.	Constellation, Name of Star, or Synonym.	Magnitude.	Right Ascension, 1850.0.	Annual Precession.	South Declination, 1850.0.	Annual Precession.	Mean year.	No. of obs.
			h. m. s.	s.	° ′ ″	″		
12 546	Gould, Z. C., 17h 1096	9.0	17 13 40.71	+ 6.015	66 0 31.5	4.03	52.60	1
12 547	Gould, Z. C., 17h 1115	9.5	42.48	6.554	69 24 4.8	.02	52.62	2
12 548	Apodis	9.5	49.68	8.967	77 28 13.8	.01	52.46	2
12 549	Gould, 23557	8.5	51.10	6.179	67 8 29.8	.01	51.65	2
12 550	Gould, Z. C., 17h 1144	9.0	13 51.45	+ 7.312	72 49 54.1	- 4.01	52.60	1
12 551	Apodis	9.5	17 14 2.24	+ 9.600	78 39 1.9	3.99	50.62	1
12 552	Apodis	10.0	2.49	9.595	78 38 32.0	.99	50.64	1
12 553	Apodis	9.0	10.22	6.893	71 4 12.0	.98	52.50	2
12 554	Apodis	9.5	17.66	7.241	72 32 58.6	.97	52.52	1
12 555	Gould, Z. C., 17h 1150	9.5	20.19	+ 6.308	68 0 2.9	− 3.97	52.67	1
12 556	Gould, Z. C., 17h 1223	9.2	17 14 21.53	+ 8.638	76 45 1.1	3.97	51.92	3
12 557	Apodis	9.5	22.28	10.907	80 32 45.3	.97	51.10	2
12 558	Aræ	9.0	27.30	5.991	65 49 3.4	.96	52.60	1
12 559	Lacaille, 7078	7.1	29.54	18.461	85 7 43.4	.96	50.96	5
12 560	Gould, Z. C., 17h 1163	9.0	39.14	+ 6.179	67 7 23.5	− 3.94	52.62	1
12 561	Apodis	9.2	17 14 49.66	+10.490	79 58 35.6	3.93	50.64	2
12 562	Octantis	9.0	49.69	13.269	82 40 22.2	.93	51.00	2
12 563	Aræ	9.5	50.63	6.225	67 25 17.2	.93	51.65	2
12 564	Apodis	10.0	14 56.47	7.676	74 6 11.0	.92	52.52	1
12 565	Apodis	9.0	15 1.94	+ 9.286	78 5 0.6	-- 3.91	52.43	1
12 566	Gould, 23590	7.0	17 15 2.22	+ 6.122	66 44 3.2	− 3.91	52.62	2
12 567	Apodis	9.8	3.88	6.774	70 29 44.4	.91	52.48	2
12 568	Apodis	9.3	10.13	6.726	70 15 12.8	.90	52.50	3
12 569	Apodis	9.5	11.48	11.082	80 42 2.1	.90	51.10	2
12 570	Gould, Z. C., 17h 1255	9.0	28.94	+ 7.121	72 2 54.0	− 3.87	52.04	2
12 571	Apodis	9.5	17 15 38.80	+ 7.451	73 19 16.0	− 3.86	52.58	1
12 572	Gould, Z. C., 17h 1295	9.0	45.75	8.232	75 43 53.4	.85	52.62	1
12 573	Gould, Z. C., 17h 1246	10.0	48.97	6.024	66 2 9.0	.84	52.60	1
12 574	Aræ	9.5	49.88	6.124	66 44 11.8	.84	52.61	1
12 575	Apodis	9.5	15 59.86	+12.538	82 6 20.3	− 3.83	50.64	1
12 576	Apodis	10.0	17 16 1.72	+ 7.065	71 48 15.8	− 3.82	52.52	1
12 577	Apodis	10.0	2.03	7.941	74 54 44.3	.82	52.53	1
12 578	Aræ	9.5	9.82	6.057	66 15 46.9	.81	52.61	1
12 579	Gould, Z. C., 17h 1326	9.0	14.62	8.079	75 18 34.6	.81	50.56	1
12 580	Gould, 23647	10.0	17.56	+ 9.553	78 32 55.3	− 3.80	50.64	1
12 581	Apodis	10.0	17 16 20.87	+ 8.049	75 13 26.2	− 3.80	50.56	1
12 582	Gould, Z. C., 17h 1314	8.8	44.97	6.313	67 56 22.4	.76	52.65	2
12 583	Apodis	9.0	49.34	6.394	68 25 42.8	.76	52.65	1
12 584	Apodis	10.0	50.71	7.070	71 48 54.8	.75	52.52	1
12 585	Gould, Z. C., 17h 1319	9.5	16 53.92	+ 6.209	67 17 2.2	3.75	50.67	1

Number.	Constellation, Name of Star, or Synonym.	Magnitude.	Right Ascension, 1850.0.	Annual Precession.	South Declination, 1850.0.	Annual Precession.	Mean year.	No. of obs.
			h. m. s.	s.	° ′ ″	″		
12 586	Apodis	9.5	17 17 15.62	+ 7.332	72 51 28.1	— 3.72	52.60	1
12 587	Apodis	9.5	16.30	9.698	78 47 1.9	.72	50.62	1
12 588	Apodis	8.2	18.41	6.654	69 51 21.2	.71	52.58	2
12 589	Apodis	9.2	20.77	7.032	71 38 35.5	.71	52.09	2
12 590	Apodis	9.2	26.57	+ 7.181	72 13 6.9	— 3.70	52.20	3
12 591	Apodis	9.0	17 17 31.46	+10.078	79 22 14.1	— 3.70	51.13	2
12 592	Herschel, 4947[1]	9.7	32.82	12.189	81 47 50.0	.69	50.58	3
12 593	Herschel, 4947[2]	9.7	37.59	12.189	81 47 47.3	.69	50.58	3
12 594	Aræ	9.5	40.31	6.113	66 37 40.6	.68	52.61	1
12 595	Aræ	9.2	44.05	+ 6.776	70 27 42.6	— 3.68	52.48	2
12 596	Gould, Z. C., 17h 1378 .	9.0	17 17 45.98	+ 6.433	68 38 22.4	— 3.68	52.52	2
12 597	Gould, Z. C., 17h 1428 .	9.0	51.94	7.546	73 37 40.0	.67	52.48	2
12 598	Apodis	9.0	17 55.57	6.547	69 16 46.9	.66	52.64	1
12 599	Apodis	9.3	18 4.96	6.735	70 15 23.5	.65	52.51	3
12 600	Apodis	8.5	5.49	+ 7.187	72 16 52.7	— 3.65	52.52	1
12 601	Apodis	9.0	17 18 5.96	+ 7.164	72 11 23.2	— 3.65	52.04	2
12 602	Aræ	9.5	9.50	6.115	66 37 46.4	.64	52.61	1
12 603	Gould, Z. C., 17h 1453 .	9.0	13.40	7.692	74 6 47.0	.64	52.52	1
12 604	Apodis	9.5	14.84	6.953	71 17 9.2	.63	52.48	1
12 605	Apodis	9.2	16.17	+ 7.028	71 36 45.0	— 3.63	52.09	2
12 606	Aræ	9.0	17 18 22.36	+ 6.090	66 27 21.8	— 3.62	52.61	1
12 607	Lacaille, 7240	7.5	29.19	7.900	74 47 47.2	.61	52.54	2
12 608	Octantis	9.7	31.81	48.364	88 20 11.8	.61	51.04	3
12 609	Gould, Z. C., 17h 1438 .	9.0	33.12	6.162	66 56 37.8	.61	52.62	1
12 610	Apodis	9.5	38.22	+ 6.949	71 15 32.1	— 3.60	52.48	1
12 611	Lacaille, 7361	9.2	17 18 45.44	+ 7.523	73 32 7.2	— 3.59	52.52	3
12 612	Octantis	10.0	46.05	15.557	83 59 10.0	.59	51.49	1
12 613	Apodis	10.0	46.15	7.766	74 20 36.2	.59	52.54	2
12 614	Apodis	9.5	48.35	9.766	78 52 54.6	.59	50.62	1
12 615	Apodis	9.2	18 59.22	+ 8.152	75 28 51.9	— 3.57	50.59	2
12 616	Gould, Z. C., 17h 1473 .	8.0	17 19 1.15	+ 6.616	69 38 50.7	— 3.57	52.60	1
12 617	Brisbane, 6087	8.5	4.55	6.012	65 53 12.1	.56	52.60	1
12 618	Aræ	9.5	5.34	6.132	66 43 50.1	.56	52.62	2
12 619	Apodis	10.0	10.68	7.447	73 15 32.3	.55	52.58	1
12 620	Apodis	10.0	12.27	+ 6.603	69 33 43.7	— 3.55	52.60	1
12 621	Lacaille, 7229	7.8	17 19 15.69	+ 9.592	78 35 19.4	— 3.55	50.63	2
12 622	Octantis	9.1	17.20	16.333	84 19 56.6	.54	51.17	4
12 623	Apodis	10.0	32.64	9.010	77 30 5.8	.52	52.46	2
12 624	Apodis	9.2	43.40	7.881	74 41 27.2	.51	52.54	2
12 625	Apodis	8.2	47.28	+ 6.365	68 12 35.2	— 3.50	52.66	2

Number.	Constellation, Name of Star, or Synonym.	Magnitude.	Right Ascension, 1850.0.	Annual Precession.	South Declination, 1850.0.	Annual Precession.	Mean year.	No. of obs.
			h. m. s.	s.	° ′ ″	″		
12 626	Apodis	9.8	17 19 49.88	+ 7.277	72 36 50.1	— 3.50	52.56	2
12 627	Gould, Z. C., 17ʰ 1552	8.0	52.46	7.362	72 56 24.6	.49	52.60	1
12 628	Octantis	9.0	19 57.96	19.309	85 21 49.6	.49	51.04	3
12 629	Apodis	10.0	20 6.63	10.113	79 24 24.1	.47	51.13	2
12 630	Apodis	9.5	6.89	+ 7.406	73 6 4.1	— 3.47	52.59	2
12 631	Apodis	9.0	17 20 9.26	+ 6.947	71 13 51.9	— 3.47	52.48	1
12 632	Gould, Z. C., 17ʰ 1540	9.5	9.62	6.140	66 46 22.2	.47	52.62	1
12 633	Octantis	8.7	10.98	19.684	85 28 1.7	.47	51.05	3
12 634	Brisbane, 6093	7.2	42.30	6.138	66 44 55.6	.42	52.62	2
12 635	Gould, Z. C., 17ʰ 1607	9.5	50.87	+ 6.739	70 14 19.8	— 3.41	52.56	2
12 636	Gould, Z. C., 17ʰ 1620	9.0	17 20 52.80	+ 7.133	72 1 23.2	— 3.41	52.04	2
12 637	Apodis	10.0	21 2.76	7.212	72 20 33.9	.39	52.46	2
12 638	Gould, Z. C., 17ʰ 1624	9.0	3.83	7.769	70 22 52.6	.39	52.56	2
12 639	Gould, Z. C., 17ʰ 1643	9.3	8.10	7.108	71 54 53.5	.38	51.91	3
12 640	Lacaille, 7290	7.5	11.19	+ 6.016	65 52 58.1	— 3.38	52.60	1
12 641	Gould, Z. C., 17ʰ 1670	8.8	17 21 13.19	+ 7.551	73 36 11.6	— 3.38	52.48	2
12 642	Apodis	10.0	16.93	7.904	74 44 31.3	.37	52.53	1
12 643	Aræ	9.0	28.80	6.251	67 28 34.6	.36	51.65	2
12 644	Apodis	10.0	29.97	6.468	68 47 15.3	.35	52.59	1
12 645	Aræ	9.5	36.79	+ 6.101	66 28 59.7	— 3.34	52.61	1
12 646	Gould, Z. C., 17ʰ 1680	8.8	17 21 40.86	+ 6.541	69 11 37.4	— 3.34	52.62	2
12 647	Aræ	9.0	45.64	6.040	66 2 57.1	.33	52.60	1
12 648	Gould, Z. C., 17ʰ 1681	8.5	52.25	6.345	68 3 36.4	.32	52.67	1
12 649	Lacaille, 7285	7.8	21 58.06	6.814	70 35 12.4	.31	52.54	2
12 650	Apodis	10.0	22 2.06	+ 7.099	71 52 1.4	— 3.31	51.66	1
12 651	Apodis	9.5	17 22 8.68	+ 9.982	79 11 42.9	— 3.30	51.64	1
12 652	Aræ	8.0	8.74	6.092	66 24 43.4	.30	52.61	1
12 653	Brisbane, 6083	7.8	8.98	14.065	83 9 30.4	.30	51.01	2
12 654	Gould, 23772	7.5	20.59	6.406	68 24 51.0	.28	52.65	1
12 655	Gould, Z. C., 17ʰ 1794	8.0	29.58	+ 8.905	77 15 25.4	— 3.27	52.56	1
12 656	Aræ	9.0	17 22 29.88	+ 6.061	66 11 10.1	— 3.27	52.61	1
12 657	Gould, Z. C., 17ʰ 1742	9.2	34.95	6.828	70 38 56.9	.26	52.54	2
12 658	Octantis	9.3	35.06	18.215	85 1 56.7	.26	50.91	3
12 659	Gould, Z. C., 17ʰ 1823	8.5	49.02	9.379	78 11 0.7	.24	51.52	2
12 660	Apodis	10.0	51.28	+ 6.942	71 10 25.3	— 3.24	52.48	1
12 661	Apodis	10.0	17 22 56.45	+ 7.693	74 3 49.5	— 3.23	52.52	1
12 662	Aræ	9.5	22 58.53	6.109	66 30 48.0	.23	52.61	1
12 663	Gould, Z. C., 17ʰ 1753	9.0	23 1.65	5.913	66 4 48.3	.22	52.60	1
12 664	Apodis	10.0	7.35	6.383	68 16 13.7	.21	52.65	1
12 665	Apodis	9.0	7.86	+ 9.039	77 31 44.6	— 3.21	52.46	2

Number.	Constellation, Name of Star, or Synonym.	Magnitude.	Right Ascension, 1850.0.	Annual Precession.	South Declination, 1850.0.	Annual Precession.	Mean year.	No. of obs.
			h. m. s.	s.	° ′ ″	″		
12 666	Apodis	9.2	17 23 20.10	+ 6.832	70 39 26.2	— 3.20	52.54	2
12 667	Gould, Z. C., 17h 1829 .	9.0	23.88	8.208	75 35 9.4	.19	52.11	4
12 668	Gould, Z. C., 17h 1783 .	9.5	31.91	6.121	66 35 18.1	.18	52.61	1
12 669	Lacaille, 7292	7.0	40.43	7.471	73 17 37.9	.17	52.58	1
12 670	Apodis	9.5	43.56	+ 7.680	74 0 42.4	— 3.16	52.05	2
12 671	Gould, Z. C., 17h 1864 .	9.2	17 23 45.19	+ 8.633	76 38 39.4	— 3.16	51.64	2
12 672	Aræ	10.0	46.49	6.116	66 33 8.3	.16	52.61	1
12 673	Apodis	9.0	48.10	6.742	70 12 38.9	.15	52.56	2
12 674	Gould, Z. C., 17h 1806 .	8.0	50.19	6.350	68 3 34.3	.15	52.67	1
12 675	Gould, Z. C., 17h 1824 .	9.0	23 53.01	+ 6.886	70 54 15.2	— 3.15	52.52	1
12 676	Apodis	9.2	17 24 5.96	+ 7.658	73 56 4.6	— 3.13	52.27	4
12 677	Apodis	9.0	12.24	6.520	69 2 33.0	.12	52.62	2
12 678	Apodis	9.0	24.00	6.622	69 35 31.5	.10	52.62	2
12 679	Apodis	9.5	42.13	6.816	70 33 39.1	.08	52.56	1
12 680	Gould, Z. C., 17h 1897 .	9.2	42.29	+ 7.536	73 30 43.9	— 3.08	52.52	3
12 681	Apodis	9.5	17 24 43.01	+ 8.792	76 59 37.2	— 3.08	52.10	2
12 682	Lacaille, 7316	6.5	56.92	6.303	67 45 16.6	.06	52.63	1
12 683	Apodis	10.0	24 57.70	10.096	79 20 39.5	.06	51.64	1
12 684	Apodis	9.0	25 8.39	8.736	76 53 41.2	.04	51.64	1
12 685	Gould, Z. C., 17h 1936 .	9.0	9.03	+ 7.967	74 53 28.9	— 3.04	52.53	1
12 686	Apodis	9.2	17 25 14.68	+12.478	82 0 4.2	— 3.03	50.60	2
12 687	Apodis	10.0	19.75	10.079	79 19 1.3	.02	51.64	1
12 688	Gould, Z. C., 17h 1962 .	9.0	20.02	8.620	76 35 56.8	.02	51.64	2
12 689	Apodis	9.5	21.34	9.039	77 30 32.0	.02	52.36	1
12 690	Apodis	9.0	23.92	+10.008	79 12 37.3	— 3.02	51.64	1
12 691	Gould, Z. C., 17h 1939 .	9.0	17 25 24.89	+ 7.387	72 58 8.2	— 3.02	52.59	2
12 692	Apodis	10.0	26.83	8.772	76 56 37.2	.01	51.64	1
12 693	Apodis	10.0	33.01	8.400	76 3 55.2	3.00	51.64	1
12 694	Apodis	10.0	44.26	7.415	73 4 4.0	2.99	52.59	2
12 695	Apodis	9.8	51.74	+ 7.404	73 3 54.1	— 2.98	52.59	2
12 696	Apodis	8.5	17 25 53.45	+ 6.503	68 55 26.5	— 2.97	52.59	1
12 697	Apodis	10.0	25 56.02	7.650	73 53 22.2	.97	52.52	1
12 698	Gould, Z. C., 17h 1967 .	9.0	26 6.03	6.997	71 22 48.0	.96	52.48	1
12 699	Apodis	10.0	14.81	8.312	75 50 7.0	.94	51.64	1
12 700	Lacaille, 7317	7.0	18.74	+ 7.176	72 8 9.9	— 2.94	51.67	1
12 701	Apodis	10.0	17 26 23.51	+ 7.110	71 51 49.4	— 2.93	51.66	1
12 702	Lacaille, 7001	5.5	30.14	35.172	87 38 25.1	.92	51.04	4
12 703	Apodis	9.5	30.62	7.930	74 46 4.1	.92	52.54	2
12 704	Apodis	9.5	35.60	6.487	68 49 31.7	.91	52.59	1
12 705	Apodis	9.2	38.38	+ 7.737	74 10 9.7	— 2.91	52.54	2

Number.	Constellation, Name of Star, or Synonym.	Magnitude.	Right Ascension, 1850.0.	Annual Precession.	South Declination, 1850.0.	Annual Precession.	Mean year.	No. of obs.
			h. m. s.	s.	° ′ ″	″		
12 706	Gould, Z. C., 17h 2019 .	7.8	17 26 39.25	+ 7.320	72 42 7.8	− 2.91	52.56	2
12 707	Aræ	10.0	48.49	6.228	67 15 21.1	.89	50.67	1
12 708	Apodis	10.0	49.11	8.329	75 52 24.9	.89	52.14	2
12 709	Gould, Z. C., 17h 2103 .	9.0	26 58.14	9.577	78 30 8.3	.88	50.63	1
12 710	Aræ	10.0	27 7.53	+ 6.193	67 1 14.2	− 2.87	52.62	1
12 711	Gould, Z. C., 17h 2036 .	8.5	17 27 11.67	+ 6.738	70 9 3.4	− 2.86	52.56	1
12 712	Gould, Z. C., 17h 2030 .	9.0	12.26	6.491	68 50 30.9	.86	52.59	1
12 713	Apodis	9.0	19.90	8.878	77 9 31.9	.85	52.10	2
12 714	Apodis	9.5	28.88	7.417	73 3 23.6	.84	52.60	1
12 715	Apodis	9.8	41.54	+ 9.049	77 30 43.8	− 2.82	52.46	2
12 716	Gould, Z. C., 17h 2074 .	9.0	17 27 42.20	+ 6.697	69 56 25.0	− 2.82	52.58	2
12 717	Apodis	10.0	43.75	6.388	68 14 29.0	.81	52.65	1
12 718	Apodis	10.0	46.77	6.286	67 36 54.6	.81	52.63	1
12 719	Apodis	9.0	47.60	6.557	69 11 47.9	.81	52.64	1
12 720	Gould, Z. C., 17h 2082 .	9.0	52.09	+ 6.511	68 56 40.2	− 2.80	52.59	1
12 721	Aræ	9.5	17 27 52.81	+ 6.234	67 16 58.0	− 2.80	50.67	1
12 722	Gould, Z. C., 17h 2109 .	8.0	27 57.88	7.333	72 44 18.2	.79	52.56	2
12 723	Apodis	10.0	28 2.81	9.706	78 42 45.9	.79	50.62	1
12 724	Gould, Z. C., 17h 2100 .	9.0	6.63	6.656	69 43 24.3	.78	52.60	1
12 725	Aræ	8.0	11.51	+ 6.239	67 18 44.7	− 2.77	50.67	1
12 726	Apodis	9.5	17 28 17.62	+ 7.175	72 6 49.5	− 2.77	51.67	1
12 727	Gould, Z. C., 17h 2098 .	8.8	18.36	6.150	66 43 16.0	.76	52.62	2
12 728	Gould, Z. C., 17h 2212 .	9.0	20.33	9.867	78 58 19.5	.76	50.62	1
12 729	Lacaille, 7275	7.5	23.21	11.817	81 23 17.8	.76	50.57	3
12 730	Gould, Z. C., 17h 2153 .	8.8	28.22	+ 7.658	73 53 29.7	− 2.75	52.33	5
12 731	Gould, Z. C., 17h 2175 .	9.5	17 28 28.98	+ 8.599	76 31 28.2	− 2.75	51.64	1
12 732	Gould, Z. C., 17h 2143 .	9.0	33.45	7.051	71 35 23.1	.74	52.07	2
12 733	Lacaille, 7319	8.0	35.10	8.434	76 7 28.3	.74	51.64	1
12 734	Apodis	9.8	40.66	7.781	74 17 22.3	.73	52.54	2
12 735	Octantis	9.0	43.06	+13.387	82 40 56.2	− 2.73	51.00	2
12 736	Apodis	10.0	17 28 48.50	+ 6.286	67 36 6.1	− 2.72	52.63	1
12 737	Gould, Z. C., 17h 2165 .	9.2	29 1.06	6.767	70 16 36.9	.70	52.56	2
12 738	Apodis	9.0	6.05	10.028	79 12 57.7	.70	51.64	1
12 739	Gould, Z. C., 17h 2190 .	10.0	26.43	6.771	70 17 24.3	.67	52.56	1
12 740	Octantis	9.0	27.38	+25.847	86 40 11.6	− 2.67	51.04	2
12 741	Aræ	9.0	17 29 27.49	+ 6.201	67 2 41.9	− 2.67	52.62	1
12 742	Gould, Z. C., 17h 2193 .	9.0	27 59	6.764	70 15 17.0	.67	52.56	2
12 743	Gould, Z. C., 17h 2260 .	9.0	31.80	8.643	76 37 9.9	.66	51.64	2
12 744	Apodis	8.0	33.56	6.448	68 34 1.0	.66	52.62	2
12 745	Apodis	10.0	35.46	+ 6.769	70 16 36.5	− 2.65	52.56	2

Number.	Constellation, Name of Star, or Synonym.	Magnitude.	Right Ascension, 1850.0.	Annual Precession.	South Declination, 1850.0.	Annual Precession.	Mean year.	No. of obs.
			h. m. s.	s.	° ′ ″	″		
12 746	Apodis	9.7	17 29 40.83	+ 8.343	75 53 12.7	— 2.65	52.30	3
12 747	Gould, Z. C., 17h 2214	9.3	46.05	6.517	68 57 25.1	.64	52.59	1
12 748	Apodis	9.0	50.82	6.928	71 1 54.0	.63	52.50	2
12 749	Lacaille, 7332	7.7	51.36	8.382	75 59 2.8	.63	52.30	3
12 750	Octantis	9.5	56.42	+16.154	84 12 50.0	— 2.62	51.07	2
12 751	Apodis	9.2	17 29 56.66	+ 9.219	77 49 49.0	— 2.62	52.46	2
12 752	Apodis	9.5	30 10.32	7.275	72 29 43.9	.60	52.52	1
12 753	Apodis	9.5	25.58	8.716	76 47 46.0	.58	51.64	1
12 754	Apodis	9.5	31.42	7.081	71 41 57.2	.57	51.66	1
12 755	Gould, Z. C., 17h 2296	8.8	32.48	+ 7.423	73 3 4.4	— 2.57	52.59	2
12 756	Octantis	10.0	17 30 35.32	+18.506	85 5 22.2	— 2.57	50.64	1
12 757	Octantis	10.0	31 12.71	14.534	83 23 52.1	.51	50.50	1
12 758	Gould, Z. C., 17h 2341	9.5	25.58	6.864	70 42 59.3	.49	52.56	1
12 759	Gould, Z. C., 17h 2321	9.0	29 27	6.056	66 1 44.8	.49	52.60	1
12 760	Gould, Z. C., 17h 2346	7.8	32.08	+ 6.618	69 29 8.2	— 2.48	52.62	2
12 761	Apodis	9.5	17 31 41.12	+ 6.401	68 16 19.5	— 2.47	52.66	3
12 762	Lacaille, 7355	7.5	43.96	6.931	71 1 25.2	.47	52.50	2
12 763	Lacaille, 7359	8.0	51.68	6.690	69 51 31.2	.46	52.58	2
12 764	Apodis	9.8	31 58.52	10.851	80 19 6.9	.45	51.15	2
12 765	Apodis	9.5	32 4.35	+ 6.495	68 48 38.5	— 2.44	52.59	1
12 766	Apodis	9.7	17 32 5.64	+ 7.769	74 13 22.7	— 2.44	52.22	3
12 767	Apodis	9.2	7.96	6.555	69 8 22.4	.43	52.62	2
12 768	Apodis	9.2	9.13	9.207	77 47 35.8	.43	52.39	2
12 769	Apodis	9.0	10.05	6.848	70 38 10.0	.43	52.52	1
12 770	Apodis	9.5	20.51	7.816	74 22 8.2	.41	52.22	3
12 771	Pavonis	8.8	17 32 24.76	+ 6.247	67 18 37.8	— 2.41	51.67	2
12 772	Apodis	8.2	29.86	6.296	67 37 17.6	.40	52.65	2
12 773	Apodis	9.5	30.15	7.284	72 30 23.7	.40	52.52	1
12 774	Apodis	9.8	31.10	6.279	67 30 46.7	.40	52.65	2
12 775	Apodis	9.0	34.28	+ 7.179	72 5 21.0	— 2.39	51.67	1
12 776	Gould, Z. C., 17h 2433	8.0	17 32 35.09	+ 6.981	71 14 36.2	— 2.39	52.48	1
12 777	Octantis	9.0	41.94	22.157	86 1 15.1	.38	51.04	3
12 778	Apodis	10.0	44.84	6.288	67 34 17.5	.38	52.65	2
12 779	Gould, Z. C., 17h 2445	8.0	51.16	6.657	69 40 45.0	.37	52.60	1
12 780	Apodis	9.5	52.45	+ 8.726	76 47 4.1	— 2.37	51.64	1
12 781	Gould, 24006	7.5	17 32 53.00	+ 6.073	66 7 58.9	— 2.37	52.60	1
12 782	Gould, Z. C., 17h 2505	8.8	54.50	9.061	77 30 2.7	.37	52.46	2
12 783	Apodis	9.5	59.35	9.937	79 3 12.1	.36	51.13	2
12 784	Brisbane, 6164	8.0	32 59.84	8.008	74 56 41.8	.36	52.53	1
12 785	Gould, Z. C., 17h 2474	9.0	33 9.63	+ 6.850	70 38 11.0	— 2.34	52.56	1

No. 12 747. Gould's declination is 1′ less.

No. 12 757. The right ascension is doubtful by 2s.

Number.	Constellation, Name of Star, or Synonym.	Magnitude.	Right Ascension, 1850.0.	Annual Precession.	South Declination, 1850.0.	Annual Precession.	Mean year.	No. of obs.
			h. m. s.	s.	° ′ ″	″		
12 786	Apodis	9.8	17 33 24.36	+ 8.716	76 45 27.6	− 2.32	51.64	2
12 787	Apodis	10.0	27.56	7.713	74 1 56.2	.32	52.52	1
12 788	Apodis	9.2	33.94	6.398	68 13 55.8	.31	52.66	2
12 789	Apodis	9.5	34.75	11.266	80 47 34.1	.31	50.56	1
12 790	Apodis	9.2	34.86	+ 7.311	72 36 19.1	− 2.31	52.56	2
12 791	Apodis	10.0	17 33 40.10	+ 6.317	67 44 16.2	− 2.30	52.68	1
12 792	Octantis	9.3	41.41	23.589	86 17 44.6	.30	51.04	3
12 793	Brisbane, 6170	8.8	49.72	8.091	75 10 21.8	.29	51.54	2
12 794	Apodis	9.5	54.19	6.294	67 35 47.2	.28	52.68	1
12 795	Apodis	9.8	54.92	+ 7.342	72 43 5.6	− 2.28	52.56	2
12 796	Apodis	10.0	17 33 59.33	+ 8.582	76 26 36.3	− 2.27	51.64	1
12 797	Octantis	9.2	34 3.68	14.053	83 6 5.4	.27	51.01	2
12 798	Gould, 24354	9.7	9.67	35.575	87 39 33.4	.26	51.04	3
12 799	Apodis	10.0	13.92	6.391	68 11 11.0	.25	52.66	2
12 800	Gould, Z. C., 17ʰ 2540	9.5	18.65	+ 7.073	71 37 40.8	− 2.24	51.66	1
12 801	Pavonis	9.5	17 34 27.04	+ 6.267	67 {25 17.1 / 24 57.4}	2.23	51.67	2
12 802	Apodis	9.0	30.06	6.366	68 1 56.9	.23	52.68	1
12 803	Gould, Z. C., 17ʰ 2530	8.7	33.73	6.162	66 43 58.4	.22	52.63	3
12 804	Apodis	9.0	40.07	6.383	68 8 0.7	.21	52.68	1
12 805	Gould, Z. C., 17ʰ 2560	8.8	49.68	+ 6.695	69 51 33.4	− 2.20	52.58	2
12 806	Octantis	10.0	17 34 52.73	+13.751	82 54 19.6	− 2.19	51.52	1
12 807	Gould, Z. C., 17ʰ 2574	9.0	53.97	6.957	71 6 59.4	.19	52.50	2
12 808	Apodis	10.0	34 55.97	8.726	76 46 20.3	.19	51.64	1
12 809	Gould, Z. C., 17ʰ 2570	9.5	35 7.96	6.214	67 4 24.4	.17	52.66	3
12 810	Apodis	9.5	23.00	+ 6.716	69 57 35.1	− 2.15	52.60	1
12 811	Lacaille, 7327	7.0	17 35 28.70	+11.914	81 26 58.4	− 2.14	50.57	3
12 812	Gould, Z. C., 17ʰ 2634	9.5	39.46	7.219	72 13 30.8	.13	52.09	2
12 813	Apodis	9.5	40.70	6.810	70 25 19.9	.12	52.56	1
12 814	Apodis	9.0	43.05	6.923	70 57 5.8	.12	52.52	1
12 815	Apodis	10.0	53.43	+ 6.626	69 29 23.2	− 2.11	52.64	1
12 816	Gould, Z. C., 17ʰ 2628	8.6	17 35 55.24	+ 6.290	67 32 53.8	− 2.10	52.16	4
12 817	Apodis	10.0	35 55.38	6.685	69 47 44.2	.10	52.60	1
12 818	Gould, Z. C., 17ʰ 2662	8.3	36 0.36	7.538	73 25 11.1	.10	52.53	3
12 819	Gould, Z. C., 17ʰ 2641	9.5	7.11	6.377	68 4 59.6	.09	52.68	1
12 820	Apodis	10.0	13.07	+ 6.519	68 58 16.9	− 2.08	52.59	1
12 821	Lacaille, 7361	6.8	17 36 13.28	+ 9.212	77 46 46.6	− 2.08	52.16	3
12 822	Apodis	9.2	15.78	6.309	67 40 3.0	.07	52.65	2
12 823	Apodis	9.5	26.78	6.382	68 6 39.8	.06	52.68	1
12 824	Apodis	10.0	27.70	6.386	68 7 52.0	.06	52.68	1
12 825	Lacaille, 7372	7.0	28.56	+ 8.459	76 7 52.4	− 2.05	51.64	1

Number.	Constellation, Name of Star, or Synonym.	Magnitude.	Right Ascension, 1850.0.	Annual Precession.	South Declination, 1850.0.	Annual Precession.	Mean year.	No. of obs.
			h. m. s.	s.	° ′ ″	″		
12 826	Gould, Z. C., 17ʰ 2706 .	9.0	17 36 47.14	+ 7.628	73 43 29.0	— 2.03	52.51	2
12 827	Apodis	9.5	51.33	6.388	68 8 32.2	.02	52.68	1
12 828	Apodis	9.5	51.51	11.103	80 35 43.2	.02	51.63	1
12 829	Apodis	9.5	36 56.43	6.964	71 7 41.0	.01	52.48	1
12 830	Gould, Z. C., 17ʰ 2749 .	9.0	37 0.90	+ 8.805	76 56 9.4	— 2.01	51.64	1
12 831	Apodis	9.2	17 37 9.88	+ 6.631	69 30 13.6	— 2.00	52.62	2
12 832	Gould, 24172	7.8	13.02	9.520	78 20 31.0	1.99	51.53	2
12 833	Apodis	9.0	16.66	6.538	68 59 54.5	.99	52.64	1
12 834	Octantis	9.0	26.36	14.568	83 23 52.8	.97	50.50	1
12 835	Gould, Z. C., 17ʰ 2762 .	9.5	29.05	+ 8.105	75 11 23.6	— 1.97	52.53	1
12 836	Apodis	9.8	17 37 32.56	+ 7.327	72 38 6.7	— 1.96	52.26	3
12 837	Apodis	9.0	39.48	7.154	71 56 37.6	.95	51.66	2
12 838	Gould, Z. C., 17ʰ 2726 .	9.2	42.14	7.160	66 41 21.9	.95	52.63	3
12 839	Gould, Z. C., 17ʰ 2749 .	9.8	42.31	8.857	77 2 44.8	.95	52.10	2
12 840	Pavonis	10.0	53.49	+ 6.060	65 59 27.1	— 1.93	52.60	1
12 841	Apodis	9.8	17 37 57.02	+ 7.425	72 59 58.3	— 1.93	52.59	2
12 842	Apodis	8.5	57.12	7.131	71 50 56.8	.93	51.66	2
12 843	Gould, Z. C., 17ʰ 2758 .	8.0	37 57.53	6.813	70 25 3.4	.93	52.56	1
12 844	Gould, Z. C., 17ʰ 2752 .	9.0	38 12.23	6.059	65 58 56.7	.90	52.60	1
12 845	Gould, 24164	8.0	13.02	+ 6.470	68 36 27.2	— 1.90	52.62	2
12 846	Apodis	9.2	17 38 14.18	+ 6.404	68 13 33.1	— 1.90	52.66	2
12 847	Pavonis	10.0	15.07	6.202	66 57 57.8	.90	52.67	1
12 848	Lacaille, 7401	8.7	15.46	7.233	72 15 48.0	.90	51.95	3
12 849	Apodis	9.2	16.80	8.315	75 45 24.0	.90	52.64	2
12 850	Apodis	9.0	29.73	+ 6.936	70 59 20.8	— 1.88	52.52	1
12 851	Apodis	10.0	17 38 33.51	+ 7.876	74 30 41.5	— 1.87	52.56	1
12 852	Gould, Z. C., 17ʰ 2817 .	8.5	44.65	6.828	70 29 4.2	.86	52.56	1
12 853	Apodis	9.8	50.71	9.514	78 19 22.3	.85	51.23	2
12 854	Apodis	9.5	38 52.48	6.838	71 31 57.1	.85	51.66	1
12 855	Pavonis	8.5	39 5.76	+ 6.040	65 50 21.1	— 1.83	52.60	1
12 856	Gould, Z. C., 17ʰ 2923	9.0	17 39 11.70	+ 9.717	78 40 7.0	— 1.82	50.63	2
12 857	Gould, Z. C., 17ʰ 2826 .	8.0	14.52	6.070	66 3 6.9	.81	52.60	1
12 858	Gould, Z. C., 17ʰ 2858 .	9.0	18.69	7.118	71 46 57.6	.81	51.66	1
12 859	Apodis	10.0	19.26	6.795	70 19 3.6	.81	52.56	1
12 860	Gould, Z. C., 17ʰ 2842 .	9.0	24.47	+ 6.211	67 0 57.8	— 1.80	52.66	3
12 861	Gould, Z. C., 17ʰ 2872 .	10.0	17 39 30.05	+ 7.234	72 15 30.6	— 1.79	51.95	3
12 862	Apodis	10.0	53.28	7.871	74 29 18.2	.76	52.56	1
12 863	Lacaille, 7415	7.5	39 53.32	6.951	71 2 53.2	.76	52.50	2
12 864	Apodis	8.8	40 0.75	7.601	73 36 38.4	.75	52.54	3
12 865	Apodis	9.5	1.06	+ 6.985	71 12 15.5	— 1.75	52.48	1

Number.	Constellation, Name of Star, or Synonym.	Magnitude.	Right Ascension, 1850.0.	Annual Precession.	South Declination, 1850.0.	Annual Precession.	Mean year.	No. of obs.
			h. m. s.	s.	° ′ ″	″		
12 866	Apodis	10.0	17 40 1.99	+ 6.984	71 11 54.5	— 1.75	52.48	1
12 867	Apodis	9.0	13.80	11.879	81 23 57.9	.73	50.57	3
12 868	Gould, Z. C., 17ʰ 2908 .	9.0	17.52	6.756	70 7 19.3	.72	52.56	1
12 869	Apodis	10.0	22.79	6.495	68 44 6.3	.71	52.59	1
12 870	Apodis	9.3	28.60	+ 6.297	67 33 16.0	— 1.71	52.66	3
12 871	Melbourne (1), 899 . . .	8.0	17 40 28.94	+ 6.242	67 12 24.8	— 1.71	50.67	2
12 872	Pavonis	10.0	30.36	6.223	67 4 52.6	.70	52.67	1
12 873	Gould, Z. C., 17ʰ 2958 .	9.0	35.47	7.604	73 37 6.3	.70	52.53	3
12 874	Pavonis	9.5	40.44	6.166	66 42 5.7	.69	52.61	1
12 875	Gould, Z. C., 17ʰ 3063 .	9.0	49.96	+10.476	79 48 55.5	1.68	50.64	2
12 876	Octantis	9.0	17 40 54.20	+17.596	84 45 26.6	— 1.67	50.78	3
12 877	Apodis	9.5	40 57.42	6.806	70 21 44.5	.66	52.66	2
12 878	Apodis	9.0	41 3.45	7.104	71 42 47.5	.66	51.66	1
12 879	Apodis	9.0	6.85	8.956	77 14 21.9	.65	52.56	1
12 880	Gould, Z. C., 17ʰ 3004 .	9.0	26.47	+ 7.208	72 8 31.8	— 1.62	51.67	1
12 881	Apodis	9.0	17 41 26.61	+ 8.081	75 6 0.3	— 1.62	51.65	1
12 882	Apodis	9.5	33.12	6.412	68 14 52.8	.61	52.68	1
12 883	Apodis	9.0	37.11	6.497	68 44 26.0	.61	52.62	2
12 884	Gould, Z. C., 17ʰ 3036 .	9.0	39.42	7.579	73 31 36.4	.60	52.53	3
12 885	Apodis	9.5	42.78	+ 6.693	69 47 34.6	— 1.60	52.60	1
12 886	Pavonis	9.0	17 41 51.90	+ 6.088	66 9 35.0	— 1.59	52.60	2
12 887	Apodis	9.5	41 57.79	6.545	69 0 17.2	.58	52.64	1
12 888	Gould, Z. C., 17ʰ 3070 .	9.0	42 7.36	7.615	73 38 53.3	.56	52.57	1
12 889	Pavonis	10.0	14.73	6.045	65 51 10.6	.55	52.60	1
12 890	Octantis	9.7	15.54	+23.333	86 14 8.1	— 1.55	51.04	3
12 891	Gould, 24283	7.8	17 42 26.02	+ 8.287	75 39 43.6	— 1.54	52.64	2
12 892	Gould, Z. C., 17ʰ 3074 .	9.0	35.27	6.787	70 10 2.7	.52	52.56	1
12 893	Apodis	9.5	42 37.38	6.375	68 1 15.3	.52	52.68	1
12 894	Apodis	9.0	43 0.10	12.166	81 39 30.0	.49	50.55	2
12 895	Pavonis	9.8	10.84	+17.627	84 45 55.7	— 1.47	50.65	2
12 896	Gould, Z. C., 17ʰ 3115 .	9.0	17 43 17.46	+ 6.560	69 4 44.6	— 1.46	52.62	2
12 897	Gould, Z. C., 17ʰ 3168 .	8.8	26.27	8.084	75 6 1.7	.45	51.54	2
12 898	Apodis	8.8	27.64	6.805	70 17 26.4	.45	52.57	3
12 899	Gould, Z. C., 17ʰ 3135 .	9.0	27.69	6.852	70 33 56.7	.45	52.56	1
12 900	Gould, Z. C., 17ʰ 3165 .	8.8	31.58	+ 7.812	74 17 13.4	— 1.44	52.07	2
12 901	Apodis	10.0	17 43 36.89	+12.348	81 49 6.7	— 1.43	50.58	3
12 902	Gould, 24295	8.0	43 39.92	6.395	68 7 50.2	.43	52.68	1
12 903	Gould, Z. C., 17ʰ 3193 .	9.0	44 0.03	7.613	73 37 59.1	.40	52.57	1
12 904	Pavonis	9.5	1.40	6.239	67 9 48.3	.40	52.16	4
12 905	Apodis	8.7	11.44	+12.376	81 50 27.6	— 1.38	50.58	3

Number.	Constellation, Name of Star, or Synonym.	Magnitude.	Right Ascension, 1850.0.	Annual Precession.	South Declination, 1850.0.	Annual Precession.	Mean year.	No. of obs.
			h. m. s.	s.	° ′ ″	″		
12 906	Gould, Z. C., 17h 3240	9.5	17 44 13.63	+ 8.642	76 32 2.7	— 1.38	51.64	1
12 907	Apodis	9.0	16.94	7.014	71 18 15.2	.37	52.48	1
12 908	Apodis	10.0	24.95	10.267	79 29 46.8	.36	51.64	1
12 909	Melbourne (1), 902	9.5	26.95	6.071	66 1 7.4	.36	52.60	1
12 910	Apodis	10.0	35.71	+10.168	79 21 10.4	— 1.35	51.64	1
12 911	Gould, Z. C., 17h 3209	8.2	17 44 37.89	+ 6.585	69 12 26.5	— 1.34	52.62	2
12 912	Apodis	9.0	44.89	7.044	71 26 8.6	.33	52.07	2
12 913	Apodis	9.8	45.30	6.655	69 34 34.4	.33	52.62	2
12 914	Apodis	10.0	45.86	9.085	77 29 25.8	.33	52.12	2
12 915	Apodis	9.5	44 56.03	+ 6.741	70 1 2.6	— 1.32	52.57	1
12 916	Apodis	10.0	17 45 3.51	+ 6.732	69 58 28.5	— 1.31	52.56	1
12 917	Apodis	9.0	3.86	7.199	72 4 59.3	.31	51.67	1
12 918	Apodis	9.2	7.52	6.874	70 39 54.0	.30	52.54	2
12 919	Apodis	9.5	13.18	9.921	78 58 34.8	.29	50.62	1
12 920	Apodis	9.5	14.07	+ 7.621	73 39 13.5	— 1.29	52.57	1
12 921	Gould, Z. C., 17h 3242	9.0	17 45 16.04	+ 6.177	66 44 57.2	— 1.29	52.64	2
12 922	Pavonis	9.0	17.99	6.156	66 35 43.1	.29	52.67	1
12 923	Pavonis	9.5	18.44	6.165	66 39 56.6	.28	52.61	1
12 924	Apodis	9.5	25.43	9.922	78 58 39.5	.27	50.62	1
12 925	Apodis	9.0	26.50	+ 6.818	70 23 28.8	— 1.27	52.57	1
12 926	Octantis	9.2	17 45 28.28	+14.026	83 3 16.4	— 1.27	51.01	2
12 927	Apodis	9.5	37.89	6.604	69 18 17.5	.26	52.64	1
12 928	Apodis	10.0	40.55	7.437	72 58 6.8	.25	52.60	1
12 929	Apodis	9.5	43.41	7.444	73 1 38.7	.25	52.59	2
12 930	Gould, Z. C., 17h 3334	10.0	46.65	+ 8.464	76 5 54.3	— 1.24	51.64	1
12 931	Gould, Z. C., 17h 3279	9.0	17 45 53.23	+ 6.125	66 23 13.5	— 1.23	52.61	1
12 932	Apodis	10.0	54.72	6.629	69 26 2.2	.23	52.59	1
12 933	Apodis	9.0	45 55.65	6.833	70 27 56.8	.23	52.56	1
12 934	Apodis	9.8	46 2.58	7.253	72 17 45.1	.22	51.67	2
12 935	Octantis	8.9	5.04	+21.044	85 45 9.8	— 1.22	51.04	4
12 936	Octantis	10.5	17 46 6.10	+21.519	85 51 43.6	— 1.22	51.05	2
12 937	Apodis	10.0	12.64	6.579	69 8 9.2	.21	52.64	1
12 938	Lacaille, 7457	8.0	15.16	7.420	72 56 2.5	.20	52.60	1
12 939	Lacaille, 7348	7.0	18.98	16.717	84 24 47.4	.20	50.67	2
12 940	Brisbane, 6251	8.5	23.10	+ 6.133	66 26 30.9	— 1.19	52.61	1
12 941	Apodis	9.0	17 46 23.16	+ 6.951	71 0 41.8	— 1.19	52.50	2
12 942	Apodis	9.5	36.31	6.855	70 33 55.5	.17	52.56	1
12 943	Gould, Z. C., 17h 3354	8.2	37.26	7.267	72 20 58.4	.17	51.67	2
12 944	Gould, Z. C., 17h 3339	10.0	37.73	6.597	69 15 38.7	.17	52.64	1
12 945	Apodis	9.0	40.42	+ 6.957	71 2 17.9	— 1.17	52.23	3

Number.	Constellation, Name of Star, or Synonym.	Magnitude.	Right Ascension, 1850.0.	Annual Precession.	South Declination, 1850.0.	Annual Precession.	Mean year.	No. of obs.
			h. m. s.	s.	° ′ ″	″		
12 946	Lacaille, 7456	7.8	17 46 45.07	+ 7.725	73 59 31.2	— 1.16	52.08	2
12 947	Apodis	10.0	54.80	8.839	76 58 3.4	.15	51.64	1
12 948	Gould, 24383	10.0	46 56.00	7.920	74 36 19.9	.14	51.65	1
12 949	Apodis	7.7	47 1.01	6.341	67 50 2.9	.14	52.66	3
12 950	Gould, Z. C., 17h 3353	8.5	1.22	+ 6.302	67 32 38.0	— 1.14	52.16	4
12 951	Gould, Z. C., 17h 3406	10.0	17 47 13.83	+ 7.746	74 3 38.0	— 1.12	51.58	1
12 952	Lacaille, 7394	7.0	17.52	13.263	82 31 11.8	.11	50.49	1
12 953	Gould, Z. C., 17h 3380	9.0	31.89	6.082	66 4 42.3	.09	52.60	1
12 954	Apodis	9.8	33.58	6.430	68 19 12.8	.09	52.66	2
12 955	Apodis	9.8	34.68	+ 6.643	69 29 58.4	1.09	52.62	2
12 956	Apodis	8.0	17 47 45.72	+ 7.728	73 59 59.1	— 1.07	51.70	1
12 957	Gould, Z. C., 17h 3418	9.0	54.85	6.660	69 35 10.4	.06	52.60	1
12 958	Apodis	9.0	47 55.18	9.935	78 59 25.5	.06	50.62	1
12 959	Lacaille, 7481	6.5	48 10.59	6.146	66 31 18.2	.03	52.61	1
12 960	Apodis	9.2	11.37	+10.822	80 13 29.5	— 1.03	51.31	3
12 961	Apodis	10.0	17 48 13.97	+ 7.747	74 3 39.9	— 1.03	51.70	1
12 962	Apodis	10.0	15.14	6.427	68 17 59.4	.03	52.66	2
12 963	Apodis	10.0	18.64	6.435	68 20 47.7	.02	52.66	2
12 964	Lacaille, 7462	8.0	19.90	8.090	75 5 53.2	.02	51.58	3
12 965	Gould, Z. C., 17h 3444	10.0	29.54	+ 6.083	66 4 59.4	— 1.01	52.60	1
12 966	Gould, Z. C., 17h 3471	8.8	17 48 37.56	+ 6.785	70 13 6.4	— 0.99	52.58	2
12 967	Apodis	9.2	43.88	6.979	71 7 38.2	.99	52.09	2
12 968	Apodis	9.0	43.97	6.956	71 1 36.1	.99	52.51	1
12 969	Gould, Z. C., 17h 3475	9.0	45.26	6.637	69 27 49.4	.98	52.62	2
12 970	Gould, 24433	9.2	46.86	+ 7.465	73 5 33.4	— 0.98	52.59	2
12 971	Apodis	9.5	17 48 50.44	+ 6.340	67 46 35.4	— 0.98	52.65	2
12 972	Apodis	9.2	54.02	9.301	77 53 58.6	.97	52.05	2
12 973	Apodis	9.5	48 54.26	6.804	70 18 34.5	.97	52.59	1
12 974	Gould, 24440	10.0	49 8.08	7.453	73 2 45.4	.97	52.60	1
12 975	Apodis	9.0	19.68	+ 9.653	78 31 42.2	— 0.93	50.64	1
12 976	Octantis	9.0	17 49 22.39	+13.867	82 56 42.8	— 0.93	51.11	2
12 977	Apodis	9.0	28.45	10.711	80 4 56.8	.92	50.67	1
12 978	Gould, Z. C., 17h 3534	9.0	34.64	6.723	69 54 21.4	.91	52.59	2
12 979	Apodis	10.0	34.81	11.194	80 39 38.6	.91	51.63	1
12 980	Octantis	10.0	49.72	+17.416	84 40 47.8	— 0.89	50.64	1
12 981	Apodis	9.5	17 49 56.63	+ 6.326	67 40 57.3	— 0.88	52.63	1
12 982	Apodis	9.5	49 57.01	9.001	77 18 11.7	.88	52.56	1
12 983	Apodis	9.0	50 1.65	7.809	74 15 12.5	.87	52.56	1
12 984	Octantis	9.5	12.15	18.516	85 3 23.7	.86	50.91	3
12 985	Apodis	9.8	13.48	+ 7.268	72 20 24.6	— 0.86	51.67	2

No. 12 970. Probably variable.

Number.	Constellation, Name of Star, or Synonym.	Magnitude.	Right Ascension, 1850.0.	Annual Precession.	South Declination, 1850.0.	Annual Precession.	Mean year.	No. of obs.
			h. m. s.	s.	° ′ ″	″		
12 986	Lacaille, 7474	8.2	17 50 14.61	+ 8.124	75 11 19.0	− 0.85	51.11	2
12 987	Gould, Z. C., 17h 3565 .	9.4	14.88	6.290	67 27 27.4	.85	52.16	4
12 988	Apodis	9.0	15.06	11.664	81 9 44.5	.85	50.59	2
12 989	Stone, 9814	9.0	15.20	8.125	75 11 34.2	.85	51.11	2
12 990	Lacaille, 7473	6.3	17.04	+ 8.383	75 52 57.0	− 0.85	52.30	3
12 991	Octantis	9.0	17 50 18.32	+14.203	83 9 23.4	− 0.85	50.91	3
12 992	Apodis	10.0	21.93	11.171	80 38 2.7	.84	51.63	1
12 993	Apodis	10.0	33.61	10.284	79 30 23.4	.84	51.64	1
12 994	Gould, Z. C., 17h 3700 .	9.0	37.24	10.066	79 11 9.2	.82	51.64	1
12 995	Brisbane, 6229	8.9	40.03	+23.634	86 15 56.6	− 0.82	51.04	4
12 996	Apodis	9.8	17 50 44.24	+ 6.325	67 40 21.6	− 0.81	52.65	2
12 997	Apodis	9.4	44.48	6.548	68 58 29.1	.81	52.62	2
12 998	Apodis	10.0	49.32	7.585	73 30 44.1	.80	52.57	1
12 999	Octantis	9.5	50 53.36	25.436	86 34 51.3	.80	51.05	2
13 000	Apodis	9.0	51 5.60	+ 9.489	78 14 19.5	− 0.78	51.53	2
13 001	Apodis	10.0	17 51 8.08	+ 7.669	73 47 39.5	− 0.78	52.57	1
13 002	Apodis	10.0	16.53	6.302	67 31 40.8	.76	52.65	2
13 003	Apodis	10.0	17.73	6.486	68 37 41.5	.76	52.65	1
13 004	Apodis	10.0	20.74	10.582	79 54 41.8	.76	50.67	1
13 005	Apodis	9.2	22.06	+ 7.358	72 41 15.6	− 0.76	52.14	2
13 006	Gould, Z. C., 17h 3669 .	8.7	17 51 32.50	+ 6.349	67 49 3.2	− 0.74	52.66	3
13 007	Apodis	10.0	37.20	6.796	70 16 8.8	.73	52.59	1
13 008	Gould, Z. C., 17h 3681 .	9.8	46.36	6.350	67 49 34.8	.72	52.65	2
13 009	Apodis	8.0	51.83	6.759	70 4 47.7	.71	52.59	1
13 010	Apodis	10.0	51 59.40	+ 6.300	67 30 45.4	− 0.70	51.99	3
13 011	Apodis	10.0	17 52 1.26	+ 8.388	75 55 13.6	− 0.70	51.14	2
13 012	Apodis	9.5	6.89	6.400	68 7 20.9	.69	52.68	1
13 013	Lacaille, 7500	8.0	9.18	7.100	71 38 48.8	.69	51.66	1
13 014	Apodis	9.0	22.84	6.802	70 17 18.7	.67	52.59	1
13 015	Apodis	9.5	30.78	+ 7.502	73 12 46.7	− 0.66	52.58	1
13 016	Gould, Z. C., 17h 3724 .	9.5	17 52 36.41	+ 6.979	71 7 4.4	− 0.65	52.09	2
13 017	Apodis	9.5	38.34	9.122	77 32 40.1	.64	52.12	2
13 018	Gould, Z. C., 17h 3737 .	9.5	47.62	6.110	66 15 25.4	.63	52.60	2
13 019	Apodis	9.0	53.42	11.843	81 20 13.6	.62	50.65	1
13 020	Apodis	9.0	53.94	+13.006	82 20 16.1	− 0.62	50.59	2
13 021	Gould, Z. C., 17h 3783 .	8.5	17 52 56.20	+ 7.338	72 36 23.2	− 0.62	52.14	2
13 022	Apodis	9.8	56.76	6.884	70 40 45.4	.62	52.13	2
13 023	Apodis	9.5	52 59.12	11.535	81 1 32.8	.61	50.59	2
13 024	Gould, Z. C., 17h 3770 .	9.8	53 3.41	6.511	68 45 36.6	.61	52.62	2
13 025	Lacaille, 7489	7.5	3.76	+ 8.263	75 33 44.8	− 0.61	51.94	3

No. 13 017. The declination may be 1 revolution = 27.9″ too large.

Number.	Constellation, Name of Star, or Synonym.	Magnitude.	Right Ascension, 1850.0.	Annual Precession.	South Declination, 1850.0.	Annual Precession.	Mean year.	No. of obs.
			h. m. s.	s.	° ′ ″	″		
13026	Apodis	9.5	17 53 6.92	+ 6.408	68 6 31.3	− 0.60	52.68	1
13027	Apodis	10.0	10.73	11.681	81 10 32.9	.60	50.56	1
13028	Lacaille, 7466	8.0	14.36	10.851	80 15 51.0	.59	51.31	3
13029	Gould, Z. C., 17h 3782 .	10.0	16.21	6.492	68 39 26.5	.59	52.65	1
13030	Gould, Z. C., 17h 3823 .	8.2	18.43	+ 7.826	74 17 50.6	− 0.59	51.95	3
13031	Apodis	9.0	17 53 19.63	+10.584	79 54 41.2	− 0.58	50.61	1
13032	Lacaille, 7486	7.5	19.89	8.903	77 5 23.8	.58	52.10	2
13033	Gould, Z. C., 17h 3799 .	9.0	22.91	6.802	70 17 20.5	.58	52.56	1
13034	Apodis	9.5	27.73	10.153	79 18 38.9	.57	51.64	1
13035	Apodis	9.5	28.34	+ 6.384	68 1 38.9	− 0.57	52.68	1
13036	Lacaille, 7507	6.0	17 53 30.70	+ 7.624	73 40 23.5	− 0.57	52.57	1
13037	Gould, 24565	8.2	32.87	7.435	72 58 11.8	.56	52.59	2
13038	Lacaille, 7511	8.8	37.75	8.718	76 40 58.4	.56	51.64	2
13039	Apodis	10.0	38.20	8.718	76 40 57.0	.56	51.64	1
13040	Gould, 24573	9.5	53 43.37	+ 7.386	72 47 7.6	− 0.55	52.60	1
13041	Gould, Z. C., 17h 3891 .	9.2	17 54 11.06	+ 7.806	74 14 0.1	− 0.51	51.85	3
13042	Apodis	8.5	21.08	7.835	74 19 31.7	.49	51.95	3
13043	Octantis	9.0	22.25	31.499	87 18 28.1	.49	51.04	4
13044	Apodis	9.5	30.47	7.044	71 24 11.2	.48	52.07	2
13045	Apodis	9.2	31.34	+ 6.965	71 { 2 23.1 / 1 59.8 }	− 0.48	52.09	2
13046	Octantis	10.2	17 54 38.82	+60.413	88 39 53.6	− 0.47	51.05	2
13047	Gould, Z. C., 17h 3874 .	7.8	40.20	6.164	66 37 25.3	.47	52.64	2
13048	Gould, Z. C., 17h 3907 .	9.0	54 48.90	6.811	70 19 37.5	.45	52.58	2
13049	Octantis	10.2	55 3.96	54.833	88 31 14.2	.43	51.05	2
13050	Apodis	9.5	19.43	+ 8.471	76 5 32.9	− 0.41	51.64	1
13051	Apodis	9.5	17 55 24.79	+ 7.087	71 35 10.4	− 0.40	51.66	1
13052	Stone, 9866	7.5	32.61	8.672	76 36 17.5	.39	51.64	2
13053	Gould, Z. C., 17h 3972 .	9.0	32.86	6.798	70 15 57.6	.39	52.58	2
13054	Apodis	9.0	40.30	7.178	71 57 57.5	.38	51.66	2
13055	Gould, Z. C., 17h 3970 .	9.0	48.04	+ 6.135	66 24 44.8	− 0.37	52.60	2
13056	Gould, Z. C., 17h 4018 .	8.5	17 55 50.00	+ 7.751	74 3 21.8	− 0.36	51.64	2
13057	Apodis	10.0	52.92	8.883	77 2 41.9	.36	52.10	2
13058	Gould, Z. C., 17h 3983 .	9.2	55 59.04	6.184	66 45 18.0	.35	52.64	2
13059	Gould, Z. C., 17h 4006 .	8.2	56 3.70	6.733	69 56 19.0	.34	52.59	2
13060	Gould, 24633	10.0	16.21	+ 7.383	72 46 18.0	− 0.33	52.60	1
13061	Apodis	9.8	17 56 42.47	+ 6.488	68 37 41.5	− 0.29	52.62	2
13062	Pavonis	9.0	44.03	6.151	66 31 57.6	.29	52.61	1
13063	Apodis	10.0	56 45.58	8.882	77 2 32.1	.28	52.10	2
13064	Apodis	10.0	57 1.29	8.476	76 6 13.0	.26	51.64	1
13065	Apodis	9.8	2.02	+ 7.257	72 16 55.4	− 0.26	51.67	2

Number.	Constellation, Name of Star, or Synonym.	Magnitude.	Right Ascension, 1850.0.	Annual Precession.	South Declination. 1850.0.	Annual Precession.	Mean year.	No. of obs.
			h. m. s.	s.	° ′ ″	″		
13 066	Gould, Z. C., 18h 29 . .	10.0	17 57 6.14	+ 8.066	75 0 47.3	− 0.25	51.65	1
13 067	Lacaille, 7532	7.0	8.98	6.905	70 46 19.8	.25	52.13	2
13 068	Apodis	9.0	13.00	7.287	72 24 12.4	.24	51.68	1
13 069	Gould, Z. C., 18h 73 . .	9.0	17.48	9.206	77 42 15.2	.24	51.68	1
13 070	Apodis	10.0	22.05	+ 8.463	76 4 16.1	− 0.23	51.64	1
13 071	Gould, 24665	9.0	17 57 35.25	+ 7.418	72 54 8.6	− 0.21	52.60	1
13 072	Apodis	8.2	35.77	6.353	67 49 41.3	.21	52.66	3
13 073	Gould, 24668	9.5	40.14	7.358	72 40 23.2	.20	52.14	2
13 074	Gould, Z. C., 18h 79 . .	9.5	47.84	7.990	74 47 16.4	.19	52.11	2
13 075	Apodis	10.0	48.90	+ 8.750	76 44 57.1	− 0.19	51.64	1
13 076	Apodis	11.0	17 57 50.86	+ 6.646	69 29 35.0	− 0.19	52.60	1
13 077	Pavonis	10.0	52.24	6.175	66 41 29.3	.19	52.61	1
13 078	Apodis	10.0	53.30	7.250	72 15 21.8	.18	51.67	2
13 079	Apodis	9.8	54.46	6.648	69 29 55.9	.18	52.62	2
13 080	Apodis	9.0	57 58.84	+ 6.699	69 45 46.5	− 0.18	52.60	1
13 081	Apodis	8.3	58 11.90	+ 6.351	67 49 12.8	− 0.16	52.66	3
13 082	Gould, Z. C., 18h 82 . .	10.0	20.04	6.889	70 41 39.5	.15	52.56	1
13 083	Apodis	10.0	27.31	8.532	76 14 26.9	.13	51.64	2
13 084	Apodis	10.0	43.56	6.630	69 24 14.5	.11	52.64	1
13 085	Gould, 24699	8.3	58 58.59	+ 6.885	70 40 19.8	− 0.09	52.28	3
13 086	Pavonis	8.8	17 59 3.37	+ 6.188	66 42 46.2	− 0.08	52.64	2
13 087	Apodis	10.0	10.96	7.578	73 28 29.2	.07	52.57	1
13 088	Pavonis	8.8	12.25	6.180	66 43 23.7	.07	52.64	2
13 089	Gould, Z. C., 18h 119 . .	8.5	14.37	6.055	65 51 32.0	.07	52.60	1
13 090	Apodis	9.5	17.69	+ 6.391	68 3 39.5	− 0.06	52.68	1
13 091	Gould, 24702	7.0	17 59 21.19	+ 6.198	66 50 26.8	− 0.06	52.67	1
13 092	Gould, Z. C., 18h 150 . .	9.0	23.49	6.769	70 7 6.5	.05	52.59	1
13 093	Apodis	9.7	25.45	6.352	67 49 17.0	.05	52.66	3
13 094	Apodis	9.5	26.83	6.429	68 17 5.6	.05	52.65	1
13 095	Gould, Z. C., 18h 158 . .	8.5	40.44	+ 6.202	66 50 31.5	.03	52.67	1
13 096	Apodis	10.0	17 59 43.84	+ 7.478	73 7 18.4	− 0.02	52.59	2
13 097	Lacaille, 7515	7.8	46.19	10.635	79 58 28.1	.02	50.64	2
13 098	Apodis	9.5	46.20	7.399	72 49 41.0	.02	52.60	1
13 099	Apodis	9.5	50.07	7.050	71 25 15.5	.01	52.48	1
13 100	Octantis	10.0	54.90	+16.462	84 17 50.7	− 0.01	50.67	2
13 101	Pavonis	10.0	17 59 57.72	+ 6.276	67 21 7.7	− 0.00	51.67	2
13 102	Apodis	9.8	17 59 59.76	9.297	77 52 43.4	− 0.00	52.03	2
13 103	Pavonis	9.5	18 0 5.20	7.336	72 35 32.8	+ 0.01	51.68	1
13 104	Gould, Z. C., 18h 179 . .	8.2	5.34	6.112	66 15 38.6	.01	52.60	2
13 105	Octantis	9.0	7.44	+12.369	81 48 55.8	+ 0.01	50.60	2

Number.	Constellation, Name of Star, or Synonym.	Magnitude.	Right Ascension, 1850.0.	Annual Precession.	South Declination, 1850.0.	Annual Precession.	Mean year.	No. of obs.
			h. m. s.	s.	° ′ ″	″		
13 106	Gould, Z. C., 18h 232 . .	9.2	18 0 15.56	+ 7.475	73 6 27.8	+ 0.02	52.59	2
13 107	Pavonis	9.5	21.00	6.964	71 2 39.2	.03	52.48	1
13 108	Gould, Z. C., 18h 228 . .	9.0	31.71	6.717	69 51 31.6	.05	52.59	2
13 109	Pavonis	10.0	42.71	6.533	68 52 38.7	.06	52.59	1
13 110	Octantis	9.8	48.53	+ 8.855	76 58 55.6	+ 0.07	52.12	2
13 111	Pavonis	9.0	18 0 49.97	+ 6.322	67 41 1.0	+ 0.07	52.63	1
13 112	Pavonis	10.0	54.96	6.067	65 56 26.2	.08	52.60	1
13 113	Octantis	10.0	0 58.37	10.018	79 6 13.0	.09	51.13	2
13 114	Pavonis	9.0	1 1.19	7.336	72 35 30.3	.09	52.60	1
13 115	Pavonis	9.0	8.61	+ 6.711	69 49 31.0	+ 0.10	52.59	2
13 116	Gould, Z. C., 18h 312 . .	9.2	18 1 16.46	+ 7.964	74 42 51.0	+ 0.11	52.11	2
13 117	Pavonis	9.5	21.98	7.692	73 51 30.9	.12	52.57	1
13 118	Pavonis	9.5	22.74	7.041	71 23 4.9	.12	52.48	1
13 119	Pavonis	9.5	24.85	6.904	70 46 8.2	.12	52.58	1
13 120	Octantis	9.5	26.96	+ 9.350	77 58 41.8	+ 0.13	52.03	2
13 121	Gould, Z. C., 18h 282 . .	9.3	18 1 29.06	+ 6.363	67 53 24.6	+ 0.13	52.66	3
13 122	Octantis	9.0	35.87	13.042	82 21 41.4	.14	50.61	3
13 123	Lacaille, 7529	8.0	37.88	10.161	79 19 12.0	.14	51.64	1
13 124	Octantis	10.0	46.14	9.597	78 25 9.2	.15	50.64	1
13 125	Pavonis	9.8	46.15	+ 7.353	72 39 18.3	+ 0.15	52.14	2
13 126	Gould, Z. C., 18h 342 . .	8.8	18 1 47.45	+ 7.865	74 24 46.5	+ 0.16	52.13	2
13 127	Octantis	9.5	54.98	20.396	85 35 12.0	.17	51.04	2
13 128	Gould, Z. C., 18h 315 . .	8.0	57.50	6.365	67 54 4.8	.17	52.66	3
13 129	Lacaille, 7574	6.8	1 57.96	6.426	68 15 53.8	.17	52.66	2
13 130	Pavonis	9.5	2 2.10	+ 6.629	69 24 4.6	+ 0.18	52.62	2
13 131	Pavonis	10.0	18 2 2.19	+ 7.653	73 43 52.6	+ 0.18	52.57	1
13 132	Pavonis	9.5	6.29	7.036	71 21 38.5	.18	52.48	1
13 133	Lacaille, 7572	7.5	6.66	6.675	69 38 12.8	.18	52.62	2
13 134	Brisbane, 6335	8.0	9.67	6.282	67 23 28.9	.19	51.67	2
13 135	Lacaille, 7525	6.8	14.69	+10.884	80 17 16.7	+ 0.20	51.14	2
13 136	Octantis	8.8	18 2 23.87	+11.924	81 24 40.2	+ 0.21	50.57	3
13 137	Pavonis	10.0	37.22	7.514	73 14 55.4	.23	52.58	1
13 138	Gould, 24801	8.2	45.14	7.178	71 57 53.5	.24	51.66	2
13 139	Octantis	9.3	47.85	17.809	84 48 58.5	.24	50.91	3
13 140	Pavonis	9.5	50.72	+ 7.442	72 59 15.9	+ 0.25	52.60	1
13 141	Pavonis	10.0	18 2 52.83	+ 6.531	68 51 52.1	+ 0.25	52.59	1
13 142	Gould, Z. C., 18h 392 . .	10.0	53.70	6.776	70 9 6.6	.25	52.59	2
13 143	Gould, 24802	8.8	2 55.98	6.772	70 8 8.0	.26	52.59	2
13 144	Octantis	10.0	3 5.50	8.393	75 53 47.0	.27	52.64	1
13 145	Pavonis	9.5	6.67	+ 7.630	73 39 13.9	+ 0.27	52.57	1

Number.	Constellation, Name of Star, or Synonym.	Magnitude.	Right Ascension, 1850.0.	Annual Precession.	South Declination, 1850.0.	Annual Precession.	Mean year.	No. of obs.
			h. m. s.	s.	° ′ ″	″		
13 146	Lacaille, 7559	6.5	18 3 24.60	+ 8.094	75 5 38.9	+ 0.30	51.11	2
13 147	Brisbane, 6340	8.5	39.10	6.279	67 22 8.5	.32	51.65	2
13 148	Pavonis	9.2	50.94	7.575	73 27 50.1	.34	52.58	2
13 149	Pavonis	9.5	3 59.02	7.494	73 10 46.2	.35	52.59	2
13 150	Pavonis	9.0	4 4.42	+ 7.767	74 6 20.1	+ 0.36	51.70	1
13 151	Gould, 24828	7.2	18 4 22.20	+ 6.176	66 41 52.2	+ 0.38	52.64	2
13 152	Gould, Z. C., 18ʰ 515 . .	8.5	30.23	7.463	73 4 9.4	.39	52.59	2
13 153	Gould, Z. C., 18ʰ 484 . .	9.5	32.43	7.143	66 28 50.3	.40	52.61	1
13 154	Octantis	9.8	40.10	10.616	79 57 4.0	.41	50.64	2
13 155	Octantis	9.8	43.75	+ 9.332	77 56 41.4	+ 0.41	51.66	2
13 156	Octantis	9.0	18 4 45.09	+ 8.831	76 55 53.8	+ 0.42	51.64	1
13 157	Brisbane, 6338	8.2	46.08	9.743	78 40 6.4	.42	50.63	2
13 158	Pavonis	9.8	4 48.74	6.113	66 16 13.0	.42	52.60	2
13 159	Gould, 24852	9.5	5 2.12	6.551	68 58 56.7	.44	52.59	1
13 160	Octantis	10.0	5.52	+14.675	83 25 42.8	+ 0.45	50.68	1
13 161	Pavonis	9.2	18 5 17.19	+ 6.282	67 23 33.6	+ 0.46	51.65	2
13 162	Octantis	9.0	19.11	9.585	78 24 4.4	.47	50.64	1
13 163	Pavonis	10.0	20.45	6.080	66 2 37.5	.47	52.60	1
13 164	Gould, Z. C., 18ʰ 540 . .	9.5	22.42	6.106	66 13 15.3	.47	52.60	2
13 165	Gould, Z. C., 18ʰ 546 . .	9.0	30.28	+ 6.099	66 10 31.2	+ 0.48	52.60	2
13 166	Octantis	10.0	18 5 39.47	+ 9.776	78 43 30.4	+ 0.50	50.62	1
13 167	Pavonis	10.0	46.04	7.089	71 35 42.2	.50	51.66	1
13 168	Octantis	9.5	57.32	9.013	77 19 13.1	.52	52.60	1
13 169	Lacaille, 7569	7.0	5 58.98	9.065	77 25 38.2	.52	52.14	2
13 170	Gould, Z. C., 18ʰ 620 . .	8.8	6 2.53	+ 7.614	73 36 4.6	+ 0.53	52.58	2
13 171	Pavonis	10.0	18 6 6.78	+ 6.077	66 1 16.8	+ 0.53	52.60	1
13 172	Gould, Z. C., 18ʰ 619 . .	10.0	18.68	6.976	71 6 11.5	.55	51.69	2
13 173	Pavonis	10.0	26.70	6.387	68 2 35.2	.56	52.68	1
13 174	Octantis	9.0	26.77	12.978	82 18 57.0	.56	50.69	1
13 175	Pavonis	9.2	33.99	+ 6.658	69 33 43.7	+ 0.57	52.62	2
13 176	Octantis	7.5	18 6 35.87	+13.935	82 59 9.7	+ 0.58	50.62	3
13 177	Lacaille, 7442	8.6	42.99	20.774	85 40 55.8	.59	51.04	4
13 178	Octantis	10.0	6 53.65	11.601	81 5 41.1	.60	50.69	2
13 179	Octantis	10.0	7 3.04	9.741	78 40 7.5	.62	50.62	1
13 180	Pavonis	8.8	17.46	+ 6.096	66 9 31.3	+ 0.64	52.60	2
13 181	Stone, 10026	9.2	18 7 19.28	+20.729	85 40 17.8	+ 0.64	51.04	4
13 182	Gould, 24911	8.8	20.68	6.249	67 11 9.5	.64	51.67	2
13 183	Gould, Z. C., 18ʰ 678 . .	9.5	27.92	6.368	67 55 47.8	.65	52.66	3
13 184	Gould, Z. C., 18ʰ 707 . .	8.8	30.69	7.483	73 8 48.2	.66	52.59	2
13 185	Pavonis	10.0	38.49	+ 6.589	69 11 37.7	+ 0.67	52.64	1

Number.	Constellation, Name of Star, or Synonym.	Magnitude.	Right Ascension, 1850.0.	Annual Precession.	South Declination, 1850.0.	Annual Precession.	Mean year.	No. of obs.
			h. m. s.	s.	° ′ ″	″		
13 186	Gould, Z. C., 18h 695 . .	9.5	18 7 42.39	+ 6.447	68 24 5.6	+ 0.67	52.65	1
13 187	Gould, Z. C., 18h 718 . .	9.2	8 11.18	6.506	68 44 27.1	.72	52.62	2
13 188	Octantis	10.0	23.80	11.226	80 41 36.2	.73	50.69	2
13 189	Octantis	9.5	24.33	8.747	76 45 6.0	.74	51.64	2
13 190	Pavonis	9.5	24.82	+ 6.925	70 52 34.6	+ 0.74	51.70	1
13 191	Gould, Z. C., 18h 735 . .	9.0	18 8 27.48	+ 6.597	69 14 29.3	+ 0.74	52.64	1
13 192	Pavonis	10.0	31.24	6.180	66 44 26.5	.75	52.64	2
13 193	Pavonis	9.5	33.10	7.933	74 37 54.2	.75	52.11	2
13 194	Brisbane, 6361[1]	9.5	39.18	7.147	71 50 52.7	.76	51.62	2
13 195	Brisbane, 6361[2]	9.5	41.53	+ 7.148	71 51 0.1	+ 0.76	51.66	2
13 196	Octantis	10.0	18 8 49.12	+10.869	80 16 33.7	+ 0.77	51.62	1
13 197	Octantis	9.8	51.84	9.517	78 17 20.2	.78	51.14	2
13 198	Pavonis	10.0	52.22	6.171	66 40 45.3	.78	52.64	2
13 199	Gould, Z. C., 18h 778 . .	9.0	53.57	6.868	70 36 36.0	.78	52.14	2
13 200	Gould, Z. C., 18h 760 . .	9.5	54.17	+ 6.459	68 28 13.2	+ 0.78	52.65	1
13 201	Gould, Z. C., 18h 756 . .	9.0	18 8 57.80	+ 6.211	66 56 41.3	+ 0.78	52.67	1
13 202	Pavonis	8.8	8 59.33	6.302	67 31 51.2	.79	51.65	2
13 203	Gould, Z. C., 18h 796 . .	9.0	9 6.31	7.052	71 26 35.6	.80	51.67	2
13 204	Pavonis	10.0	11.68	6.203	66 53 33.0	.80	52.67	1
13 205	Gould, 24970	9.5	23.49	+ 6.456	68 27 23.7	+ 0.82	52.65	1
13 206	Gould, Z. C., 18h 836 . .	8.2	18 9 37.69	+ 7.612	73 36 14.7	+ 0.84	52.58	2
13 207	Brisbane, 6365	9.2	40.32	7.172	71 57 16.2	.85	51.66	2
13 208	Gould, Z. C., 18h 838 . .	9.0	45.63	7.421	72 55 26.2	.85	52.60	1
13 209	Lacaille, 7628	7.2	9 50.80	6.659	69 34 29.1	.86	52.62	2
13 210	Pavonis	9.0	10 2.49	+ 6.993	71 11 16.0	+ 0.88	51.68	1
13 211	Gould, Z. C., 18h 841 . .	9.5	18 10 12.93	+ 6.457	68 27 52.5	+ 0.89	52.65	1
13 212	Octantis	9.8	20.04	8.228	75 28 35.7	.90	51.60	2
13 213	Gould, Z. C., 18h 855 . .	9.5	30.33	6.468	68 31 50.4	.92	52.65	1
13 214	Pavonis	9.0	30.40	6.389	68 4 0.0	.92	52.68	2
13 215	Lacaille, 7636	7.2	30.68	+ 6.635	69 26 56.2	+ 0.92	52.62	2
13 216	Pavonis	9.8	18 10 30.96	+ 6.870	70 37 28.4	+ 0.92	52.14	2
13 217	Octantis	9.8	34.37	8.233	75 29 27.6	.92	51.60	2
13 218	Gould, Z. C., 18h 856 . .	9.0	37.29	6.176	66 43 22.4	.93	52.64	2
13 219	Pavonis	9.0	43.09	6.043	65 47 35.8	.94	52.60	1
13 220	Brisbane, 6369	8.5	49.12	+ 7.147	71 51 14.8	+ 0.95	51.66	2
13 221	Pavonis	9.0	18 10 59.69	+ 6.059	65 54 49.8	+ 0.96	52.60	1
13 222	Gould, 25050	8.5	11 8.40	10.467	79 43 0.9	.97	50.61	1
13 223	Gould, Z. C., 18h 900 . .	9.0	13.39	6.812	70 21 4.7	.98	52.59	2
13 224	Gould, Z. C., 18h 925 . .	9.5	15.72	7.685	73 51 19.7	.99	52.08	2
13 225	Octantis	10.0	17.09	+ 9.817	78 48 1.5	+ 0.99	50.62	1

No. 13 186. Gould's declination is 11.5″ greater.

Number.	Constellation, Name of Star, or Synonym.	Magnitude.	Right Ascension, 1850.0.	Annual Precession.	South Declination, 1850.0.	Annual Precession.	Mean year.	No. of obs.
			h. m. s.	s.	° ′ ″	″		
13 226	Lacaille, 7562	6.5	18 11 20.84	+12.463	81 54 22.9	+ 0.99	50.58	3
13 227	Pavonis	9.5	23.04	7.011	71 16 22.7	1.00	51.69	1
13 228	Pavonis	9.2	34.37	6.488	68 39 7.1	.02	52.62	2
13 229	Stone, 10005	7.5	36.36	6.622	69 23 7.5	.02	52.62	2
13 230	Pavonis	9.2	39.78	+ 7.550	73 23 39.0	+ 1.02	52.58	2
13 231	Gould, Z. C., 18h 907 . .	9.0	18 11 39.92	+ 6.112	66 17 12.3	+ 1.02	52.60	2
13 232	Pavonis	9.2	12 0.14	6.490	68 39 51.8	.05	52.62	2
13 233	Pavonis	10.0	3.29	10.636	79 59 21.7	.06	50.61	1
13 234	Brisbane, 6374	9.0	3.54	7.079	71 34 23.9	.06	51.66	1
13 235	Pavonis	9.0	25.32	+ 6.066	65 58 4.2	+ 1.09	52.60	1
13 236	Octantis	8.6	18 12 25.44	+22.972	86 9 42.4	+ 1.09	51.04	4
13 237	Pavonis	10.0	32.38	7.310	72 30 49.6	.10	51.68	1
13 238	Pavonis	9.5	12 44.69	6.147	66 31 55.1	.11	52.61	1
13 239	Gould, 25061	8.0	13 11.56	6.539	68 56 38.3	.15	52.62	2
13 240	Lacaille, 7642	7.0	38.08	+ 7.741	74 2 48.8	+ 1.19	51.64	2
13 241	Gould, 25075	9.0	18 13 47.39	+ 6.533	68 54 54.2	+ 1.21	52.59	1
13 242	Gould, 25073	9.0	48.32	6.317	67 38 34.1	.21	52.63	1
13 243	Lacaille, 7548	8.5	53.13	15.125	83 40 52.5	.21	50.68	1
13 244	Pavonis	9.5	57.10	6.162	66 38 34.0	.22	52.64	2
13 245	Gould, Z. C., 18h 1056 .	9.5	13 58.74	+ 6.443	68 24 9.7	+ 1.22	52.65	1
13 246	Gould, Z. C., 18h 1087 .	9.2	18 14 2.32	+ 7.495	73 12 37.6	+ 1.23	52.59	2
13 247	Gould, Z. C., 18h 1091 .	8.0	9.26	7.376	72 46 22.6	.24	52.14	1
13 248	Gould, Z. C., 18h 1136 .	9.5	14.58	9.048	77 24 43.8	.25	51.44	2
13 249	Gould, Z. C., 18h 1069 .	9.0	16.36	6.204	66 55 5.6	.25	52.67	2
13 250	Pavonis	8.5	23.04	+ 6.697	69 47 26.3	+ 1.26	52.59	2
13 251	Octantis	9.5	18 14 25.51	+ 8.324	75 44 40.0	+ 1.26	52.64	1
13 252	Pavonis	10.0	36.18	7.933	74 39 5.7	1.28	51.66	1
13 253	Octantis	9.5	51.21	8.401	75 56 35.9	.30	51.64	1
13 254	Melbourne (1), 932 . . .	7.8	14 53.39	6.218	67 1 4.9	.30	52.00	3
13 255	Pavonis	9.0	15 0.94	+ 6.784	70 13 57.9	+ 1.31	52.59	2
13 256	Pavonis	9.3	18 15 21.32	+ 7.863	74 26 22.4	+ 1.34	51.97	3
13 257	Pavonis	10.0	24.49	6.872	70 39 28.2	.35	52.14	2
13 258	Gould, Z. C., 18h 1139 .	9.8	29.20	6.342	67 48 18.5	.35	52.65	2
13 259	Brisbane, 6394	8.5	38.02	6.380	68 2 27.5	.37	52.68	1
13 260	Gould, Z. C., 18h 1177 .	8.5	45.46	+ 7.379	72 47 27.6	+ 1.38	52.60	1
13 261	Lacaille, 7573	7.8	18 15 49.04	+14.670	83 26 20.6	+ 1.38	50.67	2
13 262	Gould, Z. C., 18h 1163 .	9.2	52.84	6.339	67 47 35.9	.39	52.65	2
13 263	Pavonis	9.0	55.52	6.774	70 11 16.7	.39	52.59	2
13 264	Gould, Z. C., 18h 1166 .	8.5	58.46	6.144	66 31 52.9	.40	52.61	1
13 265	Gould, Z. C., 18h 1174	9.0	58.74	+ 6.644	69 31 21.0	+ 1.40	52.62	2

Number.	Constellation, Name of Star, or Synonym.	Magnitude.	Right Ascension, 1850.0.	Annual Precession.	South Declination, 1850.0.	Annual Precession.	Mean year.	No. of obs.
			h. m. s.	s.	° ′ ″	″		
13 266	Lacaille, 7666	7.2	18 15 59.80	+ 7.142	71 51 33.2	+ 1.40	51.66	2
13 267	Gould, Z. C., 18h 1173	9.0	15 59.97	6.446	68 25 57.6	.40	52.65	1
13 268	Pavonis	10.0	16 10.19	7.756	74 6 19.1	.41	51.70	1
13 269	Gould, Z. C., 18h 1237	9.5	14.79	8.795	76 52 55.7	.42	51.64	1
13 270	Pavonis	10.0	18.06	+ 6.091	66 10 6.2	+ 1.43	52.62	1
13 271	Gould, Z. C., 18h 1260	9.2	18 16 23.86	+ 9.495	78 16 12.9	+ 1.43	51.14	2
13 272	Pavonis	9.4	37.95	7.833	74 21 4.6	.45	51.87	4
13 273	Gould, Z. C., 18h 1218	10.0	38.84	7.049	71 27 54.5	.45	51.67	2
13 274	Octantis	9.0	50.57	13.711	82 51 21.0	.47	50.69	1
13 275	Pavonis	10.0	16 56.32	+ 6.080	66 5 26.1	+ 1.48	52.62	1
13 276	Gould, Z. C., 18h 1286	9.0	18 17 19.00	+ 8.385	75 54 49.3	+ 1.51	52.14	2
13 277	Gould, 25190	8.0	19.76	8.668	76 36 1.2	.51	51.64	2
13 278	Octantis	9.2	24.72	11.349	80 50 56.6	.52	50.69	2
13 279	Pavonis	9.0	31.38	6.734	69 59 49.8	.53	52.59	1
13 280	Lacaille, 7678	7.8	31.98	+ 7.037	71 25 0.4	+ 1.53	51.67	2
13 281	Gould, Z. C., 18h 1259	9.2	18 17 41.93	+ 6.354	67 53 52.0	+ 1.55	52.65	2
13 282	Octantis	9.5	47.16	10.188	79 23 23.0	.55	51.13	2
13 283	Gould, Z. C., 18h 1337	10.0	17 51.29	8.807	76 54 57.8	.56	51.64	1
13 284	Pavonis	9.2	18 0.68	7.872	74 28 45.9	.57	51.68	2
13 285	Pavonis	10.0	0.68	+ 7.781	74 11 46.5	+ .57	51.64	2
13 286	Pavonis	10.0	18 18 1.41	+ 6.244	67 12 32.3	+ 1.58	50.67	1
13 287	Gould, 25186	9.0	2.20	6.531	68 55 37.4	.58	52.61	2
13 288	Gould, 25226	9.5	11.48	8.846	77 0 4.8	.59	52.12	2
13 289	Octantis	9.5	11.94	8.366	75 52 11.7	.59	52.14	2
13 290	Gould, Z. C., 18h 1312	10.0	13.34	+ 7.025	71 22 0.9	+ 1.59	51.69	1
13 291	Pavonis	9.5	18 18 15.24	+ 6.255	67 16 53.5	+ 1.60	50.67	1
13 292	Octantis	9.5	15.64	17.020	84 32 28.8	.60	50.66	1
13 293	Lacaille, 7679	7.5	17.86	7.402	72 53 26.6	.60	52.60	1
13 294	Gould, Z. C., 18h 1326	9.5	28.64	6.943	71 0 13.5	.61	51.70	1
13 295	Octantis	10.0	39.32	+13.453	82 41 6.5	+ 1.63	50.69	1
13 296	Pavonis	10.0	18 18 39.44	+ 6.625	69 26 29.1	+ 1.63	52.60	1
13 297	Pavonis	9.0	39.69	6.795	70 18 25.8	.63	52.59	1
13 298	Pavonis	9.5	40.11	6.783	70 14 50.8	.63	52.58	1
13 299	Pavonis	9.5	41.22	7.499	73 14 46.3	.63	52.58	1
13 300	Pavonis	9.0	49.71	+ 6.740	70 2 12.6	+ 1.65	52.59	1
13 301	Gould, Z. C., 18h 1336	9.0	18 18 51.50	+ 6.327	67 44 9.3	+ 1.65	52.63	1
13 302	Pavonis	9.8	52.19	6.620	69 24 49.5	.65	52.62	2
13 303	Lacaille, 7697	7.2	18 59.12	6.118	66 22 31.8	.66	52.62	2
13 304	Pavonis	10.0	19 11.81	6.063	65 59 14.8	.68	52.62	1
13 305	Gould, Z. C., 18h 1442	9.8	18.36	+ 9.518	78 19 19.8	+ 1.69	51.14	2

Number.	Constellation, Name of Star, or Synonym.	Magnitude.	Right Ascension, 1850.0.	Annual Precession.	South Declination, 1850.0.	Annual Precession.	Mean year.	No. of obs.
			h. m. s.	s.	° ′ ″	″		
13 306	Gould, Z. C., 18h 1443 .	9.5	18 19 24.15	+ 9.353	78 1 26.6	+ 1.70	51.63	1
13 307	Gould, Z. C., 18h 1379 .	9.0	32.07	6.309	67 38 3.3	.71	52.63	1
13 308	Pavonis	9.0	34.88	6.071	66 3 1.0	.71	52.62	1
13 309	Gould, Z. C., 18h 1402 .	9.8	36.62	7.040	71 26 31.6	.71	51.67	2
13 310	Lacaille, 7664	7.5	36.99	+ 9.438	78 10 49.5	+ 1.71	51.14	2
13 311	Gould, Z. C., 18h 1395 .	9.5	18 19 44.11	+ 6.607	69 21 0.1	+ 1.72	52.64	1
13 312	Gould, 25245	8.2	44.76	6.729	69 59 12.2	.73	52.59	2
13 313	Gould, Z. C., 18h 1391 .	8.8	51.98	6.102	66 16 25.5	.74	52.62	2
13 314	Pavonis	10.0	56.58	7.435	73 1 {1.6 / 29.5}	.74	52.59	2
13 315	Brisbane, 6412	9.0	19 58.81	+ 7.191	72 4 51.3	+ 1.75	51.67	1
13 316	Pavonis	9.0	18 20 1.76	+11.876	81 23 54.0	+ 1.75	50.59	2
13 317	Brisbane, 6415	8.0	6.55	6.575	69 10 58.6	.76	52.62	3
13 318	Gould, Z. C., 18h 1445 .	8.5	29.38	6.857	70 37 3.8	.79	52.14	2
13 319	Pavonis	9.5	52.90	7.628	73 44 34.5	.82	52.57	1
13 320	Pavonis	10.0	20 54.48	+ 7.031	71 24 39.0	+ 1.83	51.67	2
13 321	Pavonis	9.8	18 21 1.87	+ 7.443	73 3 25.6	+ 1.84	52.59	2
13 322	Octantis	10.0	3.49	18.346	85 1 5.0	.84	50.64	1
13 323	Gould, Z. C., 18h 1541 .	8.5	5.39	10.151	79 20 55.9	.84	51.13	2
13 324	Gould, 25278	8.0	7.95	6.207	66 59 32.7	.85	52.67	2
13 325	Gould, Z. C., 18h 1521 .	9.5	12.26	+ 8.497	76 12 38.0	+ 1.85	51.64	2
13 326	Gould, Z. C., 18h 1487 .	9.2	18 21 28.28	+ 6.556	69 5 30.1	+ 1.88	52.62	3
13 327	Pavonis	10.0	29.77	7.982	74 49 48.9	.88	51.65	1
13 328	Gould, Z. C., 18h 1498 .	9.0	32.47	6.935	70 59 15.0	.88	51.70	1
13 329	Octantis	10.0	34.34	9.511	78 19 10.7	.88	51.63	1
13 330	Octantis	9.8	34.60	+11.345	80 51 32.1	+ 1.89	50.69	2
13 331	Octantis	9.5	18 21 35.61	+35.100	87 37 12.1	+ 1.89	51.04	1
13 332	Lacaille, 7615	6.8	35.76	14.878	83 34 0.8	.89	50.67	2
13 333	Pavonis	9.5	38.68	7.314	72 34 38.9	.89	52.60	1
13 334	Lacaille, 7706	7.5	46.20	6.798	70 20 42.8	.90	52.59	2
13 335	Octantis	9.5	47.17	+22.625	86 6 21.4	+ 1.90	51.04	3
13 336	Gould, Z. C., 18h 1513 .	9.0	18 21 50.92	+ 6.658	69 38 12.3	+ 1.91	52.60	1
13 337	Lacaille, 7707	7.0	56.04	6.918	70 54 50.4	.92	51.70	1
13 338	Gould, Z. C., 18h 1506 .	9.0	21 59.55	6.046	65 52 19.5	.92	52.62	1
13 339	Gould, Z. C., 18h 1572 .	8.5	22 3.76	9.025	77 23 47.2	.93	52.14	2
13 340	Pavonis	10.0	5.72	+ 6.447	68 28 55.3	+ 1.93	52.65	1
13 341	Octantis	9.2	18 22 34.36	+10.882	80 19 52.0	— 1.97	51.14	2
13 342	Pavonis	9.5	39.50	7.897	74 35 1.0	1.98	51.66	1
13 343	Gould, Z. C., 18h 1568 .	9.0	51.78	6.938	71 0 33.6	2.00	51.69	2
13 344	Gould, Z. C., 18h 1610 .	10.0	55.44	8.706	76 42 54.4	.00	51.64	2
13 345	Brisbane, 6421	9.0	22 58.87	+ 7.212	72 11 16.4	+ 2.01	51.67	1

No. 13 321. Declination possibly 39.7″, one observation ½ rev. wrong. No. 13 323. Gould's declination is 79° 20′ 39.0″.

Number.	Constellation, Name of Star, or Synonym.	Magnitude.	Right Ascension, 1850.0.	Annual Precession.	South Declination, 1850.0.	Annual Precession.	Mean year.	No. of obs.
			h. m. s.	s.	° ′ ″	″		
13 346	Lacaille, 7612	8.0	18 23 12.72	+15.930	84 5 35.9	+ 2.03	50.68	1
13 347	Gould, 25343	9.5	18.82	7.044	71 28 46.7	.04	51.66	1
13 348	Gould, Z. C., 18h 1601	8.8	20.31	7.251	72 20 33.8	.04	51.67	2
13 349	Pavonis	10.0	35.31	7.859	74 28 25.2	.06	51.66	1
13 350	Pavonis	9.8	23 43.45	+ 6.625	69 28 45.2	+ 2.07	52.62	2
13 351	Pavonis	10.0	18 24 6.12	+ 6.378	68 5 33.6	+ 2.11	52.68	1
13 352	Gould, Z. C., 18h 1647	8.8	7.55	7.958	74 46 40.4	.11	51.65	2
13 353	Pavonis	10.0	16.48	7.204	72 9 43.2	.12	51.67	1
13 354	Gould, 25366	9.5	22.70	7.988	74 51 57.8	.13	51.65	1
13 355	Gould, 25363	10.0	27.73	+ 6.821	70 28 35.7	+ 2.14	52.58	1
13 356	Gould, 25364	8.5	18 24 30.09	+ 6.814	70 26 33.9	+ 2.14	52.58	1
13 357	Pavonis	9.8	31.71	8.518	76 16 57.8	.14	51.64	2
13 358	Brisbane, 6440	9.8	32.74	7.316	72 36 24.8	.14	52.14	2
13 359	Pavonis	9.8	32.86	6.628	69 30 3.8	.14	52.62	2
13 360	Octantis	10.0	35.00	+10.568	79 56 36.5	+ 2.15	50.61	1
13 361	Octantis	9.2	18 24 38.18	+ 9.657	78 35 14.4	+ 2.15	50.63	2
13 362	Pavonis	9.5	39.98	6.303	67 38 24.1	.15	52.63	1
13 363	Gould, 25357	9.0	41.86	6.058	66 0 22.7	.16	52.62	1
13 364	Lacaille, 7700	8.0	42.45	9.329	78 0 20.5	.16	51.63	1
13 365	Pavonis	10.0	46.39	+ 6.296	67 35 41.6	+ 2.16	52.63	1
13 366	Octantis	9.8	18 24 47.81	+ 9.120	77 36 19.0	+ 2.17	52.14	2
13 367	Pavonis	9.5	24 49.22	6.669	69 43 18.2	.17	52.60	1
13 368	Octantis	10.0	25 0.21	8.712	76 44 23.6	.18	51.64	2
13 369	Gould, Z. C., 18h 1702	9.0	0.22	8.409	76 0 56.5	.18	51.64	1
13 370	Gould, Z. C., 18h 1706	9.0	6.07	+ 8.426	76 3 33.6	+ 2.19	51.64	1
13 371	Pavonis	9.5	18 25 10.88	+ 6.717	69 58 11.3	+ 2.20	52.59	2
13 372	Octantis	9.5	19.90	9.792	78 48 50.3	.21	50.62	1
13 373	Octantis	10.0	20.01	9.396	78 7 55.8	.21	51.63	1
13 374	ζ Pavonis	4.2	29.15	7.054	71 32 45.0	.23	51.67	2
13 375	Gould, Z. C., 18h 1694	8.3	32.87	+ 6.562	69 9 18.2	+ 2.23	52.62	3
13 376	Pavonis	9.5	18 25 33.00	+ 6.739	70 4 55.8	+ 2.23	52.59	1
13 377	Lacaille, 7740	7.5	33.04	6.809	70 25 35.1	.23	52.58	1
13 378	Gould, 25387	8.5	42.34	7.023	71 24 44.0	.24	51.67	2
13 379	Pavonis	9.5	42.92	6.856	70 39 4.8	.25	52.14	2
13 380	Octantis	10.0	45.76	+ 8.826	76 59 54.2	+ 2.25	52.60	1
13 381	Pavonis	9.5	18 25 46.35	+ 6.668	69 43 15.1	+ 2.25	52.60	1
13 382	Gould, Z. C., 18h 1704	8.7	48.74	6.562	69 9 41.9	.25	52.62	3
13 383	Pavonis	9.0	52.99	6.033	65 50 7.1	.26	52.62	1
13 384	Pavonis	10.0	25 57.26	6.888	70 48 25.0	.27	51.70	1
13 385	Lacaille, 7749	7.8	26 11.80	+ 6.627	69 30 43.6	+ 2.29	52.62	2

Number.	Constellation, Name of Star, or Synonym.	Magnitude.	Right Ascension, 1850.0.	Annual Precession.	South Declination, 1850.0.	Annual Precession.	Mean year.	No. of obs.
			h. m. s.	s.	° ′ ″	″		
13 386	Octantis	9.8	18 26 12.82	+15.269	83 47 4.7	+ 2.29	50.68	2
13 387	Octantis	10.8	18.33	15.272	83 47 10.5	.30	50.68	2
13 388	Octantis	10.0	26.40	9.385	78 7 10.8	.31	51.63	1
13 389	Octantis	9.5	29.58	8.377	75 56 45.0	.31	52.14	2
13 390	Octantis	10.0	37.37	+10.262	79 32 6.4	+ 2.32	51.13	2
13 391	Gould, Z. C., 18h 1773 .	9.5	18 26 50.66	+ 7.318	72 38 8.5	+ 2.34	52.14	2
13 392	Octantis	9.2	52.52	10.243	79 30 33.4	.35	51.13	2
13 393	Octantis	9.8	55.80	9.943	79 3 45.3	.35	51.13	2
13 394	Brisbane, 6440	8.5	26 56.13	7.306	72 35 12.5	.35	52.14	2
13 395	Pavonis	10.0	27 16.23	+ 7.939	74 44 29.7	+ 2.38	51.66	1
13 396	Octantis	10.0	18 27 27.38	+ 9.504	78 20 22.2	+ 2.40	51.14	2
13 397	Pavonis	9.5	27.72	6.340	67 53 31.2	.40	52.68	1
13 398	Pavonis	9.8	34.88	6.150	66 40 27.2	.41	52.64	2
13 399	Pavonis	10.0	35.08	6.510	68 53 19.5	.41	52.62	1
13 400	Gould, Z. C., 18h 1801 .	8.0	39.19	+ 6.586	69 18 27.8	+ 2.41	52.64	1
13 401	Pavonis	9.5	18 27 40.52	+ 6.773	70 16 21.5	+ 2.42	52.59	1
13 402	Gould, Z. C., 18h 1795 .	9.5	42.74	6.175	66 50 39.7	.42	52.68	1
13 403	Pavonis	9.0	48.66	6.114	66 25 37.8	.43	52.61	1
13 404	Pavonis	10.0	51.82	7.852	74 28 58.0	.43	51.66	1
13 405	Lacaille, 7752	7.0	54.46	+ 7.398	72 56 44.3	+ 2.44	52.64	2
13 406	Octantis	10.0	18 27 58.20	+10.885	80 21 36.6	+ 2.44	51.14	2
13 407	Lacaille, 7699	8.5	28 0.18	11.337	80 52 41.8	.44	50.69	2
13 408	Octantis	10.0	24.93	15.796	84 2 48.5	.48	50.68	1
13 409	Octantis	9.0	36.86	10.858	80 19 46.7	.50	51.14	2
13 410	Gould, Z. C., 18h 1844 .	9.0	38.51	+ 6.468	68 39 52.0	+ 2.50	52.62	3
13 411	Octantis	10.0	18 28 40.33	+ 9.461	78 16 11.4	+ 2.50	51.63	1
13 412	Pavonis	9.5	43.58	6.088	66 15 20.3	.51	52.61	1
13 413	Octantis	9.5	44.55	26.878	86 48 38.1	.51	51.04	1
13 414	Gould, Z. C., 18h 1922 .	9.0	28 48.66	10.115	79 20 3.3	.51	51.64	1
13 415	Octantis	10.0	29 4.01	+10.557	79 57 6.1	+ 2.54	50.61	1
13 416	Gould, Z. C., 18h 1954 .	9.0	18 29 12.98	+10.592	79 59 49.3	+ 2.55	50.64	2
13 417	Gould, Z. C., 18h 1879 .	9.5	15.06	7.122	71 54 34.2	.55	51.66	1
13 418	Pavonis	9.5	24.24	7.548	73 29 52.2	.57	52.58	1
13 419	Gould, Z. C., 18h 1895 .	9.0	27.46	7.226	72 17 48.1	.57	51.67	2
13 420	Pavonis	10.0	28.68	+ 7.207	67 4 17.8	2.57	51.67	2
13 421	Octantis	10.0	18 29 40.16	+ 9.457	78 16 12.3	+ 2.59	50.64	1
13 422	Pavonis	10.0	41.59	6.204	67 3 32.7	.59	51.67	2
13 423	Gould, Z. C., 18h 1898 .	10.0	43.18	6.755	70 12 15.7	.59	52.59	1
13 424	Octantis	10.0	46.10	12.030	81 34 54.2	.60	50.62	1
13 425	Octantis	9.0	57.14	+10.915	80 24 23.1	+ 2.61	51.14	2

Number.	Constellation, Name of Star, or Synonym.	Magnitude.	Right Ascension, 1850.0.	Annual Precession.	South Declination, 1850.0.	Annual Precession.	Mean year.	No. of obs.
			h. m. s.	s.	° ′ ″	″		
13426	Gould, 25487	9.0	18 29 59.66	+ 6.351	67 59 16.0	+ 2.62	52.65	2
13427	Octantis	9.2	30 4.75	13.451	82 43 19.6	.62	50.69	2
13428	Octantis	9.0	24.01	22.347	86 3 59.2	.65	51.04	1
13429	Pavonis	9.0	24.04	7.545	73 29 50.3	.65	52.57	1
13430	Gould, Z. C., 18h 1933 .	9.5	26.92	+ 6.528	69 1 15.0	+ 2.66	52.63	2
13431	Pavonis	9.7	18 30 28.87	+ 6.086	66 16 8.1	+ 2.66	52.63	3
13432	Gould, Z. C., 18h 1952	10.0	32.06	7.276	72 30 21.8	.66	51.68	1
13433	Pavonis	9.0	32.97	6.069	66 8 43.0	.67	52.65	2
13434	Gould, 25509	8.0	34.36	6.738	70 7 33.6	.67	52.59	1
13435	Pavonis	10.0	38.81	+ 6.357	68 2 9.7	+ 2.67	52.68	1
13436	Pavonis	10.0	18 30 42.46	+ 6.472	68 42 23.7	+ 2.68	52.65	1
13437	Gould, Z. C., 18h 1950 .	9.0	43.31	6.743	70 9 12.1	.68	52.59	1
13438	Gould, 25518	8.8	46.64	7.433	73 6 1.4	.69	52.59	2
13439	Octantis	9.5	47.85	10.990	80 29 57.2	.69	51.62	1
13440	Pavonis	10.0	30 48.53	+ 6.335	67 54 11.0	+ 2.69	52.68	1
13441	Brisbane, 6453	8.0	18 31 0.81	+ 8.907	77 12 29.7	+ 2.70	52.60	1
13442	Lacaille, 7771	7.0	14.20	7.444	73 8 34.0	.72	52.59	2
13443	Octantis	10.0	16.63	9.116	77 38 17.9	.73	51.68	1
13444	Pavonis	10.0	26.07	6.461	68 39 9.3	.74	52.63	2
13445	Pavonis	9.5	28.05	+ 6.755	70 13 22.2	+ 2.74	52.58	1
13446	Pavonis	10.0	18 31 32.10	+ 7.884	74 36 46.3	+ 2.75	51.66	1
13447	Gould, Z. C., 18h 1996 .	9.8	44.58	6.542	69 6 43.3	.77	52.63	2
13448	Gould, Z. C., 18h 2014 .	8.5	31 48.01	7.355	72 49 8.0	.77	52.60	1
13449	Brisbane, 6457	9.0	32 6.44	7.416	73 3 5.3	.80	52.59	2
13450	Pavonis	9.5	14.03	+ 7.561	73 33 58.3	+ 2.81	52.57	1
13451	Gould, Z. C., 18h 2020 .	9.2	18 32 17.87	+ 6.540	69 6 22.8	+ 2.82	52.63	2
13452	Gould, Z. C., 18h 2034 .	9.5	27.31	6.753	70 13 24.2	.83	52.59	1
13453	Pavonis	9.5	31.00	6.079	66 14 27.2	.84	52.63	3
13454	Gould, Z. C., 18h 2117 .	9.5	37.92	10.018	79 12 45.1	.85	51.64	1
13455	Pavonis	10.0	32 56.95	+ 8.980	77 22 32.5	+ 2.87	52.60	1
13456	Pavonis	9.8	18 33 5.30	+ 6.379	68 11 32.6	+ 2.89	52.66	2
13457	Octantis	12.0	6.28	13.684	82 53 34.0	.89	50.69	1
13458	Octantis	10.0	8.80	8.782	76 57 13.6	.89	51.64	1
13459	Pavonis	9.5	9.09	6.681	69 52 3.8	.89	52.60	1
13460	Pavonis	9.8	12.84	+ 6.068	66 10 29.2	+ 2.90	52.65	2
13461	Gould, Z. C., 18h 2098 .	10.0	18 33 14.58	+ 7.482	73 18 4.3	+ 2.90	52.58	1
13462	Gould, Z. C., 18h 2102 .	7.0	15.35	7.614	73 45 40.7	.90	52.57	1
13463	Pavonis	9.2	20.24	7.218	72 20 25.0	.91	51.67	2
13464	Octantis	9.8	28.77	13.448	82 44 5.4	.92	50.69	2
13465	Gould, Z. C., 18h 2126 .	9.0	29.16	+ 8.381	75 58 37.0	+ 2.92	52.14	2

Number.	Constellation, Name of Star, or Synonym.	Magnitude.	Right Ascension, 1850.0.	Annual Precession.	South Declination, 1850.0.	Annual Precession.	Mean year.	No. of obs.
			h. m. s.	s.	° ′ ″	″		
13466	Brisbane, 6465	9.5	18 33 30.31	+ 7.461	73 13 36.0	+ 2.92	52.58	1
13467	Gould, Z. C., 18^h 2170 .	9.0	31.26	10.072	79 17 59.3	.92	51.18	2
13468	Lacaille, 7781	8.0	44.76	8.048	75 6 57.1	.94	51.11	2
13469	Octantis	8.5	33 51.38	13.361	82 40 39.4	.95	50.69	2
13470	Pavonis	10.0	34 1.58	+ 6.186	67 0 38.7	+ 2.97	52.68	1
13471	Octantis	9.3	18 34 3.35	+18.185	84 59 57.5	+ 2.97	50.91	3
13472	Octantis	9.5	4.92	15.847	84 5 26.2	.97	50.68	1
13473	Gould, Z. C., 18^h 2115	10.0	6.99	6.196	67 3 42.4	.97	52.68	1
13474	Gould, Z. C., 18^h 2146	9.2	8.89	7.229	72 21 17.5	.98	51.67	2
13475	Lacaille, 7789	7.5	18.38	+ 7.619	73 47 16.6	+ 2.99	52.14	2
13476	Gould, 25590	8.0	18 34 23.03	+ 6.254	67 26 19.8	+ 3.00	51.65	2
13477	Pavonis	10.0	24.11	6.752	70 14 33.4	.00	52.58	1
13478	Pavonis	10.0	26.89	6.098	66 24 3.2	.00	52.61	1
13479	Gould, Z. C., 18^h 2187 .	9.0	39.90	8.372	75 59 51.6	.02	52.14	2
13480	Octantis	10.0	40.89	+ 8.755	76 54 22.6	+ 3.02	51.64	1
13481	Octantis	10.0	18 34 51.07	+10.832	80 20 6.0	+ 3.04	50.67	1
13482	Pavonis	10.0	35 0.97	6.750	70 14 18.7	.05	52.58	1
13483	Octantis	10.0	1.72	9.710	78 44 28.5	.05	50.62	1
13484	Lacaille, 7705	8.0	26.81	14.986	83 40 19.8	.09	50.68	1
13485	Pavonis	10.0	30.66	+ 6.363	68 7 48.9	+ 3.09	52.68	1
13486	Pavonis	9.5	18 35 37.03	+ 6.435	68 33 20.2	+ 3.10	52.65	1
13487	Gould, Z. C., 18^h 2191 .	9.5	38.47	6.174	66 56 1.7	.11	52.68	1
13488	Octantis	10.0	38.57	9.435	78 16 20.0	.11	51.14	2
13489	Octantis	9.5	39.48	8.339	75 55 15.2	.11	52.14	2
13490	Gould, 25652	8.8	41.00	+ 8.386	76 0 55.8	+ 3.11	52.14	2
13491	Pavonis	10.0	18 35 42.83	+ 7.680	74 0 24.4	+ 3.11	51.64	2
13492	Gould, Z. C., 18^h 2211 .	9.0	50.81	6.573	69 19 51.6	.12	52.64	1
13493	Gould, Z. C., 18^h 2246 .	9.0	35 57.37	7.946	74 50 21.7	.13	51.65	1
13494	Octantis	9.0	36 7.35	9.147	77 44 12.9	.15	51.68	1
13495	Octantis	9.0	13.38	+16.638	84 26 26.6	+ 3.16	50.66	1
13496	Gould, Z. C., 18^h 2231 .	8.0	18 36 13.61	+ 6.714	70 4 21.1	+ 3.16	52.59	1
13497	Octantis	10.0	25.84	8.583	76 31 42.5	.17	51.64	1
13498	Pavonis	10.0	37.21	6.983	71 21 4.2	.18	51.69	1
13499	Pavonis	9.8	38.78	7.306	72 41 6.6	.19	52.14	2
13500	Octantis	9.2	39.53	+12.468	82 0 12.0	+ 3.19	50.62	3
13501	Octantis	9.2	18 36 44.16	+26.108	86 44 5.2	+ 3.20	51.04	3
13502	Gould, 25673	8.0	37 6.66	6.696	69 59 41.4	.23	52.59	2
13503	Pavonis	9.2	9.92	6.602	69 30 13.6	.24	52.62	2
13504	Octantis	8.8	14.45	43.443	88 7 41.1	.24	51.04	4
13505	Octantis	8.5	29.96	+ 9.741	78 48 32.6	+ 3.25	50.62	1

Number.	Constellation, Name of Star, or Synonym.	Magnitude.	Right Ascension, 1850.0.	Annual Precession.	South Declination, 1850.0.	Annual Precession.	Mean year.	No. of obs.
			h. m. s.	s.	° ′ ″	″		
13 506	Gould, Z. C., 18h 2360	8.2	18 37 30.30	+10.118	79 23 46.3	+ 3.25	51.03	3
13 507	Octantis	9.5	30.53	9.809	78 58 18.2	.25	50.62	1
13 508	Gould, 25682	8.5	30.56	6.502	68 57 50.4	.25	52.62	1
13 509	Gould, Z. C., 18h 2307	8.8	37 53.22	6.310	67 50 15.6	.30	52.65	2
13 510	Pavonis	8.0	38 10.51	+ 6.828	70 39 31.1	+ 3.33	51.70	1
13 511	Octantis	9.5	18 38 11.94	+ 9.583	78 33 6.2	+ 3.33	50.64	1
13 512	Gould, 25704	8.0	17.48	6.953	71 14 12.5	.34	51.69	1
13 513	Gould, Z. C., 18h 2345	9.5	18.14	7.389	73 1 1.8	.34	52.59	2
13 514	Pavonis	10.0	19.24	6.012	65 50 46.7	.34	52.67	1
13 515	Octantis	10.0	23.43	+16.314	84 18 52.5	+ 3.34	50.67	2
13 516	Gould, Z. C., 18h 2358	9.5	18 38 29.10	+ 7.582	73 42 20.8	+ 3.35	52.57	1
13 517	Pavonis	10.0	38 57.91	6.573	69 22 20.0	.39	52.64	1
13 518	Lacaille, 7822	7.0	39 8.44	8.192	75 34 7.8	.41	51.60	2
13 519	Gould, Z. C., 18h 2374	8.0	10.78	6.826	70 39 32.0	.41	52.58	1
13 520	Gould, Z. C., 18h 2398	10.0	12.82	+ 7.974	74 57 18.9	+ 3.41	51.65	1
13 521	Pavonis	9.0	18 39 16.66	+15.529	83 57 44.7	+ 3.42	50.68	1
13 522	Gould, Z. C., 18h 2378	8.5	19.74	6.686	69 57 36.2	.42	52.59	2
13 523	Pavonis	10.0	39.12	6.175	67 0 17.0	.45	52.68	1
13 524	Gould, Z. C., 18h 2404	8.2	43.71	7.002	71 28 19.1	.46	51.67	2
13 525	Octantis	9.0	45.69	+11.155	80 44 44.9	+ 3.46	51.00	3
13 526	Pavonis	9.5	18 39 49.04	+ 6.482	68 52 52.7	+ 3.47	52.62	1
13 527	Pavonis	9.5	40 30.73	6.205	67 12 10.7	.53	50.67	1
13 528	Gould, Z. C., 18h 2478	9.0	42.15	8.674	76 46 49.2	.54	51.64	1
13 529	Pavonis	9.5	42.64	6.694	70 1 49.5	.54	52.59	1
13 530	Gould, Z. C., 18h 2439	8.5	40 49.84	+ 6.304	67 50 46.1	+ 3.55	52.65	2
13 531	Gould, Z. C., 18h 2468	9.0	18 41 15.54	+ 6.891	70 59 32.8	+ 3.59	51.70	1
13 532	Gould, Z. C., 18h 2508	9.0	21.10	8.694	76 49 53.0	.60	51.64	1
13 533	κ Pavonis	4.5	27.28	6.233	67 24 43.6	.61	51.65	2
13 534	Pavonis	9.5	32.29	6.671	69 55 20.8	.62	52.59	1
13 535	Pavonis	9.5	33.52	+ 6.619	69 39 17.3	+ 3.62	52.60	1
13 536	Pavonis	10.0	18 41 37.12	+ 7.144	72 5 31.6	+ 3.62	51.67	1
13 537	Pavonis	9.5	42.78	6.645	69 47 34.6	.63	52.60	1
13 538	Pavonis	10.0	52.94	7.842	74 35 29.7	.64	51.66	1
13 539	Pavonis	9.5	41 55.03	7.342	72 53 7.5	.65	52.60	1
13 540	Lacaille, 7864	8.0	42 5.38	+ 6.097	66 30 35.8	+ 3.66	52.61	1
13 541	Pavonis	10.0	18 42 9.04	+ 7.346	72 54 13.4	+ 3.67	52.60	1
13 542	Lacaille, 7848	7.0	14.28	7.145	72 6 59.1	.67	51.67	1
13 543	Octantis	9.2	14.52	8.786	77 2 38.2	.67	52.12	2
13 544	Gould, 25868	8.0	16.36	11.994	81 37 12.2	.68	50.64	2
13 545	Pavonis	9.0	19.46	+ 6.314	67 55 49.2	+ 3.68	52.65	2

Number.	Constellation, Name of Star, or Synonym.	Magnitude.	Right Ascension, 1850.0.	Annual Precession.	South Declination, 1850.0.	Annual Precession.	Mean year.	No. of obs.
			h. m. s.	s.	° ′ ″	″		
13 546	Octantis	10.0	18 42 21.56	+ 9.643	78 41 22.4	+ 3.69	50.62	1
13 547	Lacaille, 7851	8.5	28.52	7.050	71 43 7.7	.70	51.66	1
13 548	Gould, Z. C., 18h 2543 .	9.2	32.07	7.738	74 16 15.5	.70	51.66	4
13 549	Brisbane, 6504	8.8	33.25	8.464	76 18 10.2	.70	51.64	2
13 550	Lacaille, 7857	7.0	34.16	+ 6.814	70 38 55.8	+ 3.70	52.14	2
13 551	Octantis	9.5	18 42 53.95	+ 9.279	78 3 1.5	+ 3.73	51.63	1
13 552	Octantis	9.5	42 58.18	11.001	80 35 37.0	.74	51.62	1
13 553	Gould, Z. C., 18h 2546	9.0	43 9.02	6.355	68 11 44.9	.76	52.66	2
13 554	Gould, Z. C., 18h 2566	9.0	15.74	7.321	72 49 30.4	.76	52.70	1
13 555	Gould, 25839	8.0	16.06	+ 6.034	66 5 19.1	+ 3.78	52.67	1
13 556	Pavonis	9.0	18 43 16.64	+ 6.034	66 5 20.8	+ 3.76	52.67	1
13 557	Octantis	9.2	17.55	11.134	80 44 54.2	.77	51.15	2
13 558	Octantis	10.0	21.53	13.140	82 34 6.8	.77	50.69	1
13 559	Gould, 25845	8.5	22.24	6.704	70 7 24.2	.77	52.59	1
13 560	Pavonis	9.0	22.61	+ 6.252	67 33 47.6	+ 3.77	52.63	1
13 561	Octantis	10.0	18 43 59.80	8.687	76 50 40.7	+ 3.83	51.64	1
13 562	Pavonis	10.0	44 1.10	7.031	71 39 39.9	.83	51.66	1
13 563	Gould, Z. C., 18h 2597 .	9.0	6.36	6.830	70 44 59.9	.84	52.14	2
13 564	Gould, Z. C., 18h 2593 .	9.2	12.36	6.448	68 45 34.4	.84	52.63	2
13 565	Pavonis	10.0	21.16	+ 6.217	67 21 5.0	+ 3.86	50.67	1
13 566	Octantis	10.0	18 44 21.23	+ 8.868	77 14 44.1	+ 3.86	52.60	1
13 567	Octantis	10.0	24.39	9.103	77 43 34.8	.86	51.68	1
13 568	Gould, Z. C., 18h 2638 .	9.0	33.88	7.894	74 46 52.3	.88	51.65	1
13 569	Octantis	9.5	36.41	11.772	81 25 34.6	.88	50.67	2
13 570	Lacaille, 7880	7.0	47.88	+ 6.138	66 50 29.8	+ 3.90	52.68	1
13 571	Gould, Z. C., 18h 2630 .	9.5	18 44 57.82	+ 6.461	68 50 56.6	+ 3.91	52.62	1
13 572	Pavonis	9.0	45 1.57	6.251	67 34 59.6	.92	52.63	1
13 573	Pavonis	9.0	2.38	7.121	72 3 23.5	.92	51.67	1
13 574	Pavonis	9.0	5.63	6.293	67 50 50.6	.92	52.65	2
13 575	Gould, Z. C., 18h 2666 .	10.0	12.02	+ 7.854	74 40 4.0	+ 3.93	51.65	2
13 576	Pavonis	9.5	18 45 25.44	+ 6.506	69 6 31.8	+ 3.95	52.63	2
13 577	Gould, Z. C., 18h 2677 .	9.2	29.42	7.500	73 30 40.2	.95	52.58	2
13 578	Gould, Z. C., 18h 2652 .	8.0	31.28	6.608	69 39 32.4	.96	52.60	1
13 579	Gould, Z. C., 18h 2655 .	9.2	32.18	6.723	70 15 9.2	.96	52.69	2
13 580	Pavonis	8.8	35.36	+ 7.250	72 35 2.0	+ 3.96	52.14	2
13 581	Lacaille, 7751	8.6	18 45 39.40	+17.932	84 57 34.2	+ 3.97	51.03	4
13 582	Pavonis	10.0	42.81	7.523	73 35 38.3	.97	52.57	1
13 583	Gould, Z. C., 18h 2656	8.5	47.88	6.100	66 35 46.8	.98	52.64	2
13 584	Gould, Z. C., 18h 2740 .	8.0	50.27	10.284	79 42 5.8	.98	50.61	1
13 585	Octantis	10.0	52.49	+15.524	83 59 32.9	+ 3.99	50.68	1

Number.	Constellation, Name of Star, or Synonym.	Magnitude.	Right Ascension, 1850.0.	Annual Precession.	South Declination, 1850.0.	Annual Precession.	Mean year.	No. of obs.
			h. m. s.	s.	° ′ ″	″		
13 586	Gould, Z. C., 18h 2669	8.2	18 45 57.41	+ 6.067	66 22 12.4	+ 3.99	52.64	2
13 587	Pavonis	10.0	46 1.07	12.481	82 4 19.4	4.00	50.65	1
13 588	Pavonis	8.8	2.41	6.222	67 24 55.4	.00	51.65	2
13 589	Pavonis	9.5	6.57	7.588	73 49 23.3	.01	52.57	1
13 590	Pavonis	10.0	8.56	+ 7.632	73 58 18.3	+ 4.01	52.57	1
13 591	Gould, Z. C., 18h 2691	8.0	18 46 9.00	+ 6.369	68 19 42.9	+ 4.01	52.66	2
13 592	Gould, Z. C., 18h 2713	8.8	18.30	7.634	73 58 51.5	.02	51.95	3
13 593	Octantis	9.8	20.94	8.160	75 33 47.7	.03	51.60	2
13 594	Octantis	9.5	33.46	8.687	76 52 20.6	.05	51.64	1
13 595	Pavonis	9.2	34.78	+ 6.044	66 13 22.2	+ 4.05	52.64	2
13 596	Pavonis	9.8	18 46 38.76	+ 7.709	74 13 56.5	+ 4.05	51.68	4
13 597	Pavonis	9.5	40.09	6.511	69 9 30.3	.06	52.63	2
13 598	Pavonis	9.5	42.34	6.086	66 31 13.3	.06	52.61	1
13 599	Pavonis	10.0	46.95	7.815	74 34 1.2	.07	51.66	1
13 600	Octantis	10.0	54.96	+ 8.696	76 53 49.1	+ 4.08	51.64	1
13 601	Octantis	9.2	18 46 59.04	+16.041	84 14 9.6	+ 4.08	50.67	2
13 602	Pavonis	10.0	47 0.95	6.222	67 26 3.2	.09	51.65	2
13 603	Pavonis	10.0	2.72	7.264	72 39 40.2	.09	52.60	1
13 604	Lacaille, 7877	7.8	16.84	7.943	74 57 43.8	.11	51.11	2
13 605	Gould, Z. C., 18h 2752	9.2	20.67	+ 7.167	72 16 45.6	+ 4.11	51.67	2.
13 606	Octantis	9.0	18 47 24.66	+ 9.374	78 16 11.2	+ 4.12	51.14	2
13 607	Pavonis	9.5	25.09	6.744	70 22 40.6	.12	52.59	2
13 608	Lacaille, 7897	6.5	25.74	6.473	68 57 20.0	.12	52.62	1
13 609	Pavonis	10.0	30.17	7.488	73 29 50.5	.13	52.58	1
13 610	Gould, Z. C., 18h 2769	9.5	43.68	+ 7.128	72 7 25.5	+ 4.15	51.67	1
13 611	Octantis	9.0	18 47 43.69	+52.422	88 28 53.7	+ 4.15	51.04	3
13 612	Pavonis	10.0	51.60	7.834	74 38 27.3	.16	51.66	1
13 613	Pavonis	10.0	56.02	6.229	67 29 50.6	.16	52.63	1
13 614	Gould, Z. C., 18h 2775	9.5	47 58.97	7.302	72 49 9.4	.17	52.60	1
13 615	Gould, Z. C., 18h 2773	9.0	48 15.53	+ 6.435	68 45 20.3	+ 4.19	52.63	2
13 616	Gould, Z. C., 18h 2786	9.5	18 48 16.89	+ 7.298	72 48 28.1	+ 4.19	52.60	1
13 617	Gould, Z. C., 18h 2841	10.0	24.77	9.977	79 16 44.5	.21	50.72	1
13 618	Pavonis	10.0	28.71	6.072	66 27 18.7	.21	52.61	1
13 619	Pavonis	10.0	32.55	6.802	70 41 14.4	.22	52.14	2
13 620	Octantis	9.0	36.72	+15.265	83 52 51.6	+ 4.22	50.68	2
13 621	Pavonis	9.5	18 48 37.39	+ 6.431	68 44 19.5	+ 4.22	52.63	2
13 622	Gould, Z. C., 18h 2788	9.0	50.62	6.019	66 5 22.1	.24	52.67	1
13 623	Gould, Z. C., 18h 2801	8.2	52.76	6.716	70 16 22.6	.25	52.59	2
13 624	Gould, Z. C., 18h 2809	9.0	59.41	7.044	71 47 24.2	.25	51.66	2
13 625	Pavonis	9.5	59.83	+ 6.016	66 3 56.6	+ 4.26	52.67	1

Number.	Constellation, Name of Star, or Synonym.	Magnitude.	Right Ascension, 1850.0.	Annual Precession.	South Declination, 1850.0.	Annual Precession.	Mean year.	No. of obs.
			h. m. s.	s.	° ′ ″	″		
13 626	Gould, Z. C., 18^{h} 2795	9.5	18 48 59.95	+ 6.075	66 29 21.7	+ 4.26	52.61	1
13 627	Pavonis	10.0	49 4.08	6.768	70 31 47.7	.26	52.58	1
13 628	Octantis	10.5	14.12	9.680	78 49 0.6	.28	50.62	1
13 629	Octantis	9.0	15.50	9.684	78 49 26.8	.28	50.62	1
13 630	Octantis	9.5	30.94	+22.422	86 8 20.2	+ 4.30	51.09	3
13 631	Pavonis	9.5	18 49 34.28	+ 6.467	68 57 54.8	+ 4.30	52.62	1
13 632	Gould, Z. C., 18^{h} 2854	9.5	38.93	7.487	73 31 21.9	.31	52.58	1
13 633	Octantis	9.0	39.25	16.974	84 38 2.5	.31	50.65	2
13 634	Lacaille, 7890	8.8	40.21	8.560	76 37 0.6	.31	51.69	4
13 635	Gould, 25977	8.0	49 46.13	+ 6.667	70 2 20.3	+ 4.32	52.59	1
13 636	Octantis	8.8	18 50 19.30	+ 8.263	75 53 35.4	+ 4.37	52.14	2
13 637	Lacaille, 7884	8.0	20.58	9.257	78 5 11.4	.37	51.63	1
13 638	Octantis	10.0	29.21	12.778	82 20 36.8	.38	50.69	1
13 639	Octantis	9.5	42.06	8.240	75 50 0.6	.40	52.14	2
13 640	Octantis	9.2	44.00	+ 9.328	78 13 21.0	+ 4.40	51.14	2
13 641	Pavonis	9.5	18 50 47.72	+ 6.525	69 18 23.9	+ 4.41	52.64	1
13 642	Gould, 26051	9.0	50 59.20	9.944	79 15 15.0	.43	50.72	1
13 643	Octantis	10.0	51 17.44	8.713	76 59 10.6	.45	52.60	1
13 644	Octantis	10.0	18.42	8.058	75 20 49.8	.45	50.56	1
13 645	Pavonis	10.0	19.63	+ 6.510	69 14 3.5	+ 4.45	52.64	1
13 646	Pavonis	11.0	18 51 20.77	+11.777	81 28 57.7	+ 4.46	50.62	1
13 647	Gould, 26020	9.8	26.48	6.338	68 14 41.4	.46	52.68	2
13 648	Octantis	9.0	27.59	29.138	87 8 13.8	.47	51.12	7
13 649	Gould, Z. C., 18^{h} 2913	9.0	27.82	6.575	69 35 15.0	.47	52.62	2
13 650	Octantis	9.5	31.52	+28.534	87 4 10.4	+ 4.47	51.22	1
13 651	Octantis	9.0	18 51 38.13	+12.386	82 2 4.6	+ 4.48	50.65	1
13 652	Gould, 26034	7.2	41.66	6.484	69 5 51.6	.49	52.63	2
13 653	Octantis	10.0	46.11	11.361	81 3 55.1	.49	50.69	2
13 654	Gould, Z. C., 18^{h} 2931	9.0	50.51	6.653	70 1 27.7	.50	52.59	1
13 655	Gould, Z. C., 18^{h} 2926	9.0	52.32	+ 6.148	67 2 27.9	+ 4.50	51.67	2
13 656	Octantis	9.5	18 51 53.24	+11.758	81 28 9.7	+ 4.50	50.67	2
13 657	Octantis	9.0	52 15.34	10.185	79 37 15.7	.53	50.61	1
13 658	Lacaille, 7906	7.0	34.01	8.839	77 16 36.9	.56	52.60	1
13 659	Gould, Z. C., 18^{h} 2968	9.0	39.90	6.946	71 25 30.8	.57	51.67	2
13 660	Lacaille, 7928	6.5	44.02	+ 7.024	71 46 13.0	+ 4.57	51.66	2
13 661	Gould, Z. C., 18^{h} 2984	9.2	18 52 56.20	+ 7.172	72 23 14.2	+ 4.59	51.67	2
13 662	Pavonis	10.0	53 0.98	6.270	67 51 17.5	.60	52.63	1
13 663	Gould, Z. C., 18^{h} 2976	9.0	13.64	6.082	66 37 29.2	.62	52.64	2
13 664	Gould, Z. C., 18^{h} 2992	8.2	21.06	6.539	69 25 58.6	.63	52.62	2
13 665	Octantis	8.8	30.37	+18.814	85 16 34.4	+ 4.64	51.11	5

Number.	Constellation, Name of Star, or Synonym.	Magnitude.	Right Ascension, 1850.0.	Annual Precession.	South Declination, 1850.0.	Annual Precession.	Mean year.	No. of obs.
			h. m. s.	s.	° ′ ″	″		
13 666	Octantis	9.5	18 53 37.52	+ 9.306	78 12 54.1	+ 4.65	51.14	2
13 667	Pavonis	10.0	40.02	6.215	67 31 24.6	.65	52.63	1
13 668	Pavonis	9.5	40.32	6.260	67 48 15.5	.65	52.68	1
13 669	Gould, 26086	8.2	51.16	6.252	67 45 40.4	.67	52.65	2
13 670	Pavonis	10.0	55.02	+ 7.544	73 47 22.1	+ 4.68	52.57	1
13 671	Lacaille, 7944	5.2	18 53 56.34	+ 6.398	68 38 43.8	+ 4.68	52.63	2
13 672	Brisbane, 6564	7.5	54 1.21	6.376	68 31 15.5	.68	52.65	1
13 673	Octantis	9.5	23.36	11.985	81 42 14.6	.72	50.72	1
13 674	Pavonis	8.2	25.06	6.702	70 18 15.6	.72	52.59	2
13 675	Octantis	9.8	29.43	+ 9.788	79 2 59.1	+ 4.72	50.67	2
13 676	Octantis	9.2	18 54 36.12	+12.070	81 46 59.4	+ 4.73	50.68	2
13 677	Octantis	9.5	41.40	11.902	81 37 48.1	.74	50.72	1
13 678	Gould, Z. C., 18h 3069 .	8.8	47.71	7.470	73 32 41.9	.75	52.58	2
13 679	Pavonis	8.2	48.60	6.536	69 26 38.5	.75	52.62	2
13 680	Octantis	9.5	49.17	+22.601	86 11 40.1	+ 4.75	51.04	1
13 681	Gould, Z. C., 18h 3059 .	9.0	18 54 51.61	6.851	71 2 0.4	+ 4.76	51.70	1
13 682	Pavonis	10.0	54 53.03	6.011	66 9 35.7	.76	52.67	1
13 683	Octantis	9.5	55 0.46	7.977	75 10 17.2	.77	50.56	1
13 684	Gould, Z. C., 18h 3090 .	9.0	0.77	8.079	75 27 28.7	.77	52.64	1
13 685	Gould, 26114	8.2	3.00	+ 6.390	68 37 34.1	+ 4.77	52.63	2
13 686	Pavonis	9.0	18 55 5.81	+ 6.001	66 5 14.7	+ 4.78	52.67	1
13 687	Pavonis	9.5	10.32	7.306	72 57 4.7	.78	52.60	1
13 688	Octantis	10.0	11.69	8.712	77 2 14.6	.78	52.12	2
13 689	Gould, Z. C., 18h 3070 .	9.5	18.15	6.375	68 32 42.2	.79	52.65	1
13 690	Octantis	9.8	27.66	+11.043	80 45 1.5	+ 4.81	51.15	2
13 691	Gould, Z. C., 18h 3092 .	9.2	18 55 49.42	+ 6.187	67 23 8.0	+ 4.84	51.65	2
13 692	Lacaille, 7935	6.5	55 55.19	8.290	76 2 12.7	.85	51.64	1
13 693	Octantis	9.0	56 1.20	16.169	84 20 35.3	.85	50.67	2
13 694	Octantis	10.0	11.36	10.085	79 31 5.3	.87	50.61	1
13 695	Gould, Z. C., 18h 3133 .	9.2	13.68	+ 7.333	73 4 7.8	+ 4.87	52.59	2
13 696	Pavonis	10.0	18 56 29.52	+ 6.192	67 26 10.1	+ 4.89	52.63	1
13 697	Octantis	9.0	30.46	10.183	79 39 58.9	.90	50.61	1
13 698	Lacaille, 7964	7.0	30.76	5.999	66 6 25.5	.90	52.67	1
13 699	Pavonis	9.0	43.46	6.511	69 20 44.3	.91	52.64	1
13 700	Gould, 26152	8.0	43.99	+ 6.306	68 9 37.8	+ 4.91	52.68	1
13 701	Pavonis	9.5	18 56 45.23	+ 5.973	65 55 39.7	+ 4.92	52.67	1
13 702	Pavonis	9.5	48.92	5.980	65 58 29.9	.92	52.67	1
13 703	Octantis	9.2	56 54.41	8.197	75 48 32.5	.93	52.14	2
13 704	Octantis	10.0	57 3.19	8.219	75 52 10.9	.94	51.64	1
13 705	Octantis	9.8	12.08	+12.066	81 48 8.4	+ 4.95	50.68	2

Number.	Constellation, Name of Star, or Synonym.	Magnitude.	Right Ascension, 1850.0.	Annual Precession.	South Declination, 1850.0.	Annual Precession.	Mean year.	No. of obs.
			h. m. s.	s.	° ′ ″	″		
13706	Pavonis	10.0	18 57 19.22	+ 6.803	70 51 6.4	+ 4.96	51.70	1
13707	Pavonis	10.0	26.17	6.920	71 23 30.5	.97	51.66	1
13708	Gould, Z. C., 19^h 4	9.0	30.06	6.250	67 49 28.8	.98	52.65	2
13709	Gould, 26171	9.0	33.81	6.170	67 18 59.5	.98	50.67	1
13710	Pavonis	10.0	37.85	+ 6.809	70 53 18.6	+ 4.99	51.70	1
13711	Gould, Z. C., 19^h 42	9.7	18 57 56.77	+ 7.636	74 9 51.5	+ 5.02	51.66	3
13712	Octantis	8.6	57 59.45	17.782	84 58 14.3	.02	51.04	5
13713	Pavonis	9.5	58 7.10	7.260	72 49 30.8	.03	52.60	1
13714	Pavonis	10.0	7.41	7.287	72 55 45.8	.03	52.60	1
13715	Octantis	8.8	12.75	+ 70.681	88 54 11.9	+ 5.04	51.13	8
13716	Gould, 26186	9.5	18 58 21.47	+ 6.183	67 25 9.0	+ 5.05	52.63	1
13717	Gould, Z. C., 19^h 37	9.0	22.27	6.267	67 57 7.5	.05	52.68	1
13718	Octantis	9.5	32.43	15.267	83 56 41.5	.07	50.68	1
13719	Gould, Z. C., 19^h 62	8.5	34.54	6.911	71 22 42.8	.07	51.67	2
13720	Gould, Z. C., 19^h 80	8.7	39.71	+ 7.758	74 34 10.8	+ 5.08	51.67	3
13721	Pavonis	9.5	18 58 42.06	+ 6.393	68 43 22.6	+ 5.08	52.63	2
13722	Lacaille, 7969	7.0	44.64	6.224	67 41 27.9	.08	52.63	1
13723	Pavonis	10.0	51.80	6.067	66 38 52.3	.09	52.64	2
13724	Pavonis	9.5	58 55.21	7.244	72 46 48.4	.10	52.60	1
13725	Pavonis	9.5	59 15.99	+ 6.680	70 17 32.9	+ 5.13	52.59	1
13726	Octantis	11.0	18 59 20.00	+ 9.019	77 44 15.5	+ 5.13	51.68	1
13727	Gould, Z. C., 19^h 87	9.5	23.50	6.295	68 8 58.9	.14	52.68	1
13728	Gould, Z. C., 19^h 92	8.8	27.08	6.303	68 12 33.2	.14	52.66	2
13729	Pavonis	9.5	31.19	6.272	68 0 9.4	.15	52.68	1
13730	Octantis	9.5	35.59	+ 9.021	77 44 44.0	+ 5.16	51.21	2
13731	Gould, Z. C., 19^h 132	9.5	18 59 39.56	+ 8.285	76 4 48.8	+ 5.16	51.64	1
13732	Pavonis	9.0	40.73	6.750	70 38 43.4	.16	52.58	1
13733	Octantis	9.0	41.20	10.790	80 29 48.2	.16	51.62	1
13734	Pavonis	9.5	18 59 52.22	7.411	73 25 19.0	.18	52.58	2
13735	Pavonis	10.0	19 0 2.35	+ 6.669	70 15 9.3	+ 5.19	52.59	1
13736	Pavonis	10.0	19 0 3.79	+ 7.192	72 35 52.1	+ 5.20	51.68	1
13737	Pavonis	9.2	5.28	6.911	71 24 36.9	.20	51.67	2
13738	Pavonis	10.0	12.21	7.052	72 1 30.1	.21	51.67	1
13739	Pavonis	10.0	15.99	7.389	73 21 1.0	.21	52.58	1
13740	π Pavonis	6.0	19.86	+ 6.513	69 26 7.2	+ 5.22	52.62	2
13741	Octantis	10.0	19 0 42.22	+ 7.958	75 12 24.3	+ 5.25	50.56	1
13742	Gould, Z. C., 19^h 139	9.0	0 44.86	6.162	67 20 14.6	.25	50.67	1
13743	Gould, Z. C., 19^h 170	8.0	1 0.07	7.255	72 51 38.0	.28	52.60	1
13744	Pavonis	11.0	0.11	6.685	70 21 6.8	.28	52.59	1
13745	Pavonis	10.0	7.44	+ 6.306	68 15 33.8	+ 5.29	52.65	1

Number.	Constellation, Name of Star, or Synonym.	Magnitude.	Right Ascension, 1850.0.	Annual Precession.	South Declination, 1850.0.	Annual Precession.	Mean year.	No. of obs.
			h. m. s.	s.	° ′ ″	″		
13 746	Pavonis	9.0	19 1 10.44	+ 6.194	67 33 8.9	+ 5.29	52.63	1
13 747	Gould, Z. C., 19h 180	7.5	12.76	7.264	72 54 3.4	.29	52.60	1
13 748	Gould, Z. C., 19h 192	9.2	15.53	7.896	75 2 3.3	.30	50.99	3
13 749	Octantis	9.2	16.61	40.695	88 2 12.7	.30	51.15	10
13 750	Lacaille, 7975	9.0	19.15	+ 7.856	74 54 51.5	+ 5.30	51.65	1
13 751	Gould, Z. C., 19h 171	9.0	19 1 26.25	+ 6.248	67 54 26.1	+ 5.31	52.63	1
13 752	Pavonis	10.0	28.78	6.390	68 46 8.5	.32	52.63	2
13 753	Gould, Z. C., 19h 178	9.0	29.07	6.557	69 41 57.2	.32	52.60	1
13 754	Gould, 26275	7.0	39.60	6.17.	67 24 52.6	.33	51.65	2
13 755	Pavonis	10.0	47.27	+ 7.023	71 56 7.2	+ 5.34	51.67	1
13 756	Octantis	11.0	19 1 49.00	+41.067	88 3 26.5	+ 5.34	51.20	1
13 757	Lacaille, 7997	6.5	2 4.94	6.094	66 54 38.1	.37	52.68	1
13 758	Octantis	9.0	21.22	11.373	81 10 52.0	.39	50.62	1
13 759	Gould, Z. C., 19h 268	9.7	22.69	9.099	77 56 20.6	.39	51.35	3
13 760	Gould, Z. C., 19h 239	9.5	42.66	+ 6.685	70 23 25.0	+ 5.42	52.59	2
13 761	Pavonis	9.2	19 2 47.72	+ 5.995	66 14 0.8	+ 5.43	52.64	2
13 762	Gould, Z. C., 19h 257	7.5	56.62	7.251	72 53 2.4	.44	52.60	1
13 763	Pavonis	9.5	2 57.57	6.799	70 56 59.6	.44	51.70	1
13 764	Gould, Z. C., 19h 288	10.0	3 5.20	8.395	76 24 39.6	.45	51.74	1
13 765	Gould, Z. C., 19h 252	8.5	8.21	+ 6.540	69 38 52.2	+ 5.46	52.60	1
13 766	Brisbane, 6588	8.2	19 3 27.92	+12.005	81 48 27.7	+ 5.48	50.68	2
13 767	Pavonis	9.5	31.72	6.127	67 10 23.8	.49	52.68	1
13 768	Pavonis	9.0	44.85	5.946	65 54 12.8	.51	52.67	1
13 769	Pavonis	10.0	47.25	6.256	68 0 38.0	.51	52.68	1
13 770	Pavonis	9.5	48.49	+ 6.220	67 47 11.0	+ 5.51	52.63	1
13 771	Lacaille, 8001	8.0	19 3 49.48	+ 6.630	70 8 2.3	+ 5.51	52.59	1
13 772	Pavonis	10.0	3 56.14	7.511	73 51 10.7	.52	52.57	1
13 773	Pavonis	10.0	4 12.49	6.440	69 7 7.9	.55	52.62	1
13 774	Gould, 26347	8.0	14.18	6.627	70 7 45.6	.55	52.59	1
13 775	Gould, 26387	9.0	33.74	+ 9.382	78 29 39.7	+ 5.58	50.64	1
13 776	Pavonis	10.0	19 4 39.19	+ 6.028	66 31 13.4	+ 5.58	52.61	1
13 777	Pavonis	9.0	43.30	5.939	65 52 26.7	.59	52.67	1
13 778	Pavonis	10.0	49.14	7.810	74 50 19.4	.60	51.65	1
13 779	Pavonis	10.0	52.46	6.829	71 8 11.5	.60	51.69	1
13 780	Pavonis	10.0	56.23	+ 6.261	68 4 19.6	+ 5.61	52.68	1
13 781	Octantis	10.0	19 4 56.65	+ 8.435	76 32 23.8	+ 5.61	51.74	1
13 782	Octantis	9.0	5 2.36	10.553	80 15 52.0	.62	51.14	2
13 783	Gould, Z. C., 19h 406	9.8	7.60	10.210	79 48 24.6	.62	50.64	2
13 784	Brisbane, 6598	7.0	14.72	12.249	82 2 21.7	.63	50.65	1
13 785	Octantis	9.6	20.07	+32.840	87 31 54.7	+ 5.64	51.15	5

No. 13 754. Gould's declination is 10″ too large.

Number.	Constellation, Name of Star, or Synonym.	Magnitude.	Right Ascension, 1850.0.	Annual Precession.	South Declination, 1850.0.	Annual Precession.	Mean year.	No. of obs.
			h. m. s.	s.	° ′ ″	″		
13786	Pavonis	9.0	19 5 22.25	+ 6.535	69 40 10.3	+ 5.64	52.60	1
13787	Gould, Z. C., 19h 358	8.5	26.98	6.744	70 44 25.2	.65	52.14	2
13788	Gould, Z. C., 19h 401	9.0	30.71	8.473	76 38 20.7	.66	51.70	3
13789	Octantis	10.0	36.20	8.626	76 59 44.6	.66	52.12	2
13790	Octantis	9.0	39.41	+15.175	83 57 3.8	+ 5.67	50.68	2
13791	Pavonis	9.8	19 5 48.05	+ 7.185	72 40 47.6	+ 5.68	52.14	4
13792	Gould, Z. C., 19h 380	9.0	50.41	6.966	71 46 31.5	.68	51.66	1
13793	Gould, Z. C., 19h 382	10.0	5 50.66	6.971	71 47 42.1	.68	51.66	1
13794	Pavonis	9.0	6 1.33	5.933	65 52 2.5	.70	52.67	1
13795	Gould, Z. C., 19h 413	8.5	21.01	+ 7.836	74 56 44.4	+ 5.73	51.20	2
13796	Gould, Z. C., 19h 402	9.0	19 6 31.44	+ 6.088	66 59 17.8	+ 5.74	52.68	1
13797	Octantis	8.6	37.87	37.215	87 51 4.8	.75	51.15	10
13798	Pavonis	10.2	41.39	7.532	73 58 31.2	.75	52.16	2
13799	Gould, Z. C., 19h 442	9.2	6 55.72	7.831	74 56 32.2	.77	51.20	2
13800	Gould, Z. C., 19h 430	9.0	7 3.46	+ 7.032	72 5 10.6	+ 5.79	51.67	1
13801	Pavonis	9.2	19 7 13.39	+ 7.154	72 35 15.2	+ 5.80	52.14	2
13802	Gould, Z. C., 19h 463	9.0	17.67	8.182	75 56 19.0	.80	52.14	2
13803	Pavonis	10.0	20.59	7.609	74 14 52.2	.81	51.68	2
13804	Pavonis	8.0	22.71	6.528	69 40 50.2	.81	52.60	1
13805	Pavonis	9.5	30.48	+ 5.973	66 12 1.2	+ 5.82	52.64	2
13806	Octantis	10.0	19 7 36.31	+ 8.571	76 54 13.4	+ 5.83	51.64	1
13807	Gould, Z. C., 19h 440	8.5	39.08	6.005	66 26 9.8	.83	52.61	1
13808	Pavonis	9.5	41.34	6.450	69 15 31.9	.84	52.64	1
13809	Octantis	9.0	46.32	11.691	81 33 34.6	.84	50.67	2
13810	Gould, Z. C., 19h 488	9.5	47.97	+ 8.228	76 4 4.4	+ 5.85	51.64	1
13811	Brisbane, 6613	7.8	19 7 48.65	+ 6.658	70 22 24.4	+ 5.85	52.59	2
13812	Gould, Z. C., 19h 456	9.5	51.82	6.448	69 15 2.8	.85	52.64	1
13813	Brisbane, 6615	8.5	53.15	6.066	66 52 11.3	.85	52.68	1
13814	Gould, Z. C., 19h 455	9.0	7 57.08	6.000	66 24 24.6	.86	52.61	1
13815	Gould, 26424	9.0	8 6.82	+ 6.331	68 34 44.5	+ 5.87	52.65	1
13816	Gould, Z. C., 19h 465	10.0	19 8 11.33	6.088	67 2 2.0	+ 5.88	52.68	1
13817	Gould, Z. C., 19h 484	9.0	12.88	7.084	72 19 46.2	.88	51.67	2
13818	Octantis	9.2	15.22	9.960	79 29 11.8	.89	50.67	2
13819	Gould, 26428	7.0	16.59	6.475	69 24 58.1	.89	52.64	1
13820	Pavonis	9.5	17.22	+ 6.042	66 42 45.4	+ 5.89	52.64	2
13821	Octantis	10.0	19 8 20.35	+13.355	82 54 55.0	+ 5.89	50.69	1
13822	Gould, Z. C., 19h 503	9.8	21.67	7.675	74 28 55.2	.89	51.68	2
13823	Lacaille, 8034	6.5	23.68	6.340	68 38 38.6	.90	52.63	2
13824	Gould, 26436	7.5	32.82	6.321	68 31 52.9	.91	52.65	1
13825	Octantis	9.0	33.86	+17.111	84 47 59.3	+ 5.91	50.64	1

No. 13803. The declination may be 74° 18′ 58.2″.
No. 13809. The declination may be 81° 32′ 38.7″.
No. 13814. Gould's declination is 9.7″ greater.
No. 13815. Gould's declination is 10″ too great.

Number.	Constellation, Name of Star, or Synonym.	Magnitude.	Right Ascension, 1850.0.	Annual Precession.	South Declination, 1850.0.	Annual Precession.	Mean year.	No. of obs.
			h. m. s.	s.	° ′ ″	″		
13 826	Octantis	9.0	19 8 39.08	+18.228	85 10 54.0	+ 5.92	51.13	6
13 827	Lacaille, 8042	7.0	41.66	5.922	65 51 2.1	.92	52.67	1
13 828	Lacaille, 8020	7.8	46.91	7.995	75 27 16.2	.93	51.69	2
13 829	Pavonis	10.0	47.02	6.258	68 9 21.5	.93	52.68	1
13 830	Gould, Z. C., 19h 531	10.0	54.28	+ 8.000	75 28 21.7	+ 5.94	52.64	1
13 831	Octantis	9.5	19 8 54.64	+ 8.949	77 44 23.3	+ 5.94	50.73	1
13 832	Pavonis	10.0	8 59.28	6.242	68 3 34.6	.95	52.68	1
13 833	Brisbane, 6620	8.5	9 10.02	6.942	71 44 34.3	.96	51.66	1
13 834	Pavonis	10.0	14.08	6.237	68 1 59.7	.97	52.68	1
13 835	Pavonis	9.5	18.78	+ 5.928	65 55 10.5	+ 5.97	52.67	1
13 836	Pavonis	10.0	19 9 20.18	+ 6.513	69 39 8.1	+ 5.98	52.60	1
13 837	Octantis	10.0	28.70	8.679	77 10 45.6	.99	52.60	1
13 838	Pavonis	10.0	28.72	6.498	69 34 14.9	.99	52.60	1
13 839	Pavonis	9.5	33.25	6.392	68 58 41.7	.99	52.62	1
13 840	Pavonis	10.0	33.48	+ 6.153	67 30 4.8	+ 5.99	52.63	1
13 841	Pavonis	9.0	19 9 33.73	+ 6.589	70 3 28.3	+ 5.99	52.59	1
13 842	Lacaille, 8036	8.5	34.91	6.940	71 44 42.6	6.00	52.66	1
13 843	Pavonis	9.0	35.68	7.105	72 26 38.3	.00	51.68	1
13 844	Pavonis	8.5	40.27	7.011	72 3 11.9	.00	51.67	1
13 845	Pavonis	10.0	41.42	+ 6.996	71 59 25.2	+ 6.01	51.67	1
13 846	Gould, Z. C., 19h 546	8.8	19 9 43.12	+ 6.726	70 45 25.0	+ 6.01	52.14	2
13 847	Pavonis	10.0	44.67	7.584	74 12 47.4	.01	51.66	1
13 848	Pavonis	9.0	49.86	6.197	67 47 46.3	.02	52.63	1
13 849	Gould, Z. C., 19h 554	8.5	55.30	6.692	70 35 28.8	.02	52.14	2
13 850	Pavonis	9.5	9 59.26	+ 5.915	65 50 31.7	+ 6.03	52.67	1
13 851	Pavonis	9.5	19 10 3.51	+ 7.140	72 35 48.8	+ 6.04	52.60	1
13 852	Lacaille, 8031	7.5	3.78	7.341	73 22 14.6	.04	52.58	1
13 853	Pavonis	9.5	9.85	6.161	67 34 37.0	.04	52.63	1
13 854	Gould, Z. C., 19h 553	9.5	13.57	5.936	66 0 22.9	.05	52.67	1
13 855	Pavonis	8.8	17.18	+ 6.040	66 45 22.3	+ 6.06	52.64	2
13 856	Pavonis	9.3	19 10 22.32	+ 7.164	72 41 58.3	+ 6.06	52.29	3
13 857	Gould, Z. C., 19h 577	9.0	24.42	6.794	71 5 54.0	.07	51.69	2
13 858	Pavonis	8.2	28.20	6.271	68 16 48.3	.07	52.66	2
13 859	Gould, Z. C., 19h 579	9.5	35.87	6.387	68 58 38.5	.08	52.62	1
13 860	Octantis	11.0	37.43	+ 8.568	76 56 55.1	+ 6.08	51.64	1
13 861	Octantis	13.0	19 10 37.57	+ 8.567	76 56 46.5	+ 6.08	51.64	1
13 862	Octantis	10.0	40.18	13.571	83 4 52.4	.09	50.67	1
13 863	Pavonis	9.8	48.80	6.144	67 28 41.0	.10	51.65	2
13 864	Pavonis	9.2	52.31	6.872	71 28 14.8	.10	51.67	7
13 865	Gould, Z. C., 19h 603	9.5	10 57.88	+ 7.252	73 3 18.5	+ 6.11	52.59	3

Number.	Constellation, Name of Star, or Synonym.	Magnitude.	Right Ascension, 1850.0.	Annual Precession.	South Declination, 1850.0.	Annual Precession.	Mean year.	No. of obs.
			h. m. s.	s.	° ′ ″	″		
13 866	Pavonis	9.0	19 11 1.73	+ 6.917	71 40 34.1	+ 6.12	51.66	1
13 867	Pavonis	9.5	4.19	6.075	67 1 20.7	.12	52.68	1
13 868	Gould, Z. C., 19h 602 . .	9.2	15.72	6.492	69 35 5.2	.14	52.62	2
13 869	Gould, 26495	8.5	32.18	5.931	66 0 19.2	.16	52.67	1
13 870	Pavonis	10.0	32.51	+ 6.569	70 0 10.8	+ 6.16	52.59	1
13 871	Pavonis	9.0	19 11 33.18	+ 6.679	70 34 12.6	+ 6.16	52.58	1
13 872	Gould, Z. C., 19h 627 . .	9.0	34.90	6.961	71 53 1.6	.16	51.66	2
13 873	Octantis	9.0	42.34	8.156	75 57 7.4	.17	52.14	2
13 874	Lacaille, 8056	6.8	55.40	6.551	69 55 15.9	.19	52.59	2
13 875	Gould, Z. C., 19h 631 . .	9.2	11 57.46	+ 6.471	69 29 14.4	+ 6.19	52.62	2
13 876	Pavonis	10.0	19 12 9.66	+ 7.387	73 34 57.0	+ 6.21	52.57	1
13 877	Pavonis	10.0	10.08	7.737	74 45 23.2	.21	51.66	1
13 878	Brisbane, 6633	8.2	18.04	7.587	74 16 33.6	.22	51.69	3
13 879	Pavonis	10.0	18.82	6.380	68 59 3.6	.22	52.62	1
13 880	Pavonis	10.0	21.56	+ 7.051	72 17 4.7	+ 6.23	51.67	1
13 881	Gould, 26521	9.5	19 12 23.42	+ 5.964	66 16 41.4	+ 6.23	52.64	2
13 882	Pavonis	10.0	24.49	7.218	72 57 29.4	.23	52.60	1
13 883	Octantis	9.5	26.12	9.668	79 5 40.3	.23	50.62	1
13 884	Lacaille, 8046	9.8	26.12	7.527	74 4 41.0	.23	51.70	2
13 885	Gould, Z. C., 19h 651 . .	8.5	28.18	+ 5.921	65 57 42.5	+ 6.24	52.67	1
13 886	Gould, Z. C., 19h 674 . .	10.0	19 12 29.35	+ 7.045	72 15 42.4	+ 6.24	51.67	1
13 887	Brisbane, 6637	8.8	34.36	7.529	74 5 22.2	.25	51.70	2
13 888	Pavonis	9.5	35.54	6.037	66 48 28.5	.25	52.68	1
13 889	Gould, Z. C., 19h 700 . .	9.0	36.27	7.906	75 16 31.7	.25	50.75	1
13 890	Gould, 26526 . .	9.0	36.28	+ 5.981	66 24 25.7	+ 6.25	52.64	2
13 891	Gould, Z. C., 19h 673 . .	8.8	19 12 37.36	+ 6.680	70 36 8.2	+ 6.25	52.14	2
13 892	Pavonis	9.0	42.12	6.244	68 10 37.1	.26	52.66	2
13 893	Gould, 26544	9.0	48.15	6.752	70 57 27.4	.26	51.70	1
13 894	Pavonis	10.0	56.33	7.720	74 43 12.3	.28	51.66	1
13 895	Pavonis	9.8	12 57.82	+ 7.342	73 26 9.9	+ 6.28	52.58	2
13 896	Gould, Z. C., 19h 703 . .	9.0	19 13 4.65	+ 6.893	71 37 16.6	+ 6.29	51.66	1
13 897	Pavonis	9.5	5.64	7.473	73 54 31.8	.29	52.14	2
13 898	Gould, Z. C., 19h 705 . .	9.0	19.73	6.364	68 55 0.7	.31	52.62	1
13 899	Gould, 26573	9.5	29.53	7.981	75 30 30.8	.32	51.69	2
13 900	Gould, Z. C., 19h 712 . .	9.0	35.93	+ 6.350	68 50 42.3	+ 6.33	52.62	1
13 901	Gould, 26580	8.8	19 13 50.91	+ 7.978	75 30 16.8	+ 6.35	51.70	2
13 902	Octantis	9.9	56.48	152.799	89 30 15.0	.36	51.13	8
13 903	Pavonis	9.5	13 57.70	7.339	73 27 3.7	.36	52.57	1
13 904	Pavonis	9.5	14 4.41	6.598	70 13 18.7	.37	52.59	1
13 905	Gould, Z. C., 19h 761 . .	8.8	7.90	+ 7.991	75 33 0.0	+ 6.37	51.70	2

Number.	Constellation, Name of Star, or Synonym.	Magnitude.	Right Ascension, 1850.0.	Annual Precession.	South Declination, 1850.0.	Annual Precession.	Mean year.	No. of obs.
			h. m. s.	s.	° ′ ″	″		
13906	Gould, Z. C., 19^h 729 . .	9.5	19 14 8.54	+ 5.917	65 58 57.5	+ 6.38	52.67	1
13907	Pavonis	9.0	10.42	6.571	70 5 13.4	.38	52.59	1
13908	Gould, 26576	8.2	20.20	6.620	70 20 29.4	.39	52.59	2
13909	Lacaille, 8059	7.5	28.80	7.431	73 47 38.2	.40	52.57	1
13910	Brisbane, 6651	8.0	31.92	+ 6.146	67 36 2.6	+ 6.41	52.63	1
13911	Pavonis	9.5	19 14 32.31	+ 6.046	66 55 27.8	+ 6.41	52.68	1
13912	Pavonis	11.0	33.59	6.144	67 35 29.1	.41	52.63	1
13913	Pavonis	9.5	34.53	6.147	67 36 21.3	.41	52.63	1
13914	Gould, Z. C., 19^h 762 . .	9.0	43.65	6.571	70 6 9.5	.42	52.59	1
13915	Octantis	9.4	44.60	+18.972	85 26 44.0	+ 6.43	51.09	4
13916	Gould, Z. C., 19^h 784 . .	9.5	19 14 47.38	+ 7.434	73 49 35.8	+ 6.43	52.14	2
13917	Octantis	9.2	47.54	16.680	84 41 58.2	.43	50.65	2
13918	Gould, Z. C., 19^h 764 . .	9.0	14 50.20	6.403	69 16 20.0	.43	52.64	1
13919	Gould, Z. C., 19^h 792 . .	8.5	15 6.95	7.014	72 11 45.1	.46	51.67	2
13920	Pavonis	9.8	9.14	+ 6.077	67 9 16.9	+ 6.46	51.67	2
13921	Gould, Z. C., 19^h 797 . .	9.5	19 15 27.19	+ 6.405	69 18 13.4	+ 6.48	52.64	1
13922	Lacaille, 8078	6.5	31.84	6.322	68 43 50.8	.49	52.63	2
13923	Octantis	7.5	39.76	13.765	83 15 14.0	.50	50.67	1
13924	Octantis	10.0	44.21	9.193	78 19 35.9	.51	50.64	1
13925	Octantis	9.2	47.66	+13.360	82 59 32.7	+ 6.51	50.68	2
13926	Gould, Z. C., 19^h 814 . .	8.5	19 15 49.08	+ 7.059	72 24 7.8	+ 6.51	51.67	2
13927	Gould, Z. C., 19^h 808 . .	9.0	50.26	6.743	70 59 33.9	.52	51.69	1
13928	Pavonis	9.0	15 50.36	6.758	71 3 49.6	.52	51.70	1
13929	Pavonis	10.0	16 1.64	7.253	73 10 35.0	.53	52.58	1
13930	Pavonis	10.0	6.83	+ 6.230	68 11 4.9	+ 6.54	52.68	1
13931	Gould, Z. C., 19^h 819 . .	9.0	19 16 10.95	+ 6.761	71 5 12.0	+ 6.54	51.69	2
13932	Gould, Z. C., 19^h 815 . .	9.0	17.24	6.065	67 6 47.3	.55	51.67	2
13933	Gould, 26616	8.0	31.06	6.027	66 51 5.8	.57	52.68	1
13934	Gould, Z. C., 19^h 857 . .	9.0	50.08	7.112	72 38 41.0	.60	52.14	2
13935	Pavonis	9.5	52.24	+ 7.228	73 6 2.9	+ 6.60	52.59	2
13936	Pavonis	9.0	19 16 59.08	+ 6.659	70 36 53.1	+ 6.61	52.14	2
13937	Gould, Z. C., 19^h 890 . .	9.0	17 17.32	7.825	75 7 58.2	.64	51.20	2
13938	Pavonis	9.5	18.41	6.362	69 1 3.5	.64	52.64	1
13939	Pavonis	10.0	46.78	6.380	69 8 5.6	.68	52.63	2
13940	Octantis	9.5	50.66	+ 8.762	77 30 22.8	+ 6.68	51.67	2
13941	Gould, Z. C., 19^h 904 . .	9.5	19 17 52.80	+ 6.966	72 3 45.5	+ 6.69	51.67	1
13942	Gould, Z. C., 19^h 906 . .	8.5	18 2.71	6.695	70 49 3.3	.70	51.70	1
13943	Pavonis	10.0	7.06	5.902	65 59 50.8	.70	52.67	1
13944	Pavonis	9.2	7.22	7.417	73 49 32.1	.70	52.17	2
13945	Gould, Z. C., 19^h 930 . .	9.0	9.22	+ 8.022	75 43 4.5	+ 6.71	52.64	1

Number.	Constellation, Name of Star, or Synonym.	Magnitude.	Right Ascension, 1850.0.	Annual Precession.	South Declination, 1850.0.	Annual Precession.	Mean year.	No. of obs.
			h. m. s.	s.	° ′ ″	″		
13 946	Gould, Z. C., 19h 944 . .	9.0	19 18 34.38	+ 7.625	74 32 26.8	+ 6.74	51.68	2
13 947	Octantis	9.8	37.78	20.362	85 50 4.6	.75	51.17	6
13 948	Gould, 26663	7.8	38.54	6.228	68 14 59.4	.75	52.66	2
13 949	Gould, Z. C., 19h 965 . .	9.5	45.94	8.276	76 24 2.1	.76	51.74	2
13 950	Gould, 26667	9.0	46.48	+ 6.310	68 45 27.9	+ 6.76	52.63	2
13 951	Pavonis	10.0	19 18 52.57	+ 6.851	71 34 42.0	+ 6.77	51.66	1
13 952	Gould, Z. C., 19h 939 . .	8.0	18 53.79	6.583	70 16 41.9	.77	52.59	2
13 953	Gould, Z. C., 19h 938 . .	9.0	19 1.84	6.165	67 51 44.9	.78	52.65	2
13 954	Octantis	9.5	13.04	9.448	78 50 16.7	.80	50.62	1
13 955	Octantis	7.5	18.87	+10.520	80 24 51.8	+ 6.80	51.14	2
13 956	Lacaille, 8096	8.0	19 19 25.24	+ 6.417	69 23 48.6	+ 6.81	52.62	2
13 957	Octantis	10.0	28.35	10.416	80 17 2.0	.82	50.67	1
13 958	Gould, 26688	9.0	35.75	6.418	69 24 19.6	.83	52.62	2
13 959	Pavonis	9.0	50.31	6.474	69 43 34.5	.85	52.60	1
13 960	Lacaille, 8076	7.5	52.36	+ 8.933	77 53 51.0	+ 6.85	51.18	2
13 961	Octantis	10.0	19 19 52.89	+ 8.493	76 56 57.2	+ 6.85	51.64	1
13 962	Octantis	9.5	19 58.60	18.283	85 16 40.0	.86	51.14	7
13 963	Gould, Z. C., 19h 986 . .	9.0	20 0.44	6.520	69 58 40.8	.86	52.59	2
13 964	Pavonis	9.8	21.30	6.154	67 50 4.0	.89	52.15	2
13 965	Octantis	10.0	36.40	+15.162	84 4 20.7	+ 6.91	50.68	1
13 966	Gould, Z. C., 19h 1016 .	9.0	19 20 43.06	+ 6.544	70 7 51.6	+ 6.92	52.59	1
13 967	Pavonis	9.5	50.20	6.157	67 51 54.2	.93	52.65	2
13 968	Gould, Z. C., 19h 1038 .	10.0	52.77	7.204	73 6 33.4	.93	52.59	1
13 969	Gould, Z. C., 19h 1022 .	9.0	20 57.58	6.280	68 38 21.4	.94	52.63	2
13 970	Gould, Z. C., 19h 1036 .	9.5	21 5.96	+ 6.500	69 54 7.3	+ 6.95	52.59	2
13 971	Gould, Z. C., 19h 1030 .	10.5	19 21 10.36	+ 5.988	66 43 45.6	+ 6.96	52.68	2
13 972	Gould, Z. C., 19h 1031 .	10.0	10.82	5.987	66 43 19.7	.96	52.68	2
13 973	Octantis	9.4	16.62	16.695	84 44 28.7	.96	50.85	4
13 974	Pavonis	9.5	19.00	6.328	68 56 30.3	.97	52.62	1
13 975	Octantis	9.0	21.05	+13.247	82 58 33.3	+ 6.97	50.68	2
13 976	Pavonis	9.0	19 21 22.18	+ 5.886	65 59 15.2	+ 6.97	52.67	1
13 977	Gould, Z. C., 19h 1070 .	9.5	31.81	8.074	75 56 4.6	.99	52.14	2
13 978	Octantis	9.5	32.71	8.083	75 57 24.8	.99	52.14	2
13 979	Octantis	10.0	36.34	9.158	78 21 55.0	6.99	51.14	2
13 980	Octantis	10.0	46.83	+10.604	80 33 17.6	+ 7.01	51.62	1
13 981	Pavonis	9.0	19 21 47.82	+ 6.098	67 30 43.9	+ 7.01	51.65	2
13 982	Octantis	8.5	49.93	15.270	84 8 9.6	.01	50.68	1
13 983	Octantis	10.0	21 52.61	21.476	86 6 22.8	.01	51.23	1
13 984	Octantis	9.2	22 5.70	14.724	83 51 57.2	.03	50.68	2
13 985	Gould, Z. C., 19h 1074 .	9.0	9.33	+ 6.952	72 6 53.4	+ 7.04	51.67	1

No. 13 968. Gould's declination is 1′ less.

Number.	Constellation, Name of Star, or Synonym.	Magnitude.	Right Ascension, 1850.0.	Annual Precession.	South Declination, 1850.0.	Annual Precession.	Mean year.	No. of obs.
			h. m. s.	s.	° ′ ″	″		
13 986	Octantis	10.0	19 22 9.96	+ 9.645	79 12 56.9	+ 7.04	50.72	1
13 987	Gould, Z. C., 19h 1062 .	9.0	10.23	5.900	66 7 3.6	.04	52.67	1
13 988	Octantis	10.0	10.81	10.444	80 21 38.8	.04	51.61	1
13 989	Brisbane, 6680	7.5	12.94	5.895	66 5 11.5	.04	52.67	1
13 990	Octantis	9.0	18.07	+ 9.312	78 39 27.0	+ 7.05	50.97	3
13 991	Octantis	8.5	19 22 19.01	+12.587	82 30 19.9	+ 7.05	50.69	1
13 992	Pavonis	10.0	27.49	6.789	71 23 40.3	.06	51.69	1
13 993	Octantis	9.8	35.82	38.223	87 57 41.3	.07	51.14	4
13 994	Octantis	9.0	38.23	11.562	81 37 1.9	.08	50.72	1
13 995	Gould, Z. C., 19h 1102 .	9.0	43.57	+ 7.075	72 38 45.6	+ 7.08	52.14	2
13 996	Gould, Z. C., 19h 1100 .	9.0	19 22 54.52	+ 6.540	70 10 20.0	+ 7.10	52.59	1
13 997	Pavonis	9.5	22 58.61	7.000	72 20 39.0	.10	51.68	1
13 998	Gould, Z. C., 19h 1108 .	9.5	23 9.38	6.643	70 42 25.0	.12	52.14	2
13 999	Octantis	9.5	10.94	10.430	80 21 34.2	.12	50.67	1
14 000	Pavonis	10.0	11.31	+ 5.992	66 49 44.8	+ 7.12	52.68	1
14 001	Octantis	9.5	19 23 13.24	+ 8.751	77 35 13.6	+ 7.12	52.60	1
14 002	Octantis	9.5	17.77	8.988	78 4 18.3	.13	51.63	1
14 003	Gould, Z. C., 19h 1112	9.5	18.14	6.522	70 5 27.1	.13	52.59	1
14 004	Gould, 26794	8.0	25.24	7.985	75 43 53.0	.14	52.64	1
14 005	Lacaille, 8114	7.0	28.49	+ 6.443	69 39 46.0	+ 7.14	52.60	1
14 006	Pavonis	10.0	19 23 30.87	+ 7.601	74 34 49.6	+ 7.15	51.66	1
14 007	Pavonis	9.0	32.35	6.155	67 56 33.6	.15	52.65	2
14 008	Lacaille, 8113	7.0	32.67	6.490	69 55 38.2	.15	52.59	2
14 009	Lacaille, 8119	6.8	51.56	5.908	66 14 17.8	.18	52.68	2
14 010	Pavonis	9.5	23 59.70	+ 7.262	73 24 42.0	+ 7.19	52.57	1
14 011	Pavonis	8.5	19 24 4.36	+ 5.995	66 52 42.3	+ 7.19	52.68	1
14 012	Pavonis	9.0	7.24	7.095	72 45 58.2	.20	52.14	2
14 013	Pavonis	9.0	16.16	6.867	71 47 57.2	.21	51.66	2
14 014	Lacaille, 8108	8.0	18.71	7.761	75 6 3.6	.21	51.20	2
14 015	Octantis	9.5	22.94	+ 8.623	77 20 3.5	+ 7.22	52.60	1
14 016	Gould, Z. C., 19h 1158 .	9.0	19 24 34.70	+ 5.924	66 22 54.8	+ 7.23	52.68	2
14 017	Pavonis	9.2	34.95	6.125	67 47 0.6	.23	52.65	2
14 018	Pavonis	10.0	50.85	6.871	71 50 3.3	.26	51.67	1
14 019	Gould, Z. C., 19h 1172 .	10.0	51.62	6.314	68 58 4.9	.26	52.62	1
14 020	Pavonis	9.5	51.83	+ 6.926	72 4 39.9	+ 7.26	51.67	1
14 021	Octantis	9.8	19 24 53.36	+17.229	84 58 6.6	+ 7.26	50.97	3
14 022	Octantis	9.0	55.15	9.525	79 4 5.6	.26	50.62	1
14 023	Lacaille, 8127	6.5	24 58.98	6.009	67 0 42.5	.27	52.68	1
14 024	Pavonis	9.5	25 6.18	7.175	73 6 30.4	.28	52.59	2
14 025	Gould, Z. C., 19h 1191 .	9.5	6.99	+ 7.025	72 30 23.3	+ 7.28	51.68	1

Number.	Constellation, Name of Star, or Synonym.	Magnitude.	Right Ascension, 1850.0.	Annual Precession.	South Declination, 1850.0.	Annual Precession.	Mean year.	No. of obs.
			h. m. s.	s.	° ′ ″	″		
14 026	Gould, Z. C., 19h 1234 .	9.5	19 25 7.39	+ 9.187	78 29 2.8	+ 7.28	50.64	1
14 027	Gould, Z. C., 19h 1198 .	9.2	8.55	7.298	73 34 22.5	.28	52.29	3
14 028	Octantis	10.0	12.55	9.042	78 12 52.2	.29	51.14	2
14 029	Pavonis	10.0	14.75	7.459	74 8 56.6	.29	51.71	1
14 030	Gould, Z. C., 19h 1200 .	9.0	17.47	+ 7.033	72 32 34.9	+ 7.29	51.68	1
14 031	Gould, Z. C., 19h 1235 .	8.7	19 25 23.08	+ 8.655	77 25 36.0	+ 7.30	51.68	3
14 032	Pavonis	9.5	26.15	6.905	72 0 8.7	.30	51.67	1
14 033	Lacaille, 8112	8.0	28.92	7.652	74 47 33.1	.31	51.65	1
14 034	Pavonis	9.5	47.86	6.047	67 18 12.7	.33	50.67	1
14 035	Octantis	9.8	49.88	+10.979	81 3 24.4	+ 7.34	50.65	2
14 036	Gould, Z. C., 19h 1244 .	10.0	19 25 56.27	+ 7.625	74 42 56.8	+ 7.35	51.65	2
14 037	Pavonis	8.5	26 1.21	6.948	72 12 32.1	.35	51.67	1
14 038	Octantis	10.0	4.10	12.215	82 15 6.8	.36	50.69	1
14 039	Octantis	9.5	5.76	12.842	82 44 40.8	.36	50.69	2
14 040	Pavonis	9.8	11.50	+ 6.777	71 26 38.9	+ 7.37	51.67	2
14 041	Gould, Z. C., 19h 1246 .	8.8	19 26 11.70	+ 7.038	72 35 29.0	+ 7.37	52.14	2
14 042	Pavonis	10.0	19.25	7.491	74 17 9.8	.38	51.66	1
14 043	Gould, Z. C., 19h 1231 .	9.5	21.54	5.986	66 53 43.9	.38	52.68	1
14 044	Octantis	10.0	28.76	8.323	76 40 55.5	.39	51.64	1
14 045	Gould, Z. C., 19h 1253 .	9.8	29.50	+ 6.683	71 0 28.6	+ 7.39	51.69	2
14 046	Pavonis	10.0	19 26 38.82	+ 7.099	72 51 1.8	+ 7.40	52.59	1
14 047	Gould, Z. C., 19h 1268 .	9.0	27 2.44	6.227	68 30 51.9	.44	52.65	1
14 048	Lacaille, 8141	6.2	2.62	5.885	66 11 8.2	.44	52.68	2
14 049	Gould, Z. C., 19h 1272 .	9.0	5.18	6.248	68 38 42.0	.44	52.63	2
14 050	Octantis	9.8	9.81	+13.391	83 8 23.0	+ 7.45	50.68	2
14 051	Octantis	10.0	19 27 17.42	+ 7.921	75 38 29.1	+ 7.46	52.64	1
14 052	Octantis	10.0	19.90	7.717	75 2 22.9	.46	50.75	1
14 053	Gould, Z. C., 19h 1286 .	9.0	24.97	6.440	69 46 22.7	.47	52.60	1
14 054	Pavonis	9.0	26.04	6.533	70 16 25.4	.47	52.59	2
14 055	Pavonis	10.0	40.25	+ 5.849	65 56 7.6	+ 7.49	52.67	1
14 056	Pavonis	9.5	19 27 46.94	+ 5.879	66 9 54.2	+ 7.50	52.67	1
14 057	Pavonis	10.0	51.60	7.044	72 39 44.4	.50	52.59	1
14 058	Gould, Z. C., 19h 1301 .	9.0	57.24	6.155	68 5 47.9	.51	52.68	1
14 059	Gould, Z. C., 19h 1340 .	9.5	27 57.49	7.782	75 15 15.6	.51	50.75	1
14 060	Octantis	9.6	28 1.32	+16.776	84 49 51.9	+ 7.52	50.92	5
14 061	Lacaille, 8094	7.0	19 28 4.03	+11.580	81 42 37.8	+ 7.52	50.72	1
14 062	Pavonis	9.5	6.90	6.405	69 36 7.7	.52	52.60	1
14 063	Gould, Z. C., 19h 1310 .	9.8	11.24	6.017	67 10 49.7	.53	50.67	2
14 064	Gould, Z. C., 19h 1313 .	10.0	11.36	6.162	68 8 40.1	.53	52.68	1
14 065	Lacaille, 8118	7.2	13.00	+ 8.552	77 15 23.4	+ 7.53	52.16	2

No. 14 042. Declination may be 74° 16′ 29.0″.

Number.	Constellation, Name of Star, or Synonym.	Magnitude.	Right Ascension, 1850.0.	Annual Precession.	South Declination, 1850.0.	Annual Precession.	Mean year.	No. of obs.
			h. m. s.	s.	° ′ ″	″		
14 066	Pavonis	9.5	19 28 14.31	+ 6.779	71 30 57.6	+ 7.53	51.66	1
14 067	Pavonis	10.0	18.71	5.849	65 57 27.2	.54	52.67	1
14 068	Gould, 26888	9.0	39.11	6.032	67 17 58.7	.57	50.67	1
14 069	Gould, Z. C., 19^h 1344	9.0	46.61	6.147	68 4 8.4	.58	52.68	1
14 070	Pavonis	9.5	52.57	+ 6.112	67 50 44.1	+ 7.58	52.63	1
14 071	Octantis	10.0	19 28 53.35	+ 8.810	77 49 17.2	+ 7.59	51.63	1
14 072	Pavonis	9.5	53.99	6.454	69 53 55.2	.59	52.59	2
14 073	Gould, Z. C., 19^h 1371 .	9.2	28 58.02	6.954	72 19 8.0	.59	51.67	2
14 074	Gould, 26893	9.0	29 3.28	6.028	67 17 16.3	.60	50.67	1
14 075	Lacaille, 8148	7.8	3.64	+ 6.539	70 21 29.0	+ 7.60	52.59	2
14 076	Octantis	9.2	19 29 5.24	+14.442	83 47 21.0	+ 7.60	50.68	2
14 077	Pavonis	10.0	9.22	8.328	76 45 18.6	.61	51.68	2
14 078	Lacaille, 8147	7.2	11.78	6.658	70 58 8.2	.61	51.69	2
14 079	Pavonis	9.8	11.96	7.470	74 17 30.6	.61	51.68	2
14 080	Gould, 26897	9.8	16.77	+ 5.934	66 37 36.2	+ 7.62	52.68	2
14 081	Brisbane, 6709	8.0	19 29 19.66	+ 5.934	66 37 50.2	+ 7.62	52.68	2
14 082	Pavonis	10.0	24.05	6.297	69 1 8.8	.63	52.64	1
14 083	Pavonis	10.0	29 46.58	6.248	68 44 19.7	.66	52.65	1
14 084	Pavonis	10.0	30 6.42	6.579	70 36 5.2	.68	52.14	2
14 085	Pavonis	9.5	7.52	+ 7.052	72 45 42.4	+ 7.69	52.59	1
14 086	Gould, Z. C., 19^h 1407 .	9.0	19 30 28.62	+ 6.040	67 25 22.9	+ 7.71	52.63	1
14 087	Lacaille, 8145	9.0	29.72	7.515	74 28 38.8	.72	51.68	2
14 088	Gould, Z. C., 19^h 1418 .	9.2	32.62	6.654	70 59 28.0	.72	51.69	2
14 089	Octantis	9.2	39.03	14.465	83 49 9.8	.73	50.68	2
14 090	Pavonis	9.8	39.33	+ 6.759	71 29 55.2	+ 7.73	51.67	2
14 091	Gould, Z. C., 19^h 1412 .	9.0	19 30 39.36	+ 5.819	65 48 48.0	+ 7.73	52.67	1
14 092	Pavonis	9.5	44.30	5.818	65 48 32.4	.74	52.67	1
14 093	Gould, Z. C., 19^h 1481 .	9.0	45.38	9.817	79 38 8.7	.74	50.61	1
14 094	Octantis	9.8	46.56	8.090	76 11 26.2	.74	51.69	2
14 095	Gould, Z. C., 19^h 1433 .	10.0	48.34	+ 6.658	71 0 31.6	+ 7.74	51.69	2
14 096	Pavonis	9.5	19 30 54.37	+ 6.047	67 29 9.1	+ 7.75	52.63	1
14 097	Pavonis	10.0	57.10	6.681	71 8 1.3	.75	51.70	1
14 098	Pavonis	9.0	30 58.97	6.069	67 38 3.9	.75	52.63	1
14 099	Octantis	9.5	31 3.29	29.475	87 19 35.0	.76	51.04	1
14 100	Gould, 26952	8.0	8.28	+ 6.032	67 23 39.6	+ 7.77	51.65	2
14 101	Gould, Z. C., 19^h 1467 .	8.8	19 31 16.18	+ 7.816	75 26 19.7	+ 7.78	51.70	2
14 102	Octantis	10.0	19.43	8.432	77 3 0.3	.78	52.60	1
14 103	Lacaille, 8151	7.8	20.73	7.335	73 52 38.0	.78	52.00	3
14 104	Gould, Z. C., 19^h 1451 .	9.5	36.11	5.906	66 30 40.1	.80	52.68	1
14 105	Gould, Z. C., 19^h 1458 .	8.2	36.91	+ 6.280	68 59 39.6	+ 7.81	52.63	2

Number.	Constellation, Name of Star, or Synonym.	Magnitude.	Right Ascension, 1850.0.	Annual Precession.	South Declination, 1850.0.	Annual Precession.	Mean year.	No. of obs.
			h. m. s.	s.	° ′ ″	″		
14 106	Octantis	10.0	19 31 37.39	+ 9.286	78 47 20.6	+ 7.81	50.62	1
14 107	Pavonis	9.8	40.80	5.868	66 13 51.5	.81	52.68	2
14 108	Pavonis	9.8	57.54	7.245	73 34 1.6	.83	52.15	2
14 109	Pavonis	9.7	57.73	7.360	73 58 56.2	.83	51.00	3
14 110	Octantis	9.8	31 59.68	+12.662	82 41 12.8	+ 7.84	50.69	2
14 111	Lacaille, 8156	7.0	19 32 2.14	+ 7.062	72 51 35.4	+ 7.84	52.17	2
14 112	Octantis	9.5	6.36	14.254	83 43 13.1	.85	50.68	1
14 113	Gould, Z. C., 19^h 1480	9.5	14.27	6.184	68 25 54.0	.86	52.65	1
14 114	Octantis	10.0	19.95	8.489	77 12 22.6	.86	52.60	1
14 115	Gould, Z. C., 19^h 1513	10.0	21.84	+ 7.401	74 8 23.2	+ 7.87	51.70	2
14 116	Gould, Z. C., 19^h 1499	9.5	19 32 32.19	+ 6.265	68 56 16.5	+ 7.88	52.62	1
14 117	Gould, 26991	9.0	33.01	6.892	72 9 26.6	.88	51.70	3
14 118	Pavonis	9.8	42.50	7.490	74 27 12.7	.89	51.68	2
14 119	Octantis	8.8	42.66	11.757	81 56 39.8	.89	50.68	2
14 120	Gould, Z. C., 19^h 1542	9.0	46.53	+ 8.208	76 32 30.6	+ 7.90	51.73	1
14 121	Pavonis	9.0	19 32 49.66	+ 6.908	72 14 11.7	+ 7.90	51.69	3
14 122	Stone, 10626	8.5	56.47	7.150	73 14 2.9	.91	52.58	1
14 123	Gould, Z. C., 19^h 1521	9.0	32 57.95	6.549	70 32 34.2	.91	52.58	1
14 124	Gould, Z. C., 19^h 1519	9.0	33 0.05	6.376	69 36 16.3	.92	52.62	2
14 125	Gould, Z. C., 19^h 1518	9.2	1.66	+ 6.284	69 4 10.4	+ 7.92	52.63	2
14 126	Octantis	9.9	19 33 18.90	+19.251	85 39 37.3	+ 7.94	51.18	7
14 127	Octantis	8.7	29.36	23.286	86 31 28.5	.96	51.17	9
14 128	Octantis	9.0	30.45	11.411	81 37 38.1	.96	50.72	1
14 129	Octantis	10.0	38.05	9.060	78 25 9.3	.97	50.64	1
14 130	Lacaille, 8168	8.0	42.25	+ 7.124	73 9 16.0	+ 7.97	52.31	3
14 131	Octantis	9.8	19 33 56.74	+25.759	86 54 21.3	+ 7.99	51.13	4
14 132	Pavonis	10.0	34 14.24	7.414	74 14 17.9	8.02	51.66	1
14 133	Lacaille, 8177	7.5	16.70	6.385	69 41 54.8	.02	52.60	1
14 134	Gould, Z. C., 19^h 1575	9.0	20.43	6.544	70 33 31.4	.02	52.58	1
14 135	Gould, 27030	8.8	42.92	+ 5.846	66 10 57.0	+ 8.05	52.68	2
14 136	Octantis	9.4	19 34 48.74	+21.662	86 13 55.0	+ 8.06	51.17	7
14 137	Brisbane, 6723	7.8	58.00	5.865	66 20 17.0	.07	52.68	2
14 138	Pavonis	9.0	58.67	6.803	71 50 17.8	.08	51.66	1
14 139	Octantis	10.0	34 58.89	8.745	77 49 36.7	.08	51.63	1
14 140	Gould, Z. C., 19^h 1628	9.5	35 2.37	+ 8.491	77 16 26.5	+ 8.08	52.60	1
14 141	Pavonis	10.0	19 35 5.49	+ 7.360	74 4 23.8	+ 8.08	51.66	1
14 142	Lacaille, 8184	7.0	6.78	5.979	67 10 30.0	.09	52.67	2
14 143	Octantis	10.0	12.81	8.793	77 55 50.2	.09	51.63	1
14 144	Gould, Z. C., 19^h 1598	9.0	14.14	5.820	66 0 25.1	.10	52.67	1
14 145	Pavonis	9.5	27.35	+ 6.448	70 5 26.8	+ 8.11	52.59	1

Number.	Constellation, Name of Star, or Synonym.	Magnitude.	Right Ascension, 1850.0.	Annual Precession.	South Declination, 1850.0.	Annual Precession.	Mean year.	No. of obs.
			h. m. s.	s.	° ′ ″	″		
14 146	Pavonis	8.5	19 35 32.49	+ 5.944	66 56 28.1	+ 8.12	52.68	1
14 147	Gould, Z. C., 19h 1626	9.5	45.58	6.627	71 1 52.6	.14	51.69	2
14 148	Gould, Z. C., 19h 1631	9.0	47.62	6.754	71 38 28.9	.14	51.66	1
14 149	Pavonis	10.0	47.74	7.544	74 42 58.7	.14	51.65	2
14 150	Pavonis	9.0	50.85	+ 5.835	66 8 41.8	+ 8.15	52.67	1
14 151	Pavonis	9.2	19 35 54.04	+ 6.397	69 49 29.5	+ 8.15	52.59	2
14 152	Pavonis	10.0	55.92	5.864	66 21 29.5	.15	52.68	1
14 153	Pavonis	10.0	58.42	6.150	68 21 9.7	.16	52.68	1
14 154	Brisbane, 6724	8.8	35 59.47	7.028	72 50 30.2	.16	52.17	2
14 155	Octantis	10.0	36 6.22	+ 8.293	76 50 9.8	+ 8.17	51.64	1
14 156	Gould, Z. C., 19h 1686	9.5	19 36 14.36	+ 8.483	77 17 3.1	+ 8.18	52.60	1
14 157	Octantis	9.0	22.82	13.402	83 15 41.6	.19	50.68	2
14 158	Gould, Z. C., 19h 1645	9.0	28.05	5.812	65 59 50.8	.20	52.68	1
14 159	Gould, Z. C., 19h 1657	9.5	28.42	6.639	71 6 44.3	.20	51.69	2
14 160	Pavonis	9.5	28.73	+ 5.972	67 10 51.7	+ 8.20	52.68	1
14 161	Lacaille, 8195	6.5	19 36 36.72	+ 5.807	65 57 50.7	+ 8.21	52.67	1
14 162	Pavonis	10.0	41.70	5.880	66 31 1.8	.21	52.68	1
14 163	Pavonis	9.5	47.14	7.053	72 58 4.9	.22	52.58	1
14 164	Gould, Z. C., 19h 1712	10.0	48.72	8.429	77 10 29.6	.22	52.12	2
14 165	Gould, Z. C., 19h 1671	8.5	36 50.69	+ 6.168	68 30 19.1	+ 8.23	52.65	1
14 166	Octantis	10.0	19 37 1.00	+11.052	81 18 50.6	+ 8.24	50.62	1
14 167	Pavonis	10.0	2.55	7.030	72 53 2.8	.24	51.74	1
14 168	Lacaille, 8172	9.0	7.62	8.652	77 40 47.4	.25	51.22	2
14 169	Lacaille, 8187	7.8	7.80	6.774	71 46 35.0	.25	51.66	2
14 170	Gould, Z. C., 19h 1697	9.0	9.43	+ 6.759	71 42 38.6	+ 8.25	51.66	1
14 171	Stone, 10652	9.0	19 37 17.29	+ 6.800	71 54 12.7	+ 8.26	51.66	2
14 172	Pavonis	9.5	28.73	5.830	66 10 16.8	.28	52.67	1
14 173	Octantis	10.0	30.44	8.828	78 3 11.0	.28	51.63	1
14 174	Octantis	10.0	34.64	7.906	75 51 46.6	.28	51.64	1
14 175	Gould, Z. C., 19h 1714	8.0	42.06	+ 6.476	70 19 22.8	+ 8.29	52.59	2
14 176	Gould, Z. C., 19h 1719	8.5	19 37 47.46	+ 6.167	68 32 4.9	+ 8.30	52.65	1
14 177	Octantis	8.0	37 52.14	13.279	83 12 2.6	.31	50.69	3
14 178	Gould, Z. C., 19h 1753	9.8	38 1.72	8.215	76 41 29.7	.32	51.68	2
14 179	Gould, Z. C., 19h 1737	9.5	13.20	6.751	71 42 27.8	.33	51.66	1
14 180	Gould, Z. C., 19h 1748	9.5	14.91	+ 7.291	73 55 17.2	+ 8.34	51.72	2
14 181	Pavonis	9.2	19 38 17.21	+ 6.397	69 54 49.4	+ 8.34	52.59	2
14 182	Pavonis	8.5	19.28	5.813	66 4 48.2	.34	52.67	1
14 183	Pavonis	10.0	25.37	5.871	66 31 21.6	.35	52.68	1
14 184	Stone, 10661	9.0	37.76	6.795	71 55 22.6	.37	51.66	2
14 185	Pavonis	9.2	42.90	+ 6.238	69 0 25.2	+ 8.37	52.63	2

No. 14 151. Right ascension possibly $19^h\ 36^m\ 43.64^s$.

Number.	Constellation, Name of Star, or Synonym.	Magnitude.	Right Ascension, 1850.0.	Annual Precession.	South Declination, 1850.0	Annual Precession.	Mean year.	No. of obs.
			h. m. s.	s.	° ′ ″	″		
14 186	Pavonis	10.0	19 38 49.79	+ 6.515	70 33 58.8	+ 8.38	51.70	1
14 187	Octantis	10.0	38 53.71	11.326	81 37 46.3	.39	50.72	1
14 188	Gould, 27122	7.5	39 1.87	6.160	68 32 0.7	.40	52.65	1
14 189	Gould, Z. C., 19^h 1771	9.0	7.12	6.977	72 43 48.5	.41	52.01	1
14 190	Gould, 27133	7.8	27.55	+ 6.173	68 38 5.8	+ 8.43	52.63	2
14 191	Octantis	10.5	19 39 29.27	+15.670	84 30 0.0	+ 8.44	50.66	1
14 192	Gould, Z. C., 19^h 1775	9.0	29.32	6.287	69 19 50.3	.44	52.64	1
14 193	Gould, 27140	8.2	36.31	6.512	70 34 57.8	.44	52.14	2
14 194	Pavonis	8.5	37.62	5.998	67 29 11.2	.45	51.65	2
14 195	Gould, Z. C., 19^h 1821	9.5	41.00	+ 8.986	78 24 48.9	+ 8.45	50.64	1
14 196	Pavonis	9.0	19 39 44.92	+ 6.021	67 39 12.2	+ 8.46	52.63	1
14 197	Pavonis	9.5	52.94	6.451	70 15 55.7	.47	52.59	1
14 198	Pavonis	9.8	39 55.21	5.985	67 24 43.2	.47	51.65	2
14 199	Gould, Z. C., 19^h 1830	9.0	40 3.68	8.688	77 49 29.8	.48	51.18	2
14 200	Lacaille, 8203	8.0	7.96	+ 6.913	72 29 40.7	+ 8.49	51.71	2
14 201	Pavonis	10.0	19 40 21.25	+ 5.980	67 23 25.7	+ 8.51	52.63	1
14 202	Pavonis	10.0	29.59	6.387	69 56 17.3	.52	52.60	1
14 203	Lacaille, 8179	9.0	37.35	9.430	79 13 21.5	.53	50.72	1
14 204	Octantis	9.0	54.16	10.749	81 2 47.9	.55	50.65	2
14 205	Octantis	8.2	56.96	+ 8.324	77 1 52.2	+ 8.55	52.12	2
14 206	Octantis	7.8	19 40 59.22	+10.705	80 59 50.0	+ 8.55	50.65	2
14 207	Pavonis	10.0	41 6.87	5.911	66 56 5.4	.56	52.68	1
14 208	Octantis	7.5	8.50	8.155	76 37 16.4	.57	51.68	2
14 209	Lacaille, 8205	7.8	8.59	7.436	74 30 57.3	.57	51.68	2
14 210	Lacaille, 8181	9.0	11.82	+ 9.393	79 10 23.7	+ 8.57	51.18	2
14 211	Gould, 27205	10.0	19 41 17.11	+ 9.397	79 10 52.0	+ 8.58	51.16	2
14 212	Pavonis	10.0	19.04	5.778	65 56 10.2	.58	52.67	1
14 213	Pavonis	9.5	25.38	6.209	68 56 1.3	.59	52.62	1
14 214	Octantis	9.0	33.51	11.817	82 8 9.9	.60	50.65	1
14 215	Gould, Z. C., 19^h 1900	9.0	41.39	+ 9.093	78 39 31.7	+ 8.61	50.84	2
14 216	Octantis	8.2	19 41 59.78	+11.493	81 50 39.6	+ 8.63	50.68	2
14 217	Pavonis	9.5	42 5.78	6.163	68 40 40.6	.64	52.65	1
14 218	Pavonis	10.0	5.86	6.107	68 19 7.9	.64	52.68	1
14 219	Lacaille, 8188	8.0	21.03	9.789	79 49 18.8	.66	50.64	2
14 220	Pavonis	10.0	22.77	+ 7.497	74 45 22.2	+ 8.66	51.65	2
14 221	Gould, Z. C., 19^h 1877	8.5	19 42 25.96	+ 5.769	65 55 0.9	+ 8.67	52.67	1
14 222	Octantis	9.2	31.84	23.443	86 36 48.1	.68	51.16	9
14 223	Pavonis	8.0	40.58	6.087	68 12 48.1	.69	52.66	2
14 224	Octantis	9.5	40.70	7.938	76 5 24.9	.69	51.64	1
14 225	Lacaille, 8213	6.8	47.78	+ 7.388	74 24 13.4	+ 8.70	51.68	2

No. 14 208. Possibly variable.

Number.	Constellation, Name of Star, or Synonym.	Magnitude.	Right Ascension, 1850.0.	Annual Precession.	South Declination, 1850.0.	Annual Precession.	Mean year.	No. of obs.
			h. m. s.	s.	° ′ ″	″		
14 226	Octantis	10.0	19 42 48.96	+15.590	84 30 9.9	+ 8.70	50.66	1
14 227	Gould, 27243	9.2	42 55.90	9.330	79 6 21.1	.71	51.18	2
14 228	Lacaille, 8212	8.0	43 0.22	7.616	75 9 24.6	.71	51.20	2
14 229	Octantis	10.0	4.37	8.172	76 42 55.9	.72	51.64	1
14 230	Octantis	10.0	4.38	+ 8.010	76 17 44.0	+ 8.72	51.73	1
14 231	Lacaille, 8224	5.8	19 43 7.22	+ 6.300	69 32 54.0	+ 8.72	52.62	2
14 232	Octantis	10.0	8.02	8.865	78 15 32.5	.72	51.63	1
14 233	ε Pavonis	5.5	8.58	7.084	73 17 49.3	.72	52.58	1
14 234	Gould, Z. C., 19h 1913 .	9.0	9.29	6.502	70 39 26.8	.73	52.14	2
14 235	Gould, Z. C., 19h 1916 .	9.0	13.30	+ 6.692	71 36 28.2	+ 8.73	51.67	2
14 236	Pavonis	10.0	19 43 13.97	+ 5.884	66 47 29.3	+ 8.73	50.68	1
14 237	Gould, Z. C., 19h 1918 .	9.0	16.83	6.698	71 38 23.0	.74	51.67	2
14 238	Pavonis	9.5	17.20	6.397	70 6 9.3	.74	52.59	1
14 239	Gould, Z. C., 19h 1915 .	9.0	18.45	6.465	70 28 20.7	.74	52.58	1
14 240	Gould, Z. C., 19h 1933 .	9.0	20.14	+ 6.681	71 33 25.2	+ 8.74	51.67	2
14 241	Octantis	9.9	19 43 22.37	+24.303	86 45 21.5	+ 8.74	51.17	5
14 242	Lacaille, 8229	6.5	30.06	6.230	69 8 57.0	.75	52.63	2
14 243	Gould, Z. C., 19h 1931 .	9.0	33.29	6.550	70 55 17.0	.76	51.70	1
14 244	Pavonis	9.0	43.75	6.003	67 41 45.6	.77	52.63	1
14 245	Gould, Z. C., 19h 1958 .	8.5	43 45.34	+ 7.805	75 44 49.3	+ 8.77	52.64	1
14 246	Pavonis	9.5	19 44 5.46	+ 6.411	70 12 37.6	+ 8.80	52.59	2
14 247	Octantis	10.0	7.23	8.825	78 12 15.0	.80	51.63	1
14 248	Octantis	8.5	9.94	12.126	82 26 29.7	.80	50.69	1
14 249	Octantis	9.0	11.88	8.940	78 25 51.6	.81	50.64	1
14 250	Octantis	11.0	28.97	+ 8.825	78 12 48.1	+ 8.83	51.63	1
14 251	Pavonis	10.0	19 44 33.57	+ 6.214	69 5 24.3	+ 8.84	52.64	1
14 252	Gould, Z. C., 19h 1960 .	10.0	36.54	5.902	67 0 56.0	.84	52.68	1
14 253	Gould, 27281	9.5	51.72	9.404	79 16 20.4	.86	50.72	1
14 254	Pavonis	10.0	52.81	6.416	70 16 9.9	.86	52.59	1
14 255	Octantis	9.0	57.29	+23.862	86 41 54.5	+ 8.87	51.15	10
14 256	Gould, Z. C., 19h 1988 .	9.8	19 44 59.19	+ 6.675	71 35 19.6	+ 8.87	51.67	2
14 257	Octantis	10.0	45 4.80	7.825	75 50 28.4	.88	52.14	2
14 258	Octantis	10.0	27.83	9.006	78 35 10.2	.91	50.64	1
14 259	Pavonis	9.5	32.01	7.417	74 35 19.0	.91	51.66	1
14 260	Pavonis	9.5	37.22	+ 6.153	68 45 31.6	+ 8.92	52.65	1
14 261	Pavonis	9.0	19 45 38.26	+ 6.516	70 49 33.5	+ 8.92	52.14	2
14 262	Pavonis	8.8	42.78	6.481	70 38 53.6	.93	52.14	2
14 263	Pavonis	10.0	43.00	7.288	74 8 45.4	.93	51.70	2
14 264	μ^1 Pavonis	6.5	43.56	5.940	67 20 17.4	.93	50.67	1
14 265	Octantis	9.5	43.70	+ 8.583	77 44 37.4	+ 8.93	50.73	1

Number.	Constellation, Name of Star, or Synonym.	Magnitude.	Right Ascension, 1850.0.	Annual Precession.	South Declination, 1850.0.	Annual Precession.	Mean year.	No. of obs.
			h. m. s.	s.	° ′ ″	″		
14 266	Pavonis	10.0	19 45 45.78	+ 7.034	73 11 8.9	+ 8.93	52.58	1
14 267	Pavonis	10.0	45 47.90	6.280	69 32 2.8	.93	52.62	2
14 268	Octantis	9.0	46 9.19	10.861	81 16 8.5	.96	50.62	1
14 269	Octantis	9.5	13.62	13.862	83 40 29.9	.97	50.68	1
14 270	Pavonis	8.0	16.58	+ 6.284	69 34 51.9	+ 8.97	52.62	2
14 271	Gould, Z. C., 19ʰ 2024 .	9.5	19 46 21.79	+ 5.922	67 14 35.1	+ 8.98	50.67	1
14 272	Pavonis	9.8	22.44	6.128	68 38 4.3	.98	52.63	2
14 273	Pavonis	10.0	26.10	6.256	69 25 20.7	.98	52.64	1
14 274	Pavonis	9.5	26.71	6.119	68 34 52.6	8.98	51.63	2
14 275	Gould, Z. C., 19ʰ 2061 .	9.2	46.56	+ 7.364	74 26 46.5	+ 9.01	51.68	2
14 276	Pavonis	9.8	19 46 54.28	+ 6.807	72 16 14.4	+ 9.02	51.67	2
14 277	Pavonis	9.5	46 58.23	5.960	67 32 17.2	.02	52.63	1
14 278	Lacaille, 8218	7.0	47 12.87	9.528	79 31 35.6	.04	51.67	2
14 279	μ^1 Pavonis	5.5	13.00	5.931	67 20 28.1	.04	50.67	1
14 280	Pavonis	10.0	36.12	+ 6.127	68 40 49.9	+ 9.07	52.63	2
14 281	Pavonis	9.5	19 47 40.10	+ 6.221	69 15 42.4	+ 9.08	52.64	1
14 282	Gould, Z. C., 19ʰ 2132 .	9.0	48.96	9.601	79 39 20.0	.09	50.61	1
14 283	Lacaille, 8236	8.2	52.64	8.448	77 30 9.8	.10	51.67	2
14 284	Octantis	9.8	54.27	20.641	86 7 11.6	.10	51.23	2
14 285	Gould, 27308	8.0	47 59.22	+ 5.805	66 29 39.5	+ 9.10	52.68	1
14 286	Pavonis	9.2	19 48 15.94	+ 5.774	66 13 24.6	+ 9.13	52.68	2
14 287	Gould, Z. C., 19ʰ 2092 .	9.5	23.71	5.889	67 5 53.2	.14	52.68	1
14 288	Octantis	10.0	24.00	24.073	86 45 22.0	.14	51.22	1
14 289	Pavonis	10.0	30.22	6.396	70 18 18.0	.14	52.59	1
14 290	Octantis	10.0	31.34	+ 9.645	79 44 17.6	+ 9.15	50.61	1
14 291	Pavonis	10.0	19 48 50.62	+ 7.272	74 11 32.8	+ 9.17	51.71	1
14 292	Pavonis	10.0	48 59.22	6.641	71 34 42.8	.18	51.66	1
14 293	Pavonis	9.0	49 10.60	6.002	67 55 41.6	.20	52.65	2
14 294	Octantis	9.8	14.51	9.452	79 27 6.6	.20	50.67	2
14 295	Pavonis	9.5	16.03	+ 6.233	69 24 12.2	+ 9.20	52.64	1
14 296	Pavonis	11.0	19 49 21.93	+ 7.503	74 59 41.7	+ 9.21	50.75	1
14 297	Pavonis	9.5	25.04	5.745	66 2 38.7	.22	52.67	1
14 298	Gould, Z. C., 19ʰ 2146 .	9.0	35.43	6.361	70 9 17.9	.23	52.59	1
14 299	Gould, Z. C., 19ʰ 2151 .	8.8	37.24	6.441	70 35 35.2	.23	52.14	2
14 300	Pavonis	10.0	58.17	+ 6.840	72 31 40.6	+ 9.26	51.68	1
14 301	Pavonis	10.0	19 49 59.52	+ 7.408	74 42 9.7	+ 9.26	51.66	1
14 302	Pavonis	10.0	50 0.69	7.095	73 34 36.8	.26	52.15	2
14 303	Lacaille, 8252	7.8	2.92	7.550	75 10 14.5	.26	51.20	2
14 304	Octantis	9.5	12.01	7.751	75 46 58.2	.28	52.64	1
14 305	Octantis	9.8	13.74	+10.025	80 19 32.4	+ 9.28	51.14	2

Number.	Constellation, Name of Star, or Synonym.	Magnitude.	Right Ascension, 1850.0.	Annual Precession.	South Declination, 1850.0.	Annual Precession.	Mean year.	No. of obs.
			h. m. s.	s.	° ′ ″	″		
14 306	Pavonis	9.5	19 50 14.56	+ 6.432	70 34 12.4	+ 9.28	52.14	2
14 307	Pavonis	10.0	17.59	7.268	74 13 38.8	.28	51.68	2
14 308	Pavonis	10.0	19.04	6.182	69 8 13.8	.29	52.63	2
14 309	Pavonis	9.5	26.33	5.728	65 57 41.4	.30	52.67	1
14 310	Pavonis	10.0	27.67	+ 5.829	66 44 53.2	+ 9.30	52.18	2
14 311	Pavonis	9.5	19 50 28.09	+ 6.135	68 51 12.2	+ 9.30	52.62	1
14 312	Lacaille, 8267	6.0	28.83	5.961	67 42 14.9	.30	52.63	1
14 313	Octantis	10.0	31.45	7.929	76 17 41.2	.30	51.69	2
14 314	Pavonis	9.0	33.38	6.466	70 45 41.7	.30	52.14	2
14 315	Octantis	9.0	35.20	+22.840	86 34 16.2	+ 9.31	51.15	10
14 316	Octantis	8.8	19 50 38.04	+23.992	86 45 35.9	+ 9.31	51.15	10
14 317	Pavonis	9.8	39.50	6.316	69 56 41.3	.31	52.59	2
14 318	Pavonis	10.0	40.21	6.847	72 34 58.4	.32	51.72	2
14 319	Gould, Z. C., 19ʰ 2197	9.0	44.01	6.552	71 12 32.4	.32	51.69	1
14 320	Pavonis	10.0	48.27	+ 6.623	71 32 54.8	+ 9.32	51.66	1
14 321	Pavonis	9.5	19 50 59.27	+ 5.718	65 54 27.2	+ 9.34	52.67	1
14 322	Lacaille, 8259	8.0	51 1.78	7.377	74 37 53.8	.34	51.65	2
14 323	Pavonis	9.8	6.19	6.442	70 39 33.2	.35	52.14	2
14 324	Gould, Z. C., 19ʰ 2223	9.0	9.52	8.563	77 50 32.8	.35	51.18	2
14 325	Gould, Z. C., 19ʰ 2213	9.2	13.86	+ 6.848	72 36 32.0	+ 9.36	51.72	2
14 326	Octantis	10.0	19 51 14.82	+12.278	82 40 46.6	+ 9.36	50.69	1
14 327	Gould, Z. C., 19ʰ 2229	9.2	14.96	8.209	77 2 8.6	.36	52.12	2
14 328	Lacaille, 8240	8.5	26.71	9.782	80 0 29.9	.37	50.67	1
14 329	Pavonis	9.5	27.21	6.126	68 50 26.8	.37	52.62	1
14 330	Lacaille, 8273	8.0	41.00	+ 6.313	69 58 12.0	+ 9.39	52.59	2
14 331	Gould, Z. C., 19ʰ 2253	9.0	19 51 44.89	+ 8.365	77 25 10.9	+ 9.40	52.60	1
14 332	Pavonis	9.5	45.40	5.953	67 42 23.3	.40	52.63	1
14 333	Gould, 27419	12.0	48.33	6.548	71 13 54.8	.40	51.69	1
14 334	Gould, 27420	9.0	51 49.89	6.548	71 13 58.2	.40	51.69	1
14 335	Gould, Z. C., 19ʰ 2239	9.0	52 3.09	+ 6.724	72 5 12.7	+ 9.42	51.67	1
14 336	Gould, 27416	7.5	19 52 3.78	+ 5.751	66 13 32.4	+ 9.42	52.68	2
14 337	Gould, Z. C., 19ʰ 2232	9.5	4.59	6.324	70 2 57.2	.42	52.59	1
14 338	Gould, Z. C., 19ʰ 2223	9.0	5.76	5.762	66 18 31.2	.42	52.68	2
14 339	Lacaille, 8270	8.0	8.17	6.698	71 58 20.2	.43	51.66	2
14 340	Octantis	10.0	8.47	+ 8.764	78 17 12.6	+ 9.43	51.14	2
14 341	Pavonis	10.0	19 52 9.43	+ 6.577	71 23 28.4	+ 9.43	51.69	1
14 342	Pavonis	10.0	12.86	7.068	73 33 1.3	.43	51.73	2
14 343	Lacaille, 8327	7.8	13.72	13.857	83 45 22.6	.43	50.86	3
14 344	Pavonis	10.0	25.80	5.872	67 9 42.6	.45	51.67	2
14 345	Octantis	10.0	34.19	+ 7.537	75 12 29.0	+ 9.46	50.75	1

Number.	Constellation, Name of Star, or Synonym.	Magnitude.	Right Ascension, 1850.0.	Annual Precession.	South Declination, 1850.0.	Annual Precession.	Mean year.	No. of obs.
			h. m. s.	s.	° ′ ″	″		
14 346	Pavonis	10.0	19 52 35.70	+ 7.250	74 14 28.9	+ 9.46	51.68	2
14 347	Pavonis	9.2	38.62	6.972	73 11 2.7	.47	52.02	3
14 348	Gould, Z. C., 19h 2264 .	9.5	51.60	6.367	70 19 31.4	.48	52.58	1
14 349	Gould, 27467	9.0	52 55.03	8.475	77 42 7.2	.49	50.73	1
14 350	Pavonis	9.5	53 4.12	+ 5.952	67 45 47.0	+ 9.50	52.63	1
14 351	Gould, Z. C., 19h 2265 .	8.8	19 53 6.94	+ 5.750	66 16 5.3	+ 9.50	52.68	2
14 352	Gould, Z. C., 19h 2275 .	8.8	13.91	6.010	68 9 59.8	.51	52.66	2
14 353	Octantis	9.5	19.08	11.547	82 5 36.6	.52	50.65	1
14 354	Gould, Z. C., 19h 2315 .	10.0	23.24	9.006	78 47 0.8	.52	51.65	2
14 355	Pavonis	9.5	23.46	+ 6.201	69 23 19.7	+ 9.52	52.62	2
14 356	Octantis	9.0	19 53 38.32	+11.493	82 2 58.9	+ 9.54	50.65	1
14 357	Pavonis	9.5	43.10	5.742	66 14 8.9	.55	52.67	1
14 358	Lacaille, 8256	8.2	48.03	9.380	79 26 34.2	.55	50.67	2
14 359	Pavonis	9.8	56.93	5.850	67 4 20.0	.57	51.67	2
14 360	d Pavonis	3.0	53 57.89	+ 5.782	66 33 25.5	+ 9.57	52.68	1
14 361	Brisbane, 6788	7.2	19 54 0.32	+ 5.810	66 46 33.2	+ 9.57	52.68	2
14 362	Pavonis	9.5	0.76	5.815	66 48 50.3	.57	52.68	1
14 363	Lacaille, 8284	8.0	8.22	6.702	72 4 5.7	.58	51.67	1
14 364	Pavonis	10.0	8.65	6.632	71 44 26.5	.58	51.66	1
14 365	Octantis	10.0	32.20	+10.724	81 16 50.4	+ 9.61	50.62	1
14 366	Pavonis	9.5	19 54 39.69	+ 5.928	67 40 1.4	+ 9.62	52.63	1
14 367	Octantis	9.0	47.70	11.130	81 43 9.4	.63	50.72	1
14 368	Octantis	9.0	48.29	7.883	76 17 57.8	.63	51.69	2
14 369	Gould, Z. C., 19h 2336 .	9.0	52.23	6.701	72 5 39.6	.64	51.67	1
14 370	Gould, Z. C., 19h 2349 .	9.0	54.39	+ 7.523	75 14 38.8	+ 9.64	50.75	1
14 371	Octantis	9.5	19 54 59.19	+13.671	83 41 18.7	+ 9.65	50.68	1
14 372	Octantis	10.0	55 20.29	10.763	81 20 28.8	.67	50.62	1
14 373	Gould, Z. C., 19h 2347 .	10.0	29.88	5.851	67 9 12.1	.68	52.68	1
14 374	Pavonis	8.5	33.04	5.956	67 54 21.8	.69	52.68	1
14 375	Pavonis	10.0	36.49	+ 7.303	74 32 15.6	+ 9.69	51.66	1
14 376	Gould, Z. C., 19h 2378 .	9.8	19 55 56.32	+ 6.938	73 10 10.4	+ 9.72	51.74	2
14 377	Gould, Z. C., 19h 2368 .	9.0	55 59.25	5.992	68 10 34.0	.72	52.66	2
14 378	Pavonis	9.7	56 3.42	7.249	74 21 37.0	.73	51.69	3
14 379	Gould, Z. C., 19h 2375 .	9.5	4.93	6.218	69 36 47.2	.73	52.60	1
14 380	Gould, Z. C., 19h 2380 .	8.2	18.08	+ 6.125	69 3 23.1	+ 9.75	52.63	2
14 381	Gould, Z. C., 20h 14 .	10.0	19 56 33.30	+ 9.608	79 52 11.6	+ 9.77	50.67	1
14 382	Gould, Z. C., 19h 2401 .	9.0	47.06	6.217	69 38 21.8	.78	52.60	1
14 383	Gould, Z. C., 19h 2413 .	9.5	50.88	6.962	73 18 9.5	.79	51.74	1
14 384	Pavonis . . .	8.8	56 56.54	6.191	69 29 19.1	.80	52.00	3
14 385	Pavonis	10.0	57 11.81	+ 6.369	70 31 26.8	+ 9.81	52.58	1

Number.	Constellation, Name of Star, or Synonym.	Magnitude.	Right Ascension, 1850.0.	Annual Precession.	South Declination, 1850.0.	Annual Precession.	Mean year.	No. of obs.
			h. m. s.	s.	° ′ ″	″		
14 386	Pavonis	10.0	19 57 18.55	+ 5.853	67 15 37.1	+ 9.82	50.67	1
14 387	Pavonis	10.0	27.93	7.126	73 57 50.5	.84	51.73	1
14 388	Octantis	10.0	33.51	7.695	75 51 20.0	.84	52.64	1
14 389	Pavonis	10.0	39.89	6.956	73 18 36.0	.85	51.74	1
14 390	Melbourne (1), 1017 . .	9.5	40.63	+ 5.780	66 43 31.6	+ 9.85	52.68	2
14 391	Octantis	10.0	19 57 45.88	+15.305	84 33 45.9	+ 9.86	51.23	1
14 392	Octantis	7.8	51.26	27.620	87 17 5.5	.87	51.15	10
14 393	Pavonis	10.0	51.94	6.611	71 47 38.3	.87	51.66	1
14 394	Pavonis	8.0	56.81	6.163	69 22 4.7	.87	52.64	1
14 395	Pavonis	9.0	58.51	+ 5.883	67 30 39.0	+ 9.87	52.63	1
14 396	Gould, Z. C., 20h 68 . .	9.0	19 57 58.72	+ 9.361	79 30 56.8	+ 9.87	50.67	2
14 397	Pavonis	10.0	58 6.09	5.798	66 53 10.0	.88	52.68	1
14 398	Gould, Z. C., 20h 80 . .	9.2	14.16	9.378	79 33 3.4	.89	50.67	2
14 399	Lacaille, 8281	7.5	18.36	9.698	80 2 48.6	.90	50.67	1
14 400	Octantis	9.8	21.88	+11.601	82 14 15.2	+ 9.90	50.67	2
14 401	Octantis	10.0	19 58 22.55	+ 7.646	75 44 17.9	+ 9.90	52.64	1
14 402	Gould, Z. C., 20h 58 . .	9.0	33.57	7.490	75 15 48.8	.92	50.75	1
14 403	Octantis	9.8	33.93	15.902	84 49 28.1	.92	50.94	2
14 404	Pavonis	10.0	36.33	6.082	68 53 32.1	.92	52.62	1
14 405	Gould, Z. C., 20h 45 . .	9.0	41.66	+ 6.465	71 5 58.4	+ 9.93	51.69	2
14 406	Pavonis	9.5	19 58 48.93	+ 5.666	65 52 46.4	+ 9.94	52.67	1
14 407	Gould, Z. C., 20h 40 . .	8.0	49.98	5.902	67 41 27.4	.94	52.63	1
14 408	Octantis	10.0	58 56.90	7.804	76 12 50.0	.95	51.69	2
14 409	Pavonis	9.8	59 2.30	5.718	66 18 51.6	.96	52.68	2
14 410	Pavonis	8.8	12.44	+ 5.929	67 53 47.4	+ 9.97	52.65	2
14 411	Pavonis	9.2	19 59 15.38	+ 5.919	67 49 42.2	+ 9.97	52.65	2
14 412	Gould, Z. C., 20h 96 . .	9.0	22.20	7.538	75 26 30.0	.98	51.70	2
14 413	Octantis	10.0	25.30	7.094	73 55 5.2	.98	51.72	2
14 414	Pavonis	10.0	33.24	7.734	76 2 15.3	9.99	51.64	1
14 415	Octantis	10.0	35.75	+ 6.166	69 27 46.0	+ 10.00	51.77	1
14 416	Gould, 27666	9.0	19 59 37.41	+13.918	83 54 0.7	+ 10.00	50.68	2
14 417	Octantis	9.5	20 0 1.15	7.922	76 34 26.9	.03	51.69	2
14 418	Octantis	10.0	0 {4.42 24.42}	35.233 35.206	87 56 18.2	.02 .06	51.22	2
14 419	Gould, Z. C., 20h 123 . .	9.0	8.36	7.756	76 7 5.4	.04	51.64	1
14 420	Lacaille, 8301	6.0	17.20	+ 9.263	79 24 46.9	+ 10.05	50.67	2
14 421	Gould, Z. C., 20h 117 . .	8.5	20 0 20.55	+ 6.770	72 37 47.8	+ 10.05	51.72	2
14 422	Melbourne (1), 1020 . .	8.5	33.25	6.065	68 52 28.3	.07	52.62	1
14 423	Gould, Z. C., 20h 119 . .	8.8	0 45.67	6.105	69 8 34.9	.09	52.63	2
14 424	Octantis	9.0	1 0.54	11.316	82 1 29.6	.10	50.65	1
14 425	Pavonis	9.5	3.84	+ 7.097	73 59 25.6	+ 10.11	51.72	2

No. 14418. There is an error of 20s in one of the transits. Each observation depends upon only one transit.

Number.	Constellation, Name of Star, or Synonym.	Magnitude.	Right Ascension, 1850.0.	Annual Precession.	South Declination, 1850.0.	Annual Precession.	Mean year.	No. of obs.
			h. m. s.	s.	° ′ ″	″		
14 426	Gould, Z. C., 20h 138 . .	8.5	20 1 8.23	+ 6.117	69 14 17.6	+ 10.11	52.64	1
14 427	Pavonis	9.5	9.04	5.805	67 5 50.0	.12	51.67	2
14 428	Octantis	9.0	12.94	14.396	84 10 45.8	.12	50.67	2
14 429	Gould, Z. C., 20h 147 . .	9.0	16.90	6.524	71 30 48.7	.13	51.67	2
14 430	Lacaille, 8306	7.5	24.53	+ 9.301	79 30 23.5	+ 10.13	50.67	2
14 431	Lacaille, 8332	8.0	20 1 25.56	+ 6.949	73 25 36.2	+ 10.14	51.73	2
14 432	Lacaille, 8335	8.0	35.46	6.906	73 15 27.6	.15	51.74	1
14 433	Gould, Z. C., 20h 171 . .	9.0	43.36	7.708	76 2 11.5	.16	51.64	1
14 434	Pavonis	10.0	1 45.87	5.951	68 10 33.7	.16	52.68	1
14 435	Pavonis	10.0	2 7.32	+ 6.341	70 35 37.4	+ 10.19	51.70	1
14 436	Octantis	9.8	20 2 19.46	+ 7.377	75 1 44.4	+ 10.20	51.20	2
14 437	Pavonis	10.0	23.91	5.977	68 23 23.4	.21	52.65	1
14 438	Gould, Z. C., 20h 194 . .	10.0	33.29	7.001	73 40 46.4	.22	51.73	1
14 439	Lacaille, 8353	6.8	35.68	5.905	67 54 2.4	.22	52.65	2
14 440	Octantis	9.5	41.01	+ 9.917	80 27 48.0	+ 10.23	51.62	1
14 441	Gould, 27667	9.0	20 2 43.23	+ 6.408	70 58 54.6	+ 10.23	51.69	2
14 442	Gould, Z. C., 20h 202 . .	9.5	50.72	7.056	73 54 16.4	.24	51.72	2
14 443	Pavonis	10.0	51.91	6.282	70 17 50.3	.24	51.74	1
14 444	Octantis	9.5	52.98	7.915	76 38 53.7	.25	51.64	1
14 445	Pavonis	9.0	2 56.98	+ 6.114	69 18 9.5	+ 10.25	52.64	1
14 446	Pavonis	9.5	20 3 1.44	+ 5.867	67 39 6.2	+ 10.26	52.63	1
14 447	Pavonis	9.0	10.69	6.055	68 56 27.2	.27	52.62	1
14 448	Gould, Z. C., 20h 201 . .	9.8	10.97	6.158	69 35 4.1	.27	52.22	2
14 449	Pavonis	9.5	12.16	5.863	67 38 8.4	.27	52.63	1
14 450	Pavonis	10.0	19.88	+ 6.916	73 22 18.9	+ 10.28	51.74	1
14 451	Octantis	9.8	20 3 26.42	+ 8.423	77 53 40.0	+ 10.29	51.18	2
14 452	Octantis	9.5	27.08	10.249	80 54 52.1	.29	50.69	1
14 453	Pavonis	9.8	32.52	5.682	66 16 18.2	.30	52.68	2
14 454	Gould, Z. C., 20h 212 . .	9.5	35.73	6.177	69 43 25.8	.30	51.77	1
14 455	Lacaille, 8342	8.5	39.09	+ 7.190	74 26 1.4	+ 10.30	51.68	2
14 456	Pavonis	8.8	20 3 42.41	+ 5.891	67 51 28.2	+ 10.31	52.65	2
14 457	Gould, Z. C., 20h 218 . .	9.0	51.12	6.203	69 53 16.7	.32	51.75	2
14 458	Octantis	9.0	3 54.20	10.465	81 11 7.1	.32	50.62	1
14 459	Gould, Z. C., 20h 239 . .	9.5	4 16.34	7.049	73 56 35.2	.35	51.72	2
14 460	Pavonis	10.0	17.56	+ 6.531	71 41 3.8	+ 10.35	51.66	1
14 461	Pavonis	9.2	20 4 20.90	+ 5.825	67 25 8.1	+ 10.36	51.65	2
14 462	Pavonis	8.8	22.66	7.665	76 0 13.2	.36	52.14	2
14 463	Pavonis	10.0	25.86	6.280	70 21 16.4	.36	52.16	2
14 464	Pavonis	10.0	27.65	6.526	71 39 50.1	.36	51.66	1
14 465	Brisbane, 6815	8.2	33.08	+ 6.128	69 28 7.6	+ 10.37	52.21	2

Number.	Constellation, Name of Star, or Synonym.	Magnitude.	Right Ascension, 1850.0.	Annual Precession.	South Declination, 1850.0.	Annual Precession.	Mean year.	No. of obs.
			h. m. s.	s.	° ′ ″	″		
14 466	Pavonis	9.0	20 4 34.47	+ 5.680	66 18 26.2	+ 10.37	52.68	2
14 467	Lacaille, 8323	8.2	47.48	9.467	79 51 34.9	.39	50.64	2
14 468	Pavonis	10.0	4 53.29	5.677	66 18 3.9	.40	52.67	1
14 469	Pavonis	9.5	5 5.87	5.656	66 8 36.4	.41	52.67	1
14 470	Pavonis	10.0	8.76	+ 6.270	70 20 23.9	+ 10.42	52.16	2
14 471	Lacaille, 8336	8.7	20 5 30.91	+ 9.070	79 13 27.0	+ 10.44	51.24	2
14 472	Gould, Z. C., 20h 298	9.4	33.72	8.764	78 39 41.3	.45	51.46	4
14 473	Pavonis	8.2	43.84	5.820	67 27 17.8	.46	51.65	2
14 474	Pavonis	10.0	45.60	5.714	66 38 38.0	.46	52.68	2
14 475	Pavonis	10.0	46.22	+ 6.734	72 41 57.4	+ 10.46	51.72	2
14 476	Lacaille, 8257	7.2	20 5 54.58	+15.846	84 53 55.4	+ 10.47	51.08	4
14 477	Octantis	9.8	5 57.98	7.603	75 52 32.9	.48	52.14	2
14 478	Pavonis	10.0	6 2.70	6.733	72 42 25.6	.48	51.72	2
14 479	Pavonis	9.5	5.03	5.958	68 26 57.8	.49	52.65	1
14 480	Pavonis	9.0	26.03	+ 6.507	71 30 32.5	+ 10.51	51.66	1
14 481	Pavonis	9.5	20 6 29.64	+ 6.102	69 24 22.6	+ 10.52	52.21	2
14 482	Gould, 27752	8.5	32.99	5.745	66 55 52.4	.52	52.68	1
14 483	Gould, Z. C., 20h 302	9.0	34.84	6.822	73 6 51.2	.52	51.74	2
14 484	Lacaille, 8328	7.8	39.38	10.273	81 1 9.0	.53	50.65	2
14 485	Octantis	9.2	39.67	+17.805	85 34 58.6	+ 10.53	51.22	6
14 486	Gould, Z. C., 20h 304	9.0	20 6 43.61	+ 6.680	72 29 45.1	+ 10.53	51.68	1
14 487	Gould, Z. C., 20h 309	9.5	6 49.41	6.918	73 31 30.8	.54	51.73	2
14 488	Gould, 27767	10.0	7 0.15	5.781	67 14 11.7	.55	50.67	1
14 489	Pavonis	10.0	5.36	5.887	68 0 35.1	.56	52.68	1
14 490	Lacaille, 8371	6.8	7.72	+ 5.854	67 46 32.1	+ 10.56	52.65	2
14 491	Pavonis	9.2	20 7 8.69	+ 6.165	69 49 24.2	+ 10.56	51.75	2
14 492	Pavonis	9.0	15.70	5.885	68 0 13.2	.57	52.68	1
14 493	Lacaille, 8374	8.0	17.66	5.776	67 12 47.3	.58	50.67	1
14 494	Pavonis	9.5	29.06	5.990	68 44 12.7	.59	52.65	1
14 495	Octantis	9.0	29.58	+11.228	82 4 28.8	+ 10.59	50.65	1
14 496	Octantis	9.0	20 7 34.77	+12.024	82 46 23.2	+ 10.60	50.69	2
14 497	Octantis	10.0	35.37	7.548	75 46 13.6	.60	52.64	1
14 498	Pavonis	10.0	37.39	5.854	67 48 19.3	.60	52.63	1
14 499	Pavonis	9.0	50.82	6.171	69 53 36.1	.62	51.75	2
14 500	Pavonis	10.0	51.65	+ 6.279	70 31 20.5	+ 10.62	52.58	1
14 501	Lacaille, 8406	6.2	20 7 56.36	+10.611	81 26 44.2	+ 10.62	50.67	2
14 502	Lacaille, 8350	8.5	8 1.92	9.120	79 22 58.4	.63	50.67	2
14 503	Pavonis	10.0	5.41	7.059	74 7 39.4	.63	51.70	1
14 504	Pavonis	9.8	5.42	6.703	72 39 49.8	.63	51.72	2
14 505	Gould, Z. C., 20h 352	10.0	7.36	+ 6.469	71 32 48.3	+ 10.64	51.67	2

Number.	Constellation, Name of Star, or Synonym.	Magnitude.	Right Ascension, 1850.0.	Annual Precession.	South Declination, 1850.0.	Annual Precession.	Mean year.	No. of obs.
			h. m. s.	s.	° ′ ″	″		
14 506	Pavonis	10.0	20 8 11.32	+ 7.256	74 50 43.1	+ 10.64	51.65	1
14 507	Pavonis	9.5	15.67	7.239	74 47 16.3	.65	51.65	1
14 508	Octantis	10.0	17.90	12.275	82 58 53.1	.65	50.70	2
14 509	Octantis	9.8	21.01	7.582	75 54 4.3	.65	52.14	2
14 510	Pavonis	9.0	23.17	+ 5.674	66 28 25.9	+ 10.66	52.68	1
14 511	Octantis	10.0	20 8 23.33	+ 7.572	75 52 19.7	+ 10.66	51.64	1
14 512	Pavonis	10.0	24.51	5.659	66 21 30.4	.66	52.67	1
14 513	Octantis	9.9	27.37	17.430	85 29 23.7	.66	51.22	6
14 514	Lacaille, 8372	7.5	34.80	6.583	72 7 38.6	.67	51.67	1
14 515	Octantis	10.0	8 44.67	+ 8.186	77 31 23.4	+ 10.68	52.60	1
14 516	Octantis	10.0	20 9 6.49	+ 9.764	80 24 43.7	+ 10.71	51.62	1
14 517	Pavonis	9.0	12.89	5.656	66 22 53.4	.72	52.68	2
14 518	Gould, Z. C., 20h 387 . .	9.0	13.29	6.782	73 3 26.6	.72	51.74	2
14 519	Octantis	9.8	22.62	18.090	85 41 54.3	.73	51.22	5
14 520	Pavonis	10.0	26.50	+ 6.291	70 40 6.8	+ 10.74	52.14	2
14 521	Gould, Z. C., 20h 424 . .	9.0	20 9 32.73	+ 8.447	78 8 12.0	+ 10.74	51.63	1
14 522	Octantis	10.0	35.32	7.312	75 5 20.7	.74	50.75	1
14 523	Gould, Z. C., 20h 434 . .	8.5	43.23	8.424	78 5 35.6	.76	51.63	1
14 524	Pavonis	10.0	44.23	6.581	72 10 23.0	.76	51.67	1
14 525	Melbourne (1), 1032 . .	8.5	46.90	+ 5.874	68 4 1.3	+ 10.76	52.68	1
14 526	Pavonis	9.2	20 9 48.38	+ 6.165	69 57 39.2	+ 10.76	51.75	2
14 527	Brisbane, 6832	8.5	55.22	5.684	66 38 54.6	.77	52.68	2
14 528	Lacaille, 8375	8.2	56.04	7.212	74 45 36.4	.77	51.65	2
14 529	Pavonis	9.5	56.90	5.637	66 16 12.0	.77	52.68	2
14 530	Gould, Z. C., 20h 409 . .	9.5	9 58.37	+ 6.544	72 0 16.0	+ 10.77	51.67	1
14 531	Gould, Z. C., 20h 444 . .	9.0	20 10 6.72	+ 8.035	77 12 8.7	+ 10.78	52.12	2
14 532	Gould, Z. C., 20h 410 . .	9.0	21.24	5.608	66 3 1.4	.80	52.67	1
14 533	Octantis	10.0	23.50	8.798	78 52 16.5	.81	51.71	2
14 534	Octantis	10.0	23.89	23.717	86 52 37.2	.81	51.22	2
14 535	Octantis	9.0	24.06	+11.105	82 1 5.9	+ 10.81	50.65	1
14 536	Pavonis	9.8	20 10 27.31	+ 6.282	70 39 52.0	+ 10.81	52.14	2
14 537	Pavonis	9.0	49.18	5.865	68 3 47.0	.84	52.68	1
14 538	Pavonis	10.0	10 56.69	6.048	69 17 45.0	.85	52.64	1
14 539	Pavonis	10.0	11 0.12	7.187	74 42 54.3	.85	51.65	2
14 540	Pavonis	10.0	5.75	+ 6.797	73 12 26.7	+ 10.86	51.74	1
14 541	Octantis	10.0	20 11 11.56	+10.788	81 42 46.5	+ 10.86	50.72	1
14 542	Pavonis	9.0	14.45	5.861	68 3 21.0	.87	52.68	1
14 543	Octantis	10.0	33.45	12.567	83 15 23.8	.89	50.70	1
14 544	Pavonis	9.5	33.53	5.596	66 1 7.7	.89	52.67	1
14 545	Gould, Z. C., 20h 459 . .	9.8	35.84	+ 6.017	69 7 58.1	+ 10.89	52.63	2

Number.	Constellation, Name of Star, or Synonym.	Magnitude.	Right Ascension, 1850.0.	Annual Precession.	South Declination, 1850.0.	Annual Precession.	Mean year.	No. of obs.
			h. m. s.	s.	° ′ ″	″		
14 546	Octantis	10.0	20 11 36.29	+11.983	82 49 12.0	+ 10.89	50.69	1
14 547	Octantis	10.0	55.36	8.515	78 21 22.8	.92	50.85	2
14 548	Lacaille, 8376	9.2	57.12	8.209	77 41 1.0	.92	51.25	2
14 549	Gould, 27910	9.8	11 59.68	8.208	77 41 2.4	.92	51.25	2
14 550	Gould, Z. C., 20h 465 . .	9.5	12 3.70	+ 5.681	66 44 51.4	+ 10.93	52.68	2
14 551	Pavonis	9.5	20 12 7.42	+ 5.779	67 30 33.4	+ 10.93	52.63	1
14 552	Pavonis	10.0	17.45	5.628	66 19 41.1	.94	52.68	1
14 553	Gould, Z. C., 20h 481 . .	8.0	22.41	6.050	69 23 8.3	.95	52.64	1
14 554	Pavonis	10.0	22.45	5.627	66 19 41.1	.95	52.68	1
14 555	Octantis	9.5	27.05	+11.063	82 1 21.4	+ 10.96	50.89	1
14 556	Lacaille, 8360	6.5	20 12 32.02	+10.826	81 47 0.2	+ 10.96	50.68	2
14 557	Lacaille, 8377	8.8	33.88	8.242	77 46 52.4	.97	51.25	2
14 558	Pavonis	10.0	35.87	6.106	69 44 48.4	.97	51.77	1
14 559	Pavonis	9.5	38.05	5.625	66 19 29.7	.97	52.68	1
14 560	Pavonis	9.5	12 40.23	+ 6.130	69 54 2.2	+ 10.97	51.75	2
14 561	Gould, 27929	8.5	20 13 0.44	+ 7.474	75 44 44.5	+ 11.00	52.64	1
14 562	Pavonis	9.5	2.50	5.770	67 29 32.8	.00	52.63	1
14 563	Pavonis	9.5	11.53	5.757	67 24 7.2	.01	50.67	1
14 564	Pavonis	10.0	16.57	6.860	73 33 55.8	.02	51.73	2
14 565	Octantis	8.5	23.06	+14.061	84 11 46.3	+ 11.02	50.67	2
14 566	Octantis	10.0	20 13 26.34	+ 7.642	76 16 7.8	+ 11.03	51.64	1
14 567	Pavonis	10.0	32.73	6.536	72 8 14.9	.04	51.67	1
14 568	Pavonis	10.0	13 53.59	6.529	72 7 8.9	.06	51.67	1
14 569	Octantis	9.8	14 7.63	18.949	85 59 4.1	.08	51.22	5
14 570	Pavonis	9.2	8.48	+ 5.996	69 8 2.8	+ 11.08	52.63	2
14 571	Brisbane, 6847	7.0	20 14 14.18	+ 5.863	68 14 29.7	+ 11.09	52.66	2
14 572	Gould, Z. C., 20h 553 . .	9.0	19.77	6.370	71 20 26.2	.09	51.69	1
14 573	Gould, Z. C., 20h 602 . .	9.2	44.49	8.779	78 58 8.3	.12	51.24	2
14 574	Gould, Z. C., 20h 582 . .	9.0	53.97	7.260	75 7 52.0	.14	51.20	2
14 575	Octantis	10.0	14 55.38	+ 9.779	80 35 11.6	+ 11.14	51.62	1
14 576	Gould, Z. C., 20h 583 . .	9.8	20 15 3.96	+ 6.839	73 33 39.9	+ 11.15	51.73	2
14 577	Pavonis	10.0	5.13	6.085	69 45 17.9	.15	51.77	1
14 578	Pavonis	9.0	12.34	5.842	68 8 41.4	.16	52.68	1
14 579	Gould, Z. C., 20h 612 . .	9.0	13.76	8.053	77 25 37.3	.16	51.70	3
14 580	Octantis	10.0	20.04	+ 9.714	80 30 27.8	+ 11.17	51.62	1
14 581	Gould, Z. C., 20h 571 . .	9.0	20 15 20.23	+ 5.573	66 3 16.8	+ 11.17	52.67	1
14 582	Pavonis	8.5	20.82	5.746	67 26 57.2	.17	51.65	2
14 583	Lacaille, 8412	7.0	25.13	6.051	69 33 19.6	.17	52.21	2
14 584	Octantis	9.8	26.10	7.381	75 33 1.8	.17	51.70	2
14 585	Pavonis	10.0	27.52	+ 5.632	66 33 29.3	+ 11.18	52.68	1

Number.	Constellation, Name of Star, or Synonym.	Magnitude.	Right Ascension, 1850.0.	Annual Precession.	South Declination, 1850.0.	Annual Precession.	Mean year.	No. of obs.
			h. m. s.	s.	° ′ ″	″		
14 586	Gould, Z. C., 20^h 599 . .	9.0	20 15 31.04	+ 6.810	73 27 44.6	+ 11.18	51.73	2
14 587	Pavonis	9.0	37.20	6.206	70 30 13.4	.19	52.58	1
14 588	Gould, Z. C., 20^h 588 . .	9.5	41.77	5.671	66 53 2.0	.19	52.68	1
14 589	Gould, Z. C., 20^h 608 . .	9.0	43.43	6.776	73 19 39.1	.20	51.74	1
14 590	Octantis	9.5	46.36	+ 8.616	78 41 11.6	+ 11.20	51.27	2
14 591	Pavonis	9.0	20 15 59.00	+ 5.749	67 30 15.8	+ 11.21	51.65	2
14 592	Octantis	10.0	16 7.15	16.241	85 11 22.3	.22	51.04	3
14 593	Gould, Z. C., 20^h 617 . .	9.0	8.75	5.886	68 30 31.3	.23	52.65	1
14 594	Pavonis	10.0	11.11	6.600	72 33 57.8	.23	51.68	1
14 595	Lacaille, 8411	7.0	11.88	+ 6.555	72 21 15.7	+ 11.23	51.67	2
14 596	Gould, Z. C., 20^h 638 . .	9.0	20 16 12.26	+ 7.290	75 17 4.9	+ 11.23	50.75	1
14 597	Gould, Z. C., 20^h 611 . .	9.8	12.98	5.601	66 20 51.8	.23	52.68	2
14 598	Pavonis	9.5	14.06	6.500	72 5 29.7	.23	51.67	1
14 599	Octantis	10.0	20.19	7.215	75 2 17.3	.24	51.65	1
14 600	Gould, Z. C., 20^h 644 . .	9.5	20.56	+ 7.282	75 15 44.2	+ 11.24	50.75	1
14 601	Octantis	9.5	20 16 33.65	+ 10.376	81 23 3.2	+ 11.26	50.62	1
14 602	Octantis	10.0	34.66	9.178	79 43 42.4	.26	50.78	1
14 603	β Octantis	7.1	34.71	133.672	89 30 52.3	.26	51.14	9
14 604	Pavonis	10.0	41.68	5.556	65 59 56.2	.27	52.67	1
14 605	Gould, Z. C., 20^h 643 . .	8.5	16 48.29	+ 5.982	69 11 36.3	+ 11.27	52.35	3
14 606	Gould, Z. C., 20^h 675 . .	9.2	20 17 6.02	+ 7.395	75 39 50.4	+ 11.29	52.11	2
14 607	Brisbane, 6853	8.0	16.47	6.434	71 48 55.9	.31	51.66	1
14 608	Pavonis	10.0	19.89	5.686	67 5 56.8	.31	52.68	1
14 609	Octantis	9.5	36.10	7.753	76 44 21.8	.33	51.64	1
14 610	Octantis	11.0	37.27	+ 12.146	83 4 7.0	+ 11.33	50.70	1
14 611	Gould, 28039	9.0	20 17 38.63	+ 6.427	71 48 5.3	+ 11.33	51.66	1
14 612	Pavonis	9.5	48.34	6.649	72 52 12.8	.35	51.75	1
14 613	Gould, Z. C., 20^h 679 . .	8.8	49.34	5.958	69 5 33.8	.35	52.35	3
14 614	Pavonis	9.2	17 58.66	6.280	71 2 33.0	.36	51.69	2
14 615	Octantis	9.0	18 3.65	+ 10.061	81 2 31.2	+ 11.36	50.65	2
14 616	Pavonis	9.5	20 18 13.81	+ 5.871	68 31 36.9	+ 11.38	52.65	1
14 617	Gould, Z. C., 20^h 692 . .	9.5	15.60	5.948	69 3 18.6	.38	52.35	3
14 618	Gould, Z. C., 20^h 691 . .	7.5	17.62	5.581	66 18 29.6	.38	52.67	3
14 619	Gould, 28052	8.0	23.76	5.919	68 52 0.8	.39	52.62	1
14 620	Pavonis	10.0	41.40	+ 5.886	68 31 14.5	+ 11.41	52.65	1
14 621	Gould, Z. C., 20^h 730 . .	10.0	20 18 43.75	+ 7.607	76 22 21.7	+ 11.41	51.69	2
14 622	Pavonis	10.0	46.80	5.864	68 30 39.9	.42	52.65	1
14 623	Lacaille, 8413	7.8	47.07	7.989	77 23 47.7	.42	51.70	3
14 624	Octantis	10.0	55.73	7.527	76 8 46.1	.43	51.64	1
14 625	Pavonis	10.0	18 56.56	+ 6.750	73 22 5.1	+ 11.43	51.74	1

Number.	Constellation, Name of Star, or Synonym.	Magnitude.	Right Ascension, 1850.0.	Annual Precession.	South Declination, 1850.0.	Annual Precession.	Mean year.	No. of obs.
			h. m. s.	s.	° ′ ″	″		
14 626	Gould, Z. C., 20h 713	9.5	20 19 4.68	+ 5.600	66 31 8.3	+ 11.44	52.68	1
14 627	Pavonis	9.5	9.73	5.727	67 31 53.5	.44	52.63	1
14 628	Pavonis	9.0	10.19	5.973	69 16 1.6	.45	52.64	1
14 629	Lacaille, 8424	6.5	28.20	6.387	71 41 26.3	.47	51.66	1
14 630	Pavonis	9.5	49.79	+ 6.160	70 27 50.2	+ 11.49	52.58	1
14 631	Octantis	9.8	20 19 49.84	+18.652	85 58 36.2	+ 11.49	51.22	3
14 632	Pavonis	9.8	54.96	6.079	69 59 4.8	.50	51.75	2
14 633	Octantis	10.0	19 59.05	13.222	83 50 28.1	.50	50.68	1
14 634	Octantis	9.5	20 3.32	7.699	76 41 8.9	.51	51.64	1
14 635	Gould, Z. C., 20h 793	10.0	5.82	+ 9.160	79 48 12.2	+ 11.51	50.73	2
14 636	Pavonis	9.4	20 20 11.86	+ 5.927	69 1 37.0	+ 11.52	52.21	4
14 637	Gould, 28097	9.0	14.18	5.681	67 14 21.1	.52	50.67	1
14 638	Gould, 28102	8.5	23.48	5.678	67 13 48.9	.53	50.67	1
14 639	Pavonis	9.5	24.81	6.092	70 5 16.8	.53	51.74	1
14 640	Lacaille, 8431	7.0	32.48	+ 6.391	71 45 53.4	+ 11.54	51.66	2
14 641	Pavonis	9.5	20 20 38.32	+ 5.799	68 9 36.4	+ 11.55	52.68	1
14 642	Pavonis	9.0	49.99	5.745	67 46 4.7	.56	52.63	1
14 643	Gould, 28113	9.2	50.02	6.350	71 34 11.6	.56	51.67	2
14 644	Octantis	9.0	53.66	11.709	82 47 33.7	.57	50.69	1
14 645	Octantis	10.0	53.80	+ 7.466	76 2 26.7	+ 11.57	51.64	1
14 646	Lacaille, 8437	7.0	20 20 53.89	+ 6.084	70 6 45.4	+ 11.57	51.74	1
14 647	Pavonis	9.0	21 1.39	6.250	71 2 19.7	.58	51.70	1
14 648	Lacaille, 8436	7.5	2.54	6.349	71 34 34.3	.58	51.67	2
14 649	Pavonis	9.6	4.68	5.940	69 9 35.3	.58	52.21	4
14 650	Pavonis	10.0	7.09	+ 5.735	67 42 49.1	+ 11.58	52.63	1
14 651	Pavonis	10.0	20 21 12.86	+ 5.931	69 6 44.7	+ 11.59	52.64	1
14 652	Octantis	9.5	36.62	7.339	75 40 23.2	.62	52.64	1
14 653	Pavonis	9.2	47.77	5.498	65 49 7.0	.63	52.67	2
14 654	Pavonis	10.0	49.81	5.650	67 6 4.7	.63	52.68	1
14 655	Gould, 28128	9.5	53.22	+ 6.077	70 4 34.8	+ 11.64	51.74	1
14 656	Lacaille, 8445	7.5	20 21 57.56	+ 6.056	69 57 23.9	+ 11.64	51.75	2
14 657	Pavonis	9.5	21 59.91	6.165	70 36 56.3	.65	52.58	1
14 658	Pavonis	9.5	22 3.96	5.544	66 14 31.4	.65	52.68	2
14 659	Brisbane, 6868	8.0	6.20	6.383	71 48 30.8	.65	51.66	2
14 660	Octantis	10.0	10.82	+14.140	84 23 11.5	+ 11.66	50.68	1
14 661	Pavonis	9.0	20 22 11.10	+ 5.740	67 49 11.7	+ 11.66	52.68	1
14 662	Pavonis	9.0	11.54	6.615	72 55 50.2	.66	51.75	1
14 663	Pavonis	9.0	17.50	6.448	72 8 44.9	.67	51.67	1
14 664	Octantis	9.3	18.78	29.232	87 37 10.4	.67	51.22	6
14 665	Gould, Z. C., 20h 829	9.0	20.03	+ 6.246	71 5 24.8	+ 11.67	51.70	1

Number.	Constellation, Name of Star, or Synonym.	Magnitude.	Right Ascension, 1850.0.	Annual Precession.	South Declination, 1850.0.	Annual Precession.	Mean year.	No. of obs.
			h. m. s.	s.	° ′ ″	″		
14 666	Octantis	10.0	20 22 26.17	+ 8.080	77 45 14.4	+ 11.68	51.76	1
14 667	Gould, Z. C., 20h 825 . .	9.0	29.71	5.676	67 20 46.6	.68	50.67	1
14 668	Octantis	9.1	33.34	29.258	87 37 25.7	.69	51.22	6
14 669	Octantis	9.8	33.86	7.645	76 37 58.7	.69	51.68	2
14 670	Octantis	9.5	40.76	+ 8.344	78 21 32.8	+ 11.70	51.71	2
14 671	Gould, Z. C., 20h 841 . .	7.5	20 22 45.16	+ 5.750	67 55 40.2	+ 11.70	52.65	2
14 672	Octantis	9.3	50.16	16.422	85 21 2.4	.71	51.14	6
14 673	Octantis	9.0	22 59.46	11.651	82 47 28.6	.72	50.69	1
14 674	Octantis	10.0	23 9.51	9.817	80 51 54.5	.73	50.69	1
14 675	μ^1 Octantis	6.7	20.20	+ 7.656	76 41 47.6	+ 11.74	51.68	2
14 676	Brisbane, 6872	7.3	20 23 25.08	+ 6.146	70 35 1.5	+ 11.75	52.29	3
14 677	Lacaille, 8455	7.2	26.21	6.331	71 36 23.6	.75	51.67	2
14 678	Lacaille, 8443	7.2	42.86	7.370	75 51 39.1	.77	51.96	3
14 679	Stone, 10969	7.7	44.35	7.368	75 51 21.6	.77	51.96	3
14 680	Octantis	9.0	46.91	+ 8.977	79 36 42.5	+ 11.77	90.73	3
14 681	Octantis	9.5	20 23 52.08	+17.006	85 33 32.9	+ 11.78	51.23	3
14 682	Pavonis	10.0	23 53.73	5.856	68 45 50.7	.78	52.65	1
14 683	Pavonis	10.0	24 5.59	7.086	74 55 22.5	.79	51.65	1
14 684	Lacaille, 8420	9.0	5.80	9.786	80 50 57.3	.80	90.69	1
14 685	Octantis	10.0	9.01	+10.482	81 41 55.9	+ 11.80	50.72	1
14 686	Octantis	9.8	20 24 12.38	+ 7.903	77 23 17.8	+ 11.80	52.18	2
14 687	Pavonis	10.0	16.35	5.723	67 49 10.3	.81	52.63	1
14 688	Lacaille, 8444	7.8	20.96	7.775	77 3 40.6	.81	52.12	2
14 689	Lacaille, 8448	8.5	29.21	7.271	75 34 28.3	.82	51.66	3
14 690	Gould, Z. C., 20h 893 . .	8.8	33.00	+ 5.620	67 2 9.1	+ 11.83	51.67	2
14 691	Pavonis	10.0	20 24 41.80	+ 6.872	74 8 37.0	+ 11.84	51.70	2
14 692	Octantis	9.5	24 49.73	12.878	83 43 17.4	.85	50.68	1
14 693	Lacaille, 8425	7.5	25 2.36	9.434	80 22 53.6	.86	51.14	2
14 694	Pavonis	9.8	5.63	6.071	70 13 43.6	.87	52.34	3
14 695	Octantis	8.5	6.00	+14.094	84 24 51.3	+ 11.87	50.67	2
14 696	Pavonis	10.0	20 25 19.73	+ 5.605	66 57 57.0	+ 11.88	52.68	1
14 697	Pavonis	9.5	26.80	6.445	72 17 43.7	.89	51.68	2
14 698	Pavonis	10.0	36.12	7.086	74 59 26.2	.90	51.20	2
14 699	Octantis	10.0	37.67	15.978	85 14 1.4	.90	50.64	1
14 700	Lacaille, 8467	9.5	39.62	+ 6.105	70 28 11.3	+ 11.91	52.64	2
14 701	Pavonis	10.0	20 25 45.40	+ 5.850	68 49 55.2	+ 11.91	52.20	2
14 702	Gould, 28204	9.3	25 55.55	6.071	70 16 48.4	.92	52.34	3
14 703	Pavonis	9.5	26 10.44	5.734	68 1 27.4	.94	52.68	1
14 704	Brisbane, 6882	9.0	19.39	6.028	70 2 24.8	.95	51.74	1
14 705	Pavonis	9.5	20.04	+ 5.756	68 12 3.7	+ 11.95	52.68	1

Number.	Constellation, Name of Star, or Synonym.	Magnitude.	Right Ascension, 1850.0.	Annual Precession.	South Declination, 1850.0.	Annual Precession.	Mean year.	No. of obs.
			h. m. s.	s.	° ′ ″	″		
14 706	Octantis	8.7	20 26 21.72	+15.972	85 14 33.9	+ 11.96	51.13	6
14 707	Pavonis	9.5	39.79	5.724	67 58 54.5	.98	52.68	1
14 708	Lacaille, 8434	8.2	42.18	9.597	80 40 1.8	.98	51.15	2
14 709	Lacaille, 8468	8.0	43.56	6.267	71 27 1.8	.98	51.67	2
14 710	Gould, Z. C., 20h 976 . .	9.0	54.50	+ 6.141	70 45 15.2	+ 11.99	52.14	2
14 711	Lacaille, 8446	8.8	20 26 56.75	+ 8.842	79 28 49.0	+ 12.00	50.73	3
14 712	Octantis	10.0	27 1.31	10.924	82 13 52.2	.00	50.65	1
14 713	Lacaille, 8469	7.8	7.64	6.458	72 27 4.3	.01	51.68	2
14 714	Pavonis	10.0	9.91	6.442	72 22 28.8	.01	51.68	1
14 715	Pavonis	9.8	19.68	+ 5.611	67 8 46.0	+ 12.02	51.67	2
14 716	Gould, Z. C., 20h 992 . .	9.3	20 27 23.49	+ 6.413	72 14 30.2	+ 12.03	51.68	3
14 717	Gould, Z. C., 20h 987 . .	8.8	29.69	5.809	68 39 27.4	.03	52.21	4
14 718	Pavonis	10.0	41.26	5.962	69 42 22.6	.05	51.77	1
14 719	Gould, Z. C., 20h 997 . .	8.5	27 50.25	5.660	67 34 17.7	.06	52.63	1
14 720	υ Pavonis	5.5	28 7.27	+ 5.622	67 17 0.5	+ 12.08	50.67	1
14 721	Pavonis	10.0	20 28 22.87	+ 6.656	73 25 43.4	+ 12.10	51.74	1
14 722	Pavonis	9.0	29.97	6.142	70 51 3.9	.11	51.70	1
14 723	Pavonis	9.8	36.32	5.773	68 28 1.8	.11	52.22	2
14 724	Pavonis	9.5	37.07	5.724	68 6 20.5	.11	52.68	1
14 725	Pavonis	9.5	41.54	+ 6.178	71 4 14.9	+ 12.12	51.69	2
14 726	Pavonis	9.5	20 28 43.20	+ 5.635	67 25 52.7	+ 12.12	52.63	1
14 727	Pavonis	9.5	28 48.54	5.640	67 28 30.6	.13	52.63	1
14 728	Gould, Z. C., 20h 1063 .	9.5	29 3.79	8.052	77 56 14.0	.14	51.28	2
14 729	Octantis	9.8	4.21	8.989	79 48 18.7	.14	50.71	3
14 730	Pavonis	9.5	4.40	+ 6.545	72 58 4.2	+ 12.15	51.77	2
14 731	Brisbane, 6891	9.0	20 29 10.81	+ 6.300	71 45 45.2	+ 12.15	51.66	1
14 732	Pavonis	10.0	16.44	6.630	73 21 50.9	.16	51.74	1
14 733	Pavonis	10.0	20.20	5.509	66 26 0.0	.16	52.68	1
14 734	Lacaille, 8467	8.8	21.35	6.055	70 22 58.7	.16	52.34	3
14 735	Pavonis	8.0	25.45	+ 5.906	69 26 29.0	+ 12.17	52.07	3
14 736	Gould, Z. C., 20h 1051 .	9.0	20 29 32.62	+ 5.853	69 5 44.8	+ 12.18	51.79	2
14 737	Octantis	9.7	44.10	15.901	85 16 9.1	.19	51.22	5
14 738	Pavonis	10.0	47.39	6.516	72 53 48.9	.19	51.75	1
14 739	Gould, Z. C., 20h 1097 .	9.0	29 49.29	8.126	78 8 25.2	.20	50.79	1
14 740	Brisbane, 6892	8.0	30 8.14	+ 6.194	71 14 39.8	+ 12.21	51.69	1
14 741	Pavonis	9.5	20 30 11.67	+ 5.712	68 7 22.5	+ 12.22	52.68	1
14 742	Pavonis	9.2	14.16	5.479	66 13 58.6	.23	52.67	2
14 743	Octantis	10.5	21.16	26.827	87 26 46.1	.23	51.25	2
14 744	Pavonis	10.0	39.88	5.692	68 0 8.7	.26	52.68	1
14 745	Gould, 28331	9.0	44.54	+ 6.424	72 28 54.9	+ 12.26	51.68	2

Number.	Constellation, Name of Star, or Synonym.	Magnitude.	Right Ascension, 1850.0.	Annual Precession.	South Declination, 1850.0.	Annual Precession.	Mean year.	No. of obs.
			h. m. s.	s.	° ′ ″	″		
14 746	Gould, Z. C., 20h 1098	10.0	20 30 44.95	+ 5.860	69 12 57.6	+ 12.26	51.78	1
14 747	Octantis . . .	10.0	45.65	8.403	78 46 24.6	.26	51.77	1
14 748	Pavonis	10.0	51.79	6.375	72 14 33.4	.27	51.67	2
14 749	Pavonis	9.2	52.46	5.567	67 1 42.9	.27	52.65	2
14 750	Gould, 28357	9.0	30 56.59	+ 8.730	79 24 52.8	+ 12.27	50.70	2
14 751	Pavonis	9.5	20 31 0.47	+ 6.582	73 14 15.6	+ 12.28	51.76	2
14 752	Lacaille, 8473	8.8	4.65	8.349	78 40 25.4	.28	51.27	2
14 753	Octantis	8.5	6.86	12.643	83 41 57.5	.29	50.68	1
14 754	Octantis	10.0	7.06	7.791	77 22 39.0	.29	52.12	2
14 755	Pavonis	10.0	21.53	+ 5.966	69 57 16.3	+ 12.30	51.77	1
14 756	β Pavonis	4.2	20 31 22.42	+ 5.528	66 44 8.2	+ 12.30	52.65	3
14 757	Gould, Z. C., 20h 1140 .	9.0	29.93	6.969	74 50 45.3	.31	51.65	1
14 758	Gould, 28344	8.2	34.50	5.609	67 24 57.4	.32	51.65	2
14 759	Lacaille, 8493	8.5	37.94	7.423	76 22 13.8	.32	51.69	2
14 760	Octantis	9.5	51.68	+23.528	87 3 0.0	+ 12.34	51.22	5
14 761	Pavonis	10.0	20 31 56.44	+ 5.942	69 50 21.2	+ 12.34	51.77	1
14 762	Gould, Z. C., 20h 1136 .	8.2	58.13	5.597	67 20 55.2	.35	51.65	2
14 763	Pavonis	10.0	31 58.44	6.325	72 3 15.8	.35	51.67	1
14 764	Gould, Z. C., 20h 1163 .	10.0	32 3.63	7.443	76 27 1.6	.35	51.73	1
14 765	Gould, Z. C., 20h 1160 .	9.0	3.67	+ 6.953	74 48 55.9	+ 12.35	51.65	1
14 766	Pavonis	7.5	20 32 25.37	+ 6.057	70 35 0.5	+ 12.38	52.69	1
14 767	Pavonis	9.3	27.41	5.515	66 42 23.4	.38	52.66	3
14 768	Gould, 28398	9.5	45.43	8.456	78 57 20.2	.40	51.77	1
14 769	Gould, 28399	9.5	45.58	8.558	79 9 18.3	.40	51.24	2
14 770	Octantis	9.0	50.10	+30.297	87 47 23.6	+ 12.40	51.22	6
14 771	Gould, Z. C., 20h 1185 .	9.2	20 32 51.86	+ 7.157	75 34 37.8	+ 12.41	51.16	2
14 772	Pavonis	10.0	54.77	6.283	71 53 12.2	.41	51.66	1
14 773	Gould, 28407	9.2	32 55.78	8.546	79 8 18.9	.41	51.24	2
14 774	Lacaille, 8502 . .	8.5	33 1.88	6.453	72 45 5.0	.42	51.72	2
14 775	Lacaille, 8501	8.2	5.28	+ 6.521	73 4 18.9	+ 12.42	51.76	3
14 776	Pavonis	8.7	20 33 11.62	+ 5.715	68 21 1.1	+ 12.43	52.37	3
14 777	Gould, 28408	9.0	36.14	7.332	76 10 49.7	.46	51.64	1
14 778	Octantis	10.0	37.13	11.626	83 0 58.7	.46	50.69	3
14 779	Octantis	9.2	33 52.42	8.164	78 23 0.8	.48	50.78	2
14 780	Pavonis	10.0	34 5.20	+ 5.406	65 51 24.3	+ 12.49	52.67	1
14 781	Pavonis	9.7	20 34 20.10	+ 8.606	73 31 27.5	+ 12.51	51.75	2
14 782	Octantis	9.0	22.78	11.323	82 46 52.8	.51	50.69	1
14 783	Lacaille, 8519	7.5	35.89	5.840	69 19 51.4	.52	51.78	1
14 784	Gould, 28436	9.5	42.78	7.215	75 51 35.4	.53	51.61	2
14 785	Octantis	10.0	50.39	+12.674	83 47 55.0	+ 12.54	50.68	1

Number.	Constellation, Name of Star, or Synonym.	Magnitude.	Right Ascension, 1850.0.	Annual Precession.	South Declination, 1850.0.	Annual Precession.	Mean year.	No. of obs.
			h. m. s.	s.	° ′ ″	″		
14 786	Pavonis	10.0	20 34 54.86	+ 5.464	66 26 13.3	+ 12.55	52.68	1
14 787	σ Pavonis	7.0	35 1.34	5.834	69 19 3.8	.56	51.78	1
14 788	Octantis	9.5	5.14	7.609	77 3 33.7	.56	51.63	1
14 789	Lacaille, 8483	9.0	5.58	9.788	81 10 38.9	.56	50.65	2
14 790	Lacaille, 8510	8.8	10.14	+ 6.764	74 14 9.4	+ 12.57	51.68	2
14 791	Pavonis	9.5	20 35 12.42	+ 5.988	70 20 6.8	+ 12.57	52.16	2
14 792	Pavonis	9.5	15.14	5.971	70 14 0.6	.57	52.16	2
14 793	Octantis	9.0	20.01	13.876	84 29 36.7	.58	50.66	1
14 794	Pavonis	10.0	23.64	5.768	68 52 56.7	.58	51.79	1
14 795	Pavonis	10.0	30.64	+ 5.851	69 27 58.7	+ 12.59	51.77	1
14 796	Pavonis	10.0	20 35 39.46	+ 5.919	69 55 35.3	+ 12.60	51.75	2
14 797	Gould, Z. C., 20h 1248	8.2	42.06	6.186	71 31 46.4	.60	51.67	2
14 798	Gould, Z. C., 20h 1263	9.5	44.57	7.292	76 9 16.0	.60	51.64	1
14 799	Pavonis	10.0	47.72	5.805	69 10 16.0	.61	51.79	2
14 800	Octantis	9.5	56.30	+ 7.672	77 16 10.5	+ 12.62	51.63	1
14 801	Gould, 28447	8.8	20 35 58.15	+ 5.676	68 14 28.0	+ 12.62	52.24	2
14 802	Pavonis	9.8	36 4.68	6.479	73 2 38.4	.63	51.76	3
14 803	Pavonis	8.c	32.30	5.602	67 43 51.2	.66	52.63	1
14 804	Pavonis	9.0	33.43	5.600	67 41 48.0	.66	52.63	1
14 805	Gould, Z. C., 20h 1267	9.5	33.52	+ 5.413	66 6 18.2	+ 12.66	52.67	1
14 806	Octantis	9.5	20 36 38.68	+ 10.229	81 45 27.4	+ 12.67	50.72	1
14 807	Pavonis	10.0	47.49	5.423	66 13 12.8	.68	52.68	1
14 808	Octantis	10.0	49.39	7.225	75 59 29.8	.68	51.64	1
14 809	Pavonis	10.0	51.73	5.617	67 51 20.4	.68	52.63	1
14 810	Gould, Z. C., 20h 1285	8.0	36 54.25	+ 5.648	68 5 42.7	+ 12.68	52.68	1
14 811	Pavonis	10.0	20 37 34.55	+ 5.860	69 40 4.8	+ 12.73	51.77	1
14 812	Octantis	10.0	37 56.20	10.571	82 9 52.4	.75	50.65	1
14 813	Gould, Z. C., 20h 1323	9.0	38 1.94	5.568	67 32 48.1	.76	52.63	1
14 814	Herschel, 5221[1]	10.0	8.30	5.417	66 15 42.4	.77	52.67	2
14 815	Herschel, 5221[2]	10.0	9.39	+ 5.417	66 15 46.0	+ 12.77	52.68	1
14 816	Gould, 28573	10.0	20 38 15.92	+ 12.684	83 52 45.2	+ 12.77	50.68	1
14 817	Pavonis	9.5	18.92	6.005	70 38 35.5	.78	52.14	2
14 818	Pavonis	9.0	22.91	6.256	72 4 14.2	.78	51.67	2
14 819	Gould, 28512	8.0	35.35	5.640	68 9 6.4	.80	52.68	1
14 820	Gould, Z. C., 20h 1337	9.0	41.30	+ 5.506	67 4 58.6	+ 12.80	51.64	2
14 821	Gould, Z. C., 20h 1357	9.2	20 38 47.19	+ 6.572	73 37 20.0	+ 12.81	51.75	2
14 822	Pavonis	9.5	38 56.65	6.443	73 2 12.9	.82	51.78	1
14 823	Pavonis	10.0	39 0.00	6.650	73 58 18.9	.82	51.72	2
14 824	Octantis	8.5	22.91	11.527	83 4 37.7	.85	50.68	2
14 825	Gould, 28561	9.0	29.04	+ 8.227	78 44 29.6	+ 12.86	51.27	2

Number.	Constellation, Name of Star, or Synonym.	Magnitude.	Right Ascension, 1850.0.	Annual Precession.	South Declination, 1850.0.	Annual Precession.	Mean year.	No. of obs.
			h. m. s.	s.	° ′ ″	″		
14 826	Pavonis	9.0	20 39 33.52	+ 5.468	66 49 12.8	+ 12.86	52.62	1
14 827	Pavonis	10.0	43.71	5.833	69 37 37.3	.87	51.77	1
14 828	Octantis	9.0	54.28	11.528	83 5 30.0	.89	50.68	2
14 829	Pavonis	10.0	54.91	6.361	72 41 51.5	.89	51.68	1
14 830	Gould, Z. C., 20^h 1383 .	9.5	39 59.12	+ 5.459	66 46 46.4	+ 12.89	52.65	2
14 831	Gould, Z. C., 20^h 1410 .	10.0	20 40 3.02	+ 7.389	76 39 17.3	+ 12.90	51.68	2
14 832	Stone, 11124	7.6	3.30	16.982	85 47 22.3	.90	51.09	7
14 833	Gould, 28552	8.5	4.78	5.368	65 57 8.7	.90	52.67	1
14 834	Gould, Z. C., 20^h 1409 .	10.0	8.85	7.111	75 46 45.5	.90	51.58	1
14 835	Gould, Z. C., 20^h 1398 .	9.0	11.54	+ 6.188	71 48 58.7	+ 12.90	51.67	3
14 836	Gould, Z. C., 20^h 1413 .	10.0	20 40 13.54	+ 7.036	75 31 31.1	+ 12.91	50.75	1
14 837	Pavonis	9.5	26.12	6.515	73 27 26.8	.92	51.78	1
14 838	Lacaille, 8536	8.5	29.60	7.358	76 34 55.0	.93	51.68	2
14 839	Octantis	9.1	31.17	25.630	87 24 18.4	.93	51.22	5
14 840	Octantis	10.0	49.80	+16.224	85 33 36.2	+ 12.95	50.74	1
14 841	Pavonis	10.0	20 40 52.09	+ 5.688	68 40 44.0	+ 12.95	51.79	1
14 842	Lacaille, 8507	8.2	41 4.69	10.305	81 58 17.1	.96	50.69	2
14 843	Octantis	11.0	11.88	8.278	78 55 2.3	.97	51.77	1
14 844	Octantis	10.0	17.92	9.541	81 2 43.0	.98	50.69	1
14 845	Gould, Z. C., 20^h 1429 .	9.0	24.60	+ 5.591	67 58 24.8	+ 12.99	52.65	2
14 846	Octantis	11.0	20 41 24.69	+ 8.277	78 55 22.3	+ 12.99	51.77	1
14 847	Pavonis	10.0	35.26	6.336	72 40 26.2	13.00	51.72	2
14 848	Octantis	9.5	41 41.50	8.054	78 27 5.4	.01	50.78	1
14 849	Gould, Z. C., 20^h 1475 .	10.0	42 8.16	6.879	75 3 15.2	.03	51.20	2
14 850	Pavonis	10.0	10.20	+ 6.336	72 42 33.8	+ 13.04	51.75	1
14 851	Pavonis	10.0	20 42 15.25	+ 6.364	72 51 15.0	+ 13.04	51.75	1
14 852	Gould, 28602	7.8	19.68	5.431	66 42 41.2	.05	52.65	2
14 853	Pavonis	10.0	23.32	5.961	70 38 21.2	.05	52.14	2
14 854	Stone, 11139	7.8	34.04	18.505	86 14 19.8	.06	51.22	6
14 855	Octantis	8.2	35.66	+18.106	86 8 22.8	+ 13.06	51.22	6
14 856	Pavonis	8.0	20 42 49.43	+ 5.348	65 59 45.7	+ 13.08	52.67	1
14 857	Pavonis	10.0	42 53.91	5.735	69 9 59.0	.09	51.79	2
14 858	Gould, Z. C., 20^h 1504 .	9.0	43 1.63	6.855	75 0 42.0	.09	51.20	2
14 859	Pavonis	9.5	9.72	6.093	71 28 4.8	.10	51.66	1
14 860	Gould, Z. C., 20^h 1511 .	9.2	15.90	+ 7.324	76 36 46.2	+ 13.11	51.68	2
14 861	Pavonis	9.5	20 43 18.49	+ 5.350	66 2 55.8	+ 13.11	52.67	1
14 862	Gould, Z. C., 20^h 1498 .	7.8	24.32	5.609	68 15 54.6	.12	52.24	2
14 863	Lacaille, 8515	7.0	31.70	10.866	82 36 28.9	.13	50.69	2
14 864	Gould, 28676	10.0	45.10	9.651	81 16 15.6	.14	50.62	1
14 865	Lacaille, 8578	7.8	46.48	+ 5.702	68 59 25.0	+ 13.14	51.79	2

No. 14 840. Minute of right ascension doubtful.

Number.	Constellation, Name of Star, or Synonym.	Magnitude.	Right Ascension, 1850.0.	Annual Precession.	South Declination, 1850.0.	Annual Precession.	Mean year.	No. of obs.
			h. m. s.	s.	° ′ ″	″		
14 866	Lacaille, 8535	7.2	20 43 47.13	+ 9.649	81 16 14.2	+ 13.14	50.66	2
14 867	Pavonis	9.5	52.20	5.393	66 29 32.5	.15	52.68	1
14 868	Octantis	10.0	55.16	11.303	83 0 28.6	.15	50.67	1
14 869	Lacaille, 8577	7.0	58.28	5.802	69 42 44.1	.16	51.77	1
14 870	Lacaille, 8573[1]	9.0	43 58.76	+ 6.001	70 59 28.2	+ 13.16	51.69	2
14 871	Lacaille, 8573[2]	8.5	20 44 0.26	+ 6.001	70 59 31.3	+ 13.16	51.69	2
14 872	Pavonis	10.0	0.43	5.964	70 45 51.3	.16	52.58	1
14 873	Pavonis	10.0	1.26	5.662	68 42 51.1	.16	51.79	2
14 874	Octantis	10.0	1.98	12.499	83 53 27.6	.16	50.68	2
14 875	Lacaille, 8562	8.0	7.55	+ 7.116	75 59 45.7	+ 13.17	51.64	1
14 876	Gould, 28666	10.0	20 44 8.65	+ 5.843	70 0 2.4	+ 13.17	51.64	1
14 877	Pavonis	9.0	11.08	5.549	67 50 43.8	.17	52.65	2
14 878	Gould, 28704	9.0	14.10	12.528	83 54 48.9	.17	50.68	2
14 879	Gould, Z. C., 20h 1523	9.5	14.88	5.715	69 7 8.8	.17	51.79	2
14 880	Octantis	10.0	19.06	+12.542	83 55 30.0	+ 13.18	50.68	1
14 881	Lacaille, 8474	9.2	20 44 30.31	+15.457	85 21 2.4	+ 13.19	50.91	3
14 882	Pavonis	8.5	35.94	5.481	67 19 2.9	.20	50.67	1
14 883	Octantis	10.0	44 52.88	10.744	82 31 53.2	.23	50.69	1
14 884	Octantis	10.0	45 15.70	6.837	75 3 55.1	.24	51.65	1
14 885	Pavonis	9.5	17.11	+ 5.754	69 28 12.4	+ 13.24	51.78	2
14 886	Octantis	9.3	20 45 19.01	+18.713	86 19 37.1	+ 13.24	51.22	5
14 887	Octantis	10.0	21.05	12.479	83 54 30.2	.25	50.68	1
14 888	Octantis	9.5	28.54	10.586	82 23 37.5	.25	50.67	2
14 889	Lacaille, 8563	8.0	30.50	7.842	78 7 21.9	.26	50.79	1
14 890	Lacaille, 8588	7.2	33.71	+ 5.849	70 8 21.7	+ 13.26	52.16	2
14 891	Octantis	10.0	20 45 34.01	+30.520	87 54 25.2	+ 13.26	51.22	2
14 892	Gould, Z. C., 20h 1569	8.8	51.58	5.870	70 17 59.4	.28	52.16	2
14 893	Octantis	8.5	45 56.88	17.018	85 53 30.8	.29	51.10	8
14 894	Pavonis	10.0	46 6.93	6.492	73 41 2.9	.30	51.73	1
14 895	Pavonis	10.0	8.24	+ 6.294	72 44 36.5	+ 13.30	51.68	1
14 896	Gould, Z. C., 20h 1581	8.5	20 46 9.50	+ 5.851	70 11 35.6	+ 13.30	52.16	2
14 897	α Octantis	6.0	20.25	7.616	77 35 15.0	.31	51.73	3
14 898	Pavonis	10.0	21.13	6.207	72 18 22.7	.31	51.68	1
14 899	Pavonis	10.0	34.56	5.772	69 41 22.6	.33	51.77	1
14 900	Octantis	9.8	45.59	+20.717	86 45 40.3	+ 13.34	51.23	2
14 901	Pavonis	9.0	20 46 50.16	+ 6.220	72 24 26.6	+ 13.34	51.68	2
14 902	Gould, Z. C., 20h 1599	9.2	51.86	5.836	70 8 36.8	.35	52.16	2
14 903	Gould, Z. C., 20h 1601	9.5	57.46	5.651	68 51 0.4	.35	51.79	1
14 904	Gould, Z. C., 20h 1609	10.0	46 58.71	6.029	71 21 33.4	.35	51.69	1
14 905	Pavonis	10.0	47 21.37	+ 5.378	66 38 27.0	+ 13.38	52.62	1

Number.	Constellation, Name of Star, or Synonym.	Magnitude.	Right Ascension, 1850.0.	Annual Precession.	South Declination, 1850.0.	Annual Precession.	Mean year.	No. of obs.
			h. m. s.	s.	° ′ ″	″		
14 906	Gould, Z. C., 20h 1640	9.5	20 47 26.68	+ 7 742	77 57 42.2	+ 13.38	51.45	3
14 907	Lacaille, 8528	8.2	27.78	12.332	83 51 44.7	.38	50.68	2
14 908	Pavonis	10.0	35.02	6.508	73 50 {69.4 / 43.4}	.39	51.72	2
14 909	Pavonis	10.0	36.33	5.467	67 26 13.1	.39	52.63	1
14 910	Gould, Z. C., 20h 1629 .	9.2	39.10	+ 6.458	73 37 23.7	+ 13.40	51.76	2
14 911	Lacaille, 8611	7.2	20 47 39.28	+ 5.635	68 47 9.2	+ 13.40	51.79	2
14 912	Pavonis	9.5	40.38	6.618	74 19 38.5	.40	51.68	2
14 913	Pavonis	9.5	47 58.93	5.315	66 6 35.3	.42	52.67	1
14 914	Gould, Z. C., 20h 1638 .	9.5	48 9.67	5.818	70 5 39.4	.43	51.74	1
14 915	Pavonis	9.5	11.76	+ 5.285	65 49 57.9	+ 13.43	52.67	1
14 916	Gould, Z. C., 20h 1647 .	10.0	20 48 29.07	+ 6.011	71 21 17.2	+ 13.45	51.69	1
14 917	Pavonis	9.8	48 45.32	5.510	67 52 55.9	.47	52.65	2
14 918	Lacaille, 8580	8.0	49 1.89	8.608	79 51 51.0	.49	50.73	2
14 919	Lacaille, 8511[1]	9.2	6.51	14.188	84 54 51.6	.49	51.03	3
14 920	Lacaille, 8511[2]	9.0	9.41	+14.186	84 54 51.3	+ 13.50	51.03	3
14 921	Pavonis	9.0	20 49 11.47	+ 5.876	70 34 9.2	+ 13.50	52.58	1
14 922	Pavonis	9.0	11.61	5.753	69 44 58.9	.50	51.77	1
14 923	Gould, Z. C., 20h 1672 .	9.0	32.78	5.942	71 0 35.4	.52	51.69	2
14 924	Pavonis	8.0	39.78	5.889	70 41 25.1	.53	52.14	2
14 925	Pavonis	9.0	41.74	+ 5.512	67 58 20.3	+ 13.53	52.65	2
14 926	Pavonis	10.5	20 49 47.82	+ 7.981	78 38 35.5	+ 13.54	50.78	1
14 927	Gould, 28749	8.0	50 0.95	5.594	68 39 15.4	.55	51.79	2
14 928	Pavonis	9.0	3.00	5.902	70 47 46.8	.55	52.14	2
14 929	Pavonis	9.3	5.30	5.351	66 37 3.9	.56	52.66	3
14 930	Octantis	9.3	11.90	+10.022	81 56 0.0	+ 13.56	51.06	3
14 931	Pavonis	10.0	20 50 13.02	+ 5.443	67 26 40.2	+ 13.56	51.65	2
14 932	Gould, Z. C., 20h 1699 .	9.8	14.46	6.687	74 45 34.9	.57	51.65	2
14 933	Pavonis	10.0	25.63	5.762	69 54 25.2	.58	51.77	1
14 934	Octantis	10.5	27.50	42.360	88 33 54.6	.58	51.23	2
14 935	Gould, Z. C., 20h 1692 .	9.0	29.08	+ 5.608	68 47 46.9	+ 13.58	51.79	1
14 936	Pavonis	9.5	20 50 32.04	+ 6.590	74 22 37.2	+ 13.58	51.68	2
14 937	Pavonis	9.0	42.52	5.523	68 8 49.3	.60	52.68	1
14 938	Gould, 28787	9.0	43.54	7.296	76 53 58.9	.60	51.64	1
14 939	Pavonis	9.5	49.77	5.542	68 18 36.5	.60	52.68	1
14 940	Pavonis	9.0	53.98	+ 5.256	65 47 2.1	+ 13.61	52.67	1
14 941	Lacaille, 8614	9.0	20 50 58.60	+ 7.184	76 34 4.0	+ 13.61	51.73	1
14 942	Pavonis	9.5	51 9.79	6.252	72 51 5.5	.62	51.75	1
14 943	Lacaille, 8615	6.0	17.03	7.254	76 48 11.5	.63	51.64	1
14 944	Stone, 11193	8.2	20.10	14.015	84 52 47.0	.64	50.94	2
14 945	Octantis	9.5	28.87	+ 6.982	75 55 44.2	+ 13.64	51.64	1

No. 14 908. One of the declinations is evidently one revolution wrong. No. 14 934. Right ascension possibly 21h 1m 47.2s.

Number.	Constellation, Name of Star, or Synonym.	Magnitude.	Right Ascension, 1850.0.	Annual Precession.	South Declination, 1850.0.	Annual Precession.	Mean year.	No. of obs.
			h. m. s.	s.	° ′ ″	″		
14 946	Pavonis . . .	9.0	20 51 29.54	+ 5.521	68 11 42.5	+ 13.65	52.68	1
14 947	Pavonis	10.0	39.13	5.781	70 7 23.2	.66	51.74	1
14 948	Pavonis	9.0	41.54	5.836	70 29 35.9	.66	52.58	1
14 949	Pavonis	9.0	41.72	5.762	69 59 46.0	.66	51.74	1
14 950	Pavonis	9.2	50.93	+ 5.439	67 32 31.0	+ 13.67	51.65	2
14 951	Pavonis	9.0	20 51 52.09	+ 5.359	66 50 24.4	+ 13.67	52.62	1
14 952	Octantis	10.0	52 3.25	12.688	84 11 40.3	.68	50.66	1
14 953	Pavonis	10.0	22.36	6.565	74 22 55.6	.70	51.68	2
14 954	Pavonis	10.0	26.24	5.653	69 16 53.3	.71	51.78	1
14 955	Pavonis	9.0	27.96	+ 5.513	68 12 19.0	+ 13.71	52.68	1
14 956	Gould, Z. C., 20h 1748 . .	9.5	20 52 35.45	+ 5.552	68 31 30.3	+ 13.72	51.80	1
14 957	Lacaille, 8627 . .	8.0	39.82	5.644	69 14 9.0	.72	51.78	1
14 958	Pavonis	10.0	40.16	5.537	68 24 49.8	.72	51.80	1
14 959	Pavonis . . .	10.0	43.72	6.519	74 12 28.8	.73	51.71	1
14 960	Gould, 28825	9.0	46.59	+ 5.779	70 11 55.0	+ 13.73	51.74	1
14 961	Pavonis	11.0	20 52 52.32	+ 6.567	74 28 18.5	+ 13.73	51.66	1
14 962	Pavonis	11.0	52 53.46	6.567	74 28 15.7	.74	51.66	1
14 963	Lacaille, 8623	8.0	53 8.43	6.464	73 59 18.3	.75	51.71	1
14 964	Gould, Z. C., 20h 1784 .	9.2	14.74	6.965	75 58 9.0	.76	51.61	2
14 965	Pavonis	10.0	26.20	+ 6.046	71 53 59.4	+ 13.77	51.65	1
14 966	Lacaille, 8625	6.0	20 53 32.06	+ 6.407	73 45 13.3	+ 13.78	51.77	2
14 967	Octantis	9.8	35.02	7.205	76 46 16.6	.78	51.68	2
14 968	Octantis	9.8	35.20	18.159	86 18 56.4	.78	51.21	2
14 969	Octantis	9.5	44.09	7.785	78 21 35.5	.79	50.79	1
14 970	Pavonis	9.0	53 49.27	+ 5.439	67 42 5.1	+ 13.79	52.63	1
14 971	Gould, Z. C., 20h 1794 .	9.0	20 54 6.11	+ 5.534	68 30 30.3	+ 13.81	51.80	1
14 972	Pavonis	10.0	15.86	6.027	71 51 57.4	.82	51.66	1
14 973	Pavonis	9.0	24.24	5.277	66 17 47.1	.83	52.68	3
14 974	Gould, Z. C., 20h 1806 .	9.0	26.20	5.707	69 49 35.1	.83	51.77	1
14 975	Gould, Z. C., 20h 1824 .	9.0	28.37	+ 7.450	77 32 1.2	+ 13.84	51.73	3
14 976	Lacaille, 8569	8.2	20 54 29.34	+12.010	83 49 10.8	+ 13.84	50.68	2
14 977	Pavonis	9.2	33.44	5.273	66 16 14.8	.84	52.68	3
14 978	Lacaille, 8592	8.2	38.11	10.746	82 48 54.8	.85	51.20	2
14 979	Pavonis	9.8	41.26	5.406	67 29 38.2	.85	51.65	2
14 980	Pavonis	9.0	50.29	+ 5.421	67 37 54.4	+ 13.86	52.63	1
14 981	Octantis	10.0	20 54 53.11	+12.107	83 53 42.2	+ 13.86	50.68	2
14 982	Gould, Z. C., 20h 1829 .	9.3	55 5.31	6.322	73 27 8.5	.88	51.77	3
14 983	Pavonis	9.5	14.92	6.079	72 12 28.4	.88	51.68	2
14 984	Lacaille, 8618	8.0	29.16	9.127	80 57 6.4	.90	50.73	2
14 985	Octantis	10.0	55 52.78	+39.974	88 30 20.6	+ 13.92	51.23	1

No. 14 985. This is probably the same as No. 15 210, differing 2 wire intervals in right ascension.

Number.	Constellation, Name of Star, or Synonym.	Magnitude.	Right Ascension, 1850.0.	Annual Precession.	South Declination, 1850.0.	Annual Precession.	Mean year.	No. of obs.
			h. m. s.	s.	° ′ ″	″		
14 986	Lacaille, 8637	6.8	20 56 7.17	+ 6.245	73 8 38.0	+ 13.94	51.77	2
14 987	Gould, 28905	7.3	7.79	5.272	66 23 40.2	.94	52.68	3
14 988	Pavonis	9.5	12.52	5.423	67 45 54.5	.95	52.63	1
14 989	Octantis	10.0	34.84	7.513	77 48 29.1	.97	50.78	1
14 990	Pavonis	10.0	40.58	+ 8.807	80 30 19.9	+ 13.97	51.62	1
14 991	Gould, Z. C., 20^h 1899	9.2	20 56 47.06	+ 7.248	77 4 12.8	+ 13.98	51.64	2
14 992	Pavonis	10.0	49.37	6.307	73 29 48.3	.98	51.75	2
14 993	Pavonis	9.2	52.62	7.376	77 26 52.6	.99	51.72	2
14 994	Octantis	10.0	56 56.76	7.056	76 28 52.3	.99	51.73	1
14 995	Octantis	10.5	57 3.63	+ 9.212	81 8 9.7	+ 14.00	50.80	1
14 996	Gould, Z. C., 20^h 1906	9.5	20 57 17.76	+ 6.317	73 34 37.2	+ 14.01	51.77	3
14 997	Octantis	10.0	24.35	8.728	80 24 7.5	.02	51.62	1
14 998	Gould, Z. C., 20^h 1903	9.2	26.60	5.799	70 41 5.3	.02	51.99	3
14 999	Pavonis	9.5	57 28.37	5.699	70 1 31.5	.02	51.74	1
15 000	Pavonis	9.2	58 0.68	+ 5.564	69 3 41.0	+ 14.06	51.79	2
15 001	Gould, 28964	8.8	20 58 7.01	+ 5.804	70 45 54.1	+ 14.06	51.99	3
15 002	Gould, Z. C., 21^h 31	9.0	12.23	7.144	76 49 48.3	.07	51.64	1
15 003	Gould, Z. C., 21^h 27	9.0	13.74	6.943	76 10 24.8	.07	51.64	1
15 004	Octantis	10.2	18.67	9.392	81 25 51.2	.08	51.11	3
15 005	Pavonis	10.0	27.30	+ 6.522	74 34 43.6	+ 14.09	51.66	1
15 006	Lacaille, 8551	7.2	20 58 30.19	+15.000	85 26 19.2	+ 14.09	51.10	8
15 007	Gould, Z. C., 21^h 19	9.3	30.29	5.233	66 13 52.6	.09	52.67	3
15 008	Pavonis	9.5	30.85	5.468	68 20 19.2	.09	52.68	1
15 009	Gould, Z. C., 21^h 34	10.0	36.46	6.367	73 53 51.3	.10	51.76	2
15 010	Pavonis	10.0	39.69	+ 5.307	66 56 52.3	+ 14.10	52.62	1
15 011	Gould, Z. C., 21^h 29	8.5	20 58 55.61	+ 5.298	66 53 14.5	+ 14.12	52.62	1
15 012	Pavonis	10.0	58 58.10	5.663	69 52 20.4	.12	51.75	2
15 013	Octantis	9.5	59 0.44	6.857	75 54 56.4	.12	51.61	2
15 014	Pavonis	10.5	3.27	5.656	69 53 40.8	.12	51.77	1
15 015	o Pavonis	5.2	10.34	+ 5.786	70 43 56.4	+ 14.13	51.99	3
15 016	Gould, 29005	9.5	20 59 10.98	+ 7.083	76 41 25.8	+ 14.13	51.68	2
15 017	Octantis	9.5	14.93	11.026	83 11 57.8	.14	50.67	1
15 018	Gould, Z. C., 21^h 64	9.2	26.68	6.614	75 1 20.2	.15	51.20	2
15 019	Gould, Z. C., 21^h 70	9.5	28.39	7.023	76 30 41.2	.15	51.73	1
15 020	Gould, Z. C., 21^h 65	9.2	34.12	+ 6.445	74 18 41.7	+ 14.16	51.68	2
15 021	Octantis	10.0	20 59 40.68	+ 8.888	80 45 12.3	+ 14.16	51.62	1
15 022	Gould, Z. C., 21^h 59	9.2	45.11	5.356	67 32 8.1	.17	51.98	3
15 023	Pavonis	10.0	45.63	5.890	71 26 15.1	.17	51.66	1
15 024	Gould, Z. C., 21^h 63	9.5	50.73	5.607	69 32 3.5	.17	51.78	2
15 025	Lacaille, 8673	8.0	52.00	+ 5.893	71 27 49.6	+ 14.17	51.67	2

Number.	Constellation, Name of Star, or Synonym.	Magnitude.	Right Ascension, 1850.0.	Annual Precession.	South Declination, 1850.0.	Annual Precession.	Mean year.	No. of obs.
			h. m. s.	s.	° ′ ″	− ″		
15 026	Octantis	10.0	20 59 52.59	+ 7.328	77 28 10.3	+ 14.17	51.72	2
15 027	Gould, Z. C., 21^{h} 67 . .	9.5	20 59 52.80	5.824	71 1 45.0	.17	51.70	1
15 028	Octantis	10.0	21 0 4.96	8.320	79 47 30.3	.19	50.67	1
15 029	Pavonis	10.0	5.17	5.793	70 50 46.1	.19	51.70	1
15 030	Pavonis	9.5	10.67	+ 6.008	72 9 52.0	+ 14.19	51.68	1
15 031	Gould, Z. C., 21^{h} 95 . .	8.5	21 0 41.43	+ 5.354	67 33 30.3	+ 14.22	51.98	3
15 032	Gould, Z. C., 21^{h} 94 . .	9.5	41.74	5.283	66 54 47.5	.23	52.62	1
15 033	Octantis	9.0	0 50.18	7.928	79 1 20.0	.23	51.24	2
15 034	Gould, 29041	9.0	1 1.99	6.750	75 38 48.3	.24	51.58	1
15 035	Lacaille, 8643	8.2	4.02	+ 8.837	80 43 47.5	+ 14.25	51.15	2
15 036	Pavonis	10.0	21 1 7.95	+ 6.381	74 7 57.6	+ 14.25	51.71	1
15 037	Octantis	9.0	14.03	11.728	83 48 0.8	.26	50.72	3
15 038	Octantis	10.0	33.81	10.338	82 38 11.4	.28	51.20	2
15 039	Gould, Z. C., 21^{h} 118 . .	9.0	34.88	5.327	67 23 37.8	.28	52.64	2
15 040	Lacaille, 8671	7.0	45.06	+ 6.824	75 57 41.8	+ 14.29	51.61	2
15 041	Gould, 29049	9.5	21 1 45.29	+ 6.113	72 51 19.2	+ 14.29	51.75	1
15 042	Pavonis	10.0	46.77	5.266	66 50 52.5	.29	52.62	1
15 043	Gould, 29050	9.0	1 53.34	5.816	71 8 4.7	.30	51.69	2
15 044	Octantis	9.5	2 17.16	6.571	75 1 30.2	.32	51.20	2
15 045	Octantis	10.0	26.56	+ 7.260	77 24 38.8	+ 14.33	51.72	2
15 046	Pavonis	10.0	21 2 26.79	+ 5.420	68 16 45.3	+ 14.33	52.68	1
15 047	Pavonis	10.0	30.76	5.626	69 53 29.6	.34	51.77	1
15 048	Octantis	9.8	45.60	9.320	81 29 50.7	.35	51.11	3
15 049	Octantis	9.2	46.34	9.072	81 9 9.6	.35	51.21	2
15 050	Pavonis	9.5	47.48	+ 5.393	68 5 10.8	+ 14.35	52.68	1
15 051	Gould, Z. C., 21^{h} 173 . .	9.0	21 2 58.45	+ 6.948	76 27 54.2	+ 14.36	51.69	2
15 052	Gould, Z. C., 21^{h} 162 . .	9.3	3 2.43	5.939	71 58 27.4	.37	51.72	3
15 053	Octantis	10.0	5.02	8.251	79 47 34.0	.37	50.78	1
15 054	Pavonis	9.5	11.38	6.631	75 19 20.2	.38	50.75	1
15 055	Lacaille, 8696	7.9	13.54	+ 6.248	73 38 40.9	+ 14.38	51.78	4
15 056	Gould, 29187	9.7	21 3 25.93	+ 15.159	85 35 41.1	+ 14.39	51.14	6
15 057	Gould, Z. C., 21^{h} 204 . .	9.0	28.36	7.408	77 52 51.1	.40	51.30	2
15 058	Gould, Z. C., 21^{h} 182 . .	9.8	34.80	6.071	72 45 48.7	.40	51.72	2
15 059	Gould, Z. C., 21^{h} 195 . .	9.0	53.26	5.870	71 37 36.1	.42	51.66	1
15 060	Octantis	9.0	54.64	+ 10.224	82 35 50.5	+ 14.52	51.20	2
15 061	Lacaille, 8696	9.8	21 3 59.06	+ 6.294	73 55 2.2	+ 14.43	51.75	3
15 062	Pavonis	10.0	4 1.98	5.145	65 52 21.9	.43	52.67	1
15 063	Pavonis	10.0	12.60	5.380	68 5 45.5	.44	52.68	1
15 064	Gould, Z. C., 21^{h} 199 . .	9.5	4 16.12	5.175	66 11 50.2	.44	52.67	3
15 065	Gould, 29126	9.5	5 3.02	+ 5.508	69 13 16.3	+ 14.49	51.78	1

Number.	Constellation, Name of Star, or Synonym.	Magnitude.	Right Ascension, 1850.0.	Annual Precession.	South Declination, 1850.0.	Annual Precession.	Mean year.	No. of obs.
			h. m. s.	s.	° ′ ″	″		
15 066	Lacaille, 8669	8.0	21 5 5.95	+ 9.121	81 18 48.2	+ 14.49	50.80	1
15 067	Lacaille, 8636	6.0	5.98	11.404	83 40 14.8	.49	50.74	2
15 068	Gould, 29222	9.9	16.99	14.695	85 27 27.3	.51	51.13	5
15 069	Pavonis	9.8	25.62	8.759	80 46 59.2	.51	51.62	2
15 070	Gould, Z. C., 21^h 239 . .	9.0	34.40	+ 5.289	67 25 6.6	+ 14.52	52.64	2
15 071	Octantis	9.5	21 5 36.63	+ 7.464	78 8 36.8	+ 14.52	50.79	1
15 072	Gould, Z. C., 21^h 246 . .	9.8	40.70	5.704	70 41 57.5	.53	51.70	2
15 073	Gould, 29165	10.0	52.35	6.499	74 57 33.2	.54	51.65	1
15 074	Gould, Z. C., 21^h 253 . .	9.3	56.54	5.889	71 53 57.3	.54	51.72	3
15 075	Pavonis	9.0	5 58.71	+ 5.116	65 45 47.6	+ 14.55	52.67	1
15 076	Gould, 29156	9.0	21 6 6.67	+ 5.239	66 59 45.6	+ 14.56	52.63	2
15 077	Octantis	10.0	10.83	7.473	78 11 50.6	.56	50.79	1
15 078	Pavonis	9.5	15.01	10.430	82 52 43.5	.56	51.71	1
15 079	Gould, 29160	9.2	17.24	5.237	66 59 34.2	.57	52.63	2
15 080	Lacaille, 8710	9.5	20.34	+ 6.381	74 28 51.4	+ 14.57	51.66	1
15 081	Pavonis	8.5	21 6 30.94	+ 5.545	69 38 7.2	+ 14.58	51.77	1
15 082	Gould, Z. C., 21^h 262 . .	9.2	33.07	5.191	66 34 51.3	.58	52.65	2
15 083	Octantis	9.9	37.97	29.826	88 2 7.5	.59	51.22	5
15 084	Octantis	10.0	41.06	7.171	77 22 57.2	.59	51.63	1
15 085	Pavonis	10.0	46.17	+ 6.269	73 59 47.0	+ 14.59	51.71	1
15 086	Pavonis	9.3	21 6 54 23	+ 6.152	73 25 48.1	+ 14.60	51.80	3
15 087	Gould, 29178	9.0	6 56.21	5.480	69 9 56.0	.60	51.79	2
15 088	Pavonis	9.5	7 0.65	5.377	68 19 50.9	.61	52.68	1
15 089	Lacaille, 8711	8.0	2.81	6.780	76 7 35.3	.62	51.64	1
15 090	Gould, Z. C., 21^h 280 . .	8.5	15.83	+ 5.318	67 52 54.3	+ 14.63	52.63	1
15 091	Pavonis	9.5	21 7 21.56	+ 5.354	68 9 25.6	+ 14.63	52.68	1
15 092	Octantis	10.0	25.14	7.262	77 41 24.7	.63	51.80	1
15 093	Octantis	10.0	36.32	12.877	84 40 37.6	.64	50.65	2
15 094	Pavonis	9.0	40.99	5.347	68 7 39.3	.65	52.68	1
15 095	Pavonis	10.0	41.49	+ 5.529	69 36 47.7	+ 14.65	51.77	1
15 096	Gould, Z. C., 21^h 298 . .	10.0	21 7 41.79	+ 5.485	69 16 13.2	+ 14.65	51.78	1
15 097	Pavonis	9.0	44.33	5.584	70 1 44.8	.65	51.73	1
15 098	Octantis	9.0	44.34	7.544	78 27 43.5	.65	50.78	1
15 099	Octantis	9.5	44.75	8.094	79 41 44.8	.65	50.78	1
15 100	Lacaille, 8729	7.2	51.35	+ 5.914	72 11 56.2	+ 14.66	51.72	4
15 101	Pavonis	10.0	21 7 54.55	+ 5.567	69 54 53.3	+ 14.66	51.75	2
15 102	Gould, Z. C., 21^h 302 . .	9.0	7 57.05	5.301	67 44 42.4	.67	52.63	1
15 103	Lacaille, 8713	8.5	8 0.44	7.072	77 9 25.7	.67	51.64	2
15 104	Gould, Z. C., 21^h 305 . .	8.5	4.11	5.313	67 51 47.6	.67	52.63	1
15 105	Octantis	10.0	22.60	+12.557	84 31 5.4	+ 14.69	50.66	1

Number.	Constellation, Name of Star, or Synonym.	Magnitude.	Right Ascension, 1850.0.	Annual Precession.	South Declination, 1850.0.	Annual Precession.	Mean year.	No. of obs.
			h. m. s.	s.	° ′ ″	″		
15 106	Pavonis	9.0	21 8 25.41	+ 5.252	67 20 45.1	+ 14.69	52.64	1
15 107	Pavonis	9.0	26.30	5.273	67 32 22.2	.69	52.64	2
15 108	Pavonis	10.0	30.33	5.273	67 32 33.9	.70	52.63	1
15 109	Octantis	9.0	30.74	7.551	78 31 8.1	.70	50.78	1
15 110	Lacaille, 8703	7.3	33.85	8.543	80 33 44.1	+ 14.70	51.63	3
15 111	Melbourne (1), 1079 . .	8.0	21 8 36.26	+ 5.245	67 17 43.1	+ 14.70	52.64	1
15 112	Gould, Z. C., 21^h 330 . .	9.0	47.73	5.472	69 16 15.5	.72	51.78	1
15 113	Lacaille, 8672	6.5	49.07	10.835	83 19 35.2	.72	50.67	1
15 114	Octantis	9.5	8 58.45	7.522	78 28 12.6	.73	50.78	1
15 115	Pavonis	9.5	9 8.63	+ 5.250	67 23 18.1	+ 14.74	52.64	2
15 116	Pavonis	9.8	21 9 8.76	+ 5.145	66 22 1.5	+ 14.74	52.68	2
15 117	Octantis	10.0	29.05	6.549	75 24 6.2	.76	50.75	1
15 118	Octantis	9.5	38.04	20.021	86 56 38.5	.77	51.23	2
15 119	Octantis	10.0	39.33	13.716	85 8 29.0	.77	50.64	1
15 120	Gould, Z. C., 21^h 392 . .	9.5	21 9 43.61	+ 7.210	77 40 16.8	+ 14.77	51.80	1
15 121	Pavonis	9.2	21 10 6.30	+ 5.768	71 28 58.6	+ 14.79	51.67	2
15 122	Octantis	10.0	10.65	9.439	81 55 53.2	.80	50.81	1
15 123	Pavonis	10.0	12.12	5.381	68 36 40.0	.80	51.79	2
15 124	Gould, Z. C., 21^h 406 . .	9.5	21.65	6.333	74 32 39.7	.81	51.66	1
15 125	Gould, Z. C., 21^h 407 . .	9.8	23.36	+ 6.355	74 38 41.2	+ 14.81	51.65	2
15 126	Pavonis	9.5	26.28	+ 5.355	68 27 13.7	+ 14.81	51.80	1
15 127	Gould, Z. C., 21^h 401 . .	10.0	31.74	5.587	70 17 32.0	.82	51.72	2
15 128	Lacaille, 8745	6.8	35.90	5.597	70 22 5.2	.82	51.72	2
15 129	Lacaille, 8746	7.5	40.51	6.060	73 13 49.0	.83	51.78	3
15 130	Pavonis	9.8	42.60	+ 6.127	73 34 51.0	+ 14.83	51.80	2
15 131	Pavonis	10.0	21 10 45.25	+ 5.829	71 55 8.9	+ 14.83	51.66	1
15 132	Octantis	10.0	47.81	17.587	86 27 8.8	.83	51.23	2
15 133	Octantis	9.5	50.50	7.534	78 36 3.8	.84	51.27	2
15 134	Pavonis	10.0	52.44	6.021	73 2 11.4	.84	51.78	1
15 135	Lacaille, 8744	6.2	10 53.31	+ 5.913	72 26 5.4	+ 14.84	51.69	3
15 136	Gould, Z. C., 21^h 413 . .	9.2	21 11 3.67	+ 5.101	66 6 41.6	+ 14.85	52.67	2
15 137	Pavonis	10.0	16.23	5.501	69 42 58.9	.86	51.77	1
15 138	Pavonis	9.5	24.49	5.189	67 1 44.7	.87	52.62	1
15 139	Lacaille, 8757	8.0	24.61	5.394	68 52 13.1	.87	51.79	1
15 140	Octantis	8.5	25.15	+ 9.668	82 15 13.0	+ 14.87	50.81	1
15 141	Lacaille, 8702	7.8	21 11 28.58	+ 9.954	82 34 28.0	+ 14.87	51.20	2
15 142	Pavonis	11.0	29.75	5.499	69 43 0.0	.88	51.77	1
15 143	Pavonis	9.0	44.12	5.187	67 2 52.9	.89	52.62	1
15 144	Gould, Z. C., 21^h 462 . .	9.8	45.43	6.863	76 42 18.2	.89	51.64	2
15 145	Gould, Z. C., 21^h 454 . .	8.5	56.31	+ 5.724	71 21 17.2	+ 14.90	51.69	1

No. 15 127. Gould's declination is 6.7″ greater.

Number.	Constellation, Name of Star, or Synonym.	Magnitude.	Right Ascension, 1850.0.	Annual Precession.	South Declination, 1850.0.	Annual Precession.	Mean year.	No. of obs.
			h. m. s.	s.	° ′ ″	″		
15 146	Octantis	8.2	21 11 58.45	+ 8.548	80 43 12.4	+ 14.90	51.40	4
15 147	Gould, Z. C., 21ʰ 452 . .	9.2	11 59.73	5.516	69 53 36.4	.90	51.75	2
15 148	Pavonis	9.3	12 0.85	5.106	66 15 38.7	.91	52.67	3
15 149	Pavonis	8.5	1.86	5.315	68 15 31.0	.91	52.24	2
15 150	Gould, Z. C., 21ʰ 459 . .	9.5	14.61	+ 5.534	70 3 14.7	+ 14.92	51.74	1
15 151	Gould, Z. C., 21ʰ 475 . .	9.2	21 12 22.01	+ 6.835	76 38 57.6	+ 14.93	51.64	2
15 152	Octantis	10.0	25.33	6.846	76 41 24.1	.93	51.64	1
15 153	Pavonis	9.5	29.74	5.189	67 8 29.3	.93	52.62	1
15 154	Pavonis	9.0	37.94	5.313	68 17 48.1	.94	52.68	1
15 155	Gould, Z. C., 21ʰ 474 . .	9.0	41.96	+ 5.905	72 31 54.6	+ 14.95	51.68	1
15 156	Pavonis	9.0	21 12 44.70	+ 5.627	70 45 53.6	+ 14.95	51.70	2
15 157	Pavonis	9.8	50.88	6.147	73 50 39.2	.95	51.76	2
15 158	Gould, Z. C., 21ʰ 486 . .	9.5	51.10	6.556	75 39 32.5	.95	51.58	1
15 159	Gould, Z. C., 21ʰ 470 . .	9.0	12 56.53	5.064	65 54 37.5	.96	52.67	1
15 160	Gould, Z. C., 21ʰ 480 . .	9.2	13 7.34	+ 5.359	68 44 21.4	+ 14.97	51.79	2
15 161	Lacaille, 8731	7.0	21 13 12.86	+ 8.495	80 41 11.2	+ 14.98	51.40	4
15 162	Pavonis	9.2	37.55	5.126	66 37 40.5	15.00	52.64	2
15 163	γ Pavonis	4.0	13 58.33	5.066	66 2 22.6	.02	52.67	1
15 164	Gould, Z. C., 21ʰ 510 . .	9.5	14 0.70	5.304	68 20 55.3	.02	52.25	2
15 165	Gould, Z. C., 21ʰ 512 . .	10.0	4.39	+ 5.320	68 29 52.4	+ 14.03	51.80	1
15 166	Pavonis	10.0	21 14 16.03	+ 5.517	70 6 21.8	+ 14.04	51.74	1
15 167	Octantis	9.8	20.96	5.290	80 22 48.7	.04	51.14	2
15 168	Octantis	10.0	21.68	8.331	80 27 13.8	.04	51.62	1
15 169	Octantis	9.5	21.86	6.706	76 19 20.4	.04	51.64	2
15 170	Octantis	9.5	22.96	+ 8.809	81 14 14.6	+ 14.04	50.80	1
15 171	Octantis	9.2	21 14 27.85	+ 12.784	84 47 58.8	+ 15.05	50.91	5
15 172	Gould, Z. C., 21ʰ 527 . .	9.2	31.00	6.098	73 43 16.4	.05	51.81	2
15 173	Octantis	9.2	31.31	8.998	81 31 6.3	.05	51.11	3
15 174	Pavonis	9.5	31.96	5.553	70 23 55.6	.05	51.70	1
15 175	Pavonis	10.0	38.96	+ 5.754	71 46 30.0	+ 15.06	51.66	1
15 176	Gould, 29431	9.5	21 14 39.45	+ 17.551	86 30 35.4	+ 15.06	51.22	6
15 177	Lacaille, 8626	9.0	42.58	17.542	86 30 30.2	.06	51.22	6
15 178	Gould, Z. C., 21ʰ 539 . .	9.2	44.38	6.081	73 38 56.8	.06	51.81	2
15 179	Lacaille, 8750	8.0	44.50	7.696	79 11 32.3	.06	51.24	2
15 180	Pavonis	12.0	48.59	+ 5.753	71 46 52.3	+ 15.07	51.66	1
15 181	Pavonis	10.0	21 14 48.90	+ 5.817	72 10 58.0	+ 15.07	51.76	2
15 182	Lacaille, 8766	8.0	51.98	6.570	75 50 56.6	.07	51.61	2
15 183	Octantis	10.0	53.25	6.790	76 39 13.9	.07	51.64	1
15 184	Gould, Z. C., 21ʰ 532 . .	8.2	57.73	5.200	67 29 28.0	.08	52.64	2
15 185	Gould, Z. C., 21ʰ 530 . .	9.0	14 58.26	+ 5.138	66 53 35.3	+ 15.08	52.62	1

Number.	Constellation, Name of Star, or Synonym.	Magnitude.	Right Ascension, 1850.0.	Annual Precession.	South Declination, 1850.0.	Annual Precession.	Mean year.	No. of obs.
			h. m. s.	s.	° ′ ″	″		
15 186	Gould, Z. C., 21h 555 . .	9.5	21 15 12.86	+ 6.322	74 50 16.4	+ 15.09	51.65	1
15 187	Pavonis	10.0	13.17	5.442	69 37 6.3	.09	51.77	1
15 188	Lacaille, 8782	6.5	14.63	5.511	70 8 54.1	.09	51.73	1
15 189	Melbourne (1), 1085 . .	7.5	16.11	5.191	67 26 14.0	.09	52.64	2
15 190	Pavonis	10.0	20.30	+ 5.962	73 3 47.8	+ 15.10	51.78	1
15 191	Gould, Z. C., 21h 569 . .	10.0	21 15 28.83	+ 6.556	75 50 2.7	+ 15.11	51.64	1
15 192	Gould, 29346	9.5	37.01	5.970	73 7 47.5	.11	51.77	2
15 193	Octantis	10.0	40.01	8.737	81 11 1.7	.12	50.80	1
15 194	Octantis	10.0	43.25	7.462	78 41 25.7	.12	51.78	1
15 195	Gould, Z. C., 21h 560 . .	9.0	45.60	+ 5.228	67 50 28.7	+ 15.12	52.65	2
15 196	Pavonis	10.0	21 15 45.87	+ 5.181	67 23 51.0	+ 15.12	52.63	1
15 197	Gould, 29360	8.8	15 53.47	6.568	75 54 21.4	.13	51.61	2
15 198	Pavonis	10.0	16 25.23	5.401	69 23 53.8	.16	51.78	1
15 199	Octantis	10.0	36.48	6.364	75 7 32.3	.17	50.75	1
15 200	Gould, Z. C., 21h 590 . .	9.5	36.54	+ 5.762	71 59 29.6	+ 15.17	51.76	2
15 201	Pavonis	10.0	21 16 38.52	+ 5.725	71 45 44.0	+ 15.17	51.66	1
15 202	Gould, Z. C., 21h 593 . .	9.2	40.94	5.762	71 59 51.8	.18	51.76	2
15 203	Octantis	10.0	44.15	10.880	83 36 48.0	.18	50.80	1
15 204	Pavonis	10.0	48.89	6.226	74 31 11.1	.18	51.66	1
15 205	Gould, Z. C., 21h 595 . .	9.5	50.24	+ 5.531	70 26 32.8	+ 15.18	51.70	1
15 206	Octantis	9.5	21 16 50.32	+ 9.275	81 59 6.0	+ 15.18	51.11	3
15 207	Pavonis	10.0	17 3.06	5.427	69 40 7.5	.20	51.77	1
15 208	Gould, Z. C., 21h 613 . .	10.0	9.56	5.603	70 59 10.7	.20	51.70	1
15 209	Lacaille, 8786	8.5	24.61	6.222	74 32 36.6	.22	51.66	1
15 210	Gould, 29629	9.5	32.68	+36.459	88 30 21.8	+ 15.22	51.23	2
15 211	Pavonis	10.0	21 17 35.90	+ 6.173	74 19 29.0	+ 15.23	51.68	2
15 212	Lacaille, 8785	8.0	39.75	6.577	76 3 51.8	.23	51.64	1
15 213	Pavonis	9.2	42.06	5.261	68 20 2.4	.23	52.24	2
15 214	Pavonis	9.5	45.16	5.063	66 24 32.8	.24	52.68	1
15 215	Gould, Z. C., 21h 631 . .	9.0	17 46.98	+ 5.886	72 49 58.5	+ 15.24	51.75	1
15 216	Gould, Z. C., 21h 634 . .	10.0	21 18 0.19	+ 5.519	70 27 46.8	+ 15.25	51.70	1
15 217	Octantis	9.2	0.82	7.111	77 52 6.4	.25	51.30	2
15 218	Pavonis	9.5	4.64	5.052	66 20 3.8	.26	52.68	2
15 219	Octantis	9.5	8.99	11.987	84 26 26.5	.26	50.66	1
15 220	Octantis	10.0	9.49	+ 8.645	81 9 2.9	+ 15.26	51.21	2
15 221	Octantis	8.0	21 18 24.11	+10.969	83 44 14.1	+ 15.27	50.74	2
15 222	Pavonis	10.0	18 29.70	5.699	71 45 19.7	.28	51.66	1
15 223	Octantis	10.0	19 4.04	8.742	81 20 25.1	.31	50.80	1
13 224	Pavonis	10.0	16.50	5.280	68 39 37.4	.32	51.79	2
15 225	Lacaille, 8806	9.0	20.90	+ 5.460	70 8 36.3	+ 15.33	51.73	1

Number.	Constellation, Name of Star, or Synonym.	Magnitude.	Right Ascension, 1850.0.	Annual Precession.	South Declination, 1850.0.	Annual Precession.	Mean year.	No. of obs.
			h. m. s.	s.	° ′ ″	″		
15 226	Gould, Z. C., 21h 677 . .	9.5	21 19 36.51	+ 5.451	70 6 15.2	+ 15.34	51.74	1
15 227	Lacaille, 8753	6.0	52.12	8.001	80 6 9.9	.36	50.67	1
15 228	Gould, 29447	8.5	58.11	4.999	65 57 21.6	.36	52.67	1
15 229	Octantis	9.7	19 59.02	6.699	76 40 3.9	.36	51.43	3
15 230	Gould, 29453	8.0	20 14.30	+ 5.042	66 27 56.5	+ 15.38	52.68	1
15 231	Gould, Z. C., 21h 699 . .	9.0	21 20 19.34	+ 5.232	68 20 3.6	+ 15.38	52.24	2
15 232	Indi	8.5	21.54	5.229	68 19 3.2	.38	52.24	2
15 233	Indi	10.0	27.50	6.027	73 49 18.2	.39	51.71	1
15 234	Gould, Z. C., 21h 719 . .	9.5	38.49	5.844	72 49 46.8	.40	51.75	1
15 235	Indi	9.0	45.37	+ 5.173	67 49 47.8	+ 15.41	52.63	1
15 236	Octantis	9.8	21 20 49.87	+27.458	87 59 24.7	+ 15.41	51.22	3
15 237	Gould, Z. C., 21h 736 . .	9.0	50.29	6.831	77 10 35.5	.41	51.09	3
15 238	Gould, Z. C., 21h 726 . .	9.0	54.91	5.489	70 30 55.5	.42	51.70	1
15 239	Indi	9.0	20 59.64	5.081	66 56 41.0	.42	52.62	1
15 240	Gould, 29447	8.8	21 33.10	+ 5.353	69 30 44.5	+ 15.45	51.78	2
15 241	Indi	10.0	21 21 41.30	+ 5.546	71 0 1.7	+ 15.46	51.69	1
15 242	Gould, Z. C., 21h 749 . .	9.0	21 50.94	5.068	66 54 9.1	.47	52.62	1
15 243	Indi	9.5	22 4.56	5.158	67 49 52.7	.48	52.63	1
15 244	Octantis	8.6	5.36	13.242	85 13 18.9	.48	51.22	4
15 245	Gould, Z. C., 21h 765 . .	9.0	18.32	+ 5.109	67 23 4.0	+ 15.49	52.65	1
15 246	Gould, Z. C., 21h 776 . .	9.2	21 22 26.66	+ 5.611	71 31 18.8	+ 15.50	51.67	2
15 247	Gould, Z. C., 21h 782 . .	9.0	27.83	6.390	75 39 26.1	.50	51.58	1
15 248	Indi	10.0	28.16	5.011	66 22 6.3	.50	52.68	1
15 249	Gould, Z. C., 21h 787 . .	9.5	39.31	6.312	75 20 24.1	.51	50.75	1
15 250	Lacaille, 8720	7.4	49.82	+14.376	85 42 58.8	+ 15.52	51.11	8
15 251	Indi	10.0	21 22 55.83	+ 5.484	70 40 10.8	+ 15.53	51.70	1
15 252	Indi	9.5	23 17.09	5.171	68 4 46.5	.55	52.68	1
15 253	Indi	9.8	20.94	5.524	70 59 43.1	.55	51.69	2
15 254	Lacaille, 8751	7.5	24.22	12.067	84 38 19.9	.55	50.89	3
15 255	Gould, Z. C., 21h 799 . .	8.5	28.39	+ 5.160	67 59 45.1	+ 15.56	52.65	2
15 256	Gould, Z. C., 21h 806 . .	9.5	21 23 45.87	+ 5.171	68 8 0.8	+ 15.57	52.68	1
15 257	Indi	10.0	52.67	6.169	74 47 29.8	.58	51.65	1
15 258	Indi	9.5	55.00	5.585	71 28 37.0	.58	51.67	2
15 259	Indi	9.8	23 58.13	5.143	67 53 18.0	.59	52.65	2
15 260	Gould, Z. C., 21h 825 . .	8.2	24 5.61	+ 6.243	75 8 45.2	+ 15.59	51.20	2
15 261	Indi	10.0	21 24 8.82	+ 5.319	69 28 59.5	+ 15.59	51.77	1
15 262	Gould, Z. C., 21h 838 . .	8.0	10.82	6.922	77 41 2.8	.60	51.81	2
15 263	Gould, Z. C., 21h 817 . .	8.5	17.29	4.943	65 48 53.9	.60	52.67	1
15 264	Indi	10.0	21.77	5.719	72 23 44.1	.61	51.68	1
15 265	Indi	9.5	22.98	+ 4.964	66 3 17.4	+ 15.61	52.67	1

Number.	Constellation, Name of Star, or Synonym.	Magnitude.	Right Ascension, 1850.0.	Annual Precession.	South Declination, 1850.0.	Annual Precession.	Mean year.	No. of obs.
			h. m. s.	s.	° ′ ″	″		
15 266	γ Octantis	4.5	21 24 34.42	+ 7.035	78 2 58.1	+ 15.62	51.47	3
15 267	Indi	9.8	36.44	5.575	71 28 35.8	.62	51.67	2
15 268	Gould, Z. C., 21ʰ 830 . .	9.5	36.90	5.239	68 50 35.8	.62	51.79	1
15 269	Octantis	10.0	53.48	7.395	79 2 17.7	.64	51.77	1
15 270	Gould, Z. C., 21ʰ 839 . .	9.8	24 56.45	+ 5.268	69 8 7.2	+ 15.64	51.79	2
15 271	Gould, 29530	7.5	21 25 3.56	+ 5.408	70 17 44.2	+ 15.64	51.72	2
15 272	Gould, Z. C., 21ʰ 846 . .	9.0	13.75	5.052	67 7 36.8	.65	52.63	2
15 273	Gould, 29529	8.5	14.71	4.947	65 58 40.3	.66	52.67	1
15 274	Octantis	9.8	16.64	15.183	86 3 10.4	.66	51.22	2
15 275	Indi	10.0	27.69	+ 5.188	68 28 24.7	+ 15.67	51.80	1
15 276	Octantis	9.5	21 25 33.23	+ 7.995	80 22 49.2	+ 15.67	51.14	2
15 277	Indi	10.0	41.98	5.212	68 42 47.3	.68	51.79	1
15 278	Lacaille, 8828	7.5	43.76	6.130	74 45 11.3	.68	51.65	2
15 279	Indi	10.0	25 57.85	5.815	73 7 7.2	.69	51.77	2
15 280	Indi	9.2	26 25.16	+ 5.811	73 7 57.0	+ 15.72	51.77	2
15 281	Lacaille, 8835	8.5	21 26 26.31	+ 5.654	72 10 15.4	+ 15.72	51.81	1
15 282	Indi	9.5	28.99	5.760	72 50 15.3	.72	51.75	1
15 283	Octantis	10.0	31.02	9.646	82 48 14.4	.72	51.71	1
15 284	Lacaille, 8797	7.8	31.84	9.942	83 6 43.6	.73	51.19	2
15 285	Gould, Z. C., 21ʰ 889 . .	9.2	26 37.72	+ 5.067	67 25 23.2	+ 15.73	52.64	2
15 286	Gould, Z. C., 21ʰ 908 . .	10.0	21 27 10.81	+ 5.256	69 15 50.5	+ 15.76	51.78	1
15 287	Gould, Z. C., 21ʰ 916 . .	9.0	11.05	5.814	73 13 3.6	.76	51.78	1
15 288	Gould, 29583	10.0	15.26	5.162	68 25 27.4	.76	51.79	1
15 289	Gould, 29584	10.0	15.50	5.158	68 23 18.6	.76	51.80	1
15 290	Octantis	5.2	17.72	+10.214	83 23 59.0	+ 15.77	50.73	2
15 291	Stone, 11436	9.0	21 27 19.67	+10.215	83 23 58.8	+ 15.77	50.67	1
15 292	Octantis	9.0	23.34	10.595	83 44 4.5	.77	50.80	1
15 293	Indi	9.5	29.84	5.736	72 47 2.4	.78	51.72	2
15 294	Lacaille, 8845	8.0	31.60	5.536	71 28 43.3	.78	51.67	2
15 295	Indi	10.0	37.17	+ 5.570	71 43 35.5	+ 15.78	51.66	1
15 296	Octantis	9.1	21 27 40.42	+16.957	86 36 3.7	+ 15.79	51.22	5
15 297	Gould, Z. C., 21ʰ 926 . .	9.0	27 43.29	5.343	70 2 47.0	.79	51.74	1
15 298	Indi	10.0	28 4.12	5.049	67 23 59.2	.81	52.63	1
15 299	Octantis	8.5	8.32	8.877	81 54 59.3	.81	51.31	2
15 300	Octantis	9.8	8.85	+19.834	87 11 27.2	+ 15.81	51.22	2
15 301	Indi	9.5	21 28 12.51	+ 5.115	68 4 44.4	+ 15.82	52.68	1
15 302	Indi	9.8	17.54	5.129	68 13 18.0	.82	52.24	2
15 303	Gould, Z. C., 21ʰ 945 . .	10.0	27.89	5.462	71 2 42.8	.83	51.70	1
15 304	Gould, Z. C., 21ʰ 962 . .	10.0	31.01	6.722	77 19 36.7	.83	50.81	1
15 305	Octantis	9.5	35.54	+ 6.401	76 9 22.8	+ 15.84	51.64	1

Number.	Constellation, Name of Star, or Synonym.	Magnitude.	Right Ascension, 1850.0.	Annual Precession.	South Declination, 1850.0.	Annual Precession.	Mean year.	No. of obs.
			h. m. s.	s.	° ′ ″	″		
15 306	Gould, Z. C., 21h 953 . .	9.0	21 28 49.24	+ 5.487	71 14 31.1	+ 15.85	51.69	1
15 307	Gould, Z. C., 21h 964 . .	10.0	28 54.19	6.062	74 40 46.7	.85	51.66	1
15 308	Gould, Z. C., 21h 979 . .	10.0	29 0.36	6.668	77 10 30.0	.86	50.82	2
15 309	Octantis	8.2	15.36	8.272	81 4 0.1	.87	51.21	2
15 310	Gould, Z. C., 21h 973 . .	9.2	21.50	+ 5.443	70 59 32.4	+ 15.88	51.69	2
15 311	Octantis	10.0	21 29 48.37	+ 6.257	75 38 51.9	+ 15.90	51.58	1
15 312	Octantis	9.8	53.02	11.334	84 22 10.8	.91	50.67	2
15 313	Octantis	9.0	29 56.08	53.223	89 4 11.3	.91	51.22	3
15 314	Gould, Z. C., 21h 1001 .	9.2	30 5.36	5.292	69 52 18.0	.92	51.75	2
15 315	Octantis	9.2	16.65	+12.608	85 7 44.3	+ 15.93	51.11	4
15 316	Gould, 29677	9.0	21 30 24.13	+10.024	83 20 14.6	+ 15.93	50.67	1
15 317	Indi	9.5	27.87	4.895	65 59 47.3	.94	52.67	1
15 318	Gould, Z. C., 21h 1010 .	10.0	31.09	5.257	69 37 34.9	.94	51.77	1
15 319	Octantis	9.0	45.86	11.365	84 24 57.2	.95	50.67	2
15 320	Gould, Z. C., 21h 1016 .	9.5	49.32	+ 4.991	67 7 4.4	+ 15.96	52.63	2
15 321	Melbourne (1), 1094 . .	8.0	21 30 57.16	+ 5.153	68 44 38.4	+ 15.96	51.79	2
15 322	Gould, Z. C., 21h 1025 .	9.2	31 3.65	5.245	69 34 40.1	.97	51.78	2
15 323	Indi	9.5	6.16	5.552	71 56 0.6	.97	51.74	2
15 324	Indi	9.8	9.93	5.024	67 29 51.7	.97	51.99	3
15 325	Gould, 29668	9.0	16.86	+ 7.277	79 7 19.4	+ 15.98	51.24	2
15 326	Octantis	10.0	28.03	+ 6.697	77 26 53.7	+ 15.99	51.80	1
15 327	Octantis	9.8	41.40	6.713	77 30 57.6	16.00	51.31	2
15 328	Octantis	10.0	55.44	9.729	83 6 18.7	.01	51.71	1
15 329	Indi	9.2	58.41	5.043	67 47 9.1	.02	52.23	2
15 330	Indi	9.5	31 58.58	+ 4.891	66 7 58.1	+ 16.02	52.67	1
15 331	Lacaille, 8860	6.0	21 32 4.24	+ 5.503	71 41 24.2	+ 16.02	51.66	1
15 332	Indi	9.0	4.38	5.478	71 30 58.7	.02	51.67	2
15 333	Indi	10.0	19.74	4.912	66 25 22.3	.04	52.68	1
15 334	Octantis	10.0	21.18	6.335	76 10 24.0	.04	51.64	1
15 335	Gould, Z. C., 21h 1107 .	9.0	23.72	+ 7.736	80 13 58.5	+ 16.04	51.10	3
15 336	Indi	10.0	21 32 33.07	+ 5.184	69 12 27.6	+ 16.05	51.78	1
15 337	Gould, Z. C., 21h 1070 .	10.0	35.06	5.278	70 1 11.8	.05	51.74	1
15 338	Gould, Z. C., 21h 1081 .	9.0	49.78	5.278	70 2 47.7	.06	51.74	1
15 339	Brisbane, 7058	9.0	50.30	5.491	71 46 3.6	.06	51.66	1
15 340	Octantis	8.2	56.00	+ 9.046	8[illegible] 22 10.4	+ 16.07	50.75	2
15 341	Gould, Z. C., 21h 1087 .	9.5	21 32 5[illegible].70	+ 5.291	70 10 8.5	+ 16.07	51.74	1
15 342	Lacaille, 8850	8.0	33 0.65	7.152	78 54 25.2	.07	51.77	1
15 343	Octantis	9.8	2.08	11.013	84 14 55.4	.07	50.67	2
15 344	Indi	8.0	4.71	10.350	83 43 47.6	.07	50.80	1
15 345	Gould, Z. C., 21h 1119 .	9.0	5.09	+ 6.823	77 57 59.5	+ 16.08	51.30	2

No. 15 340. Mean of two observations differing 9.5″.

Number.	Constellation, Name of Star, or Synonym.	Magnitude.	Right Ascension, 1850.0.	Annual Precession.	South Declination, 1850.0.	Annual Precession.	Mean year.	No. of obs.
			h. m. s.	s.	° ′ ″	″		
15 346	Indi	9.5	21 33 5.94	+ 4.947	66 54 29.8	+ 16.08	52.62	1
15 347	Indi	7.8	6.06	5.030	67 47 11.7	.08	52.38	3
15 348	Gould, Z. C., 21h 1113 .	9.8	34.93	5.352	70 43 43.2	.10	51.70	2
15 349	Indi	9.5	36.61	5.044	67 59 26.4	.10	52.68	1
15 350	Lacaille, 8738	8.6	40.66	+19.321	87 11 32.7	+ 16.11	51.22	5
15 351	Gould, Z. C., 21h 1117 .	8.0	21 33 43.44	+ 5.188	69 22 5.2	+ 16.11	51.78	1
15 352	Gould, Z. C., 21h 1112 .	9.0	43.52	4.860	65 59 6.9	.11	52.67	1
15 353	Lacaille, 8869	8.0	53.19	5.736	73 22 2.5	.12	51.78	1
15 354	Indi	10.0	33 57.79	5.256	69 59 29.6	.12	51.77	1
15 355	Indi	10.0	34 1.05	+ 5.606	72 34 47.6	+ 16.12	51.68	1
15 356	Gould, 29721	7.7	21 34 13.47	+ 5.066	68 16 50.9	+ 16.13	52.11	3
15 357	Gould, Z. C., 21h 1137 .	10.0	22.11	5.510	71 58 13.4	.14	51.81	1
15 358	Gould, Z. C., 21h 1131 .	9.0	23.03	4.854	65 59 35.0	.14	52.67	1
15 359	Gould, Z. C., 21h 1141 .	10.0	25.56	5.608	72 38 51.9	.15	51.68	1
15 360	Gould, Z. C., 21h 1146 .	9.8	36.20	+ 5.489	71 52 1.7	+ 16.15	51.74	2
15 361	Octantis	10.0	21 34 56.08	+ 6.503	77 1 12.0	+ 16.17	50.82	2
15 362	Octantis	9.0	35 11.29	6.808	78 3 54.8	.18	50.79	1
15 363	Gould, Z. C., 21h 1163 .	9.8	13.76	5.487	71 53 44.2	.19	51.74	2
15 364	Indi	10.0	24.95	5.039	68 8 57.4	.20	52.68	1
15 365	Octantis	9.5	25.18	+ 7.579	80 4 26.4	+ 16.20	50.67	1
15 366	Octantis	9.5	21 35 25.83	+ 6.460	76 53 46.0	+ 16.20	50.82	1
15 367	Gould, Z. C., 11h 1191 .	10.0	26.16	7.256	79 19 32.3	.20	50.72	1
15 368	Gould, 29738 . . .	8.2	30.32	4.974	67 29 29.4	.20	51.83	2
15 369	Octantis	9.5	31.06	18.442	87 3 51.4	.20	51.22	2
15 370	Indi	10.0	31.56	+ 4.907	66 45 47.9	+ 16.20	52.68	1
15 371	Octantis	10.0	21 35 35.13	+48.265	89 0 4.4	+ 16.21	51.23	2
15 372	Gould, Z. C., 21h 1187 .	8.5	40.72	6.227	75 59 18.8	.21	51.61	2
15 373	Octantis	8.7	51.03	11.477	84 39 12.9	.22	50.73	2
15 374	Indi	10.0	54.97	5.563	72 28 44.0	.22	51.68	1
15 375	Octantis	8.5	35 56.26	+ 9.920	83 27 6.8	+ 16.22	50.73	2
15 376	Indi	10.0	21 36 1.54	+ 5.677	73 12 55.4	+ 16.23	51.78	1
15 377	Gould, 29755	8.8	1.99	6.102	75 27 53.1	.23	51.16	2
15 378	Lacaille, 8864	8.0	9.17	7.217	79 16 24.7	.23	50.72	1
15 379	Octantis	9.5	12.08	8.432	81 40 5.8	.24	51.81	1
15 380	Gould, Z. C., 21h 1213 .	9.0	14.82	+ 6.930	78 30 9.4	+ 16.24	50.78	1
15 381	Octantis	9.0	21 36 32.69	+ 8.054	81 3 30.8	+ 16.25	51.21	2
15 382	Octantis	10.0	41.70	12.882	85 26 18.8	.26	51.23	2
15 383	Brisbane, 7075	7.0	51.19	4.844	66 11 35.7	.27	52.63	2
15 384	Octantis	10.0	58.27	6.203	75 59 16.6	.28	51.64	1
15 385	Gould, Z. C., 21h 1218 .	10.0	36 58.35	+ 5.477	72 0 15.4	+ 16.28	51.81	1

Number.	Constellation, Name of Star, or Synonym.	Magnitude.	Right Ascension, 1850.0.	Annual Precession.	South Declination, 1850.0.	Annual Precession.	Mean year.	No. of obs.
			h. m. s.	s.	° ′ ″	″		
15 386	Indi	10.0	21 37 2.42	+ 5.594	72 47 37.8	+ 16.28	51.75	1
15 387	Octantis	9.1	8.35	25.345	87 59 34.6	.28	51.22	5
15 388	Octantis	9.0	9.55	6.772	78 5 27.2	.29	50.79	1
15 389	Indi	10.0	12.62	5.316	70 49 50.1	.29	51.70	1
15 390	Indi	9.5	13.28	+ 5.655	73 11 43.4	+ 16.29	51.78	1
15 391	Lacaille, 8894	8.0	21 37 31.78	+ 5.592	72 49 59.7	+ 16.30	51.75	1
15 392	Indi	10.0	32.13	5.193	69 50 42.7	.30	51.77	1
15 393	Octantis	8.7	34.04	7.683	80 24 58.5	.31	51.03	3
15 394	Gould, Z. C., 21^h 1238 .	10.0	41.25	5.472	72 2 24.7	.31	51.81	1
15 395	Indi	9.5	44.80	+ 5.760	73 51 46.2	− 16.32	51.76	2
15 396	Gould, 29783	9.5	21 37 58.61	− 5.032	68 23 8.8	+ 16.33	51.79	1
15 397	υ Indi	6.5	38 0.10	5.243	70 19 23.8	.33	51.72	2
15 398	Indi	9.5	0.10	5.220	70 7 36.5	.33	51.74	1
15 399	Indi	10.0	1.82	5.324	70 59 2.9	.33	51.69	1
15 400	Indi	10.0	8.33	+ 5.896	74 38 20.7	− 16.34	51.65	1
15 401	Octantis .	13.0	21 38 10.21	+ 6.095	75 36 28.2	+ 16.34	50.75	1
15 402	Octantis	10.0	12.61	6.091	75 35 46.1	.34	50.75	1
15 403	Gould, Z. C., 21^h 1283 .	9.0	13.60	6.867	78 27 8.2	.34	50.78	1
15 404	Indi	10.0	14.83	5.724	73 42 17.6	.34	51.81	1
15 405	Gould, Z. C., 21^h 1253 .	8.8	15.67	+ 5.068	68 46 39.2	+ 16.34	51.79	2
15 406	Gould, Z. C., 21^h 1261 .	9.8	21 38 17.34	+ 5.433	71 49 3.3	+ 16.34	51.74	2
15 407	Octantis	9.4	18.88	28.281	88 14 21.1	.34	51.22	5
15 408	Indi	9.2	27.30	5.686	73 30 6.5	.35	51.79	2
15 409	Gould, Z. C., 21^h 1264 .	8.8	28.58	5.113	69 10 53.2	.35	51.79	2
15 410	Indi	10.0	29.60	+ 4.983	67 56 50.0	+ 16.35	51.83	2
15 411	Octantis	10.0	21 38 32.56	+ 6.030	75 20 30.3	+ 16.36	50.75	1
15 412	Gould, Z. C., 21^h 1300 .	10.0	34.52	7.164	79 17 33.4	.36	50.72	1
15 413	Indi .	10.0	34.85	5.072	68 50 53.0	.36	51.79	1
15 414	Indi	8.0	35.66	5.927	74 50 20.3	.36	51.65	1
15 415	Lacaille, 8885	8.8	35.97	+ 6.720	78 1 41.3	+ 16.36	51.30	2
15 416	Gould, Z. C., 21^h 1281	7.8	21 38 49.27	+ 5.323	71 2 41.1	+ 16.37	51.69	2
15 417	Indi	9.0	52.31	5.645	73 16 40.6	.37	51.78	1
15 418	Gould, 29828	9.0	52.88	6.411	76 57 56.7	.37	50.82	1
15 419	Octantis	9.5	54.36	8.499	81 53 42.4	.37	51.31	2
15 420	Indi	10.0	38 57.23	+ 5.427	71 50 18.3	+ 16.38	51.66	1
15 421	Gould, Z. C., 21^h 1301	9.0	21 39 6.79	+ 5.837	74 25 6.3	+ 16.38	51.66	1
15 422	Gould, 29827	7.8	17.36	5.341	71 14 47.6	.39	51.69	2
15 423	Lacaille, 8879	7.5	17.96	7.635	80 25 10.7	.39	51.03	3
15 424	Gould, Z. C., 21^h 1324	8.5	21.66	6.586	77 38 25.5	.40	51.81	1
15 425	Gould, Z. C., 21^h 1304 .	9.0	29.31	+ 5.204	70 9 37.1	+ 16.40	51.74	1

Number.	Constellation, Name of Star, or Synonym.	Magnitude.	Right Ascension, 1850.0.	Annual Precession.	South Declination, 1850.0.	Annual Precession.	Mean year.	No. of obs.
			h. m. s.	s.	° ′ ″	″		
15 426	Indi	9.5	21 39 33.66	+ 5.703	73 42 17.6	+ 16.41	51.81	1
15 427	Indi	10.0	38.36	4.818	66 14 17.1	.41	52.68	1
15 428	Gould, Z. C., 21ʰ 1329 .	9.2	45.89	5.812	74 20 37.3	.42	51.72	2
15 429	Indi	10.0	39 48.67	— 4.974	68 1 3.8	.42	51.83	1
15 430	Indi	9.2	40 2.59	+ 5.859	74 37 8.2	+ 16.43	51.65	2
15 431	Indi	9.8	21 40 14.78	+ 5.396	71 45 56.6	+ 16.44	51.70	2
15 432	Indi	8.8	20.30	5.573	72 59 21.0	.45	51.77	2
15 433	Indi	10.0	22.01	5.085	69 11 6.6	.45	51.79	2
15 434	Gould, 29861 . .	9.0	37.42	5.793	74 19 8.4	.46	51.68	2
15 435	Indi	9.2	39.56	+ 4.890	67 12 35.8	+ 16.46	52.22	2
15 436	Indi	9.0	21 40 51.29	+ 7.276	79 42 54.4	+ 16.47	50.78	1
15 437	Lacaille, 8910 . . .	8.0	40 55.79	5.752	74 7 14.9	.48	51.71	1
15 438	Indi	10.0	41 4.22	4.952	67 56 44.1	.48	51.83	2
15 439	Gould, Z. C., 21ʰ 1367 . .	9.0	12.39	6.459	77 20 5.7	.49	50.81	1
15 440	Octantis . .	9.0	23.05	+ 9.594	83 21 2.3	+ 16.50	50.67	1
15 441	Indi	9.5	21 41 33.75	+ 4.926	67 43 20.1	+ 16.51	51.83	1
15 442	Octantis	10.0	37.00	8.302	81 44 27.3	.51	51.81	1
15 443	Indi	9.7	41.79	4.831	66 41 13.6	.51	52.66	3
15 444	Lacaille, 8909	7.0	55.46	6.332	76 54 56.6	.52	50.82	1
15 445	Octantis	9.6	41 59.44	+12.973	85 37 31.7	+ 16.53	51.25	5
15 446	Indi . . .	9.0	21 42 0.02	+ 5.011	68 40 21.4	+ 16.53	51.79	2
15 447	Indi	9.0	5.94	4.783	66 8 37.3	.53	52.67	1
15 448	Indi	9.0	18.29	4.782	66 9 37.4	.54	52.67	1
15 449	Indi	10.0	21.35	5.131	69 50 58.1	.55	51.77	1
15 450	Indi	10.5	30.67	+ 5.113	69 42 7.9	+ 16.55	51.77	1
15 451	Brisbane, 7088	6.8	21 42 39.56	+ 5.240	70 49 2.8	+ 16.56	51.70	2
15 452	Indi . .	9.0	44.36	+ 4.855	67 { 5 32.0 / 4 53.4 }	.57	52.22	2
15 453	Octantis	9.5	45.65	7.346	80 0 1.3	.57	50.67	1
15 454	Gould, Z. C., 21ʰ 1397 .	9.5	45.98	5.111	69 42 46.2	.57	51.77	1
15 455	Octantis	9.5	46.05	+ 7.345	79 59 53.7	+ 16.57	50.78	1
15 456	Lacaille, 8925	6.8	21 42 47.49	+ 5.232	70 46 12.4	+ 16.57	51.70	2
15 457	Indi	9.0	55.25	4.899	67 36 13.1	.57	51.83	1
15 458	Gould, Z. C., 21ʰ 1404 .	8.5	57.97	4.989	68 33 47.5	.58	51.80	1
15 459	Stone, 11521	9.0	42 58.21	5.288	71 14 22.2	.58	51.69	2
15 460	Octantis	10.0	43 1.70	+ 8.798	82 30 51.9	+ 16.58	50.69	1
15 461	Indi	9.7	21 43 10.43	+ 4.782	66 16 34.3	+ 16.59	52.68	3
15 462	Octantis	10.0	11.88	89.947	89 30 15.6	.59	51.22	2
15 463	Indi	10.0	24.98	6.960	79 4 25.3	.60	51.77	1
15 464	Gould, Z. C., 21ʰ 1429	9.0	43 55.33	4.822	66 51 35.9	.62	51.22	1
15 465	Octantis	9.5	44 3.94	+ 6.040	75 52 4.8	+ 16.63	51.64	1

No. 15 452. Assumed to be same star; right ascensions identical; no error in declinations can be found.
No. 15 454. Gould's declination is 8.8″ greater.

Number.	Constellation, Name of Star, or Synonym.	Magnitude.	Right Ascension, 1850.0.	Annual Precession.	South Declination, 1850.0.	Annual Precession.	Mean year.	No. of obs.
			h. m. s.	s.	° ′ ″	″		
15 466	Indi	10.0	21 44 7.09	+ 5.199	70 38 41.9	+ 16.63	51.70	1
15 467	Gould, Z. C., 21^{h} 1445 .	8.8	19.17	5.466	72 41 15.4	.64	51.72	2
15 468	Gould, Z. C., 21^{h} 1439 .	9.2	19.90	4.843	67 9 30.9	.64	52.22	2
15 469	Gould, Z. C., 21^{h} 1460 .	8.5	41.52	5.880	75 9 20.5	.66	51.20	2
15 470	Indi	10.0	44 49.58	+ 4.994	68 50 43.0	+ 16.67	51.79	1
15 471	Gould, Z. C., 21^{h} 1484 .	9.2	21 45 37.48	+ 5.142	70 20 11.8	+ 16.71	51.72	2
15 472	Gould, Z. C., 21^{h} 1431 .	9.5	38.46	4.991	68 55 7.6	.71	51.79	1
15 473	Octantis	9.0	42.82	7.925	81 20 18.6	.71	50.80	1
15 474	Indi	10.0	51.19	4.857	67 31 15.1	.72	51.83	1
15 475	Gould, Z. C., 21^{h} 1493 .	8.8	51.88	+ 4.885	67 49 52.9	+ 16.72	51.83	2
15 476	Gould, Z. C., 21^{h} 1506 .	9.2	21 45 55.78	+ 5.753	74 36 6.8	+ 16.72	51.65	2
15 477	Lacaille, 8927	7.0	58.55	6.659	78 22 27.9	.72	50.78	2
15 478	Gould, Z. C., 21^{h} 1510 .	9.5	59.49	5.792	74 49 18.3	.72	51.65	1
15 479	Indi	9.5	45 59.84	5.407	72 27 31.4	.72	51.68	1
15 480	Indi	10.0	46 4.16	+ 5.090	69 55 5.4	+ 19.73	51.75	2
15 481	Octantis	9.3	21 46 6.96	+10.921	84 37 58.4	+ 16.73	50.76	3
15 482	Indi	10.0	11.85	5.408	72 29 26.2	.73	51.68	1
15 483	Gould, Z. C., 21^{h} 1508	9.5	15.57	4.988	68 58 14.3	.74	51.79	1
15 484	Octantis	9.5	17.27	11.989	85 16 40.1	.74	51.53	2
15 485	Indi	9.5	29.06	+ 4.724	65 59 42.6	+ 16.75	52.67	1
15 486	Indi	10.0	21 46 33.48	+ 7.208	79 55 15.1	— 16.75	50.78	1
15 487	Indi	9.5	34.69	4.840	67 25 34.6	.75	51.83	1
15 488	Octantis	9.0	41.27	7.476	80 31 46.4	.76	50.81	1
15 489	Octantis	9.0	47.20	9.371	83 21 9.8	.76	50.67	1
15 490	Gould, 30003	10.5	47.55	+ 6.533	78 1 34.0	+ 16.76	51.30	2
15 491	Lacaille, 8935	8.8	21 46 48.88	+ 6.532	78 1 25.3	+ 16.76	51.30	2
15 492	Indi	10.0	50.57	4.741	66 16 11.1	.77	52.68	1
15 493	Indi	10.0	52.56	4.894	68 3 52.1	.77	51.83	1
15 494	Indi	10.0	58.31	5.405	72 33 20.1	.77	51.68	1
15 495	Indi	10.0	58.53	— 5.082	69 57 38.2	+ 16.77	51.77	1
15 496	Lacaille, 8897	7.5	21 46 59.05	+10.141	84 4 44.2	+ 16.77	50.68	1
15 497	Gould, Z. C., 21^{h} 1536 .	9.2	47 13.87	4.737	66 16 10.6	.78	52.68	2
15 498	Gould, Z. C., 21^{h} 1543 .	9.8	14.60	5.076	69 56 7.0	.78	51.75	2
15 499	Lacaille, 8942	6.8	27.66	6.091	76 23 29.5	.79	51.10	3
15 500	Gould, Z. C., 21^{h} 1548 .	8.2	28.04	+ 4.776	66 47 38.2	+ 16.80	52.66	3
15 501	Gould, Z. C., 21^{h} 1569 .	9.5	21 47 39.53	+ 5.717	74 34 30.7	+ 16.80	51.66	1
15 502	Gould, 30013	8.8	43.18	5.026	69 31 34.4	.81	51.78	2
15 503	Gould, 30018	9.0	45.73	5.217	71 13 11.3	.81	51.69	1
15 504	Gould, Z. C., 21^{h} 1582 .	8.0	47 59.34	5.831	75 12 43.6	.82	50.75	1
15 505	Indi	10.0	48 2.08	+ 4.931	68 36 39.2	+ 16.82	51.79	2

No. 15 478. Gould's declination is 11.9″ greater.

Number.	Constellation, Name of Star, or Synonym.	Magnitude.	Right Ascension, 1850.0.	Annual Precession.	South Declination, 1850.0.	Annual Precession.	Mean year.	No. of obs.
			h. m. s.	s.	° ′ ″	″		
15 506	Lacaille, 8946	7.0	21 48 10.89	+ 6.181	76 49 52.4	+ 16.83	50.82	1
15 507	Octantis	9.5	22.91	6.024	76 10 42.6	.84	51.64	1
15 508	Gould, Z. C., 21h 1588	9.5	25.83	5.215	71 17 4.5	.84	51.69	1
15 509	Gould, Z. C., 21h 1612	9.5	26.15	6.771	78 53 43.8	.84	51.77	1
15 510	Indi	8.8	39.52	4.778	66 59 0.2	+ 16.85	52.22	2
15 511	Octantis	9.1	21 48 46.02	+18.987	87 23 37.2	+ 16.86	51.22	4
15 512	Indi	9.0	51.33	5.078	70 9 33.0	.86	51.74	1
15 513	Octantis	10.0	48 53.65	6.113	76 36 39.4	.86	50.83	1
15 514	Gould, Z. C., 21h 1609	10.0	49 1.62	5.249	71 37 37.5	.87	51.73	1
15 515	Gould, Z. C., 21h 1616	8.0	2.54	+ 5.802	75 9 51.3	+ 16.87	50.75	1
15 516	Indi	8.8	21 49 6.94	+ 4.736	66 31 31.8	+ 16.87	52.68	3
15 517	Indi	10.0	11.02	4.902	68 27 38.8	.88	51.80	1
15 518	Indi	10.0	36.64	5.468	73 16 14.5	.90	51.78	1
15 519	Gould, 30056	9.0	43.80	5.020	69 43 13.0	.90	51.77	1
15 520	Octantis	10.0	48.50	+10.037	84 6 13.3	+ 16.91	50.68	1
15 521	Indi	9.0	21 49 50.97	+ 5.324	72 17 21.0	+ 16.91	51.68	1
15 522	Gould, Z. C., 21h 1636	9.8	49 53.22	5.318	72 14 55.8	.91	51.75	2
15 523	Octantis	9.0	50 2.93	5.783	75 9 50.8	.92	51.65	1
15 524	Octantis	10.0	16.39	6.163	76 56 28.4	.93	50.82	1
15 525	Indi	10.0	24.85	+ 4.986	69 28 54.9	+ 16.93	51.77	1
15 526	Gould, 30087	9.0	21 50 37.24	+ 5.480	73 27 33.2	+ 16.94	51.79	2
15 527	Lacaille, 8970	7.5	37.87	4.675	65 57 12.1	.95	52.67	1
15 528	Octantis	10.0	40.78	9.191	83 20 9.3	.95	50.67	1
15 529	Gould, 30088	9.2	50 54.24	4.941	69 5 35.3	.96	51.79	2
15 530	Octantis	9.8	51 0.11	+ 4.709	66 26 51.8	+ 16.96	52.68	2
15 531	Indi	9.7	21 51 11.76	+22.127	87 51 31.6	+ 16.97	51.42	3
15 532	Gould, Z. C., 21h 1690	10.0	32.68	4.901	68 46 39.0	.99	51.79	2
15 533	Octantis	10.0	36.61	7.595	81 4 25.5	.99	51.62	1
15 534	Octantis	10.0	37.80	5.587	74 13 52.0	16.99	51.68	2
15 535	Indi	10.0	51 53.25	+ 4.952	69 20 9.2	+ 17.00	51.78	1
15 536	Octantis	7.8	21 52 25.58	+24.066	88 4 20.9	+ 17.03	51.22	6
15 537	Octantis	9.0	34.20	8.544	82 39 3.4	.04	51.08	3
15 538	Gould, Z. C., 21h 1722	9.2	52 37.48	4.781	67 34 11.7	.04	51.83	2
15 539	Octantis	9.8	53 8.17	8.250	82 16 7.7	.06	50.75	2
15 540	Indi	10.0	9.12	+ 5.597	74 26 38.3	+ 17.06	51.66	1
15 541	Gould, Z. C., 21h 1762	9.0	21 53 33.02	+ 5.653	74 48 34.2	+ 17.08	51.65	1
15 542	Gould, Z. C., 21h 1758	10.0	37.72	5.220	71 55 57.4	.08	51.76	2
15 543	Indi	10.0	41.42	5.530	74 6 6.9	.09	51.71	1
15 544	Gould, Z. C., 21h 1782	9.0	46.86	6.331	77 53 1.7	.09	51.30	2
15 545	Indi	10.0	49.93	+ 5.366	73 2 42.5	+ 17.09	51.75	1

Number.	Constellation, Name of Star, or Synonym.	Magnitude.	Right Ascension, 1850.0.	Annual Precession.	South Declination, 1850.0.	Annual Precession.	Mean year.	No. of obs.
			h. m. s.	s.	° ′ ″	″		
15 546	Indi	10.0	21 53 51.94	+ 5.179	71 38 28.7	+ 17.09	51.73	1
15 547	Gould, 30154	9.2	53 54.56	5.200	71 48 57.2	.10	51.76	2
15 548	Gould, Z. C., 21h 1773 .	9.5	54 1.43	4.985	69 56 29.8	.10	51.74	1
15 549	Indi	10.0	2.79	5.043	70 29 30.3	.10	51.70	1
15 550	Octantis	10.0	7.32	+ 6.513	78 32 4.1	+ 17.11	50.78	1
15 551	Indi	9.0	21 54 9.20	+ 4.633	65 54 46.7	+ 17.11	52.67	1
15 552	Octantis	9.8	16.49	8.201	82 15 24.4	.11	50.75	2
15 553	Lacaille, 8988	8.0	19.12	4.933	69 27 51.4	.12	51.78	2
15 554	Octantis	8.7	25.49	15.922	86 53 34.8	.12	51.32	6
15 555	Gould, 30167	9.2	33.45	+ 5.055	70 40 6.0	+ 17.13	51.70	2
15 556	Brisbane, 7115	7.8	21 54 38.61	+ 4.663	66 23 30.0	+ 17.13	52.68	3
15 557	Gould, Z. C., 21h 1795 .	9.5	45.38	5.256	72 20 57.6	.14	51.75	2
15 558	Octantis	9.0	50.99	8.657	82 54 48.4	.14	51.71	1
15 559	Gould, Z. C., 21h 1798	9.5	54 58.21	5.038	70 33 33.3	.15	51.70	1
15 560	Octantis	9.7	55 1.11	+16.425	87 1 36.8	+ 17.15	51.35	5
15 561	Indi	10.0	21 55 3.79	+ 4.681	66 41 32.0	+ 17.15	52.62	1
15 562	Indi	9.2	5.14	4.772	67 49 9.5	.15	51.83	2
15 563	Gould, Z. C., 21h 1804 .	8.0	16.88	4.613	65 48 22.3	.16	52.67	1
15 564	Indi	9.2	19.54	4.917	69 26 42.6	.16	51.78	2
15 565	Lacaille, 8994	6.8	28.28	+ 5.081	71 0 45.3	+ 17.17	51.69	2
15 566	Octantis	9.0	21 55 41.69	+ 8.300	82 27 46.2	+ 17.18	50.69	1
15 567	Indi	10.0	42.95	4.940	69 43 24.5	.18	51.77	1
15 568	Gould, 30194	7.5	45.58	4.846	68 44 52.6	.18	51.79	2
15 569	Indi	10.0	47.54	4.814	68 23 49.8	.18	51.80	1
15 570	Indi	9.5	49.06	+ 4.754	67 42 51.5	+ 17.18	51.83	1
15 571	Octantis	9.2	21 55 49.52	+20.433	87 43 32.1	+ 17.18	51.32	6
15 572	Octantis	10.0	49.81	6.536	78 44 37.8	.18	51.77	1
15 573	Indi	9.2	55 51.78	4.662	66 36 36.4	.19	52.65	2
15 574	Indi	9.0	56 9.08	4.651	66 27 27.3	.20	52.68	1
15 575	Indi	10.0	12.97	+ 4.942	69 48 56.5	+ 17.20	51.77	1
15 576	Gould, Z. C., 21h 1843 .	9.5	21 56 15.67	+ 4.961	70 0 22.5	+ 17.20	51.74	1
15 577	Lacaille, 8991	7.0	21.48	5.957	76 36 39.4	.21	50.83	1
15 578	Indi	9.5	22.48	4.673	66 46 49.6	.21	52.62	1
15 579	Gould, 30205	7.5	28.60	4.601	65 49 43.3	.21	52.67	1
15 580	Octantis	9.0	29.11	+11.613	85 24 22.4	+ 17.21	51.23	4
15 581	Gould, Z. C., 21h 1854 .	9.0	21 56 30.76	+ 5.093	71 15 3.3	+ 17.21	51.69	1
15 582	Indi	9.5	32.79	4.814	68 30 23.1	.22	51.80	1
15 583	Indi	9.0	36.96	4.705	67 13 25.7	.22	51.82	1
15 584	Octantis	10.0	47.96	6.892	79 50 24.6	.23	50.80	2
15 585	Lacaille, 9002	9.0	54.54	+ 5.132	71 38 0.5	+ 17.23	51.73	1

No. 15 584. Declination may be 79° 49′ 56.7″.

Number.	Constellation, Name of Star, or Synonym.	Magnitude.	Right Ascension, 1850.0.	Annual Precession.	South Declination, 1850.0.	Annual Precession.	Mean year.	No. of obs.
			h. m. s.	s.	° ′ ″	″		
15 586	Gould, Z. C., 21h 1887	9.2	21 56 57.76	+ 6.255	77 52 32.3	+ 17.23	51.30	2
15 587	Indi	10.0	57 5.70	4.751	67 51 22.9	.24	51.83	1
15 588	Gould, Z. C., 21h 1889	9.2	8.78	5.978	76 46 38.8	.24	50.83	2
15 589	Indi	10.0	20.30	4.916	69 40 53.7	.25	51.77	1
15 590	Lacaille, 8996	7.0	34.25	+ 5.985	76 50 48.6	+ 17.26	50.83	2
15 591	Gould, Z. C., 21h 1891	9.5	21 57 41.38	+ 4.812	68 38 50.4	+ 17.27	51.79	2
15 592	Indi	9.0	51.15	5.295	73 0 19.9	.27	51.77	2
15 593	Gould, Z. C., 21h 1898	9.5	57 52.57	4.916	69 46 48.6	.28	51.77	1
15 594	Indi	10.0	58 3.59	5.280	72 55 20.9	28	51.75	1
15 595	Gould, Z. C., 22h 23	10.0	5.39	+ 6.283	78 4 32.3	+ 17.29	50.79	1
15 596	Gould, Z. C., 22h 13	8.8	21 58 7.08	+ 5.250	72 42 25.6	+ 17.29	51.72	2
15 597	Octantis	10.0	14.87	8.718	83 9 31.8	.29	51.71	1
15 598	Indi	10.0	17.10	4.674	67 5 35.6	.29	51.82	1
15 599	Gould, Z. C., 22h 39	8.8	22.34	6.575	79 3 58.3	.30	51.11	3
15 600	Indi	9.5	38.50	+ 4.587	65 58 16.6	+ 17.31	52.67	1
15 601	Indi	9.5	21 58 41.14	+ 4.955	70 16 28.8	+ 17.31	51.72	2
15 602	Brisbane, 7128	9.2	43.28	5.101	71 36 1.6	.31	51.71	2
15 603	Lacaille, 9016 .	7.2	47.71	4.835	69 3 33.0	.32	51.79	2
15 604	Octantis	10.0	48.42	6.823	79 48 25.1	.32	50.80	2
15 605	Gould, Z. C., 22h 59	9.0	51.47	+ 6.931	80 5 25.2	+ 17.32	50.82	1
15 606	Indi	10.0	21 58 59.80	+ 4.632	66 38 59.1	+ 17.33	52.68	1
15 607	Gould, Z. C., 22h 44	10.0	59 10.86	4.827	69 1 49.6	.33	51.78	2
15 608	Indi	9.8	23.23	4.598	66 15 1.7	.34	52.68	2
15 609	Octantis	8.6	24.83	18.990	87 35 4.0	.34	51.22	7
15 610	Octantis	9.5	27.92	+ 8.850	83 22 19.2	+ 17.35	50.67	1
15 611	Indi	10.0	21 59 32.11	+ 5.036	71 8 42.6	+ 17.35	51.70	1
15 612	Indi	10.0	36.69	4.858	69 25 8.6	.35	51.77	1
15 613	Octantis	10.0	39.49	11.204	85 17 17.7	.35	51.43	3
15 614	Gould, Z. C., 22h 53	8.8	40.04	4.636	67 4 23.5	.35	52.22	2
15 615	Gould, Z. C., 22h 81	9.0	48.16	+ 6.137	77 40 43.2	+ 17.36	51.81	1
15 616	Indi	9.5	21 59 49.66	+ 4.642	66 54 32.1	+ 17.36	52.62	1
15 617	Octantis	9.8	52.28	6.285	78 14 4.2	.36	50.78	2
15 618	Gould, Z. C., 22h 70	10.0	53.47	5.372	73 46 54.5	.36	51.81	1
15 619	Octantis	9.6	54.95	15.951	87 1 35.7	.37	51.38	4
15 620	Gould, Z. C., 22h 77	9.8	55.32	+ 5.511	74 40 0.4	+ 17.37	51.65	2
15 621	Gould, Z. C., 22h 66	9.8	21 59 56.86	+ 4.650	67 2 6.1	+ 17.37	52.22	2
15 622	Indi	9.0	22 0 0.75	5.050	71 19 33.1	.37	51.69	1
15 623	Octantis	9.9	9.79	7.273	80 58 9.2	.38	51.62	1
15 624	Gould, Z. C., 22h 72	9.5	11.32	4.782	68 40 26.9	.38	51.78	2
15 625	Octantis	9.5	15.76	+ 7.380	81 11 46.2	+ 17.38	50.80	1

No. 15 595. Gould's declination is 7.9″ greater.

No. 15 604. Declination may be 79° 47′ 57.2″.

Number.	Constellation, Name of Star, or Synonym.	Magnitude.	Right Ascension, 1850.0.	Annual Precession.	South Declination, 1850.0.	Annual Precession.	Mean year.	No. of obs.
			h. m. s.	s.	° ′ ″	″		
15 626	Gould, Z. C., 22h 97 . .	9.5	22 0 21.06	+ 6.123	77 40 21.1	+ 17.38	51.81	1
15 627	Octantis	10.0	38.22	7.334	81 7 42.1	.40	50.80	1
15 628	Octantis	10.0	41.55	5.766	76 8 6.3	.40	50.84	1
15 629	Octantis	10.0	45.13	6.170	77 53 30.2	.40	51.81	1
15 630	Gould, Z. C., 22h 95 . .	10.0	53.09	+ 4.755	68 28 5.5	+ 17.41	51.80	1
15 631	Indi	10.0	22 0 54.78	+ 4.998	70 58 58.7	+ 17.41	51.70	1
15 632	Octantis	9.5	56.77	7.632	81 43 9.1	.41	51.81	1
15 633	Tucanæ	10.0	56.93	4.585	66 18 57.2	.41	52.68	2
15 634	Indi	9.5	57.05	4.938	70 25 16.2	.41	51.70	1
15 635	Gould, Z. C 22h 105 . .	9.5	0 59.48	+ 5.143	72 14 21.8	+ 17.41	51.75	2
15 636	Gould, Z. C., 22h 113 . .	9.5	22 1 3.43	+ 5.407	74 8 48.3	+ 17.42	51.71	1
15 637	Gould, Z. C., 22h 117 . .	9.0	6.17	5.679	75 44 1.3	.42	51.58	1
15 638	C Octantis	6.5	10.00	14.643	86 43 20.4	.42	51.32	6
15 639	Indi	10.0	15.34	5.152	72 20 43.8	.42	51.75	2
15 640	Octantis	10.0	25.94	+ 6.481	79 1 21.6	+ 17.43	51.78	1
15 641	Gould, Z. C., 22h 121 . .	9.0	22 1 31.92	+ 4.833	69 26 30.6	+ 17.44	51.78	2
15 642	Indi	9.2	37.07	4.945	70 34 51.3	.44	51.70	2
15 643	Indi	9.8	53.45	5.042	71 30 10.1	.45	51.71	2
15 644	Gould, Z. C., 22h 125 . .	8.5	1 58.70	4.580	66 24 24.2	.46	52.68	2
15 645	Melbourne (1), 1121 . .	7.0	2 2.38	+ 4.774	68 51 47.6	+ 17.46	51.79	1
15 646	Gould, Z. C., 22h 157 . .	9.0	22 2 13.17	+ 6.508	79 10 13.8	+ 17.47	51.11	3
15 647	Indi	8.0	13.76	4.660	67 31 11.0	.47	51.83	2
15 648	Tucanæ	9.0	14.64	4.645	67 19 48.2	.47	51.82	1
15 649	Indi	9.5	50.79	4.683	67 54 26.1	.49	51.83	1
15 650	ε Octantis	6.0	51.89	+ 7.189	81 10 52.6	+ 17.49	51.21	2
15 651	Tucanæ	9.8	22 2 52.16	+ 4.627	67 11 17.2	+ 17.49	52.22	2
15 652	Gould, Z. C., 22h 160 . .	9.0	2 53.27	4.996	71 13 47.0	.49	51.69	1
15 653	Lacaille, 9022	6.8	3 1.82	6.213	78 15 13.5	.50	50.78	2
15 654	Lacaille, 9035	8.0	3.84	5.011	71 23 26.8	.50	51.71	2
15 655	Gould, Z. C., 22h 169 . .	9.0	17.32	+ 4.583	66 40 6.8	+ 17.51	52.65	2
15 656	Octantis	10.0	22 3 28.71	+ 7.370	81 23 23.3	+ 17.52	51.81	1
15 657	Octantis	9.8	38.98	6.216	78 19 3.7	.53	50.78	2
15 658	Gould, 30373	10.0	43.16	7.267	81 11 49.6	.53	51.21	2
15 659	Gould, Z. C., 22h 178 . .	9.0	44.57	4.794	69 20 5.2	.53	51.78	1
15 660	Gould, 30376	10.0	46.87	+ 7.266	81 11 54.7	+ 17.53	50.71	2
15 661	Octantis	9.0	22 3 52.30	+ 5.616	75 41 35.5	+ 17.54	51.58	1
15 662	Lacaille, 9023	8.2	4 20.18	6.754	80 2 27.1	.56	50.80	2
15 663	Lacaille, 8998	8.8	33.17	9.500	84 16 0.8	.57	50.76	2
15 664	Octantis	10.0	42.24	8.507	83 14 10.8	.57	50.67	1
15 665	Indi .	10.0	56.24	+ 5.081	72 14 32.5	+ 17.58	51.68	1

N 15 662. There is some confusion in the recorded micrometer revolutions; both observations are unreliable.

Number.	Constellation, Name of Star, or Synonym.	Magnitude.	Right Ascension, 1850.0.	Annual Precession.	South Declination, 1850.0.	Annual Precession.	Mean year.	No. of obs.
			h. m. s.	s.	° ′ ″	″		
15 666	Gould, Z. C., 22h 210 . .	8.8	22 4 57.61	+ 4.755	69 5 15.0	+ 17.58	51.79	2
15 667	Tucanæ	10.0	5 20.51	4.536	66 20 23.6	.60	52.67	1
15 668	Indi	10.0	21.86	5.133	72 42 58.1	.60	51.72	2
15 669	Gould, Z. C., 22h 248 . .	9.5	28.58	6.707	80 0 17.0	.60	50.80	2
15 670	Indi	10.0	29.38	+ 4.718	68 43 53.0	+ 17.60	51.79	2
15 671	Gould, Z. C., 22h 234 . .	9.5	22 5 32.02	+ 5.177	73 4 36.4	+ 17.61	51.78	1
15 672	Indi	10.5	34.26	5.131	72 43 58.6	.61	71.72	2
15 673	Octantis	8.5	41.67	5.896	77 14 20.2	.61	50.81	1
15 674	Octantis	9.0	5 49.21	5.811	76 52 4.2	.62	50.82	1
15 675	Gould, Z. C., 22h 260 . .	9.0	6 2.96	+ 6.118	78 10 26.2	+ 17.63	50.78	2
15 676	Gould, Z. C., 22h 251 . .	10.0	22 6 19.69	+ 4.720	68 53 42.2	+ 17.64	51.79	1
15 677	Gould, Z. C., 22h 256 . .	9.2	32.68	4.639	67 56 15.9	.65	51.83	2
15 678	Octantis	10.5	34.88	6.376	79 7 11.3	.65	50.78	2
15 679	Octantis	11.5	36.24	6.375	79 7 14.8	.65	50.78	2
15 680	Indi	10.0	40.99	+ 4.908	70 56 6.6	+ 17.65	51.70	1
15 681	Indi	10.0	22 6 42.13	+ 4.841	70 16 5.3	+ 17.66	51.74	1
15 682	Octantis	10.0	43.11	7.260	81 23 17.3	.66	51.[illegible]1	1
15 683	Gould, Z. C., 22h 261 . .	7.0	45.34	4.480	65 45 45.8	.66	52.66	1
15 684	Gould, Z. C., 22h 277 . .	9.8	6 56.14	5.146	73 1 38.6	.66	51.77	2
15 685	Gould, Z. C., 22h 290 . .	9.5	7 11.33	+ 5.151	73 5 50.8	+ 17.68	51.77	2
15 686	Gould, Z. C., 22h 288 . .	10.0	22 7 11.84	+ 5.034	72 9 6.5	+ 17.68	51.81	1
15 687	Indi	9.0	15.00	4.616	67 45 16.7	.68	51.83	1
15 688	Lacaille, 9062	7.8	16.68	4.721	69 3 5.5	.68	51.79	2
15 689	Octantis	10.0	28.62	11.381	85 39 43.2	.69	50.74	1
15 690	Octantis	10.0	32.20	+ 5.739	76 42 33.7	+ 17.69	50.83	1
15 691	Gould, Z. C., 22h 304 . .	9.5	22 7 35.80	+ 5.135	73 1 27.3	+ 17.69	51.77	2
15 692	Octantis	11.0	47.13	7.266	81 28 22.8	.70	51.81	1
15 693	Octantis	10.0	50.16	5.418	75 0 8.2	.70	51.20	2
15 694	Lacaille, 9055	8.0	50.60	5.541	75 42 58.0	.70	51.58	1
15 695	Octantis	10.0	7 54.37	+ 5.604	76 4 0.7	+ 17.70	50.84	1
15 696	Tucanæ	9.8	22 8 13.26	+ 4.476	65 57 1.1	− 17.72	52.66	2
15 697	Indi	10.0	21.81	4.597	67 41 5.2	.72	51.83	1
15 698	Octantis	10.0	27.53	9.409	84 21 50.6	.73	50.76	2
15 699	Lacaille, 9049	8.5	29.19	6.603	79 57 30.8	.73	50.80	2
15 700	Gould, Z. C., 22h 334 . .	8.0	47.60	+ 4.587	67 37 57.4	+ 17.74	51.83	2
15 701	Tucanæ	8.0	22 8 54.36	+ 4.536	66 56 45.1	+ 17.75	52.22	2
15 702	Gould, Z. C., 22h 340 . .	10.0	8 56.03	4.617	68 2 28.8	.75	51.83	1
15 703	Indi	9.5	9 2.14	4.785	70 1 56.2	.75	51.74	1
15 704	Gould, Z. C., 22h 349 . .	10.0	13.85	4.580	67 36 21.5	.76	51.83	2
15 705	Gould, 30632	9.9	24.88	+29.540	88 39 25.9	+ 17.77	51.34	5

Number.	Constellation, Name of Star, or Synonym.	Magnitude.	Right Ascension, 1850.0.	Annual Precession.	South Declination, 1850.0.	Annual Precession.	Mean year.	No. of obs.
			h. m. s.	s.	° ′ ″	″		
15 706	Gould, Z. C., 22h 357	9.5	22 9 29.66	+ 4.694	69 4 15.2	+ 17.72	51.79	2
15 707	Gould, Z. C., 22h 356 . .	8.2	35.10	4.462	65 58 55.1	.77	52.66	2
15 708	Gould, Z. C., 22h 373 . .	9.8	45.90	4.913	71 25 30.4	.78	51.71	2
15 709	Lacaille, 9070	8.5	9 48.43	5.886	77 36 17.6	.78	51.81	1
15 710	Indi	10.0	10 10.30	+ 4.855	70 55 57.4	+ 17.80	51.76	1
15 711	Indi	9.5	22 10 20.12	+ 4.801	70 24 17.7	+ 17.80	51.70	1
15 712	Indi	10.0	29.35	5.277	74 25 11.9	.81	51.66	1
15 713	Octantis	9.5	36.89	5.639	76 32 43.0	.82	50.83	1
15 714	Octantis	9.5	42.25	5.564	76 9 52.6	.82	50.84	1
15 715	Indi	9.5	44.33	+ 4.986	72 13 32.1	+ 17.82	51.75	2
15 716	Tucanæ	10.0	22 10 48.42	+ 4.517	67 0 41.0	+ 17.82	52.62	1
15 717	Gould, 30478	8.5	48.70	4.577	67 50 6.4	.82	51.83	2
15 718	Indi	10.0	10 53.57	5.127	73 23 43.9	.83	51.78	1
15 719	Indi	9.5	11 0.16	5.235	74 11 44.0	.83	51.71	1
15 720	Indi	10.0	0.70	+ 5.209	74 1 8.7	+ 17.83	51.71	1
15 721	Indi	10.0	22 11 7.37	+ 4.854	71 3 30.2	+ 17.84	51.73	2
15 722	Tucanæ	10.0	10.06	4.508	66 56 14.0	.84	52.62	1
15 723	Stone, 11700	9.0	10.67	5.143	73 33 23.4	.84	51.79	2
15 724	Gould, 30495	8.8	15.42	5.142	73 33 24.2	.84	51.79	2
15 725	Gould, Z. C., 22h 423 . .	9.5	26.16	+ 5.007	72 30 25.4	+ 17.85	51.68	1
15 726	Gould, Z. C., 22h 421 . .	9.5	22 11 34.72	+ 4.564	67 47 28.5	+ 17.85	51.83	2
15 727	ν Indi	5.8	37.22	5.061	72 58 52.3	.86	51.77	2
15 728	Gould, Z. C., 22h 445 . .	9.0	11 57.71	6.280	79 17 2.1	.87	50.72	1
15 729	Gould, 30506	10.0	12 3.08	5.128	73 35 26.6	.87	51.81	1
15 730	Tucanæ	9.5	13.45	+ 4.426	65 54 12.0	+ 17.87	52.66	1
15 731	Indi	10.0	22 12 19.87	+ 5.002	72 35 29.6	+ 17.88	51.68	1
15 732	Octantis	9.0	20.80	6.656	80 24 58.5	.88	50.81	1
15 733	Gould, Z. C., 22h 454 . .	10.0	35.97	5.131	73 39 20.8	.89	51.81	1
15 734	Tucanæ	9.5	39.66	4.445	66 16 22.8	.90	52.67	2
15 735	Lacaille, 9090	6.8	40.71	+ 5.452	75 46 22.7	+ 17.90	51.09	3
15 736	Octantis	9.0	22 12 41.62	+ 5.400	75 28 36.9	+ 17.90	51.16	2
15 737	Gould, 30511	9.0	42.82	4.703	69 42 2.4	.90	51.77	1
15 738	Brisbane, 7155 . .	9.0	46.12	5.450	75 46 19.5	.90	51.09	3
15 739	Indi	9.0	48.59	4.540	67 40 44.6	.90	51.83	1
15 740	Octantis	10.0	54.18	+ 5.725	77 12 34.8	+ 17.91	50.81	1
15 741	Octantis	10.0	22 12 55.12	+ 5.580	76 29 49.3	+ 17.91	50.83	1
15 742	Octantis	10.0	13 0.17	5.585	76 31 58.9	.91	50.83	1
15 743	Gould, 30518	7.5	4.69	4.429	66 6 31.1	.91	52.66	1
15 744	Lacaille, 9085	9.0	5.89	6.020	78 28 19.6	.91	50.78	1
15 745	Gould, 30527	8.8	17.08	+ 4.487	67 1 3.6	+ 17.92	52.22	2

Number.	Constellation, Name of Star, or Synonym.	Magnitude.	Right Ascension, 1850.0.	Annual Precession.	South Declination, 1850.0.	Annual Precession.	Mean year.	No. of obs.
			h. m. s.	s.	° ′ ″	″		
15 746	Lacaille, 9099	6.0	22 13 20.99	+ 4.833	71 11 7.2	+ 17.92	51.73	4
15 747	Indi	9.0	42.79	4.524	67 37 2.4	.94	51.83	2
15 748	Indi	9.5	45.07	5.076	73 23 19.5	.94	51.78	1
15 749	Melbourne (1), 1131 .	7.8	46.76	4.439	66 22 38.6	.94	52.67	2
15 750	Indi	9.5	13 50.15	+ 5.047	73 10 9.8	+ 17 94	51.77	2
15 751	Tucanæ	9.5	22 14 1.17	+ 4.469	66 53 15.4	+ 17.95	52.62	1
15 752	Octantis	9.7	3.17	14.518	87 1 3.6	.95	51.31	7
15 753	Indi	10.0	7.73	5.132	73 51 51.7	.95	51.75	2
15 754	Indi	9.8	20.16	4.791	70 54 54.0	.96	51.73	2
15 755	Gould, Z. C., 22h 503 . .	9.0	29.54	+ 4.645	69 19 15.8	+ 17.97	51.78	1
15 756	Tucanæ .	9.5	22 14 35.41	+ 4.432	66 25 39.3	+ 17.97	52.68	1
15 757	Octantis	9.8	42.66	7.018	81 27 21.4	.98	51.30	2
15 758	Octantis . .	10.0	14 58.29	7.395	82 12 30.7	.99	50.81	1
15 759	Octantis	9.5	15 3.24	5.987	78 32 17.9	.99	50.78	1
15 760	Gould, Z. C., 22h 517 . .	10.0	6.74	+ 4.479	67 13 8.2	+ 17.99	51.82	1
15 761	Lacaille, 9095 . . .	8.0	22 15 20.04	+ 6.098	78 58 27.1	+ 18.00	50.83	1
15 762	Gould, 30576	9.2	36.17	4.518	67 52 7.7	.01	51.83	2
15 763	Indi	9.0	44.92	4.727	70 27 50.1	.02	51.70	1
15 764	Octantis	9.8	15 50.13	6.027	78 46 0.0	.02	50.80	2
15 765	Tucanæ .	9.5	16 10.98	+ 4.431	66 42 42.2	+ 18.03	52.62	1
15 766	Gould, Z. C., 22h 557 . .	10.0	22 16 19.72	+ 5.152	74 18 42.8	+ 18.04	51.68	2
15 767	Indi	9.5	26.39	4.672	69 57 11.1	.04	51.74	1
15 768	Indi	10.0	28.98	5.232	74 53 5.6	.04	51.65	1
15 769	d Tucanæ	5.0	35.96	4.365	65 43 34.2	.05	52.66	1
15 770	Gould, 30597	9.0	37.24	+ 5.006	73 13 57.9	+ 18.05	51.78	1
15 771	Tucanæ	9.5	22 16 41.01	+ 4.433	66 50 31.5	+ 18.05	52.62	1
15 772	Lacaille, 9106 . . .	8.5	47.12	5.685	77 26 33.4	.06	51.31	2
15 773	Octantis	10.0	47.29	6.932	81 25 36.9	.06	51.30	2
15 774	Octantis	9.5	55.12	5.476	76 24 31.7	.06	50.83	1
15 775	Octantis	8.8	55.69	+11.755	86 10 22.8	+ 18.06	51.31	7
15 776	Indi	10.0	22 16 55.75	+ 4.777	71 10 40.5	+ 18.06	51.78	2
15 777	Gould, Z. C., 22h 571 . .	9.5	16 56.81	4.664	69 57 6.7	.06	51.77	1
15 778	Octantis	9.0	17 2.24	6.390	80 4 21.6	.07	50.82	1
15 779	Octantis	9.5	26.29	8.542	83 57 45.8	.08	50.74	2
15 780	Octantis	8.0	27.65	+11.978	86 16 55.8	+ 18.08	51.37	8
15 781	Lacaille, 9117	5.5	22 17 30.43	+ 4.518	68 14 54.1	+ 18.08	51.81	2
15 782	Gould, Z. C., 22h 601 . .	9.2	30.66	5.763	77 52 29.0	.08	51.30	2
15 783	Octantis	10.0	30.84	5.491	76 33 28.8	.08	50.83	1
15 784	Octantis	10.0	36.62	7.551	82 39 6.7	.09	51.71	1
15 785	Gould, Z. C., 22h 599 . .	8.2	55.19	+ 4.672	70 12 2.3	+ 18.10	51.72	2

Number.	Constellation, Name of Star, or Synonym.	Magnitude.	Right Ascension, 1850.0.	Annual Precession.	South Declination, 1850.0.	Annual Precession.	Mean year.	No. of obs.
			h. m. s.	s.	° ′ ″	″		
15 786	Indi	9.8	22 17 58.75	+ 4.467	67 34 54.8	+ 18.10	51.83	2
15 787	Indi	10.0	59.00	4.798	71 33 19.1	.10	51.80	1
15 788	Gould, Z. C., 22h 602 . .	10.0	17 59.91	4.516	68 16 4.5	.10	51.80	1
15 789	Octantis	9.9	18 0.94	14.278	87 3 35.8	.10	51.37	4
15 790	Gould, Z. C., 22h 603 . .	9.5	1.28	+ 4.558	68 49 33.7	+ 18.10	51.79	1
15 791	Octantis	10.0	22 18 29.72	+ 8.099	83 29 49.6	+ 18.12	50.73	2
15 792	Tucanæ	9.2	38.90	4.419	66 59 9.1	.13	52.22	2
15 793	Indi	10.0	44.76	4.554	68 54 21.8	.13	51.79	1
15 794	Gould, Z. C., 22h 625 . .	9.0	18 47.02	4.645	70 2 5.9	.13	51.74	1
15 795	Indi	10.0	19 0.18	+ 4.528	68 36 55.8	+ 18.14	51.79	1
15 796	Gould, Z. C., 22h 629 . .	10.0	22 19 0.89	+ 4.831	72 2 31.2	+ 18.14	51.81	1
15 797	Indi	10.0	7.73	4.509	68 22 28.6	.14	51.80	1
15 798	Tucanæ	10.0	9.19	4.395	66 42 57.2	.15	52.68	2
15 799	Gould, Z. C., 22h 632 . .	9.8	9.54	4.747	71 13 26.4	.15	51.78	2
15 800	Octantis	10.0	9.92	+ 5.281	75 32 43.2	+ 18.15	50.75	1
15 801	Tucanæ	9.0	22 19 19.71	+ 4.358	66 9 9.9	+ 18.15	52.66	1
15 802	Gould, Z. C., 22h 637 . .	9.5	22.15	4.821	72 0 5.1	.15	51.81	1
15 803	Tucanæ	10.0	24.84	4.392	66 42 42.6	.16	52.68	2
15 804	Indi	10.0	29.70	4.772	71 32 36.0	.16	51.80	1
15 805	Gould, 30642	8.8	35.21	+ 4.418	67 9 20.2	+ 18.16	52.22	2
15 806	Gould, Z. C., 22h 646 . .	10.0	22 19 35.57	+ 5.111	74 27 23.8	+ 18.16	51.66	1
15 807	Octantis	10.0	37.48	5.709	77 51 53.4	.16	51.30	2
15 808	Octantis	10.0	41.75	7.120	82 1 51.7	.17	50.81	1
15 809	Octantis	9.8	46.45	9.343	84 50 22.1	.17	51.31	2
15 810	Octantis	9.5	48.13	+ 6.499	80 37 16.6	+ 18.17	51.21	2
15 811	Melbourne (1), 1137 . .	9.0	22 19 51.11	+ 4.477	68 4 12.4	+ 18.17	51.83	1
15 812	Octantis	10.0	51.34	6.154	79 35 44.6	.17	50.78	1
15 813	Gould, Z. C., 22h 669 . .	9.0	19 59.92	5.710	77 54 35.8	.18	51.30	2
15 814	Gould, Z. C., 22h 655 . .	9.3	20 3.49	4.578	69 26 46.9	.19	51.76	3
15 815	Octantis	9.0	3.92	+ 9.407	84 54 21.8	+ 18.19	51.31	2
15 816	Octantis	10.0	22 20 11.14	+ 5.496	76 53 54.9	+ 18.19	50.82	1
15 817	Gould, Z. C., 22h 664 . .	10.0	18.94	4.601	69 46 23.1	.19	51.75	2
15 818	Indi	10.0	24.32	4.481	68 13 34.6	.19	51.81	2
15 819	Octantis	10.0	36.00	6.243	79 57 11.0	.20	50.78	1
15 820	Octantis	10.5	36.28	+ 10.510	85 40 48.1	+ 18.20	51.23	2
15 821	Lacaille, 9105	8.3	22 20 37.67	+ 8.068	83 35 6.9	+ 18.20	50.77	3
15 822	Octantis	9.6	39.88	11.670	86 15 47.7	.20	51.31	7
15 823	Indi	9.5	43.86	4.590	69 42 31.1	.20	52.68	2
15 824	Indi	10.0	46.48	4.727	71 16 46.7	.21	51.76	1
15 825	Lacaille, 9102	7.0	48.73	+ 8.429	84 1 26.2	+ 18.21	50.68	1

No. 15 791. The declination may be 83° 29′ 21.7″.

Number.	Constellation, Name of Star, or Synonym.	Magnitude.	Right Ascension, 1850.0.	Annual Precession.	South Declination, 1850.0.	Annual Precession.	Mean year.	No. of obs.
			h. m. s.	s.	° ′ ″	″		
15 826	Lacaille, 9134	8.2	22 20 52.26	+ 4.601	69 52 24.6	+ 18.21	51.75	3
15 827	Octantis	9.3	53.65	5.247	75 33 26.2	.21	51.06	3
15 828	Lacaille, 9122	6.0	54.17	6.107	79 32 28.0	.21	50.75	2
15 829	Lacaille, 9124	8.0	54.17	6.061	79 22 58.4	.21	50.72	1
15 830	Octantis	9.3	20 57.75	+ 5.227	75 26 1.7	+ 18.21	51.06	3
15 831	Octantis	10.5	22 21 11.67	+10.258	85 33 14.3	+ 18.22	51.23	2
15 832	Gould, Z. C., 22h 706 . .	9.5	12.48	6.019	79 16 2.4	.22	50.12	1
15 833	Indi	10.0	16.88	5.094	74 29 56.1	.22	51.66	1
15 834	Octantis	9.5	23.58	5.451	76 48 27.8	.23	50.83	2
15 835	Gould, Z. C., 22h 689 . .	9.0	24.61	+ 4.380	66 55 21.8	+ 18.23	52.62	1
15 836	Gould, Z. C., 22h 694 . .	9.5	22 21 31.50	+ 4.439	67 50 3.5	+ 18.23	51.83	3
15 837	Octantis	10.0	21 44.51	6.877	81 41 19.4	.24	51.81	1
15 838	Octantis	10.2	22 1.20	10.646	85 48 50.9	.25	51.07	3
15 839	Indi	9.5	5.67	4.531	69 12 20.2	.25	51.78	1
15 840	Gould, Z. C., 22h 711 . .	8.8	6.43	+ 4.327	66 11 37.8	+ 18.25	52.66	2
15 841	Indi	10.0	22 22 19.43	+ 4.488	68 40 46.0	+ 18.26	51.79	2
15 842	Gould, Z. C., 22h 720 . .	9.5	21.87	4.394	67 19 31.4	.26	51.82	1
15 843	Octantis	9.8	22.20	8.546	84 14 12.1	.26	50.77	2
15 844	Octantis	9.5	23.49	5.343	76 19 42.3	.26	50.84	1
15 845	Indi	10.0	24.12	+ 5.038	74 18 32.0	+ 18.27	51.71	1
15 846	Octantis	10.0	22 22 30.80	+ 6.695	81 20 29.5	+ 18.27	50.80	1
15 847	Gould, Z. C., 22h 736 . .	10.0	46.67	4.383	67 13 54.9	.28	51.82	1
15 848	Indi	9.5	22 53.27	5.109	74 53 48.7	.28	51.65	1
15 849	Octantis	10.0	23 4.97	5.243	75 49 12.5	.29	50.84	1
15 850	Octantis	9.0	5.64	+ 6.617	81 12 5.2	+ 18.29	51.21	2
15 851	Gould, Z. C., 22h 751 . .	9.8	22 23 12.15	+ 4.498	68 58 20.8	+ 18.29	51.79	2
15 852	Lacaille, 9163	8.5	22.54	4.898	73 19 32.6	.30	51.78	1
15 853	Octantis	10.0	23 24.45	8.496	84 14 30.4	.30	50.84	1
15 854	Octantis	9.0	24 17.65	6.765	81 38 50.2	.33	51.81	1
15 855	Lacaille, 9155	7.8	25.92	+ 4.898	73 29 1.2	+ 18.34	51.79	2
15 856	Tucanæ	9.0	22 24 29.93	+ 3.277	65 48 42.8	+ 18.34	52.66	1
15 857	Octantis	10.0	33.24	5.492	77 24 10.8	.34	51.31	2
15 858	Lacaille, 9158	8.5	38.77	4.709	71 44 12.0	.35	51.73	1
15 859	Octantis	10.0	42.48	5.351	76 40 9.4	.35	50.83	1
15 860	Gould, Z. C., 22h 789 . .	9.0	46.60	+ 4.336	66 53 19.8	+ 18.35	52.62	1
15 861	Indi	10.0	22 24 52.90	+ 4.454	68 41 30.0	+ 18.36	51.79	2
15 862	Indi	10.0	24 58.34	4.884	73 26 38.1	.36	51.78	1
15 863	Gould, Z. C., 22h 798 . .	9.2	25 0.87	4.294	66 11 27.5	.36	52.67	2
15 864	Octantis	10.5	2.86	9.679	85 20 34.8	.36	51.23	2
15 865	Gould, Z. C., 22h 802 .	10.0	5.27	+ 4.471	68 57 57.3	+ 18.36	51.79	1

No. 15 860. Gould's declination is 4′ greater.

Number.	Constellation, Name of Star, or Synonym.	Magnitude.	Right Ascension, 1850.0.	Annual Precession.	South Declination, 1850.0.	Annual Precession.	Mean year.	No. of obs.
			h. m. s.	s.	° ′ ″	″		
15 866	Octantis	9.5	22 25 6.29	+ 5.221	75 58 47.3	+ 18.36	50.84	1
15 867	Octantis	9.8	10.25	17.216	87 49 23.8	.36	51.22	3
15 868	Gould, 30767	10.0	17.58	4.763	72 23 24.2	.37	51.73	1
15 869	Gould, Z. C., 22h 824 . .	9.8	27.63	4.982	74 19 41.1	.37	51.68	2
15 870	Gould, Z. C., 22h 816 . .	9.8	30.57	+ 4.378	67 41 22.8	+ 18.38	51.83	2
15 871	Gould, Z. C., 22h 820 . .	9.0	22 25 38.06	+ 4.325	66 52 59.0	+ 18.38	52.62	1
15 872	Gould, Z. C., 22h 827 . .	8.8	39.08	4.963	74 12 16.4	.38	51.68	2
15 873	Octantis	9.8	42.83	9.306	85 5 56.5	.38	51.08	3
15 874	Octantis	10.0	47.16	8.597	84 36 3.6	.39	50.85	1
15 875	Gould, Z. C., 22h 848 . .	9.0	49.47	+ 5.786	78 52 12.6	+ 18.39	50.83	1
15 876	Tucanæ	10.0	22 25 57.52	+ 4.265	65 59 13.2	+ 18.39	52.66	1
15 877	Indi	9.5	26 4.64	4.457	68 57 58.2	.40	51.78	1
15 878	Lacaille, 9123	10.0	15.67	8.616	84 31 19.2	.41	50.85	1
15 879	Gould, Z. C., 22h 852 . .	9.8	25.57	4.821	73 7 4.9	.41	51.77	2
15 880	Tucanæ	8.0	33.20	+ 4.299	66 37 55.8	+ 18.41	52.65	2
15 881	Octantis	9.0	22 26 35.89	+ 6.712	81 42 57.4	+ 18.41	51.81	1
15 882	Gould, Z. C., 22h 860 . .	9.0	48.22	4.529	70 3 23.8	.42	51.74	1
15 883	Indi	9.8	54.13	4.371	67 52 10.9	.42	51.83	3
15 884	Gould, Z. C., 22h 867 . .	10.0	26 55.76	4.436	68 49 55.4	.42	51.79	1
15 885	Gould, 30799	8.0	27 3.00	+ 4.276	66 20 27.4	+ 18.43	52.67	2
15 886	Indi	10.0	22 27 3.17	+ 4.366	67 49 22.3	+ 18.43	51.83	1
15 887	Octantis	9.2	8.16	6.062	80 0 17.0	.43	50.80	2
15 888	Gould, 30848	10.2	22.63	10.119	85 43 58.4	.44	51.07	4
15 889	Octantis	12.0	28.86	10.137	85 44 53.8	.44	51.23	1
15 890	Indi	9.8	32.03	+ 4.626	71 20 9.1	+ 18.45	51.78	2
15 891	Indi	10.5	22 27 44.91	+ 4.541	70 22 42.2	+ 18.45	51.70	1
15 892	Gould, Z. C., 22h 908 . .	9.8	49.36	4.988	74 43 35.4	.46	51.65	2
15 893	Indi	10.0	50.36	4.785	73 0 39.2	.46	51.77	2
15 894	Gould, Z. C., 22h 904 . .	9.0	27 57.71	4.518	70 8 9.7	.46	51.74	1
15 895	Octantis	10.0	28 6.76	+ 7.029	81 44 50.3	+ 18.47	51.81	1
15 896	Indi	10.0	22 28 11.27	+ 4.812	73 19 6.6	+ 18.47	51.78	1
15 897	Gould, Z. C., 22h 921 . .	9.0	16.81	5.141	75 52 4.5	.47	50.84	2
15 898	Gould, 30829	8.0	28 41.16	4.412	68 50 13.3	.49	51.79	1
15 899	Gould, Z. C., 22h 9[illegible]7 . .	9.0	29 32.43	5.891	79 38 20.3	.51	50.78	1
15 900	Gould, Z. C., 22h 961 . .	10.0	33.52	+ 4.908	74 21 36.8	+ 18.51	51.68	2
15 901	Tucanæ	9.5	22 29 35.64	+ 4.264	66 40 41.5	+ 18.52	52.64	2
15 902	Indi	10.0	44.67	4.599	71 25 48.2	.52	51.80	1
15 903	Gould, Z. C., 22h 960 . .	9.5	45.67	4.444	69 30 21.7	.52	51.76	2
15 904	Gould, Z. C., 22h 985 . .	9.0	45.72	5.697	78 56 44.3	.52	50.83	1
15 905	Gould, Z. C., 22h 962 . .	8.5	53.60	+ 4.253	66 33 27.6	+ 18.53	52.68	1

Number.	Constellation, Name of Star, or Synonym.	Magnitude.	Right Ascension, 1850.0.	Annual Precession.	South Declination, 1850.0.	Annual Precession.	Mean year.	No. of obs.
			h. m. s.	s.	° ′ ″	″		
15 906	Octantis	9.0	22 29 57.48	+ 6.312	81 1 14.2	+ 18.53	51.71	2
15 907	Gould, Z. C., 22h 970 .	8.5	30 4.06	4.247	66 29 3.6	.53	52.68	1
15 908	Tucanæ	8.7	4.49	4.233	66 13 44.1	.53	52.67	3
15 909	Indi	10.0	4.72	4.893	74 19 7.2	.53	51.66	1
15 910	Indi	10.0	6.45	+ 4.332	68 3 53.7	+ 18.53	51.83	1
15 911	Rümker, 571	8.0	22 30 7.93	+ 4.367	68 28 0.1	+ 18.53	51.79	1
15 912	Gould, Z. C., 22h 980 . .	9.5	10.69	4.424	69 19 56.9	.54	51.78	1
15 913	β Octantis	5.2	22.03	6.776	82 9 52.5	.54	50.82	2
15 914	Indi	10.0	31.17	4.837	73 55 20.0	.55	51.71	1
15 915	Gould, 30866 . .	9.5	36.20	+ 4.666	72 18 27.8	+ 18.55	51.77	2
15 916	Tucanæ	10.0	22 30 36.24	+ 4.290	67 21 8.3	+ 18.55	51.82	1
15 917	Indi	10.0	45.28	4.508	70 32 9.3	.55	51.72	3
15 918	Gould, Z. C., 22h 1014	9.5	45.67	5.635	78 48 3.8	.56	50.83	1
15 919	Gould, 30872	9.0	46.12	4.843	74 0 43.2	.56	51.71	1
15 920	Indi	11.0	49.67	+ 4.507	70 32 29.9	+ 18.56	51.72	3
15 921	Gould, Z. C., 22h 994 . .	8.7	22 30 52.37	+ 4.228	66 19 5.1	+ 18.56	52.67	3
15 922	Gould, Z. C., 22h 1001	9.0	30 57.34	4.469	70 4 41.4	.56	51.74	1
15 923	Gould, 30903	10.0	31 6.73	9.068	85 10 45.1	.57	50.79	1
15 924	Octantis	10.0	10.25	6.138	80 38 29.6	.57	51.62	1
15 925	Gould, Z. C., 22h 1013	9.3	17.78	+ 4.225	66 21 34.0	+ 18.57	52.07	3
15 926	Octantis	9.5	22 31 18.19	+ 6.457	81 31 16.8	+ 18.57	51.81	1
15 927	Gould, 30908 .	10.0	20.92	8.894	85 2 53.6	.57	51.01	2
15 928	Indi	9.7	21.16	4.499	70 32 4.8	.57	51.72	3
15 929	Lacaille, 9192	8.2	22.77	4.398	69 10 49.3	.58	51.79	2
13 930	Indi	10.0	30.92	+ 4.711	72 55 14.1	+ 18.58	51.75	1
15 931	Indi	9.8	22 31 32.32	+ 4.498	70 33 52.2	+ 18.58	51.72	3
15 932	Tucanæ	10.0	36.58	4.263	67 6 33.9	.58	52.22	2
15 933	Indi	9.0	31 37.38	4.356	68 36 52.3	.58	51.79	2
15 934	Gould, 30925	10.2	32 9.07	9.661	85 39 35.6	.60	51.25	5
15 935	Indi	10.0	12.70	+ 4.724	73 10 12.1	+ 18.60	51.78	1
15 936	Lacaille, 9191	7.8	22 32 16.86	+ 5.017	75 36 9.4	+ 18.61	50.80	2
15 937	Gould, Z. C., 22h 1050	9.5	18.86	4.644	72 23 11.8	.61	51.77	2
15 938	Gould, Z. C., 22h 1057 . .	10.0	32 37.92	4.569	71 37 41.5	.62	51.73	1
15 939	Indi	10.0	33 3.11	4.535	71 18 34.3	.63	51.76	1
15 940	Gould, Z. C., 22h 1077 .	9.0	27.01	+ 4.433	70 5 11.1	+ 18.64	51.74	1
15 941	Octantis	9.0	22 33 28.64	+ 5.782	79 41 33.1	+ 18.64	50.78	1
15 942	Gould, Z. C., 22h 1084 .	9.8	39.75	4.466	70 33 43.2	.65	51.74	2
15 943	Tucanæ . . .	9.0	42.50	4.531	67 0 32.2	.65	52.22	2
15 944	Gould, Z. C., 22h 1095 .	9.0	50.02	4.979	75 33 28.7	.66	50.80	2
15 945	Indi	10.0	53.34	+ 4.872	74 45 12.7	+ 18.66	51.38	3

Number.	Constellation, Name of Star, or Synonym.	Magnitude.	Right Ascension, 1850.0.	Annual Precession.	South Declination, 1850.0.	Annual Precession.	Mean year.	No. of obs.
			h. m. s.	s.	° ′ ″	″		
15 946	Octantis	10.0	22 33 58.20	+ 5.063	76 10 22.6	+ 18.66	50.83	2
15 947	Indi	11.5	34 4.40	4.867	74 44 36.2	.66	51.65	2
15 948	Octantis	9.8	9.65	12.850	87 8 9.6	.67	51.53	4
15 949	Lacaille, 9203	8.5	12.75	5.063	76 12 28.9	.67	50.83	2
15 950	Indi	10.0	12.90	+ 4.308	68 24 55.1	+ 18.67	51.80	1
15 951	Gould, Z. C., 22ʰ 1116	9.0	22 34 47.55	+ 4.213	66 56 6.9	+ 18.69	52.62	1
15 952	Gould, Z. C., 22ʰ 1121	9.8	55.45	4.269	67 56 21.4	.69	51.83	3
15 953	Indi	10.0	34 57.20	4.474	70 55 16.8	.69	51.76	1
15 954	Octantis	9.5	35 0.02	7.687	84 0 19.7	.69	50.68	1
15 955	Gould, Z. C., 22ʰ 1141	9.0	4.60	+ 5.400	78 15 5.7	+ 18.70	50.78	2
15 956	Indi	10.0	22 35 7.07	+ 4.291	68 21 23.7	+ 18.70	51.81	2
15 957	Lacaille, 9214	8.5	9.06	4.335	69 2 26.2	.70	51.79	2
15 958	Octantis	8.5	13.23	7.310	83 29 36.3	.70	50.77	3
15 959	Melbourne (1), 1156	7.5	19.99	4.203	66 53 3.5	.70	52.62	1
15 960	Tucanæ	11.0	30.58	+ 4.260	67 55 30.7	+ 18.71	51.83	1
15 961	Tucanæ	9.8	22 35 42.98	+ 4.260	67 58 51.1	+ 18.72	51.83	2
15 962	Indi	10.0	46.41	4.291	68 29 20.5	.72	51.80	1
15 963	Octantis	9.8	56.41	7.030	83 5 36.7	.72	51.28	2
15 964	Gould, Z. C., 22ʰ 1155	9.8	35 57.53	4.819	74 39 15.0	.72	51.20	2
15 965	Octantis	9.5	36 1.00	+ 6.450	81 55 41.6	+ 18.72	51.31	2
15 966	Indi	10.0	22 36 2.58	+ 4.349	69 26 45.9	+ 18.73	51.74	1
15 967	Lacaille, 9202	6.0	5.49	6.065	80 54 46.7	.73	51.62	1
15 968	Stone, 11852	8.2	7.74	4.408	70 18 49.7	.73	51.75	2
15 969	Gould, 31023	10.0	16.06	9.758	85 54 48.3	.73	51.83	1
15 970	Tucanæ	10.0	21.37	+ 4.157	66 14 51.3	+ 18.74	52.67	1
15 971	Gould, 31129	8.8	22 36 28.21	+25.188	88 45 55.7	+ 18.74	51.69	7
15 972	Gould, Z. C., 22ʰ 1185	9.0	35.10	5.766	79 58 50.0	.74	50.80	2
15 973	Indi	9.5	38.35	4.575	72 26 12.1	.74	51.73	1
15 974	Lacaille, 9220	7.5	36 49.86	4.395	70 15 47.1	.75	51.75	2
15 975	Lacaille, 9216	7.0	37 1.02	+ 5.270	77 50 26.9	+ 18.76	51.30	2
15 976	Octantis	9.5	22 37 4.10	+ 5.201	77 27 48.6	+ 18.76	50.75	2
15 977	Indi	10.0	8.88	4.450	71 1 16.2	.76	51.78	2
15 978	Gould, Z. C., 22ʰ 1208	8.8	19.10	5.736	79 57 18.6	.77	50.80	2
15 979	Lacaille, 9227	7.0	26.45	4.148	66 20 53.2	.77	52.67	2
15 980	Gould, Z. C., 22ʰ 1206	9.0	26.69	+ 5.136	77 8 20.6	+ 18.77	50.82	2
15 981	Lacaille, 9232	9.0	22 37 35.24	+ 4.496	71 43 19.9	+ 18.77	51.73	1
15 982	Octantis	10.0	38.40	7.963	84 30 32.4	.78	50.85	1
15 983	Indi	9.3	41.11	4.211	67 35 17.6	.78	51.82	3
15 984	Gould, 31079	9.2	49.72	12.568	87 10 17.6	.78	51.58	9
15 985	Gould, Z. C., 22ʰ 1209	10.0	52.85	+ 4.519	72 3 7.1	+ 18.78	51.81	1

No. 15 976. Declination may be 77° 28′ 16.5″.

Number.	Constellation, Name of Star, or Synonym.	Magnitude.	Right Ascension, 1850.0.	Annual Precession.	South Declination, 1850.0.	Annual Precession.	Mean year.	No. of obs.
			h. m. s.	s.	° ′ ″	″		
15 986	Indi	9.5	22 37 53.45	+ 4.297	69 3 31.2	+ 18.78	51.79	2
15 987	Indi	11.0	55.40	4.295	68 59 13.9	.78	51.78	1
15 988	Indi	10.0	37 58.80	4.244	68 13 1.8	.79	51.81	2
15 989	Gould, Z. C., 22^h 1210 .	9.0	38 2.76	4.132	66 10 22.2	.79	52.67	1
15 990	Indi	9.8	12.93	+ 4.202	67 32 54.2	+ 18.79	51.83	2
15 991	Octantis	10.0	22 38 13.91	+ 5.677	79 50 20.5	+ 18.79	50.82	1
15 992	Tucanæ	9.2	19.84	4.175	67 4 58.5	.80	52.22	2
15 993	Gould, 31021	9.5	21.57	4.325	69 35 35.0	.80	51.74	1
15 994	Lacaille, 9230	6.5	29.35	4.435	71 8 21.0	.80	51.78	3
15 995	Indi	10.0	32.39	+ 4.634	73 24 49.7	+ 18.80	51.78	1
15 996	Gould, Z. C., 22^h 1250 .	9.0	22 38 35.24	+ 5.184	77 34 57.7	+ 18.80	51.31	2
15 997	Indi	10.0	41.32	4.642	73 31 21.3	.81	51.78	1
15 998	Gould, Z. C., 22^h 1249 .	10.0	49.00	4.498	71 59 17.6	.81	51.77	2
15 999	Octantis	9.0	38 53.45	7.591	84 8 44.8	.81	50.68	1
16 000	Indi	9.5	39 7.38	+ 4.449	70 56 27.8	+ 18.82	51.77	2
16 001	Gould, Z. C., 22^h 1265 .	10.0	22 39 16.75	+ 5.160	77 32 31.1	+ 18.83	51.31	2
16 002	Gould, Z. C., 22^h 1266 .	9.0	41.83	4.146	66 51 25.8	.84	52.62	1
16 003	Indi	10.0	43.25	4.682	74 5 55.3	.84	51.71	1
16 004	Gould, Z. C., 22^h 1280 .	9.0	53.72	4.912	76 2 16.8	.84	50.84	1
16 005	Indi	9.5	57.56	+ 4.618	73 30 52.8	+ 18.85	51.79	2
16 006	Gould, Z. C., 22^h 1272 .	7.5	22 39 58.08	+ 4.141	66 50 27.0	+ 18.85	52.62	1
16 007	Lacaille, 9247	8.5	40 12.54	4.494	72 13 2.3	.85	51.77	2
16 008	Gould, Z. C., 22^h 1284 .	9.0	16.04	4.309	69 46 0.1	.86	51.74	1
16 009	Indi	9.2	22.46	4.372	70 42 25.5	.86	51.77	2
16 010	Indi	9.5	22.56	+ 4.564	73 2 32.4	+ 18.86	51.77	2
16 011	Gould, Z. C., 22^h 1308 .	9.0	22 40 27.88	+ 5.275	78 20 20.8	+ 18.86	50.78	2
16 012	Gould, Z. C., 22^h 1295 .	10.0	37.27	4.206	68 11 13.7	.87	51.83	1
16 013	Gould, Z. C., 22^h 1296 .	8.5	40 39.15	4.116	66 30 33.3	.87	52.68	1
16 014	Indi	10.0	41 7.97	4.494	72 28 36.6	.88	51.73	1
16 015	Gould, Z. C., 22^h 1319 .	9.2	13.16	+ 4.517	72 40 43.2	+ 18.88	51.74	2
16 016	Lacaille, 9228	8.0	22 41 23.56	+ 5.763	80 30 47.6	+ 18.89	50.84	2
16 017	Indi	9.8	26.40	4.294	69 48 14.7	.89	51.74	2
16 018	Octantis	9.8	39.55	5.267	82 0 55.0	.90	51.31	2
16 019	Gould, Z. C., 22^h 1346 .	9.0	47.10	5.671	80 13 57.7	.90	50.83	3
16 020	Octantis	10.0	48.25	+ 4.794	75 26 14.7	+ 18.90	50.75	1
16 021	Gould, 31093	8.0	22 41 55.88	+ 4.111	66 44 24.9	+ 18.90	52.65	2
16 022	Indi	10.0	41 57.72	4.200	68 24 42.9	.90	51.80	1
16 023	Octantis	10.0	42 6.66	5.634	80 7 57.5	.91	50.82	1
16 024	Gould, Z. C., 22^h 1350 .	9.0	37.53	4.142	67 31 12.9	.92	51.82	3
16 025	Indi	10.0	49.04	+ 4.280	69 54 45.3	+ 18.93	51.74	1

Number.	Constellation, Name of Star, or Synonym.	Magnitude.	Right Ascension, 1850.0.	Annual Precession.	South Declination, 1850.0.	Annual Precession.	Mean year.	No. of obs.
			h. m. s.	s.	° ′ ″	″		
16026	Octantis	9.2	22 42 51.26	+ 6.177	81 54 19.4	+ 18.93	51.31	2
16027	Gould, Z. C., 22h 1360 .	8.5	52.95	4.632	74 12 13.7	.93	51.41	3
16028	Lacaille, 9262	7.5	42 53.61	4.799	75 39 37.9	.93	50.85	1
16029	Tucanæ	9.5	43 8.67	4.058	65 55 7.2	.94	52.67	1
16030	Octantis	10.0	17.08	+ 8.010	84 55 1.6	+ 18.94	50.79	1
16031	Indi	10.0	22 43 17.57	+ 4.376	71 22 55.9	+ 18.94	51.76	1
16032	Octantis	10.0	29.19	5.562	80 1 12.9	.95	50.82	1
16033	Gould, Z. C., 22h 1373 .	9.0	33.13	4.502	72 58 50.9	.95	51.77	2
16034	Octantis	10.0	35.59	7.984	84 54 34.3	.95	51.23	1
16035	Octantis	10.0	38.46	+ 4.734	75 15 23.3	+ 18.95	50.75	1
16036	Indi	10.0	22 43 39.58	+ 4.237	69 25 39.7	+ 18.95	51.76	2
16037	Indi	9.8	45.00	{4.356 / 4.361}	71 {12 53.4 / 17 9.0}	.96	51.78	2
16038	Indi	9.5	43 58.62	4.233	69 26 29.4	.96	51.76	2
16039	Gould, Z. C., 22h 1388 .	9.2	44 3.60	4.575	73 51 56.8	.97	51.76	2
16040	ρ Indi	6.8	8.49	+ 4.325	70 52 23.0	+ 18.97	51.77	2
16041	Gould, Z. C., 22h 1392 .	8.0	22 44 20.27	+ 4.039	65 52 8.5	+ 18.97	52.67	1
16042	Indi	10.0	21.94	5.423	79 33 37.9	.97	50.78	1
16043	Indi	10.0	23.46	4.147	68 3 47.2	.97	51.83	1
16044	Lacaille, 9273	8.0	25.34	4.782	75 47 31.4	.98	50.84	2
16045	Gould, 31132	8.5	27.17	+ 4.165	68 24 25.6	+ 18.98	51.79	1
16046	Indi	9.8	22 44 27.92	+ 4.233	69 33 59.8	+ 18.98	51.76	2
16047	Gould, Z. C., 22h 1408 .	9.5	34.78	5.037	77 35 51.4	.98	51.31	2
16048	Lacaille, 9279	8.5	39.10	4.252	69 53 59.9	.98	51.74	2
16049	Indi	9.5	44 42.47	4.208	69 12 51.3	.98	51.78	1
16050	Gould, 31142	7.5	45 0.38	+ 4.043	66 7 38.0	+ 18.99	52.67	1
16051	Indi	9.2	22 45 10.54	+ 4.461	72 45 51.6	+ 19.00	51.74	2
16052	Gould, Z. C., 22h 1415 .	9.0	20.42	4.056	66 30 36.3	.00	52.68	1
16053	Indi	10.0	23.37	4.139	68 9 48.1	.00	51.83	1
16054	Tucanæ	9.8	30.44	4.100	67 26 56.6	.01	51.82	3
16055	Gould, Z. C., 22h 1429 .	9.0	35.26	+ 4.854	76 32 18.5	+ 19.01	50.83	1
16056	Indi	10.0	22 45 36.89	+ 4.636	74 45 28.0	+ 19.01	51.13	3
16057	Indi	9.8	38.36	4.106	67 36 42.3	.01	51.82	2
16058	Gould, Z. C., 22h 1428 .	8.5	42.65	4.490	73 16 58.3	.01	51.78	1
16059	Gould, Z. C., 22h 1425 .	9.0	44.89	4.270	70 26 26.2	.01	51.77	1
16060	Gould, Z. C., 22h 1438 .	9.5	45 54.16	+ 4.850	76 33 39.7	+ 19.02	50.83	1
16061	Indi	10.0	22 46 2.05	+ 4.172	68 55 9.6	+ 19.02	51.79	1
16062	Gould, 31195	10.5	9.28	8.264	85 20 16.1	.02	51.38	5
16063	Lacaille, 9291	8.5	10.89	4.209	69 34 56.9	.03	51.76	2
16064	Gould, 31197	10.5	15.21	8.258	85 20 20.6	.03	51.38	5
16065	Gould, Z. C., 22h 1443	9.2	18.27	+ 4.249	70 15 8.5	+ 19.03	51.75	2

No. 16037. One of these observations is one wire interval wrong.

Number.	Constellation, Name of Star, or Synonym.	Magnitude.	Right Ascension, 1850.0.	Annual Precession.	South Declination, 1850.0.	Annual Precession.	Mean year.	No. of obs.
			h. m. s.	s.	° ′ ″	″		
16 066	Indi	10.0	22 46 21.34	+ 4.116	67 57 23.0	+ 19.03	51.83	1
16 067	Lacaille, 9260	8.2	46 53.54	6.755	83 30 15.9	.04	50.82	2
16 068	Indi	9.8	47 7.54	4.257	70 34 42.2	.05	51.77	2
16 069	Gould, Z. C., 22^h 1468	9.0	13.21	4.397	72 31 3.0	.05	51.73	1
16 070	Indi	9.0	16.51	+ 4.311	71 23 52.8	+ 19.05	51.77	3
16 071	Lacaille, 9293	8.5	22 47 18.32	+ 4.584	74 34 41.4	+ 19.06	51.46	3
16 072	Tucanæ	10.0	27.51	4.045	66 50 43.5	.06	52.62	1
16 073	Gould, Z. C., 22^h 1473	9.0	31.17	4.230	70 14 59.4	.06	51.75	2
16 074	Gould, Z. C., 22^h 1474	9.0	33.76	4.232	70 17 59.6	.06	51.75	2
16 075	Octantis	9.5	47 34.15	+ 5.121	78 32 3.9	+ 19.06	50.78	1
16 076	Gould, Z. C., 22^h 1484	8.0	22 48 6.84	+ 4.050	67 8 10.6	+ 19.08	52.22	2
16 077	Indi	10.0	21.02	4.365	72 21 21.4	.08	51.77	2
16 078	Octantis	10.0	27.86	4.808	76 41 46.6	.09	50.83	2
16 079	Gould, 31203	8.0	30.19	3.981	66 10 13.7	.09	52.67	1
16 080	Gould, Z. C., 22^h 1492	9.8	32.00	+ 4.707	75 55 1.8	+ 19.09	50.84	2
16 081	Gould, 31205	9.0	22 48 36.83	+ 4.005	66 19 4.4	+ 19.09	52.67	2
16 082	Gould, 31208	8.0	37.71	4.382	72 38 13.1	.09	51.74	2
16 083	Tucanæ	9.5	39.88	4.023	66 43 8.3	.09	52.68	1
16 084	Gould, Z. C., 22^h 1495	8.8	43.27	4.525	74 16 28.8	.09	51.29	2
16 085	Indi	10.0	47.41	+ 4.499	74 1 5.0	+ 19.10	51.71	1
16 086	Gould, Z. C., 22^h 1509	9.0	22 48 47.58	+ 5.319	79 41 24.8	+ 19.10	50.78	1
16 087	Indi	9.5	51.66	4.093	68 11 57.6	.10	51.83	1
16 088	Octantis	9.8	55.41	5.561	80 41 27.1	.10	51.21	2
16 089	Indi	10.0	48 55.88	4.127	68 51 11.0	.10	51.79	1
16 090	Gould, Z. C., 22^h 1499	9.2	49 2.96	+ 4.133	69 0 30.3	+ 19.10	51.79	2
16 091	Indi	9.0	22 49 3.48	+ 4.098	68 20 42.2	+ 19.10	51.81	2
16 092	Indi	10.0	37.87	4.456	73 37 3.1	.12	51.81	1
16 093	Gould, 31322	9.0	41.55	17.261	88 22 11.9	.12	51.54	12
16 094	Tucanæ	9.0	41.89	3.971	65 49 50.1	.12	52.67	1
16 095	Indi	10.0	44.09	+ 4.135	69 13 25.1	+ 19.12	51.78	1
16 096	Gould, Z. C., 22^h 1534	9.5	22 49 48.52	+ 4.723	76 16 34.0	+ 19.12	50.83	2
16 097	Tucanæ	10.0	47.77	4.032	67 13 37.8	.12	51.82	1
16 098	Indi	10.0	58.06	4.122	69 2 24.6	.13	51.79	2
16 099	Tucanæ	10.0	49 59.55	4.025	67 7 26.7	.13	51.82	1
16 100	Tucanæ	10.0	50 0.62	+ 4.033	67 18 10.6	+ 19.13	51.82	1
16 101	Gould, Z. C., 22^h 1536	8.8	22 50 6.11	+ 4.047	67 37 57.6	+ 19.13	51.82	2
16 102	Tucanæ	9.8	25.86	4.017	67 5 32.6	.14	52.22	2
16 103	Indi	10.0	29.60	4.076	68 18 54.3	.14	51.81	2
16 104	Octantis	8.5	35.51	5.355	80 5 30.6	.14	50.82	1
16 105	Gould, Z. C., 22^h 1553	9.2	44.55	+ 4.612	75 30 58.5	+ 19.15	50.80	2

No. 16 088. Mean of two declinations differing 10.4″.

Number.	Constellation, Name of Star, or Synonym.	Magnitude.	Right Ascension, 1850.0.	Annual Precession.	South Declination, 1850.0.	Annual Precession.	Mean year.	No. of obs.
			h. m. s.	s.	° ′ ″	″		
16 106	Gould, 31281	9.7	22 50 45.10	+ 8.347	85 41 10.9	+ 19.15	51.38	10
16 107	Gould, Z. C., 22h 1551 .	8.5	50 52.95	3.984	66 28 2.4	.15	52.68	1
16 108	Octantis	9.0	51 19.38	4.870	77 37 17.2	.16	51.31	2
16 109	Tucanæ	9.5	39.18	3.997	66 59 28.6	.17	52.62	1
16 110	Octantis	9.5	42.64	+ 5.242	79 44 46.8	+ 19.17	50.78	1
16 111	Gould, Z. C., 22h 1580 .	9.0	22 51 44.28	+ 4.341	72 49 16.1	+ 19.17	51.75	1
16 112	Gould, Z. C., 22h 1583 .	9.0	48.76	4.239	71 25 47.7	.18	51.77	3
16 113	Gould, 31266	9.5	53.44	4.026	67 40 54.4	.18	51.82	2
16 114	Gould, 31270	11.0	51 58.86	4.024	67 40 34.2	.18	51.83	1
16 115	Indi	9.0	52 0.35	+ 4.175	70 29 8.1	+ 19.18	51.77	1
16 116	Indi	10.0	22 52 8.17	+ 4.027	67 46 40.0	+ 19.18	51.83	1
16 117	Gould, 31275	8.2	16.58	3.981	66 49 6.0	.19	52.65	2
16 118	Indi	9.8	21.66	4.162	70 21 49.6	.19	51.75	2
16 119	Indi	9.8	23.58	4.175	70 35 10.4	.19	51.77	2
16 120	Octantis	9.0	31.20	+ 4.557	75 21 48.7	+ 19.19	50.75	1
16 121	Tucanæ	9.5	22 52 32.60	+ 3.860	65 50 7.5	+ 19.19	52.67	1
16 122	Octantis	10.0	52 54.74	5.250	79 57 14.2	.20	50.78	1
16 123	Octantis	10.0	53 8.64	5.121	79 22 12.2	.21	50.72	1
16 124	Indi	10.0	15.94	4.345	73 13 17.9	.21	51.78	1
16 125	Gould, Z. C., 22h 1631 .	8.0	17.93	+ 4.837	77 44 7.6	+ 19.21	51.81	1
16 126	Octantis	9.0	22 53 19.00	+ 5.127	79 25 40.8	+ 19.21	50.75	2
16 127	Tucanæ	9.0	21.03	3.927	65 51 18.6	.21	52.67	1
16 128	Octantis	10.0	27.52	5.795	81 59 44.6	.22	51.31	2
16 129	Indi	10.0	32.94	4.116	69 51 14.4	.22	51.74	2
16 130	Indi	11.0	38.98	+ 4.338	73 13 42.7	+ 19.22	51.78	1
16 131	Indi	10.0	22 53 45.39	+ 4.044	68 35 23.7	+ 19.22	51.80	1
16 132	Octantis	10.0	53 57.12	6.086	82 48 47.5	.23	51.71	1
16 133	Indi	9.0	54 8.16	4.067	69 9 13.6	.23	51.79	2
16 134	Indi	10.0	9.19	4.202	71 28 13.8	.23	51.80	1
16 135	Tucanæ	9.5	9.91	+ 3.961	66 54 51.5	+ 19.23	52.62	1
16 136	Indi	10.0	22 54 25.26	+ 4.273	72 33 46.9	+ 19.24	51.73	1
16 137	Octantis	9.5	40.50	5.771	82 4 8.6	.25	50.81	1
16 138	Indi	9.8	41.78	4.061	69 11 8.7	.25	51.79	2
16 139	Gould, 31343	10.0	50.71	6.216	83 11 55.3	.25	51.71	1
16 140	Lacaille, 9337	6.2	53.44	+ 4.081	69 37 45.3	+ 19.25	51.76	2
16 141	Gould, 31347	10.0	22 54 56.67	+ 6.218	83 12 41.1	+ 19.25	51.71	1
16 142	Indi	9.8	54 57.81	4.032	68 42 29.1	.25	51.79	2
16 143	Octantis	9.0	55 5.72	4.524	75 34 23.3	.26	50.80	2
16 144	Gould, Z. C., 22h 1669 .	9.2	6.08	3.994	67 56 11.8	.26	51.83	2
16 145	Octantis	9.5	13.60	+ 5.843	82 21 10.0	+ 19.26	50.82	2

Number.	Constellation, Name of Star, or Synonym.	Magnitude.	Right Ascension, 1850.0.	Annual Precession.	South Declination, 1850.0.	Annual Precession.	Mean year.	No. of obs.
			h. m. s.	s.	° ′ ″	″		
16 146	Octantis	10.0	22 55 18.94	+10.239	87 1 32.6	19.26	51.87	2
16 147	Indi	10.0	20.06	4.171	71 17 40.0	.26	51.78	2
16 148	Indi	9.8	20.38	4.005	68 15 17.2	.26	51.81	2
16 149	Gould, Z. C., 22ʰ 1697	9.5	33.21	5.001	79 6 51.8	.27	50.78	2
16 150	Octantis	9.2	40.09	+ 7.124	84 46 35.8	+ 19.27	51.17	4
16 151	Octantis	10.0	22 55 41.18	+ 5.217	80 12 32.5	+ 19.27	50.81	2
16 152	Gould, Z. C., 22ʰ 1704	8.5	41.78	5.063	79 28 7.7	.27	50.75	2
16 153	Octantis	9.0	44.71	5.413	81 1 5.1	.27	51.21	2
16 154	Gould, Z. C., 22ʰ 1696	9.2	45.94	4.343	73 46 46.6	.28	51.18	3
16 155	Lacaille, 9332	6.0	55 57.32	+ 5.227	80 17 16.7	+ 19.28	50.81	2
16 156	Tucanæ	10.0	22 56 0.60	+ 3.928	66 42 19.3	+ 19.28	52.68	1
16 157	Gould, Z. C., 22ʰ 1706	9.2	4.74	4.159	71 17 44.7	.28	51.78	2
16 158	Gould, 31376	9.6	11.16	8.102	85 49 15.8	.28	51.38	10
16 159	Indi	10.0	20.15	4.020	68 51 6.4	.29	51.79	1
16 160	Gould, Z. C., 22ʰ 1712	9.0	20.92	+ 3.901	66 8 52.4	+ 19.29	52.67	1
16 161	Octantis	9.5	22 56 28.24	+ 4.521	75 49 33.9	+ 19.29	50.84	2
16 162	Tucanæ	10.0	41.33	3.940	67 13 10.1	.30	51.82	1
16 163	Indi	10.0	56 57.85	4.412	74 50 27.9	.30	50.86	1
16 164	Tucanæ	10.0	57 2.44	3.937	67 15 1.1	.30	51.82	1
16 165	Gould, Z. C., 22ʰ 1735	9.5	3.22	+ 4.096	70 29 49.6	+ 19.30	51.77	1
16 166	Tucanæ	9.5	22 57 8.93	+ 3.915	66 44 22.4	+ 19.31	52.68	1
16 167	Indi	9.8	24.05	4.153	71 33 28.2	.31	51.77	2
16 168	Octantis	9.8	28.66	5.497	81 32 59.3	.31	51.30	2
16 169	Octantis	10.0	44.55	6.254	83 34 1.8	.32	50.82	2
16 170	Indi	10.0	49.44	+ 3.978	68 26 17.6	+ 19.32	51.80	1
16 171	Octantis	10.0	22 57 54.73	+ 5.890	82 45 47.5	+ 19.33	51.71	1
16 172	Gould, Z. C., 22ʰ 1753	9.0	57 58.54	4.101	70 50 56.2	.33	51.77	2
16 173	Lacaille, 9358	6.8	58 3.05	4.350	74 23 48.2	.33	50.87	2
16 174	Indi	10.0	5.05	4.082	70 32 27.3	.33	51.77	1
16 175	Tucanæ	9.0	5.68	+ 3.931	67 27 39.0	+ 19.33	51.82	2
16 176	Octantis	10.0	22 58 7.76	+ 4.576	76 39 32.9	+ 19.33	50.83	2
16 177	Gould, Z. C., 23ʰ 10	8.5	9.44	5.009	79 34 26.4	.33	50.75	2
16 178	Gould, Z. C., 23ʰ 6	8.8	9.70	4.580	76 42 26.8	.33	50.83	2
16 179	Gould, 31413	10.0	26.87	7.300	85 12 14.0	.34	51.33	2
16 180	Octantis	9.5	38.09	+ 4.792	78 22 55.9	+ 19.34	50.78	1
16 181	Indi	9.7	22 58 40.69	+ 4.199	72 35 23.5	+ 19.34	51.74	3
16 182	Indi	10.0	43.89	3.974	68 37 53.2	.34	51.79	2
16 183	Tucanæ	10.0	58 46.32	3.923	67 28 16.8	.34	51.83	1
16 184	Indi	10.0	59 8.39	3.994	69 10 4.5	.35	51.78	1
16 185	Lacaille, 9355	6.5	22.34	+ 5.476	81 43 30.7	+ 19.36	51.81	1

No. 16 151. One record ambiguous; declination possibly 80° 12′ 10.3″.

Number.	Constellation, Name of Star, or Synonym.	Magnitude.	Right Ascension, 1850.0.	Annual Precession.	South Declination, 1850.0.	Annual Precession.	Mean year.	No. of obs.
			h. m. s.	s.	° ′ ″	″		
16 186	Gould, Z. C., 23^h 28 . .	8.8	22 59 24.77	+ 3.896	67 1 28.9	+ 19.36	52.22	2
16 187	Octantis	9.8	25.74	4.417	75 25 14.4	.36	50.80	2
16 188	Octantis	10.0	28.10	4.777	78 25 41.0	.36	50.78	1
16 189	Lacaille, 9362	8.0	42.86	4.611	77 16 33.1	.37	50.81	1
16 190	Octantis	10.0	49.51	+ 4.658	77 39 16.8	+ 19.37	51.81	1
16 191	Gould, Z. C., 23^h 38 . .	9.8	22 59 52.58	+ 4.109	71 30 56.7	+ 19.37	51.76	2
16 192	Tucanæ	9.5	22 59 57.94	3.901	67 19 29.7	.37	51.82	1
16 193	Lacaille, 9374	7.0	23 0 4.08	3.958	68 41 7.8	.37	51.79	2
16 194	Gould, Z. C., 23^h 45. . .	8.0	6.00	3.897	67 17 1.6	.38	51.82	1
16 195	Gould, Z. C., 23^h 47 . .	9.0	9.66	+ 3.950	68 32 8.0	+ 19.38	51.80	1
16 196	Lacaille, 9375	6.5	23 0 32.47	+ 3.907	67 40 14.3	+ 19.38	51.83	1
16 197	Indi	9.5	33.37	3.999	69 39 51.0	.39	51.74	1
16 198	Octantis	9.5	34.82	4.771	78 35 47.8	.39	50.80	2
16 199	Indi	10.0	42.33	4.103	71 39 19.8	.39	51.73	1
16 200	Gould, Z. C., 23^h 62 . .	9.0	49.35	+ 3.881	67 7 0.5	+ 19.39	52.62	1
16 201	Tucanæ	9.5	23 0 51.05	+ 3.844	66 9 7.6	+ 19.39	52.67	1
16 202	Tucanæ	9.5	55.15	3.864	66 42 42.8	.39	52.65	2
16 203	Gould, 31440	9.8	0 57.07	5.338	81 26 54.4	.39	51.30	2
16 204	Octantis	11.0	1 6.33	14.112	88 14 13.1	.40	51.64	3
16 205	Gould, Z. C., 23^h 72 . .	9.0	10.85	+ 4.206	73 20 17.9	+ 19.40	51.78	1
16 206	Indi	9.2	23 1 12.18	+ 3.996	69 50 36.3	+ 19.40	51.74	2
16 207	Indi	10.0	25.61	4.310	74 43 56.4	.40	50.86	2
16 208	Indi	10.0	39.56	3.933	68 58 57.7	.41	51.78	1
16 209	Gould, Z. C., 23^h 100 . .	8.5	51.86	4.824	79 9 29.1	.41	50.78	2
16 210	Gould, Z. C, 23^h 94 . .	9.0	1 57.28	+ 4.184	73 14 9.0	+ 19.42	51.78	1
16 211	Indi	10.0	23 2 5.49	+ 4.037	70 53 40.7	+ 19.42	51.76	1
16 212	Gould, Z. C., 23^h 107 . .	10.0	7.43	4.828	79 13 34.7	.42	50.72	1
16 213	Octantis	9.0	13.46	4.654	78 5 31.9	.42	50.79	1
16 214	Indi	10.0	14.38	4.079	71 41 20.3	.42	51.73	1
16 215	Indi	8.8	22.80	+ 4.034	70 55 36.4	+ 19.43	51.77	2
16 216	Tucanæ	9.5	23 2 34.88	+ 3.829	66 21 17.5	+ 19.43	52.68	2
16 217	τ Octantis	5.6	44.77	14.231	88 18 11.5	.43	51.54	14
16 218	Octantis	10.0	45.74	4.314	75 8 20.8	.43	50.75	1
16 219	Indi	10.0	47.49	4.069	71 40 34.5	.43	51.73	1
16 220	Indi	10.0	49.49	+ 4.249	74 20 19.6	+ 19.44	50.87	1
16 221	Indi	10.0	23 2 51.64	+ 4.011	70 38 28.8	+ 19.44	51.77	1
16 222	Indi	10.0	3 3.65	3.948	69 26 4.9	.44	51.74	1
16 223	Octantis	9.5	12.45	4.580	77 43 55.4	.45	51.81	1
16 224	Gould, Z. C., 23^h 130 . .	9.2	15.22	4.775	79 6 30.2	.45	50.78	2
16 225	Indi	10.0	18.00	+ 3.859	69 44 14.8	+ 19.45	51.74	1

Number.	Constellation, Name of Star, or Synonym.	Magnitude.	Right Ascension, 1850.0.	Annual Precession.	South Declination, 1850.0.	Annual Precession.	Mean year.	No. of obs.
			h. m. s.	s.	° ′ ″	″		
16 226	Octantis	10.0	23 3 23.60	+ 4.545	77 29 18.8	+ 19.45	51.31	2
16 227	Gould, Z. C., 23^h 133 . .	9.2	29.83	4.774	79 9 3.4	.45	50.78	2
16 228	Octantis	8.8	34.46	5.272	81 34 29.4	.45	51.30	2
16 229	Gould, Z. C., 23^h 138 . .	8.8	3 56.97	4.198	73 57 53.8	.46	50.87	2
16 230	Lacaille, 9378	8.0	4 5.38	+ 5.190	81 19 52.3	+ 19.46	50.84	2
16 231	Octantis	10.0	23 4 7.34	+ 5.228	81 29 7.4	+ 19.46	51.30	2
16 232	Indi	10.0	7.38	4.098	72 32 9.3	.46	51.75	1
16 233	Indi	9.0	7.96	4.066	72 1 28.7	.46	51.81	1
16 234	Octantis	10.0	16.27	4.401	76 23 16.1	.47	50.84	1
16 235	Gould, 31486 .	8.2	17.26	+ 3.915	69 6 21.4	+ 19.47	51.79	2
16 236	Lacaille, 9390	7.5	23 4 17.84	+ 4.000	70 52 32.6	+ 19.47	51.77	2
16 237	Gould, Z. C., 23^h 156 . .	9.0	19.38	4.203	74 7 45.8	.47	50.87	1
16 238	Gould, Z. C., 23^h 164 . .	9.0	26.32	4.680	78 42 33.7	.47	50.80	2
16 239	Indi	10.0	26.58	4.070	72 10 8.1	.47	51.81	1
16 240	Gould, 31491	9.0	29.44	+ 4.011	71 9 11.2	+ 19.47	51.78	3
16 241	Tucanæ	9.0	23 4 31.78	+ 3.808	66 28 20.2	+ 19.47	52.68	1
16 242	Octantis	9.5	46.98	4.262	75 0 0.5	.48	50.80	2
16 243	Octantis	9.0	4 58.76	4.294	75 20 25.6	.48	50.75	1
16 244	Gould, Z. C., 23^h 181 . .	8.5	5 17.00	3.818	67 1 52.8	.49	52.22	2
16 245	Stone, 12057	8.8	23.90	+ 5.112	81 12 34.3	+ 19.49	51.10	3
16 246	Octantis	9.8	23 5 28.16	+ 6.979	85 23 17.4	+ 19.49	51.84	7
16 247	Gould, 31509	8.2	29.06	3.793	66 23 46.9	.49	52.68	2
16 248	Indi	10.0	31.81	4.095	72 53 44.8	.49	51.75	1
16 249	Gould, 31575	10.0	32.36	14.406	88 24 33.3	.49	51.61	5
16 250	Gould, Z. C., 23^h 193 . .	9.0	40.42	+ 4.219	74 41 31.4	+ 19.50	50.86	2
16 251	Octantis	9.5	23 5 40.64	+ 4.363	76 20 22.4	+ 19.50	50.83	2
16 252	Gould, Z. C., 23^h 191 . .	9.0	41.39	4.009	71 38 37.6	.50	51.73	1
16 253	Lacaille, 9389	8.2	47.14	4.896	80 15 3.5	.50	50.81	2
16 254	Lacaille, 9392	8.5	5 52.90	5.102	81 14 30.9	.50	51.10	3
16 255	Indi	10.0	6 4.26	+ 3.857	68 20 30.8	+ 19.50	51.81	2
16 256	Indi	9.8	23 6 5.28	+ 3.851	68 11 47.4	+ 19.50	51.82	2
16 257	Tucanæ	10.0	8.28	3.813	67 12 51.6	.51	51.82	1
16 258	Octantis	9.2	11.53	4.353	76 21 0.2	.51	50.83	2
16 259	Gould, 31520	9.0	14.16	3.830	67 42 19.2	.51	51.83	1
16 260	Gould, Z. C., 23^h 209 . .	8.5	17.26	+ 3.840	67 59 16.7	+ 19.51	51.82	3
16 261	Octantis	9.0	23 6 18.34	+ 4.502	77 45 2.7	+ 19.51	51.81	1
16 262	Indi	9.5	22.41	3.822	67 32 33.4	.51	51.83	1
16 263	Octantis	9.0	26.66	4.232	75 3 14.0	.51	50.75	1
16 264	Indi	10.0	28.20	3.892	69 19 5.6	.51	51.78	1
16 265	Gould, Z. C., 23^h 216 . .	9.2	28.34	+ 3.835	67 56 20.8	+ 19.51	51.82	3

Number.	Constellation, Name of Star, or Synonym.	Magnitude.	Right Ascension, 1850.0.	Annual Precession.	South Declination, 1850.0.	Annual Precession.	Mean year.	No. of obs.
			h. m. s.	s.	° ′ ″	″		
16 266	Indi	9.0	23 6 30.76	+ 3.956	70 43 14.4	+ 19.51	51.77	2
16 267	Gould, Z. C., 23h 218	9.0	31.50	3.768	66 4 28.8	.51	52.67	1
16 268	Indi	9.5	33.77	4.079	72 57 14.7	.51	51.75	1
16 269	Tucanæ	9.5	35.61	3.780	66 26 38.7	.52	52.68	1
16 270	Octantis	10.0	41.07	+ 4.535	78 5 58.3	+ 19.52	50.79	1
16 271	Lacaille, 9402	9.0	23 6 56.36	+ 3.966	71 4 4.4	+ 19.52	51.79	2
16 272	Octantis	9.5	23 7 21.83	5.457	82 43 35.6	.53	51.26	2
16 273	Tucanæ	10.0	34.12	3.780	66 49 8.9	.53	52.62	1
16 274	Indi	10.0	36.86	3.889	69 39 14.2	.53	51.74	1
16 275	Gould, Z. C., 23h 253	8.5	37.19	+ 3.766	66 32 5.3	+ 19.53	52.68	1
16 276	Lacaille, 9399	7.2	23 7 39.22	+ 4.840	80 17 30.5	+ 19.54	50.81	2
16 277	Gould, Z. C., 23h 256	9.2	43.07	3.785	67 1 50.0	.54	52.22	2
16 278	Indi	9.5	43.85	3.807	67 39 34.3	.54	51.83	1
16 279	Indi	9.0	43.96	4.091	73 28 36.7	.54	51.78	1
16 280	Gould, Z. C., 23h 259	9.0	48.78	+ 3.809	67 44 20.2	+ 19.54	51.83	1
16 281	Gould, Z. C., 23h 261	9.0	23 7 52.75	+ 3.809	67 43 11.3	+ 19.54	51.83	1
16 282	Indi	9.5	23 8 4.11	3.802	67 38 44.3	.54	51.83	1
16 283	Gould, 31553	10.0	14.50	3.876	69 34 54.4	.55	51.76	2
16 284	Gould, 31556	9.5	16.21	3.902	70 10 48.1	.55	51.74	1
16 285	Lacaille, 9408	7.8	16.52	+ 4.215	75 20 3.5	+ 19.55	50.75	1
16 286	Gould, 31557	9.3	23 8 18.33	+ 3.874	69 34 0.8	+ 19.55	51.77	3
16 287	Tucanæ	9.5	24.42	3.742	65 58 53.2	.55	52.67	1
16 288	Indi	10.0	26.97	3.899	70 10 9.0	.55	51.74	1
16 289	Octantis	10.0	32.05	6.893	85 26 28.0	.55	51.83	1
16 290	Octantis	9.5	34.32	+ 4.524	78 25 28.0	+ 19.55	50.78	1
16 291	Indi	10.0	23 8 38.96	+ 4.036	72 52 57.9	+ 19.56	51.75	1
16 292	Octantis	9.4	45.32	3.701	85 19 46.0	.56	51.70	7
16 293	Gould, Z. C., 23h 290	9.5	54.15	3.732	65 52 31.2	.56	52.67	1
16 294	Lacaille, 9418	7.0	8 59.46	3.813	68 17 25.6	.56	51.80	2
16 295	Gould, 31576	8.0	9 35.60	+ 3.731	66 7 16.2	+ 19.57	52.67	1
16 296	Octantis	9.1	23 9 51.18	+ 8.664	87 2 33.1	+ 19.58	51.86	6
16 297	Gould, Z. C., 23h 314	9.2	51.64	3.921	71 8 26.0	.58	51.78	2
16 298	Gould, Z. C., 23h 317	9.5	52.36	3.956	71 49 25.0	.58	51.77	2
16 299	Gould, Z. C., 23h 318	9.0	9 57.84	4.032	73 12 48.5	.58	51.78	1
16 300	Indi	9.8	10 17.22	+ 3.841	69 29 56.2	+ 19.59	51.80	3
16 301	Lacaille, 9427	8.0	23 10 22.29	+ 4.264	76 27 12.6	+ 19.59	50.83	1
16 302	Indi	10.0	34.82	4.088	74 16 39.6	.59	50.86	1
16 303	Gould, Z. C., 23h 336	9.0	50.90	3.741	66 57 11.6	.60	52.25	2
16 304	Gould, Z. C., 23h 339	9.0	10 50.90	3.781	68 8 36.0	.60	51.81	1
16 305	Tucanæ	9.8	11 26.16	+ 3.725	66 43 3.7	+ 19.61	52.40	3

No. 16 281. Gould's declination is 9.6″ greater.

No. 16 289. Record confused.

Number.	Constellation, Name of Star, Synonym.	Magnitude.	Right Ascension, 1850.0.	Annual Precession.	South Declination. 1850.0.	Annual Precession.	Mean year.	No. of obs.
			h. m. s.	s.	° ′ ″	″		
16 306	Gould, Z. C., 23ʰ 361	9.0	23 11 27.26	+ 3.848	70 5 20.9	+ 19.61	51.74	1
16 307	Gould, 31626	9.0	12 0.28	3.948	72 23 40.9	.62	51.75	1
16 308	Gould, Z. C., 23ʰ 370	9.0	1.24	4.076	74 31 48.3	.62	50.86	1
16 309	Octantis	9.5	1.64	— 4.393	78 7 33.2	.62	50.79	1
16 310	Tucanæ	9.5	2.38	+ 3.697	66 1 33.2	+ 19.62	52.67	1
16 311	Gould, Z. C., 23ʰ 375	9.0	23 12 5.74	+ 4.206	74 59 13.3	+ 19.62	50.80	2
16 312	Octantis	9.9	9.21	5.884	84 22 20.3	.62	50.82	4
16 313	Indi	10.0	27.64	3.850	70 31 21.4	.63	51.77	1
16 314	Octantis	10.0	28.91	4.295	77 18 58.4	.63	50.81	1
16 315	Lacaille, 9434	8.5	29.29	+ 4.438	78 36 33.6	+ 19.63	50.80	2
16 316	Octantis	10.5	23 12 31.77	+ 4.145	75 37 53.7	+ 19.63	50.85	1
16 317	Octantis	10.0	32.91	4.112	75 11 59.3	.63	50.75	1
16 318	Octantis	9.5	34.31	5.150	82 27 15.8	.63	50.83	1
16 319	Octantis	9.5	12 40.21	4.209	76 26 22.5	.63	50.83	1
16 320	Octantis	9.5	13 1.57	+ 4.471	78 59 35.4	+ 19.64	50.78	2
16 321	Lacaille, 9401	7.7	23 13 17.67	+ 7.539	86 31 59.1	+ 19.64	51.87	7
16 322	Gould, Z. C., 23ʰ 398	8.0	22.56	3.681	66 4 59.8	.64	52.67	1
16 323	Octantis	11.0	26.29	4.128	75 40 1.2	.64	50.85	1
16 324	Gould, Z. C., 23ʰ 411	8.8	48.56	4.701	66 57 52.7	.65	52.25	2
16 325	Octantis	9.0	13 53.64	+ 4.824	81 19 55.6	+ 19.65	50.80	1
16 326	Indi	9.5	23 14 10.28	+ 3.956	73 17 20.7	+ 19.66	51.78	1
16 327	Octantis	9.5	10.28	4.986	82 6 6.4	.66	50.81	1
16 328	Octantis	9.8	10.56	4.704	80 45 32.2	.66	51.21	2
16 329	Indi	10.0	25.37	3.920	72 42 26.7	.66	51.75	1
16 330	Gould, Z. C., 23ʰ 438	9.2	39.76	+ 4.000	74 11 13.0	+ 19.67	50.87	2
16 331	Octantis	10.0	23 14 41.18	+ 4.067	75 13 6.0	+ 19.67	50.75	1
16 332	Lacaille, 9450	7.0	45.92	3.929	72 59 8.9	.67	51.77	2
16 333	Octantis	10.0	48.45	4.072	75 19 22.4	.67	50.78	1
16 334	Indi	10.0	51.03	3.858	71 35 1.3	.67	51.73	1
16 335	Octantis	10.0	14 59.74	+ 4.146	76 21 28.3	+ 19.67	50.84	1
16 336	Gould, Z. C., 23ʰ 457	8.5	23 15 4.68	+ 4.193	76 55 37.2	+ 19.67	50.82	1
16 337	Indi	9.8	11.86	3.769	69 34 54.2	.67	51.81	2
16 338	Indi	10.0	12.37	3.730	68 28 24.8	.67	51.80	1
16 339	Indi	9.5	16.02	3.755	69 12 50.8	.68	51.78	1
16 340	Octantis	9.5	19.91	+ 5.172	82 58 10.3	+ 19.68	51.71	1
16 341	Gould, Z. C., 23ʰ 460	8.5	23 15 20.34	+ 3.891	72 27 1.0	+ 19.68	51.75	1
16 342	Gould, 31692	7.5	26.67	3.978	74 5 50.0	.68	50.87	1
16 343	Indi	[illegible].2	15 35.18	4.505	79 48 55.0	.68	50.80	2
16 344	Gould, 31704	9.5	16 0.06	3.845	71 43 59.7	.69	51.73	1
16 345	Octantis	9.0	4.26	+ 4.706	81 8 29.6	+ 19.69	51.21	2

Number.	Constellation, Name of Star, or Synonym.	Magnitude.	Right Ascension, 1850.0.	Annual Precession.	South Declination, 1850.0.	Annual Precession.	Mean year.	No. of obs.
			h. m. s.	s.	° ′ ″	″		
16 346	Gould, Z. C., 23h 485 . .	9.5	23 16 5.33	+ 3.748	69 21 47.2	+ 19.69	51.78	1
16 347	Tucanæ	10.0	19.91	3.671	67 5 14.3	.69	52.62	1
16 348	Indi	10.0	22.54	3.997	74 42 42.0	.69	50.86	1
16 349	Octantis	9.0	37.99	4.451	79 39 50.2	.70	50.78	1
16 350	Octantis	9.5	50.25	+ 4.124	77 36 52.2	+ 19.70	51.81	1
16 351	Indi	9.5	23 16 59.07	+ 3.867	72 35 10.6	+ 19.70	51.75	1
16 352	Lacaille, 9459	8.2	17 15.96	3.975	74 39 39.4	.70	50.86	2
16 353	Gould, Z. C., 23h 507 . .	9.5	17.74	3.712	68 51 28.8	.71	51.79	1
16 354	Gould, Z. C., 23h 511 . .	10.0	24.55	3.800	71 15 40.4	.71	51.80	1
16 355	Indi	10.0	31.74	+ 3.741	69 47 41.1	+ 19.71	51.74	1
16 356	Gould, Z. C., 23h 519 . .	9.2	23 17 32.20	+ 4.126	76 50 39.5	+ 19.71	50.85	2
16 357	Octantis	9.0	35.37	4.231	78 1 11.9	.71	50.79	1
16 358	Indi	10.0	39.72	3.713	69 2 6.3	.72	51.79	1
16 359	Indi	8.0	39.98	3.912	73 42 22.0	.72	50.87	1
16 360	Gould, Z. C., 23h 522 . .	9.0	44.16	+ 4.073	76 14 42.9	+ 19.72	50.83	2
16 361	Octantis	9.0	23 17 55.74	+ 5.165	83 21 0.7	+ 19.72	50.84	1
16 362	Gould, Z. C., 23h 525 . .	9.2	17 56.78	3.789	71 12 30.8	.72	51.79	2
16 363	Octantis	9.0	18 0.02	4.198	77 47 22.2	.72	51.30	2
16 364	Octantis	9.2	2.54	4.985	82 44 59.6	.72	51.27	2
16 365	Lacaille, 9469	6.2	10.70	+ 3.655	67 24 17.4	+ 19.72	51.82	2
16 366	Indi	10.0	23 18 33.64	+ 3.722	69 43 0.8	+ 19.73	51.81	2
16 367	Gould, Z. C., 23h 555 . .	9.5	40.57	4.048	76 11 58.5	.73	50.83	2
16 368	Gould, Z. C., 23h 553 . .	8.8	49.20	3.619	66 23 23.0	.73	52.67	3
16 369	Indi	9.0	18 58.03	3.909	74 6 15.1	.74	50.87	1
16 370	Gould, 31758	8.0	19 8.30	+ 3.650	67 44 0.4	+ 19.74	51.83	1
16 371	Indi	10.0	23 19 35.34	+ 3.759	71 9 51.5	+ 19.75	51.79	2
16 372	Octantis	9.0	38.22	4.068	76 46 6.0	.75	50.85	3
16 373	Gould, Z. C., 23h 575 . .	8.8	45.78	3.620	66 54 38.2	.75	52.25	2
16 374	Octantis	10.0	48.40	4.081	76 59 7.1	.75	50.81	1
16 375	Gould, Z. C., 23h 582 . .	8.5	52.07	+ 4.093	77 9 0.3	+ 19.75	50.85	2
16 376	Octantis	9.2	23 19 54.10	+ 4.315	79 23 52.2	+ 19.75	51.30	2
16 377	Gould, Z. C., 23h 583 . .	9.0	19 58.02	3.856	73 29 19.2	.75	51.14	3
16 378	Octantis	9.8	20 21.54	3.990	75 55 42.2	.76	50.84	2
16 379	Gould, Z. C., 23h 597 .	9.5	25.27	3.642	68 2 34.4	.76	51.81	1
16 380	Gould, Z. C., 23h 603 . .	9.5	34.92	+ 3.797	72 29 17.1	+ 19.76	51.75	1
16 381	Octantis	9.8	23 20 45.90	+ 4.044	76 49 11.0	+ 19.76	50.85	2
16 382	Tucanæ	9.8	20 50.30	3.600	66 44 44.2	.76	52.28	2
16 383	Tucanæ	11.0	21 5.18	3.714	70 36 20.2	.77	51.77	1
16 384	Lacaille, 9475	7.5	7.47	4.611	81 39 19.4	.77	51.81	1
16 385	Tucanæ	10.0	12.74	+ 3.711	70 34 58.6	+ 19.77	51.77	1

No. 16 366. Declination possibly 69° 42′ 46.8″.

Number.	Constellation, Name of Star, or Synonym.	Magnitude.	Right Ascension, 1850.0.	Annual Precession.	South Declination, 1850.0.	Annual Precession.	Mean year.	No. of obs.
			h. m. s.	s.	° ′ ″	″		
16386	Lacaille, 9487	8.2	23 21 36.93	+ 3.901	74 57 42.4	+ 19.78	50.80	2
16387	Octantis	9.8	42.93	4.639	81 55 {49.8 / 30.7}	.78	51.31	2
16388	Octantis	9.0	52.93	4.970	83 20 49.7	.78	50.84	1
16389	Tucanæ	9.5	55.55	− 4.797	73 2 17.4	.78	51.78	1
16390	Gould, 31799	9.8	56.66	+ 3.676	69 54 26.6	+ 19.78	51.82	3
16391	Lacaille, 9492	7.5	23 21 58.71	+ 3.675	69 53 53.9	+ 19.78	51.79	3
16392	Tucanæ	10.0	22 2.95	3.714	71 3 51.6	.78	51.80	1
16393	Octantis	9.0	23.62	4.542	81 32 55.2	.79	51.30	2
16394	Tucanæ	9.8	31.82	4.565	66 11 5.0	.79	52.67	2
16395	Gould, Z. C., 23^h 652	9.5	37.50	+ 3.674	70 10 2.5	+ 19.79	51.74	1
16396	Gould, 31817	10.0	23 22 41.86	+ 3.585	67 6 19.0	+ 19.79	51.85	2
16397	Octantis	9.0	50.05	4.267	79 45 46.3	.79	50.78	1
16398	Lacaille, 9493	8.2	57.84	3.876	75 1 39.4	.80	50.80	2
16399	Tucanæ	10.0	22 58.22	3.625	68 44 55.2	.80	51.79	2
16400	Tucanæ	9.8	23 0.17	+ 3.560	66 14 36.6	+ 19.80	52.67	2
16401	Lacaille, 9494	6.5	23 23 30.46	+ 4.088	78 12 48.5	+ 19.80	50.78	2
16402	Octantis	9.5	36.02	4.836	83 9 55.0	.80	51.28	2
16403	Gould, Z. C., 23^h 680	10.0	38.32	3.676	70 44 3.3	.80	51.77	1
16404	Octantis	9.0	47.36	4.230	79 42 30.4	.81	50.78	1
16405	Octantis	9.1	55.67	+ 16.641	89 6 54.2	+ 19.81	51.86	6
16406	Gould, Z. C., 23^h 694	9.0	23 23 59.95	+ 3.891	75 40 56.8	+ 19.81	50.85	1
16407	Octantis	9.0	24 0.96	4.381	80 55 36.2	.81	51.73	2
16408	Lacaille, 9464	8.2	1.49	7.387	87 13 37.7	.81	51.87	7
16409	Octantis	10.0	2.54	4.300	80 20 37.6	.81	50.82	1
16410	Tucanæ	9.5	9.75	+ 3.542	66 7 37.3	+ 19.81	52.66	1
16411	Tucanæ	9.5	23 24 20.13	+ 3.584	67 57 17.7	+ 19.81	51.83	1
16412	Octantis	10.0	26.60	3.928	76 25 37.4	.82	50.83	1
16413	Lacaille, 9505	8.5	34.65	3.817	74 33 48.8	.82	50.86	1
16414	Gould, Z. C., 23^h 706	10.0	38.34	3.651	70 28 27.1	.82	51.77	1
16415	Gould, Z. C., 23^h 711	9.2	47.67	+ 3.736	72 55 24.5	+ 19.82	51.32	2
16416	Octantis	10.0	23 24 49.45	+ 4.014	77 45 49.3	+ 19.82	51.81	1
16417	Tucanæ	10.0	25 3.38	3.656	70 49 34.4	.82	51.77	1
16418	Gould, Z. C., 23^h 725	9.2	3.64	4.213	79 54 24.0	.82	50.80	2
16419	Octantis	9.2	5.50	3.850	75 23 12.2	.82	50.80	2
16420	Octantis	9.2	8.47	+ 5.242	84 39 59.7	+ 19.82	51.24	5
16421	Octantis	10.0	23 25 24.65	+ 3.998	77 45 4.3	+ 19.83	51.81	1
16422	Lacaille, 9511	9.0	34.36	3.680	71 47 40.2	.83	51.73	1
16423	Gould, Z. C., 23^h 732	10.0	35.26	3.571	68 9 56.9	.83	51.81	1
16424	Gould, Z. C., 23^h 740	9.0	49.82	3.645	70 53 40.4	.83	51.77	1
16425	Tucanæ	10.0	25 56.84	+ 3.762	73 59 51.4	+ 19.84	50.87	1

No. 16400. Declination possibly 66° 15′ 4.5″. No. 16407. Declination possibly 80° 51′ 22.5″

Number.	Constellation, Name of Star, or Synonym.	Magnitude.	Right Ascension, 1850.0.	Annual Precession.	South Declination, 1850.0.	Annual Precession.	Mean year.	No. of obs.
			h. m. s.	s.	° ′ ″	″		
16426	Octantis	10.0	23 26 8.06	+ 7.983	87 42 17.1	+ 19.84	51.88	3
16427	Tucanæ	9.0	46.20	3.519	66 39 28.5	.85	52.41	3
16428	Tucanæ	10.0	26 48.28	3.735	73 46 42.8	.85	50.87	1
16429	Gould, Z. C., 23h 767 .	9.5	27 3.91	3.638	71 19 21.1	.85	51.80	1
16430	Gould, Z. C., 23h 771 . .	9.5	11.09	+ 3.610	70 29 47.8	+ 19.85	51.77	1
16431	Octantis	9.0	23 27 12.31	+ 4.609	82 55 41.2	+ 19.85	51.84	2
16432	Octantis	10.0	16.74	6.299	86 37 44.3	.85	51.88	2
16433	Tucanæ	10.0	37.84	3.495	66 2 13.1	.86	52.66	1
16434	Gould, Z. C., 23h 787 . .	9.5	38.60	3.534	67 53 23.8	.86	51.82	2
16435	Gould, Z. C., 23h 790 . .	10.0	41.14	+ 3.600	70 27 18.8	+ 19.86	51.77	1
16436	Gould, Z. C., 23h 793 . .	9.5	23 27 43.97	+ 4.036	78 59 4.3	+ 19.86	50.83	1
16437	Octantis	10.0	46.08	4.239	80 52 47.2	.86	51.80	2
16438	Octantis	10.0	50.00	4.032	78 58 19.1	.86	50.84	1
16439	Tucanæ	9.3	53.20	3.525	67 36 12.6	.86	51.85	3
16440	Octantis	10.0	27 56.29	+ 4.190	80 31 52.5	+ 19.86	50.81	1
16441	Octantis	10.1	23 28 0.96	+ 5.362	85 21 24.3	+ 19.86	51.68	6
16442	Tucanæ	9.5	7.84	3.506	67 15 31.6	.86	51.82	1
16443	Octantis	10.0	24.56	3.891	77 21 49.5	.87	50.81	1
16444	Gould, Z. C., 23h 808 . .	9.0	30.02	3.600	70 52 3.0	.87	51.77	1
16445	Octantis	10.0	34.55	+ 3.907	77 39 24.8	+ 19.87	51.81	1
16446	Tucanæ	10.0	23 28 37.58	+ 3.576	70 5 58.7	+ 19.87	51.74	1
16447	Octantis	9.5	41.54	4.382	82 5 21.1	.87	50.81	1
16448	Gould, Z. C., 23h 818 . .	10.0	42.03	3.739	74 44 29.9	.87	50.86	1
16449	Lacaille, 9525	7.0	28 57.22	3.900	77 41 56.9	.87	51.81	1
16450	Octantis .	9.8	29 9.90	+ 4.155	80 35 44.8	+ 19.88	51.33	2
16451	Tucanæ . . .	10.0	23 29 16.43	+ 3.534	68 50 28.4	+ 19.88	51.79	1
16452	Lacaille, 9531 . . .	9.0	21.43	3.612	71 44 3.3	.88	51.73	1
16453	Gould, Z. C., 23h 835	9.3	23.36	3.510	67 52 39.4	.88	51.84	3
16454	Tucanæ	9.8	23.90	3.671	73 27 15.4	.88	50.88	2
16455	Tucanæ .	9.0	26.46	+ 3.474	66 9 45.2	+ 19.88	52.66	1
16456	Tucanæ	10.0	23 29 35.33	+ 3.530	68 53 39.2	+ 19.88	51.79	1
16457	Tucanæ	10.0	42.55	3.505	67 50 40.1	.88	51.83	1
16458	Tucanæ	10.0	47.48	3.533	69 9 8.0	.88	51.79	2
16459	Octantis	10.0	29 56.74	3.834	77 4 36.6	.88	50.85	2
16460	Octantis	10.0	30 20.63	+ 5.701	86 14 45.2	+ 19.89	51.88	1
16461	Gould, Z. C., 23h 872 . .	9.5	23 30 25.10	+ 3.821	77 3 45.4	− 19.89	50.82	3
16462	Gould, Z. C., 23h 869 .	10.0	25.23	3.514	68 44 29.6	.89	51.79	2
16463	Lacaille, 9537 . . .	8.2	40.24	3.649	73 31 20.2	.89	50.88	2
16464	Gould, Z. C., 23h 882 . .	8.5	43.04	3.806	76 56 35.9	.89	50.89	1
16465	Octantis	10.5	47.74	+ 4.099	80 36 23.6	+ 19.89	51.34	2

No. 16432. The two observations differ by 1 wire interval in right ascension; one has been changed from E to F. If, however, the wrong one has been changed, the position will be right ascension 23h 27m 16.74s, declination 86° 37′ 50.1″.

Number.	Constellation, Name of Star, or Synonym.	Magnitude.	Right Ascension, 1850.0.	Annual Precession.	South Declination, 1850.0.	Annual Precession.	Mean year.	No. of obs.
			h. m. s.	s.	° ′ ″	″		
16 466	Octantis	10.0	23 30 49.22	+ 3.756	76 4 24.5	+ 19.89	50.84	1
16 467	Gould, Z. C., 23^h 887 . .	10.0	56.50	3.706	75 3 52.0	.90	50.75	1
16 468	Gould, Z. C., 23^h 889 . .	10.0	58.05	3.944	79 2 28.1	.90	51.33	2
16 469	Gould, Z. C., 23^h 884 . .	9.0	30 58.26	3.536	70 1 25.3	.90	51.74	1
16 470	Melbourne (1), 1200 . .	7.0	31 9.31	+ 3.469	67 4 59.7	+ 19.90	51.87	3
16 471	Gould, 31984	10.0	23 31 20.08	+ 3.584	71 56 49.0	+ 19.90	51.77	2
16 472	Gould, 31985	11.0	21.08	3.583	71 56 55.6	.90	51.77	2
16 473	Gould, Z. C., 23^h 896 . .	9.0	21.94	3.467	67 7 26.7	.90	51.87	3
16 474	Octantis	9.8	23.72	4.081	81 22 51.3	.90	51.30	2
16 475	Tucanæ	9.5	24.22	+ 3.452	66 20 16.3	+ 19.90	52.66	2
16 476	Gould, Z. C., 23^h 902 . .	9.0	23 31 36.11	+ 3.526	70 1 31.5	+ 19.90	51.74	1
16 477	Octantis	10.0	43.50	3.735	76 4 24.5	.90	50.84	1
16 478	Octantis	9.8	44.00	4.387	82 52 13.8	.90	51.79	2
16 479	Tucanæ	10.0	46.10	3.673	74 42 17.3	.90	50.86	1
16 480	Tucanæ	9.8	48.50	+ 3.553	70 47 4.4	+ 19.90	51.77	2
16 481	Gould, Z. C., 23^h 912 . .	10.0	23 31 56.30	+ 3.912	78 59 48.9	+ 19.91	51.82	1
16 482	Octantis	10.0	32 1.39	4.105	81 2 45.9	.91	51.51	4
16 483	Octantis	10.0	6.61	5.832	86 37 50.1	.91	51.88	2
16 484	Gould, Z. C., 23^h 920 . .	9.0	21.04	3.435	66 6 25.9	.91	52.66	1
16 485	Octantis	8.8	21.66	+ 4.593	83 57 51.5	+ 19.91	51.19	3
16 486	Tucanæ	9.2	23 32 23.06	+ 3.438	66 18 26.5	+ 19.91	52.66	2
16 487	Octantis	9.0	23.23	4.663	84 13 55.2	.91	50.68	1
16 488	Tucanæ	10.0	33.06	3.589	72 49 19.8	.91	50.88	1
16 489	Tucanæ	9.5	41.20	3.649	74 36 43.5	.91	50.86	2
16 490	Octantis	9.0	54.57	+ 3.846	78 29 0.1	+ 19.92	50.78	1
16 491	Octantis	10.0	23 32 55.90	+ 3.880	78 59 48.1	+ 19.92	50.83	1
16 492	Octantis	10.0	32 55.98	3.691	75 43 35.0	.92	50.85	1
16 493	Gould, Z. C., 23^h 955 . .	8.5	33 27.57	3.503	70 16 6.8	.92	51.75	2
16 494	Gould, Z. C., 23^h 959 . .	8.5	29.78	3.466	68 36 5.6	.92	51.79	2
16 495	Tucanæ	10.0	43.25	+ 3.541	71 56 33.8	+ 19.92	51.81	1
16 496	Octantis	10.0	23 33 43.38	+ 3.665	75 33 8.6	+ 19.92	50.80	2
16 497	Octantis	10.5	54.89	4.959	85 24 0.0	.93	51.88	2
16 498	Lacaille, 9546	10.0	33 58.70	4.369	83 20 12.5	.93	50.84	1
16 499	Tucanæ	10.0	34 22.59	3.517	71 29 32.4	.93	51.77	2
16 500	Gould, Z. C., 23^h 981 . .	9.2	28.84	+ 3.812	78 38 24.7	+ 19.93	50.80	2
16 501	Tucanæ	9.5	23 34 29.29	+ 3.429	67 25 51.1	+ 19.93	51.85	3
16 502	Octantis	9.5	32.10	3.794	78 23 34.2	.93	50.78	1
16 503	Gould, 32035	9.0	33.34	3.499	70 52 31.7	.93	51.77	1
16 504	Tucanæ	10.0	34.53	3.554	72 56 5.0	.93	50.88	1
16 505	Tucanæ	10.0	37.81	+ 3.462	69 15 41.9	+ 19.93	51.78	1

Number.	Constellation, Name of Star, or Synonym.	Magnitude.	Right Ascension, 1850.0.	Annual Precession.	South Declination, 1850.0.	Annual Precession.	Mean year.	No. of obs.
			h. m. s.	s.	° ′ ″	″		
16 506	Lacaille, 9562	9.0	23 34 39.71	+ 3.517	71 39 8.8	+ 19.93	51.73	1
16 507	Lacaille, 9558	8.0	41.62	3.624	75 2 52.0	.93	50.80	2
16 508	Octantis	10.0	48.70	3.677	76 23 31.3	.94	50.83	2
16 509	Gould, Z. C., 23h 987 . .	8.5	50.76	3.822	78 57 6.7	.94	50.83	1
16 510	Gould, Z. C., 23h 990 . .	9.5	56.67	+ 3.777	78 18 55.7	+ 19.94	50.80	2
16 511	Tucanæ	10.0	23 34 58.60	+ 3.441	68 28 28.1	+ 19.94	51.80	1
16 512	Octantis	8.8	34 59.22	4.218	82 45 33.5	.94	51.47	3
16 513	Tucanæ	9.5	35 13.56	3.442	68 41 47.8	.94	51.79	2
16 514	Gould, 32052	8.5	17.40	3.399	66 15 46.5	.94	52.66	2
16 515	Lacaille, 9560	5.0	25.11	+ 3.852	79 37 25.6	+ 19.94	50.78	1
16 516	Tucanæ	9.5	23 35 25.34	+ 3.449	69 15 46.8	+ 19.94	51.78	1
16 517	Octantis	9.8	27.86	4.229	82 57 43.9	.94	51.79	2
16 518	Octantis	10.5	28.34	3.711	77 24 0.6	.94	51.31	2
16 519	Gould, 32059	8.8	36.74	3.399	66 19 18.0	.94	52.66	2
16 520	Lacaille, 9566	6.5	46.94	+ 3.489	71 19 30.1	+ 19.94	51.80	1
16 521	Tucanæ	10.0	23 35 48.39	+ 3.516	72 25 17.4	+ 19.94	51.75	1
16 522	Octantis	10.0	55.88	3.934	80 45 49.5	.95	51.32	2
16 523	Tucanæ	10.0	35 56.85	3.560	73 59 25.8	.95	50.87	1
16 524	Octantis	10.0	36 16.92	3.717	77 54 29.6	.95	51.81	1
16 525	Octantis	9.0	28.88	+ 3.600	75 28 13.6	+ 19.95	50.80	2
16 526	Tucanæ	9.3	23 36 35.07	+ 3.401	67 30 51.1	+ 19.95	51.85	3
16 527	Tucanæ	10.0	36 43.64	3.428	69 14 42.6	.95	51.79	2
16 528	Gould, 32080	9.3	37 8.30	3.427	69 27 5.3	.96	51.80	3
16 529	Gould, 32091	10.0	15.60	3.446	70 30 58.7	.96	51.77	1
16 530	Gould, Z. C., 23h 1058 .	9.0	18.88	+ 3.715	78 24 9.3	+ 19.96	50.78	1
16 531	Lacaille, 9563	8.0	23 37 23.19	+ 4.490	84 41 44.8	+ 19.96	51.33	4
16 532	Gould, Z. C., 23h 1066 .	10.0	40.29	3.689	78 5 44.2	.96	50.85	1
16 533	Melbourne (1), 1206 . .	8.5	41.85	3.388	67 41 5.2	.96	51.86	2
16 534	Tucanæ	10.0	46.46	3.468	71 31 19.8	.96	51.77	2
16 535	Lacaille, 9580	9.8	37 59.81	+ 3.464	71 54 22.5	+ 19.96	51.77	2
16 536	Gould, 32104	10.0	23 38 0.34	+ 3.486	72 48 36.9	+ 19.96	50.88	1
16 537	Gould, 32106	10.0	4.34	3.485	72 48 42.5	.96	50.88	1
16 538	Gould, Z. C., 23h 1075 .	9.0	15.75	3.400	68 53 21.5	.97	51.79	1
16 539	Octantis	9.5	20.54	4.136	83 14 39.9	.97	50.84	1
16 540	Gould, Z. C., 23h 1086 .	9.5	26.75	+ 3.666	78 4 6.5	+ 19.97	50.85	1
16 541	Lacaille, 9581	8.2	23 38 31.48	+ 3.506	73 55 55.6	+ 19.97	50.87	2
16 542	Octantis	10.0	54.66	3.735	79 30 33.1	.97	51.26	2
16 543	Lacaille, 9584	8.5	38 55.79	3.453	72 8 14.4	.97	51.81	1
16 544	Lacaille, 9588	7.0	39 3.08	3.394	69 13 34.4	.97	51.78	1
16 545	Tucanæ	9.0	8.94	+ 3.389	69 1 57.5	+ 19.97	51.79	2

Number.	Constellation, Name of Star, or Synonym.	Magnitude.	Right Ascension, 1850.0.	Annual Precession.	South Declination, 1850.0.	Annual Precession.	Mean year.	No. of obs.
			h. m. s.	s.	° ′ ″	″		
16 546	Gould, Z. C., 23h 1105 .	8.5	23 39 11.36	3.386	68 33 15.3	+ 19.97	51.80	1
16 547	Tucanæ	10.0	12.03	3.409	70 12 32.4	.97	51.77	1
16 548	Octantis	10.0	15.43	3.733	79 55 58.4	.97	50.82	1
16 549	Lacaille, 7327	7.5	26.93	3.342	66 4 26.7	.98	52.66	1
16 550	Octantis	10.0	30.41	+ 4.012	82 45 40.2	+ 19.98	51.35	2
16 551	Gould, Z. C., 23h 1114	10.0	23 39 31.04	+ 3.443	72 10 38.1	+ 19.98	51.81	1
16 552	Lacaille, 9592	7.0	44.00	3.355	67 24 5.6	.98	51.85	3
16 553	Gould, Z. C., 23h 1123	9.8	39 52.64	3.586	77 9 50.6	.98	50.85	2
16 554	Tucanæ	9.2	40 6.53	3.403	70 41 57.0	.98	51.77	2
16 555	Tucanæ	9.5	17.06	+ 3.394	70 22 23.2	+ 19.98	51.75	2
16 556	Gould, Z. C., 23h 1128	10.0	23 40 19.17	+ 3.358	68 10 38.6	+ 19.98	51.81	1
16 557	Octantis	10.0	29.46	3.917	82 20 27.5	.98	50.81	1
16 558	Gould, 32152	8.8	38.00	3.328	66 13 47.0	.98	52.66	2
16 559	Octantis	9.8	40 51.72	3.669	79 24 48.4	.99	51.30	2
16 560	Octantis	9.5	41 9.10	+ 4.671	86 4 14.3	+ 19.99	51.87	3
16 561	Gould, 32165	7.2	23 41 13.10	+ 3.331	67 5 17.0	+ 19.99	51.86	2
16 562	Gould, Z. C., 23h 1153	9.2	18.40	3.552	68 46 29.4	.99	51.79	2
16 563	Tucanæ	10.0	18.88	3.318	66 11 9.0	.99	52.66	1
16 564	Gould, Z. C., 23h 1159	9.5	33.56	3.483	75 21 58.6	.99	50.82	2
16 565	Octantis	10.3	34.78	+ 4.823	86 29 37.8	+ 19.99	51.86	3
16 566	Octantis	10.0	23 41 38.84	+ 4.127	84 12 36.9	+ 19.99	50.89	1
16 567	Gould, Z. C., 23h 1161 .	9.8	41.16	3.385	71 11 36.8	.99	51.79	2
16 568	Octantis	10.0	43.13	4.141	84 18 25.4	.99	50.86	2
16 569	Gould, 32176 . .	7.0	45.81	3.316	66 28 27.1	.99	52.67	1
16 570	Octantis	9.8	46.93	+ 4.437	85 33 13.1	+ 19.99	51.50	3
16 571	Gould, Z. C., 23h 1176 .	9.0	23 41 57.22	+ 3.374	70 47 51.3	+ 19.99	51.77	1
16 572	Gould, Z. C., 23h 1177 .	9.5	41 59.07	3.427	73 31 38.2	19.99	50.88	2
16 573	Octantis	9.8	42 7.85	4.863	86 40 19.6	20.00	51.87	6
16 574	Lacaille, 9602	8.8	9.79	3.709	80 43 57.6	.00	51.56	4
16 575	Lacaille, 9596	8.6	18.82	+ 4.876	86 43 49.4	+ 20.00	51.87	7
16 576	Lacaille, 9608	8.0	23 42 19.12	+ 3.354	69 55 30.8	+ 20.00	51.81	2
16 577	Gould, Z. C., 23h 1187 .	9.5	22.69	3.364	70 39 59.2	.00	51.77	2
16 578	Gould, Z. C., 23h 1195 .	10.0	45.61	3.380	71 57 33.2	.00	51.73	1
16 579	Gould, Z. C., 23h 1197 .	9.5	51.28	3.433	74 32 16.2	.00	50.86	1
16 580	Octantis	9.5	42 51.32	+ 5.120	87 12 22.6	+ 20.00	51.87	1
16 581	Gould, Z. C., 23h 1207 .	8.8	23 43 2.56	+ 3.506	77 11 14.9	+ 20.00	50.85	2
16 582	Brisbane, 7335	8.0	2.76	3.295	66 8 53.4	.00	52.65	1
16 583	γ¹ Octantis	5.5	7.56	3.856	82 51 9.4	.00	51.88	1
16 584	Tucanæ	10.0	9.46	3.304	67 7 10.9	.00	51.90	1
16 585	Octantis	10.0	9.70	+ 5.434	87 37 28.2	+ 20.00	51.87	2

Nos. 16 580 and 16 591. These two observations probably belong to the same star.

Number.	Constellation, Name of Star, or Synonym.	Magnitude.	Right Ascension, 1850.0.	Annual Precession.	South Declination, 1850.0.	Annual Precession.	Mean year.	No. of obs.
			h. m. s.	s.	° ′ ″	″		
16 586	Octantis	9.2	23 43 13.52	+ 3.626	79 59 43.1	+ 20.00	50.80	2
16 587	Gould, Z. C., 23^h 1216 .	9.5	22.86	3.513	77 36 31.0	.00	51.81	1
16 588	Octantis	10.0	24.27	3.863	83 1 45.3	.00	51.88	1
16 589	Gould, Z. C., 23^h 1214 .	9.0	24.74	3.409	74 1 22.3	.00	50.87	1
16 590	Gould, Z. C., 23^h 1217 .	10.0	27.04	+ 3.363	71 42 16.0	+ 20.00	51.73	1
16 591	Octantis	10.0	23 43 52.14	+ 4.985	87 11 20.8	+ 20.01	51.87	1
16 592	Gould, Z. C., 23^h 1233 .	9.5	43 53.87	3.441	75 44 27.1	.01	50.85	1
16 593	Tucanæ	9.5	44 8.80	3.295	67 29 23.0	.01	51.82	2
16 594	Gould, Z. C., 23^h 1252	8.0	30.70	3.306	68 51 32.1	.01	51.79	1
16 595	Lacaille, 9614	6.6	33.79	+ 3.906	83 50 32.2	+ 20.01	50.85	2
16 596	Lacaille, 9627	8.0	23 44 34.83	+ 3.369	73 14 2.5	+ 20.01	50.88	1
16 597	Tucanæ	10.0	39.82	3.282	66 56 13.3	.01	51.90	1
16 598	Gould, 32222	9.2	46.38	3.398	74 45 52.9	.01	50.86	2
16 599	Lacaille, 9621	9.0	54.56	3.638	81 10 31.9	.01	51.37	3
16 600	Tucanæ	10.0	55.94	+ 3.330	71 13 35.2	+ 20.01	51.80	1
16 601	Gould, Z. C., 23^h 1263 .	9.0	23 44 57.87	+ 3.268	65 57 45.2	+ 20.01	52.66	1
16 602	Octantis	8.7	44 58.71	4.517	86 31 56.2	.01	51.87	6
16 603	Tucanæ	10.0	45 3.93	3.320	70 39 56.5	.01	51.77	1
16 604	Gould, Z. C., 23^h 1265 .	9.3	7.19	3.281	67 28 40.1	.01	51.85	3
16 605	Tucanæ	9.2	14.40	+ 3.298	69 10 54.0	+ 20.01	51.79	2
16 606	Gould, 32236	10.0	23 45 22.12	+ 3.351	73 1 12.2	+ 20.02	50.88	2
16 607	Gould, 32237	10.0	24.09	3.351	73 1 7.4	.02	50.88	2
16 608	Octantis	9.8	28.32	3.419	76 18 5.2	.02	50.83	2
16 609	Tucanæ	10.0	28.98	3.291	68 53 8.9	.02	51.79	1
16 610	Brisbane, 7341	7.0	39.38	+ 3.267	66 47 6.8	+ 20.02	52.29	2
16 611	Gould, Z. C., 23^h 1282 .	9.2	23 45 40.64	+ 3.322	71 30 47.9	+ 20.02	51.77	2
16 612	Octantis	11.0	41.21	3.753	83 0 57.1	.02	51.88	1
16 613	Octantis	10.0	45 51.46	3.472	78 22 12.9	.02	50.78	1
16 614	Gould, Z. C., 23^h 1293 .	8.5	46 7.02	3.384	75 27 29.0	.02	50.80	2
16 615	Octantis	8.0	10.32	+ 3.421	77 0 1.7	+ 20.02	50.85	2
16 616	Octantis	10.0	23 46 19.20	+ 3.825	83 57 33.8	+ 20.02	50.85	2
16 617	Gould, Z. C., 23^h 1298 .	9.5	21.96	3.364	74 47 16.5	.02	50.86	2
16 618	Octantis	9.7	24.66	3.572	81 0 7.9	.02	51.38	3
16 619	Gould, 32250	9.2	26.94	3.347	73 57 45.4	.02	50.87	2
16 620	Octantis	9.5	30.32	+ 3.535	80 21 28.8	+ 20.02	50.81	2
16 621	Lacaille, 9635	8.5	23 46 30.86	+ 3.489	79 20 16.1	+ 20.02	51.82	1
16 622	Tucanæ	10.0	43.92	3.266	68 19 11.4	.02	51.80	2
16 623	Tucanæ	10.0	45.35	3.355	74 44 42.4	.02	50.86	1
16 624	Gould, Z. C., 23^h 1312 .	9.0	52.49	3.289	70 34 17.4	.02	51.77	1
16 625	Tucanæ	10.0	53.28	+ 3.262	68 4 32.8	+ 20.02	51.81	1

Nos. 16 580 and 16 591. These two observations probably belong to the same star.

Number.	Constellation, Name of Star, or Synonym.	Magnitude.	Right Ascension, 1850.0.	Annual Precession.	South Declination, 1850.0.	Annual Precession.	Mean year.	No. of obs.
			h. m. s.	s.	° ′ ″	″		
16 626	Gould, Z. C., 23h 1319	9.8	23 46 59.80	+ 3.384	76 20 55.0	+ 20.02	50.83	2
16 627	Tucanæ	9.5	47 11.71	3.248	67 3 42.4	.03	51.90	1
16 628	Tucanæ	9.5	22.85	3.238	66 10 49.6	.03	52.66	1
16 629	Tucanæ	10.0	47 51.36	3.255	68 53 29.5	.03	51.79	1
16 630	Gould, Z. C., 23h 1339	9.5	48 3.17	+ 3.251	68 45 23.6	+ 20.03	51.79	2
16 631	Lacaille, 9614	7.5	23 48 6.20	+ 3.712	83 49 9.8	+ 20.03	50.85	2
16 632	Tucanæ	10.0	8.00	3.257	69 31 4.6	.03	51.84	2
16 633	Tucanæ	10.0	17.11	3.278	71 42 13.8	.03	51.73	1
16 634	Tucanæ	10.0	18.86	3.271	71 5 5.3	.03	51.79	2
16 635	Tucanæ	9.2	30.88	+ 3.227	66 36 16.6	+ 20.03	52.29	2
16 636	Tucanæ	10.0	23 48 31.88	+ 3.267	71 5 10.2	+ 20.03	51.79	2
16 637	Tucanæ	9.5	40.45	3.283	72 38 20.4	.03	51.32	2
16 638	Octantis	10.0	46.83	3.487	81 2 22.6	.03	51.31	2
16 639	Octantis	10.0	47.17	4.446	87 16 25.5	.03	51.88	2
16 640	Octantis	9.2	50.05	+ 3.742	85 9 59.4	+ 20.03	51.64	4
16 641	Tucanæ	10.0	23 48 50.60	+ 3.247	69 36 29.6	+ 20.03	51.84	2
16 642	Gould, Z. C., 23h 1366	9.0	49 1.05	3.238	68 25 53.6	.03	51.80	1
16 643	Gould, Z. C., 23h 1369	8.5	5.68	3.311	75 8 13.8	.03	50.75	1
16 644	γ² Octantis	6.5	8.96	3.587	83 0 16.4	.03	51.36	2
16 645	Gould, Z. C., 23h 1380	9.5	30.38	+ 3.423	80 6 7.5	+ 20.04	50.82	1
16 646	Brisbane, 7351	8.5	23 49 35.57	+ 3.212	66 36 34.8	+ 20.04	51.29	2
16 647	Tucanæ	10.0	52.70	3.260	72 36 47.1	.04	51.75	1
16 648	Tucanæ	10.0	49 55.40	3.214	67 29 25.2	.04	51.83	1
16 649	Tucanæ	7.8	50 0.17	3.215	67 50 25.5	.04	51.86	4
16 650	Gould, Z. C., 23h 1393	10.0	1.35	+ 3.226	68 54 39.9	+ 20.04	51.79	1
16 651	Octantis	9.3	23 50 6.94	+ 7.023	89 9 51.5	+ 20.04	51.86	5
16 652	Tucanæ	9.8	14.56	3.234	70 45 51.8	.04	51.77	2
16 653	Gould, Z. C., 23h 1403	9.0	27.12	3.266	74 1 27.6	.04	50.87	1
16 654	Tucanæ	10.0	28.49	3.226	70 9 52.7	.04	51.74	1
16 655	Octantis	10.0	47.11	+ 3.452	81 57 26.9	+ 20.04	51.81	1
16 656	Tucanæ	10.0	23 50 51.52	+ 3.225	70 46 27.8	+ 20.04	51.77	2
16 657	Gould, Z. C., 23h 1415	9.2	50 51.96	3.221	70 20 22.6	.04	51.75	2
16 658	Gould, Z. C., 23h 1424	10.0	51 5.80	3.264	74 53 28.6	04	50.86	1
16 659	Tucanæ	10.0	7.78	3.197	67 38 17.3	.04	51.83	1
16 660	Gould, Z. C., 23h 1426	8.5	8.97	+ 3.217	70 29 20.7	+ 20.04	51.77	1
16 661	Tucanæ	10.0	23 51 8.98	+ 3.229	71 47 15.8	+ 20.04	51.73	1
16 662	Tucanæ	10.0	16.18	3.210	69 46 29.8	.04	51.89	1
16 663	Tucanæ	10.0	22.30	3.192	67 12 43.6	.04	51.82	1
16 664	Tucanæ	10.0	26.92	3.212	69 59 58.5	.04	51.74	1
16 665	Brisbane, 7357	8.8	29.52	+ 3.185	66 21 27.2	+ 20.04	52.66	2

No. 16 627. The right ascension may be 23h 47m 21.71s.

Number.	Constellation, Name of Star, or Synonym.	Magnitude.	Right Ascension, 1850.0.	Annual Precession.	South Declination, 1850.0.	Annual Precession.	Mean year.	No. of obs.
			h. m. s.	s.	° ′ ″	″		
16 666	Gould, Z. C., 23^h 1435	10.0	23 51 36.86	+ 3.252	74 48 32.2	+ 20.04	50.86	1
16 667	Tucanæ	9.8	38.10	3.191	67 48 59.0	.04	51.84	3
16 668	Gould, Z. C., 23^h 1436	9.0	39.21	3.212	70 53 7.5	.04	51.77	1
16 669	Tucanæ	10.0	56.52	3.219	72 20 20.3	.04	51.75	1
16 670	Gould, Z. C., 23^h 1444	10.0	51 57.15	+ 3.227	73 10 23.1	+ 20.04	50.88	1
16 671	ε Tucanæ	4.8	23 52 4.93	+ 3.178	66 24 41.8	+ 20.04	51.41	3
16 672	Tucanæ	10.0	14.80	3.181	67 33 48.2	.04	51.82	2
16 673	Octantis	9.8	17.45	3.374	81 30 31.6	.05	51.30	2
16 674	Gould, Z. C., 23^h 1460	8.8	20.46	3.322	79 53 38.6	.05	50.80	2
16 675	Tucanæ	10.0	38.49	+ 3.190	69 58 26.3	+ 20.05	51.89	1
16 676	Gould, Z. C., 23^h 1473	9.0	23 52 40.36	+ 3.270	77 51 19.8	+ 20.05	51.38	3
16 677	Gould, Z. C., 23^h 1474	10.0	43.34	3.211	73 2 16.6	.05	50.88	2
16 678	Gould, 32356	10.5	45.67	3.174	67 31 1.7	.05	51.91	1
16 679	Gould, 32357	7.9	46.21	3.174	67 31 3.2	.05	51.86	4
16 680	Gould, Z. C., 23^h 1480	9.0	53.10	+ 3.189	70 30 28.3	+ 20.05	51.77	1
16 681	Gould, Z. C., 23^h 1482	9.0	23 52 56.24	+ 3.208	73 9 13.4	+ 20.05	50.88	2
16 682	Octantis	9.8	53 4.30	3.288	79 24 22.1	.05	51.30	2
16 683	Brisbane, 7363	9.0	4.61	3.164	66 28 55.1	.05	52.67	1
16 684	Tucanæ	10.0	15.21	3.211	74 11 56.3	.05	50.86	1
16 685	Octantis	9.2	23.86	+ 3.461	84 21 28.4	+ 20.05	50.86	2
16 686	Gould, 32369	11.0	23 53 28.28	+ 3.183	71 1 13.6	+ 20.05	51.77	1
16 687	Gould, Z. C., 23^h 1496	8.5	30.70	3.171	69 12 18.7	.05	51.78	1
16 688	Gould, Z. C., 23^h 1499	9.0	33.13	3.181	70 57 38.9	.05	51.77	1
16 689	Gould, 32371	12.0	33.43	3.181	71 1 30.1	.05	51.77	1
16 690	Tucanæ	10.0	42.20	+ 3.166	68 46 40.0	+ 20.05	51.79	1
16 691	Gould, Z. C., 23^h 1506	9.2	23 53 43.14	+ 3.195	73 26 28.0	+ 20.05	50.88	2
16 692	θ Octantis	5.5	48.76	3.240	77 53 36.5	.05	51.47	3
16 693	Tucanæ	10.0	50.30	3.177	71 9 33.7	.05	51.80	1
16 694	Octantis	10.0	53 53.38	3.289	80 40 22.1	.05	51.80	2
16 695	Gould, Z. C., 23^h 1516	9.5	54 0.53	+ 3.163	68 56 18.0	+ 20.05	51.79	1
16 696	Octantis	9.5	23 54 8.14	+ 3.319	82 7 7.2	+ 20.05	50.81	1
16 697	Tucanæ	10.0	15.41	3.158	68 42 39.2	.05	51.79	2
16 698	Gould, Z. C., 23^h 1524	9.0	17.02	3.164	70 6 18.1	.05	51.74	1
16 699	Gould, Z. C., 23^h 1525	9.5	18.55	3.159	69 6 25.2	.05	51.79	2
16 700	Tucanæ	10.0	26.72	+ 3.153	68 21 37.7	+ 20.05	51.80	2
16 701	Tucanæ	10.0	23 54 53.44	+ 3.147	68 28 8.8	+ 20.05	51.80	1
16 702	Octantis	9.9	55 5.50	3.537	86 28 43.4	.05	51.86	4
16 703	Octantis	9.5	5.78	3.182	75 25 52.4	.05	50.85	1
16 704	Octantis	10.0	32.93	3.583	87 5 29.6	.05	51.88	1
16 705	Gould, Z. C., 23^h 1562	9.0	48.83	+ 3.128	66 39 50.6	+ 20.05	52.16	3

Number.	Constellation, Name of Star, or Synonym.	Magnitude.	Right Ascension, 1850.0.	Annual Precession.	South Declination, 1850.0.	Annual Precession.	Mean year.	No. of obs.
			h. m. s.	s.	° ′ ″	″		
16706	Octantis	10.0	23 55 55.75	− 3.228	81 20 48.0	+ 20.05	50.80	1
16707	Gould, Z. C., 23h 1569 .	9.8	56 0.72	3.134	69 31 33.2	.05	51.84	2
16708	Tucanæ	10.0	4.19	3.153	74 9 11.9	.05	50.87	1
16709	Tucanæ	9.5	13.76	3.123	66 41 32.0	.05	52.67	1
16710	Lacaille, 9704	8.0	17.75	+ 3.112	81 13 51.5	+ 20.05	51.29	2
16711	Tucanæ	10.0	23 56 34.30	+ 3.118	66 41 34.9	+ 20.05	51.90	1
16712	Gould, Z. C., 23h 1580 .	9.8	36.26	3.140	73 51 57.2	.05	50.87	2
16713	Octantis	9.2	49.78	3.157	77 42 30.3	.05	51.78	2
16714	Lacaille, 9708	8.0	51.09	3.135	73 43 50.3	.05	50.87	1
16715	Gould, Z. C., 23h 1587 .	9.5	53.41	+ 3.133	73 29 46.3	+ 20.05	50.88	1
16716	Tucanæ	9.5	23 56 54.42	+ 3.138	74 41 16.8	+ 20.05	50.86	2
16717	Gould, Z. C., 23h 1588 .	9.2	54.82	3.151	77 10 56.8	.05	50.85	2
16718	Tucanæ	10.0	55.71	3.129	72 43 44.1	.05	51.75	1
16719	Tucanæ	9.5	56.10	3.129	72 31 47.5	.05	51.75	1
16720	Tucanæ	8.5	56 58.48	+ 3.124	71 14 20.9	+ 20.05	51.80	1
16721	Octantis	8.5	23 57 1.69	+ 3.198	82 10 52.9	+ 20.05	50.81	1
16722	Lacaille, 9710	6.2	1.70	3.126	72 16 16.8	.05	51.78	2
16723	Gould, Z. C., 23h 1596 .	9.5	4.81	3.118	69 40 27.8	.05	51.89	1
16724	Tucanæ	10.0	7.08	3.130	73 58 57.5	.05	50.87	1
16725	Octantis	9.8	10.36	+ 3.160	79 24 1.6	+ 20.05	51.30	2
16726	Tucanæ	8.2	23 57 19.02	+ 3.117	71 2 37.5	+ 20.05	51.79	2
16727	Tucanæ	9.0	20.74	3.117	71 19 35.1	.06	51.80	1
16728	Octantis	9.8	25.37	3.165	80 47 40.7	.06	51.47	3
16729	Gould, Z. C., 23h 1612 .	8.8	34.16	3.104	66 11 9.4	.06	52.28	2
16730	Gould, Z. C., 23h 1618 .	9.0	43.02	+ 3.106	68 38 42.4	+ 20.06	51.79	2
16731	Gould, Z. C., 23h 1620 .	9.0	23 57 47.40	+ 3.146	80 5 15.5	+ 20.06	50.82	1
16732	Melbourne (1), 1226 . .	8.0	57 49.07	3.105	68 47 42.6	.06	51.79	2
16733	Tucanæ	9.5	58 2.69	3.099	67 20 33.0	.06	51.82	1
16734	Tucanæ	10.0	16.51	3.099	69 23 7.2	.06	51.78	1
16735	Gould, Z. C., 23h 1631 .	9.5	20.32	+ 3.098	69 38 51.2	+ 20.06	51.89	1
16736	Tucanæ	10.0	23 58 34.60	+ 3.093	68 59 42.9	+ 20.06	51.79	1
16737	Octantis	10.0	38.27	3.115	79 29 29.9	.06	50.78	1
16738	Octantis	8.8	43.84	3.155	84 56 1.4	.06	51.66	5
16739	Gould, Z. C., 0h 2 . . .	9.2	47.86	3.092	70 43 54 4	.06	51.77	2
16740	Octantis	10.0	53.50	+ 3.096	75 14 11.8	+ 20.06	50.84	2
16741	Octantis	9.8	23 58 54.13	+ 3.110	80 36 11.5	+ 20.06	51.47	3
16742	Octantis	9.5	59 1.53	3.118	83 1 51.6	.06	51.36	2
16743	Gould, Z. C., 0h 10 . . .	8.2	13.74	3.085	71 25 29.6	.06	51.77	2
16744	Lacaille, 9727	8.3	42.53	3.080	77 33 58.7	.06	51.46	3
16745	Gould, Z. C., 0h 29 . . .	9.5	43.72	+ 3.078	74 46 16.6	+ 20.06	50.86	2

No. 16730. Gould's declination is 9.7″ greater.

Number.	Constellation, Name of Star, or Synonym.	Magnitude.	Right Ascension, 1850.0.	Annual Precession.	South Declination, 1850.0.	Annual Precession.	Mean year.	No. of obs.
			h. m. s.	s.	° ′ ″	″		
16 746	Tucanæ	9.5	23 59 50.35	+ 3.075	71 0 1.3	+ 20.06	51.77	1
16 747	Octantis	9.8	50.63	3.086	86 12 44.0	.06	51.87	4
16 748	Octantis	9.5	54.88	+ 3.075	82 19 17.9	+ 20.06	50.82	2

0